SPRINGER
STUDY
EDITION

W0036171

H. Herken   F. Hucho   (Eds.)

Selective
Neurotoxicity

With 116 Figures and 73 Tables

Springer-Verlag
Berlin Heidelberg New York London Paris
Tokyo Hong Kong Barcelona Budapest

Professor Dr. med. HANS HERKEN
Institut für Pharmakologie der Freien Universität
Thielallee 69/73
14195 Berlin, FRG

Professor Dr. rer. nat. FERDINAND HUCHO
Institut für Biochemie der Freien Universität
Thielallee 63
14195 Berlin, FRG

ISBN 3-540-57815-3 Springer-Verlag Berlin Heidelberg New York
ISBN 0-387-57815-3 Springer-Verlag New York Berlin Heidelberg

Second Printing 1994 (originally published in the Handbook of
Experimental Pharmacology, Volume 102, ISBN 3-540-54654-5, 1992)

Library of Congress Cataloging-in-Publication Data.
Selective neurotoxity / H. Herken, F. Hucho, eds – Springer study ed. p. cm. Originally published in
the Handbook of experimental pharmacology, vol. 102. Berlin; New York: Springer-Verlag, 1992.
Includes bibliographical references and index. ISBN 0-387-57815-3. – ISBN 3-540-57815-3 (Berlin):
DM 158.00 1. Neurotoxicology. I. Herken, Hans. II. Hucho, Ferdinand, 1939– . [RC347.5.S45   1994]
616.8 ' 047–dc20   94-7806 CIP

This work is subject to copyright. All rights are reserved, wether the whole or part of the material is
concerned, specifically the rights of translation, reprinting, reuse of illustrations, recitation, broadcasting,
reproduction on microfilms or in other ways, and storage in data banks. Duplication of this publication
or parts thereof is only permitted under the provisions of the German Copyright Law of September 9,
1965, in its current version and a copyright fee must always be paid. Violations fall under the prosecu-
tion act of the German Copyright Law.

© Springer-Verlag Berlin Heidelberg 1992, 1994
Printed in Germany

The use of registered names, trademarks, etc in this publication does not imply, even in the absence of a
specific statement, that such names are exempt from the relevant protective laws and regulations and
therefore free for general use.

Product liability: The publisher can give no guarantee for information about drug dosage and appication
thereof contained in this book. In every individual case the respective user must check its accuracy by
consulting other pharmaceutical literature.

Typesetting: Best-set Typesetter Ltd., Hong Kong
27/3130 – 5 4 3 2 1 0 – Printed on acid-free paper

List of Contributors

AKTORIES, K., Institut für Pharmakologie und Toxikologie der Universität des Saarlandes, W-6650 Homburg, Saar, FRG

ALBUQUERQUE, E.X.,. Department of Pharmacology and Experimental Therapeutics, University of Maryland School of Medicine, Bressler Research Building, 655 W. Baltimore Street, Baltimore, MD 21201, USA (permanent address) and Laboratory of Molecular Pharmacology II, Institute of Biophysics "Carlos Chagas Filno", Federal University of Rio de Janeiro, Ilha do Fundao, CEP 21941, Rio de Janeiro, Brazil

BAUMGARTEN, H.G., Institut für Anatomie (FB1/WE1), Freie Universität Berlin, Königin-Luise-Straße 15, W-1000 Berlin 33, FRG

BECKER, C.-M., Zentrum für Molekulare Biologie und Neurologische Universitätsklinik, Universität Heidelberg, Im Neuenheimer Feld 282, W-6900 Heidelberg, FRG

BECKER, S., Max-Planck-Institut für Biophysik, Abteilung für Molekulare Membranbiologie, Heinrich-Hofmann-Str. 7, W-6000 Frankfurt/Main, FRG

DOLLY, J.O., Department of Biochemistry, Imperial College of Science, Technology and Medicine, London SW7 2AY, Great Britain

GIERSCHIK, P., Abteilung Molekulare Pharmakologie Deutsches Krebsforschungszentrum, Im Neuenheimer Feld 280, W-6900 Heidelberg, FRG

GLOSSMANN, H., Institut für Biochemische Pharmakologie, Peter-Mayr-Str. 1, A-6020 Innsbruck, Austria

GORDON, R.D., ICI Pharmaceuticals, Biotechnology Department, Mereside, Alderley Park, Macclesfield, Cheshire SK10 4TG, Great Britain

HANIN, I., Department of Pharmacology and Experimental Therapeutics, Loyola University Chicago, Stritch School of Medicine, 2160 South First Avenue, Maywood, IL 60153, USA

HERKEN, H., Institut für Pharmakologie der Freien Universität, Thielallee 69/73, W-1000 Berlin 33, FRG

HÖRTNAGL, H., Institut für Biochemische Pharmakologie der Medizinischen Fakultät, Universität Wien, Borschkegasse 8a, A-1090 Wien, Austria

HOLZER, P., Institut für experimentelle und klinische Pharmakologie, Universität Graz, Universitätsplatz 4, A-8010 Graz, Austria

HUCHO, F., Freie Universität Berlin, Fachbereich Chemie, Institut für Biochemie, Thielallee 63, W-1000 Berlin 33, FRG

JÄRV, J., Laboratory of Bioorganic Chemistry, Tartu University, Jakobi. 2, 202400 Tartu, Estonia

JAKOBS, K.H., Pharmakologisches Institut, Universitätsklinikum Essen, Hufelandstraßess, W-4300 Essen, FRG

JOHANSSON, B.B., Department of Neurology, Lund University Hospital, S-221 85 Lund, Sweden

KOPIN, I.J., Division of Intramural Research, National Institute of Neurological Disease and Stroke, National Institutes of Health, 9000 Rockville Pike, Bldg. 10, Rm. 5N214, Bethesda, MD 20892, USA

KRIEGLSTEIN, J., Institut für Pharmakologie und Toxikologie, Fachbereich Pharmazie und Lebensmittelchemie, Philipps-Universität, Ketzerbach 63, 3550 Marburg, FRG

LUMMIS, S.C.R., Molecular Neurobiology Unit, Medical Research Council Centre, Hills Road, Cambridge CB2 2QH, Great Britain

MARTIN, I.L., Molecular Neurobiology Unit, Medical Research Council Centre, Hills Road, Cambridge CB2 2QH, Great Britain

MEVES, H., I. Physiologisches Institut, Medizinische Fakultät der Universität des Saarlandes, W-6650 Homburg/Saar, FRG

NUGLISCH, J., Institut für Pharmakologie und Toxikologie, Fachbereich Pharmazie und Lebensmittelchemie, Philipps-Universität, Ketzerbach 63, W-3550 Marburg, FRG

OCHS, S., Department of Physiology and Biophysics, Indiana University School of Medicine, VanNuys Medical Science Building 374, 635 Barnhill Drive, Indianapolis, IN 46202-5120, USA

OLSSON, Y., Laboratory of Neuropathology, Institute of Pathology, University Hospital, S-751 85 Uppsala, Sweden

STRIESSNIG, J., Department of Pharmacology SJ-30, University of Washington, Seattle, WA 98195, USA. *Present address*: Institut für Biochemische Pharmakologie, Peter-Mayr-Straße 1, A-6020 Innsbruck, Austria

SWANSON, K.L., Department of Pharmacology and Experimental Therapeutics, University of Maryland School of Medicine, Bressler Research Building, 655 W. Baltimore Street, Baltimore, MD 21201, USA

TEICHBERG, V.I., Department of Neurobiology, The Weizmann Institute of Science, P.O. Box 26, 76100 Rehovot, Israel

WELLHÖNER, H.H., Zentrum Pharmakologie und Toxikologie, Abteilung Toxikologie, Medizinische Hochschule Hannover, Postfach 610180, W-3000 Hannover 61, FRG

ZELLER, W.J., Institut für Toxikologie und Chemotherapie, Deutsches Krebsforschungszentrum, Im Neuenheimer Feld 280, P. 101949, W-6900 Heidelberg 1, FRG

ZIMMERMANN, B., Institut für Anatomie, Freie Universität Berlin, FB1, Königin-Luise-Straße 15, W-1000 Berlin 33, FRG

Preface

The volume "Selective Neurotoxicity" focuses on the molecular mechanisms of action of various neurotoxins. Recent research on their targets at the nerve cell and its subcellular components is discussed.

Because of the vast number of neurotoxins a certain degree of selection was necessary, of course of a subjective nature. On the other hand, not only are natural (plant and microbial) toxins included, but also some representative artificial "chemical" toxins and other mechanisms of neurotoxicity.

Special emphasis has been placed on the selective vulnerability sites in the nervous system. The first few chapters therefore deal with the topography of these sites, with protective barriers in the peripheral nervous system, and with the blood-brain barrier in the CNS. A special chapter is devoted to the kinetics and metabolic disorders of axoplasmatic transport induced by neurotoxins. An additional mechanism of neurotoxicity is introduced in the topic of selective vulnerability: the toxicity of energy deficits (induced by metabolic disorders), which characteristically affect adrenergic, dopaminergic and cholinergic systems. A further source of toxicity is presented in a chapter dealing with biosynthesis of neurotoxic compounds as a result of enzymatic errors. The diversity of topics related to the subject is further broadened with the chapter on neurotropic carcinogenesis.

The major part of the volume deals with neurotoxicity in a more conventional sense: mostly natural toxins affecting receptors and ion channels are discussed. This part of the book can be viewed as a success report on the application of neurotoxins as tools for elucidating central functional mechanisms of nerve activity. Evolution optimized various molecules which serve certain organisms for defense and for hunting and killing prey. These molecules have been used by researchers to guide them to their targets in the nervous system. These targets must by definition be key elements of, for example, the axonal and pre- and postsynaptic membranes. Various receptors, ion channels, and components of transmitter release sites were first identified and biochemically investigated using these neurotoxins. In combination with the then new immunochemical and recombinant DNA techniques, a wealth of information became available. This information is reviewed in chapters on receptors for glutamate, GABA, glycine, dopamine, and acetylcholine, calcium channels, and voltage-dependent sodium and potassium channels.

Here again we transcended the classical definition of a neurotoxin by including cholera and pertussis toxins. Of course, the G proteins of nerve cells are not the only target of these ADP-ribosylating proteins, but they are so essential in elucidating mechanisms of signal transduction in neurons, and we have learned so much in recent years about their targets, that we did not want to omit them from this volume. The reader may see them as another variation on the theme of "toxins as tools in neurochemistry."

After all, what can be counted among the neurotoxins? "Wenn ihr jedes Gifft recht wollt ausslegen, was ist das nit Gifft ist? Alle ding sind Gifft und nichts ohn Gifft, allein die Dosis macht, dass ein Ding kein Gifft ist" says the 16th century physician Paracelsus (1537/38; Third Defension II, 170): "If you want to clarify every poison correctly, what there, that is not poison? All things are poison and nothing is without poison: the dosis alone makes a thing not poison."

The molecular and mechanistic approach to defining toxins and their targets became most prominent during the last century in Claude Bernard's work. The "toxins as tools" motif is evident in his "Leçons sur les effets des substances toxiques et médicamenteuses," published in 1857: He describes their application "comme des espèces d'instruments physiologiques plus délicats que les moyens mécaniques et destinés à disséquer pour ainsi dire, une à une, les propriétés des éléments anatomiques de l'organisme vivant" ("as a type of physiological instrument that is more delicate than mechanical methods and whose purpose is, so to speak, to dissect one by one the properties and anatomical elements in the living organism") The toxic substance provides a logical link between the interaction at the molecular level and the tissue damage or functional disorder in the whole animal. Claude Bernard proved this point with his work on d-tubocurarine, which initiated the elucidation of the mechanism of the neuromuscular impulse transmission and of the neuromuscular endplate. This volume aims, on the one hand, to document the broad scope of the topic and, on the other, to show the depth of the molecular mechanistic approach of our present-day research. Selective neurotoxicity is beginning to be understandable, not only at the cellular to subcellular level, but even at atomic resolution.

We would like to express our gratitude to Springer-Verlag and especially to Ms. Doris Walker for her excellent support during the production of this volume. We also thank Ms. Mary Wurm for valuable secretarial help.

Berlin, October 1991   H. HERKEN · F. HUCHO

Contents

CHAPTER 3

Protective Barriers in the Nervous System Against Neurotoxic Agents: The Blood-Brain Barrier

CHAPTER 4

**Kinetic and Metabolic Disorders of Axoplasmic Transport Induced
by Neurotoxic Agents**

CHAPTER 5

Metabolic Disorders as Consequences of Drug-Induced Energy Deficits
J. KRIEGLSTEIN and J. NUGLISCH. With 10 Figures 111

CHAPTER 6

Neurotoxic Synthesis by Enzymatic Error
H. HERKEN. With 12 Figures

CHAPTER 7

Neurotropic Carcinogenesis
W.J. ZELLER 193

CHAPTER 8

Neurotoxic Phenylalkylamines and Indolealkylamines
H.G. Baumgarten and B. Zimmermann. With 10 Figures   225

CHAPTER 9

Toxins Affecting the Cholinergic System
H. HÖRTNAGL and I. HANIN. With 2 Figures 293

CHAPTER 10

**Mechanisms of 1-Methyl-4-Phenyl-1,2,3,6-Tetrahydropyridine
Induced Destruction of Dopaminergic Neurons**

CHAPTER 11

Tetanus and Botulinum Neurotoxins

CHAPTER 12

Capsaicin: Selective Toxicity for Thin Primary Sensory Neurons

CHAPTER 13

Excitotoxins, Glutamate Receptors, and Excitotoxicity

CHAPTER 14

Convulsants and Gamma-Aminobutyric Acid Receptors

CHAPTER 15

Convulsants Acting at the Inhibitory Glycine Receptor

CHAPTER 16

Peptide Toxins Acting on the Nicotinic Acetylcholine Receptor

CHAPTER 17

Nicotinic Acetylcholine Receptors and Low Molecular Weight Toxins

CHAPTER 23

ADP-Ribosylation of Signal-Transducing Guanine Nucleotide
Binding Proteins by Cholera and Pertussis Toxin
P. GIERSCHIK and K.H. JAKOBS. With 2 Figures 807

CHAPTER 24

Cellular and Subcellular Targets of Neurotoxins: The Concept of Selective Vulnerability

H.G. Baumgarten and B. Zimmermann

A. Introduction

Vogt and Vogt (1937) recognized that the brain is capable of reacting to generalized forms of trauma such as hypoxia/ischemia in a nonuniform manner, i.e., with topologically restricted and cell-selective neuropathology ("pathoklise") (pyramidal neurons of the hippocampus, neocortical neurons in lamina III, V, and VI, striatal neurons, and cerebellar Purkinje cells), suggesting that specific intrinsic properties expressed by certain nerve cells are responsible for their selective vulnerability and thus for the regional patterning of neuropathology. Similar patterns of neuropathology are also seen after subjecting the brain to other forms of generalized trauma, e.g., to hypoglycemia (loss of neocortical neurons and dentate granule cells) and generalized epilepsy (Ammon's horn sclerosis). Early attempts to uncover the pathophysiology common to these forms of brain damage focussed on depletion of energy stores by critical restriction of the oxygen and glucose supply (cf. Siesjö 1981). Data which showed that incomplete ischemia can cause more extensive cell loss than complete ischemia (which results in near total depletion of ATP and phosphocreatine) were difficult to reconcile with the energy deficit concept as the common cause of selective vulnerability. Kass and Lipton (1982) and Rothman (1983) focussed on the role of synaptic transmission in the pathogenesis of hypoxic neuron damage by showing that suppression of transmission with an inhibitor of electro-secretory coupling, e.g., magnesium, attenuated hypoxic damage in hippo-campal neurons. Subsequent findings pointed to excitatory amino acids (glutamate, aspartate) as key element in the selective vulnerability of neurons subjected to hypoxia, convulsion, and hypoglycemia (Rothman 1984; Collins 1987; Choi 1988a,b, 1990; Olney 1990). The patterning of neuropathology which is specific for each condition can be replicated experi-mentally by administration of excitatory amino acids (or their synthetic analogues) to the target structures and by intense electrical stimulation of the amino acid-containing pathway which innervates susceptible target areas such as the hippocampus (Sloviter 1983). Furthermore, the development of condition-specific neuropathology can be prevented by transsectioning the glutamatergic input to target structures (Johansen et al. 1986; Onodera et al. 1986) or by pretreatment with excitatory amino acid receptor anta-

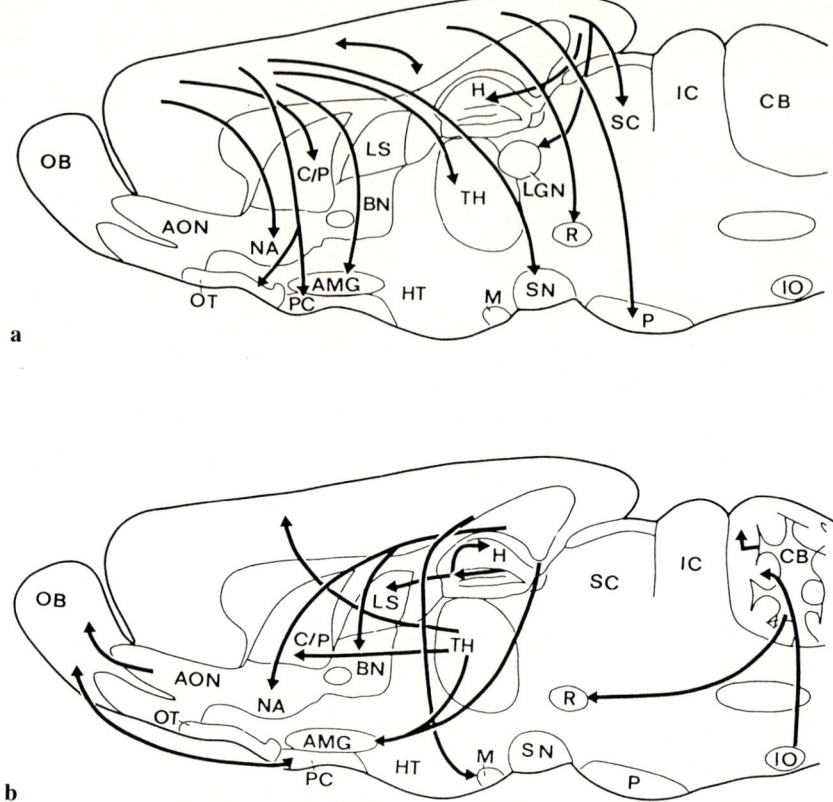

Fig. 1a,b. Excitatory amino acid-using pathways in the rat brain (after COTMAN et al. 1987): **a** corticofugal pathways; **b** allocortical and subcortical pathways. *AMG*, amygdala; *AON*, anterior olfactory nucleus; *BN*, bed nucleus of the stria terminalis; *C/P*, caudate/putamen; *CB*, cerebellum; *H*, hippocampus; *HT*, hypothalamus; *IC*, inferior colliculus; *IO*, inferior olivary complex; *LGN*, lateral geniculate nucleus; *LS*, lateral septum; *M*, mammillary bodies; *NA*, nucleus accumbens; *OB*, olfactory bulb; *OT*, olfactory tubercle; *P*, pontine nucleus; *PC*, pyriform cortex; *R*, red nucleus; *SC*, superior colliculus; *SN*, substantia nigra; *TH*, thalamus

gonists (ROTHMAN 1984; SIMON et al. 1984). These findings suggest that those neuronal targets in the brain which receive important functional excitatory amino acid input (Fig. 1a,b) and express the corresponding postsynaptic dendrosomal receptor molecules [the *N*-methyl-D-aspartate (NMDA), quisqualate, and kainic acid receptor subtypes; cf. COTMAN et al. 1987] are the potential victims of excitotoxicity. Factors which modify the release and the extracellular disposition of endogenous amino acids or their interaction with the postsynaptic receptors influence the extent of hypoxic neuron damage (e.g., destruction of the substantia nigra attenuates but destruction of the locus coeruleus enhances toxic injury; cf. GLOBUS et al. 1987; DAVIS et al. 1987; antagonists at the NMDA and non-NMDA receptors partially protect against hypoxic damage, cf. WATKINS et al. 1990; LODGE and JOHNSON 1990).

B. Central Glutamatergic and Aspartatergic Pathways as Mediators of Excitotoxicity

Several methods have been combined to reveal the existence and identity of glutaminergic and aspartatergic pathways such as immunohisto/cytochemistry of antibodies against albumin-conjugated glutamate, high affinity uptake of L-[^{3}H]-glutamate and D-[^{3}H]aspartate combined with tract lesion experiments, and retrograde transport of D-[^{3}H]aspartate and electrical release of amino acids from nerve terminals before and after tract lesions. Using these methods, a number of amino acid containing projections have been proposed (COTMAN et al. 1987) which include major efferent pathways from the neocortex and hippocampus to the basal ganglia and thalamus, to pontine relay nuclei and to motor neurons of the brainstem and spinal cord (Fig. 1a,b). The origin of many of these pathways in pyramidal-type executive neurons implies that corticocortical associational systems also make use of excitatory amino acids (EAA) as do well-known links between the neocortex and allocortex (e.g., the perforant entorhinohippocampal path). Granule-type excitatory neurons intrinsic to the hippocampal formation (mossy fibers from the dentate gyrus to CA3 pyramidal cells) and intrinsic to the cerebellum (parallel fibers) appear to release aspartate and/or glutamate. Important links within the central motor system circuitries which interconnect various modulators of the basal ganglia to the thalamus and cortex have been shown to employ excitatory amino acids, e.g., the thalamostriatal, thalamo-cortical, and subthalamopallidal projections. Major efferent pathways from the hippocampus to the subcortical motor, non-motor, and limbic-related basal ganglia (n. accumbens, septum, amygdaloid complex, and n. basalis Meynert) take up excitatory amino acids. Finally, the well-known climbing fiber projections from the inferior olive to the cerebellum have been shown to transport D-aspartate.

Current evidence suggests that excessive, prolonged, and poorly controlled release of glutamate and/or aspartate from nerve terminals onto susceptible dendrosomatic receptors and insufficient glial clearance of these amino acids from the extracellular space and breakdown of surrounding GABA-ergic inhibition transform the excitatory transmitters into potential toxins (OLNEY 1990). The nature of the receptors and their concentration in the postsynaptic membrane of target neurons is an important determinant of the vulnerability of neurons innervated by glutamatergic and/or aspartatergic systems. This is illustrated by the distribution of NMDA and non-NMDA (i.e., quisqualate and kainate) binding sites in the hippocampal formation: NMDA receptors are enriched in the stratum oriens and radiatum of CA1

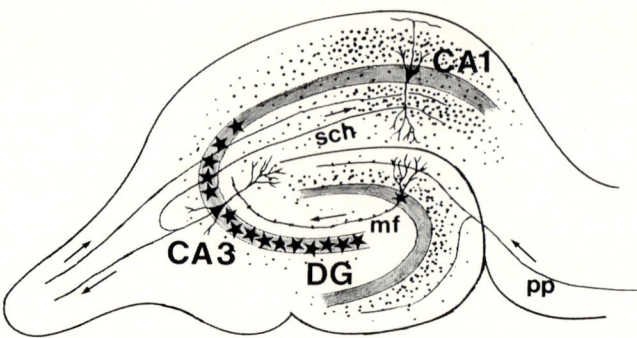

Fig. 2. Structures of the hippocampal formation and subfields CA1 and CA3 and transitional area (hilus) to the dentate gyrus (*DG*). Pyramidal neurons of the hippocampus are outlined; their intrahippocampal axon collaterals (Schaffer collateral, *sch*) extend through CA1 and CA3; they overlap with terminal fields of extrinsic afferents from the entorhinal area (perforant path, *pp*), which use excitatory amino acid transmitters. Also shown are the granule cells of the dentate gyrus and the mossy fiber (*mf*) terminal domains in CA3 and CA2. The granule-type dentate neurons use glutamate as transmitter. High-density areas of *N*-methyl-D-aspartate-sensitive sites (*dotted areas*) and of kainate-sensitive sites (*stars*) are shown. (Modified from COTMAN et al. 1987)

(site of termination of the Schaffer collaterals and the commissural projections), quisqualate receptors are confined to CA1 pyramidal cells, whereas kainate-sensitive sites are concentrated within the CA3 termination zone of the dentate granule cell projections (COTMAN et al. 1987; Fig. 2). This regional selectivity in the concentration of receptor subtypes may have relevance to the preferential vulnerability patterns of hippocampal neurons subjected to hypoxia/hypoglycemia (preferential damage of CA1 pyramidal neurons) and to certain forms of epilepsy (preferential destruction of CA3 pyramidal neurons in rodents; cf. COLLINS 1987). Such differences may vanish with time when repeated epileptic insults cause a loss of the entire pyramidal neuron population of CA1 and CA3, resulting in the well-known Ammon's horn sclerosis; under such conditions, the spread of epileptic activity to other limbic structures (via glutamatergic mechanisms) may cause extensive neuron damage (COLLINS 1987). The spread of damage beyond the termination fields of primarily affected neurons to synaptically linked neurons and their projection territories reflects the potential risk of transsynaptic degeneration of neurons interconnected in functional circuits, a mirror of "system degeneration" (Fig. 3) in the human (e.g., in Parkinson's dementia, Alzheimer's disease, progressive supranuclear palsy, and olivopontocerebellar atrophy; cf. OLNEY 1990; MELDRUM and GARTHWAITE 1990; SAPER et al. 1987). The induction of provoked release of glutamate/aspartate by NMDA, kainic acid, quisqualate, ibotentate, and quinolinic acid, which in addition to being releasers also directly stimulate postsynaptic EAA receptors, is a well-established experimental method for

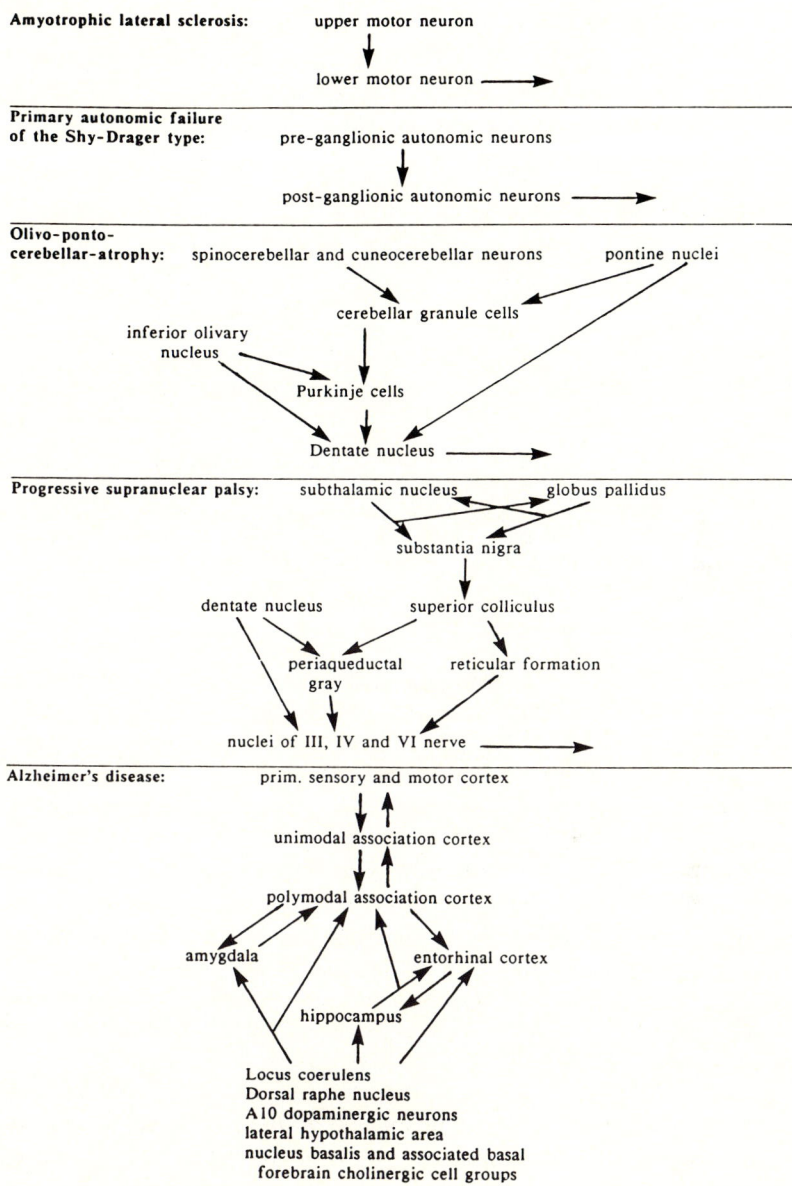

Fig. 3. Schematic representation of some connections of brain structures involved in system degeneration. (Based on SAPER et al. 1987)

region-selective destruction of neurons and their connections ("animal models of disease").

Cells overstimulated by excitotoxic input react to the depolarization of their membranes by Na^+ and H_2O influx which results in acute swelling

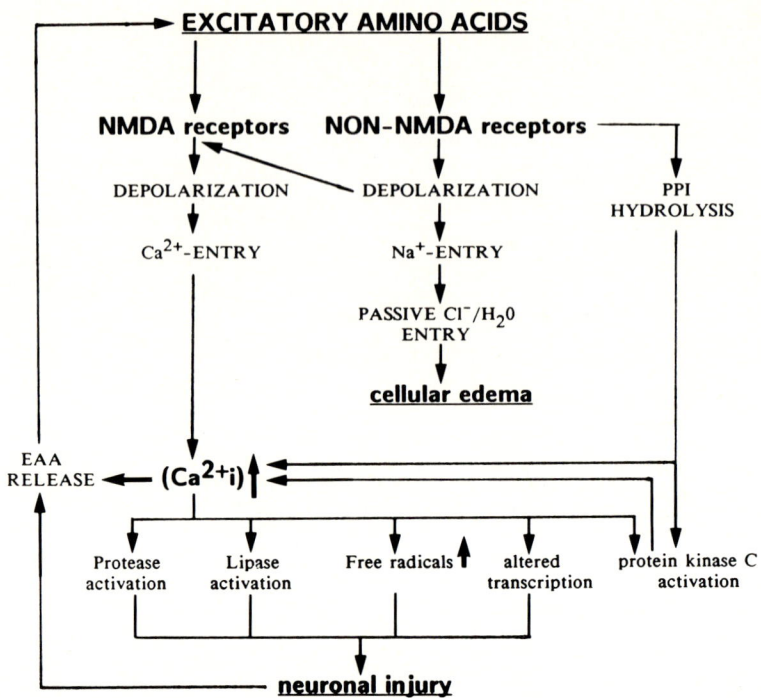

Fig. 4. Pathogenesis of excitotoxic cell damage (modified from McDonald and Johnston 1990). *EAA*, excitatory amino acids; *NMDA*, N-methyl-D-aspartate *PPI*, phosphatidylinositol

(edema); subsequently, the cells are exposed to deleterious consequences of cytoplasmic free Ca^{2+} concentration increase due to channel-mediated influx, mobilization of Ca^{2+} from internal stores (endoplasmic reticulum, mitochondria) due to activation of second messenger systems, and problems in Ca^{2+} clearance due to depletion of energy reserve or problems of ATP resynthesis (Fig. 4). Increased levels of cytoplasmatic free Ca^{2+} are known to activate autolytic enzymes which excessively degrade membrane phospholipids, the metabolism of which generates eicosanoids and free radicals. Free radical attack on protective SH pools alters cellular redox potentials and impairs mitochondrial ATP generation. In the final phase, cytoskeletal components and membranes are degraded. Prolonged oxidative stress episodes associated with depletion of the energy reserve will compromise clearance of excitotoxic amino acids in nervous tissue and may thus be considered a common cause of neurotoxicity under a wide variety of pathological conditions. Therefore, the second phase of excitotoxic injury represents an unspecific response of neurons to various types of central nervous system (CNS) trauma (Choi 1988a,b; McDonald and Johnston 1990).

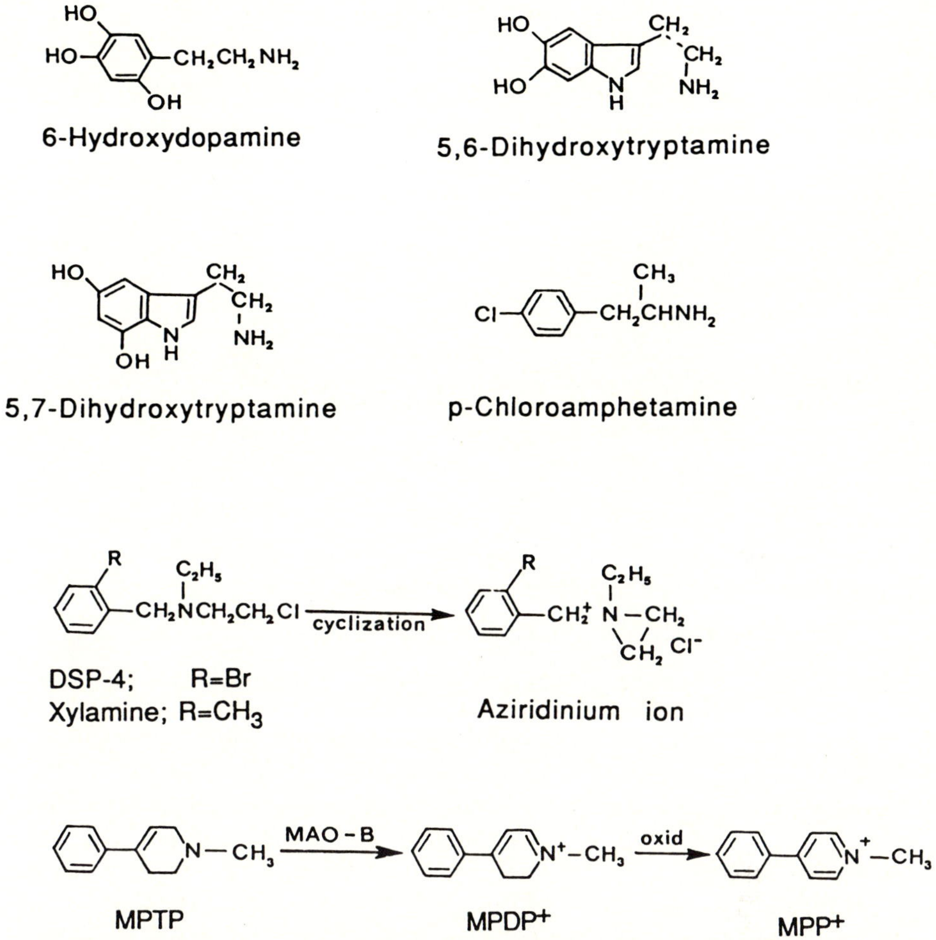

6-Hydroxydopamine

5,6-Dihydroxytryptamine

5,7-Dihydroxytryptamine

p-Chloroamphetamine

DSP-4; R=Br
Xylamine; R=CH$_3$

Aziridinium ion

MPTP

MPDP$^+$

MPP$^+$

Fig. 5. Chemical structure of neurotoxic transport ligands (for details, see text). *DSP-4*, β-haloethylamine derivative of benzylamine; *MPTP*, 1-methyl-4-phenyl-1, 2, 3,6-tetrahydropyridine; *MPDP$^+$*, 1-methyl-4-phenyl-2,3-dihydropyridine; *MPP$^+$*, 1-methyl-4-phenylpyridinium ion; MAO-β, monoamine oxidase β

C. Role of Active Transport Systems as Mediators of Selective Neurotoxicity

More selective and restricted injury is, however, documented in neurons that inactivate their natural transmitter by means of a membrane-bound, high-affinity, energy- and sodium-dependent, monoamine transporter (Ross 1987) and that are exposed to structural analogues (Fig. 5) which compromise cellular antioxidant defense, ionic homeostasis, and/or energy balance (for details, see BAUMGARTEN and ZIMMERMANN, this volume).

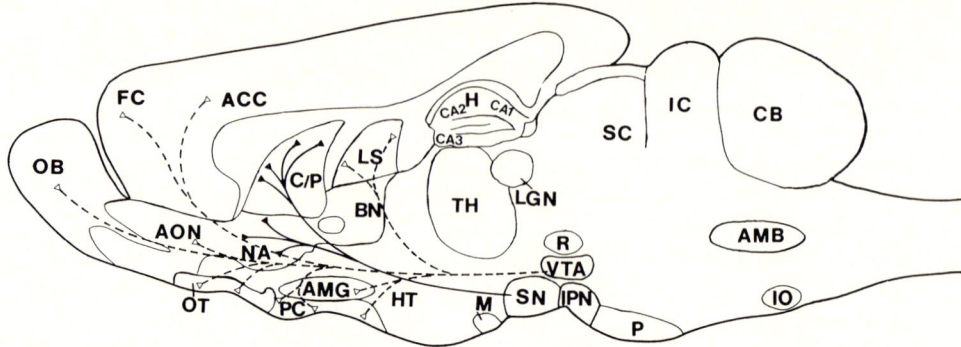

Fig. 6. Schematic representation of the major ascending dopaminergic projections: the nigrostriatal pathway originating in the substantia nigra and the mesolimbic/ mesocortical pathway originating in the ventromedial tegmental area of Tsai (based on LINDVALL and BJÖRKLUND 1974). Abbreviations cf. Figs. 1, 7, 9. *FC*, frontal cortex; *ACC*, anterior cingulate cortex

D. The Central Dopaminergic Systems as Targets of Neurotoxins

The ascending dopaminergic projections originating in the substantia nigra pars compacta and in the ventromedial tegmental area of Tsai, the nigro-striatal, the mesolimbic, and mesocortical pathways (UNGERSTEDT 1971; LINDVALL and BJÖRKLUND 1974; Fig. 6), are sensitive to intracerebrally microinjected 6-OH-dopamine (6-OH-DA; in rodents and in primates), whereas the neurodegenerative toxicity of 1-methyl-4-phenyl-1,2,3,6-tetrahydropyridine (MPTP) in the primate affects only the nigrostriatal system (BANKIEWICZ et al. 1986; BURNS et al. 1983). The nigrostriatal DA system is also more sensitive than the mesolimbic and -cortical system to intraventricularly or intracerebrally applied 6-OH-DA (SACHS and JONSSON 1975). These findings point to clinically relevant sensitivity differences of dopaminergic subsystems to neurotoxins which are most easily explained by differences in the affinity of the DA carriers in these neurons to the toxins or, alternatively, differences in the number of carrier sites per neuron type. It is now acknowledged that there is only limited anatomical separation of either DA system because some mesocortical and mesolimbic DA neurons reside in the medial portions of the substantia nigra, and some nigrostriatal neurons are located in the tegmental area of Tsai. These latter cells project to the n. accumbens (which is the limbic-related ventral striatum), whereas the bulk of nigral DA neurons project to the putamen and n. caudatus with some collaterals to the globus pallidus. The mesolimbic system densely innervates the olfactory tubercle (an important interface between the orbitofrontal neocortex and the olfactory systems), the lateral septal nucleus, bed nucleus of the stria terminalis, and the central and basal nuclei of the amygdaloid complex. The mesocortical DA projections innervate the medial part of the frontal lobe, the prepiriform and piriform, entorhinal, and anterior cingulate cortex (in rodents) which are components of limbic system circuitry. In primates and man, the mesocortical DA system is more expansive and supplies the motor, premotor, and supplementary motor areas and also (less densely) much of the primary sensorimotor and parietotemporal association areas (BERGER et al. 1991), some of which receive no DA afferents in rodents (motor and parietotemporal assoc. cortex). The DA subsystems may also differ with respect to certain pharmacological properties: The mesocortical DA pathways have been claimed to lack autoreceptors and DA transport sites and to have a higher DA turnover rate than the remaining DA subsystems (BANNON et al. 1981; DE KEYSER et al. 1990).

The exclusive neurodegenerative toxicity of MPTP in the nigrostriatal DA system with sparing of the remaining DA systems has made this compound the model drug of choice for imitating the neuropathology and symptomatology in humans affected by idiopathic Parkinson's disease. The time course of uptake and transport of radio-labelled MPTP within the primate brain in relationship to the temporal and regional evolvement of toxicity in terminals, nonterminal axons, and perikarya of the nigrostriatal DA system clearly indicates that uptake of the charged MAO-B-metabolite of MPTP, (1-methyl-4-phenylpyridinium ion, MPP^+) occurs mainly at the nerve terminals; subsequently, the radiolabel is retrogradely transported towards the cell bodies over several days concomitant with rapid loss of terminals and retrogradely progressing neuropathology in the nonterminal axons (giant axonal dilatations) with seemingly unaffected structure of the neuronal perikarya during the first 10 days after systemic administration of labelled MPTP (HERKENHAM et al. 1991). The cells undergo final degeneration within weeks to months after drug treatment; a substantial portion of the axon is lost by a dying-back process, and the neuron fails to maintain a minimal axonal volume by collateral sprouting. Previous postulates on the role of neuromelanin as a redox cycling system which generates radicals have not been substantiated (cf. KOPIN, this volume), but recent evidence from studies in cell cultures and in mitochondrial preparations point to inhibition of mitochondrial respiration at the rotenone-sensitive site (complex I) as a key mechanism of intracellular action of the charged metabolite MPP^+ which is readily trapped by positively charged molecules in the inner space of the mitochondria (RAMSAY and SINGER 1986; SAYRE et al. 1990).

E. The Central Noradrenergic Systems as Targets of Neurotoxins

The original proposal by UNGERSTEDT (1971) that the central noradrenergic (NA) systems are composed of two complementary, overlapping, and

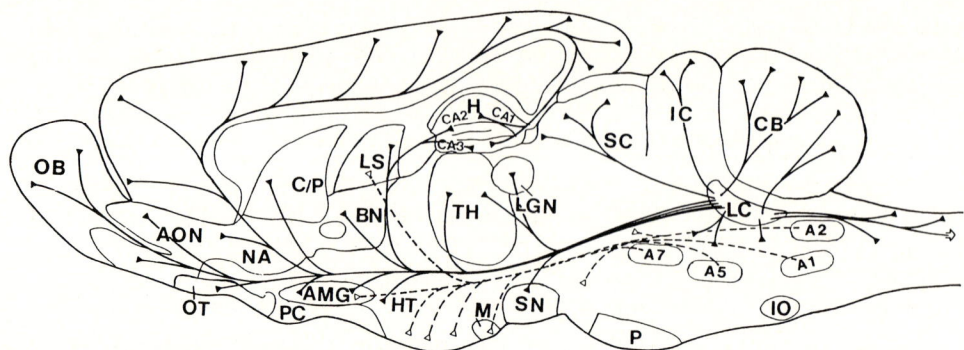

Fig. 7. The dorsal noradrenergic (NA) projection pathway (*solid lines*) originating in the NA cell group A6 (locus coeruleus) is the primary target of systemically administered DSP-4, systemically administered 6-hydroxy-dopamine (6-OH-DA) given perinatally, or intraventricularly administered 6-OH-DA. After systemic administration of DSP-4, the projections of the ventral tegmental NA system are unaffected (for details, see text). *A1, A5, A7*, lateral tegmental NA cell group; *A2*, dorsal medullary NA cell group (origin of the ventral NA projection system); *AMG*, amygdala; *AON*, anterior olfactory nucleus; *BN*, bed nucleus of the stria terminalis; *C/P*, caudate/putamen; *CB*, cerebellum; *H*, hippocampus; *HT*, hypothalamus; *IC*, inferior colliculus; *IO*, inferior olivary complex; *LC*, nucleus locus coeruleus NA cell group (origin of the dorsal NA projection system); *LGN*, lateral geniculate nucleus; *LS*, lateral septum; *M*, mammillary bodies; *NA*, nucleus accumbens; *OB*, olfactory bulb; *OT*, olfactory tubercle; *P*, pontine nucleus; *PC*, piriform cortex; *SC*, superior colliculus; *SN*, substantia nigra; *TH*, thalamus

yet distinct projection systems [a *dorsal system* originating in the n. locus coeruleus (LC) A$_6$, subcoeruleus area, and subependymal cell group A$_4$ and a *ventral system* originating in the dorsal medullary (solitarian) cell group A$_2$ and in the lateral tegmental cell groups A$_1$, A$_5$, and A$_7$; cf. Fig. 7 for details] has received elegant support by recent studies revealing exclusive and almost total disintegration of the LC-complex projections with sparing of the ventral pathway and its terminations following N-(2-chloroethyl)-N-ethyl-2-bromobenzylamine (DSP-4) administration in the adult rat (Grzanna et al. 1989; Fritschy et al. 1990); the reason for this remarkable subsystem selectivity is the higher affinity of the dorsal NA neuron monoamine transporter for this neurotoxic β-haloalkylamine derivative, thus testifying to the existence of pharmacological differences between separate systems of NA neurons. The LC NA projections are also more sensitive to the toxic actions of intraventricularly administered 6-OH-DA than the axonal projections of the ventral NA pathway.

By means of multiple axon collaterals and polarized ascending and descending main axon branches, some LC NA neurons provide projections to multiple targets along the entire neuraxis. Other neurons of the LC are less collateralized and provide more restricted projections to functionally interrelated targets. Major termination sites of this expansive reticular type neuron system comprise the neocortex, hippocampus, almost all areas of the dorsal thalamus, the anterior, ventral, and lateral nuclear complex and the geniculate bodies, the central and basal nuclei of the amygdaloid complex, the cortical and subcortical areas of the olfactory system and limbic-related grisea (diagonal band, medial septal nucleus, and bed nucleus of stria terminalis), brainstem nuclei involved in sensory integration, the raphé serotonergic nuclei, the cerebellar cortex, and deep nuclei and neurons in the ventral parts of the dorsal horn, the intermediate zone (sparing the intermediolateral column), and the ventral horn of the spinal cord. By contrast, the projections of the caudal medullary and lateral tegmental NA cell groups are less expansive and mainly confined to basal parts of the telodiencephalon providing most of the NA input to hypothalamic and preoptic nuclei, lateral septum, and periventricular thalamus. Visceromotor brainstem nuclei (general and special visceromotor nuclei) receive fibers from the noncoerulean neurons as do the preganglionic sympathetic neurons of the spinal cord; vagal and glossopharyngeal viscerosensory integration centers are also targets of the noncoerulean NA cells.

The ventral NA projection pathway is sensitive to the neurotoxic action of high doses of 6-OH-DA injected into the CSF (lateral or IVth ventricle) and directly into the brain parenchyma. In contrast to the dopaminergic neurons which are sensitive to retrograde damage and disintegration following substantial injury to their axons (after MPTP or 6-OH-DA), the NA neurons appear less sensitive to the loss of their parent axon terminals (e.g., after DSP-4 or intraventricular 6-OH-DA), possibly because the extensive branching of axons favors reestablishment of the axonal tree by collateral sprouting (JONSSON 1983). The immature LC NA neurons have, however, been found sensitive to direct toxic effects of high doses of 6-OH-DA (Fig. 8) and also to retarded retrograde degeneration following primary loss of substantial portions of their axon volume (SIEVERS et al. 1983, 1985). This acute toxic degeneration may be caused by covalent modification of functionally essential macromolecules in the cells by electrophilic quinoidal intermediates of 6-OH-DA autoxidized in the cells following active uptake by the NA carrier and by peroxidative and free radical damage at the external face of the cell membrane (for details, see BAUMGARTEN and ZIMMERMANN, this volume).

F. The Central Serotonergic Systems as Targets of Neurotoxins

Two parallel, complementary, and extensively overlapping and collateralized projection systems which originate in the dorsal and median raphé nuclei (Fig. 9) provide serotonergic afferents to almost all telodiencephalic target centers (STEINBUSCH 1981; AZMITIA and GANNON 1986; TÖRK 1990; BAUMGARTEN and LACHENMAYER 1985); the median raphé projections pre-

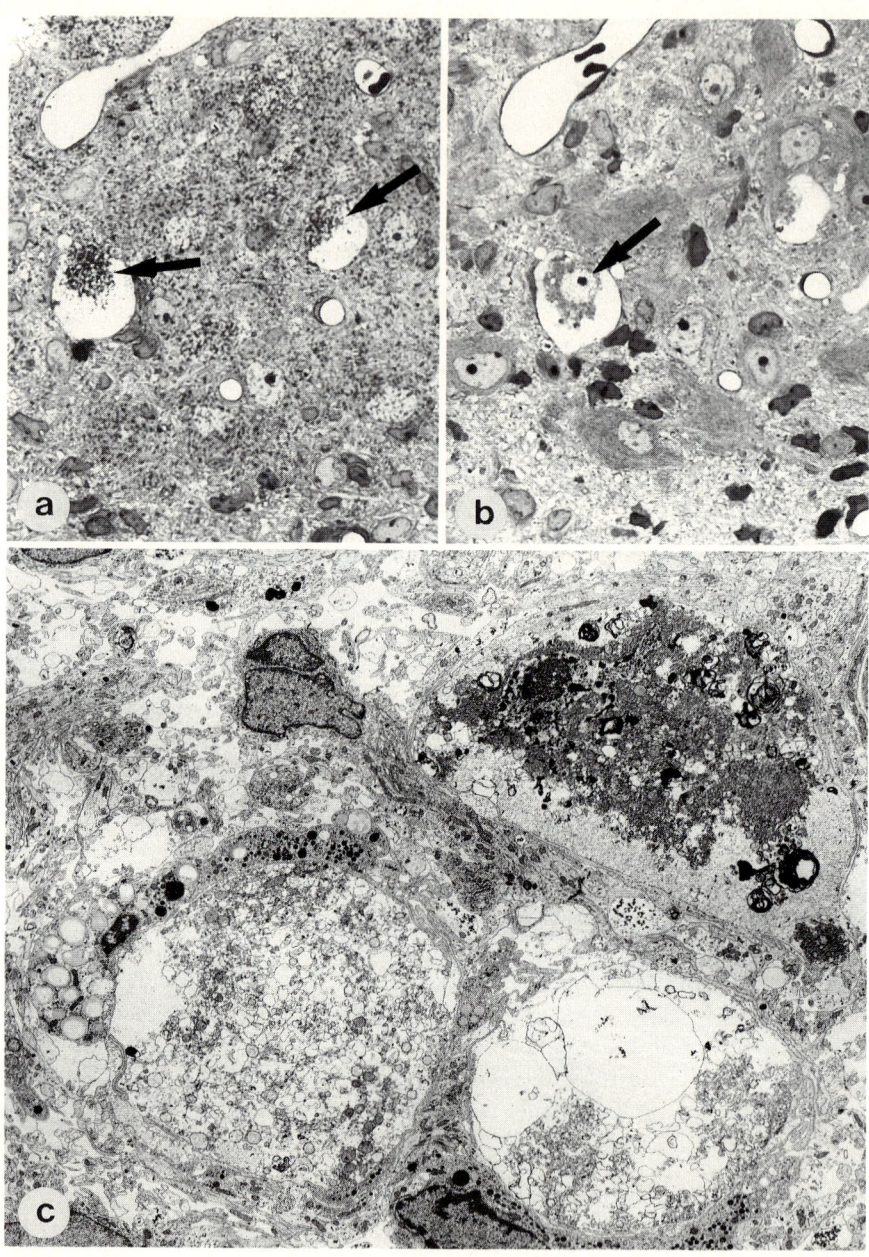

Fig. 8a–c

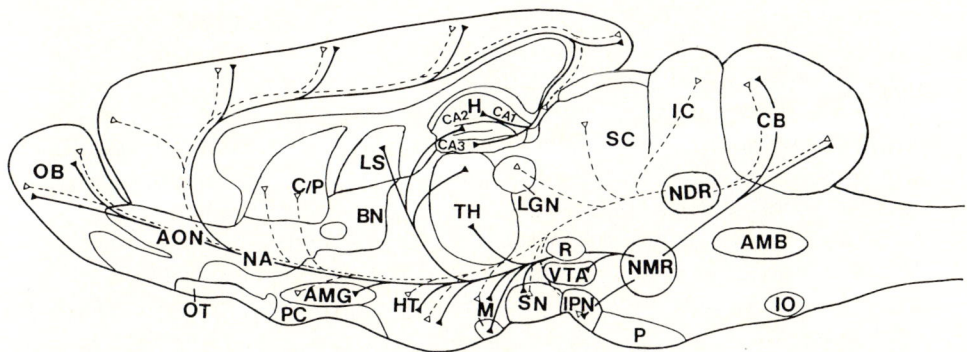

Fig. 9. Schema illustrating the two ascending reticular serotonergic pathways originating in the dorsal and median raphé nucleus of the rat which provide overlappping (complementary) innervation to telodiencephalic targets. While established "serotonin neurotoxins" like 5,7-dihydroxytryptamine do not discriminate between the two systems, the substituted amphetamines like methylenedioxymethamphetamine, methylenedioxyamphetamine, and p-chloroamphetamine affect only the dorsal raphé projection system (*dotted lines*), thereby causing long-lasting reductions in target-region 5-HT concentrations (for details, see text). *AMB*, nucleus ambiguus; *AMG*, amydala; *AON*, anterior olfactory nucleus; *BN*, bed nucleus of the stria terminalis; *C/P*, caudate/putamen; *CB*, cerebellum; *H*, hippocampus; *HT*, hypothalamus; *IC*, inferior colliculus; *IO*, inferior olivary complex; *IPN*, interpeduncular nucleus; *LGN*, lateral geniculate nucleus; *LS*, lateral septum; *M*, mammillary bodies; *NA*, nucleus accumbens; *NDR*, dorsal raphé nucleus; *NMR*, medial raphé nucleus; *OB*, olfactory bulb; *OT*, olfactory tubercle; *P*, pontine nucleus; *PC*, piriform cortex; *R*, red nucleus; *SC*, superior colliculus; *SN*, substantia nigra; *TH*, thalamus; *VTA*, ventral tegmental area

ferentially innervate basal telodiencephalic areas (preoptic, septal, and olfactory areas), the superficial laminae of the neocortex, and the hippocampal formation. The dorsal raphé projections supply the entire neocortex (deep laminae), basal ganglia, thalamic nuclei, and amygdaloid complex.

Fig. 8. a Autoradiogram of frontal section through the locus coeruleus (LC) of a newborn rat, 6 h after intracisternal (IVth ventricle) injection of [³H]6-OH-DA; note intense labelling of noradrenergic perikarya within the ventral part of the LC (*arrows* point to large vacuoles which reveal clustered radiolabel in an excentric position). As revealed in the consecutive unexposed section stained with toluidine blue/pyronine red **b**, remnants of nucleus and cytoplasm (*arrow*) are located in those regions of the vacuoles which are heavily labelled in Fig. 1a. As evident in the accompanying electron micrograph **c**, these morphological alterations reflect toxic damage and degeneration of 6-OH-DA-treated noradrenergic neurons. One cell shows large accumulations of electrondense material, myelinoid inclusions, and a pyknotic nuclues; two neuronal perikarya are fragmented into irregular assemblies of membranes and vacuoles. **a** and **b** are taken from SIEVERS et al. (1983) (×880); **c** is taken from SIEVERS et al. (1980) (×3185)

Tracing studies reveal that clusters of neurons within these two major cranial raphé nuclei have restricted terminal fields within functionally related centers throughout the telodiencephalon, giving serotonergic subsystems topological specificity. The laminar selectivity of the terminal innervation within the neocortex by either projection system suggests some specificity in the relationships of either system to innervated targets: the median raphé neurons appear to influence nonspecific thalamic afferences and various types of cortical intrinsic inhibitory neurons (stellate, bipolar, and granule cells), whereas the dorsal raphé neurons seem to influence specific thalamic afferences and pyramidal type neurons. Both systems differ in their sensitivity to the toxic actions of certain substituted amphetamines: methylenedioxymethamphetamine (MDMA), methylenedioxyamphetamine (MDA), and p-chloroamphetamine (p-CA) preferentially or exclusively affect the telodendritic terminals of the dorsal projection system in the rodent (rat) and primate (New and Old World monkey and human) brain while sparing those of the overlapping median raphé nucleus (Mamounas and Molliver 1988; O'Hearn et al. 1988; Insel et al. 1989). The reason for the preferential vulnerability of the finely varicose axons of the dorsal system by MDMA, MDA, or p-CA may be the higher affinity of the dorsal raphé axon serotonin (5-HT) transporter for this class of compounds. The observation of prolonged postdrug psychopathology in humans following consumption of high doses of MDMA (confusion, depression, severe anxiety, blurred vision, paranoia, and problems in social adjustment behavior; Barnes 1988) may be a direct reflection of the psychological and behavioral deficits following selective, temporary ablation of the neocortical and limbic allocortical and subcortical serotonergic afferences from the dorsal raphé nucleus and thus mirror the important regulatory dimensions of 5-HT in affective tone, mood, perception, and cognition, i.e., functions which are difficult to evaluate and measure in laboratory rodents and which are expressed in species-specific patterns of complex behaviour. Both systems do not significantly differ in their vulnerability to the powerful neurotoxin 5,7-dihydroxytryptamine (Baumgarten et al. 1978), which has reasonably high affinity to the 5-HT transporter of either system (for details, see Baumgarten and Zimmermann, this volume). There is evidence that deficits in serotonin transmission (pre- and postsynaptic markers) may contribute to the psychopathology, test performance, and complex behavior deficits in brain organic disease (in, e.g., ageing and Alzheimer's disease; cf. Gottfries 1991).

G. The Central Cholinergic Systems as Targets of Neurotoxins

The reticular-type, multipolar neurons of the brainstem (pontomesencephalic tegmental cell groups CH 5 and CH 6; parabrachialis-pedunculopontine complex) provide projections to the thalamus (sensory and motor relay nuclei, limbic-related, associational and reticular and intralaminar nuclei), superior colliculi, pretectal area, n. interpeduncularis, and ventral tegmental area of Tsai; they are part of the ascending reticular activating system (Fig. 10). The large multipolar cholinergic neurons of the basal forebrain (the medial forebrain cell groups CH 4 within the n. basalis of Meynert, CH 3 in the vertical and CH 2 in the horizontal limb of the diagonal band n. of Broca, and CH 1 in the medial septal nucleus) give rise to topographically

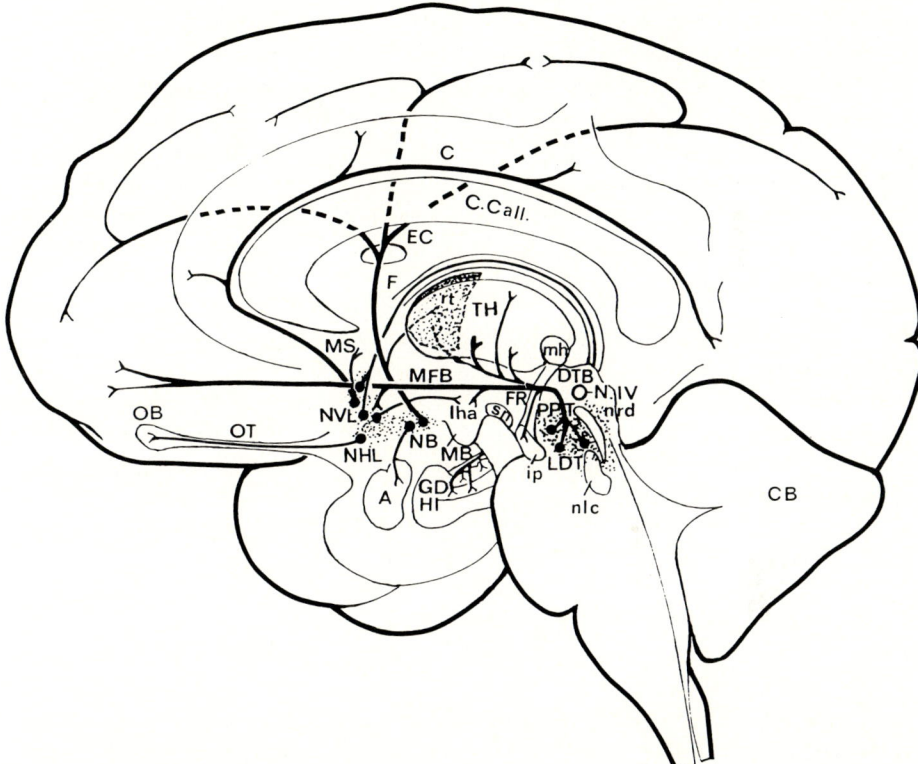

Fig. 10. The major ascending cholinergic projections arising in the parabrachialis-pedunculopontine complex (CH5 and CH6) of the upper brainstem and in the basal forebrain cell groups CH4 (n. basalis of Meynert), CH3 and CH2 within the diagonal band nuclei, and CH1 in the medial septal nucleus projected onto a mediosagittal plane of the human brain. Note overlapping of terminal fields from either system in the reticular thalamic nucleus (*rt*). *A*, amygdala; *C*, cingulum; *CB*, cerebellum; *C. Call*, corpus callosum; *DTB*, dorsal tegmental bundle; *EC*, external capsule; *F*, formix; *FR*, fasciculus retroflexus; *GD*, gyrus dentatus; *HI*, hippocampus; *ip*, interpeduncular nucleus; *LDT*, laterodorsal tegmental nucleus; *lha*, lateral hypothalamic area; *MB*, mammillary body; *MFB*, medial forebrain bundle; *mh*, medial habenula; *MS*, medial septum; *N.IV*, n. nervi trochlearis; *NB*, nucleus basalis; *NHL, NVL*, nucleus of the horizontal and vertical limb of the diagonal band of Broca; *nlc*, locus coeruleus; *nrd*, nucleus raphé dorsalis; *OB*, olfactory bulb; *OT*, olfactory tract; *PPT*, pedunculopontine tegmental nucleus; *sn*, substantia nigra; *TH*, thalamus

organized cortical projections, provide cholinergic input to the reticular
thalamus, septum, habenula, and hippocampus, and should be considered
as the telodiencephalic equivalent of the ascending reticular activating
system (Semba and Fibiger 1989), the loss of which results in learning and
memory deficits and problems in neocortical arousal-dependent activation of
adaptive, intelligent behavior due to insufficient suppression of the reticular
thalamic oscillator which deactivates neocortical neurons (Buzsáki et al.
1988; Vanderwolf 1988). It is thus not surprising that disease-related
degeneration of the basal cholinergic neurons in man are associated with
cognitive and psychometric test performance deficits (e.g., in Alzheimer's
disease and Parkinson's dementia syndrome; Bartus et al. 1982; Whitehouse
et al. 1981). The cholinergic basal forebrain neurons are sensitive to
excitotoxic destruction by local kainate or ibotenate administration (Buzsáki
et al. 1988; Stewart et al. 1984) and to alkylating mustards (Walsh et al.
1984), which are choline analogues (e.g., AF64A), taken up into cholinergic
neurons via the choline carrier, i.e., most efficiently at the presynaptic
cholinergic nerve terminals.

H. Cholinoceptive Mechanisms as Mediators of CNS Toxicity

Acetylcholine (ACh) binding sites on cholinergic nerve terminals or sites
postsynaptic to cholinergic neurons (nicotine receptors) are well-known
targets of different classes of neurotoxins:

– presynaptic: some bacterial and venom toxins (e.g., botulinum toxin,
  black widow spider venom)
– postsynaptic: muscarine, antiacetylcholinesterases (physostigmine,
  organophosphates, carbamates, nerve gases), atropine, most snake and
  spider venoms)

Several members of the postsynaptic class of neurotoxins are snake venoms,
so-called α-neurotoxins, which all act via the same mechanism, i.e., block
of the neuromuscular function by binding to the nicotinic acetylcholine
receptor (nAChR). These toxins are proteins forming three loops, the
tertiary structure of which has been elucidated recently (reviewed by Endo
and Tamiya 1987). Examples of this type of toxin comprise erabutoxin, α-
cobratoxin, and α-bungarotoxin. They bind to the α-subunit of the whole
receptor complex which carries the ACh binding site (Low and Corfield
1986, 1988). Binding of these toxins to the presynaptic (auto)receptor has
been suggested based on the demonstration of an increase in end-plate
potential concomitant with continuation of the tetanic fade shortly after
washout of the toxin (Chang and Hong 1987).

   Not all of the α-neurotoxins that block the ACh receptor at the neuromuscular junction produce the same functional deficit in the CNS; for some

of these, much higher concentrations of the toxin are required to inhibit ACh-mediated transmission in the CNS. Very specific but different froms of toxin-receptor interaction with any of these neurotoxins may be the reason for species differences in toxin sensitivity. Species differences may also depend on regional differences in brain neurochemistry and architecture (CHIAPPINELLI 1985). In addition, there exist certain types of neurotoxins which bind only to the central ACh receptor, e.g., the k-subtype of bungarotoxin (CHIAPPINELLI 1985; MEBS 1988).

Other naturally occurring neurotoxins with targets in the CNS are widely distributed in the animal kingdom, and nearly every possible target

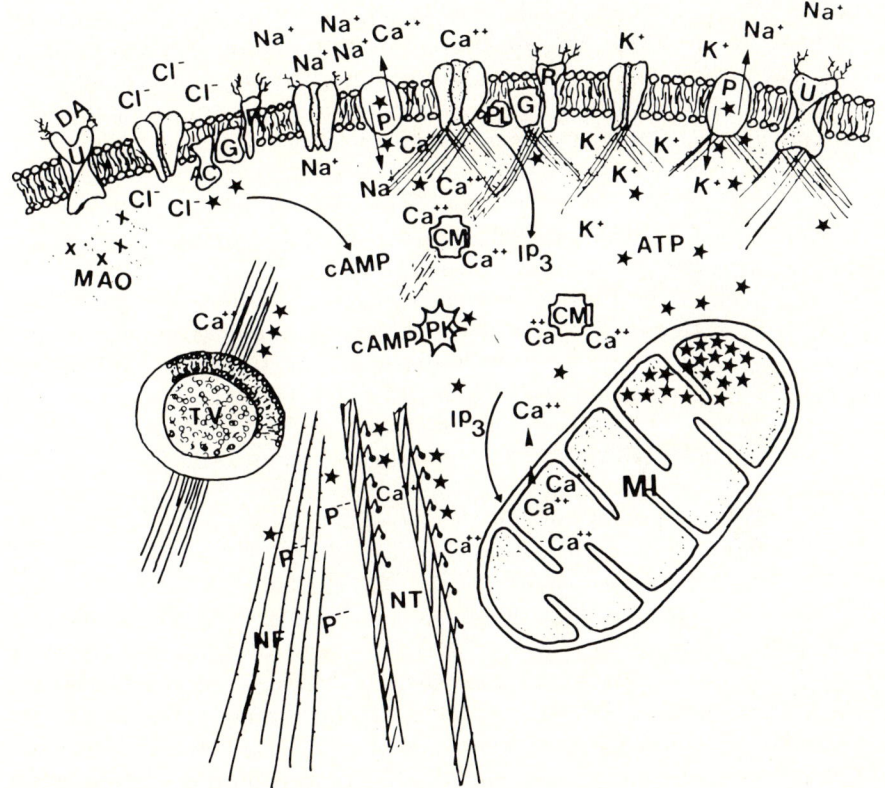

Fig. 11. Schematic presentation of some of the cellular tatgets for neurotoxins. Channel proteins in the phospholipid bilayer for Cl^-, Na^+, Ca^{2+}, and K^+, with anchorage cytoskeletal filaments (actin, α-actinin); *R*, receptor protein; *U*, uptake receptor protein; *P*, ionic pumps; *G*, guanosine triphosphate-containing protein; *PL*, phospholipase; *AC*, adenylate cyclase; *DA*, dopamine (as an example for transmitter uptake); *MAO*, monoamine oxidase; *cAMP*, cyclic adenosine monophosphate; *CM*, calmodulin; IP_3, inositol triphosphate; *PK*, phosphokinase; *NF*, neurofilaments; *NT*, neurotubules; P^{2-}, phosphate groups; *MI*, mitochondrium; *stars* represent *ATP*, adenosine triphosphate; *TV*, transport vesicle

has been chosen during evolution (e.g., ion channels, receptors, and uptake/ transport structures) (Fig. 11). These include voltage-sensitive sodium channels (Couraud et al. 1986; Brown 1988) and voltage-dependent, Ca^{2+}- triggered potassium channels (Seagar et al. 1984, 1988; Harvey and Anderson 1985; Halliwell et al. 1986).

J. Vulnerability of the CNS to Antiniacin Compounds

Toxic effects of antimetabolites on the nervous system become manifest as severe neurological disorders (irreversible paralysis, seizures) following treatment of experimental animals with 6-aminonicotinamide (6-AN) (Wolf et al. 1959; Herken et al. 1973; Gaitonde et al. 1987). 6-AN has distinct cytotoxic effects on glial cells, resulting in cell necroses in the CNS (Herken et al. 1973; Politis 1989; Aikawa et al. 1986), but neurons are also affected (Huck and Stumpf 1985). It blocks the pentose phosphate pathway, which is probably most important in brain energy metabolism, by inhibiting 6-phosphogluconate dehydrogenase, leading to drastic accumulation of 6-phosphogluconate; 6-AN shares the synthesis pathway for nicotinamide-adenine dinucleotide (NAD and NADP), resulting in the formation of 6-ANAD and 6-ANADP, which compete with the natural nucleotides (Herken et al. 1969; Kolbe et al. 1976). 6-ANADP, formed in dopaminergic neurons, impairs dopa formation by inhibiting the synthesis of the natural cofactor for tyrosine hydroxylase (biopterin) and alters and function of the cofactor regenerating enzyme dihydropteridin reductase (quinoid form of BH_2 to BH_4); in concent, these effects of 6-ANADP cause parkinsonian symptoms (Herken 1990), i.e., a specific neurological disorder.

   3-Acetylpyridine (3-AP) has been found to cause selective neuronal degeneration. After intraperitoneal application of high doses (close to the LD_{50}), almost all neurons within the inferior olivary nucleus (ION) are destroyed followed by degeneration of axon fibers in the tractus olivo-cerebellaris and the axon terminals of the climbing fibers (CF) in the lamina molecularis of the cerebellum (Desclin and Escubi 1974; Desclin 1976; Desclin and Colin 1980; Herken 1968). This apparently specific lesion of the ION and the concomitant degeneration of the CF in the cerebellum led to the conclusion that the CF originate solely in the ION (Desclin 1976; Desclin and Colin 1980). By the use of 3-AP, the role of the ION in motor learning was elucidated by Llinas et al. (1975). However, neurons in other CNS regions are also sensitive to 3-AP, but to a variable extent, e.g., in the nucleus ambiguus, facial and hypoglossal nuclei, substantia nigra, and nuclei of the dorsal raphé (Desclin and Escubi 1974). Balaban (1985) has reinvestigated the topological pattern of neuronal degeneration caused by 3-AP and has reviewed the results of other authors. He found additional degeneration in the dorsal motor nucleus of the vagal nerve, the nucleus intercalatus, the medial terminal nucleus, the interpeduncular nucleus, the ventral tegmental area, the hippocampus, the nucleus of the diagonal band,

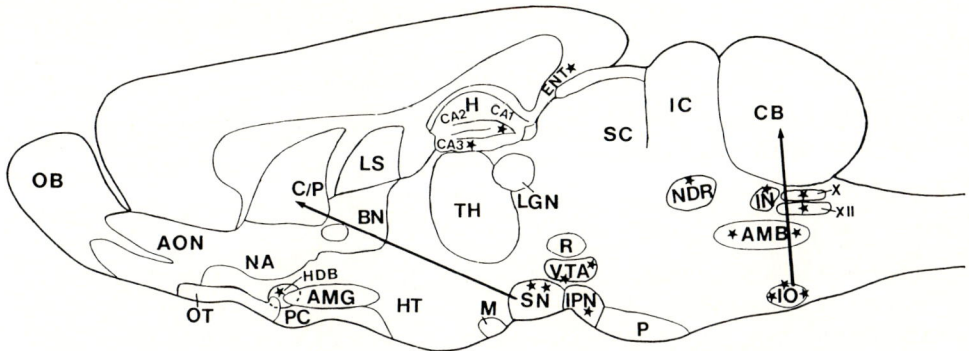

Fig. 12. Schematic representation of sites of neuronal degeneration in the rat brain after 3-acetylpyridine intoxication. *Stars*, nuclei with degenerated neurons (*** = almost complete degeneration, ** = severe degeneration, * = moderate degeneration; *arrows* = axon and terminal degeneration); *AMB*, nucleus ambiguus; *AMG*, amygdala; *AON*, anterior olfactory nucleus; *BN*, bed nucleus of the stria terminalis; *C/P*, caudate/putamen; *CB*, cerebellum; *ENT*, entorhinal cortex; *H*, hippocampus; *HDB*, horizontal limb of the nucleus of the diagonal band; *HT*, hypothalamus; *IC*, inferior colliculus; *IN*, nucleus intercalatus; *IO*, inferior olivary complex; *IPN*, interpeduncular nucleus; *LGN*, lateral geniculate nucleus; *LS*, lateral septum; *M*, mammillary body; *NA*, nucleus accumbens; *NDR*, dorsal raphé nucleus; *OB*, olfactory bulb; *OT*, olfactory tubercle; *P*, pontine nucleus; *PC*, piriform cortex; *R*, red nucleus; *SC*, superior colliculus; *SN*, substantia nigra; *TH*, thalamus; *VTA*, ventral tegmental area; *X*, dorsal motor nucleus of vagal nerve; *XII*, hypoglossal nucleus

and the lateral entorhinal cortex in layers II-V. Furthermore, degeneration of terminals originating in the substantia nigra was found in the striatum. The lesions in the hippocampus were restricted to the dentate gyrus and the area CA3/CA4. The anatomical distribution of the 3-AP-sensitive neurons is shown in Fig. 12 (BALABAN 1985). The reason for the selective vulnerability of various neurons may depend on differences in the affinity of neuronal uptake to 3-AP. With higher doses, neuron degeneration is also seen in the red nucleus (SZALAY et al. 1979) and in the amygdaloid complex (HICKS 1955). The biochemical mechanism of the degeneration by 3-AP may be similar to that by 6-AN, although the electron microscopical appearance of the lesions varies. The olivocerebellar CF which elicit complex burst spiking in Purkinje neurons (ECCLES et al. 1966) appear to use EAA (aspartate, derivatives of glutamate) as transmitters (WIKLUND et al. 1982; FOSTER and ROBERTS 1983).

K. Vulnerability of the CNS to Drugs Affecting Oxidation-Phosphorylation Coupling

Many diseases of oxidative metabolism cause encephalopathy or encephalomyopathy, indicating that nerve (and muscle) cells are particularly sensitive to disturbance of mitochondrial functions (cf. DI MAURO and DE VIVO 1989). An established toxin-sensitive portion of the energy-generating intracellular catalytic system is the coupling of NADH oxidation and AMP/ADP phosphorylation; model substances for uncoupling oxidation and phosphorylation comprise drugs like p-nitrophenol and triiodothyronine, which were used many years ago for weight reduction because they effectively inhibit ATP formation. Besides the development of hyperthermia following ingestion of these compounds, severe CNS disturbances develop (GAULTIER et al. 1974). Another drug, acetylsalicylic acid, which has been widely used for more than 100 years as an antipyretic and analgesic, also acts as an uncoupler of oxidative phosphorylation at concentrations of $1-2\,mM$ (BRODY 1956; SMITH 1959; WHITEHOUSE and HASLAM 1962; BLATTNER 1965). Such plasma levels can be reached in the treatment of rheumatoid arthritis (using $8-10\,g/day$), and neurotoxic side effects (nausea, vomiting, sedation, convulsions) have been reported after ingestion of high dosages (FREIER et al. 1957).

L. Vulnerability of Neurons Due to the Interaction of Toxins with the Cytoskeleton

Energy-dependent, bidirectional transport processes are necessary for bridging the distance between the perikaryon and the axon terminals of neurons because functioning and survival of the axon depend on perikaryal synthesis of proteins and their translocation to sites of usage and feedback transport of proteins regulating cellular m-RNA expression (e.g., by nerve growth factors). Three types of cytoskeletal elements in all cell types are involved in transport processes, axonal motility, and structural stability: microtubules, a large group of intermediate filaments, and actin. These are structures for mechanical compartmentation and distribution of cell organelles, sites of enzyme complexes, and transport structures (LENK et al. 1977; CERVERA et al. 1981; NIELSEN et al. 1983; TOMINGA and KAZIRO 1983). They play a crucial role in translocation of vesicles through the cytoplasm and mediate exocytosis of secretory products (CASE 1978; DAVIS and TAI 1980; ZIMMERMANN 1989). They have considerable importance for cell movement, division, and morphology. In neurons, specific types of intermediate filaments (neurofilaments) and microtubuli (neurotubuli) are necessary for axon growth, structural stability, and antero- and retrograde axonal transport (ALLEN et al. 1985; TOOZE and BURKE 1987; MATUS 1987). Neurofilaments and neurotubuli are sensitive to a variety of toxins which cause axonal dystrophy or degeneration.

Well-known compounds that affect cytoskeletal elements are colchicine and vincristine, which disturb microtubular integrity and function (BERGEN and BORISY 1983; BENNETT et al. 1984), and several cytochalasins, which interfere with actin polymerization (MARUYAMA et al. 1980; LIN et al. 1980; BRITCH and ALLEN 1981). Systemically applied colchicine does not induce severe neurological symptoms, but once injected into the brain it destroys neurons. Neurons at specific sites in the brain are more vulnerable than those in other areas (STEWARD et al. 1984). Other poisons, such as β, β-imidodipropionitrile, are known to affect specifically neurofilaments, leading to altered phosphorylation of polypeptide subunits and disturbance of axonal transport (DYCK et al. 1986; BRADLEY and WILLIAMS 1973; GRIFFIN et al. 1978; WATSON et al. 1989). Industrially used solvents such as toluene or hexane disturb cytoskeletal systems, thus resulting in centroperipheral neuropathies (DE CAPRIO 1985; DE CAPRIO et al. 1988; ABOU-DONIA et al. 1988). Organophosphorus compounds affect axonal transport by altering the phosphorylation status of the filament peptides (ABOU-DONIA et al. 1988). CNS symptoms caused by aluminum intoxication are due to alterations of brain neurofilaments (BIZZI and GAMBETTI 1986; OTEIZA et al. 1989).

Whereas alterations of the cytoskeleton in the PNS lead to peripheral neuropathic lesions, some toxicants like tetanus toxin or viruses, which have access to the nervous system mainly through peripheral terminals, use the retrograde transport systems to reach the perikaryon and, transsynaptically, interconnected neurons, thus producing system degeneration, a pathogenetic principle thought to be responsible for amyotrophic lateral sclerosis, Alzheimer's disease, olivopontocerebellar atrophy, and others (reviewed by SAPER et al. 1987; Fig. 3).

In conclusion, nerve cells are selectively vulnerable because they express specific proteins which mediate essential functions as e.g. neurotransmitter uptake (noradrenaline, dopamine and serotonin neurons), neurotransmitter precursor uptake (choline uptake by cholinergic neurons) or neurotransmitter receptors (NMDA- and non-NMDA receptors) which have limited substrate selectivity and therefore recognize and bind structural analogues; the structural analogues are capable of interfering with essential cellular functions (Ca^{2+} homeostasis, ionic balance, control of lysosomal enzyme activities, antiradical defense, maintenance of ATP (phosphocreatine pool) thereby mediating structural damage and cell death. Excessive, prolonged and poorly controlled release of excitatory amino acids from glutamatergic and/or aspartatergic nerve terminals combined with problems in glia-assisted removal of the released acids from the extracellular space transforms transmitters into toxins at susceptible neurons (having high dendrosomal densities of NMDA or non-NMDA receptors). The exquisite sensitivity of certain populations of nerve cells to excititoxicity by endogenous glutamate/apartate may be the cause of region- and cell-selective patterns of brain neuropathology in response to generalized forms of CNS trauma such as e.g. hypoxia, hypoglycaemia and epilepsy ("pathoklise" of VOGT and VOGT

1937). Excitotoxic damage may spread transneuronally, a potential substrate of "system degeneration". Ageing- or disease-related deficiencies in neurotrophic support may contribute to increased endo- or ektotoxin sensitivity of neurons. Neurons are particularly sensitive to toxin-induced dysfunction of voltage and Ca^{2+}-dependent ion channels, to antimetabolites, to drugs/toxins affecting oxidation-phosphorylation coupling and to anticytoskeletal toxins which impair/disrupt bidirectional axonal transport mechanisms. The application of neurotoxins to experimental animals by systemic (e.g. MPTP, substituted amphetamines, DSP-4) or by region-specific intracerebral approach (e.g. NMDA, kainate, ibotenate; 6-OH-DA; 5,7-DHT; AF64A) may provide useful animal models of human disease.

References

Abou-Donia MB, Lapadula DM, Suwita E (1988) Cytoskeletal proteins as targets for organophosphorous components and aliphatic hexacarbon-induced neurotoxicity. Toxicology 49:469–477

Aikawa H, Sobayashi S, Suzuki K (1986) Aqueductal lesions in 6-aminonicotinamide-treated suckling mice. Acta Neuropathol (Berl) 71:243–250

Allen RD, Weiss PS, Hayden JH, Brown DT, Fujiwake H, Simpson M (1985) Gliding movement of a bidirectional organelle transport along single native microtubules of squid axoplasm: evidence for an active role of microtubules in cytoplasmic transport. J Cell Biol 100:1736–1752

Azmitia EC, Gannon PJ (1986) The primate serotonergic system: a review of human and animal studies and a report on *Macaca fascicularis*. Adv Neurol 43:407–468

Balaban CD (1985) Central neurotoxic effects of intraperitoneally administered 3-acetylpyridine, harmaline and niacinamide in Sprague-Dawley and Long-Evans rats: a critical review of central 3-acetylpyridine neurotoxicity. Brain Res Rev 9:21–42

Bankiewicz KS, Oldfield EH, Chiueh CC, Doppman JL, Jacobowitz DM, Kopin IJ (1986) Hemiparkinsonism in monkeys after unilateral carotid artery infusion of 1-methyl-4-phenyl-1,2,3,6-tetrahydropyridine (MPTP). Life Sci 39:7–16

Bannon MJ, Michaud RL, Roth RH (1981) Mesocortical dopamine neurons: lack of autoreceptors modulating dopamine synthesis. Mol Pharmacol 19:270–275

Barnes DM (1988) New data intensify agony over ecstasy. Science 239:864–866

Bartus RT, Dean RL, Beer B, Lippa AS (1982) The cholinergic hypothesis of geriatric dysfunction. Science 217:408–417

Baumgarten HG, Lachenmayer L (1985) Anatomical features and physiological properties of central serotonin neurons. Pharmacopsychiatry 18:180–187

Baumgarten HG, Klemm HP, Lachenmayer L, Björklund A, Lovenbergy W, Schlossberger HG (1978) Mode and mechanism of action of neurotoxic indoleamines: a review and progress report. Ann NY Acad Sci 305:3–24

Bennett G, Parson S, Carlet E (1984) Influence of colchicine and vinblastine on the intracellular migration of secretory and membrane glycoproteins. I. Inhibition of glycoprotein migration in various rat cell types as shown by light microscope radioautography after injection of ^{3}H-fucose. Am J Anat 170:521–530

Bergen LG, Borisy GG (1983) Tubulin-colchicine complex inhibits microtubule elongation at both plus and minus ends. J Biol Chem 258:4190–4194

Berger B, Gaspar P, Verney C (1991) Dopaminergic innervation of the cerebral cortex: unexpected differences between rodents and primates. Trends Neurosci 14:21–27

Bizzi A, Gambetti P (1986) Phosphorylation of neurofilaments is altered in aluminium intoxication. Acta Neuropathol (Berl) 71:154–158

Blattner RJ (1965) Comments on current literature. Teratogenic action of salicylates in mice. Effect on mucopolysaccharide synthesis. J Pediatr 66:1102–1104

Bradley WG, Williams MH (1973) Axoplasmatic flow in axonal neuropathies. Brain 96:235–246

Britch M, Allen TD (1981) The effects of cytochalasin B on the cytoplasmic contractile network revealed by whole-cell transmission electron microscopy. Exp Cell Res 131:121–172

Brody TM (1956) Action of sodium salicylate and related compounds on tissue metabolism in vitro. J Pharmacol Exp Ther 117:39–51

Brown GB (1988) Batrachotoxin, a window on the allosteric nature of the voltage-sensitive sodium channel. Int Rev Neurobiol 29:77–116

Burns RS, Chiueh CC, Markey SP, Ebert M, Kopin IJ (1983) A primate model of parkinsonism; selective destruction of dopaminergic neurons in the pars compacta of the substantia nigra by N-methyl-4-phenyl-1,2,3,6-tetrahydropyridine. Proc Natl Acad Sci USA 80:4546–4550

Buzsáki G, Bickford RG, Ponomareff G, Thal LJ, Mandel R, Gage FH (1988) Nucleus basalis and thalamic control of neocortical activity in the freely moving rat. J Neurosci 8:4007–4026

Case RM (1978) Synthesis, intracellular transport and discharge of exportable proteins in the pancreatic acinar cells and other cells. Biol Rev 53:211–354

Cervera M, Dreyfuss G, Denman S (1981) Messenger RNA is translated when associated with the cytoskleletal framework in normal and VSV-infected HeLa cells. Cell 23:113–120

Chang CC, Hong SJ (1987) Dissociation of the end-plate potential run-down and the tetanic fade from the postsynaptic inhibition of acetylcholine receptor by alpha-neurotoxins. Exp Neurol 98:509–517

Chiappinelli VA (1985) Actions of snake venom toxins on neuronal nicotinic receptors and other neuronal receptors. Pharmacol Ther 31:1–32

Choi DW (1988a) Glutamate neurotoxicity and diseases of the nervous system. Neuron 1:623–634

Choi DW (1988b) Calcium-mediated neurotoxicity: relationship to specific channel types and role in ischemic damage. Trends Neurosci 11:465–469

Choi DW (1990) Cerebral hypoxia: some new approaches and unanswered questions. J Neurosci 10:2493–2501

Collins RC (1987) Neurotoxins and the selective vulnerability of brain. In: Jenner P (ed) Neurotoxins and their pharmacological implications. Raven, New York, pp 1–17

Cotman CW, Monaghan DT, Ottersen OP, Storm-Mathisen J (1987) Anatomical organization of excitatory amino acid receptors and their pathways. Trends Neurosci 10:273–280

Couraud F, Martin-Moutot N, Koulakoff A, Berwald-Netter Y (1986) Neurotoxin-sensitive sodium channels in neurons developing in vivo and in vitro. J Neurosci 6:192–196

Davis BD, Tai P-C (1980) The mechanism of protein secretion across membranes. Nature 283:433–438

Davis JN, Nishino K, Moore K (1987) Noradrenergic regulation of delayed neuronal death after transient forebrain ischemia. In: Ginsberg MD, Dietrich WD (eds) Cerebrovascular diseases. Raven, New York, pp 109–116

DeCaprio AP (1985) Molecular mechanisms of diketone neurotoxicity. Chem Biol Interact 54:257–270

DeCaprio AP, Briggs RG, Jackowski SJ, Kim JC (1988) Comparative neurotoxicity and pyrrole-forming potential of 2,5-hexanedione and perdeuterio-2,5-hexanedione in the rat. Toxicol Appl Pharmacol 92:75–85

De Keyser J, Herregodts P, Ebinger G (1990) The mesoneocortical dopamine neuron system. Neurology 40:1660–1662

Desclin JC (1976) Early terminal degeneration of cerebellar climbing fibers after destruction of the inferior olive in the rat. Synaptic relationships in the molecular layer. Anat Embryol (Berl) 149:87–112

Desclin JC, Colin F (1980) The olivocerebellar system. II. Some ultrastructural correlates of inferior olive destructions in the rat. Brain Res 187:28–46

Desclin JC, Escubi J (1974) Effects of 3-acetylpyridine on the central nervous system of the rat, as demonstrated by silver methods. Brain Res 77:349–364

Di Mauro S, de Vivo DC (1989) Diseases of carbohydrate, fatty acid, and mitochondrial metabolism. In: Siegel G, Agranoff B, Albers RW, Molinoff P (eds) Basic neurochemistry. Molecular, cellular, and medical aspects, 4th edn. Raven, New York, pp 647–670

Dyck PJ, Karnes JL, O'Brien P, Okazaki H, Lais A, Engelstad J (1986) The spatial distribution of fiber loss in diabetic polyneuropathy suggests ischemia. Ann Neurol 19:440–449

Eccles JC, LLinás R, Sasaki K (1966) The excitatory synaptic action of climbing fibers of the Purkinje cells of the cerebellum. J Physiol (Lond) 182:269–296

Endo T, Tamiya N (1987) Current view on the structure-function relationship of postsynaptic neurotoxins from snake venoms. Pharmacol Ther 34:403–451

Foster GA, Roberts PJ (1983) Neurochemical and pharmacological correlates of inferior olive destruction in the rat: attenuation of the events mediated by an endogenous glutamate-like substance. Neuroscience 8:277–284

Freier S, Neal BW, Nisbet HIA, Rees GJ, Wilson F (1957) Salicylate intoxication treated with intermittent positive-pressure respiration. Br Med J 1:1333–1335

Fritschy J-M, Geffard M, Grzanna R (1990) The response of noradrenergic axons to systemically administered DSP-4 in the rat: an immunohistochemical study using antibodies to noradrenaline and dopamine-β-hydroxylase. J Chem Neuroanat 3:309–321

Gaitonde MK, Jones J, Evans G (1987) Metabolism of glucose into glutamate via the hexose monophosphate shunt and its inhibition by 6-aminonicotinamide in rat brain in vivo. Proc R Soc Lond (Biol) 231:71–90

Gaultier M, Gervais P, Conso F (1974) Intoxication aigue par le dinitroorthocresol. J Eur Toxicol 7:9–11

Globus MYT, Ginsberg MD, Dietrich WD, Busto R, Scheinberg P (1987) Substantia nigra lesion protects against ischemic damage in the striatum. Neurosci Lett 80:251–256

Gottfries CG (1991) Disturbance of the 5-hydroxytryptamine metabolism in ageing and in Alzheimer's and vascular dementias. In: Sandler M, Coppen A, Harnett S (eds) 5-Hydroxytryptamine in psychiatry. Oxford University Press, Oxford, pp 309–323

Griffin JW, Hoffman PN, Clark AW, Carroll PT, Price DL (1978) Slow axonal transport of neurofilament proteins: impairment by beta, beta-iminodipropionitrile administration. Science 202:633–635

Grzanna R, Berger U, Fritschy J-M, Geffard M (1989) Acute action of DSP-4 on central norepinephrine axons: biochemical and immunohistochemical evidence for differential effects. J Histochem Cytochem 37:1435–1442

Halliwell JV, Othman IB, Pelchen-Matthews A, Dolly JO (1986) Central action of dendrotoxin: selective reduction of a transient K conductance in hippocampus and binding to localized acceptors. Proc Natl Acad Sci USA 83:493–497

Harvey AL, Anderson AJ (1985) Dendrotoxins: snake toxins that block potassium channels and facilitate neurotransmitter release. Pharmacol Ther 31:33–55

Herken H (1968) Functional disorders of the brain induced by synthesis of nucelotides containing 3-acetylpyridine. Z Klin Chem 6:357–367

Herken H (1990) Neurotoxin-induced impairment to biopterin synthesis and function: initial stage of a parkinson-like dopamine deficiency syndrome. Neurochem Int 17:223–238

Herken H, Lange K, Kolbe H (1969) Brain disorders induced by pharmacological blockade of the pentose phosphate pathway. Biochem Biophys Res Commun 36:93–100

Herken H, Keller K, Kolbe H, Lange K, Schneider H (1973) Experimentelle Myelopathie—Biochemische Grundlagen ihrer cellulären Pathgenese. Klin Wochenschr 51:644–657

Herkenham M, Littler MD, Bankiewicz K, Yang S-C, Markey SP, Johannessen JN (1991) Selective retention of MPP$^+$ within the monoaminergic systems of the primate brain following MPTP administration: an in vivo autoradiographic study. Neuroscience 40:133–158

Hicks SP (1955) Pathologic effects of antimetabolites. I. Acute lesions in the hypothalamus, peripheral ganglia, and adrenal medulla caused by 3-acetylpyridine. Am J Pathol 31:189–197

Huck S, Stumpf C (1985) Failure of 6-aminonicotinamide in selectively damaging astrocytes in vitro. Drug Res 35:333–336

Insel TR, Battaglia G, Johannessen JN, Marra S, de Souza EB (1989) 3,4-methylenedioxymethamphetamine ("ecstasy") selectively destroys brain serotonin terminals in rhesus monkeys. J Pharmacol Exp Ther 249:713–720

Johansen FF, Jorgensen MB, Diemer NH (1986) Ischemic CA-1 pyramidal cell loss is prevented by preischemic colchicine destruction of dentate gyrus granule cells. Brain Res 377:344–347

Jonsson G (1983) Chemical lesioning techniques: monoamine neurotoxins. In: Björklund A, Hökfelt T (eds) Handbook of chemical neuroanatomy, vol 1. Elservier, Amsterdam, pp 463–507

Kass IS, Lipton P (1982) Mechanisms involved in irreversible anoxic damage to the in vitro rat hippocampal slice. J Physiol (Lond) 332:459–472

Kolbe H, Keller K, Lange K, Herken H (1976) Metabolic consequences of drug-induced inhibition of the pentose phosphate pathway in neuroblastoma and glioma cells. Biochem Biophys Res Commun 73:378–382

Lenk R, Ransom L, Kaufmann Y, Penman S (1977) A cytoskeletal structure with associated polyribosomes obtained from HeLa cells. Cell 10:67–78

Lin DC, Tobin KD, Grumet M, Lin S (1980) Cytochalasins inhibit nuclei-induced actin polymerization by blocking filament elongation. J Cell Biol 84:455–460

Lindvall O, Björklund A (1974) The organization of the ascending catecholamine neuron system in the rat brain. Acta Physiol Scand [Suppl]412:1–48

Llinas R, Walton K, Hillman DE (1975) Inferior olive: its role in motor learning. Science 190:1230–1231

Lodge D, Johnson KM (1990) Noncompetitive excitatory amino acid receptor antagonists. Trends Pharmacol Sci 11:81–86

Low BW, Corfield PWR (1986) Erabutoxin b. Structure/function relationships following initial refinement at 0,140 resolution. Eur J Biochem 161:579–587

Low BW, Corfield PWR (1988) The acetylcholine receptor; identification of prime α-toxin binding site. In: Dolly JO (ed) Neurotoxins in neurochemistry. Horwood, Chichester, pp 79–99

Mamounas LA, Molliver ME (1988) Evidence for dual serotonergic projections to neocortex: axons from the dorsal and medial raphé nuclei are differentially vulnerable to the neurotoxin p-chloroamphetamine (PCA). Exp Neurol 192:23–36

Maruyama K, Hartwig JH, Stossel TP (1980) Cytochalasin B and the structure of actin gels. II. Further evidence for the splitting of F-actin by cytochalasin B. Biochim Biophys Acta 626:494–500

Matus A (1987) Putting together the neuronal cytoskeleton. Trends Neurosci 10:186–188

McDonald JW, Johnston MV (1990) Physiological and pathophysiological roles of excitatory amino acids during central nervous system development. Brain Res Rev 15:41–70

Mebs D (1988) Snake venom toxins; structural aspects. In: Dolly JO (ed) Neurotoxins in neurochemistry. Horwood, Chichester, pp 3–12

Meldrum B, Garthwaite J (1990) Excitatory amino acid neurotoxicity and neurodegenerative disease. Trends Pharmacol Sci 11:379–387

Nielsen P, Goelz S, Trachsel H (1983) The role of the cytoskeleton in eukaryotic protein synthesis. Cell Biol Int Rep 7:245–254

O'Hearn, E, Battaglia G, de Souza EB, Kuhar MJ, Molliver ME (1988) Methylenedioxyamphetamine (MDA) and methylenedioxymethamphetamine (MDMA) cause selective ablation of serotonergic axon terminals in forebrain: immunocytochemical evidence for neurotoxicity. J Neurosci 8:2788–2803

Olney JW (1990) Excitotoxic amino acids and neuropsychiatric disorders. Ann Rev Pharmacol Toxicol 30:47–71

Onodera H, Sato G, Kogure K (1986) Lesions to Schaffer collaterals prevent ischemic death of CA1 pyramidal cells. Neurosci Lett 68:169–174

Oteiza PI, Goupul MS, Gershwin ME, Donald JM, Keen CL (1989) The influence of high dietary aluminium on brain microtubule polymerization in mice. Toxicol Lett 47:279–285

Politis MJ (1989) 6-Aminonicotinamide selectively causes necrosis in reactive astroglial cells in vivo. I Neurol Sci 92:71–79

Ramsay RR, Singer TP (1986) Energy-dependent uptake of N-methyl-4-phenylpyridinum, the neurotoxic metabolite of 1-methyl-4-phenyl-1,2,3,6-tetrahydropyridine, by mitochondria. J Biol Chem 261:7585–7587

Ross SB (1987) Pharmacological and toxicological exploitation of amine transporters. Trends Pharmacol Sci 8:227–231

Rothman SM (1983) Synaptic activity mediates death of hypoxic neurons. Science 220:536–537

Rothman SM (1984) Synaptic release of excitatory amino acid neurotransmitter mediates anoxic neuronal death. J Neurosci 4:1884–1891

Sachs C, Jonsson G (1975) Mechanism of action of 6-hydroxydopamine. Biochem Pharmacol 24:1–8

Saper CB, Wainer BH, German DC (1987) Axonal and transneuronal transport in the transmission of neurological diseases: potential role in system degenerations, including Alzheimer's disease. Neuroscience 23:389–398

Sayre LM, Singh MP, Arora PK, Wang F, McPeak RJ, Hoppel CL (1990) Inhibition of mitochondrial respiration by analogues of the dopamenergic neurotoxin 1-methyl-4-phenylpyridinium: structural requirements for accumulation-dependent enhanced inhibitory potency of intact mitochondria. Arch Biochem Biophys 280:274–283

Seagar MJ, Granier C, Couraud F (1984) Interactions of the neurotoxin apamin with a Ca^{++}-activated K^+ channel in primary neuronal cultures. J Biol Chem 259:1491–1496

Seagar MJ, Marqueze B, Couraud F (1988) Characterization of the apamin-binding protein assoziated with a Ca^{2+} activated K^+ channel. In: Dolly JO (ed) Neurotoxins in neurochemistry. Horwood, Chichester, pp 178–190

Semba K, Fibiger HC (1989) Organization of central cholinergic systems. Prog Brain Res 79:37–63

Siesjö BK (1981) Cell damage in the brain: a speculative synthesis. J Cereb Blood Flow Metab 1:155–185

Sievers H, Sievers J, Baumgarten HG, König N, Schlossberger HG (1983) Distributions of tritium label in the neonate rat brain following intracisternal or subcutaneous administration of (^{3}H)6-OHDA. An autoradiographic study. Brain Res 274:23–45

Sievers J, Klemm HP, Jenner S, Baumgarten HG, Berry M (1980) Neuronal and extraneuronal effects of intracisternally administered 6-hydroxydopamine on the developing rat brain. J Neurochem 34:765–771

Sievers J, Pehlemann FW, Baumgarten HG, Berry M (1985) Selective destruction of meningeal cells by 6-hydroxydopamine: a tool to study meningeal-neuroepithelial interaction in brain development. Dev Biol 110:127–135

Simon RP, Swan JH, Griffiths T, Meldrum BS (1984) Blockade of *N*-methyl-D-aspartate receptors may protect against ischemic damage in the brain. Science 226:850–852

Sloviter RS (1983) "Epileptic" brain damage in rats induced by sustained electrical stimulation of the perforant path. I. Acute electrophysiological and light micorscopic studies. Brain Res Bull 10:675–697

Smith MJH (1959) Salicylates and metabolism. J Pharm Pharmacol 11:705–720

Steinbusch HWM (1981) Distribution of serotonin-immunoreactivity in the central nervous system of the rat. Neuroscience 4:557–618

Steward O, Goldschmidt RB, Sutula T (1984) Neurotoxicity of colchicine and other tubulin-binding agents: a selective vulnerability of certain neurons to the disruption of microtubules. Life Sci 35:43–51

Stewart DJ, MacFabe DF, Vanderwolf CH (1984) Cholinergic activation of the electrocorticogram: role of substantia innominata and effects of atropine and quinuclidinyl benzylate. Brain Res 322:219–232

Szalay J, Boti Z, Temesvari P, Bara D (1979) Alterations of the nucleus ruber in 3-acetylpyridine intoxication: a light microscopic and electron microscopic study. Exp Pathol 17:271–279

Tominaga S, Kaziro Y (1983) Adenosine triphosphate associated with bovine brain microtubules. I. Properties of ATPases I and II. J Biochem 93:1093–1100

Tooze J, Burke B (1987) Accumulation of adrenocorticotropin secretory granules in the midbody of telophase AtT20 cells: evidence that secretory granules move anterogradely along microtubules. J Cell Biol 104:1047–1057

Törk I (1990) Anatomy of the serotonergic system. Ann NY Acad Sci 600:9–35

Ungerstedt U (1971) Stereotaxic mapping of the monoamine pathways in the rat brain. Acta Physiol Scand [Suppl]367:1–48

Vanderwolf CH (1988) Cerebral activity and behavior: control by central cholinergic and serotonergic systems. Int Rev Neurobiol 30:225–340

Vogt C, Vogt O (1937) Sitz und Wesen der Krankheiten im Lichte der topistischen Hirnforschung und des Variierens der Tiere. J Psychol Neurol 47:237–457

Walsh TJ, Tilson HA, de Haven DL, Mailman RB, Fisher A, Hanin I (1984) AF64A, a cholinergic neurotoxin, selectively depletes acetylcholine in hippocampus and cortex, and produces long-term passive avoidance and radial arm deficits in the rat. Brain Res 321:91–102

Watkins JC, Krogsgaard-Larsen P, Honoré T (1990) Structure-activity relationships in the development of excitatory amino acid receptor agonists and competitive antagonists. Trends Pharmacol Sci 11:25–33

Watson DF, Griffin JW, Fittro KP, Hofman PN (1989) Phosphorylation-dependent immunoreactivity of neurofilaments increases during axonal maturation and β, β-iminodipropionitrile. J Neurochem 53:1818–1829

Whitehouse MW, Haslam JM (1962) Ability of some antirheumatic drugs to uncouple oxidative phosphorylation. Nature 196:1323–1324

Whitehouse PJ, Price DL, Clark AW, Coyle JT, DeLong MR (1981) Alzheimer disease: evidence for selective loss of cholinergic neurons in the nucleus basalis. Ann Neurol 10:122–126

Wiklund L, Toddenburger G, Cuénod M (1982) Aspartate: possible neurotransmitter in cerebellar climbing fibers. Science 216:78–80

Wolf A, Cowan D, Geller LM (1959) The effects of an antimetabolite, 6-aminonicotinamide, on the central nervous system. Trans Am Neurol Assoc 45:140–145

Zimmermann B (1989) Secretion of lamellar bodies in type II pneumocytes in organoid culture: effects of colchicine and cytochalasin B. Exp Lung Res 15:31–47

CHAPTER 2

Protective Barriers in the Peripheral Nervous System to Neurotoxic Agents

Y. OLSSON

A. Toxic Injury and Protective Barriers of the Nervous System

The blood-brain barrier (BBB) has several important implications for the normal function of the brain, regulating the passage of various substances from the blood into the parenchyma. Furthermore, there are many examples of toxic compounds which are unable to pass the BBB under normal conditions, i.e., the barrier can protect the brain parenchyma from neurotoxic substances which may be present in the blood. This aspect of the BBB has been reviewed repeatedly in the past (RAPOPORT 1976; KATZMAN 1976; BRADBURY 1979; GOLDSTEIN and BETZ 1986; CRONE 1986; NEUWELT 1988).

Over the past few years, more and more scientists have become aware of the fact that protective barriers in the vessels and the perineurium of the peripheral nervous system (PNS) profoundly and in many different ways influence the course of neurotoxic diseases afficting this part of the nervous system (cf. JACOBS 1980; DYCK et al. 1984; RECHTHAND and RAPOPORT 1987; OLSSON 1990). This chapter therefore concentrates on these barriers under normal conditions and their significance in neurotoxic disorders of the PNS.

B. Microenvironment of Peripheral Nerves

Neurotoxic agents which in some way have entered the intrafascicular compartment of a nerve or a peripheral ganglion reach their target cells. Obviously, the extracellular fluid of the PNS is essential for transfer of the toxic compounds. Let us therefore first consider the fluid microenvironment of the PNS under normal circumstances.

The neural components of peripheral nerves, i.e., the Schwann cells and the myelinated and unmyelinated axons, are all surrounded by an extracellular fluid and a matrix, forming the "microenvironment" of the PNS. This can best be considered as a fiber-reinforced gel containing various ions, large molecular substances, and fluid. This environment receives material continuously from the blood, and other compounds are removed by resorption into the vessels.

The PNS is widely distributed in the body, and the compostition of the fluid environment outside peripheral nerve branches varies in different tissues. In order to maintain a constant internal milieu of such a widely distributed "organ," the nerves in almost all parts of the body are surrounded by a specialized structure, the perineurium, which forms a barrier between the endoneurium and the surrounding tissue.

The perineurium has important regulatory functions for the fluid composition of nerve fascicles. It forms a "wall" around the endoneurium, so that the fascicles can maintain homeostasis even when there are variations in the fluid composition around the nerves. This occurs frequently, for instance under local inflammatory conditions. However, some parts of the PNS lack perineurium, and functional changes may develop as a result of diseases or injuries to the perineurium.

The extracellular fluid of the PNS also communicates with the serum in the blood vessels. i.e., the vasa nervorum. These vessels play a crucial role in the regulation of the chemical environment of the PNS, and vascular abnormalities have a profound effect on the composition of the fluid. Severe changes in the amount and composition of the endoneurial fluid as seen in peripheral nerve edema can in turn aggravate toxic neuropathies.

The blood vessels provide Schwann cells and other cells of the PNS with nutritional products, whereas the axons receive their supplies to a large extent by axonal transport mechanisms. The exchange of nutrients and metabolites between the blood and the nerve parenchyma occurs across the vascular walls where the endothelium can influence the fluid composition of the fascicles. Pathological alterations in the vessels may influence the nutrition of a nerve, and metabolic waste products may accumulate in some neuropathies to such an extent that they reach neurotoxic concentrations.

I. Intrafascicular Structural Components

Each nerve fascicle is, according to the terminology proposed by KEY and RETZIUS (1873, 1876), surrounded by a perineurial sheath which forms the border between the endoneurium and the epineurium (Fig. 1). The term "endoneurium" should be restricted to the connective tissue of the fascicles. The term "intrafascicular" refers to all components inside the perineurium, i.e., connective tissue, blood vessels, and nerve fibers. Apart from blood vessels, nerve fibers, and Schwann cells there are fibroblasts, macrophages, and mast cells in the intrafascicular compartment. Furthermore, there is a substantial amount of extracellular matrix.

Connective tissue forms the so-called endoneurial matrix which surrounds each nerve fiber and fills all available spaces in the intrafascicular compartment. Biochemically, the major groups of extracellular matrix molecules are the collagens, fibronectins, other glycoproteins, and proteoglycans (CARBONETTO 1984). The endoneurium thus contains a substantial amount of collagen, which forms a most important connective tissue skeleton of nerve fascicles. Neurotoxic lesions afflicting the endoneurial collagen obviously may affect the normal organization of a peripheral nerve.

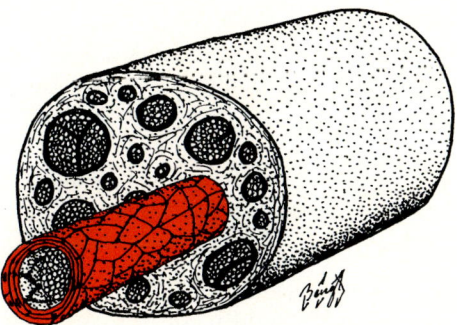

Fig. 1. Cross section of a multifascicular peripheral nerve. Each fascicle is surrounded by a dense region called perineurium. Perineurial cells form a barrier between the intra- and the extrafascicular compartments. The connective tissue binding the various fascicles together is termed epineurium and the connective tissue inside the fascicles, endoneurium. Terminology proposed by KEY and RETZIUS (1873, 1876)

The collagens belong to a family of structural proteins with some common components but also structural differences which determine their functional properties. SALONEN (1987) came to the conclusion that both the epineurium and the endoneurium have a co-localization of collagen types I and III, but biochemical investigations indicate that collagen type I is the dominating fraction in the endoneurium.

Collagens type I and III are so-called fibrillar collagens. Procollagen is considered to be secreted from the collagen-producing cells by exocytosis, and the assembly into fibrils probably takes place in compartments surrounded by cell surfaces. It is known that collagens are reconstructed particularly during growth and in various diseases and that they become progressively stronger and less resistant to degradation. Also, collagen is susceptible to various toxic influences, particularly to a group of enzymes called collagenases. During the course of chronic peripheral nerve disease, the possibility thus exists that there is not only formation but also degradation of endoneurial collagen.

II. Intrafascicular Fluid

The extracellular space of nerve fascicles continuously receives fluid components from the endoneurial vessels. This fluid is partly resorbed into the venous part of the microvessels, but there is a small over-production in the perivascular region, and in order to reach a steady state condition the fluid must be displaced in the extracellular spaces of the fascicles. In other regions of the body, overflow of extracellular fluid is resorbed into the lymphatics, but as is well known, such vessels are lacking in peripheral nerve fascicles. The turnover of endoneurial fluid has several important implications in neurotoxic lesions of the PNS. Toxic compounds which in some way have entered the intrafascicular compartment can be spread widely by this route to reach their target cells. Furthermore, such agents will be diluted in the fluid, which to a certain extent may limit their toxic actions.

Rather little is published about the mechanisms regulating the formation, resorption, and movements of endoneurial fluid. However, Low (1984) has presented an excellent review on this topic.

1. Composition

Direct sampling of endoneurial fluid has been carried out by at least three different methods in order to characterize it. By micropipettes inserted into the subperineurial space (Low et al. 1982, 1983; MYERS et al. 1983) and by similar pipettes used to remove extruded endoneurial fluid, small samples can be obtained, but the technique is time-consuming and technically demanding (PODUSLO et al. 1985; MIZISIN et al. 1986).

Other investigators (Low et al. 1977) have implanted a capsule in the fascicles through which endoneurial fluid is collected, but this technique causes slight contamination by serum components. Finally, a promising, so-called elution procedure has been introduced by PODUSLO et al. (1985).

The intrafascicular fluid is composed of water, various ions, and plasma proteins as well as other soluble compounds. Analysis of endoneurial fluid samples by the drop technique (MIZISIN et al. 1986) from normal rat sciatic nerves revealed a sodium concentration of 152 mEq/l, a chloride concentration of 122 mEq/l, and a potassium concentration of 16 mEq/l.

The endoneurial fluid also contains serum protein fractions, particularly albumin, as revealed by immunocytochemical and biochemical methods (cf. PODUSLO et al. 1985; LIEBERT et al. 1985; SEITZ et al. 1985a,b; NEUEN et al. 1987; MATA et al. 1987). However, earlier tracer studies indicated that the barriers of the PNS were impermeable to such compounds, and it was therefore considered that the endoneurial fluid lacked or had a very low concentration of serum proteins (OLSSON 1984, 1990).

2. Formation and Resorption

The endoneurial fluid originates from the microvessels, and two major forces determine its formation, namely net hydrostatic pressure and net osmotic pressure. The capillary hydrostatic pressure of endoneurial vessels has not yet been measured, but determinations carried out in vessels from other organs have revealed a pressure of 10–40 mmHg with a gradient of 2–20 mmHg between the arteriolar and venous ends of the capillary (Low 1984).

It is technically very difficult to estimate endoneurial fluid pressure. However, direct measurement with a micropipette (MYERS et al. 1978; POWELL et al. 1980b) and analysis with the implanted capsule method (Low et al. 1977; Low 1984) showed that there is a positive endoneurial pressure in contrast to other connective-tissue-rich regions, for example subcutaneous tissue. Low et al. (1977) recorded pressures of 1.2 mmHg with a SD of 0.7. Marked changes occur in the endoneurial pressure in various neurotoxic lesions.

There are indications that an "overproduction" of fluid takes place under normal conditions in the fascicles and that the excess of fluid moves away from the perivascular region. As will be evident from the following paragraphs, the fluid can either pass in a centrifugal or centripetal direction along the fascicles or possibly through the perineurium to enter the connective tissue spaces of the epineurium.

3. Intrafascicular Flow

The possibility that endoneurial fluid may move in a centrifugal (proximo-distal) direction along peripheral nerves has been discussed for a long period of time and is now considered as the major route for translocation. This opinion is to a large extent based on experiments in which colored or radioactive compounds were microinjected into nerve fascicles after which their spread was followed by various recording methods (WEISS et al. 1945). Toxic compounds which have entered the fascicles can thus be translocated by this normally occurring flow of fluid.

At which sites does the endoneurial fluid escape from the fascicles? This question cannot yet be fully answered. However, we know that peripheral nerve branches are often open-ended, i.e., they lack a perineurial sheath at the terminals (BURKEL 1967; SAITO and ZACHS 1969; KERJASCHKI and STOCKINGER 1970; MALMGREN and OLSSON 1980), and at such endings fluid might escape into the extracellular spaces outside the fascicies. Another possibility is the preterminal region where the perineurium normally has fewer layers as compared with more proximal regions of a nerve (STOCKINGER 1965; OLSSON and REESE 1969, 1971).

The possibility that endoneurial fluid normally can pass towards the nerve roots, i.e., in a centripetal direction, has also been considered. It is not known whether it occurs under normal conditions, but experiments have revealed that compounds microinjected into fascicles easily spread in a centripetal direction (cf. WEISS et al. 1945; FRENCH and STRAIN 1947). Compounds entering the extracellular spaces of nerve roots may then enter the cerebro-spinal fluid (CSF) since the sheaths of the nerve roots do not restrict fluid movements to the same extent as the perineurium of peripheral nerve fascicles.

Most likely, the centripetal mode of endoneurial fluid movement is not as pronounced as the centrifugal flow under normal conditions. However, in edematous states with abnormal pressure conditions, this mode of fluid removal might become important.

There is also a third theoretical possibility for removal of endoneurial fluid and that is a transperineurial route. If it exists, the fluid may diffuse into the large extracellular spaces of the epineurium and reach the surrounding

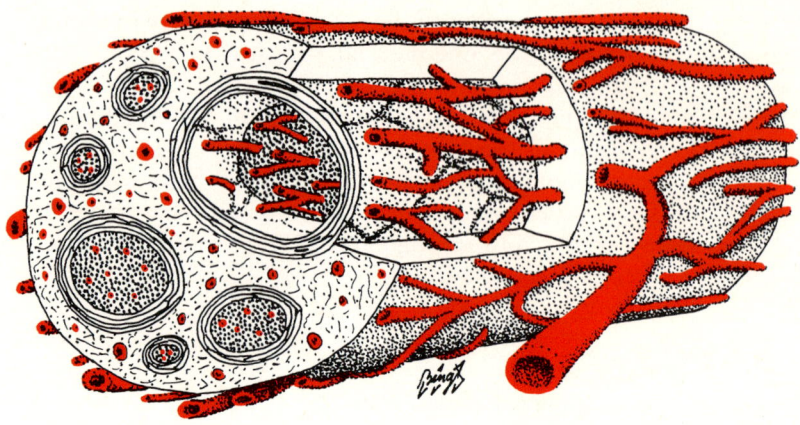

Fig. 2. Blood supply of peripheral nerve. Oblique section showing vascular plexa in the epineurium and in the endoneurium. The two plexa communicate with each other through anostomoses passing the perineurium

soft tissues to be resorbed into blood vessels or lymphatics. Normally, however, the intact perineurium restricts the passage of fluid, and therefore this route is probably of minor importance. The chances for transperineurial transfer would be highest where blood vessels pierce the perineurium and in the preterminal region with their thin perineurium.

C. Barrier in the Vasa Nervorum of Peripheral Nerves

The blood vessels in peripheral nerves have many important functions such as the formation of endoneurial fluid, exchange of gases, nutrients, and metabolic waste products and the maintenance of the blood-nerve barrier. Such functions are all of great importance under neurotoxic conditions.

I. Blood Supply

The origin and distribution of blood vessels in different peripheral nerves have been reviewed repeatedly in the past (ADAMS 1942, 1943; SUNDERLAND 1968; OLSSON 1972, 1990; LUNDBORG 1988), and two very significant papers in this field of research have recently been published (BELL and WEDELL 1984a,b). Cardinal features of these vessels are the richness of anastomoses and the presence of microvascular networks or plexuses (Fig. 2). Nutrient arteries from adjacent large vessels give off many small branches, which by anastomosis form the epineurial and perineurial vascular plexuses. Numerous vessels pierce the perineurium to join the endoneurial vascular network. This is composed of many small vessels running mainly in the longitudinal direction along the nerves.

Small terminal nerve branches lack endoneurial blood vessels, and interestingly they have a perineurial sheath, which is somewhat permeable to compounds present in the epineurium (MALMGREN and OLSSON 1980; OLDFORS 1981). Conceivably, thin nerves also receive nutrients by this route and not entirely from the endoneurial vessels.

II. Vascular Permeability

The term "vascular permeability" refers to the exchange of all kinds of substances between blood and a tissue. It is well known that the chemical nature and the electrical charge are of great importance in such a transfer of substances and that the passage across the vascular wall may occur in quite different ways, including diffusion, pinocytosis, and mediated and active transport (RAPOPORT 1976; KATZMAN 1976; BRADBURY 1979; OLSSON 1984, 1990; GOLDSTEIN and BETZ 1986; CRONE 1986).

With regard to the PNS, several quantitative studies have been carried out, but in most of them it has been difficult or impossible to sample selectively endoneurial and epineurial tissue. Several investigators have observed that different radiolabeled ions, such as ^{24}Na, ^{36}Cl, and [^{35}S] thiourea, do not penetrate as rapidly into peripheral nerves as into other soft tissues (MANERY and BALE 1941; DAINTY and KRNJEVIC 1955; WELCH and DAVSON 1972). Since glucose is the principal metabolic substrate in peripheral nerve, its transfer from the blood into the nerves is of great interest. Recently, RECHTHAND et al. (1985) showed that this transfer occurs by means of facilitated transport.

In studies carried out by BRADBURY and CROWDER (1976) on mammals, several polar nonelectrolytes were used. They found that ethylenediaminetetraacetic acid, [^{51}Cr] EDTA, and [^{14}C] sucrose equilibrated with the interstitial fluid in peripheral nerves with a rate constant about five times greater than that observed for [^{14}C] insulin. The penetration took place in two phases. The rapidly equilibrating space constituted 7%–20% of the tissue weight, and the slowly equilibrating phase was related to the penetration of substances into some remaining larger compartment. Most probably, the rapid phase corresponds to the penetration into the intrafascicular compartment.

WEERASURYIA and RAPOPORT (1986) recently reported that in frog sciatic nerve the endoneurial capillaries form an efficient part of the blood-nerve barrier to water-soluble nonelectrolytes (cf. MICHEL et al. 1984).

Most earlier studies on vascular permeability relied on so-called tracer techniques (OLSSON and KRISTENSSON 1979). The tracers included acid dyes with great affinity for serum albumin, such as trypan blue and Evans blue, albumin labeled with a fluorescent or radioactive marker (STEINWALL and KLATZO 1966; BRIGHTMAN et al. 1970), sodium fluorescein (MALMGREN and OLSSON 1980), as well as diaminoacridine and fluorochrome-labeled dextrans of various molecular sizes (AKER 1972; HULTSTRÖM et al. 1983).

Such compounds are given by intravenous injection at a predetermined interva before the animals are killed, and the localization of the tracer is determined, for instance by fluorescence microscopy.

A major advance in tracer techniques was made when Graham and Karnovsky (1966) and Karnovsky (1967) developed methods for light and electron microscopic visualization of enzymes with peroxidase activity. For instance, the fine structural localization of horseradish peroxidase (HRP) can be determined with great accuracy if the technique described by Graham and Karnovsky (1966) is applied. HRP is then used as a protein tracer, which after fixation with a glutaraldehyde-paraformaldehyde mixture and appropriate incubation can be localized both by light and electron microscopy.

However, there are only a few studies in which other compounds with peroxidase activity, such as cytochrome C, catalase, and lactoperoxidase, have been used. The refined histochemical methods for detection of HRP (Malmgren and Olsson 1978; Mesulam 1982) have with few exceptions not yet been applied in studies on the vascular permeability of the PNS. More subtle changes in vascular permeability may well be recognized with such procedures as compared with the standard method.

Recently, by means of immunohistochemistry several important observations have been made regarding the distribution of serum proteins in peripheral nerves (e.g., van Liss and Jennekens 1977; Liebert et al. 1985; Seitz et al. 1985a,b; Neuen et al, 1987; Mata et al. 1987). The technique demonstrates available antigenic sites in the tissue at a given moment, and the results provide indirect information regarding the blood-nerve barrier.

1. Endoneurial Vessels

Numerous studies on the BBB have shown that trypan blue and Evans blue injected intravenously cause a blue staining of all parenchymatous organs except the brain and spinal cord (Broman 1949; Brightman et al. 1970; Goldstein and Betz 1986). The distribution of the dye in the PNS, however, differs from that in the brain and is influenced by a number of factors such as the topographic site, the area within the nerve (epineurium, endoneurium), and the animal species (Fig. 3). For instance, the epineurium will be intensely stained, whereas the endoneurium is either completely unstained or shows only a slight bluish coloration (Doinikow 1913; Tschetschujeva 1929; Waksman 1961; Olsson 1966, 1990).

Observations by fluorescence microscopy have provided further information on vascular permeability in peripheral nerves of mammals. After intravenous injection of albumin or gamma globulin labeled with the fluorescent marker Evans blue or fluorescein isothiocyanate, these proteins can be traced to the lumen of peripheral nerve blood vessels almost immediately. After a brief delay there are signs of a normally occurring extravasation in the epineurium (Olsson 1966, 1990).

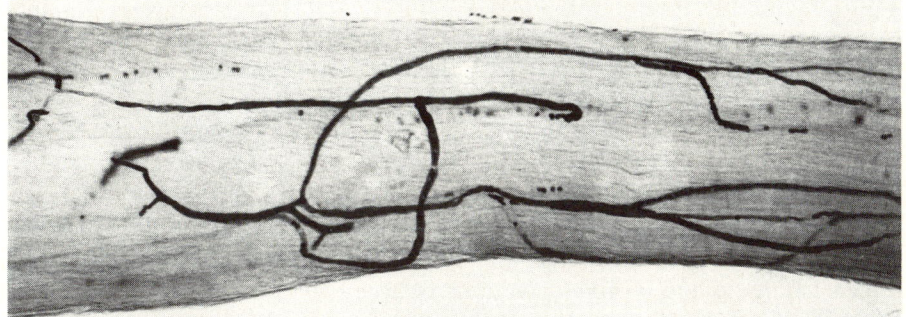

Fig. 3. The "blood-nerve barrier" is composed of two parts, the perineurial barrier and the barrier in endoneurial blood vessels. This longitudinal section from a peripheral nerve fascicle derives from an experiment in which a macromolecular tracer has been injected intravenously. The tracer (*dark stain*) remains in the lumen of the vessels and does not freely pass into the nerve

The amount of extravasated tracer in the endoneurium is usually considerably smaller than that in the epineurium, and there are variations between large nerves (like the sciatic) and intramuscular nerve branches (MALMGREN and BRINK 1975; MALMGREN and OLSSON 1980). There are also differences among animal species. For instance, in the mouse and rat sciatic nerve, no or insignificant amounts of Evans blue-labeled albumin can be detected outside endoneurial vessels after intravenous injection (OLSSON 1966; OLSSON and REESE 1971).

In many other vertebrates (e.g., guinea pig, rabbit, cat, monkey), however, passage of intravenously injected protein tracers can be observed, but the amount differs considerably among individuals and among fascicles in a multifascicular nerve (WAKSMAN 1961; OLSSON 1966, 1990). For instance, some of the fascicles in the sciatic nerve of the rabbit and the guinea pig contain extravasated Evans blue-labeled albumin, whereas other fascicles in the same animals do not show any signs of tracer outside the vessels (OLSSON 1984). More recent experiments based on immunohistochemistry have shown that serum proteins including albumin are present in the endoneurium of peripheral nerves. This indicates that the barriers in peripheral nerves are not as complete as in the CNS (LIEBERT et al. 1985; SEITZ et al. 1985a,b; NEUEN et al. 1987).

Some conflicting data appear to exist concerning the permeability of endoneurial vessels in the rabbit, which is probably due to differences in the methods applied. Thus, DOINIKOW (1913) noticed that only minimal amounts of trypan blue could pass into rabbit nerve parenchyma. WAKSMAN (1961) also was of the opinion that the parenchyma of the rabbit sciatic nerve remained completely unstained by parenterally administered trypan blue and did not contain radiolabeled albumin or diphtheria toxin or toxoid after intravenous injection. In these experiments the presence of extravasated tracer was judged macroscopically, by light microscopy, or by autoradiography. Using the sensitive fluorescence microscope technique, LUNDBORG (1970) detected extravasated fluorochrome-labeled albumin in the endoneurium of rabbit sciatic and tibial nerves, thus proving that the barrier function of the endoneurial vessels to proteins in this species is not absolutely efficient.

Applying the peroxidase technique, OLSSON and REESE (1971) studied vascular permeability of endoneurial vessels in mouse sciatic nerve. Intravenously injected horseradish peroxidase remained in the lumen of the vessels of most animals, and the extracellular diffusion of this protein tracer was restricted by tight junctions between the endothelial cells. These junctions were similar to those in the brain parenchyma, where they are considered to be of major importance for the BBB phenomenon with this protein tracer (REESE and KARNOVSKY 1967). Fenestrations of the endothelial cells were not detected, and protein transport by pinocytosis across the endothelial lining appeared to be insignificant since the covering basal lamina did not contain peroxidase (OLSSON and REESE 1971).

Later, ARVIDSON (1977) showed that the mouse endoneurial capillaries in fact must be somewhat permeable to horseradish peroxidase, since numerous macrophages with uptake of the tracer can be seen as early as 5 min after leakage. Uptake into endoneurial macrophages (OLDFORS 1980a,b) has also been reported in the sciatic and autonomic nerves of the rat (JACOBS 1977; DYCK et al. 1980). However, they were not reported by AHMED and WELLER (1979), and the ultrastructural basis for the passage of horseradish peroxidase from the blood into some fascicles is unclear. ARVIDSON (1984), on the basis of an electron microscopic study, concluded that vesicular transfer of horseradish peroxidase is one important form of vascular leakage in endoneurial vessels of normal mouse sciatic nerves.

As was mentioned earlier, thin intramuscular nerves differ from large nerves like the sciatic in their barrier functions after local or intravenous injection of macromolecular tracers. Substances such as cytochrome c and sodium fluorescein invariably leak into the endoneurium of intramuscular nerve branches, but it is not known how they gain access to this compartment in the periphery (MALMGREN and BRINK 1975; MALMGREN and OLSSON 1980). One route might be across the perineurium (OLDFORS 1981), or they may diffuse from the neuromuscular junctions where the perineurium is open-ended (BURKEL 1967; SAITO and ZACKS 1969; KERJASCHKI and STOCKINGER 1970; BROADWELL and BRIGHTMAN 1976). Finally, the possibility exists that the endoneurial vessels in fact are more permeable in this region than in large nerves.

The extracellular environment in terminal nerve branches therefore most probably differs from that in large nerve trunks with respect to the accessibility of blood-borne substances. This aspect should be considered in studies on the effects of toxic substances on peripheral nerves and in interpretations of dying-back phenomena.

2. Epineurial Vessels

As was mentioned earlier, intravenously injected Evans blue causes a blue staining of the epineurium in contrast to the endoneurium of peripheral nerves (Doinikow 1913; Tschetschujewa 1929; Waksman 1961; Olsson 1966, 1990). Observations by fluorescence microscopy after intravenous injection of albumin or gamma globulin labeled with a fluorescent marker have shown that the tracer is present in the lumen of epineurial blood vessels almost immediately. After a brief delay there are signs of the labeled proteins in the connective tissue around the vessels (Olsson 1966, 1990).

The proteins in the extracellular spaces can migrate to the innermost parts of the perineurium, but their further diffusion into the endoneurium is prevented by the barrier function of the perineurium (Olsson 1966, 1990). This pattern of spread has been observed in all species thus far examined, and a similar extravasation has also been described for other tracers, such as [131I]albumin, ferritin, HRP, cytochrome C, dextrans, and sodium fluorescein (Waksman 1961; Olsson and Reese 1971; Boddingius et al. 1972; Malmgren and Brink 1975; Sparrow and Kiernan 1979; Dyck et al. 1980).

The routes by which protein tracers leak out of epineurial vessels have been elucidated with the peroxidase method. Horseradish peroxidase injected intravenously passed out of the vessels by diffusion between adjacent endothelial cells where the junctions are of the open variety. These junctions are similar to those found in some permeable vessels, such as cardiac and skeletal muscle, lung, and liver.

A few fenestrated vessels are also present in the epineurium, and therefore the proteins may, in part, escape from the circulation through such pores (Olsson and Reese 1971). The same holds true for pinocytotic transfer of the proteins through the endothelial cells. Following passage through the endothelium protein tracers can easily penetrate the vascular basal lamina and then freely diffuse into the extracellular spaces of the epineurium (Olsson and Reese 1971).

The epineurial blood vessels thus share with many other blood vessels in the body the ability to allow the passage of serum proteins across their walls. In such tissues, there is normally a slow flow of serum proteins from the blood vessels to the extracellular spaces. The extravascular proteins are then resorbed into lymphatics and later conveyed back to the blood. In this connection it should be recalled that lymphatics are present in the epineurium but are absent from the endoneurium of nerve fascicles. For some tracers, particularly HRP, there is a marked uptake and accumulation in macrophages, but this tendency is much less pronounced for tracers like fluorochrome-labeled albumin, dextrans, cytochrome C, and sodium fluorescein.

D. Barrier in the Perineurium of Peripheral Nerves

The homeostasis of peripheral nerve fascicles is maintained by the joint actions of endoneurial blood vessels and perineurium. The perineurium is of great importance in many disease processes afflicting the peripheral nerves and is a key structure influencing the localization and extent of neurotoxic injuries.

I. Structure

The nervous system is surrounded by a membranous covering which varies in composition and function in different regions. Peripheral nerve fascicles are bordered by the perineurium (Shantha and Bourne 1968; Low 1976; Thomas and Olsson 1984), but at some places it is lacking, for instance in the pulp of the teeth (Obst 1971; Bishop 1982). The structural arrangement of spinal nerve root sheaths is special, and some terminations lack a complete perineurial sheath.

The perineurium was discovered by histologists at the end of the nineteenth century (Henle 1841; Ranvier 1871, 1872; Key and Retzius 1873, 1876), and the terminology used today was introduced by Key and Retzius (1873, 1876). Early electron microscopical investigations showed that the cellular lamellae of the perineurium are formed by concentric sleeves of flattened cells with a polygonal outline (Röhlich and Knoop 1961; Shanthaveerappa and Bourne 1963; Thomas 1963; Gamble 1964; Cravioto 1966; Waggener and Beggs 1967; Burkel 1967; Lieberman 1968; Akert et al. 1976). The number of lamellae varied depending on the size of the fascicle; the thicker the fascicle, the greater the number of lamellae. Up to 15 layers were described around mammalian nerve fascicles.

Each perineurial cell is thin but has a very large surface. Both the endoneurial and the epineurial sides are covered by a basal lamina which at some points may be lacking (Thomas and Olsson 1984). The cytoplasm contains bundles of filaments similar in appearance to myofilaments of smooth muscle. The idea has therefore been proposed that the perineurial cells may have contractile properties (Ross and Reith 1969) which might be of importance for the flow of endoneurial fluid. A very prominent feature is the presence of pinocytotic vesicles which often have openings to the cell surfaces (cf. Akert et al. 1976; Shinowara et al. 1982; Latker et al. 1985).

Freeze fracture examinations and transmission electron microscopy study have elucidated the contacts which exist between the cells in the perineurial lamellae (Olsson and Reese 1971; Reale et al. 1975; Akert et al. 1976). There are extensive zonulae occludentes (tight junctions) between the cells, which most likely are essential components of the perineurial diffusion barrier. There are also at some places punctate gap junctions between cells (Akert et al. 1976).

The perineurial lamellae are separated by extracellular clefts. At various places the cleft may be obliterated by joining surfaces of perineurial cells from different lamellae, and sometimes there are processes from perineurial cells seeking a focal contact with other cells. This implies that the shape of the perineurial space is complex.

The perineurial extracellular spaces contain collagen fibrils forming a lattice-like pattern with circular, longitudinal, and oblique (AKERT et al. 1976; THOMAS and OLSSON 1984). The diameter of such fibrils is considerable smaller than that of epineurial collagen (THOMAS 1963). There are also a few elastic fibers, and occasionally one can find a fibroblast and a mast cell.

The perineurium is traversed by blood vessels which at many places form communicants (cf. MYERS et al. 1986a) between the epineurial blood vessels and the endoneurial vasculature. The vessels carry a "sleeve" of perineurial cells (THOMAS 1963; BURKEL 1967) into the endoneurium, but these sleeves do not form a closed contact with the vessel walls. Therefore, these areas might be of importance as communication routes of fluid material including toxic agents between the epineurium and the endoneurium.

The perineurium follows the nerves towards the terminals, but the number of perineurial cell layers diminishes successively. The smallest branches will have only one layer of perineurial cells, the so-called sheath of Henle. This is continuous with the capsules of muscle spindles and encapsulated end organs (SHANTHAVEERAPPA and BOURNE 1962). At unencapsulated endings and at the neuromuscular junctions, the perineurium terminates with an open end (BURKEL 1967). At such junctions the possibility exists that material in the extracellular spaces can come in direct contact with the nerve terminals.

Histochemical studies mainly carried out by SHANTHAVEERAPPA and BOURNE (1962, 1963, 1964) have shown that the perineurial cells contain a wide range of dephosphorylating enzymes in the cytoplasm and a high activity of, for example, ATPase, 5-nucleotidase, glycerophosphatase, and creatine phosphatase. The cells are therefore equipped with enzymes compatible with a function as a metabolically active diffusion barrier.

The connective tissue layers around the spinal roots differ from those of a peripheral nerve which may very well have important implications in diseasestates. Thus, the layer corresponding to the perineurium in nerves is much looser in composition, and macromolecules present in the CSF have a direct communication with the endoneurial spaces of the nerve roots. Pathogenic substances in the CSF may thus influence the roots, and conversely substances present in the roots may enter the CSF. As an example, serum proteins which leak out of the root vessels under inflammatory conditions can easily enter the CSF.

II. Properties

As has already been emphasized, an important property of the perineurium in large nerves is its capacity to act as a perifascicular diffusion barrier to many compounds including neurotoxic substances (MARTIN 1964; SHANTHA and BOURNE 1968; OLSSON 1966; OLSSON and KRISTENSSON 1979; THOMAS and OLSSON 1984). This implies that certain substances, if present in the extracellular spaces around the nerves, are unable to diffuse into the endoneurium of the fascicles or do so only at a greatly reduced speed.

The existence of a perifascicular diffusion barrier can be demonstrated by both morphologic and neurophysiologic methods. In the typical morphologic experiment, a substance is injected around a large nerve such as the sciatic nerve. After an appropriate delay and fixation of the tissue, the distribution of the substance in the nerve is determined by microscopic methods.

Neurophysiologically, a diffusion barrier can be shown in the following way. An agent with the capacity to reduce either the action potential or the resting potential is applied to the surface of a nerve. By comparing the time required to block these potentials in normal and in desheathed nerves, a conclusion can be drawn about the barrier function of the nerve sheaths (MARTIN 1964). Obviously, this type of experiment is restricted to the demonstration of certain slowly diffusible substances for which the nerve sheaths reduce but do not totally abolish their diffusion into the endoneurium.

1. Barrier to Various Compounds

A perifascicular diffusion barrier has been shown to exist in normal large peripheral nerves of various animals for many but not all substances. MARTIN (1964), in reviewing the literature, listed previously tested substances with regard to their ability to pass into the endoneurium from the surface of a peripheral nerve trunk.

For one group of agents, the nerve sheaths act as a diffusion barrier, so that the substance usually cannot be demonstrated in the endoneurium of the fascicles but only in the perifascicular tissue. This group comprises various substances, such as lithium carmine (WEISS and RÖHLICH 1954), typan blue (EMIROUGLU 1955; CLARA and ÖZER 1960), Evans blue (OLSSON and REESE 1971), Congo red (EMIROUGLU 1955), methyl blue, and methylene blue (MARTIN 1964). In addition, sodium fluorescein (MALMGREN and OLSSON 1980) and India ink behave in the same way (KRISTENSSON and OLSSON 1971), as do several protein tracers, such as ferritin (WAGGENER et al. 1965), fluorochrome-labeled serum proteins (OLSSON 1966), HRP (OLSSON and REESE 1969, 1971; KLEMM 1970; TOWFIGHI and GONATAS 1977), cytochrome C (MALMGREN and BRINK 1975), and microperoxidase (TOWFIGHI and GONATAS 1977).

For another group of substances, diffusion through the nerve sheath is only partially restricted. These substances can therefore reach the endoneurium after a certain delay. According to MARTIN (1964), the molecular weight of such substances is lower than that of those in the first group. This class of agents includes several ionic species, such as potassium, barium, and calcium chloride, choline chloride, glucose, and many others (FENG and LIU 1949; CRESCITELLI 1951; CRESCITELLI and GEISSMAN 1951; KRNJEVIC 1954a,b).

WEERASURIYA et al. (1980) reported that in the frog there is no evidence of active Na or K transport across the perineurium. They considered that the intercellular routes in the perineurium exhibit size-dependent selective permeability properties regulating the distribution of these ions in the nerve. Finally, a group of substances with low molecular weight (KRNJEVIC 1954a,b) can diffuse into the endoneurium without any significant restriction (oxygen, carbon dioxide, ethanol).

2. Site of the Barrier

In the past, the morphologic identity of the perifascicular diffusion barrier has been extensively discussed (FENG and GERARD 1930; CRESCITELLI 1951; CAUSEY and PALMER 1953; KRNJEVIC 1954a,b; MARTIN 1964). By the use of fluorochrome-labeled albumin as a tracer it was finally shown that the perineurium acted as the barrier (OLSSON and REESE 1971; KRISTENSSON and OLSSON 1971; SIMA and SOURANDER 1974). The application of HRP as a permeability tracer showed that it fills the spaces between the outermost perineurial cell layers in both mouse and rat sciatic nerves (OLSSON and REESE 1969, 1970; KLEMM 1970; TOWFIGHI and GONATAS 1977). However, there are always one or two concentric layers of perineurial cells that serve as the final complete barrier (OLSSON and REESE 1969, 1971).

The relationship between the peroxidase tracer and the perineurial cell surfaces appears to be the same in sciatic nerves of both the mouse and the rat (KLEMM 1970; OLSSON and REESE 1969, 1971; TOWFIGHI and GONATAS 1977). In both species, the tracer passes into the basal lamina and is seen in numerous pits and vesicles of the perineurial cells. To what extent pinocytosis of peroxidase actually occurs was previously open to question because lanthanum injected after fixation also fills many of the vesicles and invaginations (OLSSON and REESE 1969, 1971; SHINOWARA et al. 1982). However, there are now studies strongly indicating that transfer of small amounts of peroxidase in fact might occur in this way across perineurial cells (OLDFORS and SOURANDER 1978; OLDFORS 1980a,b, 1981). As in the cerebral vascular endothelium (REESE and KARNOVSKY 1967), the basis of the low permeability to peroxidase appears to be a lack of significant transport of the tracer by pinocytosis across the perineurial cell cytoplasm. Furthermore, intercellular diffusion between adjacent cells of the complete layers of peri-neurium is restricted by the tight junctions connecting them (OLSSON and REESE 1969, 1971; KLEMM 1970). Basically, similar findings have also been made in the rat sciatic nerve by WAGGENER et al. (1965), who employed ferritin as an electron microscopic marker. It therefore seems justified to assume that the location of the barrier is in the same perineurial cell layers for other protein tracers, such as fluorochrome-labeled albumin.

The perineurial barrier at neuromuscular junctions also appears to be deficient, since HRP injected intramuscularly can diffuse in and reach the terminal axons in this area (ZACKS and SAITO 1969). Small nerves with a thin perineurium are considered to be somewhat more permeable than large ones such as the sciatic nerve (MALMGREN and BRINK 1975).

E. Barriers in Peripheral Ganglia

Peripheral ganglia are particularly susceptible to the actions of many neurotoxic agents. The question therefore arises whether there are any differences in the barrier properties of the perineurium surrounding ganglia and in the blood vessels as compared with the corresponding properties of nerve fascicles.

I. Blood Supply

Human sensory ganglia receive their arterial supply from the spinal branches of each segmental artery. Small vessels enter the ganglia both from the poles and from branches passing the capsule at different points. Communicants are present between these two sources within the ganglia, and the ganglionic vessels are mainly capillaries.

In human and rat spinal ganglia, almost every nerve cell is surrounded by a capillary loop (BERGMANN and ALEXANDER 1941; ANDRES 1961). The grey matter of the ganglia, i.e., the regions in which the nerve cell bodies are located, are much richer in capillaries than the white matter of the ganglia or peripheral nerves (BERGMANN and ALEXANDER 1941; SZABO and BÖLÖNYI 1955; DONAT and MUNKACSI 1954). This is in line with the concept that the greater the density of nerve cell bodies the greater the number of microvessels in the nervous system. The nutritional demand of the primary sensory neurons is probably great since they have very long cell processes which must by supported by material synthesized in the perikaryon.

The walls of the microvessels in the ganglia are made up of a thin endothelium, a basal lamina which at some points encloses pericytes. The extracellular spaces outside the vessels are wide.

The endothelial cells have fenestrations in their cytoplasm, but these are rather few in the sensory ganglia (SIEGRIST et al. 1968; MATTHEWS and RAISMAN 1969; JACOBS 1977; JOHNSON 1977; ARVIDSON 1979a). They have also a rather high frequency of pinocytotic vesicles (JACOBS et al. 1976; ARVIDSON 1979a), and both tight and open junctions have been observed between the endothelial cells (ARVIDSON 1979a). There are thus several routes by which substances in the blood can pass the endothelium and diffuse into the extracellular space of ganglia.

II. Vascular Permeability

Ganglia of the peripheral somatic nervous system, like those of the dorsal roots and the trigeminal nerve, differ with regard to normal vascular permeability from those in the CNS (Fig. 4). In the ganglia there is always leakage of tracers, such as cytochrome C (ARVIDSON 1979a), HRP (OLSSON 1972; JACOBS et al. 1976; ARVIDSON 1979a), ferritin (ROSENBLUTH and WISSIG 1964; ARVIDSON 1979a), and Evans and trypan blue (DOINIKOW 1913;

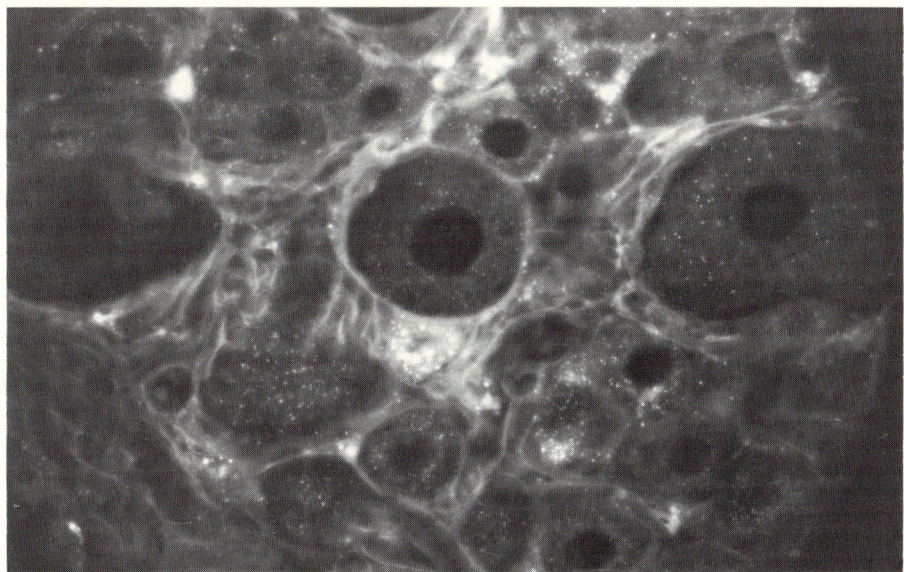

Fig. 4. Sensory ganglia do not have a vascular diffusion barrier of the same kind as in the brain or nerve fascicles. Substances from the blood are much more free to diffuse into the ganglia than in many other parts of the nervous system. One reason is that the endothelial cells of ganglionic vessels are fenestrated. This figure shows a section from a ganglion in which a fluorescent tracer (*white*) has been injected intravenously. It has passed out of the vessels and spreads in the spaces between nerve cell bodies

BEHNSEN 1926; TSCHETSCHUJEWA 1929; WAKSMAN 1961; OLSSON 1990). Radiolabeled tracers like [^{131}I]erythromycin also pass into such ganglia, and the same holds true for sodium fluorescein (MALMGREN and OLSSON 1980) and dextrans of such varying molecular weights as 3000 to 150 000 (HULTSTRÖM et al. 1983).

With the use of morphologic tracers it has been shown that leakage from the vessels occurs particularly in the areas where the nerve cell bodies are located, but adjacent parts of the nerve roots also contain extravascular tracer under normal conditions (OLSSON 1990). The leakage of macromolecular tracers from the circulation occurs presumably over fenestrations in the endothelial cells and through clefts between adjacent endothelial cells (OLSSON 1972; JACOBS et al. 1976; JOHNSON 1977; JACOBS 1977; ARVIDSON 1979a). It is not known to what extent vesicular transport of tracers occurs in the ganglionic vessels.

III. Extracellular Spaces

The extracellular space in ganglia extends from the inner part of the capsule all the way to the surface of the neurons (LIEBERMAN 1976). However, one must differentiate between a perineuronal and a perisatellite part of this space. The perineuronal compartment is demarcated by the satellite cells, which on their external side are covered by a basement lamina. The perikarya of neurons thus do not come directly into contact with connective tissue present in the extracellular space outside the satellite cells. The perisatellite portion of the extracellular space is extensive and contains collagen fibrils as well as a few fibroblasts (LIEBERMAN 1976) and surprisingly many macrophages (ARVIDSON 1979a). The composition of its matrix and fluid is unknown, but most likely there are differences between the extracellular matrix and fluid in ganglia and the endoneurium of peripheral nerves. One difference is that a substantial amount of macromolecules from the blood will form part of the fluid.

The two extracellular compartments in ganglia communicate with each other through narrow clefts present between adjacent satellite cells. Thus, it is known that substances as large as ferritin, injected intravenously in laboratory animals, can spread in the interstitial spaces and arrive as far as the plasma membrane of the neurons. Therefore, compounds originating from the blood can influence the ganglionic neurons, and in fact tracers which have arrived at the nerve cell plasma membrane will to some extent be internalized my micropinocytosis (ROSENBLUTH and WISSIG 1964; ARVIDSON 1979a).

There is a constant production of fluid from the microvessels in ganglia, and there must be a balance between production and consumption. The fluid movements in the ganglia are unknown, but we can speculate about different possibilities. Since the extracellular spaces in ganglia communicate with the spaces in the dorsal root and the peripheral nerve, part of the fluid may thus reach into these regions. Resorption into the venous part of the microvessels is another possibility. Finally, some of the fluid components may be taken up and degraded by macrophages. Experiments with HRP have revealed that these cells are very active in accumulating this compound after intravenous injection (ARVIDSON 1979a).

IV. Capsule

The capsule which surrounds the peripheral sensory ganglia is an extension of the membranous covering of peripheral nerves (Fig. 5) (WAGGENER and BEGGS 1967; LIEBERMAN 1968, 1976). The outer part is continuous with the epineurium, and the inner one is an extension of the perineurium (SHANTHAVEERAPPA and BOURNE 1962; SHANTHA and BOURNE 1968). However, some cranial ganglia, for instance the trigeminal ares, have a rather thick outer portion because of the adjacent dura.

Personal observations on mouse and rat dorsal root ganglia have revealed that their perineurium is essentially the same as in peripheral nerves as is the ultrastructure of the perineurial cells. REALE et al. (1975) examined the perineurium of rabbit spinal ganglia by freeze fracturing. Ex-

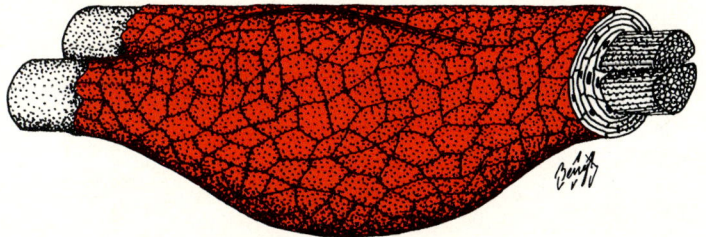

Fig. 5. Sensory ganglia are covered by a multilayered perineurium of the same kind as in peripheral nerve fascicles. It serves as an efficient diffusion barrier under normal conditions. The fluid environment of ganglia can with its help be kept independent from that in the periganglionic region

tensive meshworks of strands, presumably representing zonulae occludentes were seen between perineurial cells in the same layer. There were also occasional gap junctions, i.e., the junctional complexes are the same as those in peripheral nerve fascicles.

The perineurial sheath around ganglia most likely can act as a diffusion barrier to various substances in the same way as the perineurium in peripheral nerve fascicles. ARVIDSON (1979b) applied HRP and ferritin around the superior cervical ganglia of mice and rats and noted that the innermost layers of the perineurium acted as a barrier to these compounds under normal conditions, but we lack corresponding information from experiments on sensory ganglia.

F. Barriers and Toxic Lesions of the PNS

There are many naturally occurring and industrial substances which may cause various types of injuries to the PNS. In addition, pharmaceutical products may occasionally induce lesions of peripheral nerves or ganglia as side effects (SPENCER and SCHAUMBURG 1980; THOMAS 1980, 1984). In order to create such lesions the substances must in some way pass the vascular or the perineurial barriers of the PNS to reach their targets. This section therefore concentrates on a group of agents with known neurotoxic capacity. A selection has been done to emphasize the role of the protective barriers for the distribution of the pathological alterations and to point out the significance of associated alterations in the microenvironment.

Obviously, the functional effects are always the result of toxic influences on the neuron or its axon, the Schwann cell, or the myelin. The influence can be direct, i.e., the toxic agent reaches and affects either one of these structures as a primary event, or indirect, induced by a primary action in other structures like the endoneurial or the ganglionic blood vessels.

I. Diphtheric Neuropathy

Diphtheric neuropathy is still a matter of great concern in certain regions of the world (McDonald and Kocen 1984). The disease is caused by the toxin from *Corynebacterium diphtheriae* which is transported via the blood to the PNS. It probably passes into the parenchyma across the endoneurial blood vessels, which as is known from experimental research do not have an absolute barrier function to large-molecule proteins. It has been shown that the toxin can produce a nerve fiber lesion even after epineurial application (Boettge 1975; Oldfors et al. 1979), but it is not clear whether this effect is caused by a direct passage of the toxin across the perineurium or if it is brought into the fascicles from the local blood circulation.

The toxin thus passes into the endoneurial fluid and will finally be bound to the Schwann cells, where it elicits a segmental demyelination both in man and experimental animals (Meyer 1881; Thomas 1984; McDonald and Kocen 1984). The lesions appear to be most extensive in the spinal nerve roots and in dorsal root ganglia (Fisher and Adams 1956), i.e., in the most permeable regions of the PNS. Peripheral nerves are much better preserved (McDonald and Kocen 1984), and in order to elicit a marked lesion in nerves the toxin should be injected into the endoneurium (cf. Jacobs et al. 1966). This is in line with the concept of a "blood-nerve barrier" which is somewhat less efficient than the BBB.

II. Neuropathies Due to Metal Intoxication

There are several metal intoxications which can cause severe lesions of the PNS, the most important ones being those caused by lead, cadmium, and mercury compounds.

1. Lead Intoxication

Lead neuropathy is now a rare condition, but nevertheless experimental lead intoxication is still a topic of great scientific interest (Krigman et al. 1980). Gombault (1880, 1881) demonstrated segmental demyelination in lead-intoxicated guinea pigs, and since then this reaction, caused by a primary Schwann cell injury, is considered to represent the essential lesion. However, in man there are mainly signs of axonal degeneration, which would indicate that the axons are also primarily involved (Windebank et al. 1984).

Apart from changes in Schwann cells, myelin, and axons, the endoneurial fluid and matrix are markedly altered in lead neuropathy. For instance, chronic lead intoxication in the rat causes enlarged extracellular spaces (Ohnishi et al. 1978), increased endoneurial fluid pressure (Low and Dyck 1977; Myers et al. 1980), abnormal vascular permeability to dextrans and HRP (Myers et al. 1980), and an elevated endoneurial albumin concentration (Poduslo et al. 1982). All these changes are various expressions of an endoneurial edema of the vasogenic type. The mechanisms leading to this are not yet known in full detail, but in experimental animals the leakage of plasma occurs at least partly through widened junctions between the endothelial cells (MYERS et al. 1980; POWELL et al. 1980b).

There is clear evidence of a profound structural and functional abnormality in the endoneurial blood vessels. However, it seems unlikely that lead neuropathy is a primary vascular injury in experimental animals for the following reasons. Accumulation of lead in the endoneurium (WINDEBANK and DYCK 1981) starts earlier than the breakdown of the blood-nerve barrier to proteins (POWELL et al. 1982; PODUSLO et al. 1982), and the Schwann cell changes show an earlier onset than the endothelial alterations (POWELL et al. 1982).

Most likely small amounts of lead can pass through the endoneurial vessels as part of their normal function. When sufficient amounts have accumulated to reach a toxic concentration, the Schwann cells will suffer, and myelin will start to break down. The vascular lesions and the formation of edema may be influenced by breakdown products from myelin, and direct toxic actions on the endothelial cells may be aggravating factors. Furthermore, the increased vascular permeability may result in additional accumulation of lead in the intrafascicular compartment. The vascular lesions in lead neuropathy are important from another point of view. They create a chronic abnormal endoneurial microenvironment which can influence the nature and the progress of the lesions.

2. Mercury Intoxication

A large number of clinical and experimental studies on mercury neurotoxicity was conducted after the massive outbreak of methylmercury poisoning in Japan in the 1950s, the so-called Minamata disease (CHANG 1980). The most important neurotoxic actions of both organic and inorganic mercury occur in the central nervous system but the PNS may suffer as well. This is indicated by signs of sensory dysfunction in intoxicated patients.

Experimental animal studies have been performed to elucidate the pathomechanisms of methylmercury toxicity of the PNS. First of all, the passage of mercury into various regions of the PNS from the blood is of basic interest. SOMJEN et al. (1973) found that chronic administration of radiolabeled methylmercury resulted in an accumulation in rat spinal dorsal root ganglia to a higher degree than the cerebral cortex and the cerebellum. Spinal cord and peripheral nerves contained much lower amounts of mercury than the dorsal root ganglia. Obviously, mercury accumulates in regions of the PNS with a normally occurring high vascular permeability, i.e., the ganglia (JACOBS et al. 1975a,b; 1976; JACOBS 1980).

Subsequent studies have shown that ganglia are the preferential targets for methylmercury intoxication and that the earliest and most severe changes occur there (CHANG 1980; KUMAMOTO et al. 1986). It is also known that the dorsal roots suffer more than the ventral ones, findings in line with the assumption of an early and marked lesion in the ganglia with subsequent axonal degeneration (Arimura et al. 1988).

Experimental investigations have revealed that an early phenomenon in the peripheral ganglia after chronic exposure to mercury chloride is the formation of vacuoles around the nerve cell bodies and subsequently disintegration of the cells. By electron microscopy it has been shown that the vacuoles in fact are extracellular and formed between the ganglion neuron and the surrounding satellite cells, i.e., in the fluid microenvironment of the ganglia. However, in methylmercury poisoning the initial vacuoles are created within the cytoplasm of the ganglionic cells and not in the extracellular space (Chang 1980).

There are also indications that the blood-nerve barrier is abnormal under some conditions of mercury intoxication. Thus, Fukuhara and Tsubaki (1974) reported on increased vascular permeability in peripheral nerves of rats intoxicated with methylmercury.

3. Cadmium Intoxication

Intoxication with cadmium may occur as a hazard in factories and as a result of uncontrolled pollution of water and food products. The most well-known example of subacute and chronic intoxication leading to human disease is the so-called itai-itai syndrome which occurred in Japan from 1939 until 1945 (Murata 1971). This outbreak was caused by polluted drinking water and from rice grown on fields covered with contaminated water. Tischner (1980) in reviewing the effects of cadmium intoxication described the main symptoms as severe pain in the bones, wadding gait, aminoaciduria, glycosuria, decreased pancreatic function, pronounced osteomalacia, and multiple pathological fractures. However, there were no reports describing pathological changes in the nervous system of intoxicated patients.

Experimental research has proved that the nervous system may suffer severe changes after systemic administration of cadmium salts to laboratory animals. Profound alterations develop in dorsal root ganglia and trigeminal ganglia and lead to hemorrhages and a protein-rich edema (Gabbiani 1966; Gabbiani et al. 1974; Schlaepfer 1971; Arvidson 1980, 1983). These lesions are thought to be caused by a primary vascular degeneration afflicting the endothelial cells of the ganglionic vessels and are similar to those which may occur in the testis of intoxicated rats (Parizek 1956, 1957; Gunn et al. 1963).

The endothelial lesions present in peripheral ganglia of cadmium-intoxicated animals start as irregular dilatations of interendothelial cell clefts and damage to the cell membrane of the endothelial cells (Gabbiani et al. 1974). Later on, there are large gaps in the endothelial lining through which bleeding and vasogenic edema arise (Gabbiani 1966; Gabbiani et al. 1974; Schlaepfer 1971; Arvidson 1980). Apparently this response is also dependent on the age of the animals, immature individuals being more resistant than adult ones (ARVIDSON 1983).

ARVIDSON (1980) found marked regional differences in the severity of cadmium-induced lesions in the PNS of mice. Endothelial cell damage was a constant phenomenon in sensory ganglia, but the sympathetic superior cervical ganglion and the sciatic nerve showed such changes in only 50% of the cases. No vascular lesions at all were detected in the celiac ganglion, and the perineurium covering ganglia and sciatic nerve fascicles remained structurally and functionally intact. Acute cadmium intoxication in the mouse is thus an example of a selective lesion to the vasculature, whereas the other component of the peripheral nerve barrier, i.e., the perineurium, is unchanged. The marked variations in the intensity of the vascular damage between different topographical regions are unknown but may well reflect hitherto unknown variations in the structure or function of different endothelial cells.

Differences in the permeability properties between the brain and peripheral ganglia of adult rodents may well account for the differences in their susceptibility to cadmium intoxication (ARVIDSON 1980). According to this hypothesis, cadmium in the blood is bound to plasma proteins (NORDBERG 1977; BREMNER 1979), and the metal-protein complex can pass the endothelial lining (ARVIDSON and TJÄLVE 1982) through intercellular junctions or through fenestrations to cover both the luminal and abluminal side of the endothelial cells. This extensive exposure of the endothelial cells may produce cellular injury by interference with sulfhydryl-containing enzymes (ARVIDSON 1983).

The endothelial changes in ganglia and peripheral nerves in cadmium intoxication thus lead to such marked environmental changes as hemorrhage and vasogenic edema. Most likely, these events and direct cadmium-induced cytotoxic effects on nerve cell bodies in ganglia cause the final destruction of neurons in intoxicated rodents (TISCHNER 1980).

III. Neuropathy Due to Industrial Agents

Many industrial agents have the capacity to induce lesions of the PNS provided that they are brought into the endoneurial compartment of peripheral nerves or into the fluid microenvironment of ganglia. In this connection one example is provided by a recently recognized condition afflicting the PNS caused by an industrial agent, the so-called toxic oil syndrome.

1. Toxic Oil Syndrome

An outbreak of a condition characterized by severely disabling peripheral neuropathy combined with respiratory distress, anorexia, and myalgia started in spring 1981 in Spain which is now called "toxic oil syndrome". Within 4 weeks 20 000 sufferers were identified. This devastating disease was associated with consumption of an illegally marketed cooking oil. The "toxic oil syndrome" was thus apparently caused by the ingestion of a denaturated rape seed oil (NORIEGA et al. 1982; PORTERA-SANCHES and DEL SER QUIJANO 1983; DEL SER QUIJANO et al. 1986).

The patients suffered from a disseminated mononeuropathy involving practically every nerve trunk, which bore similarities with systemic vasculitis. Severe inflammatory changes were also present in the epi- and perineurium, and there was a high degree of perineurial fibrosis (NORIEGA et al. 1982). About 75% of the cases had such inflammatory signs with involvement of the epineurial vessels. Severe degeneration of fibers occurred in the fascicles of the affected nerves.

The components in the denaturated oil producing the vascular and inflammatory lesions have not been identified (PORTERA-SANCHES and DEL SER QUIJANO 1983), and there is no good explanation for the disseminated distribution of various peripheral nerve trunks and why sensory ganglia and spinal nerve roots are spared. The mechanisms leading to axonal degeneration and myelin changes are also unknown, but it has been proposed that the vasculitis in combination with the marked perineurial fibrosis in some way are related to the nerve fiber changes (PORTERA-SANCHES and DEL SER QUIJANO 1983).

The toxic oil syndrome is an example of a neurotoxic condition with marked vascular and inflammatory changes which in a unique way show a predilection for the epi- and the perineurium. It illustrates the great repertoire of pathological changes which may occur in neurotoxicological conditions of the PNS.

IV. Neuropathy Due to Drugs

Three examples are given of neuropathies induced by drugs, all of them with different pathogenesis. The barriers in the PNS appear to play a crucial role in their development.

1. Hexachlorophene Neuropathy

Hexachlorophene is used as an antimicrobial agents in soaps and liquid detergents to quite an extent and might be dangerous to animals and man after repeated exposure in high concentrations (TOWFIGHI 1980). Its neurotoxicity and characteristic pathological alterations have been extensively investigated in the past (KIMBROUGH 1976; TOWFIGHI and GONATAS 1976; TOWFIGHI 1980). Cell membranes are the primary targets for instance in bacteria and in red blood cells. Since myelin represents a specialization of the plasma membrane, it may suffer as well.

The most marked myelin changes occur in the brain, but peripheral myelin may be injured as well (TOWFIGHI et al. 1973; PERSSON et al. 1976, 1978). The characteristic lesion is an intramyelinic edema with a splitting of myelin lamellae between the intraperiod lines. The changes might be reversible, and for unknown reasons the nerve roots will be more damaged than peripheral nerves (TOWFIGHI et al. 1973).

Hexachlorophene intoxication is an example of an intramyelinic edema with a low protein content. Vascular permeability to proteins in the blood is normal (TOWFIGHI 1980). The myelin vacuoles remain inaccessible to protein tracers injected into the extracellular spaces of the damaged nerves (TOWFIGHI and GONATAS 1977). The edema results in an elevated endoneurial fluid pressure to such a degree that the blood flow in the nerves may be impeded (MYERS et al. 1982).

2. Isoniazid Neuropathy

Peripheral neuropathy may occur as a complication in excessive use of the antituberculosis drug isoniazid (KLINGHARDT 1954; BLAKEMORE 1980). Experimental studies in laboratory animals have shown that the essential primary event is a degeneration of axons followed by myelin breakdown (JACOBS et al. 1979). Apparently, the lesions start as multifocal vacuoles located between axon and Schwann cells (JACOBS et al. 1979). It might be a direct effect of isoniazid since this compound can easily penetrate into a nerve due to the fact thay only about 40% is protein bound (ALLEN et al. 1975). Otherwise, a major mechanism appears to be interference with the metabolism of vitamin B_6, and experimentally, addition of pyridoxine will prevent the development of the neuropathy (ROSS 1958; BLAKEMORE 1980).

There are also permeability disturbances of the endoneurial vessels which start at the same time as nerve fiber lesions become manifest (OLSSON 1990). By using intravenously injected fluorochrome-labeled serum albumin as a tracer in intoxicated rats, it was shown that there was a patchy albumin-rich endoneurial edema in the damaged areas of the nerves. Intraneural edema has also been noticed by SCHRÖDER (1975) with electron microscopy.

3. Doxorubicin Neurotoxicity

One of the most striking examples of selective neurotoxicity to ganglia is that caused by the drug doxorubicin in experimental animals (CHO et al. 1980). Doxorubicin is used in the treatment of malignant tumors and can pass into regions of the brain and the PNS which lack a barrier between the blood and the parenchyma. For this reason permeable regions like the ganglia and circumventricular brain areas will be damaged as shown in many experimental studies (CHO 1977; CHO et al. 1980; SAHENK and MENDELL 1979; BIGOTTE and OLSSON 1982a,b; BIGOTTE et al. 1982b; CAVANAGH et al. 1987). The brain parenchyma, on the other hand, will not be damaged, protected as it is by the BBB.

The mechanisms of doxorubicin cytotoxicity in sensory ganglia have been a matter of discussion. Most authors agree that the major action on cells is a direct effect on the chromosomal DNA due to the intercalation of the compounds into the DNA double helix with breakage of the nucleic acid strands. This mechanism has consequences for the protein synthesis which has been demonstrated in various types of cells, including neurons (CHO 1977; CHO et al. 1980; JORTNER and CHO 1981; CAVANAGH et al. 1987).

It is generally accepted that systemically injected doxorubicin does not cause any changes in peripheral nerves apart from those which are secondary to lesions in the ganglia and that this absence of lesions is due to the low permeability of the blood-nerve barrier (cf. BIGOTTE and OLSSON 1982a,b, 1983, 1984). However, if the blood-nerve barrier is bypassed by direct microinjections into the sciatic nerve fascicles of the rat, a delayed subacute demyelination secondary to focal Schwann cell degeneration will take place (ENGLAND et al. 1988a). Fluorescent microscopic tracing of doxorubicin showed in these experiments that almost immediately after the intraneural injection the drug became bound to the nucleus of the Schwann cells.

Doxorubicin can also damage nerve cell bodies in the ganglia, spinal cord, and brain stem after retrograde axonal transport from terminals. Local injection of the compound will lead to a delayed degeneration of both sensory (BIGOTTE and OLSSON 1987; KONDO et al. 1987) and motor neurons (BIGOTTE and OLSSON 1982, 1983; YAMAMOTO et al. 1984; ENGLAND et al. 1988b). These observations have two major implications. By using doxorubicin so-called suicide degeneration of neurons can be achieved without the use of very potent toxic agents like ricins. Secondly, degeneration of motor neurons after retrograde axonal transport of doxorubicin has similarities to motor neuron disease and might be useful for studies on certain aspects of this disease.

Epirubicin is an analogue of doxorubicin which is less cardiotoxic than its parent drug (ARCAMONE et al. 1978; GIUILANI and KAPLAN 1980). After intravenous administration epirubicin leaks into dorsal root ganglia, trigeminal and superior cervical sympathetic ganglia. Peripheral ganglia have fenestrated blood vessels which are permeable to doxorubicin and other exogenous compounds. It is therefore most likely that epirubicin will pass into the ganglia by the same mechanisms.

In the PNS the distribution and toxicity of epirubicin were thus, in general terms, similar to those observed earlier with doxorubicin (CHO et al. 1980; JORTNER and CHO 1981; BIGOTTE and OLSSON 1983).

4. Neurotoxicity of Local Anesthetics

New local anesthetic drugs obviously require extensive trials before clinical use. Previously, most tests carried out in experimental animals were limited to changes taking place in the nerve fibers. However, MYERS et al. (1986) and KALICHMAN et al. (1988) approached this problem from a much broader view. Apart from examinations of the nerve fibers, they also investigated the fluid and cells of the endoneurium.

Local anesthetic agents such as 2-chloroprocaine, tetracaine, etidocaine, and mepivacaine can cause a number of alterations in the endoneurial cells and fluid (MYERS et al. 1986b; KALICHMAN et al. 1988). Endoneurial edema may arise, mast cells may become degranulated, and the end result can be endoneurial and perineurial fibrosis of varying degrees of severity. The permeability function of the perineurium may also be disturbed, facilitating the passage of compounds into the fascicles.

G. Concluding Remarks

It is well-known that the BBB has several important implications for the normal function of the brain, regulating the passage of various substances from the blood into the parenchyma and that it has a role in protecting the brain and spinal cord from many substances with neurotoxic capacity which may be present in the blood (RAPOPORT 1976; KATZMAN 1976; BRADBURY 1979; GOLDSTEIN and BETZ 1986; CRONE 1986; NEUWELT 1988).

Over the past few years more and more scientists have become aware of the fact that also in the PNS protective barriers profoundly and in many different ways influence the course of neurotoxic diseases afflicting this part of the nervous system (cf. JACOBS 1980; DYCK et al. 1984; RECHTHAND and RAPOPORT 1987; OLSSON 1990).

Peripheral nerves are complex structures, and the composition of the extracellular matrix and fluid differs between the intra- and the extrafascicular regions. The epineurium is similar to the connective tissue in other parts in the body, and its extracellular fluid is kept separate from that in the fascicles by the barrier properties of the perineurium.

The microenvironment of peripheral nerve fascicles is normally maintained by the joint action of the endoneurial vessels and the perineurium. Endoneurial vascular permeability has certain similarities to the vascular permeability in the CNS, but compared with the BBB the blood-nerve barrier is less efficient. This would imply that toxic and infectious agents as well as some pharmaceutical products may have easier access to the parenchyma in nerves than to the brain parenchyma.

The lack of lymphatics in the nerve fascicles indicates that removal of toxic substances in the endoneurial fluid meets with some difficulties. Unlike the CNS, peripheral nerves lack the fluid-sink action exerted by the CSF and might therefore be susceptible to injury. Toxic products which in some way have passed into or accumulated in the fascicles may spread in the fluid environment of nerve fascicles in both centripetal and centrifugal directions and cause widespread injury.

At some places the fascicles of a peripheral nerve lack the efficient barrier function of the perineurium. This is the situation in terminal nerve branches and at the myoneural junctions. Toxic substances present in such regions can easily gain access to the interior of the fascicles and cause injury after they have reached their target structures. Important examples are biological toxins such as in tetanus and diphtheria.

Peripheral ganglia are covered by a perineurium which acts as an efficient diffusion barrier and prevents substances from gaining access to the ganglia from their immediate surroundings. However, ganglionic vessels lack an efficient vascular barrier to many substances. This is one important pathogenetic mechanism in toxic and inflammatory lesions with a preferential localization to the ganglia, for instance intoxication with doxorubicin, lead, mercury, and cadmium.

The protective barriers may themselves be influenced by a pathological process and then respond with increased permeability. This may then lead to the formation of edema in the PNS, i.e., one of the cardinal features of many diseases in the PNS of toxic nature. As is well-known, the essential function of the PNS is to convey information to and from the CNS by means of electrical and chemical signals. Such complicated mechanisms require for optimal performance a microenvironment of balanced ionic and osmotic composition. Barrier lesions can easily cause profound changes in the fluid microenvironment of the PNS, which in turn may lead to functional and structural abnormalities.

References

Adams WE (1942) The blood supply of nerves. I. Historical review. J Anat 76:323–341

Adams WE (1943) The blood supply of nerves. II. The effect of exclusion of its regional sources of supply on the sciatic nerve of the rabbit. J Anat 77:243–250

Ahmed AM, Weller RO (1979) The blood-nerve barrier and reconstitution of the perineurium following nerve grafting. Neuropathol Appl Neurol 5:469–475

Aker FD (1972) A study of hematic barriers in peripheral nerves of albino rabbits. Anat Rec 174:469–475

Akert K, Sandri C, Weibel ER, Peper K, Moor H (1976) The fine structure of the perineural endothelium. Cell Tissue Res 165:281–295

Allen BW, Ellard GA, Gammon PT, McDougall AC, Rees RJW, Weddell AGM (1975) The penetration of dapsone, rifampicin, isoniazid and pyraninamide into peripheral nerves. Br J Pharmacol 55:151–155

Andres KH (1961) Untersuchungen über den Feinbau von Spinalganglien. Z Zellforsch 55:1–48

Arcamone F, di Marco A, Casazza AM (1978) Chemistry and pharmacology of new antitumor anthracyclines. In: Umezawa H (ed) Advances in cancer chemotherapy. Japan Science Society Press, Tokyo, pp 297–312

Arimura K, Murai Y, Rosales RL, Izumo S (1988) Spinal roots of rats poised with methylmercury: physiology and pathology. Muscle Nerve 11:762–768

Arvidson B (1977) Cellular uptake of exogenous horseradish peroxidase in mouse peripheral nerve. Acta Neuropathol (Berl) 37:35–43

Arvidson B (1979a) Distribution of intravenously injected protein tracers in peripheral ganglia of adult mice. Exp Neurol 63:388–410

Arvidson B (1979b) A study of the perineurial diffusion barrier of a peripheral ganglion. Acta Neuropathol (Berl) 46:139–145

Arvidson B (1980) Regional differences in severity of cadmium-induced lesions in the peripheral nervous system in mice. Acta Neuropathol (Berl) 49:213–221

Arvidson B (1984) Evidence for vesicular transport of horseradish peroxidase across endoneurial vessels of the sciatic nerve in normal mice. Acta Neuropathol (Berl) 64:1–11

Arvidson B, Tjälve H (1982) Distribution of ^{109}Cd in the nervous system of rats after intravenous injection. J Neurotoxicol Neurobiol 1:1–2

Behnsen G (1926) Über die Farbstoffspeicherung im zentralen Nervensystem der weißen Maus in verschiedenen Alterszuständen. Z Zellforsch Mikrosk Anat 4:515–572

Bell MA, Wedell AGM (1984a) A morphometric study of intrafascicular vessels of mammalian sciatic nerve. Muscle Nerve 7:524–534

Bell MA, Wedell AGM (1984b) A descriptive study of the blood vessels of the sciatic nerve in the rat, man and other mammals. Brain 107:871–898

Bergmann L, Alexander L (1941) Vascular supply of the spinal ganglia. Arch Neurol Psychiatry 46:761–782

Bigotte L, Olsson Y (1983) Cytotoxic effects of adriamycin on mouse hypoglossal neurons following retrograde axonal transport from the tongue. Acta Neuropathol (Berl) 61:161–168

Bigotte L, Olsson Y (1984) Cytotoxic effects of adriamycin on the central nervous system of the mouse – cytofluorescence and electron-microscopic observations after various modes of administration. Acta Neurol Scand [Suppl 100] 70:55–67

Bigotte L, Olsson Y (1987) Degeneration of trigeminal ganglion neurons caused by retrograde axonal transport of doxorubicin. Neurology 37:985–992

Bigotte L, Arvidson B, Olsson Y (1982a) Cytofluorescence localization of adriamycin in the nervous system. I. Distribution of the drug in the central nervous system of normal adult mice after intravenous injection. Acta Neuropathol (Berl) 57:121–129

Bigotte L, Arvidson B, Olsson Y (1982b) Cytofluorescence localization of adriamycin in the nervous system. II. Distribution of the drug in the somatic and autonomic peripheral nervous system of normal adult mice after intravenous injection. Acta Neuropathol (Berl) 57:130–136

Bishop MA (1982) A fine-structural investigation on the extent of perineurial invest-ment of the nerve supply to the pulp in rat molar teeth. Arch Oral Biol 27:225–234

Blakemore WF (1980) Isoniazid. In: Spencer PS, Schaumburg HH (eds) Experi-mental and clinical neurotoxicology, vol 1. Williams and Wilkins, Baltimore, pp 476–490

Boddingius J, Rees RJW, Wedell AG (1972) Defects in the blood-nerve barrier in mice with leprosy neuropathy. Nature [New Biol] 237:190–195

Boettge R (1975) Elektronenmikroskopische Untersuchungen über die experimentelle diphtherische Neuropathie nach epineuraler Applikation des Diphtherietoxins bei neugeboreenen Ratten. Dissertation University of Frankfurt

Bradbury MWB (1979) The concept of a blood-brain barrier. Wiley, New York

Bradbury MWB, Crowder J (1976) Compartments and barriers in the sciatic nerve of the rabbit. Brain Res 103:515–522

Bremner I (1979) Mammalian absorption, transport and excretion of cadmium. In: Webb M (ed) The chemistry, biochemistry and biology of cadmium, vol 1. Elsevier, Amsterdam, pp 178–179

Brightman M, Klatzo I, Olsson Y, Reese TS (1970) The blood-brain barrier to proteins under normal and pathological conditions. J Neurol Sci 10:215–222

Broadwell RD, Brightman MW (1976) Entry of peroxidase into neurons of the central and peripheral nervous systems from extracerebral and cerebral flow. J Comp Neurol 166:257–284

Broman T (1949) The permeability of the cerebrospinal vessels in normal and pathological conditions. Munksgaard, Copenhagen

Burkel WE (1967) The histological fine structure of perineurium. Anat Rec 158: 177–190

Carbonetto S (1984) The extracellular matrix of the nervous system. Trends Neurosci 7:382–387

Causey G, Palmer E (1953) The epineurial sheath of a nerve as a barrier to the diffusion of phosphate ions. J Anat 87:30–36

Cavanagh JB, Tomiwa K, Munro PMG (1987) Nuclear and nucleolar damage in adriamycin-induced toxicity to rat sensory ganglion cells. Neuropathol Appl Neurobiol 13:23–38

Chang LW (1980) Mercury. In: Spencer PS, Schaumburg HH (eds) Experimental and clinical neurotoxicology, vol 1. Williams and Wilkins, Baltimore, pp 508–526

Cho E-S (1977) Toxic effects of adriamycin in the ganglia of the peripheral nervous system – a neuropathological study. J Neuropathol Exp Neurol 136:907–915

Cho E-S, Spencer PS, Jortner BS, Schaumburg HH (1980) A single intravenous injection of doxorubicin (adriamycin) induces sensory neuropathy in rats. Neurotoxicology 1:583–591

Clara M, Özer N (1960) Untersuchungen über die sogenannte Nervenscheide. Acta Neuroveg 20:1–15

Cravioto H (1966) The perineurium as a diffusion barrier – ultrastructural correlates. Bull Los Angeles Neurol Soc 31:196–211

Crescitelli F (1951) Nerve sheath as a barrier to the action of certain substances. Am J Physiol 166:229–236

Crescitelli F, Geissman TA (1951) Certain effects of antihistamines and related compounds on frog nerve fibers. Am J Physiol 164:509–514

Crone C (1986) The blood-brain barrier; a modified tight epithelium. In: Suckling AJ, Rumsby MG, Bradbury MWB (eds) The blood-brain barrier in health and disease. Harward, Chichester, pp 17–40

Dainty J, Krnjevic K (1955) The rate of exchange of Na24 in cat nerves. J Physiol (Lond) 128:489–496

Del Ser Quijano T, Esteban García A, Martínez Martín M, Morales Otal MA, Pondal Sordo M, Pérez Vergara P, Portera Sánches y A (1986) Evolución de la afección neuromuscular en el síndromo del aceite tóxico. Rev Med Clín 87:231–236

Doinikow B (1913) Histologische und histopathologische Untersuchungen am peripheren Nervensystem mittels vitaler Färbung. Folia Neurobiol (Leipz) 7:731–744

Donath T, Munkacsi I (1954) Morphologische Beiträge zum Kreislauf des Gasserschen Ganglions. Acta Morphol 5:275–288

Dyck PJ, Windebank AJ, Low PA, Baumann WJ (1980) Blood-nerve barrier in rat and cellular mechanisms of lead-induced segmental demyelination. J Neuropathol Exp Neurol 39:700–714

Dyck PJ, Thomas PK, Lambert EH, Bunge R (Eds) (1984) Peripheral Neuropathy Second edition. Saunders, Philadelphia pp 1773–1810

Emirouglu F (1955) The permeability of the peripheral nerve sheath in frogs. Arch Int Physiol Biochim 63:161–172

England JD, Rhee EK, Said G, Sumner AJ (1988a) Schwann cell degeneration induced by doxorubicin (adriamycin). Brain 111:901–913

England JD, Asbury AK, Rhee EK, Sumner AJ (1988b) Lethal retrograde axoplasmic transport of doxorubicin (adriamycin) to motor neurons. Brain 111:915–926

Feng TP, Gerard RW (1930) Mechanisms of nerve asphyxiation: with a note on the nerve sheath as a diffusion barrier. Proc Soc Exp Biol Med 27:1073–1078

Feng TP, Liu YM (1949) The connective tissue sheath of the nerve as effective diffusion barrier. J Cell Comp Physiol 34:1–21

Fisher CM, Adams RD (1956) Diphtheric polyneuritis – a pathologic study. J Neuropathol Exp Neurol 15:243–256

French JD, Strain WH (1947) Experimental contrast radiography of peripheral nerves. J Neuropathol Exp Neurol 4:401–407

Fukuhara N, Tsubaki T (1974) Increased vascular permeability in the peripheral nerves of rats intoxicated with methyl mercury. Rinsho Shinkeigaku 14:604–612

Gabbiani G (1966) Action of cadmium chloride on sensory ganglia. Experientia 22:261–262

Gabbiani G, Badonnel MC, Mathewson SM, Ryan GB (1974) Acute cadmium intoxication. Early selective lesions of endothelial clefts. Lab Invest 30:686–695

Gamble HJ (1964) Comparative electron-microscopic observations on the connective tissue of a peripheral nerve and a spinal nerve root in the rat. J Anat 98:17–25

Giuliani FC, Kaplan NO (1980) New doxorubicin analogs active against doxorubicin-resistant colon tumor xenografts in the nude mouse. Cancer Res 40:4682–4687

Goldstein GW, Betz AL (1986) The blood-brain barrier. Sci Am 225:70–79

Gombault A (1880) Contribution à l'étude anatomique de la névrite parenchymateuse subaiguë et chronique: névritre segmentaire périaxiale.Arch Neurol (Paris) 1:11–38

Gombault A (1881) Contribution à l'étude anatomique de la névrite parenchymateuse subaiguë et chronique: névritre segmentaire périaxiale. Arch Neurol (Paris) 1:177–190

Graham RC, Karnovsky MJ (1966) The early stages of absorbtion of injected horseradish peroxidase in the proximal tubules of mouse kidney: ultrastructural cytochemistry by a new technique. J Histochem Cytochem 14:391–398

Gunn SA, Gould TC, Anderson WAD (1963) The selective injurious response of testicular and epididymal blood vessels to cadmium and its prevention by zinc. Am J Pathol 42:695–702

Henle J (1841) Allgemeine Anatomie. Voss, Leipzig

Hultström D, Malmgren L, Gilstring D, Olsson Y (1983) FITC-dextrans as tracers for macromolecular movements in the nervous system. Acta Neuropathol (Berl) 59:53–58

Jacobs JM (1977) Penetration of systemically injected horseradish peroxidase into ganglia and nerves of the autonomic nervous system.J Neurocytol 6:607–611

Jacobs JM (1980) Vascular permeability and neural injury. In: Spencer PS, Schaumburg HH (eds) Experimental and clinical neurotoxicology, vol 1. William & Wilkins, Baltimore, pp 102–117

Jacobs JM, Cavanagh JB, Mellick RS (1966) Intraneural injections of diphtheria toxin. Br J Exp Pathol 47:507–512

Jacobs JM, Cavanagh JB, Carmichael N (1975a) The effect of chronic dosing with mercuric chloride on dorsal root and trigeminal ganglia of rats. Neuropathol Exp Neurobiol 3:321–337

Jacobs JM, Carmichael N, Cavanagh JB (1975b) Ultrastructural changes in the dorsal root and trigeminal ganglia of rats poisoned with methyl mercury. Neuropathol Exp Neurol 1:1–19

Jacobs JM, MacFarlane RM, Cavanagh JB (1976) Vascular leakage in the dorsal root ganglia of the rat studied with horseradish peroxidase. J Neurol Sci 29:95–107

Jacobs JM, Miller RH, Whittle A, Cavanagh JB (1979) Studies on early changes in acute isoniazid neuropathy in the rat. Acta Neuropathol (Berl) 47:85–92

Johnson PC (1977) Fenestrated endothelium in the human peripheral nervous system. J Neuropathol Exp Neurol 36:607–608

Jortner BS, Cho ES (1981) Neurotoxicity of adriamycin in rats. A low dose effect. Cancer Treatment Reports 64:257–263

Kalichman MW, Powell HC, Myers RR (1988) Pathology of local anesthetic-induced nerve injury. Acta Neuropathol (Berl) 75:583–589

Karnovsky MJ (1967) The ultrastructural basis of capillary permeability studied with peroxidase as a tracer. J Cell Biol 35:213–223

Katzman R (1976) Blood-brain-CSF barriers. In: Siegel GJ, et al (eds) Basic neurochemistry, 2nd edn. Little, Brown, Boston

Kerjaschki D, Stockinger L (1970) Zur Struktur und Funktion des Perineuriums. Die Endigungsweise des Perineuriums vegetativer Nerven. Z Zellforsch 110:386–400

Key A, Retzius G (1873) Studien in der Anatomie des Nervensystems. Arch Mikrosk Anat 9:308–386

Key A, Retzius G (1876) Studien in der Anatomie des Nervensystems und des Bindgewebes. Samson and Wallin, Stockholm

Kimbrough RD (1976) Hexachlorophene. Toxicity and use as an antibacterial agent. Essays Toxicol 7:99–112

Klemm H (1970) Das Perineurium als Diffusionsbarriere gegenüber Peroxydase bei epi- und endoneuraler Applikation. Z Zellforsch Mikrosk Anat 108:431–444

Klinghardt EW (1954) Experimentelle Nervenfaserschädigungen durch Isonicotinsaurehydrazide und ihre Bedeutung für die Klinik. Verh Dtsch Ges Inn Med 60:764–785

Kondo A, Ohnishi A, Nagara H, Tateishi J (1987) Neurotoxicity in primary sensory neurons of adriamycin administered through retrograde axoplasmic transport in rats. Neuropathol Exp Neurobiol 13:177–192

Krigman M, Bouldin TW, Mushak P (1980) Lead. In: Spencer PS, Schaumburg HH (eds) Experimental and clinical neurotoxicology, vol 1. Williams and Wilkins, Baltimore, pp 490–507

Kristensson K, Olsson Y (1971) The perineurium as a diffusion barrier to protein tracers. Differences between mature and immature animals. Acta Neuropathol (Berl) 17:127–135

Krnjevic K (1954a) The connective tissue of frog sciatic nerve. Q J Exp Physiol 39:55–64

Krnjevic K (1954b) Some observations on perfused frog sciatic nerves. J Physiol (Lond) 123:338–345

Kumamoto T, Fukuhara N, Miyatake T, Araki K, Takahashi Y, Araki S (1986) Experimental neuropathy induced by methyl mercury compounds: autoradiographic study of GABA uptake by dorsal root ganglia. Eur Neurol 25:269–277

Lampert PW (1969) Mechanism of demyelination in experimental allergic neuritis. Lab Invest 20:127–135

Latker CH, Lynch KJ, Shinowara NL, Rapoport SI (1985) Vesicular profiles in frog perineurial cells preserved by rapid-freezing and freeze-substitution. Brain Res 345:170–175

Lieberman AR (1968) The connective tissue elements of the mammalian nodose ganglion. An electron microscope study. Z Zellforsch Mikrosk Anat 89:95–106

Lieberman AR (1976) Sensory ganglia. In: London DN (ed) The peripheral nerve, Chapman and Hall, London, pp 188–278

Liebert UG, Seitz RJ, Weber T, Wechsler W (1985) Immunocytochemical studies of serum proteins and immunoglobulins in human sural nerve biopsies. Acta Neuropathol (Berl) 68:39–47

Low FN (1976) The perineurium and connective tissue of peripheral nerve. In: London DN (ed) The peripheral nerve. Chapman and Hall, London, pp 159–187

Low PA (1984) Endoneurial fluid pressure and microenvironment of nerve. In: Dyck PJ, Thomas PK, Lambert EH, Bunge R (eds) Peripheral neuropathy, vol 1, 2nd edn. Saunders, Philadelphia, pp 599–617

Low PA, Dyck PJ (1977) Increased endoneurial fluid pressure in experimental lead neuropathy. Nature 269:427–428

Low PA, Marchand G, Knox F, Dyck PJ (1977) Measurement of endoneurial fluid pressure with polyethylene matrix capsule. Brain Res 122:373–376

Low PA, Yao JK, Poduslo JF, Donald DE, Dyck PJ (1982) Peripheral nerve microenvironment: collection of endoneurially enriched fluid. Exp Neurol 77:208–214

Low PA, Poduslo JF, Dyck PJ (1983) Characterization of mammalian endoneurial fluid. Ann Neurol 14:120–121

Lundborg G (1970) Ischemic nerve injury. Scand Plast Reconstr Surg [Suppl]6:11–35

Lundborg G (1988) Nerve injury and repair. Livingstone, Edinburgh

Malmgren LT, Brink JJ (1975) Permeability barriers to cytochrome-C in nerves of adult and immature rats. Anat Rec 181:755–760

Malmgren LT, Olsson Y (1978) A sensitive method for histochemical demonstration of horseradish peroxidase in neurons following retrograde axonal transport, Brain Res 148:279–294

Malmgren LT, Olsson Y (1980) Differences between the central and the peripheral nervous system in permeability to sodium fluorescein. J Comp Neurol 191: 103–123

Malmgren LT, Olsson Y, Olsson T, Kristensson K (1978) Uptake and retrograde axonal transport of various exogenous macromolecules in normal and crushed hypoglossal nerves. Brain Res 153:477–493

Manery JF, Bale WF (1941) The penetration of radioactive sodium and phosphorus into the extra- and intracellular phases of tissues. Am J Physiol 1132:215–233

Martin KH (1964) Untersuchungen über die perineurale Diffusionsbarriere an gefriergetrockneten Nerven. Z Zellforsch Mikrosk Anat 64:404–415

Mata M, Staple J, Fink DJ (1987) The distribution of serum albumin in rat peripheral nerve. J Neuropathol Exp Neurol 46:485–494

Matthews MR, Raisman G (1969) The ultrastructure and somatic efferent synapses of small granule-containing cells in the superior cervical ganglion. J Anat 105: 255–282

McDonald WI, Kocen RS (1984) Diphtheric neuropathy. In: Dyck PJ, Thomas PK, Lambert EH, Bunge RWB (eds) Peripheral neuropathy, vol 2. Saunders, Philadelphia, pp 2010–2016

Mesulam MM (ed) (1982) Tracing neural connections with horseradish peroxidase. Wiley, New York

Meyer P (1881) Anatomische Untersuchungen über diphtheritische Lähmung. Virchows Arch Pathol Anat 85:181–191

Michel ME, Shinowara NL, Rapoport SI (1984) Presence of a blood-nerve barrier within blood vessels of frog sciatic nerve. Brain Res 299:25–30

Mizisin AP, Myers RR, Powell HC (1986) Endoneurial sodium accumulation in galactosemic rat nerves. Muscle Nerve 9:440–444

Murata J (1971) Chronic entero-osteo-nephropathy due to cadmium ("itai-itai" disease). J Jpn Med Assoc 65:15–24

Myers RR, Powell HC, Costello MC, Lampert PW, Zweifach WB (1978) Endoneurial fluid pressure: direct measurement with micropipettes. Brain Res 148: 510–522

Myers RR, Powell HC, Shapiro HM, Costello ML, Lampert PW (1980) Changes in endoneurial fluid pressure, permeability and peripheral nerve ultrastructure in experimental lead neuropathy. Ann Neurol 8:392–401

Myers RR, Mizisin AP, Powell HC, Lampert PW (1982) Reduced nerve blood flow in hexachlorophene neuropathy: relationship to elevated endoneurial fluid pressure. J Neuropathol Exp Neurol 40:391–399

Myers RR, Heckman HM, Powell HC (1983) Endoneurial fluid is hypertonic. Results of microanalysis and its significance in neuropathy. J Neuropathol Exp Neurol 42:217–224

Myers RR, Murakami H, Powell HC (1986a) Reduced nerve blood flow in edematous neuropathies: a biomechanical mechanism. Microvasc Res 32:145–151

Myers RR, Kalichman MW, Reisner LS, Powell HC (1986b) Neurotoxicity of local anesthetics: altered perineurial permeability, edema and nerve fiber injury. Anesthesiology 64:29–35

Neuen E, Seitz RJ, Langenbach M, Wechsler W (1987) The leakage of serum proteins across the blood-nerve barrier in hereditary and inflammatory neuropathies. Acta Neuropathol (Berl) 73:53–61

Neuwelt EA (1988) Implications of the blood-brain barrier and its manipulation, vol 1. Plenum, New York

Nordberg M (1977) Studies on metallothionein and cadmium. Thesis, Karolinska Institutet, Stockholm

Noriega AR, Gômez-Reinio J, Lôpez-Encuentra A, Martin-Escribano P, Solis-Herruzo JA, Valle-Gutiérrez F (1982) Toxic epidemic syndrome, Spain 1981. Lancet 2:697–702

Obst T (1971) Über das Endgebiet des Perineuriums und den Zahnnerven der Ratte. Z Zellforsch 14:515–531

Ohnishi A, Schilling K, Brimijoin WS, Lambert E, Fairbanks VF, Dyck PJ (1978) Lead neuropathy. Morphometry, nerve conduction, choline acetyltransferase transport: new finding of endoneurial edema associated with segmental demyelination. J Neuropathol Exp Neurol 37:499–512

Oldfors A (1980a) Functional aspects of the perineurial barrier in normal and protein-deprived rats. Thesis, University of Gothenburg

Oldfors A (1980b) Macrophages in peripheral nerve. An ultrastructural and histochemical study on rats. Acta Neuropathol (Berl) 49:43–48

Oldfors A (1981) Permeability of the perineurium of small nerve fascicles: an ultrastructural study using ferritin in rats. Neuropathol Appl Neurobiol 7:183–191

Oldfors A, Sourander P (1978) Barriers of peripheral nerve towards exogenous peroxidase in normal and protein deprived rats. Acta Neuropathol (Berl) 43:129–138

Oldfors A, Engvall J, Sourander P (1979) Effects of local extraneural application of diphtheria toxin on the sciatic nerves of normal and protein deprived rats. Acta Neuropathol (Berl) 45:141–145

Olsson Y (1966) Studies on vascular permeability in peripheral nerves. I. Distribution of circulating fluorescent serum albumin in normal, crushed and sectioned rat sciatic nerve. Acta Neuropathol (Berl) 7:1–16

Olsson Y (1972) The involvement of vasa nervorum in diseases of peripheral nerves. In: Vinken PJ, Bruyn GW (eds) Handbook of clinical neurology, vol 12. North-Holland, Amsterdam, pp 644–664

Olsson Y (1984) Vascular permeability in the peripheral nervous system. In: Dyck PJ, Thomas PK, Lambert EH, Bunge R (eds) Peripheral neuropathy, vol 1, 2nd edn. Saunders, Philadelphia, pp 579–598

Olsson Y (1990) Microenvironment of the peripheral nervous system under normal and pathological conditions. Crit Rev Neurobiol 5:265–311

Olsson Y, Kristensson K (1979) Recent applications of tracer techniques to neuropathology, with particular reference to vascular permeability and axonal flow. Recent Adv Neuropathol 2:1–26

Olsson Y, Reese TS (1969) Inaccessibility of the endoneurium in mouse sciatic nerve to exogenous proteins. Anat Rec 163:318–319

Olsson Y, Reese TS (1971) Permeability of vasa nervorum and perineurium in mouse sciatic nerve studies by fluorescence and electron microscopy. J Neuropathol Exp Neurol 30:105–116

Parizek J (1956) Effect of cadmium salts on testicular tissue. Nature 117:1036–1037

Parizek J (1957) The destructive effect of cadmium ion on testicular tissue and its prevention by zinc. J Endocrinol 15:56–63

Persson LA, Wingren U, Kristensson K (1976) Hexachlorophene-induced lesions in the developing nervous system of mice. Neuropathol Appl Neurobiol 2:167–175

Persson LA, Norlander B, Kristensson K (1978) Studies on hexachlorophene-induced myelin lesions in the trigeminal root transitional region in developing and adult mice. Acta Neuropathol (Berl) 42:115–131

Poduslo JF, Low PA, Windebank AJ, Dyck PJ, Berg CT, Schmeizer JD (1982) Altered blood-nerve barrier in experimental lead neuropathy assessed by changes in endoneurial albumin concentration. J Neurosci 2:1507–1514

Poduslo JF, Low PA, Nickander KK, Dyck PJ (1985) Mammalian endoneurial fluid: collection and protein analysis from normal and crushed nerves. Brain Res 332:91–102

Portera-Sánches YA, Franch O, del Ser Quijano T (1983) Neuromuscular manifestations of the toxic oil syndrome: a recent outbreak in Spain. In: Portera-Sàndres YA (ed) Clinical and biological aspects of peripheral nerve diseases, vol 1. Liss, New York, pp 171–181

Powell HC, Myers RR, Costello ML (1980a) Increased endoneurial fluid pressure following injection of histamine and compound 48–80 into rat peripheral nerve. Lab Invest 43:564–574

Powell HC, Myers RR, Lampert PW (1980b) Edema in neurotoxic injury. In: Spencer PS, Schaumburg HH (eds) Experimental and clinical neurotoxicology, vol 1. William and Wilkins, Baltimore, pp 118–138

Ranvier M (1871) Recherches sur l'histologie et la physiologie des nerfs. Arch Physiol Norm Pathol 4:427–446

Rapoport SI (1976) The blood-brain barrier in physiology and medicine. Raven, New York

Reale E, Luciano L, Spitznas M (1975) Freeze-fracture faces of the perineurial sheath of the rabbit sciatic nerve. J Neurocytol 4:261–270

Rechthand E, Rapoport SI (1987) Regulation of the microenvironment of peripheral nerve: role of the blood-nerve barrier. Prog Neurobiol 28:303–343

Rechthand E, Smith QR, Rapoport SI (1985) Facilitated transport of glucose from blood into peripheral nerve. J Neurochem 45: 957–964

Reese TS, Karnovsky MJ (1967) Fine structural localization of a blood-brain barrier to exogenous peroxidase. J Cell Biol 34:207–218

Röhlich P, Knoop A (1961) Elektronenmikroskopische Untersuchungen an den Hüllen des N. Ischiadicus der Ratte. Z Zellforsch Mikrosk Anat 53:299–312

Rosenbluth J, Wissig SL (1964) The distribution of exogenous ferritin in toad spinal ganglia and the mechanism of its uptake by neurons. J Cell Biol 23: 307–325

Ross MH, Reith EJ (1969) Perineurium: evidence for contractile elements. Science 165:604–614

Ross RR (1958) Use of pyridoxine hydrochloride to prevent isonoazid toxicity. JAMA 168:73–82

Sahenk Z, Mendell JR (1979) Analysis of fast axoplasmic transport in nerve ligation and adriamycin-induced neuronal perikaryon lesions. Brain Res 171:41–53

Saito A, Zacks SJ (1969) Ultrastructure of Schwann and perineurial sheaths at the mouse neuromuscular junction. Anat Rec 164:379–384

Salonen V (1987) Connective tissue reactions to peripheral nerve injury. Ann Univ Turku 28:1–55

Schlaepfer W (1971) Sequential study of endothelial changes in acute cadmium intoxication. Lab Invest 15:556–564

Schröder JM (1975) Zur Pathogenese der Isonizid-Neuropathie. I. Eine Feinstrukturelle Differenzierung gegenüber der Wallerschen Degeneration. Acta Neuropathol (Berl) 16:301–315

Seitz RJ, Neuen E, Wechsler W (1985a) The blood-nerve barrier in polyneuritis: immunocytochemical and electron microscopical studies on the human sural nerve. J Neurol [Suppl]232:244

Seitz RJ, Heininger K, Schwendemann G, Toyka KV, Wechsler W (1985b) The mouse blood-brain barrier and blood-nerve barrier for IgG: a tracer study by use the avidin-biotin system. Acta Neuropathol (Berl) 68:15–21

Shantha TR, Bourne GH (1968) The perineural epithelium – a new concept. In: Bourne GH (ed) The structure and function of nervous tissue, vol 1. Academic Press, New York, pp 379–384

Shanthaveerappa TR, Bourne GH (1962) The "perineural epithelium", a metabolically active continuous protoplasmic cell barrier surrounding peripheral nerve fascicles. Am J Anat 112:527–538

Shanthaveerappa TR, Bourne GH (1963) Electron microscope demonstration of the perineural epithelium in rat peripheral nerve. Acta Anat (Basel) 52: 193–199

Shanthaveerappa TR, Bourne GH (1964) The effects of transection of the nerve trunk on the perineral epithelium with special reference to its role nerve degeneration and regeneration Anat Rec 150:35–45

Shinowara NL, Michel ME, Rapoport SI (1982) Morphological correlates of permeability in the frog perineurium: vesicles and "transcellular channels". Cell Tissue Res 227:11–22

Siegrist G, Dolivo M, Dunant Y, Foroglou-Kerameus C, de Ribaupierre F, Rouiller C (1968) Ultrastructure and function of the chromaffin cells in the superior cervical ganglion of the rat. J Ultrastruct Res 25:381–407

Sima A, Sourander P (1974) The permeability of perineurium to peroxidase after early malnutrition. An ultrastructural study on rat sciatic nerve. Acta Neuropathol (Berl) 28:15–24 ·

Somjen GG, Herman SP, Klein R, Brubaker PE, Briner WH, Goodrich JK, Krigman MR, Maseman JK (1973) The uptake of methylmercury in different tissues related to its neurotoxic effects. J Pharmacol Exp Ther 187:602–613

Sparrow JR, Kiernan JA (1979) Uptake and retrograde transport of proteins by regenerating axons. Acta Neuropathol (Berl) 47:39–44

Spencer PS, Schaumburg HH (eds) (1980) Experimental and clinical neurotoxicology, vol 1. William and Wilkins, Baltimore

Steinwall O, Klatzo I (1966) Selective vulnerability of the blood-brain barrier in chemically induced lesions. J Neuropathol Exp Neurol 25:52–66

Stockinger L (1965) Nervenabschnitte ohne Perineurium. Acta Anat (Basel) 60:244–252

Sunderland S (1968) Nerves and nerve injuries. Livingstone, Edinburgh

Szabó Z, Bölönyi F (1955) Blood supply of the ganglia. Acta Morph Acad Sci Hung 5:165–170

Thomas PK (1963) The connective tissue of peripheral nerve: an electron microscope study. J Anat 97:35–45

Thomas PK (1980) The peripheral nervous system as a target for toxic substances. In: Spencer PS, Schaumburg HH (eds) Experimental and clinical neurotoxicology, vol 1. Williams and Wilkins, Baltimore, pp 35–47

Thomas PK (1984) The peripheral nervous system as a target for toxic substances. Acta Neurol Scand [Suppl 100] 70:21–26

Thomas PK, Olsson Y (1984) Microscopic anatomy and function of the connective tissue components of peripheral nerve. In: Dyck PJ, Thomas PK, Lambert EH, Bunge R (eds) Peripheral neuropathy, vol 1, 2nd edn. Saunders, Philadephia, pp 97–120

Tischner K (1980) In: Experimental and clinical neurotoxicology, vol 1. Spencer PS, Schaumburg HH (eds) Cadmium. Williams and Wilkins, Baltimore, pp 348-355

Towfighi J (1980) In: Experimental and clinical neurotoxicology, vol 1. Spencer PS, Schaumburg HH (eds) Hexachlorophene. Williams and Wilkins, Baltimore, pp 440–456

Towfighi J, Gonatas NK (1976) Hexachlorophene and the nervous system. Prog Neuropathol 3:101–113

Towfighi J, Gonatas N (1977) The distribution of peroxidases in the sciatic nerves of normal and hexachlorophene intoxicated developing rats. J Neurocytol 6:39–48

Towfighi J, Gonatas NK, McCree L (1973) Hexachlorophene neuropathy in rats. Lab Invest 29:428–435

Tschetschujeva T (1929) Über die Speicherung von Trypanblau in Ganglien des Nervensystems. Ges Exp Med 69:208–223

Van Lis MJ, Jennekens FGI (1977) Plasma proteins in human peripheral nerve. J Neurol Sci 34:329–341

Waggener JP, Beggs J (1967) The membranous coverings of neural tissues; an electron microscope study. J Neuropathol Exp Neurol 26:412–421

Waggener JP, Bunn SM, Beggs J (1965) The diffusion of ferritin within the peripheral nerve sheath: an electron microscopy study. J Neuropathol Exp Neurol 24:430–436

Waksman BH (1961) Experimental study of diphtheric polyneuritis in the rabbit and guinea pig. III. The blood-nerve barrier in the rabbit. J Neuropathol Exp Neurol 20:34–52

Weerasuriya A, Rapoport SI (1986) Endoneurial capillary permeability to (^{14}C) sucrose of frog sciatic nerve. Brain Res 375:150–156

Weerasuriya A, Rapoport SI, Taylor RE (1980) Ionic permeabilities of the frog perineurium. Brain Res 191:405–415

Weiss M, Röhlich P (1954) Significance of the interstice of the peripheral nerve. Acta Morphol Acad Sci Hung 4:309–319

Weiss P, Wang H, Taylor AC, Edds MV (1945) Proximodistal fluid convection in the endoneurial spaces of peripheral nerves, demonstrated by color and radioactive isotope tracers. Am J Physiol 143:521–5447

Welch K, Davson H (1972) The permeability of capillaries of the sciatic nerve of the rabbit to several materials. J Neurosurg 36:21–29

Windebank AJ, Dyck PJ (1981) Kinetics of ^{210}Pb entry into the endoneurium. Brain Res 225:67–73

Windebank AJ, Dyck PJ (1984) Lead intoxication as a model of primary segmental demyelination. In: Dyck PJ, Thomas PK, Lambert EH, Bunge R (eds) Peripheral Neuropathy, 2nd edn. Saunders, Philadelphia pp 650–665

Windebank AJ, McCall JT, Hunder HG, Dyck PJ (1980) The endoneurial content of lead related to the onset and severity of segmental demyelination. J Neuropathol Exp Neurol 39:692–699

Yamamoto T, Iwasaki Y, Konno H (1984) Retrograde axoplasmic transport of adriamycin: an experimental form of motor neuron disease. Neurology (NY) 34:1299–1304

Zacks SJ, Saito A (1969) Uptake of exogenous horseradish peroxidase by coated vesicles in mouse neuromuscular junctions. J Histochem Cytochem 17:161–166

Protective Barriers in the Nervous System Against Neurotoxic Agents: The Blood-Brain Barrier

B.B. JOHANSSON

A. Introduction

The blood-brain barrier (BBB) is a dynamic interface between blood and the central nervous system (CNS) enabling the brain to keep an optimal internal environment. The current concept of the BBB includes a number of morphological and functional characteristics that restrict or facilitate the passage of substances from blood to brain. Our knowledge of BBB physiology has advanced impressively during the past few decades (BRADBURY 1979, 1984; CRONE 1987; PARDRIDGE 1983, 1986; GOLDSTEIN and BETZ 1986). The role of the BBB in neurotoxicology has recently been reviewed by CREMER (1990).

B. Morphology of Brain Endothelial Cells

The anatomical basis of the BBB has been extensively reviewed by BRIGHTMAN (1989). In short, the brain endothelial cells are sealed together by tight junctions (REESE and KARNOVSKY 1967), which are extremely complex structures. The brain endothelial cells also contain few pinocytotic vesicles. The importance of the low endocytotic activity for the various BBB functions is controversial. Studies from other vascular beds indicate that the majority of what appears to be independent vesicles in the endothelial cytoplasm may in fact be part of membrane invaginations that communicate either with the blood or with the perivascular space (BUNDGAARD et al. 1979). No single channel seems to open to both blood and interstitial spaces simultaneously.

The mitochondrial content of brain capillary endothelial cells amounts to 8%–11% of the cytoplasmic volume, a much higher proportion than in other vascular beds (OLDENDORF et al. 1977). The high mitochondrial content is thought to reflect the high metabolic activity needed to maintain ion differentials between blood plasma and brain extracellular fluid and to maintain the unique characteristics of CNS capillaries (Fig. 1). Like epithelial cells, the brain endothelial cells have a very high electrical resistance, reflecting their low ion permeability. It has, therefore, been suggested that the BBB may be looked upon as a modified tight epithelium (CRONE 1987).

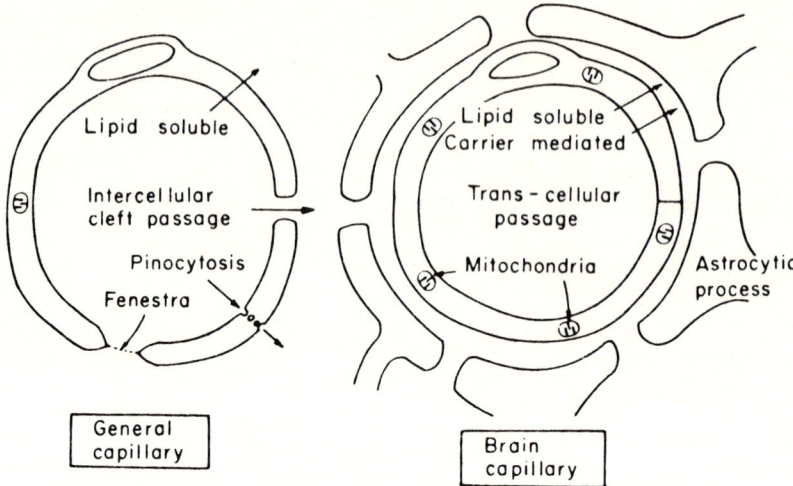

Fig. 1. The main differences between a general capillary and brain capillary. (From Oldendorf (1977), with permission from Academic Press)

The transport characteristics and polarity of cerebral endothelial cells are consistent with this concept (see below).

C. Endothelial Cell Polarity

A cell is said to be polar when the two opposing plasma membranes have different properties. Cellular polarity provides the basis for transcellular transport of many solutes and permits a cell to take up and metabolize a substance from one side while maintaining a barrier that prevents the substrate from crossing to the other side. Like epithelial cells, the endothelial cells of brain capillaries are polar. Na/K-ATPase and 5-nucleotidase are present predominantly on the abluminal side, whereas γ-glutamyl transpeptidase is mainly found on the luminal side (Betz et al. 1980). The distribution of receptors on the two sides varies, allowing a different response to agonists on the luminal and abluminal sides.

D. Possible Role of Perivascular Astrocytes

Although there is no doubt that the main site of the BBB is the endothelial cells, there is evidence that the astrocytes can modify and perhaps induce the BBB. Thus, whereas tight junctions are formed early in fetal life, the barrier to molecules like sucrose and inulin tightens later, e.g., in the rat between 4 and 9 days after birth, reaching adult degree of tightness by 2–3 weeks (FERGUSON and WOODBURY 1969; BRADBURY 1979). The closure of the BBB to small molecules seems to be related to the maturation of the glial foot processes surrounding the larger part of the endothelial cell surface and the capillary basement membrane. Based on studies using cell cultures, it has been suggested that perivascular glial cells induce some of the BBB functions such as polarity in brain endothelial cells (DE BAULT and CANCILLA 1980; BECK et al. 1984). Direct evidence that purified cultured astrocytes can induce BBB-like permeability characteristics in invading endothelial cells was provided by JANZER (1987).

E. Passage Across the Endothelial Cell Membrane

I. Role of Lipid Solubility

The capacity of a given substance to enter the brain can largely be predicted by its octanol:water partition coefficient (OLDENDORF 1974; BRADBURY 1979) although substances that have access to a carrier mechanism do not follow this role. However, larger molecules may have difficulty in entering the brain regardless of lipophilicity; cyclosporin is a good example (CEFALU and PARDRIGE 1985). Some examples of lipophilic substances that are neurotoxic are given below.

1. Hexacarbons

Organic solvents have a high penetration into the brain and tend to accumulate in brain lipids. Among the most studied solvents experimentally are *n*-hexane and methyl *n*-butyl ketone. The common metabolite 2, 5-hexanedione is thought to be responsible for most of the neurological effects associated with repetitive and prolonged occupational exposure to these solvents. Deliberate inhalation of solvents or glue thinners containing *n*-hexane in combination with various other solvents has been another source of neurological complications. Although peripheral neuropathy is the most common result of toxic exposure, distal breakdown of spinocerebellar and optic tracts have been observed after exposure in experimental animals (SPENCER et al. 1980).

2. Carbon Disulfide

Occupational exposure to carbon disulfide may occur in the production of viscose rayon. A number of clinical symptoms have been described although, with the present threshold, symptoms are rare and mild. This is a highly lipid soluble solvent with rapid penetration into the brain. It has been proposed that damage to brain microvessels may also contribute to the neurotoxicity (SEPPÄLÄINEN and HALTIA 1980).

3. Organophosphorus Compounds

Several lipophilic organophosphorus compounds are highly neurotoxic and have been used as "nerve gases," insecticides, petroleum additives, and modifiers of plastics. Many of these compounds, including nerve gases and insecticides, act by inhibiting acetylcholinesterase. Signs of acute intoxication include involuntary urination, lacrimation, muscular twitching, weakness, and convulsions. Respiratory paralysis may cause death. Other organophosphorus compounds may act by phosphorylating other enzymes and induce delayed neurotoxic lesions. A detailed account on the neurotoxic effect of various organophosphorus compounds is given by Davis and Richardson (1980).

4. Methyl Mercury

Methyl mercury is a potent neurotoxin that penetrates the brain because of its lipophilicity (Chang 1980).

5. Triethyltin

Triethyltin is soluble in organic solvents and alcohol but insoluble in water, and is highly neurotoxic. Exposure to triethyltin causes a severe edema in the brain white matter without damaging the BBB. Triethyltin uncouples oxidative phosphorylation and inhibits mitochondrial ATPase. It is believed that inhibition of the Na/K-dependent ATPase of the cell membrane contributes to the edema formation.

II. Transport Mechanisms

1. The Glucose Carrier

D-Glucose, the main fuel for the brain, is a water-soluble substance that enters the brain via a saturable, stereospecific, facilitated diffusion (Crone 1965; Harik et al. 1988). Brain microvessels have a higher density of the glucose transporter than any other tissue except human erythrocytes (Kalaria et al. 1987). The same carrier can be used by 2-deoxy-D-glucose, a substance that enters the neurons and remains there without being phosphorylated, making radiolabelled analogues very useful in experimental studies on the glucose mechanism.

Whether or not the glucose transport is "down regulated" in chronic hyperglycemia is controversial. Some earlier studies indicate that adaptation to the altered plasma concentration is achieved by reducing transport, which might lead to relative hypoglycemia if the blood glucose level is lowered too rapidly (De Fronzo et al. 1980; Gjedde and Crone 1981). However, some recent studies do not support this concept (Harik et al. 1988).

Specific impairment of glucose transport has been proposed to be present in some forms of infective encephalopathies in children (AIYATHURAI et al. 1983).

2. Carriers for Amino Acids

Neutral, basic, and decarboxylic amino acids enter the brain via separate stereospecific and saturable carriers (OLDENDORF 1977). Within the same group, the amino acids compete for the carrier. Neurological signs seen in patients with phenylketonuria are probably related to insufficient passage of other essential amino acids through the barrier because of the high serum levels of phenylalanine. Likewise, large doses of L-dopa may influence the passage of other large neutral amino acids. The transport mechanism is bidirectional; thus, single-passage studies usually show larger entry into the brain than net uptake studies. Among those amino acids that enter the brain poorly are those with supposed transmitter functions such as glutamate and aspartate.

3. Peptides

The extent of the passage of peptides across the BBB is controversial (PARDRIDGE 1983, 1986; ERMISCH et al. 1985; BANKS and KASTIN 1988). It is generally considered that they enter the brain poorly. However, it has recently been demonstrated that some peptides are capable of transversing the BBB by receptor-mediated transcytosis. Thus, receptors have been identified for insulin, insulin-like growth factor, and transferrin (PARDRIDGE 1986).

4. The Monocarboxylic Acid Transporter

The monocarboxylic acid transporter has an affinity for acetate, lactate, pyruvate, and β-hydroxybutyrate (OLDENDORF 1977). At least in the rat, the transport capacity of the monocarboxylic acid carrier is high in the neonatal and early postnatal period, when the glucose transporter is not fully developed. During this period, lactate and β-hydroxybutyrate are the main substrates for the brain (CREMER et al. 1979; CORNFORD et al. 1982). The monocarboxylic acid carrier increases its capacity in response to low plasma glucose levels (GJEDDE and CRONE 1981). During starvation in adult life, the ketone bodies β-hydroxybutyrate and acetoacetate are the main energy substrates for the brain.

Valproate, a drug used in the treatment of epilepsy, has been shown to compete with the monocarboxylic acids pyruvate and butyrate. A rapid efflux of valproate from brain to blood has been shown (CORNFORD et al. 1985; CREMER 1990), and it has been suggested that there is a carrier in normal brain that facilitates the rapid egress of monocarboxylic acids from brain tissue.

5. Organic Acids

There is some evidence that organic acids such as 5-hydroxyindoleacetic acid (5-HIAA) and homovanillic acid (HVA) can be transported from blood to brain at the brain capillary level by a transport mechanism similar to the well-studied transport from cerebrospinal fluid (CSF) to blood in the choroid plexus of the lateral and fourth ventricles (Ashcroft et al. 1968).

6. Electrolytes

In agreement with the high electrical resistance of brain capillaries, the permeability to electrolytes is very low (Crone 1987). However, carrier-mediated transport of chloride from blood to brain has been demonstrated by Smith and Rapoport (1984). Whereas the permeability of the BBB for potassium is very low in the direction from blood to brain, a Na/K-ATPase-dependent efflux of potassium from the brain has been demonstrated (Bradbury 1979). For details on the passage of electrolytes across the BBB, see Smith and Rapoport (1984), Betz (1986), and Crone (1987).

7. Others

Several other transport mechanisms have been described in the brain capillaries, such as transporters for choline and thyroid hormone.

F. Enzymatic Barriers

The enzymatic barrier mechanisms for neurotransmitter monoamines and their precursors at the blood-brain interface have recently been reviewed by Hardebo and Owman (1990). The monoamine precursors l-dopa and l-5-HTP (5-hydroxytryptamine) enter the brain by a facilitated transport sharing the same carrier as other neutral amino acids (see Sect. D). Nevertheless, the entry into the brain is restricted because the amino acids are decarboxylated in the endothelial cells by an aromatic l-amino acid decarboxylase. The enzyme capacity is limited, and given in high doses l-dopa will leak into the brain parenchyma. By combining l-dopa with a decarboxylase inhibitor in the treatment of patients with Parkinson's syndrome, the dosage and the side effects can be reduced and the clinical response improved.

All substances that can act as transmitters in the brain exhibit poor penetration through the BBB. Although the entry of monoamines like dopamine from the blood is low, the amines can be formed in the microvessels from their precursor amino acids as discussed above. However, since monoamine oxidase and catechol-O-methyltransferase are present in the endothelial cells, entry into the brain will be prevented or at least reduced. Angiotensin-converting enzyme, enkephalin-degrading aminopeptidase, and aminopeptidase A-degrading angiotensin II are other enzymes present in the brain endothelial cells.

The potential role of the enzymatic BBB in neurotoxicology can be illustrated by the studies on MPTP (1-methyl-4-phenyl-1,2,3,6-tetrahydropyridine) neurotoxicity by HARIK et al. (1990). MPTP causes a clinical, pathological, and neurochemical state that closely resembles Parkinson's disease. Oxidation by the enzyme monoamine oxidase B (MAO-B) seems necessary for neurotoxicity since pretreatment with MAO-B inhibitors, such as pargyline and Deprenyl, prevents neurotoxicity (HEYKKYLA et al. 1984; LANGSTON et al. 1984). The oxidation product N-methyl-4-phenylpyridine (MPP+) is a substrate for the high-affinity dopamine uptake membrane pump that concentrates MPP+ in dopaminergic neurons (JAVITCH et al. 1985), which probably explains the selective neurotoxicity of MPTP at dopaminergic neurons.

There is a striking species difference in the neurotoxicity of MPTP. Unlike primates, rats are resistent to systemic MPTP toxicity, even when the substance is injected into the internal carotid artery. Nevertheless, when infused directly into the rat substantia nigra it causes destruction of dopaminergic neurons, which strongly suggests that the BBB plays an important role in preventing MPTP neurotoxicity in the rat. This is probably related to a very high content of MAO-B activity in rat cerebral microvessels compared with other species, including man (HARIK et al. 1990).

Isolated rat brain microvessels contain relatively high activities of some enzymes involved in the metabolism of lipophilic xenobiotics, i.e., cytochrome P-450-linked monooxygenases, epoxide hydrolase, NADPH: cytochrome P-450 reductase, and 1-naphthol UDP-glycuronosyl transferase, which restrict the entry of lipophilic xenobiotics into the brain (GHERSI-EGEA et al. 1988).

G. Spontaneous and Induced Fluctuations of Blood-Brain Barrier Function

Most biological systems show rhythmic variations. Diurnal and other rhythmic variations in the BBB have so far not been extensively studied. However, regional BBB permeability may change with the adrenocorticotropic hormone (ACTH) plasma levels (SAIJA et al. 1987). A circadian variation has been observed in the permeability of the BBB during adrenaline-induced hypertension (JOHANSSON and MARTINSSON 1980), a variation that seems to be under noradrenergic influence from the locus coeruleus (JOHANSSON and ISAKSSON 1982). Anesthesia and drugs may also alter the BBB (JOHANSSON 1978, 1981). The possible influence of neurotransmitters on cerebrovascular function including permeability is of growing interest (see SEYLAZ and SERCOMBE 1989).

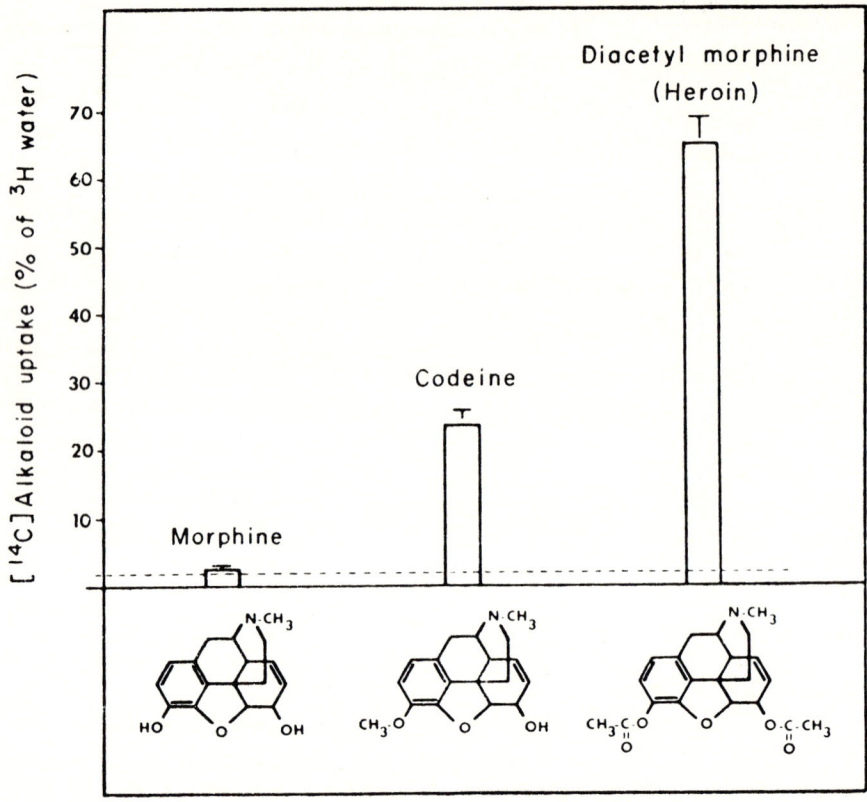

Fig. 2. Because of the highly polar hydroxyl groups, morphine penetrates the BBB poorly. Codein and heroin are more lipophilic and can therefore more easily enter the brain. (From OLDENDORF (1977), with permission of Academic Press)

H. Methods to Facilitate Passage of Substances from Blood to Brain

I. Increasing Lipid Solubility

So far, the most commonly used pharmacological principle has been to convert water-soluble substances into lipid-soluble prodrugs. Thus, morphine penetrates the BBB poorly because of two highly polar hydroxyl groups. By substituting the hydroxyl groups with two acetyl groups, heroin, which is much more lipophilic, is formed. After entering the brain, the acetyl groups are rapidly removed by pericapillary cholinesterase, forming morphine (Fig. 2).

II. Cationization and Glycosylation

Cationization of proteins enhances passage across the BBB (HOUTHOFF et al. 1984). The same is true for glycosylation of proteins, which is probably one reason for the enhanced penetration of proteins in the blood vessels of patients with diabetes. Both principles can be used for the delivery of substances such as monoclonal antibodies through the BBB.

III. Coupling to Substances Entering by Receptor-Mediated Transport

As mentioned above, there is receptor-mediated transport of some peptides across the BBB. Chimeric peptides are formed by coupling a nontransportable peptide to one with access to receptor-mediated transport such as insulin or transferrin (PARDRIDGE et al. 1987). If the coupling of the two peptides is disulfide-based, the nontransportable peptide can be cleaved from the peptide vector by brain enzymes such as glutathione-dependent disulfide reductases. Chimeric peptides can also be formed by coupling to a cationized albumin.

IV. Liposome Entrapment

Liposomes are tiny cells composed of concentrated lipid bilayers that can be made to entrap various substances such as drugs, hormones, or viruses. Their permeability can be enhanced by using liposomes with specific glycolipid ligands. However, liposomes are recognized as foreign particles by the immune system and may be phagocytosed by the reticular endothelial cells. For a recent review of the current state and the possible future development of liposomes to allow entry into the brain of various substances, see FISHMAN and CHAN (1990).

V. Methods to Open the Blood-Brain Barrier Experimentally

A number of experimental approaches have been used to alter the BBB. A discussion of them is outside the scope of this review. However, it should be stressed that the BBB can be transiently altered during an increase in blood pressure (JOHANSSON et al. 1970), particularly if the pressure increase is rapid, the vessels are dilated, or the animals are anesthetized (JOHANSSON and LINDER 1978; JOHANSSON 1989). Thus, in experiments involving the administration of vasoactive substances, the blood pressure should be monitored and a tracer for BBB permeability used whenever an intact barrier is essential for the experiments. With current immunohistochemical techniques, the endogenous serum albumins can be used if exogenous tracers are undesirable for the experimental protocol (FREDRIKSSON et al. 1985; SOKRAB et al. 1988; SALAHUDDIN et al. 1988). Likewise, immunoelectrophoresis of, e.g., endogenous albumin can be used if quantitative techniques are required (Westergren and Johansson 1990).

J. Long-Term Consequences of Blood-Brain Barrier Alterations

In spite of the importance of an intact BBB for brain homeostasis and function, the possible long-term effects of an altered brain environment induced by a transient or permanent dysfunction of the BBB have been little studied. It has recently been shown that even short-term opening of the BBB induced by acute hypertension or osmotic opening of the BBB may occasionally lead to permanent morphological changes (Sokrab et al. 1988; Salahuddin et al. 1988). However, this is not found consistently, and under what circumstances damage occurs needs to be further elucidated. Aggravating as well as protective mechanisms may be involved. Long-term consequences of a transitory BBB dysfunction may be related to alteration of the neuronal environment by extravasated normal plasma constituents, toxic substances, or drugs and/or the metabolic activity of the brain (Johansson et al. 1990).

K. Substances that Enter the Brain by Primarily Damaging the Blood-Brain Barrier

Mercury ions impair the BBB and thereby gain access to the nerve cells from the blood stream. The cerebral endothelial cells may be the primary target for lead toxicity (Goldstein 1984). However, both the passage of lead into the brain and its clearance are probably complex and there is evidence that the choroid plexus may also play a role in the control of CSF lead (see Deane and Bradbury 1990).

L. Retrograde Intraaxonal Transport in Motor Nerves

Locally produced tetanus toxin gains access to the CNS via motor nerves. Other substances that have been shown to enter through the same route are mercury, doxorubicin, and various viruses. This is probably a nonspecific route that can be used by many substances.

M. Areas Lacking a Blood-Brain Barrier

The choroid plexus and the circumventricular organs (area postrema, median eminence, pituitary neural lobe, organum vasculosum of the lamina terminalis, pineal gland, subcommissural organ, the subfornical organ) lack a BBB (see BIGOTTE et al. 1982). Thus, toxic substances that cannot enter the brain may damage these structures. So far, possible clinical consequences of damage to these areas have not been evaluated. Structures lacking a BBB do have a barrier to the CSF (BIGOTTE and OLSSON 1982). Thus, whereas there is no barrier between the CSF and the brain extracellular fluid, substances cannot enter the circumventricular organs from the CSF. Since circumventricular organs have neuroendocrine functions, the special permeability characteristics may be particularly suitable for passage from brain to blood.

N. Summary

The endothelial cells of the brain capillaries are unique epithelial like cells that are fused by tight junctions and have low pinocytotic activity. Perivascular astrocytes may have an important role in inducing and upholding some barrier functions. The entry of specific substances mainly depends on their lipid solubility and whether or not they have access to specific carriers in the endothelial cells. Enzymatic degradation in the endothelial cells can also prevent entry into the brain of substances that do enter the endothelial cells. A substance that diffusely or selectively alters some of the BBB functions can be neurotoxic through its own actions and/or by allowing other toxic substances to enter the brain.

References

Aiyathurai JEJ, Wong HB, Quak SH, Jacob E, Chio LF, Sothy SP (1983) The significance of type B hyperlactataemia in infective encephalopathy. Ann Acad Med Singapore 12:115–125

Ashcroft GW, Dow RC, Moir ATB (1968) The active transport of 5-hydroxyindol-3-ylacetic acid and 3-methoxy-4-hydroxyphenyl acetic acid from a recirculating perfusion system. J Physiol (Lond) 199:397–425

Beck DW, Vinters HV, Hart MN, Cancilla PA (1984) Glial cells influence polarity of the blood-brain barrier. J Neuropathol Exp Neurol 43:219–224

Betz AL (1986) Transport of ions across the blood-brain barrier. Fed Proc 45:2050–2054

Betz AL, Firth JA, Goldstein GW (1980) Polarity of the blood-brain barrier: distribution of enzymes between the luminal and antiluminal membranes of brain capillary endothelial cells. Brain Res 192:17–28

Bigotte L, Olsson Y (1982) Cytofluorescence localization of adriamycin in the nervous system III. Distribution of the drug in the brain of normal adult mice after intraventricular and arachnoidal injections. Acta Neuropathol (Berl) 58:193–202

Bigotte L, Arvidson B, Olsson Y (1982) Cytofluorescence localization of adriamycin in the nervous system I. Distribution of the drug in the central nervous system of normal adult mice after intravenous injection. Acta Neuropathol (Berl) 57:121–128

Bradbury MBW (1979) The concept of a blood-brain barrier. Wiley, New York

Bradbury MBW (1984) The structure and function of the blood-brain barrier. Fed Proc 43:186–190

Brightman MW (1989) The anatomic basis of the blood-brain barrier. In: Neuwelt E (ed) Implications of the blood-brain barrier and its manipulation. Plenum, New York, pp 53–83 (Basic science aspects, vol 1)

Bundgaard M, Frokjaer-Jensen J, Crone C (1979) Endothelial plasmalemmal vesicles as elements in a system of branching invaginations from the cell surface. Proc Natl Acad Sci USA 76:6439–6442

Cefalu WT, Pardridge WM (1985) Restrictive transport of a lipid-soluble peptide (cyclosporin) through the blood-brain barrier. J Neurochem 45:1954–1956

Chang LW (1980) Mercury. In: Spencer PS, Schaumburg HH (eds) Experimental and clinical neurotoxicology. Williams and Wilkins, Baltimore, pp 508–526

Cornford EM, Braun LD, Oldendorf WH (1982) Developmental modulation of BBB permeability as an indicator of changing nutrition requirement in the brain. Pediatr Res 16:324–328

Cornford EM, Diep CP, Partridge WM (1985) Blood-brain barrier transport of valproic acid. J Neurochem 44:1541–1550

Cremer JE (1990) The blood-brain barrier in neurotoxicology: an overview. In: Johansson BB, Owman C, Widner H (eds) The pathophysiology of the blood-brain barrier. Elsevier, Amsterdam, pp 243–253 (Fernström Foundation series, vol 14)

Cremer JE, Cunningham WM, Partridge WM, Braun LD, Oldendorf WH (1979) Kinetics of blood-brain barrier transport of puruvate, lactate and glucose in suckling, weanling and adult rats. J Neurochem 33:439–445

Crone C (1965) Facilitated transfer of glucose from blood into brain. J Physiol (Lond) 181:103–113

Crone C (1987) The blood-brain barrier as a tight epithelium: where is information lacking? Ann N Y Acad Sci 481:174–185

Davis CS, Richardson (1980) In: Spencer PS, Schaumburg HH (eds) Organophosphorus compounds. Williams and Wilkins, Baltimore, pp 527–544

Deane R, Bradbury MWB (1990) Lead and blood-brain barrier. In: Johansson BB, Owman C, Widner H (eds) The pathophysiology of the blood-brain barrier. Elsevier, Amsterdam, pp 279–290 (Fernström Foundation series, vol 14)

DeBault LE, Cancilla PA (1980) Gamma-glutamyl transpeptidase in isolated brain endothelial cells: in vivo induction by glial cells. Science 207:653–655

DeFronzo RA, Hendler R, Christensen N (1980) Stimulation of counterregulatory hormonal responses in diabetic man by a fall in glucose concentration. Diabetes 29:125–131

Ferguson RK, Woodbury DM (1969) Penetration of ^{14}C-inulin and ^{14}C-sucrose into brain, cerebrospinal fluid and skeletal muscle in developing rats. Exp Brain Res 7:181–194

Fishman RA, Chan PH (1990) Liposome entrapment of drugs and enzymes to enable passage across the blood-brain barrier. In: Johansson BB, Owman C, Widner H (eds) The pathophysiology of the blood-brain barrier. Elsevier, Amsterdam, pp 231–239 (Fernström Foundation series, vol 14)

Fredriksson K, Auer RN, Kalimo H, Nordborg C, Olsson Y, Johansson BB (1985) Cerebrovascular lesions in stroke-prone spontaneously hypertensive rats. Acta Neuropathol (Berl) 68:284–294

Ghersi-Egea JF, Minn A, Siest G (1988) A new aspect of the protective functions of the blood-brain barrier: activities of four drug-metabolizing enzymes in isolated rat brain microvessels. Life Sci 42:2515–2523

Gjedde A, Crone C (1981) Blood-brain glucose transfer: repression in chronic hyperglycemia. Science 214:456–457

Goldstein GW (1984) Brain capillaries: a target for inorganic lead poisoning. Neurotoxicology 5:167–176

Goldstein GW, Betz AL (1986) The blood-brain barrier. Sci Am 225:70–79

Hardebo JE, Owman C (1990) Enzymatic barrier mechanisms for neurotransmitter monoamines and their precursors at the blood-brain interface. In: Johansson

BB, Owman C, Widner H (eds) The pathophysiology of the blood-brain barrier. Elsevier, Amsterdam, pp 41–55 (Fernström Foundation series, vol 14)

Harik SI, Gravina SA, Kalaria RN (1988) Glucose transporter of the blood-brain barrier and brain in chronic hyperglycemia. J Neurochem 51:1930–1933

Harik SI, Riachi NI, Kalaria RN, LaManna JC (1990) The role of the blood-brain barrier in preventing MPTP neurotoxicity. In: Johansson BB, Owman C, Widner H (eds) The pathophysiology of the blood-brain barrier. Elsevier, Amsterdam, pp 255–263 (Fernström Foundation series, vol 14)

Heikkila RE, Manzino L, Cabbat FS, Duvoisin RC (1984) Protection against the dopaminergic neurotoxicity of 1-methyl-4-phenyl-1,2,5,6-tetrahydro-pyridine by monoamine oxidase inhibitors. Nature 311:467–469

Houthoff HJ, Moretz Rennke HG, Wisniewski HM (1984) The role of molecular charge in the extravasation and clearance of protein tracers, blood-brain impairment and cerebral edema. In: Go KG, Baethmann A (eds) Recent progress in the study and therapy of brain edema. Plenum, New York, pp 67–79

Janzer RC, Raff MC (1987) Astrocytes induce blood-brain barrier properties in endothelial cells. Nature 325:253–257

Javitch JA, D'Amato RJ, Stritmatter SM, Snyder SH (1985) Parkinsonism-inducing neurotoxin N-methyl-4-phenyl-1,2,3,6-tetrahydropyridine: uptake of the metabolite N-methyl-4-phenylpyridine by dopamine neurons explains selective toxicity. Proc Natl Acad Sci USA 82:2173–2177

Johansson BB (1978) Effect of an acute increase of the intravascular pressure on the blood-brain barrier. A comparison between conscious and anesthetized rats. Stroke 9:588–590

Johansson BB (1981) Pharmacological modification of hypertensive blood-brain barrier opening. Acta Pharmacol Toxicol 48:242–247

Johansson BB (1989) Hypertension and the blood-brain barrier. In: Neuwelt E (ed) Implications of the blood-brain barrier and its manipulation, vol 2. Plenum, New York, pp 389–410

Johansson BB, Linder LE (1978) The cerebrovascular permeability to protein in the rat during nitrous oxide anaesthesia at various blood pressure levels. Acta Anaesthesiol Scand 22:463–466

Johansson BB, Martinsson L (1980) The blood-brain barrier in adrenaline-induced hypertension. Circadian variations and modification by beta-adrenoreceptor antagonists. Acta Neurol Scand 62:96–102

Johansson BB, Li CL, Olsson Y, Klatzo I (1970) The effect of acute arterial hypertension on the blood-brain barrier to protein tracers. Acta Neuropathol (Berl) 15:117–124

Johansson BB, Nordborg C, Westergren I (1990) Neuronal injury after a transient opening of the blood-brain barrier: modifying factors. In: Johansson BB, Owman C, Widner H (eds) The pathophysiology of the blood-brain barrier. Elsevier, Amsterdam, pp 145–157 (Fernström Foundation series, vol 14)

Kalaria RN, Mitchell MJ, Harik SI (1987) Correlation of 1-methyl-4-phenyl-1,2,3,6-tetrahydropyridine neurotoxicity with blood-brain barrier monoamine oxidase activity. Proc Natl Acad Sci USA 84:3521–3525

Kalaria RN, Gravina SA, Schmidley JW et al. (1988) The glucose transporter of the human brain and blood-brain barrier. Ann Neurol 24:757–764

Langston JW, Irwin I, Langston EB, Forno LS (1984) Pargyline prevents MPTP-induced parkinsonism in primates. Science 225:1480–1482

Oldendorf WH (1974) Lipid solubility and drug penetration of the blood-brain barrier. Proc Soc Exp Biol Med 147:813–816

Oldendorf WH (1977) The blood-brain barrier. Exp Eye Res [Suppl]25:177–190

Oldendorf WH, Cornford ME, Brown WJ (1977) The large apparent work capability of the blood-brain barrier: a study of the mitochondrial content of capillary endothelial cells in brain and other tissues of the rat. Ann Neurol 1:409–417

Pardridge WM (1983) Neuropeptides and the blood-brain barrier. Annu Rev Physiol 45:73–82

Pardridge WM (1986) Receptor-mediated peptide transport through the blood-brain barrier. Endocr Rev 7:314–330

Pardridge WM, Kumagai AK, Eisenberg JB (1987) Chimeric peptides as a vehicle for peptide pharmaceutical delivery through the blood-brain barrier. Biochem Biophys Res Commun 146:307–313

Reese TS, Karnovsky MJ (1967) Fine structural localization of a blood-brain barrier to exogenous peroxidase. J Cell Biol 34:207–217

Saija A, Princi P, de Pasquale R, Costa G (1987) Regional blood-brain barrier permeability changes in response to modifications of ACTH plasma levels. Bull Mol Biol Med 12:97–100

Salahuddin TS, Kalimo H, Johansson BB, Olsson Y (1988) Structural changes in the rat brain after carotid infusion of hyperosmolar solutions. An electronmicroscopic study. Acta Neuropathol 77:5–13

Seppäläinen AM, Haltia M (1980) Carbon Disulfide. In: Spencer PS, Schaumburg HH (eds) Carbon disulfide. Williams and Wilkins, Baltimore, pp 356–373

Seylaz J, Sercombe R (1989) Neurotransmission and cerebrovascular function, vol 2. Excerpta Med Int Congr Ser 870

Smith QR, Rapoport SI (1984) Carrier-mediated transport of chloride across the blood-brain barrier. J Neurochem 42:754–763

Sokrab TEO, Johansson BB, Kalimo H, Olsson Y (1988) A transient hypertensive opening of the blood-brain barrier can lead to brain damage. Acta Neuropathol (Berl) 75:557–565, 1988

Spencer PS, Schaumburg HH (1980) Experimental and Clinical Neurotoxicology. Williams and Wilkins, Baltimore

Spencer PS, Couri D, Schaumburg HH (1980) n-Hexane and methyl n-butyl ketone. In: Spencer PS, Schaumburg HH (eds) Experimental and Clinical Neurotoxicology. Williams & Wilkins, Baltimore, pp 456–475

Watanabe I (1980) Organotins (Triethyltin). In: Spencer PS, Schaumburg HH (eds) Experimental and clinical neurotoxicology. Williams and Wilkins, Baltimore, pp 545–557

Westergren I, Johansson BB (1990) Albumin content in brain and CSF after intracarotid infusion of protamine sulphate. A longitudinal study. Exp Neurol 107:192–196

CHAPTER 4

Kinetic and Metabolic Disorders of Axoplasmic Transport Induced by Neurotoxic Agents

S. OCHS

A. Introduction

Proteins and other materials synthesized in the cell bodies are transported into the axons and dendrites to support their functions: conduction of nerve impulses, release of neurotransmitters from terminals, and transduction in terminals. The structures of the axon continually turn over and are themselves maintained by transport: the cytoskeleton microtubules and neurofilaments, the membrane and its ion channels and pumps, endoplasmic reticulum, and other membrane-bound organelles (MBO) including transmitter vesicles, etc. It is apparent that toxicants affecting axonal transport can have profound effects on the function and structure of the neuron. A synopsis of transport properties and mechanisms is given in Sect. B to set the stage for the discussion of toxicants known to involve transport in Sect. C. The discussion is restricted for the most part to studies carried out in peripheral nerves taking a wider view of neurotoxicity, including the actions of toxicants on nerves in vitro as well as those following systemic administration. Transport is commonly held to consist of separate fast and slow mechanisms. Slow transport is associated with the maintenance of the cytoskeletal organelles. Their dramatic alteration by some key toxicants is usually accounted for by a selective action on the slow transport mechanism. In this presentation the alternate concept of a unitary mechanism for both slow and fast transport is advanced and neurotoxic actions discussed on that basis.

B. The Transport System

I. Characteristics of Anterograde Transport

The main transport routes in the axon and their target sites are diagrammed in Fig. 1. Following injection of a labeled amino acid such as [³H]-leucine into a region of cell bodies, e.g., the dorsal root (DR) ganglion, the precursor is quickly taken up into the cell bodies and incorporated into a wide range of proteins and small polypeptides which enter the fibers and advance in them with a front moving linearly at a rate close to 410 mm/day (OCHS

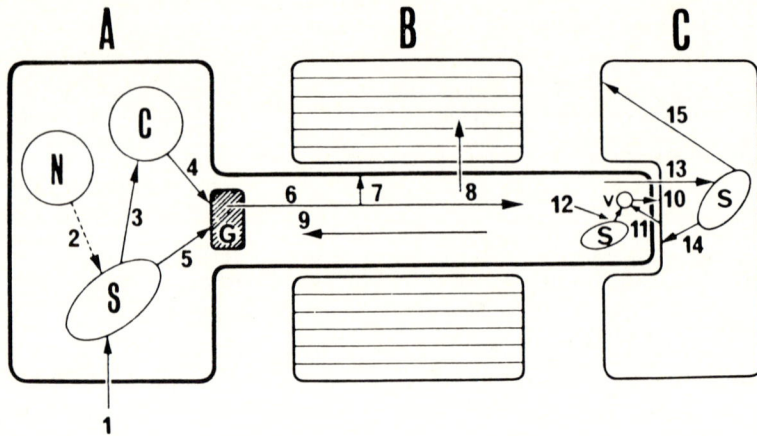

Fig. 1. Diagrammatic representation of the function of a neuron related to transport. Cell body (*A*), nerve fiber (*B*), and terminal regions (*C*) of a schematized motoneuron are shown. Precursors enter (*1*) the cell with the nucleus (*N*) controlling synthesis (*2*). *Arrow* (*3*) shows the compartmentalization (*C*) of synthesized materials which later enter the axon (*4,5*) through a gate (*G*), possibly the Golgi apparatus. Anterograde transport (*6*) supplies components to fiber, some to the membrane (*7*) and the Schwann cell (*8*). Retrograde transport (*9*) carries spent vesicles (*V*) from the terminals (*10*). Uptake of components (*11*) into the terminal, some for reconstitution within vesicles (*12*) and others to be carried by retrograde transport, is shown. Muscle processes are indicated by (*13*), (*14*), and (*15*). (From Ochs 1974c)

1972). The same rate is seen in motor and sensory myelinated fibers of all calibers and in unmyelinated fibers (Ochs and Jersild 1974). In addition, adrenergic, cholinergic, and serotinergic transmitters are fast transported at this rate within transmitter vesicles (Dahlström 1986; Grafstein and Forman 1980; Brimijoin 1982). A similar fast transport of particles and MBOs, some with the dimension of transmitter vesicles, was visualized in axoplasm by the technique introduced by Robert Allen of video-enhanced contrast differential interference contrast (AVEC-DIC) microscopy (Allen et al. 1982, 1985). The MBOs move along tracks identified as microtubules, confirming earlier evidence gained from the use of tubulin-binding agents (Sect. C, II 4) that microtubules play an essential role in transport.

Fast axoplasmic transport is closely dependent on a supply of ATP. In nerves made anoxic or exposed to agents interfering with oxidative metabolism, it is blocked within approximately 20 min following a fall in the level of ATP and phosphocreatine (CrP) (Ochs and Hollingsworth 1971; Leone and Ochs 1978). The need for ATP to maintain transport was verified by observation of particle movement in giant fiber axoplasm with AVEC-DIC microscopy. The movement stopped when ATP was removed and resumed when it was resupplied (Adams 1982; Brady et al. 1982).

Calcium is also required to maintain transport in vertebrate nerve fibers. With calcium deleted from the medium, transport in vitro is blocked in desheathed nerves and in nerves briefly exposed to Triton X-100 to increase the permeability of their perineurial sheaths (CHAN et al. 1980a; LAVOIE et al. 1979; LARIVIERE and LAVOIE 1982; KANJE et al. 1982; KANJE 1986). A similar block in vivo was seen following injection of the calcium chelator ethylenediamine tetraacetic acid (EDTA) into the endoneurial compartment of a nerve (GAN et al. 1986). In the invertebrate giant axon and in some vertebrate preparations, magnesium appears able to substitute for calcium (SMITH 1982), a finding requiring further study.

Slow transport is less well understood. Transport of labeled proteins at rates ranging from 0.2 up to 410 mm/day have been reported (GRAFSTEIN and FORMAN 1980; OCHS 1982; WILLARD et al. 1974; TASHIRO and KOMIYA 1987; WILLARD and HULEBAK 1977; MORI et al. 1979; KARLSSON and SJÖSTRAND 1971). Using SDS-PAGE, a number of labeled proteins were associated with various transport rates (WILLARD et al. 1974; WILLARD and HULEBAK 1977). Group I representing fast transport with velocities of at least 200 mm/day includes membrane and transmitter components. Group II with a velocity of 40 mm/day consists of fodrin and mitochondrial components. Groups III and IV [also termed slow component b (SCb)] with velocities of 2–8 mm/day contain actin, myosin-like proteins, glycolytic enzymes, calmodulin, clathrin, and some additional fodrin. Group V [also referred to as slow component a (SCa)], the slowest group with a rate of approximately 1 mm/day, includes the cytoskeleton proteins; neurofilament proteins (NFPs) characterized by their molecular weights of 200, 165 and 68 kDa, and proteins of 57 and 55 kDa corresponding to the α and β tubulins (HOFFMAN and LASEK 1975).

II. Retrograde Transport and Turnaround

In fibers whose transport was blocked by a local constriction (SMITH 1980) or by cooling (TSUKITA and ISHIKAWA 1980; FAHIM et al. 1985), membranous tubulovesicular structures carried by retrograde transport were seen to accumulate below the block. AVEC-DIC microscopy has also shown a retrograde movement of particles on microtubules, some even on the same microtubules carrying anterogradely transported particles (GELLES et al. 1988).

A portion of the labeled proteins carried down by anterograde transport to a distal ligation or to the nerve endings undergoes the process of turnaround, to be carried back to the cell bodies by retrograde transport. This was shown by collecting labeled proteins being carried back by retrograde transport to accumulate below a crush or an upper ligation later placed on the nerve (LASEK 1967; EDSTRÖM and HANSON 1973; OCHS 1975; BISBY and BULGER 1977; BISBY 1980) or by the use of a position-sensitive counter (SNYDER 1986).

Spent neurotransmitter vesicles are returned to the cell bodies by retrograde transport (BRIMIJOIN and HELLAND 1976; BRIMIJOIN 1982) along with membranes and substances contributed by endocytosis of the terminal membrane or taken up into the terminals through special channels or carriers (KRISTENSSON 1978). In addition to nerve growth factor (NGF), which normally plays a major role in the early development of sensory and adrenergic neurons (SCHWAB et al. 1982; LEVI-MONTALCINI 1976; MOBLEY et al. 1977), foreign substances such as tetanus toxin (ERDMANN et al. 1975; PRICE et al. 1975; SCHWAB and THOENEN 1978; SCHWAB et al. 1977), herpes simplex (KRISTENSSON et al. 1971), and (among other viruses and toxins) enzymes such as horseradish peroxidase and dyes taken up by the terminals gain access to the cell bodies via retrograde transport (KRISTENSSON 1978; PRICE et al. 1975). Retrograde transport, like anterograde transport, is dependent on ATP, and it is blocked by tubulin-binding agents (Sect. C, II 4). It is inhibited to a greater degree than anterograde transport by 9-erythro-2-(hydroxy-3-nonyl) adenine (EHNA) (FORMAN et al. 1983b) and vanadate (FORMAN 1982; FORMAN et al. 1983).

III. Transport Models

1. The Fast Transport Mechanism

Taking into account the participation of microtubules in transport, the requirement for ATP, and the need for Ca^{2+}, the transport filament hypothesis was advanced (OCHS 1971, 1980). In this model ATP is utilized by a Ca/Mg-ATPase associated with side arms of the microtubules which cyclically attach to carriers binding the various transported materials to drive them along the microtubules. The free calcium binds to calmodulin, enabling it in turn to activate the Ca/Mg-ATPase associated with the side arms. Study of the movement of organelles visualized in squid giant axon axoplasm with AVEC-DIC microscopy has allowed further insights into the transport mechanism (ALLEN et al. 1982, 1985; ALLEN 1987; VALE 1987; VALLEE et al. 1989; SHEETZ et al. 1989). A protein, kinesin, was found to be necessary to support anterograde transport (REESE 1987; VALE et al. 1985). Kinesin has ATPase activity which is induced when it binds to microtubules, a property fitting it for the role of a transport "motor" (REESE 1987; VALE 1987; VALE et al. 1985). Inhibitors of kinesin-activated ATPase such as vanadate, 5-adenylylimidodiphosphate (AMP-PNP), and ATP in the absence of magnesium block transport (COHN et al. 1987). Another protein, cytoplasmic dynein (MAP-1C), was found to be required for retrograde transport (VALLEE et al. 1989). The agent EHNA which selectively blocks retrograde transport (FORMAN et al. 1983b) also inhibits the activity of cytoplasmic dynein (SHPETNER et al. 1988).

How these force-generating motors actually produce the movement of the MBOs is as yet speculative (VALE 1987). The organelles appear to be "walked" down along the microtubules by a cyclic action of cross-bridges between them (SHPETNER et al. 1988). The manner in which kinesin and cytoplasmic dynein move organelles has been envisioned in a model in which, in addition to the motor cross-bridges, other factors activate one or other of the motors to control the direction of movement (SHEETZ et al. 1989).

2. Slow Transport and the Unitary Hypothesis

The major group of slow transported proteins, NFPs and the tubulins, were proposed in the "structural hypothesis" to be assembled in the cell bodies into neurofilaments and microtubules and moved down within the axon as part of an interconnected matrix at the SCa rate of approximately 1 mm/day (LASEK and HOFFMAN 1976; BLACK and LASEK 1980; LASEK et al. 1984) to be degraded in the terminals by a calcium-activated protease (LASEK and HOFFMAN 1976; LASEK and KATZ 1987). A strong challenge to this concept was the finding that the SCa rate decreases by as much as 50% distally in the nerve (WATSON et al. 1989). This would result in a piling up of the filaments proximally. In fibers taken following SCa downflow of labeled protein into them and beaded to observe the constrictions wherein the cytoskeleton organelles are compacted (OCHS and JERSILD 1987), autoradiography has shown few or no grains of radioactivity present in the constrictions (OCHS et al. 1989). Those studies indicated that the labeled cytoskeletal proteins in the fibers were in soluble form and had not yet been assembled into neurofilaments and microtubules. Biochemical studies using a "window" technique have shown the neurofilaments to be stationary in the fibers (NIXON and LOGVINENKO 1985; NIXON 1991). Further strong evidence that the microtubules are stationary was provided by studies of cultured neurites in which tubulin subunits were shown to add to microtubules in the terminals rather than in the cell bodies (BAMBURG et al. 1986; SAMMAK et al. 1987; BAMBURG 1988; SAMMAK and BORISY 1988; LIM et al. 1989; OKABE and HIROKAWA 1990). Addition also occurs all along the length of the neurite (OKABE and HIROKAWA 1988; LIM et al. 1990).

In an alternative model to account for slow transport as well as intermediate and fast transport rates, the "unitary hypothesis" (OCHS 1974a, 1975) was proposed in which components are considered to be bound to carriers from which they drop off in the fibers. The slow-transported components fall off earlier than the fast-transported components. Kinetic models have shown how the various transport rates can result from differing drop-off rates (RUBINOW and BLUM 1980; MUÑOZ-MARTÍNEZ 1982; TAKENAKA and GOTOH 1984). The tubulins and NFPs which drop off early on can then turn over in their respective cytoskeletal organelles or reattach to the carriers to be carried further out in the fiber for a more distal incorporation (OCHS 1975). The carriers for the various soluble proteins transported have yet to be identified. Possibly MBOs or kinesin and cytoplasmic dynein themselves can serve in this role as well as mediating fast transport. Further evidence for the hypothesis comes from the observation that the microtubule-associated protein MAP-1A moves at a different rate than tubulin and that a large portion becomes stationary as it is incorporated into the cytoskeleton (NIXON et al. 1990).

C. Neurotoxins

Nerve fibers acted on by toxic agents may show a variety of morphological features: an increase of neurofilaments and/or microtubules, tubulovesicular membranous organelles, and myelin inclusions or, alternatively, decreases of neurofilaments, microtubules, and other organelles with or without shrinkage and degeneration (SPENCER and SCHAUMBURG 1980; DYCK et al. 1984; GRIFFIN and WATSON 1988). The changes may be generalized throughout the length of the fiber or appear at different sites along its length. Those appearing at first proximally and progressing distally are termed proximo-distal axonopathies, while those appearing at first distally to ascend towards the cell bodies constitute the "dying back" or distal-proximal axonopathies. When toxicants cause structural changes in association with an altered axoplasmic transport, the problem is to determine their causal relation, namely, whether the structural changes result from a defect in transport or the structures interfere with transport. Agents which act on the cell bodies can cause a defect of components carried down which are needed to maintain transport and/or interfere with turnover of components in axonal structures. Categorizing toxicants on the basis of where structural changes first appear in the cell, proximally in the fibers or distally, is useful for exposition and to underline similarities of toxicant action. However, as will be noted below, this can only be partially successful in that toxicants often have multiple sites of action and differing spatiotemporal patterns of onset and progression.

I. Cell Body Actions

Protein synthesis in the cell bodies is suppressed by the protein synthesis blocking agents puromycin and cycloheximide (OCHS et al. 1967; McEWEN and GRAFSTEIN 1968; SJÖSTRAND and KARLSSON 1969; OCHS et al. 1970; NICHOLS et al. 1982). Injected into the DR ganglia just before the labeled precursor, they effectively block the synthesis and downflow of labeled proteins (OCHS et al. 1970). They were ineffective when injected 10–20 min afterwards, indicating that synthesis is completed by then and that these agents have no other evident effects on transport.

   Doxorubicin when systemically injected is rapidly taken up into cell bodies to form a complex with DNA (MENDELL and SAHENK 1980) with neuropathic changes seen in sensory DR ganglion neuron cell bodies (CHO et al. 1977). Degenerative changes are subsequently found in the proximal nerve fibers with a "dying forward" pattern. In the early stages of toxicity, transport is maintained at its normal rate of close to 410 mm/day. This was determined by the difference in the positions of the fronts of the fast transported labeled proteins measured at 3 and 4 h following injection of the precursor into their ganglia (MENDELL and SAHENK 1980). There was, however, a fivefold decrease in the time of egress from the cell bodies, from 0.08 to 0.4 h. This may be due to a defect in mechanisms postulated to control the "export" of substances from the cells into the fibers (HAMMERSCHLAG and STONE 1986; HAMMERSCHLAG and BRADY 1989). Using the same technique of assessing transport rates at 3 and 4 h after injection of labeled precursor, methyl mercury dicyandiamide was shown to have a similar predilection of action on the cell body followed by a defect in transport and axonopathy (MENDELL and SAHENK 1980).

Application of toxicants to the central end of cut nerves effects a more selective targeting of DR ganglia. Ricin taken up into the nerve fibers is carried by retrograde transport to their cell bodies where it blocks synthesis and causes wallerian degeneration, a process termed "suicide axoplasmic transport" (WILEY et al. 1982; YAMAMOTO et al. 1984a,b). This use of a toxicant is of considerable clinical interest in that it could control the intractable pain arising from neuromas (NENNESMO and KRISTENSSON 1986; BRANDNER et al. 1989). However, ricin and similar acting agents such as modeccin and volkensin (WILEY and STIRPE 1988) have undesirable side effects, making them unsuitable for clinical use. Doxorubicin injected in the territory of sensory nerve terminals is taken up and carried by retrograde transport to the ganglia to cause a selective degeneration of the cell bodies (BIGOTTE and OLSSON 1987; KATO et al. 1988), and it has been used clinically with success in several cases for the control of pain (SMITH et al. 1989). A somewhat similar approach has been taken with vincas and formyl-leurosin (KNYIHÁR-CSILLIK et al. 1982). Iontophoretically applied, they pass into the skin to sensory fiber terminals where they are taken up and retrogradely transported to the cells to cause their degeneration.

II. Generalized Actions on Fibers

1. Agents Acting on the Axolemma

a) Tetrodotoxin, Batrachotoxin

Block of membrane excitability by tetrodotoxin (TTX), which obstructs Na^+ channels, has no effect on anterograde axoplasmic transport (OCHS and HOLLINGSWORTH 1971; LAVOIE et al. 1976; PESTRONK et al. 1976; BRAY et al. 1979), or on retrograde transport (BOEGMAN and RIOPELLE 1980). Another highly effective neurotoxin blocking excitability, batrachotoxin (BTX), acts by keeping open the Na^+ channels in the membrane (NARAHASHI et al. 1971; ALBUQUERQUE and DALY 1976). In sheathed nerves BTX blocks transport in vitro in concentrations as low as 0.2 mM, the time of block decreasing with

increasing concentration (Ochs and Worth 1975). Retrograde transport of
[^{125}I]–NGF is also blocked by BTX in low concentrations (Boegman and
Riopelle 1980). In sheathed nerves the BTX block appears to be independent of its action on Na$^+$ channels, as an identical block was seen when
sodium was deleted from the medium (Ochs and Worth 1975). However,
Na$^+$ is retained in the endoneural space of sheathed nerves, from which it
can enter the fibers treated with BTX. In the desheathed nerve preparation
transport was inhibited with lower concentrations of $17-190$ nM, the block
depending on the presence of Na$^+$ in the medium. Transport was normal in
nerves exposed to the same range of BTX concentrations in a Na$^+$ -free
medium (Worth and Ochs 1982). The block by BTX of particulate movement observed microscopically in neuroblastoma cells was also found to
depend on Na$^+$ (Forman and Shain 1981). The increased entry of Na$^+$ into
the axon brought about by BTX may effect block of transport through the
release of Ca^{2+} sequestered in the mitochondria and other storage sites.
Exposure of isolated mitochondria to levels of Na$^+$ as low as 10 mM gives
rise to a release of Ca^{2+} (Carafoli and Crompton 1976, 1978), the elevated
Ca^{2+} concentration in turn effecting the block (Sect. C, II 3).

b) Local Anesthetics

The local anesthetic procaine can completely block membrane excitability
without affecting transport (Ochs and Hollingsworth 1971). However, it
and other local anesthetics can block transport when higher concentrations
are used (Byers et al. 1973; Edström et al. 1973; Bisby 1975; Lavoie 1982).
The potency of these agents is related to the degree to which they can enter
fibers. Important factors conditioning entry are their pK and the pH of the
medium, entry being favored by alkalinity which promotes the uncharged
state of the molecule (Lavoie 1982; Lavoie et al. 1989). The transport block
produced by these agents is not due to depression of ATP production or to a
disassembly of microtubules (Lavoie et al. 1989).

2. Metabolic Blockers

When ATP production is interfered with by making nerves anoxic, block of
transport occurs within approximately 20 min (Leone and Ochs 1978). A
similar decrease of ATP production and block time was seen with the
metabolic blocking agents azide, 2,4-dinitrophenol, and sodium cyanide
acting at the mitochondrial level (Sabri and Ochs 1972; Ochs 1974b).
Blocking glycolysis by inhibition of glyceraldehyde phosphate dehydrogenase
(GPDH) with iodoacetic acid (IAA) takes a longer time of $1.5-2$ h for
transport to fail (Ochs and Smith 1971; Sabri and Ochs 1971; Edström and
Mattsson 1976). A block of the citric acid cycle with fluoroacetate (FA)
also causes a later failure of transport, within an hour or so (Ochs 1974b).
The longer survival times are considered due to the utilization of metab-
olites other than glucose. This was also indicated in studies showing that

after glucose was depleted from a nerve, ATP and CrP levels still remained high, and only after several hours did they fall with a block of transport (GAZIRI and OCHS 1983).

When blood glucose levels are elevated in rats to 400 gm/dl or greater following streptozotocin (STZ) treatment, little effect on fast anterograde transport was seen (SIDENIUS and JAKOBSEN 1987). On the other hand, retrograde transport of glycoprotein was decreased when using labeled fucose or N-acetyl neuraminic acid as precursors (JAKOBSEN and SIDENIUS 1979). Slow transport is also decreased in the STZ-treated rat (SIDENIUS 1982; SIDENIUS and JAKOBSEN 1987; TAKENAKA et al. 1982) along with an alteration of cytoskeletal proteins (MACIOCE et al. 1989). A defect of turn-around was seen (SCHMIDT et al. 1986). That some of these effects may be due to the hyperglycemia was indicated by their reversal when insulin was given (JAKOBSEN et al. 1981). The defect of retrograde transport, however, was not restored by lowering glucose levels (SCHMIDT et al. 1986).

Gossypol, a polycyclic compound isolated from cotton seed oil and used as a contraceptive by abolishing spermatozoa motility, inhibits dynein ATPase and blocks axoplasmic transport in nerves (KANJE et al. 1986). The microtubules were not affected nor was ATP production inhibited until higher concentrations of the toxicant were applied.

3. Calcium and Calmodulin Blocking Agents

High levels of calcium blocks transport in vitro when the perineurial sheath is slit or removed (CHAN et al. 1980a) or the nerves pretreated with Triton X-100 to make the sheath permeable (KANJE et al. 1982). In sheathed nerves, the low permeability of the perineurial sheath keeps the level of Ca^{2+} relatively low in the endoneurial compartment in the face of high concentrations in the medium. The low permeability of the axolemma to Ca^{2+} and mechanisms regulating Ca^{2+} act to keep its level in the axon low. In the desheathed nerve preparation exposed in vitro to concentrations of calcium even as high as $50-60 \, mM$, the regulatory mechanisms operate until, after 2.5 h, they are overcome, and a complete block of transport ensues (CHAN et al. 1980). When transferred back to a normal Ringer's solution, transport could resume unless exposure times were too prolonged (CHAN et al. 1980). The microtubules then disassemble along with other signs of proteolytic damage (SCHLAEPFER 1974; SCHLAEPFER and HASLER 1979). When nerves are exposed to calcium ionophores such as A23187, Ca^{2+} can enter the axons from its normally lower external concentration to block transport (ESQUERRO et al. 1980; KANJE et al. 1981; LEES 1985).

Calcium is normally present in the axoplasm at submicromolar levels and binds to calmodulin, which regulates a number of enzymes, including an ATPase that may participate in transport (IQBAL and OCHS 1980a; OCHS and IQBAL 1983; IQBAL 1986). An intermediation of calmodulin in transport was indicated by the block produced by trifluoperazine (TFP) and other anti-calmodulin agents (Ekström et al. 1987; Lavoie and Tiberi 1986; Iqbal 1986). Both fast and slow anterograde transport are blocked (Ekström et al. 1987) and retrograde transport as well (Tiberi and Lavoie 1985). Amitriptyline, desipramine, and imipramine are equipotent inhibitors of calmodulin at a concentration of 0.2 mM (Lavoie and Tiberi 1986). Higher levels of 1–3 mM TFP were required to block transport fully in cat nerves in vitro with less than 1% entering the fibers (Iqbal 1986). Anticalmodulins do not effect their block of transport by disassembly of microtubules or by inhibition of ATP production (Lavoie and Tiberi 1986).

4. Tubulin-Binding Agents

Colchicine classically causes a disassembly of the microtubules forming the mitotic spindles (Dustin 1984), and it disassembles microtubules and blocks transport in nerves (Dahlström 1968; Kreutzberg 1969). A number of other agents have since been found to block transport in a similar fashion (Hanson and Edström 1978; Dustin 1978). A potent class of these toxicants, the vinca alkaloids-vincristine (VCR), vinblastine (VLB), and an amide derivative, vindesine (VDS) (Gerzon 1980a), are used clinically in the treatment of neoplastic diseases (Weiss et al. 1974). One of their untoward side effects limiting usefulness in treatment is the production of neuropathy. Using the time of transport block as the measure of their neuronal effectiveness, VCR was shown to be the most potent of this group (Chan et al. 1980b). Its greater effectiveness in blocking transport was paralleled by its greater relative uptake into nerves compared with the other vincas, as shown by the use of ^{3}H-labeled forms (Iqbal and Ochs 1980b). Their differing uptake is not related to their partition coefficients (Gerzon et al. 1979), indicating that their binding in the axon is the leading factor.

Microtubule disassembly was proposed to come about by these agents binding to tubulin which is in equilibrium with the microtubules, this shifting tubulin from the microtubules (Inoue and Sato 1967). While the block of axoplasmic transport by tubulin-binding agents is generally related to a decrease of microtubules (Banks and Till 1975; Dustin 1984), there have been variations in this relationship reported, even of a block of transport with apparently no decrease of microtubules (Byers et al. 1973), or conversely a continuation of transport with few microtubules remaining (Brady et al. 1980). To correlate the block of transport with the microtubules it is necessary to assess the microtubule number in the larger myelinated fibers where the bulk of transported proteins is carried (Ochs 1972). In such studies using maytansine (Ghetti and Ochs 1978) and vinca alkaloids (Chan et al. 1980b; Ghetti et al. 1982), microtubular density was found to be reduced to approximately half when transport was fully blocked.

Those cases reported in which microtubules were little decreased in quantity when transport was blocked require explanation. One possibility is that the tubulin-binding agents may cause selective defects at intervals along the length of the microtubules (PAULSON and McCLURE 1975; OCHS 1982). If these occur at infrequent intervals, cross-sections taken through the nerve fibers to determine microtubular numbers would not likely pass through those regions, and little decrease would be seen. Such blocks may occur at the ends of the microtubule segments (CHALFIE and THOMSON 1979; HEILDEMANN and McINTOSH 1980; BURTON 1987).

Some observations made on particle movement in fibers in vitro using dark field microscopy are of interest in this regard (HAMMOND and SMITH 1977). A reduction in the range of particle movement appeared 1–4 h after exposure to colchicine or vinblastine followed by a complete and irreversible cessation of movement. When the particles moved, they did so at their normal rates with no gradual slowing as the agent took effect. The particles remained stationary more frequently and for progressively longer periods of time. Somewhat similar observations were made of particle movement in neurites of cultured DR ganglia neurons with Nomarski optics microscopy. Colchicine and vinblastine produced regions of swelling in which the microtubule content was reduced and transport into those regions blocked (HORIE et al. 1981).

5. Microtubule "Stabilizing" Agents

Dimethylsulfoxide (DMSO), a polar solvent, "stabilizes" the microtubules. It lowers the temperature at which polymerization of tubulin occurs in vitro, and it increases the resistance of microtubules to disassembly by colchicine or in high ionic strength media (DULAK and CRIST 1974). Transport in vitro was fully blocked in nerves exposed to a concentration of DMSO of 10% (DONOSO et al. 1977). Even after 2.5 h a substantial, though not complete, recovery of axoplasmic transport was seen when the nerves were transferred to a DMSO-free medium. Taxol, which in high concentrations causes a neuropathy (ROYTTA and RAINE 1985), also has a stabilizing action at lower concentrations. This was indicated by its promotion of the assembly of microtubules (HORWITZ et al. 1986) and its ability to counter the action of colchicine to dissemble microtubules and block axoplasmic transport (HORIE et al. 1987). Similarly, the effects of estramustine phosphate to inhibit the assembly of microtubules and cause an arrest of axoplasmic transport, were blocked by a low concentration of taxol (KANJE et al. 1985).

6. Sulfhydryl Blockers

Agents which oxidize the sulfhydryl groups of tubulin reduce their ability to polymerize and form microtubules (MELLON et al. 1976), and they can block transport (EDSTRÖM and MATTSSON 1976). It is unknown whether they do so by a disassembly of microtubules. Some of these agents act on sulfhydryl-containing enzymes to reduce the production of ATP, as does IAA, which has its major effect on the glycolytic enzyme glyceraldehyde-3-phosphate dehydrogenase (GAPD) (Sabri and Ochs 1971). It has a lesser effect on other sulfhydryl enzymes and the sulfhydryl groups of tubulin. This was indicated by the almost complete reversal of the block of transport brought about by IAA when pyruvate or lactate was supplied in addition, these allowing a continued production of ATP (Ochs and Smith 1971).

Metallic ions act on the sulfhydryl groups of a number of enzymes (Mildvan 1970). The metals Zn^{2+}, Cd^{2+}, Hg^{2+}, and Cu^{2+} effect a block of transport. Conversely, at lower concentrations, Cd^{2+} and Zn^{2+} increase the amount of material fast-transported (Edström and Mattsson 1975). A possible explanation of this phenomenon is that at the lower concentrations the metals prevent the drop-off of fast-transported materials which normally takes place along the length of the fibers (Gross and Beidler 1975; Ochs 1975; Muñoz-Martínez et al. 1981). The organometal compound methyl-Hg acts on sulfhydryls and blocks transport (Abe et al. 1975) as do the sulfhydryl blocking compounds N-ethylmaleimide (NEM) and p-chloromercuribenzene sulfonic acid (PCMBS) (Edström and Mattsson 1976). At lower concentrations, NEM and PCMBS increase the amount of labeled proteins fast-transported in the fibers (Edström and Mattsson 1976). The explanation advanced above for Cd^{2+} and Zn^{2+} may apply to NEM and PCMBS when present at low concentrations, namely, they do this by decreasing the rate of drop-off of materials along the fibers.

III. Localized Actions of Toxicants on Fibers

1. Proximal Axonopathies

a) β,β'-Iminodipropionitrile

Chronic administration of β,β'-iminodipropionitrile (IDPN) produces an increase of neurofilaments in the intraspinal part of motoneuron axons with later the formation of a giant balloon (Chou and Hartmann 1965). The interpretation of these changes, based on the model that all the contents of the axon flow down at a constant rate of 1–3 mm/day (Weiss and Hiscoe 1948), was that the toxicant impeded that flow, such an "axostasis" causing the ballooning seen (Chou and Hartmann 1965). However, chronic treatment with IDPN, while it causes a reduction in the transport of NFPs into the distal fibers, leaves the microtubules unaffected in the shrunken distal fibers (Griffin et al. 1978; Clark et al. 1980). The neurofilaments become segregated in the axons where they are displaced to the periphery of the axon with the microtubules repositioned centrally (Griffin et al. 1983; Papasozomenos 1986). Fast transport is not impeded in these nerves (Griffin et al. 1978). In nerves examined following fast transport, autoradiographs showed grains of radioactivity associated with the centrally located microtubules (Papasozomenos et al. 1981; 1982b; Griffin et al. 1983).

Those results led to a modification of the axostasis hypothesis, namely that IDPN causes a selective block of the movement of neurofilaments in the proximal part of the fibers, with fewer left to move into the distal portions. However, considering the recent evidence (Sect. B, III) that the neurofilaments and the microtubules are stationary in the axons while their subunits are carried down in SCa, another explanation can be advanced on that basis: IDPN causes the NFP subunits to drop off more readily from their carriers in the proximal part of the axons, the subunits assembling there to give rise to the increased number of neurofilaments. The lesser amount of subunits transported distally results there in a decreased neurofilament assembly. Neurofilament changes can be seen as early as 1 day after injection of IDPN into the endoneurial compartment of the nerve (GRIFFIN et al. 1983) or 4 days after an intraperitoneal injection (PAPASOZOMENOS et al. 1981), suggesting a direct action of the agent on axonal processes of drop-off and neurofilament assembly.

When IDPN was administered as early as 20 min after injection of labeled precursor into the DR ganglia, it caused a small but significant decrease in the rate of fast transport, at a time well before signs of neurofilament increase could be seen (SICKLES 1989b). IDPN also leads to block of retrograde transport as assessed by the use of $[^3H]$ N-succinimidyl propionate to label retrogradely transported proteins (FINK et al. 1986). These findings further point to actions of IDPN which cannot be accounted for on the basis of a proximal stasis of slow-moving neurofilaments moving down in an assembled form.

b) Aluminum

A proliferation of neurofilaments in cells is produced by aluminum phosphate when injected into the brain or spinal cord (KLATZO et al. 1965; WISNIEWSKI et al. 1979) or applied in a slurry (BUGIANI and GHETTI 1982; TRIARHOU et al. 1985). The overproduction of neurofilaments in peripheral nerve fibers has little effect on fast transport (LIWNICZ et al. 1974). The transport of labeled proteins in aluminum-treated nerves exhibited its normal form and a rate of 440 mm/day, one close to the 410 mm/day value in control nerves.

Al causes an impairment of the slow transport of NFPs. This was seen following injection of Al into the region of the hypoglossal cell bodies (BIZZI et al. 1984). At 32 days after injection of labeled precursor, SDS-PAGE analysis showed a much greater than normal amount of NFPs in proximal segments of fibers up to 6–9 mm from the cells with a decrease distally in segments 12–21 mm from the cells. Still more distally, the fibers had their normal diameter and NF content. The proximal increase of NFPs matched with the swelling and increased content of neurofilaments seen in the proximal segments and the NFP decrease just distally with fewer neurofilaments and the shrunken fibers seen in those segments. The interpretation of these effects follows from the unitary hypothesis: Al causes a greater drop-off of NFPs proximally, leading to the increased assembly and augmentation of neurofilaments there while the diminished amount transported into the distal segments leaves fewer NFPs available for assembly into neurofilaments.

2. Distal Axonopathies

a) Acrylamide

Acrylamide produces an accumulation of neurofilaments and smooth vesicular membrane structures which appear early on in the distal regions of nerve fibers (Asbury and Brown 1980); these changes spread proximally, the pattern characteristic of a "dying-back" neuropathy (Cavanagh 1964). Earlier reports of acrylamide causing defective fast transport have been inconclusive, most indicating little change in the pattern or rate of anterograde fast transport (Sidenius and Jakobsen 1983). However, in a recent study in which the agent was administered systemically 20 min after the DR ganglia were injected with labeled precursor, the rate of fast transport was found to be reduced from 9.3% to 20.8% and the amount of labeled proteins transported reduced by 42.4%−51.3% (Sickles 1989a). The agent does not cause a global block of synthesis by the cell bodies as does cycloheximide, suggesting that it produces some alteration of the transport system.

Acrylamide reduces slow transport (Gold et al. 1985). Following a single high dose, the outflow of the leading edge of the slow component was depressed along with an accumulation of neurofilaments and swelling of the proximal regions of the axons. Following its chronic administration, the proportion of NFP contributed by SCa to the distal axons was reduced along with a decrease in fiber caliber, changes similar to those seen following exposure to IDPN. In the rat optic system, both fast and slow transport rates were reduced before overt evidence of morphological changes were seen (Sabri and Spencer 1990).

A decrease of retrograde transport was observed following a single dose (Jakobsen and Sidenius 1983; Miller et al. 1983; Miller and Spencer 1984). Either a failure of turnaround in the nerve terminals or a delay of transport into the distal terminals could account for the defect (Sahenk and Mendell 1981). An action on the distal terminals was indicated by their failure to take up [^{125}I] tetanus toxin, with a subsequent decrease in its retrograde transport (Price et al. 1975). The effect was temporary, with retrograde transport returning to normal in 48 h (Moretto and Sabri 1988). These studies suggest that the early impairment of uptake and retrograde transport contributes to the later development of axonopathy.

The pathogenesis of the changes produced by acrylamide following block of some components carried to the cells could then result in a decreased supply in the cells of some essential metabolic components needed by the fibers (Jakobsen et al. 1983; Miller and Spencer 1985;

MORETTO and SABRI 1988). A failure of glycolysis was proposed based on the irreversible reaction of acrylamide in vitro with the sulfhydryls of GPDH (SPENCER et al. 1979; MILLER and SPENCER 1985). However, in vivo the effect of GPDH on the enzyme is weak; the levels of ATP and CrP in nerve do not diminish enough to support that hypothesis (BRIMIJOIN and HAMMOND 1985). Pyruvate, on the other hand, can reverse the neurotoxic effect of acrylamide (SABRI et al. 1989b), suggesting some metabolic involvement in its toxicity. In any event, the defective supply of some needed component(s) to the distal portions of the fibers remains a possibility. In accord with this concept a greater than normal fall-off in the height of the crest and diminished slope of the advancing front of glycoproteins is observed following administration of acrylamide (HARRY et al. 1989), a fall-off which normally is greater for glycoproteins than for other proteins (ABE et al. 1973; MUÑEZ-MARTÍNEZ et al. 1981; OCHS 1982).

b) Hexacarbons

The hexacarbons, normal hexane (n-hexane), methyl n-butyl ketone (MBK), and 2,5-hexanedione (2,5-HD), produce focal swellings with an accumulation of neurofilaments in them, seen distally in the larger myelinated fibers and spreading proximally as a "dying back" neuropathy (SPENCER et al. 1980). The swellings, often located proximal to the nodes of Ranvier, contain whorls of neurofilaments peripherally placed in the axon surrounding centrally located microtubules (CAVANAGH 1985). Three weeks after injection of 2,5-HD, fast transport of labeled proteins in motor fibers was impeded with an increased labeling seen in the focal swellings (GRIFFIN et al. 1977). This was interpreted to be due to a slowing down of the rapidly transported organelles passing through the swollen neurofilament-filled regions to cause an accumulation of materials in the swellings and eventually degeneration of the fibers. 2,5-HD increases the slow rate of NFPs (MONACO et al. 1989). This was indicated by their distal shift in the nerve in fluorograms of SDS-PAGE preparations. A similar shift was seen with the toxicant carbon disulfide. The explanation given for the increase in the slow rate based on the structural hypothesis was that these agents free neurofilaments from the neurofilament-microtubule meshwork, allowing them to move faster (GAMBETTI et al. 1986). Passage of the neurofilaments through the nodes of Ranvier where the axon is narrower and neurofilaments few in number (BERTHOLD 1982), poses a problem for this hypothesis in view of the increase of NF-containing enlargements seen distally with fewer proximally, the "longitudinal" redistribution described by PAPPOLLA et al. (1987). The unitary hypothesis offers an explanation for these findings: These toxicants by reducing the rate of drop-off of NFPs proximally can therefore allow more of these proteins to be carried distally to drop off and assemble there to give rise to the increased number of neurofilaments found. 2,5-HD given 20 min after injecting labeled precursor into the DR ganglia reduced the rate of fast transport by more than 20% with an even greater decrease in the total outflow of labeled protein in the fibers (Sickles 1989). This could be ascribed to its direct action on the transport system to effect a reduction in its transport capacity.

The effects of these agents can vary greatly depending on their molecular structure. An analogue of 2,5-HD, 3,4-dimethyl-2,5-hexanedione (DMHD), causes an impairment of the slow transport of NFPs by as much as 75%–90% (Griffin et al. 1984). The decrease of downflow of NFPs was similar, though slightly less severe, than that produced by IDPN.

MBK produces focal swellings with increased amounts of neurofilaments displaced peripherally and the microtubules aggregated into groups, the animals showing partial paralysis and impairment of fast transport (Mendell et al. 1977). The difference between the fronts of the transport outflows at 3 and 4 h after injection of precursor gave a calculated fast transport rate of 283 mm/day, one significantly lower than the control rate of 417 mm/day. A decrease in the rate of egress of proteins from the cell bodies was also seen.

On the hypothesis that the local swellings produced by MBK cause an obstruction of fast transport, that in partially ligated nerves was examined (Sahenk and Mendell 1979b). The rate of fast transport, calculated from the difference in the fronts of outflow of labeled proteins after 3 and 4 h, was indeed slowed. Nevertheless, this may not be sufficient to prove that the decreased rate of fast transport following MBK treatment is due to an obstruction per se. Beading of fibers which produces a marked compaction of microtubules and neurofilaments in the constrictions (Ochs and Jersild 1987) causes very little change in the rate and pattern of fast transport of labeled proteins through them (Ochs and Jersild 1985), and as noted above (Liwnicz et al. 1974), the increased content of neurofilaments in the proximal portions of the fibers produced by IDPN and by Al does not hinder the pattern or rate of fast transport.

c) p-Bromophenylacetylurea

The nerve fiber changes produced by p-bromophenylacetylurea (BPAU) are those characteristic of a dying-back neuropathy. A single oral dose in rats results, after a delay of 8 days, in a loss of position sense and hindlimb weakness, progressing to near total paralysis over the next 2 weeks (Cavanagh et al. 1968). The fibers show an increased content of tubulovesicular membranes and then go on to degenerate. As early as 2 days after administration of the toxicant and before the first symptoms of weakness appear, the turnaround of transported material is delayed or reduced (Jakobsen and Brimijoin 1981). This observation suggests that the failure of retrograde transport causes the accumulation of the tubulovesicular membranes. Also to be considered is that the defect of some retrograde transport signal results in a reduced synthesis by the cell bodies of a needed component supplied to their fibers the lack of which results in the accumulation of the tubulovesicular structures (JAKOBSEN et al. 1983). An indication that BPAU could act on metabolic processes in the cell bodies was suggested by the increased rate of egress of labeled proteins from the cell bodies following its administration (OKA and BRIMIJOIN 1990).

d) Zinc Pyridinethione

Zinc pyridinethione (ZPT) causes a dying-back neuropathy, seen as an accumulation of tubulovesicular membranes in the motor nerve terminals followed by their degeneration (SAHENK and MENDELL 1979a). At an early stage of ZPT intoxication, a defect of fast transport of labeled protein in sensory fibers was seen (SAHENK and MENDELL 1980). This was determined by taking the difference in the positions of the fronts of labeled protein outflow at 3 and 4 h after injection of the ganglia with labeled precursor. A rate of transport of 324 mm/day was calculated, one lower than the normal rate of close to 410 mm/day. A peculiarity of the toxicant's action is that the reduction of transport rate was not related to dosage nor to the degree of hind limb paralysis (MENDELL and SAHENK 1980).

ZPT inhibits turnaround in the terminals, correlated with the swellings and increases of tubulovesicular profiles in the fiber terminals. These changes were proposed to come about by the binding of zinc to the sulfhydryl groups of thiol proteases necessary for turnaround (SAHENK and LASEK 1988). As evidence, the protease inhibitors leupeptin and E-64 applied locally to the central ends of cut peripheral nerves decreased the amount of labeled proteins carried by retrograde transport to the cells, and the axon tips at the cut ends were seen to be distended with 40–80 nm membranous tubules. These structures were proposed to be transient intermediates normally involved in the process of turnaround, their accumulation being due to a failure to convert them into the retrogradely transported organelles. Whether the membranous structures are normal intermediates in turnaround or are abnormally generated by the action of the toxicant or the protease inhibitors is an open question. These agents could interfere with a number of processes in the terminals, leading to the accumulation of the membrane structures and independently cause a defective of turnaround and retrograde transport. The complexity of effects leupeptin may produce is indicated by an augmentation of neurofilaments in nerve terminals (ROOTS 1983) and lack of effect on organelle reversal (SMITH and SNYDER 1991).

e) Organophosphates

The organophosphates diisopropylfluorophosphate (DFP) injected systemically in cats gives rise to a dying-back neuropathy (BOULDIN and CAVANAGH 1979b). Varicosities appear in the fibers; these evolve into focal degenerations to create a "chemical transection" (BOULDIN and CAVANAGH

1979b). Before that stage, anterograde transport appears to be little affected (PLEASURE et al. 1969; JAMES and AUSTIN 1970; BRADLEY and WILLIAMS 1973). Systemic injection of hens with the organophosphate di-*n*-butyl dichlorvos (DBDCP) results in a defect of retrograde transport. This was shown by injection of ^{125}I-labeled tetanus toxin into the gastrocnemius muscle for its uptake by fiber endings and transport to the cell bodies of the spinal cord and the DR ganglia (MORETTO et al. 1987). The neuropathy increases with time to reach a maximum in 7 days, the temporal evolution of neuropathology known as "organophosphate-induced delayed polyneuropathy." This appears to be due to the phosphorylation of a neurotoxic enzyme responsible for the ensuing neuropathic changes (JOHNSON 1975).

f) Tullidora (Buckthorn) Toxin

Ingestion of the toxins present in the endocarp of the fruit of *Karowinskia humboldtidiana* is followed in humans and experimental animals by a flaccid quadraparesis beginning distally in the hind limbs and progressively ascending (WELLER et al. 1980). Nerve biopsies taken at an early stage show a segmental demyelination with subsequently focal axonal swellings and later wallerian degeneration (WELLER et al. 1980). In tissue culture, the toxins were seen to act primarily on the axons, with a widening of the periaxonal space and a redistribution of axonal organelles to a marginal position associated with stacks of smooth ER, the central regions largely occupied by neurofilaments (HEATH et al. 1982). Following a single oral dose of the toxicant and an incubation period in the cat, transport was found to be impaired, more so in motor fibers than in sensory fibers (MUÑOZ-MARTÍNEZ et al. 1984). This was shown by injecting labeled precursor into the L7 DR ganglion on one side and the L7 ventral horn on the other and finding a depressed fast transport outflow in the motor fibers with a normal pattern of outflow in the sensory fibers.

This raises the question as to why motoneurons are more sensitive to the action of the toxicant. This may be due to a greater sensitivity of their cell bodies to the toxicant with a decreased synthesis of some needed component(s). Their failure to move into the farther reaches of the long nerve fibers of the hind limbs could lead to the pattern of neuropathology seen. Alternatively, the transport mechanism itself in the motor fibers may be more susceptible to the toxicant with, as a result, failure to supply critically needed components to the distal regions of the nerve fibers.

g) Alcohol

A diminished anterograde transport was shown in the sensory neurons of chronic alcohol-fed rats (McLANE 1987). Retrograde transport was also seen to be affected adversely when using [^{3}H] *N*-succinimidyl propionate [^{3}H]-NSP) to alkylate retrogradely transported proteins carried to the sensory DR ganglia (McLANE 1990). Rats maintained on a thiamine-deficient diet show no such defects nor signs of a dying-back neuropathy, indicating that alcohol itself produces the neuropathy, in an as yet unknown manner.

D. Conclusions

While toxicants can have a direct effect on the mechanism of transport and in turn cause the structural and functional defects seen in the nerve fibers, defects of transport can also be produced secondarily to neuropathic changes.

It is intuitively appealing to consider that where dense accumulations of neurofilaments or tubulovesicular membrane structures are produced in excess, they might pose a physical impediment to transport and lead to further pathological defects. This concept, however, requires a more searching examination. It has been noted that neither the overproduction of neurofilaments produced by IDPN or Al nor the marked compaction of cytoskeletal organelles in the constrictions of beaded fibers hinders fast transport. Agents such as acrylamide and hexacarbons which eventually produce such accumulations can bring about a decrease in the rate of fast transport and the amount of material carried well before such structural changes appear.

In opposition to an interruption of the slow movement of microtubules and neurofilaments as such, it is considered that agents acting on the transport mechanism as delineated in the unitary hypothesis, in which soluble subunits are dropped off in the fibers to turn over in the cytoskeleton and other structures in the fiber, can better account for a wide range of neurotoxic effects. Further clarification of the intervening stages in the process, however, is needed. Specifically, what are the carriers for the transport of soluble components? What determines the kinetics of the drop-off of components? By what means are the dropped-off components steered to their target sites, inserted into their target sites, and then later removed and disposed of? What determines the overall structure of the fiber, the numbers and relative positions of cytoskeletal organelles to one another in the axon? Answers to questions such as these will no doubt lead to a deeper understanding of neurotoxic action. It may then be possible to determine on a molecular basis exactly how a given toxicant interferes with these transport-associated processes to produce a neuropathic state.

References

Abe T, Haga T, Kurokawa M (1973) Rapid transport of phosphatidylcholine occurring simultaneously with protein transport in the frog sciatic nerve. Biochem J 136:731–740

Abe T, Haga T, Kurokawa M (1975) Blockage of axoplasmic transport and depolymerisation of reassembled microtubules by methyl mercury. Brain Res 86:504–508

Adams RJ (1982) Organelle movement in axons depends on ATP. Nature 297:327–329

Albuquerque EX, Daly JW (1976) Batrachotoxin, a selective probe for channels modulating sodium conductances in electrogenic membranes. In: Cuatrecasas P (ed) The specificity and action of animal bacterial and plant toxins. Chapman and Hall, London, pp 297–338 (Receptors and recognition, series B, vol 1)

Allen RD (1987) The microtubule as an intracellular engine. Sci Am 256:42–49

Allen RD, Travis JL, Hayden JH, Allen NS, Breuer AC, Lewis LJ (1982) Cytoplasmic transport: moving ultrastructural elements common to many cell types revealed by video-enhanced microscopy. Cold Spring Harbor Symp Quant Biol 46:85–87

Allen RD, Weiss DG, Hayden JH, Brown DT, Fujiwake H, Simpson M (1985) Gliding movement of and bidirectional transport along single native microtubules from squid axoplasm: evidence for an active role of microtubules in cytoplasmic transport. J Cell Biol 100:1736–1752

Asbury AK, Brown MJ (1980) The evolution of structural changes in distal axonopathies. In: Spencer PS, Schaumburg HH (eds) Neurotoxicology. Williams and Wilkins, Baltimore, pp 179–192

Bamburg JR (1988) The axonal cytoskeleton: stationary or moving matrix? Trends Neurosci 11:248–249

Bamburg JR, Bray D, Chapman K (1986) Assembly of microtubules at the tip of growing axons. Nature 321:788–790

Banks P, Till R (1975) A correlation between the effects of anti-mitotic drugs on microtubule assembly in vitro and the inhibition of axonal transport in noradrenergic neurones. J Physiol (Lond) 252:283–294

Berthold C-H (1982) Some aspects of the ultrastructural organization of peripheral myelinated axons in the cat. In: Weiss DG (ed) Axoplasmic transport. Springer, Berlin Heidelberg New York, pp 40–54

Bigotte L, Olsson Y (1987) Degeneration of trigeminal ganglion neurons caused by retrograde axonal transport of doxorubicin. Neurology 37:985–992

Bisby MA (1975) Inhibition of axonal transport in nerves chronically treated with local anesthetics. Exp Neurol 47:481–489

Bisby MA (1980) Retrograde axonal transport. Adv Cell Neurobiol 1:69–117

Bisby MA, Bulger VT (1977) Reversal of axonal transport at a nerve crush. J Neurochem 29:313–320

Bizzi A, Crane RC, Autilio-Gambetti L, Gambetti P (1984) Aluminum effect on slow axonal transport: a novel impairment of neurofilament transport. J Neurosci 4:722–731

Black MM, Lasek RJ (1980) Slow components of axonal transport: two cytoskeletal networks. J Cell Biol 86:616–623

Boegman RJ, Riopelle RJ (1980) Batrachotoxin blocks slow and retrograde axonal transport in vivo. Neurosci Lett 18:143–147

Bouldin T, Cavanagh J (1979a) Organophosphorous neuropathy I. A teased-fiber study of the spatio-temporal spread of axonal degeneration. Am J Pathol 94:241–252

Bouldin T, Cavanagh J (1979b) Organophosphorous neuropathy II. A fine-structural study of the early stages of axonal degeneration. Am J Pathol 94:253-270

Bradley WG, Williams MH (1973) Axoplasmic flow in axonal neuropathies I. Axoplasmic flow in cats with toxic neuropathies. Brain 96:235–246

Brady ST, Chrothers SD, Nosal C, McClure WO (1980) Fast axonal transport in the presence of high Ca^{2+}: evidence that microtubules are not required. Proc Natl Acad Sci USA 77:5909–5913

Brady ST, Lasek RJ, Allen RD (1982) Fast axonal transport in extruded axoplasm from squid giant axon. Science 218:1129–1131

Brandner MD, Buncke HJ, Campagna-Pinto D (1989) Experimental treatment of neuromas in the rat by retrograde axoplasmic transport of ricin with selective destruction of ganglion cells. J Hand Surg [Am] 14:710–714

Bray JJ, Hubbard JI, Mills RG (1979) The trophic influence of tetrodotoxin-inactive nerves on normal and reinnervated rat skeletal muscles. J Physiol (Lond) 297:479–491

Brimijoin S (1982) Axonal transport in autonomic nerves: views on its kinetics. In: Kalsner S (ed) Trends in autonomic pharmacology, vol 2. Urban and Schwarzenberg, Baltimore, pp 17–42

Brimijoin S, Helland L (1976) Rapid retrograde transport of dopamine-β-hydroxylase as examined by the stop-flow technique. Brain Res 102:217–228

Brimijoin WS, Hammond PI (1985) Acrylamide neuropathy in the rat: effects on energy metabolism in sciatic nerve. Mayo Clin Proc 60:3–8

Bugiani O, Ghetti B (1982) Progressing encephalomyelopathy with muscular atrophy, induced by aluminum powder. Neurobiol Aging 3:209–222

Burton PR (1987) Microtubules of frog olfactory axons: their length and number/axon. Brain Res 409:71–78

Byers MR, Fink BR, Kennedy RD, Middaugh ME, Hendrickson AE (1973) Effects of lidocaine on axonal morphology microtubules and rapid transport in rabbit vagus nerve in vitro. J Neurobiol 4:25–43

Carafoli E, Crompton M (1976) Calcium ions and mitochondria. Symp Soc Exp Biol 30:89–115

Carafoli E, Crompton M (1978) The regulation of intracellular calcium by mitochondria. Ann NY Acad Sci 307:269–284

Cavanagh J, Chen F, Kyu M, Ridley A (1968) The experimental neuropathy in rats caused by p-bromophenylacetylurea. J Neurol Neurosurg Psychiatry 31:471–478

Cavanagh JB (1964) The significance of the "dying back" process in experimental and human neurological disease. Int Rev Exp Pathol 3:219–267

Cavanagh JB (1985) Peripheral nervous system toxicity: a morphological approach. In: Blum K, Manzo L (eds) Neurotoxicology. Dekker, New York, pp 1–44

Chalfie M, Thomson JN (1979) Organization of neuronal microtubules in the nematode Caenorhabditis elegans. J Cell Biol 82:278–289

Chan SY, Ochs S, Worth RM (1980a) The requirement for calcium ions and the effect of other ions on axoplasmic transport in mammalian nerve. J Physiol (Lond) 301:477–504

Chan SY, Worth R, Ochs S (1980b) Block of axoplasmic transport in vitro by vinca alkaloids. J Neurobiol 11:251–264

Cho E, Schaumburg H, Spencer P (1977) Adriamycin produces ganglioradiculopathy in rats. J Neuropathol Exp Neurol 36:597

Chou S-M, Hartmann HA (1965) Electron microscopy of focal neuroaxonal lesions produced by β,β'-iminodipropionitrile (IDPN) in rats. Acta Neuropathol (Berl) 4:590–603

Clark AW, Griffin JW, Price DL (1980) The axonal pathology in chronic IDPN intoxication. J Neuropathol Exp Neurol 39:42–55

Cohn SA, Ingold AL, Scholey JM (1987) Correlation between the ATPase and microtubule translocating activities of sea urchin egg kinesin. Nature 328:160–163

Dahlström A (1968) Effect of colchicine on transport of amine storage granules in sympathetic nerves of rat. Eur J Pharmacol 5:111–113

Dahlström A (1986) Axonal transport of neurotransmitter organelles in adrenergic, cholinergic, and peptidergic neurons. In: Iqbal Z (ed) Axoplasmic transport. CRC Press, Boca Raton, pp 119–146

Donoso JA, Illanes J-P, Samson F (1977) Dimethylsulfoxide action on fast axoplasmic transport and ultrastructure of vagal axons. Brain Res 20:287–301

Dulak L, Crist RD (1974) Microtubule properties in dimethylsulfoxide. J Cell Biol 63:90–98

Dustin P (1984) Microtubules, 2nd edn. Springer, Berlin Heidelberg New York, pp 1–482

Dyck PJ, Thomas PK, Lambert EH, Bunge R (1984) Peripheral neuropathy, vol 1, 2nd edn. Saunders, Philadelphia, pp 1–1165

Edström A, Hanson M (1973) Retrograde axonal transport of proteins in vitro in frog sciatic nerves. Brain Res 6:311–320

Edström A, Mattsson H (1975) Small amounts of zinc stimulate rapid axonal transport in vitro. Brain Res 86:162–167

Edström A, Mattsson H (1976) Inhibition and stimulation of rapid axonal transport in vitro by sulfhydryl blockers. Brain Res 108:381–395

Edström A, Hansson H-A, Norstrom A (1973) Inhibition of axonal transport in vitro in frog sciatic nerves by chlorpromazine and lidocaine. Z Zellforsch Mikrosk Anat 143:53–69

Ekström P, Kanje M, McLean WG (1987) The effects of trifluoperazine on fast and slow axonal transport in the rabbit vagus nerve. J Neurobiol 18:283–293

Erdmann G, Wiegand H, Wellhoner HH (1975) Intraaxonal and extraaxonal transport of ^{125}I-tetanus toxin in early local tetanus. Naunyn-Schmiedebergs Arch Pharmacol 290:357–373

Esquerro E, Garcia AG, Sanchez-Garcia P (1980) The effects of the calcium ionophore A23187 on the axoplasmic transport of dopamine-β-hydroxylase. Br J Pharmacol 70:375–381

Fahim MA, Lasek FJ, Brady ST, Hodge AJ (1985) AVEC-DIC and electron microscopic analyses of axonally transported particles in cold-blocked squid giant axons. J Neurocytol 14:689-704

Fink DJ, Purkiss D, Mata M (1986) Beta,beta'-Iminodipropionitrile impairs retrograde axonal transport. J Neurochem 47:1032–1038

Forman D (1982) Vanadate inhibits saltatory organelle movement in a permeabilized cell model. Exp Cell Res 141:139–147

Forman DS, Shain WG Jr (1981) Batrachotoxin blocks saltatory organelle movement in electrically excitable neuroblastoma cells. Brain Res 211:242–247

Forman D, Brown K, Livengood D (1983a) Fast axonal transport in permeabilized lobster giant axons is inhibited by vanadate. J Neurosci 3:1279–1288

Forman DS, Brown KJ, Promersberger ME (1983b) Selective inhibition of retrograde axonal transport by erythro-9 [3-(2-hydroxy-nonyl)] adenine. Brain Res 272:194–197

Gambetti P, Monaco S, Autilio-Gambetti L, Sayre LM (1986) Chemical neurotoxins accelerating axonal transport of neurofilaments. In: Clarkson TW, Sager PR, Syversen TLM (eds) The cytoskeleton: a target for toxic agents. Plenum, New York, pp 129–142

Gan S-D, Fan M-M, He G-P (1986) The role of microtubules in axoplasmic transport in vivo. Brain Res 369:75–82

Gaziri LCJ, Ochs S (1983) Increased duration of axoplasmic transport in vitro with added glucose or β-hydroxybutyrate. Soc Neurosci Abst 9:150

Gelles J, Schnapp BJ, Sheetz MP (1988) Tracking kinesin-driven movements with nanometre-scale precision. Nature 331:450–453

Gerzon K (1980) Dimeric catharanthus alkaloids. In: Cassady JM, Douros JD (eds) Anticancer agents based on natural product models. Academic, New York, pp 271–317

Gerzon K, Ochs S, Todd GC (1979) Polarity of vincristine (VCR), vindesine (VDS), and vinblastine (VLB) in relation to neurological effects. Proc Am Assoc Cancer Res Abstr 20:46

Ghetti B, Ochs S (1978) On the relation between microtubule density and axoplasmic transport in nerves treated with maytansine. In: Canal N, Pozza G (eds) Peripheral neuropathies. Elsevier, Amsterdam, p 177

Ghetti B, Alyea C, Norton J, Ochs S (1982) Effects of vinblastine on microtubule density in relation to axoplasmic transport. In: Weiss DG (ed) Axoplasmic transport. Springer, Berlin Heidelberg New York pp 322–327

Gold BG, Griffin JW, Price DL (1985) Slow axonal transport in acrylamide neuropathy: different abnormalities produced by single-dose and continuous administration. J Neurosci 5:1755–1768

Grafstein B, Forman DS (1980) Intracellular transport in neurons. Physiol Rev 60:1167–1283

Griffin J, Watson D (1988) Axonal transport in neurological disease. Ann Neurol 23:3–13

Griffin J, Price D, Spencer P (1977) Fast axonal transport through giant axonal swellings in hexacarbon neuropathies. J Neuropathol Exp Neurol 36:603

Griffin JW, Hoffman PN, Clark AW, Carroll PT, Price DL (1978) Slow axonal transport of neurofilament proteins: impairment by β,β′-iminodipropionitrile administration. Science 202:633–635

Griffin JW, Fahnestock KE, Price DL, Hoffman PN (1983) Microtubule-neurofilament segregation produced by beta, beta'-iminodipropionitrile: evidence for the association of fast axonal transport with microtubules. J Neurosci 3:557–566

Griffin JW, Anthony DC, Fahnestock KE, Hoffman PN, Graham DG (1984) 3,4-Dimethyl-2,5-hexanedione impairs the axonal transport of neurofilament proteins. J Neurosci 4:1516–1526

Gross GW, Beidler LM (1975) A quantitative analysis of isotope concentration profiles and rapid transport velocities in the C-fibers of the garfish olfactory nerve. J Neurobiol 6:213–232

Hammerschlag R, Brady ST (1989) Axonal transport and the neuronal cytoskeleton. In: Siegel G, Agranoff B, Albers RW, Molinoff P (eds) Basic neurochemistry, 4th edn. Raven, New York, pp 457–478

Hammerschlag R, Stone GC (1986) Prelude to fast axonal transport: sequence of events in the cell body. In: Iqbal Z (ed) Axoplasmic transport. CRC Press, Boca Raton, pp 21–34

Hammond GR, Smith RS (1977) Inhibition of the rapid movement of optically detectable axonal particles by colchicine and vinblastine. Brain Res 128:227–242

Hanson M, Edström A (1978) Mitosis inhibitors and axonal transport. Int Rev Cytol [Suppl]7:373–402

Harry GJ, Goodrum JF, Bouldin TW, Toews AD, Morell P (1989) Acrylamide-induced increases in deposition of axonally transported glycoproteins in rat sciatic nerve. J Neurochem 52:1240–1247

Heath J, Ueda S, Bornstein M, Daves G, Raine C (1982) Buckthorn neuropathy in vitro: evidence for a primary neuronal effect. J Neuropathol Exp Neurol 2:204–220

Heidemann SR, McIntosh JR (1980) Visualization of the structural polarity of microtubules. Nature 286:517–519

Hoffman PN, Lasek RJ (1975) The slow component of axonal transport. Identification of major structural polypeptides of the axon and their generality among mammalian neurons. J Cell Biol 66:351–366

Horie H, Takenaka T, Inomata K (1981) Effects of antimitotic drugs on axoplasmic transport in tissue cultured nerve cells. Soc Neurosci Abstr 7:485

Horie H, Takenaka T, Ito S, Kim U (1987) Taxol counteracts colchicine blockade of axonal transport in neurites of cultured dorsal root ganglion cells. Brain Res 420:144–146

Horwitz S, Lothstein L, Manfredi J, Mellado W, Parness J, Roy S, Schiff P, Sorbara L, Zeheb R (1986) Taxol: mechanisms of action and resistance. Ann NY Acad Sci 466:733–744

Inoue S, Sato H (1967) Cell mobility by labile association of molecules. The nature of mitotic spindle fibers and their role in chromosome movement. J Gen Physiol 50:259–292

Iqbal Z (1986) Calmodulin and its role in axoplasmic transport. In: Iqbal Z (ed) Axoplasmic transport. CRC Press, Boca Raton, pp 45–55

Iqbal Z, Ochs S (1980a) Calmodulin in mammalian nerve. J Neurobiol 11:311–318

Iqbal Z, Ochs S (1980b) Uptake of vinca alkaloids into mammalian nerve and its subcellular components. J Neurochem 34:59–68

Jakobsen J, Sidenius P (1979) Decreased axonal flux of retrogradely transported glycoproteins in early experimental diabetes. J Neurochem 33:1055–1060

Jakobsen J, Brimijoin S (1981) Axonal transport of enzymes and labeled proteins in experimental neuropathy induced by p-bromophenylacetylurea. Brain Res 229:103–122

Jakobsen J, Sidenius P (1983) Early and dose-dependent decrease of retrograde axonal transport in acrylamide-intoxicated rats. J Neurochem 40:447–454

Jakobsen J, Brimijoin S, Skau K, Sidenius P (1981) Retrograde axonal transport of transmitter enzymes, fucose labelled proteins and nerve growth factor in streptozotocin-diabetic rats. Diabetes 30:797–803

Jakobsen J, Brimijoin S, Sidenius P (1983) Axonal transport in neuropathy. Muscle Nerve 6:164–166

James KAC, Austin L (1970) The effect of DFP on axonal transport of protein in chicken sciatic nerve. Brain Res 18:192–194

Johnson M (1975) Organophosphorus esters causing delayed neurotoxic effects. Arch Toxicol 34:259–288

Kanje M (1986) Ionic requirements for fast axonal transport. In: Iqbal Z (ed) Axoplasmic transport. CRC Press, Boca Raton pp 35–43

Kanje M, Edström A, Hanson M (1981) Inhibition of rapid axonal transport in vitro by the ionophores X-537 A and A 23187. Brain Res 204:43–50

Kanje M, Edström A, Edstrom P (1982) Divalent cations and fast axonal transport in chemically desheathed (Triton X-treated) frog sciatic nerve. Brain 241:67–74

Kanje M, Deinum J, Wallin M, Ekström P, Edström A, Hartley-Asp B (1985) Effect of estramustine phosphate on the assembly of isolated bovine brain microtubules and fast axonal transport in the frog sciatic nerve. Cancer Res 45:2234–2239

Kanje M, Ekström P, Deinum J, Wallin M (1986) The effect of gossypol on fast axonal transport and microtubule assembly. Biochim Biophys Acta 856: 437–442

Karlsson J-O, Sjöstrand J (1971) Synthesis migration and turnover of protein in retinal ganglion cells. J Neurochem 18:749–767

Kato S, Yamamoto S, Iwasaki Y, Niizuma H, Nakamura T, Suzuki J (1988) Experimental retrograde adriamycin trigeminal sensory ganglionectomy. J Neurosurg 69:760–765

Klatzo I, Wisniewski H, Streicher E (1965) Experimental production of neurofibrillary degeneration. I. Light microscopic observations. J Neuropathol Exp Neurol 24:187–199

Knyihár-Csillik E, Szücs A, Csillik B (1982) Iontophoretically applied microtubule inhibitors induce transganglionic degenerative atrophy of primary central nociceptive terminals and abolish chronic autochthonous pain. Acta Neurol Scand 66:401–412

Kreutzberg GW (1969) Neuronal dynamics and axonal flow IV. Blockage of intra-axonal enzyme transport by colchicine. Proc Natl Acad Sci USA 62:722–728

Kristensson K (1978) Retrograde transport of macromolecules in axons. Annu Rev Pharmacol Toxicol 18:97–110

Kristensson K, Lycke E, Sjöstrand J (1971) Spread of herpes simplex virus in peripheral nerves. Acta Neuropathol (Berl) 19:44–53

Lariviere L, Lavoie P-A (1982) Calcium requirement for fast axonal transport in frog motoneurons. J Neurochem 39:882–886

Lasek RJ (1967) Bidirectional transport of radioactively labelled axoplasmic components. Nature 216:1212–1214

Lasek RJ, Hoffman PN (1976) The neuronal cytoskeleton, axonal transport and axonal growth. In: Goldman R, Pollard T, Rosenbaum J (eds) Cell motility (book C). Cold Spring Harbor Laboratory, Cold Spring Harbor, pp 1021–1049

Lasek RJ, Katz MJ (1987) Mechanisms at the axon tip regulate metabolic processes critical to axonal elongation. Prog Brain Res 71:49–60

Lasek RJ, Garner JA, Brady ST (1984) Axonal transport of the cytoplasmic matrix. J Cell Biol 99:212–221

Lavoie P-A (1982) Inhibition of fast axonal transport in vitro by tetracaine: an increase in potency at alkaline pH and no change in potency in calcium-depleted nerves. Can J Physiol Pharmacol 60:1715–1720

Lavoie P-A, Tiberi M (1986) Inhibition of fast axonal transport in bullfrog nerves by dibenzazepine and dibenzocycloheptadiene calmodulin inhibitors. J Neurobiol 17:681–695

Lavoie P-A, Collier B, Tenenhouse A (1976) Comparison of alpha-bungarotoxin binding to skeletal muscles after inactivity or denervation. Nature 260:349–350

Lavoie P-A, Bolen F, Hammerschlag R (1979) Divalent cation specificity of the calcium requirement for fast transport of proteins in axons of desheathed nerves. J Neurochem 32:1745–1751

Lavoie P-A, Khazen T, Filion PR (1989) Mechanisms of the inhibition of fast axonal transport by local anesthetics. Neuropharmacology 28:175–181

Lees GJ (1985) Inhibition of the retrograde axonal transport of dopamine-beta-hydroxylase antibodies by the calcium ionophore A23187. Brain Res 345:62–67

Leone J, Ochs S (1978) Anoxic block and recovery of axoplasmic transport and electrical excitability of nerve. J Neurobiol 9:229–245

Levi-Montalcini R (1976) The nerve growth factor: its role in growth differentiation and function of the sympathetic adrenergic neuron. Prog Brain Res 45:235–258

Lim S-S, Sammak PJ, Borisy GG (1989) Progressive and spatially differentiated stability of microtubules in developing neuronal cells. J Cell Biol 109:253–263

Lim S-S, Edson KJ, Letourneau PC, Borisy GG (1990) A test of microtubule translocation during neurite elongation. J Cell Biol 111:123–130

Liwnicz BH, Kristensson K, Wisniewski HM, Shelanski ML, Terry RD (1974) Observations on axoplasmic transport in rabbits with aluminum-induced neurofibrillary tangles. Brain Res 80:413–420

Macioce P, Filliatreau G, Figliomeni B, Hassig R, Thiery J, Di Giamberardino L (1989) Slow axonal transport impairment of cytoskeletal proteins in streptozocin-induced diabetic neuropathy. J Neurochem 53:1261–1267

McEwen BS, Grafstein B (1968) Fast and slow components in axonal transport of protein. J Cell Biol 38:494–508

McLane JA (1987) Decreased axonal transport in rat nerve following acute and chronic ethanol exposure. Alcohol 4:385–389

McLane JA (1990) Retrograde axonal transport in chronic ethanol-fed and thiamine-deficient rats. Alcohol 7:103–106

Mellon M, Rhebun L, Rosenbaum J (1976) Studies on the accessible sulfhydryls of polymerizable tubulin. In: Goldman R, Pollard T, Rosenbaum J (eds) Cell motility (book C). Cold Spring Harbor Laboratory, Cold Spring Harbor, pp 1149–1163

Mendell JR, Sahenk Z (1980) Interference of neuronal processing and axoplasmic transport by toxic chemicals. In: Spencer PS, Schaumburg HH (eds) Neurotoxicology. Williams and Wilkins, Baltimore, pp 139–160

Mendell JR, Sahenk Z, Saida K, Weiss HS, Savage R, Couri D (1977) Alterations of fast axoplasmic transport in experimental methyl n-butyl ketone neuropathy. Brain Res 133:107–118

Mildvan AS (1970) Metals in enzyme catalsis vol II. In: Boyer PD (ed) The enzymes, 3rd edn. Academic, New York, pp 445–536

Miller MS, Spencer PS (1984) Single doses of acrylamide reduce retrograde transport velocity. J Neurochem 43:1401–1408

Miller MS, Spencer PS (1985) The mechanisms of acrylamide axonopathy. Annu Rev Pharmacol Toxicol 25:643–666

Mobley WC, Sever AC, Ishii DN, Riopelle RJ, Shooter EM (1977) Nerve growth factor. N Engl J Med 297:1096–1104

Monaco S, Jacob J, Jenich H, Patton A, Autilio-Gambetti L, Gambetti P (1989) Axonal transport of neurofilament is accelerated in peripheral nerve during 2,5-hexanedione intoxication. Brain Res 491:328–334

Moretto A, Sabri MI (1988) Progressive deficits in retrograde axon transport precede degeneration of motor axons in acrylamide neuropathy. Brain Res 440: 18–24

Moretto A, Lotti M, Sabri MI, Spencer PS (1987) Progressive deficit of retrograde axonal transport is associated with the pathogenesis of di-*n*-butyl dichlorvos axonopathy. J Neurochem 49:1515–1522

Mori H, Komiya Y, Kurokawa M (1979) Slowly migrating axonal polypeptides. Inequalities in their rate and amount of transport between two branches of bifurcating axons. J Cell Biol 82:74–84

Muñoz-Martínez EJ (1982) Axonal retention of transported material and the lability of nerve terminals. In: Weiss DG (ed) Axoplasmic transport. Springer, Berlin Heidelberg New York, pp 267–274

Muñoz-Martínez EJ, Núñez R, Sanderson A (1981) Axonal transport: a quantitative study of retained and transported protein fraction in the cat. J Neurobiol 12:15–26

Muñoz-Martínez EJ, Massieu D, Ochs S (1984) Depression of fast axonal transport produced by Tullidora. J Neurobiol 15:375–392

Narahashi T, Albuquerque EX, Deguchi T (1971) Effects of batrachotoxin on membrane potential and conductance of squid giant axons. J Gen Physiol 58:54–70

Nennesmo I, Kristensson K (1986) Effects of retrograde axonal transport of *Ricinus communis* agglutinin I on neuroma formation. Acta Neuropathol (Berl) 70:279–283

Nichols TR, Smith RS, Snyder RE (1982) The action of puromycin and cycloheximide on the initiation of rapid axonal transport in amphibian dorsal root neurones. J Physiol (Lond) 332:441–458

Nixon RA (1991) Axonal transport of cytoskeletal proteins. In: Burgoyne, R (ed) The Neuronal Cytoskeleton, Wiley-Liss, New York, pp 283–307

Nixon R, Logvinenko KB (1985) Multiple fates of newly synthesized neurofilament proteins: evidence for a stationary neurofilament network distributed nonuniformly along axons of retinal ganglion cell neurons. J Cell Biol 102:647–658

Nixon RA, Fischer I, Lewis SE (1990) Synthesis, axonal transport and turnover of the high molecular weight microtubule associated protein MAP 1A in mouse retinal ganglion cells: tubulin and MAP 1A display distinct transport kinetics J Cell Biol 110:437–448.

Ochs S (1971) Characteristics and a model for fast axoplasmic transport in nerve. J Neurobiol 2:331–345

Ochs S (1972) Rate of fast axoplasmic transport in mammalian nerve fibres. J Physiol (Lond) 227: 627–645

Ochs S (1974a) Retention and redistribution of fast and slow axoplasmic transported proteins in mammalian nerve. 3rd international congress on muscle disease. Abstr Excerpta Med Int Congr Ser 334:26

Ochs S (1974b) Energy metabolism and supply of ∼ P to the fast axoplasmic transport mechanism in nerve. Fed Proc 33:1049–1058

Ochs S (1974c) Systems of material transport in nerve fibers (axoplasmic transport) related to nerve function and trophic control. Ann NY Acad Sci 228:202–223

Ochs S (1975) Retention and redistribution of proteins in mammalian nerve fibres by axoplasmic transport. J Physiol (Lond) 253:459–475

Ochs S (1980) Calcium requirement for axoplasmic transport and the role of the perineurial sheath. In: Jewett DL, McCarroll HR Jr (eds) Nerve repair and regeneration. Mosby, St Louis, pp 77–88

Ochs S (1982) Axoplasmic transport and its relation to other nerve functions, 1st edn. Wiley-Interscience, New York, pp 1–462

Ochs S, Hollingsworth D (1971) Dependence of fast axoplasmic transport in nerve on oxidative metabolism. J Neurochem 18:107–114

Ochs S, Iqbal Z (1983) The role of calcium in axoplasmic transport in nerve. In: Cheung WY (ed) Calcium and cell function, vol 3. Academic, New York, pp 325–354

Ochs S, Jersild RA Jr (1974) Fast axoplasmic transport in nonmyelinated mammalian nerve fibers shown by electron microscopic radioautography. J Neurobiol 5:373–377

Ochs S, Jersild RA Jr (1985) Beading of nerve fibers and fast axoplasmic transport. Soc Neurosci Abstr 11:1135

Ochs S, Jersild RA Jr (1987) Cytoskeletal organelles and myelin structure of beaded nerve fibers. Neuroscience 22:1041–1056

Ochs S, Smith CB (1971) Fast axoplasmic transport in mammalian nerve in vitro after block of glycolysis with iodoacetic acid. J Neurochem 8:833–843

Ochs S, Worth R (1975) Batrachotoxin block of fast axoplasmic transport in mammalian nerve fibers. Science 87:87–89

Ochs S, Johnson J, Ng M-H (1967) Protein incorporation and axoplasmic flow in motoneuron fibres following intra-cord injection of labeled leucine. J Neurochem 4:3–7

Ochs S, Sabri MI, Ranish N (1970) Somal site of synthesis of fast transported materials in mammalian nerve fibers. J Neurobiol 1:329–344

Ochs S, Jersild RA, Jr Li J-M (1989) Slow transport of freely movable cytoskeletal components shown by beading partition of nerve fibers. Neuroscience 33:421–430

Oka N, Brimijoin S (1990) Premature onset of fast axonal transport in bromophenylacetylurea neuropathy: an electrophoretic analysis of proteins exported into motor nerve. Brain Res 509:107–110

Okabe S, Hirokawa N (1988) Microtubule dynamics in nerve cells: Analysis using microinjection of biotinylated tubulin into PC12 cells. J Cell Biol 107:651–664

Okabe S, Hirokawa N (1990) Turnover of fluorescently labelled tubulin and actin in the axon. Nature 343:479–482

Papasozomenos SC (1986) Reorganization of axonal cytoskeleton following b,b'-iminodipropionitrile (IDPN) intoxication. In: Clarkson TW, Sager PR, Syversen TLM (eds) The cytoskeleton – a target for toxic agents. Plenum, New York, pp 67–82

Papasozomenos SC, Autilio-Gambetti L, Gambetti P (1981) Reorganization of axoplasmic organelles following β,β'-iminodipropionitrile administration. J Cell Biol 91:866–871

Papasozomenos SC, Autilio-Gambetti L, Gambetti P (1982a) The IDPN axon: rearrangement of axonal cytoskeleton and organelles following β,β'-iminodipropionitrile (IDPN) intoxication. In: Weiss DG (ed) Axoplasmic transport. Springer, Berlin Heidelberg New York, pp 241–250

Papasozomenos SC, Yoon M, Crane R, Autilio-Gambetti L, Gambetti P (1982b) Redistribution of proteins of fast axonal transport following administration of β,β'-iminodipropionitrile: a quantitative autoradiographic study. J Cell Biol 95:672–675

Pappolla M, Penton R, Weiss HS, Miller CH Jr, Sahenk Z, Autilio-Gambetti L, Gambetti L (1987) Carbon disulfide axonopathy. Another experimental model characterized by acceleration of neurofilament transport and distinct changes of axonal size. Brain Res 424:272–280

Paulson JC, McClure WO (1975) Inhibition of axoplasmic transport by colchicine, podophyllotoxin and vinblastine: an effect on microtubules. Ann NY Acad Sci 253:517–527

Pestronk A, Drachman DB, Griffin JW (1976) Effect of muscle disuse on acetylcholine receptors. Nature 260:352–353

Pleasure DE, Mishler KC, Engle WK (1969) Axonal transport of proteins in experimental neuropathies. Science 66:524–525

Price DL, Griffin J, Young A, Peck K, Stocks A (1975) Tetanus toxin: direct evidence for retrograde intraaxonal transport. Science 88:945–947

Reese TS (1987) The molecular basis of axonal transport in the squid giant axon. In: Kandel ER (ed) Molecular neurobiology in neurology and psychiatry. Raven Press, New York, pp 89–102

Roots BI (1983) Neurofilament accumulation induced in synapses by leupeptin. Science 221:971–972

Roytta M, Raine C (1985) Taxol-induced neuropathy: further ultrastructural studies of nerve fibre changes in situ. J Neurocytol 14:157–175

Rubinow SI, Blum JJ (1980) A theoretical approach to the analysis of axonal transport. Biophys J 30:137–148

Sabri MI, Ochs S (1971) Inhibition of glyceraldehyde-3-phosphate dehydrogenase in mammalian nerve by iodoacetic acid. J Neurochem 8:1509–1514

Sabri MI, Ochs S (1972) Relation of ATP and creatine phosphate to fast axoplasmic transport in mammalian nerve. J Neurochem 9:2821–2828

Sabri MI, Spencer PS (1990) Acrylamide impairs fast and slow axonal transport in rat optic system. Neurochem Res 15:603–608

Sabri M, Dairman W, Fenton M, Juhasz L, Ng T, Spencer P (1989) Effect of exogenous pyruvate on acrylamide neuropathy in rats. Brain Res 483:1–11

Sahenk Z, Lasek RJ (1988) Inhibition of proteolysis blocks anterograde-retrograde conversion of axonally transported vesicles. Brain Res 460:199–203

Sahenk Z, Mendell J (1979a) Ultrastructural study of zinc pyridinethione-induced peripheral neuropathy. J Neuropathol Exp Neurol 58:532–550

Sahenk Z, Mendell JR (1979b) Analysis of fast axoplasmic transport in nerve ligation and adriamycin-induced neuronal perikaryon lesions. Brain Res 171:41–53

Sahenk Z, Mendell JR (1980) Axoplasmic transport in zinc pyridinethione neuropathy: evidence for an abnormality in distal turn-around. Brain Res 186:343–353

Sahenk Z, Mendell JR (1981) Acrylamide and 2,5-hexanedione neuropathies: abnormal bidirectional transport rate in distal axons. Brain Res 219:397–405

Sammak PJ, Borisy GG (1988) Direct observation of microtubule dynamics in living cells. Nature 332:724–726

Sammak PJ, Gorbsky GJ, Borisy GG (1987) Microtubule dynamics in vivo: a test of mechanisms of turnover. J Cell Biol 104:395–405

Schlaepfer WW (1974) Calcium-induced degeneration of axoplasm in isolated segments of rat peripheral nerve. Brain Res 69:203–215

Schlaepfer WW, Hasler MB (1979) Characterization of the calcium-induced disruption of neurofilaments in rat peripheral nerves. Brain Res 68:299–309

Schmidt RE, Grabau GG, Yip HK (1986) Retrograde axonal transport of [^{125}I] nerve growth factor in ileal mesenteric nerves in vitro: effect of streptozotocin diabetes. Brain Res 378:325–336

Schwab ME, Thoenen H (1978) Selective binding, uptake and retrograde transport of tetanus toxin by nerve terminals in the rat iris. J Cell Biol 77:1–13

Schwab ME, Agid Y Glowinski J, Thoenen H (1977) Retrograde axonal transport of ^{125}I-tetanus toxin as a tool for tracing fiber connections in the central nervous system: connections of the rostral part of the rat neostriatum. Brain Res 126:211–224

Schwab ME, Heumann R, Thoenen H (1982) Communication between target organs and nerve cells: retrograde axonal transport and site of action of nerve growth factor. Cold Spring Harbor Symp Quant Biol 46:125–134

Sheetz MP, Steuer ER, Schroer TA (1989) The mechanism and regulation of fast axonal transport. Trends Neurosci 12:474–478

Shpetner HS, Paschal BM, Vallee RB (1988) Characterization of the microtubule-activated ATPase of brain cytoplasmic dynein (MAP 1C). J Cell Biol 107:1001–1009

Sickles DW (1989a) Toxic neurofilamentous axonopathies and fast anterograde axonal transport. I. The effects of single doses of acrylamide on the rate and capacity of transport. Neurotoxicology 10:91–102

Sickles DW (1989b) Toxic neurofilamentous axonopathies and fast anterograde axonal transport. II. The effects of single doses of neurotoxic and non-neurotoxic diketones and beta,beta'-iminodipropionitrile (IDPN) on the rate and capacity of transport. Neurotoxicology 10:103–111

Sidenius P (1982) The axonopathy of diabetic neuropathy. Diabetes 31:356–363

Sidenius P, Jakobsen J (1983) Anterograde axonal transport in rats during intoxication with acrylamide. J Neurochem 40:697–704

Sidenius P, Jakobsen J (1987) Axonal transport in human and experimental diabetes. In: Dyck PJ, Thomas PK, Asbury AK, Winegrade AI, Porte D Jr (eds) Diabetic neuropathy. Saunders, Philadelphia, pp 260–265

Sjöstrand J, Karlsson J-O (1969) Axoplasmic transport in the optic nerve and tract of the rabbit: a biochemical and radioautographic study. J Neurochem 6:833–844

Smith FP, Kato S, Yamamoto I, Iwasaki Y (1989) Retrograde transport of adriamycin. J Neurosurg 70:819–820 (abstract)

Smith RS (1980) The short term accumulation of axonally transported organelles in the region of localized lesions of single myelinated axons. J Neurocytol 9:39–65

Smith RS (1982) Axonal transport of optically detectable particulate organelles. In: Weiss DG (ed) Axoplasmic transport. Springer, Berlin Heidelberg New York, pp 181–192

Smith RS, Snyder RE (1991) Reversal of rapid axonal transport at a lesion: leupeptin inhibits reversed protein transport, but does not inhibit reversed organelle transport. Brain Res 552:215–227

Snyder RE (1986) The kinematics of turnaround and retrograde axonal transport. J Neurobiol 17:637–647

Spencer PS, Schaumburg HH (1980) Experimental and clinical neurotoxicology, 1st edn. Williams and Wilkins, Baltimore, pp 1–929

Spencer PS, Sabri MI, Schaumburg HH, Moore CL (1979) Does a defect of energy metabolism in the nerve fiber underlie axonal degeneration in polyneuropathies? Ann Neurol 5:501–507

Spencer PS, Couri D, Schaumburg HH (1980) n-Hexane and methyl-n-butyl ketone. In: Spencer PS, Schaumburg HH (eds) Experimental and clinical neurotoxicology. Williams and Wilkins, Baltimore, pp 456–475

Takenaka T, Gotoh H (1984) Simulation of axoplasmic transport. J Theor Biol 107:579–601

Takenaka T, Inomata K, Horie H (1982) Slow axoplasmic transport of labeled protein in sciatic nerves of streptozotocin-diabetic rats and methylcobalamin treated rats. In: Goto Y, Horiuchi A, Kogure K (eds) Diabetic neuropathy. Excerpta Medica, Amsterdam, pp 99–103

Tashiro T, Komiya Y (1987) Organization of cytoskeletal proteins transported in the axon. In: Smith RS, Bisby MA (eds) Axonal transport. Liss, New York, pp 201–221

Tiberi M, Lavoie P-A (1985) Inhibition of the retrograde axonal transport of acetylcholinesterase by the anti-calmodulin agents amitriptyline and desipramine. J Neurobiol 16:245–248

Triarhou LC, Norton J, Bugiani O, Ghetti B (1985) Ventral root axonopathy and its relation to the neurofibrillary degeneration of lower motor neurons in aluminum-induced encephalomyelopathy. Neuropathol Appl Neurobiol 11:407–430

Tsukita S, Ishikawa H (1980) The movement of membranous organelles in axons. Electron microscopic identification of anterogradely and retrogradely transported organelles. J Cell Biol 84:513–530

Vale RD (1987) Intracellular transport using microtubule-based motors. Annu Rev Biol 3:347–378

Vale RD, Schnapp BJ, Mitchison T, Steuer E, Reese TS, Sheetz MP (1985) Different axoplasmic proteins generate movement in opposite directions along microtubules in vitro. Cell 43:623–632

Vallee RB, Shpetner HS, Paschal BM (1989) The role of dynein in retrograde axonal transport Trends Neurosci 12:66–70

Watson DF, Hoffman PN, Fittro KP, Griffin JW (1989) Neurofilament and tubulin transport slows along the course of mature motor axons. Brain Res 477:225–232

Weiss HD, Walker MD, Wiernik PH (1974) Neurotoxicity of commonly used antineoplastic agents. N Engl J Med 291:127–133

Weiss P, Hiscoe HB (1948) Experiments on the mechanism of nerve growth. J Exp Zool 107:315–395

Weller RO, Mitchell J, Daves GD Jr (1980) Buckthorn (*Karwinskia humboldtiana*) toxins. In: Spencer PS Schaumburg HH (eds) Experimental and clinical neurotoxicity, vol 1. Williams and Wilkins, Baltimore, pp 336–347

Wiley RG, Stirpe F (1988) Modeccin and volkensin but not abrin are effective suicide transport agents in rat CNS. Brain Res 438:145–154

Wiley RG, Blessing WW, Reis DJ (1982) Suicide transport: destruction of neurons by retrograde transport of ricin, abrin and modeccin. Science 216:889–890

Willard MB, Hulebak KL (1977) The intra-axonal transport of polypeptide H: evidence for a fifth (very slow) group of transported proteins in the retinal ganglion cells of the rabbit. Brain Res 36:289–306

Willard M, Cowan WM, Vagelos PR (1974) The polypeptide composition of intra-axonally transported proteins: evidence for four transport velocities. Proc Natl Acad Sci USA 71:2183–2187

Wisniewski HM, Sturman JA, Shek JW (1979) Aluminum chloride induced neurofibrillary changes in the developing rabbit: a chronic animal model. Ann Neurol 8:479–490

Worth RM, Ochs S (1982) Dependence of batrachotoxin block of axoplasmic transport on sodium. J Neurobiol 13:537–549

Yamamoto T, Iwasaki Y, Konno H (1984a) Experimental sensory ganglionectomy by way of suicide axoplasmic transport. J Neurosurg 60:108–114

Yamamoto T, Iwasaki Y, Konno H (1984b) Retrograde axoplasmic transport of adriamycin: an experimental form of motor neuron disease? Neurology 34:1299–1304

CHAPTER 5

Metabolic Disorders as Consequences of Drug-Induced Energy Deficits

J. Krieglstein and J. Nuglisch

A. Introduction

The human brain accounts for only 2% of the total body weight but requires more than 15% of the cardiac output. This clearly indicates that the metabolic rate of brain tissue is higher than that of most other organs and tissues. Glucose is the main substrate for brain energy metabolism under physiological conditions, but some amino acids make a minor contribution; they are probably metabolized by glial cells. Other substrates are used for energy production under pathological conditions, e.g., ketone bodies in hypoglycemia or lactate after cerebral ischemia. Free fatty acids and glycerol are metabolized only in very small amounts (Hawkins and Mans 1983).

The sensitivity of the brain to oxygen or glucose deficiency has been elegantly demonstrated in studies on the isolated perfused rat brain (Krieglstein et al. 1972; Dirks et al. 1980; Hanke and Krieglstein 1982). Interruption of the blood supply to the brain produces significant changes in the electroencephalogram (EEG) after just 3–5 s. EEG silence ensues after 20–30 s of complete cerebral ischemia. However, when the isolated rat brain is perfused with an oxygen-free or glucose-free medium, the EEG activity of the isolated brain lasts more than 1 min after anoxic perfusion (no oxygen but normal glucose concentration in the medium). After aglycemic perfusion (normal oxygen but no glucose in the medium) spontaneous EEG could be recorded for up to 15 min (Fig. 1). Interestingly, the EEG activity ceases before the ATP level in the brain tissue begins to decrease (Dirks et al. 1980).

Clearly then, a normal ATP level is a prerequisite but not a regulator of neuronal function.

B. Brain Energy Metabolism

ATP plays a fundamental role in mammalian brain function. Under normoxic conditions, ATP is synthesized almost exclusively by oxidative phosphorylation in the mitochondria, and only a small portion comes from glycolysis (Siesjö 1978). The steady-state concentrations of high-energy phosphates are essentially the same in all the various brain regions. In contrast, the rates of ATP production vary widely, with more

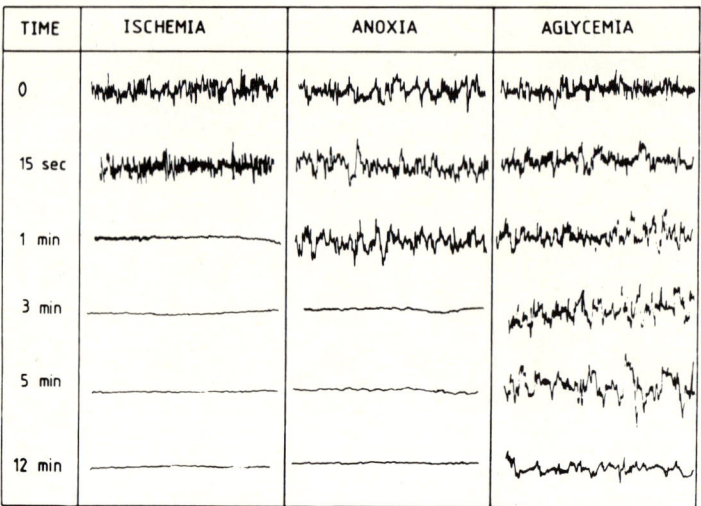

Fig. 1. Velocity of electroencephalogram (EEG) decrease in the isolated perfused rat brain during ischemia, anoxia, or hypoglycemia. EEG silence was achieved after about 30 s of ischemia, after about 1 min of anoxia, or after about 15 min of aglycemia (perfusion medium without glucose but equilibrated with 95% O_2 and 5% CO_2)

rapid synthesis in areas of higher functional activity, because ATP is predominantly utilized for ionic pumps (ERECIŃSKA and SILVER 1989).

Brain energy metabolism used to be assessed by measuring the levels of metabolic substrates and intermediates, such as high-energy phosphates, glucose, pyruvate, or lactate. However, these parameters are of limited value for characterizing brain metabolism and function. It was, therefore, an important breakthrough when SOKOLOFF et al. (1977) published their method for determining local cerebral glucose utilization (CMR_{glu}) using [^{14}C] 2-deoxy-D-glucose. This technique can be used to measure the metabolic rate in anatomically defined brain areas in vivo and not just the steady-state concentrations of energy substrates (Fig. 2).

The functional activity of brain tissue can also be characterized by this technique because glucose consumption correlates with neuronal function. The deoxyglucose technique has also been used for testing the action of drugs on distinct brain regions in vivo (SOKOLOFF et al. 1990).

Several animal species, e.g., rats, mice, rabbits, gerbils, cats, dogs, and monkeys, have been used to study disturbances in brain energy metabolism (SIESJÖ 1978; GINSBERG and BUSTO 1989). In particular, the rat and gerbil models have been extensively developed, and a large database of neurochemical, histopathological, pharmacological, and toxicological information exists for these species. The Mongolian gerbil seems to be particularly

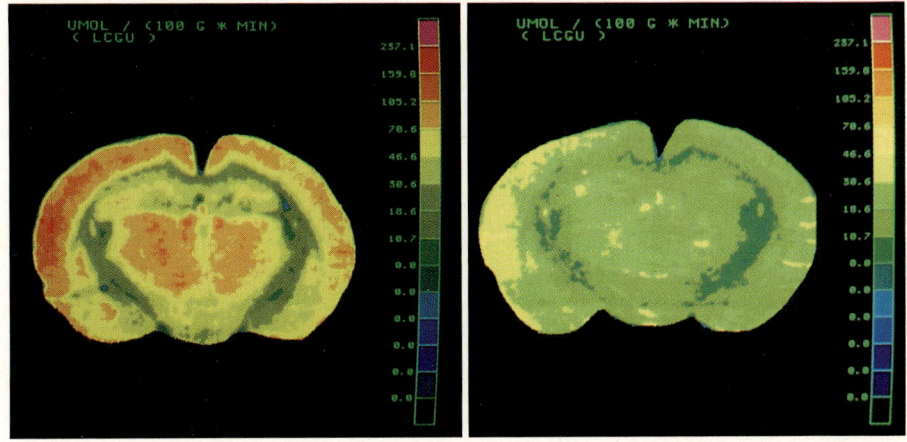

Fig. 2a,b. Autoradiographic measurement of local cerebral glucose utilization (LCGU) in the rat according to SOKOLOFF et al. (1977). **a** LCGU of the untreated conscious rat. **b** Rats treated with pentobarbital (50 mg/kg i.v.). LCGU of the color-coded pictures is given in μmol × 100 g^{-1} × min^{-1}. Pentobarbital significantly reduced LCGU in nearly all brain structures

convenient because it has an incomplete circle of Willis, lacking connections between the vertebral and the carotid arteries. Nevertheless, the occasional animal has communicating vessels, and some develop seizures. Gerbil models may therefore lack some degree of experimental reliability. This may be why rats have been used even more frequently than gerbils for studies on oxygen deficiency of the brain, i.e., cerebral ischemia and hypoxia.

C. Energy Deficits

Pathological events (ischemia, hypoxia, hypoglycemia, seizures, etc.) or toxic effects of drugs may disrupt the homeostasis of brain energy metabolism by hindering energy production or enhancing energy consumption. It was once thought that different conditions of energy deficits give rise to similar if not identical brain damage. The rationale of that thinking was that both limited energy production and pathologically enhanced energy consumption cause energy failure, leading to disorders of brain metabolism. However, it has become clear that changes in cellular energy metabolism per se do not provide common mechanisms of brain damage (SIESJÖ 1981; AUER 1986; AUER and SIESJÖ 1988; GINSBERG 1990).

I. Oxygen Deficiency

Oxygen deficiency (ischemia, hypoxia, anoxia, drug effects) leads to depletion of glucose and glycogen stores and impaired mitochondrial respiration.

Cells metabolize residual stores of glucose anaerobically to lactic acid instead of CO_2 and water. ATP is still used, but its production is decreased, leading to an increase in the proportionate amount of AMP. The adenylate energy charge, EC = [ATP] + 0.5[ADP]/[ATP + ADP + AMP], is a measure of the ratio of high-energy phosphate bond production to consumption (Atkinson 1968).

ATP-requiring processes are deprived of their energy supplies under conditions of oxygen deficiency. In particular, the loss of ATP impairs the plasma membrane ATP-linked Na^+/K^+-pump, causing cells to lose K^+ and to take up Na^+ and Cl^-. Extracellular K^+ (approximately 15 µmol/g) depolarizes voltage-dependent Ca^{2+} channels, and extracellular Ca^{2+} enters the cell (Hansen 1985).

The brain ATP level is a reliable indicator of those events, because, while it may decline within a few seconds after disturbances of energy metabolism, it is also rapidly restored after normalization of brain metabolism. However, the increase in the lactate level which results from oxygen lack is even more pronounced than the drop in ATP (Siesjö 1978; Siesjö and Bengtsson 1989; Hanke and Krieglstein 1982; Krieglstein and Weber 1986). The level of this intermediate can rise rapidly to ten times its normal concentration in brain tissue after a few minutes of ischemia or hypoxia, and these pathological values are restored more slowly than the ATP levels (Fig. 3). Since the transport of lactate across the blood-brain barrier (BBB) is very limited (Cremer et al. 1979), most of the newly formed lactate has to be metabolized in the brain tissue before returning to the normal concentration. It is argued that the increased lactate level contributes to brain damage. Lactic acid accumulation and increased PCO_2 can cause severe acidosis, giving a pH value in the tissue of approximately 6. The severity of tissue damage during an ischemic insult and the recirculation phase has been shown to be related to the degree of lactic acidosis (Siesjö 1981). Neuronal dysfunction occurs within a few seconds without prompt reoxygenation of brain tissue. Depending on the duration and extent of the oxygen deficiency, neuronal damage may be transient, reversible to some extent, or sufficient to cause cell death (necrosis).

II. Substrate Deficiency

The lactate level does not increase under conditions of substrate failure (i.e., hypoglycemia) and is not of significance for brain damage. A fall in the blood glucose level to 2.0–2.5 mmol/l causes slowing of the EGG. The ionic pumps cannot remain functional, and ions follow their concentration gradients. The aspartate-glutamate reaction is markedly shifted towards aspartate, which is released into the extracellular space. Neuronal death does not occur in hypoglycemia unless the EEG becomes isoelectric. Brain damage may occur when electrocerebral silence has lasted several minutes (Auer 1986). Regional inhomogeneities in the glucose levels of brain tissue may, at least in

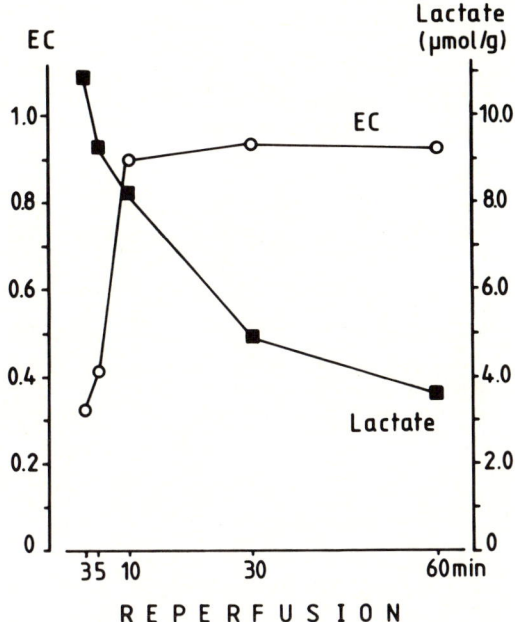

Fig. 3. Restoration of the energy charge of the adenylate pool (*EC*) and the lactate level in the isolated perfused rat brain after 30 min of ischemia. EC again has achieved the normal range (>0.90) after about 10 min of reperfusion, but the lactate level remains elevated even after 1 h of reperfusion (normal level 1 µmol/g)

part, explain differences in the vulnerability of different brain structures (Paschen et al. 1986).

D. Metabolic Disorders as Consequences of Energy Deficit

Neurons may die even after periods of time in which the energy metabolism has been normalized (Kirino 1982). It is the inability of the cells to re-establish vital functions disrupted by transient energy failure that causes gradual deterioration (Siesjö 1981). The key biochemical determinants of irreversible cell damage are still unknown. However, the interrelated factors of energy failure, lactic acidosis, disturbed protein metabolism, calcium homeostasis, neurotransmission, and membrane lipid metabolism, as well as formation of free oxygen radicals are all considered to be critical steps leading to neuronal damage or necrosis (Krieglstein 1989a; Krieglstein and Oberpichler 1990).

Many data have accumulated on the biochemical and histopathological changes which occur after brain ischemia (Siesjö and Bengtsson 1989).

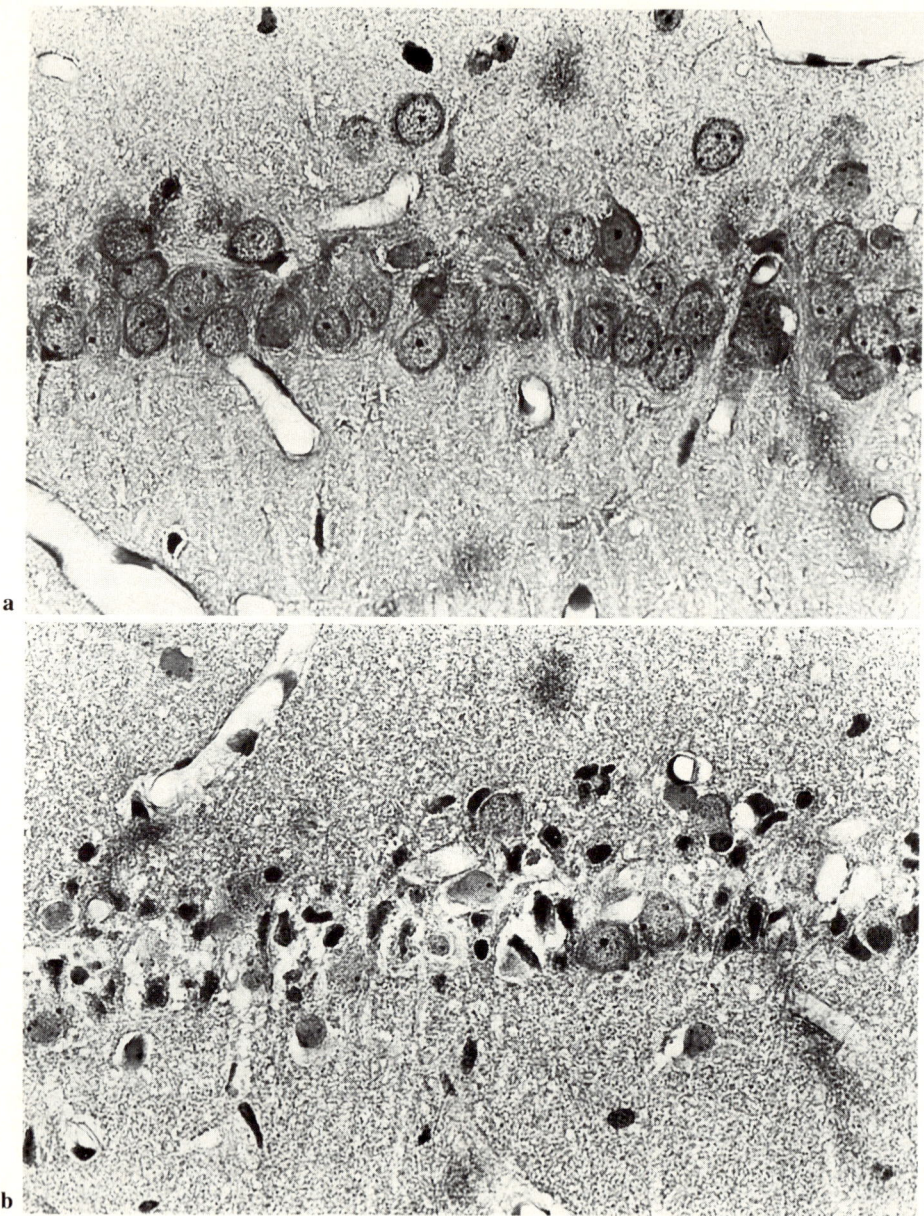

Fig. 4a,b. Histological demonstration of neuronal damage in the CA1 subfield of sham-operated (**a**) and ischemic (**b**) rats. Global ischemia was induced for 10 min, and the histological evaluation was performed 7 days after ischemia

These studies clearly show the selective vulnerability of neurons. The pyramidal neurons in the CA1 subfield of the hippocampus and the small and medium neurons of the striatum suffer more damage after short periods of global ischemia than do other regions of the brain. The Purkinje cells of the cerebellum and neurons of cortical layers 3, 5, and 6 are also vulnerable.

A sufficient supply of oxygen is undoubtedly an essential prerequisite for neuronal brain function, but it is now generally agreed that biochemical alterations occur before histological changes can be detected (KIRINO 1982). Neurons show signs of degeneration earlier than do astrocytes or endothelial cells. Mitochondrial swelling and disorganization of the cristae, as shown by microvacuoles in the cytoplasm, are early signs of neuronal damage. These changes can be reversed by prompt reperfusion. Otherwise, the neurons shrink, the nuclei become displaced, and the cells stain more darkly (Fig. 4). The histological changes caused by varying energy deficits have some of these features in common, especially in the later stages of neuronal damage.

I. Calcium Homeostasis

Calcium normally acts as a second messenger and plays an important role in membrane stabilization, metabolic regulation, and transmitter release. However, it can also cause cell death under anoxic or toxic conditions (SIESJÖ and BENGTSSON 1989; HALLAM and RINK 1989; CHOI 1990; CHOI and ROTHMAN 1990). This is because calcium can fulfill its physiological functions only when its cytosolic concentration, $[Ca^{2+}]_i$, is maintained at a low level of about $0.1\,\mu mol/l$, whereas its extracellular level is very much higher (ca. $1\,mmol/l$). Neurons can keep the $[Ca^{2+}]_i$ constant by rigorously controlling Ca^{2+} entry and intracellular storage. However, the cells must have enough energy available to transport Ca^{2+} out of the cell or sequester it in intracellular stores (SILVER and ERECINSKA 1990).

1. Routes of Ca^{2+} Entry and Means of Ca^{2+} Release

Some years ago, voltage-sensitive calcium channels (VSCCs) were considered to be the main routes of calcium entry into neurons, and the channels were demonstrated to be opened after depolarization of the cell membrane. Peripheral neurons were reported to contain three different VSCCs: L (long-lasting), T (transient) and N (neither L nor T). The N-type channels were localized on the presynaptic membrane. The entry of Ca^{2+} through these channels causes transmitter release. The L-type channels are restricted to postsynaptic membranes, and they contribute to postsynaptic Ca^{2+} influx (Figs. 5, 6; CARBONE and LUX 1984; NOWYCKY et al. 1985).

However, most of the Ca^{2+} enters postsynaptic neurons through receptor-operated calcium channels (ROCCs). These channels lie on dendrites, and the most important transmitters are glutamate and related amino acids. In recent years, an increasing number of investigations have provided new

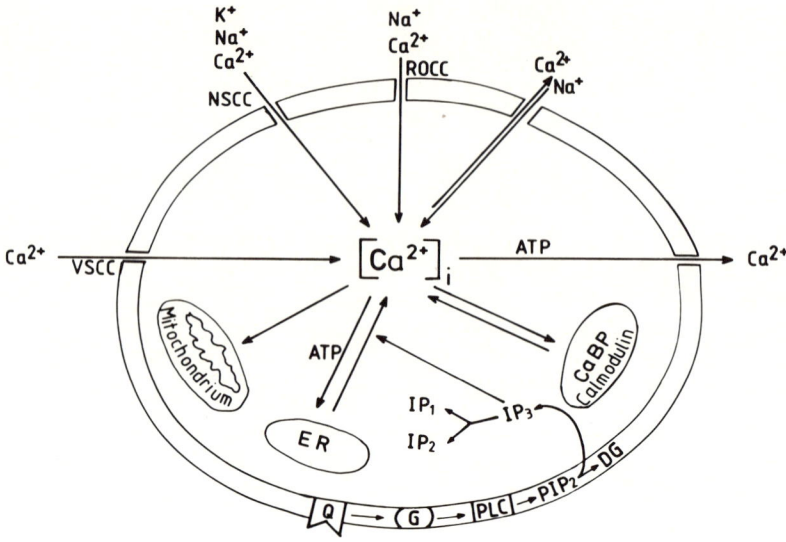

Fig. 5. Neuronal calcium homeostasis. The cytosolic Ca^{2+} concentration, $[Ca^{2+}]_i$, is low under normoxic conditions (0.1 μmol/l) and may increase tremendously after energy deficiency. Ca^{2+} may enter the neuron through voltage-sensitive channels (*VSCC*), receptor-operated channels (*ROCC*), a nonspecific cation conductance (*NSCC*), and by the Na^+/Ca^{2+} antiporter. Under conditions of energy failure, Ca^{2+} can be released by the endoplasmic reticulum (*ER*). *Q*, quisqualate-preferring metabotropic receptor; *PLC*, phospholipase C; *PIP$_2$*, phosphatidylinositol-4,5-bisphosphate; *IP$_3$*, inositol-1,4,5-triphosphate; *IP$_2$*, inositol-4,5,-biphosphate; *IP$_1$* inositol-5-phosphate. *DG*, diacylglycerol; *G*, G protein

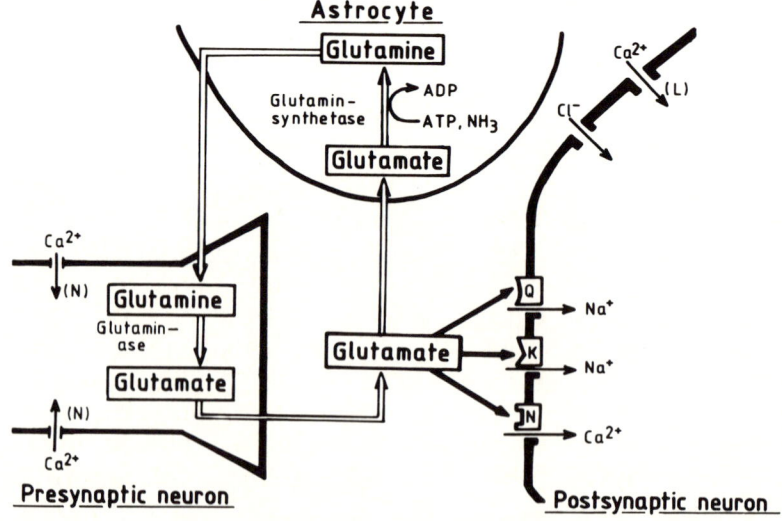

Fig. 6. Subtypes of glutamate receptors: *N*, NMDA receptor; *K*, kainate receptor; *Q*, quisqualate receptor; (*L*), L-type calcium channel; (*N*), N-type calcium channel

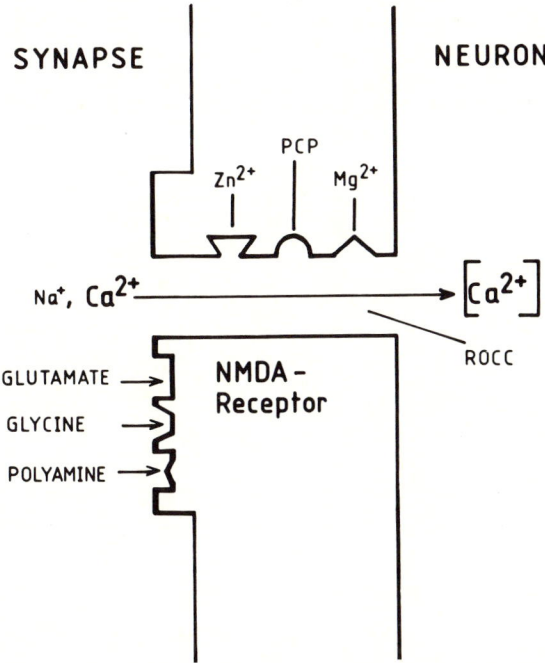

Fig. 7. *N*-methyl-D-aspartate (*NMDA*) receptor-channel complex. Stimulation of the receptor by glutamate is enhanced by glycine or polyamines. Ca^{2+} entry is blocked by Mg^{2+}, phencyclidine (*PCP*), Zn^{2+}, or various drugs binding to these specific sites within the channel. *ROCC*, receptor-operated channel

insight into glutamate receptor functions. Three types of glutamate receptors have been characterized (Fig. 6; CHOI 1990; SIESJÖ and BENGTSSON 1989; KRIEGLSTEIN 1989b). They are typified by the action of selective stimulators, i.e., *N*-methyl-D-aspartate (NMDA), kainate, and quisqualate (Q). The most important one for calcium homeostasis and the best defined is the NMDA receptor linked to a calcium channel. The hippocampus has a high density of NMDA receptors. Stimulation of the NMDA receptor opens a cation channel, allowing mainly Ca^{2+}, and to a smaller degree Na^+, to enter the cell.

Glutamate activation of this receptor and the ensuing Ca^{2+} influx are strongly potentiated by low concentrations of glycine and polyamines such as spermine or spermidine. The channel is blocked by a physiological concentration of Mg^{2+}; this block is relieved by membrane depolarization. The Ca^{2+} influx through the channel operated by this receptor is also blocked by Zn^{2+} (Fig. 7; CHRISTINE and CHOI 1990).

The other two glutamate receptors, activated by kainate and Q, are linked to channels which are indiscriminately permeable to monovalent cations (Na^+, K^+, probably H^+). These channels allow K^+ to leave and

Na^+ to enter the cell. This Na^+ influx could also indirectly induce Ca^{2+} entry via several other routes, including the most important VSCCs. Na^+ in excess of K^+ can only permeate into the cell if accompanied by an anion. Thus, when Na^+ and Cl^- accumulate within the neuron, they cause swelling due to the osmotic uptake of water. Ca^{2+} may also enter the cell via the reversed operating Na^+/Ca^{2+}-antiporter when the Na^+ gradient has collapsed. In addition, some nonspecific cation channels (NSCCs) can be activated, or channels may come into existence through membrane stretch (Fig. 5; Choi 1990).

Glutamate also activates a Q-preferring metabotropic receptor, linked via G protein to phospholipase C (Fig. 5; Dolphin 1990). Stimulation of this receptor induces hydrolysis of phosphatidylinositol-4,5-bisphosphate (PIP_2) to release the second messenger inositol-1,4,5-triphosphate (IP_3) and diacylglycerol (DG). DG, in concert with Ca^{2+}, activates protein kinase C which catalyzes the activation of many proteins, including receptor proteins. IP_3 serves as a water-soluble messenger causing release of Ca^{2+} from the endoplasmic reticulum (ER).

2. Disturbances of Neuronal Calcium Homeostasis

The Ca^{2+} entering the neuron from the extracellular space or released within the cell constitutes the Ca^{2+} signal, which is transformed into various cell responses (Rasmussen and Waisman 1983; Carafoli 1987). Membrane permeability to and extrusion of Ca^{2+} is precisely controlled at the expense of metabolic energy. Failure of ATP, influx of Na^+, and membrane depolarization all accelerate Ca^{2+} uptake and augment $[Ca^{2+}]_i$. There are several mechanisms for restricting the rising $[Ca^{2+}]_i$. When the energy state is normal, Ca^{2+} is transported out of the cell via the Na^+/Ca^{2+}-antiporter and Ca^{2+}-activated ATPase. A large fraction of the Ca^{2+} within the cell is buffered by Ca^{2+}-binding proteins such as calbindin (Rami et al. 1990) or is sequestered into the ER (Fig. 5).

However, when there is an energy deficit, $[Ca^{2+}]_i$ may rise manifold and then induce a multitude of metabolic reactions. Proteases and lipases are activated, leading to the formation of free fatty acids and destruction of various cell structures. Increased endonuclease activity hydrolyzes DNA (Katz and Messineo 1981; Siesjö 1990). The activated protein kinase C alters phosphorylation of proteins, even integral parts of receptors and ion channels, thereby accelerating neuronal damage. Mitochondria start to accumulate Ca^{2+} by an electrophoretic drag only at pathologically raised $[Ca^{2+}]_i$. An increased mitochondrial Ca^{2+} level activates rate-limiting enzymes such as pyruvate dehydrogenase, isocitrate dehydrogenase, and oxyglutamate dehydrogenase, altering the metabolic flux in the citric acid cycle. Some of the metabolic changes help adjust energy production to the metabolic demands. Thus, phosphorylase A is activated, triggering the breakdown of glycogen. Glucose is degraded anaerobically to lactate, causing cell acidosis. This could aggravate the disturbances of Ca^{2+} homeostasis, because Ca^{2+} and H^+ compete for the same anionic binding sites. Thus, acidosis curtails the Ca^{2+} buffering capacity of the cell and contributes to the increase in $[Ca^{2+}]_i$. On the other hand, it has been shown in vitro, using cultured hippocampal neurons (TOMBAUGH and SAPOLSKY 1990) or cultured cortical neurons (GIFFARD et al. 1990), that moderate acidosis may decrease neuronal vulnerability to hypoxic damage, and it need not be viewed only as a damaging component.

3. Calcium-Related Neuronal Damage

When calcium homeostasis fails, as in energy-deprived states, cell viability is threatened because calcium-activated reactions run out of control. Lipases, proteases, and endonucleases are overactivated. The cytoskeleton and membrane structures are disrupted, and the function of receptors and ion channels are adversely altered. Mitochondria are overloaded with Ca^{2+}, thereby blocking oxidative phosphorylation; the metabolic flux in the citric acid cycle is altered by overactivation of dehydrogenases (see above), causing reduced ATP synthesis. This series of devasting reactions is triggered whenever the $[Ca^{2+}]_i$ rises unduly because of excessive Ca^{2+} influx and failure of extrinsic mechanisms. Ca^{2+} may enter the neurons through various gates, with ROCCs forming a major route. The different types of glutamate receptors may act in concert to bring about Ca^{2+} influx. There is a tremendous surge in extracellular glutamate level during a period of energy failure; for instance, it rises from $5\,\mu mol/l$ to $500\,\mu mol/l$ after $30\,min$ of ischemia, and at this concentration it is rapidly toxic to cultured cortical neurons. Glutamate and related excitatory amino acids are believed to play an important role in neuronal damage caused by energy failure (OLNEY 1969; CHOI 1988, 1990; CHOI and ROTHMAN 1990). Glutamate may cause two types of neuronal damage: an immediate osmotic type and a delayed calcium-related form. However, energy depletion per se can lead to cell death without any rise in $[Ca^{2+}]_i$ and may produce catabolic changes similar to those induced by Ca^{2+}. Even though other factors than an increase in $[Ca^{2+}]_i$ may contribute to irreversible neuronal damage, one cannot ignore the potentially adverse effects triggered by Ca^{2+}. There is mounting evidence for a link between Ca^{2+} influx and neuronal necrosis (SIESJÖ and BENGTSSON 1989; MARTZ et al. 1989).

II. Oxygen Radicals

Oxygen radicals are formed in very small amounts by normal cellular metabolism and are usually kept under control by several defense mechanisms. However, metabolic disorders, such as those caused by oxygen deficiency and tissue reoxygenation, may lead to such an increased production of oxygen radicals that cellular defense mechanisms are overwhelmed. Free

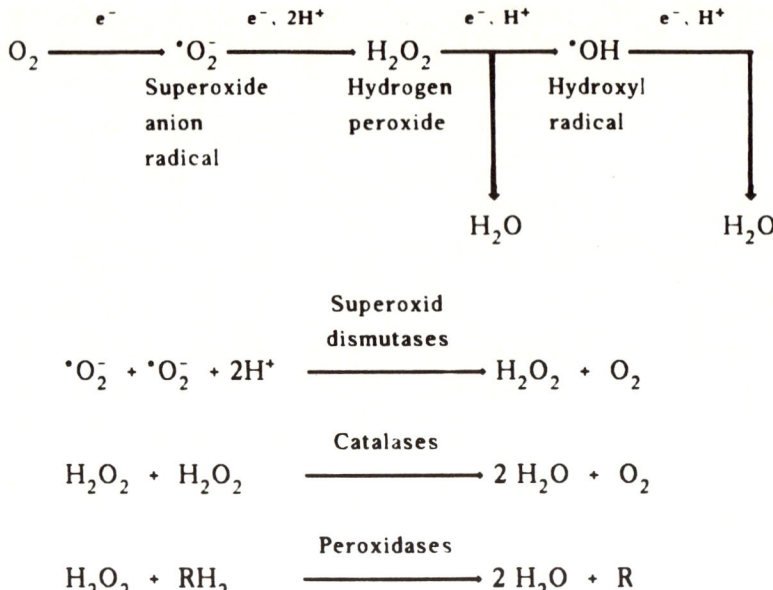

Fig. 8. Generation of oxygen radicals and enzymatic defence mechanisms

radicals may damage viable cells by peroxidation of lipids or oxidation of sulfhydryl groups, leading to perturbation of membranes and enzyme functions. The CNS is particularly sensitive to oxygen radical damage because of its high lipid content (Siesjö et al. 1989).

A free radical is defined as an agent with an unpaired electron in one of its shells. Univalent reduction of oxygen produces the superoxide anion, hydrogen peroxide, and the hydroxyl radical (Fig. 8). Superoxide and hydroxyl radical are true oxygen radicals because they contain oxygen with an unpaired electron. Hydrogen peroxide is not a radical, although it is very reactive, and it is frequently produced together with the other two agents (Halliwell 1984).

The three major types of antioxidative enzymes are superoxide dismutases, catalases, and peroxidases, but the activity of these enzymes is relatively low in brain tissue. Nonenzymatic endogenous radical scavengers include α-tocopherol (vitamin E), ascorbic acid (vitamin C), β-carotene, and compounds with SH-groups like glutathione or dihydrolipoic acid (Siesjö et al. 1989).

1. Formation of Superoxide

a) Mitochondria

Only a small portion of the oxygen reduced through the electron transport system of the mitochondria is converted to oxygen radicals. They are primarily formed by autoxidation of reduced compounds of the electron transport chain (FRIDOVICH 1978, 1979; CHANCE et al. 1979). Superoxide formation can be expected to increase under hypoxic conditions because the concentration of reduced equivalents rises under those conditions. However, superoxide will not escape from mitochondria in vivo because superoxide dismutase catalyzes the formation of hydrogen peroxide. Mitochondrial hydrogen peroxide originates entirely from this dismutation reaction of superoxide anion.

b) Oxidative Enzymes

Xanthine oxidase is present in brain tissue, primarily in the endothelium of blood vessels (BETZ 1985). It can generate substantial local concentrations of superoxide and hydrogen peroxide (BECKMAN et al. 1986). Energy failure leads to the breakdown of ATP and to the accumulation of AMP, which is further degraded by dephosphorylation and deamination to adenosine, inosine, and hypoxanthine (SIESJÖ et al. 1989); the last one is converted by xanthine oxidase to uric acid. However, in oxygen deficiency xanthine dehydrogenase is converted to an oxidase form, probably by limited proteolysis (BECKMAN et al. 1986, 1987). Thus, in contrast to the normal enzyme function, the oxidized form donates a single electron to O_2 and thereby produces $\cdot O_2^-$.

c) Degradation of Arachidonic Acid and Other Unsaturated Fatty Acids

The hydrolysis of membrane phospholipids is accelerated under anoxic conditions, and the level of arachidonic acid (AA) increases. AA is metabolized by cyclooxygenase, lipoxygenases, and cytochrome P-450 oxygenases. All three pathways are capable of producing superoxide (Fig. 9; CHAN and FISHMAN 1985). The cyclooxygenase forms endoperoxides from AA and also produces superoxide via this reaction. During the peroxidase reaction, the enzyme molecule is converted to the radical form with the unpaired electron (LAMBEIR et al. 1985). This intermediate reacts with reducing agents, such as reduced nicotinamide-adenine dinucleotide (NADH) or NADPH, which then react with oxygen to produce superoxide (LAND and SWALLOW 1971). Lipoxygenase was also shown to produce superoxide in the presence of NADH or NADPH (HEIKKILA and COHEN 1973). The cytochrome P-450 oxygenase oxidizes AA to epoxides (KUTHAN and ULLRICH 1982). The oxygen radicals themselves activate phospholipase A, which is further stimulated to release AA (Fig. 9).

d) Autoxidation

Superoxide may be produced by autoxidation of compounds such as norepinephrine, epinephrine, ascorbate, semiquinone, and hemoglobin (MISRA and FRIDOVICH 1972; COHEN 1982). The tendency toward autoxidation

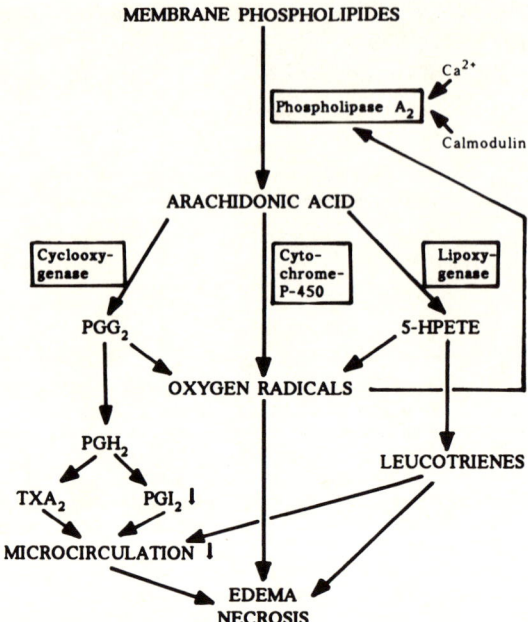

Fig. 9. Degradation of arachidonic acid and formation of free oxygen radicals. *5-HPETE*, 5-hydroxyperoxyeicosatetraenoic acid; *PG*, prostaglandin; *TX*, thromboxane

may be responsible for the cell toxicity of certain compounds, e.g., 6-hydroxydopamine, which destroys monoaminergic neurons following its uptake into the cell (HEIKKILA and COHEN 1973).

e) Polymorphonuclear Leukocytes

Polymorphonuclear leukocytes can invade tissue under various pathological conditions, including tissue injury, but invasion of the CNS appears to be slower than in other tissues (KONTOS et al. 1985). Activated polymorphonuclear leukocytes produce superoxide by NADPH oxidase (McDORD and ROY 1982). As in mitochondria, superoxide is converted to hydrogen peroxide. However, superoxide anions are also secreted directly into the extracellular space.

2. Formation of Hydroxyl Radical

Hydroxyl radical is formed according to the superoxide-driven

Fenton reaction:

$$\cdot O_2^- + H_2O_2 \rightarrow OH^- + \cdot OH + O_2 \tag{1}$$

However, this reaction only proceeds slowly unless it is catalyzed by transitional metals, usually Fe^{2+} but also Cu^{2+}.

According to the so-called metal-catalyzed *Haber-Weiss reaction*, hydrogen peroxide oxidizes Fe^{2+} and generates hydroxyl radical (HABER and WEISS 1934) according to the equation:

$$H_2O_2 + Fe^{2+} \rightarrow {}^{\cdot}OH + OH^- + Fe^{3+} \qquad (2)$$

Superoxide can now reduce Fe^{3+}, thereby making possible the repetition of Eqn. (2):

$$Fe^{3+} + {}^{\cdot}O_2^- \rightarrow Fe^{2+} + O_2 \qquad (3)$$

The summation of Eqns. (2) and (3) gives Eqn. (1).

Metals bound to protein, as for instance ferritin-bound Fe^{3+}, are at their most weakly active in catalyzing $^{\cdot}OH$ production. However, acidosis could release Fe^{3+} from ferritin, which would then trigger $^{\cdot}OH$ formation. A decrease in pH has been shown to accelerate radical production in brain tissue homogenate in vitro, and this effect was blocked by desferrioxamine (REHNCRONA et al. 1979).

3. Brain Damage and Free Radicals

Hydrogen peroxide is lipid-soluble and can cross cell membranes. Superoxide, on the other hand, is water-soluble, and its effects are likely to occur close to its site of generation. However, superoxide can cross membranes through the anion channel and enter the extracellular space (KONTOS et al. 1985; HALLIWELL and GUTTRIDGE 1984). Hydroxyl radicals are produced whenever H_2O_2 comes into contact with Fe^{2+} or Cu^{2+}. They never cross membranes and react with whatever membrane component they encounter first. Free radicals bind covalently to oxidizable groups of macromolecules and thus attack not only proteins, receptors, ion channels, and enzymes but also DNA and lipids. Enzymes, receptors, and ion channels may be destroyed and their normal function changed. Even the scavenging enzymes can be attacked, thereby prolonging the half-life and increasing the reactivity of the oxygen radicals (BRAUGHLER and HALL 1989; BAST and HAENEN 1990).

Reoxygenation of brain tissue after periods of oxygen failure leads to an explosive generation of free oxygen radicals (KIRSCH et al. 1987; HALLIWELL 1984; KONTOS and WEI 1986). Thus, the level of free radicals in brain tissue increases tremendously after cerebral ischemia, hypoxia, or intoxication. The defence mechanisms are overwhelmed, and the free radicals attack cellular constituents, causing cellular damage. Seizures are one of the cardinal CNS symptoms resulting from free radical attacks on neuronal membranes. Superoxide production has been observed in cerebral vessels of cats subjected to prolonged seizures. Drug-induced seizures lead to a large increase in cerebral levels of free fatty acids, including AA (CHAN and FISHMAN 1985). It seems likely that superoxide may arise from the metab-

Table 1. Targets of drug/toxin-induced energy deficit

Respiratory chain inhibitor	ATP synthetase inhibitor	Uncoupler	Increased lipid peroxidation	Increased oxygen consumption/ convulsives	Enzyme/ membrane defect	Energy deficit caused by detoxification
Cyanide CO Rotenone TIQ MPTP/MPP$^+$	Oligomycin	DNP TEL TET Barbiturates 6-OHDA Detergents/ LCFA	Ethanol 6-OHDA Cyanide	Organo- phosphates EAA Theophylline	MDMA *Para*-chlor- amphetamine Fenfluramine Methylmercury Ethanol	Ammonia

MPTP, 1-methyl-4-phenyl-1,2,3,6-tetrahydropyridine; MPP$^+$, 1-methyl-4-phenylpyridiniumion; TIQ, tetrahydroisoquinoline; DNP, 2,4-dinitrophenol; TEL, triethyllead; TET, triethyltin; 6-OHDA, 6-hydroxydopamine; LCFA, long chain fatty acids; EAA, excitatory amino acids; MDMA, 3,4-methylenedioxymethamphetamine.

olism of AA. Oxygen radicals also have pronounced effects on cerebral vessels. They diminish the vascular tone and reactivity and enhance the permeability of the BBB. They are capable of producing morphologically and biochemically demonstrable vascular injury (KONTOS 1989; SIESJÖ et al. 1989).

E. Drug-Induced Energy Deficits

Various drugs or toxins mediate their toxicity by causing an energy deficit. The energy failure can be generated by various mechanisms (Table 1). Several compounds have been shown to interfere with oxidative phosphorylation by inhibiting the respiratory chain or ATP synthetase and by uncoupling oxidative phosphorylation. Other compounds are toxic because they increase lipid peroxidation or oxygen consumption. An energy deficit can also be induced by an enzyme/membrane defect or as a consequence of detoxification.

I. Inhibitors of Oxidative Phosphorylation

1. Respiratory Chain Inhibitors

Cyanide inhibits cytochrome oxidase, the terminal enzyme of the mitochondrial electron transport chain (SCHUBERT and BRILL 1968). This inhibition leads to a massive failure of energy substrates in the brain. For instance, exposure of neuronal cultures derived from chick cerebral embryo hemispheres to cyanide produces a rapid decrease in the energy charge of neurons, but the energy state of the cells is quickly restored when cyanide is removed (Fig. 10; KRIEGLSTEIN et al. 1988). In contrast to the energy charge, the protein content of the cultures, which indicates cell growth and viability, did not increase during the 3-day recovery period after the hypoxic insult (AHLEMEYER and KRIEGLSTEIN 1989).

Various authors have found decreases in glycogen, ATP, creatine phosphate (CP), and pH and increases in inorganic phosphate and lactate upon administering sublethal doses of KCN in vivo (BENABID et al. 1987; PERES et al. 1988; MACMILLAN 1989). The hypoxic and energy-depleted condition after cyanide exposure leads to an accumulation of $[Ca^{2+}]_i$ (MADUH et al. 1988). In the presence of oxygen, xanthine oxidase catalyzes the formation of superoxide radicals, which initiate lipid peroxidation (see Sect. D II 1 b). Increased lipid peroxidation and activation of xanthine oxidase after cyanide intoxication have been described by JOHNSON et al. (1987) in mice. ARDELT et al. (1989) noted that cyanide intoxication caused an increase in lipid peroxidation simultaneously with decreased levels of the antioxidant enzymes catalase, glutathione peroxidase, glutathione reductase, and superoxide dismutase.

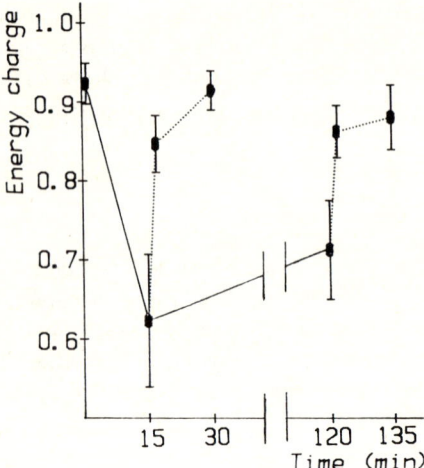

Fig. 10. The energy charge of the adenylate pool in neurons during cyanide exposure and recovery. Cyanide intoxication for 15 or 120 min (*solid line*) and recovery (*dotted line*) after 7 days in culture.
Note: The energy charge returns to the control level after 15 min of recovery

The characteristic CNS symptoms of cyanide toxicity in humans include hyperventilation, EEG irregularities, tremor, and seizures.

Hemoglobin (Hb) has about a 200-fold greater affinity for *carbon monoxide* (CO) than for oxygen. CO can thus cause toxicity by binding to Hb and reducing the oxygen-carrying capacity of the blood (OLSON 1984).

This produces hypoxia, which can lead to pathological events in those tissues most sensitive to oxygen deprivation, such as the brain and heart. Histological studies of the human brain after CO intoxication have shown pathological alterations in the white matter, globus pallidus, and hippocampus (BRUCHER 1967; LAPRESLE and FARDEAU 1967). There is also evidence that CO can poison neurons independently of its interaction with Hb, acting directly on cerebral energy metabolism by reducing cytochrome *b* (PIANTADOSI et al. 1985, 1988). Therefore, in addition to its action on Hb, a direct effect of CO on the respiratory chain cannot be excluded.

Rotenone is an inhibitor of NADH dehydrogenase. Inhibition leads to an increase in anaerobic glycolysis and lower ATP and CP levels in synaptosomes (DAGANI et al. 1989; KAUPPINEN et al. 1988a). Rotenone inhibits the energy-consuming, Ca^{2+}-dependent release of glutamate but increases its Ca^{2+}-independent release, correlating with a decline in plasma membrane potential. There is also a moderate increase in $[Ca^{2+}]_i$ after rotenone intoxication (KAUPPINEN et al. 1988a).

Other neurotoxins acting on respiratory chain of dopaminergic neurons include *1-methyl-4-phenyl-1,2,3,6-tetrahydropyridine* (MPTP), its oxidative metabolite 1-methyl-4-phenylpyridinium ion (MPP$^+$), and *tetrahydroisoquinoline* (TIQ). MPP$^+$ and TIQ were shown to inhibit the activity of NADH-ubiquinone reductase (SUZUKI et al. 1989), an enzyme-protein complex of the electron transport system. MPP$^+$ also inhibits the synthesis of ATP in mouse brain (MIZUNO et al. 1987). TIQ is now thought to be one of the endogenous substances that may be toxic to nigrostriatal dopaminergic neurons (NAGATSU and HIRATA 1987).

2. ATP Synthetase Inhibitor

Oligomycin inhibits ATP synthetase (JUNG and BRIERLEY 1983) and hence decreases the ATP and ADP content of synaptosomes (VERITY et al. 1983). It also appears to inhibit protein synthesis in synaptosomes (VERITY et al. 1983).

3. Uncouplers

2,4-Dinitrophenol (DNP). There must be adequate rates of electron flow and oxygen delivery as well as efficient coupling between electron transport and oxidative phosphorylation to maintain a balance between energy production and utilization. The increase in electron transport caused by a collapse of the H$^+$ gradient due to DNP-stimulated respiration far exceeds the available oxygen supply. The hyperactive respiration leads to tissue hypoxia, reduction of cytochrome aa_3 and cytochrome c activities, and a decrease in levels of high-energy phosphates. KARIMAN et al. (1986) found that ATP and CP were almost completely lost from the isolated perfused rat brain after DNP was added to the perfusion medium.

Triethyllead (TEL) and *triethyltin* (TET) both stimulate ATPase activity, inhibit oxidative phosphorylation, cause swelling of mitochondria (ALDRIDGE and STREET 1964), and are known to produce neuronal death in various areas of the brain (SEAWRIGHT et al. 1984). TEL administration causes neuronal damage in the neocortex (layers 3 and 5), brain stem, medulla oblongata, and most regions of the hippocampus, whereas intoxication with TET affects the pyriform cortex, amygdaloid body, and hippocampus. TEL and TET, at doses of 20 µmol/l, reduce the ATP and ADP contents of synaptosomes by a mechanism which is consistent with uncoupling of oxidative phosphorylation (KAUPPINEN et al. 1988b). TEL also inhibits pyruvate metabolism in the rat cortex in vivo and in vitro (REGUNATHAN and SUNDARESAN 1984) and enhances lipid peroxidation in the frontal cortex of the rat brain (ALI and BONDY 1989). KAUPPINEN et al. (1988b) used the fura-2 technique to determine the Ca^{2+} content of synaptosomes. They showed that 20µmol TEL or TET significantly increased the [Ca^{2+}]$_i$ of synaptosomes. The main mechanisms for this increase were uncoupling of oxidative phosphorylation in intrasynaptosomal mitochondria and inhibition of energy production (KAUPPINEN et al. 1988b).

Neuronal damage may be the direct result of inhibiting the aerobic energy metabolism of brain mitochondria, producing a marked increase in $[Ca^{2+}]_i$ (see Sect. D II 1).

Barbiturates are sedative-hypnotic drugs; their dose-response curve ranges from a mild pharmacologic to a maximal toxic effect. Toxic doses of barbiturates inhibit oxidative phosphorylation; acute barbiturate intoxication depresses the functional activity of the CNS (Sampson 1983). They have also been shown to dissociate the processes of oxidation and phosphorylation in mitochondria (Brody and Bain 1951, 1954). This uncoupling effect is qualitatively similar to that of DNP. Quantitatively, the uncoupling effect of thiobarbiturates such as thiopental is more powerful than that of oxygen barbiturates (Brody and Bain 1954). Uncoupling phosphorylation from oxidation could be one reason for the toxic effects of barbiturates. It has also recently been reported that barbiturates and other anesthetics can displace mitochondrially bound hexokinase activity and may thereby inhibit energy metabolism (Krieglstein and Bielicki 1976; Bielicki and Krieglstein 1977; Bruns et al. 1978; Dirks et al. 1980; Krieglstein et al. 1981).

6-Hydroxydopamine (6-OHDA) accumulates in catecholamine-containing neurons and exhibits selective toxicity towards them. Two mechanisms for its toxicity have been proposed. Firstly, the 6-OHDA which accumulates in neurons generates oxygen radicals and 6-OHDA-paraquinone (Sachs and Jonsson 1975; Thakar and Hassan 1988). The autoxidation of 6-OHDA (see Sect. D II 1 d) produces superoxide, hydroxyl radicals, and hydrogen peroxide. These reactive products then oxidize the SH-groups of enzymes or membrane proteins and cause lipid peroxidation as described above (Sect. D II 3) Secondly, 6-OHDA uncouples mitochondrial oxidative phosphorylation (Wagner 1971).

Detergents, long chain fatty acids. Brain mitochondria can be uncoupled by a small change in the free fatty acid composition of the mitochondrial membrane. Ogawa et al. (1988) demonstrated that oleic acid added to rat brain slices could uncouple oxidative phosphorylation. This effect could not be reversed by the addition of fatty-acid-free bovine serum albumin. The oxidative phosphorylation of brain mitochondria thus appears to depend on the precise control of the free fatty acid composition of mitochondrial membranes.

II. Ethanol

A disturbance of the oxidative processes in brain tissue is one of the early acute toxic actions of ethanol. Acute, high doses of ethanol decrease the activity of enzymes such as lactate, malate, and succinate dehydrogenases but do not appear to alter the metabolism of brain energy-rich phosphates.

Thus, alcohol does not uncouple oxidative phosphorylation in the brain but seems to inhibit the utilization of ATP.

Oxygen transport from the blood to the tissues becomes impaired after chronic exposure to ethanol, resulting in an oxygen deficiency which restricts the resynthesis of ATP (TEWARI and SYTINSKY 1983). The activity of some other enzymes, such as superoxide dismutase, is decreased by chronic ethanol treatment. The levels of the antioxidants glutathione, ascorbate, and α-tocopherol are also reduced (ROUACH et al. 1987; UYSAL et al. 1989). This drop seems to reflect consumption of the antioxidants in quenching free radicals. Acute ethanol administration also produces an increase in lipid peroxidation (ROUACH et al. 1987; UYSAL et al. 1989), which contributes to neuronal damage (see Sect. D II 3). The generation of free radicals could be one of the reasons why ethanol is toxic to brain tissue.

III. Convulsives

Organophosphate (OP) compounds are used as insecticides and chemical warfare agents. They rapidly inhibit acetylcholinesterase more or less irreversibly. A few minutes after exposure to OP, animals exhibit dyspnea, cyanosis (hypoxic effect), and generalized convulsions (McLEOD et al. 1984; DREWES and SINGH 1985). Increases in oxygen consumption (DREWES and SINGH 1985) and in the CMR_{glu} of various brain structures were seen in animals, with epileptic-like activity after soman treatment (McDONOUGH et al. 1983; SAMSON et al. 1984; MILLER and MEDINA 1986), whereas animals not suffering seizures show no changes in CMR_{glu}. Glucose utilization was above normal, as was that of lactate, but the concentration of CP was below normal (MILLER and MEDINA 1986). The cerebral concentration of ATP was only 50% of normal in strongly affected animals. Paroxon depresses the rate of glucose use in several brain regions without causing consistent changes in ATP or CP levels. The brains of rats treated with soman contain neuronal lesions in the neocortical layer III, pyriform cortex, thalamus, and all subfields of the hippocampus (McLEOD et al. 1984).

Neurotoxic concentrations of *excitatory amino acids* like glutamate and aspartate cause large increases in $[Ca^{2+}]_i$ (CHOI 1988). These toxic levels of $[Ca^{2+}]_i$ mediate reactions which lead to neuronal damage (Sect. D I 3). There are three types of receptors for excitatory amino acids, NMDA, kainate, and Q receptors (see Sect. D I 1). Compounds that have an agonistic action on NMDA receptors stimulate Ca^{2+} entry into neurons. Agonists acting on kainate or Q receptors mainly induce Na^+ movement into cells, but Ca^{2+} may also enter neurons indirectly. Agonists of these three receptors are listed in Table 2 (COTMAN and IVERSEN 1987).

Theophylline enhances brain activity; intoxication results in centrally triggered convulsions. The compound inhibits the 5′-nucleotidase (TSUZUKI and NEWBURGH 1975; JENSEN et al. 1984), an enzyme that catalyzes the forma-

Table 2. Agonists acting on NMDA, quisqualate, or kainate receptors

Receptor:	NMDA	Quisqualate	Kainate
Agonist:	NMDA	Quisqualate	Kainate
	Aspartate	AMPA	Domoate
	Ibotenate	L-Glutamate	L-Glutamate
	Kynurenate		

NMDA, N-methyl-D-aspartate; AMPA, α-amino-3-hydroxy-5-methyl-4-isoxazolepropionate.

tion of adenosine, which has an inhibitory and vasodilatory effect on the brain. Theophylline is also an adenosine receptor antagonist and inhibits the binding of adenosine to its receptor (FREDHOLM 1980). These disturbances of the inhibitory action of adenosine may lead to a toxic predominance of excitatory inputs to the neurons.

IV. Compounds Causing Energy Deficits by Enzyme/Membrane Defects

3,4-Methylenedioxymethamphetamine (MDMA), also known as "Ecstasy", is a phenylisopropylamine derivative related to amphetamines and hallucinogens. It has acute and long-term effects on serotonergic neurons (for review, see McKENNA and PEROUTKA 1990). The most pronounced effect is the decrease in cerebral 5-hydroxytryptamine (5-HT). It also reduces tryptophan hydroxylase activity and decreases 5-HT nerve terminal density. There is now evidence that the neurotoxic effect of MDMA is mediated by one of its metabolite or an endogenous substance. Dopamine (DA) has been implicated as a possible mediator of the neurotoxicity produced by MDMA (STONE et al. 1988). Oxidative processes inside the 5-HT terminal could produce cytotoxic DA metabolites capable of reacting covalently with nucleophilic cell components, such as enzymes and cell membranes (SCHMIDT et al. 1985). This could effect the neurodegenerative changes which are reported to appear in the nerve terminals of the striatum, hippocampus, and somatosensory cortex (RICAURTE et al. 1985; COMMINS et al. 1987).

Other neurotoxins influencing serotonergic neurons include the halogenated phenethylamines *para*-chloroamphetamine and fenfluramine (SANDERS-BUSH et al. 1975). These compounds also seem to be converted to a reactive intermediate within the serotonergic neurons, thereby leading to their damaging activity.

Methylmercury (Met-Hg) is a lipophilic mercurial neurotoxicant causing severe human poisoning. It inhibits the respiration of brain mitochondria and synaptosomes (VERITY et al. 1975), thereby reducing the cellular energy levels of brain cells in vitro (GRUNDT and BAKKEN 1986). KAUPPINEN et al.

(1989) also found that Met-Hg influenced synaptosomal adenylate concetrations. It was reported that $30\,\mu$mol and $100\,\mu$mol of Met-Hg reduced the ATP and ADP levels of synaptosomes, but the molecular mechanism by which it inhibits mitochondrial energy transduction is not known. It has been speculated that the toxic effect of mercurial compounds is due to their affinity for SH-groups (CLARKSON 1972). Mercury compounds have been shown to denature and inactivate SH-containing enzymes, the electron and substrate carriers of the energy transduction pathway (NICHOLLS 1982). The energy failure and a possible injury of the cell membrane (BANO and HASAN 1989) lead to increased Ca^{2+} uptake into nerve terminals. At high concentrations of Met-Hg ($100\,\mu$mol), when synaptosomes are severely energy-depleted, depolarization and a nonspecific leakage of ions through the plasma membrane produce a large increase in $[Ca^{2+}]_i$ (KAUPPINEN et al. 1989), which could cause the death of neurons. Energy failure followed by membrane depolarization and a high intraneural Ca^{2+} concentration seem to be the main reasons for the toxicity of Met-Hg.

V. Ammonia

High concentrations of ammonia in the brain may result under a number of pathological conditions, including hepatic encephalopathy or Reye's syndrome. Ammonia can interfere with astrocyte morphology and influence the integrity of the BBB, neurotransmitter function, and various biochemical pathways (COOPER and LAI 1987). The precise mechanism of its neurotoxic effect is uncertain. One hypothesis is based on the detoxification of ammonia via glutamine synthesis from glutamate. This energy-requiring process may lead to decreased concentrations of metabolites available for energy production. In acute intoxication with low ammonia levels, the CP content of the brain is lowered, whereas the adenylate pool remains unchanged (HINDFELT 1983). Similar results were obtained by McCANDLESS et al. (1985) in the primate *Tupaia glis*, in which precoma was induced with ammonia. The concentrations of ATP, CP, glucose, and glycogen all fell at higher doses of ammonia (1–3 mmol/kg brain tissue) and in coma, whereas the lactate concentration (see Sect. C I) rose (HINDFELT 1983; McCANDLESS et al. 1985). These ammonia-induced energy deficits occur particularly in brainstem structures and might be the reason for its neurotoxic effect.

F. Concluding Remarks

Disorders of cerebral energy metabolism play an important role in drug-induced toxicity. Thus, the effects of drugs may be evaluated by the way they cause an energy deficit, i.e., inhibition or uncoupling of oxidative phosphorylation, the disturbance of oxygen consumption, or an enzyme or membrane defect. These processes disturb energy metabolism by causing an oxygen or substrate deficit. Both deficits lead to metabolic disorders such as Ca^{2+} entry into neurons or the formation of oxygen free radicals and are able to cause neuronal damage.

References

Ahlemeyer B, Krieglstein J (1989) Testing drug effects against hypoxic damage of cultured neurons during long-term recovery. Life Sci 45:835–842

Aldridge WN, Street BW (1964) Oxidative phosphorylation. Biochemical effects and properties of trialkyltins. Biochem J 91:287-297

Ali SF, Bondy SC (1989) Triethyllead-induced peroxidative damage in various regions of the brain. J Toxicol Environ Health 26:235–242

Ardelt BK, Borowitz JL, Isom GE (1989) Brain lipid peroxidation and antioxidant protectant mechanisms following acute cyanide intoxication. Toxicology 56:147–154

Atkinson DE (1968) The energy charge of the adenylate pool as a regulatory parameter. Interaction with feedback modifiers. Biochemistry 7:4030–4034

Auer RN (1986) Progress review: hypoglycemic brain damage. Stroke 17:699–708

Auer RN, Siesjö BK (1988) Biological differences between ischemia, hypoglycemia and epilepsy. Ann Neurol 24:699–707

Bano Y, Hasan M (1989) Mercury induced time-dependent alterations in lipid profiles and lipid peroxidation in different body organs of cat-fish *Heteropneustes fossilis*. J Environ Sci Health [B]24:145–166

Bast A, Haenen GRMM (1990) Receptor function in free radical mediated pathologies. In: Claassen V (ed) Trends in drug research. Elsevier, Amsterdam, pp 273–285

Beckman JS, Campbell GA, Hannan CJ Jr, Karfias CS, Freeman BA (1986) Involvement of superoxide and xanthine oxidase with death due to cerebral ischemia-induced seizures in gerbils. In: Rotilio G (ed) Superoxide and superoxide dismutase in chemistry, biology and medicine. Elsevier, Amsterdam, pp 602–607

Beckmann JS, Marshall PA, Freeman BA (1987) Xanthine dehydrogenase to oxidase activity conversion in ischemic gerbil brain. Fed Proc 46:417

Benabid AL, Decorps M, Remy C, LeBas JF, Confort S, Leviel JL (1987) ^{31}P nuclear magnetic resonance in vivo spectroscopy of the metabolic changes induced in the awake rat brain during KCN intoxication and its reversal by hydroxocobalamine. J Neurochem 48:804–808

Betz AL (1985) Identification of hypoxanthine transport and xanthine oxidase activity in brain capillaries. J Neurochem 44:574–579

Bielicki L, Krieglstein J (1977) Solubilization of brain mitochondrial hexokinase by thiopental. Naunyn-Schmiedebergs Arch Pharmacol 298:61–65

Braughler JM, Hall ED (1989) Central nervous system trauma and stroke: I. Biochemical considerations for oxygen radical formation and lipid peroxidation. Free Radic Biol Med 6(3):289–301

Brody TM, Bain JA (1951) Effects of barbiturates on oxidative phosphorylation. Proc Soc Med 77:50–53

Brody TM, Bain JA (1954) Barbiturates and oxidative phosphorylation. J Pharm 110:148–156

Brucher JM (1967) Neuropathological problems posed by carbon monoxide poisoning and anoxia. Prog Brain Res 24:75–100

Bruns H, Krieglstein J, Wever K (1978) Narkose und intrazelluläre Verteilung der zerebralen Hexokinase. Anaesthesist 27:557–561

Carafoli E (1987) Intracellular calcium homeostasis. Annu Rev Biochem 56:395–433

Carbone E, Lux HD (1984) A low voltage activated, fully inactivating calcium channel in vertebrate sensory neurons. Nature 310:501–502

Chan PH, Fishman RA (1985) Free fatty acids, oxygen free radicals and membrane alterations in brain ischemia and injury. In: Plum F, Pulsinelli W (eds) Cerebrovascular diseases. Raven, New York, pp 161–167

Chance B, Sies H, Boveris A (1979) Hydroperoxide metabolism in mammalian organs. Physiol Rev 59:527–605

Choi DW (1988) Glutamate neurotoxicity and diseases of the neurons system. Neuron 1:623–634

Choi DW (1990) Methods for antagonizing glutamate neurotoxicity. Cerebrovasc Brain Metab Rev 2:105–147

Choi DW, Rothmann SM (1990) The role of glutamate neurotoxicity in hypoxic-ischemic neuronal death. Annu Rev Neurosci 13:171–182

Christine CW, Choi DW (1990) Effect of zinc on NMDA receptor-mediated channel currents in cortical neurons. J Neurosci 10:108–116

Clarkson TW (1972) The pharmacology of mercury compounds. Annu Rev Pharmacol 12:375–406

Cohen G (1982) Oxygen radicals, hydrogen peroxide, and Parkinson's disease. In: Autor AP (ed) Pathology of oxygen. Academic, New York, pp 115–126

Commins DL, Vosmer G, Virus RM, Woolverton WL, Schuster CR, Seiden LS (1987) Biochemical and histological evidence that methylendioxymethylamphetamine (MDMA) is toxic to neurons in the rat brain. J Pharmacol Exp Ther 241:338–345

Cooper AJL, Lai JCK (1987) Cerebral ammonia metabolism in normal and hyperammonemic rats. Neurochem Pathol 6:67–95

Cotman CW, Iversen LL (1987) Excitatory amino acids in the rat brain – focus on NMDA receptors. TINS 10:263–265

Cremer JE, Cunningham VJ, Partridge WM, Braun LD, Oldendorf WH (1979) Kinetics of blood-brain barrier transport of pyruvate, lactate and glucose in suckling, weanling and adult rats. J Neurochem 33:439–455

Dagani F, Feletti F, Canevari L (1989) Effects of diltiazem on bioenergetics, K^+ gradients, and free cytosolic Ca^{2+} levels in rat brain synaptosomes submitted to energy metabolism inhibition and depolarization. J Neurochem 53:1379–1389

Dirks B, Hanke J, Krieglstein J, Stock R, Wickop G (1980) Studies on the linkage of energy metabolism and neuronal activity in the isolated perfused rat brain. J Neurochem 35:311–317

Dolphin AC (1990) G protein modulation of calcium currents in neurons. Annu Rev Physiol 52:243–255

Drewes LR, Singh AK (1985) Transport, metabolism and blood flow in brain during organophosphate induced seizures. Proc West Pharmacol Soc 28:191–195

Erecinska M, Silver IA (1989) ATP and brain function. J Cereb Blood Flow Metab 9:2–19

Fredholm BB (1980) Theophylline actions on adenosine receptors. Eur J Respir Dis 61 [Suppl 109]:29–36

Fridovich I (1978) The biology of oxygen radicals. Science 201:875–880

Fridovich I (1979) Hypoxia and oxygen toxicity. Adv Neurol 26:255–266

Giffard RG, Monyer H, Christine CW, Choi DW (1990) Acidosis reduces NMDA receptor activation, glutamate neurotoxicity, and oxygen-glucose deprivation neuronal injury in cortical cultures. Brain Res 506:339–342

Ginsberg MD (1990) Local metabolic responses to cerebral ischemia. Cerebrovasc Brain Metab Rev 2:58–93

Ginsberg MD, Busto R (1989) Rodent models of cerebral ischemia. Stroke 20:1627–1642

Grundt IK, Bakken AM (1986) Adenine nucleotides in cultured brain cells after exposure to methyl mercury and triethyl lead. Acta Pharmacol Toxicol 59:11–16

Haber F, Weiss J (1934) The catalytic decomposition of hydrogen peroxide by iron salts. Proc R Soc Lond 147:332–351

Hallam TJ, Rink TJ (1989) Receptor-mediated Ca^{2+} entry: diversity of function and mechanism. TIPS 10:8–10

Halliwell B (1984) Oxygen radicals: a commonsense look at their nature and medical importance. Med Biol 62:71–77

Halliwell B, Guttidge JMC (1984) Oxygen toxicity, oxygen radicals, transition metals and disease. Biochem J 219:1–14

Hanke J, Krieglstein J (1982) Über die Mechanismen der protektiven Wirkung von Methohexital auf den cerebralen Energiestoffwechsel. Arzneimittelforsch 6:620–625

Hansen AJ (1985) Effect of anoxia on ion distribution in the brain. Physiol Rev 65:101–148

Hawkins RA, Mans AM (1983) Intermediary metabolism of carbohydrates and other fuels. In: Lajhta A (ed) Metabolism in the nervous system, 2nd edn. Plenum, New York, pp 259–294 (Handbook of neurochemistry, vol 3)

Heikkila RE, Cohen G (1973) 6-Hydroxydopamine: evidence for superoxide radical as an oxidative intermediate. Science 83:456–464

Hindfeldt B (1983) Ammonia intoxication and brain energy metabolism. In: Kleinberger G, Deutsch E (eds) New aspects of clinical nutrition. Karger, Basel, pp 474–484

Jensen MH, Jorgensen S, Nielsen H, Sanchez R, Andersen PK (1984) Is theophylline-induced seizures in man caused by inhibition of cerebral 5′-nucleotidase activity? Acta Pharmacol Toxicol 55:331–334

Johnson JD, Conroy WG, Burris KD, Isom GE (1987) Peroxidation of brain lipids following cyanide intoxication in mice. Toxicology 46:21–28

Jung DW, Brierly GP (1983) Oxidative phosphorylation. In: Lajhta A (ed) Alterations of metabolites in the nervous system, 2nd edn. Plenum, New York, pp 295–319 (Handbook of neurochemistry, vol 9)

Kariman K, Chance B, Burkhart DS, Bolinger LA (1986) Uncoupling effects of 2,4-dinitrophenol on electron transfer reactions and cell bioenergetics in rat brain in situ. Brain Res 366:300–306

Katz AM, Messineo FC (1981) Lipid-membrane interactions and the pathogenesis of ischemic damage in the myocardium. Circ Res 48:1–16

Kauppinen RA, McMahon HT, Nicholls DG (1988a) Ca^{2+}-dependent and Ca^{2+}-independent glutamate release, energy status and cytosolic free Ca^{2+} concentration in isolated nerve terminals following metabolic inhibition: possible relevance to hypoglycemia and anoxia. Neuroscience 27:175–182

Kauppinen RA, Komulainen H, Taipale HT (1988b) Chloride-dependent uncoupling of oxidative phosphorylation by triethyllead and triethyltin increases cytosolic free calcium in guinea pig cerebral cortical synaptosomes. J Neurochem 51:1617–1625

Kauppinen RA, Komulainen H, Taipale HT (1989) Cellular mechanisms underlying the increase in cytosolic free calcium concentration induced by methylmercury in cerebrocortical synaptosomes from guinea pig. J Pharmacol Exp Ther 248:1248–1254

Kirino T (1982) Delayed neuronal death in the gerbil hippocampus following ischemia. Brain Res 239:57–69

Kirsch JR, Phelan AM, Lange DG, Traystman JR (1987) Free radicals detected in brain during reperfusion from global ischemia. Fed Proc 46:799

Kontos HA (1989) Oxygen radicals in CNS damage. Chem Biol Interact 72:229–255

Kontos HA, Wei HP (1986) Superoxide production in experimental brain injury. J Neurosurg 64:803–807

Kontos HA, Wei HP, Jenkins LW, Povlishock JT, Rowe GT, Hess ML (1985) Appearance of superoxide anion radical in cerebral extracellular space during increased prostaglandin synthesis in cats. Circ Res 57:142–151

Krieglstein G, Krieglstein J, Stock R (1972) Suitability of the isolated perfused rat brain for studying effects on cerebral metabolism. Naunyn-Schmiedebergs Arch Pharmacol 275:124–134

Krieglstein J (1989a) Pharmacology of cerebral ischemia 1988. Wissenschaftliche Verlagsgesellschaft/CRC Press, Stuttgart/Boca Raton

Krieglstein J (1989b) Excitatory amino acids, adenosine and neuronal calcium homeostasis – an overview. In: Seylaz J, Sercombe R (eds) Neurotransmission and cerebrovascular function II. Elsevier, Amsterdam, pp 441–453

Krieglstein J, Bielicki L (1976) The influence of thiopental on the intracellular distribution of hexokinase in rat brain. Naunyn-Schmiedebergs Arch Pharmacol 293:R11

Krieglstein J, Oberpichler H (1990) Pharmacology of cerebral ischemia 1990. Wissenschaftliche Verlagsgesellschaft/CRC Press, Stuttgart/Boca Raton

Krieglstein J, Weber J (1986) Calcium entry blockers protect brain energy metabolism against ischemic damage. In: Longmuir JS (ed) Oxygen transport to tissue VIII. Plenum, New York, pp 243–251

Krieglstein J, Sperling G, Twietmeyer G (1981) Effects of thiopental on regulatory mechanisms of brain energy metabolism. Naunyn-Schmiedebergs Arch Pharmacol 318:56–61

Krieglstein J, Brungs H, Peruche B (1988) Cultured neurons for testing cerebroprotective drug effects in vitro. J Pharmacol Methods 20:39–46

Kuthan H, Ullrich V (1982) Oxidase and oxygenase function of the microsomal cytochrome P-450 monooxygenase system. Eur J Biochem 126:583–588

Lambeir AM, Markey CM, Dunford HB, Marnett LJ (1985) Spectral properties of the higher oxidation states of prostaglandin H synthase. J Biol Chem 260:14894–14896

Land EJ, Swallow AJ (1971) One-electron reactions in biochemical systems as studied by pulse radiolysis IV. Oxidation of dihydronicotinamide-adenine dinucleotide. Biochim Biophys Acta 234:34–42

Lapresle J, Fardeau M (1967) The central nervous system and carbon monoxide poisoning. II. Anatomical study of brain lesions following intoxication with carbon monoxide (22 cases). Prog Brain Res 24:31–74

MacMillan VH (1989) Cerebral energy metabolism in cyanide encephalopathy. J Cereb Blood Flow Metab 9:156–162

Maduh EU, Johnson JD, Ardelt BK, Borowitz JL, Isom GE (1988) Cyanide-induced neurotoxicity: mechanisms of attenuation by chlorpromazine. Toxicol Appl Pharmacol 96:60–67

Martz D, Rayos G, Schielke GP, Betz AL (1989) Allopurinol and dimethylthiourea reduce brain infarction following middle cerebral artery occlusion in rats. Stroke 20:488–494

McCandless DW, Looney GA, Modak AT, Stavinoha WB (1985) Cerebral acetylcholine and energy metabolism changes in acute ammonia intoxication in the lower primate *Tupaia glis*. J Lab Clin Med 106:183–186

McCord JM, Roy RS (1982) The pathophysiology of superoxide: roles in inflammation and ischemia. Can J Physiol Pharmacol 60:1346–1352

McDonough JH, Hackley BE, Cross R, Samson F, Nelson S (1983) Brain regional glucose use during soman-induced seizures. Neurotoxicology 4:203–210

McKenna DJ, Peroutka SJ (1990) Neurochemistry and neurotoxicity of 3,4-methylendioxymethamphetamine (MDMA, 'Ecstasy'). J Neurochem 54:14–22

McLeod CG, Singer AW, Harrington DG (1984) Acute neuropathology in soman-poisoned rats. Neurotoxicology 5:53–58

Miller AL, Medina MA (1986) Cerebral metabolic effects of organophosphorus anticholinesterase compounds. Metab Brain Dis 1:147–156

Misra HP, Fridovich I (1972) The role of superoxide anion in the autoxidation of epinephrine and a simple assay for superoxide dismutase. J Biol Chem 247:3170–3175

Mizuno Y, Suzuki K, Sone N, Saitoh T (1987) Inhibition of ATP synthesis by 1-methyl-4-phenylpyridinium ion (MPP^+) in isolated mitochondria. Neurosci Lett 81:204–208

Nagatsu T, Hirata Y (1987) Inhibition of the tyrosine hydroxylase system by MPTP, 1-methyl-4-pyridinium ion (MPP^+) and the structurally related compounds in vitro and in vivo. Eur Neurol 26 [Suppl]:11–15

Nicholls DG (1982) Bioenergetics: an introduction to the chemiosmotic theory. Academic, New York

Nowycky MC, Fox AP, Tsien RW (1985) Three types of neuronal calcium channels with different calcium agonist sensitivity. Nature 316:440–443

Ogawa M, Yoshida S, Ogawa T, Shimada T, Takeshita M (1988) Effect of oleic acid on mitochondrial oxidative phosphorylation in rat brain slices. Biochem Int 17:773–782

Olney JW (1969) Brain lesion, obesity and other disturbances in mice treated with monosodium glutamate. Science 164:719–721

Olson KR (1984) Carbon monoxide poisoning: mechanisms, presentation, and controversies in management. J Emerg Med 1:233–243

Paschen W, Siesjö BK, Ingvar M, Hossmann KA (1986) Regional differences in brain glucose content in graded hypoglycemia. Neurochem Pathol 5:131–142

Peres M, Meric P, Barrere B, Pasquier C, Beranger G, Beloeil JC, Lallemand JY, Seylaz J (1988) In vivo ^{31}P nuclear magnetic resonance (NMR) study of cerebral metabolism during histotoxic hypoxia in mice. Metab Brain Dis 3:37–48

Piantadosi CA, Sylvia AL, Saltzman HA, Jobsis-Vandervliet FF (1985) Carbon monoxide-cytochrome interactions in the brain of the fluorocarbon-perfused rat. J Appl Physiol 58:665–672

Piantadosi CA, Lee PA, Sylvia AL (1988) Direct effects of CO on cerebral energy metabolism in bloodless rats. J Appl Physiol 65:878–887

Rami A, Rabie A, Thomasset M, Krieglstein J (1990) Calbindin-D_{28k} and ischemic damage of pyramidal cells in rat hippocampus. J Neurosci Res (in press)

Rasmussen H, Waisman DM (1983) Modulation of cell function in calcium messenger system. Rev Physiol Biochem Pharmacol 95:111–148

Regunathan S, Sundaresan R (1984) Pyruvate metabolism in brain of young rats intoxicated with organic lead. J Neurochem 43:1346–1351

Rehncrona W, Mela L, Siesjö BK (1979) Recovery of brain mitochondrial function in the rat after complete and incomplete cerebral ischemia. Stroke 10:437–446

Ricaurte G, Bryan G, Strauss L, Seiden L, Schuster C (1985) Hallucinogenic amphetamine selectively destroys brain serotonin nerve terminals. Science 229–988

Rouach H, Park MK, Orfanelli MT, Janvier B, Nordmann R (1987) Ethanol-induced oxidative stress in the rat cerebellum. Alcohol Alcohol [Suppl]1:207–211

Sachs C, Jonsson G (1975) Mechanisms of action of 6-hydroxydopamine. Biochem Pharmacol 24:1–8

Sampson I (1983) Barbiturates. Mt Sinai J Med (NY) 50:283–288

Samson F, Pazdernik TL, Cross R, Giesler MP, Mewes K, Nelson S, McDonough JH (1984) Soman induced changes in brain regional glucose use. Fundam Appl Toxicol 4:S173–S183

Sanders-Bush E, Bushing JA, Sulser F (1975) Long-term effects of p-chloroamphetamine and related drugs on central serotonergic mechanisms. J Pharmacol Exp Ther 192:33–41

Schmidt CJ, Ritter JK, Sonsalla PK, Hanson GR, Gibb JW (1985) Role of dopamine in the neurotoxic effects of methamphetamine. J Pharmacol Exp Ther 233:539–544

Schubert J, Brill WA (1968) Antagonism of experimental cyanide toxicity in relation to the in vivo activity of cytochrome oxidase. J Pharmacol Exp Ther 162:352–359

Seawright AA, Brown AW, Ng JC, Hrdlicka J (1984) Experimental pathology of short-chain alkyllead compounds. In: Grandjean P, Grandjean EC (eds) Biological effects of organolead compounds. CRC Press, Boca Raton, pp 177–206

Siesjö BK (1978) Brain energy metabolism. Wiley, Chichester

Siesjö BK (1981) Cell damage in the brain: a speculative synthesis. J Cereb Blood Flow Metab 1:155–185

Siesjö BK (1990) Calcium in the brain under physiological and pathological conditions. Eur Neurol 30 [Suppl 2]:3–9

Siesjö BK, Bengtsson F (1989) Calcium fluxes, calcium antagonists, and calcium-related pathology in brain ischemia, hypoglycemia, and spreading depression: an unifying hypothesis. J Cereb Blood Flow Metab 9:127–140

Siesjö BK, Agardh CD, Bengtsson F (1989) Free radicals and brain damage. Cerebrovasc Brain Metab Rev 1:165–211

Silver IA, Erecinska M (1990) Intracellular and extracellular changes of $[Ca^{2+}]$ in hypoxia and ischemia in rat brain in vivo. J Gen Physiol 95:837–866

Sokoloff L, Reivich M, Kennedy C, Des Rosiers MH, Patlak CS, Pettigrew KD, Sakurada O, Shinhara M (1977) The ^{14}C-deoxyglucose method for the measurement of local cerebral glucose utilization: theory, procedure, and normal values in the conscious and anaesthetized albino rat. J Neurochem 28:897–916

Sokoloff L, Dienel GA, Cruz NF, Mori K (1990) Autoradiographic methods in pathology and pathophysiology. In: Krieglstein J, Oberpichler H (eds) Pharmacology of cerebral ischemia 1990. Wissenschaftliche Verlagsgesellschaft/CRC Press, Stuttgart/Boca Raton, pp 3–21

Stone DM, Johnson M, Hanson GR, Gibb JW (1988) Role of endogenous dopamine in the central serotonergic deficits induced by 3,4-methylendioxymethamphetamine. J Pharmacol Exp Ther 247:79–87

Suzuki K, Mizuno Y, Yoshida M (1989) Selective inhibition of complex I of the brain electron transport system by tetrahydroisoquinoline. Biochem Biophys Res Commun 162:1541–1545

Tewari S, Sytinsky IA (1983) Alcohol. In: Lajhta A (ed) Alterations of metabolites in the nervous system, 2nd edn. Plenum, New York, pp 219–261 (Handbook of neurochemistry, vol 9)

Thakar JH, Hassan MN (1988) Effects of 6-hydroxydopamine on oxidative phosphorylation of mitochondria from rat striatum, cortex and liver. Can J Physiol Pharmacol 66:376–379

Tombaugh GC, Sapolsky RM (1990) Mild acidosis protects hippocampal neurons from injury induced by oxygen and glucose deprivation. Brain Res 506:343–345

Tsuzuki J, Newburgh RW (1975) Inhibition of 5′-nucleotidase in rat brain by methylxanthines. J Neurochem 25:895–896

Uysal M, Kutalp G, Özdemirler G, Aykac G (1989) Ethanol-induced changes in lipid peroxidation and glutathione content in rat brain. Drug Alcohol Depend 23:227–230

Verity MA, Brown WJ, Cheung M (1975) Organic mercurial encephalopathy: in vivo and in vitro effects of methyl mercury on synaptosomal respiration. J Neurochem 25:759–766

Verity MA, Brown WJ, Cheung M (1983) Failure of atractyloside to inhibit synaptosomal mitochondrial energy transduction. Neurochem Res 8:159–166

Wagner K (1971) Uncoupling of oxidative phosphorylation by 6-hydroxydopamine. In: Malmfors T, Thoenen H (eds) 6-Hydroxydopamine and catecholamine neurons. North-Holland, Amsterdam, pp 277–278

CHAPTER 6

Neurotoxic Synthesis by Enzymatic Error

H. HERKEN

A. Introduction

The analysis of CNS functions by the use of pharmacological methods has led to the significant result that there are marked differences in each of the regions of the CNS as far as their sensitivity to chemical substances is concerned. Depending on the chemical structure and efficacy of the chemical compounds, gradually increasing doses not only produce therapeutically useful reactions but frequently can also produce reversible or irreversible damage within the same cells. For this reason, the selective vulnerability of certain cerebral regions is of special theoretical and practical significance because the examination of chemically induced neurological disturbances can furnish important information for the recognition of the pathogenesis of neurological diseases.

The accumulation of a chemical substance within a certain cerebral region does not implicitly mean that this substance will produce its greatest activity there. Accumulation gains special interest whenever it affects the metabolism of the cells concerned and gives rise to enzymic dysfunctions; these in turn are the real cause of more serious damage. The best example is still given by the elucidation of the irreversible inactivation of acetylcholinesterase by organophosphorus compounds. Studies of the chemical basis of the pharmacological actions of drugs have also led to the discovery of a peculiar lack of specificity of certain enzymes, thus enabling living cells to synthesize harmful substances (HERKEN 1965, 1970). This is not in accordance with the general opinion on the strongly specific function of enzymes.

B. Lethal Synthesis

The first observation of the lack of specificity of important enzymes in intermediary metabolism was contributed by PETERS (1952) in the course of his studies on the toxic action of fluoroacetate, which passes the BBB easily. The term "lethal synthesis" used by Peters in his Croonian Lecture before the Royal Society to define this process is very well justified. The animals died from metabolic disturbances within the mitochondria of the brain, after developing generalized convulsions and apnea.

PETERS commented in 1952 thusly: "My main theme is the biochemistry and physiology of the poison fluoroacetic acid; this is, I think, the first substance of "lethal synthesis" in the sense that fluoroacetic acid becomes "lethal" only where it has been transformed by the actions of tissue enzymes. It is the biochemistry which makes it lethal, the poison being a biochemical product of the tissue which is poisoned." This historic explanation has achieved fundamental importance in defining a neurotoxic synthesis of this kind, in which the basic similarity is due to the existence of an "enzymatic error," even though the intoxication is not always lethal.

I. Toxicity of Fluoroacetic Acid

Fluoroacetic acid is the toxic ingredient in the South African poison plant *Dichapetalum cymosum* and in other *Dichepetalum* species, which are used as poison against rodents (SWARTS 1896; MARAIS 1944).

Fluoroacetic acid ($F \cdot CH_2COOH$) is a very stable substance. It differs from halogenated compounds such as iodo- or bromoacetic acid in the stability of its CF bond and in its pharmacokinetics (SAUNDERS and STACEY 1948). The toxicity has been studied in detail by CHENOWETH and GILMAN (1946), CHENOWETH and ST JOHN (1947), CHENOWETH (1949), PATTISON and PETERS (1966) and CHENOWETH et al. (1951).

The acute toxicity of mono-fluoroacetate for animals is very high (CHENOWETH 1949; PETERS 1953). There are considerable differences in the sensitivity – dogs are very sensitive, followed by cats and rabbits, while monkeys are less sensitive.

Several comparative studies were carried out using fluoro-fatty acids of varying chain lengths to discover structure-function relationships. SAUNDERS and STACEY (1948) as well as PATTISON and SAUNDERS (1949) found while examining the toxicity of a series of co-fluorocarboxylic acids of increasing chain length and derivation that those with an even number of carbon atoms were toxic and those with an odd number were not. 2-Monofluoroacetic acid was active, 3-monofluoropropionic acid was inactive, and 4-monofluorobutyric acid was very active. These results seem to indicate that both the fluoro-fatty acids and the physiologic fatty acids with an even number of carbon chains were metabolized and thereby split to C_2 fragments, according to KNOOP and MARTIUS' principle of β-oxidation (see review CHENOWETH 1949; PATTISON and PETERS 1966).

II. Molecular Basis of Intoxication

In 1949, LIEBECQ and PETERS discovered a considerable drop in O_2 usage and a definite accumulation of citrate in kidney homogenates after adding fluoroacetic acid. This metabolic disturbance was observed in all organs of the rat during fluoroacetic acid intoxication (BUFFA and PETERS 1949) (Table 1). The levels in the brain were, however, lower than those in other organs

Table 1. Fluoroacetate poisoning; accumulation of citrate in the rat

	Citric acid (µg/g wet tissue)	
	Control	Poisoned
Kidney	14	1036
Heart	25	677
Spleen	0	413
Stomach	37	386
Small intestine	36	368
Large intestine	21	248
Lung	9	257
Brain	21	166
Blood	3	50
Liver	0.8	31
Diaphragm	0	400
Uterus (virgin)	217	207

Animals killed 1 h after an intraperitoneal injection of NaFlAc, 5 mg/kg body weight (BUFFA and PETERS 1949).

such as the kidney, even though the recognizable intoxication symptoms – opisthotonus, toxic cramps, and spastic paresis – seem to point clearly to damage in the CNS (HENDERSHOT and CHENOWETH 1955; HASTINGS et al. 1955).

The in vitro accumulation of citrate due to fluoroacetic acid poisoning was also reported by KALNITZKY (1948) and KALNITZKY and BARON (1948). MARTIUS observed the same results in 1949. The mechanism of toxic activity was at first unclear. BARTHLETT and BARON (1947) found no reaction of fluoroacetic acid with enzymes or compounds containing thiol groups, nor did they find any inhibition with enzymes of the Krebs cycle. LIEBECQ and PETERS (1949) obtained similar negative results. BARTLETT and BARRON (1947) suspected that fluoroacetate competitively inhibits the oxidation of acetate and that this could explain the poisoning effect.

This conflicts with conclusions from experimental results that CHENOWETH et al. reached in 1949 in studies of fluoroacetate poisoning using acetate and acetic acid derivates. They found that these compounds were effective antitoxins in fluoroacetate poisoning and that they increased the animals' survival rate. Most effective was monoacetin, the monoglycerolester of acetic acid. TAITELMAN et al. came to similar conclusions in 1983. Monoacetin has protective effects against fluoroacetate intoxication, if it is introduced very early (CHENOWETH et al. 1951).

In 1949, MARTIUS postulated that fluoroacetate condenses to fluorocitric acid, responsible for the inhibition of citrate oxidation and the subsequent accumulation of citric acid.

PETERS considered the following to be the most likely synthesis path for the inhibitor: activated fluoroacetate is guided into the citric acid cycle (just like acetate) and enzymatically condensed with oxaloacetate to form a fluorotricarboxylic acid. This presumes an enzymatic error during the synthesis of this inhibitor, and this was, in fact, later proved to be the case. The isolation and crystallization of monofluorocitric acid was first accomplished after incubation of kidney homogenates from rabbits, guinea pigs, and dogs initially reinforced with ATP and Mg^{2+} after the application of fluoroacetate (Buffa et al. 1951; Peters and Wakelin 1953b). There seems to be a competitive antagonism between acetate and fluoroacetate concerning an enzyme system for synthesizing the postulated inhibitor. An increase in acetate concentration in the homogenate inhibits the synthesis of fluorocitric acid. This conflicts with Bartlett and Barrons's observations and confirms those made by Chenoweth et al. (1949) concerning antagonism between acetate and fluoroacetate in animals.

1. Synthesis of Fluorocitric Acid, Inhibitor of Aconitase

The molecular mechanism that leads to the lethal synthesis of fluorocitric acid has been described almost completely. Brady reported in 1955 on the preparation and properties of fluoroacetyl coenzyme A and its consequence on the function of various enzymes. "The formation of fluorocitrate is presumed to be due to the activation of fluoroacetic acid to fluoroacetyl S-CoA. F-acetyl-S-CoA reacts with oxaloacetate in the presence of crystalline condensing enzymes (EC 4.1.3.7, citrate synthase) described by Ochoa et al. (1951) to form fluorocitrate. The initial velocity of their reaction was much more rapid with acetyl-S-CoA than with F-acetyl-S-CoA. However, the reactions proceeded to approximately the same extent. In subsequent experiments with condensing enzymes, the condensing reaction proceeded further with fluoroacetyl coenzyme A than with acetylCoA." Fluoroacetyl S-coenzyme A is a potent competitive inhibitor of acetyl S-CoA in the latter system (Brady 1955). The unspecificity of coenzyme A and the condensing citrate synthase allows the lethal synthesis of fluorocitric acid (Fig. 1).

A high sensitivity for aconitase (aconitate isomerase) was observed in studies on the effect of fluorocitric acid on the activity of various enzymes of the citric acid cycle (Morrison and Peters 1954). Aconitase is a complex that catalyzes the reversible transformation of citrate into isocitrate (Peters and Shorthouse 1970). Fluorocitrate inhibited the enzyme's activity by 50% at low concentrations (10^{-8} to 10^{-9} M) (Peters 1963). This causes the accumulation of citrate in the various organs. Lotspeich et al. (1952) and Peters (1952, 1963, 1972) were able to prove that fluorocitrate does not inhibit isocitrate dehydrogenase. The inhibitor reacts with the inner membrane of mitochondria and binds to it in an irreversible manner (Kirsten et al. 1978; Kun et al. 1978). This leads to a long-lasting disturbance of energy production in the mitochondria of various organs. Extramitochondrial aconitase is inhibited to a lesser extent by fluorocitrate than mitochondrial aconitase (Peters 1972). Four isomers of fluorocitrate have

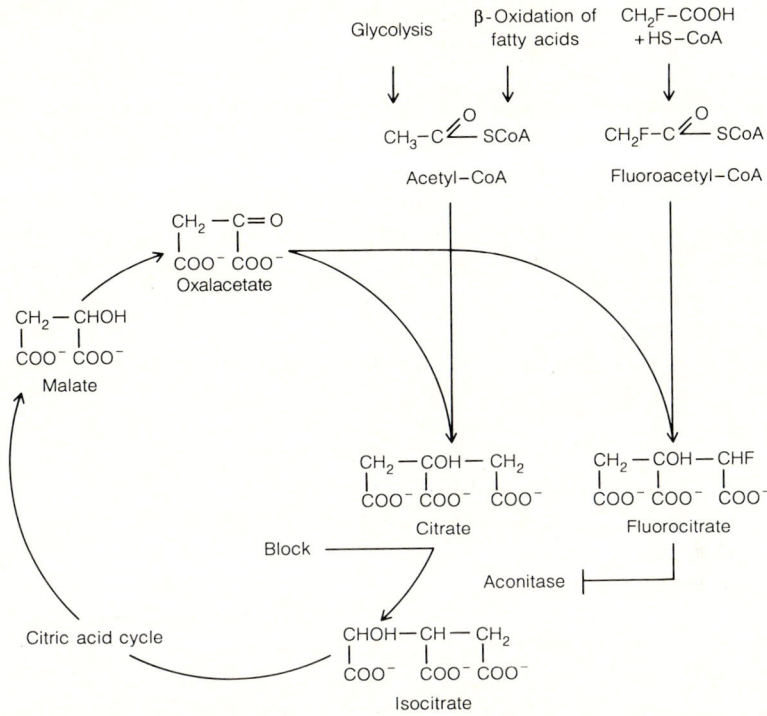

Fig. 1. Pathway of synthesis of fluoroacetate to fluorocitrate. Fluorocitrate blocks the metabolism of the C_6 acids at the aconitase stage of the citric acid cycle. The reactions of these are: citrate $\rightleftharpoons$ cis-aconitate $\rightleftharpoons$ isocitrate. It is to be noted that the increasing citrate concentration also blocks the enzyme phosphofructokinase. According to PETERS (1952, 1963), BRADY 1955

been synthesized and isolated (see discussion of the molecular mechanism of fluorocitrate action: KUN 1972; EANES and KUN 1971; DUMMEL and KUN 1969; PETERS 1972). The preferred disturbance in the brain is understandable because of the fact that the brain is almost totally dependent on glucose for its nutrition and function. The central position of coenzyme A in intermediary cell metabolism as a transient carrier of acetyl or acyl groups allows fluoroacetate to be incorporated into various compounds. However, no additional neurotoxically effective synthesis has been discovered (PETERS 1972). Furthermore, the much-discussed hypothesis that fluoroacetylcholine could be synthesized and that it could cause toxic effects by competitive displacement of the natural transmitter acetylcholine from the synaptic receptor was not confirmed.

The uptake of acetate and fluoroacetate in nerve and glia cells is catalyzed through a Na^+-dependent carrier system (GONDA and QUASTEL 1966; QUASTEL 1974).

The high neurotoxicity of fluoroacetate seems to be based not only on a blockade of the Krebs cycle by fluorocitric acid, it is rather strengthened by the inhibition of glutamine synthesis (Lahiri and Quastel 1963; Quastel 1974). Fluoroacetate is known to increase ammonia concentrations in the CNS. The ammonia ion, a common metabolite in cellular metabolism, can be extremely toxic if it accumulates in the brain after inhibition of glutamine synthetase by fluoroacetate. This intoxication leads to a disturbance of the synaptic function mediated by the transmitter of cortical inhibition, γ-aminobutyric acid (GABA) (Raabe 1981). In addition to the disturbance in glutamine synthesis, the metabolism is also affected by glutamate, aspartate, and GABA (Clarke et al. 1970).

Fluoroacetate also reduced the uptake of synaptically released glutamate. It leads to an increase in the stimulation-induced overflow of excitatory acids from hippocampal slices (Szerb and O'Regan 1988). The reason for the inhibition of glial glutamate uptake by fluoroacetate is probably a decrease in the activity of the sodium pump resulting from an insufficient production of ATP or as a consequence of the inhibition of cellular energy production (Szerb and Issekutz 1987).

III. Citrate Accumulation – A Marker in Fluoroacetate Intoxication

Tissue citrate accumulation has been found in all species intoxicated with fluoroacetic acid or fluorocitric acid, as proof of the inhibition of aconitase (Gall et al. 1956; Liebecq and Liebecq-Hutter 1958; Buffa et al. 1973; Liang 1977). A clear dose-response relationship between the accumulation of citrate in the brain and the intensity of the neurological disorders in vivo could not be found. Fluor can be enzymatically cleared from fluoroacetate, which is thereby detoxified (Soiefer and Kostyniak 1983).

It was, however, possible to prove the synthesis of fluorocitric acid in brain tissue in vitro or in homogenates using certain animal species. In 1966, Peters and Shorthouse found an accumulation of citrate only after adding fluorocitric acid and not after adding fluoroacetate.

Morselli et al. (1968) studied the neurotoxic effects of fluoroacetate and fluorocitrate after intracerebral microinjection into various brain regions of Sprague-Dawley rats under ether anesthesia. Fluorocitric acid was at least 100 times more toxic than fluoroacetic acid under various experimental conditions.

From the major differences in neurotoxicity of the two fluoro compounds, the authors came to the conclusion that the brain has a limited ability to synthesize fluorocitric acid from fluoroacetate, or that the synthesis occurs very slowly. The high citrate values found in other organs such as the kidney in fluoroacetate poisoning (see Table 2) seem to indicate that larger amounts of fluorocitric acid are synthesized outside of the brain. They reached the brain via the bloodstream after passing the BBB. The amounts of fluorocitrate in the brain leading to neurotoxic symptoms or death are very small. Intracerebral microinjection of 5 μg of the crude synthetic fluorocitrate killed all rats of 200 g body weight. The 5-μg dose was equivalent to 0.56 μg of the active sodium fluorocitrate isomer (MORSELLI et al. 1958).

As a consequence of the synthesis of fluorocitric acid and the blockade of the Krebs cycle that it produces, many disturbances in intermediary metabolism have been described. BOSAKOWSKI and LEWIN (1986) provide a review of intoxication syndromes. They reported a positive correlation between serum citrate levels and the rise of intoxication. They conclude that the determination of serum citrate content is a readily accessible and accurate peripheral indicator for determining the intensity of fluoroacetate toxicity and, therefore, lethal synthesis of fluorocitrate.

C. Biosynthesis of Neurotoxic Dinucleotides

I. The NAD(P) Glycohydrolase-Transferase Exchange Reaction

The NAD(P) glycohydrolase (EC 3.2.2.5) localized in the endoplasmic reticulum of nerves, neuroglia cells, and other organ cells was discovered by ZATMAN et al. in 1953. This enzyme cleaves the coenzymes NAD and NADP between the pyridine nitrogen of the nicotinic acid amide and the ribose in the adenosine ribose diphosphate part. This enzyme, besides being capable of hydrolyzing, also has a transferring activity, which, however, is not specific (ZATMAN et al. 1953, 1954; KAPLAN and CIOTTI 1956; KAPLAN et al. 1957). After hydrolysing NAD or NADP the transferase catalyzes the synthesis of abnormally structured nucleotides by an unspecific reaction, if a sufficient concentration of antimetabolites such as 3-acetylpyridine or 6-aminonicotinamide are available to the enzyme, instead of nicotinamide. Since then, many analogous nucleotides have been enzymatically synthesized in this way. Nicotinamide in these nucleotides has variously been replaced by imidazole derivative, (see ALIVISATOS and WOOLLEY 1955; ALIVISATOS 1959; ALIVISATOS et al. 1960), thiadiazoles (CIOTTL et al. 1960), by variously structured pyridine derivatives (ANDERSON and KAPLAN 1959; ANDERSON et al. 1959; KAPLAN and CIOTTI 1956; KAPLAN 1960), and aminochinoline (SCHABAREK 1967; see reviews: COPER 1966; COPER and HERKEN 1963; CHRIST 1975). Most of these nucleotides containing foreign compounds were produced in vitro (WALTER and KAPLAN 1963) with enzyme preparations from rat microsomes. More than 40 analogous nucleotides are known, mostly derivatives of NAD. Very few could be synthesized under in vivo conditions or isolated from the brains of those animals that had been treated with the substances mentioned above. Of special interest in this latter group are the neurotoxic antimetabolites of the nicotinic acid amide, 3-acetylpyridine (3-AP) and 6-aminonicotinic amide (6-AN), which have a convincing structural similarity to nicotinamide (Fig. 2).

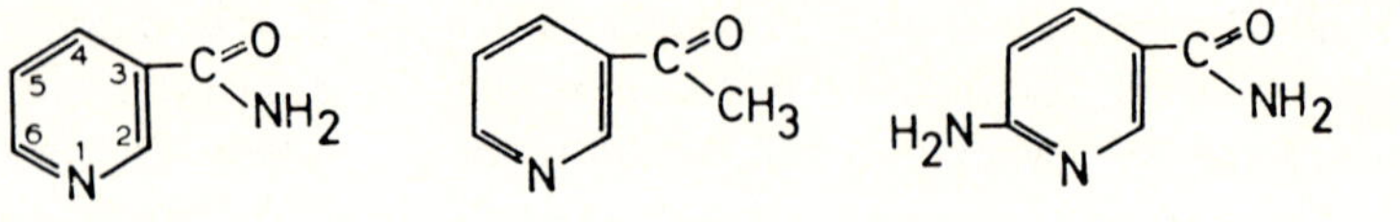

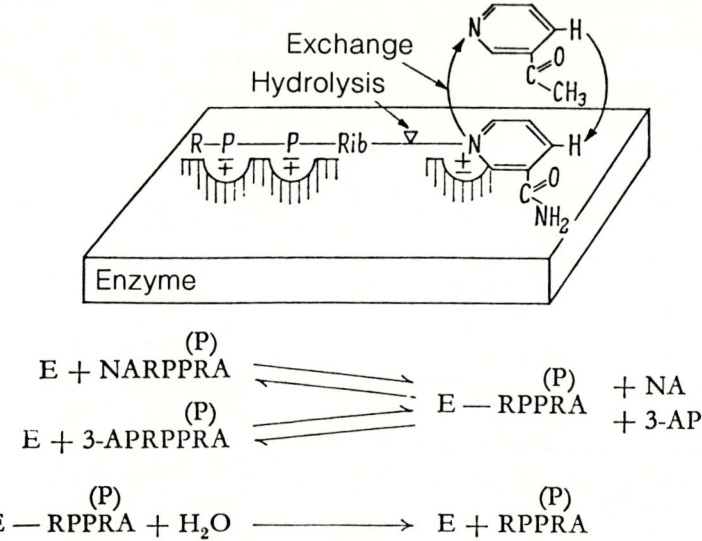

Nicotinamide 3-Acetylpyridine 6-Aminonicotinamide

Fig. 2. Formulae of nicotinamide and its antimetabolites

E + NARPPRA (P) ⇌ E — RPPRA (P) + NA

E + 3-APRPPRA (P) ⇌ E — RPPRA (P) + 3-AP

E — RPPRA (P) + H_2O ⟶ E + RPPRA (P)

Fig. 3. Hydrolysis of NAD(P) and synthesis of 3-APAD(P) by NAD(P) (glycohydrolase) from the endoplasmic reticulum according to ZATMANN et al. (1953), KAPLAN (1960, 1963), BRUNNEMANN and COPER (1965b), and HERKEN (1968a), modified for 3-acetylpyridine. *NAD*, nicotinamide-adenine-dinucleotide; *NADP*, nicotinamide-adenine-dinucleotide phosphate; *3-APAD(P)*, 3-acetylpyridine-adenine-dinucleotide (*P*); *NA*, nicotinamide; *ARPPR*, adenosine-diphosphate-ribose (*P*)

The glycohydrolase cleaves only the oxidised pyridine nucleotides between the ribose and the nitrogen of the pyridine derivative. The reduced compounds are not split, as first shown by MCILWAIN and RODNIGHT (1949). The glycohydrolase from pig brain was separated from the structure of the endoplasmic reticulum, solubilized, and purified. However, the tranferase could not be removed or isolated (WALLENFELS et al. 1960; WINDMUELLER and KAPLAN 1962; DICKERMAN et al. 1962).

The erroneous action (ZATMAN et al. 1953; ALIVISATOS and WOOLLEY 1955; KAPLAN et al. 1956; KAPLAN 1960, 1963) (Fig. 3) can be schematically represented using the exchange of nicotinic amide for 3-AP in NAD or NADP, respectively.

The mechanism for the synthesis of nucleotides containing 3-AP operates in the following manner: the hydrolysis of the nucleotide NAD(P) to nicotinamide and adenosine diphosphate ribose (P) initiates the exchange reaction. Nicotinamide is an inhibitor of this reaction, so that the hydrolysis is slowed down with subsequent growing concentration of free nicotinamide. The exchange reaction will remain untouched, and adenosine diphosphate ribose can serve as an acceptor for 3-AP or 6-AN. The concentration of 3-AP or 6-AN should be at least 10 times higher than that of nicotinamide for synthesis to occur. The competitive antagonism explains why the onset of toxic symptoms can be prevented in vivo by the early injection of nicotinamide. The lack of specificity of the NAD(P) glycohydrolase also allows the synthesis of abnormally structured nucleotides, presuming that unchanged NAD or NADP are available. The final hydrolytic cleavage of adenosine diphosphate ribose from the enzyme ends the transfer reaction at the same time. This breakdown and resynthesis pathway has been called the pyridine nucleotide cycle (GHOLSON 1966). The resynthesis of NAD(P) or 3-acetylpyridine adenine dinucleotide phosphate (3-APADP) can affect the process of ADP ribosylation reactions (OLIVERA and FERRO 1982).

The glycohydrolase activity varies considerably in the different regions of the brain (KAPLAN 1960; COPER 1961). This applies to the hydrolyzing reaction as well as to the transferring reaction. There is remarkable agreement between the localization of high glycohydrolase activity and cell lesions after application of 3-AP and 6-AN, which seems to suggest that there is a close connection between the synthesis of abnormally structured nucleotides and cell destruction (BRUNNENMANN and COPER 1964a, 1965a).

The biosynthesis of the nucleotides containing 6-AN derived from NAD or NADP operates in the same way, but the symptoms of intoxication and the changes in intermediary metabolism caused by the antimetabolites show distinct differences between 3-AP and 6-AN.

This justifies a separate discussion of the effects of 3-AP and 6-AN on the CNS. The toxicity is not limited to the CNS. Both antimetabolites have a teratogenic effect on embryos (CHAMBERLAIN 1972; CAPLAN 1971). 6-AN has a cytotoxic effect on carcinoma cells, but because of its high neurotoxicity, it is of no use in carcinoma therapy (DIETRICH 1975). The action of 6-AN on cell growth, poly (ADP-ribose) synthesis and nucleotide metabolism was investigated by HUNTING et al. (1985). 3-AP inhibits myelin synthesis in the developing rat brain (NAKASHIMA and SUZUE 1984).

II. 3-Acetylpyridine and 3-Acetylpyridine-Adenine-Dinucleotides

1. Neurotoxic Symptoms

3-AP was injected into Sprague-Dawley male rats at a dosage of 80 mg/kg. After a latent period of 2–3 h, intoxication gradually increased, and during this time the effect on the CNS became markedly evident. After a period of paralysis a condition of excitement gradually developed involving constant rolling cramps and a massive loss of equilibrium. At this stage some of the animals died. Among the survivors, chronic effects were evident, which may be seen as hyperkinesis, disturbance of muscular coordination with ataxia, and, at intervals, tonic cramps. In addition, between the cramp attacks, which appear spontaneously without visible cause, the animals exhibit abnormal behavior. Heavy tremor of the front legs with muscle fibrillation may occur. Many animals retained a hypersensitivity, which manifested itself in spontaneous convulsions for more than a year. One can distinguish between symptoms which appear as attacks at intervals and those that occur continuously with losses of general nervous function (HERKEN 1968b).

Electroencephalographic (EEG) studies reveal definite pathological changes. Some 4 h after application of 80 mg/kg into male rats of 3-acetylpyridine, a general increase in action potentials of the EEG with spikes and waves is found, which show rhythmic and generalized seizure patterns as signs of convulsive activity (HERKEN 1968a).

2. Molecular Basis of Intoxication

a) Synthesis of Nucleotides Containing 3-Acetylpyridine in the Brain

All neurotoxic symptoms became evident after a latent period of several hours, which implies that the effector compound is synthesized by the metabolism of organ cells. The question arose whether a dysfunction of dehydrogenases and reductases after biosynthesis of derivatives of NAD(P) containing 3-AP could be the cause (KAPLAN et al. 1957; KAPLAN 1960, 1963). A 3-AP derivative of NAD was first isolated from tumors by KAPLAN et al. (1954). The isolation and identification of NAD containing 3-AP from the brain of rats which had received 80 mg/kg of 3-AP was accomplished in 1962 (COPER and HERKEN 1963; BRUNNEMANN et al. 1962, 1963). This substance was purified by chromatography and identified by spectrophotometric and enzymatic procedures. However, the amount was so small that it was impossible to ascribe the loss of neurological functions to the activity of this substance. In addition, the substance acted as a hydride ion acceptor and donor with various NAD-dependent dehydrogenases, so that a severe inhibition of metabolism by this nucleotide is unlikely. With isolated brain microsomes containing high NAD(P) glycohydrolase activity, the synthesis of the derivative of NADP which contains 3-AP (3-APADP) is greater than that of the NAD derivative (3-APAD) in the same period and under similar conditions (BRUNNEMANN et al. 1962, 1963). The quantity of 3-APADP in the brain cells of the intoxicated animals was difficult to estimate: The concentration of the coenzyme NADP in the brain is 15–20 times lower than that of NAD.

A spectrofluorometric analysis was developed which has a sensitivity that is one order of magnitude higher than the method published by LOWRY et al. (1957). It allows an accurate differentiation of the natural coenzymes

NAD and NADP from their derivatives containing 3-AP. Even the presence of other derivatives of nicotinamide did not interfere with this method (HERKEN and NEUHOFF 1963). The excitation spectra of NAD and 3-APAD (3-APADP) show typical differences between the natural and the abnormally structured products. For those nucleotides which contain 3-AP, there is a second characteristic maximum occurring at a wavelength of 265 nm. The maximum emission is at 500 nm. This emission is used for the determination of the concentration of the compound in the brain. The high sensitivity of this method made it possible to demonstrate the presence of nucleotides containing 3-AP down to $2 \cdot 10^{-12}$ mol present in 1 mg of wet weight of brain cells.

3. Distribution of 3-APADP in the Brain Region

The derivative of NAD containing 3-AP, which is of minor interest, and the natural coenzyme were destroyed after previous enzymic hydrogenation with alcohol dehydrogenase by subsequent treatment with perchloric acid. During this reaction, the oxidized part of NADP and also of 3-APADP was retained in the extract, which made possible a quantitative determination of these nucleotides by the spectrofluorometric method.

Extraction of the whole brain furnished evidence that nearly 40% of naturally occurring NADP is converted into an abnormally formed compound containing 3-AP. NEUHOFF (1964) found that 4 h after injection the concentration of 3-AP in the brain is 5–10 times higher than the concentration of nicotinamide. The establishment of a concentration gradient against the extracellular space indicates an active transport of 3-AP through the cell membrane, and the formation of abnormally structured nucleotides reaches its maximal level at about the same time. Using the spectrofluorometric method, it is possible to determine the biosynthesis of 3-APADP in 9 different regions of the brain. In all of them, the abnormally structured nucleotide was, however, proved to be present in varying concentrations (WILLING et al. 1964).

The exchange of 3-AP with nicotinamide reaches the maximum value in the NADP of the hippocampus (Fig. 4). As is well-known, the hippocampus plays a dominating role in the regulation of excitation processes in the brain and is a part of the limbic system.

While examining cell fractions, it was discovered that only small amounts of 3-APADP were formed in the mitochondria, while more than 90% was found in the cytoplasm of the brain cells which did not include the mitochondria. Since, according to KLINGENBERG (1963), nearly 35% of the total NADP of the brain cells is present in the mitochondria, and this is not converted, the proportion of 3-APADP to NADP in the remaining part of the cell increases correspondingly.

The preference for the hippocampus can also be confirmed by autoradiography (KORANSKY 1965). After injection of ^{14}C-labelled 3-AP, the

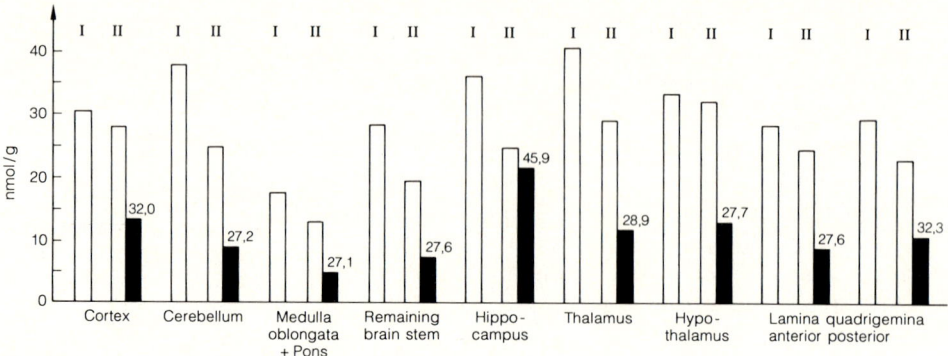

Fig. 4. Nicotinamine-adenine dinucleotide phosphate (NADP, *open bars*) and 3-acetylpyridine adenine dinucleotide phosphate (3-APADP, *black bars*) content in different brain regions. *I*, control; *II*, 6-7h after 100 mg/kg 3-AP; *Numbers*, 3-APADP in % of 3-APADP + NADP. After Willing et al. (1964)

strongest radiation was found in certain defined regions of the hippocampus. The method used does not, however, permit more detailed differentiation within the cells.

A comparison with the topographic anatomy of the hippocampus clearly shows that the stratum pyramidale and perhaps the marginal areas of the stratum oriens and the stratum radiatum in the regions CA2 and CA3 are obviously involved. These are areas which are commonly assumed to be metabolically highly active (Lowry et al. 1954; Brunnemann and Coper 1964a; Balazs et al. 1973; McLean 1963).

4. Dysfunction of Oxidoreductases with Nucleotides Containing 3-Acetylpyridine as Coenzyme

What are the consequences of the accumulation of 3-acetylpyridine dinucleotide phosphate (3-APAP) for the metabolism of the brain cell? Warburg and Christian (1936) stated: "Die Wirkung der Pyridinnucleotide in der lebendigen Substanz beruht auf der Fähigkeit ihres Pyridinanteiles, zwei Wasserstoffatome zu addieren und diesen Wasserstoff wieder abzugeben, also auf der Fähigkeit, reversible Dihydrophyridinverbindungen zu bilden." This function is disturbed. The replacement of nicotinamide by 3-AP leads to a disturbance of hydride ion transfer occurring between enzymes and substrates using pyridine nucleotides as co-factors. Of special interest is the behavior of 3-APADP in these systems. Kaplan et al. (1955) found that the isocitrate dehydrogenase (EC 1.1.1.42) obtained from pig heart reacts significantly more slowly with this unnatural pyridine nucleotide than with the natural coenzyme. Coper and Neubert (1963, 1964) and Herken and Timmler (1965) prepared 3-APADP with the help of rat microsomal enzymes from rat brain and studied the kinetics of hydride ion transfer with different dehydrogenases in vitro. The Michaelis constant for NADP with the isocitrate dehydrogenase was, at approximately $2 \times 10^{-5} M$, in the same range as that published in the literature, with an indicated value of $1.25 \times 10^{-5} M$. The Michaelis constant for 3-APADP was one order of magnitude higher, at $K_m = 2 \times 10^{-4} M$. COPER and NEUBERT (1964) were able to confirm the observations made by KAPLAN et al. in 1955. The conversion of dihydrofolic acid (FH_2) into tetrahydrofolic acid (FH_4) by dihydrofolate reductase (EC 1.5.1.3) (which is slowed down in presence of 3-APADP) is probably important for the understanding of the long-lasting effects of 3-AP (HERKEN and TIMMLER 1965). The evaluation of Lineweaver–Burk diagrams showed a significant retardation of the reaction velocity of the enzymic reduction of folic acid with 3-APADP as coenzyme. The velocity of the FH_2 conversion to FH_4 slows down to approximately one-half its normal value. During their investigations concerning the velocity of the hydrogen transfer between substrate and NADP-dependent enzymes, NEUBERT and COPER (1965), while studying the malic enzyme (EC 1.1.1.40) which had been discovered by Ochoa, obtained some results which may contribute toward the understanding of the neurotoxicity of 3-AP. The measurement of the kinetics showed that the reaction from malate to pyruvate, with 3-APADP acting as hydrogen acceptor, was faster than that with the natural coenzyme NADP. As yet, this has not been found with any other NADP-dependent enzyme. The reversal of the reaction, pyruvate to malate, in the presence of 3-APADPH was extremely slow.

Studies of the reaction kinetics of various NADP-dependent oxidoreductases showed that the K_m values for 3-APADP were, in general, one order of magnitude higher than those for NADP, as calculated from Lineweaver–Burk diagrams (KAPLAN et al. 1955; COPER and NEUBERT 1963, 1964; HERKEN and TIMMLER 1965).

However, it must be mentioned that most observations were not obtained from experiments with enzymes from brain cells. There are no indications that the kinetics of brain enzymes act fundamentally differently. This was proved, for example, by investigations done on the NADPH-aldehyde reductase (EC 1.1.1.2) with nucleotides containing 3-AP from cattle brain (JOCHHEIM 1981). It was of particular interest to note that the first two NADP-dependent steps of the pentose phosphate cycle are involved, and these are important for the production of reduction equivalents for many syntheses in the cells. The enzymes concerned are glucose-6-phosphate dehydrogenase (EC 1.1.1.49) and 6-phosphogluconic acid dehydrogenase (EC 1.1.1.44). If the first enzymic step of the Warburg–Dickens–Horecker cycle is reduced to 1/6 of the original turnover after conversion of NADP into the nucleotide containing 3-AP, it will have obvious unfavorable consequences on the following reactions (COPER and NEUBERT 1964).

The study revealed that all NADP-dependent enzymes have a retarded velocity for hydride ion transfer if they are incubated with 3-APADP instead of with the natural coenzyme. KAPLAN and CIOTTI (1956) thought that the retardation of enzymatic reactions by exchange of NAD(P) for analogous pyridine nucleotides in oxidoreductase reactions was the reason for the intoxication (Fig. 5) (Kaplan 1960, 1963).

The specific activity of NADP-dependent enzymes taking part in the pentose phosphate cycle is not the same in different brain regions (Lowry et al. 1954). Brunnemann and Coper (1964a) found that the hippocampus holds a special position in having the highest activity of glucose-6-phosphate dehydrogenase.

The investigation of different enzymes with 3-APADP as coenzyme showed that each reaction which leads to a change in the redox quotient in favor of the oxidized state of pyridine nucleotides will also influence the hydride ion acceptor systems. There is, e.g., a very strong inhibition of glutathione reductase (EC 1.6.4.2) at different concentrations of the substrates (Coper and Neubert 1964).

The diminution of the reduction equivalents by a slow hydride ion transfer in presence of 3-APADP as coenzyme of the dehydrogenase indirectly leads to a lowering of the efficiency of reductases (Hotta 1962). This, however, presupposes that the enzymes mentioned are directly connected with each other in one compartment of the cell.

3-APADP can act as an inhibitor in the presence of NADP, which entails that the catalysis can also be slowed down in vivo if more than 50% of the natural coenzyme is present as a derivative of 3-AP. This has been proved, for example, in the hippocampus (Willing et al. 1964).

5. Predilection Sites of Functional Disorders and Lesions in the Central Nervous System

Considering the intensity and severity of recognizable symptoms of intoxication, it would be logical to expect a definite destruction of the morphologic structure in certain zones of the brain. The hippocampus and the olivary nucleus area seem to be the preferred sites of cell lesion (Wolf et al. 1962; Balaban 1985).

The hippocampus belongs to the metabolically active areas of the CNS that are especially damaged by 3-AP. The NAD(P) glycohydrolase transferase develops high activity here (Brunnemann and Coper 1965a). Hicks (1955) and also Coggeshall and McLean (1958) reported on recognizable lesions in rat and mice hippocampus. Fuse and Koikegami (1961) and Montgomery and Christian (1976) reconfirmed that the most prominent lesion was in the hippocampus. Cytomorphological alteration of mossy fiber boutons of the hippocampus produced by 3-AP (Niklowitz 1969, 1972) and increased 3-AP sensitivity of the hippocampus induced by glucocorticoid treatment have been noted (Sapolsky 1985).

During experiments on local glucose utilization in the brain, Shimoda et al. (1984), working on 3-AP intoxication of mice, observed a loss of pyramidal cells and a gliosis in the dorsal hippocampus that was limited almost exclusively to the CA3 sector. Autoradiography results seem to

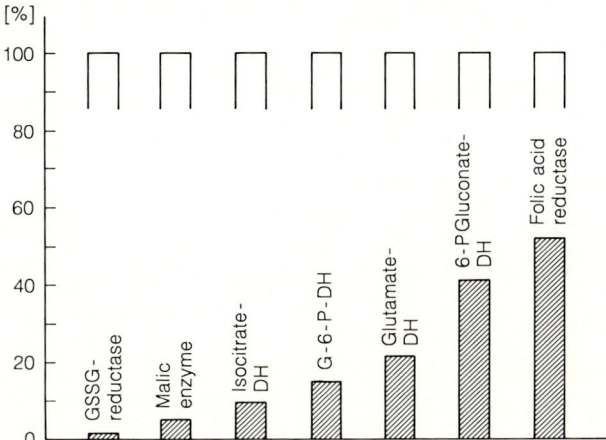

Fig. 5. Velocity of reactions of different dehydrogenases (*Dtt*) with 3-APADP(H) as cofactor [NADP(H) = 100%]. *GSSG*, Oxidized glutathione; *G-6-P*, glucose-6-phosphate

indicate a glucose utilization disturbance. The effects of 3-AP on glucose utilization, axonal transport, and function of the olivary inferior-cerebellar system (all of which were also heavily damaged) were investigated by BERNARD et al. (1984).

OZAKI et al. (1983) reported a specific sort of functional disturbance after a lesion of the CA3 area of the hippocampus, which they accomplished by single introperitoneal injection of a high dose of 3-AP (260 mg/kg) in mice. Two weeks after this treatment, learning an avoidance task was tested with a training apparatus called a jump-box. From the results, the authors concluded that there was a connection between the degree of destruction of the pyramidal cells in the CA3 area of the hippocampus and the learning deficits. DENK et al. reported similar results in 1968.

BALABAN (1985) described in a critical review the central neurotoxic effects of intraperitoneally administered 3-AP on multiple lesions in different regions of the brain. The results demonstrated that there is a strong hippocampal 3-AP neurotoxicity in rats. The studies confirm previous reports of degenerating CA3–4 regions and dentate gyrus granule cells. These are the same regions in which the highest amounts of 3-APADP were found (WILLING et al. 1964; HERKEN 1968b). BALABAN pointed out the significance of this nucleotide for setting off the neurotoxic effects and the lack of substrate specificity of NADP glycohydrolase transferase as the cause of the neurotoxic synthesis of 3-APADP.

Many studies confirmed that 3-AP produces considerable lesions of the inferior olivary nucleus in rats and mice and hence the disturbed function of the climbing fiber to the cerebellar cortex. They may be responsible for the continuing disturbance of movement (ANDERSON and FLUMERFELT 1984;

Balaban 1985; Desclin and Escubi 1974; Ito et al. 1978; Umetami 1980; Woodhams et al. 1978; Simantov et al. 1976; Batini 1985).

The effects of damage to the olivary inferior nucleus by 3-AP were studied using neurophysiologic methods by Lopiano and Savio (1986). They involve the varied behavior of Purkinje's and Deiters' cells. A reduced inhibitory efficacy of Purkinje's cells as a consequence of climbing fiber deprivation is assumed to be responsible for the long-lasting disturbance of mobility. Other authors consider the long-term effects of the inferior olive degeneration on red nucleus activity to be the cause of the mobility disturbances (Billard et al. 1988; Szalay et al. 1979).

The 3-AP-induced degeneration of the inferior olive and cerebellum is similar to the pathology seen in olivopontocerebellar atrophies (OPCA). OPCA are a heterogenous group of disorders characterized by loss of inferior olivary neurons, degeneration of the basis pontis, and cerebellar atrophy (Koeppen and Barron 1984).

Deutsch et al. (1989) reported on 3-AP-induced degeneration of the nigrostriatal dopamine system. Six weeks after 3-AP treatment, decreases in both striatal dopamine content and the activity of tyrosine hydroxylase were observed. The anatomical findings were consistent with a relatively selective effect of 3-AP on the nigrostriatal system rather than on the mesocortical or mesolimbic dopaminergic systems. Deutsch et al. (1989) concluded from these results that the profound loss of the climbing fiber input to the cerebellum (induced by 3-AP) derived from the inferior olivary nucleus and the degeneration of nigrostriatal dopamine neurons may provide a useful model of olivopontocerebellar atrophy-associated parkinsonism.

Other reports deal with the effects of climbing fiber deprivation on the release of endogenous aspartate, glutamate, and homocysteine in slices of rat cerebellar hemispheres (Vollenweider et al. 1990), putative neurotransmitter amino acids in the cerebellum and medulla of the rat (Nadi et al. 1977; Perry et al. 1976; Rea et al. 1980; Gottesfeld and Fonnum 1977), and reduction of GABA receptor binding induced by climbing fiber degeneration in the rat cerebellum after application of 3-AP (Kato and Fukuda 1985), which are connected to the mobility disturbances.

6. Metabolites of 3-Acetylpyridine

Laboratory examinations of the urine from 3-AP-treated animals have proven that no free AP is eliminated (Beher and Anthony 1953; Beher et al. 1953, 1959).

Kanig et al. (1964) isolated pyridine-3-methylcarbinol (from rat brain) as a main metabolite of 3-AP. They used tritium-labelled 3-AP, so that accurate statements could be made about the amounts that were formed in the brain. Injection of pyridine-3-methylcarbinol caused intoxication symptoms at the same dosages as 3-AP. Kaplan and Ciotti (1956) proved the exchange of pyridine-3-methylcarbinol in NAD(P) for nicotinamide.

After incubation of rat brain microsomes with tritium-labelled pyridine-3-methylcarbinol, NAD or NADP were transformed into the abnormally structured pyridine nucleotides containing pyridine-3-methylcarbinol. These nucleotides are unable to act as hydride ion carriers in oxidoreductase systems. Furthermore, they are inhibitors of dehydrogenases (BRUNNEMANN and COPER 1964b).

NEUHOFF and KÖHLER (1966) showed that during metabolism in vivo numerous compounds are derived from 3-AP. The main product of metabolism in vivo seems to be pyridine-3-methylcarbinol, produced by reduction of 3-AP; according to NEUHOFF and KÖHLER's results (1966) this product is very quickly glucuronized in the organism.

Of all the pyridine compounds hitherto isolated, only pyridine-3-methylcarbinol possesses a pharmacological activity which approximately corresponds to that of 3-AP. Among the further compounds which – due to their structure – can be exchanged for the glycohydrolase transferase, 3-AP-6-pyridone and 3-methylcarbinol-6-pyridone were entirely nontoxic.

Other acetylpyridines are not toxic. 2-AP is not transferred to NADP by the glycohydrolase of the microsomes of the brain, either in vitro or in vivo. In vitro, 4-AP is enzymatically exchanged for nicotinamide faster than 3-AP, so considerable quantities of this nucleotide could be isolated. As was to be expected from the structure, this compound was ineffective as a hydride carrier in dehydrogenase or reductase systems. A completely contrary observation was the fact that the toxicity of 4-AP is very low. No nucleotide containing 4-AP could be found in the brain (BRUNNEMANN and COPER 1964c).

III. 6-Aminonicotinamide and 6-Aminonicotinamide-Adenine-Dinucleotides

1. Neurotoxicity

6-AN, an antimetabolite of nicotinamide, differs from the naturally occurring compound by the amino group at position C6 of the pyridine ring. The remarkable pharmacological and toxicological properties of this substance were revealed in experiments with different species of animals. The effects concern mainly the function of the CNS.

JOHNSON and McCOLL (1955) were the first authors to report that 6-AN produces a spastic paresis in rats, rabbits, dogs, and cats. These results were confirmed and extended by STERNBERG and PHILIPPS (1958), WOLF and COWEN (1959), and WOLF et al. (1962). These authors also reported on the damage done to various central areas of the brain such as, among others, the hippocampus, corpus striatum, and lower olivary nucleus. SCHNEIDER and COPER (1968) extended and defined these results in exact detail using electron microscopy (SCHNEIDER and CERVOS NAVARRO 1974).

Unlike 3-AP intoxication, a single injection (intraperitoneal) of 6-AN 25 mg/kg in rats (Stamm-Sprague-Dawley) produces an uncharacteristic EEG, dysrhythmic and depressed in frequency, with scattered sharp waves and spikes (Coper and Herken 1963).

2. Biochemical Basis of Action

a) Synthesis of Nucleotides Containing 6-Aminonicotinamide
by NAD(P) Glycohydrolase

First results for understanding the mechanism of action came from the experiments by Dietrich, Friedland, and Kaplan (1958) of the enzymatic transfer of 6-AN in vivo and in vitro to analogues of the pyridine nucleotides NAD and NADP. They postulate that it produces its neurotoxic in vivo effects only after incorporation and conversion to nonphysiological nucleotide analogues, which in turn are incapable of entering into the enzymatic reactions in which pyridine nucleotides participate.

The biochemical basis of all pharmacological and toxicological effects of the antimetabolite 6-AN is found (as is also the case for 3-AP) in the unspecificity of the glycohydrolase localized in the endoplasmic reticulum of cells, which has hydrolyzing and transferring activity (Kaplan and Ciotti 1956). The exchange reaction of 6-AN for nicotinamide in NAD or NADP occurs according to the same mechanism as described for 3-AP (Fig. 6), the principle of competitive antagonism. The concentration of 6-AN in the cell must be considerably higher than that of nicotinamide for synthesis to occur. The exchange reaction results in the formation of the neurotoxic analogues 6-ANAD(P) which was isolated from the brains of rats treated with 6-AN (Coper and Herken 1963; Brunnemann et al. 1964; Brunnemann and Coper 1964c; Herken 1971). The symptoms of intoxication can be prevented in vivo by the early injection of high doses of nicotinamide. Studies of the biosynthesis of nucleotides containing 6-AN in vivo as well as studies of the kinetics of different NADP-dependent enzymes focused attention on the derivatives of NADP which contain 6-AN (Herken and Neuhoff 1964). After application of 6-AN 50 mg/kg it was found, for instance, that in the kidney the concentration of 6-ANADP is almost 8 times that of the natural coenzyme NADP. These differences in concentrations are probably due to the slow hydrolysis of the nucleotide containing 6-AN by the glycohydrolase. De novo synthesis may also be involved. In the brain, the same preference for the synthesis of derivatives of NADP containing 6-AN was found with albino rats after application of different doses of 6-AN (Coper et al. 1966). Compared with the available quantity of NAD, the conversion of NAD into a compound containing 6-AN was so negligible that it could not be considered as the cause of the disturbance of enzymic reactions (Brunnemann et al. 1964).

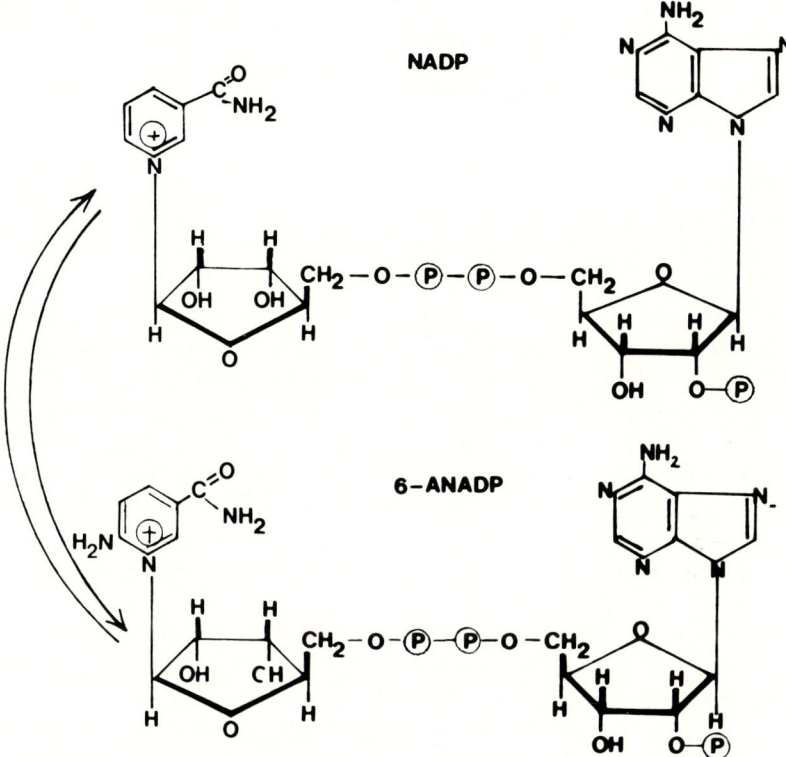

Fig. 6. Exchange reaction of 6-aminonicotinamide (*6-AN*) for nicotinamide in nicotinamide-adenine dinucleotide phosphate (*NADP*) in the endoplasmic reticulum of brain cells by glycohydrolase

On the other hand, the concentration of the coenzyme NADP in the brain is 15–20 times lower than that of NAD, so that the biosynthesis of 6-ANADP was of much greater importance for the development of the different functional disorders.

Enzyme kinetic studies with nucleotides containing 6-AN synthesized by means of microsomal enzymes (BRUNNEMANN et al. 1963, 1964; HERKEN and NEUHOFF 1964) confirmed the findings of DIETRICH et al. (1958) that these abnormally structured nucleotides are unable to act as hydride ion carriers in oxidoreductase systems.

This, however, is not the main cause of the functional disorders in the different organs, as was originally supposed. From studies of the ratios of the concentration of oxidized and reduced NAD and NADP in the kidney, it appeared that the redox status was unchanged (HERKEN and NEUHOFF 1964).

3. 6-ANADP, Inhibitor of Oxidoreductases, and the Blockade
of the Hexose Monophosphate Pathway

Of greater importance still was the result obtained during experiments on
different enzymes that the derivative of NADP containing 6-AN was able to
inhibit the activities of dehydrogenases and reductases also in the presence
of the natural coenzymes. As was shown by the evaluation of Lineweaver–
Burk diagrams, it is the machanism of a competitive inhibition (COPER and
NEUBERT 1964). The sensitivity of the enzymes differs widely.

The brain is largely dependent for nutrition on its supply of glucose.
Because the BBB is easily penetrated by glucose, the brain tissue has a
potential capacity for metabolizing glucose in different pathways which
is needed not only for energy production but also for the synthesis of numerous components in cellular structures. Because many NADP-dependent
enzymes contribute to this metabolism, it seemed to be rather difficult to
find out which metabolic reaction is most severely impaired by the derivative
of NADP containing 6-AN. Some experiments with 1-$[^{14}C]$ D-glucose and a
hyaloplasmic fraction free from nuclei, mitochondria, and microsomes from
the brain of rats which had received 6-AN 35 or 70 mg/kg show a considerable decrease of CO_2 production from 1-$[^{14}C]$ D-glucose (compared with the
untreated control animals). This led to the conclusion that either one or
both initial NADP-dependent steps of the pentose phosphate pathway are
affected in vivo (HERKEN and LANGE 1969).

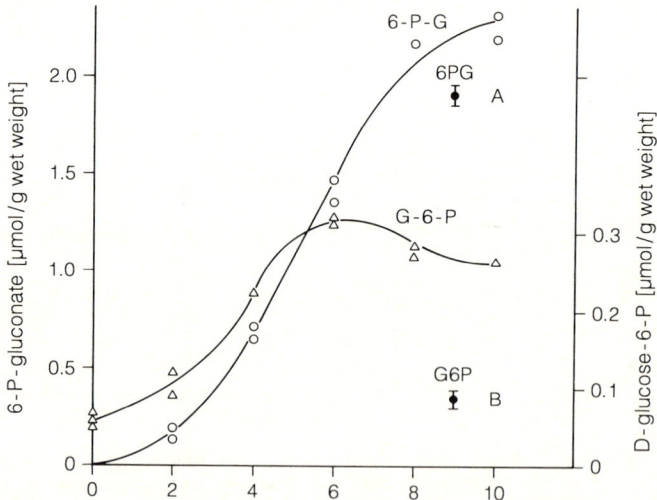

Fig. 7. 6-Phosphogluconate (*6-P-G*) and glucose-6-phosphate (*G-6-P*) levels in the
brains of rats treated with 6-aminonicotinamide (35 and 70 mg/kg). Animals killed by
decapitation. Concentrations of *A* and *B* in the brains of rats killed by deep-freezing
in Freon at −190°C (HERKEN and LANGE 1969; HERKEN et al. 1969)

This was confirmed by studies on the brains of rats which had received 6-AN by intraperitoneal injection. The animals were killed by decapitation after different periods of time following application. The brains were homogenized within 30 s in cold perchloric acid and subsequently centrifuged. The diluted clear supernatant was used for the enzymic determination of two substrates of the oxidative pentose phosphate pathway. Glucose-6-phosphate dehydrogenase and 6-phosphogluconate dehydrogenase were taken as the enzymes, and the substrate concentrations in the supernatant fraction of the brain were determined by spectrophotometrically measuring the NADPH produced. The results in Fig. 7 show a strong accumulation of 6-phosphogluconate in the brain. These concentrations showed distinct differences depending on the time which elapsed before decapitation. After 10 h, the concentration of 6-phosphogluconate reached a peak in the brain cells amounting to almost 200 times the initial value (HERKEN and LANGE 1969; HERKEN et al. 1969). The normal brain values were near the lower limit of sensitivity of this method of determination, but they correspond well with the values KAUFFMAN et al. (1969) found in the brains of mice.

The highest concentration of D-glucose-6-phosphate was found after 6 h; it diminished slightly in the further course of the experiments. These changes in the range of glucose-6-phosphate are, however, found only if the metabolism continues for about 30 s after the animals have been decapitated. If anesthetized animals in toto are frozen at $-190°C$, in which case the metabolism stops almost immediately, the value for glucose-6-phosphate is considerably lower, while under the same conditions those for 6-phosphogluconate remain unchanged (A and B on the right side of Fig. 7). These results prove that the function of the 6-phosphogluconate dehydrogenase is preferentially inhibited by 6-ANADP (HERKEN et al. 1969; HERKEN 1971). The marked differences in the accumulation of the metabolites seem to indicate that the inhibition of the 6-phosphogluconate dehydrogenase is a decisive reaction leading to disturbances of the brain metabolism. Until now, no drug was know to cause such an effective inhibition of 6-phosphogluconate dehydrogenase in the brain, blocking the oxidative pentose phosphate pathway in this manner.

The results in Table 2 show that the inhibition constant K_1 for 6-ANADP of 6-phosphogluconate dehydrogenase in the brain is almost 10 times lower than that of glucose-6-phosphate dehydrogenase. Kinetic studies of the NADP-dependent enzymes confirmed that 6-ANADP is the active inhibitor. This is also proved by the low inhibition constant, $K_1 = 1.3 \times 10^{-7} M$. The nucleotide is therefore extremely effective, because the value is within the same order of magnitude as that found for the inhibition of acetylcholinesterase by eserine and neostigmine. With isolated enzymes, 6-AN was ineffective as an inhibitor in concentrations of $10^{-3} M$ and more. These findings, in particular the inhibition of 6-phosphogluconate dehydrogenase in the brains of rats treated with 6-AN, were confirmed by several authors (KAUFFMAN and JOHNSON 1974; KRIEGLSTEIN and STOCK

Table 2. Reaction characteristics of NADP-dependent enzymes

Enzyme	Origin	K_M (NADP) (M)	K_i (6-ANADP) (M)	Conditions of reaction	Reference
Glucose 6-phosphate dehydrogenase	Yeast	$0.5-1.0 \times 10^{-5}$	$2-3 \times 10^{-5}$	pH 7.4	Coper and Neubert 1964
	Yeast	2.9×10^{-5}	2.8×10^{-5}	pH 7.6	Herken and Lange 1969
	Brain (rat)	3.4×10^{-6}	4.8×10^{-6}	pH 7.6	Herken and Lange 1969
	Embryo (rat)	1.4×10^{-5}	7.0×10^{-5}		Köhler et al. 1970
	Liver (rat)	2.0×10^{-5}	1.1×10^{-5}		Köhler et al. 1970
	Brain (guinea-pig)	2.8×10^{-6}			Lowry et al. 1957
6-Phosphogluconate dehydrogenase	Yeast	2.5×10^{-5}	4.7×10^{-7}	pH 7.6	Herken and Lange 1969
	Brain (rat)	5.3×10^{-6}	1.3×10^{-7}	pH 7.6	Herken and Lange 1969
	Embryo (rat)	1.1×10^{-5}	2.0×10^{-7}		Köhler et al. 1970
	Liver (rat)	1.3×10^{-5}	2.0×10^{-7}		Köhler et al. 1970
Malic enzyme (reaction malate → pyruvate)	Pigeon liver	1.5×10^{-5}	1.8×10^{-5}	pH 7.5	Neubert and Coper 1965

Table 3. Concentration of 6-phosphogluconate (nmol/g wet weight) in different regions of brain 9 h after i. p. application of 6-AN 35 mg/kg. After decapitation, the brains of the animals were prepared, partitioned, and frozen immediately in liquid nitrogen. The areals of 7 animals were used for each value (KELLER et al. 1972)

Region	Consisting of	Concentration (n moles/g wet weight)
Cortex cerebri	Frontal and parietal neopallium and about 20% of the hippocampus (cornu ammonis – archipallium)	1498 ± 37.5
Hippocampus	Main parts of the hippocampus (about 80%) (cornu ammonis – archipallium), corpus amygdaloideum, palaeopallium, and small temporal parts of the neopallium	1624 ± 41.8
Basal nuclear complex and thalamus	Main mass of nucleus caudatus, putamen, globus pallidus, main part of commissurae and fornix, main mass of thalamus and epithalamus, pedunculus cerebri	1934 ± 38.0
Hypothalamus		1899
Mesencephalon	Colliculus inferior and colliculus superior, tegmentum, parts of pedunculus cerebri, and small remains of pons	3040 ± 29.6
Pons		3506 ± 63.5
Medulla oblongata		4458 ± 45.2

1975; BIELICKI and KRIEGLSTEIN 1976; JANSSON et al. 1977; DESHPANDE et al. 1978; HOTHERSALL et al. 1981; GAITONDE et al. 1981, 1987, 1989). The estimation of 6-phosphogluconate content in different regions of the CNS revealed great differences in its concentration (KELLER et al. 1972; HERKEN et al. 1973) (Table 3). The concentration of 6-phosphogluconate is a marker for the extent of 6-ANADP synthesis. It is noteworthy that the highest values are found in the regions in which the greatest morphologically identifiable damage has been done to the cells (SCHNEIDER and COPER 1968). The strong inhibition of 6-phosphogluconate dehydrogenase must affect the formation of pentosephosphate. This disturbance of metabolism can therefore also influence the synthesis of nucleic acids. KAUFFMAN and JOHNSON (1974) found a decrease in ribulose-5-phosphate. The observed changes (HERKEN et al. 1969) in the metabolism of RNA in the brain cell nuclei of 6-AN-treated rats could possibly explain the cytostatic effect of the anti-metabolites (DIETRICH 1975).

6-AN has a profound influence on brain development. In exploring its biochemical mechanism of action, MORRIS et al. (1985) found an inhibition

Table 4. Kinetic constants of phosphoglucose isomerase reactions in the brains of control and 6-aminonicotinamide (6-AN)-treated rats

Kinetic constant	Forward reaction (G-6-P → F-6-P)		Reverse reaction (F-6-P → G-6-P)	
	Control	6-AN	Control	6-AN
V_{max}	$2{,}291 \pm 61$	$2{,}378 \pm 51$	$2{,}035 \pm 98$	$2{,}233 \pm 74$
K_m	0.593 ± 0.031	0.595 ± 0.022	0.095 ± 0.013	0.114 ± 0.010
K_i	0.048 ± 0.005	0.042 ± 0.002	0.042 ± 0.004	0.048 ± 0.003

Data are mean $\pm$ SEM values; V_{max} is given in nmol of the substrate utilized/min per mg of whole brain protein; K_m and K_i are given in mM; G-6-P, glucose-6-phosphate; F-6-P, fructose-6-phosphate (Gaitonde et al. 1989).

of ornithine decarboxylase. This enzyme is involved in cellular replication and differentiation of developing brain cells and is the first enzyme in the biosynthetic pathway for polyamines (Slotkin 1979). Administration of 6-AN to neonatal rats produced an inhibition of ornithine decarboxylase activity by two-thirds.

4. Inhibition of Phosphoglucose Isomerase by Phosphogluconate

In what way does 6-phosphogluconate change or limit the intermediary metabolism? It is possible that the concentration of 6-phosphogluconate in vivo leads to an inhibition of phosphoglucose isomerase (D-glucose-6-phosphate ketol isomerase, EC 5.3.1.9) as has been shown in vitro by Parr (1956) and by Salas et al. (1965). During their studies of the function of phosphoglucose isomerase and particularly of the adjustment of the equilibrilium between glucose-6-phosphate and fructose-6-phosphate, Kahana et al. (1960) found that 6-phosphogluconate is a strong inhibitor of this enzyme. The inhibitor constant K_1 amounted to $5 \times 10^{-6} M$.

Gaitonde et al. (1989) have determined the kinetic activity of phosphoglucose isomerase and the effect of 6-phosphogluconate on its activity in the forward (glucose-6-phosphate/fructose-6-phosphate) and the reverse (fructose-6-phosphate/glucose-6-phosphate) reactions in adult brain in vitro (see Table 4). No significant disorder of the enzyme kinetic were found in the brain of 6-AN intoxicated rats compared with control values. Phosphoglucose isomerase activity was intensely and competitively inhibited by 6-phosphogluconate. This confirmed the results achieved by the authors mentioned above. The highest level of accumulated 6-phosphogluconate found in the brains of animals treated with 6-AN was approximately $10^{-3} M$, which is considerably greater than the K_1 value for phosphoglucose isomerase (Herken et al. 1974). This corresponds to the data presented by Hothersall et al. (1981) and Gaitonde et al. (1983). The influence of 6-phosphogluconate on the nonoxidative part of the pentose phosphate pathway during the

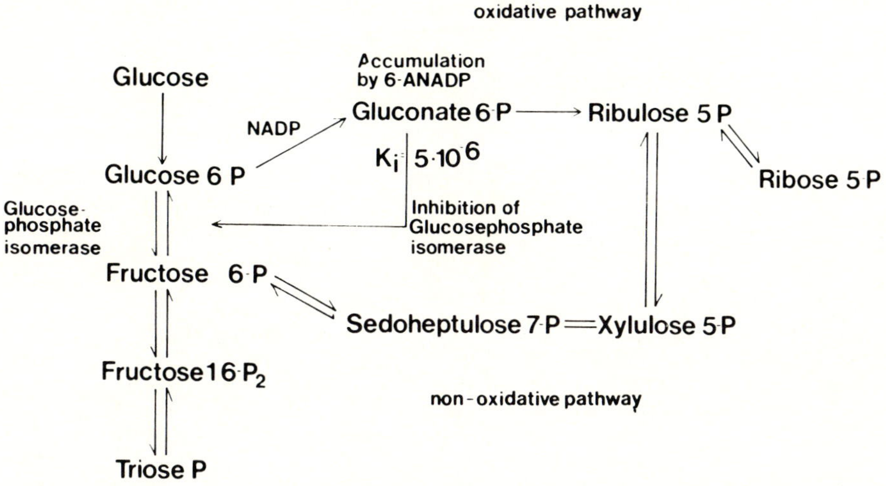

oxidative pathway

Glucose

Accumulation
by 6-ANADP

NADP

Gluconate 6 P ⟶ Ribulose 5 P

K_i $5 \cdot 10^6$

Glucose 6 P

Ribose 5 P

Glucose-
phosphate
isomerase

Inhibition of
Glucosephosphate
isomerase

Fructose 6 P

Sedoheptulose 7 P ⟹ Xylulose 5 P

Fructose 1 6 P$_2$

non-oxidative pathway

Triose P

Fig. 8. Schematic representation of the pentose phosphate pathway (adapted from HORECKER 1968). Properties of 6-phosphoglucose isomerase from rabbit brain. Inhibitor: 6-P-gluconate; direction of reaction: F6P → G6P; K_i value $5 \times 10^{-6} M$, pH 8, KAHANA et al. (1960)

synthesis of ribulose phosphate must be taken into account. The reverse reaction has also been found to be intensely inhibited by erythrose-4-phosphate and sedoheptulose-7-phosphate (GRAZI et al. 1960; HORECKER 1968; VENKATARAMAN and RACKER 1961). Figure 8 is a schematic representation of the reactions involved in the pentose phosphate pathway.

The enzymic determination of the concentration of glucose-6-phosphate and fructose-6-phosphate revealed a ratio of about 3–4:1 in the brains of normal rats. This is in agreement with the values found by KAHANA et al. (1960). In animals treated with 6-AN, this mass action ratio shifted to 16:1 in favor of glucose-6-phosphate, thus indicating that the accumulation of 6-phosphogluconate slows down the adjustment of the normal equilibrium (Table 5). HOTHERSALL et al. (1981) found a lower value in determining the ratio of glucose-6-phosphate to fructose-6-phosphate in brain tissue slices.

5. Carbohydrate Metabolism in the Brain After Inhibition of Phosphoglucose Isomerase

The enzymatic determination of the concentrations of further products of the intermediary carbohydrate metabolism in the brain offered the following results (KELLER et al. 1972).

Apart from the already mentioned deviations in the concentrations of glucose-6-phosphate and fructose-6-phosphate, the most significant change in the brains of the rats treated with 6-AN was found in the decrease in

Table 5. Average values of the substrate concentrations in µmol/g wet weight measured in the brain 6 h after i.p. injection of 6-aminonicotinamide 6-AN 35 mg/kg in rats

	6-Phosphogluconate (6-PG)	Glucose-6-phosphate (G-6-P)	Fructose-6-phosphate (F-6-P)	$\dfrac{\text{6-PG}}{\text{G-6-P}}$	$\dfrac{\text{G-6-P}}{\text{F-6-P}}$
Controls (I)	0.012 ± 0.0026	0.0607 ± 0.0014	0.0156 ± 0.0011	0.198	3.9
(n)	(5)	(5)	(5)		
6-AN (35 mg/kg rat)	1.37 ± 0.12	0.232 ± 0.014	0.0149 ± 0.0025	5.9	16.3
(n)	(5)	(5)	(5)		
P	< 0.0002	< 0.0002	< 0.54		

Enzymic determination with D-glucose-6-phosphate: NADP oxidoreducate (EC 1.1.1.49), D-glucose-6-phosphate ketol isomerase (EC 5.3.1.9), and 6-phospho-D-gluconate: NADP oxidoreductase (decarboxylating) (EC 1.1.1.44) (Herken et al. 1969).

lactate concentration. Pyruvate, too, was clearly diminished. The concentrations of citrate, malate, and 2-oxoglutarate from the Krebs cycle were also significantly decreased in the animals treated with 6-AN as compared with controls (LANGE et al. 1970). This effect is interpreted as a consequence of the slowed-down flow of substrate through glycolysis.

The assays furnished no results indicating that a further enzyme in the intermediate metabolism is inhibited by 6-ANADP as intensively as phosphogluconate dehydrogenase (LANGE et al. 1970; HOTHERSALL et al. 1981). The considerable prolongation of a pharmacologically caused anesthesia (different drugs) after pretreatment of the animals with 6-AN (REDETZKY and ALVAREZ-O'BOURKE 1962; COPER and HERKEN 1963) is not produced by an inhibition of the catabolism of the drugs. 6-AN has neither hypnotic nor narcotic properties. Also, the long-lasting decrease of body temperature (COPER et al. 1966) which has a CNS cause may be associated with the proven disturbance of brain metabolism.

The decreased flow rate through glycolysis in the CNS of 6-AN-treated rats was further discussed by KELLER et al. (1972), HOTHERSALL et al. (1981), and GAITONDE et al. (1989).

To gather more information on the glycolytic flux rate, it is necessary to measure the changes in the concentrations of the metabolites during ischemia. Tissue under ischemic conditions represents a closed biochemical system in which the glycolytic rate may be estimated by the accumulation of lactate. Under these anaerobic conditions, lactate represents the final glycolytic product. Furthermore, as the glycolytic flux increases during ischemia, limiting reactions can become more clearly visible (LOWRY et al. 1964; LOWRY and PASSONNEAU 1964). This becomes evident at the step inhibited by 6-phosphogluconate. While the glucose-6-phosphate/fructose-6-phosphate ratio under steady-state conditions after application of 6-AN amounts to 11:1, it rises during ischemia up to 40:1, indicating an activation of the initiating step of the Embden–Meyerhof pathway.

The production of lactate under the influence of 6-AN did not decrease compared with control values (KELLER et al. 1972). The concentration pattern of the phosphorylated intermediates shows that a stimulation of phosphofructokinase was the cause of this effect. The glucose-6-phosphate isomerase inhibition by 6-phosphogluconate does not cause a decrease of the glycolytic flux during ischemia; this seems to be the case under aerobic conditions only.

In GAITONDE et al. (1983), a direct method was used for evaluating the rate of glucose utilization via the hexose monophosphate shunt by measuring the accumulation of 6-phosphogluconate and the dephosphorylated metabolite gluconate. The method is based on the observation that 6-phosphogluconate dehydrogenase in rat brain is selectively inhibited by administering 6-AN (HERKEN and LANGE 1969; HERKEN et al. 1969).

A dose-dependent increase in the concentration of free glucose and glucose-6-phosphate was discovered, after accumulation of 6-phosphogluconate, in the brains of those animals that had been treated with 6-AN (Gaitonde et al. 1983). This confirmed results from Kolbe, who in 1975 discovered a considerable increase in the pool of free glucose in the brains of animals treated with 6-AN; an inhibition of hexokinase could not be proved.

The utilization rate of glucose via the hexose monophosphate shunt showed a value of 16.5 nmol of glucose utilized per min per g brain and represented approximately 2.3% of the overall utilization of glucose in the brain (730 nmol/min per g brain). By 4 h after treatment with 6-AN, there was a decrease of 16% in the overall utilization of glucose, a decrease in the labelling of amino acids after injection of [U^{14}C]glucose (Gaitonde and Evans 1982; Gaitonde et al. 1984), and a decrease (10%–13%) in the concentration of GABA and glutamic acid. The blockade of the hexose monophosphate shunt by the nucleotide containing 6-ANADP that led to a decrease of approximately 2.3% in glucose utilization in the brain cannot explain the decrease of 16%–23% in the overall utilization of glucose in the brain.

Gaitonde and his coworkers (1983) believed that there is no firm evidence to support the suggestion that the inhibition of the phosphoglucose isomerase by 6-phosphogluconate is a possible cause for the changed glucose utilization in vivo.

Hothersall et al. (1981) extensively tested the effect of 6-AN on the activities of alternative pathways of glucose metabolism in the brains of 20-day-old rats. Metabolism was evaluated by measurements of yields of $^{14}CO_2$ from glucose labelled with ^{14}C on carbons 1, 2, 3, and 4 or 6 and with uniformly labelled glucose in the various substrates. This technique (Hothersall et al. 1979) had been used to study the effect of 6-AN on the pattern of glucose utilization in the brain. At the highest dose of 6-AN used (35 mg/kg body weight), there was a significant decrease in the $^{14}CO_2$ yields via the pentose phosphate pathway, the glycolytic route, the glutamate-GABA pathway, and the tricarboxylic acid cycle. These calculations produce the following results: The most sensitive pathway was the pentose phosphate pathway, either the actual flux or the maximally stimulated activity, which was acutely inhibited even at the lowest dose of 6-AN used. The sequence of inhibition sensitivity of the other major metabolic routes appeared to be (1) the formation of lactate (decreased by 28%, resulting from the inhibition of phosphoglucose isomerase); (2) the glycolytic pathway plus pyruvate dehydrogenase, which are significantly inhibited at dose levels of 6-AN 25 mg/kg; (3) the glutamate-GABA route, which is inhibited at a dose level of 6-AN 30 mg/kg; and finally (4) the TCA cycle, which is inhibited only when the dose level is 6-AN 35 mg/kg.

The results confirm previous reports that 6-AN treatment diminishes the rate of glycolysis, as shown by decreased lactate production (Lange et al. 1970; Kauffman and Johnson 1974; Krieglstein and Stock 1975). There is good agreement among the various research groups concerning the blockade of the pentose phosphate pathway by 6-ANADP, even though the inhibition of phosphoglucose isomerase by 6-phosphogluconate in vitro has been confirmed. There are differences in the opinion about the function of phosphoglucose isomerase in 6-AN intoxication in vivo.

GAITONDE et al. (1989) studied the kinetics and were especially interested in the enzyme's behavior in vivo after accumulating 6-phosphogluconate. In contrast to the findings in vitro, the results of the studies on the inhibition of the enzyme in vivo are not clear. This applies particularly to the effects on the glycolytic flow. The following experimental approach was used by GAITONDE et al. (1989) to clarify these differences: The hexose monophosphate shunt's and the Embden–Meyerhof pathway's involvement in the metabolic flow of the C14 of glucose into amino acids in the brain can be differentiated using 6-phosphogluconate for the blockade of the pentose phosphate pathway by 6-AN (GAITONDE et al. 1987) and the Embden–Meyerhof pathway by 2-deoxyglucose-6-phosphate (WICK et al. 1957; HORTON et al. 1973). Both substances are competitive inhibitors of phosphoglucose isomerase, but 2-deoxyglucose-6-phosphate ($K_i = 5\,mM$) (HORTON et al. 1973) as compared with 6-phosphogluconate ($K_i = 0.042 - 0.048\,mM$) is considerably less effective.

In evaluating their experiments, the authors came to the following conclusions: The extent of inhibition of phosphoglucose isomerase in each compartment of the brain would depend upon the relative concentration of 2-deoxyglucose-6-phosphate to glucose-6-phosphate. The absence of a large decrease in the utilization of glucose in vivo via the Embden–Meyerhof pathway in the presence of an increased concentration of 6-phosphogluconate, in contrast to 2-deoxyglucose-6-phosphate, suggested its compartmentalization when it accumulates in the brains of rats after treatment with 6-AN. These considerations implied, therefore, the presence of distinct compartments of the two metabolic pathways, viz., the Embden–Meyerhof pathway and the hexose monophosphate shunt for glucose, in adult rat brain in vivo.

6. Inhibition of Glucose Utilization in the Glutamate γ-Aminobutyric Acid Route

The flow of glucose carbon into the neuropharmacologically active amino acids glutamate and GABA is an important route in cerebral metabolism (BERL and CLARKE 1979).

The decrease in efficiency of the glucose-glutamate-GABA route in the brain after application of 6-AN was confirmed by HOTHERSALL et al. (1979, 1981), BIELICKI and KRIEGLSTEIN (1976), and GAITONDE et al. (1981, 1987).

The metabolism of glucose in the rat brain, via both the Embden–Meyerhof pathway and the hexose monophosphate shunt, was investigated by injection of [U-^{14}C]glucose or [2-^{14}C]glucose and via the hexose monophosphate shunt alone by injecting [3,4-^{14}C]glucose. The results supplied direct evidence that the two putative transmitters glutamate and GABA in the brain were formed by metabolism of glucose via the hexose mono-phosphate shunt as well as via the Embden–Meyerhof pathway (Gaitonde et al. 1987).

The formation of radioactively labelled amino acids, particularly glutamate and GABA, after injection of [3,4-^{14}C]glucose was independent of the activity of phosphoglucose isomerase. The observed decrease in the radioactive yield of glutamate and GABA in 6-AN-treated rats, as com-pared with control animals, may be based on the inhibition of glucose utilization via the hexose monophosphate shunt. After injection of [U-^{14}C]- or [2-^{14}C]glucose, greater differences were found between their radioactive yields from controls and from 6-AN-treated animals. This could be due to the fact that not only the hexose monophosphate shunt but also the function of phosphoglucose isomerase are blocked by the increased concentration of 6-phosphogluconate, which can lead to a decrease in the glycolytic flux in the Embden–Meyerhof pathway. (Concerning the problem of the labelling pattern of carbon atoms by glucose metabolism via the Embden–Meyerhof pathway as compared with the hexose monophosphate shunt, see Gaitonde et al. 1987.)

No further enzymes (other than in the inhibition of phosphogluconate dehydrogenase by 6-AN) were found that could be responsible for the inhi-bition of the utilization of ^{14}C-labelled glucose into amino acids. The same inhibition of the pentose phosphate pathway also occurs in the synaptosomes fraction, prepared from the cerebral hemispheres by Gray and Whittaker's method (1962). These were subsequently incubated with variously labelled glucose, as described by Baquer et al. (1975) in studies of metabolic path-ways in developing brain. The experiments demonstrated that both the basal and fully activated pentose phoshate pathway are diminished in the synaptosomes prepared from the cerebral hemispheres of 20-day-old rats treated with a single dose (35 mg/kg body weight) of 6-AN (Hothersall et al. 1981).

The blockade of the pentose phosphate pathway by the antimetabolite 6-AN or by the nucleotide 6-ANADP and the resulting decrease of the formation of certain neurotransmitters can cause the multiple neurological coordination disturbances, as suggested by Gaitonde et al. (1987).

D. Lesions of Neuroglia Cells

Some aspects in these studies of distinct compartmentation of the pentose phosphate pathway and the Emden–Meyerhof pathway (glycolysis) in an organ that consists of neuroglia and extremely differentiated nerve cells were not considered. No notice was taken of the fact that large differences in the accumulation of 6-phosphogluconate, the inhibitor of phosphoglucose isomerase, exist in the various regions of the brain (Lange et al. 1970;

KELLER et al. 1972; HERKEN et al. 1973). The differences must have a very specific, strong effect on the glycolytic flux in the cells. This confirms the special vulnerability of glia cells in experiments with clonal cell lines in culture (KOLBE et al. 1976). The assumption that an uneven distribution of the two metabolic processes mentioned in the brain of rats treated with 6-AN could not be confirmed, not even in previous comparative experiments using 2-deoxyglucose-6-phosphate (KELLER et al. 1972).

6-AN administration leads to a considerable accumulation of 6-phosphogluconate, which is 3 times higher in C-6 glial cells than in C-1300 neuroblastoma cells. Dephosphorylation of the accumulated 6-phosphogluconate causes a rise of intracellular gluconate, which can be released from cells into the culture medium. The gluconate efflux, which in neuroblastoma cells is about twice that observed in glioma cells, points to the most conspicuous changes in the contents of hexose monophosphate pathway metabolites. Conversely, in neuroblastoma cells with much lower 6-phosphogluconate values, gluconate amounts to 8 times the value of 6-phosphogluconate and 4 times the gluconate content observed in glioma cells. This could be explained by more effective dephosphorylation of 6-phosphogluconate in the neuroblastoma cells (KOLBE 1976).

The reduced glycolytic flux and the diminished ATP content of the glioma cells as well as their lower dephosphorylating capacity indicate differences between both cell types as far as the inhibition of the hexose monophosphate pathway and its influence on glycolysis are concerned. With phase contrast microscopy only the glioma cells showed edematous alterations. The dephosphorylation of 6-phosphogluconate to gluconate is a detoxification reaction. Gluconate has no inhibitory effect on the phosphoglucose isomerase. Animals treated with 6-AN produce considerable amounts of gluconate, which is eliminated in the urine (HERKEN et al. 1975).

Electron microscopic studies showed fundamental differences in the sensitivity of nerve and glial cells to the antimetabolite in the brains of animals. The preferential lesion of the neuroglial cells especially in reactive astroglia cells (POLITIS 1989) was particularly conspicuous. SCHNEIDER and CERVOS-NAVARRO (1974) called it an acute gliopathy. The nerve cells were affected only secondarily if at all.

The considerable ballooning of the endoplasmic reticulum and a conspicuous enlargement of the perinuclear cisterns gave the impression of an intracellular edema in animals treated with 6-AN (BAETHMANN et al. 1968; SCHNEIDER and CERVOS-NAVARRO 1974). Changes such as these are also found in the glial cells and in the interneurons of the spinal cord, in which very high 6-phosphogluconate values (4550 ± 125 nmol/g wet weight) were measured (HERKEN et al. 1973; MEYER-ESTORFF et al. 1973).

Electron microscopic studies have shown that about 90% of the capillary surface is covered by the end feet of glial cells (WOLFF 1965). The astroglial projections as well as the laminae of these cells are closely associated with

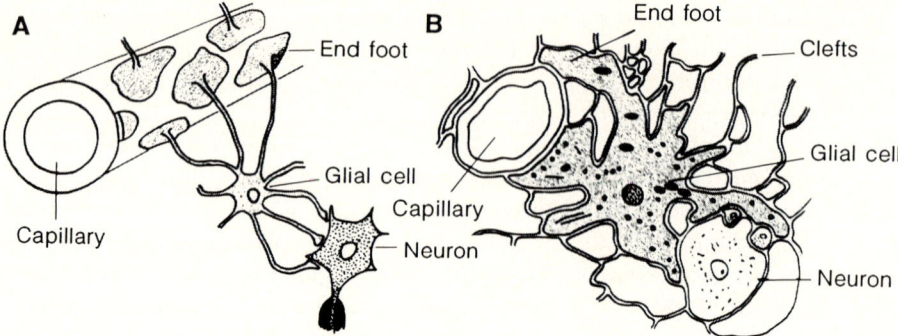

Fig. 9. A Schematic diagram of neuron-glia-capillary relationship in the vertebrate brain as seen under the light microscope. Capillary walls are closely invested by 'end feet' of astrocytes, which also make contact with neurons. Only one astrocyte is shown bridging the space between a neuron and a capillary. This type of intimate relationship served as the basis for Golgi's concept of a glia being a channel for the distribution of nutritive materials between the blood and neurons. **B** Sketch of neuron-glia-capillary relationship as seen under the electron microscope. Again, only one astrocyte (*shaded*) is shown, interposed between the endothelium of the capillary and the neuron. All the cells, axons, dendrites, and astrocytes are tightly packed with narrow clefts (their dimensions greatly exaggerated in this scheme) of about 15 nm between them (Kuffler and Nicholls 1966)

the neurons, so that a functional unit can be established (Fig. 9). The macroglia is involved in excitatory processes and in the nutrition of nerve cells (Kuffler and Nichols 1966).

The topography can also explain the preferential vulnerability of astrocytes. From the blood vessel the antimetabolite can diffuse into the glial cell, where it accumulates. The biosynthesis of the abnormal nucleotide occurs there also. Here, the astroglia seems to fulfill a barrier function which protects the nerve cells from the direct action of toxic compounds. It seems that 6-AN diffuses into the oligodendrocytes and the neurons from the extracellular space. The nucleotides containing 6-AN influence ion transport processes across biological membranes (Herken et al. 1966; Herken 1968c). 6-AN inhibits the insulin-mediated diffusion of glucose in fat cells (von Bruchhausen and Herken 1966). An inhibition of protein synthesis is involved in cell destruction by 6-AN (Zeitz et al. 1978).

E. 6-Aminonicotinamide Induces a Parkinson-Like Syndrome

A pathologically interesting disturbance concerns the nigrostriatal system with its neurotransmitter dopamine. 6-AN or 6-ANADP produces parkinsonism-like symptoms in rats (Kehr et al. 1978; Loos et al. 1979). The degree of neuropathy is determined by the destruction of nerve cells in the compact layers of the substantia nigra. Parkinson's disease is a striatal dopamine deficiency syndrome revealed by increased muscular rigidity and further characteristic disturbances of motion (HORNYKIEWICZ 1962, 1989; HASSLER 1972, 1984; BERNHEIMER et al. 1973; JELLINGER 1989). The breakdown of the dopaminergic pathway had disadvantageous consequences for other neurotransmitter systems.

In studies on the molecular basis of Parkinson's syndrome, the limitations of the function of tyrosine hydroxylase play a very important role. This enzyme catalyzes the first rate-limiting step in the biosynthesis to dopamine and norepinephrine (UDENFRIEND 1966; BLASCHKO 1959, 1972; LEVINE 1981; NAGATSU 1983; MASSERANO et al. 1989) and needs the cofactor L-erythro-tetrahydrobiopterin, which was discovered by KAUFMAN (1967, 1975; NELSON and KAUFMAN 1987). The cofactor is also required for the reaction of phenylalanine and of tryptophan hydroxylase (BRENNEMAN and KAUFMAN 1964; MANDELL 1978).

Administration of 10 mg/kg 6-AN in rats led to a significant decrease of 37% in dopamine content in the corpus striatum. As compared with controls, 6-AN prevented the disappearance of the remaining dopamine after inhibition of catecholamine synthesis by α-methyl-p-tyrosinemethylate HCl applied 90 and 180 min before the animals were sacrificed. 6-AN impedes dopamine utilization (Fig. 10) (KEHR et al. 1978). 6-AN drastically diminishes the concentration of epinephrine and norepinephrine in the adrenal glands (SCHACHT et al. 1966).

The determination of the concentration of dopa in the animals treated with 10 mg/kg 6-AN 7 days after its administration over a period of 30 min after decarboxylase inhibition by 3-hydroxybenzoylhydrazine showed a reduced catecholamine synthesis (of 25%) in the dopamine-rich part of the limbic system ($p < 0.05$) and a decrease in the corpus striatum. These data also clearly reveal the interference with the function of the dopamine synthesizing pathway (CARLSSON et al. 1972).

Particularly remarkable was the occurrence of the symptom of a persisting muscular rigidity, which was registered electromyographically as permanent activity in the gastrocnemius muscle. Seven days after administration of 6-AN, the pathological effect was fully developed. The decrease in dopamine in the brain region normally rich in dopamine occurred parallel to the development of the pathological muscular rigidity. The increased activity registered on the electromyogram remained constant for days and even weeks and months and was the most important pathological finding (KEHR et al. 1978). 6-AN's effect does not selectively concern a defined cell type, but the damage of the dopaminergic cells caused the most prominent sign of the neurotoxic action.

The pathologically increased action potentials in the rigid musculature induced by 6-AN in rats could be temporarily abolished by administration of L-dopa in combination with benzerazide or by administration of the dopamine agonists α-bromocriptine and lisuride. A test system with quantitative evaluation of the electromyographic findings was developed (Loos et

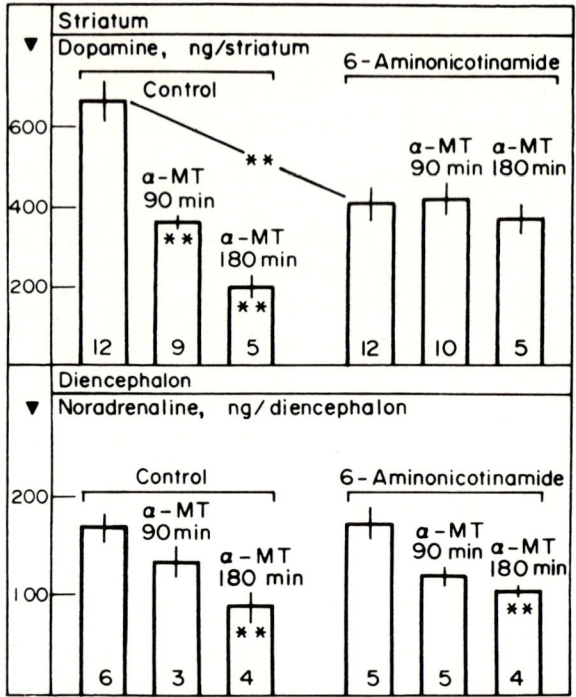

Fig. 10. Effect of 6-aminonicotinamide (6-AN) on catecholamine utilization in corpus striatum and diencephalon of the rat. 6-AN (10 mg/kg) was injected i.p. 7 days prior to α-methyl-p-tyrosine methylester HCl (*a-MT*), 250 mg/kg i.p., and the animals were killed after another 90 or 180 min. Shown are the means ±SEM. The group size is indicated within the columns. Statistics: ANOVA followed by Scheffé test. **P < 0.01 (Kehr et al. 1978)

al. 1977). In another series, muscular rigidity was produced by the i.p. administration of a high dose (5 mg/kg) of reserpine.

Both α-bromocriptine (Loew et al. 1976; Vigouret et al. 1977) and lisuride (Horowski and Wachtel 1976; Kehr 1977) are dopamine agonists with a main site of action in the nigrostriatal system. The values obtained with the reserpine model after application of α-bromocriptine correlate well with those reported by Vigouret et al. (1977). The comparative study of both substances with the 6-AN as well as the reserpine model revealed that lisuride was far more effective than α-bromocriptine in rats (Table 6). Both dopamine agonists also normalize the pathologically increased muscular rigidity in patients with Parkinson's disease. The lack of a depressant effect of the serotonin antagonist, methysergide, another lysergic acid derivative, on the electromyogram in both models excluded the possibility that the effect of lisuride was due to this antiserotoninergic action (Votava and Lamplova 1961).

Table 6. Potencies of lisuride and α-bromocriptine in reducing the pathologically increased electromyo-graphic activity caused by 6-aminonicotinamide (6-AN) or reserpine

	ED_{50}	
	Reserpine	6-AN
Lisuride	2.65 µg/kg	23.07µg/kg
α-Bromocriptine	220.7 µg/kg	1793 µg/kg
Lisuride	5.84×10^{-9} mol/kg	5.09×10^{-8} mol/kg
α-Bromocriptine	2.94×10^{-7} mol/kg	2.39×10^{-6} mol/kg

ED_{50} values were determined from the different regressions for lisuride and α-bromocriptine and defined as a percentage of the maximum inhibition of initial activity (Loos et al. 1979).

In order to compensate for the muscular rigidity, higher doses of lisuride and α-bromocriptine were required in 6-AN-induced neuropathy than in the reserpine model. The reason for the quantitative differences in the efficient dose of dopamine agonists may be due to the different gravity of the disorders induced by 6-AN or reserpine.

I. Decrease of Tetrahydrobiopterin Content and Dopa Production in PC$_{12}$ Clonal Cell Line Induced by 6-Aminonicotinamide

Various tests have confirmed that nerve cells that synthesize dopamine and norepinephrine are also able to synthesize the cofactor tetrahydrobiopterin (of tyrosine hydroxylase) de novo from guanosine triphosphate as precursor via a long chain of intermediates (KAPATOR et al. 1982; for review, see CURTIUS et al. 1983; NICHOL et al. 1985). Little is as yet known about the susceptibility of this system to lesions or diseases possibly leading to a Parkinson's syndrome by disturbance of dopa and dopamine synthesis. NAGATSU (1983) reported that there was a considerable decrease in the concentration of the biopterin cofactor of tyrosine hydroxylase in the nucleus caudatus of patients with Parkinson's disease post-mortem. LEVINE et al. (1979) found a decrease in cofactor activity in the cerebrospinal fluid of patients with Parkinson's disease. LEWITT et al. (1984) confirmed these findings. Different NADP-dependent reductases are involved in the synthesis of biopterin. Studies on the enzymatic processes in the pentose phosphate pathway gave rise to the assumption that the site of action of the antimetabolite or the nucleotide containing 6-AN might be these reductases.

Studies dealing with the evaluation of neurotropic drug action on tyrosine hydroylase activity and dopamine metabolism in clonal cell lines (HERKEN 1983; BRÄUTIGAM et al. 1985) showed that the pheochromocytoma clone (PC-12) is particularly suitable for the elucidation of this question. This clone was obtained by GREENE and TISCHLER (1976) by cloning cells from a

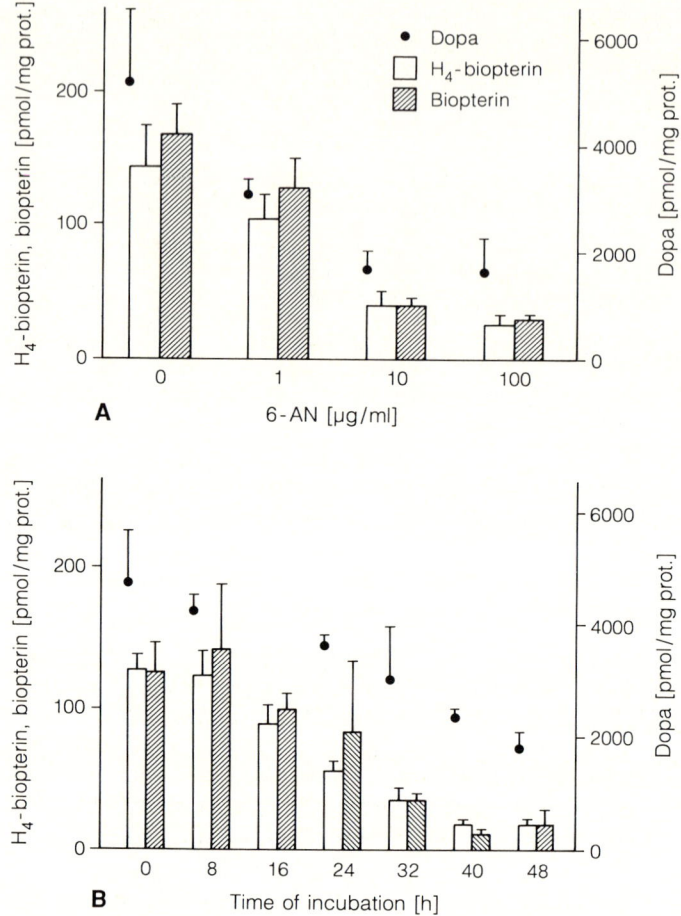

Fig. 11A,B. 6-Aminonicotinamide (6-AN)-dependent intracellular content of tetrahydrobiopterin and total biopterin (*bars*) and cellular dopa production within 1 h (*solid circles*), the latter measured in the medium. **A** Dose response after 48 h of incubation with 6-AN (*n* = 3–12; mean ±SD) (Jung and Herken 1989). **B** Time response with 6-AN 100 µg/ml medium (*n* = 3–4; mean ±SD). 6-AN was added to the culture medium. Cells were incubated under growth conditions as described by (Bräutigam et al. 1984; Jung and Herken 1989). The medium was changed 1 h before the end of the experiment. *m*-Hydrobenzoylhydrazine (final concentration 5 × 10⁻⁴ *M*) was added to the fresh medium for final incubation of 1 h. Assay procedure for simultaneous determination of the metabolites from Fukushima and Nixon (1980) and Bräutigam et al. (1982, 1984, 1985)

rat pheochromocytoma. The cell line possesses the capability of identical reproduction through numerous subcultivations. These cells contain all the enzymes for the synthesis and degradation of catecholamines as well as vesicles for their storage. Considerable quantities of tetrahydrobiopterin were shown to be present which sufficed as cofactor for tyrosine hydroxylase in dopa production of the cells (BRÄUTIGAM et al. 1984). The cells are able to synthesize biopterin de novo from GTP as precursor (BUFF and DAIRMAN 1975).

Determination of tetrahydrobiopterin and total biopterin levels confirmed that the biopterin fraction consisted almost entirely of the cofactor tetrahydrobiopterin (BUFF and DAIRMAN 1975; FUKUSHIMA and NIXON 1980; BRÄUTIGAM et al. 1982, 1984).

PC-12 cells possess the NAD(P) glycohydrolase transferase that catalyzes the exchange of nicotinamide for 6-AN to 6-ANADP. This proves that the accumulation of 6-phosphogluconate and gluconate in PC-12 cells depends on the concentration of 6-AN applied (JUNG and HERKEN 1989). The amounts of 6-phosphogluconate that accumulated in 6-AN treated PC_{12} cells at 24 h were significantly increased by treatment of the cells with nerve growth factor (DAVIS and KAUFMAN 1987). The increase in 6-phosphogluconate content is limited by enzymatic dephosphorylation to gluconate, which is released into the culture medium (KOLBE et al. 1976).

The synthesis of 6-ANADP is responsible for the initiation of further effects concerning the behavior of biopterin synthesis and dopa production in the PC-12 cells. After incubation with 6-AN added *to the* culture medium in different doses, a dose-dependent decrease in total biopterin and tetrahydrobiopterin levels was found in the cells. According to the time-dependent accumulation of 6-phosphogluconate, the effect of the anti-metabolite occurs after a latency period of several hours. After 32 h of incubation with 6-AN 100 µg/ml medium, the biopterin content in the cells is reduced to less than one-third compared with controls (JUNG and HERKEN 1989). The decrease of intracellular tetrahydrobiopterin and total biopterin resulted in reduced dopa production. The analogue of NADP containing 6-AN leads to a long-lasting impairment of biopterin synthesis. The decrease in dopa production is evidently a consequence of this process (Fig. 11).

II. De Novo Synthesis of Biopterin and the Salvage Pathway of Synthesis

The synthesis of biopterin proceeds over numerous intermediate steps in which NADP-dependent reductases are involved. Figure 12 shows a schematic representation of the synthesis. The first and rate-limiting enzyme in the de novo synthesis of tetrahydrobiopterin from guanosine triphosphate is the GTP cyclohydrolase (EC 3.5.4.16) which catalyzes the opening of the imidazole ring followed by the formation of a pyrazine ring. The first pterin derivative is D-erythro-7,8-dihydroneopterintriphosphate. Synthesis up to 6-

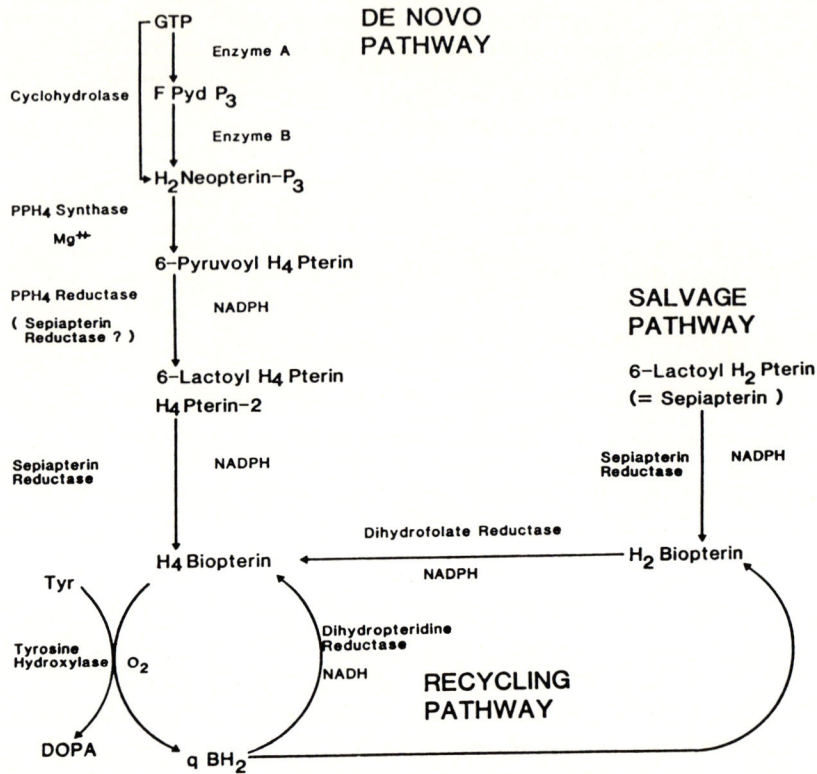

Fig. 12. Pathway of tetrahydrobiopterin biosynthesis. Working model according to
Kaufman (1967), Gal (1982), Kapatos et al. (1982), Nichol et al. (1986), Takikawa
et al. (1986)

pyruvoyltetrahydropterin does not require NADP (Gal et al. 1978; Gal
1982; Kapatos et al. 1982; Takikawa et al. 1986).

The second phase does require NADPH for the reduction of the two
keto groups in the side-chain of 6-pyruvoyltetrahydropterin. In any case,
sepiapterin reductase (EC 1.1.1.153) catalyzes the last step in the de novo
synthesis, the reduction of the second keto group (Katoh and Sueoka
1984). In the so-called salvage pathway, the same enzyme catalyzes the
reduction of lactoyldihydropterin to dihydrobiopterin, which in turn is
reduced to tetrahydrobiopterin by the NADPH-dependent dihydrofolate
reductase (Milstien and Kaufmann 1983, 1985; Nichol et al. 1983; Smith
and Nichol 1984). The decreased biopterin cofactor synthesis was compensated for by the addition of the precursor sepiapterin, indicating that the
NAD(P)-dependent reductases in biopterin synthesis were not inhibited by
the antimetabolite nucleotide 6-ANADP, as suspected by Jansson et al.
(1977), who confirmed the blockade of the hexose monophosphate shunt by
6-AN. Even after this blockage there is enough NADPH available for the

Table 7. Effect of sepiapterin on intracellular content of tetrahydrobiopterin and total biopterin (pmoles/mg protein) and on cellular dopa production ($n = 4$; means ± SD)

	Tetrahydrobiopterin	Total biopterin	dopa
No 6-AN	133.8 ± 20.9	149.7 ± 13.2	2976 ± 25*
+ sepiapterin	462.1 ± 62.3**	614.8 ± 40.8**	5698 ± 968*
6-AN 1 µg/ml	74.1 ± 11.3	99.1 ± 7.7	1869 ± 303
+ sepiapterin	462.4 ± 10.0	582.1 ± 11.9	3460 ± 119
6-AN 10 µg/ml	30.1 ± 2.1	34.4 ± 2.6	1206 ± 50*
+ sepiapterin	513.8 ± 32.2	604.2 ± 27.2	2490 ± 219*
6-AN 100 µg/ml	19.2 ± 1.3	16.6 ± 1.3	997 ± 71
+ sepiapterin	398.1 ± 27.6**	546.8 ± 25.9**	2194 ± 117

Cell monolayers were incubated with 6-aminonicotinamide (6-AN) for 48 h. The medium was changed after 47 h. Sepiapterin (final concentration $2 \times 10^{-5} M$) was added to the fresh medium containing m-hydroxybenzylhydrazine (final concentration $5 \times 10^{-4} M$) and 6-AN concentration as before. Final incubation with sepiapterin for 1 h. Biopterin values represent intracellular concentration, dopa was measured in the medium. Groups marked with the same figure were compared according to Student's t-test: $*P < 0.05$ = significant, $**P \geq 0.05$ = not significant. JUNG and HERKEN (1989).

enzymic reduction of sepiapterin to tetrahydrobiopterin (Table 7). Utilization of sepiapterin for tetrahydrobiopterin synthesis in the salvage pathway requires two NADPH-dependent reductases, sepiapterin reductase and dihydrofolate reductase (see Fig. 12). Therefore, the simultaneous decrease of tetrahydrobiopterin and total biopterin levels to less than one-third of controls, depending on the concentration of 6-AN, can only be the result of a direct blockage in the first phase of cofactor synthesis. The latency period of this reaction, correlating with the accumulation of 6-phosphogluconate, implies that the effect is caused by the inhibitor, 6-ANADP. Since there is no indication of a blockage of 6-pyruvoyltetrahydropterin synthase, the formation of GTP to D-erythro-7,8-dihydroneopterintriphosphate seems to be the step that is inhibited by the antimetabolite (GAL et al. 1978; GAL and WHITACRE 1981; GÁL 1982). The subunit of the cyclohydrolase described as enzyme b seems to be particularly sensitive to the action of the neurotoxic antimetabolite.

III. Biopterin Recycling Pathway and Dopa Production

Tetrahydrobiopterin as cofactor of hydroxylases provides electrons to reduce molecular oxygen and is, in turn, oxidized to the quinoid dihydrobiopterin. The activation of molecular oxygen results in the formation of 4-a-hydroperoxytetrahydropterin, which participates in the hydroxylation of the substrate tyrosine (KAUFMAN 1975; NICHOL et al. 1985; DIX et al. 1987). The catalytic process requires the participation of a recycling system for the regeneration of tetrahydrobiopterin (GAL and WHITACRE 1981). Experiments with sepiapterin as a precursor of tetrahydrobiopterin synthesis added to 6-AN-treated cells showed that the compensation of the cofactor content was not followed by a compensation of dopa production. This might indicate that the recycling system is inhibited by 6-AN. As shown in Fig. 12, the reduction of quinoid dihydrobiopterin is catalyzed by dihydrobiopterin reductase, which is NADH-dependent. Comparison of the dopa values of treated and untreated cells revealed that the increase of dopa production, particularly as found in 6-AN-treated cells after application of high concentrations of NADH, may be explained by a competitive antagonism versus the presumed inhibitor 6-ANAD (Jung and Herken 1989). The results explain the reason for the decrease in dopa content and its utilization in the corpus striatum of rat brain, resulting in a Parkinson-like dopamine deficiency syndrome induced by the antimetabolite 6-AN. Restriction of the synthesis of the biopterin cofactor might be a pathogenetically important disorder in the initial phase of Parkinson's disease (Herken 1990).

References

Alivisatos SGA (1959) Mechanism of action of diphosphopyridine nucleotidases. Nature 183:1034–1037

Alivisatos SGA, Woolley DW (1955) Synthesis of 5-amino-4-carboxamidoimidazole-AD. J Am Chem Soc 77:1065–1066

Alivisatos SGA, Ungar F, Lukacs L, La Mantia L (1960) Imidazolytic processes. J Biol Chem 235:1742–1750

Anderson WA, Flumerfelt BA (1984) Sensitivity of rat inferior olivary neurons to 3-acetylpyridine. Brain Res 314:285–291

Anderson BM, Kaplan NO (1959) Enzymatic studies with analogues of diphosphopyridine nucleotide. J Biol Chem 234:1226–1232

Anderson BM, Ciotti CI, Kaplan NO (1959) Chemical properties of 3-substituted pyridine analogues of diphosphopyridine nucleotides. J Biol Chem 234: 1219–1225

Annison EF, Hill KJ, Lindsay DB, Peters RA (1960) Fluoroacetate poisoning in sheep. J Comp Pathol Ther 70:145–155

Bachelard HS (1971) Specificity and kinetic properties of monosaccharide uptake into guinea pig cerebral cortex in vitro. J Neurochem 18:213–222

Baethmann H, Reulen HJ, Brendel W (1968) Die Wirkung des Antimetaboliten 6-Aminonikotinamid (6-ANA) auf Wasser- und Elektrolytgehalt des Rattenhirns und ihre Hemmung durch Nikotinsäure. Z Ges Exp Med 146:226–240

Balaban CD (1985) Central neurotoxic effects of intraperitoneally administered 3-acetylpyridine, harmaline and niacinamide in Sprague-Dawley and Long-Evans rats: a critical review of central 3-acetylpyridine neurotoxicity. Brain Res Rev 9:21–42

Balazs R, Patel AJ, Richter D (1973) Metabolic compartments in the brain: their properties and relation to morphological structures. In: Balazs R, Cremer JE (eds) Metabolic compartmentation in the brain. MacMillan, London, pp 167–183

Baquer NZ, McLean P, Greenbaum AL (1975) System relationships and the control of metabolic pathways in developing brain. In: Hommes FA, van den Berg CJ (eds) Normal and pathological development of energy metabolism. Academic, New York, pp 109–132

Bartlett GR, Barron ESG (1947) The effect of fluoroacetate on enzymes and on tissue metabolism. Its use for the study of the oxidative pathway of pyruvate catabolism. J Biol Chem 170:67–82

Batini C, Billard JM, Daniel H (1985) Long term modification of cerebellar inhibition after inferior olive degeneration. Exp Brain Res 59:404–409

Beher WT, Anthony WL (1953) Studies of antimetabolites: III. The metabolism of 3-acetylpyridine in vivo. J Biol Chem 203:895–898

Beher WT, Baker GD, Madoff M (1959) Studies of antimetabolites: V. Metabolism in vivo of 3-acetylpyridine. J Biol Chem 234:2388–2390

Berl S, Clarke DD (1979) Compartmentation of amino acid metabolism. In: Lajtha A (ed) Handbook of neurochemistry, vol 2. Plenum, New York, pp 447–472

Bernard JF, Buisseret-Delmas C, Laplante S (1984) Inferior olivary neurons: 3-acetylpyridine effects on glucose consumption, axonal transport, electrical activity and harmaline-induced tremor. Brain Res 322:382–387

Bernheimer H, Birkmayer W, Hornykiewicz O, Jellinger K, Seitelberger F (1973) Brain dopamine and the syndromes of Parkinson and Huntington. Clinical, morphological and neurochemical correlations. J Neurol Sci 20:415–455

Bielicki L, Krieglstein J (1976) Decreased GABA and glutamate concentration in rat brain after treatment with 6-aminonicotinamide. Naunyn Schmiedebergs Arch Pharmacol 294:157–160

Billard JM, Batini C, Daniel H (1988) The red nucleus activity in rats deprived of the inferior olivary complex. Behav Brain Res 28:127–130

Blaschko H (1959) The development of current concepts of catecholamine formation. Pharmacol Rev 11:307–316

Blaschko H (1972) Introduction. Catecholamines 1922–1971. In: Blaschko H, Muscholl E (eds) Catecholamines. Springer, Berlin Heidelberg New York, pp 1–15 (Handbook of experimental Pharmacology, vol 33)

Bosakowski T, Levin AA (1986) Serum citrate as a peripheral indicator of fluoroacetate and fluorocitrate toxicity in rats and dogs. Toxicol Appl Pharmacol 85:428–436

Brady RO (1955) Fluoroacetyl coenzyme A. J Biol Chem 217:213–224

Bräutigam M, Dreesen R, Herken H (1982) Determination of reduced biopterins by high pressure liquid chromatography and electrochemical detection. Hoppe Seylers Z Physiol Chem 363:341–343

Bräutigam M, Dreesen R, Herken H (1984) Tetrahydrobiopterin and total biopterin content of neuroblastoma (N1E-115, N2A) and pheochromocytoma (PC-12) clones and the dependence of catecholamine synthesis on tetrahydrobiopterin concentration. J Neurochem 42:390–396

Bräutigam M, Kittner B, Herken H (1985) Evaluation of neurotropic drug actions on tyrosine hydroxylase activity and dopamine metabolism in clonal cell lines. Arzneimittelforschung 35:277–284

Brennemann AR, Kaufman S (1964) The role of tetrahydropteridin in the enzymatic conversion of tyrosine to 3,4-hydroxyphenylalanine. Biochem Biophys Res Commun 17:177–183

Bruchhausenvon F, Herken H (1966) Wirkung des 6-Aminonicotin-säureamids auf die insulinabhängige Glucose-Aufnahme in das epididymale Fettgewebe. Naunyn Schmiedebergs Arch Pharmakol Exp Pathol 254:388–400

Brunnemann A, Coper H (1964a) Die Aktivität NAD-und NADP-abhängiger Enzyme in verschiedenen Teilen des Rattengehirns. Naunyn Schmiedebergs Arch Exp Pathol Pharmakol 246:493–503

Brunnemann A, Coper H (1964b) Untersuchungen zur Wirkungsweise des Pyridin-3-methylcarbinol. Naunyn Schmiedebergs Arch Exp Pathol Pharmakol 247:315

Brunnemann A, Coper H (1964c) Vergleichende Untersuchungen über die Biosynthese der NAD(P)-Analogen des 3-Acetylpyridins, 4-Acetylpyridins, 6-Aminonikotinamids und Isonikotinsäurehydrazids. Naunyn Schmiedebergs Arch Pharmakol 248:514–520

Brunnemann A, Coper H (1965a) Nucleosidaseaktivität im Gehirn verschiedener Tierarten. Naunyn Schmiedebergs Arch Exp Pathol Pharmakol 250:242

Brunnemann A, Coper H (1965b) Hydrolyse von NAD(P) und Biosynthese von 3-APAD(P) durch Hirnmikrosomen verschiedener Tierarten. Naunyn Schmiedebergs Arch Exp Pathol Pharmakol 250:469–478

Brunnemann A, Coper H, Herken H (1962) Isolierung von 3-Acetylpyridin-adenindinucleotid (3-APAD) aus dem Gehirn 3-Acetylpyridin-vergifteter Ratten. Naunyn Schmiedebergs Arch Exp Pathol Pharmakol 244:223–236

Brunnemann A, Coper H, Herken H (1963) Biosynthese von 3-Acetylpyridin-adenindinucleotidphosphat (3-APADP) aus Nikotinamid-adenindinucleotid-phosphat. Naunyn Schmiedebergs Arch Exp Pathol Pharmakol 245:541–550

Brunnemann A, Coper H, Neubert D (1964) Biosynthese und Wirkung des 6-Aminonikotinamiddinucleotids (6-ANAD). Naunyn Schmiedebergs Arch Exp Pathol Pharmakol 46:437–451

Buff K, Dairman W (1975) Biosynthesis of biopterin of two clones of mouse neuroblastoma. Mol Pharmacol 11:87–93

Buffa P, Peters RA (1949) The in vivo formation of citrate induced by fluoroacetate and its significance. J Physiol (Lond) 110:488–500

Buffa P, Guarriera-Bobyleva V, Costa-Tiozzo R (1973) Metabolic effects of fluoroacetate poisoning in animals. Fluoride 6:224–227

Buffa PGF, Peters RA, Wakelin RW (1951) Biochemistry of fluoracetate poisoning. Isolation of an active tricarboxylic acid fraction from poisoned kidney homogenates. Biochem J 48:467–477

Caplan AI (1971) The teratogenic action of the nicotinamide analogues 3-acetylpyridine and 6-aminonicotinamide on developing chick embryos. J Exp Zool 178:351–357

Carlsson A, Davis JN, Kehr W, Lindqvist M, Atack CV (1972) Simultaneous measurement of tyrosine and tryptophan hydroxylase activities in brain in vivo using an inhibitor of the aromatic amino acid decarboxylase. Naunyn Schmiedebergs Arch Pharmacol 275:153–168

Chamberlain JG (1972) 6-aminonicotinamide (6-AN)-induced abnormalities of the developing ependyma and choroid plexus as seen with the scanning electron microscope. Teratology 6:281–288

Chenoweth MB (1949) Monofluoroacetic acid and related compounds. Pharmacol Rev 1:383–424

Chenoweth MB, Gilman A (1946) Studies on the pharmacology of fluoroacetate I. Species response to fluoroacetate. J Pharmacol Exp Ther 87:90–103

Chenoweth MB, St John EF (1947) Studies on the pharmacology of fluoroacetate III. Effects on the central nervous system of dogs and rabbits. J Pharmacol Exp Ther 89:76–82

Chenoweth MB, Scott EB, Sati SL (1949) Prevention of fluoroacetate poisoning by acetate donor. Fed Proc 8:280

Chenoweth MB, Kandel A, Johnson LB, Benett DR (1951) Factors influencing fluoracetate poisoning. Practical treatment with glycerol monoacetate. J Pharmacol Exp Ther 102:31–54

Christ W (1975) Zur Biochemie und Pharmakologie von Nikotinamid und NAD(P)-Analogen. Untersuchungen über Struktur-Wirkungsbeziehungen. Habilitations-schrift. Freie Universität Berlin

Ciotti MM, Humphreys STR, Venditti JM, Kaplan NO, Goldin A (1960) The antileukemic action of two thiadiazole derivatives. Cancer Res 20:1195

Clarke DD, Nicklas WJ, Berl S (1970) Tricarboxylic acid-cycle metabolism in the brain. Effect of fluoroacetate and fluorocitrate on the labeling of glutamate, aspartate, glutamine and γ-aminobutyrate. Biochem J 120:345–351

Coggeshall RE, McLean PD (1958) Hippocampal lesions following administration of 3-acetylpyridine. Proc Soc Exp Biol Med 98:687–688

Coper H (1961) Die Aktivität der DPN- und TPN-Nucleosidase des Gehirns nach Einwirkung verschiedener Pharmaka. Naunyn Schmiedebergs Arch Exp Pathol Pharmakol 242:24–33

Coper H (1966) Stoffwechselstörungen durch Antimetaboliten des Nicotinamids. In: Kress HV, Blum K-U (eds) B-Vitamine. Klinische und physiologisch-chemische Probleme. Schattauer, Stuttgart, pp 275–291

Coper H, Herken H (1963) Schädigung des Zentralnervensystems durch Antimetaboliten des Nikotinsäureamids. Ein Beitrag zur Molekularpathologie der Pyridinnukleotide. Dtsch Med Wochenschr 88:2025–2036

Coper H, Neubert D (1963) Reaktionsgeschwindigkeit NADP-abhängiger Oxydoreduktasen am Rattengehirn mit 3-Acetylpyridin-Adenin-dinucleotidphosphat. Reaction rate of NADP-dependent oxidoreductases in rat brain with 3-acetylpyridine-adenine-dinucleotidphosphate. J Neurochem 10:513–522

Coper H, Neubert D (1964) Einfluß von NADP-Analogen auf die Reaktionsgeschwindigkeit einiger NADP-bedürftiger Oxidoreduktasen. Biochim Biophys Acta 89:23–32

Coper H, Schabarek A (1966) Enzymatischer Austausch von 5-Fluornikotinamid gegen Nikotinamid in Pyridinnukleotiden. Naturwissenschaften 53:225

Coper H, Hadass H, Lison H (1966) Untersuchungen zum Mechanismus zentralnervöser Funktionsstörungen durch 6-Aminonicotinsäureamid. Naunyn Schmiedebergs Arch Pharmakol Exp Pathol 255:97–106

Curtius HC, Pfleiderer W, Wachter H (eds) (1983) Biochemical and clinical aspects of pteridines, vol 2. De Gruyter, Berlin, p 435

Davis LH, Kauffman FC (1987) Metabolism via the pentosephosphate pathway in rat pheochromocytoma PC12 cells: effects of nerve growth factors and 6-aminonicotinamide. Neurochem Res 37:291–295

Denk H, Haider M, Kovac W, Studynka G (1968) Verhaltensänderung und Neuropathologie bei der 3-Acetylpyridinvergiftung der Ratte. Acta Neuropathol 10:34–44

Desclin JC, Escubi J (1974) Effects of 3-acetylpyridine on the central nervous system of the rat as demonstrated by silver methods. Brain Res 77:349–364

Deshpande SS, Albuquerque EX, Kauffman FC, Guth L (1978) Physiological, biochemical and histological changes in skeletal muscle, neuromuscular junction and spinal cord of rats rendered paraplegic by subarachnoidal administration of 6-aminonicotinamide. Brain Res 140:89–109

Deutsch AY, Rosin DL, Goldstein M, Roth RH (1989) 3-Acetylpyridine-induced degeneration of the nigrostriatal dopamine system: an animal model of olivopontocerebellar atrophy-associated Parkinsonism. Exp Neurol 105:1–9

Dickermann HW, Sanpietro A, Kaplan NO (1962) Pig-spleen pyridine trans-glycosidase. I. Purification and properties. Biochim Biophys Acta 62:230–244

Dietrich LS (1975) Cytotoxic analogs of pyridine nucleotide coenzymes. In: Sartorelli AC, Johns DG (eds) Antineoplastic and immunosuppressive agents. Springer, Berlin Heidelberg New York, pp 539–542, 891 (addendum) (Handbook of experimental pharmacolology, vol 37/2)

Dietrich LS, Friedland IM, Kaplan LA (1958) Pyridine nucleotide metabolism: mechanism of action of the niacin antagonist, 6-aminonicotinamide. J Biol Chem 233:964–958

Dix TA, Kuhn DM, Bencovi I (1987) Mechanism of oxygen activation of tyrosine hydroxylase. Biochemistry 26:3354–3361

Dummel RI, Kun E (1969) Studies with specific enzyme inhibitors: XII. Resolution of d, l-erythrofluorocitric acid into optically active isomers. J Biol Chem 244:2966–2969

Eanes RZ, Kun E (1971) Separation and characterization of aconitate hydratase isoenzymes from pig tissues. Biochim Biophys Acta 227:204–210

Fukushima T, Nixon JC (1980) Analysis of reduced forms of biopterin in biological tissues and fluids. Anal Biochem 102:176–188

Fuse S, Koikegami H (1961) On the degenerative changes of the hippocampus and other brain structures induced by administration of 3-acetyl-pyridine (3-AP). Acta Anat Nippon 36:330–331

Gaitonde MK, Evans GM (1982) The effect of inhibition of hexose-monophosphate shunt on the metabolism of glucose and function in rat brain in vivo. Neurochem Res 7:1163–1178

Gaitonde MK, Lewis LP, Evans G, Clapp A (1981) The effect of 6-aminonicotinamide on the levels of brain amino acids and glucose and their labelling with ^{14}C after injection of [U-^{14}C] glucose. Neurochem Res 6:1153–1161

Gaitonde MK, Evison E, Evans GM (1983) The rate of utilization of glucose via hexosemonophosphate shunt in brain. J Neurochem 41:1253–1260

Gaitonde MK, James MD, Evans GM (1984) Decreased labelling of amino acids by inhibition of the utilization of [^3H, ^{14}C]glucose via the hexosemonophosphate shunt in rat brain in vivo. Neurochem Res 9:367–385

Gaitonde MK, Jones J, Evans G (1987) Metabolism of glucose into glutamine via the hexose monophosphate shunt and its inhibition by 6-aminonicotinamide in rat brain in vivo. Proc R Soc Lond [Biol] 231:71–90

Gaitonde MK, Murray E, Cunningham VI (1989) Effect of 6-phosphogluconate on phosphoglucose-isomerase in rat brain in vitro and in vivo. J Neurochem 52:1348–1352

Gal EM (1982) Biosynthesis and function of unconjugated pterins in mammalian tissues. Adv Neurochem 4:83–148

Gal EM, Whitacre DH (1981) Biopterin VII. Inhibition of synthesis of reduced biopterins and its bearing on the function of cerebral tryptophan 5-hydroxylase in vivo. Neurochem Res 6:233–241

Gal EM, Peters RA, Wakelin RW (1956) Some effects of synthetic fluoro compounds on the metabolism of acetate and citrate. Biochem J 64:161–168

Gal EM, Nelson SM, Sherman AD (1978) Biopterin III. Purification and characterization of enzymes involved in the cerebral synthesis of 7,8-dihydrobiopterin. Neurochem Res 3:69–88

Gholson RK (1966) The pyridine nucleotide cycle. Nature 212:933–935

Gonda O, Quastel JH (1966) Transport and metabolism of acetate in rat brain cortex in vitro. Biochem J 100:83–94

Gottesfeld Z, Fonnum F (1977) Transmitter synthesizing enzymes in the hypoglossal nucleus and cerebellum – effect of acetylpyridine and surgical lesions. J Neurochem 28:237–239

Gray EG, Whittaker VP (1962) The isolation of nerve endings from the brain: an electron microscopy study of cell fragments by homogenization and centrifugation. J Anat 96:79–88

Grazi E, De Flora A, Pontremoli S (1960) The inhibition of phosphoglucose isomerase by D-erythrose 4-phosphate. Biochem Biophys Res Commun 2:121–125

Greene LA, Tischler AS (1976) Establishment of a noradrenergic clonal line of rat adrenal pheochromocytoma cells which respond to nerve growth factor. Proc Natl Acad Sci USA 73:2424–2428

Hassler R (1972) Physiopathology of rigidity. In: Siegfried J (ed) Parkinson's disease, vol 1. Huber, Bern, pp 19–45

Hassler RG (1984) Role of pallidum and its transmitters in the therapy of Parkinsonian rigidity and akinesia. Adv Neurol 40:1–14

Hastings AB, Peters RA, Wakelin RW (1955) The effect of subarachnoid injection of fluorocitrate into cerebrospinal fluid of rabbits. Q J Exp Physiol 40:258–268

Hendershot LC, Chenoweth MB (1955) Fluorocitrate and fluorobutyrate convulsions in the isolated cerebral cortex of the dog. J Pharmacol Exp Ther 113:160–168

Herken H (1965) Pharmakologische Untersuchungen zum Problem der selektiven Vulnerabilität des Gehirns. Arzneimittelforschung 15:707–718

Herken H (1968a) Drug-induced pathobiotic effects. In: Raikova H (ed) Mechanisms of drug toxicity. Proceedings of the 3rd international congress of pharmacol meeting, vol 4. Pergamon, Oxford, pp 3–12

Herken H (1968b) Functional disorders of the brain induced by synthesis of nucleotides containing 3-acetylpyridine. Z Klin Chem Klin Biochem 6:357–367

Herken H (1968c) Biosynthesis and action of dinucleotide containing 6-aminonicotinamide on membrane processes. Arzneimittelforschung 18:1235–1245

Herken H (1970) Pharmacological blockade of the pentose phosphate pathway and the problem of the selective vulnerability of the central nervous system. G Accad Med (Torino) 133:1–12

Herken H (1971) Antimetabolic action of 6-aminonicotinamide on the pentose phosphate pathway in the brain. In: Aldridge WN (ed) Mechanisms of toxicity. Macmillan, London, pp 189–203

Herken H (1983) Clonale Nervenzellen in der Kultur – Modelle zum Studium molekularer Grundlagen neuropharmakologischer Wirkungen. Klin Wochenschr 61:1–16

Herken H (1990) Neurotoxin-induced impairment of biopterin synthesis and function: initial stage of a Parkinson-like dopamine deficiency syndrome. Neurochem Int 17:223–238

Herken H, Lange K (1969) Blocking of pentose phosphate pathway in the brain of rats by 6-aminonicotinamide. Naunyn Schmiedebergs Arch Pharmakol Exp Pathol 263:496–499

Herken H, Neuhoff V (1963) Mikroanalytischer Nachweis von 3-Acetylpyridin-adenindinucleotid und Acetylpyridin-adenindinucleotidphosphat im Gehirn. Hoppe Seylers Z Physiol Chem 331:85–94

Herken H, Neuhoff V (1964) Spektrofluorometrische Bestimmung des Einbaus von 6-Aminonicotinsäureamid in die oxydierten Pyridinnucleotide der Niere. Naunyn Schmiedebergs Arch Exp pathol Pharmakol 247:187–201

Herken H, Timmler R (1965) Kinetik der Dihydrofolsäurereduktase aus Rattenleber und -gehirn mit 3-APADPH als Wasserstoffdonator. Naunyn Schmiedebergs Arch Exp Pathol Pharmakol 50:293–301

Herken H, Lange K, Kolbe H (1969) Brain disorders induced by pharmacological blockade of pentose phosphate pathway. Biochem Biophys Res Commun 36:93–100

Herken H, Keller K, Kolbe H, Lange K, Schneider H (1973) Experimentelle Myelopathie – Biochemische Grundlagenforschung ihrer cellulären Pathogenese. Klin Wochenschr 51:644–657

Herken H, Lange K, Kolbe H, Keller K (1974) Antimetabolic action of the pentose phosphate pathway in the central nervous system induced by 6-aminonicotinamide. In: Genazzani E, Herken H (eds) Central nervous system – studies on the metabolic regulation and function. Springer, Berlin Heidelberg New York, pp 41–54

Herken H, Keller K, Kolbe H, Lange K (1975) Renal gluconate excretion after 6-AN. Naunyn Schmiedebergs Arch Pharmacol 291:213–216

Hicks SP (1955) Pathologic effects of antimetabolites: I. Acute lesions in the hypothalamus, peripheral ganglia and adrenal medulla caused by 3-acetylpyridine and prevented by nicotinamide. Am J Pathol 31:189–199

Horecker B (1968) In: Dickens F, Randle PJ, Whelan WJ (eds) Carbohydrate metabolism and its disorders, vol 1. Academic Press, London, pp 139–167

Hornykiewicz O (1962) Dopamin (3-Hydroxytyramin) im Zentralnervensystem und seine Beziehung zum Parkinson-Syndrom des Menschen. Dtsch Med Wochenschr 87:1807–1810

Hornykiewicz O (1989) The neurochemical basis of the pharmacology of Parkinson's disease. In: Calne DB (ed) Drugs for the treatment of parkinson's disease. Springer, Berlin Heidelberg New York, pp 185–204 (Handbook of experimental pharmacology, vol 88)

Horowski R, Wachtel H (1976) Direct dopaminergic action of lisuride hydrogenmaleate, an ergot derivate, in mice. Eur J Pharmacol 36:373

Horton RW, Meldrum BS, Bachelard HS (1973) Enzymic and cerebral metabolic effects of 2-deoxy-D-glucose. J Neurochem 21:507–520

Hothersall JS, Baquer NZ, Greenbaum AL, McLean P (1979) Alternative pathways of glucose utilisation in brain. Changes in the pattern of glucose utilisation in brain during development and the effects of phenazine methosulphate on the integration of metabolic routes. Arch Biochem Biophys 198:478–492

Hothersall JS, Zabairu S, McLean P, Greenbaum AL (1981) Alternative pathways of glucose utilization in brain; changes in the pattern of glucose utilization in brain resulting from treatment of rats with 6-aminonicotinamide. J Neurochem 37:1484–1496

Hotta SS (1962) Glucose metabolism in brain tissue: the hexose monophosphate shunt and its role in glutathione reduction. J Neurochem 9:43–51

Hunting D, Gowans B, Henderson JF (1985) Effects of 6-aminonicotinamide on cell growth, poly (ADP-ribose) synthesis and nucleotide metabolism. Biochem Pharmacol 34:3999–4003

Ito M, Orlov I, Shimoyama I (1978) Reduction of the cerebellar stimulus effect on rat Deiters neurons after chemical destruction of the inferior olive. Exp Brain Res 33:143–146

Jansson SE, Gripenberg J, Härkönen M (1977) The effect of 6-aminonicotinamide blockade of the pentose phosphate pathway on catecholamines in the rat adrenal medulla, superior cervical ganglion, hypothalamus and synaptosome fractions. Acta Physiol Scand 99:467–475

Jellinger K (1989) Pathology of Parkinson's syndrome. In: Calne DB (ed) Drugs for the treatment of Parkinson's disease. Spinger, Berlin Heidelberg New York, pp 47–94 (Handbook of experimental pharmacology, vol 88)

Jochheim U (1981) Die NADPH-abhängige Aldehyd-Reduktase (EC 1.1.1.2) aus Rindergehirn. Untersuchungen zum Mechanismus, zur Coenzym- und Substratspezifität, sowie zur Hemmbarkeit des Enzyms. Dissertation. Freie Universität Berlin

Johnson WJ, McColl JD (1955) 6-Aminonicotinamide – a potent nicotinamide antagonist. Science 122:834

Jung W, Herken H (1989) Inhibition of biopterin synthesis and DOPA production in PC-12 pheochromocytoma cells induced by aminonicotinamide. Naunyn Schmiedebergs Arch Pharmacol 339:424–432

Kahana SE, Lowry OH, Schulz DW, Passonneau JV, Crawford EJ (1960) The kinetics of phosphoglucoisomerase. J Biol Chem 235:2178–2184

Kalnitzky G (1948) The effect of $BaCl_2$, $MgCl_2$ and fluoroacetate on the formation and utilization of citrate. Arch Biochem 17:403–411

Kalnitzky G, Barron ESG (1948) The inhibition by fluoroacetate and fluorobutyrate of fatty acids and glucose oxidation produced by kidney homogenate. Arch Biochem 19:75–87

Kanig K, Koransky W, Münch G, Schulze PE (1964) Isolierung eines Metaboliten des 3-Acetylpyridins aus dem Gehirn und seine Identifizierung als Pyridin-3-methylcarbinol. Naunyn Schmiedebergs Arch Exp Pathol Pharmakol 249:43–53

Kapatos G, Katoh S, Kaufman S (1982) Biosynthesis of biopterin by rat brain. J Neurochem 39:1152–1162

Kaplan NO (1960) Coenzyme metabolism of the brain. In: Brady RO, Tower DB (eds) Neurochemistry of nucleotides and aminoacids. Wiley, New York, pp 70–88

Kaplan NO (1963) Analogs of pyridine coenzymes. In: Sinakian NM (ed) Proceedings of the 5th international congress of biochemistry, Moscow 1961, vol 4. Pergamon, Oxford, pp 295–309

Kaplan NO, Ciotti MM (1956) Chemistry and properties of the 3-acetylpyridine analog of diphosphopyridine nucleotide. J Biol Chem 221:823–832

Kaplan NO, Ciotti MM, Stolzenbach FE (1955) Reaction of pyridine nucleotide analogues with dehydrogenases. J Biol Chem 221:833–844

Kaplan NO, Ciotti MM, Stolzenbach FE (1957) Studies on the interactions of diphosphopyridine nucleotide analogues with dehydrogenases. Arch Biochem Biophys 69:441–457

Kato K, Fukuda H (1985) Reduction of $GABA_B$ receptor binding induced by climbing fiber degeneration in the rat cerebellum. Life Sci 37:279–288

Katoh S, Sueoka T (1984) Sepiapterin reductase exhibits a NADPH-dependent dicarbonyl reductase. Biochem Biophys Res Commun 118:859–869

Kauffman FC, Johnson EC (1974) Cerebral energy reserves and glycolysis in neural tissue of 6-aminonicotinamide-treated mice. J Neurobiol 5:379–392

Kauffman FC, Brown JG, Passonneau JV, Lowry OH (1969) Effects of changes in brain metabolism on levels of pentose phosphate pathway intermediates. J Biol Chem 244:3646–3653

Kaufman S (1967) Pteridine cofactors. Ann Rev Biochem 36:171–184

Kaufman S (1975) Studies on the mechanism of phenylalanine hydroxylase: detection of an intermediate. In: Pfleiderer W (ed) Chemistry and biology of Pteridines. De Gruyter, Berlin, pp 291–301

Kehr W (1977) Effect of lisurid and other ergot derivates on monoaminergic mechanisms in brain. Eur J Pharmacol 41:261–273

Kehr W, Halbhübner K, Loos D, Herken H (1978) Impaired dopamine function and muscular rigidity induced by 6-aminonicotinamide in rats. Naunyn Schmiedebergs Arch Pharmacol 304:317–319

Keller K, Kolbe H, Lange K, Herken H (1972) Behaviour of the glycolytic system of rat brain and kidney in vivo after inhibition of the glucose phosphate isomerase: I. Kinetic studies on rat brain glucose phosphate isomerase. Hoppe Seylers Z Physiol Chem 353:1389–1400

Kirsten E, Sharma MI, Kun E (1978) Molecular toxicology of (−)-erythro-fluorocitrate. Selective inhibition of citrate transport in mitochondria and the binding of fluorocitrate to mitochondrial proteins. Mol Pharmacol 14:172–184

Klingenberg M (1963) Struktur und funktionelle Biochemie der Mitochondrien: II. Die funktionelle Biochemie der Mitochondrien. In: Karlson P (ed) Funktionelle und morphologische Organisation der Zelle. Springer, Berlin Heidelberg New York, pp 69–85

Köhler E, Barrach HJ, Neubert D (1970) Inhibition of NADP dependent oxidoreductases by the 6-aminonicotinamide analogue of NADP. FEBS Lett 6:225–228

Koeppen AH, Barron KD (1984) The neuropathology of olivopontocerebellar atrophy. In: Duvoisin RC, Plaitakis A (eds) The olivopontocerebellar atrophies. Raven, New York, pp 13–38

Kolbe H (1977) Möglichkeiten und Grenzen der Anwendung von ^{14}C-Glucose zur Untersuchung von Hirnstoffwechselveränderungen durch zentralwirksame Substanzen am lebenden, wachen Tier, dargestellt am Beispiel der 6-Aminonicotinsäureamidwirkung. Dissertation. Freie Universität Berlin

Kolbe H, Keller K, Lange K, Herken H (1976) Metabolic consequences of drug-induced inibition of the pentose phosphate pathway in neuroblastoma and glioma cells. Biochem Biophys Res Commun 73:378–382

Koransky W (1965) Autoradiographic studies of the distribution of centrally acting drugs in the brain. In: Roth LY (ed) Isotopes in experimental pharmacology. University of Chicago, pp 91–98

Krieglstein J, Stock R (1975) Decreased glycolytic flux rate in the isolated perfused rat brain after pretreatment with 6-aminonicotinamide in rats. Naunyn Schmiedebergs Arch Pharmacol 290:323–327

Kuffler SW, Nicholls IG (1966) The physiology of neuroglial cells. In: Kramer E et al. (eds) Ergebnisse der Physiologie, vol 57. Springer, Berlin Heidelberg New York, pp 1–90

Kun E (1972) Discussion on fluoracetate and its metabolism. In: Carbo-fluorine compounds. pp 71–76 (Ciba foundation symposium, new series, vol 2)

Kun E, Kirsten E, Sharma ML (1978) Catalytic mechanism of citrate transport through the inner mitochondrial membrane: enzymatic synthesis and hydrolysis of glutathione-citric acid thioester. In: Azzone GF et al. (eds) The proton and calcium pumps. Elsevier, Amsterdam, pp 285–295

Lahiri S, Quastel JH (1963) Fluoroacetate and the metabolism of ammonia in brain. Biochem J 89:157–163

Lange K, Kolbe H, Keller K, Herken H (1970) Der Kohlenhydratstoffwechsel des Gehirns nach Blockade des Pentose-Phosphat-Weges durch 6-Aminonicotinsäureamid. Hoppe Seylers Z Physiol Chem 351:1241–1252

Levine RA, Williams AC, Robinson DS, Calne DB, Lovenberg W (1979) Analysis of hydroxylase cofactor activity in the cerebrospinal fluid of patients with Parkinson's disease. In: Poirier LJ, Sourkes TL, Bédard PJ (eds) Advances in neurology. Raven, New York, pp 303–307

Levine RA, Miller LP, Lovenberg W (1981) Tetrahydrobiopterin in striatum: localization in dopamine terminals and role in catecholamine synthesis. Science 214:919–921

Le Witt PA, Miller LP, Newman RP, Burns RS, Insel T, Levine RA Lovenberg W, Calne DB (1984) Tyrosine hydroxylase cofactor (tetrahydrobiopterin) in parkinsonism. In: Hassler RG, Christ JF (eds) Advances in neurology, vol 40. Raven, New York, pp 459–462

Liang CS (1977) Metabolic control of circulation: effects of iodoacetate and fluoroacetate. J Clin Invest 60:61–69

Liébecq C (1989) Commentary on Liébecq C, Peters RA (1949) The toxicity of fluoroacetate and the tricarboxylic acid cycle (Biochim Biophys Acta 3:215–230) Biochim Biophys Acta 1000:251–253

Liébecq C, Peters RA (1949) The toxicity of fluoroacetate and the tricarboxylic acid cycle, Biochim Biophys Acta 3:215–230

Liébecq C, Liébecq-Hutter S (1958) Effet du fluoroacetate et des rayons X sur la teneur en acide critique du foie de la souris. Experientia 14:216–217

Loew DM, Vigouret JM, Jaton AL (1976) Neuropharmacological investigations with two ergot alkaloids, hydergine and bromocriptine. Postgrad Med J 52 (Suppl 1):40–46

Loos D, Halbhübner K, Herken H (1977) Lisuride, a potent drug in the treatment of muscular rigidity in rats. Naunyn Schmiedebergs Arch Pharmacol 300:195–198

Loos D, Halbhübner K, Kehr W, Herken H (1979) Action of dopamine agonists on Parkinson-like muscle rigidity induced by 6-aminonicotinamide. Neuroscience 4:667–676

Lopiano L, Savio T (1986) Inferior olive lesion induces long-term modifications of cerebellar inhibition on Deiters nuclei. Neurosci Res 4:51–61

Lotspeich WD, Peters RA, Wilson TH (1952) The inhibition of aconitase by "inhibitor fractions" isolated from tissues poisoned with fluoroacetate. Biochem J 51:20–25

Lovenberg W, Levine RA, Robinson DS, Ebert M, Williams AC, Calne DB (1978) Hydroxylase cofactor activity in cerebrospinal fluid of normal subjects and patients with Parkinson's disease. Science 204:624–626

Lowry OH, Passonneau JV (1964) The relationships between substrates and enzymes of glycolysis in brain. J Biol Chem 239:31–42

Lowry OH, Roberts NR, Leiner KY, Wu M, Farr AL, Albers RW (1954) The quantitative histochemistry of brain: III. Ammons horn. J Biol Chem 207:39–49

Lowry OH, Passonneau JV, Hasselberger FX, Schulz DW (1964) Effect of ischemia on known substrates and cofactors of glycolytic pathway in brain. J Biol Chem 239:18–30

Lowry OH, Roberts NR, Kapphahn JJ (1957) The fluorometric measurement of pyridine nucleotides. J Biol Chem 224:1047–1064

Mandell AJ (1978) Redundant mechanisms regulating brain tyrosine and tryptophan hydroxylase. Ann Rev Pharmacol Toxicol 18:461–493

Marais JSC (1944) Monofluor acetic acid, the toxic principle of "Giftblaa" Dichapetalum cymosum? (Hook) England Onderstepoort. J Vet Sci Animal Ind 20:67–73

Martius C (1949) Über die Unterbrechung des Citronensäurecyclus durch Fluoressigsäure. Liebigs Ann 561:227–232

Masserano IM, Vulliet PR, Tank AW, Weiner N (1989) The role of tyrosine hydroxylase in the regulation of catecholamine synthesis. In: Trendelenburg U, Weiner N (eds) Catecholamines. Springer, Berlin Heidelberg New York, pp 427–469 (Handbook of experimental pharmacology, vol 90/II)

McIlwain H, Rodnight R (1949) Breakdown of cozymase by a system from nervous tissue. Biochem J 44:470

McLean PD (1963) Comments on the selective vulnerability of the hippocampus. In: Selective vulnerability of the brain in hypoxaemia. Blackwell, Oxford, p 177

Meyer-Esdorf G, Schulze PE, Herken H (1973) Distribution of ^{3}H-labelled 6-aminonicotinamide and accumulation of 6-phosphogluconate in the spinal cord. Naunyn Schmiedebergs Arch Pharmacol 276:235–241

Milstien S, Kaufman S (1983) Tetrahydro-sepiapterin is an intermediate in BH_4 synthesis. Biochem Biophys Res Commun 115:888–893

Milstien S, Kaufman S (1985) Synthesis of tetrahydrobiopterin: conversion of dihydroneopterin triphosphate to tetrahydrobiopterin intermediates. Biochem Biophys Res Commun 128:1099–1107

Montgomery RL, Christian EL (1976) Pathologic effects of antimetabolites on the hippocampus of diet controlled mice. Brain Res Bull 1:255–259

Morris G, Nadler JV, Nemeroff CB, Slotkin TA (1985) Effects of neonatal treatment with 6-aminonicotinamide on basal and isoproterenol-stimulated ornithine decarboxylase activity in cerebellum of the developing rat. Biochem Pharmacol 34:3281–3284

Morrison JF, Peters RA (1954) Biochemistry of fluoroacetate poisoning. The effect of fluorocitrate on purified aconitase. Biochem J 58:473–479

Morselli PL, Garattini S, Marcucci F, Mussini E, Rewersky W, Valzelli L, Peters RA (1968) The effect of injections on fluorocitrate into the brains of rats. Biochem Pharmacol 17:195–202

Nadi NS, Kanter D, McBride WJ, Aprison MH (1977) Effects of 3-acetylpyridine on several putative neurotransmitter amino acids in the cerebellum and medulla of the rat. J Neurochem 28:661–662

Nagatsu T (1983) Biopterin cofactor and monoamine-synthesizing monooxygenases. Neurochem Int 5:27–38

Nakashima Y, Suzue R (1984) Effect of 3-acetylpyridine on the content of myelin in the developing rat brain. J Nutr Sci Vitaminol (Tokyo) 30:441–451

Nelson TI, Kaufman S (1987) Activation of rat caudate tyrosine hydroxylase phosphatase by tetrahydropterin. J Biol Chem 262:16470–16475

Neubert D, Coper H (1965) Die Geschwindigkeit der "malic enzyme" Reaktion in Gegenwart von NADP-Analogen. Biochem Z 341:485–494

Neuhoff V (1964) Pharmakologische Untersuchungen zur lokalisierten Vulnerabilität des Gehirns. Habilitationsschrift Freie Universität Berlin

Neuhoff V, Köhler F (1966) Biochemische Analyse des Stoffwechsels von 3-Acetylpyridin. Isolierung and Identifizierung von 6-Metaboliten. Naunyn Schmiedbergs Arch Pharmacol 254:301–326

Nichol CA, Lee CL, Edelstein MP, Chao JY, Churd DS (1983) Biosynthesis of tetrahydrobiopterin by de novo and salvage pathways in adrenal medulla extracts, mammalian cell cultures and rat brain in vivo. Proc Natl Acad Sci USA 80:1546–1550

Nichol CA, Smith GK, Duch DS (1985) Biosynthesis and metabolism of tetrahydrobiopterin and molybdopterin. Ann Rev Biochem 54:729–764

Niklowitz WJ (1969) Cytomorphological alterations of mossy fiber boutons of the hippocampus produced by 3-acetylpyridine, methoxypyridoxine, reserpine, and iproniazid. Z Zellforsch Mikrosk Anat 101:192–202

Ochoa S, Stern IR, Schneider MC (1951) Enzymatic synthesis of citric acid: II. Crystalline condensing enzymes. J Biol Chem 193:691–702

Olivera BM, Ferro AM (1982) Pyridine nucleotide metabolism and ADP-ribosylation. In: Hayaishi O, Ueda K (eds) ADP-Ribosylation reactions. Biology and medicine. Academic Press, New York, pp 19–40

Ozaki HS, Murakami TH, Shimada M (1983) Learning deficits on avoidance task and hippocampal lesions in area CA-3 following intraperitoneal administration of 3-acetylpyridine. J Neurosci Res 10:425–436

Parr CW (1956) Inhibition of phosphoglucose isomerase. Nature 78:1401

Pattison FLM, Peters RA (1966) Monofluoro-aliphatic compounds. In: Smith FA (ed) Pharmacology of fluorides. Springer, Berlin Heidelberg New York, pp 387–458 (Handbook of experimental pharmacology, vol XX/1)

Pattison FLM, Saunders BC (1949) Toxic fluorine compounds containing the C-F link. Part VII. Evidence for the β-oxidation of ω-fluorocarboxylic acids in vivo. J Chem Soc ●:2745–2749

Perry TL, MacLean I, Lijo T, Hansen S (1976) Effects of 3-acetylpyridine on putative neurotransmitter amino acids in rat cerebellum. Brain Res 109:632–635

Peters RA (1952) Lethal synthesis (Croonian lecture). Proc R Soc Lond [Biol] 139:10–170

Peters RA (1957) Mechanism of the toxicity of the active constituent of Dichapetalum cymosum and related compounds. Adv Enzymol 18:113–159

Peters RA (1963) Biochemical lesions and lethal synthesis. In: Alexander P, Bacq ZM (eds) Carbon-fluorine compounds. Pergamon, Oxford, pp 88–13

Peters RA (1972) Some metabolic aspects to fluoroacetate especially related to fluorocitrate. In: Carbo-fluorine compounds. Elsevier, Amsterdam, pp 55–70 (Ciba foundation symposium, new series, vol 2)

Peters RA, Shorthouse M (1966) Note upon the behaviour of rat tissue treated with fluoroacetate in vitro. Biochem Pharmacol 15:2130–2131

Peters RA, Shorthouse M (1970) Some new observations with especial relation to its use as a test system for fluorocitrate, together with studies on the electrophoretic separation of components of the crude enzyme. Biochem Biophys Acta 220:569–579

Peters RA, Wakelin RW (1953b) Biochemistry of fluoroacetate poisoning: the isolation and some properties of the tricarboxylic acid inhibitor of citrate metabolism. Proc R Soc Lond [Biol] 140:497–507

Politis MJ (1989) 6-aminonicotinamide selectively causes necrosis in reactive astroglia cells in vivo. Preliminary morphological observation. J Neurol Science 92:71–79

Quastel JH (1974) Metabolic compartmentation in the brain and effect of metabolic inhibitors. In: Berl S, Clarke DD, Schneider D (eds) Metabolic compartmentation and neurotransmission. Plenum, New York, pp 337–361

Raabe WA (1981) Ammonia and disinhibition in cat motor cortex by ammonium acetate, monofluoroacetate and insulin-induced hypoglycemia. Brain Res 210:311–322

Rea MA, McBride WI, Rhode BH (1980) Regional and synaptosomal levels of amine neurotransmitter in 3-acetylpyridine deafferented rat cerebellium. J Neurochem 34:1106–1108

Redetzki HM, Alvarez-O'Burke (1962) 6-Aminonicotinamide, a central nervous system depressant. J Pharmacol Exp Ther 137:173–178

Salas M, Vinuela E, Sols A (1965) Spontaneous and enzymatically catalyzed anomerization of glucose 6-phosphate and anomeric specifity of related enzymes. J Biol Chem 240:561–568

Sapolsky RM (1985) Glucocorticoid toxicity in the hippocampus: temporal aspects of neuronal vulnerability. Brain Res 359:300–305

Sasaki S (1982) Brain edema and gliopathy induced 6-aminonicotinamide intoxication in the central nervous system of rats. Am J Vet Res 43:1691–1695

Schabarek A (1967) Untersuchungen Über die Reaktion von Pyridin-, Piperidin-, Piperazin-, Phenazin-, Phenthiazin-, Chinolin und Acridinverbindungen mit der NAD(P) Glycohydrolase. Ein Beitrag zur enzymatischen Synthese von NAD(P) Analogen. Dissertation. Freie Universität Berlin

Schacht U, Schultz G, Senft G (1966) Catecholamin- und ATP-Gehalt der Nebennieren nach Gabe von 6-Aminonicotinsäureamid. Naunyn Schmiedebergs Arch Exp Pathol Pharmakol 253:355–363

Schneider H, Coper H (1968) Morphologische Befunde am Zentralnervensystem nach Vergiftung mit Antimetaboliten des Nicotinamids (6-Aminonicotinsäureamid und 3-Acetylpyridin) und einem Chinolinderivat (5-Nitro-8-hydroxychinolin). Arch Psychol 211:138–154

Schneider H, Cervos-Navarro I (1974) Acute gliopathy in spinal cord and brainstem induced by 6-aminonicotinamide. Acta Neuropathol 27:11–23

Shimada M, Ozaki HS, Murakami TH, Imahayashi T (1984) Effects of 3-acetylpyridine on local cerebral glucose utilization. Neurosci Res 1:357–362

Simantov R, Snyder SH, Oster-Granite ML (1976) Harmaline-induced tremor in the rat abolition by 3-acetylpyridine destruction of cerebellar climbing fibers. Brain Res 114:144–151

Slotkin A (1979) Ornithine decarboxylase as a tool in developmental neurobiology. Life Sci 24:1623–1630

Smith GK, Nichol CA (1984) Two new tetrahydropterin intermediates in the adrenal medullary de novo biosynthesis of tetrahydrobiopterin. Biochem Biophys Res Commun 120:761–766

Soiefer AI, Kostyniak PJ (1983) The enzymatic defluorination of fluoroacetate in mouse liver cytosol: the separation of defluorination activity from several glutathione-S-transferases of mouse liver. Arch Biochem Biophys 225:928–935

Spencer RF, Lowenstein JM (1967) Citrate content of liver and kidney of rat in various metabolic states and in fluoroacetate poisoning. Biochem J 103: 342–348

Sternberg SS, Philips FS (1958) 6-Aminonicotinamide and acute degenerative changes in the central nervous system. Science 127:644–646

Swarts F (1896) Sur l'acide fluoroacétique. Bull Soc Chim Paris (Série 3) 15:1134

Szalay J, Boti Z, Temesvari P, Bara D (1979) Alteration of the nucleus ruber in 3-acetylpyridine intoxication: a light microscopic and electron microscopic study. Exp Pathol 17:271–279

Szerb JC, Issekutz B (1987) Increase in the stimulation-induced overflow of glutamate by fluoroacetate, a selective inhibitor of the glial tricarboxylic cycle. Brain Res 410: 116–120

Szerb JC, O'Regan PA (1988) Increase in the stimulation-induced overflow of excitatory amino acids from hippocampal slices: interaction between low glucose concentration and fluoroacetate. Neurosci Lett 86:207–212

Taitelman U, Roy (Shapira) A, Raikhlin-Eisenkraft B, Hoffer E (1983) The effect of monoacetin and calcium chloride on acid-base balance and survival in experimenal sodium fluoroacetate poisoning. Arch Toxicol [Suppl] 6:222–227

Takikawa SI, Curtius HC, Redweik U, Leimbacher W, Ghisla S (1986) Biosynthesis of tetrahydrobiopterin. Purification of 6-pyruvoyl tetrahydrobiopterin synthase from human liver. Eur J Biochem 161:295–302

Udenfriend S (1966) Tyrosine hydroxylase. Pharmacol Rev 18:43–51

Umetani T (1980) The degeneration of the inferior olive of the rat induced by administration of 3-acetylpyridine: I. The degeneration of the perikaryons. Med J Kobe Univ 41:33–40

Venkataraman R, Racker E (1961) Mechanism of action of transaldolase: I. Crystallization and properties of yeast enzyme. J Biol Chem 236:1876–1882

Vigouret IM, Loew DM, Jaton AL (1977) Effects of bromocriptine on and rigidity in rats in comparison to apomorphine and amphetamine. Naunyn Schmiedebergs Arch Pharmacol 297 [Suppl II]:R54

Vollenweider FX, Cuenod M, Do KQ (1990) Effect of climbing fiber deprivation on release of endogenous aspartate, glutamate, and homocysteate in slices of rat cerebellar hemispheres and vermis. J Neurochem 54:1533–1540

Votáva Z, Lamplová I (1961) Antiserotonin activity of some ergolenyl and isoergolenyl derivates in comparison with LSD and the influence of monoamine inhibition on this antiserotonin effect. In: Rothlin E (ed) Neuropsychopharmacology, vol 2. Elsevier, Amsterdam, pp 68–73

Wallenfels K, Stathakos D, Dieckmann (1960) A new procedure for the solubilization of beef spleen diphosphopyridine nucleotidase. Biochem Biophys Res Commun 3:632–634

Walter P, Kaplan NO (1963) Substituted nicotinamide analogues of nicotinamide adenine dinucleotide. J Biol Chem 238:2823–2830

Warburg O, Christian W (1936) Pyridin, der wasserstoffübertragende Bestandteil von Gärungsfermenten. Biochem Z 287:291–328

Wick AN, Drury DR, Nakada HI, Wolfe JB (1957) Localization of the primary metabolic block produced by 2-deoxyglucose. J Biol Chem 224:963–969

Wiklund L, Toggenburger G, Cuenod M (1984) Selective retrograde labelling of the rat olivocerebellar climbing fiber system with D-[^{3}H]aspartate. Neuroscience 13:441–468

Williams JF, Blackmore PF (1983) Non-oxidative synthesis of pentose 5-phosphate from hexose 6-phosphate and triose phosphate by the L-type pentose pathway. Int J Biochem 15:797–816

Willing H, Neuhoff V, Herken H (1964) Der Austausch von 3-Acetylpyridin gegen Nicotinsäureamid in den Pyridinnucleotiden verschiedener Hirnregionen. Naunyn Schmiedebergs Arch Exp Pathol Pharmakol 247:254–266

Windmueller HG, Kaplan NO (1962) Solubilization and purification of diphosphopyridine nucleotidase from pig brain. Biochim Biophys Acta 56:388–391

Wolf A, Cowen D (1959) Pathological changes in the central nervous system produced by 6-aminonicotinamide. Bull NY Acad Med 35:814–817

Wolf A, Cowen D, Geller LM (1962) Structural and functional effects of 6-aminonicotinamide and other antimetabolites on the central nervous system. In: Jacob H (ed) Proceedings of the 6th international congress of neuropathology, vol III. Thieme, Stuttgart, pp 447–453

Wolf I (1965) Elektronenmikroskopische Untersuchungen über Struktur and Gestalt von Astrocytenfortsätzen. Z Zellforsch 66:811–828

Woodhams P, Rodd R, Balazs R (1978) Age dependent susceptibility of inferior olive neurons to 3-acetylpyridine in the rat. Brain Res 153:194–198

Zatman LJ, Kaplan NO, Colowick SP (1953) Inhibition of spleen diphosphopyridine nucleosidase by nicotinamide, an exchange reaction. J Biol Chem 200:197–212

Zatman LJ, Kaplan NO, Colowick SP, Ciotte MM (1954) The isolation and properties of the isonicotinic acid hydrazide analogue of diphosphopyridine nucleotide. J Biol Chem 209:467–484

Zeitz M, Lange K, Keller K, Herken H (1978) Effects of 6-aminonicotinamide on growth and acetylcholinesterase activity during differentiation of neuroblastoma cells in vitro. Naunyn Schmiedebergs Arch Pharmacol 305:117–121

CHAPTER 7

Neurotropic Carcinogenesis

W.J. ZELLER

A. Introduction

Tumors of the nervous system can be induced in experimental animals by chemical compounds from different classes. With regard to the subject of this survey, methods of inducing brain tumors in experimental animals by intracranial (topical) administration of carcinogenic agents will not be discussed in detail. As shown in Table 1, different measures, e.g., administration of aromatic hydrocarbons, of RNA- and DNA-tumor viruses, or of radioactive elements like ^{60}Co or ^{241}Am, have been successfully used to induce brain tumors upon topical administration (ZIMMERMAN 1969, 1982; BIGNER and SWENBERG 1977).

Some aromatic hydrocarbons were found to induce neurogenic tumors also with systemic application. However, this was only possible in the fetal brain by transplacental administration; 7,12-dimethylbenzanthracene (7,12-DMBA) was the most active compound of this class (NAPALKOV and ALEXANDROV 1974; RICE et al. 1978).

The discovery of the carcinogenic potential of N-nitroso compounds (MAGEE and BARNES 1956) initiated the synthesis and biological investigation of an increasing number of analogues (Table 2) (DRUCKREY et al. 1967b; PREUSSMANN and STEWART 1984).

While N-nitrosodialkylamines (nitrosamines), hydrazo-, azo-, and azoxyalkanes as well as 1-aryl-3,3-dialkyltriazenes require metabolic activation mediated by microsomal enzymes predominantly in the liver (DRUCKREY et al. 1967b; LIJINSKY et al. 1968; PREUSSMANN et al. 1969a,b; PREUSSMANN and STEWART 1984), acylalkylnitrosamines (syn. alkylnitrosoureas or nitrosamides) release alkyl cations spontaneously without the requirement of metabolic activation (GARETT et al. 1965; SWANN 1968; SCHEIN 1969; SWANN and MAGEE 1971). Especially candidates of the latter group of compounds proved to be highly active neurotropic carcinogens following systemic application both in adult animals and in prenatal or postnatal phases of development. The present survey will focus on selected aspects of experimental neurotropic carcinogenesis in different phases of development; special emphasis will be placed on the activity of nitrosamides. (For detailed review, see the comprehensive presentations by DRUCKREY et al. 1967b, 1969, 1972; MENNEL and IVANKOVIC 1975; IVANKOVIC 1975, 1979;

Table 1. Induction of brain tumors in rodents upon topical (intracranial) administration

---

Aromatic hydrocarbons
  20-Methylcholanthrene
  3-Methylcholanthrene
  3,4-Benzopyrene
  1,2-Benzopyrene
  3,4,8,9-Dibenzopyrene
  1,2,5,6-Dilbenzanthracene
  1,2,6-Dibenzanthracene
  9,10-Dimethyl-1,2-benzanthracene
  6, 9, 10-Trimethyl-1, 2-benzanthracene
Viruses
  RNA−oncornaviruses
    ASV, MSV, SSV
  DNA−papovaviruses
    Murine polyoma
    Simian SV40
    Human papova JC
  DNA−adenoviruses
    Human adenovirus type 12
    Simian SA-7
    Simian SV 20
    Avian chicken embryo lethal
      orphan(CELO)
Radiation
  ^{60}Co, ^{241}Am

(Zimmermann 1969; Bigner and Swenberg 1977; Maltoni et al. 1982; Takakura et al. 1987).
ASV, Avian Sarcoma Virus; MSV, Murine Sarcoma Virus; SSV, Simian Sarcoma Virus.

Kleihues et al. 1976a; Jänisch and Schreiber 1969, 1974; Bigner and Swenberg 1977; Maekawa and Mitsumori 1990.)

B. Neurotropic Carcinogenicity of Alkylnitrosoureas

I. Activity in Adult Animals

1. Methylnitrosourea, Ethylnitrosourea

Druckrey and coworkers observed brain tumors after chronic application of methylnitrosourea (MNU); intravenous injection of MNU 5 mg/kg per week induced brain tumors in about 70% of rats (Druckrey et al. 1965a). This neurotropic effect of MNU was later confirmed by several other investigators (Thomas et al. 1967, 1968; Swenberg et al. 1972a, 1975a; Ogiu et al. 1977;

Table 2. N-Nitroso compounds inducing neurogenic tumors following systemic administration

Chemical designation	Abbreviation	Chemical formula	Animal species
N-Methyl-N-nitrosourea	MNU	$ON-N{\overset{CH_3}{\underset{CO-NH_2}{}}}$	Rat, rabbit, mouse, dog
N-Ethyl-N-nitrosourea	ENU	$ON-N{\overset{CH_2-CH_3}{\underset{CO-NH_2}{}}}$	Rat, hamster, mouse, monkey, opossum
N-(n-Propyl)-N-nitrosourea	n-propyl-NU	$ON-N{\overset{(CH_2)_2-CH_3}{\underset{CO-NH_2}{}}}$	Rat
N-(n-Butyl)-N-nitrosourea	n-butyl-NU	$ON-N{\overset{(CH_2)_3-CH_3}{\underset{CO-NH_2}{}}}$	Rat, hamster
N-(n-Pentyl)-N-nitrosourea	n-pentyl-NU	$ON-N{\overset{(CH_2)_4-CH_3}{\underset{CO-NH_2}{}}}$	Rat
N-(n-Hexyl)-N-nitrosourea	n-hexyl-NU	$ON-N{\overset{(CH_2)_5-CH_3}{\underset{CO-NH_2}{}}}$	Rat
N-(n-Heptyl)-N-nitrosourea	n-heptyl-NU	$ON-N{\overset{(CH_2)_6-CH_3}{\underset{CO-NH_2}{}}}$	Rat
N-(n-Octyl)-N-nitrosourea	n-octyl-NU	$ON-N{\overset{(CH_2)_7-CH_3}{\underset{CO-NH_2}{}}}$	Rat
N,N'-Dimethyl-N-nitrosourea		$ON-N{\overset{CH_3}{\underset{CO-NH-CH_3}{}}}$	Rat
N,N',N'-Trimethyl-N-nitrosourea		$ON-N{\overset{CH_3}{\underset{CO-N{\overset{CH_3}{\underset{CH_3}{}}}}{}}}$	Rat
N-Methyl-N',N'-diethyl-N-nitrosourea		$ON-N{\overset{CH_3}{\underset{CO-N{\overset{C_2H_5}{\underset{C_2H_5}{}}}}{}}}$	Rat
N-Methyl-N-nitrosobiuret		$ON-N{\overset{CH_3}{\underset{CO-NH-CO-NH_2}{}}}$	Rat
N-Ethyl-N-nitrosobiuret		$ON-N{\overset{CH_2-CH_3}{\underset{CO-NH-CO-NH_2}{}}}$	Rat
N-Methyl-N-nitroso-N'-acetylurea		$ON-N{\overset{CH_3}{\underset{CO-NH-CO-CH_3}{}}}$	Rat

JÄNISCH and SCHREIBER 1969; for review, see BIGNER and SWENBERG 1977; KLEIHUES et al. 1976a).

Whereas the chronic application of MNU showed this distinct neurotropism, its single application yielded a broader spectrum of tumor types including, however at a lower incidence, neurogenic tumors. Thus, JÄNISCH and SCHREIBER (1969) observed only two brain tumors following a single application of MNU 70 mg/kg to 28 adult rats.

Peroral or subcutaneous application of MNU in rats was less effective than intravenous administration. MNU was also highly active in inducing neurogenic tumors in other species than in the rat. In rabbits between 56% and 70% (STAVROU 1969; SCHREIBER et al. 1969; KLEIHUES et al. 1970) and in dogs between 44% and 50% neurogenic tumors (SCHNEIDER and WARZOK 1972; STAVROU and HAGLID 1972) were observed following multiple application in adult animals. In mice, postnatal application of MNU was considerably less effective; in C3HeB/FeJ mice, DENLINGER et al. (1974) observed only 10% neurogenic tumors following chronic application of MNU.

Chronic administration of ethylnitrosourea (ENU) 10 mg/kg per week to rats resulted in the development of leukemias in >50% of the animals and in the development of neurogenic tumors in about 30% (DRUCKREY et al. 1967b). Comparable to MNU, single doses of ENU yielded a lower incidence of neurogenic tumors and a broader spectrum of tumors in other organ systems. Single intravenous doses of ENU 60 mg/kg in adult rats resulted in 6% neurogenic tumors; an increase in the single dose of ENU to a subtoxic dose level (140–200 mg/kg) yielded a higher incidence of neurogenic tumors in adult rats (45%–60%) (IVANKOVIC and DRUCKREY1968).

In mice, ENU is a weak neurotropic carcinogen following single-dose application in postnatal phases (WECHSLER et al. 1979).

2. Neurotropic Carcinogenicity of Alkylnitrosoureas with Increasing Length of the Alkyl Substituent

In adult Wistar/Furth rats, continuous oral application of *n*-butylnitrosourea (BNU) led to the development of neurogenic tumors in 23% of the animals (TAKIZAWA and NISHIHARA 1971). A comparative investigation of the carcinogenicity of single doses of the members of the homologous series of alkylnitrosoureas (MNU, ENU, *n*-propyl-NU, BNU, *n*-pentyl-NU, *n*-hexyl-NU, *n*-heptyl-NU, *n*-octyl-NU) in adult BD rats (single doses ranging between $<LD_{10}$ and $\geq LD_{50}$) revealed that compounds with longer alkyl chains also had a marked neurotropic carcinogenic effect. The yield of neurogenic tumors was dependent on the route of administration. Following peroral application, the yield of neurogenic tumors ranged between four tumors in 23 rats (*n*-octyl-NU) and 14 tumors in 29 rats (MNU), following subcutaneous application, between 4 tumors in 50 rats (*n*-heptyl-NU) and 5 tumors in 20 rats (*n*-propyl-NU). With regard to nonneurogenic tumors, subcutaneous administration increased the incidence of tumors located in

the subcutis whereas peroral administration increased the incidence in the gastrointestinal tract. Tumors observed in the brain and spinal cord were mainly oligodendrogliomas, astrocytomas, and ependymomas, whereas neurogenic tumors on cranial nerves and in the peripheral nervous system were predominantly malignant neurinomas (ZELLER et al. 1982).

II. Transplacental and Perinatal Activity of Alkylnitrosoureas

1. Transplacental Activity of Ethylnitrosourea in Rats

Stimulated by the work of IVANKOVIC et al. (1966) and DRUCKREY et al. (1966a), transplacental and perinatal carcinogenesis by nitrosamides was investigated in detail by several groups. IVANKOVIC and DRUCKREY (1968) reported that single doses of ENU ranging between 5 and 80 mg/kg body weight induced malignant tumors of the central and peripheral nervous system in practically all offspring when the compound was given prenatally in the second half of pregnancy (days 13–23). Application of ENU to pregnant rats before day 12 of gestation, on the other hand, did not result in the formation of neurogenic tumors in the progeny. A total of 821 neurogenic tumors was observed in 543 offspring whose mothers had been treated with ENU during the second half of pregnancy (IVANKOVIC and DRUCKREY 1968); 207 tumors (about 25%) were located in the brain, 197 on cranial nerves, 107 in the spinal cord, and 310 in the peripheral nervous system. The brain tumors had an average latency period of 250 days; they were mainly located in the temporal region around the lateral ventricles. The majority of tumors of cranial nerves was located on the nervi trigemini; their average latency period was 190 days. Tumors on other cerebral nerves (nervi facialis, hypoglossus, acusticus, vagus) were observed only occasionally; tumors of the bulbus olfactorius were not observed following transplacental application of ENU. The average latency period of peripheral nerve tumors was 215 days; main localizations were the plexus lumbosacralis, the plexus brachialis, the nervus ischiadicus, and the nervus femoralis. A comparable latency period (210 days) was observed for tumors located in the spinal cord; here, practically all regions were affected.

a) Comparative Sensitivity of the Nervous System to Ethylnitrosourea in Different Phases of Development: Interspecies Comparisons

The sensitivity of rat fetal brain tissue to ENU is about 50 times higher compared with adult nervous tissue (IVANKOVIC and DRUCKREY 1968). A linear regression was observed in the prenatal dose-response curve; the median effective dose (dose resulting in neurogenic tumor formation in 50% of the offspring) was calculated to be ENU 3.2 mg/kg body weight (less than 2% of the LD_{50} in adult rats) (IVANKOVIC and DRUCKREY 1968). The con-

Table 3. Tumor incidence and survival time of Sprague-Dawley rats transplacentally exposed to ethylnitrosourea

Single dose[a] (i.v., mg/kg)	Rats with neuroectodermal tumors/total no. of rats	Rats with nonneural tumors/total no. of rats	Mean no. of neuroectodermal tumors/rats with neural tumors	Mean survival time (days)
1	5/41	31/41	1.0	655 ± 128
5	19/24	6/24	1.6	477 ± 192
20	17/17	5/17	2.4	288 ± 108
50	25/25	6/25	4.1	211 ± 70

(From SWENBERG et al. 1972b).
[a] Application on the 20th day of gestation to pregnant rats.

clusion that even smaller doses than 3.2 mg/kg should also induce neurogenic tumors was subsequently confirmed. In BD IX rats, ENU 1 mg/kg corresponding to 0.4% of the LD_{50} resulted in 16% neurogenic tumors in the offspring (IVANKOVIC 1979). The incidence of neuroectodermal tumors following transplacental exposure to ENU is similar for BD IX, Sprague-Dawley, and Fischer rats (KOESTNER et al. 1979). Neurogenic tumors in the offspring of rats are also inducible by simultaneous peroral application of ethylurea plus nitrite, the precursors of ENU (IVANKOVIC and PREUSSMANN 1970), or by percutaneous administration of ENU (GRAW et al. 1974) underlining the high activity of this compound in prenatal phases of development.

SWENBERG et al. (1972b) showed in Sprague-Dawley rats that the mean number of tumors per animal rises and that the survival time of the offspring shortens following increasing doses of transplacentally administered ENU on the 20th day of gestation (Table 3).

In mice, the incidence of neurogenic tumors following transplacental and perinatal exposure to ENU is much lower than in the rat (VESSELINOVITCH et al. 1974, 1977; WECHSLER et al. 1979). Following prenatal administration of different single doses of ENU, the incidence of neurogenic tumors was <10%; postnatal application (15–42 days after birth) of various single doses of ENU was even less effective, resulting in an incidence of <1% neurogenic tumors in different strains of mice (WECHSLER et al. 1979). Thus, comparison of neurogenic tumor incidence in prenatal, neonatal, and postnatal phases of development in mice shows a decrease in sensitivity with increasing age; the sensitivity in the postnatal phase is decreased by a factor of approximately 20 vs. prenatal exposure. On the other hand, alveologenic adenomas of the lung and hepatomas are the predominating tumors following transplacental and perinatal exposure to ENU in mice (RICE 1969; KOESTNER et al. 1979).

Table 4 gives the localization and classification of neurogenic tumors in mice exposed perinatally and postnatally to ENU; in the brain oligodendrogliomas and in the peripheral nervous system neurinomas were the most

Table 4. Localization and classification of neurogenic tumors induced by ethylnitrosourea in mice by time of treatment

Tumor type	No. of tumors	
	Perinatal[a] application (1828 animals)	Postnatal[b] application (2460 animals)
Brain		
Oligodendroglioma	11	2
Medulloblastoma	4	0
Astrocytoma	3	0
Oligoastrocytoma	1	0
Gangliocytoma	1	0
Chorioid plexus papilloma	0	1
Unclassified neuroectodermal	6	0
Nerves		
Neurinoma	71	5
Other	3	0
Meninges		
Meningioma	2	0
Myxofibroma	2	0
Sarcoma	5	0
Total	109	8

(From WECHSLER et al. 1979).
[a] Prenatal (days 12–18 of gestation) plus neonatal (day <1 after birth).
[b] Days 15–42 after birth.

common types. Interestingly, in mice in contrast to rats, medulloblastomas, a characteristic brain tumor in humans, are inducible by neurotropic carcinogens. On the other hand, ependymomas, a common tumor type in rats following transplacental exposure to ENU, were not observed in mice (WECHSLER et al. 1979).

Comparing the anatomic distribution of neurogenic tumors induced transplacentally in rats and mice by ENU uncovers some further distinctions between the two species (Table 5): In mice, the ratio of tumors in the peripheral nervous systems to those in the CNS is double that in rats (2.4 vs. 1.2). No tumors were observed in the spinal cord and on the cauda equina of mice while this localization was frequent in rats transplacentally exposed to ENU (WECHSLER et al. 1979). In mice, the neurotropic carcinogenicity of ENU is largely strain-dependent. This is expressed by the considerable range of neurogenic tumor incidence given in Table 6, which additionally shows that in hamsters only tumors of the peripheral nervous system were induced by transplacental administration of ENU (MENNEL and ZÜLCH 1972).

Table 5. Comparison of the localization of neurogenic tumors induced transplacentally by ethylnitrosourea in rats and mice

Localization	No. of tumors	
	BD IX Rats ($n = 612$)	Mice[a] ($n = 1652$)
Total	1021	98
CNS		
Cerebrum	267	15
Optic nerve	0	2
Pons and medulla oblongata	21	0
Cerebellum	23	3
Spinal cord	130	0
Meninges	18	9
PNS		
Cranial nerves	204	54
Peripheral nerves	319	15
Cauda equina	39	0
Ratio PNS/CNS	1.2	2.4

(From WECHSLER et al. 1979).
CNS, Central nervous system; PNS, peripheral nervous system.
[a] Different strains (for details see WECHSLER et al. 1979).

WECHSLER and coworkers (1979) compared the sensitivity of different prenatal phases of development in mice and rats to ENU. While rats exhibited the highest sensitivity of the nervous system towards the end of gestation (from day 16) and had only a marginal sensitivity on day 12, the mouse on the other hand showed the highest sensitivity between days 12 and 14 of gestation; thereafter, the sensitivity declined.

Comparison of latency periods of neurogenic tumors induced by ENU in rats and mice revealed that for all sites latency in mice is approximately double that in rats (WECHSLER et al. 1979). The short latency period of neurogenic tumors in rats is apparently the reason for the fact that in this species nonneurogenic tumors are rather the exception following higher transplacental dosages of ENU. Vice versa, the high incidence of non-neurogenic tumors in mice attributable to transplacental ENU administration (RICE 1969; KOESTNER et al. 1979; RICE and WARD 1982) appears at least in part explicable by the longer manifestation time of neurogenic tumors in this species. That the relative incidence of neurogenic and non-neurogenic tumors in rats following transplacental exposure to ENU is essentially determined by the manifestation time of neurogenic tumors is documented by the observation that a decrease of the transplacental dose of ENU is associated with a significant increase in survival time of offspring and a concomitant increase in the rate of nonneurogenic tumors, which have longer latency periods (Table 3) (SWENBERG et al. 1972b).

Table 6. Relative incidence of different tumor types in rats, mice and hamsters following perinatal application of ethylnitrosourea

Organ	Rats (%)	Mice (%)	Hamsters (%)
Brain	27–69	1–3	0
Spinal cord	4–24	0	0
Cranial nerve	15	0.5–29	25
Peripheral nerve	8–30	1	36
Total neurogenic	100	2.5–32	61
Lymphopoietic ⎫ Hematopoietic ⎭	0	6.5	0
Stomach	0	3.2	0
Lung	0	54.8	0
Liver	0	38.7[a]	0
Other	4 (renal)	0	18
Total nonneurogenic	4	80	18

(From KOESTNER et al. 1979).
[a] Similar incidence in controls.

Table 7. Carcinogenicity of ethylnitrosourea for the fetal, infant, and adult nervous system in 5 species

	Transplacental exposure	Neurogenic tumors[a]/animals at risk	
		Early postnatal administration	Administration during adolescent or adult life
Rat	1021/612	131/72	6/102
Mouse	98/1652	7/222	3/540
Rabbit	8/12	–	0/8
Opossum	–	12/198	0/46
Patas monkey	4/51	–	0/70

(From RICE and WARD 1982).
[a] Including tumors of brain, spinal cord, meninges, and peripheral nerves.

The age-dependence of susceptibility to the induction of neurogenic tumors by ENU is not confined to rats or mice. As shown in Table 7, in three other species (rabbit, opossum, and patas monkey) the nervous system in prenatal or early postnatal phases of development is significantly more sensitive than in later phases (RICE and WARD 1982). Since results observed in nonhuman primates appear to be of special importance for the human situation, these data will be addressed below in a separate chapter.

Table 8. Incidence and localization of tumors in Wistar rats following a single administration of n-butylnitrosourea 120 mg/kg: comparison of transplacental application, direct application to fetuses after surgical delivery, and neonatal application

Group	Application	Rats with tumor(s), no./total(%)	Total no. of tumors	Localization (no.)				
				Brain	Spinal cord	Cranial nerves	PNS	Others
1	Transplacental, day 22, p.c.							
a	p.o	10/15 (66)	12	6	1	1	2	2
b	i.p.	9/15 (60)	11	3	2	1	1	4
c	s.c.	7/21 (33)	10	3	2	1	1	3
2	Surgical delivery day 22, p.c., s.c.	27/50 (54)	31	7	5	9	7	3
3	Neonatal day 1 or 2, s.c.	26/30 (87)	27	6	5	5	10	1

(From ZELLER et al. 1978).
p.o., peroral; s.c., subcutaneous; p.c., post conceptionem; PNS, peripheral nervous system.

2. Transplacental and Postnatal Activity of Further Compounds of the Homologous Series of Alkylnitrosoureas in Rats

Investigation of other compounds of the homologous series of alkylnitrosoureas in prenatal phases of development revealed a distinctly weaker activity of MNU. ALEXANDROV (1969) observed only 13 animals with malignant tumors including 6 animals with tumors of the nervous system in 33 offspring of rats that had been treated with MNU 20 mg/kg on the 21st day of gestation. The investigation of *n*-propyl-NU in rats revealed a prenatal neurotropic carcinogenic activity comparable with ENU (IVANKOVIC and ZELLER 1972). The prenatal carcinogenic action of BNU lies approximately between those of MNU and ENU (ZELLER et al. 1978). Depending on the route of administration, 33%–66% of the progeny developed tumors (predominantly neurogenic) after diaplacental application of BNU on day 22 of gestation; following direct application of the substance to the fetuses after surgical delivery on the same day with subsequent feeding by nurses, 54% of the animals developed tumors (predominately neurogenic) (Table 8), thus demonstrating that the maternal metabolism is without importance for the transplacental carcinogenic action of this compound. Application of BNU 1 or 2 days later increased the yield of neurogenic tumors (Table 8) (ZELLER and IVANKOVIC 1972; ZELLER et al. 1978). MAEKAWA et al. (1977) observed neurogenic tumors in the offspring when sodium nitrite and butylurea, the precursors of BNU, were given simultaneously perorally to pregnant rats. BNU has also a neurotropic carcinogenic action if administered to newborn rats via maternal milk (MAEKAWA and ODASHIMA 1975).

Compounds of the homologous series of alkylnitrosoureas with longer alkyl chains, such as *n*-pentyl-, *n*-hexyl-, *n*-heptyl-, or *n*-octyl-NU, did not show any carcinogenic effect in prenatal experiments with rats (IVANKOVIC and ZELLER 1972; IVANKOVIC 1979). The carcinogenic activity of the homologous series of alkylnitrosoureas in different postnatal phases of development of BD IX rats (days 1, 10, 30, and 60 after birth) is given in Table 9; the data on ENU originate from an investigation by DRUCKREY et al. (1970d). In each experimental group, a single dose, equivalent to 10% of the LD_{50} (in adult rats) was administered subcutaneously, except for ENU, which (only on day 30) was injected intravenously. The long-chain compounds also revealed a marked carcinogenic effect; the highest neurotropic carcinogenic effect of all compounds was observed following administration on days 1 or 10 of life, whereas treatment on day 60 produced fewer malignant tumors in the nervous system (IVANKOVIC et al. 1981; ZELLER et al. 1982).

III. Diagnosis and Histology of Neurogenic Tumors in Rodents

Depending on the localization of brain tumors in rats, spontaneous death occurred either rapidly or more or less delayed; increased irritability, apathy, uncontrolled movements, and refusal to take food were typical

Table 9. Neurotropic carcinogenesis following administration of single doses of compounds of the homologous series of alkylnitrosoureas in different phases of postnatal development in BD rats

Compound[a], day of application after birth	No. of rats with tumor/total (%)	Total no. of tumors (= 100%)	No. of neurogenic tumors (%)	No. of nonneurogenic tumors (%)
MNU				
1	15/26 (58)	18	6 (33)	12 (67)
10	15/30 (50)	16	7 (44)	9 (56)
30	18/35 (51)	21	7 (33)	14 (67)
60	12/16 (75)	12	3 (25)	9 (75)
ENU				
1	16/16 (100)	35	34 (97)	1 (3)
10	16/18 (89)	31	30 (97)	1 (3)
30	18/29 (62)	23	18 (78)	5 (22)
n-Propyl-NU				
1	19/23 (83)	20	17 (85)	3 (15)
10	10/25 (40)	11	9 (82)	2 (18)
30	13/29 (45)	15	9 (60)	6 (40)
60	6/21 (29)	6	1 (17)	5 (83)
BNU				
1	21/29 (72)	25	24 (96)	1 (4)
10	18/24 (75)	23	18 (78)	5 (22)
30	16/26 (62)	19	6 (32)	13 (68)
60	7/26 (27)	9	3 (33)	6 (67)
n-Pentyl-NU				
1	12/25 (48)	14	10 (71)	4 (29)
10	13/24 (54)	16	12 (75)	4 (25)
30	15/25 (60)	16	9 (56)	7 (44)
60	4/25 (16)	4	2 (50)	2 (50)
n-Hexyl-NU				
1	13/20 (65)	13	9 (69)	4 (31)
10	12/24 (50)	13	6 (46)	7 (54)
30	11/26 (42)	11	5 (45)	6 (55)
60	5/28 (18)	5	1 (20)	4 (80)
n-Heptyl-NU				
1	5/25 (20)	5	1 (20)	4 (80)
10	11/17 (65)	11	6 (55)	5 (45)
30	4/25 (16)	4	1 (25)	3 (75)
60	3/22 (14)	3	1 (33)	2 (67)
n-Octyl-NU				
1	10/22 (45)	11	9 (82)	2 (18)
10	5/21 (24)	6	1 (17)	5 (83)
30	8/18 (44)	8	1 (12.5)	7 (87.5)
60	9/22 (41)	10	—	10 (100)

(From ZELLER et al. 1982).
[a] Dose: 10% of the LD_{50}
NU, nitrosourea; M, methyl; E, ethyl; B, n-butyl.

"clinical" symptoms before the animals fell into coma (IVANKOVIC 1975). Comparable symptoms caused by increasing brain pressure were described in animals with trigeminal nerve tumors. Tumors of the peripheral nervous system were usually early diagnosable by increasing weakness of the respective extremity. "Clinical" symptoms of spinal cord tumors ranged from motor weakness in earlier stages to monoplegia, paraplegia, or tetraplegia in advanced stages of development (IVANKOVIC 1975).

The histology of neurogenic tumors induced in prenatal and postnatal phases of development by nitrosamides was described in detail by WECHSLER et al. (1969), ZÜLCH and MENNEL (1971), as well as by DRUCKREY et al. (1972). Comprehensive reviews were given by KLEIHUES et al. (1976a) and MENNEL and IVANKOVIC (1975). Except for single observations suggesting in isolated cases neoplastic transformation of neurons also (STROOBANDT and BRUCHER 1968; KLEIHUES et al. 1970; JONES et al. 1973), analysis of a large number of tumors revealed that mainly glial cells and Schwann cells were targets of the carcinogenic action (KLEIHUES et al. 1976a).

Accordingly, in the CNS predominantly oligodendrogliomas, astrocytomas, mixed gliomas, and ependymomas were observed, while in the peripheral nervous system malignant neurinomas (schwannomas) were found. With regard to a predominance of fibrillar or reticular cell populations, so-called Antony type A or type B neurinomas, respectively, were distinguished. As an example, Table 10 gives a classification and percentage distribution of 1731 tumors of the central and peripheral nervous systems as reported by MENNEL and IVANKOVIC (1975). It should, however, be noted that slightly different classifications were suggested by other authors.

Table 10. Classification of 1731 tumors of the nervous system

Tumor type	Percentage
Oligodendrogliomas	
isomorphic	8.3
polymorphic	0.1
Astrocytomas	
isomorphic	5.0
polymorphic	0.2
Mixed gliomas	
isomorphic	17.5
polymorphic	3.9
Ependymomas	
isomorphic	11.8
polymorphic	1.3
Malignant neurinomas	42
Esthesioneuroepitheliomas/medulloepitheliomas	1.3
Hypophyseal adenomas	2.9
Meningeal tumors	1.2
Unclassified and others	4.4

(MENNEL and IVANKOVIC 1975).

IV. Neurotropic Carcinogenesis in Nonhuman Primates

Rice and coworkers (1979, 1989) investigated the effects of ENU both in adult nonhuman primates and in their offspring following transplacental administration. In patas monkeys transplacentally exposed to ENU early in gestation, neurogenic tumors were observed exclusively in the brain (cerebrum, cerebellum, brainstem); schwannomas of the peripheral nervous system were not found (Rice et al. 1989). Besides brain tumors, a variety of other tumors including nephroblastomas, mesenchymomas, vascular tumors, etc. was found. Early stages of prenatal development in the patas monkey, which correspond to the first trimester of human pregnancy, represent a period of distinct susceptibility to transplacental carcinogenesis (Rice et al. 1989). While Jaenisch et al. (1977) noted no transplacental carcinogenic effects of MNU and ENU in rhesus monkeys, Rice et al. (1989) saw tumors in the offspring of rhesus monkeys following transplacental exposure to ENU. The difference between both investigations was that the latter administered ENU early in gestation and gave repeated dosage while the former administered not more than two doses towards the end of gestation. These results confirm that in nonhuman primates early stages of gestation represent vulnerable phases of prenatal developement (Rice et al. 1989). Repeated intravenous administration of ENU to pregnant patas monkeys caused gestational choriocarcinoma in the dams; this is the only animal model for this rapidly proliferating tumor type (Rice et al. 1989). Certain parallels to observations in rats are discernible where in the dams an increased incidence of ovarian tumors was observed following administration of ENU during pregnancy (Ivankovic 1969). The conclusion by Rice et al. (1989) that gestational choriocarcinoma in the mother may serve as an epidemiologically exploitable marker for human populations in which transplacental carcinogenesis appears to be likely deserves special attention.

V. Further *N*-Nitroso Compounds

Application of N,N'-dimethyl-NU (3 or 6 mg/kg daily) in the drinking water resulted in the development of neurogenic tumors in 73% and 100% of the animals (Druckrey et al. 1967b). Following administration of N,N',N'-trimethyl-N-NU in the drinking water, Ivankovic et al. (1965) observed neurogenic tumors in 57% of treated rats (8 out of 14).

Chronic application of N-methyl-N-nitrosobiuret in the drinking water at a daily dosage of 5 or 10 mg/kg resulted in neurogenic tumors in 2 out of 10 and 9 out of 16 rats, respectively (Druckrey et al. 1971).

N-Ethyl-N-nitrosobiuret was investigated in adult animals and in perinatal phases of development. Single s.c. administration (40 or 80 mg/kg) to 10-day-old rats resulted in the development of a high yield of neurogenic tumors (Druckrey and Landschütz 1971). Following transplacental exposure, the majority of offspring (>90%) developed tumors of the central and

Table 11. Azo-, azoxy-, and hydrazocompounds inducing neurogenic tumors

Chemical designation	Chemical formula	Animal species
Azoethane	$C_2H_5-N=N-C_2H_5$	Rat
Azoxymethane	$CH_3-N=N-CH_3$ $\downarrow$ O	Rat
Azoxyethane	$C_2H_5-N=N-C_2H_5$ $\downarrow$ O	Rat
Methylazoxymethanol (aglycone of cycasin)	$CH_3-N=N-CH_2OH$ $\downarrow$ O	Rat
1,2-Diethylhydrazine	$C_2H_5-NH-NH-C_2H_5$	Rat
1-Methyl-2-benzylhydrazine	$CH_3-NH-NH-CH_2-$⬡	Rat
Procabarzine	$CO-NH-CH-(CH_3)_2$ ⬡ $CH_2-NH-NH-CH_3$	Rat

peripheral nervous systems. In adult BD rats, oral application of ethyl-nitrosobiuret resulted in the development of carcinomas of the stomach (DRUCKREY and LANDSCHÜTZ 1971).

Continuous oral application of N-methyl-N-nitroso-N'-acetylurea resulted both in the development of stomach tumors and in neurogenic tumors (DRUCKREY et al. 1970a; MAEKAWA et al. 1976). Some chemotherapeutically active 2-chloroethylnitrosoureas were shown to have a proven but limited potential to induce neurogenic tumors in rats (ZELLER and SCHMÄHL 1979; HABS 1981; ZELLER et al. 1982).

VI. Azo-, Azoxy-, and Hydrazocompounds

Subcutaneous application of azoethane (Table 11) in adult rats induced brain tumors in 10% (DRUCKREY et al. 1965b); following inhalation of azoethane during pregnancy, nearly all offspring developed tumors of the central and peripheral nervous systems (DRUCKREY et al. 1968). While trans-placental application of 1,2-dimethylhydrazine and azoxymethane on day 15 of pregnancy in rats did not induce tumors in the offspring, application of 1,2-diethylhydrazine as well as azo- and azoxyethane on the same day of gestation resulted in the development of neurogenic tumors in almost all offspring (DRUCKREY 1973). Towards the end of pregnancy (day 22) a low

Table 12. Induction of neurogenic tumors in rats by 1-phenyl-3,3-dimethyltriazene (PDMT) and 1,2-diethylhydrazine

Compound, route of administration	n	No. of rats with tumors in		
		Brain	Spinal cord	Peripheral nervous system
PDMT, p.o.	12	8	–	1
PDMT, s.c.	21	9	2	3
Diethylhydrazine, s.c.	45	8	1	–

p.o., peroral; s.c. subcutaneous.
(From Druckrey et al. 1966b, 1967a).

incidence of neurogenic tumors was also observed following transplacental exposure to azoxymethane (Druckrey 1973). Like dialkylnitrosamines and aryldialkyltriazenes, compounds of this group require enzymatic activation to form the "proximate" carcinogen. Following subcutaneous administration of 1,2-diethylhydrazine to rats, 20% of the animals developed neurogenic tumors, predominantly of the brain (Druckrey et al. 1966b) (Table 12). Cycasin (β-D-glycosyloxyazoxymethane), a naturally occurring azoxyalkane, was shown to possess a limited potential to induce neurogenic tumors (Laqueur and Spatz 1973). 1-Methyl-2-benzylhydrazine induces tumors of the central and peripheral nervous systems in rats following subcutaneous or peroral treatment (Druckrey 1970).

VII. Aryldialkyltriazenes

A further group of "indirect" alkylating carcinogens studied in perinatal phases of development and in adult animals are aryldialkyltriazenes (Table 13) (Druckrey et al. 1967a; Preussmann et al. 1969a,b; Imhof and Ivankovic 1983). As given in Table 12, 1-phenyl-3,3-dimethyltriazene (PDMT) showed a significant neurotropic carcinogenic activity when given to adult animals. Comparison of the activity of PDMT and 1-phenyl-3,3-diethyltriazene showed that the dimethyl compound was not carcinogenic to the rat fetus following transplacental administration on day 15 of pregnancy; however, towards the end of gestation (23rd day), a dose of 75 mg/kg of PDMT resulted in the development of 12 neurogenic tumors in 40 surviving offspring (Druckrey 1973). The corresponding diethyl compound was transplacentally already highly active on day 15 of pregnancy, inducing 11 neurogenic tumors in 12 surviving offspring; the same is valid for 1-pyridyl-3,3-diethyltriazene (Druckrey 1973). Diaplacental administration of the cytostatic drug DTIC [5-(3,3-dimethyl-1-triazeno)-imidazole-4-carboxamide] at the end of pregnancy in BD IX rats gave rise to the development of 10 malignant tumors in 9 out of 39 offspring. The majority (6 of 10) were

Table 13. Aryldialkyltriazenes inducing neurogenic tumors

Chemical designation	Chemical formula	Animal species
1-Phenyl-3,3-dimethyltriazene		Rat
1-Phenyl-3,3-diethyltriazene		Rat
1-(*o*-Methylphenyl)-3,3-dimethyltriazene		Rat
1-(*m*-Methylphenyl)-3,3-dimethyltriazene		Rat
1-(*p*-Chlorphenyl)-3,3-dimethyltriazene		Rat
1-(*p*-Nitrophenyl)-3,3-dimethyltriazene		Rat
1-Pyridyl-3,3-dimethyltriazene		Rat
1-(Pyridyl-3-*N*-oxid)-3,3-dimethyltriazene		Rat
1-Pyridyl-3,3-diethyltriazene		Rat
1-(5-Imidazole-4-carboxamide)-3,3-dimethyltriazene		Rat

neurogenic (malignant neurinomas) (ZELLER 1980). Chronical peroral application of 1-pyridyl-3,3-diethyltriazene to adult rats resulted in the formation of malignant tumors in the heart which were of neurogenic origin (IVAN-KOVIC et al. 1972). Several other triazenes have been synthesized and tested for carcinogenicity (PREUSSMANN et al. 1974).

Table 14. Other agents with neurotropic carcinogenic action

Chemical designation	Chemical formula	Animal species
Methylmethanesulfonate	$CH_3SO_2OCH_3$	Rat
Ethylmethanesulfonate	$CH_3SO_2OC_2H_5$	Rat
Dimethylsulfate	$(CH_3)_2SO_4$	Rat
Diethylsulfate	$(CH_2-CH_3)_2SO_4$	Rat
Propanesultone		Rat
Propyleneimine		Rat
Acrylonitrile	$CH_2=CHCN$	Rat
Vinyl chloride		Rat
Ethylene oxide		Rat
Bis(chloromethyl)ether	$ClCH_2OCH_2Cl$	Rat

VIII. Other Alkylating Agents and Miscellaneous Compounds

Other alkylating agents with proven neurotropic carcinogenic action are methylmethanesulfonate, ethylmethanesulfonate, dimethylsulfate, diethylsulfate, propanesultone, and propyleneimine. Their structures are shown in Table 14. Following transplacental administration of methylmethanesulfonate, approximately 20% of the offspring developed neurogenic tumors (KLEIHUES et al. 1972). In adult rats, there is only limited evidence of a neurotropic carcinogenic action of methylmethanesulfonate and ethylmethanesulfonate (SWANN and MAGEE 1969).

Following transplacental or postnatal administration of dimethylsulfate, neurogenic tumors were observed in the minority of exposed animals (<5%)

(DRUCKREY et al. 1970b). Transplacental administration of diethylsulfate resulted in neurogenic tumors in 2 out of 30 offspring (DRUCKREY et al. 1970b).

Propanesultone showed a marked neurotropic carcinogenic activity both following administration in adult rats (ULLAND et al. 1971) and after transplacental administration (DRUCKREY et al. 1970c). Propyleneimine induced brain tumors in approximately 10% of rats (10 out of 104) following repeated administration to adult rats (ULLAND et al. 1971).

Compounds with limited neurotropic carcinogenic action following postnatal administration in rats are 2-acetylaminofluorene (OYASU et al. 1970), lead subacetate (OYASU et al. 1970), elaiomycin (SCHOENTAL 1969), and pyrrolizidine alkaloids (SCHOENTAL and CAVANAGH 1972).

Industrial chemicals known to produce brain tumors in experimental animals include vinyl chloride (MALTONI and LEFEMINE 1975; MALTONI 1977; MALTONI et al. 1982), acrylonitrile (MALTONI et al. 1977; EPA 1980; BIGNER et al. 1986), bis(chloromethyl)ether (KUSCHNER et al. 1975), and ethylene oxide (GARMAN et al. 1986). For further compounds and combinations producing a low incidence of neurogenic tumors in rats following systemic administration, see KLEIHUES et al. (1976a). Finally, it has to be noted that a variety of N-nitrosodialkylamines (nitrosamines) is able to induce esthesioneuroepitheliomas mainly located in the nasal cavity (GARCIA et al. 1970; GARCIA and LIJINSKY 1972; REZNIK et al. 1975; REZNIK-SCHÜLLER 1983; reviews, LIJINSKY 1987; MAEKAWA and MITSUMORI 1990).

IX. Mechanisms Involved in Neurooncogenesis

Chemical modification of DNA bases by alkylation or aralkylation is considered to be a crucial event in the initiation of malignant transformation (review, LAWLEY 1990). At least 13 sites are known for alkylation of DNA by the reaction of alkylating agents under physiological conditions (PEGG 1990). Compounds reacting predominantly at ring N-atoms of DNA bases are in general less carcinogenic but strongly cytotoxic, while agents that preferentially alkylate oxygen positions usually have a higher carcinogenic potential (PEGG 1977). To uncover possible differences in alkylation between brain (target tissue) and nontarget tissues, two approaches may be chosen: determination of the extent of alkylation at different positions of DNA bases or determination of the persistence of different alkyl adducts.

The extent of alkylation of DNA in target and nontarget tissues produced by neurooncogenic agents was found to be essentially dependent on the metabolism of these compounds. Since, in contrast to N-nitrosodialkylamines and dialkyltriazenes, nitrosamides do not require enzymatic activation but decompose spontaneously, levels of DNA alkylation following nitrosamide administration are similar in brain, liver, and other tissues (KLEIHUES and MAGEE 1973). Accordingly, following transplacental administration of ENU (GOTH and RAJEWSKY 1972) or methylmethanesulfonate (KLEIHUES et al.

1974), no significant differences were found in DNA alkylation rates between target and nontarget tissues. On the other hand, following administration of dialkylaryltriazenes to adult animals, DNA alkylation in the liver was higher than in the brain, since the latter, although being the principal target organ, is unable to activate these compounds (KLEIHUES et al. 1976b). After a single administration of [^{14}C]-3,3-dimethyl-1-phenyltriazene (DMPT) to pregnant rats, the extent of DNA methylation was similar in fetal liver and brain (KLEIHUES et al. 1979), since during prenatal development both liver and brain DNA are transplacentally methylated by a proximate carcinogen produced in the maternal organism. Administration of DMPT in the neonatal and postnatal phases was followed by a significant decrease of the brain to liver ratio of 7-methylguanine and O^6-methylguanine; in 30-day-old rats this ratio was only 0.12 for 7-methylguanine and 0.3 for O^6-methylguanine, reflecting the maturation of the hepatic drug-metabolizing enzyme system (KLEIHUES et al. 1979).

In contrast to, e.g., N^3- and N^7-alkylguanines which are lost spontaneously by chemical (nonenzymic) depurination, O^6-alkylguanine is chemically stable; this alkyl adduct, however, can be removed by the repair protein O^6-alkylguanine-DNA alkyltransferase (AGT) (reviews, WIESTLER et al. 1985, PEGG 1990). Investigations of the persistence of O^6-alkylguanine in DNA of brain and other organs exposed to ENU and MNU uncovered an inverse relationship between the carcinogenicity of both compounds and the capacity for repair of their adducts (GOTH and RAJEWSKY 1974; KLEIHUES and MARGISON 1974). Following administration of a single dose of ENU to 10-day-old rats, a significantly slower elimination rate from DNA of O^6-ethylguanine in the brain (half-life about 220 h) than in the liver (half-life about 30 h) was observed, suggesting that this insufficient repair capacity might be a determinant in the nervous system-specific neoplastic transformation by ENU, e.g., by enhancing the probability of "fixation" of structural DNA alterations during subsequent replication (GOTH and RAJEWSKY 1974). This reduced repair is not restricted to O^6-ethylation and is also to be noted in adult animals; for example, in adult BD IX rats, O^6-methylguanine was removed less rapidly from DNA of the brain, the principal target organ of MNU carcinogenicity, than from DNA of the liver (KLEIHUES and MARGISON 1974) (Table 15). Following a single dose of MNU, a rapid decrease of O^6-methylguanine was observed during the first hours after administration. Up to 14 days, O^6-methylguanine was reduced to 50% of the initial concentration; thereafter, the rate of removal decreased, so that 25% of the initial concentration was still detectable in cerebral DNA after 6 months. Extrapolation of these results suggested that complete removal of O^6-methylguanine would require approximately 1 year (KLEIHUES and BÜCHELER 1977).

One week after a single administration of ENU (75 mg/kg) or MNU (10 mg/kg) to 10-day-old BD IX rats, the brain-to-liver ratio of O^6-ethylguanine and O^6-methylguanine was about 20 and 90, respectively, i.e.,

Table 15. Persistence of methylated purines in DNA of brain and liver following single-dose administration of methylnitrosourea (MNU) 80 mg/kg in rats

Organ	Time after [³H]-MNU pulse (h)	Methylated bases (mol % × 10⁴ of the parent base)			Ratio O⁶-MeG/7-MeG
		3-Methyladenine	7-Methylguanine	O⁶-Methylguanine	
Liver	6	47.5	1470.9	122.9	0.084
	78	4.7 (10)[a]	742.5 (51)	19.9 (16)	0.027
Brain	6	36.8	1293.4	163.4	0.126
	78	3.6 (10)	565.2 (44)	109.3 (67)	0.193

(From KLEIHUES and MARGISON 1974).
[a] Numbers in parentheses express the amount at 78 h as percentage of the amount at 6 h.

Table 16. Accumulation of O^6-methylguanine in rat brain DNA following repetitive administration of methylnitrosourea (MNU)[a]

Organ	$10^4 \times$ mol % of guanine		Ratio O^6-MeG/7-MeG
	O^6-Methylguanine	7-Methylguanine	
Brain	59.70	88.42	0.675
Kidney	13.93	61.35	0.227
Small intestine	2.79	22.25	0.125
Liver	0.45	43.72	0.010

(From MARGISON and KLEIHUES 1975).
[a] Five injections of [^{3}H]-MNU 10 mg/kg each at weekly intervals; determination 1 week after the final dose.

by the end of 1 week the O^6-ethylguanine and O^6-methylguanine concentrations in the brain were 20 and 90 times higher, respectively, than in the liver. On the other hand, the brain-to-liver ratio of 7-ethylguanine and 7-methylguanine increased only slightly during the first week following administration of both compounds, and this ratio remained <1 (GOTH and RAJEWSKY 1974; KLEIHUES et al. 1979).

Since in adult rats selective induction of neurogenic tumors by MNU requires repetitive administration of low doses, MARGISON and KLEIHUES (1975) investigated concentrations of O^6-methylguanine in DNA of various rat tissues following repetitive application of MNU. After five weekly applications of MNU 10 mg/kg each, the brain-to-liver ratio of O^6-methylguanine was 133 (brain-to-kidney ratio: 4.3; brain-to-spleen ratio: 12.3; brain-to-small-intestine ratio: 21.4) (Table 16). An accumulation of 3-methylguanine was observed only in brain DNA and not in DNA of other organs (MARGISON and KLEIHUES 1975). While investigations in rats suggest that O^6-alkylation of guanine in DNA plays a decisive role for neurotropic carcinogenesis by alkylating agents, studies in other species do not support this hypothesis and have left some questions open. As detailed above, neurotropic carcinogenesis by nitrosamides in mice is significantly less pronounced as compared with rats; repair of O^6-methylguanine from brain DNA of mice, however, is also lower than in other organs, e.g., liver, kidney, or intestine. Investigations in different strains of mice with different susceptibilities to the neurotropic carcinogenic effect of MNU uncovered a similar rate of O^6-methylguanine repair (KLEIHUES et al. 1982). Of special interest are the results obtained in gerbils. In this species, a single administration of ENU (50 mg/kg) given postnatally (days 7 or 20) resulted in the development of cutaneous melanomas in 43% of the animals (KLEIHUES et al. 1978), while administration of MNU to adult gerbils led predominantly to carcinomas of the midventral gland (HAAS et al. 1975). In both investigations, the development of neurogenic tumors, which are characteristic for these compounds when given to other species, was not observed; thus, the gerbil brain is apparently not susceptible to the neurotropic carcinogenic potential of nitrosamides. Despite this, following a single i.v. administration of N-[^{14}C]-MNU 10 mg/kg to adult female gerbils, the persistence of O^6-methylguanine in brain DNA was much more pronounced than in other organs. For instance, on day 30, 80% of the initial amount of O^6-methylguanine was still detectable in the brain (for comparison: lung 40%, kidney 18%, liver not calculable trace amounts); 6 months following this treatment, in brain DNA of gerbils 40% of the initial O^6-methylguanine concentration was still present, while concentrations in other organs were not calculable trace amounts (lung, kidney) or were not detectable (liver) (KLEIHUES et al. 1980).

Altogether, these results emphasize that the formation and persistence of O^6-alkylguanine may constitute only one event for the initiation of carcinogenesis by monofunctional alkylating agents in the brain.

Tumors induced by nitrosamides in rats are as a rule of glial origin. Various possibilities were discussed as an underlying cause for the differing sensitivity of glial and neuronal cells to neurotropic carcinogenic agents. In order to exclude possible variations in alkylation between both cell types, KLEIHUES et al. (1973) investigated the question of whether neuronal cells are alkylated by nitrosamides to the same extent as glial cells. They showed that methylation of neuronal cell nuclear DNA was not less extensive than that of glial cell DNA. These results support the view that the failure of tumor induction in neuronal cells is caused by the fact that these cells do not divide, so that miscoding during the subsequent cell cycle cannot occur.

X. Malignant Transformation of Rat Neural Cells In Vitro and In Vivo

Transplacental exposure of BD IX rats to a single dose of ENU results in a high incidence of neurogenic tumors in the offspring. Sequential morphological and biochemical changes following exposure to ENU have been studied by several groups (for details, see LANTOS 1972; ZÜLCH and MENNEL 1973; SWENBERG et al. 1972c, 1975b; PILKINGTON and LANTOS 1979; RAJEWSKY 1980; reviews LANTOS 1986; PILKINGTON 1989; BILZER et al. 1989). Increased cellularity of the trigeminal nerve was already found 20 days following transplacental exposure to ENU. One month after exposure, more than 50% of the trigeminal nerves exhibited increased cellularity. Transplantation of grossly normal trigeminal nerves from approximately 1-, 2- and 3-month-old prenatally ENU-exposed rats developed into neurinomas at the site of transplantation, so the observed increased cellularity was an expression of early neoplastic proliferation (SWENBERG et al. 1975b).

Laerum and coworkers investigated cultures of rat fetal brain cells which had been exposed transplacentally to single doses of ENU (LAERUM and RAJEWSKY 1975; LAERUM et al. 1979, 1982). They observed malignant transformation in vitro during a time period similar to that of tumor development in vivo; after 200–270 days, ENU-pretreated cultured cells formed tumors following subcutaneous injection of these cells into 5–10-day-old rats. Malignancy of these cultured cells could also be assessed in vitro by infiltration into and destruction of chicken embryo heart muscle fragments (LAERUM et al. 1982).

In order to elucidate the time course of early events in neoplastic transformation, KOESTNER et al. (1979) exposed pregnant rats intravenously to ENU (50 mg/kg). Following removal of the fetuses 1, 4, or 24 h after treatment, the brains of the offspring of each time group were pooled and transplanted subcutaneously into 3-day-old recipients. In the recipients receiving the 1-h and 4-h collections, 2 out of 11 and 1 out of 9 animals, respectively, developed subcutaneous mixed glial neoplasms.

Up to now several approaches have been taken to elucidate the histogenesis of neurogenic tumors induced by monofunctional alkylating agents like ENU. The highest incidence of ENU-induced brain tumors was observed around the lateral ventricles, suggesting that most gliomas originate from the mitotically active, undifferentiated stem cells of the subependymal plate. This was confirmed by the detection of abnormal cell clusters by 8 weeks of age following transplacental ENU exposure (LANTOS and PILKINGTON 1979) and the observation of microtumors in this area in 16-week-old rats (PILKINGTON and LANTOS 1979). Cells in the subependymal layer continue to proliferate throughout adult life (SCHULTZE-MAURER 1978).

KLEIHUES and coworkers (1989) grafted suspensions from fetal forebrains, that were exposed transplacentally 24 h before on day 14 of gestation to ENU 50 mg/kg, stereotaxically into the caudoputamen of adult rats. No tumors were observed in these recipients if no further treatment was given. If the recipients received one intravenous dose of ENU 50 mg/kg 8 days after grafting or two intravenous doses of ENU 50 mg/kg each 8 days and 9 weeks after grafting, the tumor incidence in the graft was 40% and 100%, respectively, thus presenting in vivo evidence for a multistep development of neurogenic tumors. Tumors within the graft were exclusively oligodendrogliomas, indicating that neoplastic transformation can occur in precursor cells committed to oligodendrocytic differentiation or in oligodendrocytes. These results further suggest that transformation of a pluripotent stem cell is obviously not necessary to induce neurogenic tumors (KLEIHUES et al. 1989). In accordance with LANTOS (1972) who suggested that targets for transformation by ENU are most likely primitive cells of the subependymal plate and oligodendrocytes, KLEIHUES et al. (1989) proposed two different histogenic pathways of brain tumor induction: transformation of embryonal matrix cells–the different histological types (oligodendrogliomas, astrocytomas, ependymomas, mixed gliomas) reflect subsequent differentiation–and transformation of oligodendrocytes or precursor cells committed to oligodendrocytic differentiation.

Sequential ultrastructural investigations of early neoplastic lesions in ENU-induced rat gliomas revealed that in the initial or very early stages of glial cell neoplastic proliferation the neoplastic cells obviously need contact with neurons for metabolic purposes; following loss of this contact, the

neoplastic cells proliferate autonomously. The labelling index increased from
2.6% in hyperplastic cells to 3.3% in foci of early neoplastic proliferation to
9.6% in the central area of microtumors (IKEDA et al. 1989). Investigation of
the role of oncogene activation for neurooncogenesis (review, COOPER 1990)
demonstrated that schwannomas consistently contain the *neu* oncogene
(BARGMANN et al. 1986; PERANTONI et al. 1987). Neu is an oncogene of the
transmembrane growth factor receptor/tyrosine kinase family that appears
to be activated by a single T-A transversion mutation. It was found in
schwannomas induced by ENU in F 344 rats (PERANTONI et al. 1987). RICE
(1990) described *neu* in 32 of 32 schwannomas of Sprague-Dawley rats and
in 16 of 16 schwannomas of Syrian hamsters. This oncogene, however, was
not found to be activated in experimentally induced gliomas or in non-tumor
tissues (PERANTONI et al. 1987; RICE 1990). In order to identify cellular
oncogenes capable of transforming neural cells after introduction into susceptible neuroectodermal tissue, WIESTLER et al. (1989) investigated the
effect of the viral *src* gene (v-*src*) on developing neural transplants. Prior to
stereotaxical injection into the caudoputamen of adult host rats, transplants
were infected with the v-*src* vector. After latent periods of 2–4 months, the
recipients developed anaplastic gliomas. Extension of these studies to other
oncogenes offers a novel experimental approach for the study of oncogenesis
in the CNS (WIESTLER et al. 1989).

References

Alexandrov VA (1969) Transplacental blastomogenic action of *N*-nitrosomethylurea
    on rat offspring (in Russian). Vopr Onkol 15:55–61
Bargmann CI, Hung M-C, Weinberg RA (1986) The *neu* oncogene encodes an
    epidermal growth factor receptor-related protein. Nature 319:226–230
Bigner DD, Swenberg JA (eds) (1977) Jänisch and Schreiber's experimental tumors
    of the central nervous system. Upjohn, Kalamazoo
Bigner DD, Bigner SH, Burger PC, Shelburne JD, Friedman HS (1986) Primary
    brain tumors in Fischer 344 rats chronically exposed to acrylonitrile in their
    drinking water. Food Chem Toxicol 24:129–137
Bilzer T, Reifenberger G, Wechsler W (1989) Chemical induction of brain tumors
    in rats by nitrosourea: molecular biology and neuropathology. Neurotoxicol
    Teratol 11:551–556
Cooper CS (1990) The role of oncogene activation in chemical carcinogenesis. In:
    Cooper CS, Grover PL (eds) Chemical carcinogenesis and mutagenesis II.
    Springer, Berlin Heidelberg New York, pp 319–352 (Handbook of experimental
    pharmacology, vol 94, part 2)
Denlinger RH, Koestner A, Wechsler W (1974) Induction of neurogenic tumors in
    C3HeB/FeJ mice by nitrosourea derivatives: observations by light microscopy,
    tissue culture, and electron microscopy. Int J Cancer 13:559–571
Druckrey H (1970) Production of colonic carcinomas by 1,2-dialkylhydrazines and
    azoxyalkanes. In: Burdette WJ (ed) Carcinomas of the colon and antecedent
    epithelium. Thomas, Springfield, pp 267–269
Druckrey H (1973) Chemical structure and action in transplacental carcinogenesis
    and teratogenesis. In: Tomatis L, Mohr U (eds) Transplacental carcinogenesis.
    IARC Sci Publ 4:45–58
Druckrey H, Landschütz C (1971) Transplacentale und neonatale Krebserzeugung
    durch Äthylnitrosobiuret (ÄNBU) an BD IX-Ratten. Z Krebsforsch 76:45–58

Druckrey H, Ivankovic S, Preussmann R (1965a) Selektive Erzeugung maligner Tumoren im Gehirn und Rückenmark von Ratten durch N-Methyl-N-Nitrosoharnstoff. Z Krebsforsch 66:389–408

Druckrey H, Preussmann R, Ivankovic S, Schmidt CH, So BT, Thomas C (1965b) Carcinogene Wirkung von Azoäthan und Azoxyäthan an Ratten. Z Krebsforsch 67:31–45

Druckrey H, Ivankovic S, Preussmann R (1966a) Teratogenic and carcinogenic effects in the offspring after single injection of ethylnitrosourea to pregnant rats. Nature 210:1378–1379

Druckrey H, Preussmann R, Matzkies F, Ivankovic S (1966b) Carcinogene Wirkung von 1,2-Diäthylhydrazin an Ratten. Naturwissenschaften 53:557–558

Druckrey H, Ivankovic S, Preussmann R (1967a) Neurotrope carcinogene Wirkung von Phenyl-dimethyl-triazen an Ratten. Naturwissenschaften 54:171

Druckrey H, Preussmann R, Ivankovic S, Schmähl D (1967b) Organotrope carcinogene Wirkungen bei 65 verschiedenen N-Nitroso-Verbindungen an BD-Ratten. Z Krebsforsch 69:103–201

Druckrey H, Ivankovic S, Preussmann R, Landschütz C, Stekar J, Brunner U, Schagen B (1968) Transplacentar induction of neurogenic malignomas by 1,2-diethyl-hydrazine, azo-, and azoxy-ethane in rats. Experientia 24:561–562

Druckrey H, Preussmann R, Ivankovic S (1969) N-Nitroso compounds in organotropic and transplacental carcinogenesis. Ann NY Acad Sci 163:676–695

Druckrey H, Ivankovic S, Preussmann R (1970a) Selektive Erzeugung von Carcinomen des Drüsenmagens bei Ratten durch orale Gabe von N-Methyl-N-nitroso-N'-acetylharnstoff (AcMNH). Z Krebsforsch 75:23–33

Druckrey H, Kruse H, Preussmann R, Ivankovic S, Landschütz C (1970b) Cancerogene alkylierende Substanzen. III. Alkylhalogenide, -sulfate, -sulfonate und ringgespannte Heterocyclen. Z Krebsforsch 74:241–273

Druckrey H, Kruse H, Preussmann R, Ivankovic S, Landschütz C, Gimmy J (1970c) Cancerogene alkylierende Substanzen. IV. 1,3-Propansulton und 1,4-Butansulton. Z Krebsforsch 75:69–84

Druckrey H, Schagen B, Ivankovic S (1970d) Erzeugung neurogener Malignome durch einmalige Gabe von Äthyl-nitrosoharnstoff (ÄNH) an neugeborene und junge BD IX-Ratten. Z Krebsforsch 74:141–161

Druckrey H, Landschütz C, Preussmann R, Ivankovic S (1971) Erzeugung von Magenkrebs und neurogenen Malignomen durch orale Gabe von Methylnitrosobiuret (MNB) and Ratten. Z Krebsforsch 75:229–239

Druckrey H, Ivankovic S, Preussmann R, Zülch KJ, Mennel D (1972) Selective induction of malignant tumors of the nervous system by resorptive carcinogens. In: Kirsch WM, Grossi-Paoletti E, Paoletti P (eds) The experimental biology of brain tumors. Thomas, Springfield, pp 85–147

Environmental Protection Agency (EPA) (1980) Carcinogen assessment of acrylonitrile. Office of Research and Development, Carcinogen Assessment Group, Washington DC, p 37

Garcia H, Lijinsky W (1972) Tumorigenicity of five cyclic nitrosamines in MRC rats. Z Krebsforsch 77:257–261

Garcia H, Keefer L, Lijinsky W, Wenyon CEM (1970) Carcinogenicity of nitrosothiomorpholine and 1-nitrosopiperazine in rats. Z Krebsforsch 74:179–184

Garman RH, Snellings WM, Maronpot RR (1986) Frequency, size and location of brain tumours in F-344 rats chronically exposed to ethylene oxide. Food Chem Toxicol 24:145–153

Garrett ER, Goto S, Stubbins JF (1965) Kinetics of solvolyses of various N-alkyl-N-nitrosoureas in neutral and alkaline solutions. J Pharm Sci 54:119–123

Goth R, Rajewsky MF (1972) Ethylation of nucleic acids by ethylnitrosourea-1-^{14}C in the fetal and adult rat. Cancer Res 32:1501–1505

Goth R, Rajewsky MF (1974) Persistence of O^6-ethylguanine in rat brain DNA: correlation with nervous system-specific carcinogenesis by ethylnitrosourea. Proc Natl Acad Sci USA 71:639–643

Graw J, Zeller WJ, Ivankovic S (1974) Enstehung von neurogenen Malignomen bei den Nachkommen von Sprague-Dawley Ratten nach perkutaner Applikation von Äthylnitrosoharnstoff (ÄNH) während der Tragezeit. Z Krebsforsch 81: 169–172

Haas H, Hilfrich J, Kmoch N, and Mohr U (1975) Specific carcinogenic effect of N-methyl-N-nitrosourea on the midventral sebaceous gland of the gerbil (*Meriones unguiculatus*). JNCI 55:637–640

Habs M (1981) Animal carcinogenicity studies of nitrosatable chemicals and drugs containing N-nitroso-structures. In: Gibson GG, Joannides C (eds) Safety evaluation of nitrosatable drugs and chemicals. Pergamon, London, pp 141–156

Imhof W, Ivankovic S (1983) Karzinogene Wirkung von 1-Phenyl- und 1-(Pyridyl-3)-3,3-dimethyl-triazen sowie 1-Phenyl- und 1-(Pyridyl-3) -3,3-diäthyl-triazen bei einmaliger prä- und postnataler Verabreichung an BD IX Ratten. Arch Geschwulstforsch 53:557–569

Ikeda T, Mashimoto H, Iwasaki K, Shimokawa I, Matsuo T (1989) A sequential ultrastructural and histoautoradiographic study of early neoplastic lesions in ethylnitrosourea-induced rat glioma. Acta Pathol Jpn 39:487–495

Ivankovic S (1969) Erzeugung von Genitalkrebs bei trächtigen Ratten. Arzneimittelforschung 19:1040–1041

Ivankovic S (1975) Pränatale Carcinogenese. In: Grundmann E (ed) Geschwülste/tumors III. Springer, Berlin Heidelberg New York, pp 941–1002 (Handbuch der allgemeinen Pathologic, vol 6, part 7)

Ivankovic S (1979) Teratogenic and carcinogenic effects of some chemicals during prenatal life in rats, Syrian golden hamsters, and minipigs. NCI Monogr 51: 103–115

Ivankovic S, Druckrey H (1968) Transplacentare Erzeugung maligner Tumoren des Nervensystems I. Äthylnitrosoharnstoff (ÄNH) an BD IX-Ratten. Z Krebsforsch 71:320–360

Ivankovic S, Preussmann R (1970) Transplazentare Erzeugung maligner Tumoren nach oraler Gabe von Äthylharnstoff und Nitrit an Ratten. Naturwissenschaften 57:460

Ivankovic S, Zeller WJ (1972) Transplazentare blastomogene Wirkung von n-Propyl-nitrosoharnstoff an BD-Ratten. Arch Geschwulstforsch 40:99–102

Ivankovic S, Druckrey H, Preussmann R (1965) Erzeugung von Tumoren im peripheren und zentralen Nervensystem durch Trimethyl-nitroso-harnstoff an Ratten. Z Krebsforsch 66:541–548

Ivankovic S, Druckrey H, Preussmann R (1966) Erzeugung neurogener Tumoren bei den Nachkommen nach einmaliger Injektion von Äthylnitrosoharnastoff an schwangeren Ratten. Naturwissenschaften 53:410

Ivankovic S, Wohlenberg H, Mennel HD, Preussmann R (1972) Erzeugung von Herztumoren durch chronische orale Gabe des Carcinogens 1-Pyridyl-3,3 -Diäthyl-Triazen an BD-Ratten. Z Krebsforsch 77:217–225

Ivankovic S, Klimpel F, Wiessler M, Preussmann R (1981) Karzinogene Wirkung von 7 Homologen der N-Nitroso-N-n-Alkyl-Harnstoffe in verschiedenen postnatalen Entwicklungsphasen an BD-IX-Ratten. Arch Geschwulstforsch 51: 187–203

Jänisch W, Schreiber D (1969) Experimentelle Geschwülste des Zentralnervensystems. VEB Gustav Fischer, Jena

Jänisch W, Schreiber D (1974) Experimental brain tumors. In: Vinken PJ, Bruyn BW (eds) Handbook of clinical neurology vol 17, part II. North-Holland, Amsterdam Elsevier, New York, pp 1–41

Jänisch W, Schreiber D, Warzok R, Scholtze P (1977) Versuche mit den Kanzerogenen Methyl- und Äthylnitrosoharnstoff bei *Macaca mulatta*. Arch Geschwulstforsch 47:123–126

Jones EL, Searle CE, Smith WT (1973) Tumours of the nervous system induced in rats by the neonatal administration of N-ethyl-N-nitrosourea. J Pathol 109: 123–139

Kleihues P, Bücheler J (1977) Long-term persistence of O^6-methylguanine in rat brain DNA. Nature 269:625–626

Kleihues P, Magee PN (1973) Alkylation of rat brain nucleic acids by N-methyl-N-nitrosourea and methyl methanesulfonate. J Neurochem 20:595–606

Kleihues P, Margison GP (1974) Carcinogenicity of N-methyl-N-nitrosourea. Possible role of excision repair of O^6-methylguanine from DNA. JNCI 53:1839–1841

Kleihues P, Zülch KJ, Matsumoto S, Radke U (1970) Morphology of malignant gliomas induced in rabbits by systemic application of N-methyl-N-nitrosourea. Z Neurol 198:65–78

Kleihues P, Mende C, Reucher W (1972) Tumors of the peripheral and central nervous system induced in BD-rats by prenatal application of methyl methane-sulphonate. Eur J Cancer 8:641–645

Kleihues P, Magee PN, Austoker J, Cox D, Mathias AP (1973) Reaction of N-methyl-N-nitrosourea with DNA of neuronal and glial cells in vivo. FEBS Lett 32:105–108

Kleihues P, Patzschke K, Margison GP, Wegner LA, Mende C (1974) Reaction of methyl methanesulfonate with nucleic acids of fetal and newborn rats in vivo. Z Krebsforsch 81:273–283

Kleihues P, Lantos PL, Magee PN (1976a) Chemical carcinogenesis in the nervous system. Int Rev Exp Pathol 15:153–232

Kleihues P, Kolar GF, Margison GP (1976b) Interaction of the carcinogen 3,3-dimethyl-1-phenyltriazene with nucleic acids of various rat tissues and the effect of a protein-free diet. Cancer Res 36:2189–2193

Kleihues P, Bücheler J, Riede UN (1978) Selective induction of melanomas in gerbils (Meriones unguiculatus) following postnatal administration of N-ethyl-N-nitrosourea. JNCI 61:859–863

Kleihues P, Cooper HK, Bücheler J, Kolar GF, Diessner H (1979) Mechanism of perinatal tumor induction by neuro-oncogenic alkylnitrosoureas and dialkyltria-zenes. NCI Monogr 51:227–231

Kleihues P, Bamborschke S, Doerjer G (1980) Persistence of alkylated DNA bases in the mongolian gerbil (Meriones unguiculatus) following a single dose of methylnitrosourea. Carcinogenesis 1:111–113

Kleihues P, Patzschke K, Doerjer G (1982) DNA modification and repair in the experimental induction of nervous system tumors by chemical carcinogens. Ann NY Acad Sc 381:290–303

Kleihues P, Shibata T, Aguzzi A, Burger PL (1989) Cell specific brain tumor induction in neural transplants: evidence for multistep carcinogenesis in the nervous system. In: Napalkov NP, Rice JM, Tomatis L, Yamasaki H (eds) Perinatal and multigeneration carcinogenesis. IARC Sci Publ 96:121–129

Koestner A, Swenberg JA, Denlinger RH (1979) Host factors affecting perinatal carcinogenesis by resorptive alkylnitrosoureas in rats. NCI Monogr 51:211–217

Kuschner MS, Laskin S, Drew RT, Cappiello V, Nelson N (1975) The inhalation carcinogenicity of alpha halo-ether. 3. Life-time and limited period inhalation study with bis(chloromethyl)ether at 0.1 ppm. Arch Environ Health 30:73–77

Laerum OD, Rajewsky MF (1975) Neoplastic transformation of fetal rat brain cells in culture after exposure to ethylnitrosourea in vivo. JNCI 55:1177–1187

Laerum OD, Haugen A, Rajewsky MF (1979) Neoplastic transformation of foetal rat brain cells in culture after exposure to ethylnitrosourea in vivo. In: Franks LM, Wigley CB (eds) Neoplastic transformation in differentiated epithelial cell systems. Academic, London, pp 189–201

Laerum OD, Rajewsky MF, de Ridder L (1982) Malignant transformation of rat neural cells in culture. Ann NY Acad Sci 381:264–273

Lantos PL (1972) The fine structure of periventricular pleomorphic gliomas induced transplacentally by N-ethyl-N-nitrosourea in BD IX-rats. J Neurol Sci 17:443–460

Lantos PL (1986) Development of nitrosourea-induced brain tumors with a special note on changes occurring during latency. Food Chem Toxicol 24:121–127

Lantos PL, Pilkington GJ (1979) The development of experimental brain tumours. A sequential light and electron microscope study of the subependymal plate. I. Early lesions (abnormal cell clusters). Acta Neuropathol (Berl) 45:167–175

Laqueur GL, Spatz M (1973) Transplacental induction of tumours and malformations in rats with cycasin and methylazoxymethanol. In: Tomatis L, Mohr U (eds) Transplacental carcinogenesis. IARC Sci Publ 4:59–64

Lawley PD (1990) N-Nitroso-Compounds. In: Cooper CS, Grover PL (eds) Chemical carcinogenesis and mutagenesis I. Springer, Berlin Heidelberg New York, pp 409–469 (Handbook of experimental pharmacology, vol 94, part 1)

Lijinsky W (1987) Structure-activity relations in carcinogenesis by N-nitroso compounds. Cancer Metastasis Rev 6:301–356

Lijinsky W, Loo J, Ross AE (1968) Mechanisms of alkylation of nucleic acids by nitrosodimethylamine. Nature 218:1174–1175

Maekawa A, Mitsumori K (1990) Spontaneous occurrence and chemical induction of neurogenic tumors in rats–influence of host factors and specifity of chemical structure. Crit Rev Toxicol 20:287–310

Maekawa A, Odashima S (1975) Induction of tumors of the nervous system in the ACI/N rat with 1-butyl-1-nitrosourea administered transplacentally, neonatally or via maternal milk. Gann 66:175–183

Maekawa A, Odashima S, Nakadate M (1976) Induction of tumors in the stomach and nervous system of the ACI/N rat by continuous oral administration of 1-methyl-3-acetyl-1-nitrosourea. Z Krebsforsch 86:195–207

Maekawa A, Ishiwata H, Odashima S (1977) Transplacental carcinogenesis and chemical determination of 1-butyl-1-nitrosourea in stomach content after simultaneous oral administration of 1-butylurea and sodium nitrite to ACI/N rats. Gann 68:81–87

Magee PN, Barnes JM (1956) The production of malignant primary hepatic tumours in the rat by feeding dimethylnitrosamine. Br J Cancer 10:114–122

Maltoni C (1977) Vinyl chloride carcinogenicity: an experimental model for carcinogenesis studies. In: Hiatt HH, Watson JD, Wingsten JA (eds) Origins of human cancer. Cold Spring Harbor Laboratory, Cold Spring Harbor, pp 119–146

Maltoni C, Lefemine G (1975) Carcinogenicity bioassays of vinyl chloride: current results. Ann NY Acad Sci 246:195–218

Maltoni C, Ciliberti A, Di Maio V (1977) Carcinogenicity bioassays on rats of acrylonitrile administered by inhalation and by ingestion. Med Lav 68:401–411

Maltoni C, Ciliberti A, Carretti D (1982) Experimental contributions in identifying brain potential carcinogens in the petrochemical industry. Ann NY Acad Sci 381:216–249

Margison GP, Kleihues P (1975) Chemical carcinogenesis in the nervous system. Preferential accumulation of O^6-methylguanine rat brain deoxyribonucleic acid during repetetive administration of N-methyl-N-nitrosourea. Biochem J 148:521–525

Mennel HD, Ivankovic S (1975) Experimentelle Erzeugung von Tumoren des Nervensystems. In: Grundmann E (ed) Geschwülste/tumors III. Springer, Berlin Heidelberg New York, pp 33–122 (Handbuch der allgemeinen Pathologie, vol 6, part 7)

Mennel HD, Zülch KJ (1972) Zur Morphologie transplacentar erzeugter neurogener Tumoren beim Goldhamster. Acta Neuropathol (Berl) 21:194–203

Napalkov NP, Alexandrov VA (1974) Neurotropic effect of 7,12-dimethylbenz[a]anthracene in transplacental carcinogenesis. JNCI 52:1365–1366

Ogiu T, Nakadate M, Furuta K, Maekawa A, Odashima S (1977) Induction of tumors of peripheral nervous system in female Donryu rats by continuous oral administration of 1-methyl-1-nitrosourea. Gann 68:491–498

Oyasu R, Battifora HA, Clasen RA, McDonald JH, Hass GM (1970) Induction of cerebral gliomas in rats with dietary lead subacetate and 2-acetylaminofluorene. Cancer Res 30:1248–1261

Pegg AE (1977) Formation and metabolism of alkylated nucleosides: possible role in carcinogenesis by nitroso compounds and alkylating agents. Adv Cancer Res 25:195–267

Pegg AE (1990) DNA repair and carcinogenesis by alkylating agents. In: Cooper CS, Grover PL (eds) Chemical carcinogenesis and mutagenesis II. Springer, Berlin Heidelberg New York, pp 103–131 (Handbook of experimental pharmacology, vol 94, part 2)

Perantoni AO, Rice JM, Reed CD, Watatani M, Wenk ML (1987) Activated *neu* oncogene sequences in primary tumors of the peripheral nervous system induced in rats by transplacental exposure to ethylnitrosourea. Proc Natl Acad Sci USA 84:6317–6321

Pilkington GJ (1989) The biology, pathogenesis and spread of malignant glioma. Strahlenther Onkol 165:235–238

Pilkington GJ, Lantos PL (1979) The development of experimental brain tumours. A sequential light and electron microscope study of the subependymal plate. II. Microtumours. Acta Neuropathol (Berl) 45:177–185

Preussmann R, Stewart BW (1984) *N*-Nitroso carcinogens. ACS Monogr 182(2): 643–828

Preussmann R, Druckrey H, Ivankovic S, von Hodenberg A (1969a) Chemical structure and carcinogenicity of aliphatic hydrazo, azo, and azoxy compounds and of triazenes, potential in vivo alkylating agents. Ann NY Acad Sci 163:697–714

Preussmann R, von Hodenberg A, Hengy H (1969b) Mechanism of carcinogenesis with 1-aryl-3,3-dialkyltriazenes. Enzymatic dealkylation by rat liver microsomal fraction in vitro. Biochem Pharmacol 18:1–13

Preussmann R, Ivankovic S, Landschütz C, Gimmy J, Flohr E, Griesbach U (1974) Carcinogene Wirkung von 13 Aryldialkyltriazenen an BD-Ratten. Z Krebsforsch 81:285–310

Rajewsky MF (1980) Molecular and cellular aspects of chemical neuro-oncogenesis. In: Evans AE (ed) Advances in neuroblastoma research. Raven, New York, pp 127–134

Reznik G, Mohr U, Krüger FW (1975) Carcinogenic effects of di-*n*-propylnitrosamine, β-hydroxypropyl-*n*-propylnitrosamine, and methyl-*n*-propylnitrosamine on Sprague-Dawley rats. JNCI 54:937–943

Reznik-Schüller HM (1983) Pathogenesis of tumors induced with *N*-nitrosomethylpiperazine in the olfactory region of the rat nasal cavity. JNCI 71:165–172

Rice JM (1969) Transplacental carcinogenesis in mice by 1-ethyl-1-nitrosourea. Ann NY Acad Sci 163:813–827

Rice JM (1990) Oncogene activation during experimental prenatal carcinogenesis. J Cancer Res Clin Oncol 116 [Suppl]:1001

Rice JM, Ward JM (1982) Age dependence of susceptibility to carcinogenesis in the nervous system. Ann NY Acad Sci 381:274–289

Rice JM, Joshi SR, Shenefelt RE, Wenk ML (1978) Transplacental carcinogenic activity of 7,12-dimethylbenz[*a*]anthracene. In: Jones PW, Freudenthal RI (eds) Carcinogenesis, vol 3: polynuclear aromatic hydrocarbons. Raven, New York, pp 413–422

Rice JM, Sly DL, Palmer AE, London WT (1979) Transplacental effects of ethyl-nitrosourea in a nonhuman primate, *Erythrocebus patas*. NCI Monogr 51: 185–192

Rice JM, Rehm S, Donovan PJ, Perantoni AO (1989) Comparative transplacental carcinogenesis by directly acting and metabolism-dependent alkylating agents in rodents and nonhuman primates. In: Napalkov NP, Rice JM, Tomatis L, Yamasaki H (eds) Perinatal and multigeneration carcinogenesis. IARC Sci Publ 96:17–34

Schein PS (1969) 1-Methyl-1-nitrosourea depression of brain nicotinamide adenine dinucleotide in the production of neurologic toxicity. Proc Soc Exp Biol Med 131:517–520

Schneider J, Warzok R (1972) Induktion von Hirntumoren durch Methylnitrosoharnstoff beim Hund. Z Gesamte Inn Med 27:580–582

Schoental R (1969) Carcinogenic action of elaiomycin in rats. Nature 221:765–766

Schoental R, Cavanagh JB (1972) Brain and spinal cord tumors in rats treated with pyrrolizidine alkaloids. JNCI 49:665–671

Schreiber D, Jänisch W, Warzok R, Tausch H (1969) Die Induktion von Hirn- und Rückenmarktumoren bei Kaninchen mit N-Methyl-N-Nitrosoharnstoff. Z Gesamte Exp Med 150:76–86

Schultze-Maurer B (1978) Proliferative properties of neural cell populations. In: Laerum OD, Bigner DD, Rajewsky MF (eds) Biology of brain tumors. UICC, Geneva, pp 39–59 (UICC technical report series, vol 30)

Stavrou D (1969) Morphology and histochemistry of experimental brain tumours in rabbits. Z Krebsforsch 73:98–109

Stavrou D, Haglid KG (1972) Experimentell induzierte Tumoren des peripheren Nervensystems beim Hund. Naturwissenschaften 59:317–318

Stroobandt G, Brucher JM (1968) Neural tumors induced by administering methylnitrosourea in rats. Neurochirurgie 14:515

Swann PF (1968) The rate of breakdown of methyl methanesulphonate, dimethyl sulphate, and N-methyl-N-nitrosourea in the rat. Biochem J 110:49–52

Swann PF, Magee PN (1969) Induction of rat kidney tumors by ethyl methanesulphonate and nervous tissue tumors by methyl methanesulphonate and ethyl methanesulphonate. Nature 223:947–948

Swann PF, Magee PN (1971) Nitrosamine-induced carcinogenesis. The alkylation of N-7 of guanine of nucleic acids of the rat by diethylnitrosamine, N-ethyl-N-nitrosourea and ethylmethanesulphonate. Biochem J 125:841–847

Swenberg JA, Koestner A, Wechsler W (1972a) The induction of tumors of the nervous system with intravenous methylnitrosourea. Lab Invest 26:74–85

Swenberg JA, Koestner A, Wechsler W, Denlinger RH (1972b) Quantitative aspects of transplacental tumor induction with ethylnitrosourea in rats. Cancer Res 32:2656–2660

Swenberg JA, Wechsler W, Koestner A (1972c) The sequential development of transplacentally induced neuroectodermal tumors. J Neuropathol Exp Neurol 31:202–203

Swenberg JA, Koestner A, Wechsler W, Brunden M, Abe H (1975a) Differential oncogenic effects of methylnitrosourea. JNCI 54:89–96

Swenberg JA, Clendenon N, Denlinger R, Gordon WA (1975b) Sequential development of ethylnitrosourea-induced neurinomas: morphology, biochemistry, and transplantability. JNCI 55:147–152

Takakura K, Inoya H, Nagashima K, Kazuhiko I, Tomonaga M, Kondo K (1987) Viral neurooncogenesis. Prog Exp Tumor Res 30:10–20

Takizawa S, Nishihara H (1971) Induction of tumors in the brain, kidney, and other extra-mammary gland organs by a continuous oral administration of N-nitrosobutylurea in Wistar/Furth rats. Gann 62:495–503

Thomas C, Sierra JL, Kersting G (1967) Hirntumoren bei Ratten nach oraler Gabe von N-Nitroso-N-methylharnstoff. Naturwissenschaften 54:228

Thomas C, Sierra JL, Kersting G (1968) Neurogene Tumoren bei Ratten nach intraperitonealer Applikation von N-Nitroso-N-methyl-harnstoff. Naturwissenschaften 55:183

Ulland B, Finkelstein M, Weisburger EK, Rice JM, Weisburger JH (1971) Carcinogenicity of industrial chemicals propylene imine and propane sultone. Nature 230:460–461

Vesselinovitch SD, Rao KVN, Mihailovich N, Rice JM, Lombard LS (1974) Development of broad spectrum of tumors by ethylnitrosourea in mice and the modifying role of age, sex and strain. Cancer Res 34:2530–2538

Vesselinovitch SD, Koka M, Rao KVN, Michailovich N, Rice JM (1977) Prenatal multicarcinogenesis by ethylnitrosourea in mice. Cancer Res 37:1822–1828

Wechsler W, Kleihues P, Matsumoto S, Zülch KJ, Ivankovic S, Preussmann R, Druckrey H (1969) Pathology of experimental neurogenic tumors chemically induced during prenatal and postnatal life. Ann NY Acad Sci 159:360–408

Wechsler W, Rice JM, Vesselinovitch SD (1979) Transplacental and neonatal induction of neurogenic tumors in mice: comparison with related species and with human pediatric neoplasms. NCI Monogr 51:219–226

Wiestler OD, Pegg, AE, Kleihues P (1985) Repair of O⁶-alkylguanine in cellular DNA: significance for tumour induction and chemotherapy. In: Voth D, Krauseneck P (eds) Chemotherapy of gliomas. Basic research, experiences and results; de Gruyter Berlin, New York pp 61–69

Wiestler OD, Aguzzi A, Williams RL, Wagner EF, Boulter CA, Kleihues P (1989) Tumour induction in fetal brain transplants exposed to the viral oncogenes polyoma middle T and v-*src*. In: Napalkov NP, Rice JM, Tomatis L, Yamasaki H (eds) Perinatal and multigeneration carcinogenesis. IARC Sci Publ 96:267–274

Zeller WJ (1980) Pränatal karzinogene Wirkung von 5-(3,3-Dimethyl-1-triazeno)-imidazol-4-carboxamid (DTIC) bei den Nachkommen von BD IX-Ratten. Arch Geschwulstforsch 50:306–308

Zeller WJ, Ivankovic S (1972) Cancerogene Wirkung einer einmaligen subcutanen Gabe von *n*-Butyl-nitroso-harnstoff an neugeborenen Ratten. Z Neurol 202: 121–127

Zeller WJ, Schmähl D (1979) Development of second malignancies in rats after cure of acute leukemia L 5222 by single doses of 2-chloroethylnitrosoureas. J Cancer Res Clin Oncol 95:83–86

Zeller WJ, Ivankovic S, Zeller J (1978) Induction of malignant tumors in Wistar and Sprague-Dawley rats by single doses of *n*-butyl-nitrosourea in perinatal and juvenile phases of development. Arch Geschwulstforsch 48:9–16

Zeller WJ, Ivankovic S, Habs M, Schmähl D (1982) Experimental chemical production of brain tumors. Ann NY Acad Sci 381:250–263

Zimmerman HM (1969) Brain tumors. Their incidence and classification in man and their experimental production. Ann NY Acad Sci 159:337–359

Zimmerman HM (1982) Production of brain tumors with aromatic hydrocarbons. Ann NY Acad Sci 381:320–324

Zülch KJ, Mennel HD (1971) Die Morphologie der durch alkylierende Substanzen erzeugten Tumoren des Nervensystems. Zentralbl Neurochir 32:225–243

Zülch KJ, Mennel HD (1973) Recent results in new models of transplacental carcinogenesis on rats. In: Tomatis L, Mohr U (eds) Transplacental carcinogenesis. IARC Sci Publ 4:29–44

CHAPTER 8

Neurotoxic Phenylalkylamines and Indolealkylamines

H.G. Baumgarten and B. Zimmermann

A. Introduction

The finding that the transport systems, storage mechanisms, and enzymes involved in the biosynthesis/metabolism of catecholamines in sympathetic adrenergic neurons have limited substrate specificity and, therefore, allow chemically related (and even structurally foreign) compounds to accumulate within adrenergic nerves and influence the storage, release, and metabolism of the natural transmitter served as the foundation of the "false transmitter" concept (cf. Thoenen 1969); this concept implies that the false transmitters may be handled within the sympathetic nerves like the natural transmitter, thereby influencing the intraneuronal disposition and release of the natural transmitter. Adrenergic neurotransmission may, in addition, be modified by differences in the affinity/activity of the substitute transmitter to/at pre- and/or postsynaptic binding sites in comparison with the natural transmitter. Working with hydroxylated positional isomers of dopamine such as 3,4,5-trihydroxyphenylethylamine (5-hydroxydopamine, 5-OH-DA) and 2,4,5-trihydroxyphenylethylamine (6-hydroxydopamine, 6-OH-DA), Tranzer and Thoenen (1967; Thoenen and Tranzer 1968) discovered that 5-OH-DA is handled by sympathetic nerves like norepinephrine (NE) but that it is less potent a releaser for NE and less potent a postsynaptic receptor stimulant than NE, although it shares the same biosynthetic and degradative pathways, thus resulting in weakening of adrenergic transmission. Due to its physicochemical properties [(in situ precipitation with glutaraldehyde and reductive reaction with osmiumtetroxide (O_sO_4)], 5-OH-DA was successfully localized by electron microscopy (EM) within the small and large granular vesicles of adrenergic profiles; when 6-OH-DA was used instead of 5-OH-DA, dramatic degenerative alterations were documented in certain adrenergic nerve fiber populations, indicating that 6-OH-DA is a potential catecholamine neurotoxin. Using fluorescence hitochemistry, Malmfors and Sachs (1968) confirmed that 6-OH-DA is neurotoxic towards adrenergic nerve fibers and pointed to "chemical axotomy" as one of the mechanisms of action of 6-OH-DA resembling ligation- or cut-induced Wallerian degeneration of terminal adrenergic fiber networks. Ungerstedt (1968), Bloom et al. (1969), and Uretsky and Iversen (1969, 1970) showed that 6-OH-DA is capable of destroying central noradrenergic and dopaminergic projections provided it is either injected directly into the brain parenchyma or into the ventricular CSF. These pioneer studies highlighted the enormous potential of 6-OH-DA as a selective tool in experimental neurobiology. The pharmacology, toxicology, methods of application, usefulness in behavioral studies, and in animal models of human disease have been reviewed previously (cf. THOENEN and TRANZER 1973; KOSTRZEWA and JACOBOWITZ 1974; BREESE 1975; SACHS and JONSSON 1975; BREESE and COOPER 1977; JONSSON 1980, 1983; SCHALLERT and WILCOX 1985; KOSTRZEWA 1988, 1989).

Shortly after the establishment of 6-OH-DA as a powerful neurotoxin in central and peripheral catecholamine neurons, BAUMGARTEN et al. (1971, 1972a, 1973a,b, 1974a, 1975a) showed that selective chemical axotomy can be accomplished in central serotonin neurons by intraventricular administration of hydroxylated analogues of serotonin, in particular of 5,6- and 5,7-dihydroxytryptamine (5,6- and 5,7-DHT). Subsequent work by different groups of authors was devoted to optimizing the conditions for applying 5,6- and 5,7-DHT via various routes of administration in adult and neonate animals, to improving the selectivity of action on 5-HT neurons, to elucidating the mechanisms of action of 5,6- and 5,7-DHT, and to extensive neurobehavioral testing (cf. reviews by BAUMGARTEN and BJÖRKLUND 1976; BAUMGARTEN et al. 1971, 1972a, 1974b, 1975a, 1978, 1982a,b; BJÖRKLUND et al. 1973a,b, 1974; JONSSON 1980, 1983; READER 1989; BREESE 1975).

The aim of the present review is to summarize current knowledge and opinions on the molecular mechanisms of action of trihydroxyphenylethylamines and dihydroxytryptamines, on factors that modify their potency and specificity, and on their potential relevance for drug- or disease-induced neurotoxicity. Finally, similarities and differences in the mode of action of structurally unrelated catecholamine and serotonin neurotoxins, N-(2-chloroethyl)-N-ethyl-2-bromobenzylamine (DSP-4) and substituted amphetamines, will be discussed.

B. Hydroxylated Analogues of Dopamine

I. False Transmitter Potential of 6-Hydroxydopamine

Following uptake into adrenergic nerves by a cocaine- and desipramine-sensitive transport (uptake I, i.e., neuronal high affinity transport mechanism), [^{3}H]-6-OH-DA can be released together with NE by electrical stimulation of the splenic nerves (THOENEN and TRANZER 1968); in these experiments, a very low dose of 6-OH-DA (0.3 mg/kg i.v.) was administered, which only partly replaces the endogenous neuronal transmitter stores. A significant fraction (up to 50%) of the total amount of [^{3}H]-6-OH-DA taken up into adrenergic nerves of mouse atria in vitro ($10^{-6} M$ 6-OH-DA) appears to accumulate within the amine storage vesicles; the remaining fraction distributes in the cytoplasmic pool (soluble fraction). The retention of 6-OH-DA in the amine storage vesicles can be partially inhibited by nialamide/reserpine pretreatment (HAMBERGER 1967). The accumulation 6-OH-DA in the cytoplasmic and the vesicular pool of adrenergic nerves results in displacement of NE from either compartment, though not necessarily in stoichiometric proportion.

6-OH-DA acts mainly as an indirect sympathomimetic agent through release of endogenous NE because the response of different adrenergically innervated targets can be eliminated or attenuated by pretreatment with reserpine, guanethidine, and 6-OH-DA, i.e., treatments known to deplete NE from the presynaptic stores. In harmony with this mechanism of action, established antagonists at postsynaptic α- and β-adrenoceptors, e.g., phentolamine and propranolol, have been found to prevent the actions of NE released by 6-OH-DA. These findings also imply that the intrinsic activity of 6-OH-DA as a direct adrenoceptor agonist must be small in comparison with NE; thus, 6-OH-DA does not substitute for NE at postsynaptic sites and mechanisms.

II. Structure-Activity Relationship of Dopamine Analogues

1. Competition for Catecholamine and Serotonin Transport Sites

6-OH-DA is less potent a competitor for uptake of [^{3}H]-NE into adrenergic nerves of the mouse heart than 5-OH-DA, DA, and metaraminol (JONSSON and SACHS 1971). At concentrations of 6-OH-DA ($10^{-6} - 10^{-5} M$), close to the K_m of NE uptake ($0.5 \times 10^{-6} M$), [^{3}H]-6-OH-DA is mainly taken up into adrenergic nerves (30%–35% is bound to extraneuronal uptake sites), but at higher concentrations ($10^{-4} M$) an increasingly larger fraction becomes bound to nonneuronal binding sites (65%–70%), indicating that (toxic) actions of 6-OH-DA on nonneuronal cells are likely to occur when high doses of 6-OH-DA are used. Repeated, high doses of 6-OH-DA (100 mg/kg, i.p. or i.v.) have been shown to cause cytopathological reactions in adrenocortical cells of lizards and rats (UNSICKER et al. 1976a,b) and destruction of muscle and gland cells in the octopus posterior salivary gland (Martin and Barlow 1975).

LUNDSTRÖM et al. (1973) studied the effects of a number of isomers of 6-OH-DA on the uptake of [^{3}H]-NE into the adrenergic nerves of mouse heart in vivo and found the following rank order of potency (5-OH-DA = 6-OH-DA = 2,3,4-tri-OH->2,3,5-tri-OH- > 2,4,6-tri-OH- > 2,3,6-tri-OH-phenylethylamine), indicating that hydroxylation of dopamine at position 5 or 6 (to give 5- or 6-OH-DA) significantly attenuates uptake site affinity but that the presence of two hydroxyl groups in the vicinity of the side chain (at positions 2 and 6) drastically diminishes affinity to the NE carrier.

The kinetic constants for inhibition of the uptake of [^{3}H]-NE into slices of rat hypothalamus (uptake into NE neurons) and rat striatum (uptake into

DA neurons) indicate that 6-OH-DA has a rather low affinity to NE transport sites compared with NE itself (100 times less affinity) and that it is (10 times) more potent a competitor for uptake into NE than into DA neurons (Iversen 1970). The difference in affinity to NE and DA transport sites appears to influence the relative potency of 6-OH-DA as a catecholamine depleting agent and a toxin in central NE and DA neurons (cf. below). 6-OH-DA is a selective inhibitor for catecholamine transport in brain slices; doses up to $10^{-4} M$ 6-OH-DA fail to interfere with γ-aminobutyric acid ([³H]-GABA) and [³H]-serotonin transport (Iversen 1970; Heikkila and Cohen 1973b). The preferential selectivity of 6-OH-DA for catecholamine neurons may, however, be compromised if the concentration exceeds $10^{-4} M$ as there is published evidence for neurodegenerative actions of high doses of 6-OH-DA in certain central serotonergic terminal fields (cf. Reader 1989; Reader and Gauthier 1984; Commins et al. 1989), "nonspecific" cytotoxicity in catecholaminergic and nonaminergic neurons after high intraventricular doses of 6-OH-DA (Butcher et al. 1975; Descarries et al. 1975) and following intracerebral administration of 6-OH-DA (Javoy et al. 1975), and morphological evidence for toxicity of 6-OH-DA in nonmonoaminergic types of neurons, in glial cells, and in meningeal fibroblasts (cf. Hedreen and Chalmers 1972; Hedreen 1976; Sievers et al. 1980, 1981, 1983, 1985, 1987; Onténiente et al. 1980; Pehlemann et al. 1987) of the adult and perinatal rat brain.

2. Acute Displacement of [³H]Norepinephrine by Hydroxylated Phenylethylamines

Rank order potencies of various substituted phenylethylamines as releasers of [³H]-NE from prelabelled cardiac adrenergic storage sites in vivo (α-methyl-6-OH-DA > 6-OH-DA = 5-OH-DA > 2,3,4-tri-OH- > 2,3,5-tri-OH- ≫ 2,4,6 > 2,3,6-tri-OH-phenylethylamine) indicate that α-methylation of the side chain carbon in 6-OH-DA increases potency due to monoamine oxidase (MAO) resistance whereas O-methylation of the hydroxyls that participate in uptake site affinity decreases potency (e.g., 2,4-di-OH-5-methoxy-phenylethylamine) (Lundström et al. 1973). The data on the inhibition of [³H]-NE uptake site affinity and on the release of [³H]-NE from adrenergic nerves by the various phenylethylamines in vivo testify to the catecholamine-like properties of these compounds but fail to provide clear-cut neurotoxicity indices (e.g., the parallel behavior of 5-OH-DA and 6-OH-DA).

3. Acute and Long-Term Depletion of Tissue Norepinephrine Content by Various Phenylethylamines

Tranzer and Thoenen (1973) investigated several hydroxylated phenylethylamines and their corresponding α-methylated, O-methylated, amino- and nitro-substituted analogues as acute and long-term depletors for NE in several organs of the rat and supplemented the obtained neurochemical data by ultramorphological findings on the integrity of adrenergic nerves, thereby providing unequivocal evidence for the neurotoxicity potential of several hydroxylated and amino-substituted phenylethylamines. Table 1 shows that only those phenylethylamines that induce significant and persistent long-term reductions in tissue levels of NE by 7 days after i.v. administration are toxic to adrenergic nerves by causing morphological signs of axon damage and destruction by 24 h after the second of two consecutive doses (20 h apart). The data also indicate that the O-methylated analogues of 6-OH-DA (2-methoxy-4,5-di-OH-) and of 2,3,5-trihydroxyphenylethylamine (3-methoxy-2,5-di-OH-) are devoid of NE depleting capacity possibly due to reduced uptake site affinity. The amino-substituted analogues of 6-OH-DA (2-amino-, 4-amino-, and 5-amino-) are potent depletors for tissue NE levels and toxic to adrenergic nerve terminals, but their high general cytotoxicity potential has precluded meaningful use as selective catecholamine neurotoxins (apart from comparative model studies). Similar arguments are valid for 2-OH-DA and its α-methyl congener. TRANZER and THOENEN (1973) concluded from their structure-activity studies that the affinities of the drugs to catecholaminergic transport sites and their redox potentials (cf. below) are the most important denominators and predictors of neuro/cytotoxicity in adrenergic axons.

4. Long-Term Effects of Trihydroxyphenylethylamines on the Uptake of [³H]-Norepinephrine In Vivo

The long-lasting reduction in [³H]-NE uptake sites has been found a reliable indicator of decreases in the number of functioning adrenergic nerves in a given tissue. The rank order potency data on the reduction of [³H]-NE uptake in mouse heart by 5 days after s.c. injection of various trihydroxyphenylethylamines (α-methyl-6-OH-DA = 6-OH-DA > 2,3,5-tri-OH- > 2,3,4,5-tetra-OH-) thus match with the structure-activity concepts formulated by TRANZER and THOENEN (1973). The data by LUNDSTRÖM et al. (1973) do suggest that the α-methylated congener of 6-OH-DA may be more potent a long-term depleting agent than 6-OH-DA; in harmony with this assumption, JONSSON et al. (1972; JONSSON and SACHS 1971) have found that the neurodegenerative action of 6-OH-DA on peripheral NE neurons can be potentiated by MAO inhibition, and the same is true for the toxic actions of centrally administered 6-OH-DA on DA neurons (BREESE 1975). Metabolism of 6-OH-DA by catechol-O-methyltransferase (COMT) appears to serve as a means of detoxification, as does metabolism by MAO (cf. JONSSON et al. 1972).

5. Time Course of Effects of 6-Hydroxydopamine on [³H] Norepinephrine Uptake and on CNS Catecholamine Levels

JONSSON and SACHS (1970) analyzed dose-related effects of 6-OH-DA on the fluorescence histochemistry of adrenergic nerves in mouse heart and iris, on endogenous NE levels, on [³H]-NE uptake in vitro, and on the subcellular

Table 1. Acute and long-term effects of substituted phenylethylamines on norepinephrine (NE) content and ultrastructural integrity of adrenergic nerves in various organs of the rat

Compound	Dose (mmol/kg)	NE content (% of control)					Electron microscopic signs of destruction of adrenergic nerves
		Heart, 4h	Heart, 7 days	Salivary gland, 7 days	Spleen, 7 days	Vas deferens, 7 days	
2,4,5-tri-OH-phenylethylamine (6-OH-DA)	2 × 0.25	7 ± 2	15 ± 6	20 ± 4	24 ± 6	20 ± 4	+
α-methyl-2,4,5-tri-OH- (α-methyl-6-OH-DA)	2 × 0.25	7 ± 2	12 ± 2	16 ± 4	8 ± 4	16 ± 4	+
2-OCH$_3$-4,5-di-OH-	2 × 0.25	89 ± 3					
2-amino-4,5-di-OH-	2 × 0.125	16 ± 3	20 ± 4	32 ± 5	25 ± 7	42 ± 5	+
4-amino-2,5-di-OH-	2 × 0.125	< 1	9 ± 3	30 ± 5	16 ± 2	30 ± 3	+
5-amino-2,4-di-OH-	2 × 0.25	7 ± 1	15 ± 1	35 ± 6	12 ± 3	37 ± 2	+
3-methyl-2,4,5-tri-OH-	2 × 0.25	97 ± 10					
3,4,5-tri-OH-phenyl-ethylamine (5-OH-DA)	2 × 0.25	10 ± 3	98 ± 4	102 ± 4	104 ± 4	81 ± 2	
2,3,5-tri-OH-phenyl-ethylamine	2 × 0.25	< 1	25 ± 3	49 ± 6	33 ± 4	77 ± 5	+
3-OCH$_3$-2,5-di-OH-	2 × 0.25	101 ± 8					

Drugs were given i.v., 20h apart. Time of determination as indicated above. Taken from Tranzer and Thoenen (1973).

distribution of [^{3}H]-NE labelled storage sites in the mouse atrium. A parallel time-course was found for those neurochemical indices which characterize toxic effects on adrenergic nerves. Using an established neurodegenerative dose of 6-OH-DA (20 mg/kg i.v.), there was a rapid depletion of the cardiac NE levels to about 20% of control by 1 h which, subsequently, fell to 6% of control by 8 h. The reduction of [^{3}H]-NE uptake into adrenergic nerves of the mouse atrium via the high-affinity neuronal transport sites (uptake I according to Iversen 1975) progressed more rapidly than the fall in NE levels and had already reached minimum values by 1 h (5% of control) with negligible recovery by 4 and 8 h, provided the assay was performed at concentrations of [^{3}H]-NE close to the K_m of the uptake system ($10^{-7} M$), at which no significant binding of NE to extraneuronal transport sites complicates the interpretation of the data. Employing these criteria, Jonsson and Sachs report dose-related, significant, rapidly developing decreases in [^{3}H]-NE uptake with increasing concentrations of 6-OH-DA in vitro that are more extensive with the lower concentration of the uptake ligand used. These reductions in uptake capacity nicely match with the loss of adrenergic nerve terminals in the tissue investigated, and recovery of uptake capacity depends on the regeneration of the 6-OH-DA lesioned neurons. 6-OH-DA caused a rapid change in the subcellular distribution of [^{3}H]-NE, i.e., time-dependent decreases of radioactivity in the microsomal P_2-fraction (synaptosomes, vesicles, mitochondria) and increases in the high-speed supernatant fraction signifying loss of [^{3}H]-NE from the reserpine-sensitive granular storage compartment.

A wealth of data confirms similar effects of intracranially administered 6-OH-DA on central NE and DA neurons, although detailed structure-activity relationships have not been studied using adult or perinatal brain tissue. It has repeatedly been documented that 6-OH-DA, given via the intraventricular route of administration, is more powerful a long-term NE than DA depletor and that certain drugs have differential actions on the percentage degeneration of NE and DA axons in various regions of the CNS. These findings have been summarized and discussed at length in previous reviews (cf. Kostrzewa and Jacobowitz 1974; Breese 1975).

III. Autoxidation and Catalyzed Oxidation of Dopamine and Trihydroxyphenylethylamines

A common characteristic of those hydroxylated phenylethylamines that demonstrate neurotoxic activity in catecholaminergic systems in vivo is their low redox potential and readiness to become autoxidized at biological pH with the resulting formation of quinoidal electrophilic systems and (variable) production of secondary reaction products, i.e., reduced oxygen species and related free radicals (H_2O_2, $OH\cdot$, O_2^- and $^1O_2$); the quinoidal (organic) radical intermediates exhibit varying degrees of nucleophilic reactivity

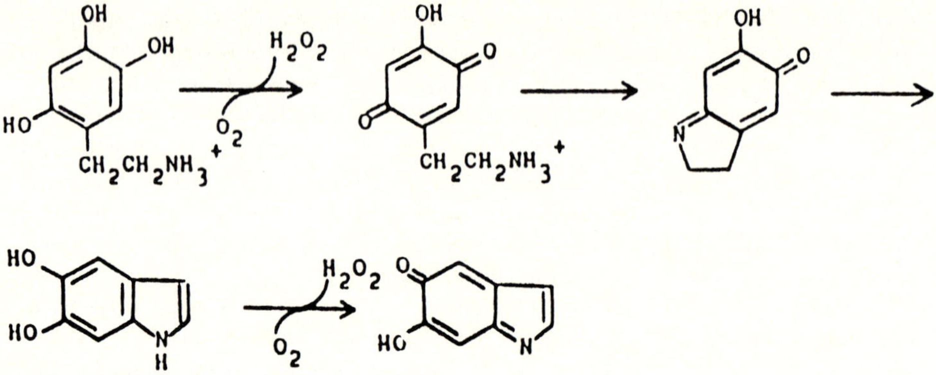

Fig. 1. Pathways and products of oxidation of 6-hydroxydopamine (6-OH-DA) in vitro. 6-OH-DA-Q, aminochrome I; 5,6-di-OH-indole, aminochrome II (Borchardt 1975)

which may result in covalent modification of SH-nucleophiles and/or NH_2-nucleophiles.

1. Pathways, Products, and Byproducts of Autoxidation

Upon exposure to air at pH 7.4 and room temperature, 6-OH-DA is instantaneously oxidized to its corresponding 2,5-*p*-quinone, said to be rapidly transformed to 4,6,7-trihydroxyindoline; this species rearranges to form 6-hydroxyindoline-4,7-dione which, upon further oxidation, generates the corresponding indole-*p*-quinone (Saner and Thoenen 1971a,b). This sequence, involving the classic 1,4-Michael type addition reaction (intramolecular side chain cyclization at position 6 of the benzene nucleus), had been proposed earlier by Senoh and Witkop (1959a,b), and Senoh et al. (1959).

An alternate oxidation pathway involving slow intramolecular cyclization of the *p*-quinone at position 2 of the benzene nucleus to give the aminochrome species "dopamine-chrome" which rearranges to form 5,6-dihydroxyindole was proposed by Powell and Heacock (1973). This reaction sequence is in harmony with that proposed earlier by Adams et al. (1972) for 6-OH-DA (Fig. 1) and, more specifically, for 6-amino-DA by Blank et al. (1972). The degree to which intramolecular cyclization of the *p*-quinone oxidation product of 6-OH-DA to 5,6-dihydroxyindole can occur in vivo depends on the presence of competing SH-nucleophiles because the readiness of the *p*-quinone to add sulfhydrylamino acids, cysteine, and GSH may be orders of magnitude higher than intramolecular cyclization readiness (Adams et al. 1972; Blank et al. 1972). McCreery et al. (1974a,b) employed cyclic voltammetry to trace the in vivo fate of 6-OH-DA and DA injected into the caudate nucleus of the rat and found that 6-OH-DA equilibrates rapidly with its *p*-quinone (with approximately 40% of 6-OH-DA in its oxidized state); the average half-life (concentration decay rate) for 6-OH-DA and its quinone was 25 min, that of DA 150 min. Sporadically, waves were monitored that conform to the time course and wave characteristics of 5,6-dihydroxyindole, the intracyclization product of either 6-OH-DA quinone or DA quinone, suggesting that the *p*-quinone of 6-OH-DA is normally cleared from brain tissue by alternate pathways, most probably through arylation of SH-nucleophiles. The chance for slow intramolecular cyclization to occur in vivo is therefore small. In harmony with this postulate, JONSSON (1976) failed to detect significant amounts of [³H]-melanin in the mouse heart following i.v. administration of [³H]-6-OH-DA.

TSE et al. (1976) investigated the oxidation of DA which proceeds at near neutral pH via the electron deficient *o*-quinone to leucochrome (5,6-dihydroxyindoline), its side-chain intracyclization product; this species is readily oxidized to the parent *o*-quinone, which rearranges at pH 7 to yield aminochrome II. This highly electron-deficient species is known to undergo rapid nucleophilic reaction with nonoxidized 5,6-dihydroxyindole, resulting in the formation of melanoid polymers. The data by TSE et al. (1976) indicate, however, that the *o*-quinone of DA reacts much more readily (1400–2100 times as fast) with SH-nucleophiles (GSH, cysteine) than with its side-chain amino function. This behavior is reflected in, e.g., the in vivo generation of cysteinyl adducts of DA (ROSENGREN et al. 1985) in the human brain, suggesting that intracyclization requires very unique, protected conditions to occur in vivo such as compartmentalization in cytosomes storing concentrates of a catalytic enzyme (tyrosinase-rich melanosomes) or a relative deficiency in SH nucleophiles. The occurrence of neuromelanin in brain indicates that such conditions are realized in certain neurons of the vertebrate brain.

2. Radical-Catalyzed Oxidation

Apart from the possibility of slow spontaneous autoxidation of DA at biological pH, there is a chance for radical-catalyzed oxidation of DA, NE, and E in vivo by a number of enzymes that produce superoxide radicals (xanthine oxidase, aldehyde oxidase, NADPH-cytochrome *c* reductase; UEMURA et al. 1977) and by redox cycling enzymes. O_2^- is a classic catalyst in the oxidation of catecholamines, resulting in aminochrome and hydrogen peroxide production according to the following reaction sequence:

$$DA + O_2^- + H^+ \longrightarrow DASQ\cdot + H_2O_2$$
$$DASQ\cdot + O_2 \longrightarrow DAQ + O_2^- + H^+$$
$$DAQ \xrightarrow{\text{cyclization}} 5,6\text{-di-OH-indole (leucochrome)}$$
$$5,6\text{-di-OH-indole} + O_2^- + H^+ \longrightarrow 5,6\text{-di-OH-indole SQ}\cdot + H_2O_2$$
$$5,6\text{-di-OH-indoleSQ}\cdot + O_2 \longrightarrow 5,6\text{-O-indoleQ} + O_2^- + H^+$$
$$5,6\text{-O-indoleQ} \longrightarrow \text{aminochrome}$$

Table 2. Autoxidation of hydroxylated phenylethylamines (spectroscopic recording of initial rates of aminochrome or p-quinone formation at pH 6.2–9.5 at room temperature)

Polyphenol (0.2 mM)	pH 6.2	pH 6.8	pH 7.2	pH 9.0	pH 9.5
6-OH-Dopamine	0.0189	0.0586	0.0759	>1	
6-OH-dopa	0.0156	0.0363	0.0559	>1	
Dopamine	0.0001	0.0002	0.0003	0.0054	0.0137
Norepinephrine				0.0008	0.0025
Dopa					0.0019
Epinephrine					0.0023

(Taken from Graham et al. 1978).

where Q = quinone; SQ· = semiquinone radical; aminochrome = imino-quinone ionic form of the indoline quinone at pH 7.

Graham (1978) and Graham et al. (1978) investigated the pH-dependent rates of autoxidation of DA, related catecholamines, and 6-OH-DA at room temperature (Table 2). The results testify to the sluggish autoxidation tendency of DA and even more so of NE at biological pH compared with 6-OH-DA and underline the importance of pH in the modification of catecholamine autoxidizability. Lundström et al. (1973) measured initial rates of autoxidation of a number of trihydroxyphenylethylamines at pH 7.0 by oxygen polarography. At a concentration of 0.5 mM, 6-OH-DA, its N,N-dimethylated analogue, 2,3,5- and 2,4,6-trihydroxyphenylethylamine undergo autoxidation at rates 10 times those of the remaining isomers. With the exception of the N,N-dimethylated analogue of 6-OH-DA, the readily autoxidized polyphenols have been shown to cause long-lasting reductions in NE levels and degeneration of sympathetic adrenergic axons in various peripheral organs of the rat (see Tranzer and Thoenen 1973). These relationships clearly suggest a role for oxygen reactivity and formation of electrophilic quinoidal intermediates in the neurotoxicity potential of hydroxylated phenylethylamines. Borchardt et al. (1977) compared the chemical and electrochemical behavior of 6-OH-DA and its 2- and 5-methylated analogue. Despite similarities in redox potentials (E°) and cyclization readiness (aminochrome formation), there were striking differences in nucleophile addition capacity of the p-quinone intermediates: only 6-OH-DA and 5-methyl-6-OH-DA reacted with GSH to form a 6-S-glutathionyl adduct (similar rate constants $t_{1/2}$). It is evident that GSH addition occurs much faster (within seconds) than aminochrome formation (within minutes).

Apart from the O_2^--catalyzed oxidation of catecholamines which occurs secondary to the production of superoxide anion radicals by enzyme activities, there is a chance of nonenzymatic oxidation of catecholamines, in particular of DA, by hydroxyl radical generating redox systems in vivo. Slivka and Cohen (1985) have shown that 2-OH-DA, 5-OH-DA, and 6-

OH-DA (in an approximate ratio of 3:2:1) are formed from DA at pH 7.2 in the presence of Fe^{2+}-EDTA/ascorbate (cyclical redox reaction) or of H_2O_2 and ferrous ethylenetriaminepentaacetate (ETPA) (Fenton-type reaction). Metal catalytic systems (Fe^{2+}/Fe^{3+}; Cu^+/Cu^{2+}) exist in various biological compartments that are capable of reacting with appropriate redox partners (e.g., peroxide) to generate hydroxyl radicals which can hydroxylate catecholamines at various positions of the benzene nucleus. Known scavengers for hydroxyl radicals such as ethanol, mannitol, and dimethyl sulfoxide antagonize the catechol hydroxylation reaction. Apart from Fe^{2+}/Fe^{3+}, the Mn^{2+}/Mn^{3+} and Cu^+/Cu^{2+} metal redox systems are also capable of enhancing the (aut) oxidation of DA (DONALDSON et al. 1980), a process associated with increased production of free radicals and other reactive species (H_2O_2, O_2^-, and OH·). The apparent similarities in the physicochemical behavior of 6-OH-DA and DA – disregarding their widely differing redox potential and the type of quinone formed – extend beyond the similarities in autoxidation pathways to interactions between antioxidant redox systems (e.g., ascorbate) and to generation of secondary reaction products. Ascorbic acid, which is enriched in brain regions having high densities of dopaminergic nerve terminals (MILBY et al. 1982), is generally believed to be an antioxidant. However, depending on the conditions prevailing (e.g., addition of L-ascorbic acid to preoxidized DA), ascorbate/dehydroascorbate may function as a prooxidant and radical and redox cycling system. Using concentrations of DA ($1-20\,mM$) and of ascorbate ($10\,mM$) thought to reflect the in vivo situation in dopaminergic striatal nerve endings, PILEBLAD et al. (1988) were able to show that ascorbate can act as a prooxidant for DA-enhancing oxygen consumption and H_2O_2 formation by as much as 640%. Redox cycling catalysis and speeding and amplification of oxidation has also been verified for 6-OH-DA autoxidation concomitant with increased production of H_2O_2 and enhancement of the damage inflicted upon catecholaminergic uptake systems in vitro (HEIKKILA and COHEN 1972); ascorbate catalyses the process of autoxidation of 6-OH-DA by continually reducing the p-quinone of 6-OH-DA back to its hydroquinone precursor.

HEIKKILA and COHEN (1971, 1973a; COHEN and HEIKKILA 1974) and HEIKKILA and CABBAT (1977) focussed attention on the role of free radicals in the cytotoxicity of 6-OH-DA and related agents which may be responsible for the dynamics and characteristics of the autoxidation process itself. COHEN and HEIKKILA (1974) investigated the oxygen consumption characteristics of 6-OH-DA, 6-amino-DA, and 6,7-DHT by polarographic techniques and found that the addition of catalase to the incubation media at the time of slowing-down of the autoxidation reaction caused the return to solution of about one-half of the consumed oxygen. This finding parallels observations by PILEBLAD et al. who studied DA oxidation characteristics in the presence of DTPA ($0.1\,mM$; $20\,mM$ DA; modified Krebs-Ringer-phosphate buffer) and found a rapid return of 44.4% of the consumed oxygen after partial oxidation of DA (expected return: 50% for the dismutation of H_2O_2 to give oxygen and water). Assuming stoichiometric conditions, the reaction can be formulated to obey the following equation:

$$QH_2 + O_2 \rightarrow Q + H_2O_2$$

in which QH symbolizes hydroquinone and Q, quinone. Heikkila and Cohen proposed an analogous reaction sequence for 6-OH-DA autoxidation:

$$6\text{-OH-DA} + O_2 \rightarrow 6\text{-Q} + H_2O_2 \tag{1}$$

A proportion of H_2O_2 formed according to Eq. (1) may compete with O_2 thereby oxidizing unchanged 6-OH-DA according to the formula:

$$6\text{-OH-DA} + H_2O_2 \rightarrow 6\text{-Q} + 2H_2O \tag{2}$$

Thus, less than stoichiometric amounts of O_2 should be recovered by adding catalase to solutions in which most of the 6-OH-DA has been allowed to oxidize. Liang et al. (1975) have presented evidence in favor of the auto-catalytic participation of H_2O_2 in the aerobic oxidation of 6-OH-DA in vitro. To unravel the potential role of H_2O_2 in the neurotoxicity caused by 6-OH-DA, Heikkila and Cohen (1971) investigated the protective role of catalase in the 6-OH-DA-mediated inhibition of [³H]-NE, DA, and 5-HT uptake into synaptosomes and slices of rat brain. Protection of the uptake mechanism against 6-OH-DA-induced damage by catalase was complete for 5-HT uptake, suggesting that H_2O_2 is involved in the inhibition of the neuronal uptake system most probably via an extracellular site of action. Protection of the catecholaminergic uptake systems by catalase was incomplete, suggesting that, due to more efficient accumulation of 6-OH-DA by the transporter, an intraneuronal component of toxicity evolves in these nerve endings that cannot be counteracted by extracellularly acting catalase.

As O_2^- may serve as an intermediate in the generation of H_2O_2 during the autoxidation of aromatic compounds, Heikkila and Cohen (1973a) evaluated the effect of superoxide dismutase (SOD) on oxygen consumption characteristics and formation of a quinoidal product during the autoxidation of 6-OH-DA. SOD effectively inhibits the rate of oxygen consumption by 6-OH-DA over a wide pH range (7.4–9.0). Simultaneously, the rate of formation of the quinone was reduced by 93%. This finding suggests that the superoxide radical is an essential mediator of the initial reaction in the autoxidation of 6-OH-DA. Adapting the theoretical reaction sequence – formulated by Misra and Fridovich (1972) for the oxidation of epinephrine – to 6-OH-DA autoxidation, Heikkila and Cohen arrived at the following sequence:

$$6\text{-OH-DA} + O_2 \rightarrow SQ\cdot + O_2^- + H^+ \tag{3}$$
$$H^+ + 6\text{-OH-DA} + O_2^- \rightarrow SQ\cdot + H_2O_2 \tag{4}$$
$$SQ\cdot + O_2 \rightarrow Q + O_2^- + H^+ \tag{5}$$
$$SQ\cdot + O_2^- + H^+ \rightarrow Q + H_2O_2 \tag{6}$$

SOD, by decomposing O_2^-, will eliminate reactions (4) and (6) which results in significant retardation of the autoxidation rate of 6-OH-DA. The results suggest that (a) the one-electron oxidation products, the semiquinones, are important initial intermediary products and driving forces in 6-OH-DA oxidation, and (b) H_2O_2 is not an essential catalytic component in the acceleration of the autoxidation of 6-OH-DA.

The presence or absence of metal redox pairs and/or other biologically important redox systems such as ascorbate significantly modify the autoxidation characteristics of 6-OH-DA. Iron-chelated EDTA functions as a catalyst in 6-OH-DA autoxidation which renders the reaction independent of O_2^- as testified by the lack of effect of SOD (SULLIVAN and STERN 1981); in this case, H_2O_2 becomes catalytic for 6-OH-DA autoxidation as demonstrated by the inhibitory actions of catalase. By removing H_2O_2, catalase may prevent the Fenton reaction between Fe^{2+} and H_2O_2 which results in the formation of the OH radical. Indirect evidence for the generation of OH· radicals during the autoxidation of 6-OH-DA was provided by COHEN and HEIKKILA (1974) through the application of the ethylene-producing system and through the use of known OH· trapping compounds in this assay system. The results obtained, i.e. that 6-OH-DA, 6-amino-DA, and 6,7-dihydroxytryptamine are potential generators of OH radicals are of general biological interest because they prove the OH· scavenging potency of catecholamines. More direct evidence for the generation of hydroxyl radicals and the semiquinone radical of 6-OH-DA was obtained by FLOYD and WISEMAN (1979) in electron spin resonance spectroscopic measurements using the spin-trapping compound DMPO (5,5-dimethyl-1-pyrroline-N-oxide), various metal chelators, catalase, and SOD by showing that the amount of trapped OH· increases linearly with the initial 6-OH-DA concentration (up to $2 mM$) and that the two iron chelators, catalase, and SOD considerably reduce the amount of trapped hydroxyl radical. Taken together, these findings indicate that 6-OH-DA autoxidation results in the generation of a variety of free radicals and reactive products (O_2^-, OH·, H_2O_2, $^1O_2$) which, apart from participating in the oxidative breakdown of 6-OH-DA, must be considered as potential cytotoxins that mediate various aspects of the in vivo neurotoxicity of 6-OH-DA.

IV. Effect of Radical Scavengers on the Axodestructive Potency of 6-Hydroxydopamine In Vivo

NE (a scavenger for superoxide radicals) and established scavengers for hydroxyl radicals [ethanol, n-butanol, methimazole, diethyldithiocarbamate, cysteamine, and 1-phenyl-3(2-thiazolyl)-2-thiourea, 1-PTTU] protect cardiac and iridical adrenergic nerves against damage by radicals and H_2O_2 generated during the autoxidation of 6-OH-DA in vivo (SACHS et al. 1975; COHEN et al. 1976). PTTU, an effective scavenger of hydroxyl radicals, provided significant protection against radical injury; its action was dependent on the relationship between the dose of 6-OH-DA and of PTTU and on the temporal relationship between the time of addmministration of either compound, i.e., full protection was guaranteed by closely spaced preadministration of PTTU (1 h before 6-OH-DA), but protection was found decremental with increasing time interval. Protection was also afforded by diethyldithiocarbamate and methimazole. While α-tocopherol (25 mg/kg, 16, 24, and 41 h before i.v. 6-OH-DA) was found ineffective as a protectant in the study of Cohen et al. (1976), nasogastric administration of all-racemic-α-tocopherol and of D-α-tocopherol (50 I.U./kg daily for 1 month prior to 6-OH-DA) was shown to protect central DA neurons and their functional capacity from damage by intrastriatally infused 6-OH-DA (20 µg) (Cadet et al. 1989). Tocopherol is an established scavenger for lipid peroxy radicals by donating hydrogen. Vitamin E thereby protects cell membranes against lipid peroxidation. These findings clearly support the free radical hypothesis of the molecular mechanism of action of 6-OH-DA and suggest that dietary vitamin E supplementation may also attenuate and thus delay the radical-mediated damage of human nigrostriatal DA neurons (Cadet 1986; Lewin 1985).

V. Role of Covalent Modification of Proteins Versus Role of Free Radicals in Toxicity of 6-Hydroxydopamine

1. Studies in Protein Model Systems

Saner and Thoenen (1971a) showed that the incubation of [^{3}H]-6-OH-DA (1–300 mM) with bovine serum albumin (BSA 0.5 mM) at pH 7.4 (38°C) resulted in covalent binding of radioactivity to protein to a concentration-dependent degree with saturation obtained at 70 mM [^{3}H]-6-OH-DA. This irreversible binding process was autoxidation-dependent and declined to 19% of control level when acetylated serum albumin was used instead of native BSA, indicating that nucleophilic groups of the protein are involved. Rotman et al. (1976b) confirmed and extended these early findings by showing that a near stoichiometric titration of free sulfhydryl groups by an autoxidation product of 6-OH-DA is responsible for the protein arylation process which finally results in extensive crosslinking of the protein monomers to polymeric high-molecular weight products. Intramolecular cyclization of the p-quinone of 6-OH-DA to 5,6-dihydroxyindole is not required for this protein arylating capacity of autoxidized 6-OH-DA. This finding is in perfect harmony with the results by Adams et al. (1972) and Tse et al. (1976) that the rate of nucleophilic reactivity of the p-quinone of 6-OH-DA towards cysteine and other sulfhydryl compounds is manifold faster than the tendency for intramolecular cyclization. Liang et al. (1975) used GSH as a model protein to investigate the interaction of the 1,4-p-quinone with thiols. GSH, when present in at least 10-fold excess, reacted with the p-quinone in a concentration-dependent manner, obeying pseudo first-order conditions over the entire pH range. Under physiological pH conditions, the p-quinone reacted rapidly with a half-life of 16.1 s with GSH at a concentration of 2 mM. According to Liang et al. (1975), the 1,4-p-quinone of 6-OH-DA adds GSH to position 6 of the benzene nucleus, and this glutathionyl adduct was subsequently isolated from rat brain and identified. By contrast, ROT-MAN et al. (1976b) proposed two electrophilic binding sites in the 1,4-*p*-quinone of 6-OH-DA capable of adding SH-nucleophiles to explain the crosslinking capacity of autoxidized 6-OH-DA and suggested an analogy to the in vivo fate of tyrosinase metabolites of dopa and-DA such as 2,5-*S*-dicysteinyldopa and DA. SH-reagents compete with protein thiols for binding to the electrophilic oxidation product of 6-OH-DA. Following unfolding of the tertiary structure in native and in acetylated BSA by 6M urea, there is a marked increase in binding of 6-OH-DA products, suggesting that some nucleophilic sites in the protein are hindered with respect to attack by the quinone product; these sites appear to be disulfide bonds which become exposed inter alia through the denaturation of the protein by the free radical species generated during the autoxidation of 6-OH-DA. The authors hypothesize that these in vitro conditions for protein arylation and crosslinking may reflect and parallel the in vivo situation in 6-OH-DA-exposed catecholaminergic nerve endings which accumulate concentrations of [^{3}H]-6-OH-DA at or above 50 mM following, e.g., i.v. administration of [^{3}H]-6-OH-DA 3 mg/kg (JONSSON 1976); at this dose level, 40% of the adrenergic nerves in the mouse heart undergo degeneration.

2. In Vivo Studies

JONSSON (1976) analyzed the actions of a known inhibitor of the neuronal high-affinity NE transporter, desmethylimipramine (DMI), and a known scavenger for radicals, PTTU, which has no effect on the carrier-mediated uptake of 6-OH-DA, on the covalent binding of radioactivity to mouse heart proteins following i.v. injection of a degeneration-inducing dose (3 mg/kg) of [^{3}H]-6-OH-DA. Irrespective of the profound differences in their mechanism of action, both drugs caused a significant and almost comparable reduction in the amount of covalently bound radioactivity, which suggests that the process of irreversible binding of products from 6-OH-DA occurs mainly within the adrenergic nerves of the heart and that radicals may be involved in catalyzing the autoxidation of 6-OH-DA, in the covalent binding of the arylating 6-OH-DA intermediates to proteins, and also direct toxic actions on bioenergetic systems, enzymes, antioxidants, and other radical-sensitive components of cells. These results are in agreement with in vivo findings by BORCHARDT et al. (1977), who used 2- and 5-methylated analogues of 6-OH-DA in order to evaluate the relative roles of quinone arylation and free radical mechanisms in the neurotoxicity of 6-OH-DA in peripheral and central NA neurons.

3. Studies in Enzyme Systems

AMAREGO and WARING (1983) have shown that the oxidation/cyclization products of DA, NE, and E, i.e., the corresponding aminochromes, inhibit

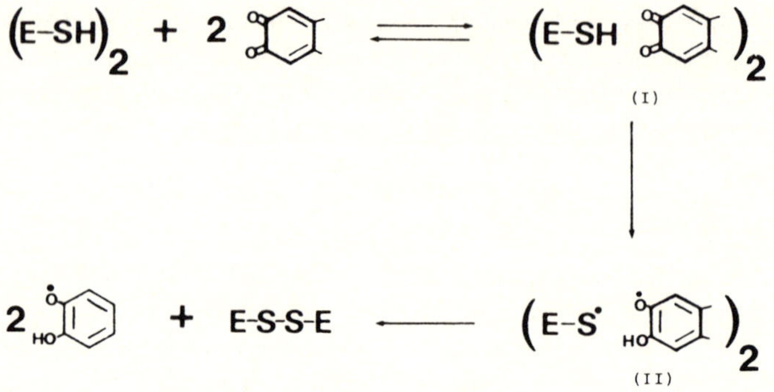

$$\left(E\text{-}SH\right)_2 \;+\; 2 \;\text{[o-quinone]} \;\rightleftharpoons\; \left(E\text{-}SH \;\text{[quinone]}\right)_2$$

(I)

$$2 \;\text{[catechol]} \;+\; E\text{-}S\text{-}S\text{-}E \;\longleftarrow\; \left(E\text{-}S^{\bullet} \;\text{[semiquinone]}\right)_2$$

(II)

Fig. 2. Molecular mechanism of inactivation of human dihydropteridine reductase by autoxidation products of dopamine (WARING 1986)

human dihydropteridine reductase (DHPR) in the presence of H_2O_2 and O_2 and that adrenochrome itself is an inhibitor for this enzyme in the presence of NADH and the quinoid dihydroform of 6-methyltetrahydropterin. The mechanism of inactivation of DHPR by aminochrome species involves intra-molecular disulphide formation near the active site of the enzyme (WARING 1986). Following transitory formation of a reversible complex (I) involving the o-quinone, internal electron/hydrogen transfer may result in the gener-ation of the one-electron reduction product of the aminochrome, i.e., the semiquinone (radical), and the one-electron oxidation product of the SH function, i.e., the thiyl radical; the thiyl radical partners in this reaction interact to establish the inactive enzyme dimer (Fig. 2). 5,6-Dihydroxyindole, the redox partner of the quinone "dopamine-chrome," has no effect on DHPR. H_2O_2, in the absence of L-dopa and peroxidase, has no effect on enzyme activity. The mechanism of DHPR inactivation which requires a preformed complex between the aminochrome and the (still functional) monomeric forms of the enzyme thus differs from that by SH-oxidizing, free radical-producing catechols like 6-OH-DA.

WICK and FITZGERALD (1981) studied the mechanism of inhibitory action of hydroquinones/quinones formed from dopa and DA by tyrosinase on the SH-dependent enzyme RNA-dependent DNA polymerase or reverse trans-criptase (RT) and found evidence for the involvement of the common intermediate(s) in both the oxidation of the dihydroxyphenols and the reduc-tion of the quinones, i.e., the semiquinones, in the inhibition/inactivation of this enzyme. For full inhibition, the presence of an autocatalytic redox system is required that (continually) produces the inhibitory species, the semiquinone. Such an active redox system is capable of generating secondary reaction products, i.e., free radicals, which may be responsible for enzyme inactivation and/or disulfide formation and breaking of nucleic acids in the assay mixture and the template. This mechanism of action resembles the redox cycling of *o*-dihydrophenol/benzoquinones by liver microsomes as characterized by KAPPUS (1985).

BORCHARDT (1975) studied the inactivation of COMT by 6-OH-DA in vitro and found that arylation of one of the (essential) sulfhydryl groups near the active site of the enzyme by 5,6-*o*-indolequinone is required. This indicates that the initial product of the autoxidation cascade of 6-OH-DA, the 1,4-*p*-quinone, is a relatively weak arylator for the SH-group near the active center of the enzyme; however, this quinone is a soft electrophile and would readily and preferably interact with nonprotein thiols (cysteine, GSH) and accessible protein thiols if these were offered to compete with the side-chain amino function of the *p*-quinone for intramolecular cyclization. Thus, artificial conditions with the exclusion of readily reactive nucleophiles are required to demonstrate the consequences of aminochrome formation from the *p*-quinone of 6-OH-DA and subsequent irreversible enzyme inactivation. The validity of such models to predict mechanisms of toxicity in vivo is thus clearly limited.

4. Studies in Cell Culture Systems

In order to mimic the in vivo conditions, several authors (cf. below) have used cell culture systems in order to provide correlations between viability indices of cells (trypan blue exclusion, thymidine incorporation, activity of DNA polymerase *a*, scavenger enzyme activities, general morphology) and aspects of the redox chemistry of 6-OH-DA and other polyphenols (nature and dynamics of formation of quinoidal products, their nucleophile reactivity, their capacity to generate radicals and reactive oxygen species). In C-1300 neuroblastoma cells, the cytotoxicity of the investigated polyphenols (thymidine incorporation, cf. Table 3) correlated with the oxidation potentials, the most cytotoxic drugs showing the most negative half-wave potential. An inverse correlation was established between the sulfhydryl reactivity of the quinone oxidation products and their ability to inhibit calf thymus DNA polymerase *a*, i.e., the most toxic drug, 6-OH-DA, was oxidized to the least reactive quinone (1,4-*p*-quinone). *N*-Acetyl-DA did not fit into this rank order because its apparent toxicity was larger than predicted by its oxidation behavior and electrochemistry; its quinone product (4-(2-*N*-acetylaminoethyl)-1,2-benzoquinone) was by far the most potent sulfhydryl reagent, and its toxicity was barely modified by NE, a scavenger for O_2^- and OH· radicals; by contrast, NE eliminated the toxicity of 6-OH-DA and attenuated that of dopa and DA. GRAHAM et al. (1978) concluded from these results that 6-OH-DA becomes cytotoxic by autoxidation-dependent production of free radicals and that dopa and DA develop toxicity by nucleophilic reactivity of their quinones and free radical mechanisms. By contrast, ROTMAN et al. (1976a) have concluded from their cytotoxicity studies in an "adrenergic" clonal line of neuroblastma cells N1E-115 (derived from the murine tumor C-1300) that 6-OH-DA oxidation products

Table 3. Autoxidation rates and cytotoxicity of dopa and dopamine derivatives (polyphenols) and sulfhydryl reactivity of their quinones as studied in C-1300 neuroblastoma cells in vitro

Polyphenol	O_2 consumption by 0.5 μM of polyphenol (μmoles/min) measured at 30°C and at pH 10.5	Half-wave potential ($E_{1/2}$) (V)	Resulting quinone	IC_{50} for [quinone] inhibition of DNA polymerase a (μM)	IC_{50} for polyphenol inhibition of [^{3}H]thymidine incorporation (μM)
6-OH-dopamine	15.90 ± 2.16	−0.09	p-quinone of 6-OH-dopamine	1700	37
Dopamine	1.67 ± 0.31	+0.24	aminochrome	150	162
N-Acetyldopamine	0.28 ± 0.05	+0.28	(ortho-quinone of N-acetyldopamine)	13	64
Dopa	0.80 ± 0.14	+0.30	dopachrome	155	196
Norepinephrine	0.33 ± 0.09	+0.37	noradrenochrome	130	>1000

(Taken from Graham et al. 1978).

Table 4. Scavenger enzyme activities (units/mg protein) in cultured cell lines

Cell line (substrates)	Catalase (H_2O_2)	GSH-S-T (1-Cl-2,4-di-nitrobenzene)	GSH-peroxidase	
			(tBHP)	(H_2O_2)
SY5Y	1.38	0.17	0.32	0.13
A_1B_1	3.53[a]	0.12[a]	0.55[a]	0.13
CHO	3.64[a]	0.37[a]	0.44[a]	0.055[a]

tBHP, tertiary butylhydroperoxide; GSH-T, GSH-transferase.
[a] Values significantly different from SY5Y data.
(From TIFFANY-CASTIGLIONI et al. 1982).

undergo extensive covalent binding to cellular proteins with concomitant protein crosslinking. Uptake of 6-OH-DA and DA into these cells is mediated by (facilitated) diffusion and not by a saturable high-affinity carrier. The amount of 6-OH-DA incorporated into these cells is, due to a lack in energy-dependent selective carriers, probably small compared with the amount accumulated in, e.g., noradrenergic nerve cells (e.g., the locus coeruleus of the developing rat, cf. SIEVERS et al. 1981, 1983) or adrenergic nerve terminal varicosities and therefore probably not a valid mirror of toxicity. As this study does not give viability data, any correlation of protein-binding data and cytotoxicity remains to be established.

Correlation of viability indices, p-quinone levels, reduced oxygen species, and the protective effects of added scavenger enzymes in 6-OH-DA-exposed human SY5Y neuroblastoma indicates that extracellular H_2O_2 and other reactive oxygen species may be responsible for the acute toxicity of 6-OH-DA and that cell-line specific patterns in catalase activity may be involved in modifying the sensitivity/resistance towards 6-OH-DA (TIFFANY-CASTIGLIONI et al. 1982). Table 4 reports scavenger enzyme activities of the three cell types compared. Catalase activity in A_1B_1 and CHO cells was 2.5 times that of SY5Y cells, GSH-transferase activity (GSG-Tr) in CHO cells was more than twice as high as in SY5Y cells, and GSH-peroxidase (GSH-Px) activity 72% higher in A_1B_1 and 30% higher in CHO than in SY5Y cells. The H_2O_2-reducing activity of GSH-Px in A_1B_1 and SY5Y cells was twice that in CHO cells. These findings suggest that H_2O_2 is the major byproduct of 6-OH-DA oxidation responsible for cell damage but that it is itself less potent than 6-OH-DA. Therefore, a portion of the cytotoxicity of 6-OH-DA should be mediated by oxidation products different from H_2O_2 (e.g., by OH· and O_2^-). By all probability, the added scavenger enzymes in this cytotoxicity assay system operate from the extracellular space and thus the generated radicals attack oxidation-sensitive components of the cell membrane. However, as stated earlier, an intracellular component in radical defence is involved in the modification of 6-OH-DA-induced radical toxicity, and this intracellular component is represented by cell-line-specific differences in radical metabolizing enzyme activities. The calculations presented

Table 5. LD_{50} of 6-OH-dopamine with or without ascorbate in mouse neuroblastoma cell lines

Cell line	LD_{50} of 6-OH-DA [A] (no ascorbate) ($\mu g/ml$)	LD_{50} of 6-OH-DA [B] (ascorbate added, 200 $\mu g/ml$)	[A]/[B]
C-1300	843 ± 157	54 ± 7	15.7
N1E-115 (adrenergic phenotype)	136 ± 12	27 ± 13	5.0
NS-20 (cholinergic phenotype)	180 ± 16	6.3 ± 1.6	28.6
N-18	36 ± 3	3.4 ± 0.2	10.5

(From Zaizen et al. 1986).

by Tiffany-Castiglioni et al. clearly show that the potency of this enzyme-mediated detoxifying principle is limited and must be balanced against the amount of radicals produced by a given dose of 6-OH-DA and of 6-OH-DA/ascorbate. As uptake of 6-OH-DA by these types of cells is generally low when compared with that of NE or DA nerve cell bodies and of nerve terminal varicosities, the conclusion seems justified that the mechanisms and the site of cytotoxic action of the oxidation products of 6-OH-DA in tumor cell lines and neurons should be different.

Zaizen et al. (1986) analyzed the sensitivity of various cloned versus wild-type (uncloned) mouse neuroblastoma cells (C-1300) and found remarkable ranges in the LD_{50} values of cloned subpopulations (Table 5). The wild-type C-1300 cells had the least sensitivity, the N-18 clone cells the highest. Ascorbate (200 $\mu g/ml$) increased the sensitivity of all types of cloned and uncloned neuroblastoma cells to 6-OH-DA. From all scavengers tested, only catalase (and catalase plus SOD) coincubation protected C-1300 cells against 6-OH-DA/ascorbate toxicity. This suggests that H_2O_2, formed rather rapidly during the autoxidation of 6-OH-DA with ascorbate as a recycler for the *p*-quinone to native 6-OH-DA (whereby its potency is enhanced), is the major toxic principle that causes cell death, most probably by an action on oxidizable groups in the external domains of the cell membrane. However, the inherent sensitivity differences of the various cloned cell populations to 6-OH-DA and 6-OH-DA/ascorbate suggest that intracellularly represented radical scavenging priciples participate in the radical defence mechanisms. This conclusion is in principal agreement with the results obtained by Tiffany-Castiglioni et al. (1982).

An intracellular component in 6-OH-DA-mediated toxicity is, however, found in cells that are endowed with a catecholamine transporter, e.g., DA neurons in culture. Mytilineou and Danias (1989) have shown that

dopaminergic neurons which have traditionally been considered more resistant to 6-OH-DA than locus coeruleus noradrenergic neurons are sensitive to 0.1 mM 6-OH-DA (after 1 h exposure in the presence of equimolar
ascorbate concentrations, 22% of the treated neurons have lost their [³H]-
DA uptake capacity and show degenerative features 24 h later) and that
these effects can be prevented by mazindol, a blocker of the high-affinity
DA carrier. Secondly, SOD, which inactivates O_2^- by producing O_2 and
H_2O_2, potentiates the toxicity of 6-OH-DA dramatically (74% lost their
[³H]-DA uptake capacity), while catalase has no effect on the potency of 6-
OH-DA. This is opposite to the findings in SY5Y tumor cells in which
catalase eliminates the toxicity while SOD had no significant long-term
effects on viability indices (see above). Moreover, SOD significantly retards
formation of the quinone in the DA neuron culture, thus facilitating uptake
of non-oxidized 6-OH-DA. Finally, an intracellular site of action of 6-OH-
DA on DA neurons and of its secondary reaction products is also suggested
by the failure of 1 mM GSH to modify the toxicity of 6-OH-DA.

However, the margin for selective neurotoxic actions on DA neurons is
rather small in view of the surprisingly low uptake site affinity of 6-OH-DA
in DA neurons in comparison with that in NE neurons. This is evidenced by
the findings of MICHEL and HEFTI (1990) that concentrations of 6-OH-DA
between 0.01 and 0.1 mM caused non-specific toxic effects in dopaminergic
and nondopaminergic neurons of cultured fetal rat mesencephalic cells. At
concentrations higher than 0.1 mM, 6-OH-DA evoked a fixative-like denaturation of cells. DA-uptake blockers which might have shifted the toxicity from DA neurons to non-DA neurons were not tested in this study.
Even more interesting than the negative specificity claims for 6-OH-DA are
the observations by MICHEL and HEFTI (1990) that DA, at concentrations
3–5 times higher than the ED_{50} for 6-OH-DA ($>100 \mu M$), is a toxin for
dopaminergic and nondopaminergic neurons and provokes fixation-like denaturation of cells, like 6-OH-DA.

Absence of selectivity of 6-OH-DA, similarities in the cytotoxicity potential of 6-OH-DA and the naturally occurring catecholamines NE, DA,
and E, and still higher sensitivities to toxic drug actions in cultured cells
were recently described by ROSENBERG (1988) in dissociated rat fetal cortical
cells (ED-15) by 24 h after plating where concentrations of the catecholamines as low as 0.025 mM were found to cause loss of viability indices
(trypan blue test) in neurons and glial cells. The catecholamine-induced
cytotoxicity could be mimicked by H_2O_2 and prevented by the addition of
catalase, indicating that autoxidation-dependent free radical mechanisms are
operative in these immature cells, which seem to have a very low potential
of radical defence mechanisms.

Collectively, these findings indicate that there is a decreasing sensitivity
to catecholamine toxicity with differentiation and maturity. Extrapolation
from culture studies to the in vivo situation is problematic and does not yield
a unifying concept on toxicity mechanisms. Further progress in our under-

standing of the extra- and intracellular sites and mechanisms of cytotoxic action of 6-OH-DA and its autoxidation products requires more detailed knowledge on the partition coefficients of polyphenols, their affinities to transport mechanisms, their scavenger properties for radicals, and their relative degrees as tissue nucleophile alkylators and radical generators in relation to intra- and extracellular autoxidant defence capacities.

C. DSP-4-Targets and Mechanism of Action

DSP-4 is a β-haloethylamine derivative of benzylamine and a close analogue of xylamine; this group of compounds includes the well-known, nonselective, irreversible α-adrenoceptor antagonists phenoxybenzamine and dibenzylchlorethamine. While DSP-4 and xylamine exhibit preferential affinity for presynaptic binding sites, phenoxybenzamine and dibenzylchlorethamine interact with postsynaptic sites. Other members include the mustards which establish covalent bonds with electrophilic groups (alkylation reaction) at or in the immediate vicinity of their binding sites. This alkylating process involves ring closure and formation of an aziridinium ion which readily undergoes nucleophilic reaction with tissue nucleophiles, e.g., amines and thiols. The action of DSP-4 on central and peripheral catecholaminergic uptake sites was discovered by Ross and Renyi in 1976 and characterized in more detail by Ross (1976). A single dose of DSP-4 (50 mg/kg, given i.p.) caused a rather selective, long-lasting reduction of [^{3}H]-NE uptake into homogenates of rat frontal cortex and hypothalamus, of NE content in whole rat brain, and of DA-β-hydroxylase activity in brain homogenates; these effects were significantly attenuated by pretreatment with desipramine, indicating that the drug acts toxically on noradrenergic axons subsequent to uptake and accumulation of DSP-4 means of the neuronal, DMI-sensitive, high-affinity NE transporter. Although the findings obtained by Ross suggested that different types of noradrenergic projections have differing sensitivities to DSP-4, and subsequent studies by Jaim-Etcheverry and Zieher (1980) and Lookingland et al. (1986) pointed to similarities in the mode of action of 6-OH-DA and, more specifically, of 6-hydroxydopa and DSP-4 on subtypes of central noradrenergic projection systems (i.e., the locus coeruleus NE system [LC]:cf. Jonsson 1980; Jonsson et al. 1982; Hallman and Jonsson 1984), it was not until reinvestigation of the topological specificity of DSP-4 action in brain by NE- and DBH-immunofluorescence (Fritschy and Grzanna 1989; Grzanna et al. 1989), by fluorescent tracer transport studies (Lyons et al. 1989), and by pharmacological characterization of the kinetics of NE transport and the kinetics of its inhibition by DSP in synaptosomes of rat cortex and hypothalamus (Zaczek et al. 1990b; Fig. 3) that the reason for the peculiar sensitivity of the LC-NE system towards DSP-4 toxicity became apparent: LC axon terminals have 2.7 times higher affinity for DSP-4 than non-LC terminals

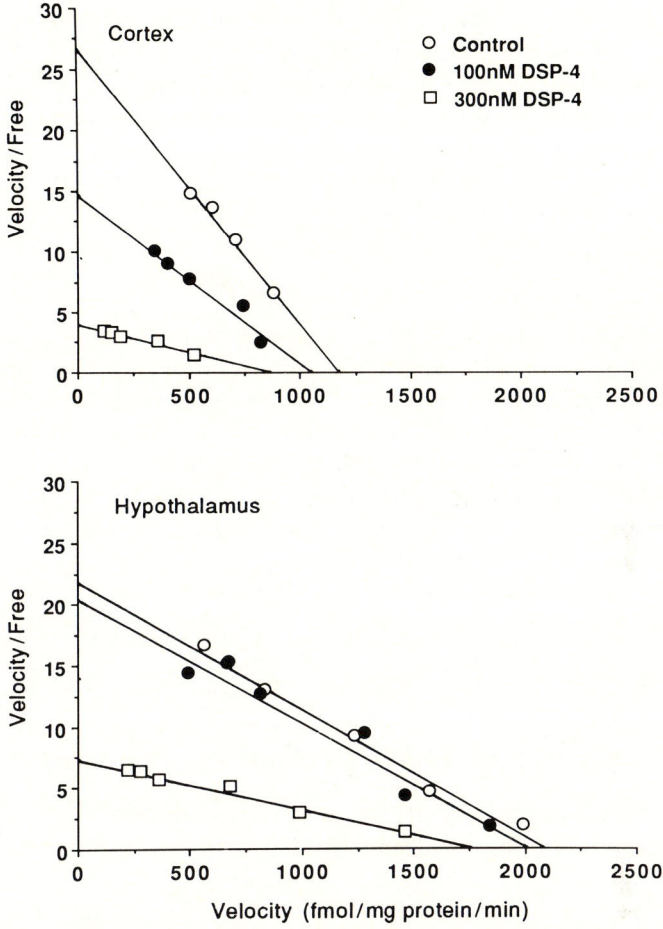

Fig. 3. Effects of DSP-4 on the kinetics of norepinephrine transport into synaptosomes from cerebral cortex and hypothalamus of the rat (From ZACZEK et al. 1990b). The micrograph copy was kindly supplied by Dr. GRZANNA

(originating in the tegmental NE cell groups), and a similar difference exists with respect to affinity towards NE itself (K_m NE cortex $39.5 \pm 7.5\,nM$, K_m hypothalamus $100 \pm 12.1\,nM$). The neurochemical profile of xylamine resembles that of DSP-4, the main variation being a slightly more potent action on cortical 5-HT terminals (GEYER et al. 1984; DUDLEY et al. 1988). Progress has been made recently in characterizing the pharmacology and cellular targets of the close analogue of DSP-4, [3H]xylamine, in PC-12 cells (KOIDE et al. 1986) and in synaptosomes (CUSHING 1988). PC-12 cells which resemble endocrine cells can be stimulated by nerve growth factor (NGF) to develop neuron-like characteristics (GREENE and TISCHLER 1982); these cells are endowed with a transport system for catecholamines which can be inhibited by DMI and cocaine. PC-12 cells accumulate xylamine (K_m 13 µM, V_{max} 36.8 pmole/min per mg protein) by a saturable, Na^+-dependent process which is inhibited by DMI (IC_{50} < 1 µM), cocaine (IC_{50} 0.6 µM), and NE (IC_{50} 1 µM). Following exposure of PC-12 cells to xylamine (10 µM) for extended periods of time, the NE transporter is irreversibly inhibited. Upon incubation in [³H]xylamine, several proteins of unstimulated or nerve growth factor-stimulated PC-12 cells become radiolabelled (Koide et al. 1986). This labelling can be prevented by known NE transport blockers (cocaine, DMI). With the exception of two (among those the transporter protein), the proteins are covalently labelled by [³H]xylamine testifying to its nucleophilic reactivity in vivo. PC-12 variants equipped with a deficient catecholamine transporter are less well labelled with [³H]xylamine than wild-type cells with a functioning transporter (Howard et al. 1990, cited from Dudley et al. 1990). Among the heavily labelled proteins are mitochondrial peptides (Cushing 1988). Xylamine appears to inhibit oxidative phosphorylation, and this blockade of ATP synthesis may result in deficient NE storage. It has not been established whether these effects on mitochondrial ATP synthesis are the main cause of its axodegenerative potential.

D. Hydroxylated Analogues of Tryptamine and Serotonin

I. False Transmitter Potential of 5,6- and 5,7-Dihydroxytryptamine

Evidence for a false transmitter-like behavior of 5,6-DHT and 5,7-DHT is indirect because quantitative studies on the uptake and release of (non-neurotoxic) doses of radiolabelled 5,6- and 5,7-DHT in relationship to the disposition of the endogenous transmitter(s) and its metabolites have not been carried out. However, it has been shown that 5,7-DHT inhibits [³H]-NE from being taken up into noradrenergic nerves of the mouse heart in vivo (IC_{50}: 15 µmoles/kg); 5,7-DHT is also capable of releasing [³H]-NE from prelabelled stores within cardiac adrenergic nerves (ED_{50} 60 µmoles/ kg) (Creveling et al. 1975). At this concentration, 5,7-DHT has no significant long-term effects on the endogenous stores of NE and no long-term inhibitory effect on the uptake of [³H]-NE into the adrenergic nerves of the heart (5 days after s.c. administration of 5,7-DHT). This finding agrees with results obtained by Baumgarten et al. (1974a) in, e.g., mouse spleen in which a high dose of 5,7-DHT (60 mg/kg i.p.) caused massive depletion of NE by 3 h; NE content recovered to control levels by day 6 after treatment, clearly testifying to the release of NE from adrenergic perivascular sympathetic nerves without concomitant neurodegenerative action. At rather high concentrations, substituted tryptamines have been shown to cause increases in the release of NE from cardiac adrenergic nerves in vitro (Paton 1973); the rank order of potency indicates that 5,6-DHT is almost equipotent a releaser for NA as 5-HT (6-HT > 7-HT > 5-HT > 5,6-DHT > 5,7-DHT > 6,7-DHT).

An interesting study was performed on the cerebral ganglion of the snail (*Helix pomatia*) which contains an identified 5-HT neuron ("giant neuron"); doses of 5,7-DHT which have been found to inhibit the uptake and accumulation of [^{3}H]-5-HT by more than 50% ($5 \times 10^{-5} M$ 5,7-DHT) and displace 5-HT from the giant neuron by about 50% ($10^{-5} M$ 5,7-DHT) fail to cause degeneration of the serotonergic axons, thus behaving like 5-HT itself (OSBORNE and PENTREATH 1976). However, 5,7-DHT produces a post-synaptic blockade of transmission from these axons by blocking the 5-HT receptors of the follower neurons. This action is reversible with time provided the tissue is washed for from 15 up to 180 min, depending on the concentration and time of exposure of 5,7-DHT used in the incubation experiments. These findings clearly indicate that 5,7-DHT has little intrinsic activity as a 5-HT receptor agonist, whereas it certainly has reasonable affinity to these 5-HT binding sites.

The false transmitter-like potential of 5,6-DHT has not been characterized as yet. However, at rather low concentrations ($0.1 \mu M$), it is apparently taken up into serotonergic nerves and synaptosomes from rat brain and releases serotonin from synaptosomal 5-HT stores without necessarily causing impairment of the uptake/storage mechanisms (WOLF and BOBIK 1988); this is mainly reflected in increases of (total) 5-hydroxyindoleacetic acid (5-HIAA) concentrations and only small decreases in intrasynaptosomal 5-HT. When MAO metabolism of 5-HT was prevented, 5,6-DHT released proportional amounts of 5-HT from the synaptosomes. At higher concentrations ($10-100 \mu M$), 5,6-DHT impaired serotonin synthesis and vesicular storage capacity. This finding agrees with results obtained by BJÖRKLUND et al. (1975) in rat brain cortical slices demonstrating rapid impairment of [^{3}H]-5-HT uptake into serotonergic nerve endings at concentrations of 10 and 100 μM 5,6-DHT. On the other hand, 5,6-DHT is a more powerful direct serotomimetic agent than 5,7-DHT. This has been shown by BAUMGARTEN et al. (1972b) in the pithed rat in which 5,6-DHT causes dose-dependent increases of blood pressure that is exclusively due to interaction with vascular 5-HT receptors (competitive inhibition by methysergide). While the affinity of 5,6-DHT to 5-HT binding sites is lower than that of 5-HT, its intrinsic activity is higher. 5-HT-like actions of 5,6-DHT have been described in various organs of the rat and guinea-pig by BARZAGHI et al. (1973) and also in platelets (DA PRADA et al. 1973).

II. Structure-Activity Relationship

1. Competition for Monoamine Uptake Sites

Table 6 lists those tryptamines for which data on the inhibition of [^{3}H]-5-HT, NE, and DA uptake in slices of rat brain or brain homogenates are

Table 6. Inhibition of serotonin uptake by tryptamines

	IC_{50} (M)
5-hydroxytryptamine (5-HT)	2.0×10^{-7}
N-methyl-5,6-DHT	3.7×10^{-7}
5,6-dihydroxytryptamine (5,6-DHT)	6.0×10^{-7}
5-hydroxy-7-methoxytryptamine	7.9×10^{-7}
6-hydroxytryptamine	8.0×10^{-7}
4,5-dihydroxytryptamine	1.7×10^{-6}
4,7-dimethyl-5,6-DHT[a]	2.0×10^{-6}
4-methyl-5,6-DHT[a]	2.3×10^{-6}
7-hydroxytryptamine	2.8×10^{-6}
5,7-dihydroxytryptamine (5,7-DHT)	4.0×10^{-6}
α-methyl-5,7-DHT	4.4×10^{-6}
6,7-dihydroxytryptamine	4.4×10^{-6}
4-hydroxytryptamine	4.5×10^{-6}
7-methyl-5,6-DHT[a]	5.2×10^{-6}
4-hydroxy-5-methoxytryptamine	5.6×10^{-6}
5-hydroxy-6-methoxytryptamine	7.0×10^{-6}
5,6-dibutyroxytryptamine	1.5×10^{-5}
N^{ω}-dimethyl-5,6-DHT	2.4×10^{-5}
5,6-diacetoxytryptamine	7.4×10^{-5}
5,6,7-trihydroxytryptamine	No inhibition of uptake at 1 or $10\,\mu M$

IC_{50}, concentration of inhibitor required to produce a 50% inhibition of the uptake of $[^3H]$-5-HT into synaptosome-rich
homogenates of rat brain in vitro (taken from: Baumgarten
et al. 1972a; Horn et al. 1973; Björklund et al. 1975; Klemm
et al. 1980). K_i value for 5,6-DHT determined by Heikkila and
Cohen (1973b), $7.1 \times 10^{-7}\,M$, in slices of rat brain tissue is in
perfect correspondence with the inhibitory constants measured in
synaptosome-rich homogenates by Horn et al. (1973).
[a] Data from Sinhababu and Borchardt 1985 obtained in neuroblastoma N-2a cells in culture; these cells are claimed to have
characteristics of 5-HT neurons (Yoffe and Borchardt 1982).

available in the literature. Comparison of the IC_{50} values for inhibition of 5-
HT uptake reveals that N-methyl-5,6-DHT ($3.7 \times 10^{-7}\,M$) and 5,6-DHT
($6 \times 10^{-7}\,M$) are only slightly less potent than 5-HT itself in competing for
uptake into serotonin neurons (K_m of 5-HT uptake: $2.0 \times 10^{-7}\,M$; cf.
Shaskan and Snyder 1970). The two nonneurotoxic analogues, 6-HT and 5-
OH-7-CH$_3$OT, have rather high affinities for serotonin transport and are
useful tools for histochemically identifying serotonergic structures provided
their accumulation into NE and DA neurons is prevented (Jonsson et al.
1969; Baumgarten et al. 1975a, 1979). Methylation of the indole carbons at
position 4, or 4 and 7 of 5,6-DHT (to give 4-methyl-, and 4,7-dimethyl-5,6-
DHT) causes only a moderate fall in uptake site affinity (3–4 times less
potent competitors for uptake) but results in acceleration of autoxidation
and changes in electrophilicity. The potency of 5,7-DHT and α-methyl-5,7-
DHT as inhibitors of 5-HT uptake is 10 times lower than that of N-methyl-

5,6-DHT and about 6 times lower than that of 5,6-DHT. Dimethylation of the side-chain nitrogen and 0,0-diacylation of the hydroxyls in 5,6-DHT result in a striking decrease of uptake site affinity, whereas the potency fall caused by methylation of one of the hydroxyls in the DHTs (to give the corresponding mono-O-methylated derivatives) is rather small. While O-methylation has thus negligibly attenuating effects on the substrate properties of the DHTs for transport into the 5-HT neurons, it profoundly alters their chemical characteristics as it abolishes their capacity to form quinoidal products by interaction with oxygen. At least for the DHTs, it is doubtful whether the IC_{50} values measured reflect the true affinity for the 5-HT transport system since the DHTs are likely to cause significant impairment of the uptake mechanism(s) during the preincubation and the coincubation period. As this occurs fairly rapidly, the affinity values for the neurotoxic DHTs are probably underestimated (cf. BJÖRKLUND et al. 1975 for details). 5,6,7-Trihydroxytryptamine (THT) does not interfere with [³H]-5-HT transport (at 1 or $10\,\mu M$); this can be explained by assuming that this analogue exists as a quinone structure in solution which lacks substrate properties for the uptake system. All serotonin analogues tested inhibit NE and DA transport (HORN et al. 1973; BJÖRKLUND et al. 1975) although – with few exceptions – less effectively than 5-HT uptake; such exceptions include, e.g., 7-HT and 5,7-DHT which have an affinity for the NE transport sites that is slightly higher than for the 5-HT uptake sites. However, this is apparently not reflected in the NE depleting capacity of 5,7-DHT (see below). The small difference in 5-HT, NE, and DA uptake site affinity measured for 5,6-DHT (IC_{50} for NE and DA uptake: $2.7 \times 10^{-6}\,M$, IC_{50} for 5-HT uptake $6.0 \times 10^{-7}\,M$, cf. BAUMGARTEN et al. 1975a) also has no correlate in its powerful long-term serotonin-depleting capacity and the lack of effect in NE neurons. The data obtained by HEIKKILA and COHEN (1973b) in chopped slices of rat brain in the presence of catalase indicate that 5,6-DHT is only 7 times more potent a competitor for 5-HT than for DA uptake [K_i for 5,6-DHT: $7.1 \times 10^{-7}\,M$ (5-HT transport) and $4.8 \times 10^{-6}\,M$ (DA transport)]. The findings by these authors confirm that 6-OH-DA is less potent an inhibitor for [³H]-DA uptake than 5,6-DHT.

2. Impairment of [³H]-Serotonin and [³H]-Norepinephrine Uptake

To obtain a quantitative measure for the acute neurotoxicity potential and the relative monoamine neuron selectivity of substituted tryptamines, brain slices were preincubated with the test substances at variable time periods and at various concentrations (10^{-7}–$10^{-4}\,M$) and, following a washout period, postincubated with [³H]-5-HT ($0.5 \times 10^{-7}\,M$) or [³H]-NA ($1 \times 10^{-7}\,M$) for 10 min (BJÖRKLUND et al. 1975). With 5,6-DHT (10^{-5}–$10^{-4}\,M$), uptake is reduced by more than 50% after as little as 5 min preincubation, demonstrating that toxicity at a sensitive membrane marker develops immediately after the onset of autoxidation of 5,6-DHT and that there is little further effect on the serotonin carrier system with time. In accordance with the data on long-term depleting capacity and selectivity of action of the various tryptamines (see below), the rank order of potency in impairing uptake (5,7-DHT $\approx$ 5,6-DHT > N-methyl-5,7-DHT > α-methyl-5,7-DHT > 6,7-DHT > 4,5-DHT > α-methyl-5,6-DHT) indicates that only 5,6-DHT, 5,7-DHT, and their N-methylated analogues are powerful serotonergic neurotoxins; of these, 5,7-DHT, 6,7-DHT, and their N-methylated derivatives cause significant damage to the NE high-affinity transport system.

3. Long-Term Effects of Dihydroxytryptamines on Cerebral Monoamine Levels and Morphology of Serotonergic Axons

Long-term cerebral serotonin and catecholamine level reductions (by 8–12 days after drug introduction into the CSF) are fairly reliable indicators of the neurodegenerative potency of substituted tryptamines (rank order of potency:5,7-DHT $\approx$ N-methyl-5,7-DHT $\approx$ 5,6-DHT > α-methyl-5,7-DHT > N-methyl-5,6-DHT > 6,7-DHT $\approx$ 4,5-DHT $\approx$ 4-OH-7-CH$_3$OT), in particular when such studies are supplemented by immunohistochemical analysis of the topological morphology of the respective neuron systems and when tracing studies are performed to testify antero- and retrograde transport capacities. By 8–12 days after neurotoxin injection, retarded anterograde degeneration of axons distal to sites of chemical axotomy can be considered complete, and regeneration efforts will not significantly influence regional brain amine levels in target areas of monoaminergic neurons (with the exception of cell-body-rich regions) (for detailed discussion of this issue, cf. BAUMGARTEN et al. 1971, 1973a,b, 1974a,b, 1975a, 1977; BJÖRKLUND et al. 1973a,b, 1974, 1975). Following intraventricular injection of 50 μg free base of the various tryptamines, only 5,6-DHT, 5,7-DHT, and their N-methylated analogues reveal themselves as rather potent long-term serotonin depleting agents, whereas α-methylation and O,O-diacylation attenuates neurotoxic activity. In accordance with the uptake site affinity and uptake impairment data, these compounds also affect central catecholamine levels although, with the exception of 5,7-DHT, to a lesser extent. A valuable, supplementary index for long-term neurodegenerative effects on monoaminergic axons and terminals is the reduction of [^{3}H]-5-HT and [^{3}H]-NE uptake capacity by 9–11 days after injection of the substituted tryptamines into the CSF (cf. BAUMGARTEN et al. 1973b, 1975a). Of the tryptamine derivatives analyzed, only 5,7-DHT has been found to produce dose-related neurodegenerative actions on cerebral serotonin systems up to 200 μg free base given as a bolus injection into the lateral or IVth ventricle of the rat (cf. BAUMGARTEN et al. 1973b, 1975a, 1978). The selectivity of action of this m-substituted DHT can be manipulated by pretreating animals with blockers of NE (e.g., desmethylimipramine) and NE and DA transport (e.g., nomifensine) (cf. BAUMGARTEN et al. 1978, 1979, 1981, 1982a,b; BJÖRKLUND et al. 1975; BREESE and COPER 1975; GERSON and BALDESSARINI 1975; READER

Table 7. Autoxidation rates of tryptamine derivatives (and 6-OH-dopamine) (initial rates of oxygen consumption measured with a Clark-type oxygen electrode)

Tryptamine derivative Final concentration of amine	Oxidation rate (n moles O_2/min)		
	$0.5\,\text{m}M^a$	$1.0\,\text{m}M^a$	$1.0\,\text{m}M^b$
5-HT	<2.0	3.7	<2.0
4,7-DHT	38.0	56.0	
5,6-DHT	11.0	22.0	2.7
5,7-DHT	16.5	36.2	33.4
6,7-DHT	40.5	59.0	
5,6,7-DHT	36.0	53.0	986.0
α-methyl-5,7-DHT	13.0		20.9
6-OH-dopamine	276 ± 13		414.0
4-methyl-5,7-DHT	(0.5 mM)	17.7 times the rate of 5,7-DHT[c]	
6-methyl-5,7-DHT	(0.5 mM)	12.8 times the rate of 5,7-DHT	
4,6-dimethyl-5,7-DHT	(0.5 mM)	178 times the rate of 5,7-DHT	

DHT, dihydroxytryptamine.
[a] Data from CREVELING et al. (1975).
[b] Data from KLEMM et al. (1980).
[c] Data from SINHABABU and BORCHARDT (1985).

1989; READER and GAUTHIER 1984; LISTON et al. 1982: amfonelic acid). These properties have rendered 5,7-DHT the most frequently applied serotonin neurotoxin.

III. Autoxidation of Dihydroxytryptamines

Data for autoxidation rates are available for only a restricted number of hydroxylated tryptamines (Table 7). At biological pH, the di- and trihydroxytryptamines react with oxygen but at considerably different velocities. For certain ones, this process can easily be followed by observing changes in the color of the solution due to the formation of conjugated quinoidal structures, by registration of changes in the UV absorption characteristics, and by polarographic measurements of oxygen consumption. Electrochemical oxidation of 5,7-DHT (see below) may result in the formation of intermediates different from those obtained in aerobic oxidation of 5,7-DHT at biological pH (SINHABABU and BORCHARDT 1985, 1988; WRONA et al. 1986). In general, the *o*- and *p*-substituted DHTs autoxidize (depending on the nature of the remaining substituents in the benzene nucleus of the indole) via semiquinone anion radicals to ortho- and paraquinones, respectively, i.e., electrophilic structures which are able to undergo Michael-type addition reactions. Provided that significant amounts of semiquinone radical intermediates are formed in the autoxidation process of, e.g., 5-HT, 5,6-DHT, or 5,7-DHT, O_2^- radicals will be generated which may participate in the further oxidation of the indole-semiquinone radicals or undergo secondary free radical reactions. If nonoxidized DHTs are partners in the nucleophilic

reaction with *o*-quinones, low or high molecular weight oligomeric or polymeric adducts may be formed. Wrona and Dryhurst (1988; Wrona et al.
1988) have shown that 4,5-DHT, generated from electrooxidized 5-HT,
tends to form di- and trimeric species due to intermolecular nucleophilic
attack at positions 2, 4, and 7 of the electrophilic *o*-quinone. Another
example is provided by the extensive intermolecular crosslinking process of
air oxidized 5,6-DHT/*o*-quinone resulting in high molecular weight,
melanoid, insoluble, polymeric complexes (Klemm et al. 1978, 1979), a
process resembling the formation of melanin species through phenoloxidase-
catalyzed oxidation of dopa/DA via aminochrome intermediates (Cromartie
and Harley-Mason 1957; Swan 1974). Furthermore, *o*-quinones may
rapidly tautomerize to form more stable *p*-quinones.

In contrast to 5,6- and 5,7-DHT, serotonin reacts only slowly with
oxygen in vitro; it has, however, been claimed that 5-HT can be oxidized by
ceruloplasmin concentrates and by cytochrome *c* (Fe^{3+}) in the presence of
NADH (Perez-Reyes and Mason 1981); this interaction is said to result in
the generation of semiquinone-imine radical intermediates of 5-HT capable
of interreacting to form dimeric or oligomeric species. This concept is
supported by the results of electrochemical oxidation studies (Wrona and
Dryhurst 1988) showing that the initial one-electron reaction ($1e^-$, $-1H^+$)
gives rise to a short-lived serotonin radical intermediate which, depending
on the pH, the oxidation potential used, and the concentration of 5-HT,

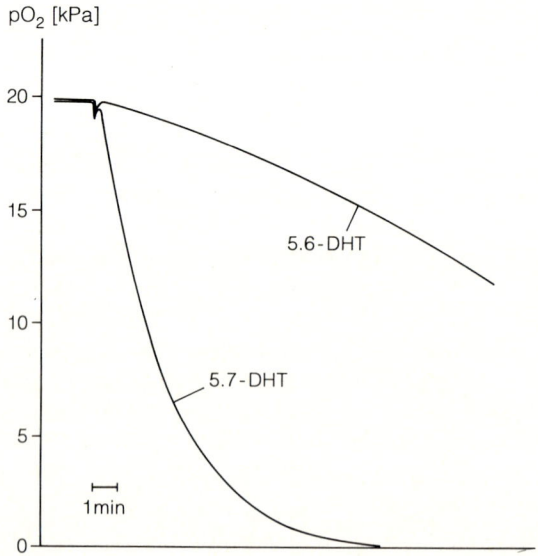

Fig. 4. Kinetics of reaction of 5,6-dihydroxytryptamine (*5,6-DHT*) (1 mM) and *5,7-DHT* (1 mM) with molecular oxygen in vitro at pH 7.2 (0.1 M phosphate buffer, 37°C). (From Klemm et al. 1980)

undergoes either intermolecular nucleophilic reactions to form dimers and/ or oligomers or further oxidation ($1e^-$, $-1H^+$) to form the p-quinone-imine of 5-HT. This quinone-imine reacts with H_2O (Michael-type addition reaction) to give 4,5-DHT; upon further oxidation ($-2e$, $-2H^+$), this DHT is transformed into its corresponding o-quinone, a highly electrophilic species which avidly undergoes addition reactions with tissue SH-nucleophiles (cf. CAI et al. 1990; CHEN et al. 1989; VOLICER et al. 1989; Ladislav Volicer, personal communication). It has been speculated that aberrant oxidation of 5-HT in the aging and/or disease-affected human brain may contribute to neurotoxic damage (cf. below).

1. Pathways, Products, Byproducts, and Biological Consequences of Autoxidation of 5,6-DHT

Oxygen electrode recordings of 5,6-DHT and 5,7-DHT oxidation are shown in Fig. 4. Initially, 5,6-DHT consumes oxygen at rather low rates (2.7 nmol O_2/min at pH 7.2); at later time points, 5,6-DHT oxidation reveals increasing reaction velocities, suggestive of autocatalytic promotion of oxidation until deoxygenation of the incubation medium. Figure 5 illustrates that there is little, if any, relationship between increasing concentrations and reaction velocity in the case of 5,6-DHT. During the consumption of oxygen, black, melanoid, insoluble polymers are formed from 5,6-DHT. Secondary reaction products are apparently responsible for the autocatalytic enhancement of 5,6-DHT oxidation since about 40% of the consumed oxygen is accounted for by H_2O_2 formation:

$$5,6\text{-DHT} + O_2 \rightarrow 5,6\text{-DHT-quinone} + H_2O_2$$
$$H_2O_2 + 5,6\text{-DHT} \rightarrow 5,6\text{-DHT-quinone} + 2H_2O$$

The latter reaction seems to be responsible for the loss of about 20% of the theoretical amount of H_2O_2 formed (see KLEMM et al. 1980). The ubiquitous presence of trace metals (e.g., Fe^{2+}) in living systems renders it likely that the fraction of H_2O_2 which escapes catalase and GSH-peroxidase breakdown will be converted to highly reactive hydroxyl radicals capable of attacking a variety of organic substrates (e.g., lipid peroxidation). At least at sites of accumulation of high concentrations of 5,6-DHT, H_2O_2 may reach critical concentrations ($10^{-4} M$), both extracellularly (near the injection cannula) and intracellularly (in heavily 5,6-DHT-accumulating serotonergic axon terminals). In vitro studies by KLEMM et al. (1980) have shown that the rate of autoxidation of 5,6-DHT can be retarded by the presence of catalase and of SOD, supporting the concept of free radicals and of other reactive species (e.g., H_2O_2) as modifiers of the autoxidation characteristics of 5,6-DHT and as potential mediators of certain aspects of its in vivo toxicity.

Alkylation of nucleophiles by autoxidation products of 5,6-DHT has been found to occur in vitro and in vivo (KLEMM et al. 1978, 1979; BAUMGARTEN et al. 1978, 1984; CREVELING and ROTMAN 1978; CREVELING et al.

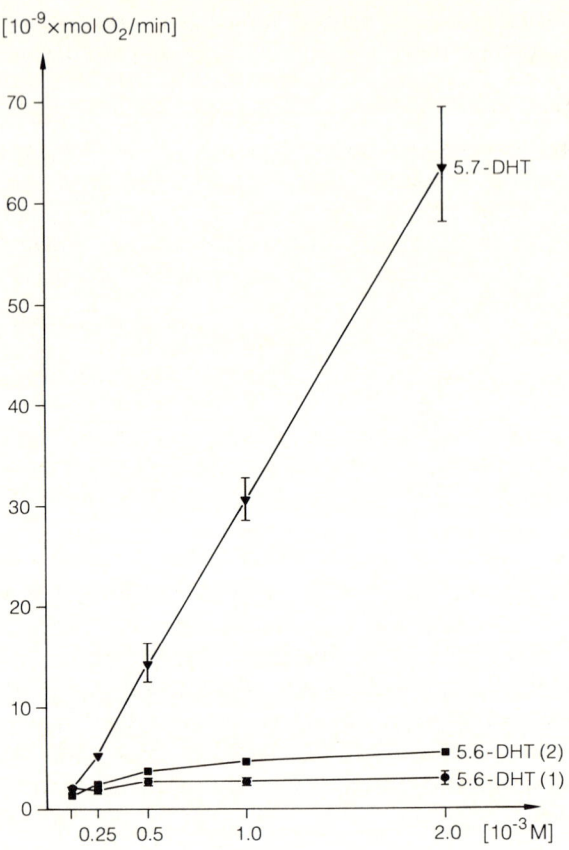

Fig. 5. Concentration-dependent increases in oxygen consumption by 5,6- and 5,7-DHT (dihydroxytryptamine) in vitro. 5,6-DHT (2) shows initial rates of oxygen consumption of 5,6-DHT, 15 min after onset of the reaction. (From KLEMM et al. 1980)

1975; ROTMAN et al. 1976b). ROTMAN et al. (1976b) studied oxygen-dependent interactions of [³H]-5,6-DHT with BSA and found time-dependent increases in the amount of irreversibly protein-bound radioactive material. The extent of protein binding with [³H]-5,6-DHT was similar to that with [³H]-6-OH-DA (12.5 versus 11.9 moles of amine/mole protein), but binding was faster after 5,6-DHT. 5,6-DHT was more potent a cross-linking agent than 6-OH-DA (94% versus 83% BSA polymerization). Binding and crosslinking capacity are oxygen-dependent and mainly due to nucleophilic attack of SH-groups at electrophilic sites in the *o*-quinone of 5,6-DHT and the 1,4-*p*-quinone of 6-OH-DA (CREVELING and ROTMAN 1978). For both 6-OH-DA and 5,6-DHT, the accumulation of secondary reaction products may assist in the alkylating process by denaturation and exposure of disulfide bonds which may become reduced by nonoxidized 5,6-DHT and 6-OH-DA.

KLEMM and BAUMGARTEN (1978) and KLEMM et al. (1979) studied the MAO-dependent metabolism of 5,6- and 5,7-DHT and covalent binding characteristics of oxidation products from either compound to homogenates of rat brain and found that both DHTs are substrates for MAO and converted to the corresponding indoleacetic acids and that 5,6-DHT, at concentrations above $10^{-4}M$, irreversibly inactivates MAO. This inactivation could be partially prevented by coincubation of 5,6-DHT with GSH or DTT (dithiothreitol) but not by catalase or SOD. The process of inactivation was paralleled by time-dependent irreversible alkylation of homogenate proteins. These findings were interpreted as indicative of autoxidation-dependent formation of highly electrophilic intermediates, most probably the corresponding *o*-quinone, which titrates SH-nucleophiles in peptides and proteins and inactivates catalytic sites of the enzyme. The autoxidation products of the major MAO/aldehyde dehydrogenase metabolite of 5,6-DHT, 5,6-dihydroxyindoleacetic acid, became covalently bound to homogenate proteins in a time- and concentration-dependent manner similar to 5,6-DHT, a process subject to inhibition by coincubation with the same sulfhydryl reagents (GSH, DTT, cysteine), suggesting that autoxidation of the amine, the aldehyde intermediate (refractory to successful isolation), and acid products all contribute to the neurotoxicity of 5,6-DHT.

5,6-DHT, 5,6-dihydroxyindole, and aminochrome species formed from catecholamines by (aut) oxidation and intramolecular cyclization are potential inhibitors/inactivators for enzymes that are dependent on accessible SH-groups near the active sites for functional/structural integrity. 5,6-Dihydroxyindole and its corresponding quinone, aminochrome II, are substrates/inhibitors/inactivators for COMT in vitro (BORCHARDT 1975) which apparently arylate an SH-group near the active site in this enzyme; aminochrome II appears to be mainly responsible for the inactivation of COMT by 6-OH-DA in vitro, and a similar mechanism of action may apply to a number of related polyphenols. These findings point to converging aspects in the mechanism of toxic action of both 6-OH-DA and dihydroxyindoles. VICTOR et al. (1974) observed a rapidly developing inhibition of tryptophan hydroxylase in ventricle-near regions of the rat CNS after lateral ventricle injection of a high dose of 5,6-DHT and also after addition of 5,6-DHT to homogenates of rat midbrain and suggested that this inhibition might have occurred through a concerted action of H_2O_2 and 5,6-DHT on this enzyme. However, apart from the possibility of direct inactivating effects of the *o*-quinone and H_2O_2 on tryptophan hydroxylase, alternative effects on its cofactor (tetrahydrobiopterin) and on the cofactor regenerating enzyme system dihydrobiopterin reductase seem likely. ARMAREGO and WARING (1983) have shown that the oxidation products of catecholamines, i.e., the corresponding aminochromes, inhibit human dihydropteridine reductase in the presence of H_2O_2 and O_2. Adrenochrome is an inhibitor of this enzyme in the presence of NADH and the quinonoid dihydroform of 6-methyl-tetrahydropterin.

The MAO/COMT/aldehyde dehydrogenase-dependent metabolism of 5,6- and 5,7-DHT was studied by Baumgarten et al. (1984) in rat brain in vivo. Both COMT and MAO were found to participate in the metabolism of intraventricularly injected [^{14}C]-5,6-DHT. Under steady-state conditions, 50% of the indoleacetic acids consisted of mono-O-methylated acids, clearly indicating that COMT is involved in the detoxification of 5,6-DHT, since the mono-O-methylated products have a negligible affinity to oxygen at biological pH and, therefore, do not readily form quinoidal intermediates. 5,6-DHT appears to have a rather high affinity to COMT (Fuller and Rush 1974), which is understandable in view of the catechol-like substitution of the hydrogens in the indole. At high concentrations, 5,6-DHT should act as a substrate/inactivator for COMT similar to 5,6-dihydroxyindole (Borchardt 1975), the intramolecular cyclization product of the p-quinone of 6-OH-DA. In addition to COMT, MAO also contributes to the detoxification of 5,6-DHT in vivo since inhibition of this enzyme prior to intraventricular 5,6-DHT administration enhances the toxicity (by decreasing the LD$_{50}$ dose). Inhibition of MAO prior to intrathecal administration of [^{14}C]-5,6-DHT by clorgyline (MAO-A) or low-dose pargyline (MAO-B) or high-dose pargyline (MAO-A plus B) only moderately decreases and retards the development of protein arylation by autoxidation products of 5,6-DHT, suggesting that MAO catalysis has no major role in the expression of the in vivo neurotoxicity by 5,6-DHT.

The role of active transport into monoaminergic neurons as a prerequisite for the expression of target-specific toxicity is much more pronounced with 5,7-DHT, and this finding is also paralleled by a significant effect of MAO-A inhibition (by clorgyline) on the protein-alkylating capacity of autoxidation products after [^{14}C]-5,7-DHT administration. Accord-

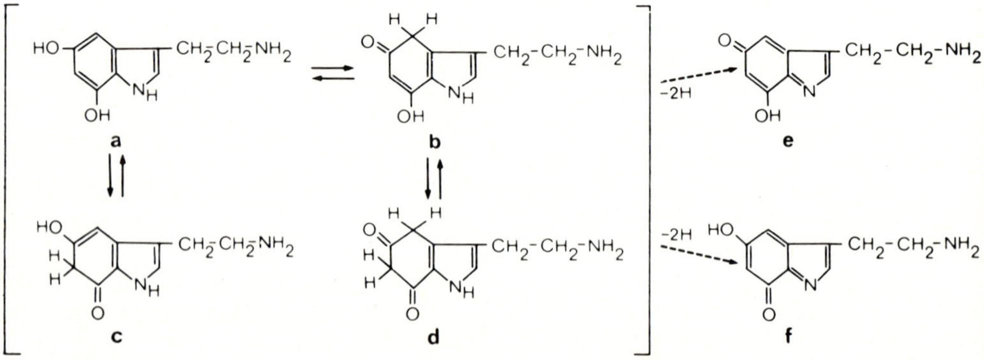

Fig. 6. Keto-enol tautomeric forms of 5,7-dihydroxytryptamine (a−d) and hypothetical oxidation products (p-and o-quinone-imine, e and f). (From Björklund et al. 1975)

ingly, MAO-A metabolism of 5,7-DHT occurs largely within the monoaminergic neurons, and the same is true for the time-dependent intraneuronal protein alkylation process which is enhanced by MAO metabolism of 5,7-DHT. Inhibition of MAO prior to intrathecal application of 5,7-DHT has indeed been found to prevent toxicity in peripheral and central NE neurons and to attenuate the long-term depleting capacity of 5,7-DHT in some but not all regions of the rat CNS (BAUMGARTEN et al. 1979). These results clearly indicate that 5,6-DHT is potentially cytotoxic to any type of cell, and this is supported by morphological analysis of brain tissue in the vicinity of the injection cannula (cf. BAUMGARTEN et al. 1972c).

2. Pathways, Products, Byproducts, and Biological Consequences of Autoxidation of 5,7-DHT

a) Keto-Enol Tautomerism

A key to understanding the unique properties of 5,7-DHT is the existence of a pH-dependent (pH 2.0–8.0) equilibrium between different interchangeable keto-enol tautomeric forms of 5,7-DHT in water (Fig. 6), and this is supported by nuclear magnetic resonance spectroscopy which indicates that the two aromatic protons in positions 4 and 6 of 5,7-DHT (creatinine sulfate) are quickly replaced by deuterium (SCHLOSSBERGER 1978). Neither 5-HT nor 5,6-DHT reveal any exchange of protons by deuterium under comparable conditions. Of the methylated analogues of 5,7-DHT (4-methyl-, 6-methyl-, and 4,6-dimethyl-5,7-DHT) studied by SINHABABU and BORCHARDT (1985, 1989), only the 6-methylated analogue exhibited keto-phenol tautomerism between pH 2–8 in the absence of oxygen similar to 5,7-DHT. The absence of isosbestic points and strong reduction in the intensity of the absorption maxima at 348 nm for 4-methyl-5,7-DHT and at 364 nm for 4,6-dimethyl-5,7-DHT indicate that the predominant keto form of 5,7-DHT and 6-methyl-5,7-DHT are those that involve proton shifts to the 4-position. Three out of eight different tautomeric forms of 5,7-DHT involve proton transfer to position 4. Of these three, only the 5-ketonized, 7-hydroxy species (cf. Fig. 6, formula b; Fig. 7, formula 2) has the maximum of 10e conjugation and is thus thermodynamically favored. From these and other spectroscopic properties of the methylated analogues, SINHABABU and BORCHARDT concluded that this form constitutes the predominant keto form of 5,7-DHT and 6-methyl-5,7-DHT at pH 7.4, which is the critical intermediate in the unique autoxidation pathway of 5,7-DHT.

b) Characteristics, Products, and Biological Consequences

5,7-DHT reacts with oxygen in a concentration-dependent manner (reaction of second order type) at least during the initial stages; the velocity of the initial, linear reaction is more than ten times higher than that of 5,6-DHT with oxygen (Fig. 4). Stable, colored and soluble products are formed

a: R=CH$_2$CH$_2$NH$_2$;   b: R=CH$_2$CH$_2$NHCO$_2$C$_2$H$_5$;   c: R=H;   d: R=CH$_2$CHO;   e: R=CH$_2$COOH

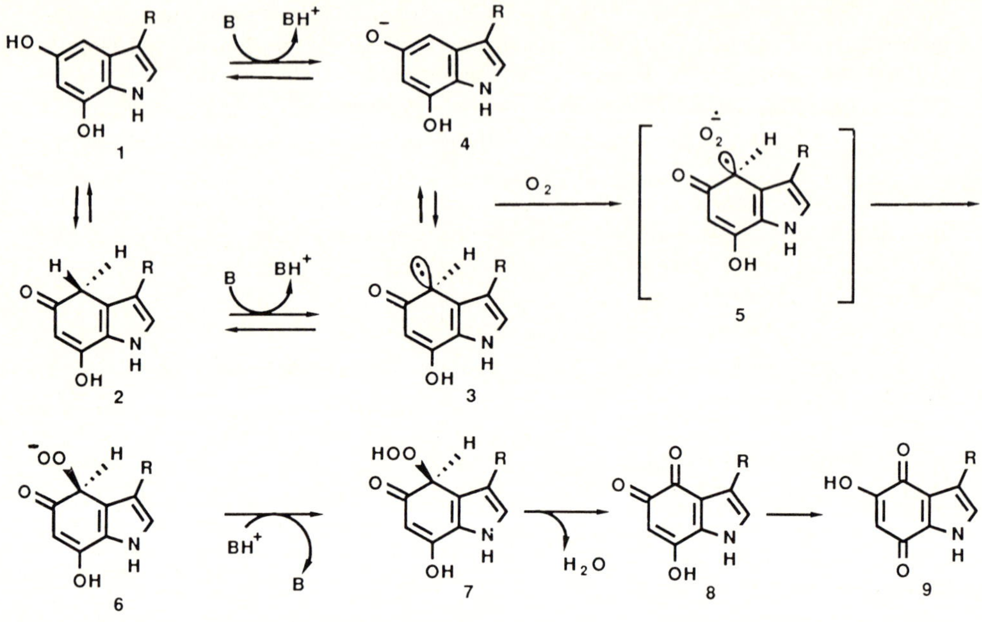

Fig. 7. Mechanism of autoxidation of 5,7-dihydroxytryptamine (*a*), of the corresponding aldehyde (*d*) and acid monoamine oxidase metabolites (*e*). The predominant keto-tautomeric form of 5,7-DHT at pH 7.4 is seen in formula (*2*). Autoxidation proceeds via the superoxide radical complex (*5*) to the hydroperoxide product (*7*) which is decomposed to the *o*-quinone (*8*) which tautomerizes to the more stable 4,7-*p*-quinone (*9*). (From Sinhababu and Borchardt 1989)

during the autoxidation of 5,7-DHT. During the aerobic oxidation of 5,7-DHT, O$_2$ is at first incorporated into the indole molecule (most probably at position 4) and subsequently released into the solution (cf. Fig. 7). That oxygen is incorporated into the indole molecule during autoxidation at position 4 was verified by Sinhababu and Borchardt (1989) through the use of ^{18}O$_2$. This finding rules out the alternative oxidation pathway for 5,7-DHT said to involve nucleophilic addition of H$_2$O to position 4 of the *p*-quinone-imine of 5,7-DHT, the originally postulated reactive quinone intermediate responsible for neurotoxicity. There is, however, evidence from electrochemical studies in favor of the transitory formation of this quinone-imine under acidic pH conditions (Sinhababu and Borchardt 1985). Upon addition of H$_2$O, a trihydroxyindole species is formed which, upon further oxidation, generates the 4,7-dione of 5-HT (Fig. 7). The initial product of autoxidation of 5,7-DHT, the hydroperoxy derivative of the keto tautomer, is subsequently decomposed to the 4,5-*o*-quinone of 7-HT which tautomerizes to form the more stable 4,7-*p*-quinone of 5-HT; this *p*-quinone is considered weakly electrophilic and, therefore, probably not significantly involved in covalent binding. In fact, as shown by KLEMM and BAUMGARTEN (1978), ROTMAN et al. (1976b), and CREVELING and ROTMAN (1978), 5,7-DHT weakly reacts with proteins in vitro and fails to produce protein-crosslinking, confirming the poor reactivity of the 4,7-p-quinone derivative of 5-HT.

However, upon introduction into living systems, radioactivity derived from [^{14}C]- or [^{3}H]-5,7-DHT becomes extensively and irreversibly bound to proteins (BAUMGARTEN et al. 1978; CREVELING and ROTMAN 1978), suggesting that catalytic mechanisms are operative which significantly enhance the reactivity and toxicity of 5,7-DHT and its major metabolites (the corresponding MAO and aldehyde dehydrogenase products). These observations parallel findings obtained by CREVELING and ROTMAN (1978) in mouse atria in vitro which reveal time-dependent increases in irreversibly bound radioactivity following incubation with $30\,\mu M$ [^{3}H]-5,7-DHT. In 1974 BAUMGARTEN et al. demonstrated that 5,7-DHT is taken up into sympathetic axons of various organs in the mouse and causes dose-dependent axodegenerative effects similar to those seen after 6-OH-DA. Subcellular distribution studies in rat brain following intraventricular injection of [^{14}C]-5,7-DHT confirm the observations by CREVELING and ROTMAN that a major percentage of radioactivity is confined to the mitochondrial fraction (BAUMGARTEN et al. 1975b) and a minor percentage to the microsomal fraction, both of which also contain synaptosomes, suggesting that metabolites and/or autoxidation products of 5,7-DHT may accumulate in mitochondria.

3. Relevance of Autoxidation for Neurotoxicity

The overall importance of intraneuronal autoxidation of 5,7-DHT in the development of neurodegenerative toxicity was elaborated by BAUMGARTEN et al. (1982a,b) by comparing uptake site affinities, serotonin-depleting capacity, and oxygen consumption rates of 5,7-DHT analogues (Table 8).

Table 8. Comparison of the initial oxygen consumption rates of various hydroxylated tryptamines with their uptake site affinity

Compound (1 mM)	Oxygen consumption rate (nmoles O_2/min)	IC$_{50}$	Forebrain 5-HT content in % of control[a]
5-OH-7-CH$_3$O-T	insignificant	7.9×10^{-7} M	90%
5,6-DHT	2.7 ± 0.5	6.0×10^{-7} M	56%
α-methyl-5,7-DHT	20.9 ± 1.8	4.4×10^{-6} M	62%
5,7-DHT	33.4 ± 2.1	4.0×10^{-6} M	48%
5,6,7-THT	986.0	no inhibition	97%

5-OH-7-CH$_3$O-T, 5-hydroxy-7-methoxytryptamine.
[a] Measured at 12 days after intraventricular injection of $50\,\mu$g of either compound.
(Taken from KLEMM et al. 1980).

These studies clearly indicated that substitution of the hydroxyl functions by methoxy groups in 5,7-DHT (5-hydroxy-7-methoxytryptamine, 5-methoxy-7-hydroxytryptamine) eliminates the toxicity by impairing oxygen reactivity despite little effect on uptake site affinity. The same study also revealed that the autoxidation process must take place subsequent to incorporation of 5,7-DHT into the monoaminergic axons since oxidation prior to uptake does not result in selective serotonin axon toxicity but rather in a dramatic increase in general brain toxicity (significantly lowered lethal dose); an example of this concept is provided by 5,6,7-THT which is instantaneously converted into a quinone species that does not accumulate in 5-HT neurons. Another example is provided by 4,5,7-THT which tautomerizes at pH 7.4 to form the more stable 4,7-dione of 5-HT; injection of electrooxidized 5,7-DHT into the CSF of mice hardly affected CNS serotonin neurons but greatly increased the unspecific toxicity, causing significant decreases in the LD_{50} compared with 5,7-DHT (Wrona et al. 1986). Finally, the electrochemical oxidation product of 5-HT, tryptamine 4,5-dione, the o-quinone of 4,5-DHT, when injected into the lateral ventricle of rats, produces ventricle close (and much less surface near) argyrophila that is most elaborated in the hippocampus; the morphological analysis indicates rather nonspecific neuronal damage of hippocampal fiber and neuron systems (Crino et al. 1989; Volicer et al. 1989). The o-quinone of 4,5-DHT is a strong electrophile which rapidly undergoes nucleophilic reactions with appropriate partners, e.g., SH-groups; in this respect, it resembles the o-quinone of 5,6-DHT which almost quantitatively reacts with available SH-groups in proteins. Provided the o-quinones of 5,6-DHT and 4,5-DHT are formed extracellularly in the brain prior to removal of the DHTs by active transport into 5-HT, DA, and NE neurons, they should be able to attack SH-groups in every cell membrane and cause impairment of membrane functions, cell damage, and eventually cell death.

IV. Potential Intracellular Catalytic Mechanisms Involved in Enhancement of Autoxidation of 5,7-DHT and Radical Formation

Compared with the straightforward, simple mechanism of toxic action of 5,6-DHT which is easily reproduced in model systems and in vitro, the mechanism of action of 5,7-DHT is remarkable and paradoxical: 5,7-DHT is poorly electrophilic when autoxidized at pH 7.4 in the presence of nucleophilic proteins or peptides, and there is no evidence for the generation of secondary reaction products during the autoxidation process. However, in vivo administration of [14C]- or [3H]-5,7-DHT results in time-dependent increases in covalent binding of reactive intermediates to brain proteins, and there is at least circumstantial evidence for the formation of radicals during the autoxidation of 5,7-DHT in vivo since pretreatment of animals with radical scavengers (nialamide, PTTU, and ethanol) lends significant protection against the degenerative action of 5,7-DHT in cardiac adrenergic nerves of mice (Cohen and Heikkila 1978; Allis and Cohen 1977). To explain these discrepancies in the physicochemical behavior of 5,7-DHT in vitro and in vivo, BAUMGARTEN et al. proposed the existence of catalytic intracellular mechanisms for enhancing the autoxidizability of 5,7-DHT and its aldehyde and acid MAO metabolites (KLEMM et al. 1980). Oxygen consumption rates of 5,6-DHT and 5,7-DHT, α-methyl-5,7-DHT (a MAO-resistant analogue of 5,7-DHT), and 5,6- and 5,7-dihydroxyindoleacetic acid were enhanced by the addition of mitochondria to the incubation medium in a dose-dependent manner; as pargyline did not modify this acceleration of autoxidation, MAO was not involved in this catalytic process. Similarly, addition of oxidized cytochrome c to the incubation mixture speeded the autoxidation rates of 5,6- and 5,7-DHT. To rule out any effect of uncoupling of phosphorylation in this acceleratory property of mitochondria, phosphorylation rates for ADP were studied under conditions of stimulated respiration in states 3 and 4 in the presence of 5,6-, 5,7-DHT, 6-OH-DA, and the acid MAO metabolites of 5,6- and 5,7-DHT. None of these compounds affected electron transport and oxidative phosphorylation. These results suggest the possibility that electron transport from 5,6- and 5,7-DHT to cytochrome c occurs within the terminal segment of the respiratory chain. The exact mechanism and significance of this mitochondria-mediated acceleration of autoxidation of 5,6- and 5,7-DHT for the in vivo neurotoxicity remains to be clarified.

SINHABABU and BORCHARDT proposed additional intracellular mechanisms to explain the unusual molar toxicity of 5,7-DHT in vivo which, at the same time, bear on the suspected production of secondary reaction products by 5,7-DHT, i.e., interaction of the hydroperoxide products of 5,7-DHT and its metabolites with glutathione peroxidase and redox cycling of the 4,7-p-quinone autoxidation products of 5,7-DHT, 5,7-DHIAA, and 5,7-dihydroxyindoleacetaldehyde (5,7-DHIA) (Fig. 8). GSH peroxidase is characterized by a low degree of substrate specificity and accepts a wide variety of bulky organic hydroperoxides which are not metabolized by catalase, such as peroxides of steroids and prostaglandins as well as synthetic hydroperoxides. By analogy, the hydroperoxides of 5,7-DHT, 5,7-DHIA, and 5,7-DHIAA may be metabolized by this enzyme which is localized in cytosolic as well as mitochondrial compartments; the peroxidase products are supposed to tautomerize rapidly to the corresponding fully aromatized trihydroxyindoles which quickly autoxidize to generate the 4,7-p-quinone species. This quinone is very toxic when injected intraventricularly (MASSOTTI et al. 1974). SINHABABU and BORCHARDT propose that this quinone, when formed intraneuronally by autoxidation of 5,7-DHT, may become a substrate of redox cycling microsomal and mitochondrial enzyme systems, i.e., mixed function oxidases which produce free radicals (O_2^-) and require GSH and reducing equivalents (cf. Fig. 9). If indole-p-quinones enter redox cycling, they might easily compromise cellular defence capacities by accumulation of free radicals, GSH consumption, depletion of GSSG reducing equivalents, and hypoxia. Hypoxia enhances the radical-induced toxicity of redox-cyclers.

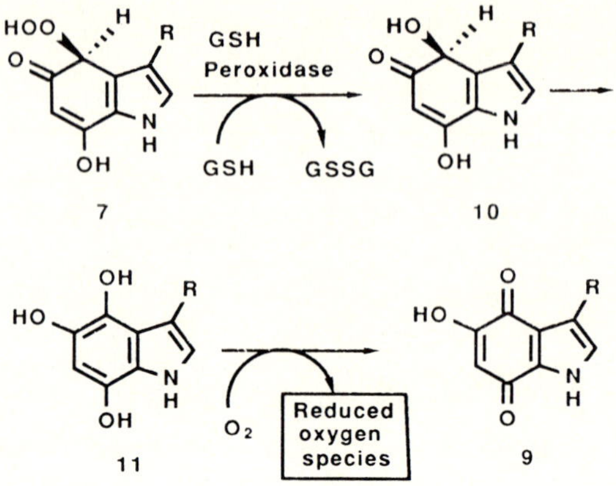

a: R=CH₂CH₂NH₂;   d: R=CH₂CHO;   e: R=CH₂COOH

Fig. 8. Role of reduced glutathione (*GSH*) peroxidase in the metabolism of hydroperoxide 7*a*, *d*, or *e*. (From Sinhababu and Borchardt 1988)

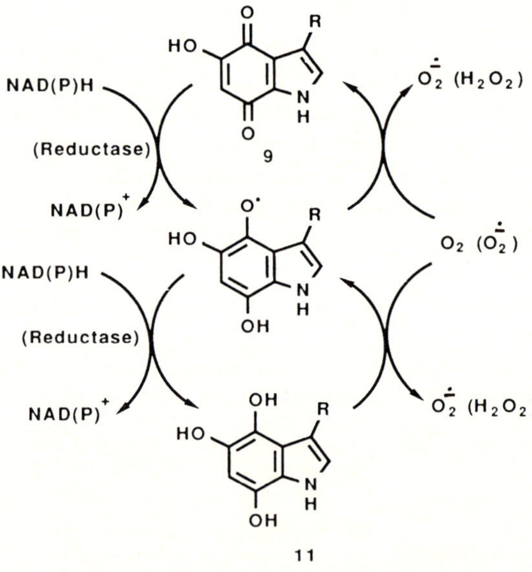

a: R=CH₂CH₂NH₂;   d: R=CH₂CHO;   e: R=CH₂COOH

Fig. 9. Redox cycling of quinones 9*a*, *d*, or *e*. (From Sinhababu and Borchardt 1988)

Finally, MAO also has been claimed to participate in the overall neurotoxicity of 5,7-DHT (Creveling and Rotman 1978; Baumgarten et al. 1982a,b). The evidence for this assumption derives from the observation that pretreatment of mice with pargyline impairs the degenerative action of

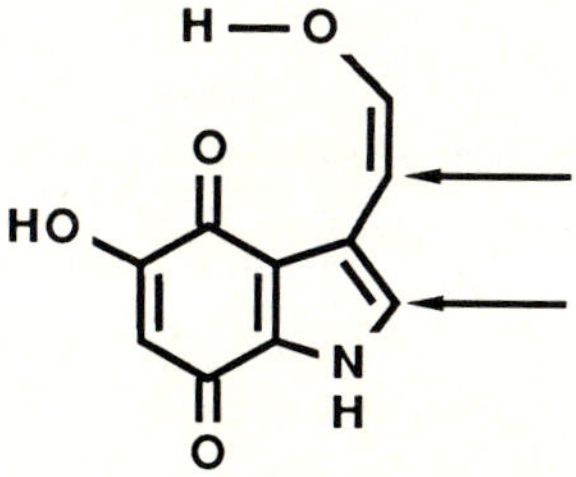

Fig. 10. Enol form of the 4,7-*p*-quinone of 5-hydroxyindole acetaldehyde (9*d* in Fig. 9). *Arrows* indicate electrophilic centers through which protein nucleophiles may undergo covalent modification (and crosslinking). (From SINHABABU and BORCHARDT 1988)

5,7-DHT in cardiac sympathetic nerves and that pretreatment of rats with various MAO inhibitors counteracts the reductions in cerebral NE levels by intraventricular 5,7-DHT (BREESE and MÜLLER 1978). However, the same treatment had little protective effect on the long-term reductions in cerebral 5-HT levels, indicating that there are considerable differences in the biochemical pharmacology/drug sensitivity/radical toxicity in NE versus 5-HT neurons that render any meaningful interpretation difficult. In the hands of BAUMGARTEN et al. (1978), α-methyl-5,7-DHT, a MAO-resistant analogue of 5,7-DHT, proved less than half as efficient a long-term cerebral 5-HT and NE depletor as 5,7-DHT. In order to more directly attack this problem, BAUMGARTEN et al. (1984) investigated the role of MAO in the covalent binding of metabolites/autoxidation products from [^{14}C]-5,7-DHT to brain proteins in vivo and found a contributing role of MAO in the extent of binding which was interpreted as due to the formation and reactivity of the aldehyde metabolite. However, since the covalent binding process after 5,7-DHT is only temporarily attenuated by MAO inhibition, other catalytic mechanisms than formation of the aldehyde metabolite are of major importance for the protein-arylating capacity of products from 5,7-DHT (see above). SINHABABU and BORCHARDT propose a role for the autoxidized 5,7-di-OH-indolacetaldehyde in the covalent modification of proteins through transformation into the corresponding *p*-quinone and enolization of the side chain aldehyde (stabilized through intramolecular hydrogen bonding with the 4-oxo function); this electrophilic structure may then react with tissue nucleophiles at the sites indicated in Fig. 10.

E. Substituted Amphetamines

I. Neurotoxicity

The long-lasting reduction of brain 5-HT, 5-HIAA, synaptosomal [^{3}H]-5-HT uptake, and tryptophan hydroxylase activity by *p*-chloroamphetamine (p-

CA; 5 mg/kg, i.p.) was discovered in 1972 by Sanders-Bush et al. who also found that this effect could be prevented by pretreatment with selective 5-HT uptake blockers (e.g., fluoxetine, Sanders-Bush and Steranka 1978), suggesting that transport of the drug into 5-HT neurons and subsequent release of 5-HT and inhibition of 5-HT uptake are required for long-term depression of functional indices in brain 5-HT neurons. The claim by Harvey et al. (1975) that 5-HT cell bodies in the midbrain of rats degenerate after p-CA treatment has not been confirmed, but there is evidence for acute morphological alteration of axon terminals in the striatum (Hattori et al. 1976) and formation of large swellings and transmitter pile-up in preterminal 5-HT axons (Lorez et al. 1976) resembling proximal stumps of axons due to disintegration of their parent terminals. Sanders-Bush and Steranka (1978) had already noted that the long-term actions of p-CA are nonuniform throughout the CNS, and they further pointed to species differences in the sensitivity of brain 5-HT neurons to the toxic actions of p-CA (little evidence for toxicity in mice in contrast to rats) focussing on the role of species-dependent differences in metabolism and pharmacokinetics in the events that render p-CA toxic or nontoxic. The idea that subsystems of 5-HT neurons may have differential sensitivity to the toxic actions of p-CA was substantiated by Mamounas and Molliver (1988) and Fritschy et al. (1988) through the use of retrograde axonal transport tracers; these authors demonstrated that the projections of the dorsal raphé nucleus, characterized by delicate, fine terminal varicosities, are preferentially affected by p-CA. These axons are also the primary target of methamphetamine, methylenedioxyamphetamine (MDA), and methylenedioxymethamphetamine (MDMA) neurotoxicity (Molliver and Molliver 1990; O'Hearn et al. 1988; Insel et al. 1989; Battaglia et al. 1987, 1988, 1989; cf. introductory review by Baumgarten and Zimmermann, this volume). The mechanism whereby these amphetamines develop neurotoxicity in axonal projections of the dorsal raphé nucleus appears to involve inhibition of the 5-HT transporter ("uptake"), enhanced release of 5-HT, inhibition of intraneuronal MAO, and formation of neurotoxic analogues of serotonin and DA, such as 5,6-DHT (Commins et al. 1987a,b) and 6-OH-DA (Seiden and Vosmer 1984) by free-radical-catalyzed hydroxylation of DA and 5-HT. Gál et al. (1975) proposed conversion of p-CA to 5,6-dihydroxyindoles to explain the neurotoxicity of p-CA, but the pathways leading to these structures have not yet been substantiated. Ames et al. (1977) showed that liver microsomes metabolize p-CA to a reactive intermediate which becomes covalently bound to microsomal proteins, but the relevance of these findings to the CNS toxicity of pCA has not been clarified. By contrast, Commins et al. (1987a,b) suggest that 5,6-DHT and/or 6-OH-DA (Seiden and Vosmer 1984) may be formed in vivo following p-CA or methamphetamine-induced release of DA and 5-HT by free-radical-catalyzed conversion of DA and/or serotonin to neurotoxic hydroxylated analogues. The evidence for nonenzymatic formation of 5,6-DHT and/or 6-OH-DA following carrier-mediated release by these amphetamines awaits confirmation by other laboratories. Long-lasting (2 weeks) depression of functional indices in serotonergic projections to the forebrain (5-HT, 5-HIAA) have been reported in the rat following parenteral (i.p.) administration of very high single doses of fenfluramine (20 and 40 mg/kg), but tryptophan hydroxylase, the enzyme responsible for 5-HT synthesis, was only significantly affected after even higher doses of fenfluramine (60 mg/kg); the potential recovery of these markers with time was not studied, however. CLINESCHMIDT et al. (1978) studied the effects of a high oral dose of fenfluramine (15 mg/kg) on brain 5-HT and 5-HIAA and showed recovery of 5-HT to near-normal levels by 64 days after treatment, indicating that fenfluramine's actions differ from the almost permanent irreversible effects of established serotonin neurotoxins, such as 5,7-DHT. These observations focus attention on the importance of the route of administration and pharmacokinetics of fenfluramine in modifying its actions on serotonin neurons.

Fenfluramine appears to differ from p-CA and methamphetamine in two further aspects, namely the absence of MAO-inhibiting properties and lack of evidence for monoamine neurotoxin (5,6-DHT, 6-OH-DA) formation; furthermore, while fenfluramine is a preferential 5-HT uptake inhibitor, its metabolite norfenfluramine is a preferential 5-HT releaser, suggesting that species-dependent differences in the kinetics of accumulation and disappearance of fenfluramine and norfenfluramine may have a considerable impact on the extent and duration of its neurochemical effects in brain. In fact, STERANKA and SANDERS-BUSH noted already in 1979 that the time-course and efficacy of action of parenterally administered fenfluramine differs widely according to the species investigated (the rat being much more sensitive to fenfluramine than the mouse) and showed that the main interspecies difference relates to the retarded clearance of original drug and its main metabolite from rat brain compared with mouse and to lower peak values of norfenfluramine in mouse compared with rat brain.

II. Importance of Pharmacokinetics and Metabolism in the Long-Term Actions of Fenfluramine

The observation that parenteral administration of moderate to high doses of fenfluramine to rats causes long-term depression of functional indices of 5-HT neurons (HARVEY and McMASTER 1975, 1977; HARVEY et al. 1975; CLINESCHMIDT et al. 1976, 1978; STERANKA and SANDERS-BUSH 1979; KLEVEN and SEIDEN 1989; KLEVEN et al. 1988; MOLLIVER and MOLLIVER 1990; APPEL et al. 1989; ZACZEK et al. 1990a) in brain, has led authors to express concern about the safety of oral fenfluramine in man. The following facts and arguments clearly indicate that extrapolation from rat data – gained by either parenteral or oral dosing regimens – to the human situation yields erroreous risk concepts because of differences in the pharmacokinetics and metabolism of fenfluramine in either species. Chronic oral d,l-fenfluramine given to rats at doses much higher than the average daily human oral anorectic dose (which is in the order of 1 mg/kg daily) appear to be marginally toxic in the rat; following 21 days of consecutive oral treatment with 10 mg/kg (5 mg b.i.d.), only transient and sporadic ("rare") signs of serotonin axon pathology were detected, which were normalized by day 30 after the onset of treatment (i.e., 7 days after finishing treatment) (Sotelo 1991). By contrast. parenteral doses of 4 mg/kg *d,l*-fenfluramine, given for only 4 days (2 mg/kg, s.c., twice daily) caused long-term reduction of 5-HT uptake sites (Zaczek et al. 1990a), showing that the route of administration has a considerable impact on the risk of *d,l*-fenfluramine to develop long-lasting effects in serotonergic axons. Dexfenfluramine (*d*-fenfluramine), the biologically active stereoisomer (average human anorectic dose 0.5 mg/kg daily), has been found to cause transient depletion of 5-HT from neocortical serotonergic axons (loss of immunoreactivity) in the rat (Kalia 1991) without evidence of axon pathology after high oral doses (up to 12 mg/kg daily for 4 days), implying high safety margins in a species considered particularly fenfluramine-sensitive in comparison with man. The reasons for the high sensitivity of the rat to (parenteral) fenfluramine actions are unfavorable pharmacokinetics and differences in metabolism. The development of neurotoxicity depends on the tendency to accumulate unproportionally high brain levels of norfenfluramine (a more potent 5-HT releaser) in addition to fenfluramine (a preferential inhibitor for the 5-HT transporter); in this situation which is apparent in the rat after 8 consecutive doses of as little as 4 mg *d,l*-fenfluramine given every 12 h, the clearance mechanisms for norfenfluramine are saturated, and the rate for elimination does not cope with the rate of formation of norfenfluramine (Zaczek et al. 1990a; Marchant et al. 1990; Campbell et al. 1986, 1990; Garattini and Caccia 1979; Caccia et al. 1982). Therefore, the rat exhibits disproportionate nondose-related rises in brain norfenfluramine and fenfluramine levels after moderate parenteral dosing regimens.

That long-term actions are dependent on critical brain concentrations of norfenfluramine (besides fenfluramine) is demonstrated by the results of a recent study employing intraventricular infusions of *d*-fenfluramine or *d*-norfenfluramine in the rat (Invernizzi et al. 1991) showing that long-term (7 days) reductions in cerebral 5-HT levels (cortex, hippocampus, striatum) require build-up of high brain concentrations of *d*-norfenfluramine (4–5 µg/g). Such high brain levels of *d*-norfenfluramine (in the face of much lower brain *d*-fenfluramine levels, 0.5 µg/g) were obtained by a 2 h continuous infusion of 500 µg *d*-fenfluramine/h or by infusion of its metabolite *d*-norfenfluramine. When *d*-norfenfluramine was infused over a 2 h period, minimum cerebral 5-HT levels were reached between 4 and 8 h, but peak norfenfluramine concentrations in the brain were already found at the end of the infusion period, indicating that prolonged presence of high concentrations of the main metabolite of *d*-fenfluramine are required for irreversible toxic effects on the brain 5-HT axon terminal system. The long-term toxicity of intracranially infused *d*-fenfluramine thus depends on the N-dealkylation to *d*-norfenfluramine in the liver and subsequent accumulation of *d*-norfenfluramine in the brain. The ratio between brain and plasma levels of *d*-norfenfluramine between 4 and 24 h after the end of the central infusion of *d*-fenfluramine (40–47) is within the range of ratios measured following systemic administration of the parent compound (CACCIA et al. 1982; GARATTINI et al. 1979; GARATTINI and CACCIA 1990).

The idea that the more potent 5-HT releaser *d*-norfenfluramine may be mainly responsible for the neurotoxicity risk of parenterally administered *d*-fenfluramine in sensitive rodent species is also supported by the recent findings of JOHNSON and NICHOLS (1990) that *d*-norfenfluramine is nearly twice as potent as *d*-fenfluramine in reducing [³H]paroxetine binding and 5-HT concentrations in cerebral cortex and hippocampus by 1 week following a single high dose of either drug (10 mg/kg, s.c.). These observations require confirmation by more selective methods as paroxetine may not only bind to the 5-HT transporter (GOBBI et al. 1990).

The rat converts the major proportion of fenfluramine to norfenfluramine, but this metabolite is poorly deaminated. The human, in contrast to the rat, metabolizes fenfluramine to norfenfluramine which is subsequently deaminated to polar compounds including the major inactive metabolite 3,4-trifluoromethylphenylpropandiol, which is excreted as the glucuronide conjugate. Consequently, man has, even at high doses, low and stable brain/plasma ratios, approximately 9:1 of both fenfluramine and norfenfluramine, but the relevant amounts of the norfenfluramine metabolite still remain low. Under overdosage conditions (48 mg/kg s.c.), the rat shows a 200–400 fold increase in peak brain fenfluramine/norfenfluramine levels (brain/plasma ratio >40). Under comparable overdosage conditions in man (>40 mg/kg, per os; cf. FLEISHER and CAMPBELL 1969) brain/plasma ratios remained at the ratio (9:1) determined following therapeutic doses of fenfluramine (6:1 at 1 mg/kg, per os). Thus, species-dependent differences in metabolism and pharmacokinetics (and species-independent differences in drug kinetics related to the route of administration) mainly determine the risk for neurotoxicity which apparently is evident in the rat upon moderate parenteral dosing of *d,l*-fenfluramine or high oral dosing of *d,l*-fenfluramine but negligible for man at doses manifold higher than the average daily anorectic dose. It is thus unlikely that oral overdosing in the human (particularly of *d*-fenfluramine which is active at half the daily dose of *d,l*-fenfluramine) could result in brain accumulation of concentrations of fenfluramine and particularly norfenfluramine required for toxicity in monoaminergic neurons.

III. Do Amphetamines Cause Acute Destruction of Serotonergic Axons Like 5,7-DHT?

Finally, it must be asked whether or not moderate or high parenteral doses of substituted amphetamines – which definitely can cause long-term depression of functional indices in central 5-HT axons within defined terminal targets (neocortex, hippocampus, and striatum) – are capable of precipitating acute axodegeneration in a manner comparable with 5,7-DHT, the best characterized "serotonin neurotoxin" (cf. reviews by Baumgarten and Björklund 1976 and Baumgarten et al. 1972c, 1974a,b, 1975a, 1977, 1978, 1981, 1982a,b).

As most of the methods that have been applied to study this issue are measures of defined functions of serotonergic neurons (serotonin synthesis, serotonin uptake, presence of the transporter protein, axonal transport, presence of immunoreactive material, etc.) but fail to provide generalized vitality indices of 5-HT neurons, a clear-cut answer is difficult to provide at present. While temporary depression of the diverse functional indices of 5-HT neurons after low or moderate parenteral doses of amphetamines is well compatible with the structural integrity and survival of axons, the axotomy-like grotesque alterations in axonal morphology and neuritic plaque-like appearance of sprouting phenomena in the vicinity of these dilated (stump-like) axons following very high parenteral doses of, e.g., MDMA or p-CA do suggest (even in the absence of electron microscopical data) that a fraction of axon terminals undergoes disintegration that are subsequently rather rapidly restored by regeneration from preterminal axons within the affected target areas which – due to the lack of injury imposed upon the remaining target structures – meet with ideal growth conditions (Molliver et al. 1990).

If one is willing to accept indirect measures of the vitality state of neurons, such as the expression of glial fibrillar acidic protein (GFAP) by glial cells in the vicinity of degenerating axons, the absence of glial activation subsequent to administration of a suspected neurotoxin would signify structural integrity. Using this criterion, O'Callaghan et al. (1990) failed to find any glial activation after rather high acute (20 mg/kg s.c.) or chronic (5–30 mg/kg, s.c., twice daily for 7 days) parenteral MDMA dosing regimens but detected GFAP increases (50%) in the cortex after extraordinary doses of MDMA (4 × 80 mg/kg, s.c.). In contrast, widespread CNS glial activation is seen after i.c.v. administered 5,7-DHT which causes less (percentage) regional 5-HT depletion but massive degeneration of serotonergic axonal profiles. This suggests that the "neurotoxicity" of moderate parenteral doses of substituted amphetamines is of a nature different from the axodestructive capacity of established serotonin neurotoxins.

IV. Relevance of 6-Hydroxydopamine- and Dihydroxytryptamine-Induced Neurotoxicity

1. Drug-Induced Toxicity in Central Monoaminergic Neurons

It has been claimed that hydroxylated analogues of DA and serotonin can be formed in the brain of rodents upon massive release of DA and/or

5-HT by substituted amphetamines (e.g., methamphetamine, p-chloroamphetamine; cf. SEIDEN and VOSMER 1984; COMMINS et al. 1987a,b). Following a single dose of methamphetamine-HCl (100 mg/kg expressed as the salt, given s.c.), 6-OH-DA was said to have been detected by HPLC and electrochemical analysis in the caudate nucleus of the rat between 30 and 120 min at a level of 0.39 ± 0.31 ng/mg wet tissue concomitant with measurable release of DA and inhibition of its MAO-catalyzed conversion to DOPAC. Given the accurateness and reproducibility of the measured signal as representing authentic 6-OH-DA, the question arises whether transitory formation of trace amounts of 6-OH-DA is sufficient to explain the long-term DA reductions after methamphetamine (50% of control by 14 days) by degeneration of dopaminergic nerve terminals within the caudate. The only study that may assist in elucidating this question is that by EVANS and COHEN (1989a,b) using 6-hydroxydopa as a tool to build up levels of 6-OH-DA in the caudate (5.55 µg/g) known to destroy NE terminals within the striatum, n. accumbens, olfactory cortex, and telencephalon (JACOBOWITZ and KOSTRZEWA 1971; KOSTRZEWA 1989). With this concentration of 6-OH-DA, there is no measurable depletion of DA in the caudate but the expected reduction of [^{3}H]-NE uptake in cortical synaptosomes (73% reduction). This raises doubt on the postulate that much lower rates of formation of 6-OH-DA should be sufficient to provoke degeneration of significant numbers of DA terminals after methamphetamine. Furthermore, ROLLEMA et al. (1986) failed to detect 6-OH-DA in the striatum of the rat after DA-releasing compounds (dexamphetamine, methylamphetamine, and MPTP). Finally, according to the findings of SLIVKA and COHEN (1985), the major products of nonenzymatic hydroxylation of DA are 2-OH- and 5-OH-DA, both of which are nontoxic to mouse heart adrenergic nerves in vivo (LUNDSTRÖM et al. 1973). Thus, the role of 6-OH-DA in amphetamine-induced DA neuron toxicity remains enigmatic.

More recently, COMMINS et al. (1987a,b) detected the transient appearance of an electrochemical signal in HPLC-separated samples of rat hippocampus having the characteristics of 5,6-DHT following treatment with either p-chloroamphetamine or methamphetamine (maximum concentration of 5,6-DHT-like signal: 0.032 ng/mg wet tissue) concomitant with significant depletion of 5-HT by 1 h and of 5-HIAA by 4 h after either amphetamine derivative. Furthermore, the authors reported sporadic appearance of a 5,6-DHT-like electrochemical peak in neocortical samples after giving methamphetamine. In order to evaluate the potential neurotoxicity of small concentrations of 5,6-DHT in structures exposed to the CSF, COMMINS et al. infused 5 µg 5,6-DHT-creatinine sulfate (calculated as free base) into either of the two lateral ventricles and detected levels of 5,6-DHT in hippocampal homogenates ranging between 0.5 ng/mg wet tissue at 4 h and 0.82 ng/mg 5,6-DHT at 5 min after termination of the infusion. Although there is a discrepancy between the trace levels of 5,6-DHT recorded after substitute amphetamine administration and after experimental administration of high amounts (10 μg) of 5,6-DHT directly into the CSF, 5,6-DHT is – compared with 6-OH-DA – a more powerful neurotoxin which has a high apparent affinity to 5-HT transport sites. Consequently, there is – even with the low concentrations measured in hippocampal and neocortical tissue after amphetamine administration – a reasonable chance for serotonin axon damage provided 5,6-DHT is formed in the brain.

In the preceding paragraphs, examples of nonenzymatic, electrochemical, and metal-catalyzed oxidation of DA and 5-HT have been presented. The products generated in these (model) reactions comprise nontoxic trihydroxyphenylethylamines (2- and 5-OH-DA) and the potentially toxic species 6-OH-DA, semiquinone radicals of dopa and DA and of 5-HT and 4,5-diketotryptamine. One would have to postulate conditions combing reduced antioxidant defence and increased radical production activity in a given neuron ("oxidative stress") in order to create a situation in which DA would be oxidized significantly to reactive semiquinone and quinone intermediates (though not necessarily 6-OH-DA and its p-quinone), causing neurotoxic "structural" damage. Indeed, it has been postulated that DA is somehow involved in the long-term toxic actions of methamphetamine in brain 5-HT neurons (Schmidt et al. 1985). DA, released from striatal dopaminergic axons by methamphetamine, may be taken up into the serotonergic nerve endings and become oxidized by unknown mechanisms, a process favored by amphetamine-induced inhibition of DA catabolism by MAO. However, although similar relationships have been invoked to understand the role of DA in serotonin nerve fiber damage by MDMA administration (Stone et al. 1988), other findings are difficult to reconcile with this concept. Thus, Hekmatpanah et al. (1989) have found that depletion of central monoamine stores by pretreatment with reserpine fails to counteract the evolvement of neurotoxicity in serotonin fiber systems (measured as reduction in the density of [³H]paroxetine-labelled 5-HT uptake sites by 2 weeks after MDMA administration). Long-term inhibition of tryptophan hydroxylase by MDMA (thought to reflect toxic damage to 5-HT axons) was counteracted by either α-methyl-para-tyrosine or reserpine pretreatment and at least attenuated by bilateral degeneration of the dopaminergic nigrostriatal projections (4 μg 6-OH-DA into the substantia nigra of either side) or by blocking uptake of DA (into 5-HT neurons?) with GBR 12909 (Stone et al. 1988). These data suggest a role of DA release in the serotonergic toxicity of MDMA.

While direct evidence for the involvement of autoxidation products from DA and/or 5-HT in amphetamine neurotoxicity is inconclusive, there is indirect evidence for the involvement of reactive alkylating species in the monoaminergic toxicity of amphetamine and p-chloroamphetamine since systemic administration of L-cysteine (a thiol replenisher) attenuates DA reductions by amphetamine and 5-HT reductions by para-chloroamphetamine (Steranka and Rhind 1987). These observations may also be explained by postulating drug-induced deficits in thiol antioxidant potential by inhibition of the disulfide reducing cellular mechanisms [NAD(P)H-dependent GSSG reductase] and/or increases in oxidative stress. As l-cysteine can interfere with the metabolism of amphetamines (INVERNIZZI et al. 1989), the formation of toxic metabolites may have been prevented. It has been suggested that the acute inhibitory actions of substituted amphetamines on striatal tryptophan hydroxylase in vivo may be the result of conformational changes in the enzyme due to disulfide formation (reversible upon addition of DTT in vitro: JOHNSON et al. 1989a,b). Intracellular oxidative stress and thiol oxidation by semiquinone/quinone products of DA (or 5-HT) formed in the extracellular space after release from and subsequent uptake into the dopaminergic and serotonergic axons might indeed account for these acute effects of amphetamines. As there is experimental evidence for a role of quinone/semiquinone redox pairs (generated through cytochrome P-450-dependent hydroxylases) in serious perturbations of intracellular Ca^{2+} homeostasis (BELLOMO et al. 1987; GANT et al. 1988), similar mechanisms may well be operative in the striatal monoaminergic nerve endings subjected to uncontrollable releasing stress by amphetamines. The prevention of amphetamine-induced monoaminergic neostriatal toxicity by postadministration of potent uptake blockers (fluoxetine, amfonelic acid) indicates that prolonged uptake of toxins and/or their immediate precursors is a prerequisite in the cascade of events that finally result in axon degeneration. Finally, none of the proposed concepts may be relevant, and the main effect of drug-induced, poorly controlled, excessive release of DA and 5-HT could involve increased toxicity of endogenous excitotoxins (e.g., glutamate) at dysregulated NMDA receptor sites, resulting in a disruption of intracellular calcium homeostasis and phospholipase- and protease-mediated cellular autolysis. In fact, circumstantial evidence points to a role of NMDA receptor mechanisms in methamphetamine-induced toxicity in striatal dopaminergic systems of mice (i.e., protection by NMDA antagonists) (SONSOLLA et al. 1989) and also in the DA-mediated toxicity of methamphetamine for striatal tryptophan hydroxylase of rats because coadministration of methamphetamine and MK-801 (the most potent noncompetitive NMDA receptor antagonist) provided partial but significant protection against the inhibition of the enzyme of serotonergic neurons. FINNEGAN et al. (1990) reported prevention of MDMA-induced long-term reductions in striatal 5-HT and DA levels by another NMDA receptor antagonist, dextrorphan. More recently, OLNEY et al. (1990) have shown that l-dopa is a weak and 6-OH-dopa a powerful excitotoxin in the chick embryo retina, imitating a quisqualate-induced toxicity pattern subject to prevention by appropriate doses of CNQX (6-cyano-7-nitroquinoxaline-2,3-dione) and thus suggesting cooperative interaction of endogenous excitatory amino acids and catecholamines or their precursor amino acids.

A role for calcium channels in the MDMA-mediated toxicity in neostriatal serotonergic neurons is suggested by the ability of flunarizine to attenuate significantly the reduction in 5-HT level and in tryptophan hydroxylase activity (Gibb et al. 1990); flunarizine failed to protect this enzyme against the toxicity induced by methamphetamine, pointing to differences in the mechanisms of action of methamphetamine and MDMA. More detailed evidence for a participation of Ca^{2+} in the serotonergic neurotoxicity of MDMA was recently provided by Azmitia et al. (1990) in cultured raphé 5-HT neurons; the potency of MDMA to inhibit the development of [^{3}H]-5-HT uptake capacity in these cultures of serotonin neurons was associated with its action on Ca^{2+}-dependent and -independent forms of 5-HT release, on 5-HT-2-receptor-mediated mobilization of Ca^{2+} from intracellular stores and on increase in Ca^{2+} influx through specific channels. These findings suggest that substituted amphetamines cause a derangement of intracellular Ca^{2+} homeostasis.

At present, the following fragmentary concept of substituted amphetamine neurotoxicity in central monoaminergic neurons can be formulated. MDMA and p-chloroamphetamine appear to release monoamines by using carrier-and noncarrier-mediated mechanisms coupled with (a) inhibition of uptake of monoamines, (b) release from (mainly) the nongranular pool, and (c) inhibition of MAO-dependent metabolism of intraneuronal monoamines. Assuming that the massive drug-induced release of DA causes oxidative stress within the dopaminergic nerve endings and a shift in the redox balance, it appears possible that fractions of DA in excess of those produced regularly undergo oxidation to form radical intermediates, e.g., the semiquinone radicals and O_2^-. In the absence of redox buffers (GSH), these intermediates may autocatalytically (by means of O_2^- and H_2O_2) promote further oxidation of the semiquinones to quinones which might then have a chance to cyclize and form 5,6-dihydroxyindoles. These may be released into the extracellular space and become substrates for the DA and 5-HT transporter. Alternatively, free radicals catalyze nonenzymatic hydroxylation of DA to generate 6-OH-DA, which enters a comparable release/reuptake/autoxidation cycle, causing injury to nerve terminals. This process may continue for extended periods of time (hours), and the cumulative effect of rather small amounts of quinones and radicals produced in fluctuating quantities is the real cause of toxicity. Given these conditions, the sporadic nature of small levels of 6-OH-DA, 5,6-dihydroxyindole, or related dopamine-derived quinoidal systems reported by Axt et al. (1990) and Marek et al. (1990) would not devaluate the concept that (at least a portion of) amphetamine neurotoxicity might be due to generation of monoaminergic neurotoxins and/or their close analogues. Progress in analytical technologies may help to resolve this controversial issue.

2. Monoamine Neurotoxins, Aging, and Disease

The idea has been expressed that chronic imbalance between radical generating and radical detoxifying cellular mechanisms has an impact upon life expectancy. Such mechanisms would be particularly important in nondividing cells that subserve essential functions such as certain neurons of the brain. Parkinson's disease (idiopathic form) may represent an example of age-related increases in the risk of radical damage because of the continuous synthesis of large amounts of autoxidizable species, e.g., semiquinone radicals of dopa and DA, the detoxification of which requires reducing equivalents, accessible SH-nucleophiles, radical defending enzyme activities (Cu, Zn-SOD, catalase, GSH peroxidase), and radical scavenging organic compounds such as α-tocopherol and ascorbic acid. Some data seem to support a relationship between increased radical activity (lipid peroxidation in the substantia nigra, cf. DEXTER et al. 1986) and reduced thiol defence potential (GSH levels in the substantia nigra and various forebrain regions, cf. PERRY et al. 1982; RIEDERER et al. 1989) in dopaminergic and non-dopaminergic areas of the human brain which correlate with the severity (stage) of the disease. However, it remains to be shown that there exist temporal and causal relationships between increases in oxidative stress, decreases in antioxidant defence capacities, increases in autoxidation rate of endogenous dopa and DA, and risk of degeneration in human nigral DA neurons. Model studies on striatal synaptosomes indicate that exposure to L-dopa increases the levels of GSSG (SPINA and COHEN 1988), suggesting that treatment of Parkinson's disease with high doses of L-dopa may actually accelerate nigral DA neuron damage and disintegration. Furthermore, increases in striatal DA turnover, evoked by reserpine treatment in mice (i.e., enhanced loss of DA and deaminated metabolite formation), are accompanied by significant rises in GSSG that can be prevented by inhibition of MAO, which is responsible for H_2O_2 production. The neostriatal dopaminergic nerve endings, therefore, are subjected to oxidative stress due to significantly enhanced turnover of its transmitter because of blockade of its vesicular storage (SPINA and COHEN 1989). If a similar condition prevails in the nigrostriatal DA neurons of patients suffering from Parkinson's disease in which the surviving neurons attempt to compensate for the loss of more than 80% of DA neurons, then chronic oxidative stress and shifting of the redox balance may signify a vulnerability factor for radical and toxin sensitivity that accelerates the disintegration of the activated DA neurons. Whether nonenzymatic oxidation of DA to 6-OH-DA occurs in the human mesolimbic and mesostriatal DA neurons is unknown at present; as outlined earlier, this is unlikely since autoxidation of DA and of dopa presumably results in the generation of semiquinone radical intermediates and the corresponding o-quinones which either undergo nucleophilic addition reactions with SH-reagents or (less likely) proceed to form cyclized aminochrome species (TSE et al. 1976; GRAHAM et al. 1978). In fact, ROSENGREN et al. (1985) have isolated 5-S-cysteinyldopa and -DA from human brain suggesting that oxidized dopa and DA can react with SH-nucleophiles. If significant amounts of 6-OH-DA were formed, its corresponding p-quinone would undergo nucleophilic thiol addition reactions at position 3 or 6 of the benzene nucleus. No such adduct has yet been isolated from brain tissue.

Finally, as pointed out earlier, radical attack on DA yields mainly nontoxic hydroxylated isomers (SLIVKA and COHEN 1985). Therefore, while autoxidation products of dopa and DA might contribute to aging-related oxidative stress in central dopaminergic neurons, there is little evidence to suggest a role of 6-OH-DA in this process.

The claim that 5-HT may be oxidized by nonenzymatic mechanisms to radical intermediates has been raised earlier (UEMURA et al. 1979, 1980; UEMURA and SHIMAZU 1980), but neither the fate of such radicals has been analyzed nor the postulated relationship to mental illness (WOOLLEY and SHAW 1954) or neurologic disease ever substantiated. However, recent studies by WRONA and DRYHURST (1987, 1988, 1989) and VOLICER and coworkers (CHEN et al. 1989; VOLICER et al. 1989; VOLICER and CRINO 1990; CAI et al. 1990; CRINO et al. 1989) have shown that the products of electrochemical oxidation of 5-HT, mainly the o-quinone of 4,5-DHT (tryptamine-4,5-dione), are highly reactive electrophiles that undergo covalent binding to tissue nucleophiles and damage a variety of neurons in a rather indiscriminate way after introduction into the ventricular CSF. There has been no description of nonenzymatic pathways resulting in the formation of 5,6-DHT from 5-HT; 5,6-dihydroxyindoles can, however, easily be generated by intramolecular cyclization from the o-quinone of DA. While nonenzymatic oxidation of 5-HT by a Fenton-type reaction and by the xanthine oxidase reaction seems feasible (Volicer, personal communication) and free radical mechanisms may evolve in the brain that oxidize serotonin to its 4,5-quinone and semiquinone radical intermediates as a result of ischemia/reperfusion cycles, the tracing of these reactive species is beset with technical difficulties because of their readiness to undergo addition reactions with sulfur and nitrogen nucleophiles. Therefore, the evaluation of the interesting neurotoxicity potential of o-quinone products formed from 5-HT under rather extreme oxidation conditions requires the isolation of quinone conjugates. No such conjugate has as yet been isolated from the human brain or CSF.

F. Conclusions

The affinity of 6-OH-DA, related polyphenols, DSP-4, and xylamine to and the rate of transport by the neuronal NE or DA transporter (uptake I) or the nonneuronal catecholamine transporter (uptake II) renders neurons and nonneuronal cells sensitive to the toxic actions of 6-OH-DA and DSP-4. During transport by the carrier, the structure and function of the transporter may become irreversibly altered, a reliable sign of cytotoxicity. The margin for selective toxic actions of 6-OH-DA is narrow because the affinity to the catecholamine transporters is low when compared with NE and DA. In the absence of such transport systems, 6-OH-DA can be toxic to almost every cell. Studies in cell cultures suggest that an as yet poorly characterized further determinant of the 6-OH-DA sensitivity or resistance of cells may be their variable capability to scavenge peroxides and radicals.

The cytotoxicity potential of hydroxylated phenylethylamines appears to depend on their redox potential, spontaneous oxygen reactivity (readiness to become autoxidized), and the nature and reactivity of their oxidation products, i.e., the semiquinone (anion) radical intermediates, the quinones (o-, p-quinones, p-quinone-imines), and in the presence of molecular oxygen on the reactivity of the one-electron reduction product of O_2, i.e., the superoxide (anion) radical (O_2^-). In the presence of the reduced form of metal redox systems (Fe^{2+}, Cu^+), the product of dismutation of O_2^- in water (H_2O_2) is transformed into the highly reactive hydroxyl ($OH\cdot$) radical. This radical avidly hydroxylates DNA, proteins, lipids, and carbohydrates and is thus mainly responsible for free radical-elicited cytotoxicity. The O_2^- radical which is constantly produced in cells operating under aerobic conditions (cytochrome aa_3) and by a variety of enzymes (tryptophan dioxygenase, xanthine oxidase, cytochrome P-450-dependent hydroxylating enzymes, phenoloxidase) catalyzes the oxidation of catecholamines; in this oxidation pathway, H_2O_2 is generated. When present in critical concentrations, ascorbate can serve as a redox cycling catalyst in the autoxidation of 6-OH-DA and DA in vitro. Autoxidation of 6-OH-DA proceeds via the semiquinone (and O_2^-) to the 2,5-p-quinone which, in the absence of competing nucleophiles, undergoes slow intramolecular cyclization (by attack of the side chain nitrogen on the 2-carbonyl group) to give the aminochrome intermediate (indoline-5,6-quinone, aminochrome I) which rearranges to form 5,6-dihydroxyindole. Via semiquinone intermediates, 5,6-dihydroxyindole autoxidizes to its parent o-quinone, a well-established intermediate in melanin formation. DA autoxidation proceeds via the corresponding semiquinones to the o-quinone which (in the absence of competing nucleophiles) undergoes slow intramolecular cyclization to form 5,6-dihydroxyindoline which tautomerizes to the more stable iminoquinone aminochrome I at pH 7 and 5,6-dihydroxyindole. This species is easily air oxidized to yield the corresponding o-quinone. 6-OH-DA and DA thus share the final steps in their autoxidation cascade.

The p-quinone of 6-OH-DA, the o-quinone of DA and the o-quinone of 5,6-dihydroxyindole are electron-deficient species which avidly react with nonprotein (cysteine, GSH) and protein thiol nucleophiles in vitro and in vivo thus yielding cysteinyl and/or glutathionyl adducts or arylation products; relatively inaccessible thiol functions in proteins or enzymes may be oxidized to disulfides by interaction with the semiquinone radical intermediates of DA and 6-OH-DA without becoming covalently bound. The mechanism of inhibition and inactivation of different enzymes by the oxidation products and byproducts of 6-OH-DA and DA differs according to the physicochemical nature, structure, and substrate properties of the enzyme investigated. 6-OH-DA appears to act extracellularly by free radical attack in cells lacking active transport mechanisms and intracellularly by free radical damage and covalent binding to nucleophiles in cells endowed with an efficient catecholamine transporter. The surprising selective toxicity of xylamine and DSP-4 in locus coeruleus NA projections exemplifies the importance of carrier properties in the shaping of the toxin sensitivity since DSP-4 and xylamine are alkylating compounds for principally every susceptible protein in all cells.

5,6-DHT and 5,7-DHT are potent competitors for transport into 5-HT neurons (IC$_{50}$ 6 × 10^{-7} M and 4 × 10^{-6} M, respectively). Among other factors (i.e., differences in the rate of transport), the small difference in uptake site affinity to 5-HT and to catecholamine transport is responsible for the restricted selectivity of the DHTs when administered in vivo. 5,6- and 5,7-DHT are metabolized by MAO and COMT (5,6-DHT) and by MAO-A/aldehyde dehydrogenase (5,7-DHT), an essential pathway for elimination of 5,7-DHT from the CNS compartment. 5,6- and 5,7-DHT are easily autoxidized at pH 7.4. Oxygen consumption rates of 5,6-DHT, 5,7-DHT, and the acid metabolites 5,6- and 5,7-DHIAA are significantly enhanced in the presence of mitochondria; the dihydroxyindoles serve as electron donors to cytochrome c in the terminal segment of the respiratory chain (complex IV). Following a lag period within which radicals and their reaction products (H$_2$O$_2$) are formed, 5,6-DHT oxidation reveals autocatalytic promotion. Via semiquinone (anion) radical intermediates, 5,6-DHT is oxidized to the parent o-quinone, a highly electrophilic species capable of attacking and crosslinking nonoxidized 5,6-DHT to yield melanoid polymer products in vitro (and in vivo). The o-quinone of 5,6-DHT is highly reactive with protein and nonprotein thiols forming cysteinyl/glutathionyl adducts and covalent modification and extensive crosslinking of proteins in vitro and in vivo. The physicochemistry of 5,7-DHT is unique due to the fact that, at pH 2–8 in water, 5,7-DHT exists as a mixture of several keto-enol tautomers of which the 5-ketonized, 7-hydroxy form predominates at pH 7.4. Via a transitory free radical superoxide complex, a hydroperoxide product is formed which is decomposed to the 4,5-dione of 7-HT which tautomerizes to yield the more stable 4,7-p-quinone of 5-HT. In order to explain the transformation of a poorly reactive quinone into a reactive species capable of arylating tissue nucleophiles, redox cycling has been proposed as a mechanism whereby small amounts of the quinone/hydroquinone are capable of producing deleterious amounts of free radicals, hypoxia, and exhaustion of reducing equivalents. Not only 5,7-DHT but also the subsequent products in the obligatory metabolic pathway, the aldehyde intermediate and its acidic product 5,7-DHIAA, autoxidize via the same cascade, all contributing to the enormous intraneuronal cytotoxicity potential of this meta-substituted DHT. The relative roles of covalent modification of proteins versus free radicals in the toxicity of 5,7-DHT remain to be determined.

The potential involvement of 6-OH-DA and 5,6-DHT – formed by nonenzymatic hydroxylation of endogenous DA and 5-HT continuously and extensively released from the nongranular and granular pool in an oxidative stress situation – in the dopaminergic and serotonergic neurotoxicity induced by various substituted amphetamines is discussed. The formation of neurotoxic analogues from physiological precursors may, however, represent only part of complex interdigitating mechanisms which could also involve loss of control over intracellular Ca^{2+} homeostasis and excitotoxic injury by glutamate acting upon dysregulated NMDA or quisqualate receptors.

Acknowledgement. We are grateful to Prof. H. Herken for constructive criticism throughout the preparation of this chapter.

References

Adams RN, Murrill E, McCreery R, Blank L, Karolczak M (1972) 6-Hydroxydopamine, a new oxidation mechanism. Eur J Pharmacol 17:287–292

Allis B, Cohen G (1977) The neurotoxicity of 5,7-dihydroxytryptamine in the mouse atrium: protection by 1-phenyl-3-(2-thiazolyl)-2-thiourea and by ethanol. Eur J Pharmacol 43:269–272

Ames MM, Nelson SD, Lovenberg W, Sasame HA (1977) Metabolic activation of *para*-chloroamphetamine to a chemically reactive metabolite. Comm Psychopharmacol 1:455–460

Appel NM, Contrera JF, De Souza EB (1989) Fenfluramine selectively and differentially decreases the density of serotonergic nerve terminals in rat brain: evidence from immunocytochemical studies. J Pharmacol Exp Ther 249(3): 928–943

Amarego WLF, Waring P (1983) Inhibition of human brain dihydropteridine reductase (EC 1.6.99.10) by the oxidation products of catecholamines, the aminochromes. Biochem Biophys Res Commun 113:895–899

Axt KJ, Commins DL, Vosmer G, Seiden LS (1990) α-Methyl-*p*-tyrosine pretreatment partially prevents methamphetamine-induced endogenous neurotoxin formation. Brain Res 515:269–276

Azmitia EC, Murphy RB, Whitaker-Azmitia PM (1990) MDMA (Ecstasy) effects on cultured serotonergic neurons: evidence for Ca^{2+}-dependent toxicity linked to release. Brain Res 510:97–103

Barzaghi F, Baumgartner HR, Carruba M, Mantegazza P, Pletscher A (1973) The 5-hydroxytryptamine-like actions of 5,6-dihydroxytryptamine. Br J Pharmacol 48:245–254

Battaglia G, Yeh SY, O'Hearn E, Molliver ME, Kuhar MJ, De Souza EB (1987) 3,4-Methylenedioxymethamphetamine and 3,4-methylenedioxyamphetamine destroy serotonin terminals in rat brain: quantification of neurodegeneration by measurement of [³H]paroxetine-labeled serotonin uptake sites. J Pharmacol Exp Ther 242:911–916

Battaglia G, Yeh SY, De Souza EB (1988) MDMA-induced neurotoxicity: parameters of degeneration and recovery of brain serotonin neurons. Pharmacol Biochem Behav 29:269–274

Baumgarten HG, Björklund A (1976) Neurotoxic indoleamines and monoamine neurons. Annu Rev Pharmacol 16:101–111

Baumgarten HG, Björklund A, Lachenmayer L, Nobin A, Stenevi U (1971) Long-lasting selective depletion of brain serotonin by 5,6-dihydroxytryptamine. Acta Physiol Scand [Suppl]373:1–15

Baumgarten HG, Evetts KD, Holman RB, Iversen LL, Vogt M, Wilson G (1972a) Effects of 5,6-dihydroxytryptamine on monoaminergic neurones in the central nervous system of the rat. J Neurochem 19:1587–1597

Baumgarten HG, Göthert M, Schlossberger HG, Tuchinda P (1972b) Mechanism of pressor effect of 5,6-dihydroxytryptamine in pithed rats. Arch Pharmacol 274:375–384

Baumgarten HG, Björklund A, Holstein AF, Nobin A (1972c) Chemical degeneration of indoleamine axons in rat brain by 5,6-dihydroxytryptamine. Z Zellforsch 129:256–271

Baumgarten HG, Lachenmayer, L, Bjöklund A, Nobin A, Rosengren E (1973a) Long-term recovery of serotonin concentrations in the rat CNS following 5,6-dihydroxytryptamine. Life Sci 12:357–364

Baumgarten HG, Björklund A, Lachenmayer L, Nobin A (1973b) Evaluation of the effects of 5,7-dihydroxytryptamine on serotonin and catecholamine neurons in the rat CNS. Acta Physiol Scand [Suppl]391:1–19

Baumgarten HG, Groth HP, Göthert M, Manian AA (1974a) The effect of 5,7-dihydroxytryptamine on peripheral adrenergic nerves in the mouse. Naunyn-Schmiedebergs Arch Pharmacol 282:245–254

Baumgarten HG, Björklund A, Horn AS, Schlossberger HG (1974b) Studies on the neurotoxic properties of hydroxylated tryptamines. In: Fuxe K, Olson L, Zotterman Y (eds) Dynamics of degeneration and grôwth in neurons. Pergamon, Oxford, pp 153–167

Baumgarten HG, Björklund A, Nobin A, Rosengren E, Schlossberger HG (1975a) Neurotoxicity of hydroxylated tryptamines: structure-activity relationships. 1. Long-term effects on monoamine content and fluorescence morphology of central monoamine neurons. Acta Physiol Scand [Suppl]429:1–27

Baumgarten HG, Björklund A, Bogdanski DF (1975b) Similarities and differences in the mode of action of 6-hydroxydopamine and neurotoxic indoleamines. In: Jonsson G, Malmfors T, Sachs Ch (eds) 6-Hydroxydopamine as a denervation tool in catecholamine research. North-Holland, Amsterdam, pp 59–66

Baumgarten HG, Lachenmayer L, Björklund A (1977) Chemical lesioning of indoleamine pathways. In: Myers RD (ed) Methods in psychobiology, vol III. Academic, London, pp 47–98

Baumgarten HG, Klemm HP, Lachenmayer L, Björklund A, Lovenberg W, Schlossberger H (1978) Mode and mechanism of action of neurotoxic indoleamines; a review and progress report. Ann NY Acad Sci 305:3–24

Baumgarten HG, Jenner S, Schlossberger HG (1979) Serotonin neurotoxins: effects of drugs on the destruction of brain serotonergic, noradrenergic and dopaminergic axons in the adult rat by intraventricularly, intracisternally or intracerebrally administered 5,7-dihydroxytryptamine and related compounds. In: Chubb IW, Geffen LB (eds) Neurotoxins, fundamental and clinical advances. Adelaide University Union Press, Adelaide, pp 221–226

Baumgarten HG, Jenner S, Klemm HP (1981) Serotonin neurotoxins: recent advances in the mode of administration and molecular mechanism of action. J Physiol (Paris) 77:309–314

Baumgarten HG, Jenner S, Björklund A, Klemm HP, Schlossberger HG (1982a) Serotonin neurotoxins. In: Osborne NN (ed) Biology of serotonergic transmission. Wiley, New York, pp 249–277

Baumgarten HG, Klemm HP, Sievers J, Schlossberger HG (1982b) Dihydroxytryptamines as tools to study the neurobiology of serotonin. Brain Res Bull 9:131–150

Baumgarten HG, Klemm HP, Schlossberger HG (1984) In-vivo-metabolism of ^{14}C-5-HT, ^{14}C-5,6-DHT and ^{14}C-5,7-DHT by MAO/COMT/aldehyde dehydrogenase in rat brain. In: Schlossberger HG, Kochen W, Linzen B, Steinhart H (eds) Progress in tryptophan and serotonin research. De Gruyter, Berlin, pp 241–249

Bellomo G, Mirabelli F, DiMonte D, Richelmi P, Thor H, Orrenius C, Orrenius S (1987) Formation and reduction of glutathione protein mixed disulfides during oxidative stress. Biochem Pharmacol 36:1313–1320

Björklund A, Nobin A, Stenevi U (1973a) The use of neurotoxic dihydroxytryptamines as tools for morphological studies and localized lesioning of central indolamine neurons. Z Zellforsch 145:479–501

Björklund A, Nobin A, Stenevi U (1973b) The use of neurotoxic dihydroxytryptamines as tools for morphological studies on central indolamine neurons. Comm Dept Anat Univ Lund 3:1–26

Björklund A, Baumgarten HG, Nobin A (1974) Chemical lesioning of central monoamine axons by means of 5,6- and 5,7-dihydroxytryptamine. Adv Biochem Psychopharmacol 10:13–33

Björklund A, Baumgarten HG, Horn AS, Nobin A, Schlossberger HG (1975) Neurotoxicity of hydroxylated tryptamines: structure-activity relationships. 2. In vitro studies on monoamine uptake inhibition and uptake impairment. Acta Physiol Scand [Suppl]429:31–60

Blank CL, Kissinger PT, Adams RN (1972) 5,6-dihydroxyindole formation from oxidized 6-hydroxydopamine. Eur J Pharmacol 19:391–394

Bloom FE, Algeri S, Gropetti A, Revuelta A, Costa E (1969) Lesions of central norepinephrine terminals with 6-OH-dopamine: biochemistry and fine structure. Science 166:1284–1286

Borchardt RT (1975) Affinity labeling of catechol O-methyltransferase by the oxidation products of 6-hydroxydopamine. Mol Pharmacol 11:436–449

Borchardt RT, Burgess SK, Reid JR, Liang YO, Adams RN (1977) Effects of 2- and/or 5-methylated analogues of 6-hydroxydopamine on norepinephrine- and dopamine-containing neurons. Mol Pharmacol 13:805–818

Breese GR (1975) Chemical and immunochemical lesions by specific neurotoxic substances and antisera. In: Iversen LL, Iversen SD, Synder SH (eds) Handbook of psychopharmacology, vol 1. Plenum, New York, pp 137–189

Breese GR, Cooper BR (1975) Behavioral and biochemical interactions of 5,7-dihydroxytryptamine with various drugs when administered intracisternally to adult and developing rats. Brain Res 98:517–527

Breese GR, Cooper BR (1977) Chemical lesioning: catecholamine pathways. In: Myers RD (ed) Methods in psychobiology, vol 3. Academic, London, pp 27–46

Breese GR, Müller RA (1978) Alterations in the neurocytotoxicity of 5,7-dihydroxytryptamine by pharmacologic agents in adult and developing rats. Ann NY Acad Sci 305:160–174

Butcher LL, Hodge GK, Schaeffer JC (1975) Degenerative processes after intraventricular infusion of 6-hydroxydopamine. In: Jonsson G, Malmfors T, Sachs C (eds) Chemical tools in catecholamine research I. North-Holland, Amsterdam, pp 83–90

Caccia S, Ballabio M, Guiso G, Rocchetti M, Garattini S (1982) Species differences in the kinetics and metabolism of fenfluramine isomers. Arch Int Pharmacodyn 258:15–28

Cadet JL (1986) The potential use of vitamin E and selenium in parkinsonism. Med Hypotheses 20:87–94

Cadet JL, Katz M, Jackson-Lewis V, Fahn S (1989) Vitamin E attenuates the toxic effects of intrastriatal injection of 6-hydroxydopamine (6-OHDA) in rats: behavioral and biochemical evidence. Brain Res 476:10–15

Cai P, Synder JK, Chen J-C, Fine R, Volicer L (1990) Preparation, reactivity, and nerotoxicity of tryptamine-4,5-dione. Tetrahedron Lett 31:969–972

Campbell DB, Richards RP, Caccia S, Garattini S (1986) Stereoselective metabolism and the fate of fenfluramine in animals and man. In: Development of drug and modern medicines. Horwood, Chichester UK, pp 298–311

Campbell DB, Ings RM, Gordon BH (1990b) The measurement of plasma and brain levels of ($\pm$) fenfluramine and ($\pm$) norfenfluramine in rats dosed for 4 days with 1 mg/kg p.o. and 5–10 and 40 mg/kg s.c. twice daily. (Unpublished data)

Chen J-C, Crino PB, Schnepper PW, To ACS, Volicer L (1989) Increased serotonin efflux by a partially oxidized serotonin: tryptamine-4,5-dione. J Pharmacol 250(1):141–148

Clineschmidt BV, Totaro JA, McGuffin JC, Pflueger AB (1976) Fenfluramine: long-term reduction in brain serotonin (5-hydroxytryptamine). Eur J Pharmacol 35:211–214

Clineschmidt BV, Zacchei AG, Totaro JA, Pflueger AB, McGuffin JC, Wishousky TI (1978) Fenfluramine and brain serotonin. Ann NY Acad Sci 305:222–241

Cohen G, Heikkila RE (1974) The generation of hydrogen peroxide, superoxide radical, and hydroxyl radical by 6-hydroxydopamine, dialuric acid, and related cytotoxic agents. J Biol Chem 249:2447–2452

Cohen G, Heikkila RE (1978) Mechanisms of action of hydroxylated phenylethylamine and indoleamine neurotoxins. Ann NY Acad Sci 305:74–84

Cohen G, Heikkila RE, Allis B, Cabbat F, Dembiec D, McNamee D, Mytilineou C, Winston B (1976); Destruction of sympathetic nerve terminals by 6-hydroxydopamine: protection by 1-phenyl-3-(2-thiazolyl)-2-thiourea, diethyldithiocarbamate, methimazole, cysteamine, ethanol and *n*-butanol. J Pharmacol Exp Ther 199:336–352

Commins DL, Axt KJ, Vosmer G, Seiden LS (1987a) 5,6-dihydroxytryptamine, a serotonergic neurotoxin, is formed endogenously in the rat brain. Brain Res 403:7–14

Commins DL, Axt KJ, Vosmer G, Seiden LS (1987b) Endogenously produced 5,6-dihydroxytryptamine may mediate the neurotoxic effects of *para*-chloroamphetamine. Brain Res 419:253-261

Commins DL, Shaughnessy RA, Axt KJ, Vosmer G, Seiden LS (1989) Variability among brain regions in the specificity of 6-hydroxydopamine (6-OHDA)-induced lesions. J Neural Transm 77:197–210

Creveling CR, Rotman A (1978) Mechanism of action of dihydroxytryptamines. Ann NY Acad Sci 305:57–84

Creveling CR, Lundström J, McNeal ET, Tice L, Daly JW (1975) Dihydroxytryptamines; effects on noradrenergic function in mouse heart in vivo. Mol Pharmacol 11:211–222

Crino PB, Vogt BA, Chen J-C, Volicer L (1989) Neurotoxic effects of partially oxidized seorotonin: tryptamine-4,5-dione. Brain Res 504:247–257

Cromartie RIT, Harley-Mason J (1957) Melanin and its precursors. Biochem J 66:713–720

Cushing SD (1988) Characterization of the binding of xylamine, an irreversible inhibitor of the catecholamine transporter and depletor of neuronal noradrenergic stores. PhD thesis, University of California, Los Angeles (212 pp)

Da Prada M, O'Brien RA, Tranzer JP, Pletscher A (1973) The effect of 5,6-dihydroxytryptamine on uptake, storage and metabolism of 5-hydroxytryptamine by blood platelets. J Pharmacol Exp Ther 186:213–219

Descarriers L, Beaudet A, De Champlain J (1975) Selective deafferentiation of rat neocortex by destruction of catecholamine neurons with intraventricular 6-hydroxydopamine. In: Jonsson G, Malmfors T, Sachs C (eds) Chemical tools in catecholamine research I. North-Holland, Amsterdam, pp 101–106

Dexter D, Carter C, Agid F, Agid Y, Lees AJ, Jenner P, Mardsen CD (1986) Lipid peroxidation as cause of nigral cell death in Parkinson's disease. Lancet II:639–640

Donaldson J, LaBella FS, Gesser D (1980) Enhanced autoxidation of dopamine as a possible basis of manganese neurotoxicity. Neurotoxicology 2:53–64

Dudley MW, Siegel BS, Ogden AM, McCarty DR (1988) A low dose of xylamine produces sustained and selective decreases in rat brain norepinephrine without evidence of neuronal degeneration. J Pharmacol Exp Ther 247:174–179

Dudley MW, Howard BD, Cho AK (1990) The interaction of the beta-haloethyl benzylamines, xylamine, and DSP-4 with catecholaminergic neurons. Annu Rev Pharmacol Toxicol 30:387–403

Evans JM, Cohen G (1989a) Studies on the formation of 6-hydroxydopamine in mouse brain after administration of 2,4,5-trihydroxyphenylalanine (6-hydroxyDOPA). J Neurochem 52:1461–1467

Evans JM, Cohen G (1989b) Can trace amounts of neurotoxins destroy dopamine neurons? Neurochem Int 15:127–129

Finnegan KT, Skratt JJ, Irwin I, Langston JW (1990) The *N*-methyl-D-asparate (NMDA) receptor antagonist, dextrorphan, prevents the neurotoxic effects of 3,4-methylenedioxymethamphetamine (MDMA) in rats. Neurosci Lett 105:300–306

Fleisher MR, Campbell DB (1969) Fenfluramine overdosage. Lancet 2:1306–1307

Floyd RA, Wiseman BB (1979) Spin-trapping free radicals in the autooxidation of 6-hydroxydopamine. Biochim Biophys Acta 586:196–207

Fritschy J-M, Grzanna R (1989) Immunohistochemical analysis of the neurotoxic effects of DSP-4 identifies two populations of noradrenergic axon terminals. Neuroscience 30:181–197

Fritschy J-M, Lyons WE, Molliver ME, Grzanna R (1988) Neurotoxic effects of p-chloroamphetamine on the serotoninergic innervation of the trigeminal motor nucleus: a retrograde transport study. Brain Res 473:261–270

Fuller RW, Rush BW (1974) 5,6-dihydroxytryptamine is a substrate for catechol-O-methyltransferase. Biochem Pharmacol 23:2208–2209

Gál EM, Christiansen PA, Yunger LM (1975) Effect of p-chloroamphetamine on cerebral tryptophan-5-hydroxylase in vivo: a reexamination. Neuropharmacology 14:31–39

Gant TW, Rao DNR, Mason PR, Cohen GM (1988) Redox cycling and sulphydryl arylation; their relative importance in the mechanism of quinone cytotoxicity to isolated hepatocytes. Chem Biol Interact 65:157–173

Garattini S, Caccia S (1979) Comparison of the plasma levels of fenfluramine in rats after a toxic dose and in man after a maximal therapeutic dose. Toxicol Lett 3:285–290

Garattini S, Caccia S (1990) Significance of fenfluramine neurotoxicity: a kinetic approach. In: Paoletti R, Vanhoutte PM, Brunello N, Maggi FM (eds) Serotonin – from cell biology to pharmacology and therapeutics. Kluwer, Dordrecht, pp 637–643

Garattini S, Caccia S, Mennini T, Samanin R, Consolo S, Ladinsky H (1979) Biochemical pharmacology of the anorectic drug fenfluramine: a review. Curr Med Res Opin 6:15–27

Gerson S, Baldessarini RJ (1975) Selective destruction of serotonin terminals in rat forebrain by high doses of 5,7-dihydroxytryptamine. Brain Res 85:140–145

Geyer MA, Gordon J, Adams LM (1984) Depletion of central norepinephrine by intraventricular xylamine in rats. Eur J Pharmacol 100:227–231

Gibb JW, Mitros K, Stone DM, Hanson GR, Johnson M (1990) Flunarizine prevents the 3,4-methylenedioxymethamphetamine-induced alteration in the serotonergic system. Abstracts, 2nd IUPHAR satellite meeting on serotonin, Basel, July 11–13, p 127

Gobbi M, Cerro L, Taddei C, Menini T (1990) Autoradiographic localization of (3H) paroxetine specific binding in the rat brain. Neurochem Intern 16:247–251

Graham DG (1978) Oxidative pathways for catecholamines in the genesis of neuromelanin and cytotoxic quinones. Mol Pharmacol 14:633–643

Graham DG, Tiffany SM, Bell WR Jr, Gutknecht WF (1978) Autoxidation versus covalent binding of quinones as the mechanism of toxicity of dopamine, 6-hydroxydopamine, and related compounds toward C1300 neuroblastoma cells in vitro. Mol Pharmacol 14:644–653

Greene LA, Tischler AS (1982) PC12 pheochromocytoma cultures in neurobiological research. Adv Cell Neurobiol 3:373–414

Grzanna R, Berger U, Fritschy J-M, Geffard M (1989) Acute action of DSP-4 on central norepinephrine axons: biochemical and immunohistochemical evidence for differential effects. J Histochem Cytochem 37:1435–1442

Hallman H, Jonsson G (1984) Pharmacological modifications of the neurotoxic action of the noradrenaline neurotoxin DSP-4 on central noredrenaline neurons. Eur J Pharmacol 103:269–278

Hamberger B (1967) Reserpine-resistant uptake of catecholamines in isolated tissues of the rat. Acta Physiol Scand [Suppl]295:1–56

Harvey JA, McMaster SE (1975) Fenfluramine: evidence for a neurotoxic action on midbrain and a long-term depletion of serotonin. Psychopharmacol Commun 1:217–228

Harvey JA, McMaster SE (1977) Fenfluramine: cumulative neurotoxicity after chronic treatment with low dosages in the rat. Comm Psychopharmacol 1:3–17

Harvey J, McMaster S, Yunger L (1975) p-Chloroamphetamine: selective neurotoxic action in brain. Science 187:841–843

Hattori T, McGeer PL, McGeer EG (1976) Synaptic morphology in the neostriatum of the rat: possible serotonergic synapse. Neurochem Res 1:451–467

Hedreen J (1975) Increased nonspecific damage after lateral ventricle injection of 6-OHDA compared with fourth ventricle injection in rat brain. In: Jonsson G, Malmfors T, Sachs C (eds) Chemical tools in catecholamine research I. North-Holland, Amsterdam, pp 91–100

Hedreen JC, Chalmers JP (1972) Neuronal degeneration in rat brain induced by 6-hydroxydopamine, a histological and biochemical study. Brain Res 47:1–36

Heikkila R, Cohen G (1971) Inhibition of biogenic amine uptake by hydrogen peroxide: a mechanism for toxic effects of 6-hydroxydopamine. Science 172:1257–1258

Heikkila RE, Cohen G (1972) Further studies on the generation of hydrogen peroxide by 6-hydroxydopamine. Mol Pharmacol 8:241–248

Heikkila RE, Cohen G (1973a) 6-Hydroxydopamine: evidence for superoxide radical as an oxidative intermediate. Science 181:456–457

Heikkila RE, Cohen G (1973b) The inhibition of ^{3}H-biogenic amine uptake by 5,6-dihydroxytryptamine: a comparison with the effects of 6-hydroxydopamine. Eur J Pharmacol 21:66–69

Heikkila RE, Cabbat FS (1977) Chemiluminescence from 6-hydroxydopamine: involvement of hydrogen peroxide, the superoxide radical and the hydroxyl radical, a potential role for singlet oxygen. Res Commun Chem Pathol Pharmacol 17:649–662

Hekmatpanah CR, McKenna DJ, Peroutka SJ (1989) Reserpine does not prevent 3,4-methylenedioxymethamphetamine-induced neurotoxicity in the rat. Neurosci Lett 104:178–182

Horn AS, Baumgarten HG, Schlossberger HG (1973) Inhibition of the uptake of 5-hydroxytryptamine, noradrenaline and dopamine into rat brain homogenates by various hydroxylated tryptamines. J Neurochem 21:233–236

Insel TR, Battaglia G, Johannessen JN, Marra S, De Souza EB (1989) 3,4-methylenedioxymethamphetamine("Ecstasy")selectively destroys brain serotonin terminals in rhesus monkeys. J Pharmacol Exp Ther 249:713–720

Invernizzi R, Fracasso C, Caccia S, DiClemente A, Garattini S, Samanin R (1989) Effect of L-cysteine on the long-term depletion of brain indoles caused by p-chloroamphetamine and d-fenfluramine in rats. Relation to brain concentrations. Eur J Pharmacol 163:77–83

Invernizzi R, Fracasso C, Caccia S, Garattini S, Samanin R (1991) Effects of intracerebroventricular d-fenfluramine and d-norfenfluramine as a single injection or 2-h infusion on brain serotonin: relation to brain drug concentrations. Neuropharmacology 30:119–123

Iversen LL (1970) Inhibition of catecholamine uptake by 6-hydroxydopamine in rat brain. Eur J Pharmacol 10:408–410

Iversen LL (1975) Uptake processes for biogenic amines. In: Iversen LL, Iversen SD, Snyder SH (eds) Handbook of psychopharmacology, vol 3. Plenum, New York pp 96–117

Jacobowitz D, Kostrzewa R (1971) Selective action of 6-hydroxydopa on noradrenergic terminals: mapping of preterminal axons of the brain. Life Sci 10:1329–1341

Jaim-Etcheverry G, Zieher LM (1980) DSP-4: a novel compound with neurotoxic effects on noradrenergic neurons of adult and developing rats. Brain Res 188:513–523

Javoy F, Agid Y, Sotelo C (1975) Specific and non-specific catecholaminergic neuronal destruction by intracerebral injection of 6-OH-DA in the rat. In: Jonsson G, Malmfors T, Sachs C (eds) Chemical tools in catecholamine research I. North-Holland, Amsterdam, pp 75–82

Johnson MP, Nichols DE (1990) Comparative serotonin neurotoxicity of the stereoisomers of fenfluramine and norfenfluramine. Pharmacol Biochem Behav 36:105–109

Johnson M, Hanson GR, Gibb JW (1989a) Characterization of acute N-ethyl-3,4-methylenedioxyamphetamine (MDE) action on the central serotonergic system. Biochem Pharmacol 38:4333–4338

Johnson M, Hanson GR, Gibb JW (1989b) Effect of MK-801 on on the decrease in tryptophan hydroxylase induced by methamphetamine and its methylenedioxy analog. Eur J Pharmacol 165:315–318

Jonsson G (1976) Studies on the mechanism of 6-hydroxydopamine cytotoxicity. Med Biol 54:406–420

Jonsson G (1980) Chemical neurotoxins as denervation tools in neurobiology. Annu Rev Neurosci 3:169–187

Jonsson G (1983) Chemical lesioning techniques: monoamine neurotoxins. In: Björklund A, Hökfelt T (eds) Methods in chemical neuroanatomy. Elsevier, Amsterdam, pp 463–507 (Handbook of chemical neuroanatomy, vol 1)

Jonsson G, Sachs C (1970) Effects of 6-hydroxydopamine on the uptake and storage of noradrenaline in sympathetic adrenergic neurons. Eur J Pharmacol 9:141–155

Jonsson G, Sachs C (1971) Uptake and accumulation of ^{3}H-6-hydroxydopamine in adrenergic nerves. Eur J Pharmacol 16:55–62

Jonsson G, Fuxe K, Hamberger B, Hökfelt T (1969) 6-Hydroxytryptamine: a new tool for monoamine fluorescence histochemistry. Brain Res 13:190–195

Jonsson G, Malmfors T, Sachs C (1972) Effects of drugs on the 6-hydroxydopamine induced degeneration of adrenergic nerves. Res Commun Chem Pathol Pharmacol 3:543–556

Jonsson G, Hallman H, Sundstrom E (1982) Effects of the noradrenaline neurotoxin DSP4 on the postnatal development of central noradrenaline neurons in the rat. Neuroscience 7:2895–2907

Kalia M (1991) Reversible, short lasting, and dose-dependent effect of d-fenfluramine on neocortical serotonergic axons. Brain Res 548:111–125

Kappus H (1985) Overview of enzymes systems involved in bio-reduction of drugs and in redox cycling. Biochem Pharmacol 35:1–6

Klemm HP, Baumgarten HG (1978) Interaction of 5,6- and 5,7-dihydroxytryptamine with tissue monoamine oxidase. Ann NY Acad Sci 305:36–56

Klemm HP, Baumgarten HG, Schlossberger HG (1979) In vitro studies on the interaction of brain monoamine oxidase with 5,6- and 5,7-dihydroxytryptamine. J Neurochem 32:111–119

Klemm HP, Baumgarten HG, Schlossberger HG (1980) Polarographic measurements of spontaneous and mitochondria-promoted oxidation of 5,6- and 5,7-dihydroxytryptamine. J Neurochem 35:1400–1408

Kleven MS, Seiden LS (1989) D-, L- and DL-fenfluramine cause long-lasting depletions of serotonin in rat brain. Brain Res 505:351–353

Kleven MS, Schuster CR, Seiden LS (1988) Effects of depletion of brain serotonin by repeated fenfluramine on neurochemical and anorectic effects of acute fenfluramine. J Pharmacol Exp Ther 246:822–828

Koide M, Cho AK, Howard BD (1986) Characterization of xylamine binding to proteins of PC12 pheochromocytoma. J Neurochem 47:1277–1285

Kostrzewa RM (1988) Reorganization of noradrenergic neuronal systems following neonatal chemical and surgical injury. Prog Brain Res 73:405–423

Kostrzewa RM (1989) Neurotoxins that affect central and peripheral catecholamine neurons. In: Boulton AA, Baker GB, Juorio AV (eds) Neuromethods 12. Humana, Clifton, pp 1–48

Kostrzewa RM, Jacobowitz DM (1974) Pharmacological actions of 6-hydroxydopamine. Pharmacol Rev 26:199–288

Lewin R (1985) Clinical trial for Parkinson's disease. Science 230:527–528

Liang Y-O, Wightman RM, Plotsky P, Adams RN (1975) Oxidative interactions of 6-hydroxydopamine with CNS constituents. In: Jonsson G, Malmfors T, Sachs C (eds) Chemical tools in catecholamine research I. North-Holland, Amsterdam, pp 15–22

Liston DR, Franz DN, Gibb JW (1982) Biochemical evidence for alteration of neostriatal dopaminergic functions by 5,7-dihydroxytryptamine. J Neurochem 38:1329–1335

Lookingland KJ, Chapin DS, McKay DW, Moore KE (1986) Comparative effects of the neurotoxins N-chloroethy-N-ethyl-N-2-bromobenzylamine hydrochloride (DSP4) and 6-hydroxydopamine on hypothalamic noradrenergic, dopaminergic and 5-hydroxytryptaminergic neurons in the male rat. Brain Res 365:228–234

Lorez H, Saner A, Richards JG, Da Prada M (1976) Accumulation of 5HT in non-terminal axons after p-chloro-N-methyl-amphetamine without degeneration of identified 5HT nerve terminals. Eur J Pharmacol 38:79–88

Lundström J, Ong H, Daly J, Creveling CR (1973) Isomers of 2,4,5-trihydroxy-phenethylamine (6-hydroxydopamine): long-term effects on the accumulation of [^{3}H]-norepinephrine in mouse heart in vivo. Mol Pharmacol 9:505–513

Lyons WE, Fritschy J-M, Grzanna R (1989) The noradrenergic neurotoxin DSP-4 eliminates the coeruleospinal projection but spares projections of the A5 and A7 groups to the ventral horn of the rat spinal cord. J Neurosci 9:1481–1489

Malmfors T, Sachs C (1968) Degeneration of adrenergic nerves produced by 6-hydroxydopamine. Eur J Pharmacol 3:89–92

Mamounas LA, Molliver ME (1988) Evidence for dual serotonergic projections to neocortex: axons from the dorsal and medial raphe nuclei are differentially vulnerable to the neurotoxin p-chloroamphetamine (PCA). Exp Neurol 102:23–36

Marchant NC, Bass S, Breen MA, Tucker FA, Richards RP, Campbell DB (1991) Species differences in the metabolism of (±)fenfluramine, In: Hlavica P, Damani LA, Gorrod JW (eds) Progress in Pharmacology and Clinical Pharmacology, vol 8/3. Fischer, Stuttgart New York, pp 23–30 Proceedings of the 4th international conference on biological oxidation of nitrogen in organic molecules, Munich 1989

Marek GJ, Vosmer G, Seiden LS (1990) The effects of monoamine uptake inhibitors and methamphetamine on neostriatal 6-hydroxydopamine (6-OHDA) formation, short-term monoamine depletions and locomotor activity in the rat. Brain Res 516:1–7

Martin R, Barlow JJ (1975) Muscle and gland cell degeneration in the octopus posterior salivary gland after 6-hydroxydopamine administration. J Ultrastruct Res 52:167–178

Massotti M, Scotti de Carolis A, Longo VG (1974) Effects of three dihydroxylated derivatives of tryptamine on behavior and on brain amine content in mice. Pharmacol Biochem Behav 2:769–775

McCreery RL, Dreiling R, Adams RN (1974a) Voltammetry in brain tissue: the fate of injected 6-hydroxydopamine. Brain Res 73:15–21

McCreery RL, Dreiling R, Adams RN (1974b) Voltammetry in brain tissue: quantitative studies of drug interactions. Brain Res 73:23–33

Mennini T, Borroni E, Samanin R, Garattini S (1981) Evidence of the existence of two different intraneuronal pools from which pharmacological agents can release serotonin. Neurochem Internat 3:289–294

Michel PP, Hefti F (1990) Toxicity of 6-hydroxydopamine and dopamine for dopaminergic neurons in culture. J Neurosci Res 26:428–435

Milby K, Oke A, Adams RN (1982) Detailed mapping of ascorbate distribution in rat brain. Neurosci Lett 28:15–20

Misra HP, Fridovich I (1972) The role of superoxide anion in autoxidation of epinephrine and a simple assay for superoxide dismutase. J Biol Chem 247(10):3170–3175

Molliver DC, Molliver ME (1990) Anatomic evidence for a neurotoxic effect of (±)-fenfluramine upon serotonergic projections in the rat. Brain Res 511:165–168

Molliver ME, Berger UV, Mamounas LA, Molliver DC, O'Hearn E, Wilson MA (1990) Neurotoxicity of MDMA and related compounds: anatomic studies. Ann NY Acad Sci 600:640–664

Mytilineou C, Danias P (1989) 6-Hydroxydopamine toxicity to dopamine neurons in culture: potentiation by the addition of superoxide dismutase and N-acetylcysteine. Biochem Pharmacol 38(11):1872–1875

O'Callaghan JP, Miller DB, Jensen KF, Schmidt CJ (1990) Serotonin depletions are not predictive of neurotoxicity: evidence from increases in glial fibrillary acidic protein induced by methylendioxymethamphetamine (MDMA) and 5,7-dihydroxytryptamine (5,7-DHT), meeting Oct 28–Nov 2. Society of neurosci, St Louis

O'Hearn E, Battaglia G, De Souza EB, Kuhar MJ, Molliver ME (1988) Methylenedioxyamphetamine (MDA) and methylenedioxymethamphetamine (MDMA) cause selective ablation of serotonergic axon terminals in forebrain: immunocytochemical evidence for neurotoxicity. J Neurosci 8:2788–2803

Olney JW, Zorumski CF, Stewart GR, Price MT, Wang G, Labruyere J (1990) Excitotoxicity of L-DOPA and 6-OH-DOPA: implications for Parkinson's and Huntington's diseases. Exp Neurol 108:269–272

Onténiente B, König N, Sievers J, Jenner S, Klemm HP, Marty R (1980) Sturctural and biochemical changes in rat cerebral cortex after neonatal 6-hydroxydopamine administration. Anat Embryol (Berl) 159:245–255

Osborne NN, Pentreath VW (1976) Effects of 5,7-dihydroxytryptamine on an identified 5-hydroxytryptamine-containing neurone in the central nervous system of the snail *Helix pomatia*. Br J Pharmacol 56:29–38

Paton DM (1973) Effects of substituted tryptamines on the efflux of noradrenaline from adrenergic nerves in rabbit atria. J Pharm Pharmacol 25:905–907

Pehlemann FW, Mohr S, Korr H, Sievers J, Beryy M (1987) Influence of meningeal cells on cell proliferation in the cerebellum. NATO ASI Ser 5:247–253

Perez-Reyes E, Mason RP (1981) Characterization of the structure and reactions of free radicals from serotonin and related indoles. J Bio Chem 256:2427–2432

Perry TL, Godin DV, Hansen S (1982) Parkinson's disease: a disorder due to nigral glutathione deficiency? Neurosci Lett 33:305–310

Pileblad E, Slivka A, Bratvold D, Cohen G (1988) Studies on the autoxidation of dopamine: interaction with ascorbate. Arch Biochem Biophys 263:447–452

Powell WS, Heacock RA (1973) The oxidation of 6-hydroxydopamine. J Pharm Pharmacol 25:193–200

Reader TA (1989) Neurotoxins that affect central indoleamine neurons. In: Boulton AA, Baker GB, Juorio AV (eds) Neuromethods 12, Humana, Clifton, pp 49–102

Reader TA, Gauthier P (1984) Catecholamines and serotonin in the rat central nervous system after 6-OHDA, 5,7-DHT and p-CPA. J Neural Trans 59:207–227

Riederer P, Sofic E, Rausch WD, Schmidt B, Reynolds GP, Jellinger K, Youdim MBH (1989) Transition metals, ferritin, glutathione, and ascorbic acid in parkinsonian brains. J Neurochem 52:515–520

Rollema H, De Vries JB, Westerink BHC, Van Putten FMS, Horn AS (1986) Failure to detect 6-hydroxydopamine in rat striatum after the dopamine releasing drugs dexamphetamine, methylamphetamine and MPTP. Eur J Pharmacol 132:65–69

Rosenberg PA (1988) Catecholamine toxicity in cerebral cortex in dissociated cell culture. J Neurosci 8:2887–2894

Rosengren E, Linder-Eliasson E, Carlsson A (1985) Detection of 5-S-cysteinyldopamine in human brain. J Neural Transm 63:247–253

Ross SB (1976) Long-term effects of N-2-chloroethyl-N-ethyl-2-bromobenzylamine hydrochloride on noradrenergic neurones in the rat brain and heart. Br J Pharmacol 58:521–527

Ross SB, Renyi AL (1976) On the long-lasting inhibitory effect of N-(2-chloroethyl)-N-ethyl-2-bromobenzylamine (DSP-4) on the active uptake of noradrenaline. J Pharm Pharmacol 28:458–459

Rotman A, Daly JW, Creveling CR, Breakefield XO (1976a) Uptake and binding of dopamine and 6-hydroxydopamine in murine neuroblastoma and fibroblast cells. Biochem Pharmacol 25:383–388

Rotman A, Daly JW, Creveling CR (1976b) Oxygen-dependent reaction of 6-hydroxydopamine, 5,6-dihydroxytryptamine, and related compounds with proteins in vitro: a model for cytotoxicity. Mol Pharmacol 12:887–899

Sachs C, Jonsson G (1975) Mechanisms of action of 6-hydroxydopamine. Biochem Pharmacol 24:1–8

Sachs C, Jonsson G, Heikkila R, Cohen G (1975) Control of the neurotoxicity of 6-hydroxydopamine by intraneuronal noradrenaline in rat iris. Acta Physiol Scand 93:345–351

Sanders-Bush E, Steranka LR (1978) Immediate and long-term effects of p-chloroamphetamine on brain amines. Ann NY Acad Sci 305:208–220

Sanders-Bush E, Bushing J, Sulser F (1972) Long-term effects of p-chloroamphetamine on tryptophan hydroxylase activity and on the levels of 5-hydroxytryptamine and 5-hydroxyindoleacetic acid in brain. Eur J Pharmacol 20:385–388

Saner A, Thoenen H (1971a) Model experiments on the molecular mechanism of action of 6-hydroxydopamine. Mol Pharmacol 7:147–154

Saner A, Thoenen H (1971b) Contributions to the molecular mechanism of action of 6-hydroxydopamine. In: Malmfors T, Thoenen H (eds) 6-Hydroxydopamine and catecholamine neurons. North-Holland, Amsterdam, pp 265–275

Schallert T, Wilcox RE (1985) Neurotransmitter-selective brain lesions. In: Boulton AA, Baker GB (eds) Neuromethods, vol 1. Humana, Clifton, pp 343–387

Schlossberger HG (1978) Synthesis and chemical properties of some indole derivatives. Ann NY Acad Sci 305:25–35

Schmidt CJ, Ritter JK, Sonsalla PK, Hanson GR, Gibb JW (1985) Role of dopamine in the neurotoxic effects of methamphetamine. J Pharmacol Exp Ther 233:539–544

Seiden LS, Vosmer G (1984) Formation of 6-hydroxydopamine in caudate nucleus of the rat brain after a single large dose of methylamphetamine. Pharmacol Biochem Behav 21:29–31

Senoh S, Witkop B (1959a) Non-enzymatic conversions of dopamine to norepinephrine and trihydroxyphenethylamines. J Am Chem Soc 81:6222–6231

Senoh S, Witkop B (1959b) Formation and rearrangements of aminochromes from a new metabolite of dopamine and some of its derivatives. J Am Chem Soc 81:6231–6235

Senoh S, Creveling CR, Udenfriend S, Witkop B (1959) Chemical, enzymatic and metabolic studies on the mechanism of oxidation of dopamine. J Am Chem Soc 81:6236–6240

Shaskan EG, Snyder SH (1970) Kinetics of serotonin accumulation into slices from rat brain; relationships to catecholamine uptake. J Pharmacol Exp Ther 178:404–418

Sievers J, Klemm HP, Jenner S, Baumgarten HG, Berry M (1980) Neuronal and extraneuronal effects of intracisternally administered 6-hydroxydopamine on the developing rat brain. J Neurochem 34:765–771

Sievers, J, Berry M, Baumgarten HG (1981) The role of noradrenergic fibres in the control of postnatal cerebellar development. Brain Res 207:200–208

Sievers H, Sievers J, Baumgarten HG, König N, Schlossberger HG (1983) Distribution of tritium label in the neonate rat brain following intracisternal or subcutaneous administration of [^{3}H]6-OHDA. An autoradiographic study. Brain Res 275:23–45

Sievers J, Pehlemann FW, Baumgarten HG, Berry M (1985) Selective destruction of meningeal cells by 6-hydroxydopamine: a tool to study meningeal-neuropithelial interaction in brain development. Dev Biol 110:127–135

Sievers J, Hartmann D, Gude S, Pehlemann FW, Berry M (1987) Influences of meningeal cells on the development of the brain. NATO ASI Ser 5:171–188

Sinhababu AK, Borchardt RT (1985) Mechanism and products of autoxidation of 5,7-dihydroxytryptamine. J Am Chem Soc 107:7618–7627

Sinhababu AK, Borchardt RT (1988) Molecular mechanism of biological action of the serotonergic neurotoxin 5,7-dihydroxytryptamine. Neurochem Int 12:273–284

Sinhababu AK, Borchardt RT (1989) Mechanism of autoxidation of 5,7-dihydroxytryptamine: ^{18}O is incorporated on C-4 during oxidation with $^{18}O_2$. J Am Chem Soc 111:2230–2233

Slivka A, Cohen G (1985) Hydroxyl radical attack on dopamine. J Biol Chem 260:15466–15472

Sonsalla PK, Nicklas WJ, Heikkila RE (1989) Role for excitatory amino acids in methamphetamine-induced nigrostriatal dopaminergic toxicity. Science 243:398

Sotelo C (1991) Immunohistochemical study of short- and long-term effects of dlfenfluramine on the serotonergic innervation of the rat hippocampal formation. Brain Res 541:309–326

Spina MB, Cohen G (1988) Exposure of school synaptosomes to L-dopa increases levels of oxidized glutathione. J Pharmacol Exp Therap 247:502–507

Spina MB, Cohen G (1989) Dopamine turnover and glutathione oxidation: implications for Parkinson disease. Proc Natl Acad Sci USA 86:1398–1400

Steranka LR, Rhind AW (1987) Effect of cysteine on the persistent depletion of brain monoamines by amphetamine, p-chloroamphetamine and MPTP. Eur J Pharmacol 133:191–197

Steranka LR, Sanders-Bush E (1979) Species differences in the rate of disappearance of fenfluramine and its effects on brain serotonin neurons. Biochem Pharmacol 28:3103–3107

Stone DM, Johnson M, Hanson GR, Gibb JW (1988) Role of endogenous dopamine in the central serotonergic deficits induced by 3,4 methylenedioxymethamphetamine. J Pharmacol Exp Ther 247:79–87

Sullivan SG, Stern A (1981) Effects of superoxide dismutase and catalase on catalysis of 6-hydroxydopamine and 6-aminodopamine autoxidation by iron and ascorbate. Biochem Pharmacol 30:2279–2285

Swan GA (1974) Structure, chemistry, and biosynthesis of the melanins. Fortschr Chem Org Naturst 31:522–585

Thoenen H (1969) Bildung und funktionelle Bedeutung adrenerger Ersatztransmitter. Exp Med Pathol Klin 27:1–85

Thoenen H, Tranzer JP (1973) The pharmacology of 6-hydroxydopamine. Annu Rev Pharmacol 13:169–180

Thoenen H, Tranzer JP (1968) Chemical sympathectomy by selective destruction of adrenergic nerve endings with 6-hydroxydopamine. Naunyn-Schmiedebergs Arch Pharmacol 261:271–288

Tiffany-Castiglioni E, Saneto RP, Proctor PH, Perez-Polo JR (1982) Participation of active oxygen species in 6-hydroxydopamine toxicity to a human neuroblastoma cell line. Biochem Pharmacol 31:181–188

Tranzer JP, Thoenen H (1967) Electronmicroscopic localization of 5-hydroxydopamine (3,4,5-trihydroxy-phenyl-ethylamine), a new "false" sympathetic transmitter. Experientia 23:743–745

Tranzer JP, Thoenen H (1973) Selective destruction of adrenergic nerve terminals by chemical analogues of 6-hydroxydopamine. Experientia 29:314–315

Tse DCS, McCreery RL, Adams RN (1976) Potential oxidative pathways of brain catecholamines. J Med Chem 19:37–40

Uemura T, Shimazu T (1980) NADPH-dependent melanin pigment formation from 5-hydroxyindolealkylamines by hepatic and cerebral microsomes. Biochem Biophys Res Commun 93:1074–1081

Uemura T, Chiesara E, Cova D (1977) Interaction of epinephrine metabolites with the liver microsomal electron transport system. Mol Pharmacol 13:196–215

Uemura T, Matsushita H, Ozawa M, Fiori A, Chiesara E (1979) Irreversible binding of 5-hydroxytryptamine and 5-hydroxytryptophan metabolites to rat liver microsomal protein. FEBS Lett 101:59–62

Uemura T, Shimazu T, Miura R, Yamano T (1980) NADPH-dependent melanin pigment formation from 5-hydroxyindoleamines by hepatic and cerebral microsomes. Biochem Biophys Res Commun 93:1074–1081

Ungerstedt U (1986) 6-hydroxydopamine induced degeneration of central monoamine neurons. Eur J Pharmacol 5:107–110

Unsicker K, Allan IJ, Newgreen DF (1976a) Extraneuronal effects of 6-hydroxydopamine and extraneuronal uptake of noradrenaline. Cell Tissue Res 173:45-69

Unsicker K, Chamley JH, McLean J (1976b) Extraneuronal effects of 6-hydroxydopamine. Cell Tissue Res 174:83–97

Uretsky NJ, Iversen LL (1969) Effects of 6-hydroxydopamine on noradrenaline-containing neurons in the rat brain. Nature 221:557–559

Uretsky NJ, Iversen LL (1970) Effects of 6-hydroxydopamine on catecholamine containing neurones in the rat brain. J Neurochem 17:269–278

Victor SJ, Baumgarten HG, Lovenberg W (1974) Depletion of tryptophan hydroxylase by 5,6-dihydroxytryptamine in rat brain – time course and regional differences. J Neurochem 22:541–546

Volicer L, Crino PB (1990) Involvement of free radicals in demential of the Alzheimer type: a hypothesis. Neurobiol Aging 11:567–571

Volicer L, Chen J-C, Crino PB, Vogt BA, Fishman J, Rubins J, Schnepper PW, Wolfe N (1989) In: Igbal K, Wisniewski HM, Winblad B (eds) Neurotoxic properties of a serotonin oxidation product: possible role in Alzheimer's disease. Alzheimer's disease and related disorders. Liss, New York, pp 453–465

Waring P (1986) The time-dependent inactivation of human brain dihydropteridine reductase by the oxidation products of L-dopa. Eur J Biochem 155:305–310

Wick MM, Fitzgerald G (1981) Inhibition of reverse transcriptase by tyrosinase generated quinones related to levodopa and dopamine. Chem Biol Interact 38:99–107

Wolf WA, Bobik A (1988) Effects of 5,6-dihydroxytryptamine on the release, synthesis, and storage of serotonin: studies using rat brain synaptosomes. J Neurochem 50:534–542

Woolley DW, Shaw E (1954) A biochemical and pharmacological suggestion about certain mental disorders. Proc Natl Acad Sci USA 40:228–231

Wrona MZ, Dryhurst G (1987) Oxidation chemistry of 5-hydroxytryptamine. 1. Mechanism and products formed at micromolar concentrations. J Org Chem 52:2817–2825

Wrona MZ, Dryhurst G (1988) Further insights into the oxidation chemistry of 5-hydroxytryptamine. J Pharm Sci 77:911–917

Wrona MZ, Dryhurst G (1989) Electrochemical oxidation of 5-hydroxytryptamine in acidic aqueous solution. J Org Chem 54:2718–2721

Wrona MZ, Lemordant D, Lin L, LeRoy Blank C, Dryhurst G (1986) Oxidation of 5-hydroxytryptamine and 5,7-dihydroxytryptamine. A new oxidation pathway and formation of a novel neurotoxin. J Med Chem 29:499–505

Wrona MZ, Humphries K, Dryhurst G (1988) Oxidation chemistry of CNS indoles. In: Dryhurst G, Niki K (eds) Redox chemistry and interfacial behavior of biological molecules. Plenum, New York, pp 425–445

Yoffe JR, Borchardt RT (1982) Characterization of serotonin uptake in cultured neuroblastoma cells. Mol Pharmacol 21:362–367

Zaczek R, Battaglia G, Culp S, Appel NM, Contrera JF, De Souza EB (1990a) Effects of repeated fenfluramine administration on indices of monoamine function in rat brain: pharmacokinetic, dose response, regional specificity and time course data. J Pharmacol Exp Ther 253:104–112

Zaczek R, Fritschy J-M, Culp S, De Souza EB, Grzanna R (1990b) Differential effects of DSP-4 on noradrenaline axons in cerebral cortex and hypothalamus may reflect heterogeneity of noradrenaline uptake sites. Brain Res 522:308–314

Zaizen Y, Nakagawara A, Ikeda K (1986) Patterns of destruction of mouse neuroblastoma cells by extracellular hydrogen peroxide formed by 6-hydroxydopamine and ascorbate. J Cancer Res Clin Oncol 111:93–97

CHAPTER 9

Toxins Affecting the Cholinergic System*

H. Hörtnagl and I. Hanin

A. Introduction

Long before the cholinergic neuronal system was identified and characterized, toxins affecting this system were available and in common use.
Extracts from belladonna plants have been used for cosmetic purposes for
many centuries and have been applied as a well-known poison. Curare has
been skilfully employed for centuries by the Indians as an arrow poison for
paralyzing animals during the hunt, prior to killing them. The betel nut,
containing arecoline, has been consumed as a euphoretic by the natives of
East India. Chewing of leaves of *Pilocarpus* plants has long been known to
cause salivation, and physostigmine was used as a therapeutic agent even in
1877 by Laqueur in the treatment of glaucoma (Gilman et al. 1985).

It is an interesting phenomenon that nature has provided various groups
of plants, bacteria, and animals with the capacity to produce a large series
of several classes of toxins which are directed more or less specifically
towards cholinergic neurons. The life-threatening toxicity of these poisons
is based on a fatal interference with the function of the cholinergic system.
Botulinus and tetanus toxins, from the anaerobic bacteria *Clostridium
botulinum* and *tetani*, belong to the most toxic compounds known. Several
plants are sources for drugs and/or toxins that affect cholinergic function,
including atropine, scopolamine, physostigmine, muscarine, arecoline,
pilocarpine, nicotine, lobeline, and curare. Anatoxin-a from fresh water
algal blooms has been responsible repeatedly for the death of livestock
and waterfowl in several countries of the world (Carmichael et al. 1975).
Various classes of cholinergic neurotoxins are contained in the venoms of
spiders, snakes, and snails, such as α-latrotoxin, α-bungarotoxin, erabutoxin b,
α-conotoxin, as well as a series of neurotoxins from snake venoms with
an intrinsic phospholipase A_2 activity, namely β-bungarotoxin, notexin,
taipoxin, and crotoxin.

This list of naturally occurring cholinergic toxins is, in addition, extended by several classes of synthetic compounds produced by mankind.
The class of the organophosphorus anticholinesterases contains the most

* A portion of the work described in this review was supported by National Institute
of Mental Health, grant no MH 42572 to IH.

potent synthetic toxic agents known, including the "nerve gases" tabun, sarin, and soman. The potential of these highly toxic organophosphorus compounds has been of considerable military and agricultural interest and has been utilized to produce chemical warfare agents and later on to develop a class of potent insecticides.

Several cholinergic and anticholinergic compounds also have been developed which are important pharmacological tools. Most recently, several substances have become available, which were designed specifically to destroy cholinergic pathways on an experimental basis (e.g., AF64A) or for the purpose of elucidating important steps in the process of synthesis, storage, and release of acetylcholine (ACh).

Many of these cholinergic toxins, whether provided by nature or synthesized by chemists, are not only of considerable pharmacological and toxicological interest but have become valuable tools in the molecular and functional investigation of various aspects of cholinergic neuronal function. The contribution of drugs and toxins to the improvement of our understanding of cholinergic function as well as their historical background have been recently described (WHITTAKER 1990). The present review will survey the various groups of cholinergic toxins according to their mechanism of action and target site on the cholinergic neuron, with particular emphasis on experimental findings pertaining to their toxic interference with the natural functioning of the cholinergic system.

B. Target Sites of Cholinergic Toxins at Various Levels of the Cholinergic Neuron

The cholinergic neuron is vulnerable at various sites along it to the entry of toxic compounds which interfere either reversibly or irreversibly with the function of the cholinergic system (Fig. 1). Several of these target sites are proteins, receptors, and components, which are specifically localized at cholinergic neurons. An intervention of a toxin here is likely to guarantee its exclusive action on the cholinergic neuron. These specific sites include choline acetyltransferase (ChAT), the enzyme essential for the synthesis of ACh, the sodium-dependent high-affinity choline transport (HAChT) system, preferentially localized in the membrane of cholinergic nerve terminals (SUSZKIW and PILAR 1976), a receptor site at cholinergic synaptic vesicles which is probably involved in the modulation of ACh transport into its storage system, the various types of pre- and postsynaptic cholinoreceptors including muscarinic and nicotinic subtypes, and acetylcholinesterase (AChE), the enzyme catalyzing the inactivation of ACh. There are, however, several toxins which affect the cholinergic neuron at sites which are common also to other populations of neurons. This is the case for toxins interfering with excitatory glutamate receptors, axoplasmic transport, the release process, or the transneuronal modulation. The selectivity of such a

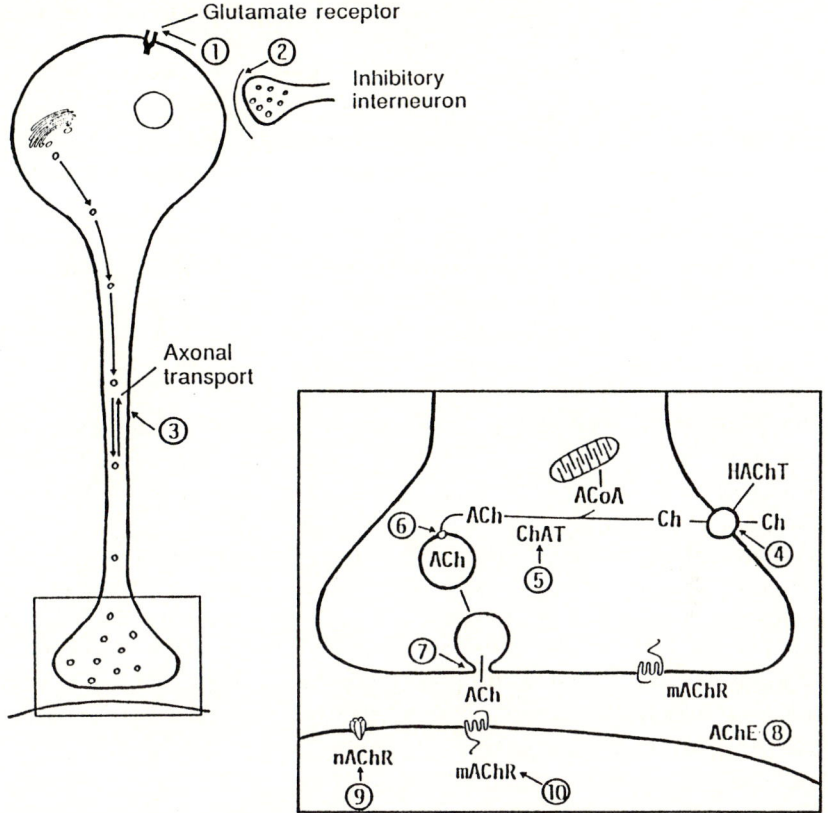

Fig. 1. Target sites of the cholinergic neuron accessible to the various classes of toxins (representative examples given):

① Interaction at the glutamate receptor: **kainic acid, ibotenic acid, quinolinic acid, quisqualic acid,** N-**methyl-D-aspartate,** β-N-methylamino-L-alanine (**L-BMAA**), other "excitotoxins"

② Blockade of inhibitory interneurons: **tetanus toxin**

③ Blockade of axonal transport: **colchicine,** ethylcholine aziridinium (**AF64A**)

④ Inhibition of the high-affinity choline transport (*HAChT*) system: **hemicholinium-3,** bis-4-methylpiperidine tertiary amine analogue (**A-4**) (reversible); hemicholinium mustard-9 (**HcM-9**), **choline mustard aziridinium, acetylcholine** (*ACh*) **mustard aziridinium, AF64A, ethoxycholine mustard aziridinium, propyl-, butyl-, and cyclopropyl- AF64A analogues, AF64A-picrylsulfonate** (irreversible)

⑤ Inhibition of ACh synthesis (*ChAT*, choline acetyltransferase; *ACoA*, acetyl-CoA): N-methyl-4-(1-naphthylvinyl) pyridinium (**NVP**), **acetyl-seco-hemicholinium, naphthoquinone, mono-, di-, or triethylcholine, diethylaminoethanol,** N-**amino-deanol, 3-hydroxypiperidinium, bromoacetylcholine**

⑥ Inhibition of vesicular ACh uptake: **vesamicol, cetiedil** (noncompetitive); **cholinergic false neurotransmitters** (competitive)

⑦ Interaction with the release of ACh: **botulinum toxin,** β-**bungarotoxin, notexin, taipoxin, crotoxin** (inhibition); α-**latrotoxin, iotrochotin,** β-**leptinotarsin-h** (stimulation)

⑧ Inhibition of AChE: **physostigmine** (reversible); **soman, sarin, tabun, tri-*o*-cresylphosphate, parathion, malathion** (irreversible)

⑨ + ⑩ Interaction at the nicotinic (*nAChR*) and muscarinic (*mAChR*) acetylcholine receptor: **reversible and irreversible nicotinic or muscarinic agonists or antagonists**

toxin would depend on the site of application, its distribution, and the presence of specific transport mechanisms. The response to a certain toxin may also differ among the various types of cholinergic pathways in the periphery and the CNS. Several different types of cholinergic pathways are known to exist. These can be distinguished according to the type of neuropeptide(s) coexisting with ACh at the nerve terminal or according to the degree of dependence of the neuron on nerve growth factor for its recovery following insult or damage.

In the following we will survey various types of toxins. They will be discussed according to their target site at the cholinergic neuron and their mechanism of action, starting at the level of the perikaryon. This will be followed by a look at the effect of selective toxins at the nerve terminal and synaptic cleft. Finally, Sect. F will be dedicated to toxins affecting axonal transport of the cholinergic neuron, as well as to toxins with miscellaneous actions.

Since extensive coverage is available elsewhere in this volume on glutamate receptors and excitotoxic substances (Chap. 12), on toxins affecting nicotinic ACh receptors (Chaps. 15, 16) and muscarinic ACh receptors (Chap. 17), and on tetanus and botulinum toxins (Chap. 11), these will only be referred to in passing in the present chapter. The reader is referred to those other chapters as well, in order to attain a highly comprehensive, overall perspective of the variety of neurotoxic agents which are presently known to affect the cholinergic system.

C. Toxins Effective at the Perikaryon of the Cholinergic Neuron

Similar to many other types of central neurons, central cholinergic neurons, including spinal motoneurons, are sensitive to the destructive effect of excitatory amino acids and their structural analogues. Focal microinjection of various excitotoxins into brain areas rich in cholinergic cell bodies such as the nucleus basalis magnocellularis (nbM), medial septum, pontomesencephalic tegmentum or striatum induce cell death due to massive overstimulation, followed by anterograde degeneration of the cholinergic projections to the target areas. As excitotoxins, kainic acid, ibotenic acid, quinolinic acid, quisqualic acid, or N-methyl-D-aspartate (NMDA) have been applied. Although lesions of cholinergic pathways can be achieved by this treatment, this effect lacks specificity, since the excitotoxic action is not directed specifically towards the cholinergic perikaryon, and noncholinergic cell bodies at the site of injection are affected as well. For example, kainic acid injections into the dorsolateral pontomesencephalic tegmentum in cats result in the destruction not only of the majority of cholinergic neurons but also of tyrosine hydroxylase-immunoreactive and other neurons in this region (Jones and Webster 1988).

Excitotoxins have been used frequently to produce cholinergic lesions in the basal forebrain in various species. Cortical cholinergic impairment and behavioral deficits including selective memory disturbances have been reported after injection of kainic acid into the nbM of rats (LERER et al. 1985; EL-DEFRAWY et al. 1985) or cats (SATO et al. 1987) and of quisqualic acid into nbM, substantia innominata, and globus pallidus of rats (ROBBINS et al. 1989; DUNNETT et al. 1989). Similar, but obviously smaller changes occur following the injection of ibotenic acid (WATSON et al. 1985; CROSS and DEAKIN 1985; HEPLER et al. 1985; KESNER et al. 1986; THAL et al. 1988) and quinolinic acid or NMDA into the nbM in rats (EL-DEFRAWY et al. 1985; METCALF et al. 1987; SCARTH et al. 1989; STEWART et al. 1986). To achieve a persistent deficit, bilateral lesions appear to be essential, since a remarkable neurochemical and behavioral recovery within 6 months has been demonstrated in rats with a unilateral ibotenic acid-induced lesion of the nbM, but not with a bilateral lesion (CASAMENTI et al. 1988). It was also shown that excitotoxic lesions in the nbM and medial septal area of rats produce qualitatively similar impairments (HEPLER et al. 1985). Only in rats with small lesions of the medial septum or nbM can a differentiation of memory deficits be achieved (KESNER et al. 1986). Since the microinjection of excitotoxins into the nbM and/or medial septum can produce some of the neurochemical changes and cognitive deficits observed in Alzheimer's disease, this approach has been employed to obtain experimental paradigms for this neurodegenerative disorder.

Of special interest is the recent finding that cholinergic spinal motoneurons also appear to be affected by a plant excitatory neurotoxin, the amino acid β-N-methylamino-L-alanine (L-BMAA), contained in the seed of the false sago palm (*Cycas circinalis*). Feeding of macaques with this amino acid for 2–12 weeks produced chromatolytic and degenerative changes of motoneurons in the spinal cord as well as in the cerebral cortex. Consequently, the possible involvement of this or similar exogenous excitotoxic amino acids in the etiology of sporadic amyotrophic lateral sclerosis and especially of the Guam-type amyotrophic lateral sclerosis – Parkinsonism – dementia complex has been hypothesized (SPENCER et al. 1987).

Tetanus toxin has an indirect effect on cholinergic motoneurons mainly via blockade of inhibitory interneurons which modulate the activity of motoneurons. In general, the activity of tetanus toxin is not specific to a particular neurotransmitter, type of synapse, region of the nervous system, or animal species. Although this toxin is also able to block cholinergic transmission to voluntary muscles and the cholinergic innervation of the iris, the major acute symptoms of clinical tetanus involve a spastic paralysis with muscle rigidity, paroxysmal muscle contraction, and seizures. This clinical picture indicates that a central block of inhibition of the cholinergic motoneurons, induced by inhibiting the release of glycine and/or GABA, is clinically the significant effect (reviewed by MELLANBY and GREEN 1981). More details about this toxin are presented in Chap. 11 of this volume.

D. Toxins Effective at the Level of the Cholinergic Nerve Terminal

I. High-Affinity Choline Transport (HAChT) System

1. Reversible Interactions

Hemicholinium-3 (α,α'-dimethylethanolamino-4,4'-bis-acetophenone; Fig. 2) originally introduced as a neuromuscular blocking agent (Long and Schueler 1954) has been shown to be a highly potent and selective inhibitor of HAChT (Guyenet et al. 1973; Haga and Noda 1973; Yamamura and Snyder 1973). The reversible inhibition of choline uptake results from its competitive interaction with choline for a common site in the choline carrier system. Although it competes for transport sites, hemicholinium-3 is not transported into synaptosomes and has a limited ability to cross the BBB (Domer and Schueler 1960; Giarman and Pepeu 1962; Freeman et al. 1982). The HAChT system is closely linked to the synthesis of ACh (Tuček 1985). Blockade of this system, therefore, results in a reduction of choline uptake and consequently in a diminished ability of the cholinergic neuron to synthesize ACh. Likewise, intracerebroventricular (i.c.v.) administration of hemicholinium-3 or its injection into various brain areas produces a reversible reduction in the levels of ACh in those specific brain areas (Gardiner 1961; Hebb et al. 1964; Slater 1968; Ansell and Spanner 1975; Freeman et al. 1975, 1979). I.c.v. application of hemicholinium-3 is followed by an acute impairment of various learning experiments in several species (Russell and Macri 1978; Freeman et al. 1979; Caulfield et al. 1981; Ridley et al. 1987; Hagan et al. 1989).

Although its effect is reversible, continuous infusion over a sufficient period of time may result in an eventual depletion of intraneuronal sources of choline and subsequently in cellular atrophy produced by prolonged blockade of choline uptake and thus prevention of a vital cell function. This phenomenon has been demonstrated by infusion of hemicholinium-3 into the rat nbM over a 14-day period. By this extended treatment an almost complete destruction of nbM cholinergic neurons is achieved concurrently with axonal and terminal field degeneration in the cortex (Hurlbut et al. 1987). Lasting hippocampal cholinergic deficit also occurs following intrahippocampal injections of large doses of hemicholinium-3 (Raaijmakers 1983).

[³H]Hemicholinium-3 selectively labels presynaptic sites associated with the HAChT system (Vickroy et al. 1984). For this reason, it has been gaining increasing importance as a specific marker for presynaptic cholinergic nerve terminals in various species, including man (Rainbow et al. 1984; Bekenstein and Wooten 1989; Pascual et al. 1989).

Several derivatives have been introduced and tested, including a series of substituted pipcridine analogues (Bhattacharyya et al. 1986; Tedford

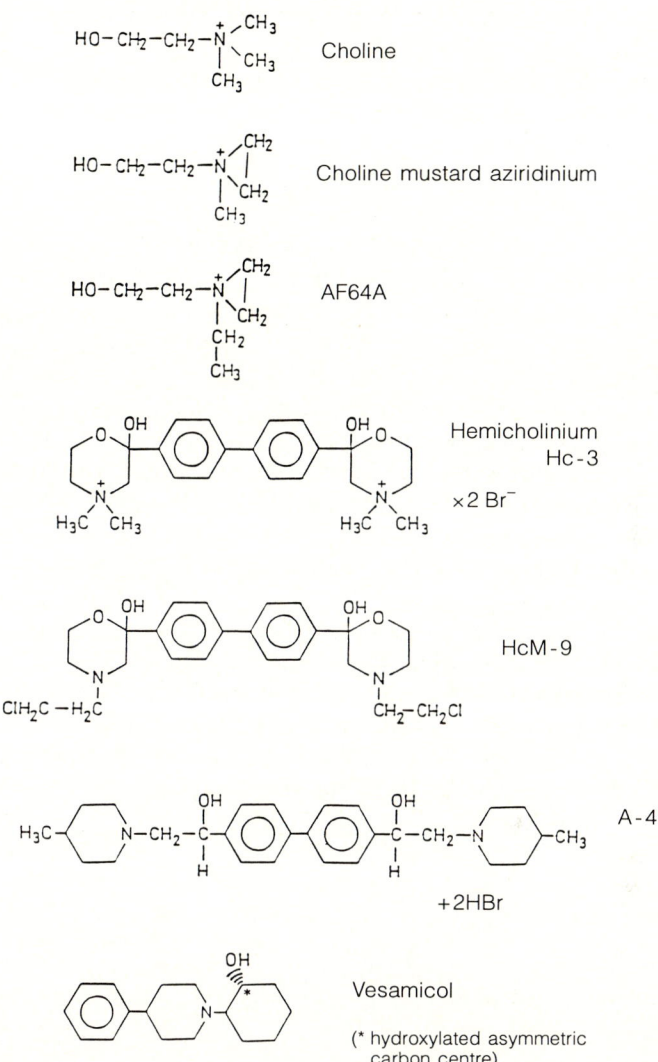

Fig. 2. Chemical structure of toxins interfering at the high-affinity choline transport (HAChT) or the vesicular acetylcholine transport system. Note the structural similarity of ethylcholine aziridinium (AF64A) and choline mustard aziridinium with choline

et al. 1986). Among them, A-4, a bis-4-methylpiperidine tertiary amine analogue of hemicholinium-3 (Fig. 2), is unique in its ability to act centrally after systemic administration (TEDFORD et al. 1987). Intraperitoneal (i.p.) injections of A-4 produce a dose-dependent reversible reduction in ACh content of several brain regions concurrent with an increase in choline content. The quarternary amine N-methyl-4-methylpiperidine analogue of hemicholinium-3 represents the most potent of all known inhibitors of sodium-dependent HAChT and acts through a noncompetitive, but reversible mechanism (Chatterjee et al. 1988).

2. Irreversible Interactions

Hemicholinium mustard derivatives, including the cyclized bis-ethylenimine form of hemicholinium-3-bromo mustard and hemicholinium mustard-9 (HcM-9; Fig. 2) have the potential to inhibit HAChT irreversibly in vitro or in vivo, respectively (Smart 1981; Tagari et al. 1986). After i.c.v. injections of HcM-9 a considerable loss of ChAT was demonstrated in the rat hippocampus, striatum, and cortex 7 days after treatment (Tagari et al. 1986).

Much more detailed knowledge is available about the effect of choline mustard aziridinium analogues. Several nitrogen mustard derivatives of choline, including choline mustard aziridinium (Fig. 2), ACh mustard aziridinium, ethylcholine aziridinium, and ethoxycholine mustard aziridinium, have been tested for their ability to inhibit irreversibly and specifically HAChT in vitro as well as in vivo. Among them, ethylcholine aziridinium (synonyms: AF64A, ECA, ECMA, ECMAz; Fig. 2) has been most widely used for over a decade both in in vitro and in vivo systems of various species, from cockroach to monkey (Morio and Yagasaki 1989; I. Hanin, personal communication). The synonym AF64A will be employed here. Since the introduction of AF64A several comprehensive reviews of this substance have been published which provide additional information (Fisher and Hanin 1986; Hanin et al. 1987; Hanin 1990).

a) In Vitro Studies

Effects on High-Affinity Choline Transport. The transport of choline into brain synaptosomes is inhibited irreversibly by choline mustard aziridinium analogues. Choline mustard aziridinium and AF64A appear to be comparable in potency as irreversible inhibitors of HAChT, whereas ACh- and ethoxycholine mustard aziridinium are much less potent (Rylett and Colhoun 1980). The choline carrier is protected by choline and hemicholinium-3 from this irreversible inhibition (Curti and Marchbanks 1984). A great selectivity for the sodium-dependent HAChT system has been demonstrated. Choline mustard aziridinium was 30 times more potent as an inhibitor of HAChT when compared with its effect on sodium-independent low-affinity choline uptake (Rylett and Colhoun 1984). AF64A does not affect choline transport in erythrocytes or kidney slices even at 20 times the concentration necessary to inhibit choline uptake into synaptosomes (Uney and Marchbanks 1987). Furthermore, the choline aziridinium analogues seem to be specific for the choline carrier since choline mustard aziridinium does not interfere with the synaptosomal uptake of serotonin, norepinephrine (NE), or GABA (Rylett 1986), and AF64A has been shown not to interfere with the synaptosomal uptake of serotonin (Mantione et al. 1983b). [^{3}H]Choline mustard aziridinium has been applied recently as a tool to purify the HAChT system from the electric organ of *Torpedo marmorata* (Rylett 1988).

Effect on Enzymes Using Choline as Substrate. Several enzymes including ChAT, choline kinase, choline dehydrogenase, and AChE are irreversibly inhibited by AF64A or choline mustard aziridinium (Rylett and Colhoun 1979; Barlow and Marchbanks 1984; Sandberg et al. 1985b). The affinity of choline mustard aziridinium as a substrate for ChAT was found to be one-third that of choline (Rylett and Colhoun 1979). The irreversible inactivation of partially purified rat brain ChAT appears at much higher concentrations than those required to decrease ChAT activity in synaptosomes incubated with this compound (Rylett and Colhoun 1985). With regard to a possible site of action of choline mustard analogues, recent evidence indicates that synaptosomal "membrane-bound" ChAT is most sensitive to inhibition by choline mustard aziridinium, suggesting that the choline mustard aziridinium transported by the high-affinity choline carrier may be directed towards a membrane-associated form of ChAT (Rylett 1989).

b) Effects in Neuronal Tissue Culture Systems

The specific cholinotoxicity of AF64A has been evaluated in various neuronal cell cultures including primary cultures prepared from whole brain, septum, or midbrain of fetal rats (Koppenaal et al. 1986; Amir et al. 1988), neuron-enriched cultures derived from 8-day old embryonic chick cerebrum (Davies et al. 1986), fetal rat reaggregate culture of whole brain (Pillar et al. 1988; Atterwill et al. 1989; Atterwill and Meakin 1990), and various neuroblastoma cell lines (Sandberg et al. 1985b; Barnes et al. 1988). In these culture systems, AF64A significantly and selectively reduced the number of AChE-stained cells and ChAT activity. In studies conducted with primary cultures, many of the surviving cholinergic cells appeared with intact somata but damaged processes, indicating a retrograde degeneration starting at the nerve terminal (Amir et al. 1988).

c) In Vivo Studies

Most in vivo studies with irreversible inhibitors of presynaptic cholinergic nerve terminal function have almost exclusively involved AF64A. To our knowledge there is only one study, in which choline mustard aziridinium ion was injected into the medial septum and dorsal hippocampus of rat brain. Since, however, in this case the resulting reduction of ChAT activity was associated with a decline in serotonin levels, a non-specific tissue damage was suggested for the action of this compound (Colhoun et al. 1986). More extensive studies with AF64A indicate that this compound does indeed have a selective central cholinotoxic effect in vivo but that this effect is highly dependent on the dose used and on the site of administration (Hanin et al. 1987; Hanin 1990).

The following mechanisms are believed to be involved in the specific cholinotoxic action of choline aziridinium compounds in vivo. AF64A appears to induce a two-component effect on the HAChT system. At low concentrations (IC_{50} values: $1.35-2.25\,\mu M$) it produces a reversible inhibition and is transported in competition with choline into the cholinergic neuron. At higher concentrations (IC_{50} values: $25-30\,\mu M$) the inhibition becomes irreversible, probably via alkylation of the choline transporter at its outward position (Pittel et al. 1987; Uney and Marchbanks 1987). The transport of choline via the HAChT system is an extremely fast process, the velocity being dependent on synaptic activity (Tuček 1985). Since alkylation (i.e., a covalent bound formation) is probably a much slower reaction than the transport process itself, it is conceivable that part of the compound is transported into the cholinergic nerve ending before it irreversibly blocks the transport system. Once inside the neuron this compound exerts its cytotoxic action, e.g., irreversibly inhibiting enzymes that use choline as substrate. Since the high-affinity choline carrier is preferentially localized on cholinergic nerve terminals but appears to be lacking on the membrane of the perikarya (Suszkiw and Pilar 1976), the choline mustard analogues are prone to accumulate mainly within cholinergic nerve terminals. The degeneration process induced by these compounds should consequently start in a retrograde direction. An uptake of AF64A or choline mustard aziridinium into intact synaptosomes followed by inhibition of ChAT has been demonstrated (Rylett and Colhoun 1980; Pedder and Prince 1984), ensuring an accumulation of the toxin within the cholinergic nerve ending in concentrations sufficient to induce intracellular DNA damage (Barnes et al. 1988), to inhibit enzyme activities irreversibly, and to block the axonal transport system in cholinergic nerve fibers (Kasa and Hanin 1985). The in vivo toxicity of choline aziridinium analogues clearly depends on an intact HAChT system, since pretreatment with reversible inhibitors of this transport system almost completely abolished the cholinergic deficit in the rat hippocampus induced by i.c.v. injections of AF64A (Chrobak et al. 1989; Potter et al. 1989). Furthermore, pretreatment with choline antagonized the AF64A-induced lethality in mice (Fisher et al. 1982) and diminished the neurotoxic changes in neuron-enriched cell cultures exposed to AF64A (Davies et al. 1986; Amir et al. 1988). Beside the important intracellular sites of action of AF64A, long-lasting blockade of the HAChT system is likely to contribute to the irreversible damage of the cholinergic neuron as well, in a manner similar to that demonstrated by prolonged exposure to hemicholinium-3, which lacks an intracellular effect at the nerve terminal (Hurlbut et al. 1987).

Peripheral Cholinergic Systems. In the cat superior cervical ganglion it has been demonstrated convincingly that AF64A produces a dose-dependent and long-lasting inhibition of cholinergic transmission after injection into the carotid artery, without interfering with adrenergic transmission or inactivating ganglionic or skeletal muscle nicotinic or smooth muscle muscarinic receptors (MANTIONE et al. 1983a). Furthermore, an acute in vivo exposure of mice to AF64A resulted in a depression of the secretion of quanta from motor nerve terminals (MCARDLE and HANIN 1986). It has also been demonstrated recently that AF64A is selectively neurotoxic for the presynaptic cholinergic neurons in the sixth abdominal ganglion of the cockroach (MORIO and YAGASAKI 1989). A selective impairment of cholinergic neuromuscular transmission in the ileum and urinary bladder and of cholinergic modulation in colon is achieved 7 days after i.p. injection of AF64A into guinea-pigs (HOYLE et al. 1986).

The effects of intravitreal injections of AF64A have also been studied in the rat, chicken, and goldfish. In all three species a profound, selective, and essentially irreversible loss of enzymatically detectable ChAT has been observed without apparent changes in markers for other neuronal systems (MILLAR et al. 1987; ESTRADA et al. 1988; VILLANI et al. 1988). AF64A, however, failed to damage the type II cholinergic amacrine cells in chicken retina, whereas in goldfish retina ChAT dropped to practically zero levels 40 days after treatment. In rats, a muscarinic receptor supersensitivity developed 3 weeks after a single intravitreal injection of AF64A (MOROI-FETTERS et al. 1990).

Central Cholinergic Systems. After i.c.v. application of ≥ 2 nmol into the lateral ventricle, AF64A has the potential to produce an irreversible cholinergic deficit in the brain of various species, which in the rat persists up to at least 1 year (LEVENTER et al. 1987). On the other hand, at lower doses administered i.c.v. (EL-TAMER et al. 1990) or intracortically (MOUTON et al. 1989), a time-dependent recovery from cholinergic hypofunction has been observed in the hippocampus and cortex, respectively. In the rat, the cholinergic pathway most affected by this toxin is the septohippocampal cholinergic pathway (HÖRTNAGL et al. 1987a,b; LEVENTER et al. 1987). A more preferential effect on the cerebral cortex can be achieved by injection of AF64A into the carotid artery following an unilateral opening of the BBB by a hypertonic treatment (PITTEL et al. 1989). In the mouse, besides the irreversible loss of cholinergic function in the hippocampus a transient cholinergic hypofunction in the cortex and striatum develops initially, lasting about 7 days (FISHER et al. 1982). With substantially higher doses of AF64A than those employed in mammals, a selective and long-lasting decrease of ChAT is achieved in the optic tectum of the goldfish (CONTESTABILE et al. 1987).

Within 4 days after treatment with AF64A, various parameters used as markers for cholinergic function dose-dependently and irreversibly decrease in the affected brain areas, including the levels of ACh, activities of ChAT and AChE, HAChT, ACh release, and hemicholinium-3 binding sites

(Fisher et al. 1982; Jarrard et al. 1984; Walsh et al. 1984; Leventer et al. 1985, 1987; Vickroy et al. 1985; Potter et al. 1986, 1989; Chrobak et al. 1987; Eva et al. 1987; Hörtnagl et al. 1987a,b, 1989; Hörtnagl and Berger 1989; Jossan et al. 1989). The levels of ACh and the activities of ChAT and AChE decrease to a comparable extent, reaching a maximal reduction of up to about 70% in the hippocampus (Fisher et al. 1982; Leventer et al. 1985, 1987; Hörtnagl et al. 1989; Hörtnagl and Berger 1989). A complete reduction in cholinergic function has not been achieved to date. Recent data also indicate subregional variations in the susceptibility of the hippocampal cholinergic innervation to the cholinotoxicity of AF64A, the CA3 subregion and the ventral portion of the hippocampus being most affected (Hörtnagl and Berger 1989; Laganiere et al. 1990). On the other hand, drug binding sites of muscarinic cholinergic receptors as revealed by high-affinity binding of $(-)$-[^{3}H]quinuclidinyl benzilate (QNB), [^{3}H]pirenzepine, and $(+)$-[^{3}H]-cis-methyldioxolane are not affected significantly by AF64A treatment (Fisher et al. 1982; Vickroy et al. 1985). Choline levels are in general also not altered significantly, although there is a slight tendency towards higher levels (Fisher et al. 1982; Eva et al. 1987; Hörtnagl et al. 1987b).

An attenuation of the AF64A-induced cholinergic lesion in vivo has been achieved by several approaches, including treatment with vitamin E prior to AF64A injection (Johnson et al. 1988) and previous depletion of brain NE by DSP-4 application (Hörtnagl et al. 1989). In contrast, pretreatment of animals with nerve growth factor or GM_1 ganglioside seems to be ineffective (Johnson et al. 1988), while basic fibroblast growth factor does exert a protective effect (Potter and Morrison 1991).

In principle, two main aspects on the consequences of the cholinergic lesion in the rat hippocampus have been followed: firstly, the effect of this lesion on specific behavioral parameters and, secondly, the impact of the cholinergic deficit towards the function of other neurotransmitter systems. The prevailing *alterations in behavior* include the development of a deficit in cognitive function. Memory deficits in a T-maze, radial arm maze, or Morris water maze task, in active and passive avoidance tests, as well as an impairment of learning acquisition occur in mice and rats (Walsh et al. 1984; Jarrard et al. 1984; Pope et al. 1985; Bailey et al. 1986; Brandeis et al. 1986; Blaker and Goodwin 1987; Chrobak et al. 1987, 1988; Ogura et al. 1987; Tateishi et al. 1987; Nakamura et al. 1988; Nakahara et al. 1989; Gower et al. 1989; Hanin 1990; Tedford et al. 1990). The improvement of these impairments by treatment with physostigmine or the M_1-muscarinic agonist AF102B has been reported (Ogura et al. 1987; Fisher et al. 1989; Nakahara et al. 1989).

In a series of experiments it has become evident that the withdrawal of cholinergic function in the rat hippocampus induced by AF64A consequently leads to *functional changes in other neurotransmitter systems*. Transient changes in the levels of various neurotransmitters and their metabolites have been described in detail (POTTER et al. 1986; EVA et al. 1987; HÖRTNAGL et al. 1987a,b, 1989, 1990; HÖRTNAGL and BERGER 1990). For the affected neurotransmitters, including NE, serotonin, somatostatin, neuropeptide Y, and glutamate, close synaptic interactions with cholinergic systems in the hippocampus have been established. The transient nature of these changes, their reversibility by pharmacological manipulations (HÖRTNAGL et al. 1989), and morphological studies (GAÁL et al. 1986) add further evidence for the interpretation of these changes as a secondary, functional sequel to the abrupt loss of the cholinergic input. The combination of a specific cholinergic deficit with secondary changes in other defined neurotransmitter systems and with the development of characteristic cognitive dysfunction provide an excellent basis for the use of AF64A to create a lucrative and appropriate animal model for the Alzheimer's disease state.

d) Structural Requirements and Question of Specificity of AF64A

Aziridinium Structure.   Aziridinium salts are known to have the potential to alkylate substrates of biological importance (ROSS 1962). This property has been utilized to develop specific neurotoxins by combining a compound which can be specifically recognized by a certain population of neurons with an aziridinium moiety or a precursor molecule. So far, this principle has been applied for the development of two classes of neurotoxins, namely choline aziridinium analogues and DSP-4 (*N*-[2-chloroethyl]-*N*-ethyl-2-bromobenzylamine hydrochloride), which is converted to an aziridinium salt in vivo (ZIEHER and JAIM-ETCHEVERRY 1980). The role of the aziridinium moiety in the neurotoxicity has been analyzed in vitro and in vivo. Opening of the aziridinium ring by either thiosulfate or alkaline hydrolysis results in a loss of the inhibitory effect on HAChT in synaptosomes in vitro, followed by a complete loss of the in vivo neurotoxic activity (COLHOUN and RYLETT 1986; HÖRTNAGL et al. 1988). In addition, the inhibition of choline dehydrogenase and ChAT was prevented by prior reaction of AF64A with thiosulfate (BARLOW and MARCHBANKS 1984; SANDBERG et al. 1985b). An inverse temperature dependence for stability of the aziridinium ring in aqueous media has also been documented (SANDBERG et al. 1985a; GOLDSTEIN et al. 1988).

Side Chain Substitution.   Substitution of one of the methyl groups of AF64A also appears to be of importance. Increasing the chain length in this position was found to decrease the potency for the inhibition of the HAChT. Propyl-, butyl-, and cyclopropyl AF64A analogues had higher IC_{50} values than AF64A (MISTRY et al. 1986). Interestingly, replacement of one of the methyl groups on the quarternary nitrogen of AF64A with a cyclopropyl moiety made the compound highly reactive toward QNB binding sites in vitro (MISTRY et al. 1986).

AF64A-picrylsulfonate.   AF64A has to be prepared by hydrolysis of acetyl-
ethylcholine mustard followed by aziridinium formation immediately prior
to injection into animals or for in vitro use (FISHER et al. 1982). AF64A-
picrylsulfonate, which already contains the active ethylcholine aziridinium
ion, has been shown to have the advantage of that it can be dissolved just
before administration without any further manipulations and thereby pre-
serving neurotoxic capacity (TONNAER et al. 1986).

Is AF64A a specific cholinergic neurotoxin? This critical question
has been asked several times since its introduction (JARRARD et al. 1984;
COLHOUN and RYLETT 1986; McGURK et al. 1987; see review by HANIN 1990)
and has mainly come up when it was injected directly into certain brain
areas including the caudate-putamen complex, substantia nigra, substantia
innominata, interpeduncular nucleus, and nbM (ASANTE et al. 1983; LEVY
et al. 1984; SPENCER et al. 1985; VILLANI et al. 1986; KOZLOWSKI and
ARBOGAST 1986; STWERTKA and OLSON 1986; McGURK et al. 1987; ALLEN
et al. 1988). In other brain areas, however, such as the hippocampus and
cerebral cortex, injections of AF64A induce specific cholinergic lesions
(MANTIONE et al. 1983b; BLAKER and GOODWIN 1987; MOUTON et al. 1988,
1989; MOUTON and ARENDASH 1990) provided the dose is carefully controlled. This evidence also applies to injections into the lateral ventricles.
Following this route of application in a very low percentage of studies has
unspecific tissue damage been claimed (JARRARD et al. 1984; EVA et al.
1987). These findings imply that in certain brain areas the range between the
specific and non-specific dose appears to be rather narrow. The conversion
of cholino-specificity to unspecific toxicity by increasing the dose of AF64A
has been convincingly demonstrated in the nbM of the rat (KOZLOWSKI and
ARBOGAST 1986) and after injection into the lateral ventricle of the rat
(GOWER et al. 1989; EL-TAMER et al. 1990). It is of interest to note that, in
contrast to specific lesions, the nonspecific tissue damage, e.g., after microinjections of AF64A into the striatum, is only marginally prevented by
concomitant administration of hemicholinium-3 (ALLEN et al. 1988).

II. Acetylcholine Synthesis, Storage, and Release

1. Inhibition of Acetylcholine Synthesis

Several inhibitors of ChAT have been introduced which act either by
alkylating or acylating a reactive nucleophilic group on the enzyme (reviewed
by HAUBRICH 1976). Analogues of styrylpyridine, including *N*-methyl-4-
(1-naphthylvinyl)pyridinium (NVP), halogenated analogues of choline,
halogenated aldehyde analogues of ACh, acetyl-seco-hemicholinium, or
quinones, primarily naphthoquinones, are effective inhibitors of ChAT in
vitro. Their in vivo use is, however, limited partly due to their inability to
cross cellular membranes, instability, and unspecificity. So far, none of the
compounds tested in vivo has the capacity of reducing the concentration of

ACh in the brain, with the exception of acetyl-seco-hemicholinium-3. There is still a need for an inhibitor of ChAT which combines the requirements of specificity, stability, potency, and the ability to cross membrane barriers.

A disturbance of ACh synthesis also occurs in the presence of the precursors of false cholinergic neurotransmitters. A number of linear and cyclic choline analogues, such as mono-, di-, or triethylcholine, diethylaminoethanol, N-aminodeanol, and 3-hydroxypiperidinium, have the potential to be transported via the HAChT system into the cholinergic nerve terminal, to be acetylated by ChAT, to interfere with the uptake into and storage of ACh in the synaptic vesicles, and, finally, to be released upon stimulation instead of ACh (HEMSWORTH et al. 1984; COLLIER and WELNER 1986; NEWTON and JENDEN 1986; JENDEN et al. 1989). The toxic effects of the compounds may, for example, result in a neuromuscular blocking activity as described, e.g., for triethylcholine (BOWMAN and RAND 1961). In general, the pharmacological activity of these cholinergic false neuro-transmitters may be similar to or lower than that of ACh and may undergo a different rate of hydrolysis (NEWTON and JENDEN 1986). Compared with ACh the acetylated 3-hydroxypiperidinium is 57 times less potent as an agonist at the nicotinic receptors of the frog rectus abdominus muscle and 162 times less potent as an agonist at the muscarinic receptors of the guinea-pig ileum (HEMSWORTH et al. 1984). As a consequence, cholinergic false neurotransmitters may have the potential of producing animal models for conditions of cholinergic hypofunction.

2. Inhibition of Acetylcholine Storage

The storage of ACh by synaptic vesicles can be disturbed by several compounds. One of the most extensively studied, vesamicol [2-(4-phenylpiperidino)cyclohexanol, AH 5183; Fig. 2) was initially shown to produce muscle paralysis after either oral or parenteral administration (BRITTAIN et al. 1969). Since muscular contraction elicited by ACh was not blocked, a predominantly prejunctional effect has been postulated. Con-sequently, MARSHALL (1970) suggested a specificity of vesamicol towards the vesicular ACh uptake system. Since then, this substance has been widely used to characterize the vesicular ACh transport system (MARSHALL and PARSONS 1987). Vesamicol impairs electrically stimulated neuromuscular transmission in a variety of nerve-muscle preparations (BRITTAIN et al. 1969; MARSHALL 1970; GANDIHA and MARSHALL 1973) and blocks the active transport of ACh into synaptic vesicles isolated from the electric organ of *Torpedo*, a model system for the mammalian cholinergic synapse (ANDERSON et al. 1983), in an ACh-synthesizing line of PC12 cells (MELAGA and HOWARD 1984), in rat cortical and striatal slices, as well as in rat cortical synaptosomes (JOPE and JOHNSON 1985; ŘÍČNÝ and COLLIER 1986; SUSZKIW and TOTH 1986).

Evidence is available that vesamicol mainly inhibits translocation of newly synthesized ACh and has little effect on previously loaded ACh

(Marshall and Parsons 1987). A block in the process of mobilization of ACh from less to more readily releasable stores by vesamicol has also been suggested (Collier et al. 1986).

The mechanism by which vesamicol inhibits vesicular ACh uptake still remains unclear. Meanwhile, a so-called vesamicol receptor has been identified and further characterized in cholinergic synaptic vesicles from the *Torpedo* electric organ, comprising a stable protein, which faces the cytoplasmic compartment of the cholinergic nerve terminal (Bahr and Parsons 1986a; Kornreich and Parsons 1988). Since vesamicol is not transported into the vesicle and the inhibition of ACh transport is noncompetitive (Anderson et al. 1983; Bahr and Parsons 1986b), the binding site for vesamicol and the ACh site involved in transport are not likely to be identical. Recent evidence indicates that vesamicol appears to be coupled to a low-affinity ACh binding site that is different from the higher-affinity transport binding site (Noremberg and Parsons 1989).

In rat brain, [³H]vesamicol labels a pharmacologically unique binding site with a distribution that closely matches that of cholinergic markers (Altar and Marien 1988). [³H]Vesamicol autoradiography may, therefore, serve as an additional tool to study subregional changes in cholinergic innervation during development, aging, and disease states.

The selectivity of vesamicol for vesicular ACh transport is indicated by its inability to affect either the HAChT, the activity of ChAT, the influx of calcium, or the vesicular content of previously loaded ACh. A second drug, cetiedil, with a similar action on the vesicular ACh transport, interacts, in addition, with the ACh release mechanism (Gandry-Talarmain et al. 1989). Inhibition of ACh storage is also achieved by a series of ACh analogues formed from suitable precursors acetylated by ChAT. The in vitro action of these compounds is characterized by a competitive inhibition, confirming that their site of action is the active transporter site. The structure-activity relationship of these compounds has been analyzed recently in detail (Rogers and Parsons 1989). In general, these compounds, particularly vesamicol, represent an excellent in vitro tool to resolve questions about ACh storage and quantal and nonquantal release. The experience of their in vivo use is still limited.

3. Acetylcholine Release

The release process at the cholinergic nerve terminal is markedly affected by various natural toxins. The evoked quantal release of ACh is almost completely inhibited by botulinum toxin, inducing a long-lasting paralysis of cholinergic transmission at all peripheral cholinergic junctions. This toxin does not interfere with the depolarization-related uptake of Ca^{2+} into the nerve terminal but rather desensitizes a subsequent Ca^{2+}-dependent process (reviewed by Gundersen 1980). The action of this toxin produced by *Clostridium botulinum* is discussed in considerably more detail in Chap. 11 of this book.

Neurotoxins from snake venoms with an intrinsic phospholipase A_2 activity, including β-bungarotoxin, notexin, taipoxin, and crotoxin, are potent presynaptic neurotoxins which interfere with the release of ACh from motor nerve terminals. The reduction of transmitter release is usually preceded by an initial facilitation. The effect is characterized by a long latency in the onset of paralysis and a continuous progress of inhibitory action after removal of the toxin. The action of these neurotoxins is accelerated by nerve impulses (LEE 1979). The neurotoxins block the choline transport by virtue of their phospholipase A_2 activity.

In contrast, α-latrotoxin, the active component (a 130-kDa single polypeptide chain) of the black widow spider venom, induces a massive discharge of ACh quanta at the frog neuromuscular junction. The toxin acts by greatly increasing the probability of vesicle fusion with the presynaptic membrane, followed by a concomitant inhibition of vesicle recycling (reviewed by MELDOLESI et al. 1986). Since apparently all types of neurotransmitters are released, the toxin seems to act via a mechanism common to the release of all transmitters (HURLBUTT and CECCARELLI 1979). It appears to interact with a specific receptor, a high molecular weight integral membrane protein, located at the frog neuromuscular junction and in the plasma membrane of mammalian brain synaptosomes. Recent data on an inhomogenous distribution of the α-latrotoxin receptor in various rat brain areas provide evidence that this receptor may not be present in all presynaptic plasma membranes in a comparable concentration range (MALGAROLI et al. 1989). Although α-latrotoxin has no selectivity for specific types of synapses and its action is not at all restricted to the cholinergic synapse, this toxin deserves mention in this chapter, since the initial studies have been worked out on the cholinergic synapse, and from a toxicological standpoint, several symptoms of the intoxication after a spider bite are preferentially referred to an effect on the cholinergic synapse.

Two other proteinaceous toxins have been described by McCLURE et al. (1986), which also apparently act at the presynaptic nerve terminal to stimulate release of ACh: iotrochotin, and β-leptinotarsin-h. While data are preliminary, nevertheless they indicate that iotrochotin acts directly on the molecular system controlling the rate of ACh release. On the other hand, it is suggested that β-leptinotarsin-h exerts its action by forcing open voltage-sensitive Ca^{2+} channels, resulting in an increased influx of ions responsible for neurotransmitter release.

III. Other Toxins at the Presynaptic Level

In an attempt to destroy basal forebrain cholinergic neurons selectively with the intention of developing a new animal model for Alzheimer's disease, several interesting strategies have been employed. One of these is treatment with AF64A, which has already been discussed in depth earlier in this chapter. Another approach has been based on the observation that cholinergic pathways in the basal forebrain differ from other cholinergic projections by the presence of, and their dependence on, nerve growth factor (for review, see Hefti et al. 1989). It should thus be possible to destroy these neurons selectively with a cytotoxic compound conjugated to nerve growth factor. The complex would then be specifically internalized by means of the nerve growth factor receptor into those cholinergic nerve endings which are provided with this receptor. This principle has been realized in the rat by injecting a nerve growth factor-diphteria toxin conjugate unilaterally into the cerebral cortex. The treatment induces a marked, specific, ipsilateral reduction of cholinergic neurons in the horizontal limb of the diagonal band and nbM (Kudo et al. 1989). Yet another strategy has been used, which is directed towards the cholinergic specific Chol-1 gangliosides, which have been recently defined (Derrington et al. 1989). An antiserum raised to *Torpedo* electromotor synaptosomal membranes which recognizes Chol-1 gangliosides has been shown to induce a cholinergic-specific immune lysis of mammalian brain synaptosomes. The question arises whether this antiserum possibly interferes specifically with cholinergic neurons also under in vivo conditions. Nevertheless, this antiserum has already been used to recognize changes in the concentration of Chol-1 as a possible marker for cholinergic neurons after degeneration and sprouting of the cholinergic innervation of the hippocampus (Derrington et al. 1989).

E. Toxins Interfering at the Level of the Synapse

Cholinergic neurotransmission can be effectively influenced by agonists or antagonists of the various subtypes of cholinergic receptors, either in the direction of potentiation or inhibition. These compounds are not only of considerable pharmacological and toxicological importance but have also contributed to the discovery and subclassification as well as purification of cholinergic receptors. The variety of these compounds is separately reviewed later in this volume.

Another potent class of toxins interfering with cholinergic function at the level of the synapse is the group of anticholinesterase agents which either reversibly or irreversibly inhibit the action of AChE, responsible for terminating the function of ACh at the cholinergic synapse. The synthesis of the first irreversible organophosphorus anticholinesterase tetraethyl pyrophosphate goes back to 1854, which was 10 years before the isolation of the naturally occurring, reversible inhibitor physostigmine from the Calabar bean. Since then, this class of cholinergic toxins has continuously attracted attention because of their wide spectrum of applicability, reaching from their importance as pharmacological tools to their worldwide use as insecticides up to their tragical misuse as chemical warfare agents. Due to the long and diverse history and tradition of these compounds, it is far beyond the limit of this chapter to present a comprehensive overview. In this respect, we would like to refer to other handbooks of this series, in which anticholinesterase agents are reviewed in detail (HOLMSTEDT 1963; GROB 1963; MOUNTER 1963; HOBBIGER 1976). Here we shall try to focus only on certain aspects of anticholinesterase activity, namely acute and delayed neurotoxicity as well as the consequences of long-term exposure.

A characteristic feature for the different types of inhibitors of AChE is their interaction at various subsites of the enzyme. The active site of AChE is composed of a catalytic (esteratic) and an anionic subsite. The sequence of the active site region has been analysed in the AChE from *Torpedo californica* (MACPHEE-QUIGLEY et al. 1985) and contains an extremely nucleophilic serine residue (in position 200), which is the target of organophosphorus inhibitors such as diisopropylfluorophosphate. Certain organophosphorus compounds (e.g., echothiophate) with high potency and selectivity interact with both the esteratic and anionic subunits (HOLMSTEDT 1963).

Besides the anionic binding site in close proximity to the esteratic subsite at the active center, one or several additional anionic subsites are located peripherally, which might allosterically regulate the center's activity. Sequences in the *Torpedo* enzyme involved in the formation of the peripheral and active site anionic subsites have been recently analyzed (WEISE et al. 1990). Propidium has been shown to be a specific ligand at the peripheral anionic site, and edrophonium interacts at the active anionic subsite, whereas decamethonium and *N*, *N*-dimethyl-2-phenylaziridinium are ligands of both active and peripheral anionic sites (WEISE et al. 1990).

For medical treatment of an acute intoxication with organophosphorus compounds, atropine sulfate and the cholinesterase reactivator pralidoxime Cl (2-PAM), originally introduced by WILSON and GINSBURG (1955), are available as antidotes. Their efficacy, however, is limited. Atropine antagonizes the actions of excess ACh at muscarinic ACh receptor sites but is virtually ineffective against the peripheral neuromuscular activation and subsequent paralysis. The efficacy of reactivation by nucleophilic agents, such as oximes, depends on the fairly rapid process of "aging" of phosphorylated AChE. The "aging" process may be related to the hydrolysis of one alkyl or alkoxy group, converting the phosphorylated AChE to the nonreactivatable monoalkyl- or monoalkoxyphosphonate enzyme (FLEISHER and HARRIS 1965), and occurs within minutes or hours depending on the structure of the organophosphorus compounds. The half-time for "aging" of, e.g., soman-inhibited blood cholinesterase in rabbits is reported to be 7.6 min (HARRIS et al. 1981), and the soman-phosphorylated AChE rapidly becomes resistant to reactivation (FLEISHER and HARRIS 1965). Because of this, pralidoxime appears to be of little value in the case of exposure to soman (LOOMIS and SALAFSKY 1963). Consequently, a number of bis-quaternary oximes have been developed to improve therapy against soman intoxication, including obidoxime. Recently, two other oximes, 1,1-methylenebis-[4-(hydroxyiminomethyl)pyridinium]-di-Cl   and   1-(2-hydroxyiminomethyl-1-pyridinio)-3-(4-carbamoyl-1-pyridinio)-2-oxapropane-di-Cl, have been shown to reactivate soman-inhibited whole blood and diaphragm cholinesterases in rabbits (Harris et al. 1990).

The organophosphorus compounds soman, sarin, and tabun, synthesized by Schrader in the years 1937–1944, belong to the most potent synthetic toxins available. These volatile toxins can kill man within minutes in submilligram doses. Besides this devastating potential, soman (pinacolyl-methylphosphonofluoridate) is increasingly utilized as an experimental tool in brain research. Within minutes following the subcutaneous injection of 120 µg/kg soman into rats, a sustained increase in brain ACh levels occurs, especially in the cerebral cortex and hippocampus (Shih 1982). Excess of ACh or muscarinic agonists, e.g., pilocarpine, are known to initiate seizure activity (reviewed by Turski et al. 1989). Likewise, in the acute phase of intoxication, using near-lethal doses of soman, the most prominent symptoms developing in the rat are repetitive convulsions which persist for several hours (McDonough et al. 1983; McLeod et al. 1984; Lemercier et al. 1983). Animals surviving this acute phase or repeatedly exposed to low doses develop neuronal lesions which are most prominent in the pyriform cortex, amygdala, hippocampus, and thalamic nuclei (Lemercier et al. 1983; McLeod et al. 1984; Pazdernik et al. 1985). Additionally, myocardial lesions progressing to fibrosis have been described (Singer et al. 1987). These pathological changes are in part linked to a cholinergically mediated phenomenon, since high doses of atropine are effective in preventing soman-induced damage (McDonough et al. 1989). Furthermore, direct microinjections of soman into the amygdala produce repetitive limbic convulsions and seizure-related brain damage (McDonough et al. 1987). Thus, besides pilocarpine, soman as well as other comparable agents, e.g., sarin, are useful tools to evaluate further the role of cholinergic systems in triggering, maintaining, and inducing spread of seizure activity (Turski et al. 1989).

Another interesting but still unsolved phenomenon in connection with some but not all of the organophosphorus compounds, such as tri-o-cresylphosphate, parathion, or malathion, is the development of a delayed neurotoxicity, which occurs in certain species, including humans. As early as 1899, it was established that humans are sensitive to delayed neurotoxicity induced by tri-o-cresylphosphate. Since then, more than 40,000 episodes of delayed neurotoxicity in response to various organophosphorus compounds have been reported (reviewed by Abou-Donia and Lapadula 1990). The initial toxicity of these compounds is the result of their ability to inhibit AChE in an irreversible manner. A single or repeated exposure to certain organophosphorus compounds may, however, additionally result in a delayed onset of neuropathological symptoms, including prolonged locomotor ataxia followed by paralysis (for review see Abou-Donia 1981; Abou-Donia and Lapadula 1990). The neurotoxic lesions, initially affecting motoneurons, are characterized by degeneration of axons with subsequent secondary degeneration of myelin in the central and peripheral nervous systems. Type I and type II compounds have been differentiated to date, according to their chemical structure. The delay before the onset of neurological deficits varies within the two types. In the cat, the delay for type II is 4–7 days and for type I, 14–21 days. In humans, the delay is 6 days to 1 month (usually 2 weeks) after exposure to type I compounds.

The mechanism of action on which this delayed neurotoxicity is based is still unknown. The involvement of a certain esterase, the so-called neurotoxic esterase, that is preferentially inhibited by these organophosphorus compounds has been discussed (JOHNSON 1982). Recent evidence, however, points to an interaction with Ca^{2+}/calmodulin kinase II, resulting in increased activity with enhanced phosphorylation of cytoskeletal proteins, which may finally lead to their disassembly (ABOU-DONIA and LAPADULA 1990). The knowledge about the delayed neurotoxic action of a certain organophosphorus compound is of extreme importance for its selection as an insecticide.

A further risk associated with the use of organophosphorus compounds as insecticides are the neurotoxic consequences of long-term low-level exposures. The neurological symptoms among exposed humans comprise headaches, giddiness, visual disturbances, paresthesia, myopia, cognitive deficits, including disturbances of memory recall and retention, and psychiatric episodes (GERSHON and SHAW 1961; METCALF and HOLMES 1969; MISRA et al. 1985). Experimental evidence indicates that in both young and old rats the long-term, intermittent exposure to low levels of organophosphate insecticides results in extensive and progressive damage of the hippocampus (VERONESI et al. 1990) and in learning and memory deficit (GARDNER et al. 1984; McDONALD et al. 1988).

F. Neurotoxins with Miscellaneous Actions, Preferentially Affecting Cholinergic Pathways

In addition to those agents which affect the cholinergic system through actions described so far, there also are a number of neurotoxins with miscellaneous actions preferentially affecting cholinergic pathways. These are by no means specific for cholinergic neuronal mechanisms and may also affect other neurotransmitter systems by similar mechanisms to those described herein. Nevertheless, we have included them here in order to provide as comprehensive an overview as possible on all types of agents known to affect cholinergic mechanisms in vivo.

I. Colchicine

Colchicine has become an important neurobiological tool because of its affinity for tubulin and its ability to block axoplasmic transport. It has been used as a neurotoxin to produce selective neuronal death, especially in the hippocampal formation. When administered directly into the hippocampus, colchicine exerts a preferential neurotoxicity towards dentate granule cells while sparing pyramidal cells (GOLDSCHMIDT and STEWARD 1980, 1982). Its neurotoxic effects, however, seem to depend on the site of application and on the differential susceptibility of various neuronal populations. The cerebral cortex has been shown to be relatively unaffected, while the dentate gyrus, olfactory bulb, and cerebellum suffer extensive damage following local injections. In the latter three areas, the disruption is most pronounced in the granule cell layers (GOLDSCHMIDT and STEWARD 1980, 1982).

Besides this preferential effect on granule cells, colchicine also interferes with cholinergic pathways in the CNS. For example, injection of colchicine into the lateral ventricle of the rat brain exerts a direct neurotoxic effect on cholinergic neurons in both the medial septum and in the adjacent striatum, without loss of granule cells in the dentate gyrus and without affecting GABAergic neurons (PETERSON and McGINTY 1988). In addition, SOFRONIEW et al. (1987) have reported that cholinergic neurons in the septum are shrunken 24–48 h after i.c.v. injection of colchicine. The disruption of the retrograde axoplasmic transport of a neurotrophic factor has been suggested as a possible mechanism by which colchicine causes cholinergic cell death (PETERSON and McGINTY 1988). In this respect, it is of special interest that hippocampal extracts prepared from colchicine-lesioned rats can enhance neurotrophic activity in cultured cholinergic neuronal cells (NAKAGAWA and ISHIHARA 1988).

Furthermore, in the rat, cholinergic neurons of the supracommissural septum projecting to the habenula and nucleus interpeduncularis are selectively killed by local injections of colchicine (FONNUM and CONTESTABILE 1984). Moreover, after bilateral injection of colchicine into the nbM of the rat, decrease in the ChAT activity and in nicotinic binding sites were observed in the frontal cortex (TILSON et al. 1989).

Cholinergic neurons are also affected, in a biphasic mode, after infusion of colchicine into the rat hippocampus. After a high dose (15 µg) a considerable reduction in hippocampal ChAT activity (down to 43% of controls) appears between 4 and 12 days later. This effect parallels in time a concurrent loss of granule cells and pyramidal cells (NAKAGAWA et al. 1987). These neuronal changes have been associated with an impairment in T-maze behavior ability. The reduction in cholinergic function in the hippocampus at early time points (~2 weeks), which is also reflected by decreased AChE staining, is followed by an increase in AChE staining and a decrease in the number of QNB binding sites over extended time periods (2–5 months; DRUST and CRAWFORD 1985; TILSON et al. 1988). This biphasic effect suggests compensatory changes in cholinergic function following intradentate administration of colchicine. In addition, 12 weeks after colchicine treatment an increase in carbachol-stimulated phosphoinositide hydrolysis was also observed in hippocampal slices, indicating an alteration in the signal transduction process for muscarinic cholinergic receptors in the colchicine-damaged hippocampus (TANDON et al. 1989).

Interestingly, beside these pronounced effects on cholinergic function, no significant changes in the hippocampal levels of serotonin, 5-HIAA, taurine, aspartate, glutamine, glutamate, glycine, or GABA were found 12 weeks after intradentate administration of colchicine (TILSON and PETERSON 1987).

II. Ethanol

It is well documented that ethanol is a neurotoxin affecting cholinergic function. Although its neurotoxicity is certainly not restricted to cholinergic neurons, the most consistent data on the effect of chronic intake of ethanol have been obtained concerning cholinergic transmission. In rodents, the cerebral cortex and hippocampus have been demonstrated to be amongst the areas most susceptible to ethanol-induced changes in ACh content, release, and turnover as well as in the number of muscarinic cholinergic receptors, in ChAT and AChE activity, and in HAChT (KALANT and GROSS 1967; KALANT et al. 1967; SMYTH and BECK 1969; ERICKSON and GRAHAM 1973; RAWAT 1974; PARKER et al. 1978; HUNT et al. 1979; TABAKOFF et al. 1979; RABIN et al. 1980; DURKIN et al. 1982; SMITH 1983; BERACOCHEA et al. 1986; HOFFMAN et al. 1986). After prolonged intake of ethanol, a loss of neurons in the rat basal forebrain cholinergic projection system has been described (ARENDT et al. 1988a,b). Transplantation of cholinergic-rich fetal basal forebrain cell suspensions into the hippocampus and cortex corrected both the cholinergic deficits and the memory abnormalities, suggesting that, at least in rats, ethanol damages directly cholinergic projection neurons and that the substantial memory losses produced by ethanol intake are associated with an impairment of cholinergic function (ARENDT et al. 1988a).

The reduction in cholinergic markers appears to be more pronounced in the basal nucleus-septum complex than in the target areas (ARENDT 1988b). This finding supports the assumption of a direct toxic effect of ethanol on cholinergic neurons in the basal forebrain, which is then reflected by a secondary degeneration of axon terminals in the cortex and hippocampus.

Ethanol has been shown to decrease synaptic ACh release both in vitro (KALANT and GROSS 1967; KALANT et al. 1967) and in vivo (ERICKSON and GRAHAM 1973; PHILLIS et al. 1980), as well as to reduce the rate of ACh turnover (RAWAT 1974; PARKER et al. 1978). An increase in the number of muscarinic cholinergic receptors in the mouse cortex and hippocampus has been repeatedly demonstrated after systemic administration of ethanol (TABAKOFF et al. 1979; RABIN et al. 1980; SMITH 1983; WITT et al. 1986; PIETRZAK et al. 1990). However, there is also evidence for a decrease (MULLER et al. 1980; SYVÄLAHTI et al. 1988) or no reported changes in muscarinic receptor density (WIGELL and OVERSTREET 1984; WITT et al. 1986) following chronic ethanol administration. Chronic ethanol treatment mainly alters M_1 muscarinic receptors and correspondingly carbachol-induced phosphotidylinositol-4,5-biphosphate hydrolysis (Hoffman et al. 1986).ACh release from rat brain cortical slices is more sensitive to inhibition by ethanol than the release of several other neurotransmitters (Carmichael and Israël 1975; Phillis et al. 1980). A dose- and time-dependent decrease in ACh content as well as in AChE and ChAT activities has been observed over the duration of prolonged ethanol intake in rats (Smyth and Beck 1969). In mice, ethanol administration is also followed by a reduction in HAChT in the hippocampus (Hunt et al. 1979; Durkin et al. 1982; Beracochea et al. 1986).

A study of the regional distribution of ethanol in the rat brain revealed that the septal area concentrates the highest levels of ethanol after chronic dosing (Erickson 1976). These differences in distribution of effective concentrations of ethanol after systemic application may at least partially account for the topographic differences in ethanol-induced cell damage and might explain the preferential involvement of the septal-diagonal band area (Arendt et al. 1988b).

In man, alcohol-induced memory impairment has been attributed to deficiences in subcortical cholinergic systems (Lishman 1986). Cholinergic neurons in the nbM have been reported to be reduced in patients with Korsakoff's syndrome, a disease in which alcohol poisoning causes apparently irreversible memory defects (Arendt et al. 1983). ChAT activity is reduced in the brains of alcoholics (Autuono et al. 1980; Nordberg et al. 1980). In addition, an acceleration of the onset of Alzheimer's disease is indicated following chronic ethanol ingestion (Maciejek et al. 1987).

III. Aluminum

Aluminum neurotoxicity has been implicated in the pathology of several neurological disorders associated with cognitive impairments (Wisniewski et al. 1985). The neurological dysfunction observed in patients with dialysis encephalopathy has been linked to toxicity induced by high aluminum concentrations in the dialysate and to the use of phosphate binding gels containing aluminum (Alfrey et al. 1976; Parkinson et al. 1981). Aluminum has also been suggested to be of pathogenetic importance in Alzheimer's disease and has been associated with both neurofibrillary tangles and senile plaques (Candy et al. 1986; Perl and Brody 1980). A possible role of aluminum in amyotrophic lateral sclerosis and Parkinsonism-dementia syndrome of Guam has also been proposed (Perl et al. 1982).

Increasing evidence indicates that the neurotoxicity of aluminum is directed against cholinergic systems, although not exclusively. In various animal species (rat, rabbit), the i.c.v. injection of $AlCl_3$ results in reductions in ChAT activity and in HAChT in the hippocampus and cerebral cortex (Zubenko and Hanin 1989; Beal et al. 1989). $AlCl_3$ shows a relative selectivity for cholinergic neurons in rat brain as compared with other neurotransmitters, such as NE (ZUBENKO and HANIN 1989). Variable decreases in other neurotransmitters like serotonin, glutamate, and substance P have, however, also been found in the rabbit in response to $AlCl_3$ administration (BEAL et al. 1989). In the rabbit, the cholinergic deficits were accompanied by neurofibrillary degeneration in AChE-positive basal forebrain and medial septal neurons (KOWALL et al. 1989). After microinjection of aluminum directly into the spinal cord of rabbits, significant reductions in ChAT activity in the sciatic nerves were accompanied by neurofibrillary degeneration in the anterior horn cells of the lumbar spinal cord (KOSIK et al. 1983). After injection of a 1% slurry of metallic aluminum powder into the cisterna magna of rabbits, decreases in ChAT activity occurred in the hippocampus and striatum (HOFSTETTER et al. 1987). After intracisternal injection of $AlCl_3$ into rabbits, a reduction of ChAT activity was found in the spinal cord and hypoglossal nucleus (YATES et al. 1980).

In rats, a slight increase in muscarinic receptor density in the hippocampus has also been reported after chronic oral administration of aluminum sulfate in the drinking water for 30 days, suggesting a slight presynaptic cholinergic deficiency, although ChAT activity was not reduced after such treatment (CONNOR et al. 1988). Oral aluminum treatment resulted also in an impairment of both consolidation and extinction of a passive avoidance task.

It seems to be of importance in which form and route aluminum is applied. ZUBENKO and HANIN (1989) have demonstrated that equivalent concentrations of aluminum citrate did not share the cholinotoxic properties of $AlCl_3$. Similarly, JOHNSON and JOPE (1987) found no significant effect of i.c.v. aluminum citrate on ChAT activity, HAChT, and ACh levels in the hippocampus, cortex, and striatum. On the other hand, dietary aluminum citrate significantly reduced choline levels in the cortex, hippocampus, and striatum (JOHNSON and JOPE 1987). Reduced availability of choline subsequently may play an important role in the destruction of cholinergic nerve endings (WURTMAN et al. 1985). The transport of choline into human erythrocytes is also inhibited by Al^{3+} (KING et al. 1983).

In neuroblastoma cell lines, $AlCl_3$ was more effective than aluminum lactate in inducing perikaryal neurofibrillar accumulations (SHEA et al. 1989). Different salts of aluminum appear to vary in the proportion of free Al^{3+} and in the complexes formed. Since only free ions are believed to exert toxicity (MARTIN 1986), differences in the relative toxicity of aluminum salts may relate to the free Al^{3+} levels achieved intracellularly following administration.

G. Concluding Remarks

In the present chapter we have attempted to summarize, in general terms, the various mechanisms by which several classes of toxins can affect the function of a cholinergic neuron. The interaction can be on a merely reversible basis but can also result in an irreversible destruction of the neuron. It is a remarkable phenomenon that cholinergic toxins of natural as well as synthetic origin (e.g., botulinum toxin and soman, respectively) belong to the most potent toxins known to date. Sophisticated approaches appear to have been followed by nature and mankind in order to interfere specifically with cholinergic systems.

One of the most specific and strategic target sites for a cholinergic toxin appears to be the sodium-dependent HAChT, which is localized on and is of vital importance for the cholinergic neuron. Several compounds affecting this target site have been introduced, including hemicholinium-3 and AF64A, as representative agents able to induce a reversible and irreversible inhibition of HAChT, respectively. Another elegant technique is the internalization of a toxin via the nerve growth factor receptor, at least into certain subpopulations of cholinergic neurons.

Beside the tremendous pharmacological and toxicological importance of the different classes of drugs and toxins affecting cholinergic function, various toxins mentioned in this chapter are also of considerable interest because of their diverse scientific input to our understanding of the function and the physiological and pathophysiological significance of cholinergic systems. These toxins have been and are still of great scientific value in elucidating various steps in the synthesis, storage, and release of ACh, in the subclassification of cholinergic receptors, and in the isolation of cholinergic receptors or the HAChT carrier. Several are invaluable tools for the identification of cholinergic pathways in the CNS. They have, furthermore, contributed to the elucidation of the role of cholinergic systems in various brain functions such as, for example, memory and learning processes or regulation of sleep. Finally, several of these cholinotoxins have been applied to the development of animal models for various neuropsychiatric diseases, such as Alzheimer's disease or epilepsy.

Acknowledgement. We wish to acknowledge the excellent secretarial help of Waltraud Krivanek.

References

Abou-Donia MB (1981) Organophosphorus ester-induced delayed neurotoxicity. Annu Rev Pharmacol Toxicol 21:511–548

Abou-Donia MB, Lapadula DM (1990) Mechanisms of organophosphorus ester-induced delayed neurotoxicity: type I and type II. Annu Rev Pharmacol Toxicol 30:405–440

Alfrey AC, LeGundre GR, Kaehny WD (1976) The dialysis encephalopathy syndrome. N Engl J Med 294:184–188

Allen YS, Marchbanks RM, Sinden JD (1988) Non-specific effects of the putative cholinergic neurotoxin ethylcholine mustard aziridinium ion in the rat brain examined by autoradiography, immunocytochemistry and gel electrophoresis. Neurosci Lett 95:69–74

Altar CA, Marien MR (1988) ^{3}H vesamicol binding in brain: autoradiographic distribution, pharmacology and effects of cholinergic lesions. Synapse 2:486–493

Amir A, Pittel Z, Shahar A, Fisher A, Heldman E (1988) Cholinotoxicity of the ethylcholine aziridinium ion in primary cultures from rat central nervous system. Brain Res 454:298–307

Anderson DC, King SC, Parsons SM (1983) Pharmacological characterization of the acetylcholine transport system in purified electric organ synaptic vesicles. Mol Pharmacol 24:48–54

Ansell GB, Spanner S (1975) The metabolism of choline in regions of rat brain and the effect of hemicholinium-3. Biochem Pharmacol 24:1719–1723

Arendt T, Bigl V, Arendt A, Tennstedt A (1983) Loss of neurons in the nucleus basalis of Meynert in Alzheimer's disease, paralysis agitans and Korsakoff's disease. Acta Neuropathol (Berl) 61:101–108

Arendt T, Allen Y, Sinden J, Schugens MM, Marchbanks RM, Lantos PL, Gray JA (1988a) Cholinergic-rich transplants reverse alcohol-induced memory deficits. Nature 332:448–450

Arendt T, Hennig D, Gray JA, Marchbanks R (1988b) Loss of neurons in the rat basal forebrain cholinergic projection system after prolonged intake of ethanol. Brain Res Bull 21:563–570

Asante JW, Cross AJ, Deakin JFW, Johnson JA, Slater HR (1983) Evaluation of ethylcholine mustard aziridinium ion (ECMA) as a specific neurotoxin of brain cholinergic neurones. Br J Pharmacol 80:573P

Atterwill CK, Meakin JM (1990) Delayed treatment with nerve growth factor (NGF) reverses ECMA-induced cholinergic lesions in rat brain reaggregate cultures. Biochem Pharmacol 39:2073–2076

Atterwill CK, Collins P, Meakin J, Pillar AM, Prince AK (1989) Effect of nerve growth factor and thyrotropin releasing hormone on cholinergic neurones in developing rat brain reaggregate cultures lesioned with ethylcholine mustard aziridinium. Biochem Pharmacol 38:1631–1638

Autuono P, Sorbi S, Bracco L, Fusco T, Amaducci L (1980) A discrete sampling technique in senile dementia of the Alzheimer type and alcoholic dementia: study of the cholinergic system. In: Amaducci L, Davison AN, Autuono P (eds) Aging of the brain and dementia. Raven, New York, p 151 (Aging, vol 13)

Bahr BA, Parsons SM (1986a) Demonstration of a receptor in *Torpedo* synaptic vesicles for the acetylcholine storage blocker L-trans-2-(4-phenyl-[3,4-³H]piperidino) cyclohexanol. Proc Natl Acad Sci USA 83:2267–2270

Bahr BA, Parsons SM (1986b) Acetylcholine transport and drug inhibition kinetics in *Torpedo* synaptic vesicles. J Neurochem 46:1214–1218

Bailey EL, Overstreet DH, Crocker AD (1986) Effects of intrahippocampal injections of the cholinergic neurotoxin AF64A on open-field activity and avoidance learning in the rat. Behav Neurol Biol 45:263–274

Barlow P, Marchbanks RM (1984) Effect of ethylcholine mustard on choline dehydrogenase and other enzymes of choline metabolism. J Neurochem 43:1568–1573

Barnes DM, Hanin I, Erickson LC (1988) Cytotoxic and DNA-damaging effects of AF64A in cholinergic and non-cholinergic human cell lines. Fed Proc 47:1749

Beal MF, Mazurek MF, Ellison, DW, Kowall NW, Solomon PR, Pendlebury WW (1989) Neurochemical characteristics of aluminum-induced neurofibrillary degeneration in rabbits. Neuroscience 29:339–346

Bekenstein JW, Wooten GF (1989) Hemicholinium-3 binding sites in rat brain: a quantitative autoradiographic study. Brain Res 481:97–105

Beracochea D, Durkin TP, Jaffard R (1986) On the involvement of the central cholinergic system in memory deficits induced by long term ethanol consumption in mice. Pharmacol Biochem Behav 24:519–524

Bhattacharyya B, Flynn JR, Cannon JG, Long JP (1986) Anticholinesterase activity and structure activity relationships of a new series of hemicholinium-3 analogs. Eur J Pharmacol 132:107–114

Blaker WD, Goodwin SD (1987) Biochemical and behavioral effects of intrahippocampal AF64A in rats. Pharmacol Biochem Behav 28:157–163

Bowman WC, Rand MJ (1961) Actions of triethylcholine on neuromuscular transmission. Br J Pharmacol 17:176–195

Brandeis R, Pittel Z, Lachman C, Heldman E, Luz S, Dachir S, Levy A, Hanin I, Fisher A (1986) AF64A-induced cholinotoxicity: behavioural and neurochemical correlates. In: Fisher A, Hanin I, Lachman C (eds) Alzheimer's and Parkinson's diseases. Strategies for research and development. Plenum, New York, p 469

Brittain RT, Levy GP, Tyers MB (1969) The neuromuscular blocking action of 2-(4-phenylpiperidino)cyclohexanol (AH 5183). Eur J Pharmacol 8:93–99

Candy JM, Klinowski J, Perry RH, Perry EK, Fairbairn A, Oakley AE, Carpenter TA, Atack JR, Blessed G, Edwardson JA (1986) Aluminosilicates and senile plaque formation in Alzheimer's disease. Lancet i:354–357

Carmichael FJ, Israël Y (1975) Effects of ethanol on neurotransmitter release by rat brain cortical slices. J Pharmacol Exp Ther 193:824–834

Carmichael WW, Biggs DV, Gorham PR (1975) Toxicology and pharmacological action of *Anabaena flosaquae* toxin. Science 187:542–544

Casamenti F, DiPatre PL, Bartolini L, Pepeu G (1988) Unilateral and bilateral nucleus basalis lesions: differences in neurochemical and behavioural recovery. Neuroscience 24:209–215

Caulfield MP, Fortune DH, Roberts PM, Stubley JK (1981) Intracerebroventricular hemicholinium-3 (HC-3) impairs learning of a passive avoidance task in mice. Br J Pharmacol 75:865p

Chatterjee TK, Long JP, Cannon JG, Bhatnagar RK (1988) Methylpiperidine analog of hemicholinium-3: a selective, high affinity non-competitive inhibitor of sodium dependent choline uptake system. Eur J Pharmacol 149:241–248

Chrobak JJ, Hanin I, Walsh TJ (1987) AF64A (ethylcholine aziridinium ion), a cholinergic neurotoxin, selectively impairs working memory in a multiple component T-maze task. Brain Res 414:15–21

Chrobak JJ, Hanin I, Schmechel DE, Walsh TJ (1988) AF64A-induced memory impairment: behavioral, neurochemical and histological correlates. Brain Res 463:107–117

Chrobak JJ, Spates MJ, Stackman RW, Walsh TJ (1989) Hemicholinium-3 prevents the working memory impairments and the cholinergic hypofunction induced by ethylcholine aziridinium ion (AF64A). Brain Res 504:269–275

Colhoun EH, Myles LA, Rylett RJ (1986) An attempt to produce cholinergic hypofunction in rat brain using choline mustard aziridinium ion: neurochemical and histological parameters. Can J Neurol Sci 13:517–520

Colhoun EH, Rylett RJ (1986) Nitrogen mustard analogues of choline: potential for use and misuse. Trends Pharmacol Sci 7:55–58

Collier B, Welner SA (1986) Synthesis, storage and release of choline analog esters. In: Hanin I (ed) Dynamics of cholinergic function. Plenum, New York, p 1161

Collier B, Welner SA, Říčný J, Aranjo DM (1986) Acetylcholine synthesis and release by a sympathetic ganglion in the presence of 2-(4-phenylpiperidino)cyclohexanol (AH 5183). J Neurochem 46:822–830

Connor JD, Jope RS, Harrell LE (1988) Chronic, oral aluminum administration to rats: cognition and cholinergic parameters. Pharmacol Biochem Behav 31:467–474

Contestabile A, Bissoli R, Saverino O, Villani L (1987) Neurochemical and anatomical effects of the presumptive cholinergic toxin, AF64A, in various areas of the goldfish brain. Comparison with the effects in mammalian brain. In: Dowdall MG, Hawthorne JN (eds) Cellular basis of cholinergic function. Horwood, Chichester, p 651

Cross AJ, Deakin JFW (1985) Cortical serotonin receptor subtypes after lesion of ascending cholinergic neurones in rat. Neurosci Lett 60:261–265

Curti D, Marchbanks RM (1984) Kinetics of irreversible inhibition of choline transport in synaptosomes by ethylcholine mustard aziridinium. J Membr Biol 82:259–268

Davies DL, Sakellaridis N, Valcana T, Vernadakis A (1986) Cholinergic neurotoxicity induced by ethylcholine aziridinium (AF64A) in neuron-enriched cultures. Brain Res 378:251–261

Derrington EA, Masco D, Whittaker VP (1989) Confirmation of the cholinergic specificity of the Chol-1 gangliosides in mammalian brain using affinity-purified antisera and lesions affecting the cholinergic input to the hippocampus. J Neurochem 53:1686–1692

Domer FR, Schueler FW (1960) Synthesis and metabolic studies of ^{14}C-labeled hemicholinium number three. J Am Pharm Assoc 49:553–558

Drust EG, Crawford IL (1985) Enhanced acetylcholinesterase staining in hippocampal area CA3 after lesion of granule cells by infusion of colchicine. Brain Res Bull 14:9–14

Dunnett SB, Rogers DC, Jones GH (1989) Effects of nucleus basalis magnocellularis lesions in rats on delayed matching and non-matching to position tasks. Eur J Neurosci 1:395–406

Durkin TP, Hashem-Zadeh H, Mandel P, Ebel A (1982) A comparative study of the acute effects of ethanol on the cholinergic system in hippocampus and striatum of inbred mouse strains. J Pharmacol Exp Ther 220:203–208

El-Defrawy SR, Coloma F, Jhamandas K, Boegman RJ, Beninger RJ, Wirsching BA (1985) Functional and neurochemical cortical cholinergic impairment following neurotoxic lesions of the nucleus basalis magnocellularis in the rat. Neurobiol Aging 6:325–330

El-Tamer A, Wülfert E, Hanin I (1990) Dose-time effect of ethylcholine aziridinium (AF64A) on cyclic AMP (cAMP) levels in rat brain: correlation with the reversibility of cholinergic disruption at low doses of AF64A. Soc Neurosci Abstr 16:1308

Erickson CK (1976) Regional distribution of ethanol in rat brain. Life Sci 19:1439–1446

Erickson CK, Graham DT (1973) Alterations of cortical and reticular acetylcholine release by ethanol in vivo. J Pharmacol Exp Ther 185:583–593

Estrada C, Triguero D, del Rio RM, Ramos PG (1988) Biochemical and histological modifications of the rat retina induced by the cholinergic neurotoxin AF64A. Brain Res 439:107–115

Eva C, Fabrazzo M, Costa E (1987) Changes of cholinergic, noradrenergic and serotonergic synaptic transmission indices elicited by ethylcholine aziridinium ion (AF64A) infused intraventricularly. J Pharmacol Exp Ther 241:181–186

Fisher A, Hanin I (1986) Potential animal models for senile dementia of Alzheimer's type, with emphasis on AF64A-induced cholinotoxicity. Annu Rev Pharmacol Toxicol 26:161–181

Fisher A, Mantione CR, Abraham DJ, Hanin I (1982) Long-term central cholinergic hypofunction induced in mice by ethylcholine aziridinium ion (AF64A) in vivo. J Pharmacol Exp Ther 222:140–145

Fisher A, Brandeis R, Pittel Z, Karton I, Sapir M, Dachir S, Levy A, Heldman E (1989) (+)-cis-2-Methyl-spiro(1,3-oxythiolane-5,3′)quinuclidine (AF102B): a new M_1 agonist attenuates cognitive dysfunctions in AF64A-treated rats. Neurosci Lett 102:325–331

Fleisher JH, Harris LW (1965) Dealkylation as a mechanism for aging of cholinesterase after poisoning with pinacolyl methylphosphonofluoridate. Biochem Pharmacol 14:641–650

Fonnum F, Contestabile A (1984) Colchicine neurotoxicity demonstrates the cholinergic projection from the supracommissural septum to the habenula and the nucleus interpeduncularis in the rat. J Neurochem 43:881–884

Freeman JJ, Choi RL, Jenden DJ (1975) The effect of hemicholinium on behaviour and on brain acetylcholine. Psychopharmacol Commun 1:15–27

Freeman JJ, Macri JR, Choi RL, Jenden DJ (1979) Studies on the behavioral and biochemical effects of hemicholinium in vivo. J Pharmacol Exp Ther 210:91–97

Freeman JJ, Kosh JW, Parrish JS (1982) Peripheral toxicity of hemicholinium-3 in mice. Br J Pharmacol 77:239–244

Gaál GY, Potter PE, Hanin I, Kakucska I, Vizi ES (1986) Effects of intracerebroventricular AF64A administration on cholinergic, serotonergic, and catecholaminergic circuitry in rat dorsal hippocampus. Neuroscience 19:1197–1205

Gandiha A, Marshall IG (1973) The effects of 2-(4-phenylpiperidino) cyclohexanol (AH 5183) on the acetylcholine content of, and output from, the chick biventer cervicis muscle preparation. Int J Neurosci 5:191–196

Gandry-Talarmain YM, Diebler M-F, O'Regan S (1989) Compared effects of two vesicular acetylcholine uptake blockers, AH 5183 and cetiedil, on cholinergic functions in *Torpedo* synaptosomes: acetylcholine synthesis, choline transport, vesicular uptake, and evoked acetylcholine release. J Neurochem 52:822–829

Gardiner JE (1961) The inhibition of acetylcholine synthesis in brain by a hemicholinium. Biochem J 18:297–303

Gardner R, Ray R, Frankenheim J, Wallace K, Loss M, Robichaud R (1984) A possible mechanism for DFP induced memory loss in rats. Pharmacol Biochem Behav 21:43–46

Gershon S, Shaw FH (1961) Psychiatric sequelae of chronic exposure to organophosphate insecticides. Lancet I:1371–1374

Giarman NJ, Pepeu G (1962) Drug induced changes in brain acetylcholine. Br J Pharmacol 19:226–234

Gilman AG, Goodman LS, Rall TW, Murad F (1985) Goodman and Gillman's the pharmacological basis of therapeutics, 7th edn. Macmillan, New York

Goldschmidt RB, Steward O (1980) Preferential neurotoxicity of colchicine for granule cells of the dentate gyrus of the adult rat. Proc Natl Acad Sci USA 77:3047–3051

Goldschmidt RB, Steward O (1982) Neurotoxic effects of colchicine: differential susceptibility of CNS neuronal depletion. Neuroscience 7:695–714

Goldstein S, Grimee R, Hanin I, Wülfert E (1988) Formation and degradation of 1-(ethyl)-1-(2-hydroxyethyl)aziridinium chloride in aqueous media – a comparative NMR study. J Neurosci Methods 23:101–105

Gower AJ, Rousseau D, Jamsin P, Gobert J, Hanin I, Wülfert E (1989) Behavioural and histological effects of low concentrations of intraventricular AF64A. Eur J Pharmacol 166:271–281

Grob D (1963) Anticholinesterase intoxication in man and its treatment. In: Koelle GB (ed) Cholinesterases and anticholinesterases. Springer, Berlin Heidelberg New York, p 989 (Handbuch der Experimentellen Pharmakologie, vol 15)

Gundersen CB (1980) The effects of botulinum toxin on the synthesis, storage and release of acetylcholine. Prog Neurobiol 14:99–119

Guyenet P, Lefresne P, Rossier J, Beaujouan JC, Glowinski J (1973) Inhibition by hemicholinium-3 of [^{14}C]acetylcholine synthesis and [^{3}H]choline high affinity uptake in rat striatal synaptosomes. Mol Pharmacol 9:630–639

Haga T, Noda H (1973) Choline uptake systems of rat brain synaptosomes. Biochim Biophys Acta 291:564–575

Hagan JJ, Jansen JHM, Broekkamp CLE (1989) Hemicholinium-3 impairs spatial learning, and the deficit is reversed by cholinomimetics. Psychopharmacology 98:347–356

Hanin I (1990) AF64A-induced cholinergic hypofunction. Prog Brain Res 84:289–299

Hanin I, Fisher A, Hörtnagl H, Leventer SM, Potter PE, Walsh TJ (1987) Ethylcholine aziridinium (AF64A; ECMA) and other potential cholinergic neuron-specific neurotoxins. In: Meltzer HY (ed) Psychopharmacology: the third generation of progress. Raven, New York, p 341

Harris LW, Stitcher DL, Heyl WC (1981) Protection and induced reactivation of cholinesterase by HS-6 in rabbits exposed to soman. Life Sci 29:1747–1753

Harris LW, Anderson DR, Lennox WJ, Woodard CL, Pastelak AM, Vanderpool BA (1990) Evaluation of several oximes as reactivators of unaged soman-inhibited whole blood acetylcholinesterase in rabbits. Biochem Pharmacol 40:2677–2682

Haubrich DR (1976) Choline acetyltransferase and its inhibitors. In: Goldberg AM, Hanin I (eds) Biology of cholinergic function. Raven, New York, p 239

Hebb CO, Ling GM, McGeer EG, McGeer PL, Perkins D (1964) Effect of locally applied hemicholinium on the acetylcholine content of the caudate nucleus. Nature 204:1309–1311

Hefti F, Hartikka J, Knusel B (1989) Function of neurotrophic factors in the adult and aging brain and their possible use in the treatment of neurodegenerative diseases. Neurobiol Aging 10:513–533

Hemsworth BA, Shreeve SM, Veitch GBA (1984) 3-Hydroxy-N,N-dimethylpiperidinium: a precursor of a false cholinergic transmitter. Br J Pharmacol 82:477–484

Hepler DJ, Olton DS, Wenk GL, Coyle JT (1985) Lesions in nucleus basalis magnocellularis and medial septal area of rats produce qualitatively similar memory impairments. J Neurosci 5:866–873

Hobbiger F (1976) Pharmacology of anticholinesterase drugs. In: Zaimis E (ed) Neuromuscular junction. Springer, Berlin Heidelberg New York, p 487 (Handbuch der Experimentellen Pharmakologie, vol 42)

Hoffman PL, Moses F, Luthin GR, Tabakoff B (1986) Acute and chronic effects of ethanol on receptor-mediated phosphatidylinositol 4,5-bisphosphate breakdown in mouse brain. Mol Pharmacol 30:13–18

Hofstetter JR, Vincent I, Bugiani O, Ghetti B, Richter JA (1987) Aluminum-induced decreases in choline acetyltransferase, tyrosine hydroxylase, and glutamate decarboxylase in selected regions of rabbit brain. Neurochem Pathol 6:177–193

Holmstedt B (1963) Structure-activity relationships of the organophosphorus anticholinesterase agents. In: Koelle GB (ed) Cholinesterases and anticholinesterases. Springer, Berlin Heidelberg New York, p 428 (Handbuch der Experimentellen Pharmakologie, vol 15)

Hörtnagl H, Berger ML (1989) Subregional differences of cholinergic deficit in rat hippocampus induced by ethylcholine aziridinium ion (AF64A). J Neurochem 52 [Suppl]:S94

Hörtnagl H, Berger ML (1990) Effect of cholinergic lesion induced by ethylcholine aziridinium on glutamatergic neurons in rat hippocampus. Naunyn-Schmiedebergs Arch Pharmacol 341 [Suppl]:R95

Hörtnagl H, Potter PE, Hanin I (1987a) Effect of cholinergic deficit induced by ethylcholine aziridinium (AF64A) on noradrenergic and dopaminergic parameters in rat brain. Brain Res 421:75–84

Hörtnagl H, Potter PE, Hanin I (1987b) Effect of cholinergic deficit induced by ethylcholine aziridinium on serotonergic parameters in rat brain. Neuroscience 22:203–213

Hörtnagl H, Potter PE, Happe K, Goldstein S, Leventer S, Wulfert E, Hanin I (1988) Role of the aziridinium moiety in the in vivo cholinotoxicity of ethylcholine aziridinium ion (AF64A). J Neurosci Methods 23:107–113

Hörtnagl H, Potter PE, Singer EA, Kindel G, Hanin I (1989) Clonidine prevents transient loss of noradrenaline in response to cholinergic hypofunction induced by ethylcholine aziridinium (AF64A). J Neurochem 52:853–858

Hörtnagl H, Sperk G, Sobal G, Maas D (1990) Cholinergic deficit induced by ethylcholine aziridinium ion (AF64A) transiently affects somatostatin and neuropeptide Y levels in rat brain. J Neurochem 54:1608–1613

Hoyle CHV, Moss HE, Burnstock G (1986) Ethylcholine mustard aziridinium (AF64A) impairs cholinergic neuromuscular transmission in the guinea-pig ileum and urinary bladder, and cholinergic neuromodulation in the enteric nervous system of the guinea-pig distal colon. Gen Pharmacol 17:543–548

Hunt WA, Majchrowicz E, Dalton TK (1979) Alterations in high-affinity choline uptake in brain after acute and chronic ethanol treatment. J Pharmacol Exp Ther 210:259–263

Hurlbut BJ, Lubar LF, Switzer R, Dougherty J, Eisenstadt ML (1987) Basal forebrain infusion of HC-3 in rats: maze learning deficits and neuropathology. Physiol Behav 39:381–393

Hurlbutt WP, Ceccarelli B (1979) Use of black widow spider venom to study the release of neurotransmitters. In: Ceccarelli B, Clementi F (eds) Neurotoxins: tools in neurobiology. Raven, New York, p 87

Jarrard E, Kant GJ, Meyerhoff JL, Levy A (1984) Behavioural and neurochemical effects of intraventricular AF64A administration in rats. Pharmacol Biochem Behav 21:273–280

Jenden DJ, Russell RW, Booth RA, Knusel BJ, Lauretz SD, Rice KM, Roch M (1989) Effects of chronic in vivo replacement of choline with a false cholinergic precursor. In: Frotscher M, Misgeld U (eds) Central cholinergic synaptic transmission. Birkhäuser, Basel, p 229

Johnson GVW, Jope RS (1987) Aluminum alters cyclic AMP and cyclic GMP levels but not presynaptic cholinergic markers in rat brain in vivo. Brain Res 403: 1–6

Johnson GVW, Simonato M, Jope RS (1988) Dose- and time-dependent hippocampal cholinergic lesions induced by ethylcholine mustard aziridinium ion: effects of nerve growth factor, GM_1 ganglioside and vitamin E. Neurochem Res 13:685–692

Johnson MK (1982) The target for initiation of delayed neurotoxicity by organophosphorus ester: biochemical studies and toxicological applications. Rev Biochem Toxicol 4:141–202

Jones BE, Webster HH (1988) Neurotoxic lesions of the dorsolateral pontomesencephalic tegmentum-cholinergic cell area in the cat. I. Effects upon the cholinergic innervation of the brain. Brain Res 451:13–32

Jope RS, Johnson GVW (1985) Quinacrine and 2-(4-phenylpiperidino)cyclohexanol (AH 5183) inhibit acetylcholine release and synthesis in rat brain slices. Mol Pharmacol 29:45–51

Jossan SS, Hiraga Y, Oreland L (1989) The cholinergic neurotoxin ethylcholine mustard aziridinium (AF64A) induces an increase in MAO-B activity in the rat brain. Brain Res 476:291–297

Kalant H, Gross W (1967) Effects of ethanol and pentobarbital on release of acetylcholine from cerebral cortex slices. J Pharmacol Exp Ther 158:386–393

Kalant H, Israël Y, Mahon MA (1967) The effect of ethanol on acetylcholine synthesis, release and degradation in brain. Can J Physiol Pharmacol 45:172–176

Kasa P, Hanin I (1985) Ethylcholine mustard aziridinium blocks the axoplasmic transport of acetylcholinesterase in cholinergic nerve fibres of the rat. Histochemistry 83:343–345

Kesner RP, Crutcher KA, Measom MO (1986) Medial septal and nucleus basalis magnocellularis lesions produce order memory deficits in rats which mimic symptomatology of Alzheimer's disease. Neurobiol Aging 7:287–295

King RJ, Sharp JA, Boura ALA (1983) The effects of Al^{3+}, Cd^{2+} and Mn^{2+} on human erythrocyte choline transport. Biochem Pharmacol 32:3611–3617

Koppenaal DW, Raizada MK, Momol EA, Morgan E, Meyer EM (1986) Effects of AF64A on [^{3}H]acetylcholine synthesis in neuron-enriched primary brain cell cultures. Dev Brain Res 30:110–113

Kornreich WD, Parsons SM (1988) Sidedness and chemical and kinetic properties of the vesamicol receptor of cholinergic synaptic vesicles. Biochemistry 27:5262–5267

Kosik KS, Bradley WG, Good PF, Rasool CG, Selkoe DJ (1983) Cholinergic function in lumbar aluminum myelopathy. J Neuropathol Exp Neurol 42:363–375

Kowall NW, Pendlebury WW, Kessler JB, Perl DP, Beal MF (1989) Aluminum-induced neurofibrillary degeneration affects a subset of neurons in rabbit cerebral cortex, basal forebrain and upper brainstem. Neuroscience 29:329–337

Kozlowski MR, Arbogast RE (1986) Specific toxic effects of ethylcholine nitrogen mustard on cholinergic neurons of the nucleus basalis of Meynert. Brain Res 372:45–54

Kudo Y, Shiosaka S, Matsuda M, Tohyama M (1989) An attempt to cause the selective loss of the cholinergic neurons in the basal forebrain of the rat: a new animal model of Alzheimer's disease. Neurosci Lett 102:125–130

Laganiere S, Marinko M, Corey J, Wulfert E, Hanin I (1990) Sector-dependent neurotoxicity of ethylcholine aziridinium (AF64A) in the rat hippocampus. Neuropharmacology 29:961–964

Lee CY (1979) Recent advances in chemistry and pharmacology of snake toxins. In: Ceccarelli B, Clementi F (eds) Neurotoxins: tools in neurobiology. Raven, New York, p 1

Lemercier G, Carpentier P, Sentenac-Roumanou H, Morelis P (1983) Histological and histochemical changes in the central nervous system of rat poisoned by an irreversible anticholinesterase organophosphorus compound. Acta Neuropathol (Berl) 61:123–129

Lerer B, Warner J, Friedman E (1985) Cortical cholinergic impairment and behavioural deficits produced by kainic acid lesions of rat magnocellular basal forebrain. Behav Neurosci 99:661–677

Leventer S, McKeag D, Clancy M, Wulfert E, Hanin I (1985) Intracerebroventricular administration of ethylcholine mustard aziridinium ion (AF64A) reduces release of acetylcholine from rat hippocampal slices. Neuropharmacology 24:453–459

Leventer SM, Wülfert E, Hanin I (1987) Time course of ethylcholine aziridinium ion (AF64A)-induced cholinotoxicity in vivo. Neuropharmacology 26:361–365

Levy A, Kant GJ, Meyerhoff JL, Jarrard LE (1984) Noncholinergic neurotoxic effects of AF64A in the substantia nigra. Brain Res 305:169–172

Lishman WA (1986) Alcoholic dementia: a hypothesis. Lancet i:1184–1186

Long JP, Schueler FW (1954) A new series of cholinesterase inhibitors. J Am Pharm Assoc (Sci Ed) 43:79–86

Loomis TA, Salafsky B (1963) Antidotal action of pyridinium oximes in anticholinesterase poisoning; comparative effects of soman, sarin, and neostigmine on neuromuscular function. Toxicol Appl Pharmacol 5:685–701

Maciejek Z, Slotala T, Nicpon K (1987) Chronic alcoholism and Alzheimer's disease. Neuroscience 22 [Suppl]:S440

MacPhee-Quigley K, Taylor P, Taylor S (1985) Primary structures of the catalytic subunits from two molecular forms of acetylcholinesterase. A comparison of NH_2-terminal and active center sequences. J Biol Chem 260:12185–12189

Malgaroli A, DeCamilli P, Meldolesi J (1989) Distribution of α-latrotoxin receptor in the rat brain by quantitative autoradiography: comparison with the nerve terminal protein, synapsin I. Neuroscience 32:393–404

Mantione CR, DeGroat WC, Fisher A, Hanin I (1983a) Selective inhibition of peripheral cholinergic transmission in the cat produced by AF64A. J Pharmacol Exp Ther 255:616–622

Mantione CR, Zigmond MJ, Fisher A, Hanin I (1983b) Selective presynaptic cholinergic neurotoxicity following intrahippocampal AF64A injection in rats. J Neurochem 41:251–255

Marshall IG (1970) Studies on the blocking action of 2-(4-phenylpiperidino)-cyclohexanol. Br J Pharmacol 38:503–516

Marshall IG, Parsons SM (1987) The vesicular acetylcholine transport system. Trends Neurosci 10:174–177

Martin RB (1986) The chemistry of aluminum as related to biology and medicine. Clin Chem 32:1797–1806

McArdle J, Hanin I (1986) Acute in vivo exposure to ethylcholine aziridinium (AF64A) depresses the secretion of quanta from nerve terminals. Eur J Pharmacol 131:119–121

McClure WO, Baxter DE, Brusca R, Crosland RD, Hsiao TH, Koenig ML, Martin JV, Yoshino JE (1986) Studies of two novel presynaptic toxins. In: Hanin I (ed) Dynamics of cholinergic function. Plenum, New York, p 1203

McDonald BE, Costa LG, Murphy SD (1988) Spatial memory impairment and central muscarinic receptor loss following prolonged treatment with organophosphates. Toxicol Lett 40:47–56

McDonough JH, Hackley BE, Cross R, Samson F, Nelson S (1983) Brain regional glucose use during soman-induced seizures. Neurotoxicology 4:203–210

McDonough JH, McLeod CG, Nipwoda T (1987) Direct microinjection of soman or VX into the amygdala produces repetitive limbic convulsions and neuropathology. Brain Res 435:123–137

McDonough JH, Jaax NK, Crowley RA, Mays MZ, Modrow HE (1989) Atropine and/or diazepam therapy protects against soman-induced neural and cardiac pathology. Fundam Appl Toxicol 13:256–276

McGurk SP, Hartgraves SL, Kelly PH, Gordon MN, Butcher LL (1987) Is ethylcholine mustard aziridinium ion a specific cholinergic neurotoxin? Neuroscience 22:215–224

McLeod CG, Singer AW, Harrington DG (1984) Acute neuropathology in soman poisoned rats. Neurotoxicology 5:53–58

Melaga WP, Howard BD (1984) Biochemical evidence that vesicles are the source of the acetylcholine released from stimulated PC12 cells. Proc Natl Acad Sci USA 81:6536–6538

Meldolesi J, Scheer H, Madeddu L, Wanke E (1986) Mechanism of action of α-latrotoxin: the presynaptic stimulatory toxin of the black widow spider venom. Trends Pharmacol Sci 7:151–155

Mellanby J, Green J (1981) How does tetanus toxin act? Neuroscience 6:281–300

Metcalf DR, Holmes JH (1969) EEG, psychological and neurological alterations in humans with organophosphorus exposure. Ann NY Acad Sci 160:357–365

Metcalf RH, Boegman RJ, Quirion R, Riopelle RJ, Ludwin SK (1987) Effect of quinolinic acid in the nucleus basalis magnocellularis on cortical high-affinity choline uptake. J Neurochem 49:639–644

Millar TJ, Ishimoto I, Boelen M, Epstein ML, Johnson CD, Morgan IG (1987) The toxic effects of ethylcholine mustard aziridinium ion on cholinergic cells in the chicken retina. J Neurosci 7:343–356

Misra UK, Nag D, Bhushan V, Ray PK (1985) Clinical and biological changes in chronically exposed organophosphorous workers. Toxicol Lett 24:187–193

Mistry JS, Abraham DJ, Hanin I (1986) Neurochemistry of aging: I. Toxins for an animal model of Alzheimer's disease. J Med Chem 29:376–380

Morio Y, Yagasaki O (1989) Effects of AF64A on cholinergic neurotransmission in the sixth abdominal ganglion of the cockroach. Comp Biochem Physiol [c] 94:121–128

Moroi-Fetters SE, Neff NH, Hadjiconstantinou M (1990) Ethylcholine aziridinium ion depletes acetylcholine and causes muscarinic receptor supersensitivity in rat retina. Neurosci Lett 109:304–308

Mounter LA (1963) Metabolism of organophosphorus anticholinesterase agents. In: Koelle GB (ed) Cholinesterases and anticholinesterases. Springer, Berlin Heidelberg New York, p 488 (Handbuch der Experimentellen Pharmakologie, vol 15)

Mouton PR, Arendash GW (1990) Atrophy of cholinergic neurons within the rat nucleus basalis magnocellularis following intracortical AF64A infusion. Neurosci Lett 111:52–57

Mouton PR, Meyer EM, Dunn AJ, Millard W, Arendash GW (1988) Induction of cortical cholinergic hypofunction and memory retention deficits through intracortical AF64A infusions. Brain Res 444:104–118

Mouton PR, Meyer EM, Arendash GW (1989) Intracortical AF64A: memory impairments and recovery from cholinergic function. Pharmacol Biochem Behav 23:841–848

Muller P, Britton RS, Seeman P (1980) The effect of long-term ethanol on brain receptors for dopamine, acetylcholine, serotonin and noradrenaline. Eur J Pharmacol 65:31–37

Nakagawa Y, Ishihara T (1988) Enhancement of neurotrophic activity in cholinergic cells by hippocampal extract prepared from colchicine-lesioned rats. Brain Res 439:11–18

Nakagawa Y, Nakamura S, Kasé Y, Noguchi T, Ishihara T (1987) Colchicine lesions in the rat hippocampus mimic the alterations of several markers in Alzheimer's disease. Brain Res 408:57–64

Nakahara N, Iga Y, Saito Y, Mizobe F, Kawanishi G (1989) Beneficial effects of FKS-508 (AF102B), a selective M_1 agonist, on the impaired working memory in AF64A-treated rats. Jpn J Pharmacol 51:539–547

Nakamura S, Nakagawa Y, Kawai M, Tohyama M, Ishihara T (1988) AF64A (ethylcholine aziridinium ion)-induced basal forebrain lesion impairs maze performance. Behav Brain Res 29:119–126

Newton MW, Jenden DJ (1986) False neurotransmitters as presynaptic probes for cholinergic mechanisms and function. Trends Pharmacol Sci 7:316–320

Nordberg A, Adolfsson R, Aquilonius SM, Marklund S, Oreland L, Winblad B (1980) Brain enzymes and acetylcholine receptors in dementia of Alzheimer type and chronic alcohol abuse. In: Amaducci L, Davison AN, Autuono P (eds) Aging of the brain and dementia. Raven, New York, p 169 (Aging, vol 13)

Noremberg K, Parsons SM (1989) Regulation of the vesamicol receptor in cholinergic synaptic vesicles by acetylcholine and an endogenous factor. J Neurochem 52:913–920

Ogura H, Yamanishi Y, Yamatsu K (1987) Effects of physostigmine on AF64A-induced impairment of learning acquisition in rats. Jpn J Pharmacol 44:498–501

Parker TH, Roberts RK, Henderson GI, Hoyumpa AM, Schmidt DE, Schenker S (1978) The effect of ethanol on cerebral regional acetylcholine concentration and utilization. Proc Soc Exp Biol Med 159:270–275

Parkinson IS, Ward MK, Kerr DNS (1981) Dialysis encephalopathy, bone disease and anemia: the aluminum syndrome during regular hemodialysis. J Clin Pathol 34:1285–1294

Pascual J, González AM, Pazos A (1989) Autoradiographic distribution of [^{3}H]hemicholinium-3 binding sites in human brain. Brain Res 505:306–310

Pazdernik TL, Cross R, Giesler M, Nelson SR, Samson FE, McDonough JH (1985) Delayed effects of soman: brain glucose use and pathology. Neurotoxicology 6:61–70

Pedder EK, Prince AK (1984) The reaction of rat brain choline acetyltransferase (ChAT) with ethylcholine mustard aziridinium ion (ECMA) and phenoxybenzamine (PB). Br J Pharmacol 83:134P

Perl DP, Brody AR (1980) Alzheimer's disease: X-ray spectrometric evidence of aluminum accumulation in neurofibrillary tangle-bearing neurons. Science 208:297–299

Perl DP, Gajdusek DC, Garruto RM, Yanagihara RT, Gibbs CJ (1982) Intraneuronal aluminum accumulation in amyotrophic lateral sclerosis and Parkinsonism-dementia of Guam. Science 217:1053–1055

Peterson GM, McGinty JF (1988) Direct neurotoxic effects of colchicine on cholinergic neurons in medial septum and striatum. Neurosci Lett 94:46–51

Phillis JW, Jiang ZG, Chelack B (1980) Effect of ethanol on acetylcholine and adenosine efflux from the in vivo rat cerebral cortex. J Pharm Pharmacol 32:871–872

Pietrzak ER, Wilce PA, Shanley BC (1990) Interaction of chronic ethanol consumption and aging on brain muscarinic cholinergic receptors. J Pharmacol Exp Ther 252:869–874

Pillar AM, Prince AK, Atterwill CK (1988) The neurotoxicity of ethylcholine mustard aziridinium (ECMA) in rat brain reaggregate cultures. Toxicology 49:115–119

Pittel Z, Fisher A, Heldman E (1987) Reversible and irreversible inhibition of high-affinity choline transport caused by ethylcholine aziridinium ion. J Neurochem 49:468–474

Pittel Z, Fisher A, Heldman E (1989) Cholinotoxicity induced by ethylcholine aziridinium ion after intracarotid and intracerebroventricular administration. Life Sci 44:1437–1448

Pope CN, Englert LF, Ho BT (1985) Passive avoidance deficits in mice following ethylcholine aziridinium chloride treatment. Pharmacol Biochem Behav 22:297–299

Potter PE, Morrison RS (1991) Basic fibroblast growth factor protects septal-hippo-campal cholinergic neurons against lesions induced by AF64A. In: Iqbal K, McLachlan DRC, Winblad B, Wisniewski HM (eds) Alzheimer's disease: Basic mechanisms, diagnosis and therapeutic strategies. John Wiley & Sons, New York, p 639

Potter PE, Hársing LG Jr, Kakucska I, Gaál G, Vizi ES (1986) Selective impairment of acetylcholine release and content in the central nervous system following intracerebroventricular administration of ethylcholine mustard aziridinium ion (AF64A) in the rat. Neurochem Int 8:199–206

Potter PE, Tedford CE, Kindel G, Hanin I (1989) Inhibition of high affinity choline transport attenuates both cholinergic and noncholinergic effects of ethylcholine aziridinium (AF64A). Brain Res 487:238–244

Raaijmakers WGM (1983) Lasting hippocampal cholinergic deficit and memory impairment in the rat. Neurosci Lett [Suppl]14:S294

Rabin RA, Wolfe DB, Dibner MD, Zahniser NR, Melchior C, Molinoff PB (1980) Effects of ethanol administration and withdrawal on neurotransmitter receptor systems in C 57 mice. J Pharmacol Exp Ther 213:491–496

Rainbow TC, Parsons B, Wieczorek CM (1984) Quantitative autoradiography of [^{3}H]-hemicholinium-3 binding sites in the rat brain. Eur J Pharmacol 102:195–196

Rawat AK (1974) Brain levels and turnover rates of presumptive neurotrans-mitters as influenced by administration and withdrawal of ethanol in mice. J Neurochem 22:915–922

Říčný J, Collier B (1986) Effect of 2(4-phenylpiperidino)cyclohexanol on acetylcholine release and subcellular distribution in rat striatal slices. J Neurochem 47:1627–1633

Ridley RM, Baker HF, Drewett B (1987) Effects of arecoline and pilocarpine on learning ability in marmosets pretreated with hemicholinium-3. Psychopharmacology 91:512–514

Robbins TW, Everitt BJ, Ryan CN, Marston HM, Jones GH, Page KJ (1989) Comparative effects of quisqualic and ibotenic acid-induced lesions of the substantia innominata and globus pallidus on the acquisition of a conditional visual discrimination: differential effects on cholinergic mechanisms. Neuroscience 28:337–352

Rogers GA, Parsons SM (1989) Inhibition of acetylcholine storage by acetylcholine analogs in vitro. Mol Pharmacol 36:333–341

Ross WCJ (1962) Biological alkylating agents. Butterworth, London

Russell RW, Macri J (1978) Some behavioural effects of suppressing choline transport by cerebroventricular injection of hemicholinium-3. Pharmacol Biochem Behav 8:399–403

Rylett RJ (1986) Choline mustard: an irreversible ligand for use in studies of choline transport mechanisms at the cholinergic nerve terminal. Can J Physiol Pharmacol 64:334–340

Rylett RJ (1988) Affinity labelling and identification of the high-affinity choline carrier from synaptic membranes of *Torpedo* electromotor nerve terminals with [^{3}H]-choline mustard. J Neurochem 51:1942–1945

Rylett RJ (1989) Synaptosomal "membrane-bound" choline acetyltransferase is most sensitive to inhibition by choline mustard. J Neurochem 52:869–875

Rylett RJ, Colhoun EH (1979) The interactions of choline mustard aziridinium ion with choline acetyltransferase. J Neurochem 32:553–558

Rylett BJ, Colhoun EH (1980) Carrier mediated inhibition of choline acetyltransferase. Life Sci 26:909–914

Rylett RJ, Colhoun EH (1984) An evaluation of irreversible inhibition of synaptosomal high-affinity choline transport by choline mustard aziridinium ion. J Neurochem 43:787–794

Rylett RJ, Colhoun EH (1985) Studies on the alkylation of choline acetyltransferase by choline mustard aziridinium ion. J Neurochem 44:1951–1954

Sandberg K, Schnaar RL, Coyle JT (1985a) Method for the quantitation and characterization of the cholinergic neurotoxin, monoethylcholine mustard aziridinium ion (AF64A). J Neurosci Methods 14:143–148

Sandberg K, Schnaar RL, McKinney M, Hanin I, Fisher A, Coyle JT (1985b) AF64A: an active site directed irreversible inhibitor of choline acetyltransferase. J Neurochem 44:439–445

Sato H, Hata Y, Tsumoto T (1987) Effects of cholinergic depletion on neuron activities in the rat visual cortex. J Neurophysiol 58:781–794

Scarth BJ, Jhamandas K, Boegman RJ, Beninger RJ, Reynolds JN (1989) Cortical muscarinic receptor function following quinolinic acid-induced lesion of the nucleus basalis magnocellularis. Exp Neurol 103:158–164

Shea TB, Clarke JF, Wheelock TR, Paskevich PA, Nixon RA (1989) Aluminum salts induce the accumulation of neurofilaments in perikarya of NB 2a/dl neuroblastoma. Brain Res 492:53–64

Shih T-M (1982) Time course effects of soman on acetylcholine and choline levels in six discrete areas of the rat brain. Psychopharmacology 78:170–175

Singer A, Jaax NK, Graham J (1987) Acute neuropathology and cardiomyopathy in soman and sarin intoxicated rats. Toxicol Lett 36:243–249

Slater P (1968) The effects of triethylcholine and hemicholinium-3 on the acetylcholine content of the rat brain. Int J Neuropharmacol 7:421–427

Smart L (1981) Hemicholinium 3-bromo mustard: a new high affinity inhibitor of sodium-dependent high affinity choline uptake. Neuroscience 6:1765–1770

Smith TL (1983) Influence of chronic ethanol consumption on muscarinic cholinergic receptors and their linkage to phospholipid metabolism in mouse synaptosomes. Neuropharmacology 22:661–663

Smyth RD, Beck H (1969) The effect of time and concentration of ethanol administration on brain acetylcholine metabolism. Arch Int Pharmacodyn 182:295–299

Sofroniew MV, Pearson RCA, Powell TPS (1987) The cholinergic nuclei of the basal forebrain of the rat: normal structure, development and experimentally induced degeneration. Brain Res 411:310–331

Spencer DG, Horvath E, Luiten P, Schuurman T, Traber J (1985) Novel approaches in the study of brain acetylcholine function: neuropharmacology, neuroanatomy and behavior. In: Traber J, Gipsen WH (eds) Senile dementia of the Alzheimer type. Springer, Berlin Heidelberg New York, p 325

Spencer PS, Nunn PB, Hugon J, Ludolph AC, Ross SM, Roy DN, Robertson RC

(1987) Guam amyotrophic lateral sclerosis-Parkinsonism-dementia linked to a plant excitant neurotoxin. Science 237:517–522

Stewart GR, Price M, Olney JM, Hartman BK, Cozzari C (1986) *N*-Methylaspartate: an effective tool for lesioning basal forebrain cholinergic neurons in the rat. Brain Res 369:377–382

Stwertka SA, Olson GL (1986) Neuropathology and amphetamine-induced turning resulting from AF64A injections into the striatum of the rat. Life Sci 38:1105–1110

Suszkiw JB, Pilar G (1976) Selective localization of a high affinity choline uptake system and its role in ACh formation in cholinergic nerve terminals. J Neurochem 26:1123–1131

Suszkiw JB, Toth G (1986) Storage and release of acetylcholine in rat cortical synaptosomes: effects of D,L-2-(4-phenylpiperidino)cyclohexanol (AH 5183). Brain Res 386:371–378

Syvälahti EKG, Hietala J, Röyttä N, Grönroos J (1988) Decrease in the number of rat brain dopamine and muscarinic receptors after chronic alcohol intake. Pharmacol Toxicol 62:210–212

Tabakoff B, Munoz-Markus M, Fields JZ (1979) Chronic ethanol feeding produces an increase in muscarinic cholinergic receptors in mouse brain. Life Sci 25:2173–2180

Tagari PC, Maysinger D, Cuello AC (1986) Hemicholinium mustard derivatives: preliminary assessment of cholinergic neurotoxicity. Neurochem Res 11:1091–1102

Tandon P, Harry GJ, Tilson HA (1989) Colchicine-induced alterations in receptor-stimulated phosphoinositide hydrolysis in the rat hippocampus. Brain Res 477:308–313

Tateishi N, Takano Y, Honda K, Yamada K, Kamiya Y, Kamiya H (1987) Effects of intrahippocampal injections of the cholinergic neurotoxin AF64A on presynaptic cholinergic markers and on passive avoidance response in the rat. Clin Exp Pharmacol Physiol 14:611–618

Tedford CE, Reed D, Bhattacharyya B, Bhalla P, Cannon JG, Long JP (1986) Evaluation of 4-methylpiperidine analogs of hemicholinium-3. Eur J Pharmacol 128:231–239

Tedford CE, Schott MJ, Flynn JR, Cannon JG, Long JP (1987) A-4, a bis tertiary amine derivative of hemicholinium-3 produces in vivo reduction of acetylcholine in rat brain regions. J Pharmacol Exp Ther 240:476–485

Tedford CE, Lorens SA, Corey JC, Lokhorst D, Kindel G, Wülfert E, Hanin I (1990) Behavioural and neurochemical effects of AF64A in young and old Fischer 344 male rats. In: Yoshida M, Fisher A, Nagatsu T (eds) Alzheimer's and Parkinson's diseases: basic and therapeutic strategies, vol 2. Plenum, New York, pp 321–334

Thal LJ, Dokla CPJ, Armstrong DM (1988) Nucleus basalis magnocellularis lesions: lack of biochemical and immunocytochemical recovery and effect of cholinesterase inhibitors on passive avoidance. Behav Neurosci 102:852–860

Tilson HA, Peterson NJ (1987) Colchicine as an investigative tool in neurobiology. Toxicology 46:159–173

Tilson HA, Harry GJ, McLamb RL, Peterson NJ, Rogers BC, Pediaditakis P, Ali SF (1988) Role of dentate gyrus granule cells in retention of a radial arm maze task and sensitivity of rats to cholinergic drugs. Behav Neurosci 102:835–842

Tilson HA, Schwartz RD, Ali SF, McLamb RL (1989) Colchicine administered into the area of nucleus basalis decreases cortical nicotinic cholinergic receptors labelled by [^{3}H]-acetylcholine. Neuropharmacology 28:855–861

Tonnaer JADM, Lammers AJJC, Wieringa JH, Steinbusch HWM (1986) Immunohistochemical evidence for degeneration of cholinergic neurons in the forebrain of the rat following injection of AF64A-picrylsulfonate into the dorsal hippocampus. Brain Res 370:200–203

Tuček S (1985) Regulation of acetylcholine synthesis in the brain. J Neurochem 44:11–24

Turski K, Ikonomidou C, Turski WA, Bortolotto ZA, Cavalheiro EA (1989) Cholinergic mechanisms and epileptogenesis. The seizures induced by pilocarpine: a novel experimental model of intractable epilepsy. Synapse 3:154–171

Uney JB, Marchbanks RM (1987) Specificity of ethylcholine mustard aziridinium as an irreversible inhibitor of choline transport in cholinergic and noncholinergic tissue. J Neurochem 48:1673–1676

Veronesi B, Jones K, Pope C (1990) The neurotoxicity of subchronic acetylcholinesterase (AChE) inhibition in rat hippocampus. Toxicol Appl Pharmacol 104:440–456

Vickroy TW, Roeske WR, Yamamura HI (1984) Sodium-dependent high affinity binding of [^{3}H]-hemicholinium-3 in rat brain: a potentially selective marker for presynaptic cholinergic sites. Life Sci 35:2335–2343

Vickroy TW, Watson M, Leventer SM, Roeske WR, Hanin I, Yamamura HI (1985) Regional differences in ethylcholine mustard aziridinium ion (AF64A)-induced deficits in presynaptic cholinergic markers for the rat central nervous system. J Pharmacol Exp Ther 235:577–582

Villani L, Contestabile A, Migani P, Poli A, Fonnum F (1986) Ultrastructural and neurochemical effects of the presumed cholinergic toxin AF64A in the rat interpeduncular nucleus. Brain Res 379:223–231

Villani L, Bissoli R, Garolini S, Guarneri T, Battistini S, Saverino O, Contestabile A (1988) Effect of AF64A on the cholinergic systems of the retina and optic tectum of goldfish. Exp Brain Res 70:455–462

Walsh TJ, Tilson HA, DeHaven DL, Mailman RB, Fisher A, Hanin I (1984) AF64A, a cholinergic neurotoxin, selectively depletes acetylcholine in hippocampus and cortex and produces long-term passive avoidance and radial-arm maze deficits in the rat. Brain Res 321:91–102

Watson M, Vickroy TW, Fibiger HC, Roeske W, Yamamura HI (1985) Effects of bilateral ibotenate-induced lesions of the nucleus basalis magnocellularis upon selective cholinergic biochemical markers in the rat anterior cerebral cortex. Brain Res 346:387–391

Weise C, Kreienkamp H-J, Raba R, Pedak A, Aaviksaar A, Hucho G (1990) Anionic subsites of the acetylcholinesterase from Torpedo californica: affinity labelling with the cationic reagent N,N-dimethyl-2-phenylaziridinium. EMBO J 9:3885–3888

Whittaker VP (1990) The contribution of drugs and toxins to understanding of cholinergic function. Trends Pharmacol Sci 11:8–13

Wigell AH, Overstreet DH (1984) Acquisition of behaviorally augmented tolerance to ethanol and its relationship to muscarinic receptors. Psychopharmacology 83:88–92

Wilson IB, Ginsburg S (1955) A powerful reactivator of alkylphosphate-inhibited acetylcholinesterase. Biochim Biophys Acta 18:168–170

Wisniewski HM, Sturman JA, Shek JW, Iqbal K (1985) Aluminum and the central nervous system. J Environ Pathol Toxicol Oncol 6:1–8

Witt ED, Mantione CR, Hanin I (1986) Sex differences in muscarinic receptor binding after chronic ethanol administration in the rat. Psychopharmacology 90:537–542

Wurtman RJ, Blusztajn JK, Maire J-C (1985) Autocannibalism of choline-containing membrane phospholipids in the pathogenesis of Alzheimer's disease – a hypothesis. Neurochem Int 2:369–372

Yamamura HI, Snyder SH (1973) High affinity transport of choline into synaptosomes of rat brain. J Neurochem 21:1355–1375

Yates CM, Simpson J, Russell D, Gordon A (1980) Cholinergic enzymes in neurofibrillary degeneration produced by aluminum. Brain Res 197:269–274

Zieher LM, Jaim-Etcheverry G (1980) Neurotoxicity of N-(2-chloroethyl)-N-ethyl-2-bromobenzylamine hydrochloride (DSP4) on noradrenergic neurons is mimicked by its cyclic aziridinium derivative. Eur J Pharmacol 65:249–256
Zubenko GS, Hanin I (1989) Cholinergic and noradrenergic toxicity of intraventricular aluminum chloride in the rat hippocampus. Brain Res 498:381–384

Mechanisms of 1-Methyl-4-Phenyl-1,2,3,6-Tetrahydropyridine Induced Destruction of Dopaminergic Neurons

I.J. KOPIN

A. Introduction

Inadvertent self-administration by drug abusers of 1-methyl-4-phenyl-1,2,3,6-tetrahydropyridine (MPTP), a side reaction product formed during the illicit synthesis of a synthetic narcotic analogue of meperidine, led to the discovery that MPTP produces in humans a movement disorder which closely resembles Parkinson's disease (DAVIS et al. 1979; LANGSTON et al. 1983). MPTP administered to monkeys was found to result in depletion of dopamine in the caudate-putamen, destruction of dopaminergic neurons in the substantia nigra, and appearance of a syndrome virtually identical to advanced Parkinson's disease in humans (BURNS et al. 1983; LANGSTON et al. 1984a; JENNER et al. 1984). Although rats, guinea pigs, and rabbits are relatively resistant to the toxic effects of MPTP (CHIUEH et al. 1984a,b; BOYCE et al. 1984), high doses were shown to be neurotoxic in mice (HALLMAN et al. 1984; HEIKKILA et al. 1984a). MYTILINEOU and COHEN (1984) demonstrated that MPTP was toxic to rat dopaminergic neurons in vitro. Vulnerability to MPTP toxicity was found to vary with age and sex, as well as with species or even strains within species; the degree of specificity of tissue damage was also found to alter among species. These differences in vulnerability to the toxin provide a challenge to understanding the mechanisms involved in producing damage to the target cells. The close resemblance of MPTP toxicity to Parkinson's disease has stimulated enormous interest in determining whether there is any relationship between the neurotoxic effects of MPTP and the pathogenesis of the spontaneously occurring movement disorder in humans.

To understand fully the mechanisms involved in producing the effects of toxin, it is necessary to determine the factors which influence its distribution and metabolism, which of its molecular interactions disturb metabolic processes, and the physiological responses to the disturbances produced by the toxin. Metabolism may be responsible for bioactivation of a protoxin into the active agent(s) or may inactivate the toxin. Limited accessibility of a toxin (e.g., inability to penetrate the BBB) to a potential target may be protective, whereas selective accumulation of the toxin can enhance toxicity and result in highly selective cellular damage. The biochemical interactions with the toxic agent(s) may inactivate essential enzymes, disrupt membrane integrity, transport, or ion sequestration, or result in the production of free radicals which attack essential cellular components. Survival of the targeted cells depends on the ability of cellular mechanisms to limit damage, repair injured cellular components, or replace irreversibly inhibited enzymes, receptors, ion channels, etc. It is the purpose of this chapter to review the current status of our understanding of the mechanisms of the neurotoxicity of MPTP and related substances.

B. Bioactivation of MPTP to MPP$^+$ by Monoamine Oxidase

After administration of [^{14}C]-MPTP to monkeys, MARKEY et al. (1984) found that its oxidized derivative, N-methyl-4-phenylpyridinium (MPP$^+$), accumulated in the brain. MPTP appeared to have entered the brain and been rapidly converted to MPP$^+$. Whereas MPP$^+$ is retained in the brains of primates, it is rapidly cleared from brains of rodents (JOHNNESSEN et al. 1985). In monkeys, high concentrations of MPP$^+$ were found in areas of brain (e.g., n. accumbens) which escape extensive destruction, indicating that the generation and accumulation of MPP$^+$ is not sufficient to ensure cell death. In squirrel monkeys, MPP$^+$ selectively accumulates in the substantia nigra, whereas in other areas of brain, its concentration declines (IRWIN and LANGSTON 1985).

In rat brain preparations, conversion of MPTP to MPP$^+$ was found to be inhibited by pargyline, a nonspecific monamine oxidase (MAO) inhibitor, and by deprenyl, a specific inhibitor of MAO type B (MAO-B), whereas MPP$^+$ formation was not affected significantly by a specific inhibitor of MAO type A (MAO-A), clorgyline (CHIBA et al. 1984). These results indicate that mainly MAO-B is responsible for metabolic oxidation of MPTP to MPP$^+$, consistent with the report by SALACH et al. (1984) that MPTP is metabolized by MAO-A at a much slower rate than by MAO-B. In monkeys, both the accumulation of MPP$^+$ and the neurotoxicity of MPTP are prevented by pretreatment with pargyline or deprenyl (MARKEY et al. 1984; LANGSTON et al. 1984b; COHEN et al. 1985). Similarly, in mice, administration of either nonselective or MAO-B selective MAO inhibitors prevents MPTP-induced depletion of striatal dopamine and the MPTP-induced diminution in the capacity of striatal synaptosomes to take up [^{3}H]-dopamine, whereas clorgyline (MAO-A selective) has no such protective effects (HEIKKILA et al. 1984b). Since deprenyl fails to protect mice against toxicity of intracerebroventricularly injected MPP$^+$, it may be concluded that MPTP itself is not toxic and that MAO-B-mediated oxidation of MPTP to MPP$^+$ is necessary to achieve toxicity; the rate of oxidation by MAO-A appears to be insufficient. Furthermore, it is clear that MPP$^+$ is the toxic agent, i.e., toxicity is not a result of the process by which MPP$^+$ is formed. MPP$^+$ is toxic to dopaminergic neurons in mesencephalic explants in culture (MYTILINEOU and COHEN 1984; MYTILINEOU et al. 1985; SANCHEZ-RAMOS et al. 1986), to hepatocytes in culture (DI MONTE et al. 1986a,b), and when administered directly into the substantia nigra, MPP$^+$ destroys dopaminergic neurons (HEIKKILA et al. 1985b).

C. Toxicity of MPTP Analogues

Studies with analogues of MPTP have supported the conclusion that formation of MPP$^+$-like compounds is necessary for toxicity. Analogues of MPTP which are poor substrates for MAO are also relatively nontoxic, although their pyridinium derivatives may be toxic when administered directly in the brain or when examined in vitro. The N-methyl group of MPTP appears to be essential for toxicity. Those analogues of MPTP and MPP$^+$ which lack this methyl group, 4-phenyl-1,2,3,6-tetrahydropyridine (PTP) and 4-phenylpyridine (4PP), as well as 4-phenylpiperidine, even when continuously infused directly into the rat substantia nigra for 4 days, appear to be nontoxic (BRADBURY et al. 1985). PTP caused only a small decrease in striatal dopamine level, and 3,4-dihydroxyphenylacetic acid (DOPAC) and the other analogues failed to modify the striatal content of dopamine or its metabolities. In C57 black mice injected repeatedly with maximal tolerated doses, 4-PP and 4-PTP failed to cause any reduction in striatal dopamine concentration (PERRY et al. 1987a). In fact, 4PP administered prior to the toxin provided significant protection against the dopamine-depleting effects of MPTP (IRWIN et al. 1987a). In vitro studies showed that 4PP may inhibit the biotransformation of MPTP to MPP$^+$, presumably because 4PP is a substrate and an inhibitor of MAO (see below).

Replacement of the N-methyl moiety of MPTP or MPP$^+$ with any of a variety of alkyl groups also diminishes or prevents toxicity. Thus, even 1-ethyl and 1-propyl derivatives of MPTP (EPTP and PPTP) were found to lack toxicity in mice (HEIKKILA et al. 1985a), although the ethyl analogue of MPP$^+$ (EPP$^+$) is toxic to dopaminergic neurons in culture (SANCHEZ-RAMOS et al. 1988). EPTP is a poor substrate for MAO, and PPTP is not oxidized at all (YOUNGSTER et al. 1989a). Similarly, although 4'-Me-MPP$^+$ is a potent toxin in vitro, 4'-Me-MPTP, which is a poor substrate for MAO, is nontoxic (YOUNGSTER et al. 1989b,c). Complete oxidation to the pyridinium derivative appears to be important for toxicity since 1,3,3-trimethyl-4-phenyl-2,3-dihydropyridinum cation (3,3-dimethyl-MPDP$^+$), a nonoxidizable analogue of MPDP$^+$, infused directly into the substantia nigra was found to be 2–3 orders of magnitude less toxic than MPP$^+$ or MPDP$^+$, which can be oxidized to MPP$^+$ (SAYRE et al. 1986). Infusion of the corresponding analogue of MPTP, 3,3-dimethyl-MPTP, had no apparent effect on striatal dopamine content (HARIK et al. 1987). Lack of 3,3-dimethyl-MPTP toxicity appears to be due to its inefficient conversion to the oxidized form, as well as to the relative impotency of 3,3-dimethyl-MPDP$^+$. The relative impotency of the oxidized molecule could be a consequence of other factors which are important for the manifestation of toxicity (see below). The 2,2-dimethyl analogue of MPDP$^+$, which is also nonoxidizable, was only slightly more toxic, presumably because it can form an nonionized species that is more lipid soluble and can more easily penetrate into the neuron. Both dimethyl analogues of MPDP$^+$ are about two orders of magnitude less potent than MPP$^+$.

Although activation of MPTP involves primarily MAO-B, some analogues of MPTP are substrates for MAO-A as well as MAO-B. HEIKKILA et al. (1988) provided the first evidence that MAO-A can support conversion of some MPTP analogues to toxic pyridinium derivatives. This led to the examination of a number of analogues of MPTP for their capacity to be oxidized by the two MAO types in crude mitochondrial preparations from mouse brain (YOUNGSTER et al. 1989a). Oxidation rates after pretreatment with deprenyl or clorgyline compared with oxidation rates in the absence of any MAO inhibitor were used to determine the rates of oxidation by MAO-A and MAO-B, respectively; also, the formation of the corresponding pyrimidines were determined by high pressure liquid chromatography (HPLC). They found that substituents on the 2'- or 3'-positions (on the

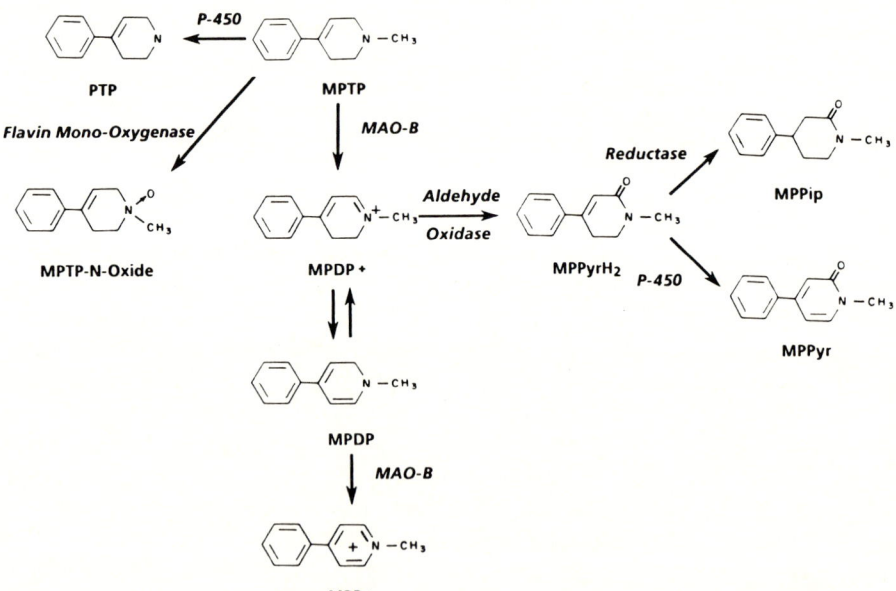

Fig. 1. Alternative metabolic pathways for 1-methyl-4-phenyl-1,2,3,6-tetrahydro-pyridine (*MPTP*). *PTP*, 4-phenyl-1,2,3,6-tetrahydropyridine; *MAO*, monoamine oxidase; *MPP$^+$*, *N*-methyl-4-phenylpyridinium; *MPDP*, 1-methyl-4-phenyl-2,3-dihydropyridinium; *MPPyrH$_2$*, 1-methyl-4-phenyl-5,6-dihydro-2(1*H*)-pyridinone; *MPPip*, 1-methyl-4-phenyl-2-piperidinone

4-phenyl group) of MPTP generally enhanced the capacity to be oxidized by MAO-A. Substituents on the 2'-position did not affect the rate of oxidation by MAO-B, except the markedly enhanced (fourfold) rate of oxidation with a 2'-methyl substituent (2'-methyl-MPTP). In mice, this compound is more toxic than MPTP (SONSALLA et al. 1987), but in marmosets it is considerably less toxic than MPTP (ROSE et al. 1990). Thus, species variations in factors other than MAO activity must be important determinants of the toxicity of MPTP and its analogues. Substituents in the 3'position either did not affect the rate of oxidation by MAO-B or decreased the rate. Significant decreases in MAO-B-catalyzed oxidation were found with 3'-methoxy- (3'-CH$_3$-MPTP) or 3'-halogen-(3'-F-, 3'-Cl-, or 3'-Br-MPTP) substituents; these resulted in 85% and 35%−65% reductions, respectively. Dihydropyridinium intermediates were formed by MAO-catalyzed oxidation of all of the MAO substrates, and the corresponding pyridinium species was the final oxidation product, except for the dihydropyridinium of 1-methyl-4-benzyl-1,2,3,6-tetrahydropyridine (M4BzTP) which was oxidized to a compound other than the corresponding pyridinium (M4BzP$^+$). Except for M4BzTP, the compounds which were substrates for MAO were neurotoxic, and their neurotoxicity was blocked by inhibiting MAO-B, MAO-A, or both together. However, not all tetrahydropyridines which were substrates for MAO-B or MAO-A were found to be neurotoxic to mice in vivo. Thus, the capacity of an MPTP analogue to be oxidized by MAO to a pyridinium appears to be necessary, but not sufficient, for in vivo neurotoxicity.

D. Alternative Routes for MPTP Metabolism

Whereas in the brain MPTP is metabolized primarily to MPP$^+$ by MOA-B, in other tissues there are several alternative metabolic routes (Fig. 1). Incubation of MPTP with hepatic microsomes from rat, mouse, or rabbit liver results in flavin monooxygenase-mediated formation of MPTP-N-oxide and cytochrome P-450-mediated N-demethylation of MPTP (CASHMAN and ZIEGLER 1986; WEISSMAN et al. 1985; CHIBA et al. 1988). Neither of these MPTP metabolites are toxic to dopaminergic neurons in intact mice (CHIBA et al. 1988). In intact liver cells, MAO (or possibly cytochrome P-450) mediated oxidation of MPTP to 1-methyl-4-phenyl-2,3,-dihydropyridinium (MPDP$^+$) appears to predominate, but further oxidation of this intermediate is mainly mediated sequentially by aldehyde oxidase to 1-methyl-4-phenyl-5,6-dihydro-2(1H)-pyridinone (MPPyrH$_2$) and by cytochrome P-450 to 1-methyl-4-phenyl-2(1H)-pyridinone (MPPyr) as the final product (WU et al. 1988). After administration of MPTP to rats, ARORA et al. (1988) identified three lactams in the tissues: MPPyrH$_2$, MPPyr, and a previously unreported metabolite, 1-methyl-4-phenyl-2-piperidinone (MPPip), the reduction product of MPPyrH$_2$. Whereas MPP$^+$ was the predominant metabolite in the brain, in the liver the lactam metabolites predominated;

there were only minimal amounts of the N-oxide and N-demethylated metabolites of MPTP. These lactam metabolites are nontoxic and appear to be mechanisms for detoxification of MPTP and MPDP$^+$ (YANG et al. 1988). CHIBA et al. (1988, 1990), however, showed that administration to mice of an alternative substrate for flavin monooxygenase increases the levels of MPTP, MPDP$^+$, and MPP$^+$ found in the brain after administration of MPTP, suggesting that in mice flavin monooxygenase, presumably by MPTP-N-oxide formation, is responsible for metabolism of a significant portion of the administered MPTP.

E. Mechanism of MPTP Oxidation by Monoamine Oxidase

The mechanisms involved in bioactivation of MPTP by MAO-B have received considerable attention. The initial oxidation results in the formation of an unstable dihydropyridinium derivative, MPDP$^+$, which can be captured by its reaction with cyanide ion (GESSNER et al. 1985; PETERSON et al. 1985). On the basis of the diminished rate of oxidation of 6-[^{2}H]-MPTP, it was suggested that oxidation at the 6-position is the rate-limiting step in MPDP$^+$ formation (GESSNER et al. 1986). In the absence of cyanide, MPDP$^+$ undergoes slow disproportionation to form MPTP and MPP$^+$, but in vivo most MPP$^+$ is probably formed by MAO-mediated oxidation of MPDP$^+$ (TREVOR et al. 1987; KRUEGER et al. 1990a). In stop-flow kinetic studies, RAMSAY et al. (1987a) demonstrated that MPTP and MPDP$^+$ can form ternary complexes with reduced MAO-B and oxygen and that the overall rate of oxidation of MPTP is dominated by reduction. During oxidation of MPTP and MPDP$^+$, several reactive intermediates are formed (PETERSON et al. 1985). These may be responsible for the pargyline- and deprenyl-sensitive irreversible binding of [^{3}H]-MPTP metabolites to rat brain protein (CORSINI et al. 1986a,b); there is a similar covalent binding in mouse and rat brain tissue in vivo (CORSINI et al. 1988). The binding sites may not be confined to MAO-B, since total binding in the liver did not increase with time, while it did in mouse brain. Covalent binding of a product of MPTP oxidation to both MAO-A and MAO-B is consistent with the progressive irreversible "suicide" or "mechanism-based" inhibition of both forms of MAO during incubation with MPTP (SALACH et al. 1984).

I. Inhibition of Monoamine Oxidase by MPTP

After administration of MPTP to monkeys, there is a rapid decrease in the cerebrospinal fluid (CSF) levels of 5-hydroxyindoleacetic acid (5-HIAA), the deamination product of serotonin, 3-methoxy-4-hydroxyphenylglycol (MHPG), the major metabolite of norepinephrine, as well as homovanillic acid (HVA), the deaminated metabolite of dopamine (BURNS et al. 1983), consistent with MAO inhibition. CSF levels of 5-HIAA and MHPG return to normal, whereas HVA levels remain depressed. Bocchetta et al. (1985) also showed that MPTP treatment diminishes oxidative metabolism of dopamine in vivo and in vitro, consistent with MAO inhibition.

Parsons and Rainbow (1984) demonstrated that MPTP was nearly as potent an inhibitor of both MAO-A and MAO-B as was pargyline. Furthermore, the distribution of [^{3}H]-MPTP binding sites in rat brain slices corresponds to those of [^{3}H]pargyline, suggesting that the toxin is associated with MAO (Rainbow et al. 1985). As indicated above, Salach et al. (1984) proposed that MPTP was a substrate and suicide inhibitor of both MAO-A and MAO-B. They showed that pure MAO-B from bovine liver oxidizes MPTP 38% as rapidly as benzylamine with a comparable K_m value and that MAO-A from human placenta catalyzes the same reaction at about 12% of the rate of oxidation of kynurenine. Inactivation of both enzymes progressed with first-order reaction kinetics. Tipton et al. (1986) estimated that the rate of oxidation of MPTP by MAO-B was about 17 000 times as rapid as the rate of irreversible inhibition of the enzyme. MAO-A inactivation proceeded at a slower rate, and its inhibition appeared to be largely reversible.

II. Cellular Localization of Monoamine Oxidases

MAO-A and MAO-B are distributed differently throughout the body tissues and in brain cells. Using monoclonal antibodies specific for MAO-A and MAO-B, respectively, Westlund et al. (1985) showed that MAO-A appeared to be localized to regions of brain containing catecholaminergic neurons (e.g., substantia nigra, locus coeruleus, and periventricular region of the hypothalamus), whereas MAO-B immunoreactivity was found primarily in serotonin regions (e.g., n. raphe dorsalis). After unilateral stereotaxic injection of kainic acid to produce selective degeneration of striatal neurons and subsequent proliferation of astrocytes, Francis et al. (1985) demonstrated a persistent small (15%–20%) decrease in MAO-A activity, whereas MAO-B activity decreased initially by 25% and then increased to more than twice the control value by 54 days after the lesions. The results obtained in these two studies are consistent with localization of immunoreactivity to MAO-B in both serotonergic neurons and astrocytes (Levitt et al. 1982).

In human brain examined by biochemical and immunohistological methods, MAO-A was found in microvessels and neurons of the locus coeruleus, whereas MAO-B was present predominantly in neuronal (presumably serotonergic) cells of the dorsal raphe nucleus (Konradi et al. 1987). In this study there was no immunohistologically detectable MAO in neurons of the substantia nigra. Astrocytes which appeared to express both subtypes of MAO were observed in all cortical areas examined and throughout the brain stem, including the substantia nigra, caudate nucleus, and putamen. MAO activity in several brain areas seemed to occur mostly in astrocytes rather than in neurons. Cultured astrocytes rapidly convert MPTP to MPP$^+$, which escapes into the culture medium and attains concentrations within the toxic range for neurons (Ransom et al. 1987). Pargyline markedly reduces this conversion of MPTP to MPP$^+$ consistent with MAO-mediated oxidation. Similar findings were reported by Schinelli et al. (1988) in cultured mesencephalic glial cells from mouse embryos. When glia and neurons remain in contact in culture, the concentration in the medium need not attain toxic levels in order for toxic effects to become manifest, presumably because astroglial-neuron interaction (perhaps by proximity) enhances exposure of the neurons to MPP$^+$ (Marini et al. 1989). Escape of positively charged MPP$^+$ from astrocytes might be unexpected, but Reinhard et al. (1990b), using molecular orbital calculations, found that the positive charge is highly delocalized throughout the pyridinium ring, making the compound less polar; this might explain its relative lipid solubility (Riachi et al. 1989) and its ability to diffuse slowly through lipid membranes.

Immunoreactivity to MAO-B is present in both neural and nonneural elements in many areas of cat brain, including the hypothalamus raphe system, dorsal tegmental nucleus, locus coeruleus, etc. (Schneider and Markham 1987). As in human brain, neurons in the cat substantia nigra pars compacta and ventral tegmental areas did not appear to contain MAO-B. Glial cells (astrocytes) stained positively for MAO-B in many regions; there was a significantly greater number of MAO-B-positive glial cells in the substantia nigra pars compacta than in other adjacent dopaminergic regions. Whereas pretreatment of mice with fluoxetine (a serotonergic uptake inhibitor) attenuated MPTP neurotoxicity, intrastriatal MPP$^+$ toxicity was unaffected (Brooks et al. 1988, 1989). Destruction of serotonergic neurons with 5,7-dihydroxytryptamine failed to diminish MPTP neurotoxicity, but even in the absence of serotonergic neurons, pretreatment with fluoxetine reduced dopaminergic neurotoxicity, suggesting that other sites (astrocytes) are a principle site of MPP$^+$ production. When MPTP was used as the MAO substrate to identify sites in mouse brain which can generate hydrogen peroxide, Vincent (1989) found that serotonergic, histaminergic, and noradrenergic, but not dopaminergic, neurons appeared to contain MAO activity. Since clorgyline inhibited the apparent MAO activity in noradrenergic neurons, MAO-A appears to have generated the hydrogen peroxide formed in these cells. In serotonergic and histaminergic neurons, MAO activity was abolished by deprenyl, indicating that MAO-B was the enzyme responsible. Using the astroglia-specific toxin L-α-aminoadipic acid (L-α-AA) and fluorescent retrograde axonal tracing, Takada et al. (1990) showed in rats that selective astroglial ablation protected against neurotoxicity of locally injected MPTP. The preventive action of L-α-AA was considerably reduced 3 days after its injection into the substantia nigra compacta, whereas at 7 days, presumably as a result of gliosis, nigrostriatal cell loss was enhanced rather than attenuated. Thus, in rats, the observations are consistent with the importance of astroglia in generating MPP$^+$ from MPTP to cause neurotoxicity, whereas in mice there appears to be a possibility that serotonergic or histaminergic neurons might be important sites for MPP$^+$ formation.

Species differences in the toxicity of MPTP have been attributed, at least in part, to variations in the ability of MPTP to penetrate into the brain as a result of its metabolism by endothelial MAO to MPP$^+$. Although rats are very resistant to toxicity after its systemic administration, MPTP causes selective dopaminergic neurotoxicity when infused directly into rat substantia nigra. KALARIA et al. (1987) attributed this to a high level of MAO and/or other enzymes in rat brain capillaries which prevent MPTP from reaching its neuronal sites of toxicity. They assessed MAO activity in isolated cerebral microvessels of humans, rats, and mice by measuring the specific binding of [^{3}H]-pargyline and by estimating the rates of MPTP and benzylamine oxidation. [^{3}H]-Pargyline binding to rat cerebral microvessels was about 10-fold higher than to human or mouse microvessels, and MPTP oxidation by rat brain microvessels was about 30-fold greater than by human microvessels; mouse microvessels yielded intermediate values. Furthermore, RIACHI et al. (1988, 1989) showed that toxicity of systemically administered MPTP is correlated with its penetration into brains and inversely related to the formation of metabolites in the brains and livers of rats and of two strains of mice. RIACHI et al. (1989) later reported, however, that MPTP and butanol, both of which have high octanol/water partition coefficients, are almost completely extracted by all regions of the brain on the first pass. MPTP is retained (with a rate constant of about $0.1\,\mathrm{min}^{-1}$ for its decline in concentration), whereas butanol is cleared much more rapidly (rate constant of $1.2\,\mathrm{min}^{-1}$). Since pargyline did not affect the rate constant for initial efflux of the MPTP tracer, retention in the brain was not due to its rapid metabolism by MAO to MPP$^+$. Thus, it appears that exclusion of MPTP from the brain by a metabolic barrier of endothelial MAO does not play a major role in accounting for species differences in vulnerability to MPTP.

III. Uptake of MPP$^+$

Since MAO-B is located primarily in cells different from those which are target sites of MPTP toxicity, the toxic oxidation product MPP$^+$ must be transferred from its site of formation to its site of action. As indicated above, MPP$^+$ formed in astrocytes or serotonergic neurons, because of its relative lipid solubility, can reach the extracellular space. JAVITCH and SNYDER (1984) showed that synaptosomes prepared from rat striatum avidly accumulate [^{3}H]-MPP$^+$, but not [^{3}H]-MPTP, and that potencies of several drugs in blocking dopamine uptake were highly correlated with the extent of their blocking uptake of MPP$^+$. Similarly, the potencies of drugs in blocking uptake of [^{3}H]-norepinephrine by cortical synaptosomes was correlated with the extent of their blocking uptake of $[^3H]$-MPP$^+$ (JAVITCH et al. 1985). Furthermore, pretreatment of mice with mazindol, which blocks uptake of dopamine and of MPP$^+$, prevents MPTP-induced damage to nigrostriatal neurons. Amfonelic acid, a specific inhibitor of dopamine uptake, also prevents MPTP-induced destruction of dopaminergic neurons in mouse striatum, whereas selective norepinephrine uptake inhibitors protect against norepinephrine depletion by MPTP without affecting its action on the dopaminergic neurons in the striatum (FULLER et al. 1985; SUNDSTROM and JONSSON 1985; PILEBLAD and CARLSSON 1985; MAYER et al. 1986). Similarly, nomifensine, another dopamine uptake inhibitor, completely prevented MPTP toxicity, whereas combined treatment with drugs which inhibit uptake of norepinephrine and serotonin had no effect on dopamine neurotoxicity (MELAMED et al. 1985). Inhibition of MPTP toxicity by drugs which block uptake of dopamine has been more difficult to demonstrate in primates. SCHULTZ et al. (1986) showed that nomifensine effectively blocked MPTP neurotoxicity in primates only when administered both before and after toxin administration; best results required giving the uptake blocker for weeks after MPTP treatment, consistent with the long persistence of MPP$^+$ in the brain of monkeys treated with MPTP (JOHANNESEN et al. 1985). Thus, after its formation in MAO-B-containing astrocytes or serotonergic neurons, MPP$^+$ escapes into the intracellular space and is concentrated in dopaminergic neurons. The selective uptake of MPP$^+$ by catecholaminergic cells may account in large part for target sites of MPTP toxicity in mice but does not explain the marked vulnerability of nigrostriatal dopaminergic neurons in primates.

F. Molecular Bases for MPP$^+$ Toxicity

When the structure of the toxic product of MPTP was identified as MPP$^+$, which had been previously known as a potential herbicide (Cyperquat, Gulf Oil Company), attention focussed on the known mechanisms of toxicity of a related herbicide, paraquat (1,1'-dimethyl-4,4'bipyridinium). Paraquat toxicity, which affects primarily the lung, liver and kidney, is a consequence of the generation of superoxide anions (O_2^-) by redox cycling (BUS et al. 1976). Formation of free radicals is a well-known and important mechanism for toxic damage, and the several cellular defence mechanisms which effectively dispose of highly reactive transient free radicals have been well described (FREEMAN and CRAPO 1982; TRUSH et al. 1982). Superoxide dismutase converts superoxide radicals to hydrogen peroxide and water; catalase rapidly decomposes hydrogen peroxide to water and oxygen. If these are allowed to accumulate, the metal (e.g., Fe^{2+}) catalyzed reaction of these two oxygen metabolites results in the formation of the hydroxyl radical (OH·), which is a powerful oxidant. Soluble reducing substances (glutathione, α-tocopherol, ascorbic acid, uric acid, etc.) can prevent accumulation of these reactive molecules, but if the protective mechanisms fail and reducing substances are depleted, damaging interactions occur with an array of sensitive biomolecules, including membrane lipids (peroxidation), DNA, proteins, cofactors, etc. The consequent disruptions in cellular metabolic processes and transport mechanisms result in cell death.

A role for redox cycling in MPP$^+$ toxicity has been supported only by indirect evidence. Transition metals (such as Fe^{2+}) involved in free radical toxicity were implicated in the vulnerability of the substantia nigra to MPTP toxicity (Poirier et al. 1986). MPTP toxicity was reported to be diminished by treatment with one or several antioxidants administered to mice before, during, and after giving the toxin. Perry et al. (1985) partially prevented the depletion of striatal dopamine 1 month after the administration of a single dose of MPTP by treatment with α-tocopherol, β-carotene, L-ascorbic acid, or N-acetylcysteine daily for 5 days beginning 2 days before the toxin dose. Wagner et al. (1985) reported that ascorbic acid reduces dopamine depletion induced by amphetamine as well as by MPP$^+$, and Yong et al. (1986) found that MPTP lowers mouse brainstem glutathione levels; the levels were normalized by antioxidant treatment. Furthermore, pretreatment of mice with diethyl dithiocarbamate (DDC), which chelates copper and inhibits superoxide dismutase, potentiates the toxicity of both paraquat (Goldstein et al. 1979) and MPTP (Corsini et al. 1985). Although formation of the free radical, MPP·, from MPP$^+$ could not be demonstrated by electron spin resonance under conditions in which paraquat free radicals were apparent, by using a spin-trapping technique, Sinha et al. (1986) did find significant stimulation by MPP$^+$ of superoxide and hydroxyl radical formation. During incubation of NADPH-cytochrome P-450 reductase and NADPH in the presence of oxygen, MPP$^+$ stimulated formation of these radicals, but the stimulation was less than that from paraquat. Finally, rats treated systemically with MPTP showed paraquat-like lung toxicity and had increased plasma levels of glutathione disulfide (Johannessen et al. 1986).

Regardless of this indirect evidence supporting a role for MPP$^+$ redox cycling with free radical formation, it appears unlikely that this mechanism is responsible for the toxic effects of MPTP and MPP$^+$. Neither Baldessarini et al. (1986) nor Martinovits et al. (1986) could demonstrate at 7 or 30 days any antioxidant protection from MPTP toxicity in mice. Maintaining glutathione levels did not protect mice from MPTP neurotoxicity (Perry et al. 1986), lipid peroxidation was not apparent in MPTP-treated mice (Corongiu et al. 1987), and marmosets were not protected from MPTP toxicity by pretreatment with antioxidants (Perry et al. 1987b).

In isolated rat hepatocytes, paraquat toxicity is enhanced by pretreatment with 1,3-bis(2-chloroethyl)-1-nitrosourea (BCNU), an inhibitor of glutathione reductase, and prevented by an antioxidant (N,N'-diphenyl-p-phenylenediamine) or chelation of iron (with desferrioxamine). Toxicity of MPP$^+$, however, was unaffected by either BCNU or agents lending protection against paraquat toxicity (Di Monte et al. 1986b). Thus, the protective effect of antioxidants has not been confirmed. Since satisfactory alternatives have been offered to explain MPTP-induced depletion of glutathione (DI MONTE et al. 1987a) and potentiation of MPTP toxicity by DDC (IRWIN et al. 1987b), it appears extremely unlikely that redox cycling is involved. Further evidence against generation of MPP· free radical formation from redox cycling of MPP$^+$ is derived from studies of the redox potentials of the required reactions. At physiological pH, one electron reduction of MPP$^+$ to MPP· requires a reduction potential of -1.07 V (LINKOUS et al. 1988). This is more negative than can be developed by NADPH and liver microsomes or purified cytochrome P-450 (FRANK et al. 1987). The unlikelihood that free radical formation by redox cycling might play a role in the toxicity of MPP$^+$ was supported also by studies of MPP$^+$ analogues. Although they have less negative reduction potentials than MPP$^+$, the 4-(4-fluorophenyl) and 4-(2-pyridyl) analogues of MPP$^+$ decrease striatal dopamine and metabolites levels but were found to be 5–10 times less potent than MPP$^+$ when infused directly in the rat substantia nigra (HARIK et al. 1987), consistent with the unimportance of redox cycling as a means of generating free radicals.

Although redox cycling may not be an important mechanism for generating free radicals, during oxidation of MPTP, MPP· and other free radicals can be formed, and these may contribute to MPTP toxicity, particularly when the protoxin is administered repeatedly and in high doses, as in rats or mice. (1) Oxidations mediated by MAO generate hydrogen peroxide, and in the presence of reduced transition metals, reactive hydroxyl free radicals are formed. (2) Mechanism-based inactivation of MAO by MPTP has been attributed to enzyme adducts with MPP· formed by one electron oxidation of MPDP$^+$ (TREVOR et al. 1988). (3) KORYTOWSKI et al. (1987) provided optical data which show that MPDP$^+$ undergoes one-electron oxidation reduction in the presence of iron chelates to form MPP. (4) Oxidation of MPDP$^+$ to MPP$^+$ by synthetic neuromelanin was found to be attended by the formation of hydrogen peroxide; hydroxyl free radical formation was detected also and was enhanced by the presence of iron chelates (KORYTOWSKY et al. 1988). (5) Most recently, HASEGAWA et al. (1990) found that inhibition of bovine heart mitochondrial particle respiration with MPP$^+$ (see below), as with rotenone, generated superoxide ions (determined by the oxidation of adrenaline to adrenochrome) and enhanced NADH-supported lipid membrane oxidation.

I. Inhibition of Mitochondrial Respiration by MPP$^+$

Reports of MPP$^+$ interference with mitochondrial function and ATP formation provided an alternative possible mechanism to explain their toxic effects. POIRIER and BARBEAU (1985) noted that MPP$^+$, but not MPTP, reversibly inhibited NADH-cytochrome c reductase, and NICKLAS et al. (1985) showed that MPP$^+$, but not MPTP, inhibited ADP-stimulated and uncoupled oxidation of NAD^+-linked substrates (e.g., pyruvate and glutamate) by mitochondria isolated from mouse brain, without affecting $NADP^+$-linked succinate oxidation. This suggested that MPP^+ blocks mitochondrial oxidation at complex I (NADH: ubiquinone oxidoreductase). Similar effects were found in mitochondria isolated from rat brain or liver. This was confirmed by Mizuno et al. (1987a) in rat brain mitochondria; they demonstrated a 50% inhibition of this enzyme complex during incubation with $50 \mu M$ MPP^+. In mouse neostriatal slices, however, MPTP as well as MPP^+ caused metabolic changes consistent with inhibition of mitochondrial oxidation. In the slices, these effects were prevented by pargyline, which blocks formation of MPP^+ from MPTP (Vyas et al. 1986). As might be expected, ATP formation is reduced markedly by MPP^+, and this was demonstrated in mouse brain mitochondria (Mizuno et al. 1987b) and in synaptosomes (Scotcher et al. 1990). The striking reduction in ATP formation is believed to be a major factor responsible for cell death.

Much higher concentrations of MPP^+ are required to inhibit oxidation of NAD^+-linked substrates in inverted mitochondria or in isolated inner mitochondrial membranes than in intact mitochondria (Ramsay et al. 1986a). This was explained by the discovery of energy-driven mitochondrial uptake of MPP^+, which results in very high concentrations of the toxin in mitochondria (Ramsay et al. 1986b). The mitochondrial uptake system for MPP^+, which concentrates MPP^+ 40-fold in 10 min, is greatly potentiated by the presence of malate plus glutamate and inhibited by respiratory inhibitors (Ramsay et al. 1986c).

Since mitochondrial uptake of MPP^+ is abolished by valinomycin plus K^+, which collapses the mitochondrial membrane electrochemical gradient, but not by agents which collapse the proton gradient, it appears likely that the uptake is energized by the transmembrane potential (Ramsay et al. 1986b,c; Ramsay and Singer 1986). Furthermore, if an uncoupler is added to mitochondria preloaded with MPP^+, efflux of the quaternary amine follows its concentration gradient. Uptake of MPP^+ has been shown using mitochondria from rat liver, whole brain, cortex, and midbrain. This uptake differs from the dopamine reuptake system since it is blocked by uncouplers and respiratory inhibitors but not by inhibitors of dopamine uptake. Although the kinetics of mitochondrial MPP^+ uptake are consistent with an energy-dependent carrier, Hoppel et al. (1987) proposed that nonspecific passive transport across the inner mitochondrial membrane, energized by the transmembrane potential, could account for the accumulation of MPP^+ in the mitochondria by an apparently saturable uptake process. The electrochemical gradient is sufficient to explain the accumulation in mitochondria of lipophilic cations (see e.g., Rottenberg 1984) such as MPP^+ without mediation of a specific carrier or transporter.

Toxicity of MPP^+ appears to be dependent upon mitochondrial concentrating mechanisms as well as upon the plasma membrane uptake system. This is evident from the comparative efficacies of inhibition of mitochondrial

respiration by charged and uncharged analogues of MPP^+ in intact mitochondria and on submitochondrial particles. In intact mitochondria, MPP^+ is
considerably more inhibitory to oxidation of NAD^+-linked substrates than is
4-phenylpyridinium (4-PP); in isolated inner mitochondrial membranes, 4-PP
is a more potent inhibitor of NADH oxidation than is MPP^+ (RAMSAY et al.
1986b). Similarly, in intact mitochondria, MPP^+ analogues were found to be
much more potent than the corresponding pyridines (lacking the N-methyl
group), whereas the neutral pyridines were more toxic than the N-methyl
pyridiniums when acting on submitochondrial particles (RAMSAY et al. 1987c;
HOPPEL et al. 1987). Upon adding uncouplers, inhibition by MPP^+ progressively diminishes, whereas the effect of 4-PP remains (RAMSAY et al.
1986b). The relatively nonpolar 4-PP, which enters the mitochondria passively
by diffusion, is unaffected, whereas collapse of the electrochemical gradient
after addition of uncouplers, as indicated above, is attended by efflux of
MPP^+ from the mitochondria, and its inhibitory actions gradually diminish.

SAYRE et al. (1990) compared the relative respiratory inhibitory potencies
of a variety of MPP^+ analogues incubated for 10 min with intact mitochondria
or for 1 min with mitochondrial electron transport particles. Differences in
the ratios of inhibitory potencies presumably reflected differences in the
rates of accumulation of the toxic compounds by the intact mitochondria
and were generally related to the degree of charge delocalization.

Acceleration (and potentiation) by tetraphenylboron (TPB) of the
inhibitory effect of MPP^+ on mitochondrial respiration provided further
evidence against carrier-mediated accumulation of MPP^+ (AIUCHI et al.
1988). This effect of TPB cannot be explained solely on the basis of facilitated penetration of the mitochondrial membrane because the magnitude as
well as the rate of inhibition are increased; TPB increases mitochondrial
concentration of MPP^+ in the matrix two- to three-fold (RAMSAY et al.
1989). But TPB, by ion-pairing, enhances lipid solubility about 10-fold,
reduces by an order of magnitude the concentration of MPP^+ required to
inhibit respiration of inverted mitochondrial membranes (RAMSAY et al.
1989), and enhances 12-fold the potency in electron transport particles
(SAYRE et al. 1989). Similar effects of TPB were found with several MPP^+
analogues (SAYRE et al. 1990). These observations suggested that in addition
to facilitating entry of MPP^+ into the mitochondrial matrix, the TPB-MPP^+
ion pair enhance accessibility of the toxin to a hydrophobic inhibition site on
NADH dehydrogenase. Complex I consists of more than 20 polypeptide
subunits along with flavin mononucleotide and several iron-sulfur cofactor
clusters (RAGAN 1987). The iron-protein fragment is transmembraneous
and surrounded by a shell of hydrophobic proteins in the lipid phase of the
membrane. Using crosslinking to study the constituent subunits of bovine
mitochondrial NADH dehydrogenase, PATEL and RAGAN (1988) demonstrated the proximity of the iron-protein and flavoprotein domains to the
hydrophobic domain of complex I.

In addition to its marked enhancement of complex I inhibition by MPP^+ and related pyridiniums in isolated mouse liver mitochondria, in intact mice administration of TPB potentiates MPTP dopaminergic neurotoxicity (HEIKKILA et al. 1990). These observations lend further support to the conclusion that TPB enhances toxicity by facilitating access of the toxic MPP^+ to its site of action. The site at which MPP^+ acts to block mitochondrial respiration appears to be at or near the region at which rotenone acts. MPP^+ and two of its analogues have been found to prevent binding of [^{14}C]rotenone to mitochondrial electron transport particles and to decrease rotenone inhibition of electron transport (KRUEGER et al. 1990b). This region is also the site at which several other agents act to interfere with mitochondrial oxidation. Interestingly, pethidine analogues are among the most potent inhibitors of NADH: ubiquinone reductase (FILSER and WERNER 1988). Furthermore, the mechanism of the protective effects of phenobarbital and of diphenylhydantoin on MPTP toxicity (MELAMED et al. 1986), which had not been easily explained, may be related to the well-known inhibitory effects of barbiturates on NAD^+-linked mitochondrial oxidation (see, e.g., CHANCE and HALLUNGER 1963). Barbiturates also appear to act at the rotenone-sensitive site but are 1000-fold less potent and might, by interfering with MPP^+ binding to its site of action, prevent the toxin from inhibiting NAD: ubiquinone oxidase in much the same way as MPP^+ blocks the actions of rotenone.

Disturbances in calcium homeostasis have also been implicated in MPP^+ toxicity (FREI and RICHTER 1986). They showed that MPP^+ in the presence of dopamine or 6-hydroxydopamine stimulated release of Ca^{2+} ions from mitochondria. Since with glutamate or malate as substrates, the respiratory rate of mitochondria from aged (2-year-old) rats is lower than in mitochondria from young (2-month-old) rats, but there is no difference when succinate is the substrate, and mitochondria from aged rats accumulate Ca^{2+} ions more slowly (VITORICA et al. 1985; VITORICA and SATRUSTEGUI 1986), age-related differences in vulnerability to MPTP may be related to differences in complex I function and capacity to reverse increases in the cytoplasmic Ca^{2+} level in aged animals.

G. Role of Neuromelanin in MPTP Toxicity

Initial observations on species differences in vulnerability to MPTP toxicity appeared to be related to the concentration of neuromelanin in the substantia nigra (BURNS et al. 1984). After the demonstration that MPP^+ binds with high affinity to neuromelanin (D'AMATO et al. 1986), it was suggested that such binding provides a store which, by its gradual release, maintains high cytoplasmic levels of the toxic metabolite and thereby selectively enhances toxicity in neuromelanin-bearing cells. Intracellular sequestration of MPP$^+$ similar to neuromelanin binding has been reported in adrenal medullary chromafin granules, but in this case resistance to MPP$^+$ toxicity was attributed to the sequestration (REINHARD et al. 1990a). In primates, neuromelanin content increases with age, and MPTP toxicity increases with age, but the effect of age on MPTP toxicity is found also in mice (JARVIS and WAGNER 1985) and rats (JARVIS and WAGNER 1990), species with almost no neuromelanin. Furthermore, sheep, which also have little neuromelanin, are about as vulnerable to MPTP toxicity as are dogs, and their vulnerability increases with age (HAMMOCK et al. 1989; BEALE et al. 1989). Thus, the association of neuromelanin with MPTP vulnerability has been weakened. It is possible, however, that neuromelanin might in some species play a role in MPTP toxicity or provide an index of vulnerability to toxic mechanisms involving inhibition of mitochondrial respiration.

H. Conclusion

There are many factors which influence MPTP toxicity, and although much progress has been made in understanding the various mechanisms involved, there remain several important incompletely answered questions. Clearly, the disposition and metabolism of MPTP and its toxic derivative, MPP$^+$, are important in determining the neuronal specificity of its toxicity. MPTP metabolism in peripheral tissues results in detoxification or formation of MPP$^+$ which, because of its positive charge, has limited access to the brain, but MPTP can easily penetrate the BBB and gain access to astrocytes and neurons which contain MAO-B. After enzymatic oxidation to MPDP$^+$ and subsequent conversion to MPP$^+$, the toxic pyridinium derivative, which is somewhat lipid soluble because of delocalization of its charge, gains access to the extracellular fluid. MAO-A is able to oxidize some MPTP analogues to the corresponding MPP$^+$ analogues. Because MPP$^+$ (or its analogue) is accumulated avidly via the dopamine uptake system, it achieves relatively high concentrations in dopaminergic neuronal cytoplasm. The pyridinium is further concentrated in the mitochondrial matrix as a result of the electro-chemical gradient across the membrane. Because of its slight lipid solubility and high concentrations, levels attained in the lipid membrane and hydrophobic shell enclosed domain of complex I are sufficient to inhibit electron transport from NADH to ubiquinone and to disrupt mitochondrial respiration. The consequences of inhibition of mitochondrial respiration include diminished ATP production, increased cytosolic Ca^{2+} levels, and formation of superoxide and other free radicals, all of which have damaging effects on the cell and threaten its survival. The apparent increase of MPTP potency with age, species differences in vulnerability to MPTP and its analogues, and specific targeting of nigrostriatal dopaminergic neurons for the most severe damage remain unsatisfactorily explained. Differences in metabolism to MPP$^+$ (or an equivalent charged ion), uptake and mitochondrial accumulation of the charged molecule, its accessibility to the lipophilic enclosed site of action, its intrinsic potency in inhibiting mitochondrial oxidative enzymes, as well as differences in the capacity of neurons to survive the metabolic insult are probably factors in target specificity. The relationship of MPTP toxicity to the etiology or pathogenesis of Parkinson's disease remains the subject of intense interest and much speculation.

References

Aiuchi T, Shirane Y, Kinemuchi H, Arai Y, Nakaya K, Nakamura Y (1988) Enhancement by tetraphenylboron of inhibition of mitochondrial respiration induced by 1 methyl-4-phenylpyridinium ion MPP$^+$. Neurochem Int 12:525–532

Arora PK, Riachi NJ, Harik SI, Sayre LM (1988) Chemical oxidation of 1-methyl-4-phenyl-1,2,3,6-tetrahydropyridine (MPTP) and its in vivo metabolism in rat brain and liver. Biochem Biophys Res Commun 152:1339–1347

Baldessarini RJ, Kula NS, Francoeur D, Finklestein SP (1986) Antioxidants fail to inhibit depletion of striatal dopamine by MPTP. Neurology 36:735

Beale AM, Higgins RJ, Work TM, Bailey CS, Smith MO, Shinka T, Hammock BD (1989) MPTP-induced Parkinson-like disease in sheep: clinical and pathologic findings. J Environ Pathol Toxicol Oncol 9:417–428

Bocchetta A, Piccardi MP, Del Zompo M, Pintus S, Corsini GU (1985) 1-methyl-4-phenyl-1,2,3,6-tetrahydropyridine: correspondence of its binding sites to monoamine oxidase in rat brain, and inhibition of dopamine oxidative deamination in vivo and in vitro. J Neurochem 45:673–676

Boyce S, Kelly E, Reavill C, Jenner P, Marsden CD (1984) Repeated administration of 1,2,5,6-tetrahydropyridine to rats is not toxic to striatal dopamine neurons. Biochem Pharmacol 33:1747–1752

Bradbury AJ, Costall B, Domeney AM, Testa B, Jenner PG, Marsden CD, Naylor RJ (1985) The toxic actions of MPTP and its metabolite MPP$^+$ are not mimicked by analogues of MPTP lacking an N-methyl moiety. Neurosci Lett 61:121–126

Brooks WJ, Jarvis MF, Wagner GC (1988) Attenuation of MPTP-induced dopaminergic neurotoxicity by a serotonin uptake blocker. J Neural Transm 71:85–90

Brooks WJ, Jarvis MF, Wagner GC (1989) Astrocytes as a primary locus for the conversion MPTP into MPP$^+$. J Neural Transm 76:1–12

Burns RS, Chiueh CC, Markey SP, Ebert MH, Jacobowitz DM, Kopin IJ (1983) A primate model of parkinsonism: selective destruction of dopaminergic neurons in the pars compacta of the substantia nigra by N-methyl-4-phenyl-1,2,3,6-tetrahydropyridine. Proc Natl Acad Sci USA 80:4546–4550

Burns RS, Markey SP, Phillips JM, Chiueh CC (1984) The neurotoxicity of 1-methyl-4-phenyl-1,2,3,6-tetrahydropyridine in the monkey and man. Can J Neurol Sci 11 [Suppl 1]:166–168

Bus JS, Cagen SZ, Olgaard M, Gibson JE (1976) A mechanism of paraquat toxicity in mice and rats. Toxicol Appl Pharmacol 35:501–513

Cashman JR, Ziegler DM (1986) Contribution of N-oxygenation to the metabolism of MPTP (1-methyl-4-phenyl-1,2,3,6-tetrahydropyridine) by various liver preparations. Mol Pharmacol 29:163–167

Chance B, Hollunger G (1963) Inhibition of electron and energy transfer in mitochondria; I. Effect of amytal, thiopental, rotenone, progesterone and methylene glycol. J Biol Chem 239:418–431

Chiba K, Trevor A, Castagnoli N Jr (1984) Metabolism of the neurotoxic tertiary amine, MPTP, by brain monoamine oxidase. Biochem Biophys Res Commun 120:574–578

Chiba K, Kubota E, Miyakawa T, Kato Y, Ishizaki T (1988) Characterization of hepatic microsomal metabolism as an in vivo detoxification pathway of 1-methyl-4-phenyl-1,2,3,6-tetrahydropyridine in mice. J Pharmacol Exp Ther 246:1108–1115

Chiba K, Horii H, Kubota E, Ishizaki T, Kato Y (1990) Effects of N-methylmercaptoimidazole on the disposition of MPTP and its metabolites in mice. Eur J Pharmacol 180:59–67

Chiueh CC, Markey SP, Burns RS, Johannessen JN, Jacobowitz DM, Kopin IJ (1984a) Neurochemical and behavioral effects of 1-methyl-4-phenyl-1,2,3,6-tetrahydropyridine (MPTP) in rat, guinea pig, and monkey. Psychopharmacol Bull 20:548–553

Chiueh CC, Markey SP, Burns RS, Johannessen JN, Pert A, Kopin IJ (1984b) Neurochemical and behavioral effects of systemic and intranigral administration of N-methyl-4-phenyl-1,2,3,6-tetrahydropyridine in the rat. Eur J Pharmacol 100(2):189–194

Cohen G, Pasik P, Cohen B, Leist A, Mytilineou C, Yahr M (1985) Pargyline and deprenyl prevent the neurotoxicity of 1-methyl-4-phenyl-1,2,3,6-tetrahydropyridine (MPTP in monkeys). Eur J Pharmacol 106:209–210

Corongiu FP, Dessi' MA, Banni S, Bernardi F, Piccardi MP, Del Zompo M, Corsini GU (1987) MPTP fails to induce lipid peroxidation in vivo. Biochem Pharmacol 36:2251–2253

Corsini GU, Pintus S, Chiueh CC, Weiss JF, Kopin IJ (1985) 1-Methyl-4-phenyl-1,2,3,6-tetrahydropyridine (MPTP) neurotoxicity in mice is enhanced by pretreatment with diethyldithiocarbamate. Eur J Pharmacol 119:127–128

Corsini GU, Pintus S, Bocchetta A, Piccardi MP, Del Zompo M (1986a) Primate-rodent ^{3}H-MPTP binding differences, and biotransformation of MPTP to a reactive intermediate in vitro. J Neural Transm [Suppl]22:55–60

Corsini GU, Pintus S, Bocchetta A, Piccardi MP, Del Zompo M T (1986b) A reactive metabolite of 1-methyl-4-phenyl-1,2,3,6-tetrahydropyridine is formed in rat brain in vitro by type B monoamine oxidase. J Pharmacol Exp Ther 238:648–652

Corsini GU, Bocchetta A, Zuddas A, Piccardi MP, Del Zompo M (1988) Covalent protein binding of a metabolite of 1-methyl-4-phenyl-1,2,3,6-tetrahydropyridine to mouse and monkey brain in vivo and in vitro. Biochem Pharmacol 37:4163–4169

D'Amato RJ, Lipman ZP, Snyder SH (1986) Selectivity of the parkinsonian neurotoxin MPTP: toxic metabolite MPP$^+$ binds to neuromelanin. Science 231:987–989

Davis GC, Williams AC, Markey SP, Ebert MH, Caine ED, Beichert CM, Kopin IJ (1979) Chronic parkinsonism secondary to intravenous injection of meperidine analogues. Psychiatry Res 1:249–254

Di Monte D, Jewell SA, Ekström G, Sandy MS, Smith MT (1986a) 1-Methyl-4-phenyl-1,2,3,6-tetrahydropyridine (MPTP) and 1-methyl-4-phenylpyridine (MPP$^+$) cause rapid ATP depletion in isolated hepatocytes. Biochem Biophys Res Commun 137:310–315

Di Monte D, Sandy MS, Ekström G, Smith MT (1986b) Comparative studies on the mechanisms of paraquat and 1-methyl-4-phenylpyridine (MPP$^+$) cytotoxicity. Biochem Biophys Res Commun 137:303–309

Di Monte D, Ekström G, Shinka T, Smith MT, Trevor AJ, Castagnoli N Jr (1987a) Role of 1-methyl-4-phenylpyridinium ion formation and accumulation in 1-methyl-4-phenyl-1,2,3,6-tetrahydropyridine toxicity to isolated hepatocytes. Chem Biol Interact 62:105–116

Di Monte D, Sandy MS, Smith MT (1987b) Increased efflux rather than oxidation is the mechanism of glutathione depletion by 1-methyl-4-phenyl-1,2,3,6-tetrahydropyridine (MPTP). Biochem Biophys Res Commun 148:153–160

Filser M, Werner S (1988) Pethidine analogues, a novel class of potent inhibitors of mitochondrial NADH: ubiquinone reductase. Biochem Pharmacol 37:2551–2558

Francis A, Pearce LB, Roth JA (1985) Cellular localization of MAO A and B in brain: evidence from kainic acid lesions in striatum. Brain Res 334:59–64

Frank DM, Arora PK, Blumer JL, Sayre LM (1987) Model study on the bioreduction of paraquat, MPP$^+$, and analogs. Evidence against a "redox cycling" mechanism in MPTP neurotoxicity. Biochem Biophys Res Commun 147:1095–1104

Freeman BA, Crapo JD (1982) Biology of disease: free radicals and tissue injury. Lab Invest 47:412–426

Frei B, Richter C (1986) N-methyl-4-phenylpyridine (MPP$^+$) together with 6-hydroxydopamine or dopamine stimulates Ca^{++} release from mitochondria. FEBS Lett 198:99–102

Fuller RW, Hemrick-Luecke SK (1985) Inhibition of types A and B monoamine oxidase by 1-methyl-4-phenyl-1,2,3,6-tetrahydropyridine. J Pharmacol Exp Ther 232:696–701

Gessner W, Brossi A, Shen RS, Abell CW (1985) Further insight into the mode of action of the neurotoxin 1-methyl-4-phenyl-1,2,3,6-tetrahydropyridine (MPTP). FEBS Lett 1985 Apr 22; 183(2):345–348

Gessner W, Brossi A, Bembenek ME, Fritz RR, Abell CW (1986) Studies on the mechanism of MPTP oxidation by human liver monoamine oxidase B. FEBS Lett 199:100–102

Goldstein BD, Rozen MG, Quintavalla JC, Amoruso MA (1979) Decrease in mouse lung and liver glutathione peroxidase activity and potentiation of the lethal effects of ozone and paraquat by the superoxide dismutase inhibitor diethyldithiocarbamate. Biochem Pharmacol 28:27–30

Hallman H, Olson L, Jonsson G (1984) Neurotoxicity of the meperidine analogue N-methyl-4-phenyl-1,2,3,6-tetrahydropyridine on brain catecholamine neurons in the mouse. Eur J Pharmacol 97:133–136

Hammock BD, Beale AM, Work T, Gee SJ, Gunther R, Higgins RJ, Shinka T, Castagnoli N Jr (1989) A sheep model for MPTP induced Parkinson-like symptoms. Life Sci 45:1601–1608

Harik SI, Schmidley JW, Iacofano LA, Blue P, Arora PK, Sayre LM (1987) On the mechanisms underlying 1-methyl-4-phenyl-1,2,3,6-tetrahydropyridine neurotoxicity: the effect of perinigral infusion of 1-methyl-4-phenyl-1,2,3,6-tetrahydropyridine, its metabolite and their analogs in the rat. J Pharmacol Exp Ther 241:669–676

Hasegawa E, Takeshige K, Oishi T, Murai Y, Minakami S (1990) 1-methyl-4-phenylpyridinium (MPP$^+$) induces NADH-dependent superoxide formation and enhances NADH-dependent lipid peroxidation in bovine heart mitochondrial particles. Biochem Biophys Res Commun 107:1049–1055

Heikkila RE, Hess A, Duvoisin RC (1984a) Dopaminergic neurotoxicity of 1-methyl-4-phenyl-1,2,5,6-tetrahydropyridine in mice. Science 224:1451–1453

Heikkila RE, Manzino L, Cabbat FS, Duvoisin RC (1984b) Protection against the dopaminergic neurotoxicity of 1-methyl-4-phenyl-1,2,5,6-tetrahydropyridine by monoamine oxidase inhibitors. Nature 311:467–469

Heikkila RE, Manzino L, Cabbat FS, Duvoisin RC (1985a) Effects of 1-methyl-4-phenyl-1,2,3,6-tetrahydropyridine (MPTP) and several of its analogues on the dopaminergic nigrostriatal pathway in mice. Neurosci Lett 58:133–137

Heikkila RE, Nicklas WJ, Duvoisin RC (1985b) Dopaminergic neurotoxicity after stereotaxic administration of 1-methyl-4-phenyl-pyridinium (MPP$^+$) to rats. Neurosci Lett 59:135–140

Heikkila RE, Kindt MV, Sonsalla PK, Giovanni A, Youngster SK, McKeown KA, Singer TP (1988) Importance of monoamine oxidase A in the bioactivation of neurotoxic analogs of 1-methyl-4-phenyl-1,2,3,6-tetrahydropyridine. Proc Natl Acad Sci USA 85:6172–6176

Heikkila RE, Hwang J, Ofori S, Geller HM, Nicklas WJ (1990) Potentiation by the tetraphenylboron anion of the effects of 1-methyl-4-phenyl-1,2,3,6-tetrahydropyridine and its pyridinium metabolite. J Neurochem 54:743–750

Hoppel CL, Grinblatt D, Kwok HC, Arora PK, Singh MP, Sayre LM, Grinblatt D (1987) Inhibition of mitochondrial respiration by analogs of 4-phenylpyridine and 1-methyl-4-phenylpyridinium cation (MPP⁺), the neurotoxic metabolite of MPTP. Biochem Biophys Res Commun 148:684–693

Irwin I, Langston JW (1985) Selective accumulation of MPP⁺ in the substantia nigra: a key to neurotoxicity? Life Sci 36:207–212

Irwin I, Langston JW, DeLanney LE (1987a) 4-Phenylpyridine (4PP) and MPTP: the relationship between striatal MPP⁺ concentrations and neurotoxicity. Life Sci 40:731–740

Irwin I, Wu EY, DeLanney LE, Trevor A, Langston JW (1987b) The effect of diethyldithiocarbamate on the biodisposition of MPTP: an explanation for enhanced neurotoxicity. Eur J Pharmacol 141:209–217

Jarvis MF, Wagner GC (1985) Age-dependent effects of 1-methyl-4-phenyl-1,2,5,6-tetrahydropyridine (MPTP). Neuropharmacology 24:581–583

Jarvis MF, Wagner GC (1990) 1-methyl-4-phenyl-1,2,3,6-tetrahydropyridine-induced neurotoxicity in the rat: characterization and age-dependent effects. Synapse 5:104–112

Javitch JA, Snyder SH (1984) Uptake of MPP⁺ by dopamine neurons explains selectivity of parkinsonism-inducing neurotoxin, MPTP. Eur J Pharmacol 106:455–456

Javitch JA, D'Amato RJ, Strittmatter SM, Snyder SH (1985) Parkinsonism-inducing neurotoxin, N-methyl-4-phenyl-1,2,3,6-tetrahydropyridine: uptake of the metabolite N-methyl-4-phenylpyridine by dopamine neurons explains selective toxicity. Proc Natl Acad Sci USA 82:2173–2177

Jenner P, Rupniak NM, Rose S, Kelly E, Kilpatrick G, Lees A, Marsden CD (1984) 1-Methyl-4-phenyl-1,2,3,6-tetrahydropyridine-induced parkinsonism in the common marmoset. Neuroscience 50:85–90

Johannessen JN, Chiueh CC, Burns RS, Markey SP (1985) Differences in the metabolism of MPTP in the rodent and primate parallel differences in sensitivity to its neurotoxic effects. Life Sci 36:219–224

Johannessen JN, Adams JD, Schuller HM, Bacon JP, Markey SP (1986) 1-Methyl-4-phenylpyridine (MPP⁺) induces oxidative stress in the rodent. Life Sci 3:743–749

Kalaria RN, Mitchell MJ, Harik SI (1987) Correlation of 1-methyl-4-phenyl-1,2,3,6-tetrahydropyridine neurotoxicity with blood-brain barrier monoamine oxidase activity. Proc Natl Acad Sci USA 84:3521–3525

Konradi C, Riederer P, Jellinger K, Denney R (1987) Cellular action of MAO inhibitors. J Neural Transm [Suppl]25:15–25

Korytowski W, Felix CC, Kalyanaraman B (1987) Evidence for the one-electron oxidation of 1-methyl-4-phenyl-2,3-diahydropyridinium (MPDP⁺). Biochem Biophys Res Commun 147:354–60

Korytowski W, Felix CC, Kalyanaraman B (1988) Oxygen activation during the interaction between MPTP metabolites and synthetic neuromelanin – an ESR-spin trapping, optical, and oxidase electrode study. Biochem Biophys Res Commun 154:781–788

Krueger MJ, McKeown K, Ramsay RR, Youngster SK, Singer TP (1990a) Mechanism-based inactivation of monoamine oxidases A and B by tetrahydropyridines and dihydropyridines. Biochem J 268:219–224

Krueger MJ, Singer TP, Casida JE, Ramsay RR (1990b) Evidence that the blockade of mitochondrial respiration by the neurotoxin 1-methyl-4-phenylpyridinium (MPP⁺) involves binding at the same site as the respiratory inhibitor, rotenone. Biochem Biophys Res Commun 169:123–128

Langston JW, Ballard P, Tetrud JW, Irwin I (1983) Chronic Parkinsonism in humans due to a product of meperidine-analog synthesis. Science 219:979–980

Langston JW, Forno LS, Rebert CS, Irwin I (1984a) Selective nigral toxicity after systemic administration of 1-methyl-4-phenyl-1,2,5,6-tetrahydropyridine (MPTP) in the squirrel monkey. Brain Res 292:390–394

Langston JW, Irwin I, Langston EB, Forno LS (1984b) Pargyline prevents MPTP-induced parkinsonism in primates. Science 225:1480–1482

Levitt P, Pintar JE, Breakefield XO (1982) Immunochemical demonstration of monoamine oxidase B in brain astrocytes and serotonergic neurons. Proc Natl Acad Sci USA 79:6385–6389

Linkous CA, Schaiac KM, Forman A, Borg DC (1988) An electrochemical study of the neurotoxin 1-methyl-4-phenyl-1,2,3,6-tetrahydropyridine and its oxidation products. Bioelectrochem Bioenerg 19:447–490

Marini AM, Schwartz JP, Kopin IJ (1989) The neurotoxicity of 1-methyl-4-phenylpyridinium in cultured cerebellar granule cells. J Neurosci 9:3665–3672

Markey SP, Johannessen JN, Chiueh CC, Burns RS, Herkenham MA (1984) Intraneuronal generation of a pyridinium metabolite may cause drug-induced parkinsonism. Nature 311:464–467

Martinovits G, Melamed E, Cohen O, Rosenthal J, Uzzan A (1986) Systemic administration of antioxidants does not protect mice against the dopaminergic neurotoxicity of 1-methyl-4-phenyl-1,2,5,6-tetrahydropyridine (MPTP). Neurosci Lett 69:192–197

Mayer RA, Kindt MV, Heikkila RE (1986) Prevention of the nigrostriatal toxicity of 1-methyl-4-phenyl-1,2,3,6-tetrahydropyridine by inhibitors of 3,4-dihydroxyphenyl-ethylamine transport. J Neurochem 47:1073–1079

Melamed E, Rosenthal J, Cohen O, Globus M, Uzzan A (1985) Dopamine but not norepinephrine or serotonin uptake inhibitors protect mice against neurotoxicity of MPTP. Eur J Pharmacol 116:179–181

Melamed E, Martinovits G, Pikarsky E, Rosenthal J, Uzzan A (1986) Diphenylhydantoin and phenobarbital suppress the dopaminergic neurotoxicity of MPTP in mice. Eur J Pharmacol 128:255–257

Mizuno Y, Saitoh T, Sone N (1987a) Inhibition of mitochondrial NADH-ubiquinone oxidoreductase activity by 1-methyl-4-phenylpyridinium ion. Biochem Biophys Res Commun 143:294–299

Mizuno Y, Saitoh T, Sone N (1987b) Inhibition of mitochondrial alpha-ketoglutarate dehydrogenase by 1-methyl-4-phenylpyridinium ion. Biochem Biophys Res Commun 143:971–976

Mytilineou C, Cohen G (1984) 1-methyl-4-phenyl-1,2,3,6-tetrahydropyridine destroys dopamine neurons in explants of rat embryo mesencephalon. Science 225:529–531

Mytilineou C, Cohen G, Heikkila RE (1985) 1-Methyl-4-phenylpyridine (MPP$^+$) is toxic to mesencephalic dopamine neurons in culture. Neurosci Lett 57:19–24

Nicklas WJ, Vyas I, Heikkila RE (1985) Inhibition of NADH-linked oxidation in brain mitochondria by 1-methyl-4-phenyl-pyridine, a metabolite of the neurotoxin, 1-methyl-4-phenyl-1,2,5,6-tetrahydropyridine. Life Sci 36:2503–2508

Parsons B, Rainbow TC (1984) High-affinity binding sites for ^{3}H-MPTP may correspond to monamine oxidase. Eur J Pharmacol 102:375–377

Patel SD, Ragan CI (1988) Structural studies on mitochondrial NADH dehydrogenase using chemical cross-linking. Biochem J 256:521–528

Perry TL, Yong VW, Clavier RM, Jones K, Wright JM, Foulks JG, Wall RA (1985) Partial protection from the dopaminergic neurotoxin N-methyl-4-phenyl-1,2,3,6-tetrahydropyridine by four different antioxidants in the mouse. Neurosci Lett 60:109–114

Perry TL, Yong VW, Jones K, Wright JM (1986) Manipulation of glutathione contents fails to alter dopaminergic nigrostriatal neurotoxicity of N-methyl-4-phenyl-1,2,3,6-tetrahydropyridine (MPTP) in the mouse. Neurosci Lett 70:261–265

Perry TL, Jones K, Hansen S, Wall RA TI (1987a) 4-phenylpyridine and three other analogues of 1-methyl-4-phenyl-1,2,3,6-tetrahydropyridine lack dopaminergic nigrostriatal neurotoxicity in mice and marmosets. Neurosci Lett 75:65–70

Perry TL, Yong VW, Hansen S, Jones K, Bergeron C, Foulks JG, Wright JM (1987b) Alpha-tocopherol and beta-carotene do not protect marmosets against the dopaminergic neurotoxicity of N-methyl-4-phenyl-1,2,3,6-tetrahydropyridine. J Neurol Sci 81:321–331

Peterson LA, Caldera PS, Trevor A, Chiba K, Castagnoli N Jr TI (1985) Studies on the 1-methyl-4-phenyl-2,3-diahydropyridinium species 2,3-MPDP$^+$, the monoamine oxidase catalyzed oxidation product of the nigrostriatal toxin 1-methyl-4-phenyl-1,2,3,6-tetrahydropyridine (MPTP). J Med Chem 28:1432–1436

Pileblad E, Carlsson A (1985) Catecholamine-uptake inhibitors prevent the neurotoxicity of 1-methyl-4-phenyl-1,2,3,6-tetrahydropyridine (MPTP) in mouse brain. Neuropharmacology 24:689–692

Poirier J, Barbeau A (1985) 1-Methyl-4-phenyl-pyridinium-induced inhibition of nicotinamide adenosine dinucleotide cytochrome c reductase. Neurosci Lett 62:7–11

Poirier J, Donaldson J, Barbeau A (1986) The specific vulnerability of the substantia nigra to MPTP is related to the presence of transition metals. Biochem Biophys Res Commun 128:25–33

Ragan CI (1987) Structure of NADH-ubiquinone reductase (complex I). Curr Top Bioenerg 15:1–35

Rainbow TC, Parsons B, Wieczorek CM, Manaker S (1985) Localization in rat brain of binding sites for parkinsonian toxin MPTP: similarities with [^{3}H]pargyline binding to monoamine oxidase. Brain Res 330:337–342

Ramsay RR, Singer TP (1986) Energy-dependent uptake of N-methyl-4-phenylpyridinium, the neurotoxic metabolite of 1-methyl-4-phenyl-1,2,3,6-tetrahydropyridine, by mitochondria. J Biol Chem 261:7585–7587

Ramsay RR, Dadgar J, Trevor A, Singer TP (1986a) Energy-driven uptake of N-methyl-4-phenylpyridine by brain mitochondria mediates the neurotoxicity of MPTP. Life Sci 39:581–588

Ramsay RR, Salach JI, Dadgar J, Singer TP (1986b) Inhibition of mitochondrial NADH dehydrogenase by pyridine derivatives and its possible relation to experimental and idiopathic parkinsonism. Biochem Biophys Res Commun 135:269–275

Ramsay RR, Salach JI, Singer TP (1986c) Uptake of the neurotoxin 1-methyl-4-phenylpyridine (MPP$^+$) by mitochondria and its relation to the inhibition of the mitochondrial oxidation of NAD$^+$-linked substrates by MPP$^+$. Biochem Biophys Res Commun 134:743–738

Ramsay RR, Koerber SC, Singer TP TI (1987a) Stopped-flow studies on the mechanism of oxidation of N-methyl-4-phenyltetrahydropyridine by bovine liver monoamine oxidase B. Biochemistry 26:3045–3050

Ramsay RR, McKeown KA, Johnson EA, Booth RG, Singer TP (1987b) Inhibition of NADH oxidation by pyridine derivatives. Biochem Biophys Res Commun 146:53–60

Ramsay RR, Mehlhorn RJ, Singer TP (1989) Enhancement by tetraphenylboron of the interaction of the 1-methyl-4-phenylpyridinium ion (MPP$^+$) with mitochondria. Biochem Biophys Res Commun 159:983–990

Ransom BR, Kunis DM, Irwin I, Langston JW (1987) Astrocytes convert the parkinsonism inducing neurotoxin, MPTP, to its metabolite, MPP$^+$. Neurosci Lett 75:323–328

Reinhard JF Jr, Carmichael SW, Daniels AJ (1990a) Mechanisms of toxicity and cellular resistance to 1-methyl-4-phenyl-1,2,3,6-tetrahydropyridine and 1-methyl-4-phenylpyridinium in adrenomedullary chromaffin cell cultures. J Neurochem 55:311–320

Reinhard JF Jr, Daniels AJ, Painter GR (1990b) Carrier-independent entry of 1-methyl-4-phenylpyridinium (MPP$^+$) into adrenal chromaffin cells as a consequence of charge delocalization. Biochem Biophys Res Commun 168:1143–1148

Riachi NJ, Harik SI, Kalaria RN, Sayre LM (1988) On the mechanisms underlying 1-methyl-4-phenyl-1,2,3,6-tetrahydropyridine neurotoxicity. II. Susceptibility among mammalian species correlates with the toxin's metabolic patterns in brain microvessels and liver. J Pharmacol Exp Ther 244:443–448

Riachi NJ, LaManna JC, Harik SI (1989) Entry of 1-methyl-4-phenyl-1,2,3,6-tetrahydropyridine into the rat brain. J Pharmacol Exp Ther 249:744–748

Rose S, Nomoto M, Jackson EA, Gibb WR, Jenner P, Marsden CD (1990) 1-Methyl-4-(2'-methylphenyl)-1,2,3,6-tetrahydropyridine (2'-methyl-MPTP) is less neurotoxic than MPTP in the common marmoset. Eur J Pharmacol 181:97–103

Rottenberg H (1984) Membrane potential and surface potential in mitochondria: uptake and binding of lipophilic cations. J Membr Biol 81:127–138

Salach JI, Singer TP, Castagnoli N Jr, Trevor A (1984) Oxidation of the neurotoxic amine 1-methyl-4-phenyl-1,2,3,6-tetrahydropyridine (MPTP) by monoamine oxidases A and B and suicide inactivation of the enzymes by MPTP. Biochem Biophys Res Commun 125:831–835

Sanchez-Ramos J, Barrett JN, Goldstein M, Weiner WJ, Hefti F (1986) 1-Methyl-4-phenyl-pyridinium (MPP+) but not 1-methyl-4-phenyl-1,2,3,6-tetrahydropyridine (MPTP) selectively destroys dopaminergic neurons in cultures of dissociated rat mesencephalic neurons. Neurosci Lett 72:215–220

Sanchez-Ramos J, Michel P, Weiner WJ, Hefti F (1988) Selective destruction of cultured dopaminergic neurons from fetal rat mesencephalon by 1-methyl-4-phenylpyridinium: cytochemical and morphological evidence. J Neurochem 50:1934–1944

Sayre LM, Arora PK, Lacofano LA, Harik SI TI (1986) Comparative toxicity of MPTP, MPP+ and 3,3-dimethyl-MPDP+ to dopaminergic neurons of the rat substantia nigra. Eur J Pharmacol 124:171–174

Sayre LM, Wang F, Hoppel CL (1989) Tetraphenylborate potentiates the respiratory inhibition by the dopaminergic neurotoxin MPP+ in both electron transport particles and intact mitochondria. Biochem Biophys Res Commun 161:809–818

Sayre LM, Singh MP, Arora PK, Wang F, McPeak RJ, Hoppel CL (1990) Inhibition of mitochondrial respiration by analogues of the dopaminergic neurotoxin 1-methyl-4-phenylpyridinium: structural requirements for accumulation-dependent enhanced inhibitory potency on intact mitochondria. Arch Biochem Biophys 280:274–283

Schinelli S, Zuddas A, Kopin IJ, Barker JL, di Porzio U (1988) 1-Methyl-4-phenyl-1,2,3,6-tetrahydropyridine metabolism and 1-methyl-4-phenylpyridinium uptake in dissociated cell cultures from the embryonic mesencephalon. J Neurochem 50:1900–1907

Schneider JS, Markham CH (1987) Immunohistochemical localization of monoamine oxidase-B in the cat brain: clues to understanding N-methyl-4-phenyl-1,2,3,6-tetrahydropyridine (MPTP) toxicity. Exp Neurol 97:465–468

Schultz W, Scarnati E, Sundstrom E, Tsutsumi T, Jonsson G TI (1986) The catecholamine uptake blocker nomifensine protects against MPTP-induced parkinsonism in monkeys. Exp Brain Res 63:216–220

Scotcher KP, Irwin I, DeLanney LE, Langston JW, Di Monte D (1990) Effects of 1-methyl-4-phenyl-1,2,3,6-tetrahydropyridine and 1-methyl-4-phenylpyridinium ion on ATP levels of mouse brain synaptosomes. J Neurochem 54:1295–1330

Sinha BK, Singh Y, Krishna G (1986) Formation of superoxide and hydroxyl radicals from 1-methyl-4-phenylpyridinium ion (MPP+): reductive activation by NADPH cytochrome P-450 reductase. Biochem Biophys Res Commun 135:583–588

Sonsalla PK, Youngster SK, Kindt MV, Heikkila RE (1987) Characteristics of 1-methyl-4-(2'-methylphenyl)-1,2,3,6-tetrahydropyridine-induced neurotoxicity in the mouse. J Pharmacol Exp Ther 242:850–857

Sundstrom E, Jonsson G (1985) Pharmacological interference with the neurotoxic action of 1-methyl-4-phenyl-1,2,3,6-tetrahydropyridine (MPTP) on central catecholamine neurons in the mouse. Eur J Pharmacol 110:293–299

Takada M, Li ZK, Hattori T (1990) Astroglial ablation prevents MPTP-induced nigrostriatal neuronal death. Brain Res 509:55–61

Tipton KF, McCrodden JM, Youdim MB TI (1986) Oxidation and enzyme-activated irreversible inhibition of rat liver monoamine oxidase-B by 1-methyl-4-phenyl-1,2,3,6-tetrahydropyridine (MPTP). Biochem J 240:379–383

Trevor AJ, Castagnoli N Jr, Caldera P, Ramsay RR, Singer TP (1987) Bioactivation of MPTP: reactive metabolites and possible biochemical sequelae. Life Sci 40:713–719

Trevor AJ, Castagnoli N, Singer TP (1988) The formation of reactive intermediates in the MAO-catalyzed oxidation of the nigrostriatal toxin 1-methyl-4-phenyl-1,2,3,6-tetrahydropyridine (MPTP). Toxicology 49:513–519

Trush MA, Mimnaugh EG, Gram TE (1982) Activation of pharmacologic agents to radical intermediates. Implications for the role of free radicals in drug action and toxicity. Biochem Pharmacol 31:3335–3346

Vincent SR (1989) Histochemical localization of 1-methyl-4-phenyl-1,2,3,6-tetrahydropyridine oxidation in the mouse brain. Neuroscience 28:189–199

Vitorica J, Satrustegui J (1986) Involvement of mitochondria in the age-dependent decrease in calcium uptake of rat brain synaptosomes. Brain Res 16:36–48

Vitorica J, Clark A, Machado A, Satrustegui J (1985) Impairment of glutamate uptake and absence of alterations in the energy-transducing ability of old rat brain mitochondria. Mech Ageing Dev 29:255–266

Vyas I, Heikkila RE, Nicklas WJ (1986) Studies on the neurotoxicity of 1-methyl-4-phenyl-1,2,3,6-tetrahydropyridine: inhibition of NAD-linked substrate oxidation by its metabolite, 1-methyl-4-phenylpyridinium. J Neurochem 46:1501–1507

Wagner GC, Jarvis MF, Carelli RM (1985) Ascorbic acid reduces the dopamine depletion induced by MPTP. Neuropharmacology 24:1261–1262

Weissman J, Trevor A, Chiba K, Peterson LA, Caldera P, Castagnoli N Jr, Baillie T (1985) Metabolism of the nigrostriatal toxin 1-methyl-4-phenyl-1,2,3,6-tetrahydropyridine by liver homogenate fractions. J Med Chem 28:997–1001

Westlund KN, Denney RM, Kochersperger LM, Rose RM, Abell CW (1985) Distinct monoamine oxidase A and B populations in primate brain. Science 230:181–183

Wu E, Shinka T, Caldera-Munoz P, Yoshizumi H, Trevor A, Castagnoli N (1988) Metabolic studies on the nigrostriatal toxin MPTP and its MAO B generated dihydropyridinium metabolite MPDP$^+$. Chem Res Toxicol 1:186–194

Yang S, Johannessen JN, Markey SP (1988) Metabolism of [^{14}C]MPTP in mouse and monkey implicates MPP$^+$, and not bound metabolites, as the operative neurotoxin. Chem Res Toxicol 1:228–233

Yong VW, Perry TL, Krisman AA (1986) Depletion of glutathione in brainstem of mice caused by N-methyl-4-phenyl-1,2,3,6-tetrahydropyridine is prevented by antioxidant pretreatment. Neurosci Lett 63:56–60

Youngster SK, McKeown KA, Jin YZ, Ramsay RR, Heikkila RE, Singer TP (1989a) Oxidation of analogs of 1-methyl-4-phenyl-1,2,3,6-tetrahydropyridine by monoamine oxidases A and B and the inhibition of monoamine oxidases by the oxidation products. J Neurochem 53:1837–1842

Youngster SK, Nicklas WJ, Heikkila RE (1989b) Structure-activity study of the mechanism of 1-methyl-4-phenyl-1,2,3,6-tetrahydropyridine (MPTP)-induced neurotoxicity. II. Evaluation of the biological activity of the pyridinium metabolites formed from the monoamine oxidase-catalyzed oxidation of MPTP analogs. J Pharmacol Exp Ther 249:829–83

Youngster SK, Sonsalla PK, Sieber BA, Heikkila RE (1989c) Structure-activity study of the mechanism of 1-methyl-4-phenyl-1,2,3,6-tetrahydropyridine (MPTP)-induced neurotoxicity. I. Evaluation of the biological activity of MPTP analogs. J Pharmacol Exp Ther 249:820–828

CHAPTER 11

Tetanus and Botulinum Neurotoxins*

H.H. Wellhöner

A. Introduction

Tetanus toxin and the botulinum toxins A, B, C1, D, E, F, and G are proteins produced by bacilli of the genus *Clostridium*. The eight toxins have a similar structure, they are translocated into neurons by adsorptive endocytosis, and they act predominantly on nerve cells. They have so many features in common (van Heyningen 1982; Mellanby 1984; Simpson 1990) that a comparative discussion not only appears to be justified but may contribute to a better understanding of the whole group of "clostridial neurotoxins" (CNTs).

Their scientific evaluation began about 120 years ago. Thousands of papers have been published since then, and reviews have been compiled from time to time, the later ones referring to their antecedents. This approach will also be followed in the present chapter, in which priority will be given to the period between 1981 and 1990. For the older literature on tetanus toxin, the reader is referred to Mellanby and Green (1981), Wellhöner (1982), van Heyningen (1986), and Habermann and Dreyer (1986). Reviews on the botulinum toxins have been published by Gundersen (1980), Sugiyama (1980), Simpson (1981, 1986), Sakaguchi (1983), Middlebrook (1986), and Sellin (1987). By the middle of the 1980s, the interest of the reviewers concentrated on the action of the CNTs at the cellular level (Middlebrook and Dorland 1984; Habermann and Dreyer 1986; Middlebrook 1986; Simpson 1986). Multiauthor textbooks covering both experimental and clinical aspects of tetanus and botulism (Veronesi 1981; Lewis 1981; Simpson 1989) and proceedings of conferences on tetanus (Anonymous 1982; Nistico et al. 1985) have been published. For clinical aspects of tetanus and botulism and for the use of botulinum toxin as a therapeutical agent, the reader is referred to Veronesi (1981) and Simpson (1989). This article will neither deal with these aspects nor include the ADP-ribosylation of actin and of GTP binding proteins by some of the botulinum toxins, because the ADP-ribosylation is not related to their typical neurotoxic action i.e., their inhibitory action on the transmitter release of nerve cells.

* Work supported by the Deutsche Forschungsgemeinschaft.

Reviews on this subject have been published by AKTORIES and WEGNER (1989) and by AKTORIES (1990).

B. Sources of Clostridial Neurotoxins

Only one serological type of tetanus toxin (Teta) is known so far. Teta is produced by *Clostridium tetani* under strictly anaerobic conditions. The gene for Teta resides on a plasmid. When strains of *C. tetani* lost this large plasmid, they are no longer toxigenic (LAIRD et al. 1980). A ^{32}P-labeled oligonucleotide coding for six contiguous amino acids at the N-terminus of Teta could be hybridized with plasmid DNA from toxigenic strains but not with plasmid DNA from nontoxigenic strains (FINN et al. 1984). The hybridization technique was also useful for the localization of the genes for the botulinum toxins (see below).

Seven serologically distinct botulinum neurotoxins (A, B, C1, D, E, F, and G) are known. They are produced by four different groups of *C. botulinum*, but BotE and BotF may also be produced by *C. butyricum* (McCROSKEY et al. 1986) and *C. baratii* (HALL et al. 1985), respectively. The genes for BotCl and BotD are carried by bacteriophages (FUJII et al. 1987; KIMURA et al. 1990). The gene for BotG resides on a plasmid (EKLUND et al. 1988), whereas those for BotA and BotE are embodied in the chromosomal DNA (BINZ et al. 1990a,b; THOMPSON et al. 1990). The gene loci for the other botulinum toxins are still unknown.

C. Production, Purification, and Structure

I. Introduction

A deductive approach to this chapter would start with the primary structures of the CNTs, calculate their molecular weights from their primary structures, find subunits, define hydrophobic and hydrophilic regions, discuss the secondary and tertiary structures with reference to the primary sturctures, and finally compare the structures of the eight forms. The historical sequence of experiments and results was not deductive in this way but was led by the availability and development of experimental techniques in protein and nucleic acid biochemistry. First, methods for the production of the CNTs were developed together with some rough enrichment and purification procedures based on filtration, precipitation, and dialysis. The yield was estimated with bioassays. Next, with the advent of chromatographic techniques, very pure samples of the CNTs were prepared. It could be shown by gel filtration and SDS gel electrophoresis that all CNTs were proteins with a molecular weight of around 150 000. At that time, apparatus became available for semiautomatic amino acid analysis, and programs were initiated to elucidate the primary structure of the CNTs with this technique. It was obvious that for amino acid analysis, the 150 000 dalton molecules had to be split into smaller subunits.

When Teta purified from the culture fluid was reduced, two subunits were obtained: an N-terminal 50-kDa subunit and a C-terminal 100-kDa subunit. The former was termed the L-chain and the latter, the H-chain by CRAVEN and DAWSON (1973). In retrospective, this first nomenclature was probably the best: Many other bacterial and plant toxins like diphtheria toxin, ricin, abrin, etc. have a similar L-chain–S-S–H-chain structure. The toxic action resides in the L-chain, the uptake and routing of the toxins are governed by structural elements residing mainly in the H-chain, and the terms L-chain and H-chain are now used with most toxins. We shall, therefore, adopt these terms for all eight CNTs in this article. Other nomenclatures will be listed later.

When Teta was retrieved directly from the clostridia and not from the culture fluid, an additional covalent link between the L-chain and H-chain had to be opened ("nicking") by mild trypsination before the chains could be separated by reduction of the disulfide bond between them. When Teta was treated with papain, a split occurred in the H-chain: The C-terminal fragment of the H-chain (H_C) and a fragment (L-H_N) consisting of the L-chain linked to the N-terminal fragment of the H-chain (H_N) were retrieved. In a final step, the L-chain was separated from the N-terminal fragment of the H-chain. Subunits of the botulinal toxins were prepared by similar methods. Frequently, the biological action of a subunit was tested first in the laboratory in which it was prepared, and a method was used that was simple and just at hand, but the results could be formed into a concept only later.

Painstaking efforts were then made to elucidate the primary structure of the CNTs by amino acid analysis of their subunits. While this work was in progress, the methods for nucleotide sequence analysis were developed and refined so fast that the complete nucleotide sequences of several CNTs together with the encoded amino acid sequences were published before the classic approach through amino acid analysis had led to final results. It should be emphasized, however, that the direct amino acid analysis greatly facilitated the analysis by nucleotide sequencing because it provided the sequences of amino acids for the synthesis of oligonucleotide probes.

Independent of investigations into the primary structure, efforts were made to gain insights into the secondary and tertiary structure of the CNTs. Up to now, spectroscopic and other methods prevail. The modern computer-assisted analysis of the spatial structure is not only difficult and time-consuming for 150-kDa molecules, it is also hampered by a lack of published crystallographic data for the CNTs.

II. Terminology of the Toxin Fragments

The historical development of the terminology for Teta led into a most confusing situation. At the Fifth European Workshop on Bacterial Protein

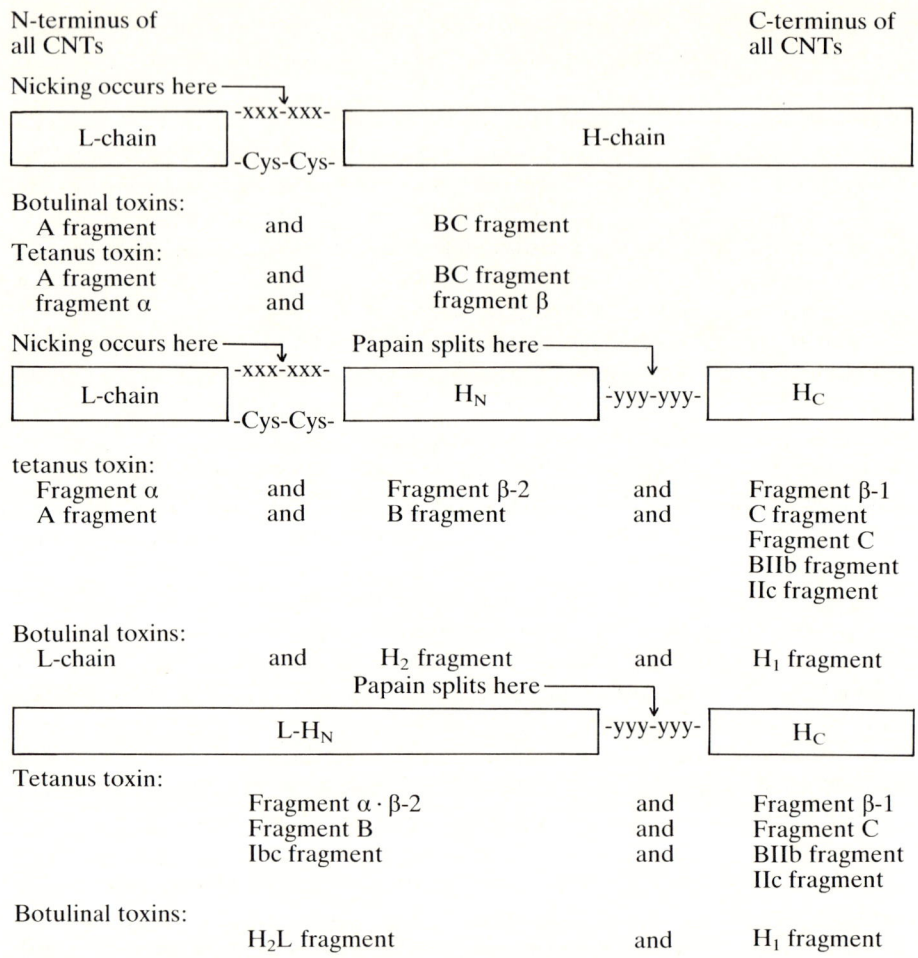

Fig. 1. Fragments of clostridial toxins and their synonyms

Toxins, a unifying terminology for all clostridial toxins was suggested and accepted. In Fig. 1, this terminology has been boxed. The symbol xxx-xxx stands for two adjacent amino acids between which a split may occur through the action of a nicking enzyme (e.g., trypsin). Papain splits between the amino acids marked yyy-yyy.

III. Production and Purification

For amino acid analysis, very pure preparations of the toxins are needed. The large-scale production and purification of Teta for industrial purposes was well-known by 1980 (cp. WELLHÖNER 1982). A modern production and purification method on the laboratory scale has been described by WELLER et al. (1988). Salt precipitates of tetanus toxin that have been produced with their method can be stored frozen over years without any loss of toxicity. Important improvements of the purification techniques for Teta were published by CHIU et al. (1982), DiMARI et al. (1982), LAZAROVICI et al. (1984), OZUTSUMI et al. (1985), and SHEPPARD et al. (1987). ROBINSON (1988) has written a review on the purification of tetanus toxin and its fragments. The purification techniques used in various combinations involve salt precipitation, size exclusion and ion exchange chromatography, affinity chromatography on solid-phase gangliosides or immunoadsorbents, and hydrophobic chromatography. Major efforts had to be made in order to remove even traces of proteolytic activities.

The techniques available by 1982 for the purification of BotA, BotB, BotC1, BotD, and BotE have been described in sufficient detail by SAKAGUCHI (1983) in his review. They resemble those used for the purification of Teta. Important improvements have been published for BotA (TSE et al. 1982), BotCl and BotD (DEJONGH et al. 1989), BotE (DASGUPTA and RASMUSSEN 1983), and for BotA, BotB, and BotE (SATHYAMOORTHY and DASGUPTA 1985b; WOODY and DASGUPTA 1988). MORIISHII et al. (1990) could separate the toxic activity of BotD from the accompanying ADP-ribosylating activity. This work was important also in a functional aspect: It helped to disprove the assumption that the inhibitory action of the Bot CNTs is mediated through the ADP-ribosylation of a protein. WADSWORTH et al. (1990) succeeded with the large-scale purification of BotF. NUKINA et al. (1988) could purify BotG.

With the progress in molecular biology, the production of CNT fragments by expression of their genes in bacteria became feasible. The interest in such techniques is great (FAIRWEATHER et al. 1987), because the active immunization with fragments of the CNTs is considered to be preferable to the immunization with toxoids. The C-terminal fragment of Teta has been expressed in *Escherichia coli* (FAIRWEATHER et al. 1986; EISEL et al. 1987; HALPERN et al. 1990a). The level of expression could be improved by removing rare codons (MAKOFF et al. 1989).

IV. Primary Structures

1. Tetanus Toxin

ROBINSON and HASH (1982) summarized the results of early attempts to sequence the N-terminus of Teta and, using two different methods of sequencing, found strong evidence for Pro as the N-terminal amino acid. By nucleotide sequencing (EISEL et al. 1986; FAIRWEATHER and LYNESS 1986), the CCA codon for Pro was identified as the second codon. In the first place, the ATG start codon of the Teta structure gene was found. It codes for a Met(1) that is removed later. No removable signal peptide or leader peptide is encoded between Met(1) and Pro(2). The Teta structure gene codes for a total of 1315 amino acids. For the following discussion, the

amino acid numeration used by Eisel et al. (1986) is adopted: Met(1),
Pro(2), . . . The disulfide bridge linking the L-chain to the H-chain is located
between Cys(439) and Cys(467) (Krieglstein et al. 1990). Obviously, nicking
may occur at several positions between these two Cys, and the nicking
between Ala(457) and Ser(458) found by Eisel et al. (1986) is only one of
several possibilities. As a consequence, the lengths of the L-chain and H-
chain are defined only within a range of <27 amino acids. In the invariable
part of the L-chain, four half-cysteines are found at positions 27, 186, 199,
and 312. A hydrophobic sequence is located between Tyr(223) and Ile(253).
It contains three acidic amino acids and four His and is of a predictable α-
helical structure. The stretch between Cys(429) and Cys(467) does not
contain half-cysteines.

Papain splits the H-chain between Ser(864) and Lys(865) (Fairweather
et al. 1986; Eisel et al. 1986; Halpern et al. 1990a). TET H_N, the N-terminal
papain fragment of the H-chain begins with a half-cysteine followed by a
hydrophilic domain between Lys(469) and Thr(476), and there is a hydrophobic sequence between Asn(660) and Ala(691). The latter sequence is
long enough to span a plasma membrane and contains three acidic amino
acids. On TET H_C, the C-terminal papain fragment of the H-chain, a disulfide
bridge is found between Cys(1077) and Cys(1093). This loop is smaller
than the Cys(439)-Cys(476) loop. Free half-cysteine residues are located at
positions 869 and 1301.

2. Botulinum A Toxin

Amino acid sequences in different regions of BotA have been described
by DasGupta and Sathyamoorthy (1984), Schmidt et al. (1984),
Sathyamoorthy and DasGupta (1985b), DasGupta et al. (1987b, 1988),
Betley et al. (1989), DasGupta and Foley (1989), and DasGupta and
Deklava (1990). The nucleotide sequence was published by Binz et al.
(1990b) and Thompson et al. (1990). BotA is synthetized with a leading Met
that is removed later, and without a removable signal peptide. The BotA
structure gene codes for 1296 amino acids. The L-chain is linked to the H-
chain by a disulfide bridge probably located between Cys(430) and Cys(454).
Nicking occurs between these two positions, and Ala(449) may become the
N-terminal amino acid of a H-chain (Sathyamoorthy et al. 1988; Thompson
et al. 1990). An attack by a protease from C. botulinum probably aiming
at Arg(432) has been reported by Deklava and DasGupta (1989,
1990).

Alignment of the BotA L-chain to the Teta L-chain shows that three of
the four His in the hydrophobic sequence between Tyr(223) and Ile(253) of
Teta are conserved in the corresponding sequence between Phe(213) and
Phe(243) of BotA. Likewise, alignment of the H-chains of the two CNTs
reveals that the N-terminal hydrophobic sequence of the Teta H-chain has
been largely conserved in BotA.

3. Botulinum B Toxin

Sequences of amino acids in different regions of BotB have been published by KLYUCHEVA et al. (1982), DASGUPTA (1984), DASGUPTA and WOODY (1984), SCHMIDT et al. (1985), and DASGUPTA and DATTA (1988). The N-terminal amino acid of BotB is Pro. BotB can be split into an L-chain and a H-chain by mild trypsination and reduction of a disulfide bond.

4. Botulinum C1 Toxin

The nucleotide sequence has been published by HAUSER et al. (1990) and KIMURA et al. (1990). The structure gene codes for 1291 amino acids. Met(1) is removed, and Pro(2) becomes the N-terminal amino acid. A removable signal peptide has not been found. the L-chain is linked to the H-chain by a disulfide bridge, probably between Cys(437) and Cys(453). Nicking occurs between these two positions.

Alignment of the L-chains of BotC1 and Teta shows that three of the four His in the hydrophobic sequence between Tyr(223) and Ile(253) of Teta are conserved in a hydrophobic sequence of BotC1 between Asp(222) and Gly(240).

5. Botulinum D Toxin

The nucleotide sequence has been published by BINZ et al. (1990a). The structure gene codes for 1276 amino acids. The homology with BotC1 is high, particularly in the N-terminal region of the H-chain.

6. Botulinum E Toxin

Amino acid sequences in different regions of BotE have been described by DASGUPTA and RASMUSSEN (1983), SATHYAMOORTHY and DASGUPTA (1985a,b), SCHMIDT et al. (1985), GIMINEZ et al. (1988), GIMINEZ and SUGIYAMA (1988), DASGUPTA and FOLEY (1989), and GIMINEZ and DASGUPTA (1990). The N-terminal amino acid of BotE is Pro. BotE retrieved from *C. botulinum* and *C. butyricum* differ in a few amino acids. In contrast to the other botulinum toxins and Teta, BotE is not nicked by bacterial proteases in the culture broth. It can be split into an L-chain and a H-chain by mild trypsination and reduction of a disulfide bond. Trypsin attacks an Arg-Lys bond. The complete amino acid sequences are unknown.

V. Higher Structures

1. Crystallography

So far, only ROBINSON and coworkers have been able to crystallize Teta and to subject the crystals to electron crystallography (CHIU et al. 1982; REIDLER and ROBINSON 1988; ROBINSON et al. 1988a). In Fig. 6 of the last paper, a gallery of instructive three-dimensional surface representations is shown. Teta has an asymmetric, three-lobed structure. Robinson et al. (1988b) have also crystallized BotA, but these crystals have not yet been used for electron crystallography study. For BotC1, Syuto and Kubo (1982) have calculated a diameter of 7.4 nm and a thickness of 4.3 nm.

2. Secondary Structure

More than 55% of the molecule of all CNTs shows an ordered structure. For Teta, the most recent study has been published by Singh et al. (1990). They found 20% ± 2.1% α-helix, 50.5% ± 2% β-sheets, no β-turns, and 29.5% random coils. At the endosomal pH of 5.5 they observed changes in the amide III frequency region. Datta and DasGupta (1988a,b) studied the secondary structure of BotA and BotE. They found 28% α-helix and 47% β-sheet for BotA at pH 6.0 and 20% α-helix and 47% β-sheet for nicked BotE. The secondary structure of BotA was not changed much by toxoidation or by chain separation (Singh and DasGupta 1989a,b,c,d). A change in Trp-containing segments occurred on separation of the BotA chains (Singh and DasGupta 1989a).

3. Tertiary Structure

Nicking increases the number of exposed Tyr in BotE (Datta and DasGupta 1988b), but not in BotA and BotB. Some 84% of all Tyr are exposed in BotA, 58% in BotB, and 61% in nicked BotE (Singh and DasGupta 1989c). Toxoidation considerably changed the number of exposed Tyr and Trp (Singh and DasGupta 1989d).

Experiments with monoclonal antibodies shed light on similarities in the tertiary structure of the CNTs. This is evident from Table 1.

For the discussion of differences between the actions of Teta and BotA (see below) it should be kept in mind that two L-chain-specific antibodies have been found that cross-react with Teta and several botulinal toxins with the exception of BotA.

Table 1. Cross-reactivity of monoclonal antibodies directed against clostridial neurotoxins. L indicates that the antibody has been raised against the L-chain

	Teta	BotA	BotB	BotC	BotD	BotE	BotF
Tzuzuki et al. (1987)	++	0	+	++	+	L++	0
Halpern et al. (1989)	L++	0	++	++		+	
Hambleton et al. (1984)		++	++			++	
Oguma et al. (1981)				++	+		
				+	++		

Polyclonal antibodies with a neutralizing action on CNTs may not pre-
cipitate. This has been shown by LIN et al. (1985) for antibodies directed
against the Teta L-chain.

VI. Binding, Uptake, Routing, and Action of Clostridial Neurotoxins as a Function of Their Subunits and Structure

1. Introduction

If the molecular biochemistry of binding, uptake, action, and routing of the
CNTs was sufficiently known, modifications of the CNT structures and CNT
substructures could be discussed in such a context. Modifications could then
provide additional proof for a particular part of a hypothesis and contribute
to a better understanding of the events at the molecular level. For several
reasons we are far from such an approach. First, the biological consequences
of modifications have been very often checked with overall assays like lethal
dose estimation. If a modification has no effect on the lethal dose, it may
provisionally be assumed that it has no influence on the binding, uptake,
strength of action, degradation, or elimination of the CNT. If, however, a
modification decreases the lethal dose, this could have several reasons.
Second, many modifications have been produced in the classic way: The
CNTs were reacted with protein reagents. Protein reagents modify more
than one amino acid on the surface of a CNT molecule. Therefore, with this
approach, there is virtually no chance of identifying the particular amino
acid whose modification was crucial for a given change in the biological
activity. Site-specific mutagenesis would be a far better approach. Third, it is
still a matter of debate what binding sites on cells mediate the action of the
CNTs. Fourth, the mechanism of action at the molecular level is unknown
for all CNTs. For all these reasons, the discussion of modifications has
remained largely descriptive. The situation is more satisfying with respect to
the CNT subunits. For a classification of the results obtained with subunits
and modifications, fitting a personal system of creative speculation, the
reader has to know only three basic facts about the binding, uptake, and
action of CNTs: (1) CNTs are bound by receptors on the plasma membrane
of nerve cells. (2) CNTs are taken up into nerve cells by receptor-mediated
endocytosis. (3) CNTs act on nerve cells by inhibiting the stimulated release
of neurotransmitters.

2. Isotoxins and Subunits of the Clostridial Neurotoxins

a) Nicking

Nicking increases the toxicity of most CNTs (Teta: WELLER et al. 1988;
AHNERT-HILGER et al. 1989a,c; BERGEY et al. 1989; BotA: STECHER et al.
1989a; BotE: SIMPSON and DASGUPTA 1983; YOKOSAWA et al. 1986; BITTNER et
al. 1989a; LOMNETH et al. 1990). The extent of this increase is dependent on

the biological test substrate, it varies for the different CNTs, but surprisingly it is virtually independent of the cleavage position in the amino acid sequence between the two Cys involved in the link between the L-chain and the H-chain (Teta: WELLER et al. 1988). At least in chromaffin cells, nicking with subsequent intraneuronal complete separation of the L-chain from the H-chain in the intraneuronal space is a prerequisite for the action of the L-chain (AHNERT-HILGER et al. 1989a; STECHER et al. 1989a), but not for the binding of the CNT (BotA: EVANS et al. 1986).

It is still an open question whether another partial enzymatic hydrolysis at an unknown site of Teta is responsible for the reported existence of its two chromatorgraphically and biologically different forms. The phenomenon was first reported by LAZAROVICI et al. (1984) and later confirmed by PARTON et al. (1989). The two Teta forms bind to neurons, synaptic membranes, and polysialogangliosides (FEDINEC et al. 1986) with varying affinities.

b) Function of the L-Chain

On extracellular application, the L-chains of the CNTs are not bound to or taken up into nerve cells, and no action has been observed (Teta: DAUZENROTH et al. 1989; MOCHIDA et al. 1989; BotA, BotB: BANDYOPADHYAY et al. 1987; BotC1: KUROKAWA et al. 1987). After injection into nerve cells or trans-location following permeabilization (digitonin, electroporation), they exhibit the action of the full toxins (Teta: AHNERT-HILGER et al. 1989c; BITTNER et al. 1989b; MOCHIDA et al. 1989; BotA: STECHER et al. 1989b; McINNES and DOLLY 1990; BotE: BITTNER et al. 1989a).

In chromaffin cells, the L-chains exhibit the action of their parent CNTs in the absence of the H-chains (Teta: AHNERT-HILGER et al. 1989c; BotA: STECHER et al. 1989b). In *Aplysia* neurons, the additional intracellular presence of the H-chain is necessary for the BotA L-chain, but not for the Teta L-chain (POULAIN et al. 1988a,b, 1989a,b; MOCHIDA et al. 1990; TAUC et al. 1990).

c) Function of the H-Chain

The H-chains mediate the binding and uptake of the CNTs (BotA: DOLLY et al. 1988; POULAIN et al. 1988b; BotB: KOZAKI et al. 1987; BotC1: AGUI et al. 1985; KUROKAWA et al. 1987; BotC1: MURAYAMA et al. 1987). They can compete with their source CNT for binding sites (Teta: SIMPSON 1984; BotA: LOMNETH et al. 1990; BotA, BotB: BANDYOPADHYAY et al. 1987; BotC: AGUI et al. 1985; MURAYAMA et al. 1984). A cross-competition with another CNT has been observed: The binding of BotD to synaptosomes could be inhibited by the H-chain of BotC1 (MURAYAMA et al. 1984).

Under most experimental conditions, the H-chains were not toxic by themselves (Teta: MOCHIDA et al. 1989; BotA: STECHER et al. 1989a; BITTNER et al. 1989a). A weak inhibition on the K^+-evoked norepinephrine (NE) release from synaptosomes was the only effect observed by WELLER et al. (1989).

H-chains are hydrophobic (Teta: Lazarovici et al. 1987) and form pores in lipid membranes (BotA: Hoch et al. 1985).

d) Functions Assigned to the C-Terminal Part of the H-Chain, H_C

The C-terminal parts of the H-chains mediate the binding, uptake, and release functions of the CNTs (Teta: Bizzini et al. 1981; Simpson and Hoch 1985; Evinger and Erichsen 1986; Büttner-Ennever et al. 1981; Meckler et al. 1990; Manning et al. 1990; and cp. Wellhöner 1982 for previous work; BotA: Shone et al. 1985; Kozaki et al. 1989; BotE: Kamata et al. 1986). The cloned and functionally expressed C-terminal part of the H-chain of Teta (Fairweather et al. 1986; Halpern et al. 1990a) displays all capabilities of the fragment produced by papain hydrolysis. Moreover, the C-terminal parts of the H-chains can compete with their parent toxins (Morris et al. 1980; Goldberg et al. 1981; Simpson 1984b, 1985; Simpson and Hoch 1985; Fishman and Carrigan 1988) and with some other CNTs (Simpson 1984a,c) for binding sites.

e) Functions Assigned to the N-Terminal Part of the H-Chain

The N-terminal part of the H-chain produces pores in lipid membranes (Teta: Matsuda et al. 1989; Simpson and Hoch 1985; BotA: Shone et al. 1987; Blaustein et al. 1987; BotE: Blaustein et al. 1988). So far, binding of the N-terminal part of the H-chain to plasma membranes has been observed only in *Aplysia* neurons. It mediates the uptake of the CNTs into these neurons (Teta: Mochida et al. 1989; BotA: Poulain et al. 1989a,b).

f) Interaction of the L-Chain with the H-Chain, H_N

The L-chain and the H-chain assist each other in obtaining the optimal structure for binding and action. The C-terminal fragment of the H-chain usually has a lower binding affinity than the parent toxin (Teta: Simpson 1984a,b; Mellanby and Leonhard 1986; Weller et al. 1986; Teta, BotA: Montecucco et al. 1988; BotA, BotB: Maisey et al. 1988).

H-chains and L-chains can reunite (Teta: Matsuda and Yoneda 1976; Mochida et al. 1989; Weller et al. 1989; BotA: Bhattacharyya et al. 1988; Lomneth et al. 1990; BotC1: Syuto and Kubo 1981), and the reconstituted molecules show virtually no loss of toxicity. The reconstitution of the disulfide linkage between the L-chain and H-chain seems to be necessary for an action of the reconstituted and externally applied CNT on mammalian nerve cells (Weller et al. 1989), but not for an action on *Aplysia* (Maisey et al. 1988). In *Aplysia*, even the BotA H-chain may mediate the entry of the Teta L-chain (Dauzenroth et al. 1989).

g) The L-Chain Linked to the C-Terminal Part of the H-Chain, $L-H_N$

The Teta $L-H_N$, unlike the H-chain, does not bind to polysialogangliosides. Whether Teta $L-H_N$ in the extracellular space has an action on neurons distinct from Teta has remained a matter of discussion. In the intact mouse, Teta L-H$_N$ produces systemic paralysis (FEDINEC et al. 1987). GAWADE et al. (1985) incubated Teta and its fragments with nonprecipitating antibodies directed either against Teta H$_C$ or against Teta L-H$_N$. Systemic injection of a high dose of Teta into mice led to a flaccid paralysis, while systemic injection of Teta incubated with H$_C$-specific Fab led to spastic paralysis, and in turn the authors could convert the spastic action of intracerebrally injected toxin to a flaccid paralysis when they preincubated Teta with L-H$_N$-specific Fab. They suggested that a second action of Teta may be encoded somewhere in fragment L-H$_N$. TAKANO et al. (1989a) injected Teta or Teta L-H$_N$ directly into the spinal cord and obtained a weaker action on the monosynaptic reflex with the fragment than with Teta. In the mouse phrenic nerve-hemidiaphragm, Teta was by at least two orders of magnitude more potent than its fragment L-H$_N$ (SIMPSON and HOCH 1985), but according to the authors, a trace contamination of the fragment with Teta could not be excluded. After injection into bovine adrenal chromaffin cells, the Teta L-H$_N$ inhibited the transmitter release as effectively as the source toxin (PENNER et al. 1986). This should be expected, because the fragment contains the L-chain of Teta. The Teta L-H$_N$ was fully active also in *Aplysia* (MOCHIDA et al. 1989).

It was outlined above that the N-terminal parts of the H-chains may produce pores in lipid membranes. The Teta L-H$_N$ also forms such pores (BOQUET and DUFLOT 1982; SIMPSON and HOCH 1985), apparently because they contain the H$_N$ part of the H-chain. BotB L-H$_N$ does not bind to polysialogangliosides (KOZAKI et al. 1987).

3. Modification of Amino Acids

Most modifications of tetanus toxin were tried before 1981 and have been reviewed by WELLHÖNER (1982). Modifications have been listed in Table 2.

Teta, BotA, and BotB are resistant against modification of the Lys residues to the extent that the modification of a few Lys remains virtually without influence on the biological activity. Therefore, the decrease of biological activity of Teta after acylation with Bolton and Hunter reagent or other N-succinimidyl derivatives may in part be due to the acylation of Tyr residues. Indeed, the modification of a few Tyr residues in Teta with tetranitromethane results in a profound reduction of toxicity. Tetranitromethane also modifies Trp residues, but this seems to be of no importance, because the modification of a few Trp with Koshland's reagent does not reduce the toxicity of Teta. The strong reduction of Teta toxicity by oxidative iodination may be due in part to the modification of Tyr but also partly to the oxidation of His. The modification of His with ethoxyformic acid profoundly decreases Teta toxicity.

In contrast to Teta, the oxidative iodination of BotA and BotF had virtually no influence on their toxicity, although as in Teta, the modification

Table 2. Modification of residues in clostridial neurotoxins

References	Toxin	Amino acids modified	Reagent or procedure
HABERMANN (1972)	Teta	Tyr (His, Trp)	Oxidative iodination
BIZZINI et al. (1980)	Teta	Tyr (Trp)	Tetranitromethane
AN DER LAHN et al. (1973)	Teta	Lys (Tyr)	Bolton and Hunter reagent
MEYER-EPPLER and WELLHÖNER (1983)	Teta	Lys (Tyr)	N-Succinimidyl-(2,3-³H)-propionate
FUJITA et al. (1988)	Teta	Lys (Tyr)	N-Succinimidyl-biotin
cf. WELLHÖNER (1982)	Teta	Lys	Reductive alkylation, amidination, maleylation, carbamylation
BIZZINI et al. (1973)	Teta	Trp	Koshland's reagent
STEIN and BIEL (1973)	Teta	His	Ethoxyformic acid
cf. WELLHÖNER (1982)	Teta	Several	Aldehydes
DASGUPTA et al. (1987a)	BotA	Lys	Methylation
SATHYAMOORTHY and DASGUPTA (1988)	BotA BotB	Lys	Reductive methylation
DEKLEVA et al. (1988, 1989)	BotA	Lys	Reductive alkylation
DASGUPTA et al. (1987a)	BotA BotB BotE	Tyr	Tetranitromethane
TERAJIMA et al. (1987)	BotC1	Tyr	Spin labeling
HABERMANN (1974)	BotA	Tyr (His, Trp)	Oxidative iodination
DOLLY et al. (1982) WILLIAMS et al. (1983)	BotA	Tyr (His, Trp)	Oxidative iodination
EVANS et al. (1986, 1988)	BotB	Tyr (His, Trp)	Oxidative iodination
DASGUPTA and SUGIYAMA (1980)	BotA BotE	Arg	1,2-Cyclohexanedione
WOODY et al. (1989)	BotA BotE	Asp Glu	
SINGH and DASGUPTA (1989d)	BotA	Several	Toxoidation
WADSWORTH et al. (1990)	BotF	Tyr (His, Trp)	Oxidative iodination

Teta, tetanus toxin; Bot, botulinum toxin.

of a few Tyr in BotA, BotB, and BotE with tetranitromethane reduced the toxicity. Such a reduction was also observed in BotC1 when spin labels were affixed to Tyr. Obviously, not enough Tyr were modified by oxidative iodination; however, this argument should also apply to Teta. The oxidation of His may be more important for the decrease in toxicity. Since oxidative iodination removes only the action of the CNTs on transmitter release and does not erase the binding and uptake capabilities of Teta (cp. WELLHÖNER 1982) or BotA (HABERMANN 1974), a nonconservative His exchange should be sought in the primary structures of the L-chains. In the section on primary structures, attention was focussed on a hydrophobic sequence between Tyr(223) and Ile(253) of Teta that contains four His. Alignment of the BotA structure of the Teta structure (THOMPSON et al. 1990) shows that three of the four His have been conserved, but the counterpart of His(250) in Teta is Asn(240) in BotA. Unfortunately, a study on the biological activity of BotA modified in its His residues has not been published yet.

BotA, BotB, and BotE are not very sensitive to a modification of their Lys residues. As one may expect, they are sensitive to a modification of the charged amino acids Arg, Asp, and Glu.

Toxoid formation with aldehydes destroys the toxicity of the CNTs. Strongly toxoided CNTs lose their toxicity and are no longer bound by nerve cells. Their immunogenicity remains virtually unimpaired. Mildly toxoided Teta is still taken up by nerve axons (WELLER et al. 1986).

D. Binding of Clostridial Neurotoxins to Cells and the Routing of the Toxins Through Neurons

I. The Three-Step Model

A great deal of the evidence for the binding of CNTs to nerve cells was available by 1980 and has been reviewed (cp. SIMPSON 1981; WELLHÖNER 1982). The idea that the CNTs do not act directly by activation of receptors on the plasma membrane but only after internalization by the nerve cells was first forwarded for BotA by SIMPSON (1980). He proposed a model consisting of three steps: binding, translocation, and intracellular processing leading to an overt action. Later, this model was generally accepted for all CNTs. Experimenting with the phrenic nerve-hemidiaphragm, Simpson had already described some essential features of the model: The binding occurred in the absence of external Ca^{2+} and nerve stimulation and had a low temperature dependence (Q_{10} about 1.6), the half-time of binding was about 12 min, and the two subsequent steps were temperature-dependent and required calcium. SCHMITT et al. (1981) investigated Teta with the same simple technique and could confirm and extend the model. Binding of Teta proceeded at temperatures as low as 0°C, translocation occurred at elevated temperatures and was moderately temperature-dependent, while the progression to paralysis was strongly dependent on temperature. Additional experiments have shown that CNTs are not taken up at low temperatures (BotA, BotB into mammalian presynaptic terminals: BLACK and DOLLY 1986b).

Electrical activity of a neuron may accelerate the uptake of CNTs without increasing the number of binding sites on their plasma membranes (Teta: WELLHÖNER et al. 1973; BotA: SIMPSON 1985; BotA, BotB: BLACK and DOLLY 1986b).

Antibodies to Teta inactivated bound toxin but could no longer neutralize translocated toxin (see below). Next, the three-step model was confirmed for BotE (SIMPSON and DASGUPTA 1983). Again, BotE-specific antibodies inactivated only bound BotE and did not neutralize translocated BotE.

The three-step model was worked out by employing a simple method of "classic" pharmacology. Subsequently, the steps were investigated in more detail and in other biological substrates by biochemical methods. The binding of Teta was temperature-independent in cultured rat cerebral and mouse spinal neurons (YAVIN et al. 1983; CRITCHLEY et al. 1985; PARTON et al. 1987), while the translocation of Teta occurred only at higher temperatures (YAVIN et al. 1982, 1983; PARTON et al. 1987). Teta disappeared from the surface of PC12 cells with a half-life of only $1-2$ min (SANDBERG et al. 1989b). By 10 min after its application, 50% of the Teta bound to N18-RE-105 cells was inaccessible for pronase (STAUB et al. 1986). Teta was seen in the endosomes of cultured mouse spinal neurons 15 min after its application (PARTON et al. 1987). The lag time between application of Teta and the reduction of transmitter release was $15-30$ min for rat particulate cortex (ALBUS and HABERMANN 1983) and 30 min for NG108-15 cells at a Teta concentration of $10^{-12} M$ (WELLHÖNER and NEVILLE 1987). Corresponding results have been obtained with BotA. It is taken up into the cholinergic nerve terminals of *Torpedo marmorata* (SOLSONA et al. 1990). The lag time of its inhibitory action on transmitter release from rat brain synaptosomes is time- and temperature-dependent (ASHTON and DOLLY 1988). The lag time for an action of BotC1 on rat brain synaptosomes is 15 min (MURAYAMA et al. 1987).

II. Binding

1. Cell Specificity

CNTs are predominantly bound to the unmyelinated parts of neuronal plasma membranes (BotA, BotB to motor nerve terminals: BLACK and DOLLY 1986a). A preference of binding to specific presynaptic terminals has been shown (BotA to cholinergic terminals: BLACK and DOLLY 1987).

Nonneural cells which have been observed to bind CNTs include glia cells (RAFF et al. 1983a,b; SCHNITZER et al. 1984), Müller cells (retinal glia cells that contain polysialogangliosides; HUBA and HOFMANN 1988), macrophages

(BLASI et al. 1990), pancreatic islet cells (EISENBARTH et al. 1982), thymic epithelium cells (HAYNES et al. 1983), and some kidney cells (HABERMANN and ALBUS 1986).

Neurons express binding sites for Teta as early as fetal day 10 (KOULAKOF et al. 1982). The sites appear in different parts of the brain at different times (PUYMIRAT et al. 1982). On Müller cells such sites have been found postnatally between day 0 and day 30 (HUBA and HOFMANN 1988). In amphibians, DUPRAT et al. (1986) found Teta binding sites in the late gastrula stage. As development advances, some neurons may lose their Teta binding sites later: In the early postnatal period, rat sensory and sympathetic neurons stain positively for Teta and negatively for neuron-specific enolase, while at a later stage, they stain negatively for Teta and positively for enolase (GROTHE and UNSICKER 1988).

Fetal, dividing, undifferentiated cells of neuronal origin may not necessarily express Teta binding sites (KOULAKOF et al. 1983). However, many neuroblastoma cells in patients may be able to express such sites, and Teta binding tests have been introduced as a method for differential diagnosis (BERLINER and UNSICKER 1985). If growing neuroblastoma cells bind Teta, toxin binding sites remain constant throughout the cell cycle (N_2AB-1 cells:

Table 3. Binding constants of clostridial neurotoxins to neuronal substrates

Toxin	Neural substrate	K_d (pM)	Reference
Teta	Rat and bovine brain membranes	1200	ROGERS and SNYDER (1981)
Teta	Rat brain membrane	2000–12 000	GOLDBERG et al. (1981)
Teta	Rat brain membrane	2000	CRITCHLEY et al. (1986)
Teta	Rat brain membrane	260–1140	PIERCE et al. (1986)
Teta	Rat brain membranes	140 and 2000	WELLER et al. (1988)
Teta	Rat cortex homogenate and gangliosides	2.3 and 200	ALBUS and HABERMANN (1983)
Teta	PC12 cells	1250	WALTON et al. (1988)
Teta	N18-RE-105 cells	620	STAUB et al. (1986)
Teta	NG108-15 cells	20	WELLHÖNER and NEVILLE (1987)
Teta	Rat brain membranes	270	PARTON et al. (1989)
BotA	Rat synaptosomes	600	WILLIAMS et al. (1983)
BotB	Rat synaptosomes	300–500	EVANS et al. (1986) PARK et al. (1990)
BotC1	Rat synaptosomes, ganglioside GT1b	79 and 35 000	AGUI et al. (1985)
BotC1	NG108-15 cells	150 and 14 500	YOKOSAWA et al. (1989)
BotF	Rat synaptosomes	150 and 20 000	WADSWORTH et al. (1990)

Teta, tetanus toxin; Bot, botulinum toxin.

NOTTER and LEARY 1985). Many types of cultured cells express the full array of their CNT binding sites only after forced differentiation, for instance with nerve growth factor (NGF) (N_2AB-1 cells: NOTTER and LEARY 1986; PC12 cells: FIGLIOMENI and GRASSO 1985; FUJITA et al. 1988; NATHAN and YAVIN 1989; WALTON et al. 1988; SANDBERG et al. 1989b; rat adrenal medullary neurons: LIETZKE and UNSICKER 1983) or after autodifferentiation forced by serum withdrawal (rat hippocampal neurons: HODSON and CURBEAM 1982; NG108-15 and NCB 20 cells: YAVIN and HABIG 1984; NG108-15, NBr-10a, and N18 cells: KALZ and WELLHÖNER 1990).

2. Affinity of Binding

Binding occurs with high affinity and may be dependent on the ionic strength and the pH of the incubation medium. Binding constants are listed in Table 3. At low ionic strengths, CNTs may bind with a comparable low affinity to sites of high capacity. At high ionic strengths, these sites virtually disappear, and binding sites with high affinity and low capacity are found.

The CNTs bind to the same substrate with different strengths: In cultured mouse brain cells, the binding strength decreases in the order BotA > BotC1 >> BotE (KUROKAWA et al. 1987).

The high-affinity binding sites have a low density on plasma membranes (450 Teta sites on one NG108-15 cell: WELLHÖNER and NEVILLE 1987; 150–650 BotA sites on one mammalian presynaptic terminal: DOLLY et al. 1984).

Binding sites of lower affinity ($K_d > 2\,nM$) have been detected in many biological substrates. Occasionally, two classes of binding sites were detected. PARTON et al. (1988, 1989) found on rat brain membranes a high-affinity, low-capacity class of neuraminidase- and protease-sensitive sites and a low-affinity, high-capacity class of sialidase-sensitive and protease-insensitive sites.

3. Chemical Nature of the Binding Sites

It was first shown by VAN HEYNINGEN (1959; cp. WELLHÖNER 1982 for later evidence) that Teta was bound to gangliosides and, with high affinity, to some polysialogangliosides. Subsequently, it turned out that all CNTs were bound by polysialogangliosides. The more recent evidence is summarized in Table 4.

Oxidation with periodate of the carbohydrate moieties of gangliosides increased the binding capacity of differentiated PC12 cells of Teta 2.5-fold (NATHAN and YAVIN 1989).

If the receptors for CNTs indeed are polysialogangliosides, substances that bind to polysialogangliosides should compete with the binding of CNTs, and a pretreatment of the biological substrate with neuraminidase should prevent the binding and action of CNTs. The evidence obtained with competing substances is very circumstantial. MARCONI et al. (1982) were able to reduce the systemic toxicity of Teta in mice by pretreating them with the lectins concanavalin A or phytohemagglutinin P. HEREDO and OJA (1985) suggested that the antagonistic action of ruthenium red on the stimulation of GABA release from rat brain

Table 4. Recent experiments demonstrating binding of CNTs to different polysialogangliosides

Toxin	Ganglioside	Reference
BotA	Preferentially to GT1b	KITAMURA et al. (1980)
BotA	GQ1b > GT1b, GD1a, not to GD1b	TAKAMIZAVA et al. (1986)
BotA	GD1a	MARXEN et al. (1989)
BotB	GT1b and GD1a	KOZAKI et al. (1987)
BotE	GT1b, GD1a, GQ1b	KAMATA et al. (1986)
BotB, BotC1, BotF	GD1a, GD1b, GT1b	OCHANDA et al. (1986)
Teta, Tetatoxoid	Solid-phase gangliosides	HABERMANN and TAYOT (1985)
Teta, BotA, BotB, BotE,	Photolabeled phospholipids	MONTECUCCO et al. (1986, 1989) SCHIAVO et al. (1990)

Teta, tetanus toxin; Bot, botulinum toxin.

slices may be due to a binding of ruthenium red to sialic acid residues. Many more experiments have been done with neuraminidase; however, even after exhaustive treatment, the results remain equivocal and are different for Teta and BotA (Table 5).

Using bovine chromaffin cells, MARXEN and BIGALKE (1989) and MARXEN et al. (1989) provided direct evidence that different gangliosides could serve not only as binding sites but as receptors for Teta. Cultured bovine chromaffin cells do not express binding sites for Teta on their plasma membrane (LIETZKE and UNSICKER 1988). MARXEN and BIGALKE loaded such cells with [^{3}H]-NE, incubated them with various gangliosides, exposed them to Teta or BotA thereafter, and finally measured the K^+-induced Ca^{2+}-dependent release of tritium. The inhibition of tritium release by Teta was maximal when the cells had been incubated with GD1b. It was 10 times less after incubation with GT1b, and 30 times less for cells incubated with either GD1a or GM1. The entry of BotA was mediated by GD1a but also by other gangliosides. These results help a great deal to explain why in many biological substrates neuraminidase pretreatment does not completely abolish the binding of Teta.

It has been suggested that glycoproteins may function as receptors, too. CRITCHLEY et al. (1986), PIERCE et al. (1986), and STAUB et al. (1986) found a reduction of Teta binding after pronase treatment at physiological salt concentrations. Trypsination destroyed >40% of the Teta binding sites on cultured mouse neurons (YAVIN and NATHAN 1986), BotA sites of high affinity (WILLIAMS et al. 1983) and BotB sites of low affinity (EVANS et al. 1986) on rat brain synaptosomes. Using trypsin or chymotrypsin, LAZAROVICI and YAVIN (1986) were able to reduce the number of Teta binding sites

Table 5. Influence of pretreatment with neuraminidase on the binding and action of clostridial neurotoxins

Toxin	Neural substrate	Findings and reference
Teta	Rat brain membranes	Binding reduced (CRITCHLEY et al. 1986)
Teta	Rat, guinea-pig, and bovine synaptosomes	Binding reduced (LAZAROVICI and YAVIN 1986)
Teta	Rat particulate cortex	ACh release impaired, long chain gangliosides below detection limit (BIGALKE et al. 1981a)
Teta	Rat particulate cortex	Release of GABA and D-Asp impaired (ALBUS and HABERMANN 1983), but binding largely removed (HABERMANN and ALBUS 1986)
Teta	NCB 20 cells	Binding decreased by 29% (YAVIN and HABIG 1984)
Teta	Cultured mouse cerebral neurons	Cerebral neurons, immunofluorescence, binding sites decreased (YAVIN et al. 1982)
Teta	Cultured mouse spinal neurons	Polysialogangliosides removed, binding sites preserved, depressing action on postsynaptic potentials preserved (BIGALKE et al. 1986)
Teta	Cultured mouse spinal neurons	Binding sites virtually not reduced, action abolished (FUJITA et al. 1988). Biotinylation of Teta may have destroyed its biological activity
Teta	Isolated chromaffin granules	Binding reduced (LAZAROVICI et al. 1989)
BotA	Cultured mouse spinal neurons	Polysialogangliosides removed, binding sites removed, action of BotA on postsynaptic potentials removed (BIGALKE et al. 1986)
BotA	Mammalian synaptosomes	Binding sites removed (WILLIAMS et al. 1983; KITAMURA and SONE 1987)
BotB	Rat brain synaptosomes	Binding sites almost completely removed (EVANS et al. 1986)
BotF	Rat brain synaptosomes	Binding sites decreased (WADSWORTH et al. 1990)

Teta, tetanus toxin; Bot, botulinum toxin; ACh, acetylcholine; GABA, γ-aminobutyric acid.

in guinea-pig synaptosomes but not in rat or bovine synaptosomes. Subsequently, however, even the most dedicated efforts to detect and demonstrate glycoproteins as Teta receptors were futile (NATHAN and YAVIN 1989). It may well be that the enzymes decrease the number of apparent ganglioside binding sites by disturbing their protein environment. Evidence for such a mechanism has been provided in experiments with erythrocytes by LAZAROVICI and YAVIN (1985a,b). Erythrocytes do not physiologically express binding sites for tetanus toxin; however, when they are incubated with gangliosides, they bind Teta, and the sites binding Teta at physiological salt concentrations are sensitive to trypsin.

The evidence available from experiments with only one CNT in one or more biological substrates is sufficient to allow the assumption that a particular class of neurons may have more than one class of binding sites for a CNT.

There is also evidence that the different CNTs may share one class of binding sites in a particular biological substrate and may compete for it (BotC1 and BotD; MURAYAMA et al. 1984) but may not share a second class present on the same or another biological substrate. BotB, but not BotA, inhibited the binding of an antibody to a ganglioside on cholinergic nerve terminals (EVANS et al. 1988). At mammalian neuromuscular synapses, BotA binding was only partially inhibited by Teta (DOLLY et al. 1984) or Teta fragments and vice versa. In rat brain synaptosomes, BotB did not inhibit the binding of BotA (WILLIAMS et al. 1983), BotA was only a weak antagonist of BotB binding (EVANS et al. 1986), and BotA, BotB, and BotE did not inhibit the binding of BotF (WADSWORTH et al. 1990). Extracellularly applied BotA, but not Teta, inhibited the stimulated transmitter release from cultured bovine chromaffin cells (MARXEN et al. 1989). This was due to the fact that the cells contained the ganglioside GD1a (but not GD1b or GT1b) and that BotA could bind to GD1a 30 times better than Teta. The binding strength of the two toxins prepared by LAZAROVICI et al. (1984) and the two toxins used by PARTON et al. (1989) differed in neurons or rat brain membranes.

The various binding sites may be of different functional importance, because it has been shown that the CNTs enter the cells by different routes of receptor-mediated endocytosis (Teta into NG108-15 cells: KALZ and WELLHÖNER 1990; BotA and BotB into motor nerve terminals: BLACK and DOLLY 1986b). There may be uptake routes that would be less likely to lead to an action on transmitter release (BLACK and DOLLY 1986b).

III. Uptake of Clostridial Neurotoxins into Endosomes of Neurons

The CNTs are taken up into cells by adsorptive endocytosis (Teta: MONTESANO et al. 1982; PARTON et al. 1987; KALZ and WELLHÖNER 1990; BotA and BotB into motor nerve terminals: BLACK and DOLLY 1986b; BotA, BotB, BotC, BotE, Teta: SIMPSON 1988). It was outlined above that CNTs are not internalized at low temperatures and that electrical activity of a neuron may accelerate the uptake of CNTs. Adsorptive endocytosis may occur through clathrin-coated or noncoated membrane invaginations. Using Teta labelled with colloidal gold, MONTESANO et al. (1982) could demonstrate the uptake of Teta-gold through noncoated invaginations into liver cells. In apparent contradiction, PARTON et al. (1987) found Teta-gold in coated pits, coated vesicles, endosomes, and tubules of cultured mouse spinal neurons. In a follow-up investigation (PARTON et al. 1988), they incubated cells devoid of polysialogangliosides first with gangliosides and then with Teta. This time they could observe an uptake of Teta through the noncoated pits. KALZ and WELLHÖNER (1990), using the procedure devised by SANDVIG et al. (1989), succeeded in reducing the uptake of Teta into NG108-15 neurohybridoma cells by preincubation with ammonium chloride. This would mean that a fraction of Teta (>40%) was taken up through clathrin-coated pits and vesicles. Ammonium chloride (and methylamine) reduced not only the uptake but also the action of Teta, BotA, BotB, BotC (SIMPSON 1983, 1988), and BotE (SIMPSON and DASGUPTA 1983). Chloroquine reduced the uptake of BotA better than of Teta (SIMPSON 1982). It appears that CNTs are taken up into neurons both by clathrin-coated and noncoated vesicles.

IV. Handling of the Clostridial Neurotoxins in the Endosomes

Inside the endosomes, the CNTs are probably protected against proteolysis (Teta: ROA and BOQUET 1985). They become more hydrophobic at the acidic endosomal pH (Teta: BOQUET et al. 1984; CABIAUX et al. 1985; BotA, BotB, BotE: MONTECUCCO et al. 1986, 1989) and possibly produce pores in the endosomal membrane. Evidence for the formation of pores by CNTs has been obtained from experiments with lipid model membranes (Teta: BOROCHOV-NEORI 1984b; MENESTRINA et al. 1989; RAUCH et al. 1990; Teta and BotB: HOCH et al. 1985; BotC: DONOVAN and MIDDLEBROOK 1986; BotE: BLAUSTEIN et al. 1988). The channels open at a positive voltage on the CNT side measured against 0 mV on the non-CNT side (+50 mV for Teta in lipid model membranes: BOROCHOV-NEORI et al. 1984a; +40 mV for BotC in lipid model membranes: DONOVAN and MIDDLEBROOK 1986). A negative voltage inactivates the channels. A low pH on the CNT side of the membrane may favor channel formation. Teta and Bot form channels in lipid model membranes preferably if gangliosides are present on the Teta side (BOROCHOV-NEORI et al. 1984b; MONTECUCCO et al. 1988). These gangliosides probably interact with the hydrophobic region of the N-terminal fragment of the H-chain (MONTECUCCO et al. 1986); for this interaction, the hydrophobic character of that region must be strengthened by an acidic pH. The Teta channels are cation-selective (BOROCHOV-NEORI et al. 1984a; HOCH et al. 1985; GAMBALE and MONTAL 1988). Their lifetime is about 1 ms (GAMBALE and MONTAL 1988). Their diameter is more than 120 nm (FINKELSTEIN 1990). MENESTRINA et al. (1989) have calculated that the channels would allow the passage of molecules with a molecular weight <700. Such channels would be too narrow for the L-chain of any CNT, but they would be wide enough to allow a part of the L-chain to protrude into the cytoplasm (FINKELSTEIN 1990) or to allow depolarization of endosomes containing CNTs. This depolarization may be an important step in the translocation of the CNTs from the endosomes into the cytoplasm.

Two lines of circumstantial evidence indicate that the CNTs are trans-located from the endosomes to the cytoplasm. Upon injecting Teta into the cytoplasm of bovine chromaffin cells, Penner et al. (1986) measured an almost immediate reduction of quantal transmitter release. The complete separation of the L-chain from the H-chain by reduction of the disulfide bond connecting them is a prerequisite for the action of Teta in chromaffin cells (Ahnert-Hilger et al. 1989a; Stecher et al. 1989a). Reductases or glutathione are not normally available in endosomes and cannot enter them through the channels created by the CNTs in the endosomal membrane. Glutathione would, of course, be small enough to pass, but it is very improbable that glutathione could reduce the interchain disulfide bond in the absence of enzymes.

What target the CNTs reach next after they have entered the cytoplasm is unknown.

V. Axonal Transport

Because of the predominantly spinal action of Teta, the question as to how it is reaches the CNS fascinated investigators as long ago as the nineteenth century. A great many papers have been published; some were written in the aggressive style of former days. Some 90 years after the early paper of Bruschettini (1892), there was no longer any doubt about Teta reaching the CNS by retrograde axonal transport through motor and vegetative efferent axons. Conclusive evidence was provided simultaneously by Erdmann et al. (1975), Price et al. (1975), and Stöckel et al. (1975), and Stöckel et al. (1975) who, by means of histoautoradiography, demonstrated the intraaxonal localization of $[^{125}I]$-Teta in motor and adrenergic (Stöckel et al. 1975) fibers. With a few exceptions, the toxin transport is retrograde. Retrograde axonal transport has also been demonstrated for BotA (Habermann 1974; Wiegand et al. 1976). Papers on axonal transport deal no longer with this question but with isotoxins, toxin fragments, and conjugates containing a CNT or a CNT fragment. Virtually no work on the molecular mechanism of axonal transport of the CNTs has been published.

It has been outlined above that the L-chain and the H-chain of Teta assist each other in obtaining the optimal structure for binding. Accordingly, the axonal transport was most efficient when derivatives of the complete toxin were studied. This was already known for Teta labeled either by oxidation or acylation (Table 2). Fedinec et al. (1986) compared the Teta isotoxins described by Lazarovici et al. (1984) and found that both were transported to the spinal cord. Weller et al. (1986) produced a toxoid by a mild treatment of Teta with formaldehyde. This toxoid retained only 0.25%–1% of Teta's toxicity, but its uptake and transport were still considerable (20%–30%). The C-terminal part of the Teta H-chain contains the binding function of Teta, and therefore an axonal transport of Teta H_C was expected and found by Bizzini et al. (1981), Büttner-Ennever et al.

(1981), EVINGER and ERICHSEN (1986), MANNING et al. (1990), and MECKLER et al. (1990). WELLER et al. (1986) found in experiments with intact animals that, as compared with Teta, the uptake and axonal transport of Teta H_C were reduced to nothing more than 1%–2%. In spite of this drawback, the Teta H_C fragment is of great interest as a constituent of conjugates. In such conjugates, the toxin or a toxin fragment carrying the selective uptake and transport capabilities of the toxin is linked to another molecule that would not enter nerve cells normally. HABIG et al. (1983) used a nonneutralizing monoclonal Teta-specific antibody in an immunocomplex with Teta and showed that Teta mediated the uptake of the complex. Conjugates containing the nontoxic C-terminal Teta H_C were synthesized by BIZZINI et al. (1980, 1984), FISHMAN and CARRIGAN (1988), FISHMAN and SAVITT (1989), FISHMAN et al. (1990), and BEAUDE et al. (1990). The conjugate of BIZZINI et al. (1980) was very similar to Teta and was also substantially transported by neurons. FISHMANN and SAVITT (1989) synthesized a conjugate that contained the Teta H_C and horseradish peroxidase. The molecular weight of the conjugate exceeded 200 kDa, one conjugate molecule containing more than 1 molecule each of fragment (100 kDa) and peroxidase (44 kDa). The conjugate was readily taken up by axons. Using the C-terminal Teta fragment and IgG, FISHMAN and SAVITT (1989) were also able to introduce IgG into axons. Of particular interest was the approach of BEAUDE et al. (1990). They produced a conjugate consisting of the C-terminal Teta fragment and the marker protein glucose oxidase. The two proteins were linked through a disulfide bridge, and such bridges may be reduced inside the cell. The conjugate was readily taken up by motor axons and accumulated in motoneurons.

VI. Elimination and Degradation

In the spinal cords of rats, Teta had a half-life of 6.5 days (HABERMANN and DIMPFEL 1973), and virtually the same half-life of 5–6 days was found in cultured mouse spinal cells (HABIG et al. 1986). HABERMANN et al. (1977) produced local tetanus with $[^{125}I]$-Teta and extracted radioactive material from the spinal cord segments that had accumulated the highest amount of radioactivity. On gel filtration, some degradation of Teta became apparent from the slight shift of the extracted radioactive material to a lower molecular weight. In cultured rat cerebral neurons, no firm evidence for a degradation of Teta (YAVIN et al. 1981) was obtained. In cultured mouse spinal cells, HABIG et al. (1986) demonstrated a degradation of the L-chain and H-chain of Teta. In NG108-15 cells we found most recently that nondegraded Teta is released from these cells (KALZ and WELLHÖNER 1990). There is, however, also a fraction of degraded Teta. The cells retained the split products for much longer than the intact toxin. MARXEN and BIGALKE (1991a) found virtually the same degradation pattern of Teta in bovine chromaffin cells.

VII. Transsynaptic Transport

Intact tetanus toxin can cross synaptic clefts. This has been demonstrated with [125I]-Teta (Schwab and Thoenen 1977) and with Teta labeled with horseradish peroxidase (Schwab et al. 1979). Erdmann et al. (1981) injected Teta into a leg muscle; some hours later and well in advance of any signs of spasticity, they injected 125I-labeled antibodies into the CSF. This injection prevented the development of local tetanus, and on histological examination the motoneurons were enveloped by a layer of precipitated immunocomplex; on leaving the motoneurons, Teta had been immobilized by the 125I-labeled antibodies and could no longer enter its effector cells, the inhibitory interneurons. A transsynaptic transport was also described for the iodinated C-terminal fragment (Büttner-Ennever et al. 1981) and later for the native C-terminal fragment (Evinger and Erichsen 1986).

E. Actions of Clostridial Neurotoxins

I. Introduction

A great deal of information about the action of CNTs was obtained from investigations in which nerve-muscle preparations were employed. In these experiments, the nerve or muscle was stimulated, and the mechanical contractions were recorded. Paralysis of the skeletal musculature, the leading clinical symptom of botulism, led early onto experimental studies of the effect of botulinum toxin on neuromuscular transmission (Dickson and Shevky 1923; Edmunds and Long 1923; Schübel 1923). These authors advanced the idea that botulinum toxins act at the presynaptic terminal of cholinergic nerve endings. Burgen et al. (1949) provided conclusive evidence for this hypothesis. It took a further 25 years until Ambache et al. (1948) described for Teta an inhibition of the cholinergic transmission at the rabbit sphincter pupillae and another 16 years until Kryzhanovsky and Kasymov (1964) published experiments with Teta on the neuromuscular transmission in cats (for experiments in other animals, cp. Wellhöner 1982). Ambache et al. (1948) as well as Kryzhanovsky and Kasymov (1964) forwarded the idea that Teta inhibits the cholinergic transmission by inhibiting the release of ACh from the presynaptic terminal.

About 1950, a rapid advance in the field of electrophysiological recording equipment took momentum, and Brooks (1954, 1956) provided evidence that BotA reduced the frequency of miniature end-plate potentials (mepps). For the next 15 years, physical methods of classic pharmacology and electrophysiological investigations into the neuromuscular synapse prevailed in the further elucidation of the locus and mode of action of the CNTs. Biochemical methods had not yet been developed to a comparable level, the culture techniques for nerve cells were in their beginnings, and accordingly, some questions had to remain open for a time. As late as 1980, Gundersen had to conduct an extensive discussion dealing with possible effects of the botulinum toxins on transmitter synthesis, storage, and reuptake. With the advent of biochemical micromethods and cell culture techniques, the situation has changed greatly. The first results of biochemical investigations using synaptosomes were published for Teta by Osborne and Bradford (1973) and for BotA by Wonnacott and Marchbanks (1976). Yet, the simple nerve-muscle preparation continues to be of great value, and electrophysiology as a tool has not fallen behind biochemistry.

For some time it had been assumed that, in every aspect, Teta and the botulinum toxins acted in the same way. Biochemical investigations on transmitter release provided only limited evidence for major variations. However, more detailed electrophysiological analyses now leave little doubt that, in spite of all similarities, the CNTs in some aspects have distinctly different modes of action.

II. The Locus and the Phenomenology of Action

1. Clostridial Neurotoxins do not Impair Conduction

Teta reaches the CNS by retrograde axonal transport (cp. Wellhöner 1982). Since it was found in motoneurons, an action on these cells was initially expected, but it could be ruled out later. All conductile parameters of the motoneurons remained unchanged in the presence of Teta (Sverdlov 1969; Wiegand and Wellhöner 1979). Accordingly, the axonal conduction was not impaired by BotA (Dickson and Shevky 1923; Burgen et al. 1949; Bishop and Bofenbrenner 1936; Guyton and MacDonald 1947) or Teta (Diamond and Mellanby 1971; Wiegand et al. 1977).

The CNTs do not affect the spread of electrical excitation to the presynaptic nerve terminals. This has been shown for BotD by Harris and Miledi (1971) and for BotA and BotB in *Aplysia* by Poulain et al. (1988a). Circumstantial evidence may be derived from the finding that the block for synaptic transmission by CNTs can be overcome by repetitive stimulation, as has been shown first for BotA by Brooks (1956) and for Teta by Mellanby and Thompson (1972) and Polgar et al. (1972). More recent evidence is presented below.

2. Ion Fluxes Through the Plasma Membranes of Presynaptic Terminals

a) Action on Calcium Fluxes Through Neuronal Plasma Membranes

The CNTs do not inhibit the influx of Ca^{2+} into rat synaptosomes (Teta and BotA: Bigalke et al. 1981a) or the stimulated influx of Ca^{2+} into presynaptic nerve terminals (Teta, BotA: Dreyer et al. 1983; BotA: Gundersen et al. 1982; Molgo and Thesleff 1984; Mallart et al. 1989). They do not interfere with the intracellular calcium sequestration processes either (Mallart et al. 1989). It appears, therefore, that the action of CNTs on the transmitter release is not a consequence of their action on the Ca^{2+} current at presynaptic terminals. At sections of the plasma membrane distant from the terminal, however, an action of CNTs on Ca^{2+} currents may occur: A rapid block of the Ca^{2+} component of the action potential by Teta has been observed in N1E-115 cells (HIGASHIDA et al. 1983; SUGIMOTO et al. 1983).

b) Influence of Elevated Extracellular Ca^{2+} Concentration, of Other Cations, and of Aminopyridines

In rat brain synaptosomes preincubated with BotA, the release of NE but not of ACh could be restored with high Ca^{2+}, Ba^{2+}, or Sr^{2+} levels (ASHTON and DOLLY 1988). High Ca^{2+} concentrations had a weak to moderate antagonistic action against the inhibition by BotA and BotE of the transmission at vertebrate neuromuscular junctions (MOLGO et al. 1990). In guinea-pig synaptosomes, the inhibition of glutamate release by BotA could be overcome with ionomycin, a $Ca^{2+}/2H^+$ ionophore (SANCHEZ-PRIETO et al. 1987), whereas raising the extracellular Ca^{2+} content remained ineffective. The authors suggest that BotA may block the entry of Ca^{2+} into a hydrophobic compartment and that ionomycin may help the Ca^{2+} ions to bypass this block.

Raising the extracellular Ca^{2+} concentration did not antagonize the action of Teta in the neuromuscular junction of mice (DUCHEN and TONGE 1973), rats (WENDON 1980), or goldfish (MELLANBY et al. 1988), in guinea-pig cortex synaptosomes (SANCHEZ-PRIETO et al. 1987), or in chromaffin cells (BITTNER and HOLZ 1988), nor the action of botulinum toxins in the vertebrate neuromuscular junction (CULL-CANDY et al. 1976; GUNDERSEN et al. 1982; DREYER et al. 1983).

La^{3+} was first used as a "super-calcium" by MELLANBY and THOMPSON (1975). In the neuromuscular synapse of the goldfish fin muscle, it antagonized the action of Teta and restored a moderate mepp frequency. The La^{3+} treatment did not deplete the presynaptic terminal of vesicles (MELLANBY et al. 1988), but vesicles accumulated in a row near the presynaptic membrane. La^{3+} was also a moderate antagonist of Teta in the rat soleus muscle (BEVAN and WENDON 1984). Moreover, it was a weak antagonist of BotE in the rat neuromuscular junction (MOLGO et al. 1989a) but could not antagonize the BotA inhibition of the frog neuromuscular junction (ANGAUT-PETIT et al. 1988).

Zinc seems to be a more effective antagonist than lanthanum against the reduction of the mepp frequency by BotA in the mouse phrenic nerve-hemidiaphragm (NISHIMURA et al. 1988). However, zinc by itself increases the mepp frequency, but the mechanism of this action is unknown.

The K^+-channel blockers 3,4-aminopyridine or 4-aminopyridine increase the intraterminal Ca^{2+} concentration. They antagonize the actions of the CNTs. The extent of antagonism depends on the CNT, the Ca^{2+} concentration, and the biological substrate. 4-Aminopyridine largely antagonized the actions of Teta and BotA at the myenteric plexus-longitudinal muscle strip of the guinea-pig ileum (BIGALKE and HABERMANN 1980; MACKENZIE et al. 1982). It also inhibited the action of BotA in the frog gastrocnemius nerve-muscle preparation (METEZEAU and DESBAN 1982) and the rat extensor digitorum longus muscle (CULL-CANDY et al. 1976). Similarly, the effects of Teta and BotA in the mouse phrenic nerve-hemidiaphragm were antagonized by 4-aminopyridine (HABERMANN et al. 1980), whereas the inhibition by BotA of stimulus-evoked transmitter release from rat brain synaptosomes was not affected (ASHTON and DOLLY 1988). On repetitive stimulation in the presence of 4-aminopyridine, only a few transmitter quanta were released from presynaptic terminals poisoned with BotA (DREYER et al. 1984) or with BotE (MOLGO et al. 1989a). For BotA, however, the release occurred without delay. In contrast, a remarkable increase of quantal release was observed for Teta after a delay of 2–3 s (DREYER et al. 1984). 3,4-Diaminopyridine, an antagonist of BotA at the neuromuscular junction (MOLGO et al. 1984), was more effective against BotA than against BotB (SELLIN et al. 1983b) or BotF (KAUFMANN et al. 1985) in the rat neuromuscular junction. In the mouse phrenic nerve-hemidiaphragm, SIMPSON (1988) found that 3,4-aminopyridine antagonized the action of BotA but not that of BotB, BotC, BotE, or Teta.

3. Actions on the Synthesis, Reuptake, Repartition, and Storage of Transmitters

a) Synthesis

Teta and BotA inhibit to some extent the synthesis of ACh from choline (mouse brain slices and synaptosomes: GUNDERSEN and HOWARD 1978; cultured mouse spinal neurons: BIGALKE et al. 1978). However, this has been considered to be not a direct action of the CNTs but a regulatory phenomenon.

b) Uptake

Under the influence of CNTs, either a small reduction in the uptake of transmitters or virtually no change has been observed (Table 6). It is an open question whether the reduction in transmitter uptake seen in some experiments represents a direct action of the CNTs or reflects only a regulatory answer of the neurons to the impairment of transmitter release by the CNTs.

c) Content

The content of transmitters remains either unchanged (Teta, rat brain slices and particles: BIGALKE et al. 1981b; BotA, synaptosomes: WONNACOTT and MARCHBANKS 1976; WONNACOTT 1980; GUNDERSEN and HOWARD 1978;

Table 6. Action of clostridial neurotoxins on the uptake of transmitters

Toxin	Action on uptake	Substrate	Reference
Teta	Uptake of GABA: small reduction	Rat brain particles	BIGALKE et al. (1981b) ALBUS and HABERMANN (1983)
Teta	Uptake of GABA: no influence	Rat hippocampal slices	COLLINGRIDGE et al. (1981)
Teta	Uptake of GABA: no influence	Rat brain synaptosomes	KRYZHANOVSKY et al. (1982)
Teta	Uptake of ACh: reduction	Rat brain synaptosomes	HABERMANN et al. (1981)
Teta	Uptake of NF: no influence	Rat brain particles	BIGALKE et al. (1981b)
BotA	Uptake of ACh: small reduction	Rat brain synaptosomes	HABERMANN et al. (1981) ASHTON and DOLLY (1988)

Teta, tetanus toxin; Bot, botulinum toxin; GABA, γ-aminobutyric acid; ACh, acetylcholine; NE, norepinephrine.

BotA, rat muscle: POLAK et al. 1981; GUNDERSEN and JENDEN 1983; BotA, *Torpedo* electric organ: DUNANT et al. 1987) or is slightly elevated (Teta, spinal cord segments: HILBIG et al. 1979). The number of synaptic vesicles remained either virtually unchanged (BotA in the *Torpedo* electric organ: DUNANT et al. 1987) or was slightly increased (Teta in goldfish: MELLANBY and THOMPSON 1975; Teta in the mammalian spinal cord: KRYZHANOVSKY et al. 1971). Again, the elevation of transmitter content as well as the increase in vesicle number may reflect a regulatory phenomenon rather than a direct action of the CNTs.

d) Compartmentalization

Studies of the compartmentalization of transmitters were conducted using cholinergic synapses. BotA had no influence on the repartition of ACh in the *Torpedo* electric organ (DUNANT et al. 1987).

4. Clostridial Neurotoxins Impair the Release of Transmitters, Neurohormones, and Neuromodulators

a) Biochemical and Biophysical Evidence

CNTs inhibit the Ca^{2+}-dependent release of transmitters from all presynaptic terminals (Tables 7 and 8), preferentially from the free pool of newly synthetized transmitter (ACh from *Torpedo* marmorata: MARSAL et al. 1988). According to STANLEY and DRACHMANN (1983), the quantal rather

Table 7. Biochemical evidence for the inhibition of transmitter release from nervous tissue by clostridial neurotoxins

Toxin	Action on release	Substrate	Reference
Teta	Release of GABA: inhibition	Rat brain synaptosomes	KRYZHANOVSKY et al. (1982)
Teta	Release of GABA: inhibition	Brain particles from rat and chicken	ALBUS and HABERMANN 1983
Teta	Release of GABA: inhibition	Cultured rat cerebellar neurons	PEARCE et al. (1983)
Teta	Release of D-Asp: inhibition	Brain particles from rat and chicken	ALBUS and HABERMANN (1983)
Teta	Release of Glu: inhibition	Guinea-pig brain synaptosomes	SANCHEZ-PRIETO et al. (1987)
Teta	Release of Glu: inhibition	Cultured cerebellar granule cells from mice	VAN VLIET et al. (1989)
Teta	Release of NE: inhibition	Rat brain particles	HABERMANN (1981)
Teta	Release of NE: inhibition	Permeabilized chromaffin cells	BITTNER and HOLZ (1988) AHNERT-HILGER et al. (1989a) LAZAROVICI et al. (1989)
Teta	Release of NA: inhibition	PC12 cells	FIGLIOMENI and GRASSO (1985)
Teta	Release of NE: inhibition	Cultured mouse brain cells	HABERMANN et al. (1988)
Teta	Release of ACh: inhibition	*Torpedo marmorata* synaptosomes	EGEA et al. (1990)
Teta	Release of ACh: inhibition	Rat brain slices	BIGALKE et al. (1981a)
Teta	Release of ACh: inhibition	NG108-15 cells	WENDON and GILL (1982) WELLHÖNER and NEVILLE (1987)
Teta	Release of ACh: inhibition	PC12 cells	SANDBERG et al. (1989b)
Teta	Release of opioids: inhibition	Rat brain particles	JANICKI and HABERMANN (1983)
Teta	Release of oxytoxin and vasopressin: inhibition	Rat pituitaries	HALPERN et al. (1990b)
BotA	Release of ACh: inhibition	Guinea-pig cortex synaptosomes	WONNACOTT and MARCHBANKS (1976) WONNACOTT (1980)
BotA	Release of ACh: inhibition	Rat brain slices	BIGALKE et al. (1981a)
BotA	Release of ACh: inhibition	*Torpedo* electric organ	DUNANT et al. (1988)

Table 7 (continued)

Toxin	Action on release	Substrate	Reference
BotA	Release of NE: inhibition	Rat brain particles	HABERMANN (1981)
BotA	Release of NE: inhibition	Permeabilized chromaffin cells	AHNERT-HILGER et al. (1989b)
BotA	Release of several transmitters: inhibition	Rat brain synaptosomes	ASHTON and DOLLY (1988)
BotA, BotB, BotD	Release of catecholamines: inhibition	Bovine chromaffin cells	KNIGHT (1986)
BotA, BotC	Release of NE: inhibition	Cultured mouse brain cells	HABERMANN et al. (1988)
BotA	Release of opioids: inhibition	Rat brain particles	JANICKI and HABERMANN (1983)

Teta, tetanus toxin; Bot, botulinum toxin; GABA, γ-aminobutyric acid; ACh, acetylcholine; NE, morepinephrine.

Table 8. Early reports on the action of botulinum toxins on mepps

Toxin	Substrate	Mepps frequency	Reference
Bot	Guinea-pig	Reduced	BROOKS (1956)
BotD	Frog	Reduced; skewed amplitude	HARRIS and MILEDI (1971)
Bot	Rat	Reduced; skewed amplitude	SPITZER (1972)
BotA	Frog	Reduced; skewed amplitude	BOROFF et al. (1974)
Bot		Reduced; skewed amplitude	CULL-CANDY et al. (1976)
BotD	Rat	Reduced; skewed amplitude	BRAY and HARRIS (1975)

than the nonquantal release is affected, but this statement has been questioned (DOLEZAL et al. 1983).

The release of substances other than transmitters, for instance ATP (WHITE et al. 1980; RABASSEDA et al. 1987; MARSAL et al. 1987, 1989), remains unimpaired.

The biochemical experiments supporting an action of the CNTs on transmitter release are listed in Table 7.

The conclusion from the list is straightforward: whatever the neuronal substrate used, whatever the transmitter investigated, and whatever the CNT used, the CNT inhibited the release of the transmitter, and this was true not only for transmitters in the narrow sense but also for neurohormones (oxytocin,vasopressin) and neuromodulators (opioids).

In biochemical experiments, qualitative differences between the CNTs occasionally became apparent when the synergistic or antagonistic action of modulator substances on the inhibition of transmitter release by CNTs were investigated. Carbachol was a partial antagonist of BotA but not of Teta (chromaffin cells; BIGALKE et al. 1989). The bulk of proof for qualitative differences between the modes of action of the CNTs has been provided by biophysical experiments at the neuromuscular junction, in which the membrane potential changes of the postsynaptic muscle cell were recorded in the absence of stimulation and in response either to a single stimulus or to repetitive stimulation of the presynaptic neuron.

In the early experiments, it was shown that the CNTs reduce the probability of both the spontaneous and the stimulus-evoked discharge of transmitter quanta. Teta reduced the number of mepps in all but one (BEVAN and WENDON 1984) of the experiments (MELLANBY and THOMPSON 1972; KRYZHANOVSKY et al. 1971; DUCHEN and TONGE 1973; HABERMANN et al. 1980; DREYER and SCHMITT 1981; BERGEY et al. 1987). The reduction of mepps frequency and evoked postsynaptic potential (EPP) was also noted for BotA (BROOKS 1954, 1956; SPITZER 1972; BOROFF et al. 1974; CULL-CANDY et al. 1976) and BotD (HARRIS and MILEDI 1971; BRAY and HARRIS 1975). Subsequently, this has been found by many authors and is now a standard part of virtually any report that deals with a particular feature of the action of CNTs on mepps or EPPs. The reduction in the number, though not in the size, of mepps has been observed not only in vertebrates but also in invertebrates (Bot A and BotB in *Aplysia*: POULAIN et al. 1988a).

With the classic recording technique in the neuromuscular synapse, one does not record events arising in the presynaptic terminal where the CNTs act: The mepps rather represent the answer of the postsynaptic cell (mepps or EPPs) to events in the presynaptic terminal (release of transmitter quanta). PENNER et al. (1986) used a new and most attractive technique for the direct measurement of the CNT action on transmitter release. Whenever a cell discharges a transmitter quantum, this event is accompanied by a very small enlargement of the cell surface. The cell may be seen as a (leaky) capacitor: the plasma membrane is the insulating sheet that separates two conductors, represented by the intracellular and the extracellular space. A small enlargement of the cell surface during the exocytosis of a transmitter quantum results in a small and transient increase of the cell capacitance. If the cell is mounted as a capacitor in a capacitance meter, the small capacitance changes can be recorded. PENNER et al. (1986) used bovine adrenal chromaffin cells and injected BotA, Teta, or Teta fragment B (the L-chain linked to the N-terminus of the H-chain) into the cytoplasm. The frequency of the capacitance changes were sharply reduced within a few minutes. This experiment provided the most important information that CNTs can reach their ultimate target from the cytoplasmic space very quickly. Therefore, if antibodies could be introduced into this space, there would be some chance that the antibodies could neutralize CNTs inside neurons.

The CNTs have different potencies at the neuromuscular synapse (see below), but this may be taken as a quantitative and not a qualitative difference.

For some time a qualitative difference in the action of the CNTs on different populations of mepps had been considered. Kriebel et al. (1976) have shown that in the absence of stimulation the mepps evoked by spontaneous release of transmitter belong to two populations: About 90% are medium-sized 'classic" mepps and form a Gaussian distribution (mean amplitude 0.5–1.2 mV); the remaining 10% are small mepps with <0.2 mV amplitude which form a skewed distribution. Teta preferentially blocks the medium-sized mepps (Dreyer and Schmitt 1983), so that a skewed distribution of small mepps is recorded from nearly completely blocked preparations (Duchen and Tonge 1973; Dreyer and Schmitt 1983). Repetitive stimulation causes some of the medium-sized mepps to reappear but has no influence on the frequency of the small-sized mepps (Teta: Duchen and Tonge 1973; BotA: Harris and Miledi 1971; Teta and BotA in comparison: Dreyer and Schmitt 1983). They can be observed, because at higher concentrations the CNTs block the physiological EPP. Dreyer and Schmitt found in mouse hemidiaphragms poisoned in vitro that, in contrast to Teta, BotA blocked both the medium-sized and the small-sized mepps. They considered this an important difference between Teta and BotA. This view has not been generally accepted because a skewed distribution of the mepps amplitudes has been found also in preparations poisoned with BotA (frog: Boroff et al. 1974; rat: Tse et al. 1982; mouse: Dolly et al. 1987) or with BotE (Molgo et al. 1989a).

The most consistent qualitative difference between the CNTs is observed when the synchronization of discharge during repetitive stimulation is recorded. In neuromuscular synapses poisoned with either BotA or BotE, the discharge is synchronous: The delays between stimulus and release are almost identical for all released quanta. In contrast, the delays are widely scattered in neuromuscular synapses poisoned with Teta, BotB, BotD, or BotF (Table 9). At present, this difference between the CNTs cannot be explained.

The differences in the synchronization of discharge become more pronounced when the CNT-poisoned synapses are cooled to room temperature (Lundh 1983; Dreyer et al. 1984) or are treated with 3,4-aminopyridine (Dreyer and Schmitt 1981; Sellin et al. 1983b). 3,4-Aminopyridine is a more effective antagonist of BotA action than of Teta action (Dreyer and Schmidt 1981) or of BotB action (Sellin et al. 1983b), and 4-aminopyridine is a better antagonist of Bot A than of BotB (Gansel et al. 1987).

Another difference between Teta and BotA was detected when Dreyer et al. (1987) used black widow spider venom to release transmitter from mouse hemidiaphragms paralyzed with CNTs. The venom was just as effective in promoting transmitter release from BotA-treated synapses as from control synapses and produced striking morphological changes: The

Table 9. Synchronous (S) and asynchronous (A) release of transmitter quanta on repetitive stimulation of presynaptic terminals of neuromuscular synapses

Teta	BotA	BotB	BotC1	BotD	BotE	BotF	Reference
A							Bevan and Wendon (1984)
A	S						Dreyer and Schmitt (1981, 1983)
A	S	A					Gansel et al. (1987)
	S	A					Sellin et al. (1983 b)
	S			A			Molgo et al. (1989b)
					S		Molgo et al. (1989a)
	S					A	Kaufmann et al. (1985)

vesicles were depleted, and the mitochondria were swollen and disrupted. When Teta-treated synapses were exposed to the venom, the increase of transmitter release was moderate, and no ultrastructural alterations could be observed. Further studies with other CNTs may reveal whether the action of black widow spider venom is strong for the BotA, BotE group and weak for the group containing the Teta-like botulinum toxins.

b) Quantitative Comparisons

The Clostridial Neurotoxin as the Variable. Two CNTs may suppress the release of a particular transmitter from a particular biological substrate with different potency. Teta was more potent than BotA as an inhibitor of NE release from rat brain particles (Habermann 1981). It was 20 times as potent as BotA in blocking the release of methionine-enkephalin-like substances from the same substrate (Janicki and Habermann 1983). Conversely, BotA was 200 times as potent as Teta at the neuromuscular junction of the mouse diaphragm (Dreyer and Schmitt 1981), and BotB exceeded the potency of Teta by a factor of 500 in the mouse triangulus sterni muscle (Gansel et al. 1987). BotE was more potent at inhibiting spontaneous mepps in the rat extensor digitorum longus muscle than BotA, BotB, or BotD (Molgo et al. 1989a). However, it was less potent than BotA in the rat tibialis anterior muscle (Sellin et al. 1983a).

The Transmitter as the Variable. A particular CNT may suppress the release of different transmitters from the same biological substrate with varying potency. BotA inhibits ACh release from rat brain synaptosomes more strongly than the release of catecholamines (Ashton and Dolly 1988) and from rat brain particles more strongly than the release of GABA (Bigalke et al. 1981a,b). Teta inhibits the transmitter release from rat brain particles in the order of Gly > GABA ≫ ACh (Bigalke et al. 1981b) and in the order of GABA > D-Asp (Albus and Habermann 1983). Correspondingly, differences in the inhibition by Teta of synaptic transmission in slices from different brain areas have been observed (CALABRESI et al. 1989).

The Function of the Synapse as the Variable.   It has been assumed for a long time that Teta acts on the inhibitory synapses at the spinal cord, whereas the botulinum toxins act at the excitatory neuromuscular synapse in the periphery. Evidence to the contrary was first obtained for Teta (see MATSUDA et al. 1982). Later it was shown that BotA disinhibited inhibitory synapses in the spinal cord (MIKHAILOV and BARASHKOV 1977; WIEGAND and WELLHÖNER 1977). Therefore, it was difficult to understand that Teta should have no influence on the classic spinal monosynaptic reflex, as has been stated repeatedly (cp. WELLHÖNER 1982). TAKANO and his group were able to show that this statement held true only for the early period of local tetanus and that the spinal monosynaptic reflex was indeed reduced at a later stage of local tetanus (TAKANO et al. 1983, 1989a,b; KANDA and TAKANO 1983). The disinhibition of the monosynaptic reflex just has a shorter delay (about 1 day) than the inhibition (about 5 days). Two arguments may be used for the explanation of the different delays. According to the first argument, inhibitory synapses are located at the soma membrane of motoneurons and may be the first to be reached by Teta, while excitatory synapses are located on the motoneuron dendrites at which Teta may arrive some time later. This thesis has not gained support from experiments involving intraspinal injections of Teta (TAKANO et al. 1989a). According to the second argument, inhibitory synapses are more sensitive to Teta than excitatory synapses. The results of several experiments are in favor of this hypothesis. Upon intraspinal injection (TAKANO et al. 1989b), the inhibitory and excitatory synapses in a spinal motoneuron pool are reached by Teta at virtually the same time, yet the disinhibition of the monosynaptic reflex occurred before its inhibition. BERGEY et al. (1987) exposed cultured mouse spinal neurons to Teta and recorded mepps, EPSPs, and IPSPs (excitatory and inhibitory postsynaptic potentials). The IPSPs were markedly reduced by Teta before the EPSPs were clearly affected.

A varying sensitivity of inhibitory and excitatory synapses has not been found for BotA (BIGALKE et al. 1985). Accordingly, the discharge pattern in the neuronal syncytium differed for cultures exposed to Teta or BotA.

The release of the inhibitory transmitter GABA from superfused rat particulate cortex was more sensitive to Teta than the release of the excitatory transmitter Asp (ALBUS and HABERMANN 1983).

The Biological Substrate as the Variable.   The order of potency for two CNTs acting on the uptake or release of a particular transmitter may vary with different biological substrates. The uptake of choline into rat synaptosomes is more sensitive to Teta than to BotA, while, vice versa, the uptake into chicken brain synaptosomes is more sensitive to BotA than to Teta (HABERMANN et al. 1981). In the investigations of RABASSEDA et al.

(1988), Teta was 100–200 times more potent than BotA as an inhibitor of ACh release from mammalian central cholinergic nerve terminals (rat striatum slices), whereas BotA was 100 times more potent than Teta at peripheral cholinergic nerve terminals (*Torpedo* electric organ).

5. Action on Postsynaptic Cells

The CNTs do not impair the responsiveness of the postsynaptic membrane at a time when the synaptic transmission is impaired already. Early evidence was provided for Teta (in rats: KAESER and SANER 1969; in goldfish: DIAMOND and MELLANBY 1971; MELLANBY and THOMPSON 1972; in the mouse diaphragm: HABERMANN et al. 1980). It was shown in cultured mouse spinal neurons by means of iontophoretic application of transmitter amino acids that Teta had no influence on the neuronal response to Gly and GABA (BERGEY et al. 1983) and that BotA did not change the neuronal response to Glu, Gly, and GABA (BIGALKE et al. 1985).

6. Late Effects

a) Damage to Neurons

ROTHMAN (1983) has found that synaptic overactivity may lead to the death of the afflicted, hyperactive neurons. For Teta, damage to neurons has repeatedly been described and must be considered the nonspecific consequence of overactivity without involving a particular Teta action. GADOTH et al. (1981) found a permanent tetraplegia in children who had suffered from tetanus neonatorum. CHOU and PAYNE (1982) published a case report on vacuolation and chromatolysis of motoneurons in tetanus. BAGETTA et al. (1990a,b) found a loss of neurons at the application site after injection of Teta into the hippocampus.

Another source of neuronal dammage, ADP-ribosylation or ribosylation of a rho-protein by BotC or BotD preparations may play an important part. Reports to this effect are available from MATSUOKA et al. (1986) and KUROKAWA et al. (1987) for BotC1.

b) Sprouting

Following the injection of a sublethal dose of BotA into a muscle, the ensuing inhibition of the stimulus-evoked transmitter release may last for weeks. After a few days, one may observe an outgrowth of motor fibers from the terminal arborization of the efferent axon. This phenomenon has been described for the mouse (DUCHEN and STRICH 1968; ANGAUT-PETIT et al. 1990), rat (DUXON and STOLKIN 1979), and frog (DIAZ et al. 1989) and has been covered in a review (THESLEFF 1989). No sprouting was observed in frogs after application of BotD (ANTONY et al. 1981). Sprouting seems to be an unspecific phenomenon, resulting from the unspecific release of a growth factor from the postsynaptic muscle cell (PAMPHLETT 1989). It may be prevented by direct stimulation or by X-irradiation of the muscle (Duchen et al. 1981).

c) Large-Sized Mepps

After injection of a sublethal dose of BotA into a muscle, the ensuing inhibition of the stimulus-evoked transmitter release may last for weeks. A few days later, spontaneous mepps of large size (up to 15 mV with a slow rise time of up to 10 ms) may occur. A few of these mepps may also be recorded from nonintoxicated neuromuscular synapses. They have been described first in BotD-treated synapses by Harris and Miledi (1971) and later for BotA by Cull-Candy et al. (1976), Sellin and Thesleff (1981), Colmeus et al. (1982), Thesleff et al. (1983), Sellin et al. (1983b, BotA and BotB), Kim et al. (1984), Lupa et al. (1987). They cannot be evoked by electrical stimuli, and a high K^+ concentration has hardly any effect either, because their release is Ca^{2+}-independent. The large-sized mepps are not correlated to sprouting, because X-irradiation of a BotA-inhibited neuromuscular synapse prevents sprouting but not the development of large-sized mepps (Duchen et al. 1981).

7. Action on Nonneuronal Cells

An inhibition by Teta of the secretion of lysosomal contents from macrophages has been found by Ho and Klempner (1985b) and by Pitzurra et al. (1989). Although Teta alters calcium homoeostasis in macrophages (Ho and Klempner 1985a), this disturbance is not directly linked to the action of Teta on lysosomal secretion (Ho and Klempner 1986).

8. Action on Multicellular Systems

a) Neuronal Systems

More than 100 papers, dealing in particular with the action of Teta on multineuronal systems in the CNS of intact animals, had been published up to 1980 (cp. Wellhöner 1982). In all these investigations, Teta inhibited the synaptic transmission. Because inhibitory synapses are more sensitive than excitatory synapses to Teta, this CNT is now used as a tool in neurophysiological studies. Experiments of this type are listed in Table 10. More recently, the action of CNTs have been investigated in cultured mammalian neurons. The results have already been discussed above.

b) Nonneuronal Systems

Teta has a toxic action on the heart (Aleksevich et al. 1983; Binah 1987; Lamanna et al. 1988). The mechanism of the depression by Teta of cardiac frequency and force of contraction is probably distinct from its action on presynaptic terminals, because it is rapidly reversible (Lamanna et al. 1988).

Table 10. Actions of tetanus toxin in neuronal systems

Investigations before 1981	cp. WELLHÖNER (1982)
Teta and BotA inhibit transmission through inhibitory and excitatory synapses in cultured neurons	BERGEY et al. (1983, 1987); BIGALKE et al. (1985)
The monosynaptic reflex may be reduced by Teta	TAKANO et al. (1983, 1989a,b)
Motoneurons without inhibitory synapses cannot be disinhibited	MIKHAILOV and SHUBIN (1987)
Removal of recurrent inhibition in the hippocampus	WIERASZKO (1985)
Generation of epileptic foci in the hippocampus	JEFFERYS and EMPSON (1990)
GABA but not GABA binding sites are reduced in the hippocampus	BAGETTA et al. (1990b)
Behavioral changes after injection into the hippocampus	BAGETTA et al. (1990a)
Inhibition of granule cells in the dentate gyrus of rats	SUNDSTROM and MELLANBY (1990)
Valproate antagonizes the actions of intracerebral Teta	FOCA et al. (1984)

III. Intracellular Mode of Action

The intracellular mode of action of the CNTs is obscure. Several bio-chemical pieces of the puzzle are known, but as yet they cannot be arranged properly to reveal the outline of the picture. One of the reasons is that we do not know whether a particular observation is closely related to the prime cause of CNT action or mirrors one of the conceivable counterregulations that may be activated in the neurons.

A reduction of creatine kinase activity, ATP, and creatine phosphate content in the *Torpedo* electric organ by BotA was reported by DUNANT et al. (1987, 1988), but this is still a matter of debate. In mammalian syn-aptosomes (prepared from guinea-pig cerebral cortex) SANCHEZ-PRIETO et al. (1987) could not find any sign of BotA action on ATP synthesis, respiratory control, or respiratory capacity.

The phosphorylation of proteins was not changed by Teta in NG108-15 cells (WENDON and GILL 1982) or by BotA in rat brain synaptosomes (ASHTON et al. 1988a,b). In synaptosomes from *Torpedo marmorata* a stimulus-related dephosphorylation of a 37-kDa protein has been reported (GUITART et al. 1987). Teta inhibited the depolarization-dependent increase in synapsin I phosphorylation (PRESEK et al. 1989).

Teta and BotA were reported to inhibit the guanylate cyclase activity in guinea-pig synaptosomes and NG108-15 cells (SMITH and MIDDLEBROOK 1985). In PC12 cells, the stimulus-related increase of cGMP was attenuated by Teta, and zaprinast, a phosphodiesterase inhibitor that raises the level of cGMP, was an effective antagonist of Teta (SANDBERG et al. 1989a). The reports on the action of 8-bromo-cGMP are conflicting: The substance antagonized the action of Teta on ACh release in PC12 cells (SANDBERG et al. 1989a), but it had no effect on the inhibition of oxytocin and vasopressin release by Teta (HALPERN et al. 1990b).

Teta did not change ADP-ribosylation in NG108-15 cells (WENDON and GILL 1982). No ADP-ribosylation was detected in synaptosomes exposed to BotA or BotB (ASHTON et al. 1988a,b) or in rabbit hippocampal slices incubated with BotA (NAKOV et al. 1989). It was mentioned in the introduction that the ADP-ribosylation observed with BotC and BotD bears no relation to the inhibitory action of the CNTs on transmitter release. The antagonistic action of nicotinamide against the block imposed by BotD on neuromuscular transmission in the rat (SHIELLS and FALK 1988) is, therefore, hard to explain in terms of ADP reconstitution.

After intraventricular injection of Teta, a transformation and relocation of protein kinase C from the inactive form in the cytosol to the active membrane-bound form was followed by a time-dependent reduction in both the amount of enzyme protein and total activity (AGUILERA and YAVIN 1990; AGUILERA et al. 1990). Correspondingly, Teta attenuated the ability of phorbol ester to mobilize cytosolic protein kinase C (NG108-15 cells; CONSIDINE et al. 1990), and a reduction of protein kinase C activity by Teta in spinal cords (from mice with general tetanus) and in macrophages has been reported by Ho and KLEMPNER (1988). Activation of protein kinase C is known to be important for the agglomeration of F-actin. MARXEN and BIGALKE (1991b) have shown that Teta and BotA inhibit the carbachol-induced F-actin agglomeration in bovine chromaffin cells. One could speculate that this may be due to their action on protein kinase C. However, the activation of protein kinase C or G proteins had no influence on the inhibition of NE release from chromaffin cells (BITTNER and HOLZ 1988; BIGALKE et al. 1989).

G. Antibodies Against Clostridial Neurotoxins as Experimental Tools

I. Monoclonal Antibodies

AHNERT-HILGER et al. (1983) raised monoclonal antibodies (MoAbs) from mice against Teta. When, however, they used Teta formol toxoid as the antigen, two of five MoAbs reacted only with the formol toxoid but not with Teta and did not protect biological systems against Teta. Therefore, a second method of immunization was employed: Mice were primed with an immunocomplex made from Teta and Teta-specific polyclonal antibody and were then boosted with Teta. MoAbs reacting with the L-chain (and with proteins containing the L-chain) and those reacting with the C-terminal of the H-chain (and with proteins containing this fragment) were obtained. Not only the C-terminal-specific but also the L-chain-specific ones neutralized Teta in vitro (rat brain membranes) and in vivo (mice). The MoAb battery of AHNERT-HILGER et al. (1983) was also used by GORETZKI and HABERMANN (1985) and by ANDERSEN-BECKH et al. (1989) for the characterization of epitopes on Teta fragments. KENIMER et al. (1983) produced another set of 14 MoAbs against Teta. Two were directed against the C-terminal of the H-chain, 3 were directed against the N-terminal of the H-chain, and 9 were directed against the L-chain. At least one MoAb of each group exhibited significant toxin neutralization activity. Two of the 10 MoAbs produced by SHEPPARD et al. (1984) reacted only with toxoid, and all four neutralizing antibodies were directed against the Teta C-terminal. A fifth MoAb against the C-terminal did not neutralize. VOLK et al. (1984), with their battery of MoAbs, could identify 20 epitopes on the Teta molecule. The neutralizing MoAbs were all directed against the H-chain, 6 against its N-terminal and 4 against the C-terminal.

MoAbs have also been raised against BotB (KOZAKI et al. 1987). One in a battery of seven neutralized BotB and was directed against the L-chain. Three of the remaining six MoAbs were neutralizing and were directed against the papain fragment that contained the L-chain and the N-terminal of the H-chain. The other three were directed against the BotB C-terminal but were not neutralizing.

MoAbs directed against the L-chain have the weakest neutralizing potential (ANDERSEN-BECKH et al. 1989); those directed against C-terminal epitopes of the H-chain are much stronger (FAIRWEATHER et al. 1987; ANDERSEN-BECKH et al. 1989), although they are still inferior to polyclonal antibodies (PABs) (MIZUGUCHI et al. 1982; AHNERT-HILGER et al. 1983; VOLK et al. 1984; ANDERSEN-BECK et al. 1989). The neutralizing power of MoAbs could be greatly increased by mixing two or more together (MIZUGUCHI et al. 1982; VOLK et al. 1984). In human serum, 25% of the PABs neutralizing Teta are directed against the L-chain (LIN et al. 1985).

Monoclonal antibodies can be used to detect similarities of the tertiary structure between CNTs, as discussed above.

II. Neutralization of Cell-Associated Clostridial Neurotoxins with Antibodies

As long as the CNTs are only bound to the plasma membrane and have not been internalized, they can be neutralized with extracellular antibodies. This has been shown in isolated preparations, which allow the temperature to be manipulated (Teta: SCHMITT et al. 1981; SIMPSON 1985; BotA: SIMPSON 1974; Teta, BotA, BotB: SIMPSON 1983; BotC1: MURAYAMA et al. 1987). Internalized CNTs can no longer be neutralized with extracellular antibodies. This has been known for a long time both from clinical and experimental (BotA: BURGEN et al. 1949; Teta: ERDMANN et al. 1981) experience.

Recently, CNT-specific antibodies were introduced into the intra-neuronal space. MARXEN et al. (1990) used bovine chromaffin cells loaded with Teta and [^{3}H]-NE. These cells, after permeabilization with digitonin, were exposed to Teta-specific F(ab')$_2$. The authors showed that the Teta-specific F(ab')$_2$ were bound by intracellular Teta. However, this had not yet had an effect on the impairment by Teta of the stimulated [^{3}H]-NE release when the cells were stimulated 30 min after exposure. An action of F(ab')$_2$ at a later point of time is virtually impossible to study in per-meabilized cells, because measurements become unreliable 1 h after per-meabilization. Therefore, in subsequent experiments (BARTELS et al. 1990), digitonin permeabilization was replaced by electroporation. Chromaffin cells that survive electroporation can be kept alive in culture for several days. BARTELS et al. (1990) showed that, after 2 days, F(ab')$_2$-treated cells resumed the release of NE upon stimulation, while the control cells did not respond to stimulation even after 6 days. They concluded that (1) Teta had irre-versibly inactivated a component essential for exocytosis, (2) Teta-specific F(ab')$_2$ neutralized Teta by interrupting its activity as an irreversible in-activator, and (3) a few days had to pass until a substance was resynthesized that had been destroyed or permanently inactivated by Teta.

Permeabilization with digitonin or streptolysin as well as electroporation inflict permanent damage on many cells in a culture. WELLHÖNER et al. (1990, 1991) succeeded in translocating Teta-specific F(ab')$_2$ into cells by linking them with an acid-labile linker or to transferrin. The conjugates were taken up by receptor-mediated endocytosis through the transferrin pathway, much of the linker was hydrolysed in the acid environment of the endo-somes, transferrin was recirculated to the surface, and F(ab')$_2$ remained in the cells. In other experiments, the authors tried using a disulfide bridge as a link or replaced transferrin with wheat germ agglutinin. All conjugates were successfully translocated into cells.

Further investigations may show whether antibodies conjugated to transferrin or other carriers can neutralize CNTs after translocation of the antibody conjugate into the intracellular space.

Acknowledgement. My sincere thanks are due to my colleague Georg Erdmann for valuable suggestions on the manuscript.

References

Agui T, Syuto B, Oguma K, Iida, Kubo S (1985) The structural relation between the antigenic determinants to monoclonal antibodies and binding sites to rat brain synaptosomes and GT1b ganglioside in *Clostridium botulinum* type C neurotoxin. J Biochem (Tokyo) 97:213–218

Aguilera J, Yavin E (1990) In vivo translocation and down-regulation of protein kinase C following intraventricular administration of tetanus toxin. J Neurochem 54:339–342

Aguilera J, Lopez LA, Yavin E (1990) Tetanus toxin-induced protein kinase C activation and elevated serotonin levels in the perinatal rat brain. FEBS Lett 263:61–65

Ahnert-Hilger G, Bizzini B, Goretzski K, Müller H, Volckers C, Habermann E (1983) Monoclonal antibodies against tetanus toxin and toxoid. Med Microbiol Immunol (Berl) 172:123–135

Ahnert-Hilger G, Bader MF, Bhakdi S, Gratzl M (1989a) Introduction of macromolecules into bovine adrenal medullary chromaffin cells and rat pheochromocytoma cells (PC12) by permeabilization with streptolysin O – inhibitory effect of tetanus toxin on catecholamine secretion. J Neurochem 52:1751–1758

Ahnert-Hilger G, Stecher B, Gratzl M (1989b) Effects of tetanus toxin and botulinum A toxin on exocytosis from permeabilized adrenal chromaffin cells. Biol Chem Hoppe Seyler 370:613

Ahnert-Hilger G, Weller U, Dauzenroth ME, Habermann E, Gratzl M (1989c) The tetanus toxin light chain inhibits exocytosis. FEBS Lett 242:245–248

Aktories K (1990) Clostridial ADP-ribosyltransferases – modification of low-molecular-weight GTP-binding proteins and of actin by clostridial toxins. Med Microbiol Immunol (Berl) 179:123–136

Aktories K, Wegner A (1989) ADP-ribosylation of actin by clostridial toxins. J Cell Biol 109:1385–1388

Albus U, Habermann E (1983) Tetanus toxin inhibits the evoked outflow of an inhibitory (GABA) and an excitatory (D-aspartate) amino acid from particulate brain cortex. Toxicon 21:97–110

Aleksevich YI, Tumanov VP, Yavorski OG, Gordii PD, Kovalyshin VI (1983) Effect of tetanus toxin on the myocardium. Bull Exp Biol Med (Russ) 95:130–133

Ambache N, Morgan RS, Wright GP (1948) The action of tetanus toxin on the rabbit's iris. J Physiol (Lond) 107:45–53

An der Lan B, Habig WH, Hardegree MC, Chrambach A (1980) Heterogenity of ^{125}I-labeled tetanus toxin in isoelectric focussing on polyacrylamide gel and polyacrylamide gel electrophoresis. Arch Biochem Biophys 200:206–215

Andersen-Beckh B, Binz T, Kurazono H, Mayer T, Eisel U, Niemann H (1989) Expression of tetanus toxin subfragments in vitro and characterization of epitopes. Infect Immun 57:3498–3505

Angaut-Petit D, Molgo J, Thesleff S (1988) Presynaptic study of frog neuromuscular junctions in vitro poisoned with botulinum A toxin. J Physiol (Lond) 406:59P

Angaut-Petit D, Molgo J, Comella JX, Faille L, Tabti N (1990) Terminal sprouting in mouse neuromuscular junctions poisoned with botulinum type A toxin – morphological and electrophysiological features. Neuroscience 37:799–909

Anonymous (1982) Sixth international conference on tetanus, 3–5 Dec 1981, Lyon. Fondation Marcel Merieux, Lyon

Antony MT, Sayers H, Stolkin C, Tonge DA (1981) Prolonged paralysis, caused by the local injection of botulinum toxin, fails to cause motor nerve terminal sprouting in skeletal muscle of the frog. Q J Exp Physiol 66:525–532

Ashton AC, Dolly JO (1988) Characterization of the inhibitory action of botulinum neurotoxin type A on the release of several transmitters from rat cerebrocortical synaptosomes. J Neurochem 50:1808–1816

Ashton AC, Edwards K, Dolly J (1988a) Action of botulinum neurotoxin A on protein phosphorylation in relation to blockade of transmitter release. Biochem Soc Trans 16 A:885–886

Ashton AC, Edwards K, Dolly JO (1988b) Lack of detectable ADP-ribosylation in synaptosomes associated with inhibition of transmitter release by botulinum neurotoxins A and B. Biochem Soc Trans 16:883–884

Bagetta G, Corasaniti MT, Nistico G, Bowery NG (1990a) Behavioral and neuropathological effects produced by tetanus toxin injected into the hippocampus of rats. Neuropharmacology 29:765–770

Bagetta G, Knott C, Nistico G, Bowery NG (1990b) Tetanus toxin produces neuronal loss and a reduction in GABAa but not GABAb binding sites in rat hippocampus. Neurosci Lett 109:7–12

Bandyopadhyay S, Clark AW, DasGupta BR, Sathyamoorthy V (1987) Role of the heavy and light chains of botulinum neurotoxin in neuromuscular paralysis. J Biol Chem 262:2660–2663

Bartels F, Marxen P, Bigalke H (1990) Translocation of tetanus toxin into intact and permeabilized chromaffin cells: a comparison. Biol Chem Hoppe Seyler 371:1035–1036

Beaude P, Delacour A, Bizzini B, Domuado D, Remy MH (1990) Retrograde axonal transport of an exogenous enzyme covalently linked to BIIb fragment of tetanus toxin. Biochem J 271:87–91

Bergey GK, MacDonald RL, Habig WH, Hardegree MC, Nelson PG (1983) Tetanus toxin: convulsant action on mouse spinal cord neurons in culture. J Neurosci 3:2310–2324

Bergey GK, Bigalke H, Nelson PG (1987) Differential effects of tetanus toxin on inhibitory and excitatory synaptic transmission in mammalian spinal cord neurons in culture: a presynaptic locus of action for tetanus toxin. J Neurophysiol 57:121–131

Bergey GK, Habig WH, Bennett JI, Lin CS (1989) Proteolytic cleavage of tetanus toxin increases activity. J Neurochem 53:155–161

Berliner P, Unsicker K (1985) Tetanus toxin labeling as a novel rapid and highly specific tool in human neuroblastoma differential diagnosis. Cancer 56:419–423

Betley MJ, Somers E, DasGupta BR (1989) Characterization of botulinum type A neurotoxin gene – delineation of the N-terminal encoding region. Biochem Biophys Res Commun 162:1388–1395

Bevan S, Wendon LMB (1984) A study of the action of tetanus toxin at rat soleus neuromuscular junctions. J Physiol (Lond) 348:1–17

Bhattacharyya SD, Sugiyama H, Rust P, Lacey D (1988) Evidence that subunits of type A botulinum toxin need not be linked by disulfide. Toxicon 26:817–825

Bigalke H, Habermann E (1980) Blockade by tetanus and botulinum A toxin of postganglionic cholinergic nerve endings in the myenteric plexus. Naunyn-Schmiedebergs Arch Pharmacol 312:255–263

Bigalke H, Dimpfel W, Habermann E (1978) Suppression of ^{3}H-acetylcholine release from primary nerve cell cultures by tetanus and botulinum-A toxin. Naunyn-Schmiedebergs Arch Pharmacol 303:133–138

Bigalke H, Ahnert-Hilger G, Habermann E (1981a) Tetanus toxin and botulinum A toxin inhibit acetylcholine release from but not calcium uptake into brain tissue. Naunyn-Schmiedebergs Arch Pharmacol 316:143–148

Bigalke H, Heller I, Bizzini B, Habermann E (1981b) Tetanus toxin and botulinum A toxin inhibit release and uptake of various transmitters, as studied with particulate preparations from rat brain and spinal cord. Naunyn-Schmiedebergs Arch Pharmacol 316:244–251

Bigalke H, Dreyer F, Bergey G (1985) Botulinum A neurotoxin inhibits non-cholinergic synaptic transmission in mouse spinal cord neurones in culture. Brain Res 360:318–324

Bigalke H, Müller H, Dreyer F (1986) Botulinum A neurotoxin unlike tetanus toxin acts via a neuraminidase sensitive structure. Toxicon 24:1065–1074

Bigalke H, Marxen P, Ahnert-Hilger G (1989) Restoration of noradrenaline release by increasing concentrations of carbachol in botulinum A neurotoxin blocked chromaffin cells in culture. Biol Chem Hoppe Seyler 370:993–994

Binah O (1987) Tetanus in the mammalian heart – studies in the shrew myocardium. J Mol Cell Cardiol 19:1247–1252

Binz T, Kurazono H, Popoff MR, Eklund MW, Sakaguchi G, Kozaki S, Krieglstein K, Henschen A, Gill DM, Niemann H (1990a) Nucleotide sequence of the gene encoding Clostridium botulinum neurotoxin type D. Nucleic Acids Res 18:5556–5556

Binz T, Kurazono H, Wille M, Frevert J, Wernars K, Niemann H (1990b) The complete sequence of botulinum neurotoxin type A and comparison with other clostridial neurotoxins. J Biol Chem 265:9153–9158

Bishop GH, Bofenbrenner JJ (1936) The site of action of botulinum toxin. Am J Physiol 117:393–404

Bittner MA, Holz RW (1988) Effects of tetanus toxin on catecholamine release from intact and digitonin-permeabilized chromaffin cells. J Neurochem 51:451–456

Bittner MA, DasGupta BR, Holz RW (1989a) Isolated light chains of botulinum neurotoxins inhibit exocytosis. Studies in digitonin-permeabilized chromaffin cells. J Biol Chem 264:10354–10360

Bittner MA, Habig WH, Holz RW (1989b) Isolated light chain of tetanus toxin inhibits exocytosis: studies in digitonin-permeabilized cells. J Neurochem 53:966–968

Bizzini B (1984) Investigation of the mode of action of tetanus toxin with the aid of hybrid molecules consisting in part of tetanus toxin-derived fragments. In: Alouf JE et al. (eds.) Bacterial protein toxins. FEMS Symposium, Academic Press, London, p 427

Bizzini B, Turpin A, Raynaud M (1973) Immunochemistry of tetanus toxin. The nitration of tyrosyl residues in tetanus toxin. Eur J Biochem 39:171–181

Bizzini B, Grob P, Glicksman MA, Akert K (1980) Use of the BIIb tetanus toxin derived fragment as a specific neuropharmacological transport agent. Brain Res 193:221–227

Bizzini B, Grob P, Akert K (1981) Papain-derived fragment IIC of tetanus toxin – its binding to isolated synaptic membranes and retrograde axonal transport. Brain Res 210:291–299

Black JD, Dolly JO (1986a) Interaction of ^{125}I-labeled botulinum neurotoxins with nerve terminals. I. Ultrastructural autoradiographic localization and quantitation of distinct membrane acceptors for types A and B on motor nerves. J Cell Biol 103:521–534

Black JD, Dolly JO (1986b) Interaction of ^{125}I-labeled botulinum neurotoxins with nerve terminals. II. Autoradiographic evidence for its uptake into motor nerves by acceptor-mediated endocytosis. J Cell Biol 103:535–544

Black JD, Dolly JO (1987) Selective location of acceptors for botulinum neurotoxin A in the central and peripheral nervous system. Neuroscience 23:767–780

Blasi E, Pitzurra F, Fuad AMB, Marconi P, Bistoni F (1990) Gamma interferon induced specific binding of tetanus toxin on the GG2EE macrophage cell line. Scand J Immunol 32:289–292

Blaustein RO, Germann WJ, Finkelstein A, DasGupta BR (1987) The N-terminal half of the heavy chain of botulinum type A neurotoxin forms channels in planar phospholipid bilayers. FEBS Lett 226:115–120

Blaustein RO, Hoch DH, DasGupta BR (1988) Channels formed by botulinum type E neurotoxin in planar lipid bilayers. FASEB J 2:1750

Boquet P, Duflot E (1982) Tetanus toxin fragment forms channels in lipid vesicles at low pH. Proc Natl Acad Sci USA 79:7614–7618

Boquet P, Duflot E, Hauttecoeur B (1984) Low pH induces a hydrophobic domain in the tetanus toxin molecule. Eur J Biochem 144:339–344

Borochov-Neori H, Delbruck T, Yavin E, Montal M (1984a) Tetanus toxin channels in ganglioside containing lipid bilayers are voltage dependent. Biophys J 45:59A

Borochov-Neori H, Yavin E, Montal M (1984b) Tetanus toxin forms channels in planner lipid bilayers containing gangliosides. Biophys J 45:83–85

Boroff DA, DelCastillo J, Evoy WH, Steinhardt RA (1974) Observations on the actions of type A botulinum toxin on frog neuromuscular junctions. J Physiol (Lond) 240:227–253

Bray JJ, Harris AJ (1975) Dissociation between nerve-muscle transmission and nerve trophic effects on rat diaphragm using type D botulinum toxin. J Physiol (Lond) 253:53–77

Brooks VB (1954) The action of botulinum toxin on motor nerve terminals. J Physiol (Lond) 123:501–515

Brooks VB (1956) An intracellular study of the action of repetitive nerve volleys and of botulinum toxin on miniature end-plate potentials. J Physiol (Lond) 134:264–277

Bruschettini A (1892) Sulla diffusione del veleno del tetano nell'organismo. Riforma Med 8:256–259, 270–273

Burgen ASV, Dickens F, Zatman LJ (1949) The action of botulinum toxin on the neuro-muscular junction. J Physiol (Lond) 109:10–24

Büttner-Ennever JA, Grob P, Akert K, Bizzini B (1981) Transsynaptic retrograde labeling in the oculomotor system of the monkey with [^{125}I]tetanus toxin BIIb fragment. Neurosci Lett 26:233–238

Cabiaux V, Lorge P, Vandenbranden M, Famalgne P, Ruysschaert JM (1985) Tetanus toxin induces fusion and aggregation of lipid vesicles containing phosphatidylinositol at low pH. Biochem Biophys Res Commun 128:840–949

Calabresi P, Benedetti M, Mercuri NB, Bernardi G (1989) Selective depression of synaptic transmission by tetanus toxin. A comparative study in hippocampal and neostriatal slices. Neuroscience 30:663–670

Chiu W, Rankert D, Cumming MA, Robinson JP (1982) Characterization of crystalline filtrate tetanus toxin. J Ultrastruct Res 79:285–293

Chou SM, Payne WN (1982) Vacuolation and chromatolysis of lower motoneurons in tetanus. A case report and review of the literature. Cleve Clin Q 49:255–264

Collingridge GL, Thompson PA, Davies J, Mellanby J (1981) In vitro effect of tetanus toxin on GABA release from rat hippocampal slices. J Neurochem 37:1039–1041

Colmeus C, Gomez S, Molgo J, Thesleff S (1982) Discrepancies between spontaneous and evoked synaptic potentials at normal, regenerating and botulinum toxin poisoned mammalian neuro-muscular junctions. Proc R Soc [B] 215:63–74

Considine RV, Bielicki JK, Simpson LL, Sherwin JR (1990) Tetanus toxin attenuates the ability of phorbol myristate acetate to mobilize cytosolic protein kinase C in NG108 cells. Toxicon 28:13–20

Craven CJ, Dawson DJ (1973) The chain composition of tetanus toxin. Biochem Biophys Acta 317:277–285

Critchley DR, Nelson PG, Habig WH, Fishman PH (1985) Fate of tetanus toxin bound to the surface of primary neurons in culture: evidence for rapid internalization. J Cell Biol 100:1499–1507

Critchley DR, Habig WH, Fishman PH (1986) Reevaluation of the role of gangliosides as receptors for tetanus toxin. J Neurochem 47:213–222

Cull-Candy SG, Lundh H, Thesleff S (1976) Effects of botulinum toxin on neuromuscular transmission in the rat. J Physiol (Lond) 260:177–203

DasGupta BR (1984) Amino acid composition of *Clostridium botulinum* type B neurotoxin. Toxicon 22:312

DasGupta BR, Datta A (1988) Botulinum neurotoxin type B (strain 657) – partial sequence and similarity with tetanus toxin. Biochimie 70:811–817

DasGupta BR, Deklava ML (1990) Botulinum neurotoxin type A: sequence of aminoacids at the N-terminus and around the nicking side. Biochimie 72:661–664

DasGupta BR, Foley J (1989) *C. botulinum* neurotoxin type A and type E – isolated light chain breaks down into 2 fragments – comparison of their amino acid sequences with tetanus neurotoxin. Biochimie 71:1193–1200

DasGupta BR, Rasmussen S (1983) Purification and amino acid composition of type E botulinum neurotoxin. Toxicon 21:535–545

DasGupta BR, Sathyamoorthy V (1984) Purification and amino acid composition of type A botulinum neurotoxin. Toxicon 22:415–424

DasGupta BR, Sugiyama H (1980) Role of arginine residues in the structure and biological activity of botulinum neurotoxin types A and E. Biochem Biophys Res Commun 93:369–375

DasGupta BR, Woody MA (1984) Amino acid composition of *Clostridium botulinum* type B neurotoxin. Toxicon 22:312–315

DasGupta BR, Bandyopadhyay S, Clark AW, Herian A (1987a) Botulinum neurotoxin – structure function and similarity to tetanus. Toxicon 25:137–138

DasGupta BR, Foley J, Niece R (1987b) Partial sequence of the light chain of botulinum neurotoxin type A. Biochemistry 26:4162

DasGupta BR, Foley J, Wadsworth C (1988) Botulinum neurotoxin type A: partial sequence of L-chain and its two fragments. FASEB J 2:A1750

Datta A, DasGupta BR (1988a) Circular dichroic and fluororescence spectroscopic study of the conformation of botulinum neurotoxin types A and E. Mol Cell Biochem 79:153–159

Datta A, DasGupta BR (1988b) Botulinum neurotoxin types A, B & E: pH induced difference spectra. Mol Cell Biochem 81:187–194

Dauzenroth ME, Dolly JO, Habermann E, Mochida S, Poulain B, Tauc L, Wadsworth JDF, Weller U (1989) Light chain of tetanus toxin can be internalized into a cholinergic synapse of *Aplysia* by the heavy chain of botulinum neurotoxin and depresses the transmitter release. J Physiol (Lond) 418:71P

Dejongh KS, Schwartzkopf CL, Howden MEH (1989) *Clostridium botulinum* type D neurotoxin – purification and detection. Toxicon 27:221–228

Dekleva ML, DasGupta BR (1989) Nicking of single chain *Clostridium botulinum* type A neurotoxin by an endogenous protease. Biochem Biophys Res Commun 162:767–772

Dekleva ML, DasGupta BR (1990) Purification and characterization of a protease from *Clostridium botulinum* type A that nicks single-chain type A botulinum neurotoxin into the di-chain form. J Bacteriol 172:2498–2503

Dekleva ML, DasGupta BR, Sathyamoorthy V (1989) Botulinum neurotoxin type A radiolabeled at either the light or the heavy chain. Arch Biochem Biophys 274:235–240

Diamond J, Mellanby J (1971) The effect of tetanus toxin in the goldfish. J Physiol (Lond) 215:727–741

Diaz J, Molgo J, Pecot-Dechavassine M (1989) Sprouting of frog motor nerve terminals after long-term paralysis by botulinum type A toxin. Neurosci Lett 96:127–132

Dickson EC, Shevky R (1923) Botulism, studies on the manner in which the toxin of *Clostridium botulinum* acts upon the body. II. The effect on the voluntary nervous system. J Exp Med 37:711–731

DIMari SJ, Cumming MA, Hash JM, Robinson JP (1982) Purification of tetanus toxin and its peptide components by preparative polyacrylamide gel electrophoresis. Arch Biochem 214:342–353

Dolezal V, Vyskocil F, Tucek S (1983) Decrease of the spontaneous nonquantal release of acetylcholine from the phrenic nerve in botulinum-poisoned rat diaphragm. Pflugers Arch 397:319–322

Dolly JO, Williams RS, Black JD, Tse CK, Hambleton P, Melling J (1982) Localization of sites for I[125]–labeled botulinum neurotoxin at murine neuromuscular junction and its binding to rat brain synaptosomes. Toxicon 20:141–148

Dolly JO, Black J, Williams RS, Melling J (1984) Acceptors for botulinum neurotoxin reside on motor nerve terminals and mediate its internalization. Nature 307:457–460

Dolly JO, Lande S, Wray DW (1987) The effects of in vitro application of purified botulinum neurotoxin at mouse motor-nerve terminals. J Physiol (Lond) 386:475–484

Dolly JO, Maisey EA, Poulain B, Tauc L, Wadsworth JDF (1988) Uptake of both light and heavy chains of botulinum neurotoxin by cholinergic neurons of *Aplysia* is mediated by the larger chain. J Physiol (Lond) 406:196

Donovan JJ, Middlebrook JL (1986) Ion-conducting channels produced by botulinum toxin in planar lipid membranes. Biochemistry 25:2872–2876

Dreyer F, Schmitt A (1981) Different effect of botulinum A toxin and tetanus toxin on the transmitter releasing process at the mammalian neuromuscular junction. Neurosci Lett 26:307–311

Dreyer F, Schmitt A (1983) Transmitter release in tetanus and botulinum A toxin-poisoned mammalian motor endplates and its dependence on nerve stimulation and temperature. Pflugers Arch 399:228–234

Dreyer F, Mallart A, Brigant JL (1983) Botulinum A toxin and tetanus toxin do not affect presynaptic membrane currents in mammalian motor nerve endings. Brain Res 270:373–375

Dreyer F, Becker C, Bigalke H, Funk J, Penner R, Rosenberg F, Ziegler M (1984) Action of botulinum A toxin and tetanus toxin on synaptic transmission. J Physiol (Paris) 79:252–258

Dreyer F, Rosenberg F, Becker C, Bigalke H, Penner R (1987) Differential effects of various secretagogues on quantal transmitter release from mouse motor nerve terminals treated with botulinum A and tetanus toxin. Naunyn-Schmiedebergs Arch Pharmacol 335:1–7

Duchen LW, Strich SJ (1968) The effects of botulinum toxin on the pattern of innervation of skeletal muscle in the mouse. Q J Exp Physiol 53:84–89

Duchen LW, Tonge DA (1973) The effects of tetanus toxin on neuromuscular transmission and on the morphology of motor end-plate in slow and fast skeletal muscle of the mouse. J Physiol (Lond) 228:157–172

Duchen LW, Gomez S, Hornsey S (1981) Effects of X-irradiation on axonal sprouting induced by botulinum toxin in skeletal muscle of the mouse. J Physiol (Lond) 312:31–32

Dunant Y, Esquerda JE, Loctin F, Marsal J, Muller D (1987) Botulinum toxin inhibits quantal acetylcholine release and energy metabolism in the *Torpedo* electric organ. J Physiol (Lond) 385:677–692

Dunant J, Loctin F, Marsal J, Muller D, Parducz A, Rabasseda X (1988) Energy metabolism and quantal acetylcholine release: effects of botulinum toxin, 1-fluoro-2, 4-dinitrobenzene, and diamide in the *Torpedo* electric organ. J Neurochem 50:431–439

Duprat AM, Gualandris L, Foulquier F, Paulin D, Bizzini B (1986) Neural induction and in vitro initial expression of neurofilament and tetanus toxin binding site molecules in amphibians. Cell Differ 18:57–64

Duxon MJ, Stolkin C (1979) Early structural changes in the motor nerve terminals of rat skeletal muscle after local injection of botulinum toxin. J Physiol (Lond) 296:12P

Edmunds CW, Long PH (1923) Contribution to the pathologic physiology of botulism. J Am Med Assoc 81:542–547

Egea G, Rabasseda X, Solsona C, Marsal J, Bizzini B (1990) Tetanus toxin blocks potassium-induced transmitter release and rearrangement of intramembrane particles at pure cholinergic synaptosomes. Toxicon 28:311–318

Eisel U, Jarausch W, Goretzki K, Henschen A, Engels J, Weller U, Hudel M, Habermann E (1986) Tetanus toxin: primary structure, expression in *E. coli*, and homology with botulinum toxins. EMBO J 5:2495–2502

Eisel U, Binz T, Niemann H (1987) Characterization of the tetanus toxin promoter and expression of nontoxic fragments in *Escherichia coli*. Biol Chem Hoppe Seyler 368:1037–1038

Eisenbarth GS, Shimizu K, Bowring MA, Wells S (1982) Expression of receptors for tetanus toxin and monoclonal antibody A2B5 by pancreatic islet cells. Proc Natl Acad Sci USA 79:5066–5070

Eklund MW, Poysky FT, Mseittif LM, Strom MS (1988) Evidence for plasmid-mediated toxin and bacteriocin production in *Clostridium botulinum* type G. Appl Environ Microbiol 54:1405–1408

Erdmann G, Wiegand H, Wellhöner HH (1975) Intraaxonal and extraaxonal transport of ^{125}I-tetanus toxin in early local tetanus. Naunyn-Schmiedebergs Arch Pharmacol 290:357–373

Erdmann G, Hanauske A, Wellhöner HH (1981) Intraspinal distribution and reaction in the grey matter with tetanus toxin of intracisternally injected anti-tetanus toxoid F(ab′)2 fragments. Brain Res 211:367–377

Evans DM, Williams RS, Shone CC, Hambleton P, Melling J, Dolly JO (1986) Botulinum neurotoxin type B – its purification, radioiodination and interaction with rat brain synaptosomal membranes. Eur J Biochem 154:409–416

Evans DM, Richardson PJ, Fine A, Mason WT, Dolly JO (1988) Relationship of accpetors for botulinum neurotoxins (type A and type B) in rat CNS with the cholinergic marker, Chol-I. Neurochem Int 13:25–36

Evinger C, Erichsen JT (1986) Transsynaptic retrograde transport of fragment C of tetanus toxin demonstrated by immunohistochemical localization. Brain Res 380:383–388

Fairweather NF, Lyness VA (1986) The complete nucleotide sequence of tetanus toxin. Nucleic Acids Res 14:7809–7812

Fairweather NF, Lyness VA, Pickard DJ, Allen G, Thomson RO (1986) Cloning, nucleotide sequencing, and expression of tetanus toxin fragment C in Escherichia coli. J Bacteriol 165:21–27

Fairweather NF, Lyness VA, Makel DJ (1987) Immunization of mice against tetanus with fragments of tetanus toxin synthesized in Escherichia coli. Infect Immun 55:2541–2545

Fedinec AA, Lazarovici P, Yavin E, Bizzini B (1986) Two ^{125}I-tetanus toxins with different affinities for gangliosides – retrograde transport in the rat sciatic nerves. J Toxicol Toxin Rev 5:191

Fedinec AA, Toth P, Bizzini B (1987) Theophylline and CGMP prolong the survival of mice paralyzed with tetanus toxin fragment Ibc. Toxicon 25:139–140

Figliomeni B, Grasso A (1985) Tetanus toxin affects the K^+-stimulated release of catecholamines from nerve growth factor-treated PC12 cells. Biochem Biophys Res Commun 128:249–256

Finkelstein A (1990) Channels formed in phospholipid bilayer membranes by diphteria, tetanus, botulinum and anthrax toxin. J Physiol (Paris) 84:188–190

Finn CW, Silver RP, Habig WH, Hardegree MC, Zon G, Garon CF (1984) The structural gene for tetanus neurotoxin is on a plasmid. Science 224:881–884

Fishman PS, Carrigan DR (1988) Motoneuron uptake from the circulation of the binding fragment of tetanus toxin. Arch Neurol 45:558–561

Fishman PS, Savitt JM (1989) Transsynaptic transfer of retrogradely transported tetanus protein peroxidase conjugates. Exp Neurol 106:197–203

Fishman PS, Savitt JM, Farrand DA (1990) Enhanced CNS uptake of systemically administered proteins through conjugation with tetanus C-fragment. J Neurol Sci 98A:311–325

Foca A, Rotirotti D, Mastroeni P, Nistico G (1984) Effects of tetanus toxin after intracerebral microinjection are antagonized by drugs enhancing GABAergic transmission in adult fowls. Neuropharmacology 23:155–158

Fujii N, Oguma K, Yokosawa N, Tsuzuki K (1987) Bacteriophages and plasmids in Clostridium botulinum type C and type D. Jpn J Med Sci Biol 40:217

Fujita K, Guroff G, Yavin E, Lazarovici P (1988) Preparation of affinity-purified, biotinylated tetanus toxin and characterization of cell-surface binding sites on nerve growth factor-treated PC12 cells. FASEB J 2:1773

Gadoth N, Dagan R, Sandbank U, Levy D, Moses SW (1981) Permanent tetraplegia as a consequence of tetanus neonatorum – evidence for widespread lower motor neuron damage. J Neurol Sci 51:273–278

Gambale F, Montal M (1988) Characterization of the channel properties of tetanus toxin in planar lipid bilayers. Biophys J 53:771–783

Gansel M, Penner R, Dreyer F (1987) Distinct sites of action of clostridial neurotoxins revealed by double-poisoning of mouse motor-nerve terminals. Pfluegers Arch 409:533–539

Gawade S, Bon C, Bizzini B (1985) The use of antibody Fab fragments specifically directed to two different complementary parts of the tetanus toxin molecule for studying the mode of action of the toxin. Brain Res 334:139–146

Gimenez JA, DasGupta BR (1990) Botulinum neurotoxin type E fragmented with endoproteinase LYS-C reveals the site trypsin nicks and homology with tetanus neurotoxin. Biochimie 72:213–217

Gimenez JA, Sugiyama H (1988) Comparison of toxins of *Clostridium butyricum* and *Clostridium botulinum* type E. Infect Immun 56:926–929

Gimenez J, Foley J, DasGupta BR (1988) Neurotoxin type E from *Clostridium botulinum* and *Clostridium butyricum* – partial sequence and comparison. FASEB J 2:1750

Goldberg RL, Costa T, Habig WH, Kohn LD, Hardegree MC (1981) Characterization of fragment C and tetanus toxin binding to rat brain membranes. Mol Pharmacol 20:565–570

Goretzki K, Habermann E (1985) Enzymatic hydrolysis of tetanus toxin by intrinsic and extrinsic proteases – characterization of the fragments by monoclonal antibodies. Med Microbiol Immunol (Berl) 174:139–150

Grothe C, Unsicker K (1988) Reciprocal age-dependent pattern of 2 neuronal markers, tetanus toxin and neuron-specific enolase, in postnatal rat sensory and sympathetic neurons. Dev Brain Res 39:1–8

Guitart X, Egea G, Solsona C, Marsal J (1987) Botulinum neurotoxin inhibits depolarization-stimulated protein phosphorylation in pure cholinergic synaptosomes. FEBS Lett 219:219–223

Gundersen CB (1980) The effects of botulinum toxin on the synthesis, storage and release of acetylcholine. Prog Neurobiol 14:99–119

Gundersen CB, Howard BD (1978) The effects of botulinum toxin on acetylcholine metabolism in mouse brain slices and synaptosomes. J Neurochem 31:1005–1013

Gundersen CB, Jenden DJ (1983) Spontaneous output of acetylcholine from rat diaphragm preparations declines after treatment with botulinum toxin. J Pharmacol Exp Ther 224:265–268

Gundersen CB, Katz B, Miledi R (1982) The antagonism between botulinum toxin and calcium in motor-nerve terminals. Proc R Soc [B] 216:369–376

Guyton AC, MacDonald MA (1947) Physiology of botulinum toxin. Arch Neurol Psychiatry 57:578–592

Habermann E (1972) Distribution of ^{125}I-tetanus toxin and ^{125}I-toxoid in rats with local tetanus, as influenced by antitoxin. Naunyn-Schmiedebergs Arch Pharmacol 272:85–88

Habermann E (1974) ^{125}I-labeled neurotoxin from *Clostridium botulinum* A: preparation, binding to synaptosomes and ascent to the spinal cord. Naunyn-Schmiedebergs Arch Pharmacol 281:47–56

Habermann E (1981) Tetanus toxin and botulinum A neurotoxin inhibit and at higher concentrations enhance noradrenaline outflow from particular brain cortex in batch. Naunyn-Schmiedebergs Arch Pharmacol 318:105–111

Habermann E, Albus U (1986) Interaction between tetanus toxin and rabbit kidney: a comparison with rat brain preparations. J Neurochem 46:1219–1226

Habermann E, Dimpfel W (1973) Distribution of ^{125}I-tetanus toxin and ^{125}I-toxoid in rats with generalized tetanus, as influenced by antitoxin. Naunyn-Schmiedebergs Arch Pharmacol 176:327–340

Habermann E, Dreyer G (1986) Clostridial neurotoxins: handling and action at the cellular and molecular level. In: Compans R et al. (eds) Current topics in microbiology and immunology, vol 129. Springer, Berlin Heidelberg New York, pp 93–179

Habermann E, Tayot JL (1985) Interaction of solid-phase gangliosides with tetanus toxin and toxoid. Toxicon 23:913–920

Habermann E, Wellhöner HH, Räker KO (1977) Metabolic fate of ^{125}I-tetanus toxin in the spinal cord of rats and cats with early local tetanus. Naunyn-Schmiedebergs Arch Pharmacol 299:187–196

Habermann E, Dreyer F, Bigalke H (1980) Tetanus toxin blocks the neuromuscular transmission in vitro like botulinum A toxin. Naunyn-Schmiedebergs Arch Pharmacol 311:33–40

Habermann E, Bigalke H, Heller J (1981) Inhibition of synaptosomal choline uptake by tetanus and botulinum A toxin. Naunyn-Schmiedebergs Arch Pharmacol 316:135–142

Habermann E, Müller H, Hudel M (1988) Tetanus toxin and botulinum A and botulinum C neurotoxins inhibit noradrenaline release from cultured mouse brain. J Neurochem 51:522–527

Habig WH, Kenimer JG, Hardegree MC (1983) Retrograde axonal transport of tetanus toxin: toxin mediated antibody transport. In: Liu TY et al. (eds) Frontiers in biochemical and biophysical studies of proteins and membranes. Elsevier, New York, pp 463–473

Habig WH, Bigalke H, Bergey GK, Neale EA, Hardegree MC, Nelson PG (1986) Tetanus toxin in dissociated spinal cord cultures: long term characterization of form and action. J Neurochem 47:930–937

Hall JD, McCroskey LM, Pincomb BJ, Hatheway CL (1985) Isolation of an organism resembling *Clostridium barati* which produces type F botulinal toxin from an infant with botulism. J Clin Microbiol 21:654–655

Halpern JL, Smith LA, Seamon KB, Groover KA, Habig WH (1989) Sequence homology between tetanus and botulinum toxins detected by an antipeptide antibody. Infect Immun 57:18–22

Halpern JL, Habig WH, Neale EA, Stibitz S (1990a) Cloning and expression of functional fragment C of tetanus toxin. Infect Immun 58:1004–1009

Halpern JL, Habig WH, Trenchard H, Russell JT (1990b) Effect of tetanus toxin on oxytocin and vasopressin release from nerve endings of the neurohypophysis. J Neurochem 55:2072–2078

Hambleton P, Shone C, Wilton-Smith P, Melling J (1984) A possible common antigen on chostridial toxins detected by monoclonal antibotulinum neurotoxin antibodies. In: Alouf JE, Fehrenbach FJ, Freer JH, Jeljaszewicz J (ed) Bacterial protein toxins. Academic London, pp 449–450

Harris AJ, Miledi R (1971) The effect of type D botulinum toxin on frog neuromuscular junctions. J Physiol (Lond) 217:497–515

Hauser D, Eklund MW, Kurazono H, Benz TH, Niemann H, Gill DM, Boquet P, Popoff MR (1990) Nucleotide sequence of *Clostridium botulinum* C1 neurotoxin. Nucleic Acids Res 18:4924

Haynes BF, Shimizu K, Eisenbarth GS (1983) Identification of human and rodent thymic epithelium using tetanus toxin and monoclonal-antibody A2B5. J Clin Invest 71:9–14

Heredero J, Oja SS (1985) Ruthenium red interferes with the tetanus toxin inhibition of potassium-stimulated GABA release from rat cerebral cortex slices. Neurochem Int 7:861–866

Higashida H, Sugimoto N, Ozutsumi K, Miki N, Matsuda M (1983) Tetanus toxin: a rapid and selective blockade of the calcium, but not sodium, component of action potentials in cultured neuroblastoma N1E-115 cells. Brain Res 279:363–368

Hilbig G, Räker KO, Wellhöner HH (1979) Local tetanus in rats; concentration of amino acids as studied in spinal cord segments, spinal roots, and dorsal root ganglia. Naunyn-Schmiedebergs Arch Pharmacol 307:287–290

Ho JL, Klempner MS (1985a) Tetanus toxin alters calcium homeostasis of human macrophages stimulated by the calcium ionophore ionomycin. Clin Res 33:405

Ho JL, Klempner MS (1985b) Tetanus toxin inhibits secretion of lysosomal contents from human macrophages. J Infect Dis 152:922–928

Ho JL, Klempner MS (1986) Inhibition of macrophage secretion by tetanus toxin is not directly linked to cytosolic calcium homeostasis. Biochem Biophys Res Commun 135:16–24

Ho JL, Klempner MS (1988) Diminished activity of protein kinase C in tetanus toxin treated macrophages and in the spinal cord of mice manifesting generalized tetanus intoxication. J Infect Dis 157:925–933

Hoch DH, Romero-Mira M, Ehrlich BE, Finkelstein A (1985) Channels formed by botulinum, tetanus, and diphteria toxins in planar lipid bilayers: relevance to translocation of proteins across membranes. Proc Natl Acad Sci USA 82:1692–1697

Hodson AK, Curbeam RV (1982) Rat hippocampal neurons maintained in serum-free media can be positively identified with monoclonal antibody A285 and tetanus toxin. Clin Res 30:899A

Huba R, Hofmann HD (1988) Tetanus toxin binding to isolated and cultured rat retinal glial-cells. GLIA 1:156–164

Janicki P, Habermann E (1983) Tetanus and botulinum toxins inhibit, and black widow spider venom stimulates the release of methionine-enkephalin-like material in vitro. J Neurochem 41:395–402

Jefferys JGR, Empson RM (1990) Development of chronic secondary epileptic foci following intrahippocampal injection of tetanus toxin in the rat. Exp Physiol 75:733–736

Kaeser HE, Saner A (1969) Tetanus toxin, a neuromuscular blocking agent. Nature 223:842

Kalz HJ, Wellhöner HH (1990) The uptake of tetanus toxin into NG108-15 neuroblastoma-glioma hybrid cells may occur both through the clathrin-dependent and the clathrin-independent path of adsorptive endocytosis. Biol Chem Hoppe Seyler 342:R17

Kamata Y, Kozaki S, Sakaguchi G, Iwamori M, Nagai Y (1986) Evidence for direct binding of *Clostridium botulinum* type E derivative toxin and its fragments to gangliosides and free fatty acids. Biochem Biophys Res Commun 140:1015–1019

Kanda K, Takano K (1983) Effect of tetanus toxin on the excitatory and the inhibitory post-synaptic potentials in the rat motoneurone. J Physiol (Lond) 335:319–333

Kaufmann JA, Way JF, Siegel LS, Selling LC (1985) Comparison of the actions of types A and F botulinum toxin at the rat neuromuscular injection. Toxicol Appl Pharmacol 79:211–217

Kenimer JG, Habig WH, Hardegree MC (1983) Monoclonal antibodies as probes of tetanus toxin structure and function. Infect Immun 42:942–948

Kim Yi, Lomo T, Lupa MT, Thesleff S (1984) Miniature end plate potentials in rat skeletal muscle poisoned with botulinum toxin. J Physiol (Lond) 356:587–599

Kimura K, Fujii M, Tsuzuki K, Murakami T, Indoh T, Yokosawa N, Takeshi K, Syuto B, Oguma K (1990) The complete nucleotide sequence of the gene coding for botulinum type C1 toxin in the C-ST phage genome. Biochem Biophys Res Commun 171:1304–1311

Kitamura M, Sone S (1987) Binding ability of *Clostridium botulinum* neurotoxin to the synaptosomes upon treatment with various kinds of enzymes. Biochem Biophys Res Commun 143:928–933

Kitamura M, Iwamori M, Nagai Y (1980) Interaction between *Clostridium botulinum* neurotoxin and gangliosides. Biochim Biophys Acta 628:328–335

Klyucheva VV, Saprykin TP, Dolgikh MS, Blagoves VA (1982) The amino-acid composition of components from *Cl. botulinum* of the B-type (in Russian). Vopr Med Khim 28:29

Knight DE (1986) Botulinum toxin types A, B and D inhibit catecholamine secretion from bovine adrenal medullary cells. FEBS Lett 207:222–226

Koulakof A, Bizzini B, Berwald-Netter Y (1982) A correlation between the appearance and the evolution of tetanus toxin binding cells and neurogensis. Dev Brain Res 5:139–147

Koulakoff A, Bizzini B, Berwald-Netter Y (1983) Neuronal acquisition of tetanus toxin binding sites: relationship with the last mitotic cycle. Dev Biol 100:350–357

Kozaki S, Ogasawara J, Shimote Y, Kamata Y, Sakaguchi G (1987) Antigenic structure of *Clostridium botulinum* type B neurotoxin and its interaction with gangliosides, cerebroside, and free fatty-acids. Infect Immun 55:3051–3056

Kozaki S, Miki A, Kamata Y, Ogasawara J, Sakaguchi G (1989) Immunological characterization of papain-induced fragments of *Clostridium botulinum* type A neurotoxin and interaction of the fragments with brain synaptosomes. Infect Immun 57:2634–2639

Kriebel ME, Llados F, Matteson DR (1976) Spontaneous subminiature endplate potentials in mouse diaphragm muscle: evidence for synchronous release. J Physiol (Lond) 262:553–581

Krieglstein K, Henschen A, Weller U, Habermann E (1990) Arrangement of disulfide bridges and positions of sulfhydryl groups in tetanus toxin. Eur J Biochem 188:39–45

Kryzhanovsky, GN, Kasymov AKh (1964) Action of tetanus toxin on neuromuscular transmission. Bull Exp Biol Med (Engl Transl) 58:1199–1203

Kryzhanovsky GN, Pozdynakov OM, D'yakonova MV, Polgar AA, Smirnova VS (1971) Disturbance of neurosecretion in myoneural junctions of muscle poisoned with tetanus toxin. Bull Exp Biol Med (Engl Transl) 72:1387–1391

Kryzhanovsky GN, Lutsenko VK, Sakharova OP, Lutsenko NG (1982) Disturbance of H-3-GABA transport in synaptosomes by tetanus toxin. Bull Exp Biol Med (Engl Transl) 94:910–914

Kurokawa Y, Oguma K, Yokosawa N, Syuto B, Fukatsu R, Yamashita I (1987) Binding and cytotoxic effects of *Clostridium botulinum* A, C1 and E toxins in primary neuron cultures from fetal mouse brains. J Gen Microbiol 133:2647–2657

Laird WJ, Aaronson W, Silver RP, Habig WH, Hardegree MC (1980) Plasmid-associated toxigenicity of *Clostridium terani*. J Infect Dis 142:623

Lamanna C, Elhage AN, Vick JA (1988) Cardiac effects of botulinal toxin. Arch Int Pharmacodyn 293:69–83

Lazarovici P, Yavin E (1985a) Tetanus toxin interaction with human erythrocytes. I. Properties of polysialoganglioside association with the cell surface. Biochim Biophys Acta 812:523–532

Lazarovici P, Yavin E (1985b) Tetanus toxin interaction with human erythrocytes. II. Kinetic properties of toxin association and evidence for a ganglioside-toxin macromolecular complex formation. Biochim Biophys Acta 812:532–542

Lazarovici P, Yavin E (1986) Affinity-purified tetanus neurotoxin interaction with synaptic membranes: properties of a protease-sensitive receptor component. Biochemistry 25:7047–7054

Lazarovici P, Tayot JL, Yavin E (1984) Affinity chromatographic purification and characterization of two iodinated tetanus toxin fractions exhibiting different binding properties. Toxicon 22:401–413

Lazarovici P, Yanai P, Yavin E (1987) Molecular interactions between micellar polysialogangliosides and affinity-purified tetanotoxins in aqueous solution. J Biol Chem 262:2645–2651

Lazarovici P, Fujita K, Contreras ML, Diorio JP, Lelkes PI (1989) Affinity purified tetanus toxin binds to isolated chromaffin granules and inhibits catecholamine release in digitonin-permeabilized chromaffin cells. FEBS Lett 253:121–128

Lewis GE (1981) Biomedical aspects of botulism. Academic, New York

Lietzke R, Unsicker K (1983) Tetanus toxin binding to different morphological phenotypes of cultured rat and bovine adrenal medullary cells. Neurosci Lett 38:233–238

Lin CS, Habig WH, Hardegree MC (1985) Antibodies against the light chain of tetanus toxin in human sera. Infect Immun 49:111–115

Lomneth R, Suszkiw JB, DasGupta BR (1990) Response of the chick ciliary ganglion iris neuromuscular preparation to botulinum neurotoxin. Neurosci Lett 113:211–216

Lundh H (1983) Antagonism of botulinum toxin paralysis by low temperature. Muscle Nerve 6:56–60

Lupa MT, Tabti N, Thesleff S, Vyskocil F, Yu SP (1987) The nature and origin of calcium-insensitive miniature end-plate potentials at rodent neuromuscular junctions. J Physiol (Lond) 381:607–618

MacKenzie I, Burnstock G, Dolly JO (1982) The effects of purified botulinum neurotoxin type A on cholinergic, adrenergic and non-adrenergic atropine resistant autonomic neuro-muscular transmission. Neuroscience 7:997–1006

Maisey EA, Wadsworth JDF, Poulain B, Shone CC, Melling J, Gibbs P, Tauc L, Dolly JO (1988) Involvement of the constituent chains of botulinum neurotoxins A and B in the blockade of neurotransmitter release. Eur J Biochem 177:683–691

Makoff AJ, Oxer MD, Romanos MA, Fairwheather NF, Ballantine S (1989) Expression of tetanus toxin fragment C in E. coli: high level expression by removing rare codons. Nucleic Acids Res 17: 10191–10202

Mallart A, Molgo J, Angaut-Petit D, Thesleff S (1989) Is the internal calcium regulation altered in type A botulinum toxin poisoned motor endings? Brain Res 479:167–171

Manning KA, Erichsen JT, Evinger C (1990) Retrograde transneuronal transport-properties of fragment C of tetanus toxin. Neuroscience 34:251–263

Marconi P, Pitzurra M, Vicchiar A, Pitzurra L, Bistoni F (1982) Resistance induced by concanavalin A and phytohemagglutinin P against tetanus toxin in mice. Annu Rev Immunol 133:15–27

Marsal J, Solsona C, Rabasseda X, Blasi J, Casanova A (1987) Depolarization-induced release of ATP from cholinergic synaptosomes is not blocked by botulinum toxin type A. Neurochem Int 10:295–302

Marsal J, Solsona C, Rabasseda X, Blasi J (1988) Botulinum neurotoxin inhibits the release of newly synthesized acetylcholine from Torpedo electric organ synaptosomes. Neurochem Int 12:439–445

Marsal J, Egea G, Solsona C, Rabasseda X, Blasi J (1989) Botulinum toxin type A blocks the morphological changes induced by chemical stimulation on the presynaptic membrane of torpedo synaptosomes. Proc Natl Acad Sci USA 86:372–376

Marxen P, Bigalke H (1989) Tetanus toxin: inhibitory action in chromaffin cells is initiated by specified types of gangliosides and promoted in low ionic strength solution. Neurosci Lett 107:261–266

Marxen P, Bigalke H (1991a) The chromaffin cell: a suitable model for investigating the actions and the metabolism of tetanus and botulinum A neurotoxins. Naunyn-Schmiedebergs Arch Pharmacol 343 [Suppl]:12–29

Marxen P, Bigalke H (1991b) Tetanus and botulinum A toxins inhibit stimulated F-actin rearrangement in chromaffin cells. Neuroreport 2:33–37

Marxen P, Fuhrmann U, Bigalke H (1989) Gangliosides mediate inhibitory effects of tetanus and botulinum A neurotoxins on exocytosis in chromaffin cells. Toxicon 27:849–859

Marxen P, Ahnert-Hilger G, Wellhöner HH, Bigalke H (1990) Tetanus antitoxin binds to intracellular tetanus toxin in permeabilized chromaffin cells without restoring Ca^{2+}-induced exocytosis. Toxicon 28:1077–1082

Matsuoka I, Syuoto B, Kurihara K, Kubo S (1986) Cytotoxic action of Clostridium botulinum type C1 toxin on neurons of central nervous system in dissociated culture. Jpn J Med Sci Biol 39:247–248

Matsuda M, Yoneda M (1976) Reconstitution of tetanus neurotoxin from two antigenically active polypeptide fragments. Biochem Biophys Res Commun 68:668–674

Matsuda M, Sugimoto N, Ozutsumi K, Hirai T (1982) Acute botulinum-like intoxication by tetanus neurotoxin in mice. Biochem Biophys Res Commun 104:799–805

Matsuda M, Lei DL, Sugimoto M, Ozutsumi K, Okabe T (1989) Isolation, purification and characterization of fragment B, the NH_2-terminal half of the heavy chain of tetanus toxin. Infect Immun 57:3588–3593

McCroskey L, Hatheway CL, Fenicia L, Pasolini B, Aureli P (1986) Characterization of an organism that procudes type E botulinal toxin but which resembles *Clostridium butyricum* from the feces of an infant with type E botulism. J Clin Microbiol 23:201–202

McInnes C, Dolly JO (1990) Ca^{2+}-dependent noradrenaline release from permeabilised PC12 cells is blocked by botulinum neurotoxin A or its light chain. FEBS Lett 261:323–326

Meckler RL, Baron R, Mclachlan EM (1990) Selective uptakes of C-fragment of tetanus toxin by sympathetic preganglionic nerve terminals. Neuroscience 36:823–829

Mellanby J (1984) Commentary: comparative activities of tetanus and botulinum toxins. Neuroscience 11:29–34

Mellanby J, Green J (1981) How does tetanus toxin act? Neuroscience 6:281–300

Mellanby J, Leonard A (1986) Effect of C-fragment of tetanus toxin on subsequent toxicity of the toxin in vivo. J Med Microbiol 22:16R

Mellanby J,Thompson PA (1972) The effect of tetanus toxin at the neuromuscular junction in the goldfish. J Physiol (Lond) 224:407–419

Mellanby J, Thompson PA (1975) The effect of lanthanum on miniature junction potentials at the goldfish neuromuscular junction after block by tetanus toxin. J Physiol (Lond) 252:81

Mellanby J, Beaumont MA,Thompson PA (1988) The effect of lanthanum on nerve terminals in goldfish muscle after paralysis with tetanus toxin. Neuroscience 25:1095–1106

Menestrina G, Forti S, Gambale F (1989) Interaction of tetanus toxin with lipid vesicles – Effects of pH, surface-charge, and transmembrane potential on the kinetics of channel formation. Biophys J 55:393–405

Molgo J, Lemeignan M, Thesleff S, Lechat P (1984) Potentiation of aminosidic antibiotics of the inhibitor effect of botulic toxin on nervous motorial insertion – antagonism by diamine 3,4 pyridine. J Pharmacol 15:508–509

Molgo J, DasGupta BR, Thesleff S (1989a) Characterization of the actions of botulinum neurotoxin type E at the rat neuromuscular junction. Acta Physiol Scand 137:497–501

Molgo J, Siegel LS, Tabti N, Thesleff S (1989b) A study of synchronization of quantal transmitter release from mammalian motor endings by the use of botulinal toxins type A and D. J Physiol (Lond) 411:195–205

Molgo J, Comella JX, Angaut-Petit D, Pecot-Dechavassine M, Tabti N, Faille L, Mallart A, Thesleff S (1990) Presynaptic actions of botulinal neurotoxins at vertebrate neuromuscular junctions. J Physiol (Paris) 84:152–166

Montecucco C, Schiavo G, Brunner J, Duflot E, Boquet P, Roa M (1986) Tetanus toxin is labeled with photoactivatable phospholipids at low pH. Biochemistry 25:919–923

Montecucco C, Schiavo G, Gao Z, Bauerlein E, Boquet P, DasGupta BR (1988) Interaction of botulinum and tetanus toxins with the lipid bilayer surface. Biochem J 251:379–383

Montecucco C, Schiavo G, DasGupta BR (1989) Effect of pH on the interaction of botulinum neurotoxin A, neurotoxin B and neurotoxin E with liposomes. Biochem J 259:47–53

Montesano R, Roth J, Robert A, Orci L (1982) Non-coated membrane invaginations are involved in binding and internalization of cholera and tetanus toxin. Nature 296:651–653

Moriishi K, Syuto B, Oguma K, Saito M (1990) Separation of toxic activity and ADP-ribosylation activity of botulinum neurotoxin D. J Biol Chem 265:16614–16616

Morris NP, Consiglio E, Kohn LD, Habig WH, Hardegree MC, Helting TB (1980) Interaction of fragments B and C of tetanus toxin with neural and thyroid membranes and with gangliosides. J Biol Chem 255:6071–6076

Metezeau P, Desban M (1982) Botulinum toxin type A: kinetics of calcium dependent paralysis of the neuromuscular junction and antagonism by drugs and animal toxins. Toxicon 20:649–654

Meyer-Eppler TBE, Wellhöner HH (1983) Tetanus toxin-[2,3-^{3}H]propionamide. Arch Toxicol 52:303–310

Middlebrook JL (1986) Cellular mechanism of action of botulinum neurotoxin. J Toxicol Toxin Rev 5:177–196

Middlebrook JL, Dorland RB (1984) Bacterial toxins: cellular mechanism of action. Microbiol Rev 48:199–221

Mikhailov VV, Barashkov GN (1977) Mechanism of disturbance of inhibitory electrogenesis in spinal *-motoneurons in experimental local botulinus poisoning. Bull Exp Biol Med (Engl Transl) 83:771–774

Mikhailov VV, Shubin AI (1987) Mechanism of paradoxical resistance of the digastric muscle to the action of tetanus toxin. Bull Exp Biol Med (Engl Transl) 103:604–606

Mizuguchi J, Yoshida T, Sato Y, Nagaoka F, Kondo S, Matuhasi T (1982) Requirements for at least two distinct monoclonal antibodies for efficient neutralization of tetanus toxin in vivo. Naturwissenschaften 69:597–598

Mochida S, Poulain B, Weller U, Habermann E, Tauc L (1989) Light chain of tetanus toxin intracellularly inhibits acetylcholine release at neuro-neuronal synapses, and its internalization is mediated by heavy chain. FEBS Lett 253:47–51

Mochida S, Poulain B, Eisel U, Binz T, Kurazono H, Niemann H, Tauc L (1990) Exogenous messenger-RNA encoding tetanus or botulinum neurotoxins expressed in *Aplysia* neurons. Proc Natl Acad Sci USA 87:7844–7848

Molgo J, Thesleff S (1984) Studies on the mode of action of botulinum toxin type A at the frog neuromuscular junction. Brain Res 293:309–316

Murayama S, Syuto B, Oguma K, Iida H, Kubo S (1984) Comparison of *Clostridium botulinum* toxins type D and C, in molecular property, antigenicity and binding ability to rat brain synaptosomes. Eur J Biochem 142:487–492

Murayama S, Umezawa J, Terajima J, Syuto B, Kubo S (1987) Action of botulinum neurotoxin on acetylcholine release from rat brain synaptosomes: putative internalization of the toxin into synaptosomes. J Biochem 102:1355–1364

Nakov R, Habermann E, Hertting G, Wurster S, Allgaier C (1989) Effects of botulinum A toxin on presynaptic modulation of evoked transmitter release. Eur J Pharmacol 164:45–53

Nathan A, Yavin E (1989) Periodate-modified gangliosides enhance surface binding of tetanus toxin to PC12 pheochromocytoma cells. J Neurochem 53:88–94

Nishimura M, Kozaki S, Sakagushi G (1988) Zinc antagonises the effect of botulinum type A toxin at the neuromuscular junction. Experientia 44:18–20

Nistico G, Mastroeni P, Pitzurra M (1985) Seventh international conference on tetanus, Copanello, Italy, 10–15 Sept 1984. Gangemi, Rome

Notter MFD, Leary JF (1985) Flow cytometric analysis of tetanus toxin binding to neuroblastoma cells. J Cell Physiol 125:476–484

Notter MFD, Leary JF (1986) Tetanus toxin binding to neuroblastoma cells differentiated by antimitotic agents. Dev Brain Res 26:59–68

Nukina M, Mochida Y, Sakaguchi S, Sakaguchi G (1988) Purification of *Clostridium botulinum* type G progenitor toxin. Zentralbl Bakteriol Mikrobiol Hyg [A] 268:220–227

Ochanda JO, Syuto B, Naiki M, Kubo S (1986) Binding of *Clostridium botulinum* neurotoxin to gangliosides. J Biochem 100:127–166

Oguma K, Syuto B, Agui T, Ilda H (1981) Homogeneity and heterogeneity of toxins produced by *Clostridium botulinum* type C and D strains. Infect Immun 34:382–388

Osborne RH, Bradford HF (1973) Patterns of amino acid release from nerve-endings isolated from spinal cord and medulla. J Neurochem 21:407–419

Ozutsumi K, Sugimoto N, Matsuda M (1985) Rapid, simplified method for production and purification of tetanus toxin. Appl Environ Microbiol 49:939–943

Pamphlett R (1989) Early terminal and nodal sprouting of motor axons after botulinum toxin. J Neurol Sci 92:181–192

Park MK, Jung HH, Yang KH (1990) Binding of *Clostridium botulinum* type B toxin to rat brain synaptosomes. FEMS Microbiol Lett 72:243–247

Parton RG, Ockleford CD, Critchley DR (1987) A study of the mechanism of internalisation of tetanus toxin by primary mouse spinal cord cultures. J Neurochem 49:1057–1068

Parton RG, Ockleford CD, Critchley DR (1988) Tetanus toxin binding to mouse spinal cord cells: an evaluation of the role of gangliosides in toxin internalization (BRE 14094). Brain Res 475:118–127

Parton RG, Davison MD, Critchley DR (1989) Comparison of the binding characteristics of 2 different preparations of tetanus toxin to rat brain membranes. Toxicon 27:127–135

Pearce BR, Gard AL, Dutton GR (1983) Tetanus toxin inhibiton of K$^+$-stimulated [^{3}H]GABA release from developing cell cultures of the rat cerebellum. J Neurochem 40:887–890

Penner R, Neher E, Dreyer F (1986) Intracellularly injected tetanus toxin inhibits exocytosis in bovine adrenal chromaffin cells. Nature 324:76–77

Pierce EJ, Davison MD, Parton RG, Habig WH, Critchley DR (1986) Characterization of tetanus toxin binding to rat brain membranes. Evidence for a high-affinity proteinase-sensitive receptor. Biochem J 236:845–852

Pitzurra L, Marconi P, Bistoni F, Blasi E (1989) Selective inhibition of cytokine-induced lysozyme activity by tetanus toxin in the GG2EE macrophage cell line. Infect Immun 57:2452–2456

Polak RL, Sellin LC, Thesleff S (1981) Acetylcholine content and release in denervated or botulinum poisoned rat skeletal muscle. J Physiol (Lond) 319:253–259

Polgar AA, Smirnova VS, Kryzhanovsky GN (1972) Activation of synaptic processes in the myoneural junction poisoned by tetanus toxin in response to repetitive nerve stimulation. Bull Exp Biol Med (Engl Transl) 73:504–508

Poulain B, Tauc L, Maisey EA, Dolly JO (1988a) Ganglionic synapses of *Aplysia* as model for the study of the mechanism of action of botulinum neurotoxins. CR Acad Sci [III] 306:483–488

Poulain B, Tauc L, Maisey EA, Wadsworth JDF, Mohan PM, Dolly JO (1988b) Neurotransmitter release is blocked intracellularly by botulinum neurotoxin, and this requires uptake of both toxin polypeptides by a process mediated by the larger chain. Proc Natl Acad Sci USA 85:4090–4094

Poulain B, Wadsworth DF, Shone CC, Mochida S, Lande S, Melling J, Dolly JO, Tauc L (1989a) Multiple domains of botulinum neurotoxin contribute to its inhibition of transmitter release in *Aplysia* neurons. J Biol Chem 264:21928–21933

Poulain B, Wadsworth JDF, Maisey EA, Shone CC, Melling J, Tauc L, Dolly JO (1989b) Inhibition of transmitter release by botulinum neurotoxin A – contribution of various fragments to the intoxication process. Eur J Biochem 185:197–203

Presek P, Jarvie PE, Dunkley PR, Dreyer F (1989) Tetanus toxin inhibits the depolarization-dependent increase in synapsin I phosphorylation. Naunyn-Schmiedebergs Arch Pharmacol 339:R18

Price DL, Griffin J, Young A, Peck K, Stocks A (1975) Tetanus toxin: direct evidence for retrograde intraaxonal transport. Science 188:945–947

Puymirat J, Faivreba A, Bizzini B, Tixiervi A (1982) Prenatal and postnatal ontogenesis of neurotransmitter synthetizing enzymes and I^{125}-labeled tetanus toxin binding capacity in the mouse hypothalamus. Dev Brain Res 3:199–206

Rabasseda X, Solsona C, Marsal J, Egea G, Bizzini B (1987) ATP release from pure cholinergic synaptosomes is not blocked by tetanus toxin. FEBS Lett 213:337–340

Rabasseda X, Blast J, Marsal J, Dunant Y, Casanova A, Bizzini B (1988) Tetanus and botulinum toxins block the release of acetylcholine from slices of rat striatum and from the isolated electric organ of torpedo at different concentrations. Toxicon 26:329–336

Raff MC, Abney ER, Cohen J, Lindsay R, Noble M (1983a) Two types of astrocytes in cultures of developing rat white matter: differences in morphology, surface gangliosides, and growth characteristics. J Neurosci 3:1289–1300

Raff MC, Miller RH, Noble M (1983b) A glial progenitor cell that develops in vitro into an astrocyte or an oligodendrocyte depending on culture medium. Nature 303:390–396

Rauch G, Gambale F, Montal M (1990) Tetanus toxin channels in phosphatidylserine planar bilayers – conductance states and pH-dependence. Eur Biophys J 18:79–83

Reidler J, Robinson JP (1988) Two-dimensional crystals of tetanus toxin. Methods Enzymol 165:389–396

Roa M, Boquet P (1985) Interaction of tetanus toxin with lipid vesicles at low pH. Protection of specific polypeptides against proteolysis. J Biol Chem 260:6827–6835

Robinson JP (1988) Purification of tetanus toxin and its major peptides. Methods Enzymol 165:85–90

Robinson JP, Hash JH (1982) A review of the molecular structure of tetanus toxin. Mol Cell Biochem 48:33–44

Robinson JP, Schmid MF, Morgan DG, Chiu W (1988a) Three-dimensional structural analysis of tetanus toxin by electron crystallography. J Mol Biol 200:367–375

Robinson JP, Chiu W, DasGupta BR (1988b) Two-dimensional crystals of botulinum toxin type A. FASEB J 2:1749

Rogers TR, Snyder SH (1981) High-affinity binding of tetanus toxin to mammalian brain membranes. J Biol Chem 256:2402–2407

Rothman SM (1983) Synaptic activity mediates death of hypoxic neurons. Science 220:536–537

Sakaguchi G (1983) *Clostridium botulinum* toxins. Pharmacol Ther 19:165–194

Sanchez-Prieto J, Sihra TS, Evans D, Ashton A, Dolly JO, Nicholls DG (1987) Botulinum toxin blocks glutamate exocytosis from guinea-pig cerebral cortical synaptosomes. Eur J Biochem 165:675–681

Sandberg K, Berry CJ, Eugster E, Rogers TB (1989a) A role for cGMP during tetanus toxin blockade of acetylcholine release in the rat pheochromocytoma (PC12) cell line. J Neurosci 9:3946–3954

Sandberg K, Berry CJ, Rogers TB (1989b) Studies on the intoxication pathway of tetanus toxin in the rat pheochromocytoma (PC12) cell line. Binding, internalization, and inhibition of acetylcholine release. J Biol Chem 264:5679–5686

Sandvig K, Olsnes S, Petersen OW, VanDeurs B (1989) Control of coated-pit function by cytoplasmic pH. Methods Cell Biol 32:365–382

Sathyamoorthy V, DasGupta BR (1985a) Partial amino acid sequences of the heavy and light chains of botulinum neurotoxin type E. Biochem Biophys Res Commun 127:768–772

Sathyamoorthy V, DasGupta RB (1985b) Separation, purification, partial characterization and comparison of the heavy and light chains of botulinum neurotoxin types A, B, and E. J Biol Chem 260:10461–10466

Sathyamoorthy V, DasGupta BR (1988) Reductive methylation of lysine residues of botulinum neurotoxin types A and B. Mol Cell Biochem 83:65–72

Sathyamoorthy V, DasGupta BR, Foley J, Niece RL (1988) Botulinum neurotoxin type A: cleavage of the heavy chain into two halves and their partial sequences. Arch Biochem Biophys 266:142–151

Schiavo G, Boquet P, DasGupta BR, Montecucco C (1990) Membrane interactions of tetanus and botulinum neurotoxins – a photolabeling study with photoactivatable phospholipids. J Physiol (Paris) 84:180–187

Schmidt JJ, Sathyamoorthy V, DasGupta BR (1984) Partial amino acid sequence of the heavy and light chains of botulinum neurotoxin type A. Biochem Biophys Res Commun 119:900–904

Schmidt JJ, Sathyamoorthy V, DasGupta BR (1985) Partial amino acid sequences of botulinum neurotoxins types B and E. Arch Biochem Biophys 238:544–564

Schmitt A, Dreyer F, John C (1981) At least three sequential steps are involved in the tetanus toxin-induced block of neuromuscular transmission. Naunyn-Schmiedebergs Arch Pharmacol 317:326–330

Schnitzer J, Kim SU, Schachner M (1984) Some immature tetanus toxin-positive cells share antigenic properties with subclasses of glial cells. An immunofluorescence study in the developing nervous system of the mouse using a new monoclonal antibody. Dev Brain Res 16:203–217

Schübel K (1923) Über das Botulinumtoxin. Arch Exp Pathol Pharmakol 96:193–259

Schwab ME, Thoenen H (1977) Selective trans-synaptic migration of tetanus toxin after retrograde axonal transport in peripheral sympathetic nerves: a comparison with nerve growth factor. Brain Res 122:459–474

Schwab ME, Suda K, Thoenen H (1979) Selective retrograde transsynaptic transfer of a protein, tetanus toxin, subsequent to its retrograde axonal transport. J Cell Biol 82:798–810

Sellin LC (1987) Botulinum toxin and the blockade of transmitter release. Asia Pac J Pharmacol 2:203–222

Sellin LC, Thesleff S (1981) Pre- and post-synaptic actions of botulinum toxin at the rat neuromuscular junction. J Physiol (Lond) 317:487–495

Sellin LC, Kaufman JA, DasGupta BR (1983a) Comparison of the effects of botulinum neurotoxin type A and type E at the rat neuromuscular junction. Med Biol 61:120–125

Sellin LC, Thesleff S, DasGupta BR (1983b) Different effects of types A and B botulinum toxin on transmitter release at the rat neuromuscular junction. Acta Physiol Scand 119:127–133

Sheppard AJ, Cussell D, Hughes M (1984) Production and characterization of monoclonal antibodies to tetanus toxin. Infect Immun 43:710–714

Sheppard AJ, Hughes M, Stephen J (1987) Affinity purification of tetanus toxin using polyclonal and monoclonal antibody immunoadsorbents. J Appl Bacteriol 62:335–348

Shiells RA, Falk G (1988) Reversal of botulinum toxin synaptic blockade by nicotinamide at the rat neuromuscular junction. Pestic Sci 23:357–358

Shone CC, Hambleton P, Melling J (1985) Inactivation of Clostridium botulinum type A neurotoxin by trypsin and purification of two tryptic fragments – proteolytic action near the COOH-terminus of the heavy subunit destroys toxin-binding activity. Eur J Biochem 151:75–82

Shone CC, Hambleton P, Melling J (1987) A 50-KDA fragment from the NH$_2$-terminus of the heavy subunit of Clostridium botulinum type A neurotoxin forms channels in lipid vesicles. Eur J Biochem 167:175–180

Simpson LL (1974) Studies on the binding of botulinum toxin type A to the rat phrenic nerve-hemidiaphragm preparation. Neuropharmacology 13:683–691

Simpson LL (1980) Kinetic studies on the interaction between botulinum toxin type A and the cholinergic neuromuscular junction. J Pharmacol Exp Ther 212:16–21

Simpson LL (1981) The origin, structure, and pharmacological activity of botulinum toxin. Pharmacol Rev 33:155–188

Simpson LL (1982) The interaction between aminoquinolines and presynaptically acting neurotoxins. J Pharmacol Exp Ther 222:43–48

Simpson LL (1983) Ammoniumchloride and methylamine hydrochloride antagonize clostridial neurotoxins. J Pharmacol Exp Ther 225:546–552

Simpson LL (1984a) Botulinum toxin and tetanus toxin recognize similar membrane determinants. Brain Res 305:177–180

Simpson LL (1984b) Fragment C of tetanus toxin antagonizes the neuromuscular blocking properties of native tetanus toxin. J Pharmacol Exp Ther 228:600–604

Simpson LL (1984c) The binding fragment from tetanus toxin antagonizes the neuromuscular blocking actions of botulinum toxin. J Pharmacol Exp Ther 229:182–187

Simpson LL (1985) Pharmacological experiments on the binding and internalization of the 50,000 Dalton carboxyterminus of tetanus toxin at the cholinergic neuromuscular junction. J Pharmacol Exp Ther 234:100–105

Simpson LL (1986) Molecular pharmacology of botulinum toxin and tetanus toxin. Annu Rev Pharmacol Toxicol 26:427–453

Simpson LL (1988) Use of pharmacologic antagonists to deduce commonalities of biologic activity among clostridial neurotoxins. J Pharmacol Exp Ther 245:867–872

Simpson LL (1989) Botulinum neurotoxin and tetanus toxin. Academic, San Diego

Simpson LL (1990) The study of clostridial and related toxins – the search for unique mechanisms and common denominators. J Physiol (Paris) 84:143–152

Simpson LL, DasGupta BR (1983) Botulinum neurotoxin type E – studies on mechanism of action and on structure-activity relationships. J Pharmacol Exp Ther 224:135–140

Simpson LL, Hoch DH (1985) Neuropharmacological characterization of fragment B from tetanus toxin. J Pharmacol Exp Ther 232:223–227

Singh BR, DasGupta BR (1989a) Changes in the molecular topography of the light and heavy chains of type A botulinum neurotoxin following their separation. Biophys Chem 34:259–267

Singh BR, DasGupta BR (1989b) Structure of heavy and light chain subunits of type A botulinum neurotoxin analyzed by circular dichroism and fluorescence measurements. Mol Cell Biochem 85:67–73

Singh BR, DasGupta BR (1989c) Molecular topography and secondary structure comparisons of botulinum neurotoxin type A, B and E. Mol Cell Biochem 86:87–95

Singh BR, DasGupta BR (1989d) Molecular differences between type A botulinum neurotoxin and its toxoid. Toxicon 27:403–410

Singh BR, Fuller MP, Schiavo G (1990) Molecular structure of tetanus neurotoxin as revealed by Fourier transform infrared and circular dichroic spectroscopy. Biophys Chem 36:155–166

Smith LA, Middlebrook JL (1985) Botulinum and tetanus neurotoxins inhibit guanylate cyclase activity in synaptosomes and cultured nerve cells. Toxicon 23:611

Solsona C, Ega A, Blasi J, Casanova C, Marsal J (1990) The action of botulinum toxin on cholinergic nerve terminals isolated from the electric organ of *Torpedo marmorata* – detection of a putative toxin receptor. J Physiol (Paris) 84:174–179

Spitzer N (1972) Miniature end plate potentials at mammalian neuromuscular junctions poisoned by botulinum toxin. Nature New Biol 237:26–27

Stanley EF, Drachman DB (1983) Botulinum toxin blocks quantal but not non-quantal release of ACH at the neuromuscular junction (technical note). Brain Res 261:172–175

Staub GC, Walton KM, Schnaar RL, Nichols T, Baichwal R, Sandberg K, Rogers TB (1986) Characterization of the binding and internalization of tetanus toxin in a neuroblastoma hybrid cell line. J Neurosci 6:1443–1451

Stecher B, Gratzl M, Ahnert-Hilger G (1989a) Reductive chain separation of botulinum toxin – a prerequisite to its inhibitory action on exocytosis in chromaffin cells. FEBS Lett 248:23–27

Stecher B, Weller U, Habermann E, Gratzl M, Ahnert-Hilger G (1989b) The light chain but not the heavy chain of botulinum A toxin inhibits exocytosis from permeabilized adrenal chromaffin cells. FEBS Lett 255:391–394

Stein P, Biel H (1973) Modification of tetanus toxin with selective chemical reagents. Z Immunitaetsforsch 145:418–431

Stöckel K, Schwab M, Thoenen H (1975) Comparison between the retrograde axonal transport of nerve growth factor and tetanus toxin in motor, sensory and adrenergic neurons. Brain Res 99:1–16

Sugimoto N, Higashida H, Ozutsumi K, Miki N, Matsuda M (1983) Tetanus toxin blocks Ca spikes in neuroblastoma clone NlE-115 cells. Biochem Biophys Res Commun 115:788–793

Sugiyama H (1980) *Clostridium botulinum* neurotoxin. Microbiol Rev 44:419–448

Sundstrom LE, Mellanby JH (1990) Tetanus toxin blocks inhibition of granule cells in the dentate gyrus of the urethane-anesthetized rat. Neuroscience 38:621–627

Svennerholm L (1970) Gangliosides. In: Laitha A (ed) Handbook of neurochemistry. Plenum, New York, pp 425–452

Sverdlov YuS (1969) Potentials of spinal motoneurons in cats with experimental tetanus. Neurophysiology 1:18–25

Syuoto B, Kubo S (1981) Separation and characterization of heavy and light chains from *Clostridium botulinum* type C toxin and their reconstitution. J Biol Chem 256:3712–3717

Syuto B, Kubo S (1982) *Clostridium botulinum* type C toxin – a sketch of the molecule. Mol Cell Biochem 48:25–32

Takamizawa K, Iwamori M, Kozaki S, Sakaguchi G, Tanaka R, Takayama H, Nagai Y (1986) TLC immunostaining characterization of *Clostridium botulinum* type A neurotoxin binding to gangliosides and free fatty acids. FEBS Lett 201:229–232

Takano K, Kirchner F, Terhaar P, Tiebert B (1983) Effect of tetanus toxin on the monosynaptic reflex. Naunyn-Schmiedebergs Arch Pharmacol 323:217–220

Takano K, Kirchner F, Gremmelt A, Matsuda M, Ozutsumi N, Sugimoto N (1989a) Blocking effects of tetanus toxin and its fragment (A-B) on the excitatory and inhibitory synapses of the spinal motoneuron of the cat. Toxicon 27:385–392

Takano K, Kirchner F, Tiebert B, Terhaar P (1989b) Presynaptic inhibition of the monosynaptic reflex during local tetanus in the cat. Toxicon 27:431–438

Tauc L, Mochida S, Poulain B (1990) *Aplysia* central synapses as models for the study of botulinum and tetanus neurotoxins. Eur J Pharmacol 183:2087–2088

Terajima J, Kuwabara M, Syuto B, Kubo S (1987) Spin labeling study of the role of tyrosyl groups of C1 toxin of *Clostridium botulinum* strain Stockholm. Jpn J Med Sci Biol 40:204–205

Thesleff S (1989) Botulinal neurotoxins as tools in studies of synaptic mechanisms. Q J Exp Physiol Med Sci 74:1003–1017

Thesleff S, Molgo J, Lundh H (1983) Botulinum toxin and 4-aminoquinoline induce a similar abnormal type of spontaneous quantal transmitter release at the rat neuromuscular-junction. Brain Res 264:89–97

Thompson DE, Brehm JK, Oultram JD, Swinfield TJ, Shone CC, Atkinson T, Melling J, Minton NP (1990) The complete amino acid sequence of the *Clostridium botulinum* type A neurotoxin, deduced by nucleotide sequence analysis of the encoding gene. Eur J Biochem 189:73–118

Tse CK, Dolly JG, Hambleton P, Wray D, Melling J (1982) Preparation and characterization of homogenous neurotoxin type A from *Clostridium botulinum*, Its inhibitory action on neuronal release of acetylcholine in the absence and presence of β-bungarotoxin. Eur J Biochem 122:493–500

Tsuzuki K, Yokosawa N, Syuto B, Ohishi I, Fujii N, Kimura K, Oguma K (1987) Establishment of a monoclonal antibody recognizing an antigenic site common to *Clostridium botulinum* type B, C1, D, and E toxins and tetanus toxin. Infect Immun 56:898–902

Van Heyningen S (1982) Similarities in the action of different toxins. In: Cohen P, VanHeyningen S (ed) Molecular action of toxins and viruses. Elsevier Biomedical, Amsterdam, pp 169–187

Van Heyningen S (1986) Tetanus toxin. In: Dorner F, Drews J (ed) Pharmacology of bacterial toxins. Pergamon, Oxford, pp 549–569

Van Heyningen WE (1959) Chemical assay of the tetanus toxin receptor in nervous tissue. J Gen Microbiol 20:301–309

Van Vliet BJ, Sebben M, Dumuis A, Gabrion J, Bockaert J, Pin JP (1989) Endogenous amino-acid release from cultured cerebellar neuronal cells – effect of tetanus toxin on glutamate release. J Neurochem 52:1229–1239

Veronesi R (1981) Tetanus. Important new concepts. Excerpta Medica, Amsterdam

Volk WA, Bizzini B, Snyder RM, Bernhard E, Wagner RR (1984) Neutralization of tetanus toxin by distinct monoclonal antibodies binding to multiple epitopes on the toxin molecule. Infect Immun 45:604–609

Wadsworth JDF, Desai M, Tranter HS, King HJ, Hambleton P, Melling J, Dolly JO, Shone CC (1990) Botulinum type F neurotoxin. Large-scale purification and characterization of its binding to rat cerebrocortical synaptosomes. Biochem J 268:123–128

Walton KM, Sandberg K, Rogers TB, Schnaar RL (1988) Complex ganglioside expression and tetanus toxin binding by PC12 pheochromocytoma cells. J Biol Chem 263:2055–2063

Weller U, Taylor CF, Habermann E (1986) Quantitative comparison between tetanus toxin, some fragments and toxoid for binding and axonal transport in the rat. Toxicon 24:1055–1064

Weller U, Mauler F, Habermann E (1988) Tetanus toxin – biochemical and pharmacological comparison between its protoxin and some isotoxins obtained by limited proteolysis. Naunyn-Schmiedebergs Arch Pharmacol 338:99–106

Weller U, Dauzenroth ME, Meyer zu Heringdorf D, Habermann E (1989) Chains and fragments of tetanus toxin – separation, reassociation and pharmacological properties. Eur J Biochem 182:649–656

Wellhöner HH (1982) Tetanus neurotoxin. Rev Physiol Biochem Pharmacol 93:1–68

Wellhöner HH, Neville DM Jr (1987) Tetanus toxin binds with high affinity to neuroblastoma x glioma hybrid cells NG108-215 and impairs their stimulated acetylcholine release. J Biol Chem 262:7374–7378

Wellhöner HH, Seib UC, Hensel B (1973) Local tetanus in cats: the influence of neuromuscular activity on spinal distribution of ^{125}I labelled tetanus toxin. Naunyn-Schmiedebergs Arch Pharmacol 276:387–394

Wellhöner HH, Bigalke H, Borcholte T, Erdmann G, Eschenhagen T, Jung KH, Marxen P, Peukert U, Neville DM Jr, Srinivasachar K (1990) Uptake of antitetanus F(ab')2 fragments into eucaryotic cells. J Physiol (Paris) 84:206–210

Wellhöner HH, Neville DM Jr, Srinivasachar K, Erdmann G (1991) Uptake and concentration of bioactive macromolecules by K562 cells via the transferrin cycle utilizing an acid-labile transferrin conjugate. J Biol Chem 266:4309–4314

Wendon LMB (1980) Action of tetanus toxin at the rat neuromuscular junction. J Physiol (Lond) 300:23

Wendon LMB, Gill DM (1982) Tetanus toxin action on cultured nerve cells – does it modify a neuronal protein? Brain Res 238:292–297

White T, Potter P, Wonnacott S (1980) Depolarization-induced release of ATP from cortical synaptosomes is not associated with acetylcholine release. J Neurochem 34:1109–1112

Wiegand H, Wellhöner HH (1977) The action of botulinum A neurotoxin on the inhibition by antidromic stimulation of the lumbar monosynaptic reflex. Naunyn-Schmiedebergs Arch Pharmacol 298:235–238

Wiegand H, Wellhöner HH (1979) Electrical excitability of motoneurones in early local tetanus. Naunyn-Schmiedebergs Arch Pharmacol 308:71–76

Wiegand H, Erdmann G, Wellhöner HH (1976) ^{125}I-labelled botulinum A neurotoxin: pharmacokinetics in cats after intramuscular injection. Naunyn-Schmiedebergs Arch Pharmacol 292:161–165

Wiegand H, Hilbig G, Wellhöner HH (1977) Early local tetanus: does tetanus toxin change the stimulus evoked discharge in afferents from injected muscle? Naunyn-Schmiedebergs Arch Pharmacol 298:189–191

Wieraszko A (1985) Attenuation of inhibitory processes in the central nervous system by tetanus toxin: an in vitro study on rat hippocampal slices. Life Sci 37:2059–2065

Williams S, Dolly JO, Ise CK, Hambleton P, Melling J (1983) Radioiodination of botulinum neurotoxin type A with retention of biological activity and its binding to brain synaptosomes. Eur J Biochem 131:437–445

Wonnacott S (1980) Inhibition by botulinum toxin of acetylcholine release from synaptosomes: latency of action and the role of gangliosides. J Neurochem 34:1567–1573

Wonnacott S, Marchbanks RM (1976) Inhibition by botulinum toxin of depolarization-evoked release of (^{14}C)acetylcholine from synaptosomes in vitro. Biochem J 156:701–712

Woody MA, DasGupta BR (1988) Fast protein liquid chromatography of botulinum neurotoxin type A, type B and type E. J Chromatogr Biomed Appl 430:279–289

Woody EMA, Herian A, DasGupta BR (1989) Modification of carboxyl groups in botulinum neurotoxin type A and type E. Toxicon 27:1143–1150

Yavin E, Habig WH (1984) Binding of tetanus toxin to somatic neural hybrid cells with varying ganglioside composition. J Neurochem 42:1313–1321

Yavin E, Nathan A (1986) Tetanus toxin receptors on nerve cells contain a trypsin-sensitive component. Eur J Biochem 154:403–407

Yavin E, Yavin Z, Habig WH, Hardegree MC (1981) Tetanus toxin association with developing neuronal cell cultures. Kinetic parameters and evidence for ganglioside-mediated internalization. J Biol Chem 256:7014–7022

Yavin Z, Yavin E, Kohn LD (1982) Sequestration of tetanus toxin in developing neuronal cell cultures. J Neurosci Res 7:267–278

Yavin E, Yavin Z, Kohn LD (1983) Temperature-mediated interaction of tetanus toxin with cerebral neuron cultures: characterization of a neuraminidase-insensitive toxin-receptor complex. J Neurochem 40:1212–1219

Yokosawa N, Tsuzuki K, Syuto B, Oguma K (1986) Activation of *Clostridium botulinum* type E toxin purified by two different methods. J Gen Microbiol 132:1981–1988

Yokosawa N, Kurokawa Y, Tsuzuki K, Syuto B, Fuji N, Kimura K, Oguma K (1989) Binding of *Clostridium botulinum* type C neurotoxin to different neuroblastoma cell lines. Infect Immun 57:272–277

CHAPTER 12

Capsaicin: Selective Toxicity for Thin Primary Sensory Neurons*

P. Holzer

A. Introduction

Capsaicin is the pungent ingredient in a wide variety of red peppers of the genus *Capsicum*. Chemically it is a derivative of vanillyl amide, 8-methyl-*N*-vanillyl-6-nonenamide (Fig. 1) with a molecular weight of 305.42. It was only by the middle of this century that N. Jancsó realized that capsaicin, in addition to its irritant effect, has a long-term sensory neuron-blocking action which can be used in the functional investigation of sensory neurons (Jancsó 1968; Szolcsányi 1982, 1984a). A large number of studies have since corroborated this discovery and established capsaicin as an important probe for sensory neuron mechanisms. With these advances it has become possible to explain the selectivity of its action on a cellular basis.

In principle, capsaicin exerts two actions on excitable cells. One action is selective for thin afferent neurons of mammalian species and expresses itself in an initial short-lasting stimulation that is followed by desensitization to capsaicin and other stimuli of sensory neurons. If the dose of capsaicin is high enough, a long-term inhibition or even ablation of thin sensory neurons is achieved. Both these stimulant and long-term inhibitory effects may arise from a common mechanism of action, i.e., activation of a cation channel in the cell membrane of mammalian sensory neurons. The other action occurs with high concentrations of capsaicin, is cell-nonselective, and is seen throughout the animal kingdom. It manifests itself in a transient depression of excitability with no long-lasting consequences for the cell.

The scope of the present chapter is to describe and discuss the phenomenology, cellular targets, and mechanisms of action of capsaicin and to present an integrated summary of its current neuropharmacology. The early work and the related advances in the physiology of capsaicin-sensitive afferent neurons have been summarized in various reviews (Nagy 1982; Szolcsányi 1982, 1984a,b, 1990; Fitzgerald 1983; Coleridge and Coleridge 1984; Russell and Burchiel 1984; Marley and Livett 1985; Buck and Burks 1986; Jancsó et al. 1987; Lundberg and Saria 1987; Chahl 1988; Holzer 1988; Lembeck 1988; Maggi and Meli 1988; Maggi

* Work in the author's laboratory was supported by the Austrian Scientific Research Funds (Grants 5552 and 7845).

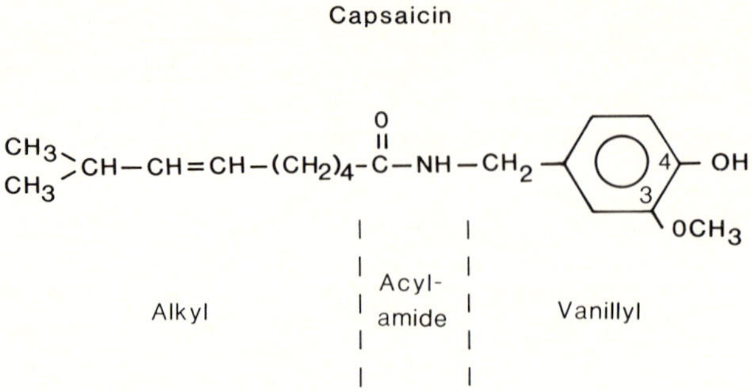

Capsaicin

$$CH_3 > CH—CH=CH—(CH_2)_4-\overset{\overset{O}{\parallel}}{C}—NH—CH_2 —\left(\bigcirc\right)^4_3— OH$$

OCH$_3$

Alkyl   Acyl-   Vanillyl
              amide

Fig. 1. Chemical structure of capsaicin

1991), to which the reader is referred. The disposition of this commentary follows the time course of the action of capsaicin on sensory neurons, proceeding from the acute to the intermediate and long-term effects of the drug. This phenomenological description is followed by a review of the mechanisms of action and an attempt to explain its selective effects on thin sensory neurons by a coherent theory.

B. Types and Targets of Action

I. Types of Primary Afferent Neurons

The classification of primary afferent neurons according to morphological, functional, bio- and histochemical criteria is dealt with here only to the extent that the target of action of capsaicin can be appreciated. The cell bodies of primary afferent neurons are located in the spinal (dorsal root) and cranial sensory ganglia and send fibers in both the central and peripheral direction. *Morphologically*, primary afferent neurons are divided into those with large light somata (A-type neurons) and those with small dark somata (B-type neurons), although the cell body diameters overlap considerably (Lawson and Harper 1984). The size of the cell bodies is grossly related to the diameter of the fibers that arise from them (Harper and Lawson 1985; Hoheisel and Mense 1986) and that can be separated into (1) thick myelinated, (2) thin myelinated, and (3) thin unmyelinated fibers. This morphological heterogeneity is closely paralleled by a functional heterogeneity with regard to projection into different layers of the spinal cord or medulla, conduction velocity, and sensory modality (Salt and Hill 1983).

Physiologically, the thick myelinated fibers have the highest conduction velocities (Aαβ-fibers) and carry nonnociceptive mechanical information from the skin and muscle. The thin unmyelinated fibers have the slowest conduction velocities (C-fibers) and are primarily nociceptors which respond to noxious mechanical, thermal, and chemical stimuli. In addition, they also comprise some specific thermonociceptors as well as some nonnociceptive mechanical, warm, and cold receptors. The small myelinated fibers conduct at intermediate velocities (Aδ-fibers) and carry both nociceptive (mechanonociceptors and polymodal nociceptors) and nonnociceptive (mechanoreceptors and cold receptors) information.

Primary afferent neurons can be further differentiated by their *ultrastructural*, *biochemical*, and *histochemical* properties. The A-type neurons in the rat can be labelled selectively with RT97, a monoclonal antibody to a neurofilament protein which is absent from B-type sensory neurons (LAWSON and HARPER 1984; LAWSON et al. 1984; KAI-KAI et al. 1986; WINTER 1987). Table 1 lists some biochemical and histochemical markers that have been found primarily in thin afferent nerve fibers. They include a number of peptides such as substance P and calcitonin gene-related peptide which play a role in the communication of primary sensory neurons with other neuronal and nonneuronal cells (SALT and HILL 1983; WEIHE 1990). However, these peptide markers are by no means exclusive for afferent neurons but also label certain neurons of the central, motor, autonomic, and enteric nervous systems.

II. Acute Excitatory Effects on Sensory Neurons

1. Targets of Action

Administration of capsaicin to the peripheral nerve endings results in depolarization and discharge of action potentials, which in turn evokes burning pain. This sensation is produced by a threshold concentration of 30 nM capsaicin on the rat eye (SZOLCSÁNYI and JANCSÓ-GÁBOR 1975) or blister base in human skin (SZOLCSÁNYI 1977), whereas the threshold concentration in the oral cavity and tongue of man is about 0.7 μM (SZOLCSÁNYI and JANCSÓ-GÁBOR 1975; SZOLCSÁNYI 1977; ROZIN et al. 1981; SIZER and HARRIS 1985). In the rat and monkey intradermal capsaicin (30 nmol) activates chemonociceptors and elicits pain (LAMOTTE et al. 1988; MARTIN et al. 1987). Injected close arterially to exteroceptive fields, doses of 0.07 nmol capsaicin in the cat (SZOLCSÁNYI 1977) and guinea-pig (SZOLCSÁNYI et al. 1986), 0.3 nmol in the rat (SZOLCSÁNYI et al. 1988), and 7 nmol in the rabbit (SZOLCSÁNYI 1987) are suprathreshold in activating polymodal nociceptors or producing pain. This high potency of capsaicin is also seen in vitro (DRAY et al. 1989a, 1990a,c) and when capsaicin is administered to the axons or somata of afferent neurons. Periaxonal concentrations as low as 30–100 nM capsaicin are capable of stimulating afferent nerve fibers in the rat isolated vagus, sciatic, and sural nerves or lumbar dorsal roots (AULT and EVANS 1980; YANAGISAWA et al. 1980; HAYES et al. 1984a; BEVAN et al. 1987; MARSH et al. 1987). Similar concentrations of capsaicin depolarize dorsal

Table 1. Some markers of capsaicin-sensitive primary afferent neurons

Marker	Selected references
Adenosine deaminase	1
Arginine vasopressin	2
Bombesin/gastrin-releasing peptide	3
Calcitonin gene-related peptide	4,5,6,7,8
Cholecystokinin[a]	8,9
Cholecystokinin receptor binding	10
Corticotropin-releasing factor[b]	11
Dynorphin	8,12
Fluoride-resistant acid phosphatase	13,14,15,16,17,18
γ-Aminobutyric acid receptor binding	19
Galanin	20
β-Glycerophosphatase	21
5-Hydroxytryptamine receptor binding	22
Lactoseries carbohydrate antigens	23
Leucine enkephalin	12
Neurokinin A	24,25
Opiate receptor binding	26,27,28
Peptide histidine methionine	29
Somatostatin	9,16,30
Substance P	9,14,16,27,30,31,32
Thiamine monophosphatase	21,33
Vasoactive intestinal polypeptide	9,11

References: (1) Nagy and Daddona 1985; (2) Kai-Kai et al. 1986; (3) Decker et al. 1985; (4) Gibbins et al. 1985; (5) Lundberg et al. 1985; (6) Skofitsch and Jacobowitz 1985b; (7) Franco-Cereceda et al. 1987; (8) Gibbins et al. 1987; (9) Jancsó et al. 1981; (10) Ladenheim et al. 1986; (11) Skofitsch et al. 1985; (12) Weihe 1990; (13) Jancsó and Knyihár 1975; (14) Jessell et al. 1978; (15) Ainsworth et al. 1981; (16) Nagy et al. 1981a, 1981b; (17) Gamse et al. 1982; (18) McDougal et al. 1983, 1985; (19) Singer and Placheta 1980; (20) Skofitsch and Jacobowitz 1985a; (21) Bucsics et al. 1988; (22) Hamon et al. 1989; (23) Kirchgessner et al. 1988; (24) Maggio and Hunter 1984; (25) Hua et al. 1985; (26) Gamse et al. 1979a; (27) Nagy et al. 1980; (28) Laduron 1984; (29) Chéry-Croze et al. 1989; (30) Gamse et al. 1981b; (31) Gamse et al. 1980; (32) Hayes and Tyers 1980; (33) Inomata and Nasu 1984.
[a] Cholecystokinin-like immunoreactivity in rat sensory neurons (Jancsó et al. 1981a) may represent calcitonin gene-related peptide (Ju et al. 1986).
[b] Corticotropin-releasing factor-like immunoreactivity in rat sensory neurons (Skofitsch et al. 1985) may represent substance P (Berkenbosch et al. 1986).

root and nodose ganglion cells of the rat in vitro or in culture (Williams and Zieglgänsberger 1982; Baccaglini and Hogan 1983; Heyman and Rang 1985; Bevan et al. 1987; Marsh et al. 1987). Behavioral evidence indicates that intrathecal, intracisternal, or intracerebroventricular injection of 33–

330 nmol capsaicin activates the central endings of nociceptive afferent nerve fibers (YAKSH et al. 1979; GAMSE et al. 1981b; JANCSÓ 1981).

The primary target of the excitatory action of capsaicin are thin primary afferent neurons. When capsaicin is administered on or into the skin of man, cat, rabbit, and rat, most if not all C-fiber chemonociceptors (SZOLCSÁNYI 1987; LANG et al. 1990), many polymodal nociceptors (SZOLCSÁNYI 1977, 1987; FOSTER and RAMAGE 1981; KENINS 1982; KONIETZNY and HENSEL 1983; MARTIN et al. 1987; SZOLCSÁNYI et al. 1988; LANG et al. 1990), some C-fiber warm receptors (FOSTER and RAMAGE 1981; KENINS 1982; SZOLCSÁNYI 1983a), and some Aδ-fiber polymodal nociceptors (MATSUMIYA et al. 1983; SZOLCSÁNYI et al. 1988) are stimulated. Other types of C- and A-fiber afferents are not affected by capsaicin. Accordingly, perineural application of capsaicin depolarizes only C-fibers in the rat vagus nerve (BEVAN et al. 1987) but also some Aδ-fibers in the cat saphenous nerve (SUCH and JANCSÓ 1986). The situation is similar with skeletal muscle and visceral afferents. Thus, capsaicin stimulates the majority of group IV (C-fiber) and some group III (Aδ-fiber) afferents from the canine skeletal muscle (KAUFMAN et al. 1982) and feline knee joint (HE et al. 1988). It likewise activates preferentially C- but also some chemosensitive Aδ-fibers in visceral nerves (COLERIDGE and COLERIDGE 1977, 1984; LONGHURST et al. 1984; SZOLCSÁNYI 1984b). When administered to the cell bodies of afferent neurons in vitro (HEYMAN and RANG 1985; BEVAN et al. 1987; MARSH et al. 1987) or in culture (BEVAN et al. 1987; WOOD et al. 1988), capsaicin stimulates only somata that are connected to C-fibers, although one report holds that the drug activates somata connected to both C- and A-fibers (WILLIAMS and ZIEGLGÄNSBERGER 1982).

Taken together, capsaicin is selective in *stimulating* primary afferent C- and Aδ-fibers, although a few other A-fibers might also respond to the drug. However, not all C- and Aδ-fibers and not all sensory neuron somata connected to these fibers are sensitive to capsaicin. Neurons other than primary afferents are not stimulated by capsaicin, the only exception to date being a group of thermosensitive neurons in the hypothalamus (see SZOLCSÁNYI 1982; HORI 1984). Thus, ventral roots of the spinal cord (AULT and EVANS 1980; MARSH et al. 1987), fibers of the optic nerve (MARSH et al. 1987), pre- and postganglionic sympathetic nerve fibers and sympathetic ganglion cells (AULT and EVANS 1980; BACCAGLINI and HOGAN 1983; SUCH and JANCSÓ 1986; BEVAN et al. 1987; MARSH et al. 1987; WOOD et al. 1988), and cerebellar neurons (SALT and HILL 1980) are not stimulated by capsaicin. There is indirect evidence that neurons of the enteric nervous system are not activated either (BARTHÓ and SZOLCSÁNYI 1978; BARTHÓ et al. 1982a; HOLZER 1984; TAKAKI and NAKAYAMA 1989). The excitatory actions of capsaicin seen, e.g., in neurons of the enteric nervous system (BARTHÓ and SZOLCSÁNYI 1978; TAKAKI and NAKAYAMA 1989), in the spinal cord (YANAGISAWA et al. 1980; J. M. CHUNG et al. 1985; URBÁN et al. 1985; HAJÓS et al. 1986b), in the medulla (SALT and HILL 1980), and in certain areas of the brain (Rabe et al. 1980; Andoh et al. 1982; Hajós et al. 1986b; Braga et al. 1987; Szikszay and London 1988) are seen as secondary consequences of sensory neuron stimulation. This argument, though, is not conclusive in all instances.

The acute actions of capsaicin are not restricted to sensory neurons and not even to neurons. These cell-nonselective effects of capsaicin include inhibition of cardiac muscle excitability (Zernig et al. 1984; Franco-Cereceda and Lundberg 1988), inhibition of visceral smooth muscle activity (Barthó et al. 1982b, 1987; Maggi et al. 1987c, 1989b), and contraction of vascular smooth muscle (Donnerer and Lembeck 1982; Duckles 1986; Bény et al. 1989; Edvinsson et al. 1990; Holzer et al. 1990b). Contradictory data suggest that capsaicin might either enhance (Juan et al. 1980; Moritoki et al. 1990) or inhibit (Flynn et al. 1986) the formation of prostanoids in vascular tissue, whereas no effect on prostanoid formation is observed in the gastric mucosa (Holzer et al. 1990a). In addition, capsaicin has been reported to inhibit platelet aggregation (Wang et al. 1984) and to influence various metabolic processes (Miller et al. 1983). Importantly, the cell-nonselective effects are usually produced by concentrations of capsaicin that are far in excess of those needed to stimulate sensory neurons; they are sustained, do not undergo desensitization, and are easily reproducible on reapplication of capsaicin (Barthó et al. 1982b, 1987; Maggi et al. 1987c, 1989b).

2. Consequences of Excitation

Given the selectivity with which capsaicin stimulates thin afferent neurons, this drug proved to be instrumental in the recognition of a "dual sensory-efferent function" (Szolcsányi 1984b) of afferent neurons. Thus, excitation of afferent nerve fibers either by capsaicin or other stimuli is followed not only by centripetal conduction of nerve activity but also by release of transmitter substances from the activated peripheral nerve endings themselves. In addition, nerve activity may travel among the peripheral branches of sensory nerve fibers and thereby account for local "axon reflexes" such as the spreading flare (vasodilatation) around a focal injury of the skin (Szolcsányi 1984b, 1988; Holzer 1988; Maggi and Meli 1988). The *afferent function* of sensory neurons enables information to be transmitted to the CNS. Capsaicin gives rise to a painful sensation and activates protective reflexes including avoidance or escape reactions (Gamse 1982; Fitzgerald 1983; Russell and Burchiel 1984; Buck and Burks 1986) or sneezing, coughing, and bronchoconstriction in response to airway irritation (Lundberg and Saria 1987). Other reflexes involve thermoregulatory (Rabe et al. 1980; Szikszay et al. 1982; Szolcsányi 1982; Donnerer and Lembeck 1983; Hayes et al. 1984b; Hori 1984; de Vries and Blumberg 1989; Szallasi and Blumberg 1989), cardiovascular (Crayton et al. 1981; Donnerer and Lembeck 1982; Jancsó and Such 1983; Ordway and

LONGHURST 1983; LONGHURST et al. 1984; SZOLCSÁNYI et al. 1986; AMANN et al. 1989a), and neuroendocrine (MUELLER 1981; WATANABE et al. 1988) control mechanisms.

The local release of peptide mediators from peripheral sensory nerve endings reflects a *local effector function* (HOLZER 1988) of these neurons, as the released peptides influence a variety of local tissue functions (SZOLCSÁNYI 1984b; LUNDBERG and SARIA 1987; CHAHL 1988; HOLZER 1988; MAGGI and MELI 1988). These include local blood flow, vascular permeability, cardiac and smooth muscle activity, tissue growth and repair, immunological processes, and regulation of activity in postganglionic sympathetic efferents. Substance P, neurokinin A, calcitonin gene-related peptide, and vasoactive intestinal polypeptide are among the identified peptides released from peripheral sensory nerve endings (HOLZER 1988; MAGGI and MELI 1988; SARIA et al. 1988; MAGGI et al. 1989d). However, capsaicin is also able to release, e.g., substance P, somatostatin, and calcitonin gene-related peptide from their central endings in the spinal cord (GAMSE et al. 1979b, 1981a; THÉRIAULT et al. 1979; YAKSH et al. 1980; HELKE et al. 1981b; JHAMANDAS et al. 1984; SARIA et al. 1986). Substance P in neurons of the central (GAMSE et al. 1979b; HELKE et al. 1981b) or enteric (HOLZER 1984) nervous system is not released by capsaicin. Likewise, the release of neurotransmitters in the spinal cord such as glutamic acid (AKAGI et al. 1980; SINGER et al. 1982), γ-aminobutyric acid (GABA), glycine (AKAGI et al. 1980), 5-hydroxytryptamine (BERGSTROM et al. 1983), vasoactive intestinal polypeptide, cholecystokinin (YAKSH et al. 1982; JHAMANDAS et al. 1984), or bombesin (MOODY et al. 1981) is not affected by capsaicin.

III. Intermediate Effects on Sensory Neurons

1. Sensitization and Desensitization

Desensitization is a characteristic sequel of capsaicin-induced excitation of sensory neurons, but if low doses of capsaicin are administered repeatedly, sensitization of the human tongue to capsaicin (GREEN 1989) and of cutaneous C-fiber polymodal nociceptors to thermal (KENINS 1982; KONIETZNY and HENSEL 1983) and mechanical (KENINS 1982) noxious stimuli may also take place. In the latter case, sensitization precedes desensitization (KENINS 1982). Sensitization could have a bearing on the capsaicin-induced hyperalgesia to thermal (SZOLCSÁNYI 1977, 1990; CARPENTER and LYNN 1981; GREEN 1986; SIMONE et al. 1987) and mechanical (SZOLCSÁNYI 1977; CULP et al. 1989; SIMONE et al. 1989) stimuli.

Typically, though, capsaicin-induced excitation of primary afferent neurons soon subsides, and the neurons become unresponsive to further applications of the drug and to other chemical, thermal, and mechanical stimuli. This desensitization is often seen as the first manifestation of the long-term neurotoxic action of capsaicin on sensory neurons. Consequently, the term desensitization has also been used to denote the chronic functional inhibition of sensory neurons which, as is clear by now, involves morphologic changes. In this commentary, such morphologic changes are used as criterion to define chronic effects as neurotoxic, whereas the term desensitization implies a readily reversible functional refractoriness in the absence of morphologic changes. Capsaicin desensitization has been observed both by recording the activity of sensory neurons and by examining the consequences of sensory neuron activation. The rapidity and extent with which desensitization to capsaicin develops is related to the dose of, and the time of exposure to, capsaicin and the time interval between consecutive dosings (SZOLCSÁNYI et al. 1975; SZOLCSÁNYI and JANCSÓ-GÁBOR 1976; SZOLCSÁNYI 1977, 1987; FOSTER and RAMAGE 1981; KAUFMAN et al. 1982; ZERNIG et al. 1984; MAGGI et al. 1988f, 1990a; DRAY et al. 1989a,b).

With low suprathreshold doses of capsaicin given at appropriate intervals, desensitization does not necessarily take place (SZOLCSÁNYI et al. 1975, 1986, 1988; SZOLCSÁNYI and JANCSÓ-GÁBOR 1976; SZOLCSÁNYI 1977, 1987; DONNERER and LEMBECK 1982; KAUFMAN et al. 1982; KENINS 1982; LONGHURST et al. 1984; AMANN et al. 1989a; GREEN 1989; DRAY et al. 1989a,b). With higher doses of capsaicin or prolonged exposure to the drug, however, desensitization ensues. Although desensitization to comparatively low doses may be specific for capsaicin (SZOLCSÁNYI 1977, 1987; BERNSTEIN et al. 1981; DRAY et al. 1989a,b), desensitization to higher doses is also associated with a loss of responsivity to other chemoirritant (SZOLCSÁNYI 1977, 1987; FOSTER and RAMAGE 1981; KENINS 1982; JANCSÓ and SUCH 1983; UEDA et al. 1984; GEPPETTI et al. 1988c; DRAY et al. 1989a), warm (SZOLCSÁNYI 1977, 1983a, 1987; FOSTER and RAMAGE 1981; KENINS 1982; WILLIAMS and ZIEGLGÄNSBERGER 1982), and high-threshold mechanical stimuli (KENINS 1982; SZOLCSÁNYI 1987) and to potassium depolarization (SARIA et al. 1983; DRAY et al. 1989b; DONNERER and AMANN 1990; MAGGI et al. 1990b). Specific taste chemoreceptors, cold receptors, and low-threshold mechanoreceptors are not inhibited by desensitization to capsaicin (SZOLCSÁNYI 1977, 1990; DRAY et al. 1989b). The duration of the desensitization appears to be a matter of a few hours to a few days (SZOLCSÁNYI et al. 1975; SZOLCSÁNYI and JANCSÓ-GÁBOR 1976; SZOLCSÁNYI 1977, 1983a, 1990; HAYES et al. 1981b, 1984b; GAMSE 1982; BITTNER and LaHANN 1985; GREEN 1989).

2. Blockade of Nerve Conduction

Within a few minutes from perineural administration of capsaicin, nerve conduction in most but not all afferent C-fibers of the rat coccygeal, saphenous, sciatic, sural, and vagus nerves and of the monkey sural nerve is blocked, polymodal C-fiber nociceptors being most often affected (PETSCHE et al. 1983; PINI 1983; WELK et al. 1983; HANDWERKER et al. 1984; LYNN et al. 1984; J.M. CHUNG et al. 1985; BARANOWSKI et al. 1986; SUCH and

JANCSÓ 1986; MARSH et al. 1987; WADDELL and LAWSON 1989). Except in the ferret in which perineural capsaicin blocks C-fibers only (BARANOWSKI et al. 1986), the conduction in afferent A-fibers (primarily Aδ-fibers) of the rat, rabbit, guinea-pig, and monkey is also reduced to a minor degree (PINI 1983; LYNN et al. 1984; J.M. CHUNG et al. 1985; BARANOWSKI et al. 1986; SUCH and JANCSÓ 1986; MARSH et al. 1987). The potency of capsaicin in blocking C- and Aδ-fibers is considerably higher than that for fast-conducting A-fibers (BARANOWSKI et al. 1986; MARSH et al. 1987; WADDELL and LAWSON 1989). While the block of Aαβ-fibers is fully reversible within 1 h, the C-fibers show only partial, if any, recovery during the first 3 days after capsaicin application (JANCSÓ and SUCH 1983; WELK et al. 1983; LYNN et al. 1984; WADDELL and LAWSON 1989). There are species differences, however, in that C-fibers in the saphenous nerve of the guinea-pig and rabbit are less sensitive than those in the rat, and full recovery from the conduction block produced by as much as 33 mM capsaicin takes place within an hour (BARANOWSKI et al. 1986). Nerve conduction in sympathetic efferent fibers, ventral roots of the spinal cord, or fibers of the optic nerve is not altered (HANDWERKER et al. 1984; MARSH et al. 1987) or only temporarily reduced (SUCH and JANCSÓ 1986).

IV. Neurotoxic Effects on Mammalian Sensory Neurons

1. Effects of Systemic Treatment in Newborn Mammals

a) Rat

Morphological Changes. It was long assumed that the persistent inhibition of sensory neurons produced by systemic administration of capsaicin to adult rats merely reflects a sustained defunctionalization of these neurons. In 1977, however, G. JANCSÓ and his associates (1977) reported that a sub-cutaneous dose of 50 mg/kg capsaicin given to newborn rats caused a life-long degeneration of B-type primary afferent neurons. This degeneration takes place within 30 min, is permanent, and involves most B-type neurons in the dorsal root and cranial sensory ganglia and their peripheral and central nerve processes (JANCSÓ et al. 1977; JANCSÓ and KIRÁLY 1980, 1981; NAGY et al. 1980, 1983; NAGY and HUNT 1983; HOLJE et al. 1983; DINH and RITTER 1987). Less than 5 mg/kg capsaicin is able to destroy some unmyelinated dorsal root fibers (NAGY et al. 1983), and the dose of 50 mg/kg capsaicin is thought to cause maximal degeneration of unmyelinated afferent neurons, although the loss of unmyelinated fibers from the rat saphenous, inferior alveolar, and mental nerves ranges from 39% (LYNN 1984), 50% (WELK et al. 1984), 58% (HOLJE et al. 1983), and 64% (SCADDING 1980) to 67% (JANCSÓ et al. 1977, 1980a). A major loss of afferent nerve fibers from the rat hypogastric and pelvic nerves is demonstrated by a marked inhibition of the retrograde transport of horseradish peroxidase (JANCSÓ and MAGGI

1987). Part of the regional differences in nerve fiber loss might result from regional differences in the proportion of small dark afferent neurons in afferent nerves (Lawson et al. 1984; Doucette et al. 1987) or in different branches of a given nerve (Holje et al. 1983).

In the dorsal roots 72% (Arvidsson and Ygge 1986) to 95% (Lawson and Nickels 1980; Nagy et al. 1981b, 1983; Holje et al. 1983) of all unmyelinated fibers are lost. The number of myelinated Aδ-fibers either is unchanged (Scadding 1980; Nagy et al. 1981b; Arvidsson and Ygge 1986) or reduced by 10% (Jancsó et al. 1980a) to 40% (Lawson and Nickels 1980; Nagy and Hunt 1983; Nagy et al. 1983), particularly if doses higher than 50 mg/kg capsaicin are used. Doses up to 25 mg/kg lead to a selective loss of up to 95% of all unmyelinated afferent nerve fibers without affecting myelinated fibers (Nagy et al. 1983). Consistent with these values, a 93% loss of type I synaptic glomeruli thought to be terminations of unmyelinated afferent fibers (Ribeiro-da-Silva and Coimbra 1984) and an absence of low-density synaptosomes containing substance P (Bucsics et al. 1984) are observed in the dorsal horn of the rat spinal cord. Axon terminal degeneration is seen in all areas of the spinal cord and brainstem receiving primary afferent C-fiber input (Jancsó et al. 1977; Jancsó and Király 1980, 1981; Nagy et al. 1980; Dinh and Ritter 1987). The loss of fine afferent nerve fibers is associated with degeneration of prevalently small B-type cell bodies from the sensory ganglia (Jancsó et al. 1977; Lawson and Nickels 1980; Otten et al. 1983; Lawson and Harper 1984; McDougal et al. 1985; Arvidsson and Ygge 1986). The loss of cell bodies from the ganglia ranges between 28% (McDougal et al. 1985), 43% (Arvidsson and Ygge 1986), and 44% (Otten et al. 1983). Degeneration of some large light A-type somata is seen only with doses of capsaicin higher than 50 mg/kg (Lawson and Harper 1984).

Neuro- and Histochemical Changes. Capsaicin leads to a depletion of markers associated with thin primary afferent neurons. Substance P was the first marker found to be depleted by capsaicin treatment of newborn rats (Gamse et al. 1980; Nagy et al. 1980; Holzer et al. 1982), the depletion being permanent as no recovery is seen within 9 months (Gamse et al. 1981b). Thus, the majority of substance P-containing afferent neurons is ablated by neonatal treatment, although primary afferent neurons sensitive to the neurotoxic action of capsaicin do not completely overlap with substance P-containing afferents. Table 1 lists a number of other markers of capsaicin-sensitive afferent neurons which include the substance P-related peptide neurokinin A, calcitonin gene-related peptide, dynorphin, leucine-enkephalin, galanin, somatostatin, and vasoactive intestinal polypeptide. The chemical coding (Costa et al. 1986) of afferent neurons is heterogeneous in that the markers listed in Table 1 are not present in all afferent neurons and that, within as well as across different species, there are several different populations of afferent neurons characterized by particular combinations of coexisting markers (Marley and Livett 1985; Buck and Burks 1986; Gibbins et al. 1987; Holzer 1988; Maggi and Meli 1988).

The neuro- and histochemical characterization of the sensory neurons sensitive to capsaicin is limited in that many of the markers of afferent neurons also occur in afferent and nonafferent neurons that are not sensitive to capsaicin. Although the majority of capsaicin-sensitive afferent neurons falls within the group of neurofilament-negative B-type neurons (Lawson and Harper 1984; Kai-Kai et al. 1986; Winter 1987; Wood et al. 1988), some of the neurofilament-containing A-type neurons are also destroyed by neonatal administration, at least at doses higher than 50 mg/kg (Lawson and Harper 1984). Furthermore, capsaicin treatment of newborn rats leads to inhibition of axoplasmic transport of organelle-specific enzymes and the retrograde transport of nerve growth factor in sensory but not sympathetic nerves of adult animals (McDougal et al. 1983). These changes can be explained by a deficit of thin primary afferent nerve fibers.

Functional Changes. As might be expected from the morphological and neurochemical changes, capsaicin treatment of newborn rats is associated with permanent sensory and functional deficits which involve both the afferent and local effector functions of sensory neurons (Nagy 1982; Szolcsányi 1982, 1990; Jancsó 1984; Buck and Burks 1986; Lundberg and Saria 1987; Chahl 1988; Holzer 1988; Lembeck 1988; Maggi and Meli 1988). Capsaicin treatment of newborn rats therefore has widely been used to explore the functional implications of capsaicin-sensitive afferent neurons, this approach being based on the assumption that it affects thin sensory neurons only. Accordingly, consistent evidence has been obtained that capsaicin-sensitive afferent neurons mediate a number of local effector functions including vasodilatation, increase in vascular permeability, changes in the activity of cardiac and visceral muscle, changes in the activity of the immune system (Lembeck and Holzer 1979; Morton and Chahl 1980; Lundberg and Saria 1983, 1987; Helme et al. 1987; Chahl 1988; Holzer 1988; Maggi and Meli 1988; Nilsson 1989), and changes in tissue growth and repair (Gamse et al. 1981b; Shimizu et al. 1984; Kjartansson et al. 1987; Maggi et al. 1987a).

Afferent functions that are impaired by neonatal capsaicin treatment comprise nociception and nociception-associated avoidance and escape reactions (see below), warmth reception and thermoregulation (Hori and Tsuzuki 1981; Donnerer and Lembeck 1983; Dib 1983; Hajós et al. 1983), cardiovascular reflexes (Bond et al. 1982; Lorez et al. 1983; Bennett and Gardiner 1984; Donnerer et al. 1989), visceral reflexes (Cervero and McRitchie 1982; Sharkey et al. 1983; Santicioli et al. 1985; Holzer and Sametz 1986; Holzer et al. 1986), neuroendocrine reflexes (Traurig et al. 1984a,b; Amann and Lembeck 1986, 1987; Bennett and Gardiner 1986; Donnerer et al. 1989; Donnerer and Lembeck 1990), and satiety (MacLean 1985). While chemonociception is permanently inhibited by neonatal administration (Jancsó et al. 1977; Faulkner and Growcott 1980; Hayes et al. 1981a; Gamse 1982; Saumet and Duclaux 1982), there is disagreement as to whether mechano- and thermonociception are also impaired (Holzer et al. 1979; Faulkner and Growcott 1980; Jancsó and Jancsó-Gábor 1980; Nagy et al. 1980; Cervero and McRitchie 1981; Hayes et al. 1981a; Gamse 1982; Saumet and Duclaux 1982; Doucette et al. 1987; Hammond and Ruda 1989). The discrepant results may be due to several factors including differences in animal strain, capsaicin sensitivity (Nagy and van der Koy 1983), secondary changes in sensory pathway-related systems (Wall et al. 1982; Nagy and Hunt 1983; Lynn et al. 1984; Welk et al. 1984; Rethélyi et al. 1986; Saporta 1986; Cervero and Plenderleith 1987), and experimental protocols and procedures (Szolcsányi 1990). In addition, there are regional differences in the proportion of afferent nerve fibers sensitive to capsaicin (Holje et al. 1983; Doucette et al. 1987) and age-related changes in capsaicin-induced antinociception (Hammond and Ruda 1989).

The lack of a consistent effect of neonatal treatment on mechanical and thermal nociception is a paradox in view of the permanent loss of unmyelinated afferent neurons. A possible explanation comes from the finding that the fiber and receptor types in afferent nerves of adult rats treated with capsaicin as neonates is the same as in vehicle-treated rats (Lynn 1984; Welk et al. 1984; Cervero and Sharkey 1988), although the number of unmyelinated afferents is diminished by about 40%–50% (Lynn 1984; Welk et al. 1984). Since acute capsaicin produces excitation (Szolcsányi 1990) and a conduction block (Welk et al. 1984) in C-fiber polymodal nociceptors, it is evident that adult rats treated with capsaicin as neonates possess capsaicin-sensitive afferent neurons, but it is not known whether these neurons escaped or survived the neonatal capsaicin treatment or evolved at a later stage. In addition, calcitonin gene-related peptide, although being depleted from the dorsal spinal cord 10 days after capsaicin treatment of newborn rats (Hammond and Ruda 1989), can be replenished over the following months to near-normal levels in the spinal cord (Diez Guerra et al. 1988; Hammond and Ruda 1989). No such replenishment is seen with substance P or neurokinin A (Diez Guerra et al. 1988), and in the sensory ganglia and peripheral targets of sensory neurons depletion of calcitonin gene-related peptide also persists (Skofitsch and Jacobowitz 1985b; Franco-Cereceda et al. 1987; Sternini et al. 1987; Su et al. 1987; Diez Guerra et al. 1988; Geppetti et al. 1988a; Varro et al. 1988). The cellular source of the replenishment of calcitonin gene-related peptide in the spinal cord has not yet been identified.

Selectivity of the Action of Capsaicin. There is no doubt that the primary target of the neurotoxic action of capsaicin in the newborn rat is a group of thin (Aδ- and C-fiber) primary afferent neurons. Other primary afferent as well as efferent motor and autonomic neurons are not affected (Jancsó et al. 1980a; Cervero and McRitchie 1982; Sharkey et al. 1983; Soukup and JANCSÓ 1987; TAKANO et al. 1988). However, substantial deafferentation is likely to have a significant impact on the whole sensory system comprising second- and higher-order afferent neurons. Indeed, the organization of sensory pathway-related systems in the spinal cord and nucleus gracilis (NAGY and HUNT 1983; RETHÉLYI et al. 1986) and of the spinothalamic tract (SAPORTA 1986) as well as the processing of sensory information in the spinal cord (WALL et al. 1982; CERVERO and PLENDERLEITH 1987) are profoundly altered by neonatal capsaicin treatment. In the periphery, it leads to sprouting of the axon terminals of surviving afferents in the cornea (OGILVY et al. 1991), to permanent changes in the morphology of the cornea (SHIMIZU et al. 1984) and lung (AHLSTEDT et al. 1986), and to increased tissue concentrations of histamine and 5-hydroxytryptamine in the skin and lung (HOLZER et al. 1981). A particular relationship exists between the peripheral endings of sympathetic neurons and afferent neurons. Capsaicin-induced elimination of the sensory neurons results in increases in transmitter content and innervation density of sympathetic neurons and vice versa (TERENGHI et al. 1986; NIELSCH and KEEN 1987; LUTHMAN et al. 1989; ABERDEEN et al. 1990). These reciprocal changes may be due to competition of sensory and sympathetic neurons for nerve growth factor, elimination of one nerve population increasing the availability of nerve growth factor for the other nerve population.

Whether all capsaicin-induced alterations of sensory pathway-related systems are secondary to the degeneration of thin primary afferent neurons is not clear. However, most neurons of the CNS are not susceptible to the neurotoxic effect of neonatal treatment, although this is not true for all central neurons. The capsaicin-induced degeneration of axon terminals in the dorsal spinal cord and brain stem (JANCSÓ et al. 1977; JANCSÓ 1978; JANCSÓ and KIRÁLY 1980, 1981; DINH and RITTER 1987) probably reflects degeneration of the central endings of primary afferent neurons only, since markers of thick afferent and central neurons such as glutamic acid, glutamic acid decarboxylase, GABA, glycine, glycine receptors, aspartic acid, taurine, choline acetyltransferase, norepinephrine (NE), 5-hydroxytryptamine, and histamine are not decreased following neonatal treatment (NAGY et al. 1980; SINGER and PLACHETA 1980; HOLZER et al. 1981; JANCSÓ et al. 1981; SINGER et al. 1982; HAJÓS et al. 1986b; HOLZER-PETSCHE et al. 1986). Furthermore, the levels of substance P, somatostatin, and other neuropeptides found both in sensory and central neurons are not altered in the ventral spinal cord and in regions of the brain above the brainstem (GAMSE et al. 1980, 1981b; NAGY et al. 1980; HELKE et al. 1981a; JANCSÓ et al. 1981; PRIESTLEY et al. 1982; PANERAI et al. 1983), whereas neuropeptides not associated with sensory neurons such as neurotensin and methionine-enkephalin remain unaffected throughout the CNS (GAMSE et al. 1981b; JANCSÓ et al. 1981; PRIESTLEY et al. 1982; SINGER et al. 1982; PANERAI et al. 1983).

There is, however, preliminary evidence that some central neurons in the forebrain degenerate after neonatal treatment (DINH and RITTER 1987) and that β-endorphin is permanently depleted from the hypothalamus but not other regions of the brain (Panerai et al. 1983). These observations are in need of further confirmation but throw doubt on the exclusive selectivity of capsaicin for primary afferent neurons. Enteric neurons also seem to be insensitive to capsaicin as gastrointestinal tissue levels of peptides associated with enteric neurons (substance P, calcitonin gene-related peptide) do not change (Holzer et al. 1980; Geppetti et al. 1988a) except in the upper gastrointestinal tract where sensory nerve endings contribute significantly to the tissue content of these peptides (Sharkey et al. 1984; Sternini et al. 1987; Su et al. 1987; Green and Dockray 1988; Geppetti et al. 1988a). However, vasoactive intestinal polypeptide is depleted from fibers but not somata of submucosal neurons, whereas a lactoseries carbohydrate antigen colocalized with the peptide is lost from both fibers and somata (Kirchgessner et al. 1988). This may indicate that some enteric neurons are sensitive to the neurotoxic effect of capsaicin, but further proof of such a target of the drug is needed.

b) Other Mammals

Capsaicin treatment of newborn mice (Scadding 1980; Jancsó et al. 1985a; Hiura and Sakamoto 1987b; Hiura and Ishizuka 1989) and dogs (Jancsó et al. 1985a) also causes degeneration of small to medium-sized afferent nerve terminals in the dorsal spinal cord. In the mouse, 50 mg/kg capsaicin leads to rapid degeneration of B-type somata sending unmyelinated fibers into the dorsal roots, whereas later on also some A-type somata with myelinated fibers degenerate (Hiura and Ishizuka 1989). Doses of 50–150 mg/kg capsaicin cause a 41%–75% reduction of unmyelinated fibers and a 6%–12% loss of myelinated fibers in the dorsal roots; the number of small dorsal root ganglion cell bodies decreases by 51%–77% and that of large somata by 14%–52% (Hiura and Sakamoto 1987b). Peripherally, unmyelinated fibers are reduced in the sural nerve (50% loss) (Scadding 1980) and cornea (Fujita et al. 1984). The neurotoxic action of capsaicin in the newborn mouse is associated with depletion of substance P from the dorsal spinal cord and cornea and with deficits in chemo- but not thermonociception (Hayes et al. 1981a; Gamse 1982; Keen et al. 1982). Vascularization of, and lesions in, the cornea (Keen et al. 1982; Shimizu et al. 1984; Fujita et al. 1984) and changes of the somatotopic maps in the cerebral cortex (Wall et al. 1982) are considered to represent changes secondary to the removal of sensory neurons.

Administration of 70 mg/kg capsaicin to newborn rabbits failed to cause any long-term depletion of substance P from the spinal cord and eye, although the miotic and hyperemic effects of acute intracameral injection of capsaicin were blocked (Tervo 1981). In contrast, administration of 200 mg/kg capsaicin to newborn cats (Fehér and Vajda 1982) was reported to result in degeneration of axons and a few cell bodies in the myenteric and submucosal plexuses of the small intestine. It has not yet been confirmed, however, whether some enteric neurons of the cat are in fact susceptible to the neurotoxic action of capsaicin.

2. Effects of Systemic Treatment in Adult Mammals

a) Rat

Morphological Changes. Unmyelinated primary afferent neurons of the rat appear to be particularly sensitive to the neurotoxic action of capsaicin at the age of 1–12 days, whereas when given at the age of 14 days or more no major degeneration of nerve fibers is appreciable in the dorsal spinal cord (JANCSÓ and KIRÁLY 1981). Although some functional deficits produced by capsaicin in the adult rat are virtually irreversible (SZOLCSÁNYI et al. 1975), most of the long-term effects (neuropeptide depletion, defunctionalization) show recovery which may take several weeks to several months (SZOLCSÁNYI and JANCSÓ-GÁBOR 1976; JANCSÓ et al. 1977; GAMSE et al. 1980, 1981b; GAMSE 1982; BITTNER and LaHANN 1985; MAGGI et al. 1987d; SOUTH and RITTER 1988; GARDINER et al. 1989; SZALLASI et al. 1989).

Ultrastructurally, subcutaneous administration of 35–300 mg/kg capsaicin to adult rats (JOÓ et al. 1969; SZOLCSÁNYI et al. 1975; CHIBA et al. 1986) leads to swelling of mitochondria in B-type sensory neurons but not in A-type afferent or sympathetic efferent neurons when examined 1–60 days posttreatment. However, another study has noted that, within a few hours, systemic administration of capsaicin gives rise to severe ultrastructural changes in some B-type somata, which are thought to reflect degeneration (JANCSÓ et al. 1985b). The dose of 100 mg/kg is maximally effective and damages 17% of the B-type cell bodies. In addition, degenerating axon terminals are seen in the spinal cord and brainstem confined to the central projection areas of thin primary afferent neurons (JANCSÓ et al. 1985b; JANCSÓ and MAGGI 1987; RITTER and DINH 1988). Six days posttreatment there is a 45% loss of unmyelinated and a 16% loss of myelinated fibers from the rat saphenous nerve. Extensive degeneration also occurs in axons of the ureter (HOYES and BARBER 1981) and trachea (HOYES et al. 1981) 24 h after subcutaneous injection of capsaicin 50 mg/kg to adult rats, degeneration being confined to axons with terminals containing mainly large, dense-cored vesicles. In the ureter, 90% of all axons degenerate, while no appreciable degeneration in the dorsal roots is seen (K. CHUNG et al. 1985). Thus, capsaicin treatment of adult rats damages some afferent B-type somata and prevalently unmyelinated afferent nerve fibers, but the extent of degeneration of somata and axons is clearly less pronounced than when capsaicin is given to the newborn rat (JANCSÓ et al. 1977, 1980a, 1985b, 1987). However, while somata and nerve fibers show moderate degeneration in the adult rat, axon terminals in the periphery are destroyed to a considerable extent (HOYES and BARBER 1981; K. CHUNG et al. 1985, 1990).

Neuro- and Histochemical Changes. Depletion of substance P from sensory nerve pathways is produced by capsaicin 50–125 mg/kg given subcutaneously or intraperitoneally to adult rats (Gamse et al. 1981b; Gamse 1982), the i.p. route being more effective than the s.c. one (Gamse et al. 1981b). Subcutaneously, the dose of 125 mg/kg appears maximally effective (Gamse et al. 1981b), making it unnecessary to use doses as high as 950 mg/kg (Jessell et al. 1978) which may give rise to nonselective neurotoxic effects, e.g., on enteric neurons (Harti 1988). The depletion of substance P and somatostatin from sensory pathways produced by capsaicin 125 mg/kg in adult Sprague-Dawley rats is less than that caused by capsaicin 50 mg/kg in newborn animals (Gamse et al. 1981b; Gamse 1982; Priestley et al. 1982). In MRC Porton and Wistar rats, though, the depletion of substance P and calcitonin gene-related peptide is the same whether capsaicin is given to newborn or adult rats (Salt et al. 1982; Geppetti et al. 1988a), and 125 mg/kg given to adult animals is no more effective than 50 mg/kg (Geppetti et al. 1988a). Opiate receptor binding in the dorsal horn of the spinal cord remains unaltered after capsaicin treatment of adult rats (Jessell et al. 1978) but is decreased by neonatal administration (Gamse et al. 1979a; Nagy et al. 1980). While following capsaicin treatment of newborn rats chemical markers of sensory neurons (Table 1) are permanently depleted from sensory pathways (see above), there is partial or total recovery following capsaicin treatment of adult rats, although the recovery rates differ with tissue and peptide (Gamse et al. 1981b; Bittner and LaHann 1985; Maggi et al. 1987d). The depletion of somatostatin is completely reversible within 4 months in all tissues (Gamse et al. 1981b). A relationship, if any, between marker replenishment and morphologic recovery is not known.

Functional Changes. Whereas capsaicin treatment of newborn rats does not change the proportion of sensory fiber and receptor types in afferent nerves (Lynn et al. 1984; Welk et al. 1984; Cervero and Sharkey 1988), there is a significant reduction in the proportion of C-fiber polymodal nociceptors in the rat saphenous nerve following treatment of adult rats (Lynn et al. 1984; Szolcsányi et al. 1988). The responsiveness of the remaining C-fiber polymodal nociceptors to topical application of capsaicin is diminished but not abolished (Szolcsányi et al. 1988). In contrast, thermonociception seems to be impaired only temporarily by systemic capsaicin treatment of adult rats (Hayes et al. 1981b; Buck et al. 1982; Gamse 1982; Bittner and LaHann 1985; Szolcsányi 1990), while permanent changes may occur after neonatal treatment (Holzer et al. 1979; Nagy et al. 1980; Gamse 1982). Chemo- and mechanonociception, though, are reduced for a long time after capsaicin treatment of adult rats (Szolcsányi et al. 1975; Szolcsányi and Jancsó-Gábor 1976; Hayes and Tyers 1980; Gamse et al. 1981b; Gamse 1982).

Some of the different effects of neonatal versus adult capsaicin treatment can be explained by different degrees of sensory nerve ablation. Thus, capsaicin-induced release of substance P from the dorsal spinal cord in vitro is inhibited to a greater extent by neonatal capsaicin pretreatment than by pretreatment of adult animals (GAMSE et al. 1981a; GAMSE 1982). Other differences, particularly in thermonociception, can also be accounted for by changes at the level of second-order neurons in the spinal cord and medulla (SALT et al. 1982). A particularly good example of pronounced differences in the long-term effects of capsaicin when given to newborn or adult rats is illustrated by the micturition reflex. Capsaicin treatment at any age impairs this reflex, probably by interfering with its afferent arc only (SHARKEY et al. 1983; HOLZER-PETSCHE and LEMBECK 1984; MAGGI et al. 1984, 1987d, 1989b; SANTICIOLI et al. 1985; MAGGI and MELI 1986). Although the depletion of substance P from the urinary bladder remains similar (HOLZER et al. 1982; MAGGI et al. 1987d; GEPPETTI et al. 1988a), neonatal capsaicin treatment leads to a permanent abolition of distension-induced micturition and to pronounced hypertrophy of the bladder (SHARKEY et al. 1983; SANTICIOLI et al. 1985; MAGGI et al. 1989b) whereas adult treatment causes only a reversible increase in the pressure threshold of micturition (MAGGI et al. 1984, 1987d, 1989b).

Onset of Long-Term Changes. Degeneration of B-type sensory neurons takes place within 1 h after s.c. administration of capsaicin to adult rats (JANCSÓ et al. 1985b), and blockade of their afferent as well as local effector functions is observed by 10 min to 1 h after treatment (HAYES et al. 1981b; LEMBECK and DONNERER 1981; BITTNER and LAHANN 1985; MAGGI et al. 1987d). In contrast, depletion of substance P from peripheral sensory nerve endings takes at least 3 h (BITTNER and LAHANN 1985; MAGGI et al. 1987d,e) or longer (LEMBECK and DONNERER 1981). The levels of substance P (LEMBECK and DONNERER 1981; GAMSE et al. 1981b; GEPPETTI et al. 1988a) and calcitonin gene-related peptide (GEPPETTI et al. 1988a) in sensory ganglia may be increased 2–4 days posttreatment.

Selectivity of the Action of Capsaicin. Levels of substance P in the ventral spinal cord and in the brain above the brainstem remain unaffected by capsaicin treatment of adult rats (HAYES and TYERS 1980; GAMSE et al. 1981b; GAMSE 1982; VIRUS et al. 1982). Likewise, levels of glutamic acid decarboxylase, choline acetyltransferase (JESSELL et al. 1978; HOLZER-PETSCHE et al. 1986), neurotensin, and methionine-enkephalin (GAMSE et al. 1981b; PRIESTLEY et al. 1982) remain unaltered in the CNS. The tissue concentrations of NE, dopamine, and 5-hydroxytryptamine in the CNS either are unaffected (HAJÓS et al. 1986b) or may even be increased (VIRUS et al. 1983). However, intraperitoneal administration of capsaicin 50–90 mg/ kg to adult rats has been found to induce nerve terminal degeneration in certain fore- and hindbrain areas not yet associated with terminations of primary afferents (RITTER and DINH 1988). Degenerating cell bodies in the ventromedial midbrain tegmentum, supramammillary nucleus, and posterior hypothalamus indicate that some central neurons can permanently be destroyed by capsaicin (Ritter and Dinh 1988).

One group of central neurons that have long been recognized as being sensitive to capsaicin lie in the preoptic region of the hypothalamus. Administration of capsaicin 30–70 mg/kg to adult rats produces ultrastructural changes in these neurons, including swelling of mitochondria (Szolcsányi et al. 1971). They are thought to be thermosensitive neurons involved in central thermoregulation (Szolcsányi 1982), but the effect of capsaicin on thermoregulation is complicated by the fact that both peripheral C-fiber warmth receptors and central thermosensitive neurons are affected (for reviews see Szolcsányi 1982; Hori 1984). While warmth sensitivity mediated by primary afferents is impaired by capsaicin treatment of both newborn and adult rats (Hajós et al. 1986a), only treatment of adult rats produces a long-term defunctionalization of central thermosensitive neurons (Hori 1981; Dib 1983; Hajós et al. 1983; Obál et al. 1983). Likewise, the increases in the synthesis rate and content of monoamines in the area of the preoptic region and hypothalamus produced by acute capsaicin administration are unchanged after systemic capsaicin pretreatment of newborn rats but abolished by pretreatment of adult animals (Hajós et al. 1986b). The enteric nervous system seems to be spared by treatment of adult rats with capsaicin. There is no direct evidence that enteric neurons are damaged by capsaicin (Hoyes and Barber 1981), and the levels of substance P and calcitonin gene-related peptide in the enteric nervous system remain unaltered by the drug (Holzer et al. 1980; Geppetti et al. 1988a). Furthermore, there is morphological (Szolcsányi et al. 1975) and functional (Stein et al. 1986) evidence that the sympathetic nervous system is not affected.

b) Guinea-Pig

Systemic administration of capsaicin to adult guinea-pigs produces extensive axon terminal degeneration in the dorsal spinal cord and brainstem (Jancsó et al. 1987) as well as in the periphery (Papka et al. 1984). Perivascular substance P-containing nerve endings deteriorate within 5 min after the s.c. injection of capsaicin 50 mg/kg (Papka et al. 1984). Depletion of substance P and ultrastructural signs of degeneration are appreciable for at least 1 year (Papka et al. 1984), which is in keeping with virtually irreversible functional deficits (Jancsó 1968; Jancsó-Gábor et al. 1970). Quantitatively, 2.5 mg/kg given s.c. appears to be suprathreshold, while 10 mg/kg is capable of producing maximal depletion of the peptide from afferent neurons 10 days posttreatment (Miller et al. 1982b; Buck et al. 1983). The loss of substance P induced by a 10 mg/kg dose can be close to 80%–90%, this level being not surpassed by doses up to 1250 mg/kg (Gamse et al. 1981c; Miller et al. 1982a,b; Buck et al. 1983). Tissue levels of vasoactive intestinal polypeptide (Buck et al. 1983; Della et al. 1983), somatostatin (Buck et al. 1983), and substance P in the ventral spinal cord and brain above the brainstem are not affected (BUCK et al. 1983). However, calcitonin gene-related peptide (GIBBINS et al. 1985, 1987; LUNDBERG et al. 1985; FRANCO-CERECEDA et al. 1987), cholecystokinin (BUCK et al. 1983; GIBBINS et al. 1987), and dynorphin (GIBBINS et al. 1987) are depleted from sensory neurons.

The sensitivity of the cornea to chemical noxious stimuli is abolished by the same doses of capsaicin that reduce the substance P content of sensory neurons (BUCK et al. 1983). Unlike in the rat, capsaicin treatment of adult guinea-pigs is followed by cutaneous insensitivity to nonnoxious and noxious heat (MILLER et al. 1982a; BUCK et al. 1983), whereas the sensitivity to nonnoxious and noxious cold and to nonnoxious and noxious mechanical stimuli remains unchanged (BUCK et al. 1983). Substance P depletion from dorsal root ganglia is preceded by insensitivity to thermal noxious stimuli (MILLER et al. 1982a) and inhibition of the retrograde transport of nerve growth factor (MILLER et al. 1982b). A selective action of capsaicin on primary afferent neurons is supported by the absence of effects of the drug on nonsensory central neurons (BUCK et al. 1983), autonomic efferent neurons (DELLA et al. 1983), and enteric neurons (GAMSE et al. 1981c; FURNESS et al. 1982; DONNERER et al. 1984; GIBBINS et al. 1985). It is not known, however, whether the corneal and cutaneous lesions produced by capsaicin 950 mg/kg (BUCK et al. 1983) reflect a trophic role of sensory neurons in these tissues or other effects of the drug. Changes in the number of substance P binding sites in the vas deferens are considered to be primary and/or secondary consequences of sensory denervation (MUSSAP et al. 1989).

c) Other Mammals

Unlike in rats, treatment of mice at the age of 2, 4, and 7 days and of adult mice with capsaicin 50 mg/kg results in increasingly pronounced inhibition of thermo- and chemonociception, the latter being almost completely abolished by treatment of 10-day-old or older mice, whereas the degree of substance P depletion from the spinal cord and sciatic nerve is the same (GAMSE 1982). A lower dose of capsaicin, 10 mg/kg, given to adult mice reduces the substance P content of the spinal cord but not the sciatic nerve and fails to cause long-term changes in thermonociception (GAMSE 1982). Morphological data on the effect of capsaicin in the mouse are not yet available except that a marked loss of axons is seen in the cornea (FUJITA et al. 1984). Subcutaneous treatment of adult cats with capsaicin 50 mg/kg is followed by extensive axon terminal degeneration in areas of the brainstem and spinal cord known to be the central projection areas of primary sensory neurons (JANCSÓ et al. 1987). Injection of capsaicin to adult rabbits does not alter the heat thresholds of primary afferent neurons in the saphenous nerve (LYNN et al. 1984) and the substance P content of the spinal cord and eye (TERVO 1981) but abolishes the acute excitatory effects of topical capsaicin application to the skin (LYNN et al. 1984) or intracameral injection of capsaicin (Tervo 1981). Administration of a cumulative dose of 50 mg/kg to pigs results in an approximately 70% depletion of calcitonin gene-related peptide from the skin and skeletal muscle and abolishes the vasodilator effect of acute intradermal injection of capsaicin (Franco-Cereceda and Lundberg 1989). Data from other mammalian species are not available.

3. Effects of Periaxonal Application

a) Rat

Morphological and Neurochemical Changes. Application of capsaicin to axons of afferent neurons is followed by long-term ablation of primary afferent C-fiber neurons. One to several days after periaxonal application of 1% (33 mM) capsaicin to the saphenous and sciatic nerves, swelling of numerous unmyelinated fibers is observed both distal and proximal to the application site, and 2–3 weeks post-treatment the structural organization of the nerves is changed (Handwerker et al. 1984; Jancsó et al. 1985a). This damage is confined to unmyelinated fibers (Handwerker et al. 1984). Quantitatively, 34% of the dorsal root ganglion neurons (Jancsó and Lawson 1990) and 32%–40% of afferent C-fibers (Lynn et al. 1987; Jancsó and Lawson 1990) degenerate in response to periaxonal capsaicin administration, while the number of A-fibers remains unaltered. Periaxonal application of capsaicin (0.33–33 mM) inhibits the orthograde axoplasmic transport of substance P and somatostatin as seen 1 day posttreatment (Gamse et al. 1982; Handwerker et al. 1984) and the retrograde transport of horseradish peroxidase (Jancsó et al. 1985a). In contrast, the axoplasmic flow of acetylcholinesterase and NE is not interrupted, indicating that efferent motor and sympathetic nerve fibers are not affected (Gamse et al. 1982). Posttreatment, substance P is lost progressively from all parts of afferent neurons but distal to the treatment site the loss is maximal within 4 days (Gamse et al. 1982). In the spinal cord not only substance P but also somatostatin, cholecystokinin-like immunoreactivity, and fluoride-resistant acid phosphatase are depleted (Ainsworth et al. 1981; Gamse et al. 1982; Gibson et al. 1982). Proximal to the treatment site, marker depletion is maximal 2 weeks posttreatment and shows various degrees of recovery during the following 4–6 months (Gamse et al. 1982; Gibson et al. 1982). Substance P in nonsensory neurons and peptides primarily present in intrinsic neurons of the spinal cord (neurotensin, neurophysin, methionine-enkephalin, bombesin) remain unchanged after perineural application (Ainsworth et al. 1981; Gibson et al. 1982).

Functional Changes. Perineural treatment inhibits chemo- and thermo-nociception in the exteroreceptive areas supplied by the treated nerve (Jancsó et al. 1980b; Fitzgerald and Woolf 1982; Gamse et al. 1982; Gibson et al. 1982; Coderre et al. 1984; McMahon et al. 1984; Szolcsányi 1990), which is seen already by 5 h posttreatment, i.e., before substance P is depleted from afferent nerve fibers (GAMSE et al. 1982). The sensitivity to noxious cold and noxious and nonnoxious mechanical stimuli is not affected (FITZGERALD and WOOLF 1982; CODERRE et al. 1984). The early onset of functional inhibition is in keeping with the prompt inhibition of nerve conduction caused by periaxonal application. Initially, nerve conduction in polymodal C- and Aδ-fiber nociceptors is inhibited at the treatment site only (PETSCHE et al. 1983; LYNN et al. 1984; J.M. CHUNG et al. 1985). While conduction in the A-fibers is soon resumed, the polymodal C-fiber nociceptors remain unresponsive to their natural sensory modalities (WELK et al. 1983; HANDWERKER et al. 1984; LYNN et al. 1984) for several months (LYNN and PINI 1985). In addition, the processing of sensory information in the spinal cord is altered (WALL and FITZGERALD 1981; FITZGERALD 1982; FITZGERALD and WOOLF 1982; MCMAHON et al. 1984). Perineural treatment not only blocks the afferent (JANCSÓ et al. 1980b; GAMSE et al. 1982; GIBSON et al. 1982; RAYBOULD and TACHÉ 1988; SOUTH and RITTER 1988) but also the local effector functions of peripheral sensory nerve endings (JANCSÓ et al. 1980b; GAMSE et al. 1982; CODERRE et al. 1984; RAYBOULD et al. 1990). The effect of perineural application on thermonociception lasts at least 3 months (SZOLCSÁNYI 1990), and cholecystokinin-induced satiety, which is attenuated by perivagal administration, requires up to 5 months for full recovery (SOUTH and RITTER 1988). Functional recovery may not take place even within 1 year posttreatment (JANCSÓ and KIRÁLY 1984).

Topical Selectivity of Periaxonal Application. The selective target of the long-term effect of periaxonal application of capsaicin is primary afferent C-fibers with polymodal nociceptors, whereas other afferent or nonsensory (efferent motor and autonomic) nerve fibers are not directly affected. Although 30 min after application of a 33 mM solution (maximal estimated dose: 1.6 µmol) to the rat sciatic nerve the drug is found not only in the treated nerve segment and the adjacent distal and proximal segments but also in the blood (GAMSE et al. 1982; PETSCHE et al. 1983), the amounts of capsaicin delivered to other tissues appear to be too small to exert a long-term neurotoxic effect on nerves other than the treated one (JANCSÓ et al. 1980b; GAMSE et al. 1982; SOUTH and RITTER 1988). A systematic study by SOUTH and RITTER (1988) has clearly shown that perineural treatment ablates afferent fibers in the treated nerve only and hence represents an important tool for functional neuroanatomy.

b) Other Mammals

As in the rat, periaxonal application of capsaicin (33 mM) to the sciatic nerves of the guinea-pig, cat, and rabbit inhibits the axoplasmic transport of substance P (GAMSE et al. 1982). Periaxonal application to somatic nerves of the guinea-pig is followed by a long-lasting inhibition of thermonociception (SZOLCSÁNYI 1990), and in the cat, perivagal administration impairs cardiovascular and respiratory chemoreflexes (Jancsó and Such 1983). In the rabbit, however, the effect of capsaicin is much weaker than in the other species (Gamse et al. 1982) and is not associated with any sign of axon degeneration, although the substance P content of the skin is diminished, and the local effector function of the peripheral sensory nerve endings is reduced (Lynn and Shakhanbeh 1988).

4. Effects of Local Application

a) Intrathecal or Intracisternal Administration

Local administration of 100–130 nmol capsaicin into the rat cisterna magna (Jancsó 1981; Gamse et al. 1984) or lumbar/cervical subarachnoid space (Palermo et al. 1981; Ribeiro-da-Silva and Coimbra 1984) leads to degeneration of axon terminals in the brainstem and spinal cord. Whereas 53%–56% of the glomerular C-type (Palermo et al. 1981) or glomerular type I (Ribeiro-da-Silva and Coimbra 1984) nerve terminals considered to be terminations of unmyelinated afferents are lost, only 2%–5% of the type II nerve terminals thought to arise from myelinated afferents disappear within 2 days posttreatment (Ribeiro-da-Silva and Coimbra 1984). The effect of capsaicin is restricted to the central endings of primary afferent neurons as no signs of degeneration in trigeminal roots, trigeminal ganglion, and maxillary nerve have been detected (Jancsó 1981; Gamse et al. 1984). The topical selectivity of intracisternal capsaicin is underlined by a selective depletion of substance P and neurokinin A from the medulla but not the trigeminal roots, trigeminal ganglion, and maxillary nerve of the rat and guinea-pig (Gamse et al. 1984, 1986). Intrathecal capsaicin (50–1000 nmol) injection at the lumbar level causes a long-term depletion of substance P and somatostatin from the rat and mouse dorsal spinal cord (Yaksh et al. 1979; Nagy et al. 1981a; Gamse 1982; Micevych et al. 1983; Jhamandas et al. 1984; South and Ritter 1988) and dorsal roots (Gamse 1982) but not from the dorsal root ganglia and sciatic nerve (Gamse 1982). Substance P levels in spinal areas remote from the injection site are also unchanged (Yaksh et al. 1979; Nagy et al. 1981a; South and Ritter 1988). Markers of nonsensory neurons such as neurotensin (Gamse 1982), NE (Yaksh et al. 1979), 5-hydroxytryptamine, methionine-enkephalin (Micevych et al. 1983), vasoactive intestinal polypeptide, cholecystokinin (Jhamandas et al. 1984), and glutamic acid decarboxylase (Yaksh et al. 1979; Nagy et al. 1981a) are grossly unaltered after intrathecal treatment .

While afferent nociceptive functions of sensory neurons are inhibited, the local effector functions of peripheral sensory nerve endings remain unaffected by intracisternal application (Jancsó 1981; Gamse et al. 1984). Chemonociception in the corresponding exteroreceptive areas is quickly lost after intracisternal, intrathecal, or epidural capsaicin treatment (Yaksh et al. 1979; Jancsó 1981; Gamse 1982; Gamse et al. 1984; Eimerl and Papir-Kricheli 1987), whereas thermonociception is either reduced (Yaksh et al.

1979; PALERMO et al. 1981; GAMSE 1982; MICEVYCH et al. 1983; JHAMANDAS et al. 1984; EIMERL and PAPIR-KRICHELI 1987) or unchanged (HAYES et al. 1981b; NAGY et al. 1981a). Mechanonociception is only transiently diminished (HAYES et al. 1981b) but unchanged in the long term (YAKSH et al. 1979). Taken together, after intrathecal or intracisternal application only the central terminals of unmyelinated afferent neurons in the vicinity of the injection site are ablated, whereas their cell bodies and peripheral processes are left intact. When doses of capsaicin shown to be effective intrathecally or intracisternally are administered i.v. or i.p., no long-term neurotoxic effects are noted (YAKSH et al. 1979; GAMSE et al. 1986).

b) Intracerebroventricular Administration

Intracerebroventricular injection of 650–1000 nmol capsaicin reduces the content of substance P but not of somatostatin and neurotensin in the medulla oblongata (GAMSE et al. 1981b; SOUTH and RITTER 1988), with no changes in other brain areas, spinal cord, and trigeminal ganglion. In the hypothalamus, however, β-endorphin is depleted (PANERAI et al. 1983). Chemonociception in the cornea (GAMSE et al. 1981b; SOUTH and RITTER 1988) but not in skin areas supplied by the lumbar spinal cord (YAKSH et al. 1979) is greatly impaired by intracerebroventricular administration, and cholecystokinin-induced satiety is also attenuated (SOUTH and RITTER 1988). No data on morphological changes are as yet available.

c) Topical Administration to Specific Brain Regions

Topical administration of capsaicin to the brainstem of the cat causes degeneration of nerve terminals in a specific region of the ventral medulla oblongata, this area being involved in the regulation of the cardiovascular and respiratory systems (JANCSÓ and SUCH 1985). Injection into the area postrema and adjacent nucleus tractus solitarii of the rat brainstem results in overconsumption of preferred food (SOUTH and RITTER 1983). No other long-term effects of capsaicin injected topically into the CNS have been reported.

d) Topical Administration to Peripheral Endings of Sensory Neurons

Repeated administration of capsaicin (1%) to the cornea of rats and guinea-pigs causes insensitivity to chemical noxious stimuli (SZOLCSÁNYI et al. 1975; SZOLCSÁNYI and JANCSÓ-GÁBOR 1976; GAMSE et al. 1981b) but does not induce gross damage such as corneal opacity or ultrastructural changes of the corneal epithelium (SZOLCSÁNYI et al. 1975). In unmyelinated nerve endings of the cornea, however, swollen mitochondria with disorganized cristae and a 90% reduction of the number of microvesicles can be observed (SZOLCSÁNYI et al. 1975). Repeated topical application of 1% capsaicin to the rat eye is followed, within 4h, by a 80% depletion of substance P which reverses over the following 3 weeks as does the corneal sensitivity to chemogenic pain (GAMSE et al. 1981b). Injection of 26 nmol dihydrocapsaicin, a congener of capsaicin, into the skin of guinea-pigs causes localized inhibition of thermonociception but does not deplete substance P from dorsal root ganglia or inhibit the retrograde transport of nerve growth factor (MILLER et al. 1982a). However, injection of a higher dose of capsaicin (8 μmol) into rat skin inhibits the retrograde transport of horseradish peroxidase predominantly in small afferent neurons (TAYLOR et al. 1985) for at least 1 year, although no sign of degeneration in the respective dorsal root ganglia has been observed (TAYLOR et al. 1984).

Topical application of $3-30\,mM$ capsaicin to the nasal mucosa of the guinea-pig results in a long-lasting depletion of substance P from the nasal mucosa, a long-lasting inhibition of protective reflexes in response to nasal irritation, and a long-lasting block of sensory-nerve-induced extravasation of plasma proteins (LUNDBLAD 1984). These changes are evident for at least 2 months posttreatment and are restricted to the treated organ as no neurochemical or functional alterations are seen in the ureter (LUNDBLAD 1984). Topical application of 1% capsaicin to the exposed urinary bladder of the rat (MAGGI et al. 1989b) depletes substance P from the bladder but not from the adjacent ureter and leaves chemonociceptive and local effector functions of sensory neurons unaltered in organs distant from the urinary bladder such as the skin and eye. Local application of capsaicin can be as effective as systemic capsaicin treatment (LUNDBLAD 1984; MAGGI et al. 1989b). Retrobulbar injection of the drug into the rabbit eye inhibits inflammation in response to ocular irritation for several weeks (CAMRAS and BITO 1980; BYNKE 1983).

Local application of capsaicin in man is followed by a long-term defunctionalization of sensory neurons supplying the treated area. Administration of capsaicin to the monkey or human skin reduces the cutaneous substance P content (ALBER et al. 1989), inhibits local effector functions of sensory neurons (BERNSTEIN et al. 1981; CARPENTER and LYNN 1981; LUNDBLAD et al. 1987; SZOLCSÁNYI 1990), and impairs chemo- and thermonociception (BERNSTEIN et al. 1981; CARPENTER and LYNN 1981; SZOLCSÁNYI 1990), while mechanonociception and perception of touch and temperature remain unaltered (BERNSTEIN et al. 1981). Likewise, repetitive administration of capsaicin to the human nasal mucosa leads to abolition of the irritating and secretory effects of the drug and may cure vasomotor rhinitis (GEPPETTI et al. 1988b; MARABINI et al. 1988; SARIA and WOLF 1988). Cutaneous capsaicin can relieve pain associated with, e.g., postherpetic neuralgia (BERNSTEIN et al. 1987) and diabetic neuropathy (ROSS and VARIPAPA 1989).

5. Effects on Sensory Neurons In Vitro

After only a 5-min exposure of the isolated rat trachea to $640\,\mu M$ capsaicin, nerve fibers in the epithelium exhibit ultrastructural changes including an increase in axon diameter, a reduction in axoplasmic density, a loss of microtubules, and a disruption of axonal membranes (HOYES et al. 1981). These degenerative alterations are confined to axons with terminals containing a high proportion of large dense-cored vesicles and only scattered small vesicles, whereas axons with terminals containing large numbers of closely packed small vesicles are not affected (HOYES et al. 1981). Axon degeneration afflicts up to 70% of all axon profiles and is seen with concentrations of $1-10\,\mu M$ capsaicin (KIRÁLY et al. 1991). In the isolated nodose ganglion of the rat, $1-10\,\mu M$ capsaicin produces ultrastructural damage within 5 min of exposure to the drug (MARSH et al. 1987). The ultrastructural alterations include a number of signs indicative of degeneration, such as swelling of mitochondria, disruption of neurofilament organization, and fiber enlargement. These and similar observations in the isolated ureter of the guinea-pig (SIKRI et al. 1981) demonstrate rapid in vitro degeneration of sensory neurons exposed to capsaicin.

The rapidity of morphological changes is paralleled by the quick onset of functional changes. Isolated muscle tissues including cardiac muscle (ZERNIG et al. 1984; FRANCO-CERECEDA and LUNDBERG 1988), visceral smooth muscle (BARTHÓ and SZOLCSÁNYI 1978; BARTHÓ et al. 1982a, 1987; CHAHL 1982; HUA and LUNDBERG 1986; LUNDBERG and SARIA 1987; MAGGI et al. 1987c), and the iris sphincter muscle (UEDA et al. 1984) promptly respond to capsaicin with contraction or relaxation. In the isolated inferior mesenteric ganglion of the guinea-pig, capsaicin elicits a slow depolarization of the principal ganglion cells (TSUNOO et al. 1982; DUN and KIRALY 1983). All these acute effects are short-lasting, and within a few minutes, the tissues become unresponsive to both capsaicin and other means of sensory nerve stimulation. Whether this state of "desensitization" reflects a mere defunctionalization of sensory nerve fibers or is due to degenerative alterations is not known because no combined morphological and functional study has as yet been carried out. A 5-min exposure of the isolated rat urinary bladder to $1-10\,\mu M$ capsaicin also leads to substantial depletion of substance P (MAGGI et al. 1987e), which does not become appreciable, however, until 3 h posttreatment.

6. Effects on Sensory Neurons in Culture

Like in vivo, two populations of neurons, large light and small dark cell bodies, can be distinguished in cultures of dorsal root ganglia from newborn or adult rats (WINTER 1987; WOOD et al. 1988). Capsaicin selectively stimulates only a proportion of the small dark ganglion cells, which conveniently can be visualized by a cobalt uptake stain (WINTER 1987; WOOD et al. 1988). About 50% of the cells are sensitive to capsaicin (WINTER 1987; WOOD et al. 1988), and overnight treatment of cultures with $2\,\mu M$ capsaicin results in the loss of most cell bodies which showed cobalt staining. Neuritic damage is observed in less than an hour after exposure (WINTER et al. 1990).

Since prolonged exposure to capsaicin also leads to a 37% reduction in the total number of cell bodies in the culture, it follows that most of the capsaicin-sensitive cells are killed (Wood et al. 1988). Thus, the extent of cell loss is very similar to that seen after systemic capsaicin treatment of newborn rats (Otten et al. 1983; McDougal et al. 1985; Arvidsson and Ygge 1986). Cultures of large light neurons and nonneuronal cells from rat dorsal root ganglia, nonneuronal cells from the rat sciatic nerve, rat sympathetic neurons, neural crest-derived neuroblastoma cells, and hybrid cells derived from embryonal carcinoma cells are not affected by capsaicin (Bevan et al. 1987; Wood et al. 1988).

7. Age, Strain, and Species Differences in Sensitivity

There are large differences in the sensitivity of sensory neurons to capsaicin with respect to age, strain, or species of the individual under study. *Age-dependent* differences in the effects of capsaicin are best documented for the newborn rat in which the cell bodies of small afferent neurons are considerably more vulnerable than in the adult animal. The *strains* of Wistar and Sprague-Dawley rats differ not only with regard to the cardiovascular and respiratory effects of acute administration (Donnerer and Lembeck 1982) but also with regard to the long-term consequences of treatment. The variations concern the weight loss posttreatment, cardiovascular changes, sensitivity to the anesthetic methohexital (Gardiner et al. 1989), and the degree of peptide depletion from sensory neurons (Gamse et al. 1981b; Geppetti et al. 1988a). Following capsaicin treatment of neonates, Charles River and Wistar-Morini rats develop persistent skin wounds, while Sprague-Dawley rats do not (Gamse et al. 1981b; Maggi et al. 1987a). Differences are also seen following capsaicin treatment of adult Wistar-Kyoto and spontaneously hypertensive rats with regard to inhibition of thermonociception (Virus et al. 1981), substance P depletion, and 5-hydroxytryptamine and NE accumulation in the spinal cord (Virus et al. 1983).

Although capsaicin acts as a stimulant of thin sensory neurons in all mammalian species investigated so far, including man, rat, mouse, guinea-pig, hamster, rabbit, cat, dog, goat (Glinsukon et al. 1980; Szolcsányi 1990), pig (Franco-Cereceda and Lundberg 1989; Matran et al. 1989; Pierau and Szolcsányi 1989), and bear (Rogers 1984), there are considerable *species* differences in the sensitivity of afferent neurons to capsaicin which have not yet been examined systematically. Afferent neurons of the adult guinea-pig (Buck et al. 1983) are slightly more sensitive to the neurotoxic action of systemically given capsaicin than those of newborn rats (Nagy et al. 1983). Inhibition of micturition (Maggi et al. 1987b) and thermonociception (Gamse 1982; Buck et al. 1983) is more marked in guinea-pigs than in rats treated at the same age. However, perineural application in the guinea-pig is less effective in blocking nerve conduction than in the rat and ferret (Baranowski et al. 1986), although peptide transport (GAMSE et al. 1982) and thermonociception in the guinea-pig are at least as much inhibited as in the rat (SZOLCSÁNYI 1990). Afferent C-fibers of the rabbit are less sensitive to periaxonal administration than those in the rat in terms of conduction block (BARANOWSKI et al. 1986), inhibition of peptide transport (GAMSE et al. 1982), and axon degeneration (LYNN et al. 1987; LYNN and SHAKHANBEH 1988). Afferent neurons in the rabbit are in general quite resistant to the long-term neurotoxic action of capsaicin, although they are acutely excited and desensitized (TERVO 1981; BARANOWSKI et al. 1986; MORITOKI et al. 1990; TRAD et al. 1990), capsaicin being about 20 times less potent than in the rat (SZOLCSÁNYI 1987). There are also species differences in the sensitivity of afferent A-fibers to capsaicin in that perineural treatment reversibly blocks conduction in Aδ-fibers of the rat, guinea-pig, rabbit, cat, and monkey (LYNN et al. 1984; J.M. CHUNG et al. 1985; BARANOWSKI et al. 1986; SUCH and JANCSÓ 1986; MARSH et al. 1987) but not ferret (BARANOWSKI et al. 1986).

Some species differences in the excitatory and long-term effects of capsaicin are not necessarily due to differences in sensory neuron sensitivity to the drug but rather to different distributions of capsaicin-sensitive afferent neurons, other neurotransmitter contents, different actions and functions of these afferent neurons, or species differences in the metabolic effects and pharmacokinetic fate of capsaicin (MAGGI et al. 1986, 1987b,c, 1989d; BARTHÓ et al. 1987; LUNDBERG and SARIA 1987b; AMANN et al. 1988). Finally, its overall toxicity when administered systemically to mammals (GLINSUKON et al. 1980) does not necessarily correlate with the species sensitivity of afferent neurons to capsaicin. By stimulating sensory neurons capsaicin not only gives rise to nociception but also to generalized vasodilatation, extravasation of plasma protein, and bronchoconstriction. Consequently, this painful treatment ought to be performed under general anesthesia and under pharmacological precautions to control bronchoconstriction, hypersecretion in the airways, and leakage of plasma protein (see GAMSE et al. 1981c; BUCK et al. 1983).

V. Acute and Long-Term Effects in Nonmammalian Species

In contrast to mammals, nonmammalian species are only poorly sensitive to capsaicin. In the opossum, capsaicin has been found to release substance P from apparent sensory nerve fibers in the muscularis mucosae of the esophagus, whereas systemic pretreatment does not change the tissue level of the peptide (DANIEL et al. 1987). Acute local or systemic administration, at concentrations of 33 mM or doses up to 600 mg/kg, either fails to evoke pain or reactions indicative of pain or is only weakly active in pigeons and other birds (MASON and MARUNIAK 1983; SZOLCSÁNYI et al. 1986; SANN et al. 1987). No long-term changes in chemonociception have been noted after systemic capsaicin treatment of pigeons (SZOLCSÁNYI et al. 1986), and thermoregulation examined in some avian species is not affected either

(Mason and Maruniak 1983; Geisthövel et al. 1986; Sann et al. 1987). Capsaicin (10 µM) is unable to release substance P from sensory nerve endings in the pigeon spinal cord (Pierau et al. 1987), although systemic treatment of pigeons with a 950 mg/kg dose gives rise to a partial depletion of substance P and calcitonin gene-related peptide from nerve fibers in the gut, ureter, and cornea (Harti 1988; Harti et al. 1989). A reduction of peptide levels in the pigeon cornea is also seen after topical application of 33 mM capsaicin to this tissue (Harti et al. 1989). Likewise, application of 330 nmol capsaicin to the sciatic nerve of the pigeon leads to a 40% depletion of substance P and calcitonin gene-related peptide from the nerve axons, whereas these peptides accumulate in the dorsal horn of the spinal cord (Harti et al. 1987). However, systemic application of 500 mg/kg to the newborn chick does not cause degeneration of dorsal root ganglion cells (Jancsó et al. 1985a). Cultured dorsal root ganglia from this species fail to respond to capsaicin in the cobalt stain assay (Bevan et al. 1987; Wood et al. 1988), whereas exposure (30 min to 28 h) of cultured chick sensory neurons to 32–160 µM capsaicin has been reported to cause a concentration-dependent retardation of neurite outgrowth which appears to be due to degeneration of neurite tips (Hiura and Sakamoto 1987a). These changes are not permanent because quick neurite regeneration may occur even in the continued presence of capsaicin (Hiura and Sakamoto 1987a). Degeneration of nerve cell bodies has not been noted. Frogs appear to be insensitive to the irritant and neurotoxic effects of capsaicin on sensory neurons (Chéry-Croze et al. 1985; Szolcsányi et al. 1986).

VI. Summary

1. Capsaicin-Sensitive Neurons

Afferent neurons which respond to capsaicin are referred to as capsaicin-sensitive sensory neurons (Szolcsányi and Barthó 1978; Szolcsányi 1982, 1984b). They are primary afferent neurons, the majority of which have small dark cell bodies that are connected to unmyelinated (C-) or thinly myelinated (Aδ-) nerve fibers. However, a minority of afferent neurons with large light cell bodies are also sensitive to capsaicin (Lawson and Harper 1984). Characteristically, capsaicin-sensitive afferent neurons contain a variety of peptides which are thought to play a transmitter or mediator role (Table 1). However, none of these markers is an exclusive constituent of capsaicin-sensitive sensory neurons. Their classification is also hampered by the ontogenetic shift in sensory neuron sensitivity as the population of sensory neurons sensitive to the neurotoxic effect of capsaicin in the newborn rat is not completely identical with that in the adult animal (Jancsó et al. 1987; Maggi and Meli 1988; Szolcsányi 1990). Another problem consists in the fact that capsaicin-sensitive afferent neurons do not completely overlap with any population of afferent neurons that have been classified according to morphological, neurochemical, or functional criteria. Thus, not all unmyelinated afferent neurons conducting in the C- and Aδ-fiber range are sensitive to capsaicin, and capsaicin-sensitive afferent neurons are heterogeneous in terms of their sensory modality and the functions they subserve. The somatic capsaicin-sensitive afferents correspond to a large degree with nociceptors sensitive to chemical, thermal, and/or mechanical stimuli and to warmth receptors. Visceral afferent neurons sensitive to capsaicin have been found to respond to noxious (COLERIDGE and COLERIDGE 1977, 1984; CERVERO and McRITCHIE 1982; LONGHURST et al. 1984; SZOLCSÁNYI 1984b; STEIN et al. 1986) as well as nonnoxious "physiological" stimuli (MacLEAN 1985; McCANN et al. 1988; RAYBOULD and TACHÉ 1988; SOUTH and RITTER 1988; RÓZSA and JACOBSON 1989; HOLZER et al. 1990a).

2. Neurons not Sensitive to Capsaicin

Afferent neurons with thickly myelinated axons conducting in the Aα and Aβ range are not sensitive to the long-term neurotoxic action of this compound. However, some thickly myelinated afferent neurons can be stimulated by acute capsaicin treatment, and subsequently conduction can transiently be blocked in them. Most of the neurons in the CNS appear to be insensitive to its acute and long-term effects, but the situation is not yet clear and calls for further examination. Exceptions are a group of thermosensitive neurons in the preoptic region of the hypothalamus and some neurons in other nuclei of the brain (DINH and RITTER 1987; RITTER and DINH 1988). Efferent motor neurons and efferent neurons of the autonomic nervous system are not sensitive to the excitatory or neurotoxic action of capsaicin (SZOLCSÁNYI et al. 1975; AULT and EVANS 1980; JANCSÓ et al. 1980a; CERVERO and McRITCHIE 1982; GAMSE et al. 1982; DELLA et al. 1983; SHARKEY et al. 1983; HANDWERKER et al. 1984; STEIN et al. 1986; SUCH and JANCSÓ 1986; BEVAN et al. 1987; MARSH et al. 1987; TAKANO et al. 1988; WOOD et al. 1988). The enteric nervous system is believed to be grossly insensitive to capsaicin (see BARTHÓ and HOLZER 1985), although a minority of enteric neurons may be sensitive to the drug (FEHÉR and VAJDA 1982; HARTI 1988; KIRCHGESSNER et al. 1988). The conclusion to be drawn is that the selectivity with which capsaicin acts on a group of unmyelinated and thinly myelinated primary afferent neurons is exceptional but certainly not absolute.

The "nonselective" effects of capsaicin can be separated into three types of action. (1) Capsaicin can affect neural and nonneural systems that are functionally related to primary afferent neurons. Capsaicin-induced changes in these systems are interpreted to be secondary to the action of capsaicin on primary afferent neurons. (2) Certain neurotoxic actions of capsaicin on neurons in the central and enteric nervous systems appear to reflect a direct effect of the drug on these neurons. (3) High concentrations of capsaicin exert a transient effect on many neural as well as non-neural systems in both vertebrate and invertebrate species, which seem to be unrelated to its excitatory action on thin primary afferent neurons.

C. Mechanisms of Action

I. Structure-Activity Relationships

The finding of Szolcsányi and Jancsó-Gábor (1975, 1976) that the effects of capsaicin are shared by structurally related analogues of the drug paved the way to the hypothesis that afferent neurons are sensitive to capsaicin because they possess a selective recognition site for capsaicinoids. As shown in Fig. 1, capsaicin consists of a vanillyl (4-hydroxy-3-methoxy-benzyl) moiety, an acylamide group, and an alkyl chain, certain features of which are required for its activity. The acutely stimulant activity is determined by the presence of a 4-hydroxy-3-methoxy substituted aromatic (phenyl) ring connected by only one CH_2 group to the acylamide group, the polar nature of the acylamide, and the apolar nature of the alkyl chain. There is some flexibility with the acylamide bond which as such is not essential for pungency since it can be replaced by an ester group without much loss of activity (Szolcsányi and Jancsó-Gábor 1975). The optimal length of the alkyl chain is 8–10 carbon atoms, although its substitution by a cycloalkyl ring retains much of the pungent activity (Szolcsányi and Jancsó-Gábor 1975). The structural requirements for an agonist to be recognized by the putative capsaicin receptor have been confirmed by a study using cultured neurons from rat dorsal root ganglia and a capsaicin-like photoaffinity probe (James et al. 1988). The noxious activity of capsaicin congeners on the rat eye (Szolcsányi and Jancsó-Gábor 1975) correlates well with their potency in eliciting the Bezold-Jarisch reflex (apnoea, bradycardia, hypotension) (Szolcsányi 1982), producing hypothermia (Szolcsányi 1982; Hayes et al. 1984b), causing contraction of the trachea (Szolcsányi 1983b), and facilitating the release of substance P from the spinal cord (Bucsics and Lembeck 1981; Jhamandas et al. 1984). On the basis of fairly rigid structural requirements for capsaicin-like activity, Szolcsányi and Jancsó-Gábor (1975) proposed a hypothetical model of a capsaicin receptor in which multiple interactions between the capsaicin molecule and the recognition site determine optimal activity.

The structural requirements for desensitization are somewhat different from those for the acutely stimulant effect of capsaicin (Szolcsányi and Jancsó-Gábor 1976; Hayes et al. 1984b). The differences in excitatory and desensitizing potency are best illustrated by the fact that there are capsaicin analogues such as zingerone that exhibit noxious activity but are devoid of desensitizing activity (Szolcsányi and Jancsó-Gábor 1976; Szolcsányi 1982; Hayes et al. 1984b) and that there are some capsaicinoids that display desensitizing activity in the absence of excitatory activity (Dray et al.

1990b). It remains to be investigated whether the sites and mechanisms of action underlying capsaicin-induced excitation and desensitization are identical or different. The inference of a common site of action for all sensory neuron-selective effects of capsaicin derives from the findings that the activity of its congeners to induce afferent neuron degeneration in the newborn rat (JANCSÓ and KIRÁLY 1981) is similar to their noxious activity (SZOLCSÁNYI and JANCSÓ-GÁBOR 1975), and that with intrathecal administration of capsaicin analogues, comparable structure-activity relationships exist for the acute release of substance P in the spinal cord, desensitization to this effect, and long-lasting antinociception and depletion of substance P from afferent nerve terminals in the spinal cord (JHAMANDAS et al. 1984).

Biochemical characterization of the cellular recognition site for capsaicin was hampered by technical problems arising from the lipophilic nature of this ligand (JAMES et al. 1988), but these problems have recently been overcome by the use of resiniferatoxin (SZALLASI and BLUMBERG 1990a,b). This drug, known for some time to be a potent irritant, is a diterpene derivative occurring naturally in the latex of certain species of *Euphorbia*. Structurally it combines characteristics of the phorbol esters and of capsaicin. Like capsaicin, it contains a 4-hydroxy-3-methoxy benzyl (vanillyl) moiety connected to an ester group which binds the vanillyl moiety to the 20-hydroxyl substituent of the complex phorbol ester structure. Unlike active phorbol esters, resiniferatoxin lacks a free 20-hydroxyl group which is essential for phorbol ester activity and thus fails to induce their typical effects including tumor promotion, binding to and stimulation of protein kinase C (SZALLASI and BLUMBERG 1989, 1990a,b; DRAY et al. 1990a).

Resiniferatoxin appears to share all the effects of capsaicin on sensory neurons, but it is 1000-fold more potent than capsaicin in producing hypothermia and increasing vascular permeability in the rat (DE VRIES and BLUMBERG 1989), whereas the potency of the two drugs in inducing corneal nociception is similar (SZALLASI and BLUMBERG 1989). When tested in an in vitro preparation of the neonatal rat tail-spinal cord, resiniferatoxin proved to be about 100 times more potent than capsaicin in activating cutaneous nociceptors (DRAY et al. 1990a) and in depolarizing cultured neurons from rat dorsal root ganglia (WINTER et al. 1990). The inference that both capsaicin and resiniferatoxin share a common target of cellular action is supported by similarities in the intermediate and long-term effects of the two drugs. In the newborn rat, administration of resiniferatoxin 300 µg/kg leads to degeneration of about 50% of the cell bodies in dorsal root ganglia, which is associated with the neurochemical and functional alterations known to occur with capsaicin (SZALLASI et al. 1990). Given to adult rats, resiniferatoxin (300 µg/kg) induces ultrastructural changes in small dark neurons of the dorsal root ganglia, characterized by swelling and disorganization of mitochondria (SZALLASI et al. 1989). Resiniferatoxin also leads to the accumulation of ionic calcium in some small sensory neurons, an effect which is more marked than with capsaicin (SZALLASI et al. 1989) and which according to Jancsó et al. (1978) could indicate that in the adult rat resiniferatoxin causes more sensory neurons to degenerate than capsaicin. In vitro exposure of cultured sensory neurons to $10-100\,nM$ resiniferatoxin results in neuritic damage of the small ganglion cells (Winter et al. 1990).

Functionally, treatment of rats with resiniferatoxin ($100-400\,\mu g/kg$) causes desensitization to resiniferatoxin-evoked hypothermia, leakage of plasma protein and pain, with cross-desensitization between resiniferatoxin and capsaicin (De Vries and Blumberg 1989; Szallasi and Blumberg 1989; Dray et al. 1990a; Szallasi et al. 1990; Szolcsányi et al. 1990; Winter et al. 1990). However, the activity of resiniferatoxin as a sensory neuron stimulant does not completely parallel its activity in cross-desensitizing against capsaicin congeners (Szallasi and Blumberg 1989; Dray et al. 1990a; Szolcsányi et al. 1990). Differences between resiniferatoxin and capsaicin with respect to potency and time course of their effects do not necessarily indicate a heterogeneity of action, because the pharmacokinetic behavior of resiniferatoxin and capsaicin may vary profoundly (Szallasi et al. 1989; Maggi et al. 1990c). The finding of cross-desensitization between capsaicin and resiniferatoxin and the observation that ruthenium red blocks the responses to both drugs is strong pharmacological evidence that resiniferatoxin might be an "ultrapotent agonist" at the putative capsaicin receptor. More direct support for this contention comes from the presence of a single class of specific saturable binding sites for (^{3}H)resiniferatoxin on membranes from dorsal root ganglia, at which binding also is displaced by capsaicin (Szallasi and Blumberg 1990a,b). These binding sites are present in dorsal root ganglia of rat, pig, cow, and sheep but not in those of the chicken, a species that is not responsive to the pungent action of capsaicin (Szallasi and Blumberg 1990a,b). No specific binding of resiniferatoxin has been detected in the preoptic region of the brain, striatum, substantia nigra, and entire spinal cord of the rat. Resiniferatoxin-induced degeneration of primary afferent neurons in the newborn rat is associated with a major loss of resiniferatoxin binding sites (Szallasi et al. 1990).

Strong pharmacological evidence for the existence of a specific recognition site for capsaicin came from the discovery of a competitive antagonist, capsazepine. This compound (2-[2-(4-chlorophenyl)ethylamino-thiocarbonyl]-7,8-dihydroxy-2,3,4,5-tetrahydro-1H-2-benzazepine) is structurally related to capsaicin and potently antagonizes all sensory neuron-selective effects of capsaicin but lacks agonistic activity (Bevan et al. 1991; Dray et al. 1991). The concentration of the antagonist needed to reduce the excitatory effects of capsaicin on sensory neurons lies between 0.01 and $5\,\mu M$, depending on the preparation under study (Dray et al. 1991). Analysis of the inhibition of capsaicin-induced ion fluxes in cultured dorsal root ganglion neurons from the rat yields linear Schild plots, whose slope is very close to 1, giving K_d estimates of $0.1-0.7\,\mu M$ (Bevan et al. 1991). The availability of a competitive antagonist of capsaicin is a significant asset in the further exploration of the sensory neuron-selective actions of capsaicin, their mechanisms, and functional significance.

II. Effects on the Cell Membrane

1. Sensory Neuron-Selective Effects

Capsaicin depolarizes both axons and somata of rat primary afferent neurons with C-fibers (AULT and EVANS 1980; YANAGISAWA et al. 1980; WILLIAMS and ZIEGLGÄNSBERGER 1982; BACCAGLINI and HOGAN 1983; HAYES et al. 1984a; HEYMAN and RANG 1985; BEVAN et al. 1987; MARSH et al. 1987; WINTER et al. 1988, 1990; BLEAKMAN et al. 1990; DOCHERTY et al. 1991), which is associated with an increase in membrane conductance and an inward current that can be followed by an outward current. Local anesthetics and particularly tetrodotoxin do not inhibit the local excitatory action of capsaicin on sensory nerve endings and axons, indicating that fast voltage-dependent sodium channels are not involved (JANCSÓ et al. 1968; GAMSE et al. 1979b; SARIA et al. 1983; SZOLCSÁNYI 1983b; 1984b; HAYES et al. 1984a; HUA et al. 1986; SANTICIOLI et al. 1987; MAGGI et al. 1989c, 1990b). However, sodium influx is enhanced in cultured sensory neurons exposed to capsaicin (WOOD et al. 1988), and capsaicin-induced depolarization is partly inhibited by removal of extracellular sodium (BEVAN et al. 1987; MARSH et al. 1987). Removal of extracellular calcium, on the other hand, does not inhibit the excitatory or depolarizing action of capsaicin on sensory neurons but, if anything, enhances it (YANAGISAWA et al. 1980; BACCAGLINI and HOGAN 1983; MARSH et al. 1987; BETTANEY et al. 1988; AMANN et al. 1989a).

Ion flux (WOOD et al. 1988; WINTER et al. 1990) and patch-clamp (BEVAN and FORBES 1988; FORBES and BEVAN 1988) studies of cultured sensory neurons indicate that capsaicin opens a nonselective cation channel that leads to an inward current carried by both sodium and calcium ions. This inward current reverses at a potential close to 0 mV (BEVAN and FORBES 1988; WINTER et al. 1990). The finding that capsaicin also acts on isolated membrane patches (FORBES and BEVAN 1988) indicates that opening of a nonselective cation channel is a basic mechanism of its action on sensory neurons. In addition, capsaicin increases the efflux of potassium, which is inhibited in the absence of extracellular calcium (WOOD et al. 1988) and thus represents probably a calcium-activated potassium outward current (MARSH et al. 1987). Chloride efflux remains unaltered by capsaicin treatment (WOOD et al. 1988). Nickel ions and other antagonists of voltage-dependent calcium channels do not inhibit the stimulant effect of capsaicin on sensory neurons (ZERNIG et al. 1984; MAGGI et al. 1988a,c,e, 1989c,e, 1990b; WOOD et al. 1988). These data underline the uniqueness of the capsaicin-activated cation channel, which is distinct from voltage-dependent sodium and calcium channels and which can be activated even in depolarized terminals of sensory neurons (DONNERER and AMANN 1990).

There seems to be some action of capsaicin on certain voltage-dependent calcium channels of sensory neurons (Petersen et al. 1989), but this probably occurs secondarily to the depolarization (Wood et al. 1989; Bevan and Szolcsányi 1990) and is not seen in cultured sensory neurons (Docherty et al. 1991). The capsaicin-evoked influx of calcium into the cell is a specific response of B-type sensory neurons of the rat and mouse (and probably other mammalian species), which is absent in the vast majority of rat A-type sensory neurons, in rat sympathetic neurons, Schwann cells, and fibroblasts, in neuronal cell lines, and in chick sensory neurons (Wood et al. 1988; Winter et al. 1990). The finding that the nonselective cation conductance is also activated in isolated membrane patches of sensory neurons exposed to capsaicin (Forbes and Bevan 1988; Dray et al. 1990c) indicates that the drug acts directly on the plasma membrane without any intervention of a second messenger system, a conclusion for which there is also biochemical evidence (Wood et al. 1989; Winter et al. 1990).

2. Cell-Nonselective Effects

Capsaicin exerts certain actions on sensory as well as nonsensory neurons and nonneuronal excitable cells, which seem to be unrelated to its selective stimulant, desensitizing, and neurotoxic effects on thin sensory neurons. Although these "cell-nonselective actions" of capsaicin have not been investigated systematically, it is possible to see common traits in them. Thus, the nonselective effects can be shown repetitively and do not exhibit desensitization, nor are they related to cell toxicity. The potency of capsaicin for the cell-nonselective effects is typically orders of magnitude higher than that for its sensory neuron-selective effects. Capsaicin influences the action potential in all sensory neurons of the rat, guinea-pig, and chick (Godfraind et al. 1981; Petersen et al. 1987; Szolcsányi 1990). In sensory neurons of the rat and chick (Godfraind et al. 1981) and in giant neurons of the snail (Erdélyi et al. 1987), it prolongs the duration of the action potential, which may be related to reductions of both the sodium inward and potassium outward current as seen in guinea-pig and chick sensory neurons (Petersen et al. 1987; Szolcsányi 1990), rat sympathetic neurons (Bevan et al. 1987), snail neurons (Erdélyi et al. 1987), and crayfish giant axon (Yamanaka et al. 1984). Blockade of a certain component of the outward potassium current, with some inhibition of the sodium inward current, is observed in myelinated nerve fibers of the frog sciatic nerve (Dubois 1982), and in snail neurons calcium inward currents are inhibited by capsaicin as well (Erdélyi et al. 1987).

It is conceivable that the inhibitory effect on the neuronal calcium conductance bears a relation to the capsaicin-induced inhibition of calcium uptake in mouse neuroblastoma cells and rat aortic smooth muscle cells (Monsereenusorn and Kongsamut 1985). The effects of capsaicin on cardiac muscle (Zernig et al. 1984; Franco-Cereceda and Lundberg 1988) and visceral (BARTHÓ et al. 1982b, 1987; MAGGI et al. 1987c, 1989b) smooth muscle may arise from a similar action. Inhibition of membrane currents is also likely to account for a capsaicin-induced transient blockade of nerve conduction as has been shown in rat A-fiber afferent neurons (BARANOWSKI et al. 1986; MARSH et al. 1987), cat sympathetic nerve fibers (SUCH and JANCSÓ 1986), and crayfish giant axon (YAMANAKA et al. 1984). While the nonselective effects on voltage-dependent ion conductances in the cell membrane are fairly consistent in vertebrate and nonvertebrate species, there is less uniformity with its effect on the resting cell membrane. Rat and chick sensory neurons respond to capsaicin with a hyperpolarization that is blocked by removal of extracellular chloride (GODFRAIND et al. 1981). In the crayfish giant axon (YAMANAKA et al. 1984) and in the giant amoeba (FOSTER et al. 1981), capsaicin is without effect on the resting potential, although in the amoeba capsaicin reduces the input resistance of the membrane (FOSTER et al. 1981). In contrast, giant neurons of the snail are depolarized by capsaicin (ERDÉLYI et al. 1987).

III. Intracellular Effects

1. Ion Accumulation, Peptide Release, and Biochemical Effects

The membrane effects of capsaicin give way to cation influx into the cell. It is estimated that the intracellular calcium concentration can rise up to $12\,mM$ (WOOD et al. 1988). Calcium is thus concentrated in the cell, and this process requires an oxidative energy supply. A considerable proportion of influxing calcium will be sequestered by intracellular organelles, but there also is a transient rise in the intracellular concentration of free calcium ions (BLEAKMAN et al. 1990; DRAY et al. 1990c). Since drugs known to interfere with the calcium binding of the endoplasmic reticulum (e.g., caffeine, isobutyl-methylxanthine) do not inhibit capsaicin-induced uptake of calcium whereas uncoupling of oxidative phosphorylation in the mitochrondria inhibits capsaicin-induced intracellular calcium uptake (WOOD et al. 1988) and calcitonin gene-related peptide release from sensory nerve endings (AMANN et al. 1990b), it has been proposed that calcium is bound primarily by mitochondria (WOOD et al. 1988). In addition, sodium ions also enter the cell through the capsaicin-activated cation channel, which together with a passive influx of chloride results in the intracellular uptake of NaCl (WINTER et al. 1990).

Intracellular accumulation of calcium has consequences for many cell functions including peptide release. The capsaicin-induced release of peptides from the central and peripheral endings of sensory neurons is inhibited by removal of extracellular calcium (GAMSE et al. 1979b; THÉRIAULT et al. 1979; YANAGISAWA et al. 1980; HELKE et al. 1981b; HUA et al. 1986; MAGGI et al. 1988d, 1989e; AMANN et al. 1989a; AMANN 1990). Electrically induced peptide release is prevented by blockers of voltage-dependent calcium channels (Holz et al. 1988; Maggi et al. 1989c, 1990b), while peptide release induced by capsaicin is not blocked by calcium channel inhibitors (Maggi et al. 1988c,e, 1989e) or by potassium depolarization (Donnerer and Amann 1990). The massive intracellular uptake of calcium (Wood et al. 1988) may explain why the requirements of extracellular calcium for capsaicin-induced release of substance P from sensory nerve endings are substantially lower than those necessary to maintain release induced by other depolarizing stimuli (Gamse et al. 1979b; Maggi et al. 1988e, 1989e). The build-up of intracellular calcium also may account for the capsaicin-induced increase in cyclic guanosine monophosphate levels in sensory neurons (Wood et al. 1989). The levels of cyclic adenosine mono-phosphate are not changed in cultured sensory neurons (Wood et al. 1989), which contrasts with reports that capsaicin can increase the levels of this phosphate in the rat brain (Horváth et al. 1979) and spinal cord (Northam and Jones 1984). This effect may be an indirect action caused by transmitter release (Wood et al. 1989).

2. Nonspecific Desensitization

The time and dose relationships between capsaicin-induced excitation and desensitization or nerve block suggest that these phenomena are consecu-tive manifestations of one common effect on sensory neurons. Excessive stimulation of sensory neurons or simple fatigue are considered unlikely to account for desensitization (Szolcsányi and Jancsó-Gábor 1976; Green 1989). In analogy to observations on dorsal root ganglia (Williams and Zieglgänsberger 1982), the initial phase of desensitization may be due to a still prevailing depolarization but does not explain the persistence of desensitization, since depolarization of the axons abates within a few minutes (Hayes et al. 1984a; Bevan et al. 1987; Marsh et al. 1987). While specific desensitization may reflect desensitization of the capsaicin receptor/ion channel complex (Dray et al. 1989a), nonspecific desen-sitization which occurs with higher doses involves processes beyond the level of a membrane receptor for capsaicin. It seems as if the excitatory action of capsaicin triggers some functional change in the axons that outlasts the initial depolarization.

Nonspecific desensitization most likely arises from the intracellular con-sequences of capsaicin-evoked excitation of sensory neurons (Docherty et al. 1991), since capsaicin-induced activation of nonselective cation channels in isolated cell membrane patches from sensory neurons does not show desensitization (Forbes and Bevan 1988; Dray et al. 1990c). Mani-festation of desensitization to capsaicin depends on the dose of the drug and the availability of calcium, because removal of extracellular calcium reduces desensitization (Santicioli et al. 1987; Wood 1987; James et al. 1988; Jin et al. 1989; Amann 1990). Intracellular NaCl accumulation may in addition contribute to desensitization (Bevan and Szolcsányi 1990). Accordingly, ruthenium red which blocks capsaicin-induced activation of cation channels and the uptake of calcium in B-type sensory neurons (Wood et al. 1988; Dray et al. 1990c) can inhibit both the excitatory and desensitizing actions of capsaicin (Maggi et al. 1988b,d, 1989a; Chahl 1989; Jin et al. 1989). Electrophysiologically, it appears that a long-lasting inhibition of voltage-dependent calcium channels plays an important role in the manifestation of nonspecific desensitization (Bleakman et al. 1990; Docherty et al. 1991). Some studies of the capsaicin-induced release of substance P from sensory neurons in, or its actions on, the rabbit iris sphincter have led to the proposal that desensitization to capsaicin could be accounted for by depletion of the releasable peptide pool (Ueda et al. 1984; Håkanson et al. 1987). However, studies in the rat and guinea-pig have ruled out this possibility (Gamse et al. 1979b; Lembeck and Donnerer 1981; Bittner and LaHann 1985; Hua et al. 1986; Maggi et al. 1987e, 1990a; Dray et al. 1989b; Donnerer and Amann 1990; Amann 1990).

3. Neurotoxicity

There is evidence that the stimulant, desensitizing, and long-term neurotoxic effects of capsaicin on thin sensory neurons arise from a common mechanism of action in which intracellular accumulation of calcium and sodium plays a central role. Excess accumulation of calcium within cells is extremely toxic and appears to be a common final process by which toxins cause degeneration and cell death (Schanne et al. 1979; Kamakura et al. 1983) as calcium-activated degradative enzymes are thought to destroy the cytoskeletal organization (Kamakura et al. 1983). As pointed out by Maggi and Meli (1988), dorsal root ganglion cells have a relative incapacity to buffer excess intracellular calcium as compared with other neurons (Jia and Nelson 1986). Consequently, the dose-dependent desensitizing and neurotoxic effect of capsaicin may be determined by different grades of calcium accumulation in the sensory neuron, and the type of reaction (desensitization or neurotoxicity) might be related to whether or not the calcium overloading of the cell can be reversed. Furthermore, intracellular NaCl accumulation is likely to cause water influx, swelling, and osmotic lysis of the cell (Winter et al. 1990). These effects are consistent with the observation that exposure of nodose ganglion cells to $1-10\,\mu M$ capsaicin leads to swelling and disruption of the microtubular and neurofilament organization within 5 min (Marsh et al. 1987). The involvement of calcium is shown by the finding that the neurotoxic action of capsaicin can be inhibited by removal of extracellular calcium but not by blockade of voltage-dependent calcium channels (Marsh et al. 1987). Accordingly, exposure of B-type sensory neurons to capsaicin or resiniferatoxin in vivo leads to the histochemically demonstrable appearance of calcium in the degenerating cell bodies (Jancsó et al. 1978; Szallasi et al. 1989), particularly at the mitochondrial level (Jancsó et al. 1984).

IV. Ruthenium Red as a Capsaicin Antagonist

The inorganic dye ruthenium red inhibits the capsaicin-induced stimulation and desensitization of sensory nerve endings. At the concentration range of $0.03-10\,\mu M$, the dye is remarkably selective for capsaicin. Thus, cardio-vascular reflexes (Amann and Lembeck 1989; Amann et al. 1989a, 1990a; Maggi et al. 1989a, 1990c), urinary bladder motor reflexes (Maggi et al. 1989a), and nociception-associated depolarization of spinal ventral roots (Dray et al. 1990c) in response to capsaicin are blocked by ruthenium red. Activation of sensory neurons by mechanical stimuli (Maggi et al. 1989a; Dray et al. 1990c), acetylcholine, bradykinin, and 5-hydroxytryptamine (Amann and Lembeck 1989; Amann et al. 1989a, 1990a; Dray et al. 1990c) is not affected by ruthenium red. Responses to noxious heat either remain unaffected (Dray et al. 1990c) or are reduced (Amann et al. 1990a). Ruthenium red also inhibits the capsaicin-induced release of peptides (substance P and calcitonin gene-related peptide) from sensory nerve endings in the guinea-pig lung (Amann et al. 1989b) and bronchi (Maggi et al. 1989c), guinea-pig and rat urinary bladder (Maggi et al. 1988d; Amann et al. 1990b), guinea-pig heart (Franco-Cereceda et al. 1989; Maggi et al. 1989c), rat trachea (Ray et al. 1990), and rabbit ear (Amann et al. 1989a, 1990a). Peptide release evoked by bradykinin, nicotine, veratridine, or potassium depolarization remains unaltered by ruthenium red (Amann et al. 1989b, 1990a; Franco-Cereceda et al. 1989; Ray et al. 1990). The motor effects which capsaicin or resiniferatoxin produce in cardiac, visceral, and vascular muscle also are prevented by ruthenium red (Maggi et al. 1988b,d, 1989a,c, 1990c; Chahl 1989; Franco-Cereceda et al. 1989; Jin et al. 1989; Amann et al. 1990b), while those to electrical field stimulation, substance P, neurokinin A, calcitonin gene-related peptide, acetylcholine, nicotine or potassium depolarization are in most cases not inhibited by ruthenium red. The capsaicin-induced increase in blood flow and vascular permeability in the rabbit skin is prevented by ruthenium red, whereas the vascular responses to bradykinin, N-formyl-methionyl-leucyl-phenylalanine, platelet-activating factor, histamine, and calcitonin gene-related peptide are spared (Buckley et al. 1990).

Since ruthenium red does not displace the binding of resiniferatoxin to membranes from dorsal root ganglia (Szallasi and Blumberg 1990a), it is not a competitive antagonist at the resiniferatoxin/capsaicin recognition site but is a functional antagonist of capsaicin. As such, ruthenium red prevents the capsaicin-induced opening of a cation channel in isolated cell membrane patches from dorsal root ganglia cultures and inhibits the capsaicin-induced increase in free intracellular calcium (Dray et al. 1990c) and intracellular calcium accumulation (Wood et al. 1988). The dye most likely interferes with the coupling mechanism that links the capsaicin recognition site to ion channel activation (Dray et al. 1990c). An intracellular site of action of ruthenium red is unlikely (Maggi et al. 1988d, 1989c; Amann et al. 1989a,

1990b; FRANCO-CERECEDA et al. 1989; DRAY et al. 1990c). Being a selective functional antagonist of capsaicin, ruthenium red is a useful tool to search for endogenous substances that utilize the same transduction mechanisms as capsaicin (BUCK and BURKS 1986; MAGGI 1991).

V. Interaction with Nerve Growth Factor

Capsaicin disrupts the retrograde transport of nerve growth factor (MILLER et al. 1982a,b) and horseradish peroxidase (TAYLOR et al. 1985), which precedes the depletion of substance P from the dorsal root ganglia (MILLER et al. 1982b). Systemic application of nerve growth factor counteracts the capsaicin-induced depletion of substance P from sensory neurons (MILLER et al. 1982b) and protects B-type sensory neurons from capsaicin-induced degeneration (OTTEN et al. 1983). The possibility that nerve growth factor supplementation circumvents the deleterious effect of capsaicin-induced inhibition of nerve growth factor transport to the dorsal root ganglia is negated by the finding that nerve growth factor also prevents the inhibitory action of capsaicin on the retrograde transport of horseradish peroxidase when both capsaicin and nerve growth factor are administered locally to the peripheral endings of sensory neurons (TAYLOR et al. 1985). The situation is complicated by the fact that sensory neuron cultures derived from dorsal root ganglia of adult rats, survival of which does not depend on the presence of nerve growth factor, lose their responsiveness to capsaicin in the absence of nerve growth factor (WINTER et al. 1988). Full sensitivity to capsaicin is restored within 4–6 days after the addition of nerve growth factor, which suggests that nerve growth factor turns on the synthesis of cellular components which determine the capsaicin sensitivity of sensory neurons (WINTER et al. 1988). A satisfactory reconciliation of the in vivo observations with the data from cultured sensory neurons is not yet possible.

VI. Sites of Action on Sensory Neurons

From the concept that the sensory-neuron-selective actions of capsaicin are due to its recognition by a capsaicin receptor coupled to a peculiar cation conductance in the cell membrane, it would follow that the distribution of these receptor/channel complexes will determine which segment of a sensory neuron responds to capsaicin. Since somata, axons, and nerve endings of B-type sensory neurons are excited by capsaicin, it would appear that the capsaicin receptor/cation channel complex is expressed in all parts of thin sensory neurons. However, local or systemic administration of capsaicin to adult rats damages many more nerve terminals than cell bodies (HOYES and BARBER 1981; HOYES et al. 1981; JANCSÓ 1981; PALERMO et al. 1981; GAMSE 1982; GAMSE et al. 1984, 1986; RIBEIRO-DA-SILVA and COIMBRA 1984; K. CHUNG et al. 1985, 1990). This may be due to the absence of a Schwann cell sheath (perineurium) around the terminals of sensory neurons, which

hinders the access of capsaicin to other parts of the neuron, or to seg-
mentally different capacities of resistance and repair. Given the reversibility
of capsaicin's long-term effects in the adult rat, it would appear that most of
the B-type somata do not degenerate in the adult rat (JANCSÓ et al. 1985b),
although they display ultrastructural changes (JOÓ et al. 1969; SZOLCSÁNYI
et al. 1975; CHIBA et al. 1986) and are depleted of neuropeptides (JESSELL
et al. 1978; GAMSE et al. 1981b; GAMSE 1982; GEPPETTI et al. 1988a).

If indeed nerve terminals of thin sensory neurons were the primary
target of capsaicin's neurotoxic action in the adult rat, either the distribution
of the capsaicin receptor/cation channel or the vulnerability by intracellular
calcium/sodium accumulation would be expected to vary along the axons of
sensory neurons. A systematic analysis of this issue is not yet available,
although there is other indirect evidence for nerve terminals being a preferred target of capsaicin's action. Thus, thermonociception and the retrograde uptake and transport of nerve growth factor is inhibited by systemic
capsaicin before substance P is depleted from the dorsal root ganglia
(MILLER et al. 1982a), and substance P depletion from the peripheral processes of sensory neurons proceeds considerably faster than after nerve
section (MAGGI et al. 1987e). A site of action distal to the soma has been
envisaged from the neurotoxic effect of intracutaneous administration,
which can be prevented by application of nerve growth factor to the peripheral nerve processes (TAYLOR et al. 1985). These data are not conclusive,
however, and investigation of the capsaicin receptor/cation channel distribution will shed more light on this issue.

VII. Summary

The information now available on the mechanism of capsaicin's action
permits the formulation of a hypothesis that explains the principle features
of the effects of this drug on thin sensory neurons (excitation, desensitization, neurotoxicity, and selectivity) by a common site and mechanism.
It is important to realize that, apart from these sensory-neuron-selective
effects, high concentrations of capsaicin also exert cell-nonselective effects
on a variety of excitable cells. These nonselective effects manifest themselves in a kind of membrane stabilization and might be related to the
lipophilicity of capsaicin which could influence membrane fluidity and ion
permeability. The sensory neuron-selective effects of capsaicin arise from a
specific membrane recognition site for capsaicin and related compounds that
is coupled to a cation conductance of the cell membrane. Evidence for a
capsaicin recognition site comes from binding studies with resiniferatoxin,
the existence of a structure-activity relationship for capsaicin-like activity,
the availability of a competitive capsaicin antagonist (BEVAN et al. 1991;
DRAY et al. 1991), the possibility to label the putative capsaicin recognition
site with structurally related photoaffinity probes (JAMES et al. 1988), the
remarkable cell selectivity of capsaicin's actions on thin sensory neurons,

and the ability of capsaicin to activate single cation channels in membrane patches from capsaicin-sensitive neurons.

The existence of a specific capsaicin recognition site, a capsaicin receptor, provides a satisfactory explanation for the cell selectivity of capsaicin's actions on thin sensory neurons. Occupation of the capsaicin recognition site leads to activation of a cation conductance in the cell membrane which admits sodium, calcium, and other cations to the cytoplasm. Ruthenium red appears to interfere with the coupling of the occupied capsaicin receptor with the cation conductance. Cation influx causes depolarization and a build-up of intracellular calcium and sodium, which give rise to nonspecific desensitization and neurotoxicity. As a consequence, ultrastructural changes take place, leading to the long-term neurotoxic effect. If the somata of sensory neurons do not degenerate, the neurotoxic effects may be slowly reversible, whereas degeneration of the somata results in an irreversible neurotoxic effect. Desensitization and neurotoxic changes take place within a couple of minutes and are associated with the manifestation of functional deficits developing with a similar time course. Depletion of peptides and other constituents of sensory neurons becomes appreciable much later and hence is a sequel of the neurotoxic action but not the cause of defunctionalization.

D. Capsaicin as a Pharmacological Tool

The sensory-neuron-selective effects of capsaicin have made this drug an important, if not indispensable, tool with which to investigate the neuroanatomical, neurochemical, and functional implications of these neurons. In the investigation of sensory neuron functions, both the acute excitatory and the long-term neurotoxic actions of capsaicin have been and can be made use of. The selectivity of capsaicin for sensory neurons, however, is only relative but not absolute, and there is a number of limitations to be considered when capsaicin is employed as a pharmacological tool.

1. The group of capsaicin-sensitive primary afferent neurons is not identical with any population of afferent neurons classified according to morphological, neurochemical, or functional criteria.
2. The degree of selectivity differs considerably for the acute and long-term actions of capsaicin. This is because the drug exerts both sensory-neuron-selective and cell-nonselective effects. The nonselective responses need be accounted for when acute administration of capsaicin is used to test for functional implications of thin afferent neurons. In many instances, it is possible to differentiate the cell-nonselective from the sensory-neuron-selective actions, since the nonselective effects are readily reproducible, do not show desensitization, and remain unaltered after ablation of capsaicin-sensitive neurons.

3. Another limitation is related to capsaicin's action on the sensory neurons themselves. Owing to the high activity of the drug, the consequences of sensory neuron stimulation can temporarily override or obscure the function of other systems, and false-positive evidence of an involvement of capsaicin-sensitive sensory neurons in these systems may be obtained.
4. There are further restrictions of the long-term neurotoxic effect of capsaicin. Thus, ablation of a group of primary afferent neurons also results in changes of cellular systems that are functionally connected to these neurons, and reorganization of sensory pathways and associated systems can take place.
5. Certain long-term actions of capsaicin on nonsensory neural systems can at present not be explained as being consequences of sensory neuron ablation. Examples for this category of "nonselective" neurotoxic action include those on thermosensitive neurons in the hypothalamus, some neurons in other brain areas, and some neurons of the enteric nervous system.
6. Special attention needs to be given to the age, strain, and species differences in the sensory neuron sensitivity to capsaicin, which may be related to different degrees of sensory neuron selectivity.

In summary, there is a number of factors that limit the selectivity and usefulness of capsaicin as a research tool for the investigation of sensory neurons. Many of these restrictions, however, can be outweighed by appropriate controls and by a careful consideration of the advantages and disadvantages of the different experimental uses of capsaicin. Systemic treatment of small rodents with neurotoxic doses of capsaicin has been the most frequent way of examining the anatomical, neurochemical, and functional implications of capsaicin-sensitive afferent neurons. Periaxonal and local application of capsaicin to sensory neurons are methods which have great potential in selectively ablating certain sensory projections and thereby avoid some of the problems associated with systemic administration.

Acknowledgements. I am indebted to Drs. Rainer Amann, Ulrike Holzer-Petsche, and Irmgard Th. Lippe for their critical reading of the manuscript. Thanks are also due to Mrs. Irmgard Russa and Mr. Wolfgang Schluet for their help in typing the manuscript and preparing the graph, respectively.

References

Aberdeen J, Corr L, Milner P, Lincoln J, Burnstock G (1990) Marked increases in calcitonin gene-related peptide-containing nerves in the developing rat following long-term sympathectomy with guanethidine. Neuroscience 35:175–184

Ahlstedt S, Alving K, Hesselmar B, Olaisson E (1986) Enhancement of the bronchial reactivity in immunized rats by neonatal treatment with capsaicin. Int Arch Allergy Appl Immunol 80:262–266

Ainsworth A, Hall P, Wall PD, Allt G, Mackenzie ML, Gibson S, Polak JM (1981) Effects of capsaicin applied locally to adult peripheral nerve. II. Anatomy and enzyme and peptide chemistry of peripheral nerve and spinal cord. Pain 11:379–388

Akagi H, Otsuka M, Yanagisawa M (1980) Identification by high-performance liquid chromatography of immunoreactive substance P released from isolated rat spinal cord. Neurosci Lett 20:259–263

Alber G, Scheuber PH, Reck B, Sailer-Kramer B, Hartmann A, Hammer DK (1989) Role of substance P in immediate-type skin reactions induced by staphylococcal enterotoxin B in unsensitized monkeys. J Allergy Clin Immunol 84:880–885

Amann R (1990) Desensitization of capsaicin-evoked neuropeptide release – influence of Ca^{2+} and temperature. Naunyn Schmiedebergs Arch Pharmacol 342:671–676

Amann R, Lembeck F (1986) Capsaicin sensitive afferent neurons from peripheral glucose receptors mediate the insulin-induced increase in adrenaline secretion. Naunyn Schmiedebergs Arch Pharmacol 334:71–76

Amann R, Lembeck F (1987) Stress induced ACTH release in capsaicin treated rats. Br J Pharmacol 90:727–731

Amann R, Lembeck F (1989) Ruthenium red selectively prevents capsaicin-induced nociceptor stimulation. Eur J Pharmacol 161:227–229

Amann R, Skofitsch G, Lembeck F (1988) Species-related differences in the capsaicin-sensitive innervation of the rat and guinea-pig ureter. Naunyn Schmiedebergs Arch Pharmacol 338:407–410

Amann R, Donnerer J, Lembeck F (1989a) Capsaicin-induced stimulation of polymodal nociceptors is antagonized by ruthenium red independently of extracellular calcium. Neuroscience 32:255–259

Amann R, Donnerer J, Lembeck F (1989b) Ruthenium red selectively inhibits capsaicin-induced release of calcitonin gene-related peptide from the isolated perfused guinea pig lung. Neurosci Lett 101:311–315

Amann R, Donnerer J, Lembeck F (1990a) Activation of primary afferent neurons by thermal stimulation. Influence of ruthenium red. Naunyn Schmiedebergs Arch Pharmacol 341:108–113

Amann R, Maggi CA, Giuliani S, Donnerer J, Lembeck F (1990b) Effects of carbonyl cyanide *para*-trichloromethoxyphenylhydrazone (CCCP) and of ruthenium red on capsaicin-evoked neuropeptide release from peripheral terminals of primary afferent neurones. Naunyn Schmiedebergs Arch Pharmacol 341:534–537

Andoh R, Sakurada S, Sato T, Takahashi M, Kisara K (1982) Potentiating effects of prostaglandin E_2 on bradykinin and capsaicin responses in medial thalamic nociceptive neurons. Nippon Yakungakuzasshi 32:81–89

Arvidsson J, Ygge J (1986) A quantitative study of the effects of neonatal capsaicin treatment and of subsequent peripheral nerve transection in the adult rat. Brain Res 397:130–136

Ault B, Evans H (1980) Depolarizing action of capsaicin on isolated dorsal root fibres of the rat. J Physiol (Lond) 306:22P–23P

Baccaglini PI, Hogan PG (1983) Some rat sensory neurons in culture express characteristics of differentiated pain sensory cells. Proc Nat Acad Sci USA 80:594–598

Baranowski R, Lynn B, Pini A (1986) The effects of locally applied capsaicin on conduction in cutaneous nerves of four mammalian species. Br J Pharmacol 89:267–276

Barthó L, Holzer P (1985) Search for a physiological role of substance P in gastrointestinal motility. Neuroscience 16:1–32

Barthó L, Szolcsányi J (1978) The site of action of capsaicin on the guinea-pig isolated ileum. Naunyn Schmiedebergs Arch Pharmacol 305:75–81

Barthó L, Holzer P, Lembeck F, Szolcsányi J (1982a) Evidence that the contractile response of the guinea-pig ileum to capsaicin is due to release of substance P. J Physiol (Lond) 332:157–167

Barthó L, Sebök B, Szolcsányi J (1982b) Indirect evidence for the inhibition of enteric substance P neurones by opiate agonists but not by capsaicin. Eur J Pharmacol 77:273–279

Barthó L, Pethö G, Antal A, Holzer P, Szolcsányi J (1987) Two types of relaxation due to capsaicin in the guinea-pig isolated ileum. Neurosci Lett 81:146–150

Bennett T, Gardiner SM (1984) Neonatal capsaicin treatment impairs vasopressin-mediated blood pressure recovery following acute hypotension. Br J Pharmacol 81:341–345

Bennett T, Gardiner SM (1986) Neonatal treatment with capsaicin impairs hormonal regulation of blood pressure in adult, water-deprived, Long-Evans but not Brattleboro rats. Neurosci Lett 63:131–134

Bény J-L, Brunet PC, Huggel H (1989) Effects of substance P, calcitonin gene-related peptide and capsaicin on tension and membrane potential of pig coronary artery in vitro. Regul Pept 25:25–36

Bergstrom L, Hammond DL, Go VLW, Yaksh TL (1983) Concurrent measurement of substance P and serotonin in spinal superfusates: failure of capsaicin and p-chloroamphetamine to co-release. Brain Res 270:181–184

Berkenbosch F, Schipper J, Tilders FJH (1986) Corticotropin-releasing factor immunostaining in the rat spinal cord and medulla oblongata: an unexpected form of cross-reactivity with substance P. Brain Res 399:87–96

Bernstein JE, Swift RM, Soltani K, Lorincz AL (1981) Inhibition of axon reflex vasodilatation by topically applied capsaicin. J Invest Dermatol 76:394–395

Bernstein JE, Bickers DR, Dahl MV, Roshal JY (1987) Treatment of chronic postherpetic neuralgia with topical capsaicin: a preliminary study. J Am Acad Dermatol 17:93–96

Bettaney J, Dray A, Forster P (1988) Calcium and the effects of capsaicin on afferent fibres in a neonatal rat isolated spinal cord-tail preparation. J Physiol (Lond) 406:37P

Bevan SJ, Forbes CA (1988) Membrane effects of capsaicin on rat dorsal root ganglion neurones in cell culture. J Physiol (Lond) 398:28P

Bevan S, Szolcsányi J (1990) Sensory neurone-specific actions of capsaicin – mechanisms and applications. Trends Pharmacol Sci 11:330–333

Bevan SJ, James IF, Rang HP, Winter J, Wood JN (1987) The mechanism of action of capsaicin – a sensory neurotoxin. In: Jenner P (ed) Neurotoxins and their pharmacological implications. Raven, New York, pp 261–277

Bevan S, Hothi S, Hughes GA, James IF, Rang HP, Shah K, Walpole CSJ, Yeats JC (1991) Development of a competitive antagonist for the sensory neurone excitant, capsaicin. Br J Pharmacol 102:77P

Bittner MA, LaHann TR (1985) Biphasic time-course of capsaicin-induced substance P depletion: failure to correlate with thermal analgesia in the rat. Brain Res 322:305–309

Bleakman D, Brorson JR, Miller RJ (1990) The effect of capsaicin on voltage-gated calcium currents and calcium signals in cultured dorsal root ganglion cells. Br J Pharmacol 101:423–431

Bond SM, Cervero F, McQueen DS (1982) Influence of neonatally administered capsaicin on baroreceptor and chemoreceptor reflexes in the adult rat. Br J Pharmacol 77:517–524

Braga PC, Biella G, Tiengo M (1987) Effects of capsaicin peripherally applied on spontaneous and evoked activity of thalamic nociceptive neurons in rat. Med Sci Res 15:1511–1512

Buck SH, Burks TF (1986) The neuropharmacology of capsaicin – review of some recent observations. Pharmacol Rev 38:179–226

Buck SH, Miller MS, Burks TF (1982) Depletion of primary afferent substance P by capsaicin and dihydrocapsaicin without altered thermal sensitivity in rats. Brain Res 233:216–220

Buck SH, Walsh JH, Davis TP, Brown MR, Yamamura HI, Burks TF (1983) Characterization of the peptide and sensory neurotoxic effects of capsaicin in the guinea pig. J Neurosci 3:2064–2074

Buckley TL, Brain SD, Williams TJ (1990) Ruthenium red selectively inhibits oedema formation and increased blood flow induced by capsaicin in rabbit skin. Br J Pharmacol 99:7–8

Bucsics A, Lembeck F (1981) In vitro release of substance P from spinal cord slices by capsaicin congeners. Eur J Pharmacol 71:71–77

Bucsics A, Mayer N, Pabst MA, Lembeck F (1984) Distribution of spinal cord nerve endings containing various neurotransmitters on a continuous density gradient. J Neurochem 42:692–697

Bucsics A, Sutter D, Jancsó G, Lembeck F (1988) Quantitative assay of capsaicin-sensitive thiamine monophosphatase and β-glycerophosphatase activity in rodent spinal cord. J Neurosci Methods 24:155–162

Bynke G (1983) Capsaicin pretreatment prevents disruption of the blood-aqueous barrier in the rabbit eye. Invest Ophthalmol Vis Sci 24:744–748

Camras CB, Bito LZ (1980) The pathophysiological effects of nitrogen mustard on the rabbit eye. II. The inhibition of the initial hypertensive phase by capsaicin and the apparent role of substance P. Invest Ophthalmol Vis Sci 19:423–428

Carpenter SE, Lynn B (1981) Vascular and sensory responses of human skin to mild injury after topical treatment with capsaicin. Br J Pharmacol 73:755–758

Cervero F, McRitchie HA (1981) Neonatal capsaicin and thermal nociception: a paradox. Brain Res 215:414–418

Cervero F, McRitchie HA (1982) Neonatal capsaicin does not affect unmyelinated efferent fibers of the autonomic nervous system: functional evidence. Brain Res 239:283–288

Cervero F, Plenderleith MB (1987) Spinal cord sensory systems after neonatal capsaicin. Acta Physiol Hung 69:393–401

Cervero F, Sharkey KA (1988) An electrophysiological and anatomical study of intestinal afferent fibres in the rat. J Physiol (Lond) 401:381–397

Chahl LA (1982) Evidence that the contractile response of the guinea-pig ileum to capsaicin is due to substance P release. Naunyn Schmiedebergs Arch Pharmacol 319:212–215

Chahl LA (1988) Antidromic vasodilatation and neurogenic inflammation. Pharmacol Ther 37:275–300

Chahl LA (1989) The effects of ruthenium red on the response of guinea-pig ileum to capsaicin. Eur J Pharmacol 169:241–247

Chéry-Croze S, Godinot F, Jourdan G, Bernard C, Chayvialle JA (1985) Capsaicin in adult frogs: effects on nociceptive responses to cutaneous stimuli and on nervous tissue concentrations of immunoreactive substance P, somatostatin and cholecystokinin. Naunyn Schmiedebergs Arch Pharmacol 331:159–165

Chéry-Croze S, Bosshard A, Martin H, Cuber JC, Charnay Y, Chayvialle JA (1989) Peptide immunocytochemistry in afferent neurons from lower gut in rats. Peptides 9:873–881

Chiba T, Masuko S, Kawano H (1986) Correlation of mitochondrial swelling after capsaicin treatment and substance P and somatostatin immunoreactivity in small neurons of dorsal root ganglion in the rat. Neurosci Lett 64:311–316

Chung JM, Lee KH, Hori Y, Willis WD (1985) Effects of capsaicin applied to a peripheral nerve on the responses of primate spinothalamic tract cells. Brain Res 329:27–38

Chung K, Schwen RJ, Coggeshall RE (1985) Ureteral axon damage following subcutaneous administration of capsaicin in adult rats. Neurosci Lett 53:221–226

Chung K, Klein CM, Coggeshall RE (1990) The receptive part of the primary afferent axon is most vulnerable to systemic capsaicin in adult rats. Brain Res 511:222–226

Coderre TJ, Abbott FV, Melzack R (1984) Behavioral evidence in rats for a peptidergic-noradrenergic interaction in cutaneous sensory and vascular function. Neurosci Lett 47:113–118

Coleridge HM, Coleridge JCG (1977) Impulse activity in afferent vagal C-fibers with endings in the intrapulmonary airways of dogs. Respir Physiol 29:125–142

Coleridge JCG, Coleridge HM (1984) Afferent vagal C fibre innervation of the lung and airways and its functional significance. Rev Physiol Biochem Pharmacol 99:1–110

Costa M, Furness JB, Gibbins IL (1986) Chemical coding of enteric neurons. Prog Brain Res 68:217–239

Crayton SC, Mitchell JH, Payne FC (1981) Reflex cardiovascular response during injection of capsaicin into skeletal muscle. Am J Physiol 240:H315–H319

Culp WJ, Ochoa J, Cline M, Dotson R (1989) Mechanical hyperalgesia induced by capsaicin – cross modality threshold modulation in human C-nociceptors. Brain 112:1317–1331

Daniel EE, Nagahma M, Sato H, Jury J, Bowker P (1987) Immunochemical studies on substance P release from muscularis mucosa of opossum esophagus. In: Henry JL, Couture R, Cuello AC, Pelletier G, Quirion R, Regoli D (eds) Substance P and Neurokinins. Springer, Berlin Heidelberg New York, pp 266–269

Dawbarn D, Harmar AJ, Pycock CJ (1981) Intranigral injection of capsaicin enhances motor activity and depletes nigral 5-hydroxytryptamine but not substance P. Neuropharmacology 20:341–346

Decker MW, Towle AC, Bissette G, Mueller RA, Lauder JM, Nemeroff CB (1985) Bombesin-like immunoreactivity in the central nervous system of capsaicin-treated rats: a radioimmunoassay and immunohistochemical study. Brain Res 342:1–8

Della NG, Papka RE, Furness JB, Costa M (1983) Vasoactive intestinal peptide-like immunoreactivity in nerves associated with the cardiovascular system of guinea pigs. Neuroscience 9:605–619

De Vries DJ, Blumberg PM (1989) Thermoregulatory effects of resiniferatoxin in the mouse: comparison with capsaicin. Life Sci 44:711–715

Dib B (1983) Dissociation between peripheral and central heat loss mechanisms induced by neonatal capsaicin. Behav Neurosci 97:822–829

Diez Guerra FJ, Zaidi M, Bevis P, MacIntyre I, Emson PC (1988) Evidence for release of calcitonin gene-related peptide and neurokinin A from sensory nerve endings in vivo. Neuroscience 25:839–846

Dinh TT, Ritter S (1987) Capsaicin induces neuronal degeneration in midbrain and forebrain structures in neonatally treated rats. Soc Neurosci Abstr 13:1684

Docherty RJ, Robertson B, Bevan S (1991) Capsaicin causes prolonged inhibition of voltage-activated calcium currents in adult rat dorsal root ganglion neurons in culture. Neuroscience 40:513–521

Donnerer J, Amann R (1990) Capsaicin-evoked neuropeptide release is not dependent on membrane potential changes. Neurosci Lett 117:331–334

Donnerer J, Lembeck F (1982) Analysis of the effects of intravenously injected capsaicin in the rat. Naunyn Schmiedebergs Arch Pharmacol 320:54–57

Donnerer J, Lembeck F (1983) Heat loss reaction to capsaicin through a peripheral site of action. Br J Pharmacol 79:710–723

Donnerer J, Lembeck F (1990) Different control of the adrenocorticotropin-corticosterone response and of prolactin secretion during cold stress, anesthesia, surgery, and nicotine injection in the rat: involvement of capsaicin-sensitive sensory neurons. Endocrinology 126:921–926

Donnerer J, Barthó L, Holzer P, Lembeck F (1984) Intestinal peristalsis associated with release of immunoreactive substance P. Neuroscience 11:913–918

Donnerer J, Schuligoi R, Lembeck F (1989) Influence of capsaicin-induced denervation on neurogenic and humoral control of arterial pressure. Naunyn Schmiedebergs Arch Pharmacol 340:740–743

Doucette R, Theriault E, Diamond J (1987) Regionally selective elimination of cutaneous thermal nociception in rats by neonatal capsaicin. J Comp Neurol 261:583–591

Dray A, Bettaney J, Forster P (1989a) Capsaicin desensitization of peripheral nociceptive fibers does not impair sensitivity to other noxious stimuli. Neurosci Lett 99:50–54

Dray A, Hankins MW, Yeats JC (1989b) Desensitization and capsaicin-induced release of substance P-like immunoreactivity from guinea-pig ureter in vitro. Neuroscience 31:479–483

Dray A, Bettaney J, Forster P (1990a) Resiniferatoxin, a potent capsaicin-like stimulator of peripheral nociceptors in the neonatal rat tail in vitro. Br J Pharmacol 99:323–326

Dray A, Bettaney J, Rueff A, Walpole C, Wrigglesworth R (1990b) NE-19550 and NE-21610, antinociceptive capsaicin analogues: studies on nociceptive fibres of the neonatal rat tail in vitro. Eur J Pharmacol 181:289–293

Dray A, Forbes CA, Burgess GM (1990c) Ruthenium red blocks the capsaicin-induced increase in intracellular calcium and activation of membrane currents in sensory neurones as well as the activation of peripheral nociceptors in vitro. Neurosci Lett 110:52–59

Dray A, Campbell EA, Hughes GA, Patel IA, Perkins MN, Rang HP, Rueff A, Seno N, Urban L, Walpole CSJ (1991) Antagonism of capsaicin-induced activation of C-fibres by a selective capsaicin antagonist, capsazepine. Br J Pharmacol 102:78P

Dubois JM (1982) Capsaicin blocks one class of K^+ channels in the frog node of Ranvier. Brain Res 245:372–375

Duckles SP (1986) Effects of capsaicin on vascular smooth muscle. Naunyn Schmiedebergs Arch Pharmacol 333:59–64

Dun NJ, Kiraly M (1983) Capsaicin causes release of a substance P-like peptide in guinea-pig inferior mesenteric ganglia. J Physiol (Lond) 340:107–120

Edvinsson L, Jansen I, Kingman TA, McCulloch J (1990) Cerebrovascular responses to capsaicin in vitro and in situ. Br J Pharmacol 100:312–318

Eimerl D, Papir-Kricheli D (1987) Epidural capsaicin produces prolonged segmental analgesia in the rat. Exp Neurol 97:169–178

Erdélyi L, Such G, Jancsó G (1987) Intracellular and voltage clamp studies of capsaicin induced effects on a sensory neuron model. Acta Physiol Hung 69:481–492

Faulkner DC, Growcott JW (1980) Effects of neonatal capsaicin administration on the nociceptive response of the rat to mechanical and chemical stimuli. J Pharm Pharmacol 32:656–657

Fehér E, Vajda J (1982) Effect of capsaicin on the nerve elements of the small intestine. Acta Morphol Acad Sci Hung 30:57–63

Fitzgerald M (1982) Alterations in the ipsi- and contralateral afferent inputs of dorsal horn cells produced by capsaicin treatment of one sciatic nerve in the rat. Brain Res 248:92–107

Fitzgerald M (1983) Capsaicin and sensory neurones – a review. Pain 15:109–130

Fitzgerald M, Woolf CJ (1982) The time course and specificity of the changes in the behavioural and dorsal horn cell responses to noxious stimuli following peripheral nerve capsaicin treatment in the rat. Neuroscience 9:2051–2056

Flynn DL, Rafferty MF, Boctor AM (1986) Inhibition of human neutrophil 5-lipoxygenase activity by gingerdione, shogaol, capsaicin and related pungent compounds. Prostaglandins Leukotrienes Med 24:195–198

Forbes CA, Bevan SJ (1988) Properties of single capsaicin-activated channels. Soc Neurosci Abstr 14:642

Foster RW, Ramage AG (1981) The action of some chemical irritants on somatosensory receptors of the cat. Neuropharmacology 20:191–198

Foster RW, Weston AH, Weston KM (1981) Some effects of chemical irritants on the membrane of the giant amoeba. Br J Pharmacol 74:333–339

Franco-Cereceda A, Lundberg JM (1988) Actions of calcitonin gene-related peptide and tachykinins in relation to the contractile effects of capsaicin in the guinea-pig and rat heart in vitro. Naunyn Schmiedebergs Arch Pharmacol 337:649–655

Franco-Cereceda A, Lundberg JM (1989) Post-occlusive reactive hyperaemia in the heart, skeletal muscle and skin of control and capsaicin-pre-treated pigs. Acta Physiol Scand 137:271–277

Franco-Cereceda A, Henke H, Lundberg JM, Petermann JB, Hökfelt T, Fischer JA (1987) Calcitonin gene-related peptide (CGRP) in capsaicin-sensitive substance P-immunoreactive sensory neurons in animals and man: distribution and release by capsaicin. Peptides 8:399–410

Franco-Cereceda A, Lou Y-P, Lundberg JM (1989) Ruthenium red differentiates between capsaicin and nicotine effects on cardiac sensory nerves. Acta Physiol Scand 137:457–458

Fujita S, Shimizu T, Izumi K, Fukuda T, Sameshima M, Ohba N (1984) Capsaicin-induced neuroparalytic keratitis-like corneal changes in the mouse. Exp Eye Res 38:165–175

Furness JB, Papka RE, Della NG, Costa M, Eskay RL (1982) Substance P-like immunoreactivity in nerves associated with the vascular system of guinea-pigs. Neuroscience 7:447–459

Gamse R (1982) Capsaicin and nociception in the rat and mouse. Possible role of substance P. Naunyn Schmiedebergs Arch Pharmacol 320:205–216

Gamse R, Holzer P, Lembeck F (1979a) Indirect evidence for presynaptic location of opiate receptors on chemosensitive primary sensory neurones. Naunyn Schmiedebergs Arch Pharmacol 308:281–285

Gamse R, Molnar A, Lembeck F (1979b) Substance P release from spinal cord slices by capsaicin. Life Sci 25:629–636

Gamse R, Holzer P, Lembeck F (1980) Decrease of substance P in primary sensory neurones and impairment of neurogenic plasma extravasation by capsaicin. Br J Pharmacol 68:207–213

Gamse R, Lackner D, Gamse G, Leeman SE (1981a) Effect of capsaicin pretreatment on capsaicin-evoked release of immunoreactive somatostatin and substance P from primary sensory neurons. Naunyn Schmiedebergs Arch Pharmacol 316:38–41

Gamse R, Leeman SE, Holzer P, Lembeck F (1981b) Differential effects of capsaicin on the content of somatostatin, substance P, and neurotensin in the nervous system of the rat. Naunyn Schmiedebergs Arch Pharmacol 317:140–148

Games R, Wax A, Zigmond RE, Leeman SE (1981c) Immunoreactive substance P in sympathetic ganglia: distribution and sensitivity towards capsaicin. Neuroscience 6:437–441

Gamse R, Petsche U, Lembeck F, Jancsó G (1982) Capsaicin applied to peripheral nerve inhibits axoplasmic transport of substance P and somatostatin. Brain Res 239:447–462

Gamse R, Jancsó G, Király E (1984) Intracisternal capsaicin: a novel approach for studying nociceptive sensory neurons. In: Chahl LA, Szolcsányi J, Lembeck F (eds) Antidromic vasodilatation and neurogenic inflammation. Akadémiai Kiadó, Budapest, pp 93–106

Gamse R, Saria A, Lundberg JM, Theodorsson-Norheim E (1986) Behavioral and neurochemical changes after intracisternal capsaicin treatment of the guinea pig. Neurosci Lett 64:287–292

Gardiner SM, Bennett T, O'Neill TP (1989) Synthetic capsaicin reversibly impairs vasopressin-mediated blood pressure recovery. Am J Physiol 257:R1429–R1435

Geisthövel E, Ludwig O, Simon E (1986) Capsaicin fails to produce disturbances of autonomic heat and cold defence in an avian species (*Anas platyrhynchos*). Pflugers Arch 406:343–350

Geppetti P, Frilli S, Renzi D, Santicioli P, Maggi CA, Theodorsson E, Fanciullacci M (1988a) Distribution of calcitonin gene-related peptide-like immunoreactivity in various rat tissues: correlation with substance P and other tachykinins and sensitivity to capsaicin. Regul Pept 23:289–298

Geppetti P, Fusco B, Marabini S, Maggi CA, Fanciullacci M, Sicuteri F (1988b) Secretion, pain and sneezing induced by the application of capsaicin to the nasal mucosa in man. Br J Pharmacol 93:509–514

Geppetti P, Maggi CA, Perretti F, Frilli S, Manzini S (1988c) Simultaneous release by bradykinin of substance P- and calcitonin gene-related peptide immunoreactivities from capsaicin-sensitive structures in guinea-pig heart. Br J Pharmacol 94:288–290

Gibbins IL, Furness JB, Costa M, MacIntyre I, Hillyard CJ, Girgis S (1985) Co-localization of calcitonin gene-related peptide-like immunoreactivity with substance P in cutaneous, vascular and visceral sensory neurons of guinea pigs. Neurosci Lett 57:125–130

Gibbins IL, Furness JB, Costa M (1987) Pathway-specific patterns of the co-existence of substance P, calcitonin gene-related peptide, cholecystokinin and dynorphin in neurons of the dorsal root ganglia of the guinea-pig. Cell Tissue Res 248:417–437

Gibson SJ, McGregor G, Bloom SR, Polak JM, Wall PD (1982) Local application of capsaicin to one sciatic nerve of the adult rat induces a marked depletion in the peptide content of the lumbar dorsal horn. Neuroscience 7:3153–3162

Glinsukon T, Stitmunnaithum V, Toskulkao C, Baranawuti T, Tangkrisanavinont V (1980) Acute toxicity of capsaicin in several animal species. Toxicon 18:215–220

Godfraind JM, Jessell TM, Kelly JS, McBurney RN, Mudge AW, Yamamoto M (1981) Capsaicin prolongs action potential duration in cultured sensory neurones. J Physiol (Lond) 312:32P-33P

Green BG (1986) Sensory interactions between capsaicin and temperature in the oral cavity. Chem Senses 11:371–382

Green BG (1989) Capsaicin sensitization and desensitization on the tongue produced by brief exposures to a low concentration. Neurosci Lett 107:173–178

Green T, Dockray GJ (1988) Characterization of the peptidergic afferent innervation of the stomach in the rat, mouse and guinea-pig. Neuroscience 25:181–193

Hajós M, Obál F, Jancsó G, Obál F (1983) The capsaicin sensitivity of the preoptic region is preserved in adult rats pretreated as neonates, but lost in rats pretreated as adults. Naunyn Schmiedebergs Arch Pharmacol 324:219–222

Hajós M, Engberg G, Elam M (1986a) Reduced responsiveness of locus coeruleus neurons to cutaneous thermal stimuli in capsaicin-treated rats. Neurosci Lett 70:382–387

Hajós M, Svensson K, Nissbrandt H, Obál F, Carlsson A (1986b) Effects of capsaicin on central monoaminergic mechanisms in the rat. J Neural Transm 66:221–242

Håkanson R, Beding B, Ekman R, Heilig M, Wahlestedt C, Sundler F (1987) Multiple tachykinin pools in sensory nerve fibres in the rabbit iris. Neuroscience 21:943–950

Hammond DL, Ruda MA (1989) Developmental alterations in thermal nociceptive threshold and the distribution of immunoreactive calcitonin gene-related peptide and substance P after neonatal administration of capsaicin in the rat. Neurosci Lett 97:57–62

Hamon M, Gallissot MC, Menard F, Gozlan H, Bourgoin S, Verge D (1989) 5-HT$_3$ receptor binding sites are on capsaicin-sensitive fibres in the rat spinal cord. Eur J Pharmacol 164:315–322

Handwerker HO, Holzer-Petsche U, Heym C, Welk E (1984) C-fibre functions after topical application of capsaicin to a peripheral nerve and after neonatal capsaicin treatment. In: Chahl LA, Szolcsányi J, Lembeck F (eds) Antidromic vasodilatation and neurogenic inflammation. Akadémiai Kiadó, Budapest, pp 57–78

Harper AA, Lawson SN (1985) Conduction velocity is related to morphological cell type in rat dorsal root ganglion neurones. J Physiol (Lond) 359:31–46

Harti G (1988) Wirkung von Capsaicin auf afferente peptiderge Nervenfasern: Immunohistochemische Untersuchungen an Ratten, Meerschweinchen und Tauben. Thesis, University of Gießen

Harti G, Sharkey KA, Rössler W, Pierau F-K (1987) Distribution of substance P (SP) and calcitonin gene-related peptide (CGRP) and the effect of capsaicin in sensory neurones of rat and pigeon. Neuroscience [Suppl] 22:S322

Harti G, Sharkey KA, Pierau F-K (1989) Effects of capsaicin in rat and pigeon on peripheral nerves containing substance P and calcitonin gene-related peptide. Cell Tissue Res 256:465–474

Hayes AG, Tyers MB (1980) Effects of capsaicin on nociceptive heat, pressure and chemical thresholds and on substance P levels in the rat. Brain Res 189:561–564

Hayes AG, Scadding JW, Skingle M, Tyers MB (1981a) Effects of neonatal administration of capsaicin on nociceptive thresholds in the mouse and rat. J Pharm Pharmacol 33:183-185

Hayes AG, Skingle M, Tyers MB (1981b) Effect of single doses of capsaicin on nociceptive thresholds in the rodent. Neuropharmacology 20:505–511

Hayes AG, Hawcock AB, Hill RG (1984a) The depolarising action of capsaicin on rat isolated sciatic nerve. Life Sci 35:1561–1568

Hayes AG, Oxford A, Reynolds M, Shingler AH, Skingle M, Smith C, Tyers MB (1984b) The effects of a series of capsaicin analogues on nociception and body temperature in the rat. Life Sci 34:1241–1248

He X, Schmidt RF, Schmittner H (1988) Effects of capsaicin on articular afferents of the cat's knee joint. Agents Actions 25:222–224

Helke CJ, DiMiccio JA, Jacobowitz DM, Kopin IJ (1981a) Effect of capsaicin administration to neonatal rats on the substance P content of discrete CNS regions. Brain Res 222:428–431

Helke CJ, Jacobowitz DM, Thoa NB (1981b) Capsaicin and potassium evoked substance P release from the nucleus tractus solitarius and spinal trigeminal nucleus in vitro. Life Sci 29:1779–1785

Helme RD, Eglezos A, Dandie GW, Andrews PV, Boyd RL (1987) The effect of substance P on the regional lymph node antibody response to antigenic stimulation in capsaicin-pretreated rats. J Immunol 139:3470–3473

Heyman I, Rang HP (1985) Depolarizing responses to capsaicin in a subpopulation of rat dorsal root ganglion cells. Neurosci Lett 56:69–75

Hiura A, Ishizuka H (1989) Changes in features of degenerating primary sensory neurons with time after capsaicin treatment. Acta Neuropathol (Berl) 78:35–46

Hiura A, Sakamoto Y (1987a) Effect of capsaicin on neurites of cultured dorsal root ganglia and isolated neurons of chick embryos. Neurosci Lett 73:237–241

Hiura A, Sakamoto Y (1987b) Quantitative estimation of the effect of capsaicin on the mouse primary sensory neurons. Neurosci Lett 76:101–106

Hoheisel U, Mense S (1986) Non-myelinated afferent fibres do not originate exclusively from the smallest dorsal root ganglion cells in the cat. Neurosci Lett 72:153–157

Holje H, Hildebrand C, Fried K (1983) Proportion of unmyelinated axons in the rat inferior alveolar nerve and mandibular molar pulps after neonatal administration of capsaicin. Brain Res 266:133–136

Holz GG, Dunlap K, Kream RM (1988) Characterization of the electrically evoked release of substance P from dorsal root ganglion neurons: methods and dihydropyridine sensitivity. J Neurosci 8:463–471

Holzer P (1984) Characterization of the stimulus-induced release of immunoreactive substance P from the myenteric plexus of the guinea-pig small intestine. Brain Res 297:127–136

Holzer P (1988) Local effector functions of capsaicin-sensitive sensory nerve endings: involvement of tachykinins, calcitonin gene-related peptide and other neuropeptides. Neuroscience 24:739–768

Holzer P, Sametz W (1986) Gastric mucosal protection against ulcerogenic factors in the rat mediated by capsaicin-sensitive afferent neurons. Gastroenterology 91:975–981

Holzer P, Jurna I, Gamse R, Lembeck F (1979) Nociceptive threshold after neonatal capsaicin treatment. Eur J Pharmacol 58:511–514

Holzer P, Gamse R, Lembeck F (1980) Distribution of substance P in the rat gastrointestinal tract – lack of effect of capsaicin pretreatment. Eur J Pharmacol 61:303–307

Holzer P, Saria A, Skofitsch G, Lembeck F (1981) Increase in tissue concentrations of histamine and 5-hydroxytryptamine following capsaicin treatment of newborn rats. Life Sci 29:1099–1105

Holzer P, Bucsics A, Lembeck F (1982) Distribution of capsaicin-sensitive nerve fibres containing immunoreactive substance P in cutaneous and visceral tissues of the rat. Neurosci Lett 31:253–257

Holzer P, Lippe IT, Holzer-Petsche U (1986) Inhibition of gastrointestinal transit due to surgical trauma or peritoneal irritation is reduced in capsaicin-treated rats. Gastroenterology 91:360–363

Holzer P, Pabst MA, Lippe IT, Peskar BM, Peskar BA, Livingston EH, Guth PH (1990a) Afferent nerve-mediated protection against deep mucosal damage in the rat stomach. Gastroenterology 98:838–848

Holzer P, Peskar BM, Peskar BA, Amann R (1990b) Release of calcitonin gene-related peptide induced by capsaicin in the vascularly perfused rat stomach. Neurosci Lett 108:195–200

Holzer-Petsche U, Lembeck F (1984) Systemic capsaicin treatment impairs the micturition reflex in the rat. Br J Pharmacol 83:935–941

Holzer-Petsche U, Rinner I, Lembeck F (1986) Distribution of choline acetyltransferase activity in rat spinal cord – influence of primary afferents? J Neural Transm 66:85–92

Hori T (1981) Thermosensitivity of preoptic and anterior hypothalamic neurons in the capsaicin-desensitized rat. Pflugers Arch 389:297–299

Hori T (1984) Capsaicin and central control of thermoregulation. Pharmacol Ther 26:389–416

Hori T, Tsuzuki S (1981) Thermoregulation in adult rats which have been treated with capsaicin as neonates. Pflugers Arch 390:219–223

Horváth KJ, Jancsó G, Wollemann M (1979) The effect of calcium on the capsaicin activation of adenylate cyclase in rat brain. Brain Res 179:401–403

Hoyes AD, Barber P (1981) Degeneration of axons in the ureteric and duodenal nerve plexuses of the adult rat following in vivo treatment with capsaicin. Neurosci Lett 25:19–24

Hoyes AD, Barber P, Jagessar H (1981) Effect of capsaicin on the intraperitoneal axons of the rat trachea. Neurosci Lett 26:329–334

Hua X-Y, Lundberg JM (1986) Dual capsaicin effects on ureteric motility: low dose inhibition mediated by calcitonin gene-related peptide and high dose stimulation by tachykinins? Acta Physiol Scand 128:453–465

Hua X-Y, Theodorsson-Norheim E, Brodin E, Lundberg JM, Hökfelt T (1985) Multiple tachykinins (neurokinin A, neuropeptide K and substance P) in capsaicin-sensitive sensory neurons in the guinea-pig. Regul Pept 13:1–19

Hua X-Y, Saria A, Gamse R, Theodorsson-Norheim E, Brodin E, Lundberg JM (1986) Capsaicin induced release of multiple tachykinins (substance P, neurokinin A and eledoisin-like material) from guinea-pig spinal cord and ureter. Neuroscience 19:313–319

Inomata K, Nasu F (1984) Effects of neonatal capsaicin treatment on thiamine monophosphatase (TMPase) activity in the substantia gelatinosa of the spinal cord. Int J Dev Neurosci 2:307–311

James IF, Walpole CSJ, Hixon J, Wood JN, Wrigglesworth R (1988) Long-lasting agonist activity produced by a capsaicin-like photoaffinity probe. Mol Pharmacol 33:643–649

Jancsó G (1978) Selective degeneration of chemosensitive primary sensory neurones induced by capsaicin: glial changes. Cell Tissue Res 195:145–152

Jancsó G (1981) Intracisternal capsaicin: selective degeneration of chemosensitive primary sensory afferents in the adult rat. Neurosci Lett 27:41–45

Jancsó G (1984) Sensory nerves as modulators of inflammatory reactions. In: Chahl LA, Szolcsányi J, Lembeck F (eds) Antidromic vasodilatation and neurogenic inflammation. Akadémiai Kiadó, Budapest, pp 207–222

Jancsó G, Jancsó-Gábor A (1980) Effect of capsaicin on morphine analgesia – possible involvement of hypothalamic structures. Naunyn Schmiedebergs Arch Pharmacol 311:285–288

Jancsó G, Király E (1980) Distribution of chemosensitive primary sensory afferents in the central nervous system of the rat. J Comp Neurol 190:781–792

Jancsó G, Király E (1981) Sensory neurotoxins: chemically induced selective destruction of primary sensory neurons. Brain Res 210:83–89

Jancsó G, Király E (1984) Regeneration of peptidergic sensory nerves in the rat skin. Acta Physiol Acad Sci Hung 63:239–240

Jancsó G, Knyihár E (1975) Functional linkage between nociception and fluoride-resistant acid phosphatase activity in the Rolando substance. Neurobiology 5:42–43

Jancsó G, Lawson SN (1990) Transganglionic degeneration of capsaicin-sensitive C-fiber primary afferent teminals. Neuroscience 39:501–511

Jancsó G, Maggi CA (1987) Distribution of capsaicin-sensitive urinary bladder afferents in the rat spinal cord. Brain Res 418:371–376

Jancsó G, Such G (1983) Effects of capsaicin applied perineurally to the vagus nerve on cardiovascular and respiratory functions in the cat. J Physiol (Lond) 341: 359–370

Jancsó G, Such G (1985) Evidence for a capsaicin-sensitive vasomotor mechanism in the ventral medullary chemosensitive area of the cat. Naunyn Schmiedebergs Arch Pharmacol 329:56–62

Jancsó G, Király E, Jancsó-Gábor A (1977) Pharmacologically induced selective degeneration of chemosensitive primary sensory neurones. Nature 270:741–743

Jancsó G, Sávay G, Király E (1978) Appearance of biochemically detectable ionic calcium in degenerating primary sensory neurons. Acta Histochem (Jena) 62: 165–169

Jancsó G, Király E, Jancsó-Gábor A (1980a) Chemosensitive pain fibres and inflammation. Int J Tissue React 2:57–66

Jancsó G, Király E, Jancsó-Gábor A (1980b) Direct evidence for an axonal site of action of capsaicin. Naunyn Schmiedebergs Arch Pharmacol 313:91–94

Jancsó G, Hökfelt T, Lundberg JM, Király E, Halász N, Nilsson G, Terenius L, Rehfeld L, Rehfeld J, Steinbusch H, Verhofstad AER, Said S, Brown M (1981) Immunohistochemcial studies on the effect of capsaicin on spinal and medullary peptide and monoamine neurons using antisera to substance P, gastrin/CCK, somatostatin, VIP, enkephalin, neurotensin and 5-hydroxytryptamine. J Neurocytol 10:963–980

Jancsó G, Karcsú S, Király E, Szebeni A, Tóth L, Bácsy E, Joó F, Párducz A (1984) Neurotoxin induced nerve cell degeneration: possible involvement of calcium. Brain Res 295:211–216

Jancsó G, Ferencsik M, Such G, Király E, Nagy A, Bujdosó M (1985a) Morphological effects of capsaicin and its analogues in newborn and adult mammals. In: Håkanson R, Sundler F (eds) Tachykinin antagonists. Elsevier, Amsterdam, pp 35–44

Jancsó G, Király E, Joó F, Such G, Nagy A (1985b) Selective degeneration by capsaicin of a subpopulation of primary sensory neurons in the adult rat. Neurosci Lett 59:209–214

Jancsó G, Király E, Such G, Joó F, Nagy A (1987) Neurotoxic effect of capsaicin in mammals. Acta Physiol Hung 69:295–313

Jancsó N (1968) Desensitization with capsaicin and related acylamides as a tool for studying the function of pain receptors. In: Lim RKS (ed) Pharmacology of pain. Pergamon, Oxford, pp 33–55

Jancsó N, Jancsó-Gábor A, Szolcsányi J (1968) The role of sensory nerve endings in neurogenic inflammation induced in human skin and in the eye and paw of the rat. Br J Pharmacol 33:32–41

Jancsó-Gábor A, Szolcsányi J, Jancsó N (1970) Irreversible impairment of thermoregulation induced by capsaicin and similar pungent substances. J Physiol (Lond) 206:495–507

Jessell TM, Iversen LL, Cuello AC (1978) Capsaicin-induced depletion of substance P from primary sensory neurones. Brain Res 152:183–188

Jhamandas K, Yaksh TL, Harty G, Szolcsányi J, Go VLW (1984) Action of intrathecal capsaicin and its structural analogues on the content and release of spinal substance P: selective action and relationship to analgesia. Brain Res 306: 215–225

Jia M, Nelson PG (1986) Calcium currents and transmitter output in cultured spinal cord and dorsal root ganglion neurons. J Neurophysiol 56:1257–1267

Jin J-G, Takaki M, Nakayama S (1989) Ruthenium red prevents capsaicin-induced neurotoxic action on sensory fibers of the guinea pig ileum. Neurosci Lett 106:152–156

Joó F, Szolcsányi J, Jancsó-Gábor A (1969) Mitochondrial alterations in the spinal ganglion cells of the rat accompanying the long-lasting sensory disturbance induced by capsaicin. Life Sci 8:621–626

Ju G, Hökfelt T, Fischer JA, Frey P, Rehfeld JF, Dockray GJ (1986) Does cholecystokinin-like immunoreactivity in rat primary sensory neurons represent calcitonin gene-related peptide? Neurosci Lett 68:305–310

Juan H, Lembeck F, Seewann S, Hack U (1980) Nociceptor stimulation and PGE release by capsaicin. Naunyn Schmiedebergs Arch Pharmacol 312:139–143

Kai-Kai MA, Anderton BH, Keen P (1986) A quantitative analysis of the interrelationships between subpopulations of rat sensory neurons containing arginine vasopressin or oxytocin and those containing substance P, fluoride-resistant acid phosphatase or neurofilament protein. Neuroscience 18:475–486

Kamakura K, Ishiura S, Sugita H, Toyokura Y (1983) Identification of calcium activated neutral protease in the peripheral nerve and its effect on neurofilament degeneration. J Neurochem 40:908–913

Kaufman MP, Iwamoto GA, Longhurst JC, Mitchell JH (1982) Effects of capsaicin and bradykinin on afferent fibers with endings in skeletal muscle. Circ Res 50:133–139

Keen P, Tullo AB, Blyth WA, Hill TJ (1982) Substance P in the mouse cornea: effects of chemical and surgical denervation. Neurosci Lett 29:231–235

Kenins P (1982) Responses of single nerve fibres to capsaicin applied to the skin. Neurosci Lett 29:83–88

Király E, Jancsó G, Hajós M (1991) Possible morphological correlates of capsaicin desensitization. Brain Res 540:279–282

Kirchgessner AL, Dodd J, Gershon MD (1988) Markers shared between dorsal root and enteric ganglia. J Comp Neurol 276:607–621

Kjartansson J, Dalsgaard C-J, Jonsson CE (1987) Decreased survival of experimental critical flaps in rats after sensory denervation with capsaicin. Plast Reconstr Surg 79:218–221

Konietzny F, Hensel H (1983) The effect of capsaicin on the response characteristics of human C-polymodal nociceptors. J Therm Biol 8:213–215

Ladenheim EE, Speth RC, Ritter RC (1986) Reduction of CCK-8 binding in the nucleus of the solitary tract of capsaicin pretreated rats. Soc Neurosci Abstr 12:828

Laduron PM (1984) Axonal transport of opiate receptors in capsaicin-sensitive neurones. Brain Res 294:157–160

LaMotte RH, Simone DA, Baumann TK, Shain CN, Alreja M (1988) Hypothesis for novel classes of chemoreceptors mediating chemogenic pain and itch. In: Dubner R, Gebhart GF, Bond MR (eds) Pain research and clinical management. Elsevier, Amsterdam, pp 529–535

Lang E, Novak A, Reeh PW, Handwerker HO (1990) Chemosensitivity of fine afferents from rat skin in vitro. J Neurophysiol 63:887–901

Lawson SN, Harper AA (1984) Neonatal capsaicin is not a specific neurotoxin for sensory C-fibres or small dark cells of rat dorsal root ganglia. In: Chahl LA, Szolcsányi J, Lembeck F (eds) Antidromic vasodilatation and neurogenic inflammation. Akadémiai Kiadó, Budapest, pp 111–116

Lawson SN, Nickels SM (1980) The use of morphometric techniques to analyse the effects of neonatal capsaicin treatment on rat dorsal root ganglia and dorsal roots. J Physiol (Lond) 303:12P

Lawson SN, Harper AA, Garson JA, Anderton BH (1984) A monoclonal antibody against neurofilament protein specifically labels a subpopulation of rat sensory neurones. J Comp Neurol 228:262–272

Lembeck F (1988) The 1988 Ulf von Euler lecture. Substance P: from extract to excitement. Acta Physiol Scand 133:435–454

Lembeck F, Donnerer J (1981) Time course of capsaicin-induced functional impairments in comparison with changes in neuronal substance P content. Naunyn Schmiedebergs Arch Pharmacol 316:240–243

Lembeck F, Holzer P (1979) Substance P as neurogenic mediator of antidromic vasodilation and neurogenic plasma extravasation. Naunyn Schmiedebergs Arch Pharmacol 310:175–183

Longhurst JC, Kaufman MP, Ordway GA, Musch TI (1984) Effects of bradykinin and capsaicin on endings of afferent fibers from abdominal visceral organs. Am J Physiol 247:R552–R559

Lorez HP, Haeusler G, Aeppli L (1983) Substance P neurones in medullary baroreflex areas and baroreflex function of capsaicin-treated rats. Comparison with other primary afferent systems. Neuroscience 8:507–523

Lundberg JM, Saria A (1983) Capsaicin induced desensitization of the airway mucosa to cigarette smoke, mechanical and chemical irritants. Nature 302:251–253

Lundberg JM, Saria A (1987) Polypeptide-containing neurons in airway smooth muscle. Annu Rev Physiol 49:557–572

Lundberg JM, Franco-Cereceda A, Hua X-Y, Hökfelt T, Fischer J (1985) Coexistence of substance P and calcitonin gene-related peptide immunoreactivities in sensory nerves in relation to cardiovascular and bronchoconstrictor effects of capsaicin. Eur J Pharmacol 108:315–319

Lundblad L (1984) Protective reflexes and vascular effects in the nasal mucosa elicited by activation of capsaicin-sensitive substance P-immunoreactive trigeminal neurons. Acta Physiol Scand [Suppl] 529:1–42

Lundblad L, Lundberg JM, Änggard A, Zetterström D (1987) Capsaicin-sensitive nerves and the cutaneous allergy reaction in man. Possible involvement of sensory neuropeptides in the flare reaction. Allergy 42:20–25

Luthman J, Stromberg I, Brodin E, Jonsson G (1989) Capsaicin treatment to developing rats induces increase of noradrenaline levels in the iris without affecting the adrenergic terminal density. Int J Dev Neurosci 7:613–620

Lynn B (1984) Effect of neonatal treatment with capsaicin on the numbers and properties of cutaneous afferent units from the hairy skin of the rat. Brain Res 322:255–260

Lynn B, Pini A (1985) Long-term block of afferent C-fibres following capsaicin treatment in the rat. J Physiol (Lond) 362:198

Lynn B, Shakhanbeh J (1988) Substance P content of the skin, neurogenic inflammation and numbers of C-fibres following capsaicin application to a cutaneous nerve in the rabbit. Neuroscience 24:769–776

Lynn B, Carpenter SE, Pini A (1984) Capsaicin and cutaneous afferents. In: Chahl LA, Szolcsányi J, Lembeck F (eds) Antidromic vasodilatation and neurogenic inflammation. Akadémiai Kiadó, Budapest, pp 83–92

Lynn B, Pini A, Baranowski R (1987) Injury of somatosensory afferents by capsaicin: selectivity and failure to regenerate. In: Pubols LM, Sessle B (eds) Effects of injury on trigeminal and spinal somatosensory systems. Liss, New York, pp 115–124

MacLean DD (1985) Abrogation of peripheral cholecystokinin-satiety in the capsaicin-treated rat. Regul Pept 11:321–333

Maggi CA (1991) Capsaicin and primary afferent neurons: from basic science to human therapy? J Auton Nerv Syst 33:1–14

Maggi CA, Meli A (1986) The role of neuropeptides in the regulation of the micturition reflex. J Auton Pharmacol 6:133–162

Maggi CA, Meli A (1988) The sensory-efferent function of capsaicin-sensitive sensory neurons. Gen Pharmacol 19:1–43

Maggi CA, Santicioli P, Meli A (1984) The effects of capsaicin on rat urinary bladder motility in vivo. Eur J Pharmacol 103:41–50

Maggi CA, Manzini S, Giuliani S, Santicioli S, Meli A (1986) Extrinsic origin of the capsaicin-sensitive innervation of rat duodenum: possible involvement of calcitonin gene-related peptide (CGRP) in the capsaicin-induced activation of intramural non-adrenergic non-cholinergic ('purinergic'?) neurons. Naunyn Schmiedebergs Arch Pharmacol 334:172–180

Maggi CA, Borsini F, Santicioli P, Geppetti P, Abelli L, Evangelista S, Manzini S, Theodorsson-Norheim E, Somma V, Amenta F, Bacciarelli C, Meli A (1987a) Cutaneous lesions in capsaicin-pretreated rats. A trophic role of capsaicin-sensitive afferents? Naunyn Schmiedebergs Arch Pharmacol 336:538–545

Maggi CA, Giuliani S, Santicioli P, Abelli L, Geppetti P, Somma V, Renzi D, Meli A (1987b) Species-related variations in the effects of capsaicin on urinary bladder functions: relation to bladder content of substance P-like immunoreactivity. Naunyn Schmiedebergs Arch Pharmacol 336:546–555

Maggi CA, Meli A, Santicioli P (1987c) Four motor effects of capsaicin on guinea-pig distal colon. Br J Pharmacol 90:651–660

Maggi CA, Santicioli P, Geppetti P, Furio M, Frilli S, Conte B, Fanciullacci M, Giuliani S, Meli A (1987d) The contribution of capsaicin-sensitive innervation to activation of the spinal vesicovesical reflex in rats: relationship between substance P levels in the urinary bladder and the sensory-efferent function of capsaicin-sensitive sensory neurons. Brain Res 415:1–13

Maggi CA, Santicioli P, Geppetti P, Giuliani S, Patacchini R, Frilli S, Grassi J, Meli A (1987e) Involvement of a peripheral site of action in the early phase of neuropeptide depletion following capsaicin desensitization. Brain Res 436:402–406

Maggi CA, Patacchini R, Giuliani S, Santicioli P, Meli A (1988a) Evidence for two independent modes of activation of the 'efferent' function of capsaicin-sensitive nerves. Eur J Pharmacol 156:367–374

Maggi CA, Patacchini R, Santicioli P, Giuliani S, Geppetti P, Meli A (1988b) Protective action of ruthenium red toward capsaicin desensitization of sensory fibers. Neurosci Lett 88:201–205

Maggi CA, Patacchini R, Santicioli P, Lippe IT, Giuliani S, Geppetti P, del Bianco E, Selleri S, Meli A (1988c) The effect of omega conotoxin GVIA, a peptide modulator of the N-type voltage sensitive calcium channels, on motor responses produced by activation of efferent and sensory nerves in mammalian smooth muscle. Naunyn Schmiedebergs Arch Pharmacol 338:107–113

Maggi CA, Santicioli P, Geppetti P, Parlani M, Astolfi M, Pradelles P, Patacchini R, Meli A (1988d) The antagonism induced by ruthenium red of the actions of capsaicin on the peripheral terminals of sensory neurons: further studies. Eur J Pharmacol 154:1–10

Maggi CA, Santicioli P, Geppetti P, Patacchini R, del Bianco E, Meli A (1988e) Calcium and capsaicin-induced substance P release from peripheral terminals of primary sensory neurons. Regul Pept 22:117

Maggi CA, Santicioli P, Patacchini R, Geppetti P, Giuliani S, Astolfi M, Baldi E, Parlani M, Theodorsson E, Fusco B, Meli A (1988f) Regional differences in the motor response to capsaicin in the guinea-pig urinary bladder: relative role of pre- and postjunctional factors related to neuropeptide-containing sensory nerves. Neuroscience 27:675–688

Maggi CA, Giuliani S, Meli A (1989a) Effect of ruthenium red on responses mediated by activation of capsaicin-sensitive nerves of the rat urinary bladder. Naunyn Schmiedebergs Arch Pharmacol 340:541–546

Maggi CA, Lippe IT, Giuliani S, Abelli L, Somma V, Geppetti P, Jancsó G, Santicioli P, Meli A (1989b) Topical versus systemic capsaicin desensitization: specific and unspecific effects as indicated by modification of reflex micturition in rats. Neuroscience 31:745–756

Maggi CA, Patacchini R, Santicioli P, Giuliani S, del Bianco E, Geppetti P, Meli A (1989c) The efferent function of capsaicin-sensitive nerves – ruthenium red discriminates between different mechanisms of activation. Eur J Pharmacol 170:167–177

Maggi CA, Santicioli P, del Bianco E, Geppetti P, Barbanti G, Turini D, Meli A (1989d) Release of VIP- but not CGRP-like immunoreactivity by capsaicin from the human isolated small intestine. Neurosci Lett 98:317–320

Maggi CA, Santicioli P, Geppetti P, Parlani M, Astolfi M, del Bianco E, Patacchini R, Giuliani S, Meli A (1989e) The effect of calcium free medium and nifedipine on the release of substance P-like immunoreactivity and contractions induced by capsaicin in the isolated guinea pig and rat bladder. Gen Pharmacol 20:445–456

Maggi CA, Astolfi M, Donnerer J, Amann R (1990a) Which mechanisms account for the sensory neuron blocking action of capsaicin on primary afferents in the rat urinary bladder? Neurosci Lett 110:267–272

Maggi CA, Giuliani S, Santicioli P, Tramontana M, Meli A (1990b) Effect of omega conotoxin on reflex responses mediated by activation of capsaicin-sensitive nerves of the rat urinary bladder and peptide release from the rat spinal cord. Neuroscience 34:243–250

Maggi CA, Patacchini R, Tramontana M, Amann R, Giuliani S, Santicioli P (1990c) Similarities and differences in the action of resiniferatoxin and capsaicin on central and peripheral endings of primary sensory neurons. Neuroscience 37:531–539

Maggio JE, Hunter JC (1984) Regional distribution of kassinin-like immunoreactivity in rat central and peripheral tissues and the effect of capsaicin. Brain Res 307:370–373

Marabini S, Ciabatti G, Polli G, Fusco BM, Geppetti P, Maggi CA, Fanciullacci M, Sicuteri F (1988) Effect of topical nasal treatment with capsaicin in vasomotor rhinitis. Regul Pept 22:121

Marley P, Livett BG (1985) Neuropeptides in the autonomic nervous system. CRC Crit Rev Clin Neurobiol 1:201–283

Marsh SJ, Stansfeld CE, Brown DA, Davey R, McCarthy D (1987) The mechanism of action of capsaicin on sensory C-type neurons and their axons in vitro. Neuroscience 23:275–290

Martin HA, Basbaum AI, Kwiat GC, Goetzl EJ, Levine JD (1987) Leukotriene and prostaglandin sensitization of cutaneous high threshold C- and A-delta mechanonociceptors in the hairy skin of rat hindlimbs. Neuroscience 22:651–659

Mason RJ, Maruniak JA (1983) Behavioral and physiological effects of capsaicin in red-winged blackbirds. Pharmacol Biochem Behav 19:857–862

Matran R, Alving K, Martling C-R, Lacroix JS, Lundberg JM (1989) Effects of neuropeptides and capsaicin on tracheobronchial blood flow of the pig. Acta Physiol Scand 135:335–342

Matsumiya T, Sawa A, Oka T (1983) The effect of capsaicin on C-fiber reflex and heat evoked discharge in the acute spinal cat. Tokai J Exp Clin Med 8:325–332

McCann MJ, Verbalis JG, Stricker EM (1988) Capsaicin pretreatment attenuates multiple responses to cholecystokinin in rats. J Auton Nerv Syst 23:265–271

McDougal DB, Yuan MJC, Dargar RV, Johnson EM (1983) Neonatal capsaicin and guanethidine and axonally transported organelle-specific enzymes in sciatic nerve and in sympathetic and dorsal root ganglia. J Neurosci 3:124–132

McDougal DB, Yuan MJC, Johnson EM (1985) Effect of capsaicin upon fluoride sensitive acid phosphatases in selected ganglia and spinal cord and upon neuronal size and number in dorsal root ganglion. Brain Res 331:63–70

McMahon SB, Wall PD, Granum SL, Webster KE (1984) The effect of capsaicin applied to peripheral nerves on responses of a group of lamina I cells in adult rats. J Comp Neurol 227:393–400

Micevych PE, Yaksh TL, Szolcsányi J (1983) Effect of intrathecal capsaicin analogues on the immunofluorescence of peptides and serotonin in the dorsal horn in rats. Neuroscience 8:123–131

Miller MS, Buck SH, Sipes IG, Burks TF (1982a) Capsaicinoid-induced local and systemic antinociception without substance P depletion. Brain Res 244:193–197

Miller MS, Buck SH, Sipes IG, Yamamura HI, Burks TF (1982b) Regulation of substance P by nerve growth factor: disruption by capsaicin. Brain Res 250:193–196

Miller MS, Brendel K, Burks TF, Sipes IG (1983) Interaction of capsaicinoids with drug-metabolizing systems – relationship to toxicity. Biochem Pharmacol 32:547–551

Monsereenusorn Y, Kongsamut S (1985) Inhibition of calcium uptake by capsaicin. Res Commun Chem Pathol Pharmacol 47:453–456

Moody TW, Thoa NB, O'Donohue TL, Jacobowitz DM (1981) Bombesin-like peptides in rat spinal cord: biochemical characterization, localization, and mechanism of release. Life Sci 29:2273–2279

Moritoki H, Takase H, Tanioka A (1990) Dual effects of capsaicin on responses of the rabbit ear artery to field stimulation. Br J Pharmacol 99:152–156

Morton CR, Chahl LA (1980) Pharmacology of the neurogenic oedema response to electrical stimulation of the saphenous nerve in the rat. Naunyn Schmiedebergs Arch Pharmacol 314:271–276

Mueller GP (1981) Beta-endorphin immunoreactivity in rat plasma: variations in response to different physical stimuli. Life Sci 29:1669–1674

Mussap CJ, Lew R, Burcher E (1989) The autoradiographic distribution of substance P binding sites in guinea-pig vas deferens is altered by capsaicin pretreatment. Eur J Pharmacol 168:337–345

Nagy JI (1982) Capsaicin: a chemical probe for sensory neuron mechanisms. In: Iversen LL, Iversen SD, Snyder SH (eds) Handbook of psychopharmacology, vol 15. Plenum, New York, pp 185–235

Nagy JI, Daddona PE (1985) Anatomical and cytochemical relationships of adenosine deaminase-containing primary afferent neurons in the rat. Neuroscience 15:799–813

Nagy JI, Hunt SP (1983) The termination of primary afferents within the rat dorsal horn: evidence for rearrangement following capsaicin treatment. J Comp Neurol 218:145–158

Nagy JI, van der Koy D (1983) Effects of neonatal capsaicin treatment on nociceptive thresholds in the rat. J Neurosci 3:1145–1150

Nagy JI, Vincent SR, Staines WMA, Fibiger HC, Reisine TD, Yamamura HI (1980) Neurotoxic action of capsaicin on spinal substance P neurons. Brain Res 186:435–444

Nagy JI, Emson PC, Iversen LL (1981a) A re-evaluation of the neurochemical and antinociceptive effects of intrathecal capsaicin in the rat. Brain Res 211:497–502

Nagy JI, Hunt SP, Iversen LL, Emson PC (1981b) Biochemical and anatomical observations on the degeneration of peptide-containing primary afferent neurons after neonatal capsaicin. Neuroscience 6:1923–1934

Nagy JI, Iversen LL, Goedert M, Chapman D, Hunt SP (1983) Dose-dependent effects of capsaicin on primary sensory neurons in the neonatal rat. J Neurosci 3:399–406

Nielsch U, Keen P (1987) Effects of neonatal 6-hydroxydopamine administration on different substance P-containing sensory neurones. Eur J Pharmacol 138:193–197

Nilsson G (1989) Modulation of the immune response in rat by utilizing the neurotoxin capsaicin. Acta Univ Ups 230:1–53

Northam WJ, Jones DJ (1984) Comparison of capsaicin and substance P induced cyclic AMP accumulation in spinal cord tissue slices. Life Sci 35:293–302

Obál F, Jancsó G, Jancsó-Gábor A, Obál F (1983) Vasodilatation on preoptic heating in capsaicin-treated rats. Experientia 39:221–223

Ogilvy CS, Silverberg KR, Borges LF (1991) Sprouting of corneal sensory fibers in rats treated at birth with capsaicin. Invest Ophthalmol Vis Sci 32:112–121

Ordway GA, Longhurst JC (1983) Cardiovascular reflexes arising from the gallbladder of the cat. Effects of capsaicin, bradykinin, and distension. Circ Res 52:26–35

Otten U, Lorez HP, Businger F (1983) Nerve growth factor antagonizes the neurotoxic action of capsaicin on primary sensory neurones. Nature 301:515–517

Palermo NN, Brown HK, Smith DL (1981) Selective neurotoxic action of capsaicin on glomerular C-type terminals in rat SG. Brain Res 207:506–510

Panerai AE, Martini A, Locatelli V, Mantegazza P (1983) Capsaicin decreases b-endorphin hypothalamic concentrations in the rat. Pharmacol Res Commun 15:825–832

Papka RE, Furness JB, Della NG, Murphy R, Costa M (1984) Time course of effect of capsaicin on ultrastructure and histochemistry of substance P-immunoreactive nerves associated with the cardiovascular system of the guinea-pig. Neuroscience 12:1277–1292

Petersen M, Pierau F-K, Weyrich M (1987) The influence of capsaicin on membrane currents in dorsal root ganglion neurones of guinea-pig and chicken. Pflugers Arch 409:403–410

Petersen M, Wagner G, Pierau F-K (1989) Modulation of calcium-currents by capsaicin in a subpopulation of sensory neurones of guinea pig. Naunyn Schmiedebergs Arch Pharmacol 339:184–191

Petsche U, Fleischer E, Lembeck F, Handwerker HO (1983) The effect of capsaicin application to a peripheral nerve on impulse conduction in functionally identified afferent nerve fibres. Brain Res 265:233–240

Pierau F-K, Szolcsányi J (1989) Neurogenic inflammation: axon reflex in pigs. Agents Actions 26:231–232

Pierau F-K, Sann H, Harti G, Gamse R (1987) Neuropeptides in sensory neurones of pigeons and the insensitivity of avians to capsaicin. In: Schmidt RF, Schaible H-G, Vahle-Hinz C (eds) Fine nerve fibres and pain. VHC, Weinheim, pp 213–223

Pini A (1983) Effects of capsaicin on conduction in a cutaneous nerve of the rat. J. Physiol (Lond) 338:60P–61P

Priestley JV, Bramwell S, Butcher LL, Cuello AC (1982) Effect of capsaicin on neuropeptides in areas of termination of primary sensory neurons. Neurochem Int 4:57–65

Rabe LS, Buck SH, Moreno L, Burks TF, Dafny N (1980) Neurophysiological and thermoregulatory effects of capsaicin. Brain Res Bull 5:755–758

Ray NJ, Jones AJ, Keen P (1990) Ruthenium red: selective, reversible inhibition of capsaicin-stimulated substance P release from primary afferent neurons in rat trachea. Br J Pharmacol 99:186P

Raybould HE, Taché Y (1988) Cholecystokinin inhibits gastric motility and emptying via a capsaicin-sensitive vagal pathway in rats. Am J Physiol 255:G242–G246

Raybould HE, Holzer P, Sternini C, Eysselein VE (1990) Selective ablation of spinal sensory neurons containing CGRP inhibits the increase in rat gastric mucosal blood flow due to acid back-diffusion. Gastroenterology 98:A198

Réthelyi M, Salim MZ, Jancsó G (1986) Altered distribution of dorsal root fibers in the rat following neonatal capsaicin treatment. Neuroscience 18:749–761

Ribeiro-da-Silva A, Coimbra A (1984) Capsaicin causes selective damage to type I synaptic glomeruli in rat substantia gelatinosa. Brain Res 290:380–383

Ritter S, Dinh TT (1988) Capsaicin-induced neuronal degeneration: silver impregnation of cell bodies, axons, and terminals in the central nervous system of the adult rat. J Comp Neurol 271:79–90

Rogers LR (1984) Reactions of free-ranging black bears to capsaicin spray repellent. Wildl Soc Bull 12:59–61

Ross DR, Varipapa RJ (1989) Treatment of painful diabetic neuropathy with topical capsaicin. N Engl J Med 321:474–475

Rozin P, Mark M, Schiller D (1981) The role of desensitization to capsaicin in chili pepper ingestion and preference. Chem Senses 6:23–31

Rózsa Z, Jacobson ED (1989) Capsaicin-sensitive nerves are involved in bile-oleate induced intestinal hyperemia. Am J Physiol 256:G476–G481

Russell LC, Burchiel KJ (1984) Neurophysiological effects of capsaicin. Brain Res Rev 8:165–176

Salt TE, Hill RG (1980) The effect of microiontophoretically applied capsaicin and substance P on single neurones in the rat and cat brain. Neurosci Lett 20:329–334

Salt TE, Hill RG (1983) Neurotransmitter candidates of somatosensory primary afferent fibres. Neuroscience 10:1083–1103

Salt TE, Crozier CS, Hill RG (1982) The effects of capsaicin pretreatment on the responses of single neurons to sensory stimuli in the trigeminal nucleus caudalis of the rat: evidence against a role for substance P as the neurotransmitter serving thermal nociception. Neuroscience 7:1141–1148

Sann H, Harti G, Pierau F-K, Simon E (1987) Effect of capsaicin upon afferent and efferent mechanisms of nociception and temperature regulation in birds. Can J Physiol Pharmacol 65:1347–1354

Santicioli P, Maggi CA, Meli A (1985) The effect of capsaicin pretreatment on the cystometrograms of urethane anesthetized rats. J Urol 133:700–703

Santicioli P, Patacchini R, Maggi CA, Meli A (1987) Exposure to calcium-free medium protects sensory fibers by capsaicin desensitization. Neurosci Lett 80:167–172

Saporta S (1986) Loss of spinothalamic tract neurons following neonatal treatment of rats with the neurotoxin capsaicin. Somatosens Res 4:153–173

Saria A, Wolf G (1988) Beneficial effect of topically applied capsaicin in the treatment of hyperreactive rhinopathy. Regul Pept 22:167

Saria A, Lundberg JM, Hua X-Y, Lembeck F (1983) Capsaicin-induced substance P release and sensory control of vascular permeability in the guinea-pig ureter. Neurosci Lett 41:167–172

Saria A, Gamse R, Petermann J, Fischer JA, Theodorsson-Norheim E, Lundberg JM (1986) Simultaneous release of several tachykinins and calcitonin gene-related peptide from rat spinal cord slices. Neurosci Lett 63:310–314

Saria A, Martling C-R, Yan Z, Theodorsson-Norheim E, Gamse R, Lundberg JM (1988) Release of multiple tachykinins from capsaicin-sensitive sensory nerves in the lung by bradykinin, histamine, dimethylphenyl piperazinium, and vagal nerve stimulation. Am Rev Respir Dis 137:1330–1335

Saumet J-L, Duclaux R (1982) Analgesia induced by neonatal capsaicin treatment in rats. Pharmacol Biochem Behav 16:241–243

Scadding JW (1980) The permanent anatomical effects of neonatal capsaicin on somatosensory nerves. J Anat 131:473–484

Schanne FAX, Kane AB, Young EE, Farber JL (1979) Calcium dependence of toxic cell death: a final common pathway. Science 206:700–702

Sharkey KA, Williams RG, Schultzberg M, Dockray GJ (1983) Sensory substance P innervation of the urinary bladder: possible site of action of capsaicin in causing urine retention in rats. Neuroscience 10:861–868

Sharkey KA, Williams RG, Dockray GJ (1984) Sensory substance P innervation of the stomach and pancreas. Demonstration of capsaicin-sensitive sensory neurons in the rat by combined immunohistochemistry and retrograde tracing. Gastroenterology 87:914–921

Shimizu T, Fujita S, Izumi K, Koja T, Ohba N, Fukuda T (1984) Corneal lesions induced by the systemic administration of capsaicin in neonatal mice and rats. Naunyn Schmiedebergs Arch Pharmacol 326:347–351

Sikri KL, Hoyes AD, Barber P, Jagessar H (1981) Substance P-like immunoreactivity in the intramural nerve plexuses of the guinea-pig ureter: a light and electron microscopical study. J Anat 133:425–442

Simone DA, Ngeow JYF, Putterman GJ, LaMotte RH (1987) Hyperalgesia to heat after intradermal injection of capsaicin. Brain Res 418:201–203

Simone DA, Baumann TK, LaMotte RH (1989) Dose-dependent pain and mechanical hyperalgesia in humans after intradermal injection of capsaicin. Pain 38:99–107

Singer EA, Placheta P (1980) Reduction of ^{3}H-muscimol binding sites in rat dorsal spinal cord after neonatal capsaicin treatment. Brain Res 202:484–487

Singer EA, Sperk G, Schmid R (1982) Capsaicin does not change tissue levels of glutamic acid, its uptake, or release in the rat spinal cord. J Neurochem 38:1383–1386

Sizer F, Harris N (1985) The influence of common food additives and temperature on threshold perception of capsaicin. Chem Senses 10:279–286

Skofitsch G, Jacobowitz DM (1985a) Galanin-like immunoreactivity in capsaicin-sensitive sensory neurons and ganglia. Brain Res Bull 15:191–195

Skofitsch G, Jacobowitz DM (1985b) Calcitonin gene-related peptide coexists with substance P in capsaicin sensitive neurons and sensory ganglia of the rat. Peptides 6:747–754

Skofitsch G, Zamir N, Helke CJ, Savitt JM, Jacobowitz DM (1985) Corticotropin releasing factor-like immunoreactivity in sensory ganglia and capsaicin sensitive neurons of the rat central nervous system: colocalization with other neuropeptides. Peptides 6:307–318

Soukup T, Jancsó G (1987) Development of muscle stretch receptors in the rat is not affected by neonatal capsaicin treatment. Acta Physiol Hung 69:527–532

South EH, Ritter RC (1983) Overconsumption of preferred foods following capsaicin pretreatment of the area postrema and adjacent nucleus of the solitary tract. Brain Res 288:243–251

South EH, Ritter RC (1988) Capsaicin application to central or peripheral vagal fibers attenuates CCK satiety. Peptides 9:601–612

Stein RD, Genovesi S, Demarest KT, Weaver LC (1986) Capsaicin treatment attenuates the reflex excitation of sympathetic activity caused by chemical stimulation of intestinal afferent nerves. Brain Res 397:145–151

Sternini C, Reeve JR, Brecha N (1987) Distribution and characterization of calcitonin gene-related peptide immunoreactivity in the digestive system of normal and capsaicin-treated rats. Gastroenterology 93:852–862

Su HC, Bishop AE, Power RF, Hamada Y, Polak JM (1987) Dual intrinsic and extrinsic origins of CGRP- and NPY-immunoreactive nerves of rat gut and pancreas. J Neurosci 7:2674–2687

Such G, Jancsó G (1986) Axonal effects of capsaicin: an electrophysiological study. Acta Physiol Hung 67:53–63

Szallasi A, Blumberg PM (1989) Resiniferatoxin, a phorbol-related diterpene, acts as an ultrapotent analog of capsaicin, the irritant constituent in red pepper. Neuroscience 30:515–520

Szallasi A, Blumberg PM (1990a) Specific binding of resiniferatoxin, an ultrapotent capsaicin analog, by dorsal root ganglion membranes. Brain Res 524:106–111

Szallasi A, Blumberg PM (1990b) Resiniferatoxin and its analogs provide novel insights into the pharmacology of the vanilloid (capsaicin) receptor. Life Sci 47:1399–1408

Szallasi A, Joo F, Blumberg PM (1989) Duration of desensitization and ultrastructural changes in dorsal root ganglia in rat treated with resiniferatoxin, an ultrapotent capsaicin analog. Brain Res 503:68–72

Szallasi A, Szallasi Z, Blumberg PM (1990) Permanent effects of neonatally administered resiniferatoxin in the rat. Brain Res 537:182–186

Szikszay M, London ED (1988) Effects of subacute capsaicin treatment on local cerebral glucose utilization in the rat. Neuroscience 25:917–923

Szikszay M, Obál F, Obál F (1982) Dose-response relationships in the thermoregulatory effects of capsaicin. Naunyn Schmiedebergs Arch Pharmacol 320:97–100

Szolcsányi J (1977) A pharmacological approach to elucidation of the role of different nerve fibres and receptor endings in mediation of pain. J Physiol (Paris) 73:251–259

Szolcsányi J (1982) Capsaicin type pungent agents producing pyrexia. In: Milton AS (ed) Pyretics and antipyretics. Springer, Berlin Heidelbeg New York, pp 437–478 (Handbook of experimental pharmacology, vol 60)

Szolcsányi J (1983a) Disturbances of thermoregulation induced by capsaicin. J Therm Biol 8:207–212

Szolcsányi J (1983b) Tetrodotoxin-resistant non-cholinergic neurogenic contraction evoked by capsaicinoids and piperine on the guinea-pig trachea. Neurosci Lett 42:83–88

Szolcsányi J (1984a) Capsaicin and neurogenic inflammation: history and early findings. In: Chahl LA, Szolcsányi J, Lembeck F (eds) Antidromic vaso-dilatation and neurogenic inflammation. Akadémiai Kiadó, Budapest, pp 7–25

Szolcsányi J (1984b) Capsaicin-sensitive chemoceptive neural system with dual sensory-efferent function. In: Chahl LA, Szolcsányi J, Lembeck F (eds) Antidromic vasodilatation and neurogenic inflammation. Akadémiai Kiadó, Budapest, pp 27–52

Szolcsányi J (1987) Selective responsiveness of polymodal nociceptors of the rabbit ear to capsaicin, bradykinin and ultra-violet irradiation. J Physiol (Lond) 388:9–23

Szolcsányi J (1988) Antidromic vasodilatation and neurogenic inflammation. Agents Actions 23:4–11

Szolcsányi J (1990) Capsaicin, irritation, and desensitization. Neurophysiological basis and future perspectives. In: Green BG, Mason JR, Kare MR (eds) Irritation. Dekker, New York, pp 141–168 (Chemical senses, vol 2)

Szolcsányi J, Barthó L (1978) New type of nerve-mediated cholinergic contractions of the guinea-pig small intestine and its selective blockade by capsaicin. Naunyn Schmiedebergs Arch Pharmacol 305:83–90

Szolcsányi J, Jancsó-Gábor A (1975) Sensory effects of capsaicin congeners. I. Relationship between chemical structure and pain-producing potency of pungent agents. Drug Res 25:1877–1881

Szolcsányi J, Jancsó-Gábor A (1976) Sensory effects of capsaicin congeners. II. Importance of chemical structure and pungency in desensitizing activity of capsaicin-type compounds. Drug Res 26:33–37

Szolcsányi J, Joó F, Jancsó-Gábor A (1971) Mitochondrial changes in preoptic neurones after capsaicin desensitization of the hypothalamic thermodetectors in rats. Nature 229:116–117

Szolcsányi J, Jancsó-Gábor A, Joó F (1975) Functional and fine structural characteristics of the sensory neuron blocking effect of capsaicin. Naunyn Schmiedebergs Arch Pharmacol 287:157–169

Szolcsányi J, Sann H, Pierau F-K (1986) Nociception in pigeons is not impaired by capsaicin. Pain 27:247–260

Szolcsányi J, Szallasi A, Szallasi A, Joo F, Blumberg PM (1990) Resiniferatoxin: an ultrapotent selective modulator of capsaicin-sensitive primary afferent neurons. J Pharmacol Exp Ther 255:923–928

Takaki M, Nakayama S (1989) Effects of capsaicin on myenteric neurons of the guinea pig ileum. Neurosci Lett 105:125–130

Takano Y, Nagashima A, Kamiya H, Kurosawa M, Sato A (1988) Well-maintained reflex responses of sympathetic nerve activity to stimulation of baroreceptors, chemoreceptors and cutaneous mechanoreceptors in neonatal capsaicin-treated rats. Brain Res 445:188–192

Taylor DCM, Pierau F-K, Szolcsányi J (1984) Long lasting inhibition of horseradish peroxidase (HRP) transport in sensory nerves induced by capsaicin pretreatment of the receptive field. Brain Res 298:45–49

Taylor DCM, Pierau F-K, Szolcsányi J (1985) Capsaicin-induced inhibition of axoplasmic transport is prevented by nerve growth factor. Cell Tissue Res 240:569–573

Terenghi G, Zhang SQ, Unger WG, Polak JM (1986) Morphological changes of sensory CGRP-immunoreactive and sympathetic nerves in peripheral tissues following chronic denervation. Histochemistry 86:89–95

Tervo K (1981) Effect of prolonged and neonatal capsaicin treatments on the substance P immunoreactive nerves in the rabbit eye and spinal cord. Acta Ophthalmol (Copenh) 59:737–746

Thériault E, Otsuka M, Jessell T (1979) Capsaicin-evoked release of substance P from primary sensory neurons. Brain Res 170:209–213

Trad KS, Harmon JW, Fernicola MT, Hakki FZ, Rao JS, Dziki AJ, Bass BL (1990) Evidence for a role of capsaicin-sensitive mucosal afferent nerves in the regulation of esophageal blood flow in rabbits. Gastroenterology 98:A139

Traurig H, Papka RE, Saria A, Lembeck F (1984a) Substance P immunoreactivity in the rat mammary nipple and the effects of capsaicin treatment on lactation. Naunyn Schmiedebergs Arch Pharmacol 328:1–8

Traurig H, Saria A, Lembeck F (1984b) The effects of neonatal capsaicin treatment on growth and subsequent reproductive function in the rat. Naunyn Schmiedebergs Arch Pharmacol 327:254–259

Tsunoo A, Konishi S, Otsuka M (1982) Substance P as an excitatory transmitter of primary afferent neurons in guinea-pig sympathetic ganglia. Neuroscience 7:2025–2037

Ueda N, Muramatsu I, Fujiwara M (1984) Capsaicin and bradykinin-induced substance P-ergic responses in the iris sphincter muscle of the rabbit. J Pharmacol Exp Ther 230:469–473

Urbán L, Willetts J, Randic M, Papka RE (1985) The acute and chronic effects of capsaicin on slow excitatory transmission in rat dorsal horn. Brain Res 330:390–396

Varro A, Green T, Holmes S, Dockray GJ (1988) Calcitonin gene-related peptide in visceral afferent nerve fibres: quantification by radioimmunoassay and determination of axonal transport rates. Neuroscience 26:927–932

Virus RM, Knuepfer MM, McManus DQ, Brody MJ, Gebhart GF (1981) Capsaicin treatment in adult Wistar-Kyoto and spontaneously hypertensive rats: effects on nociceptive behavior and cardiovascular regulation. Eur J Pharmacol 72:209–217

Virus RM, McManus DQ, Gebhart GF (1982) Capsaicin treatment in adult Wistar-Kyoto and spontaneously hypertensive rats: effects on substance P contents of peripheral and central nervous system. Eur J Pharmacol 81:67–73

Virus RM, McManus DQ, Gebhart GF (1983) Capsaicin treatment in adult Wistar-Kyoto and spontaneously hypertensive rats: neurochemical effects in the spinal cord. Eur J Pharmacol 92:1–8

Waddell PJ, Lawson SN (1989) The C-fibre conduction block caused by capsaicin on rat vagus nerve in vitro. Pain 39:237–242

Wall PD, Fitzgerald M (1981) Effects of capsaicin applied locally to adult peripheral nerve. I. Physiology of peripheral nerve and spinal cord. Pain 11:363–377

Wall PD, Fitzgerald M, Nussbaumer JC, van der Loos H, Devor M (1982) Somatotopic maps are disorganized in adult rodents treated neonatally with capsaicin. Nature 295:691–693

Wang J-P, Hsu M-F, Teng C-M (1984) Antiplatelet effect of capsaicin. Thromb Res 36:497–507

Watanabe T, Kawada T, Kurosawa M, Sato A, Iwai K (1988) Adrenal sympathetic efferent nerve and catecholamine secretion excitation caused by capsaicin in rats. Am J Physiol 255:E23–E27

Weihe E (1990) Neuropeptides in primary afferent neurons. In: Zenker W, Neuhuber WL (eds) The primary afferent neuron. Plenum, New York, pp 127–159

Welk E, Petsche U, Fleischer E, Handwerker HO (1983) Altered excitability of afferent C-fibres of the rat distal to a nerve site exposed to capsaicin. Neurosci Lett 38:245–250

Welk E, Fleischer E, Petsche U, Handwerker HO (1984) Afferent C-fibres in rats after neonatal capsaicin treatment. Pflugers Arch 400:66–71

Williams JT, Zieglgänsberger W (1982) The acute effects of capsaicin on rat primary afferents and spinal neurons. Brain Res 253:125–131

Winter J (1987) Characterization of capsaicin-sensitive neurones in adult rat dorsal root ganglion cultures. Neurosci Lett 80:134–140

Winter J, Forbes CA, Sternberg J, Lindsay RM (1988) Nerve growth factor (NGF) regulates adult rat cultured dorsal root ganglion neuron responses to the excitotoxin capsaicin. Neuron 1:973-981

Winter J, Dray A, Wood JN, Yeats JC, Bevan S (1990) Cellular mechanism of action of resiniferatoxin: a potent sensory neuron excitotoxin. Brain Res 520:131–140

Wood JN (1987) Reversible desensitization to capsaicin induced calcium uptake in cultured rat sensory neurones. Neuroscience [Suppl] 22:S247

Wood JN, Winter J, James IF, Rang HP, Yeats J, Bevan S (1988) Capsaicin-induced ion fluxes in dorsal root ganglion cells in culture. J Neurosci 8:3208–3220

Wood JN, Coote PR, Minhas A, Mullaney I, McNeill M, Burgess GM (1989) Capsaicin-induced ion fluxes increases cyclic GMP but not cyclic AMP levels in rat sensory neurones in culture. J Neurochem 53:1203–1211

Yaksh TL, Farb DH, Leeman SE, Jessell TM (1979) Intrathecal capsaicin depletes substance P in the rat spinal cord and produces prolonged thermal analgesia. Science 206:481–483

Yaksh TL, Jessell TM, Gamse R, Mudge AW, Leeman SE (1980) Intrathecal morphine inhibits substance P release from mammalian spinal cord in vivo. Nature 286:155–157

Yaksh TL, Abay EO, Go VLW (1982) Studies on the location and release of cholecystokinin and vasoactive intestinal peptide in rat and cat spinal cord. Brain Res 242:279–290

Yamanaka K, Kigoshi S, Muramatsu I (1984) Conduction-block induced by capsaicin in crayfish giant axon. Brain Res 300:113–119

Yanagisawa M, Nakano S, Otsuka M (1980) Capsaicin-induced depolarization of primary afferent fibers and the release of substance P from isolated rat spinal cord. Biomed Res [Suppl] 1:88–90

Zernig G, Holzer P, Lembeck F (1984) A study of the mode and site of action of capsaicin in guinea-pig heart and uterus. Naunyn Schmiedebergs Arch Pharmacol 326:58–63

CHAPTER 13

Excitotoxins, Glutamate Receptors, and Excitotoxicity

V.I. TEICHBERG

A. Introduction

The increasing awareness in the various fields of the neurosciences of the dual role of glutamate (Glu) as a neurotransmitter and toxin justifies the inclusion of a chapter on excitotoxins and Glu receptors in a book devoted to neurotoxins. Recent reviews dealing with the topic are available particularly with regard to the involvement of Glu and Glu receptors in CNS ontogeny (McDONALD and JOHNSTON 1990), physiology (COLLINGRIDGE and LESTER 1989), and pathology (CHOI and ROTHMAN 1990). Rather than reiterating the issues discussed in detail in these reviews, emphasis is placed here on the molecular aspects related to the first established steps in the mechanisms of excitotoxicity and more particularly to the intimate properties of the excitotoxin receptors, i.e., the Glu receptors.

B. Excitotoxins and Excitotoxicity

Excitotoxins represent a group of neuroactive compounds characterized by their combined affinity to any of the brain Glu receptor subtypes and ability to excite and kill neurons. The connection between the neurotoxicity and neuronal excitatory potency of Glu analogues was first demonstrated by OLNEY et al. (1981) who introduced the terms "excitotoxic amino acids" and "excitotoxins" as a reference to those excitatory amino acids and Glu analogues which mimic both the excitation and toxic effects of Glu on CNS neurons. The neology was extended to include the terms "excitotoxicity" and "excitotoxicology" to refer, respectively, to the toxicity exerted by excitotoxins and its study. For a more extensive insight into the historical aspects of the discovery of excitatory amino acids and of excitotoxins, the reader can refer to the reviews written by WATKINS (1986) and OLNEY (1982).

C. Excitotoxins and Glutamate Receptors

Excitotoxins can be separated into two main classes: the endogenous ones such as Glu, aspartate (Asp), quinolinate, homocysteate, etc. which are integral components of the biochemical machinery of the brain, and the exogenous ones which are Glu analogues, often synthetic, that gain access to the brain either under experimental conditions or accidentally. Excitotoxic effects can also be produced by substances which do not interact with Glu receptors but prevent either the metabolism of endogenous excitotoxins and/or their cellular uptake or increase their release. Excitotoxins can be further defined according to the subtypes of Glu receptors through which they may exert their toxicity.

I. The *N*-Methyl-D-Aspartate Receptor

1. Pharmacological and Electrophysiological Characterization

The *N*-methyl-D-aspartate (NMDA) receptor is well characterized pharmacologically since it possesses a set of highly specific agonists, antagonists, allosteric regulators, and channel blockers acting at separate, easily identifiable binding sites. Table 1 lists NMDA receptor ligands classified according to the receptor sites with which they interact. The occupation of these by the relevant ligands exerts either a pro- or an antiexcitotoxic effect.

The agonist site of the NMDA receptor shows generally a preference for either cyclic conformationally constrained compounds or amino acids with up to a 2-carbon long side chain, both with side-chain acidic function in the following order $CO_2H > SO_3H$, $SO_2H > PO_3H_2$ (Watkins and Olverman 1987). The same site accommodates antagonists with either cyclic structures or aliphatic 5–7-carbon long D-amino acids displaying side chain acidic function in the order $PO_3H_2 > CO_2H$.

The glycine site accommodates preferentially glycine but also D-serine and 1-aminocyclopropanecarboxylate, that is, amino acids with up to 4 carbons. Larger analogues are likely to be antagonists.

The polyamine site binds N,N'-bis-(3-aminopropyl)-1,' 3-propanediamine > spermine > spermidine. It also binds putrescine, but the latter apparently blocks the action of the polyamines (Lodge and Johnson 1990).

In addition to these sites, the NMDA receptor possesses a site for dissociative anesthetic compounds such as ketamine, benzomorphans, and phencyclidine (PCP, angel dust) and its MK-801 analogue (Collingridge and Lester 1989). All the latter compounds block, in a noncompetitive fashion, the response to NMDA receptor agonists by binding in a voltage- and use-dependent manner to a site within the NMDA receptor channel. The latter harbors two additional but distinct binding sites, one for Mg^{2+} (Nowak et al. 1984), the other for Zn^{2+} ions (Westbrook and Mayer 1987), which, once bound, block the channel.

NMDA receptor ligands cause the opening of a channel of 50 pS with a Q_{10} of 1.6 (Nowak et al. 1984; Ascher et al. 1988). The opening of the channel is blocked by Mg^{2+} ions in a highly voltage-dependent manner (Nowak et al. 1984). The block is very significant at negative membrane potentials and decreases with depolarization. The opening of the NMDA

Table 1. Pharmacological properties of the *N*-methyl-D-aspartate (NMDA) receptor

---

Agonists and excitotoxins acting at the agonist/antagonist site
NMDA, glutamate, aspartate, quinolinate, homocysteate, trans-2,3-PDA, D-HCSA, CPAA, homoquinolate, cis-ACPD, BMAA

Antagonists competing at the agonist/antagonist site
D-AP5, D-AP7, D-CPP, D-CPPene, CGS 19755, CGP 37849

Agonists acting at the glycine allosteric site
Glycine, D-serine, ACPC

Antagonists acting at the glycine allosteric site
HA-966, 7-C1KYN, kynurenate, ACPC, 5-CLIAA

Agonists acting at the polyamine site
N,*N'*-bis-(3-aminopropyl)-1,3-propanediamine, spermine, spermidine, ifenprodil, SL-82.0715, HU-211

Antagonists acting at the polyamine site
Arcaine, putrescine, cadaverine

Antagonists acting at the ion channel site
Mg^{2+}, MK-801, phencyclidine, TCP, ketamine, SKF10047

Trans-2,3-PDA, *trans*-2,3 piperidinedicarboxylate; D-HCSA, D-homocysteine-sulfinate; CPAA, *trans*-2-carboxy-3-pyrrolidine acetate; cis-ACPD, *cis*-1-amino-1,3-cyclopentanedicarboxylate; BMAA, β-*N*-methylamino-L-alanine; D-AP5, 2-amino-5-phosphonopentanoate; D-AP7, 2-amino-7-phosphonoheptanoate; D-CPP, 3((±)-2-carboxypiperazin-4-yl)propyl-1-phosphonate; D-CPPene, (E)-4-(3-phosphonoprop-2-enyl)piperazine-2-carboxylate; CGS 19755, *cis*-4-(phosphonomethyl)piperidine-2-carboxylate; CGP 37849, (E)-2-amino-4-methyl-5-phosphono-3-pentanoate; ACPC, 1-amino-cyclopropanecarboxylate; HA-966, 1-hydroxy-3-aminopyrrolidone; MK-801, (+)-5-methyl-10,11-dihydro-5*H*-dibenzo[*a,d*]cyclohepten-5,10-iminemaleate; TCP, *N*-[1,2(thienyl)cyclohexyl]-piperidine; 7C1KYN, 7-chlorokynurenate; ACPC, 1-aminocyclopropanecarboxylate; arcaine, 1,4-diguanidinobutane; 5-CLIAA, 5-chloro-indole-2-carboxylate; HU-211, 7-hydroxy-Δ^6-tetrahydrocannabinol-1,1-dimethyl-heptyl; SKF10047, *N*-allylnormetacozine.

---

receptor/channel is thus conditional on a prior depolarization of the cell membrane to remove the Mg^{2+} block. Na^+, K^+, and Ca^{2+} ions have the ability to transit through the NMDA receptor channel. Its permeability to Ca^{2+} ions is quite appreciable: It is estimated that Ca^{2+} ions contribute 5%–10% of the total NMDA-gated current (PUMAIN et al. 1987). Na^+ and K^+ ions are likely to be equipermeant since the NMDA responses reverse at membrane potentials around 0 mV (NOWAK et al. 1984).

The occupation of the glycine site by glycine causes a very significant potentiation of the NMDA responses to the point that in some cell preparations NMDA is unable to produce an ionic current in the absence of glycine (KLECKNER and DINGLEDINE 1988). It has been proposed that this potentiation is due to a glycine-induced increase in the recovery rate from desensitization of the NMDA response (MAYER et al. 1989). However, this conclusion has been recently questioned following the finding that glycine can potentiate the NMDA response under conditions in which it does not regulate desensitization (SATHER et al. 1990). Possibly these conflicting data can be reconciled, beyond the more trivial methodological differences, on the basis of a heterogeneity of NMDA receptors (MONAGHAN et al. 1988). That glycine causes an increase in the binding affinity of NMDA remains also a matter of debate (YOUNG and FAGG 1990; KESSLER et al. 1989).

The question of the physiological and possibly pathological situations in which glycine would play its potentiating role needs to be addressed since the glycine site is likely to be permanently saturated by the relatively large glycine concentrations present in the extracellular environment. In agreement with this conclusion are the findings that in several preparations, glycine has no potentiating effect on NMDA responses unless it competes with its antagonists such as HA-966 or kynurenate, which depress the NMDA response. Since kynurenate is naturally present in the brain, it is a possibility that the occupancy of the glycine site can be regulated by kynurenate or by still unknown endogenous ligands to allow a regulation of the NMDA receptors by glycinergic nerve inputs.

The effects of polyamines on the NMDA receptor was studied using the *Xenopus* oocyte expression system (McGURK et al. 1990). Spermine produces an increase of up to 100% in the current induced by saturating concentrations of NMDA and glycine by acting at a site distinct from the NMDA or glycine site. Spermine also reduced the glycine-dependent desensitization of the NMDA/glycine-induced current in agreement with the concept that spermine is a regulator of the glycine site.

2. Biochemical and Structural Characterization

The NMDA receptor present on brain membranes has been extensively studied using various radiolabeled ligands. Some have helped in establishing the brain tissue localization of the NMDA receptors (MONAGHAN et al. 1984, 1988) and their developmental patterns (McDONALD and JOHNSTON 1990).

The agonist/antagonist binding site can be labeled with [^{3}H]-Glu, [^{3}H]-D-APV, [^{3}H]-APH, 3((I)-2-carboxypiperazin-4-yl)propyl-1-phosphonate ([^{3}H]-CPP), and *cis*-4-(phosphonomethyl)piperidine-2-carboxylate ([^{3}H]-CGS 19755) (Table 1). [^{3}H]-NMDA itself, however, is not a very good ligand as it binds with low affinity (FOSTER and FAGG 1987).

As PCP and its analogues bind to the NMDA receptor/channel in a voltage- and use-dependent manner, (+)-5-methyl-10,11-dihydro-5*H*-dibenzo [*a,d*]cyclohepten-5,10-iminemaleate ([^{3}H]-MK-801) and *N*-[1,2(thienyl) cyclohexyl]-piperidine ([^{3}H]-TCP) have been used as probes of channel function (KLOOG et al. 1988; REYNOLDS and MILLER 1988; JOHNSON et al. 1988; JAVITT and ZUKIN 1989) to provide information as to the modalities of opening of the channel, binding to the channel, rates of association and dissociation, and their temperature dependence (LAMDANI-ITKIN et al. 1990). Recently, some new tritiated analogues of PCP have been synthesized, including the azido derivative, and used to characterize further their binding site within the NMDA receptor/channel. The binding domain was suggested to prefer ligands forming charge transfer interactions (NADLER et al. 1990).

The strychnine-insensitive glycine sites can be labeled with [³H]-glycine (SNELL et al. 1988; KLOOG et al. 1990), allowing direct studies of the steric requirements for the activation or blockade of the NMDA receptor/channel.

The function of the polyamine site has been explored by following the binding of [³H]-MK-801 (RANSOM and STEC 1988) and [³H]-TCP (SACAAN and JOHNSON 1989) to the NMDA receptor/channel. Spermine and spermidine but not putrescine increase the binding of these channel blockers while ifenprodil, a neuroprotective agent, and its congener SL 82.0715 block this increase by acting as NMDA receptor antagonists at the polyamine site (CARTER et al. 1989). Recently, a nonpsychotropic cannabinoid HU-211 (7-hydroxy-Δ^6-tetrahydrocannabinol-1,1-dimethylheptyl) was found to act as a functional NMDA receptor blocker by binding to a site distinct from the agonist or glycine sites (FEIGENBAUM et al. 1989). It remains to be established whether HU-211 acts at the polyamine site.

On the whole, the binding studies confirm the basic principles governing the activation of the NMDA receptor/channel by agonists and excitotoxins as established by the electrophysiological studies. The binding of blockers to the channel is increased on the one hand by the combined presence of a NMDA receptor agonist, glycine and a polyamine that cause the opening of the channel, and is blocked on the other by the antagonists of NMDA, glycine and polyamines. Once bound, the channel blockers dissociate extremely slowly from the closed channel, but their rate of dissociation increases in the presence of NMDA receptor agonists as the latter open and keep the channel in an open conformation.

Recently, the NMDA receptor has been solubilized and purified 3700-fold by affinity chromatography on amino-PCP-agarose (IKIN et al. 1990). The purified protein has a pharmacological profile similar to that of membranal NMDA receptors and is composed of four polypeptides of 67, 57, 46, and 33 kDa, adding up to a total M_r of 203, a value which closely corresponds to the M_r of 209 estimated by the radiation inactivation method (HONORE 1989).

II. The Kainate Receptor

1. Pharmacological and Electrophysiological Characterization

The KA receptor has a set of specific agonists, KA, domoate, and acromelate (MARUYAMA and TAKEDA 1989), but no antagonist with an exclusive selectivity has yet been identified (Table 2). Antagonists are available to distinguish clearly the activation of the KA receptor versus the NMDA receptor, but they lack specificity when it comes to a distinction between

Table 2. Pharmacological properties of kainate and AMPA receptors

	Kainate	AMPA
Specific agonists:	Kainate, domoate, acromelate	AMPA, ATPA, 5-HPCA, 7-HPCA, Br-HIBO
Specific antagonists:	Quisqualate	NBQX, philanthotoxin-435
Common agonists:	5-bromowillardiine	
Common antagonists:	CNQX, DNQX, pCB-PzDA, pBB-PzDA, kynurenate, γDGG, GAMS, JSTX	

AMPA, α-amino-3-hydroxy-5-methyl-4-isoxazolepropionate; ATPA, α-amino-3-hydroxy-5-tert-butyl-4-isoxazolpropionate; 5-HPCA, 3-hydroxy-4,5,6,7-tetrahydroisoxazolo[5,4-c]pyridine-5-carboxylate; 7-HPCA, 3-hydroxy-4,5,6,7-tetrahydroisoxazolo[5,4-c]pyridine-7-carboxylate; Br-HIBO, α-amino-4-bromo-3-hydroxy-5-isoxazolepropionate; NBQX, 6-nitro-7-sulfamoylbenzo(f)quinoxaline-2,3,-dione; CNQX, 6-cyano-7-nitroquinoxaline-2,3-dione; DNQX, 6,7-dinitroquinoxaline-2,3-dione; pCB-PzDA, 1-(p-chlorobenzoyl)piperazine-2,3-dicarboxylate; pBB-PzDA, 1-(p-bromobenzoyl)piperazine-2,3-dicarboxylate; γDGG, γ-D-glutamylglycine; GAMS, γ-D-glutamylaminomethylusulfonate; JSTX, Joro spider toxin.

the so-called non-NMDA receptors, i.e. the KA and α-amino-3-hydroxy-5-methyl-4-isoxazolepropionate (AMPA) receptors. In contrast to the NMDA receptor, the KA receptor has not yet been found to be positively regulated by allosteric effectors with the exception of a phosphorylation reaction which increases the current flow through the KA receptor/channels, at least in the white perch horizontal cells (Liman et al. 1989) and cultured Bergmann glia (Ortega et al. 1991). However, the KA receptor harbors apparently a site allowing negative regulation since quisqualate (Q), at low but not at high concentrations, inhibits the KA receptor activation (Ishida and Neyton 1985; O'Brien and Fischbach 1986; Verdoorn and Dingledine 1988; Pin et al. 1989; Gallo et al. 1989), possibly by desensitizing the KA receptors or because of its partial agonist nature when acting on the KA receptors.

The properties of the KA receptor/channels resemble very much those of the AMPA receptor/channel (also known as Q ionotropic receptor) to the extent that both Q and KA were thought to activate the same receptor (Cull-Candy and Usowicz 1987; Jahr and Stevens 1987). KA, however, activates predominantly channels with conductances different from those displayed by the Q receptor, i.e., mainly 2 pS but also 15–30 pS (Ascher and Nowak 1988).

The molecular cloning of genes encoding various Glu receptors sensitive to both KA and AMPA has provided the explanation for the similarity of actions of AMPA and KA. Using the *Xenopus* oocyte expression system, Hollmann et al. (1989) were able to clone a rat cDNA encoding a functional Glu receptor of 99.8 kDa with a preferential sensitivity to KA. Activation by KA of this receptor, called GluR-K1, causes a depolarization of the oocyte membrane. Domoate is less potent than KA, while AMPA, Q, and

Glu produce smaller responses and NMDA, none at all (BOULTER et al. 1990). Up to five additional cDNAs encoding polypeptides of similar length to GluR-K1 and sharing about 70% sequence identity have been isolated (KEINANEN et al. 1990; BOULTER et al. 1990; NAKANISHI et al. 1990; BETTLER et al. 1990). Although all these Glu receptors are preferentially activated by KA, they bind tritiated AMPA with an affinity in the nanomolar range, larger by three orders of magnitude than that of KA. On the basis of this finding and of the regional expression patterns of the various mRNAs which correspond to the regional distribution of AMPA binding sites, it was concluded that these Glu receptors belong to the family of AMPA receptors (KEINANEN et al. 1990) rather than to that of KA receptors. Thus, the AMPA receptors are sensitive to both KA and AMPA, but while the latter is the preferential ligand in terms of binding affinity, the former is the preferential agonist producing much larger responses than AMPA (NAKANISHI et al. 1990).

2. Biochemical and Structural Characterization

$[^3H]$-KA has been used in binding studies with brain membranes to characterize the structural requirements of the binding sites, their affinities (FOSTER and FAGG 1984), and density in ontogeny and phylogeny (LONDON et al. 1980). In addition, $[^3H]$-KA has been applied in autoradiographic studies to establish the brain tissue distribution of the binding sites. One of the most striking observations made was the very significant difference (more than $100\times$) in $[^3H]$-KA binding site density between the brain of mammals and that of birds and fish (LONDON et al. 1980; HENKE et al. 1981; HENLEY et al. 1990). The functional or evolutionary significance of this difference remains to be established.

The high density of KA binding sites and the relative abundance of brain tissues from frog, pigeon, and chick have been dominant factors in the success in isolating, in pure form, KA binding proteins (KBP) from these species (HAMPSON and WENTHOLD 1988; KLEIN et al. 1988; GREGOR et al. 1988). All KBPs consist of polypeptide chains with apparent M_r of 48–49.

The frog KBP consists of several (up to 8) isoelectric forms with pI values from 5.5–6.3 (HAMPSON et al. 1989). It harbors a high ($K_d = 5.5$ nM) and low ($K_d = 34$ nM) affinity site for $[^3H]$-KA. Antibodies to the frog KBP react on immunoblots with a 99-kDa rat brain polypeptide which could be equivalent to GluR-K1. This immunoreactive polypeptide comigrates with an unreduced form of frog KBP (HAMPSON et al. 1989). Localization in frog brain tissues of KBP-specific immunoreactivity shows an association mostly with extrasynaptic neuronal membranes but occasionally with subsynaptic membranes, suggesting that KBP could be involved in synaptic neurotransmission (DECHESNE et al. 1990).

The purified chick KBP interacts with KA, domoate, Glu, and Q but not with AMPA (GREGOR et al. 1988; ORTEGA and TEICHBERG unpublished observations). It migrates on sodium dodecylsulfate gel electrophoresis as a relatively broad band with an average apparent M_r of 49 (Gregor et al. 1988). When analyzed by MonoQ ion exchange chromatography, the chick KBP appears to be composed of at least four isoelectric forms, each being able to bind [^{3}H]-KA (Gregor unpublished data). In contrast with the frog, the chick KBP has only a low affinity ($K_d = 255$ nM) for [^{3}H]-KA. A further difference is that the frog KBP is not a substrate for a cAMP-dependent protein kinase (Hampson and Hullebroeck 1990), whereas the chick KBP is phosphorylated with up to 2 moles of phosphate/mole of protein. The phosphorylation of the chick KBP is prevented by KA at the same concentrations as those that activate the KA receptor. This finding has been interpreted as suggesting that the KA-induced change in conformation which leads to the opening of the receptor channel causes as well the burial of the two amino acids sensitive to phosphorylation reactions (Ortega and Teichberg 1990). Antibodies to the chick KBP react on immunoblots of chick brain proteins with polypeptides of 49, 93, and 115 kDa. The function of these cross-reactive polypeptides has not yet been established. It is a possibility that the 115-kDa polypeptide corresponds to the AMPA receptor (see below). Histological studies, performed using light and electron micro-scopy (Gregor et al. 1988; Somogyi et al. 1990), demonstrate that the chick KBP-specific immunoreactivity is localized in the cerebellar molecular layer and associated exclusively with the Bergmann glia.

The cDNAs encoding the neuronal KBP from frog brain and the glial KBP from chick cerebellum have been isolated (Wada et al. 1990; Gregor et al. 1989). The derived amino acid sequences show the frog KBP to be composed of 470 amino acids and the chick KBP, of 464 amino acids. The alignment of the two amino acid sequences reveals 56% identical residues and 72% similar residues (identical plus functionally conserved). For com-parison, the α-, γ-, and δ-subunits of the nicotinic acetylcholine receptor in frog versus chick are 69%–82% identical. Thus, the extent of identity between frog and chick KBPs is lower than expected, suggesting that in spite of their similarities in function, M_r, and sequence (but differences in binding affinity to [^{3}H]-KA), they may not be each species' version of the same protein. This suggestion is in line with their neuronal versus glial localization. Analysis of the hydropathy profiles of these proteins demon-strates the presence of four putative transmembrane domains in a con-figuration very similar to that found in the ligand-gated receptor/channels. This result suggests that the KBPs functions as ion channels, either on their own or as subunits of an oligomeric channel complex. However, transfection experiments with the frog KBP cDNA show the expression of a de novo synthesized [^{3}H]-KA binding protein, but the latter is not able to form a functional ion channel. It is likely that the formation of a KA-gated channel requires additional subunits for physiological activity. The existence of several isoelectric forms of frog KBP and of two related mRNAs on Northern blot analysis, together with the fact that cells transfected with

KBP cDNA display solely a high affinity to [³H]-KA, rather than the high and low affinity of the pure KBP, lend credence to this hypothesis.

The chick and frog KBPs share 36% identical and 72% similar residues with the C-terminal half of GluR-K1. The shared sequences contain the four putative transmembrane domains at similar positions. Accordingly, it can be assumed that the chick and frog KBPs and the rat AMPA receptor/channel derive from a common gene pool.

III. The α-Amino-3-Hydroxy-5-Methyl-4-Isoxazolepropionate Receptor

1. Pharmacological and Electrophysiological Characterization

The AMPA receptor is characterized by specific agonists (Table 2). Its antagonists show a mixed AMPA/KA receptor specificity which may originate from the fact that KA and AMPA do not possess an exclusive specificity for their respective receptors (see above).

Activation of the AMPA receptor produces a membrane current that reverses at 0 mV (CRUNELLI et al. 1984; NAKANISHI et al. 1990). Both Na^+ and K^+ ions thus flow through the AMPA receptor/channel. This, in contrast to the NMDA receptor/channel, shows neither a significant permeability to Ca^{2+} ions (MAYER and WESTBROOK 1987) nor voltage-dependence. The channels activated by Q display various conductances from 8 pS (ASCHER and NOWAK 1988) to 35 pS (TANG et al. 1989). They all tend to open for 2–3 ms and then desensitize.

2. Biochemical and Structural Characterization

The biochemical and structural properties of the AMPA receptor have emerged from molecular cloning studies (HOLLMANN et al. 1989; BOULTER et al. 1990; KEINANEN et al. 1990; NAKANISHI et al. 1990). Some of their properties have been described above. The AMPA receptor has several forms encoded by mRNAs which are differentially distributed in brain. It is not yet clear what combination of isoforms give rise to the AMPA receptors present in situ in neuronal membranes. As a step towards the elucidation of the structure of native AMPA receptors, an AMPA binding protein was isolated from rat brain by immunoaffinity chromatography using antibodies to a synthetic peptide corresponding to the C-terminal portion of the Glur-K1. It was found to migrate on SDS-PAGE as a 114-kDa polypeptide (WENTHOLD et al. 1990) and to bind AMPA with nanomolar affinity but not KA. A KA/AMPA binding protein from *Xenopus* brain (HENLEY et al. 1990; AMBROSINI et al. 1990a) has also been isolated but shown to be composed of polypeptides of 42, 48, 52, and 55 kDa. The purified preparation harbors KA and AMPA sites in a 1:1 ratio. The rank order potency of competitive ligands in inhibiting the binding of either [³H]-KA or [³H]-AMPA is the same: domoate > KA > AMPA > Glu = 6-

cyano-7-nitroquinoxaline-2,3-dione (CNQX). Once reconstituted in artificial lipid vesicles and incorporated into planar lipid bilayers, the KA/AMPA binding protein produces Glu agonist-activated currents, suggesting that it represents a functional Glu receptor/channel. Interestingly, polyclonal antibodies directed against the purified pigeon KBP cross-react with the *Xenopus* 42-kDa polypeptide (AMBROSINI et al. 1990b), while the latter displays 65% amino acid sequence homology with the frog KBP (BARNARD unpublished observation). The sequence homologies between the KBPs, the KA/AMPA binding protein, and the AMPA receptors suggest that they all belong to the same gene family.

IV. The 2-Amino-4-Phosphonobutyrate Receptor

The 2-amino-4-phosphonobutyrate (AP4) receptor is very poorly character- ized pharmacologically, and other than AP4, very few agonist analogues are known. The existence of this receptor was inferred from the observation that AP4 depresses certain excitatory pathways in the brain and spinal cord (KOERNER and COTMAN 1981; DAVIES and WATKINS 1982). In spite of the fact that the activation of the AP4 receptor results in a blockade of synaptic excitation, this receptor is considered a subtype of the Glu receptor because L-AP4 is thought to act as an agonist at presynaptic Glu autoreceptors by an action blocked by γ-D-glutamylglycine and *cis*-2,3-piperidinedicarboxylate, two broad spectrum antagonists of excitatory amino acid receptors. Because its relevance to the excitotoxic process is moot, its properties will not be discussed further; they are described in detail elsewhere (JOHNSON and KOERNER 1988).

V. The *Trans*-1-Amino-1,3-Cyclopentanedicarboxylate Receptor

1. Pharmacological and Electrophysiological Characterization

The *trans*-1-amino-1,3-cyclopentanedicarboxylate (trans-ACPD) receptor is included in the classification of Glu receptors because it is activated by excitatory amino acids such as ibotenate (IBO) and Q (RECASENS et al. 1990). However, the activation of this receptor does not lead to an exci- tation of the cell but to the formation of inositolphosphates (IP). AP4 and aminophosphonopropionate are the major antagonists of the trans-ACPD receptor, but their respective potency varies with the maturation of the nervous system. Although the trans-ACPD receptor is activated by Q, it is not blocked by CNQX or any other AMPA receptor antagonists. The activation by Q is at the origin of some confusion in receptor nomenclature, which has not yet reached a state of consensus. The literature refers to this receptor as the Q metabotropic receptor (SUGIYAMA et al. 1988), the G_{P2} metabolotropic receptor (COSTA et al. 1988), the IBO-Q-sensitive receptor

(RECASENS et al. 1990), and the trans-ACPD receptor (MONAGHAN et al. 1989).

The electrophysiological consequences of the activation of this receptor were investigated using *Xenopus* oocytes injected with brain RNA (SUGIYAMA et al. 1988). In this system, studied under voltage-clamp conditions, Q elicits an oscillatory current response which reverses at the equilibrium potential of Cl^- ions. The oscillatory current response is blocked by pretreatment with pertussis toxin and intracellular injections of ethylene glycol-bis(β-amino ethyl Ether)N,N,N',N'-tetraacetic acid (EGTA), indicating the involvement of both a G protein and intracellular Ca^{2+} ions. The latter are thought to activate the oocyte's own Cl^- ion channels. The Q response is clearly mediated by inositol-1,4,5-trisphophate (IP_3), since intracellular injections of IP_3 not only mimic but also cause the desensitization of the oscillatory Q response. Recently, the activation of the metabotropic receptor in hippocampal neurons dialyzed with the nonhydrolyzable GTP thiol derivative, guanosine S'-triphosphate (GTPγS), was found to cause an irreversible depression of the Ca^{2+} current only when the latter ion was present either in the internal or external medium (LESTER and JAHR 1990). This result suggests that the G-protein-coupled metabotropic receptor can stimulate phosphoinositol turnover and depress voltage-dependent Ca^{2+} currents.

A metabotropic Glu receptor has recently been characterized following the isolation of its cDNA (MASU et al. 1991). It consists of a polypeptide of 1199 amino acids displaying at least 8 putative transmembrane segments and two large hydrophilic domains, one of 570 amino acids at the N-terminus, likely to be located extracellularly, and the other of about 360 amino acids at the C-terminus, presumably located intracellularly. Both the amino acid sequence and structural architecture of this Glu receptor differ from those of the conventional G-protein-coupled receptors. There is, however, some small but statistically significant similarity with some segments of the AMPA receptor. Once expressed in *Xenopus* oocytes, the receptor is sensitive to the following agonists: Q > IBO > Glu > ACPD, which all elicit the formation of IP_3. This response, which is blocked by pertussis toxin and by intracellularly applied EGTA, manifests itself by a chloride current as described above.

D. Mechanisms of Action of Excitotoxic Substances

The toxicity of excitotoxins, i.e., of glutamatergic agonists, is clearly mediated by their interactions with Glu receptors, since it can be prevented, from the onset of the excitotoxin exposure, by glutamatergic antagonists (OLNEY et al. 1981; ROTHMAN and OLNEY 1987; OLNEY 1990). Thus, an understanding of the mechanisms leading to cell death must be based first on the comprehension of the biochemical events taking place in the cell following the activation of Glu receptors (Table 3).

Table 3. Summary of properties of glutamate receptors

	NMDA	Kainate	AMPA
Ionic selectivity	Na, K, Ca	Na, K	Na, K
Voltage dependence	Yes	No	No
Channel conductance	50 pS	2 pS, 15–30 pS	8 pS, 35 pS
Desensitization	Yes	Yes	Yes
Phosphorylation	Probable	Yes	?
Divalent ion binding	Mg, Zn	?	?
Subunit composition	67, 57, 46, 33 kDa	Oligomers of 99 kDa only? 49 kDa only?	114 kDa 42, 48, 52, 55 kDa (AMPA/kainate receptor)
Native molecular weight	203 kDa	300–400 kDa	180 kDa
Allosteric regulators	Glycine, Polyamines	Quisqualate, glutamate?	?
Heterogeneity	Yes	Yes	Yes

NMDA, N-methyl-D-aspartate; AMPA, α-amino-3-hydroxy-5-methyl-4-isoxazole-propionate.

I. N-Methyl-D-Aspartate-Induced Excitotoxicity

The very first events relevant to NMDA excitotoxicity consist of influxes of Na^+ and Ca^{2+} ions through the receptor/channel. The role of these ions in excitotoxicity has been investigated by ion substitution experiments carried out with cultured neurons (ROTHMAN 1985; CHOI 1985, 1987), isolated chick embryo retinas (OLNEY et al. 1986; DAVID et al. 1988), and rat cerebellar slices (GARTHWAITE et al. 1986). The early disagreements on the identity of the offending ions were soon accounted for by methodological differences linked to the time of exposure of the cells to the excitotoxins, the nature of the preparation (age of animals, cultured cells versus brain slices, etc.), and the Ca^{2+} concentrations used. It is now established that the NMDA-mediated excitotoxic process can be separated into two phases, an early phase dependent on extracellular Na^+ and Cl^- ions and a late phase dependent on Ca^{2+} ions.

1. Early Events: Role of Sodium/Chloride Ions

The influx of Na^+ ions is the major factor responsible for the acute toxicity not only of excitatory amino acids but also of depolarizing agents such as veratridine and elevated extracellular K^+ ions (ROTHMAN 1985). This Na^+ influx causes the cell depolarization which entails a passive influx of Cl^- ions, since the distribution of Cl^- ions across the cell membrane follows the changes of membrane potential. The build-up of a NaCl intracellular pool leads to a movement of water, swelling of the cell, and acute osmotic damage. Deleting either Na^+ or Cl^- ions from the extracellular medium prevents the toxic process (ROTHMAN 1985); it, however, is not affected by changes in the extracellular Ca^{2+} ion concentration. There is some evidence, at least in the retina, that the Cl^- ion influx is not entirely passive but may take place through the Cl^-/bicarbonate channel since the excitotoxicity can be blocked by the anion channel blocker 4,4'-diisothiocyano-2,2'-stilbene disulphonate (DIDS) (ZEEVALK et al. 1989).

2. Late Events: Role of Calcium Ions and "Downstream Processes"

CHOI (1985, 1987) made the striking observation that short exposures of cultured cortical neurons to Glu (5 min) produce a reversible perikaryal swelling. However, if Ca^{2+} ions are present in the extracellular fluid during that period, the cells degenerate after a delay of several hours, during most of which they show a normal appearance. Exposure of the cells to the calcium ionophore A23187 does not produce acute cell swelling but leads to late neuronal death, indicating that the increase in intracellular Ca^{2+} ions is at the origin of the toxic process.

The phenomenon of Ca^{2+}-mediated cell death is not unique to the nervous system and not restricted to a specific cell system or organ. Therefore, the understanding of the role of Ca^{2+} ions in excitotoxicity has benefited from the numerous toxicological studies that have been performed on the susceptibility of tissues such as the liver or heart to environmental toxins, chemicals, and drugs (ORRENIUS et al. 1989). Several biochemical mechanisms have already been identified that appear to be generally involved in Ca^{2+}-dependent cell injury and death. The latter are mediated by either one or the combination of the following processes: perturbation of cytoskeletal organization, phospholipase activation, protease activation, endonuclease activation, impaired mitochondrial function, protein kinase C activation, xanthine oxidase activation, depletion of energy reserves.

Although not all of these processes have been investigated in the excitotoxicity context, an increasing body of evidence has accumulated to suggest that they do occur in neurons exposed to excitotoxins, and ways to prevent their occurrence have been reported to provide some degree of protection against the excitotoxic insult.

a) Perturbation of Cytoskeletal Organization

Excitatory amino acids have been found to activate, in rat hippocampal neurons, Ca^{2+}-dependent neutral proteases called calpains, causing selective loss of brain spectrin and microtubule-associated protein MAP2, thus inducing structural breakdown (SIMAN and NOSZEK 1988). Glu also causes a Ca^{2+}-dependent increase in tau and ubiquitin immunostaining in cultured hippocampal neurons (MATTSON 1999). Tau is a microtubule associated protein often present in the neurofibrillary tangles found in the brain of people suffering from neurodegenerative disorders such as Alzheimer's disease. Ubiquitin is believed to bind to cellular proteins including cyto-skeletal proteins and target them for degradation by specific proteases.

b) Phospholipase Activation

In differentiated striatal neurons in culture, the associative stimulation of the NMDA and trans-ACPD receptors triggers the release of arachidonic acid, AA (BOCKAERT et al. 1990), from phosphatidylinositol or phosphatidylcholine by activating a Ca^{2+}-dependent phospholipase A_2. Being a cellular second messenger, AA produces a plethora of effects (AXELROD et al. 1988), several of them being possibly relevant to the excitotoxic process. For example, it can elevate cytosolic calcium levels, thus amplifying the effects of both NMDA and IP_3. AA and its metabolites are known to play a role in brain injury and cerebral ischemia (WOLFE and PAPPIUS 1984), i.e., under pathological conditions that can be triggered by excessive release of Glu and blocked by glutamatergic antagonists (CHOI and ROTHMAN 1990). AA was also found to induce a prolonged inhibition of Glu uptake into glial cells (BARBOUR et al. 1989), a fact which contributes to an exacerbated excitotoxicity (SUGIYAMA et al. 1989).

c) Endonuclease Activation

There is no direct information as to the activation of Ca^{2+}-dependent endonucleases leading to DNA fragmentation in the excitotoxic process, but the latter is well-known to cause major disturbances in RNA and protein synthesis (COYLE et al. 1978) and to affect chromatin structure (OLNEY 1978). Glu and KA produce an increase in protein synthesis in cultured rat cerebellar granule cells and cause neuronal degeneration. Protein synthesis inhibitors, and to some extent RNA synthesis inhibitors, were reported to limit the neuronal damage (MORANDI et al. 1990), suggesting that the excitotoxicity is translation-dependent and possibly requires protein synthesis. It is of interest to mention that transcription factors such as c-*fos* and c-*jun* are expressed at high levels in brain neurons exposed to either excitatory amino acids or convulsants such as metrazole and picrotoxin (POPOVICI et al. 1988; SONNENBERG et al. 1989). The nature of the proteins synthesized de novo following exposure of neurons to excitotoxins remains to be established.

d) Protein Kinase C Activation

Glu causes the translocation of the Ca^{2+}/phospholipid-dependent protein kinase (PKC) from the cytoplasm to the membrane where the enzyme possibly phosphorylates proteins involved in the cell degeneration process.

The latter conclusion can be drawn from the findings that both gangliosides which prevent PKC translocation (VACCARINO et al. 1987) and phorbol esters which cause PKC down-regulation decrease Glu excitotoxicity without affecting the Glu-induced activation of the various Glu receptors (FAVARON et al. 1988, 1990).

e) Xanthine Oxidase Activation

Ca^{2+}-activated serine proteases cause the irreversible conversion of xanthine deshydrogenase into xanthine oxidase which generates cytotoxic superoxide and hydroxyl radicals (McCORD 1985). The KA-induced death of cultured cerebellar neurons is partly mediated by the activation of xanthine oxidase, since a significant protecting effect can be obtained by treatment of the cell cultures with allupurinol (a specific inhibitor of xanthine oxidase), mannitol (a hydroxyl radical scavenger), superoxide dismutase (a superoxide radical scavenger), and aprotinin (a serine protease inhibitor) (DYKENS et al. 1987).

II. Kainate- and Quisqualate-Induced Excitotoxicity

There is some debate as to the ionic events leading to KA- and Q-induced excitotoxicity, possibly because in some biological preparations, KA and Q effects are mediated by released Glu acting at NMDA receptors.

As described above, Q is an agonist of both the AMPA receptor and the trans-ACPD receptor. Therefore, its excitotoxicity is complex and needs to be analyzed in parallel with that of AMPA.

A 5-min exposure of cultured cortical neurons to high concentrations of Q (0.5 mM) produces acute neuronal swelling and late neuronal loss, while AMPA causes none of these changes (KOH et al. 1990). The effect of Q is blocked by APV, indicating that they are mediated by NMDA receptors. However, 24-h exposure to micromolar concentrations of Q and AMPA causes neuronal loss in a process unaffected by APV.

In young rat hippocampal slices, Q (0.1 mM) leads to, within 20 min, a process of "dark cell degeneration" which is unaffected either by the presence of APV or the reduction of extracellular Ca^{2+} and Cl^- ions and is thus likely to be due to the activation of the trans-ACPD metabotropic receptor and subsequent elevation of free intracellular Ca^{2+} levels (GARTHWAITE and GARTHWAITE 1989). In the same preparation, NMDA toxicity is Ca^{2+}-dependent and Cl^--independent.

A 5-min exposure of cultured cortical neurons to high concentrations of KA (0.5 mM) produces an acute but transient neuronal swelling resistant to APV but no late neuronal loss (KOH et al. 1990). However, 24-h exposure to micromolar concentrations of KA destroys neurons. It is suggested that the latter process does require Ca^{2+} ions, which gain access into the cell possibly via voltage-gated channels and the reverse operation of the Na^+/Ca^{2+} exchanger (KOH et al. 1990). In contrast, KA (0.1 mM), applied to cultured cerebellar granule cells, causes a rapid cell death which can be prevented by high concentrations of Ca^{2+} and the omission of Cl^- ions (McCoslin and Smith 1988). The studies of KA toxicity with brain slices produce equally conflicting results with respect to the Ca^{2+} dependence. In slices from immature rat hippocampus, the KA neurotoxicity is prevented by reduction of extracellular Cl^- ions but exacerbated by Ca^{2+} omission (Lehmann 1990), while it is abolished in rat cerebellar Golgi cells under Ca^{2+}-free conditions (Garthwaite et al. 1986) and unaffected by Cl^- ion omission.

These studies strongly suggest that the ionic mechanisms of excitotoxity not only differ from excitotoxin to excitotoxin but also vary between cell populations according to their age, biochemical make-up, and respective representation and density to Glu receptor subtypes. Indeed, NADPH-diaphorase-containing neurons and GABAergic cortical neurons are resistant to NMDA receptor-mediated injury but susceptible to KA and Q (Tecoma and Choi 1989; Koh et al. 1990), while striatal acetylcholine-containing neurons are relatively immune to KA (Koh and Choi 1988a). Also, the various cerebellar cells show a clear but distinct and graded susceptibility to KA (Hajos et al. 1986). Glial cells, though harboring Glu receptors (Backus et al. 1989), are not destroyed when exposed (even when prolonged) to Glu but disintegrate following exposure to A23187 (Choi et al. 1987). With the advances in the molecular cloning of Glu receptor genes, cytotoxicity studies using transfected nonneuronal cells can be expected to clarify the toxic mechanisms exerted by individual Glu receptors.

E. Endogenous Mechanisms of Protection from Excitotoxicity

The brain levels of Glu and Asp are extremely high (0.011 and 0.003 mmole/g, respectively) and in nerve terminals, the Glu concentration can reach the value of 66 mM (Fonnum 1984). Since Glu and Asp are released into the extracellular medium at high concentrations during synaptic transmission, protective mechanisms are imperative to prevent the excitotoxic process and dispose of the excitotoxins. Whatever their nature, these mechanisms are effective and protect the brain from the endogenous excitotoxins during its normal physiological operations but often fail during pathological states such as ischemia and hypoglycemia, leading to excessive and prolonged exposures to Glu (Choi and Rothman 1990). Moreover, they are apparently inefficient when challenged with exogenous excitotoxins such as β-N-methylamino-L-alanine and domoate. The latter have been suggested to cause, respectively, the Guam amyotrophic lateral sclerosis/ parkinsonism/ dementia complex and anterograde amnesia (Olney 1990). Probably all exogenous excitotoxins are able to exert their deleterious effects because of the failure of the protective mechanisms.

I. Receptor Desensitization

Agonist-dependent desensitization of excitotoxin receptors is the most basic mechanism protecting from cytotoxic excessive ionic imbalances. Prolonged agonist exposures cause desensitization, i.e., a closure of the receptor-associated ion channel, and may produce down-regulation, i.e., a decrease in receptor synthesis or its increased degradation. NMDA and AMPA receptors readily undergo desensitization (COLLINGRIDGE and LESTER 1989), whereas the reality of KA receptor desensitization is still being debated. In numerous instances, KA responses are found not to desensitize, but in others they do, as in frog oocytes injected with rat brain mRNA (RASSENDREN et al. 1990), rat dorsal root ganglion neurons (HUETTNER 1989), and chick motoneurons (DUDEL et al. 1990). Desensitization does not prevent the lethal excitotoxicity exerted by exogenous excitotoxins, but it remains to be demonstrated whether it protects from endogenous excitotoxins.

II. Cellular Uptake of Glutamate

The cellular uptake of Glu plays an important protective role against cytotoxic insults both in vivo and in vitro. The removal, by deafferentation of the corticostriatal pathway or blockade with DL-threo-3-hydroxyaspartate, of this uptake system increases markedly the vulnerability of striatal neurons to Glu exposure (MCBEAN and ROBERTS 1985). The presence within hippo-campal cultures of glial cells harboring Glu uptake mechanisms significantly contributes to the survival of neurons grown in a culture medium containing normally neurotoxic concentrations ($0.1\,mM$) of Glu and Asp (SUGIYAMA et al. 1989), thus emphasizing the protective role of glia.

III. Calcium Homeostasis

The cellular mechanisms regulating calcium homeostasis are likely to play a very significant role in the containment of endogenous excitotoxins. However, the individual contributions of the various components such as the Na^+/Ca^{2+} exchanger and the several Ca^{2+}-binding proteins and ATP-driven calcium pumps have not been yet analyzed in detail.

IV. Built-In Mechanisms

The presence and occupation of allosteric sites on Glu receptors able to tone down the agonist response may provide a built-in mechanism of protection against excessive excitation by endogenous excitotoxins. Thus, the occupation by Glu of the Q site present on the KA receptor may create an appropriate safety barrier. Likewise, the occupation by kynurenate of the NMDA receptor glycine site (see above) or that of the Zn site by synaptically release Zn (KOH and CHOI 1988b) may help in attenuating NMDA receptor-mediated neuronal damage.

F. Mechanisms Contributing to Excitotoxicity

I. Role of Afferences

The toxicity of intrastriatally injected KA depends on the integrity of the glutamatergic corticostriatal afferences (BIZIERE and COYLE 1978). Decortication does not decrease the density of KA binding sites, nor does it affect the excitatory effects of iontophoretically applied KA (COYLE 1982). Co-injections of Glu and KA, neither of which are neurotoxic in the deafferented striatum, restore to some extent KA neurotoxicity. This paradigm, which represents possibly the only exception to the correlation between excitation and excitotoxicity, has not yet received a convincing explanation.

II. Cellular Metabolism

Exhaustion of energy stores as a result of hyperexcitation, oxygen, glucose and/or ATP shortages and all other conditions impairing the cellular ionic balance and calcium homeostasis will exacerbate the deleterious effects of endogenous excitotoxins which otherwise are counterbalanced by the protective mechanisms. The formation of necrotic secondary foci in epilepsy are likely to be due to the combination of enhanced exposure to endogenous excitotoxins and impaired cellular metabolism.

G. Excitotoxicology: New Perspectives

During the past 20 years, the field of excitotoxicology has seen a tremendous development from the dark ages of widespread unawareness (1969–1980) to a renaissance of basic findings and explanations (1980–1990) reaching now the golden age of consolidation as a crucial link in the neuroscience web. Excitotoxicology is nowadays intertwined with neurochemistry, electrophysiology, and neuropathology. Its relation to the development of the nervous system architecture, to epigenesis and necrobiosis, and to brain plasticity is only beginning to emerge. Excitotoxicology is thus bound to keep its excitement.

References

Ambrosini A, Prestipino G, Henley JM, Barnard EA (1990a) A purified functional non-NMDA glutamate receptor reconstituted in artificial lipid bilayers. Neurochem Int [Suppl1] 16:26

Ambrosini A, Streit P, Barnard EA (1990b) Relation between pigeon and *Xenopus* oocytes kainate receptors: an immunological study. Neurochem Int [Suppl 1] 16:26

Ascher P, Nowak L (1988) Quisqualate- and kainate-activated channels in mouse central neurons in cultures. J Physiol (Lond) 399:227–246

Ascher P, Bregetovski P, Nowak L (1988) *N*-Methyl-D-aspartate activated channels in mouse central neurons in magnesium-free solution. J Physiol (Lond) 339:207–226

Axelrod J, Burch RM, Jelsema CL (1988) Receptor-mediated activation of phospholipase A2 via GTP-binding proteins: arachidonic acid and its metabolites as second messengers. Trends Neurosci 11:117–123

Backus KH, Kettenmann H, Schachner M (1989) Pharmacological characterization of the glutamate receptor in cultured astrocytes. J Neurosci 22:274–282

Barbour B, Szetkowski M, Ingledew N, Attwell D (1989) Arachidonic acid induces a prolonged inhibition of glutamate uptake into glial cells. Nature 342:918–920

Bettler B, Boulter J, Hermans-Borgmeyer I, O'Shea-Greenfield A, Deneris ES, Moll C, Borgmeyer U, Hollmann M, Heinemann S (1990) Cloning of a novel glutmate receptor subunit, GluR5: expression in the nervous system during development. Neuron 5:583–595

Biziere K, Coyle JT (1978) Influence of cortico-striatal afferents on striatal kainic acid neurotoxicity. Neurosci Lett 8:303–310

Bockaert J, Pin JP, Sebben M, Dumuis A (1990) Associative stimulation of NMDA or metabotropic quisqualate receptors and depolarization are needed to trigger arachidonic acid release. Neurochem Int [Suppl 1] 16:16

Boulter J, Hollmann M, O'Shea-Greenfield A, Hartley M, Deneris E, Maron C, Heinemann S (1990) Molecular cloning and functional expression of glutamate receptor subunit genes. Science 249:1033–1037

Carter C, Rivy JP, Thuret F, Lloyd KG, Scatton B (1989) Ifenprodil and SL 82.0715 are antagonists at the polyamine modulatory site of the NMDA receptor complex. Soc Neurosci Abstr 15:133.8

Choi DW (1985) Glutamate neurotoxicity in cortical cell cultures is calcium dependent. Neurosci Lett 58:293–297

Choi DW (1987) Ionic dependence of glutamate neurotoxicity. J Neurosci 7:369–379

Choi DW, Rothman SM (1990) The role of glutamate neurotoxicity in hypoxic-ischemic neuronal death. Annu Rev Neurosci 13:171–182

Choi DW, Maulucci-Gedde M, Kriegstein AR (1987) Glutamate neurotoxicity in cortical cell cultures. J Neurosci 7:357–368

Collingridge GL, Lester RAJ (1989) Excitatory amino acid receptors in the vertebrate central nervous system. Pharmacol Rev 40:145–210

Costa E, Fadda E, Kozikowski AP, Nicoletti F, Wroblewski JT (1988) Classification and allosteric modulations of excitatory amino acid signal transduction in brain slices and primary cultures of cerebellar neurons. In: Ferrandelli JA, Collins RC, Johnson EM (eds) Neurobiology of amino acids, peptides and trophic factors. Kluwer Academic, Boston, p 35

Cotman CW, Monaghan DT, Ottersen OP, Storm-Mathisen J (1987) Anatomical organization of excitatory amino acid receptors and their pathways. Trends Neurosci 10:273–280

Coyle JT (1982) Excitatory amino acid neurotoxins. In: Iversen LL, Iversen SD, Snyder SM (eds) Handbook of psychopharmacology, vol 15. Plenum, New York, p 237

Coyle JT, McGeer EG, McGeer PL, Schwarcz R (1978) Neostriatal injections: a model for Huntington's chorea. In: McGeer EG, Olney JW, McGeer PL (eds) Kainic acid as a tool in neurobiology. Raven, New York, p 139

Cull-Candy SG, Usowicz MM (1987) Multiple conductance channels activated by excitatory amino acids in cerebellar neurons. Nature 325:525–527

Crunelli V, Forda S, Kelly JS (1984) The reversal potential of excitatory amino acid action on granule cells of the dentate gyrus. J Physiol (Lond) 341:627–640

David P, Lusky M, Teichberg VI (1988) Involvement of excitatory neurotransmitters in the damage produced in chick retinas by anoxia and extracellular high potassium. Exp Eye Res 46:657–662

Davies J, Watkins JC (1982) Action of D and L forms of 2-amino-5-phosphonovalerate and 2-amino-4-phosphonobutyrate in the cat spinal cord. Brain Res 235:378–386

Dechesne CJ, Oberdorfer MD, Hampson DR, Wheaton KD, Nazaroli AJ, Hoping G, Wenthold RJ (1990) Distribution of a putative kainic acid receptor in the frog central nervous system determined by monoclonal and polyclonal antibodies: evidence for synaptic and extra synaptic localization. J Neurosci 10:479–490

Dudel J, Franke C, Hatt H, Rosenheimer JL, Smith DO, Zufall F (1990) Desensitization and resensitization of glutamatergic channels of motoneurons. Neurochem Int [Suppl 1] 16:75

Dykens JA, Stern A, Trenkner E (1987) Mechanism of kainate toxicity to cerebellar neurons in vitro is analogous to reperfusion tissue injury. J Neurochem 49:1222–1228

Favaron M, Manev H, Alho H, Bertolino M, Ferret B, Guidotti A, Costa E (1988) Gangliosides prevent glutamate and kainate neurotoxicity in primary neuronal cultures of neonatal rat cerebellum and cortex. Proc Natl Acad Sci USA 85:7351–7355

Favaron M, Manev H, Siman R, Bertolino M, Szekely AM, DeErausquin, Guidotti G, Costa E (1990) Down-regulation of protein kinase C protects cerebellar granule neurons in primary culture from glutamate-induced neuronal death. Proc Natl Acad Sci USA 87:1983–1987

Feigenbaum JJ, Bergmann F, Richmond SA, Mechoulam R, Nadler V, Kloog Y, Sokolovsky M (1989) Nonpsychotropic cannabinoid acts as a functional N-methyl-D-aspartate receptor blocker. Proc Natl Acad Sci USA 86:9584–9587

Fonnum F (1984) Glutamate: a transmitter in mammalian brain. J Neurochem 42:1–11

Foster AC, Fagg GE (1984) Acidic amino acid binding sites in mammalian neuronal membranes: their characteristics and relationship to synaptic receptors. Brain Res Rev 7:103–164

Foster AC, Fagg GF (1987) Comparison of L-[^{3}H]glutamate, D-[^{3}H]aspartate, DL-[^{3}H]AP5 and [^{3}H]NMDA as ligands for NMDA receptors in crude post-synaptic densities from rat brain. Eur J Pharmacol 133:291–300

Gallo V, Giovannini C, Levi G (1989) Quisqualic acid modulates kainate responses in cultured cerebellar granule cells. J Neurochem 52:10–16

Garthwaite G, Garthwaite J (1989) Differential dependence on Ca^{++} of N-methyl-D-aspartate and quisqualate neurotoxicity in young rat hippocampal slices. Neurosci Lett 97:316–322

Garthwaite G, Hajos F, Garthwaite J (1986) Ionic requirements for neurotoxic effect of excitatory amino acid analogues in rat cerebellar slices. Neuroscience 18:437–447

Gregor P, Eshhar N, Ortega A, Teichberg VI (1988) Isolation, immunochemical characterization and localization of the kainate subclass of glutamate receptor from chick cerebellum. EMBO J 7:2673–2679

Gregor P, Mano I, Maoz I, McKeown M, Teichberg VI (1989) Molecular structure of the chick cerebellar kainate binding subunit of a putative glutamate receptor. Nature 342:689–692

Hajos F, Garthwaite G, Garthwaite J (1986) Reversible and irreversible neuronal damage caused by excitatory amino acid analogues in rat cerebellar slices. Neuroscience 18:417–436

Hampson DR, Hullebroeck MF (1990) Phosphorylation of a kainic acid binding protein. Soc Neurosc Abstr 16:230.6

Hampson DR, Wenthold RJ (1988) A kainic acid receptor from frog brain purified using domoic acid affinity chromatography. J Biol Chem 263:2500–2505

Hampson DR, Wheaton KD, Dechesne CJ, Wenthold RJ (1989) Identification and characterization of the ligand binding subunit of a kainic acid receptor using monoclonal antibodies and peptide mapping. J Biol Chem 264:13329–13335

Henke H, Beaudet A, Cuenod M (1981) Autoradiographic localization of specific kainic acid binding sites in pigeon and rat cerebellum. Brain Res 219: 95–105

Henley JM, Ambrosini A, Krogsgaard-Larsen P, Barnard EA (1989) Evidence for a single glutamate receptor of the ionotropic kainate/quisqualate type. New Biol 1:153–158

Hollmann M, O'Shea-Greenfield A, Rogers SW, Heinemann S (1989) Cloning by functional expression of a member of the glutamate receptor family. Nature 342:643–648

Honoré T (1989) Excitatory amino acid subtypes and specific antagonists. Med Res Rev 9:1–23

Huettner JE (1989) Current gated by kainate in rat dorsal root ganglion neurons: conA abolishes desensitization. Soc Neurosci Abstr 16:461.5

Ikin AF, Kloog Y, Sokolovsky M (1990) N-Methyl-D-aspartate/phencyclidine receptor complex of rat forebrain: purification and biochemical characterization. Biochemistry 29:2290–2295

Ishida M, Neyton J (1985) Quisqualate and L-glutamate inhibit retinal horizontal cell response to kainate. Proc Natl Acad Sci USA 82:1837–1841

Ishida M, Shinozoki H (1988) Acromelic acid is a much more potent excitant than kainic acid or domoic acid in the isolated rat spinal cord. Brain Res 474:386–389

Jahr CE, Stevens CF (1987) Glutamate activates multiple single channel conductances in hippocampal neurons. Nature 325:522–525

Javitt DC, Zukin SR (1989) Interaction of [^{3}H]MK-801 with multiple states of the N-methyl-D-aspartate complex of rat brain. Proc Natl Acad Sci USA 86:740–744

Johnson KM, Sacaan AI, Snell LD (1988) Equilibrium analysis of [^{3}H]TCP binding: effects of glycine, magnesium and N-methyl-D-aspartate agonists. Eur J Pharmacol 152:141–146

Johnson RL, Koerner JF (1988) Excitatory amino acid neurotransmission. J Med Chem 32:2057–2066

Keinanen K, Wisden W, Sommer B, Werner P, Herb A, Verdoorn TA, Sackmann B, Seeburg PH (1990) A family of AMPA-selective glutamate receptors. Science 249:556–560

Kessler M, Terramani T, Lynch G, Baudry M (1989) A glycine site associated with N-methyl-D-aspartic acid receptors: characterization and identification of a new class of antagonists. J Neurochem 52:1319–1328

Kleckner NW, Dingledine R (1988) Requirement for glycine in activation of NMDA receptors expressed in *Xenopus* oocytes. Science 241:835–837

Klein AV, Niederoest B, Winterholter KH, Cuenod M, Streit P (1988) A kainate binding protein in pigeon cerebellum: purification and localization by monoclonal antibody. Neurosci Lett 95:359–364

Kloog Y, Haring R, Sokolovsky M (1988) Kinetic characterization of the phencyclidine-N-methyl-D-aspartate receptor interaction: evidence for a steric blockade of the channel. Biochemistry 27:843–848

Kloog Y, Lamdani-Itkin H, Sokolovsky M (1990) The glycine site of the N-methyl-D-aspartate receptor channel: differences between the binding of HA-966 and 7-chlorokynurenic acid. J Neurochem 54:1576–1583

Koerner JF, Cotman CW (1981) Micromolar L-2-amino-4-phosphonobutyric acid selectively inhibits perforant path synapses from lateral entorhinal cortex. Brain Res 216:192–198

Koh J-Y, Choi DW (1988a) Cultured striatal neurons containing NADPH-diaphorase or acetylcholinesterase are selectively resistant to injury by NMDA receptor agonists. Brain Res 446:374–378

Koh J-Y, Choi DW (1988b) Zinc alters excitatory amino acid neurotoxicity on cortical neurons. J Neurosci 8:2164–2171

Koh J-Y, Goldberg M, Hartley DM, Choi DW (1990) Non-NMDA receptor-mediated neurotoxicity in cortical culture. J Neurosci 10:693–705

Lamdani-Itkin H, Kloog Y, Sokolovsky M (1990) Modulation of glutamate-induced uncompetitive blocker binding to the NMDA receptor by temperature and by glycine. Biochemistry 29:3987–3993

Lehmann A (1990) Kainic acid neurotoxicity in slices from immature rat hippocampus: protection by chloride reduction and exacerbation by calcium omission. Neurosci Res Commun 6:27–36

Lester RAJ, Jahr CE (1990) Quisqualate receptor-mediated depression of calcium currents in hippocampal neurons. Neuron 4:741–749

Liman ER, Knapp AG, Dowling JE (1989) Enhancement of kainate-gated currents in retinal horizontal cells by cyclic AMP-dependent protein kinase. Brain Res 481:399–402

Lodge D, Johnson KM (1990) Noncompetitive excitatory amino acid receptor antagonists. Trends Pharmacol Sci 11:81–86

London ED, Klemm N, Coyle JT (1980) Phylogenetic distribution of [^{3}H]kainic acid receptor binding sites in neuronal tissue. Brain Res 192:463–476

Maruyama M, Takeda K (1989) Effects of acromelic acid A on the binding of [^{3}H]glutamic acid and [^{3}H]kainic acid to synaptic membranes and on the depolarization at the frog spinal cord. Brain Res 504:328–331

Masu M, Tanabe Y, Tsuchida K, Shigemoto R, Nakanishi S (1991) Sequence and expression of a metabotropic glutamate receptor. Nature 349:760–765

Mattson MP (1990) Antigenic changes similar to those seen in neurofibrillary tangles are elicited by glutamate and Ca^{++} influx in cultured hippocampal neurons. Neuron 2:105–117

Mayer ML, Westbrook GL (1987) Permeation and block of N-methyl-D-aspartic acid receptor channels by divalent cations in mouse cultured central neurons. J Physiol (Lond) 394:501–527

Mayer ML, Vyklicky L Jr, Clements J (1989) Regulation of NMDA receptor desensitization in mouse hippocampal neurons by glycine. Nature 338:425–427

McBean GJ, Roberts PJ (1985) Neurotoxicity of L-glutamate and DL-thio-3-hydroxyaspartate in the rat striatum. J Neurochem 44:247–254

McCord JM (1985) The role of superoxide in postischemic tissue injury. In: Oberley LW (ed) Superoxide dismutase, vol 3. CRC, Boca Raton, p 143

McCoslin PP, Smith TG (1988) Quisqualate, high calcium concentration and zero-chloride prevent kainate-induced toxicity of cerebellar granule cells. Eur J Pharmacol 152:341–346

McDonald JW, Johnston MV (1990) Physiological and pathophysiological roles of excitatory amino acids during central nervous system development. Brain Res Rev 15:41–70

McGurk JF, Zukin RS, Bennett MVL (1990) Effects of polyamines on excitatory amino acid receptor induced currents in oocytes. Neurochem Int [Suppl 1] 16:53

Monaghan DT, Yoo D, Olverman HJ, Watkins JC, Cotman CW (1984) Autoradiography of D-2-[^{3}H]amino-5-phosphonopentanoate binding sites in rat brain. Neurosci Lett 52:253–258

Monaghan DT, Olverman HJ, Ngujen L, Watkins JC, Cotman CW (1988) Two classes of N-methyl-D-aspartate recognition sites: differential distribution and differential regulation by glycine. Proc Natl Acad Sci USA 85:9836–9840

Monaghan DT, Bridges RJ, Cotman CW (1989) The excitatory amino acid receptors: their classes, pharmacology, and distinct properties in the function of the central nervous system. Annu Rev Pharmacol Toxicol 29:365–402

Morandi A, Facci L, Leon A, Toffano G, Skaper SD (1990) Inhibition of protein synthesis protects against excitatory amino acid neurotoxicity. Neurochem Int [Suppl 1] 16:56

Murphy DE, Snowhill EW, Williams M (1987) Characterization of quisqualate recognition sites in rat brain tissue using [^{3}H]AMPA and a filtration assay. Neurochem Res 187:113–127

Nadler V, Kloog Y, Sokolovsky M (1990) Distinctive structural requirement for the binding of uncompetitive blockers (phencyclidine-like drugs) to the NMDA receptor. Eur J Pharmacol 188:97–104

Nakanishi N, Shneider NA, Axel R (1990) A family of glutamate receptor genes: evidence for the formation of heteromultimeric receptors with distinct channel properties. Neuron 5:569–581

Nowak L, Bregetovski P, Ascher P, Herbert A, Prochiantz A (1984) Magnesium gates glutamate-activated channels in mouse central neurons. Nature 307:462–465

O'Brien RJ, Fischbach GD (1986) Characterization of excitatory amino acid receptors expressed by embryonic chick motoneurons in vitro. J Neurosci 6:3275–3283

Olney JW (1978) neurotoxicity of excitatory amino acids. In: McGeer EG, Olney JW, McGeer PL (eds) Kainic acid as a tool in neurobiology. Raven, New York, p 95

Olney JW (1982) The toxic effects of glutamate and related compounds in the retina and the brain. Retina 2:341–359

Olney JW (1990) Excitatory amino acids and neuropsychiatric disorders. Annu Rev Pharmacol Toxicol 30:47–71

Olney JW, Labruyère J, Collins JF, Curry K (1981) D-Aminophosphonovalerate is 100-fold more powerful than D-α-aminoadipate in blocking N-methyl-aspartate neurotoxicity. Brain Res 221:207–210

Olney JW, Price M, Samson L, Labruyère J (1986) The role of specific ions in glutamate neurotoxicity. Neurosci Lett 65:65–71

Orrenius S, McConkey DJ, Bellomo G, Nicotera P (1989) Role of Ca^{++} in toxic cell killing. Trends Pharmacol Sci 10:281–285

Ortega A, Teichberg (1990) Phosphorylation of the 49-kDa putative subunit of the chick cerebellar kainate receptor and its regulation by kainatergic ligands. J Biol Chem 265:21404–21406

Ortega A, Eshhar N, Teichberg VI (1990) Cyclic AMP prevents kainate receptor desensitization and increases kainate mediated responses in cultured Bergmann glia. Neurochem Int [Suppl 1] 16:100

Ortega A, Eshhar N, Teichberg VI (1991) Properties of kainate receptor channels on cultured Bergmann glia. Neuroscience 41:335–349

Pin JP, van Vliet BJ, Bockaert J (1989) Complex interaction between quisqualate and kainate receptors as revealed by GABA release measurements from striatal neurons in primary culture. Eur J Pharmacol 72:81–91

Popovici T, Barbin G, Ben Ari Y (1988) Kainic acid-induced seizures increase c-fos-like protein in hippocampus. Eur J Pharmacol 150:405–406

Pumain R, Kurcewicz I, Louvel J (1987) Ionic changes induced by excitatory amino acids in the rat cerebral cortex. Can J Physiol Pharmacol 65:1067–1077

Ransom RW, Stec NL (1988) Cooperative modulation of [³H]MK-801 binding to the N-methyl-D-aspartate receptor-ion channel complex by L-glutamate, glycine and polyamines. J Neurochem 51:830–836

Rassendren FA, Lory P, Nargeot J (1990) Desensitization of the kainate responses in oocytes injected with brain RNA. Neurochem Int [Suppl 1] 16:61

Recasens M, Mayat E, Guiramand J (1990) Excitatory amino acid receptors and phosphoinositide breakdown: facts and perspectives. In: Osborne NM (ed) Current aspects of the neurosciences. MacMillan, London (in press)

Reynolds IJ, Miller RJ (1988) Multiple sites for the regulation of the N-methyl-D-aspartate receptor. Mol Pharmacol 33:581–584

Rothman SM (1985) The neurotoxicity of excitatory amino acids is produced by passive chloride fluxes. J Neurosci J 5:1483–1489

Rothman SM, Olney JW (1987) Excitotoxicity and the NMDA receptor. Trends Neurosci 10:299–302

Sacaan AI, Johnson KM (1989) Characterization of polyamine effects on [³H]TCP and NMDA specific [³H]glutamate binding. Soc Neurosci Abstr 15:200

Sather W, Johnson JW, Henderson G, Ascher P (1990) Glycine insensitive desensitization of NMDA responses in cultured mouse embryonic neurons. Neuron 4:725–731

Siman R, Noszek JC (1988) Excitatory amino acids activate calpain I and induce structural protein breakdown in vivo. Neuron 1:279–287

Snell LS, Morter RS, Johnson KM (1988) Structural requirements for activation of the glycine receptor that modulates the N-methyl-D-aspartate operated ion channel. Eur J Pharmacol 156:105–110

Somogyi P, Eshhar, N, Teichberg VI, Roberts JDB (1990) Subcellular localization of a putative kainate receptor in Bergmann glial cells using a monoclonal antibody in the chick and fish cerebellar cortex. Neuroscience 35:9–30

Sonnenberg JL, Mitchelmore C, MacGregor-Leon PF, Hempstead J, Morgan JI, Curran T (1989) Glutamate receptor agonists increase the expression of Fos, Fra and AP-1 DNA binding activity in the mammalian brain. J Neurosci Res 24:72–80

Stevens CF (1987) Channel families in the brain. Nature 328:198–199

Sugiyama H, Ito I, Okada D, Hirono C, Ohmori H, Shiyemoto T, Furuya S (1988) Functional and pharmacological properties of glutamate receptors linked to inositol phospholipid metabolism. In: Cavalheiro EA, Lehmann J, Turski L (eds) Frontiers in excitatory amino acid research. Liss, New York, p 21

Sugiyama K, Brunori A, Mayer ML (1989) Glial uptake of excitatory amino acids influences neuronal survival in cultures of mouse hippocampus. Neuroscience 32:779–791

Tecoma ES, Choi DW (1989) GABA$_{ergic}$ neocortical neurons are resistant to NMDA receptor mediated injury. Neurology 39:676–682

Tang GM, Dichter M, Morad M (1989) Quisqualate activates a rapidly inactivating high conductance ionic channel in hippocampal neurons. Science 243:1474–1477

Vaccarino F, Guidotti A, Costa E (1987) Ganglioside inhibition of glutamate mediated protein kinase C translocation in primary cultures of cerebellar neurons. Proc Natl Acad Sci USA 84:8707–8711

Verdoorn TA, Dingledine K (1988) Excitatory amino acid receptor expressed in Xenopus oocytes: agonist pharmacology. Mol Pharmacol 34:298–307

Wada K, Dechesne CJ, Shimasaki S, King RG, Kusano K, Buonanno A, Hampson DR, Banner C, Wenthold RJ, Nakatani Y (1989) Sequence and expression of a frog brain complementary DNA encoding a kainate-binding protein. Nature 342:684–689

Watkins JC (1986) Twenty-five years of excitatory amino acid research. In: Roberts PJ, Storm-Mathisen J, Bradford HF (eds) Excitatory amino acids. MacMillan, London, p 1

Watkins JC, Olverman HJ (1987) Agonists and antagonists for excitatory amino acid receptors. Trends Neurosci 10:265–272

Wenthold RJ, Hunter C, Wada K, Dechesne CJ (1990) Antibodies to a C terminal peptide of the rat brain glutamate receptor subunit, GluR-A, recognize a subpopulation of AMPA binding sites but not kainate sites. FEBS Lett 276:147–150

Westbrook GL, Mayer ML (1987) Micromolar concentrations of Zn^{++} antagonize NMDA and GABA responses of hippocampal neurons. Nature 328:640–643

Wolfe LS, Pappius HM (1984) Arachidonic acid metabolites in cerebral ischemia and brain injury. In: Bes A, Braquet P, Paolitti R, Siesjo BK (eds) Cerebral ischemia. Elsevier, Amsterdam, p 223

Young AB, Fagg GE (1990) Excitatory amino acid receptors in the brain: membrane binding and receptor autoradiographic approaches. Trends Pharmacol Sci 11:126–133

Zeevalk GD, Hyndman AG, Nicklas WJ (1989) Excitatory amino acid induced toxicity in chick retina: amino acid release, histology and effects of chloride channel blockers. J Neurochem 53:1610–1619

CHAPTER 14

Convulsants and Gamma-Aminobutyric Acid Receptors

S.C.R. Lummis and I. L. Martin

A. Introduction

Gamma-aminobutyric acid (GABA) is the major inhibitory transmitter of the mammalian CNS and has been implicated in the aetiology of a number of clinically important disorders (Bartholini et al. 1985), including epilepsy (Gale 1989). The complexity of neuronal integration in the brain has precluded any detailed understanding of these conditions, though much information has been gleaned from the consequences of intervention with a variety of pharmacological agents.

The amino acid GABA is widely distributed in the vertebrate CNS, where it has a differential topographical distribution (Ottersen and Storm-Mathison 1984). It is localised in nerve terminals (Neal and Lversen 1969) from which it is released in a calcium-dependent manner by depolarizing stimuli (Bradford 1970). It is synthesised almost exclusively from glutamic acid through decarboxylation by the enzyme glutamic acid decarboxylase (GAD; EC 4.1.1.15). This step is rate limiting, and inhibition of GAD, by a variety of agents, leads to a decrease in the tissue concentration of GABA; if the inhibition is significant, convulsions ensue (Tapia et al. 1975). High-affinity neuronal uptake systems are responsible for the removal of the transmitter from the synaptic cleft subsequent to its release (Iversen and Neal 1968), although there is also evidence that it can be taken up into glia (Iversen and Kelly 1975). GABA is degraded by transamination, the enzyme 4-aminobutyrate: 2-oxoglutarate aminotransferase (EC 2.6.1.19) being responsible. Inhibition of this enzyme leads to an increase in the concentration of the transmitter in the CNS (Anlezark et al. 1976; Fowler and John 1972; Palfreyman et al. 1981), and an irreversible inhibitor of this enzyme has recently found clinical use as an anticonvulsant (Schechter 1989; Remy and Beaumont 1989). Thus, although manipulation of the metabolic machinery can clearly lead to decreases in the availability of GABA and subsequent loss of tonic inhibitory control, it is not our intention to discuss this further here.

Over the past 20 years there have been significant advances in our understanding of the receptor-mediated events which occur subsequent to GABA release. GABA interacts with three pharmacologically distinct receptors in the mammalian CNS. $GABA_A$ receptors are defined by their sensitivity to the competitive antagonist (+)-bicuculline; they are responsible for the activation of fast inhibitory post-synaptic potentials which are due to the increase in chloride conductance of the neuronal membrane. $GABA_B$ receptors are not sensitive to bicuculline; indeed, the characterisation of the functional importance of this receptor has been hampered by the lack of potent antagonists. Its activation leads to a decrease in calcium conductance and an increase in potassium conductance, both mediated by interaction with G proteins (Wojcik et al. 1989). Both $GABA_A$ and $GABA_B$ receptors exhibit a distinct topographical distribution in the mammalian CNS (Bowery et al. 1987). The third type of receptor is the autoreceptor (Mitchell and Martin 1978; Snodgrass 1978; Arbilla et al. 1979). Its function is to control the release of GABA from the pre-synaptic terminal; its pharmacology remains ill-defined, although delta-amino-laevulinic acid is a more potent agonist at the autoreceptor than it is at the A-subtype (Brennan and Cantrill 1979; Brennan et al. 1981). Excepting the GAD inhibitors, other convulsant agents thought to mediate their effects through interference with GABA-mediated transmission appear to modify the interaction of the natural transmitter with the $GABA_A$ receptor. It is these substances to which the remainder of this chapter will be devoted.

B. Pharmacology and Structure of the $GABA_A$ Receptor

It has been known for many years that activation of the $GABA_A$ receptor leads to an increase in chloride conductance of the cell membrane (Curtis et al. 1968; Kelly et al. 1969). A number of compounds structurally related to GABA such as isoguvacine, muscimol, 3-aminopropanesulphonic acid (3APS), piperidine-4-sulphonic acid (P4S) and 4,5,6,7-tetrahydro-[5,4-c]-pyridin-3-ol (THIP) (Fig. 1) are all agonists at this receptor (Krogsgaard-Larsen et al. 1984). Bicuculline acts as a competitive antagonist (Fig. 2; Johnston 1976; Frere et al. 1982; Nowak et al. 1982). Its effects are stereo-specific, and the structure-activity relationships within this chemical class have already been investigated (Kardos et al. 1984). In addition, some pyridazinyl derivatives of GABA, such as SR95531 (Fig. 2; Heaulme et al. 1986; Wermuth and Biziere 1986) and pitrazepin (Fig. 2; Gahwiler et al. 1984), appear to antagonise the effects of GABA competitively at this site. Picrotoxin is also a potent antagonist but acts in a noncompetitive manner (Fig. 3; Simmonds 1980; Barker et al. 1980). There are a number of other compounds which antagonise GABA-mediated transmission at sites distinct from the amino acid recognition site including anisatin (Kudo et al. 1981), a series of bicyclophosphates (Fig. 4) (Bowery et al. 1976, 1977), tetramethylenedisulphotetramine (Bowery et al. 1975), bemegride and pentylenetetrazole (Simmonds 1978) and penicillin (Fig. 3) (Pickles and Simmonds 1980). All these antagonists are convulsant.

GABA
Isoguvacine
Muscimol
3APS
P4S
THIP

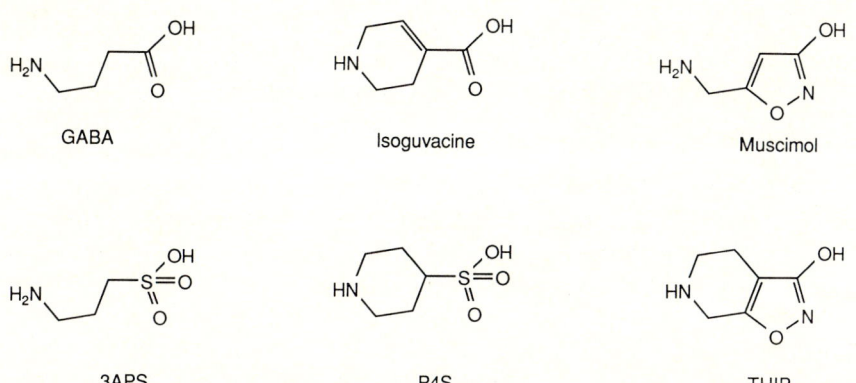

Fig. 1. Structures of $GABA_A$, receptor agonists. *GABA*, γ-aminobutyric acid; *3APS*, 3-aminopropanesulphonic acid; *P4S*, piperidine-4-sulphonic acid; *THIP*, 4,5,6,7-tetrahydro-[5,4-*c*]-pyrin-3-ol

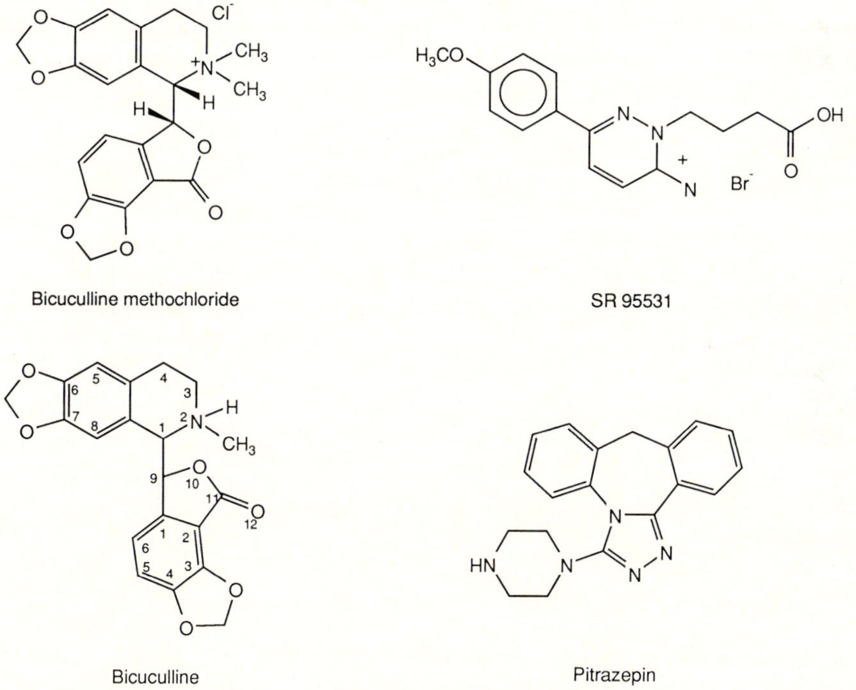

Bicuculline methochloride
SR 95531
Bicuculline
Pitrazepin

Fig. 2. Structures of $GABA_A$ receptor competitive antagonists

There are additional allosteric recognition sites on the receptor complex, the most notable of which is that for the benzodazepine class of minor tranquillisers (BOSMANN et al. 1977; MOHLER and OKADA 1977b; SQUIRES and BRAESTRUP 1977). The interaction of ligands with this site is complicated by

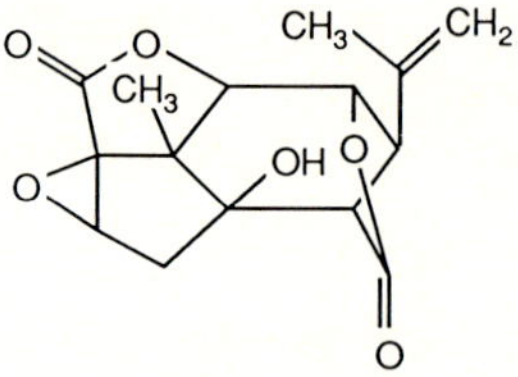

Picrotoxinin

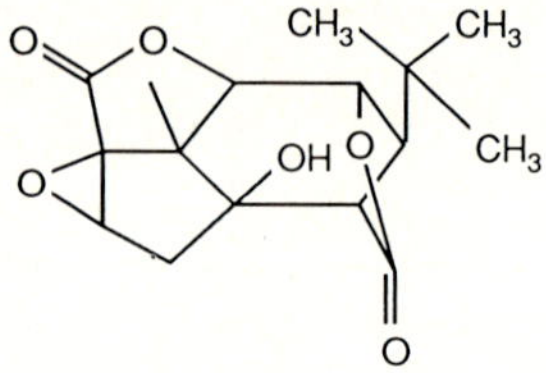

Picrotin

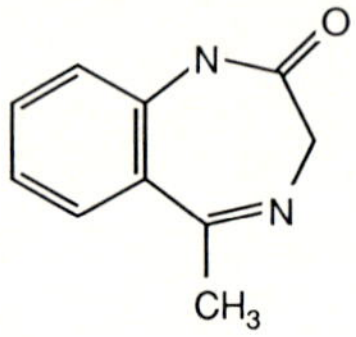

Ro 5-3663

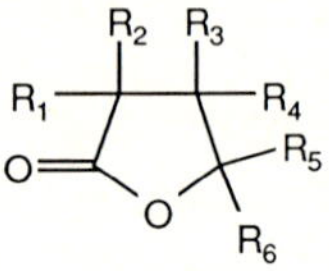

γ-Butyrolactone

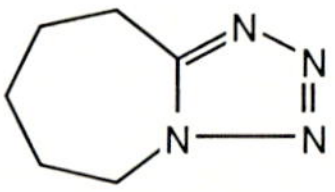

Pentylenetetrazole

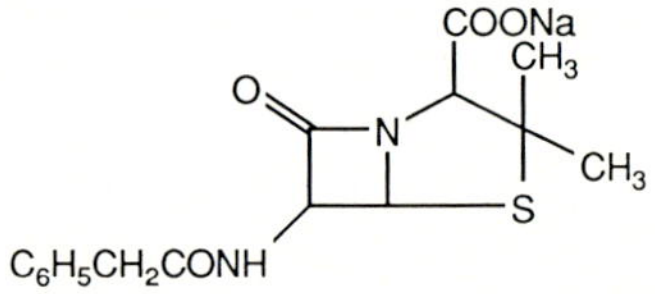

Penicillin

Fig. 3. Structures of picrotoxin site ligands

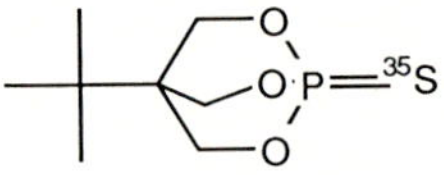

$[^{35}S]$TBPS

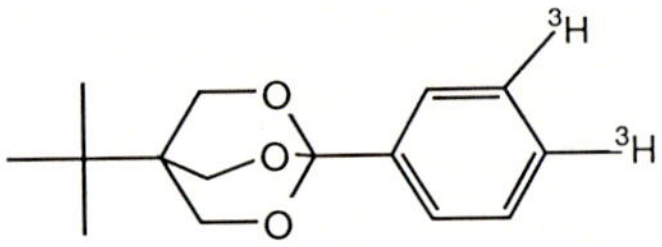

$[^{3}H]$TBOB

Fig. 4. Structures of bicyclophosphates. *TBPS*, *t*-butylbicyclophorothionate; *TBOB*, *t*-butylbicycloorthobenzoate

the fact that while the so-called agonist benzodiazepines (the anxiolytics) appear to augment the actions of a given GABA stimulus (Schmidt et al. 1967; Polc et al. 1974), other compounds, which appear to interact specifically with the benzodiazepine recognition site, are able to produce diametrically opposed effects (Tenen and Hirsch 1980; Cowan et al. 1981). These have been termed inverse agonists, partial inverse agonists being pro-convulsant and the full inverse agonists, overtly convulsant.

The GABA$_A$ receptor is a member of the family of ligand-gated ion channels, the archetypal member of which is the nicotinic acetylcholine receptor (nAChR; see Changeux et al. 1984). The GABA$_A$ receptor is a hetero-oligomeric structure in which the chloride ion channel itself is formed from portions of the constituent subunits as they pass through the lipid bilayer of the cell membrane. The recognition sites are thought to be located on the segment of the protein exposed to the extracellular environment. The mature protein is glycosylated, and there are consensus phosphorylation sites present on the predicted intracellular portion of the receptor protein (BARNARD et al. 1987). Molecular cloning studies initially identified two classes of peptide subunit, α and β, which when expressed in *Xenopus* oocytes exhibited GABA-activated currents (SCHOFIELD et al. 1987). Subsequent studies have provided evidence that a further two classes of subunit, γ and δ, are also expressed in the mammalian CNS (PRITCHETT et al. 1989; SHIVERS et al. 1989). There are multiple subtypes within the α, β and γ classes which are important in the definition of the precise pharmacology of the functional hetero-oligomer (see OLSEN and TOBIN 1990). Comparison of these sequences with those of the nAChR suggests that there is significant amino acid homology between their peptide subunits. The physical properties of the amino acids within these sequences indicate that there is likely to be considerable structural similarity between the nAChR and the GABA$_A$ receptor (see UNWIN 1989). There is no significant experimental evidence to address either the subunit composition of any naturally occurring GABA$_A$ receptor or the number of subunits comprising such a complex. However, by sequence homology with the nAChR, it is probable that all the subunits have a large amino-terminal segment, presumably located extracellularly, with four subsequent stretches which are hydrophobic (putative membrane-spanning segments). These membrane-spanning segments are each connected to short loops of amino acids with the exception of that between membrane-spanning segments 3 and 4, which is probably a large intracellular loop, containing, in the case of the β-subunits, consensus phosphorylation sites. In addition, there is in the extracellular amino-terminal region a Cys-Cys loop. There is little evidence concerning the precise location of the recognition sites for any of the ligands known to interact with the GABA$_A$ receptor protein. However, consensus opinion would place the GABA recognition site on the β-subunit (DENG et al. 1986; CASALOTTI et al. 1986; KIRKNESS and TURNER 1988; KIRKNESS et al. 1989) with that for the benzodiazepines on the α-subunit (MOHLER et al. 1980; SIGEL et al. 1983; STAUBER et al. 1987; SWEETNAM et al. 1987). Elucidation here will undoubtedly benefit from the development of antibodies with defined epitope specificity, which is in progress (EWART et al. 1990; DUGGAN and STEPHENSON 1990).

C. Competitive Antagonists at the GABA$_A$ Receptor

It is now generally accepted that the convulsant alkaloid, bicuculline, is a potent, competitive antagonist at the GABA$_A$ receptor. The initial electrophysiological studies with this compound, in fact, provided convincing evidence that GABA was an inhibitory transmitter in the mammalian CNS (CURTIS et al. 1970). There was some difficulty in reproducing these early observations, but this was almost certainly due to the fact that bicuculline is relatively insoluble in aqueous solution and unstable at physiological pH, as it is rapidly hydrolysed to the inactive bicucine (OLSEN et al. 1975). These problems have now been overcome by the use of the quaternary salts of bicuculline such as the methochloride or the methiodide (PONG and GRAHAM 1972; JOHNSTON et al. 1972).

[^{3}H]-(+)-Bicuculline methohalides have been used in radioligand binding experiments, although the interpretation of the data has proved somewhat complex. It is clear that both GABA and a number of the analogues shown in Fig. 1 can be used to label GABA$_A$ receptor sites (ENNA and SNYDER 1975; BEAUMONT et al. 1978) in a mutually competitive manner. The binding isotherms for these compounds to brain membranes are nonlinear upon Scatchard transformation (FALCH and KROGSGAARD-LARSEN 1982; JORDAN et al. 1982; Quast and BRENNER 1983). The analysis suggested that the receptor existed in multiple conformers exhibiting different affinity constants for the agonist ligands (KROGSGAARD-LARSEN et al. 1981). [^{3}H]-(+)-Bicuculline methiodide binds to brain membranes in a saturable manner, and the interaction appears to take place with a homogeneous population of sites (MOHLER and OKADA 1977a, 1978); the binding is completely displaceable by GABA. The specific radioligand binding of the methohalide salts of (+)-bicuculline is markedly enhanced in the presence of chaotropic ions, while the binding of agonists is not so affected (MOHLER and OKADA 1977a, 1978; OLSEN and SNOWMAN 1983). Comparison of the affinity constants of agonists and antagonists for the GABA$_A$ receptor reveals that the antagonists bind with higher affinity to the sites labelled with lower affinity by the agonists. Similar hypotheses for the existence of agonist- and antagonist-preferring conformations have been advanced for the nicotinic (SUGIYAMA and CHANGEAUX 1975) and muscarinic acetylcholine receptors (BIRDSALL et al. 1980).

The observation that the autoradiographic distribution of high affinity muscimol and bicuculline methochloride binding sites is distinct in certain brain regions (OLSEN et al. 1984, 1990) implies that the equilibrium between these conformers must vary, and there is some evidence to support this contention (FALCH and KROGSGAARD-LARSEN 1982; OLSEN and SNOWMAN 1982). However, recent evidence from molecular cloning studies (see Sect. B) suggests that the precise oligomeric composition of the GABA$_A$ receptor in different brain areas may vary. In addition, it is clear that desensitisation will lead to a higher affinity for agonists (see CASH and SUBBARAO 1987), which has some experimental support (BRISTOW and MARTIN 1989). The fact that distinct oligomeric complexes of the receptor exhibit distinct rates of desensitisation (VERDOORN et al. 1990) may provide an explanation of the radioligand binding data obtained at the GABA$_A$ receptor, but its validation must await further experimental results.

Following the early electrophysiological experiments with bicuculline in whole animal experiments (Curtis et al. 1971), the competitive nature of the antagonism of GABA responses was demonstrated by Simmonds (1978, 1980) using the rat cuneate slice preparation in vitro. Antagonism by bicuculline has also been demonstrated in mouse spinal cord neurons in culture (Barker et al. 1983). Here it was shown that bicuculline affected neither the conductance nor the average open time of the GABA-activated channels. In bullfrog dorsal root ganglion cells, bicuculline proved an effective antagonist of GABA responses only when applied to the outside of the ganglion, in keeping with the competitive nature of its antagonism (Akaike et al. 1985).

Structure-activity studies carried out with bicuculline and a number of its analogues have shown that addition of an N-methyl substituent at position 2 does not appear to detract from the activity of the compounds as GABA$_A$ receptor antagonists unless they are given systemically (Johnston et al. 1972). The poor access of these compounds to the CNS is probably the explanation for this observation (Pong and Graham 1972; Johnston et al. 1972). The methylenedioxy bridge across positions 6 and 7 can be replaced by methoxy substituents at the same position without loss of activity. However, the stereochemistry around positions 1 and 9 is essential: the required stereochemistry is erythro-1S,9R (Hill et al. 1974; Collins and Hill 1974). Calculations would then suggest that the N-2 to C-11/0-12 distance is about 0.56 nm (Kier and George 1973). The structure-activity relationships of 45 bicuculline-related phthalideisoquinolines have been described (Kardos et al. 1984), of which (+)-bicuculline was the most potent.

SR95531 (Fig. 2) is essentially GABA linked via its amino function to a phenylaminopyridazine heterocycle. It is a potent displacer of GABA agonist binding to the GABA$_A$ site, being about 20 times more effective than bicuculline (Heaulme et al. 1987). The binding characteristics of [^{3}H]-SR95531 appear to suffer from complex dissociation kinetics, some of which are extremely rapid. However, the compound exhibits a considerable degree of specificity in both radioligand binding (Wermuth and Biziere 1986) and electrophysiological experiments (Chambon et al. 1985; Mienville and Vicini 1987). The properties are shared by some of its analogues (Desarmenian et al. 1987).

Pitrazepin (Fig. 2) also exhibits apparent competitive antagonism at the GABA$_A$ site, though the compound shows some affinity for the benzodiazepine recognition site (Gahwiler et al. 1984; Kemp et al. 1985). In addition, R5135 displays GABA$_A$ antagonism, but here the specificity of the compound is compromised by its interaction with both benzodiazepine and glycine receptors (Hunt and Clements-Jewery 1981).

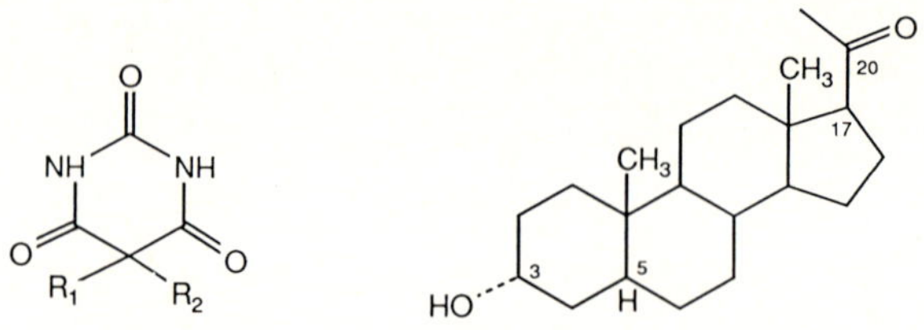

Barbiturates   5β-Pregnan-3α-ol-20-one

Fig. 5. Structures of barbiturates and steroids

D. Convulsant Agents Acting Through the Picrotoxin/Convulsant Recognition Site

The convulsant effects of picrotoxin have been known for many years, with some of its properties being described over a century ago (BENTLEY and TRIMEN 1875). It is a plant-derived polycyclic lactone consisting of two compounds, picrotin and picrotoxinin, the latter being 50-fold more potent than the former (JARBOE et al. 1968). The structures of these two compounds are shown in Fig. 3, the only difference being hydration of the isopropenyl group in picrotin.

Picrotoxin has been shown to block $GABA_A$-induced responses in both vertebrate and invertebrate preparations (TAKEUCHI and TAKEUCHI 1969; GALINDO 1969). Picrotoxin acts in a non-competitive manner, and there is a reduction in its inhibitory effect at high chloride concentrations, properties which led to the proposal that the specific binding site identified for picrotoxin is in or close to the $GABA_A$-activated chloride ion channel (TICKU et al. 1978; SIMMONDS 1980). The existence of a picrotoxin recognition site on the $GABA_A$ receptor complex has provided a potential site for the interaction of a variety of other compounds known to affect GABA responses including (1) many convulsants, such as bicyclophosphate esters, pentylenetetrazole (PTZ), penicillin G, the benzodiazepine Ro5-3663 and the γ-butyrolactones (Fig. 3); (2) the barbiturates, both convulsant and sedative/hypnotic/anti-convulsant (Fig. 5); (3) certain steroids (Fig. 5); (4) ethanol; and (5) various insecticidally active compounds (Fig. 6). The structure of these groups of compounds is very diverse, and it is now apparent that they do not all interact with picrotoxin in a competitive manner. However, many convulsant classes of ligand do appear to mediate their actions via the picrotoxin binding site, which is frequently referred to in the literature as "the convulsant binding site".

The first radioligand used to study this site was an analogue of picrotoxin, α-dihydropicrotoxinin (TICKU et al. 1978) (Table 1). The specific binding of this convulsant toxin was not directly displaced by GABA or benzo-

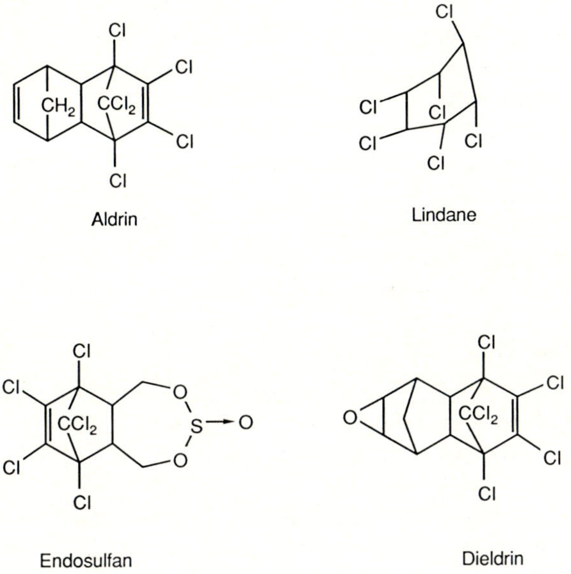

Fig. 6. Structures of insecticides

Table 1. Ligands used to study the picrotoxin binding site

Ligand	K_d	B_{max}
[3H]-α-dihydropicrotoxinin	$1-2\,\mu M$	5 pmol/mg protein
[35S]-t-butylbicyclophosphorothionate	$17\,nM$	6 pmol/mg protein
[3H]-t-butylbicycloorthobenzoate	$61\,nM$	1.6 pmol/mg protein

diazepines, but α-dihydropicrotoxinin binding was inhibited by convulsant analogues of picrotoxin and bicyclophosphate esters, with relative potencies consistent with their pharmacological activity. Other inhibitors of this binding included PTZ, the convulsant benzodiazepine Ro5-3663 and the pyrethroid insecticides, suggesting that this site is important in the action of a wide range of convulsants (OLSEN 1983; TICKU and MAKSAY 1983). There was also evidence that α-dihydropicrotoxinin binding sites represented the recognition site for barbiturates; sedative-hypnotic and anticonvulsant barbiturates competitively inhibited the binding of radiolabelled α-dihydropicrotoxinin to brain membranes (TICKU and OLSEN 1978). However, this ligand was not ideal as a probe for the picrotoxin binding site: it bound with a relatively low affinity $(1-2\,\mu M)$ and exhibited high non-specific binding. Thus, conclusions drawn from data obtained with this ligand must be viewed with caution.

More recent studies of the picrotoxin site have been greatly facilitated by the development of new radioligands. The most widely used of these is [^{35}S]-t-butylbicyclophorothionate (TBPS, Fig. 4) (Squires et al. 1983). [^{3}H]-t-Butylbicycloorthobenzoate (TBOB, Fig. 4) has also proved a suitable ligand, although it binds with slightly lower affinity (Lawrence et al. 1985). There are also reports of characterisation of the picrotoxin binding site using other ligands including [^{3}H]-n-propylbicyclophosphate (PrPO) and [^{3}H]-p-cyano-s-butylbicycloorthobenzoate (BuOB) (Casida et al. 1988), but these are not commercially available and have not been widely studied. [^{35}S]-TBPS binds with high affinity and has low non-specific binding to brain membrane preparations; it is competitively displaced by picrotoxin and bicyclophosphate esters. Its interactions with GABA and benzodiazepines suggest that the binding sites are constituents of the GABA$_A$ receptor complex, GABA and GABA$_A$ agonists inhibit binding in a non-competitive fashion, although there are also reports of enhancement at low concentrations. These interactions with GABA are not always fully considered when analysing radioligand binding data from membrane preparation, as the washing procedure and hence the concentration of GABA remaining in the preparation will affect the data. Apparently inconsistent data obtained from [^{35}S]-TBPS binding experiments between different groups may be due to this phenomenon. Benzodiazepine agonists will enhance binding, whereas inverse agonists will inhibit both TBPS binding and muscimol inhibition of TBPS binding. There are some noticeable differences between TBPS and α-dihydropicrotoxinin binding: (1) the optimal temperature for TBPS binding is 21°C, as compared with 0°C, (2) the binding is dependent on the presence of appropriate ions, and (3) potent inhibition by GABA or GABA$_A$ agonists will only occur in the presence of Eccles anions (Squires et al. 1983; Supavilai and Karobath 1984). [^{3}H]-TBOB has binding characteristics essentially identical to those of [^{35}S]-TBPS, although there are differences in the effect of temperature, with [^{3}H]-TBOB having similar binding characteristics at 0°, 25° and 37°C (Lawrence et al. 1985).

There are other subtle differences in the binding characteristics of TBPS and TBOB which are not yet understood but perhaps can be partially explained by recent studies suggesting that these two compounds represent different subclasses of convulsant ligands which bind to the picrotoxin site. Palmer and Casida (1988) have examined the effects of a wide variety of cage convulsants on toxicity, GABA-stimulated ^{36}Cl$^-$ flux and [^{35}S]-TBPS binding, and their data propose two classes of cage convulsants, possibly differing in their sites or types of action at the GABA$_A$ receptor chloride channel. In general, larger compounds including the orthobenzoates (such as TBOB), orthocarboxylates and picrotoxin fall into one class (type A). These are more potent at inhibiting both chloride flux and [^{35}S]-TBPS binding than type B compounds (which include smaller molecules such as the phosphorothionates) when the comparisons are based on compounds with a similar mouse LD$_{50}$. The authors hypothesise that the two types of compound could perhaps interact with a single convulsant site but differ in their efficiency in blocking the GABA response; alternatively, they may act at distinct sites that have different pharmacological characteristics.

Studies on various convulsants have shown that many of them will inhibit [^{35}S]-TBPS binding at low concentrations. As has been reported for dihydropicrotoxinin binding, picrotoxin and bicyclophosphate esters are potent, competitive inhibitors of [^{35}S]-TBPS binding, and both convulsant and anticonvulsant γ-butyrolactones, γ-thiobutyrolactones and succinimides will also competitively displace [^{35}S]-TBPS binding (WEISSMAN et al. 1984; LEVINE et al. 1985; HOLLAND et al. 1990a), implying that all these compounds act at the same site as picrotoxin. The convulsants anisatin, Ro5-3663 and PTZ have an IC$_{50}$ of 0.1, 22 and 700 nM, respectively, against [^{35}S]-TBPS binding, but their precise modes of action are still unclear. PTZ, for example, is a member of a group of compounds, the tetrazoles, which are reported to act competitively at both the benzodiazepine and [^{35}S]-TBPS binding sites (REHAVI et al. 1982; RAMANJANEYULU and TICKU 1984). This apparent controversy may be partly explained by the description that PTZ is a potent inhibitor of benzodiazepine binding in astrocytes, cells which possess mitochondrial or peripheral-type benzodiazepine receptors which are not linked to a GABA receptor (BENDER and HERTZ 1988). This information, coupled with the fact that the dissociation profile of [^{35}S]-TBPS binding in the presence of PTZ is identical to that in the presence of picrotoxin, indicates that the convulsive properties of PTZ are probably due to an action at the picrotoxin binding site (MAKSAY and TICKU 1984).

There has been a suggestion from the study of SQUIRES et al. (1983) that [^{35}S]-TBPS binding studies may be useful in distinguishing convulsant and anticonvulsant compounds. Detailed studies with TBPS have shown that R5135, a GABA$_A$ antagonist, will reverse the inhibitory effects of anticonvulsant, anxiolytic and sedative hyponotic substances (barbiturates, benzodiazepines and ethanol) whilst having no effect or potentiating the inhibitory effects of convulsants. This has led to the hypothesis that the occupation of GABA$_A$ receptors by R5135 induces conformational changes in TBPS binding sites such that the affinities for convulsants are unchanged but the affinities for anticonvulsant anxiolytic ligands are decreased.

The non-competitive block of GABA$_A$-mediated chloride conductance by picrotoxin was first described in the 1960s (TAKEUCHI and TAKEUCHI 1969; GALINDO 1969). Many subsequent studies have used picrotoxin to demonstrate a specific action of GABA$_A$-mediated events, but the molecular details of its mechanism of action are still not fully understood. It was proposed many years ago that picrotoxin acts on the chloride ion channel of the GABA$_A$ receptor complex, and a report by AKAIKE et al. (1985) showing block with both internally and externally applied picrotoxin in bullfrog dorsal root ganglion cells supports this idea. However, in dissociated sympathetic ganglion neurons picrotoxin was only effective when applied externally (CULL-CANDY et al. 1988), suggesting that a "plug-in-hole" model of picrotoxin action is probably too simplistic. The effect of picrotoxin at the single-channel level has also been difficult to clarify, although in a recent study it was reported to alter $GABA_A$ receptor channel bursting properties by reducing both mean open time and the average number of openings per burst, suggesting it may destabilize the open state (Twyman et al. 1989). Single-channel analysis of the effects of TBPS and picrotoxin in non-$GABA_A$ activated (spontaneous) chloride channels revealed that these compounds prolong the longer closed times, resulting in a decrease in opening probability (Hamann et al. 1990). Not all studies, however, have shown similar mechanisms of action of TBPS and picrotoxin. In studies using vertebrate (chick) $GABA_A$ receptors, both TBPS and β-propyl-γ-butyrolactone were described as mixed, rather than noncompetitive, inhibitors (Van Rentergehm et al. 1987; Clifford et al. 1989), although in rat hippocampal neurones β-substituted γ-butyrolactones were reported to have a non-competitive action on GABA responses (Holland et al. 1990a). The explanation for these apparently conflicting results is not yet clear, although it has been proposed to be due to the different preparations. Certainly there are quantitative differences between the responses to alkyl-substituted γ-butyrolactones and thiobutyrolactone in the chick and rat preparations (Clifford et al. 1989; Holland et al. 1990a), and in another preparation, the crayfish neuromuscular junction, TBPS does appear to act in the same manner as picrotoxin (Gammon and Casida 1983). PTZ (Simmonds 1982) and Ro5-3663 (Harrison and Simmonds 1983) have been found to act in a similar manner to picrotoxin. Less evidence is available for the mechanism of action of penicillin, although it has been suggested that it shortens the duration of the open $GABA_A$ channel (Mathers 1987).

A recent detailed study on the action of γ-butyrolactones and γ-thiobutyrolactones in rat hippocampal neurones has revealed some apparently novel actions of compounds at the picrotoxin site. Whilst β-substituted γ-butyrolactones block $GABA_A$ currents, the anticonvulsant α-alkyl-substituted thiobutyrolactones, which act competitively at the $[^{35}S]$-TBPS binding site, enhance $GABA_A$ currents (Holland et al. 1990a). Thus, extrapolating from the benzodiazepine site nomenclature, β-substituted γ-butyrolactones could be regarded as "picrotoxin site agonists" and α-alkyl-substituted thiobutyrolactones as "picrotoxin site inverse agonists". This series of compounds also may possess members with "partial agonist" activity: convulsant thiobutyrolactones which only partially block $GABA_A$ currents and a possible mixed inverse agonist/antagonist. This group (Holland et al. 1990b) has also reported that cyclopentanones and cyclohexanones have several properties in common with the α-butylactones and thiobutyrolactones. Cyclopentanones and cyclohexanones exhibit either anticonvulsant or convulsant activity depending upon the location and size of their alkyl substitution. Anticonvulsant compounds, which could be considered as "picrotoxin site inverse agonists", will act competitively at the $[^{35}S]$-TBPS binding site, with their potency at displacing specific $[^{35}S]$-TBPS binding correlating with their ability to prevent PTZ-induced seizures. "Picrotoxin site inverse agonists" are therefore anticonvulsants, in contrast to benzodiazepine inverse agonists, which are convulsants. In this regard, it is interesting that while many potential, but not yet proven, candidates have been proposed as endogenous ligands for the benzodiazepine site, the most promising ones appear to behave as benzodiazepine site inverse agonists. The search for an endogenous picrotoxin site ligand is much less advanced, but the most promising candidates that have emerged again appear to act as "picrotoxin site inverse agonists": steroids, which enhance rather than inhibit $GABA_A$ function.

Anticonvulsant and anaesthetic barbiturates potentiate the electrophysiological response to GABA (BROWN and CONSTANTI 1978; NICOLL 1978; EVANS 1979; SIMMONDS 1981) and enhence $GABA_A$-stimulated chloride efflux (WONG et al. 1984; HARRIS and ALLAN 1985; YANG and OLSEN 1987). Analysis of membrane noise showed that this effect is due to an increase in the average open time of $GABA_A$-activated channels (BARKER and McBURNEY 1979; STUDY and BARKER 1981). Patch clamp recordings have revealed more detail: barbiturates increase the relative frequency of occurrence of long bursts of opening of the $GABA_A$-activated channels, combined with a decrease in the relative proportion of bursts which have shorter openings (TWYMAN et al. 1989). Radiolabelled-ligand binding studies have shown that the barbiturates interact with the $GABA_A$ and benzodiazepine binding sites in an allosteric manner; for example, they enhance $GABA_A$ and benzodiazepine receptor agonist binding to their respective sites in a chloride-dependent manner (WILLOW and JOHNSTON 1981b; ASANO and OGASAWARA 1981; OLSEN and SNOWMAN 1982). The displacement of α-dihydropicrotoxinin binding by various barbiturates led to the proposal that it is the picrotoxin site which mediates the modulatory actions of these compounds at $GABA_A$ receptors (TICKU and OLSEN 1978). However, more detailed studies examining the dissociation kinetics of [^{35}S]-TBPS binding initiated by anticonvulsant and anaesthetic barbiturates have shown that these interactions are not competitive, and these non-convulsant barbiturates are now thought to act at a separate binding site which is closely associated, but not identical to, the picrotoxin binding site (TRIFILLETTI et al. 1984).

The convulsant barbiturates, which include CHEB [5-(2-cyclohexylideneethyl)-5-ethyl barbituric acid], MPPB [(1-methyl-5-phenyl-5-propyl)-5-ethylbarbituric acid], DMBB [5-(1,3-dimethyl-butyl-)5-ethyl barbituric acid] and 3M2B [5-ethyl-5-(3-methyl-but-3-enyl)-barbituric acid], may have a site of action distinct from their non-convulsant counterparts, perhaps a surprising proposal considering that most of those mentioned above have only one isomer with convulsant properties, whilst its racemate may be a potent sedative, hypnotic and/or anaesthetic barbiturate. It has been proposed that the convulsant properties of these compounds could be explained by effects unrelated to $GABA_A$ receptors (JOHNSTON 1983), but it is clear that they do interact with the GABA$_A$ receptor chloride channel complexes at physiological concentrations. The convulsant barbiturates will displace binding of ligand to the picrotoxin binding site and are more potent than the anaesthetic barbiturates (TICKU and OLSEN 1978); they are less effective in stimulating diazepam binding (LEEB-LUNDBERG and OLSEN 1982; SKOLNICK et al. 1982). There is some controversy about whether these convulsant barbiturates interact competitively with the picrotoxin site. A study comparing the convulsant (+)MPPB with the depressant (−)MPPB reported that in the presence of either isomer the rate of dissociation of [^{35}S]-TBPS was faster compared with that in the presence of picrotoxin, but there was a greater difference for the depressant isomer. A more recent study using [^{35}S]-TBPS binding in the presence of GABA showed that GABA enhanced the displacing potencies of a variety of depressants, including barbiturates, alcohol derivates and etazolate, but the IC$_{50}$ of picrotoxin, PTZ and (+)MPPB was not affected (MAKSAY and TICKU 1988). Thus, although it is clear that the site of action of depressant barbiturates and probably other depressant compounds, such as etomidate, etazolate, cyclopyrrolones and ethanol, differs from that of convulsant barbiturates, it is not yet apparent whether or not the latter have a site of action distinct from other convulsants such as picrotoxin and PTZ. If, however, these compounds interact specifically with the GABA$_A$ receptor complex, it seems probable that they do so at the picrotoxin binding site, and the apparently anomalous results described above are due to incomplete separation of the convulsant and non-convulsant isomers.

There is now considerable interest in the interaction of steroids with the GABA$_A$ receptor complex (see GEE 1988, for a recent review). Electrophysiological and radioligand binding studies have shown that the effects of anaesthetic steroids on GABA$_A$ receptors are similar to those of barbiturates: they potentiate the inhibitory effect of GABA, stimulate ^{36}Cl$^-$ uptake, increase muscimol binding and inhibit [^{35}S]-TBPS binding (HARRISON and SIMMONDS 1984; MAJEWSKA and SCHWARTZ 1987). However, detailed studies have shown that steroids and barbiturates do not act through a common binding site (PETERS et al. 1988). Certain steroids, including pregnenolone sulphate and some glucocorticoids, will enhance rather than inhibit [^{35}S]-TBPS binding, although the physiological significance of this is not yet understood; studies on glucocorticoids using GABA-stimulated ^{36}Cl$^-$ flux revealed no effects of these compounds (MAJEWSKA 1987). Whereas steroids and their metabolites may play an important role in modulation of CNS excitability via the GABA$_A$ receptor, they do not appear to act at the picrotoxin binding site and thus will not be discussed further here.

The effects of ethanol have many similarities to the sedative benzodiazepines and barbiturates that enhance GABA$_A$ action, and withdrawal from ethanol can lead to the onset of convulsions. It is therefore not surprising that some years ago ethanol was proposed to have a specific effect on the GABA$_A$ receptor complex. There is now considerable evidence to support this, both from biochemical studies such as the enhancement of $GABA_A$-stimulated $^{36}Cl^-$ flux in neurosynaptosome preparations and cultured spinal cord neurones and from physiological work in which, for example, ethanol decreased the potency of various convulsants that act on the $GABA_A$ receptor complex. More recently, the effect of GABA in two groups of mice (LS and SS) that differ in their genetic susceptibilities to ethanol have been studied in mRNA-injected oocytes. Ethanol facilitated GABA responses in oocytes injected with mRNA from LS mice but antagonised responses in oocytes injected with SS mouse mRNA (WAFFORD et al. 1990). Thus, there is a clear indication that ethanol affects the $GABA_A$ receptor complex. More studies are required to determine its precise site of action.

There is proof that a variety of insecticides, some of which are known potent convulsant agents, act at the $GABA_A$ receptor complex. Picrotoxin itself was used as an insecticide over 100 years ago (BENTLEY and TRIMEN 1875), and more recent electrophysiological and ligand binding data have demonstrated that insect GABA receptors are probably the primary site of action of certain groups of insecticides and may also be involved in the mechanism of action of a number of others (LUMMIS 1990). Type II (α-cyanophenoxybenzyl) pyrethroids, for example, are capable of producing convulsions in vertebrates, and the toxicity of several of these compounds has been shown to be directly correlated with the potency of inhibition of $[^{35}S]$-TBPS binding (LAWRENCE and CASIDA 1983). The onset of type II pyrethroid action has been observed to be delayed in the presence of diazepam and barbiturates, and certain type II pyrethroids will also inhibit $^{36}Cl^-$ influx in rat brain microsacs (ELDEFRAWI and ELDEFRAWI 1987). However, although there is a clear indication that these compounds do affect the $GABA_A$ receptor complex, the concentration required to elicit these effects is much greater than that required for insecticidal action, and it is now clear that the primary target site for these compounds is the sodium channel (NARAHASHI 1985).

The insecticidal and anthelmintic macrocyclic lactone, avermectin, has also been reported to have a variety of effects on $GABA_A$ receptor chloride ion channels (see WRIGHT 1985) and displays a biphasic influence on $[^{35}S]$-TBPS binding at nanomolar concentrations that resembles the actions of etazolate and pentobarbital. However, the ability of this compound to open non-GABA-linked chloride channels throws doubt on the proposal that the $GABA_A$ receptor is its main site of action.

The most convincing evidence that insecticides do act at the $GABA_A$ receptor site has been obtained for certain chlorinated hydrocarbon hydro-carbons including lindane, toxaphene and the cyclodienes, many of which are convulsants. The toxicity to mice of various of these polychlorinated hydrocarbons shows a good correlation with their potency in inhibiting $[^{35}S]$-TBPS binding (COLE and CASIDA 1986). Both the convulsant effect and the GABA receptor complex interaction appear to be stereo-specific: the γ-isomer of hexachlorocyclohexane (lindane) will stimulate seizures in amygdala-kindled animals, whereas the α-isomer has no effect, and the β-isomer acts as an anticonvulsant (Stark et al. 1986). [^{35}S]-TBPS binding data reflect these different potencies: the IC_{50} value is $1.7\,\mu M$ for γ-hexachlorocyclohexane, whereas for the α- and β-isomers it is greater than $10\,\mu M$ (Lawrence and Casida 1984). Other chlorinated hydrocarbon insecticides that are potent, competitive and stereo-specific inhibitors of vertebrate [^{35}S]-TBPS binding are compounds of the toxaphene and aldrin/dieldrin (cyclodienes) types, which have an IC_{50} in the nanomolar to micromolar range (Abalis et al. 1985). Indeed, the IC_{50} values for the polychlorinated convulsant insecticides α-endosulphan and endrin showed that they were more potent than a range of convulsants including TBPS. Examination of a series of these two groups of compounds revealed that the extent of inhibition of [^{35}S]-TBPS binding is closely related to mammalian toxicity. Thus, it seems likely that the convulsant actions of these compounds is mediated through the picrotoxin binding site.

E. Convulsant Agents Acting Through the Benzodiazepine Recognition Site

The 1,4-benzodiazepines were introduced into clinical practice in 1960 as anxiolytics, anticonvulsants, sedative/hypnotics and muscle relaxants (see Randall 1982). Their mechanism of action was unclear until the mid-1970s when it became apparent that they were able to augment the effects of a given GABA stimulus by some ill-defined interaction with the $GABA_A$ receptor (Schmidt et al. 1967; Polc et al. 1974). In 1977, specific high-affinity binding sites for the benzodiazepines were identified in the mammalian CNS using radioligand binding techniques (Bosmann et al. 1977; Squires and Braestrup 1977; Mohler and Okada 1977b); the distribution of these binding sites closely paralleled those for GABA (Schoch et al. 1985). The recent demonstration that transfection of cells with the cDNA of α-, β- and γ-subunits, cloned from human material, imparts GABA responses to the recipient cells which respond in the expected fashion to the benzodiazepines (Pritchett et al. 1989; Pritchett and Seeburg 1990) indicates that the recognition site for the benzodiazepines is an integral part of the $GABA_A$ receptor.

Three years after the identification of recognition sites for the benzodiazepines in the mammalian CNS, Braestrup et al. (1980) found that certain esters of the β-carboline-3-carboxylic acids (Fig. 8) were able to displace completely the benzodiazepines from their binding sites in the mammalian CNS with high affinity. However, the ethyl ester (β-CCE), when given to experimental animals, appeared to be pro-convulsant, i.e. it produced effects opposite to those of the classic benzodiazepines which were potent anticonvulsants (Tenen and Hirsch 1980; Cowan et al. 1981). The

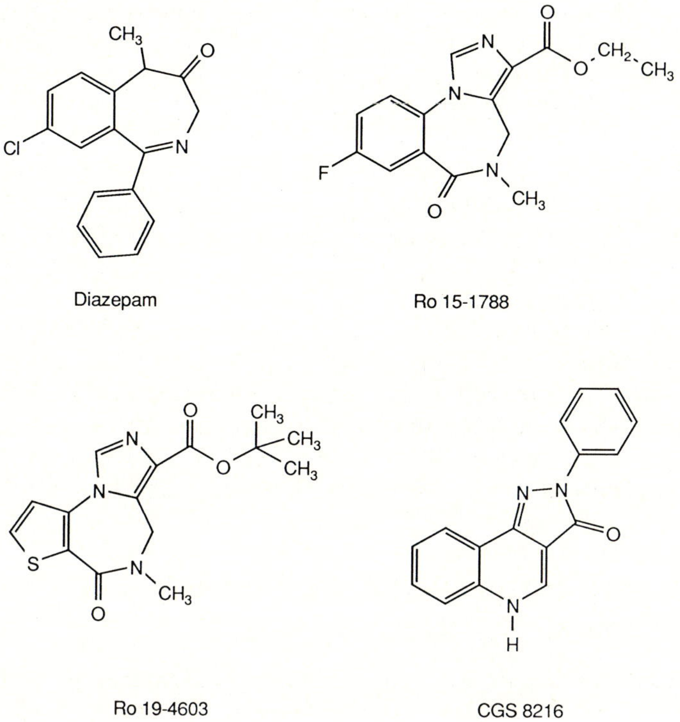

Diazepam   Ro 15-1788

Ro 19-4603   CGS 8216

Fig. 7. Structures of benzodiazepine recognition site ligands

amide analogue of this compound, FG7142, also produced marked anxiety in human volunteers (DOROW et al. 1983). Other analogues of β-CCE such as the methyl ester (β-CCM) and methyl-6,7-dimethoxy-4-ethyl-β-carboline-3-carboxylate (DMCM), are able to produce overt convulsions in laboratory animals (BRAESTRUP et al. 1982), suggesting that they have a greater efficacy than the pro-convulsants; these have been classified as full inverse agonists, with the pro-convulsants termed partial inverse agonists.

Interestingly, while the classic (agonist) benzodiazepines are both sedative and amnesic, certain β-carbolines appear to enhance learning and memory tasks (VENAULT et al. 1986), suggesting that compounds with this profile of activity may be useful in the treatment of senile dementia (SARTER and STEPHENS 1989). However, a degree of caution is required prior to the exploitation of this observation. It is only at low levels of receptor occupancy that these effects are seen; at higher occupancy, convulsions may ensue (PETERSEN et al. 1983; POTIER et al. 1988). An additional complication is the observation of chemical kindling with these compounds (LITTLE and NUTT 1984; LITTLE et al. 1984). This is a phenomenon first observed in this chemical series with FG7142, in which small doses of such compounds

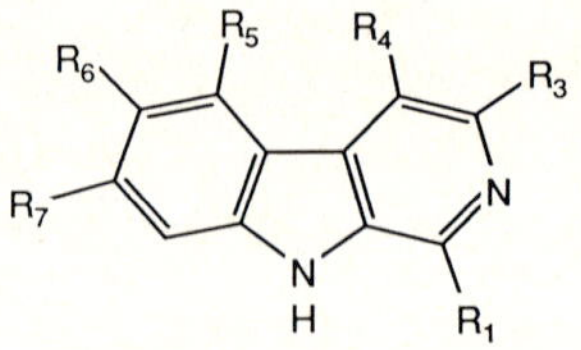

β-carbolines

	R	R₃	R₄	R₅	R₆	R₇
βCCM	H	CO_2Me	H	H	H	H
βCCE	H	CO_2Et	H	H	H	H
FG7142	H	$CONH_2$	H	H	H	H
DMCM	H	CO_2Me	Et	H	OMe	OMe
ZK93426	H	CO_2Et	Me	OⁱPr	H	H
ZK91296	H	CO_2Et	CH_2OMe	OBzy	H	H
ZK93423	H	CO_2Et	CH_2OMe	H	OBzy	H

OBzy = $O.CH_2.C_6H_5$

Fig. 8. Structures of β-carboline analogues. *OBZY*, $O \cdot CH_2 \cdot C_6H_5$

produce no overt effects initially but when given repeatedly over a period of 10–14 days begin to initiate overt convulsions in the animals. The effect is very long-lasting; dosage may cease for several months, during which period the animals appear to be completely normal, but when challenged again with the same dose of the compound, they exhibit the same overt convulsive behaviour. The phenomenon is very poorly understood at present.

Inverse agonist activity is not an inherent property of the β-carboline analogues, as antagonists (ZK93426), partial (ZK91296) and full (ZK93423) agonists have also been identified in this series (see e.g. Stephens et al. 1986) (Fig. 8). Conversely, inverse agonists, such as Ro19-4603 (Fig. 7), can be found in the benzodiazepine series (Pieri 1988). Inverse agonists, such as CGS 8216 (Fig. 7), can also be found in other chemical series (Czernik et al. 1982; Yokoyama et al. 1982). It has not yet proved possible to define the relative importance of the chemical modifications necessary, even within individual chemical series, to differentiat agonist, inverse agonist and antagonist ligands for this site, although a number of hypotheses have been put forward (Codding and Muir 1985; Fryer et al. 1986; Borea et al. 1987).

These observations have been rationalised in terms of the "three-state model" of the benzodiazepine site (Nutt et al. 1982; Polc et al. 1982; Jensen et al. 1983; Prado de Carvalho et al. 1983). The site is envisaged as existing in three conformational states. Occupation of the first one results in no effect being observed on the GABA activation of chloride channel opening: the antagonist state. The second, when occupied, allows the receptor to relax into such a conformation that an increased coupling between the GABA stimulus and its effector mechanism results; this is to so-called agonist state and leads to the overt actions produced by the classic benzodiazepines (e.g. diazepam; Fig. 7). The third state is obtained when the receptor is occupied by a compound such as β-CCM, with behavioural actions opposite to those of the benzodiazepines; it induces a conformational change which produces a decrease in the coupling between the GABA stimulus and its effector sequence; this is termed the inverse agonist state. The antagonists, such as Ro15-1788 (Fig. 7), are able to block the effects of both the agonists and the inverse agonists (NUTT et al. 1982; BRAESTRUP et al. 1983). Recently, KEMP et al. (1987) carried out an electrophysiological analysis of the actions of a series of agonists, antagonists and inverse agonists at the benzodiazepine site using isoguvacine as the $GABA_A$ agonist in the hippocampal slice preparation. They conclude that their data are consistent with the model of positive and negative modulation of $GABA_A$ responses by these agents.

The precise molecular mechanisms by which ligands for the benzodiazepine receptor produce their effects on the GABA receptor is not clear. Application of GABA to the receptor complex results in an increased permeability of the cell membrane to chloride ions (CURTIS et al. 1968; KRNJEVIC and SCHWARTZ 1967; BARKER and RANSOM 1978). This is due to the increased probability of opening of the integral chloride ion channel of the $GABA_A$ receptor as a result of the presence of the amino acid. This ion channel exhibits multiple conductance states (HAMILL et al. 1983), and there has been a detailed electrophysiological study of the characteristics of this channel (BORMANN et al. 1987). It is clear that the agonist benzodiazepines are able to enhance the GABA-induced conductance in cells, but they do not increase the maximal response which can be obtained with GABA (CHOI et al. 1981; SKERRITT and MACDONALD 1984a). Noise analysis suggests that the agonist benzodiazepines increase the frequency of this channel opening in response to a given GABA stimulus (STUDY and BARKER 1981). The inverse agonists appear to decrease this GABA-induced current (SKERRITT and MACDONALD 1984b; VICINI et al. 1986); however, there is no detailed characterisation of the effects of benzodiazepine recognition site ligands on single-channel activity. It is clear that the action of the benzodiazepines is simply an allosteric effect on GABA-mediated channel opening; the benzodiazepines are not able to open the channel in the absence of a GABA stimulus. The dose-response curve to GABA is shifted, but the maximum response is not affected.

It is not certain upon which of the peptide subunits in the receptor heterooligomer the benzodiazepine recognition site is located. Initial reports of the molecular cloning of this receptor suggested that RNA transcribed in vitro from the cloned cDNA of the α- and β-subunits of the bovine GABA$_A$ receptor, when expressed in *Xenopus* oocytes, exhibited a facilitation of the GABA-induced current with flurazepam (Schofield et al. 1987). There has been some considerable difficulty in reproducing this observation (Levitan et al. 1988). Responses to benzodiazepines and inverse agonist β-carbolines have been seen in oocytes in which only α- and β-subunits from the rat are expressed (Malherbe et al. 1990), but in this case both the agonists and inverse agonists produced an increase in the GABA-activated currents, indicating atypical receptor properties. However, transfection experiments in human embryonic kidney cells imply that the ability of the agonist benzodiazepines to angment GABA-mediated currents in these cells is critically dependent on the co-expression of α-, β- and γ-subunits (Pritchett et al. 1989). It is only under such conditions that the cells express binding sites for the benzodiazepines. Whether the γ-subunit is simply performing a structural role in allowing expression of the benzo-diazepine site in the cells or is an integral part of the recognition site for these ligands is not clear at this stage.

The subunit localisation of the benzodiazepine recognition site cannot readily be addressed with protein chemistry. Initial experiments using photolabelling of the receptor protein with flunitrazepam (Mohler et al. 1980) suggested that only a single band was labelled on SDS-PAGE, the α-subunit (Sigel et al. 1983); at that time it was thought that the receptor oligomer consisted of only two subunits (α and β). Subsequent work from Sieghart (1986; Sieghart and Drexler 1983) has shown that flunitrazepam labels multiple bands on high resolution SDS-PAGE. This group recently suggested that all the gel bands labelled with flunitrazepam were α-subunits (Fuchs and Sieghart 1989). However, it is clear that the gel separation of γ-and certain α-subunits is problematic, and it is thus difficult to state conclusively that the γ-subunit(s) are not also labelled. The development of subunit-specific antibodies with defined epitope specificity will un-doubtedly assist in resolving this uncertainty (Duggan and Stephenson 1990; Stephenson et al. 1990; Ewert et al. 1990).

F. Conclusions

It is thought that all neurons in the mammalian CNS express receptors for GABA. In some brain regions, more than 50% of the neurons use GABA as their primary neurotransmitter. It is thus not surprising that any action taken to compromise the effects of GABA produce pronounced overt effects. In this review, we have discussed the recognition sites present on the GABA$_A$ receptor which are thought to mediate the convulsant effects. The mechanisms by which these drugs subvert GABA-mediated transmission vary from apparently simple competitive antagonism at the GABA recog-nition site to the non-competitive and clearly complex interactions which take place at or close to the picrotoxin site and the modulatory effects of benzodiazepine inverse agonists. The multiplicity of genes which are present within the mammalian CNS could be responsible for the construction of an enormous variety of subtly different GABA$_A$ receptors. Currently, we do not know which of the oligomeric receptor complexes possible are actually assembled in vivo. The elucidation of this question will rely on a multi-disciplinary effort from molecular neurobiology. The prize is considerable for it will involve a greater understanding of those clinical conditions whose aetiology is currently ill-defined and the development of drugs with an increased specificity for their treatment.

References

Abalis IM, Eldefrawi ME, Eldefrawi AT (1985) High affinity stereo-specific binding of cyclodiene insecticides and γ-hexachlorocyclohexane to γ-aminobutyric acid receptors of rat brain. Pestic Biochem Physiol 24:95–105

Akaike N, Hattori K, Oomura Y, Carpenter DO (1985) Bicuculline and picrotoxin block gamma-aminobutyric acid-gated Cl⁻ conductance by different mechanisms. Experientia 41:70–71

Anlezark G, Horton RW, Meldrum BS, Swaya MCB (1976) Anti-convulsant action of ethanolamine-O-sulphate and di-n-propylacetate and the metabolism of gamma-aminobutyric acid (GABA) in mice with audiogenic seizures. Biochem Pharmacol 25:413–417

Arbilla S, Kamal L, Langer Z (1979) Presynaptic GABA auto receptors on GABAergic nerve endings of the rat substantia nigra. Eur J Pharmacol 57:211–217

Asano T, Ogasawara N (1981) Chloride-dependent stimulation of GABA and benzodiazepine receptor binding by pentobarbital. Brain Res 225:212–216

Barker JL, McBurney RN (1979) Phenobarbitone modulation of postsynaptic GABA receptor function on cultured mammalian neurones. Proc R Soc Lond [Biol] 206:319–327

Barker JL, Ransom BR (1978) Amino acid pharmacology of mammalian central neurons grown in tissue culture. J Physiol (Lond) 280:331–354

Barker JL, McBurney RN, Mathers DA, Vaughan W (1980) The actions of convulsant agents on ion-channels activated by inhibitory amino acids on mouse neurons in cell culture. J Physiol (Lond) 308:18P

Barker JL, McBurney RN, Mathers DA (1983) Convulsant-induced depression of amino acid responses in cultured mouse spinal neurones studied under voltage clamp. Br J Pharmacol 80:619–629

Barnard EA, Darlison MG, Seeburg P (1987) Molecular biology of the GABA$_A$ receptor: the receptor channel super family. Trends Neurosci 10:502–509

Bartholini G, Scatton B, Zivkovic B, Lloyd KG, Depoortere H, Langer SZ, Morselli PL (1985) GABA receptor agonists as a new therapeutic class. In: Bartholini G, Bossi L, Lloyd KG, Morselli PL (eds) Epilepsy and GABA receptor agonists. Raven, New York, pp 1–30

Beaumont K, Chilton W, Yamamura HI, Enna SJ (1978) Muscimol binding in rat brain: association with synaptic GABA receptors. Brain Res 148:153–162

Bender AS, Hertz L (1988) Evidence for involvement of the astrocytic benzodiazepine receptor in the mechanism of action of convulsant and anticonvulsant drugs. Life Sci 43:477–484

Bentley R, Trumen H (1875) *Anamirta paniculata*. In: Medicinal plants being descriptions with original figures of the principal plants employed in medicine – an account of their properties and uses, part 2. Churchill, London, p 5

Birdsall NMJ, Berrie CP, Burgen ASV, Hulme EC (1980) Modulation of the binding properties of muscarinic receptors: evidence of receptor-effector coupling. In: Pepeu GC, Kuhar MJ, Enna SJ (eds) Receptors for neurotransmitters and peptide hormones. Raven, New York, pp 107–116

Bloomquist JR, Soderland DM (1985) Neurotoxic insecticides inhibit GABA-dependent chloride uptake by mouse brain vesicles. Biochem Biophys Res Comm 133:37–43

Borea PA, Gilli G, Bertolasi V, Ferretti V (1987) Stereochemical features controlling binding and intrinsic activity properties of benzodiazepine receptor ligands. Mol Pharmacol 31:334–344

Bormann J, Clapham DE (1985) Gamma-aminobutyric acid receptor channels in adrenal chromaffin cells: a patch-clamp study. Proc Natl Acad Sci USA 82:2168–2172

Bormann J, Hamill OP, Sakmann B (1987) Mechanism of anion permeation through channels gated by glycine and gamma-aminobutyric acid in mouse cultured spinal neurones. J Physiol (Lond) 385:243–286

Bosmann HB, Case KR, Distefano P (1977) Diazepam receptor characterisation: specific binding of a benzodiazepine to macromolecules in various areas of rat brain. FEBS Lett 82:368–372

Bowery NG, Brown DA, Collins JF (1975) Tetramethylenedisulphotetramine: an inhibitor of gamma-aminobutyric acid induced depolarisation of the isolated superior cervical ganglion of the rat. Br J Pharmacol 53:422–424

Bowery NG, Collins JF, Hill RG, Pearson S (1976) GABA antagonism as a possible basis for the convulsant action of a series of bicyclic phosphorus esters. Br J Pharmacol 57:435–436

Bowery NG, Collins JF, Hill RG, Pearson S (1977) t-Butyl bicyclophosphate: a convulsant GABA antagonist more potent than bicuculline. Br J Pharmacol 60:275–276

Bowery NG, Hudson AL, Price GW (1987) $GABA_A$ and $GABA_B$ receptor site distribution in the rat central nervous system. Neuroscience 20:365–383

Bradford HF (1970) Metabolic response of synaptosomes to electrical stimulation: release of amino acids. Brain Res 19:239–247

Braestrup C, Nielsen M (1980) Multiple benzodiazepine receptors. Trends Neurosci 3:301–303

Braestrup C, Nielsen M, Olsen CE (1980) Urinary and brain beta-carboline-3-carboxylates as potent inhibitors of brain benzodiazepine receptors. Proc Natl Acad Sci USA 77:2288–2292

Braestrup C, Schmiechen R, Neef G, Nielsen M, Petersen EN (1982) Interaction of convulsive ligands with benzodiazepine receptors. Science 216:1241–1243

Braestrup C, Nielsen M, Honore T, Jensen LH, Petersen EN (1983) Benzodiazepine receptor ligands with positive/negative efficacy. Neuropharmacology 22:1451–1457

Brennan MJW, Cantrill RC (1979) Delta-aminolaevulinic acid is a potent agonist for GABA autoreceptors. Nature 280:514–515

Brennan MJW, Cantrill RC, Krogsgaard-Larsen P (1981) GABA autoreceptors: structure activity relationships for agonists. Adv Biochem Psychopharmacol 26:157–167

Bristow DR, Martin IL (1989) GABA preincubation of rat brain sections increases [^{3}H]-GABA binding to a $GABA_A$ receptor and compromises the modulatory interactions. Eur J Pharmacol 73:65–73

Brown DA, Constanti A (1978) Interaction of pentobarbitone γ-aminobutyric acid on mammalian sympathetic ganglion cells. Br J Pharmacol 63:217–224

Casalotti SO, Stephenson FA, Barnard EA (1986) Separate subunits of agonist and benzodiazepine binding to the gamma-aminobutyric acid-A receptor oligomer. J Biol Chem 261:15015–15016

Cash DJ, Subbarao K (1987) Desensitisation of gamma-aminobutyric acid receptor from rat brain: two distinguishable receptors on the same membrane. Biochemistry 26:7556–7562

Casida JE, Nicholson RA, Palmer CJ (1988) Trioxabicyclooctanes as probes for the convulsant site of the GABA-operated chloride channel in mammals arthropods. In: Lunt GG (ed) Neurotox 88: molecular basis of drug pesticide action. Elsevier, Amsterdam pp 125–144

Chambon JP, Feltz P, Heulme M, Restle S, Schlichter R, Biziere K, Wermuth CG (1985) An arylaminopyridazine derivative of gamma-aminobutyric acid (GABA) is a selective and competitive antagonist at the $GABA_A$ receptor site. Proc Natl Acad Sci USA 82:1832–1836

Chang L-R, Barnard EA (1982) The benzodiazepine/GABA receptor complex: molecular size in brain synaptic membranes in solution. J Neurochem 39:1507–1518

Changeux J-P, Devilliers-Thiery A, Chemouilli P (1984) Acetylcholine receptor: an allosteric protein. Science 225:1335–1345

Choi DW, Farb DH, Fischbach GD (1981) Chlordiazepoxide selectively potentiates GABA conductance of spinal cord sensory neurons in cell culture. J Neurophysiol 45:621–631

Clifford DB, Baker K, Yang J, Covey DF, Zorumski CF (1989) Convulsant gamma-butyrolactones block GABA currents in cultured chick spinal cord neurones. Brain Res 484:102–110

Codding PW, Muir AKS (1985) Molecular structure of Ro15-1788: a model for the binding of benzodiazepine receptor ligands. Mol Pharmacol 28:178–184

Cole LM, Casida JE (1986) Polychlorocycloalkane insecticide-induced convulsions in mice in relation to disruption of the GABA-regulated chloride ionophore. Life Sci 39:1855–1862

Collins JF, Hill RG (1974) (+) and (−) bicuculline methochloride as optical isomers of a GABA antagonist. Nature 249:845–846

Cowan PJ, Green AR, Nutt DJ, Martin IL (1981) Ethyl beta-carboline carboxylate lowers seizure threshold and antagonises flurazepam induced sedation in rats. Nature 290:54–55

Cull-Candy SG, Mathie A, Newland CF (1988) Influence of picrotoxin on ion channels activated by γ-aminobutyric acid in dissociated sympathetic neurones of the rat. J Physiol (Lond) 398:89P

Curtis DR, Hosli L, Johnston GAR, Johnston IH (1968) Pharmacological study of the depression of spinal neurones by glycine related amino acids. Exp Brain Res 6:1–18

Curtis DR, Duggan AW, Felix D, Johnston GAR (1970) GABA, bicuculline and central inhibition. Nature 226:1222–1224

Curtis DR, Duggan AW, Felix D, Johnston GAR (1971) Antagonism between bicuculline and GABA in the cat brain. Brain Res 33:57–73

Czernik AJ, Petrack B, Kalinsky HJ, Psychoyos S, Cash WD, Tsai C, Rinehart RK, Granat FR, Lovell RA, Brundish DE, Wade R (1982) CGS 8216; receptor binding characteristics of a potent benzodiazepine antagonist. Life Sci 30:363–372

Deng L, Ransom RW, Olsen RW (1986) [^{3}H]-muscimol photolabels the GABA receptor site on a peptide subunit distinct from that labelled by a benzodiazepine. Biochem Biophys Res Commun 138:1308–1314

Desarmenian N, Desaulles E, Feltz P, Hamann M (1987) Electrophysiological study of SR42641, a novel aminopyridazine derivative of GABA: antagonist properties and receptor selectivity of $GABA_A$ versus $GABA_B$ responses. Br J Pharmacol 90:287–298

Dorow R, Horowski R, Paschelke G, Amin M, Braestrup C (1983) Severe anxiety induced by FG7142, a beta-carboline ligand for benzodiazepine receptors. Lancet 2:98–99

Duggan MJ, Stephenson FA (1990) Biochemical evidence for the existence of γ-aminobutyrate$_A$ receptor iso-oligomers. J Biol Chem 265:3831–3835

Eldefrawi AT, Eldefrawi ME (1987) Receptors for γ-aminobutyric acid and voltage dependent chloride channels as targets for drugs and toxicants. FASEB J 1:262–271

Enna SJ, Snyder SH (1975) Properties of gamma-aminobutyric acid (GABA) binding in rat brain synaptic membranes. Brain Res 100:81–97

Evans RH (1979) Potentiation of the effect of GABA by pentobarbitone. Brain Res 17:113–120

Ewert M, Shivers BD, Luddens H, Mohler H, Seeburg P (1990) Subunit selectivity epitope characterisation of mAbs directed against the $GABA_A$/benzodiazepine receptor. J Cell Biol 110:2043–2048

Falch E, Krogsgaard-Larsen P (1982) The binding of the GABA agonist [^{3}H]-THIP to rat brain synaptic membranes. J Neurochem 38:1123–1129

Fowler LJ, John RA (1972) Active-site directed irreversible inhibition of rat brain 4-amino-butyrate aminotransferase by ethanolamine-O-sulphate in vitro and in vivo. Biochemistry 130:569–573

Frere RC, MacDonald RL, Young AB (1982) GABA binding and bicuculline in spinal cord cortical membranes from adult rat mouse neurons in cell culture. Brain Res 244:145–153

Fryer RI, Cook C, Gilman NW, Walser A (1986) Conformational shifts at the benzodiazepine receptor related to the binding of agonists antagonists inverse agonists. Life Sci 39:1947–1957

Fuchs K, Sieghart W (1989) Evidence for the existence of several different α- and β-subunits of the GABA/benzodiazepine receptor complex from rat brain. Neurosci Lett 97:329–333

Gahwiler BH, Maurer R, Wuthrich HJ (1984) Pitrazepin, a novel $GABA_A$ antagonist. Neuroci Lett 45:311–316

Gale K (1989) GABA in epilepsy: the pharmacologic basis. Epilepsia 30:S1–S11

Galindo A (1969) GABA-picrotoxin interaction in the mammalian nervous system. Brain Res 14:763–767

Gammon D, Casida JE (1983) Pyrethroids of the most potent class antagonize GABA action at the crayfish neuromuscular junction. Neurosci Lett 40:163–168

Gee KW (1988) Steroid modulation of the GABA/benzodiazepine receptor linked chloride ionophore. Mol Neurobiol 2:291–317

Hamann M, Desarmenien M, Vanderhayden P, Piguet P, Feltz P (1990) Electrophysiological study of tert-butylbicyclophosphorothionate-induced block of spontaneous chloride channels. Mol Pharmacol 37:578–582

Hamill OP, Borman J, Sakman B (1983) Activation of multiple conductance state chloride channels in spinal neurons by glycine GABA. Nature 305:805–808

Harris RA, Allen AM (1985) Functional coupling of γ-aminobutyric acid receptors to chloride channels in brain membranes. Science 228:1108–1110

Harrison NL, Simmonds MA (1983) The picrotoxin-like action of a convulsant benzodiazepine Ro5-3663. Eur J Pharmacol 87:287–292

Harrison NL, Simmonds MA (1984) Modulation of the GABA receptor complex by a steroid anaesthetic. Brain Res 323:287–292

Heaulme M, Chambon JP, Leyris R, Wermuth CG, Biziere K (1986) Specific binding of a phenyl-pyridazinium derivative endowed with $GABA_A$ receptor antagonist activity. Neuropharmacology 25:1279–1283

Heaulme M, Chambon J-P, Leyris R, Wermuth CG, Biziere K (1987) Characterisation of the binding of [^{3}H]SR95531, a $GABA_A$ antagonist, to rat brain membranes. J Neurochem 48:1677–1686

Hill RG, Simmonds MA, Straughan DW (1974) Effects of synaptic inhibition in the cuneate nucleus produced by stereoisomers of bicuculline methochloride. J Physiol (Lond) 239:122P–123P

Holland KD, Ferrendelli JA, Covey DF, Rothman SM (1990a) Physiological regulation of the picrotoxin receptor by γ-butyrolactones γ-thiobutyrolactones in cultured hippocampal neurons. J Neurosci 10:1719–1727

Holland KD, Narioku DK, McKeon AC, Ferrendelli JA, Covey DF (1990b) Convulsant and anticonvulsant cyclopentanones and cyclohexanones. Mol Pharmacol 37:98–103

Hunt P, Clements-Jewery S (1981) A steroid derivative, R5135, antagonises the GABA/benzodiazepine receptor interaction. Neuropharmacology 20:357–361

Iversen LL, Kelly JS (1975) Uptake and metabolism of gamma-aminobutyric acid by neurones and glial cells. Biochem Pharmacol 24:933–938

Iversen LL, Neal MJ (1968) The uptake of ^{3}H-GABA by slices of rat cerebral cortex. J Neurochem 25:1141–1149

Jarboe CH, Porter LA, Buckler RT (1968) Structural aspects of picrotoxinin action. J Med Chem 11:729–731

Jensen LH, Petersen EN, Braestrup C (1983) Audiogenic seizures in DBA/2 mice discriminate sensitively between low efficacy benzodiazepine receptor agonists and inverse agonists. Life Sci 33:393–399

Johnston GAR (1976) Physiologic pharmacology of GABA and its antagonists in vertebrate nervous system. In: Roberts E, Chase TN, Tower DB (eds) GABA in nervous system function. Raven, New York pp 395–411

Johnston GAR (1983) Regulation of GABA receptors by barbiturates and by related sedative-hypnotic and anticonvulsant drugs. In Enna SJ (ed) The GABA receptors. Humana, Clifton, pp 107–128

Johnston GAR, Beart PM, Curtis DR, Game CJA, McCulloch RM, Maclachan RM (1972) Bicuculline methochloride as a GABA antagonist. Nature [New Biol] 240:219–220

Jordan CC, Matus AI, Piotrowski W, Wilkinson D (1982) Binding of [^{3}H]-gamma-aminobutyric acid and [^{3}H]-muscimol in purified rat brain synaptic plasma membranes and the effects of bicuculline. J Neurochem 39:52–58

Kardos J, Blasko G, Kerekes P, Kovacs I, Simonyi M (1984) Inhibition of [^{3}H]-GABA binding to rat brain synaptic membranes by bicuculline related alkaloids. Biochem Pharmacol 22:3537–3545

Kelly JS, Krnjevic K, Morris ME, Yim GKW (1969) Anionic permeability of cortical neurones. Exp Brain Res 7:11–31

Kemp JA, Marshall GR, Wong EHF, Woodruff GN (1985) Pharmacological studies on pitrazepin, a GABA$_A$ receptor antagonist. Br J Pharmacol 85:237P

Kemp JA, Marshall GR, Wong EHF, Woodruff GN (1987) The affinities, potencies and efficacies of some benzodiazepine receptor agonists antagonists and inverse agonists at rat hippocampal GABA$_A$ receptors. Br J Pharmacol 91:601–608

Leeb-Lundberg F, Olsen RW (1982) Interactions of barbiturates of various pharmacological categories with benzodiazepine receptors. Mol Pharmacol 21:320–328

Levine JA, Ferrendelli JA, Covey DF (1985) Convulsant and anticonvulsant gamma-butyrolactones bind at the picrotoxinin-t-butylbicyclophosphorothionate (TBPS) receptor. Biochem Pharmacol 34:4187–4190

Levitan ES, Schofield PR, Burt DR, Rhee LM, Wisden W, Kohler M, Fujita N, Rodriguez HF, Stephenson FA, Darlison MG, Barnard EA, Seeburg PH (1988) Structure and functional basis for GABA$_A$ receptor heterogeneity. Nature 335:76–79

Little HJ, Nutt DJ (1984) Benzodiazepine contragonists cause kindling. Br J Pharmacol 81:28P

Little HJ, Nutt DJ, Taylor SC (1984) Acute chronic effects of the benzodiazepine receptor ligand FG7142: pro-convulsant properties and kindling. Br J Pharmacol 83:951–958

Lummis SCR (1990) GABA receptors in insects. Comp Biochem Physiol 95C:1–8

Majewska MD (1987) Steroids and brain activity: essential dialogue between body and mind. Biochem Pharmacol 36:3781–3788

Majewska MD, Schwartz RD (1987) Pregnenolone sulfate: an endogeneous antagonist of the γ-aminobutyric acid receptor complex in brain. Brain Res 404:355–360

Majewska MD, Harrison NL, Schwartz RD, Baker JL, Pane SM (1986) Steroid hormone metabolites are barbiturate-like modulators of the GABA receptor. Science 232:1004–1007

Maksay G, Ticku MK (1984) Dissociation of [^{35}S]-t-butylbicyclophosphorothionate binding differentiates convulsant depressant drugs that modulate GABAergic transmission. J Neurochem 44:480–486

Maksay G, Ticku MK (1988) CNS depressants accelerate the dissociation of ^{35}S-TBPS binding, GABA enhances their displacing potencies. Life Sci 43:1331–1337

Kier LB, George JM (1973) Molecular orbital studies on the conformations of bicuculline and beta-hydroxy-GABA. Experientia 29:501–503

Kirkness EF (1989) Steroid modulation reveals further complexity of GABA$_A$ receptor. TIPS 10:6–7

Kirkness EF, Turner AJ (1988) Antibodies directed against a nonapeptide sequence of the gamma-aminobutyrate (GABA)/benzodiazepine receptor alpha-subunit. Biochem J 256:291–294

Kirkness EF, Bovenkerk CF, Ueda T, Turner AJ (1989) Phosphorylation of gamma-aminobutyrate (GABA)/benzodiazepine receptors by cyclic AMP-dependent protein kinase. Biochem J 259:613–616

Krnjevic K, Schwartz S (1967) The action of gamma-aminobutyric acid on cortical neurones. Exp Brain Res 3:320–336

Krogsgaard-Larsen P, Snowman A, Lummis SCR, Olsen RW (1981) Characterisation of the binding of the GABA agonist [^{3}H]-piperidine-4-sulphonic acid to bovine brain synaptic membranes. J Neurochem 37:401–409

Krogsgaard-Larsen P, Falch E, Jacobsen P (1984) GABA agonists structural requirements for interaction with the GABA-benzodiazepine receptor complex. In: Bowery NG (ed) Actions and interactions of GABA benzodiazepines. Raven, New York, pp 109–132

Kudo Y, Oka JI, Yamada K (1981) Anisatin, a potent GABA antagonist isolated from *Illicium anisatum*. Neurosci Lett 25:83–88

Lawrence LJ, Casida JE (1983) Stereospecific actions of pyrethroid insecticides on the γ-aminobutyric acid receptor ionophore complex. Science 221:1339–1401

Lawrence LJ, Casida JE (1984) Interactions of lindane, toxaphene and cyclodienes with brain specific t-butylbicyclophosphorothionate receptor. Life Sci 35:171–178

Lawrence LJ, Palmer CJ, Gee KW, Wang X, Yamamura HI, Casida JE (1985) t-[^{3}H]Butylbicycloorthobenzoate: a new radioligand probe for the γ-aminobutyric acid-regulated chloride ionophore. J Neurochem 45:798–804

Malherbe P, Draguhn A, Multhaup G, Beyreuther K, Mohler H (1990) GABA$_A$-receptor expressed from rat brain α- and β-subunit cDNAs displays potentiation by benzodiazepine receptor ligands. Mol Brain Res 8:199–208

Mathers DA (1987) The GABA$_A$ receptor: new insights from single channel recording. Synapse 1:96–101

Mienville J-M, Vicini S (1987) A pyridazinyl derivative of gamma-aminobutyric acid (GABA), SR 95531, is a potent antagonist of Cl$^-$ channel opening regulated by GABA$_A$ receptors. Neuropharmacology 26:779–783

Mitchell PR, Martin IL (1978) Is GABA release modulated by presynaptic receptors? Nature 274:904–905

Mohler H, Okada T (1977a) GABA receptor binding with [^{3}H]-bicuculline methiodide in rat CNS. Nature 267:65–67

Mohler H, Okada T (1977b) Benzodiazepine receptor: demonstration in the central nervous system. Science 198:849–851

Mohler H, Okada T (1978) Properties of gamma-aminobutyric acid receptor binding with (+)-[^{3}H]-bicuculline methiodide in rat cerebellum. Mol Pharmacol 14:256–265

Mohler H, Battersby MK, Richards JG (1980) Benzodiazepine receptor protein identified and visualised in brain tissue by a photo affinity label. Proce Natl Acad Sci USA 77:1666–1670

Narahashi T (1985) Nerve membrane channels as the primary target of pyrethroids. Neurotoxicology 6:3–22

Neal MJ, Iversen LL (1969) Subcellular distribution of endogenous ^{3}H-GABA in rat cerebral cortex. J Neurochem 16:1245–1252

Nicoll RA (1978) Pentobarbital: differential postsynaptic actions on sympathetic ganglion cells. Science 199:451–452

Nowak LM, Young B, MacDonald RL (1982) GABA bicuculline actions on mouse spinal cord cortical neurones in cell culture. Brain Res 244:155–164

Nutt DJ, Cowen PJ, Little HJ (1982) Unusual interactions of benzodiazepine receptor antagonists. Nature 295:436–438

Olsen RW (1981) GABA-benzodiazepine-barbiturate interactions. J Neurochem 37:1–13

Olsen RW (1983) Biochemical properties of GABA$_A$ receptors. In: Enna SJ (ed) The GABA receptors. Human a Clifton

Olsen RW, Snowman AM (1982) Chloride-dependent enhancement by barbiturates of gamma-aminobutyric acid receptor binding. J Neurosci 2:1812–1823

Olsen RW, Snowman AM (1983) [^{3}H]-Bicuculline methochloride binding to low affinity GABA receptors. J Neurochem 41:1653–1663

Olsen RW, Tobin AJ (1990) Molecular biology of GABA$_A$ receptors. FASEB J 4:1469–1480

Olsen RW, Ban M, Miller T, Johnston GAR (1975) Chemical instability of the GABA antagonist bicuculline under physiological conditions. Brain Res 98:383–387

Olsen RW, Snowhill EW, Wamsley JK (1984) Autoradiographic localisation of low affinity GABA receptors with [^{3}H]-bicuculline methochloride. Eur J Pharmacol 99:247–248

Olsen RW, McCabe RT, Wamsley JK (1990) GABA$_A$ receptor subtypes: autoradiographic comparison of GABA, benzodiazepine and convulsant binding sites in the rat central nervous system. J Chem Neuroanat 3:59–76

Ottersen OP, Storm-Mathison J (1984) Neurons containing or accumulating transmitter amino acids. In: Bjorklund A, Hokfelt T, Kuhar MJ (eds) Handbook of chemical neuroanatomy, vol 3. Elsevier, Amsterdam, pp 141–246

Palfreyman MG, Schechter P, Buckett WR, Tell GP, Koch-Weser J (1981) The pharmacology of GABA-transaminase inhibitors. Biochem Pharmacol 30:817–824

Palmer CJ, Casida JE (1988) Two types of cage convulsant action at the GABA-gated chloride channel. Toxicol Lett 42:117–122

Peters JA, Kirkness EF, Callachan H, Lambert JJ, Turner AJ (1988) Modulation of the GABA$_A$ receptor by depressant barbiturates and pregnane steroids. Br J Pharmacol 94:1257–1269

Petersen EN, Jensen LH, Honore T, Braestrup C (1983) Differential pharmacological effects of benzodiazepine receptor inverse agonists. In: Biggio G, Costa E (eds) Benzodiazepine recognition site ligands: biochemistry and pharmacology. Raven, New York, pp 57–64

Petersen EN, Jensen LH, Honore T, Braestrup C, Kehr W, Stephens DN, Wachtel H, Seidelman D, Schmiechen R (1984) ZK91296 a partial agonist at benzodiazepine receptors. Psychopharmacology (Berlin) 82:260–264

Pickles HG, Simmonds MA (1980) Antagonism by penicillin of gamma-aminobutyric acid depolarisations at presynaptic sites in rat olfactory cortex and cuneate nucleus in vitro. Neuropharmacology 19:35–38

Pieri L (1988) Ro19-4603: a benzodiazepine receptor partial inverse agonist with prolonged pro-convulsant action in rodents. Br J Pharmacol 95:477P

Polc P, Mohler H, Haefely W (1974) The effects of diazepam on spinal cord activities: possible sites and mechanisms of action. Naunyn Schmiedebergs Arch Pharmacol 284:319–337

Polc P, Bonetti EP, Schaffner R, Haefely W (1982) A three state model of the benzodiazepine receptor explains the interactions between the benzodiazepine antagonist Ro 15-1788, benzodiazepine tranquilisers, beta-carbolines, and phenobarbitone. Naunyn Schmiedebergs Arch Pharmacol 321:260–26

Pong SF, Graham LT (1972) N-Methylbicuculline, a convulsant more potent than bicuculline. Brain Res 42:486–490

Potier M-C, Prado de Carvalho L, Dodd RH, Besseliever R, Rossier J (1988) In vivo binding of beta-carbolines in mice: regional differences and correlation of occupancy to pharmacological effects. Mol Pharmacol 34:124–128

Prado de Carvalho L, Grecksch G, Chapouthier G, Rossier J (1983) Anxiogenic and non-anxiogenic benzodiazepine antagonists. Nature 301:64–66

Pritchett DB, Seeburg PH (1990) γ-Aminobutyric acid$_A$ receptor α_5-subunit creates novel type II benzodiazepine receptor pharmacology. J Neurochem 54:1802–1804

Pritchett DB, Sontheimer H, Shivers BD, Ymer S, Kettenmann H, Schofield PR, Seeburg PH (1989) Importance of a novel GABA$_A$ receptor subunit for benzodiazepine pharmacology. Nature 338:582–585

Quast U, Brenner O (1983) Modulation of [^{3}H]-muscimol binding in rat cerebellar and cerebral cortical membranes by picrotoxin, pentobarbitone and etomidate J Neurochem 41:418–425

Ramanjaneyulu R, Ticku MK (1984) Binding characteristics and interactions of depressant drugs with [^{35}S]-t-butylbicyclophosphorothionate, a ligand that binds to the picrotoxinin site. J Neurochem 42:221–229

Randall LO (1982) Discovery of benzodiazepines. In: Usdin E, Skolnick P, Tallman JF, Greenblatt D, Paul SM (eds) Pharmacology of benzodiazepine. Macmillan, London, pp 15–22

Rehavi MP, Skolnick P, Paul SM (1982) Effects of tetrazole derivatives on [^{3}H]diazepam binding in vitro: correlation with convulsant potency. Eur J Pharmacol 78:353–358

Remy C, Beaumont D (1989) Efficacy and safety of vigabatrin in the long term treatment of refractory epilepsy. Br J Clin Pharmacol 27:125S–129S

Sarter M, Stephens DN (1989) Disinhibitory properties of beta-carboline antagonists of benzodiazepine receptors: a possible therapeutic approach for senile dementia. Biochem Soc Trans 17:81–83

Schechter P (1989) Clinical pharmacology of vigabatrin. Br J Clin Pharmacol 27:19S–22S

Schmidt RF, Vogel E, Zimmermann M (1967) Die Wirkung von Diazepam auf die prä-synaptische Hemmung und die Rückenmarksreflexe. Naunyn Schmiedebergs Arch Pharmacol 258:69–82

Schoch P, Haring P, Takacs B, Stahli C, Mohler H (1984) A GABA$_A$/benzodiazepine receptor complex from bovine brain. Purification, reconstitution and immunological characterisation. J Recept Res 4:189–200

Schoch P, Richards JG, Haring P (1985) Co-localisation of GABA$_A$ receptors and benzodiazepine receptors in the brain shown by monoclonal antibodies. Nature 314:168–171

Schofield PR, Darlison MG, Fujita N, Burt DR, Stephenson FA, Rodriguez H, Rhee LM, Ramachandran J, Glencorse TA, Seeburg PH, Barnard EA (1987) Sequence and functional expression of the GABA$_A$ receptor shows a ligand-gated receptor super-family. Nature 328:221–227

Shivers B, Killisch I, Sprengel R, Sontheimer H, Kohler M, Schofield PR, Seeburg P (1989) Two novel GABA$_A$ receptor subunits exist in distinct neuronal subpopulations. Neuron 3:327–337

Sieghart W (1986) Comparison of benzodiazepine receptors in cerebellum and inferior colliculus. J Neurochem 47:920–923

Sieghart W, Drexler G (1983) Irreversible binding of ^{3}H-flunitrazepam to different proteins in various brain regions. J Neurochem 41:47–55

Sigel E, Stephenson FA, Mamalaki C, Barnard EA (1983) A gamma-aminobutyric acid/benzodiazepine receptor complex of bovine cerebral cortex. J Biol Chem 258:6965–6971

Simmonds MA (1978) Presynaptic actions of gamma-aminobutyric acid and some antagonists in a slice preparation of cuneate nucleus. Br J Pharmacol 63:495–502

Simmonds MA (1980) Evidence that bicuculline and picrotoxin act at separate sites to antagonise gamma-aminobutyric acid in the rat cuneate nucleus. Neuropharmacology 19:39–45

Simmonds MA (1981) Distinction between the effects of barbiturates, benzodiazepines and phenytoin on responses to γ-aminobutyric acid receptor activation and antagonism by bicuculline and picrotoxin. Br J Pharmacol 73:739–747

Simmonds MA (1982) Classification of some GABA antagonists with regard to site of action and potency in slices of rat cuneate nucleus. Eur J Pharmacol 80: 347–358

Skerritt JH, MacDonald RL (1984a) Benzodiazepine receptor ligands actions on GABA responses. Benzodiazepines, CL218872, zopiclone. Eur J Pharmacol 101:127–134

Skerritt JH, MacDonald RL (1984b) Benzodiazepine receptor ligands actions on GABA responses. Beta-carbolines and purines. Eur J Pharmacol 101:135–141

Skolnick P, Rice KC, Baker JL, Paul SM (1982) Interaction of barbiturates with benzodiazepine receptors in the central nervous system. Brain Res 233:143–146

Snodgrass SR (1978) Use of ^{3}H-muscimol for GABA receptor studies. Nature 273:392–294

Squires RF, Braestrup C (1977) Benzodiazepine receptors in rat brain. Nature 266:732–734

Squires RF, Casida JE, Richardson M, Saederup E (1983) [^{35}S]-t-butylbicyclophosphorothionate binds with high affinity to brain specific sites coupled to γ-aminobutyric acid at ion recognition sites. Mol Pharmacol 23:326–336

Stark LG, Albertson TE, Joy RM (1986) Effects of hexachlorocyclohexane isomers on the acquisition of kindled seizures. Neurobehav Toxicol Teratol 8:487–491

Stauber GB, Ransom RW, Dilber AI, Olsen RW (1987) The gamma-aminobutyric acid-benzodiazepine receptor protein from rat brain; large-scale purification and preparation of antibodies. Eur J Biochem 167:125–133

Stephens DN, Kehr W, Duka T (1986) Anxiolytic and anxiogenic beta-carbolines: tools for the study of anxiety mechanisms. In: Biggio G, Costa E (eds) GABAergic transmission anxiety. Raven, New York, pp 91–106

Stephenson A, Duggan MJ, Pollard S (1990) The $γ_2$ subunit is an integral component of the GABA$_A$ receptor but the α polypeptide is the principle site of the agonist benzodiazepine photoaffinity labelling reaction. J Biol Chem 265:21160–21165

Study RE, Barker JL (1981) Diazepam and (−)pentobarbital: fluctuation analysis reveals different mechanisms for potentiation of γ-aminobutyric acid responses in cultures of central neurones. Proc Natl Acad Sci USA 78:7180–7184

Sugiyama H, Changeaux J-P (1975) Interconversion between different states of affinity for acetylcholine of the cholinergic receptor protein from Torpedo marmorata. Eur J Biochem 55:505–515

Supavilai P, Karobath M (1984) [^{35}S]-t-butylbicyclophosphorothionate binding sites are constituents of the γ-aminobutyric acid benzodiazepine receptor complex. J Neurosci 4:1193–1200

Suzdak PD, Paul SM, Crawley JN (1988) Effects of Ro15-4513 and other benzodiazepine receptor inverse agonists on alcohol induced intoxication in the rat. J Pharmacol Exp Ther 245:88–886

Sweetnam P, Nestle E, Gallombardo P, Brown S, Duman R, Bracha HS, Tallman J (1987) Comparison of the molecular structure of GABA/benzodiazepine receptor purified from rat and human cerebellum. Mol Brain Res 2:223–233

Takeuchi A, Takeuchi N (1969) A study of the action of picrotoxin on the inhibitory neuromuscular junction of the crayfish. J Physiol (Lond) 205:377–391

Tallman JF, Thomas JW, Gallager DW (1978) GABAergic modulation of benzodiazepine binding site sensitivity. Nature 274:383–385

Tapia R, Sandoval ME, Contreras P (1975) Evidence for role of glutamate decarboxylase activity as a regulatory mechanism of cerebral excitability. J Neurochem 24:1283–1285

Tenen SS, Hirsch JD (1980) Antagonism of diazepam activity by beta-carboline-3-carboxylic acid ethyl ester. Nature 288:609–610

Ticku MK (1986) Convulsant binding sites on the benzodiazepine GABA receptor. In: Olsen RW, Venter JC (eds) Benzodiazepine/GABA receptors of chloride channels. Structural and functional properties. Liss, New York, pp 195–207

Ticku MK, Maksay G (1983) Convulsant/depressant site of action at the allosteric benzodiazepine-GABA receptor-ionophore complex. Life Sci 33:2363–2375

Ticku MK, Olsen RW (1978) Interaction of barbiturates with dihydropicrotoxinin binding sites related to the GABA-receptor-ionophore system. Life Sci 22:1643–1652

Ticku MK, Ban M, Olsen RW (1978) Binding of [^{3}H]a-dihydropicrotoxinin, a γ-aminobutyric acid synaptic antagonist to rat brain membranes. Mol Pharmacol 14:391–402

Trifiletti RR, Snowman AM, Snyder SH (1984) Barbiturate recognition site on the GABA/benzodiazepine receptor complex is distinct from the picrotoxinin/TBPS recognition site. Eur J Pharmacol 106:441–447

Twyman RE, Rogers CJ, Macdonald RL (1989) Pentobarbital and picrotoxin has reciprocal actions on single GABA$_A$ receptor channels. Neurosci Lett 96:89–95

Unwin N (1989) The structure of ion channels in membranes of excitable cells. Neuron 3:665–676

Van Renterghem C, Bilbe G, Moss S, Smart T, Constanti A, Brown DA, Barnard EA (1987) GABA receptors induced in *Xenopus* oocytes by chick brain mRNA: evaluation of the TBPS as a use dependent channel blocker. Mol Brain Res 2:21–31

Venault P, Chapouthier G, Prado de Carvalho L, Simiand J, Morre M, Dodd RH, Rossier J (1986) Benzodiazepine impairs and beta-carboline enhances performance in learning and memory tasks. Nature 321:864–866

Verdoorn TA, Draguhn A, Ymer S, Seeburg PH, Sakmann B (1990) Functional properties of recombinant rat GABA$_A$ receptors depend upon subunit composition. Neuron 4:919–928

Vicini S, Alho H, Costa E, Mienville J-M, Santi MR, Vaccarino FM (1986) Modulation of gamma-aminobutyric acid mediated inhibitory synaptic currents in dissociated cell cultures. Proc Natl Acad Sci USA 83:9269–9273

Wafford KA, Burnett DM, Dunwiddie TV, Harris HA (1990) Genetic differences in the ethanol sensitivity of GABA$_A$ receptors expressed in *Xenopus* oocytes. Science 249:291–293

Weissman BA, Burke TR, Rice KC, Skolnick P (1984) Alkyl-substituted γ-butyrolactones inhibit [^{35}S]TBPS binding to a GABA-linked chloride ionophore. Eur J Pharmacol 105:195–196

Wermuth CG, Biziere K (1986) Pyridazinyl GABA derivatives: a new class of synthetic GABA$_A$ antagonists. Trends Pharmacol Sci 7:421–424

Willow M, Johnson GAR (1981a) Dual action of pentobarbitone on GABA binding. Role of binding site integrity. J Neurochem 37:1291–1294

Willow M, Johnston GAR (1981b) Enhancement of GABA binding by pentobarbitone. Neurosci Lett 18:323–327

Wojcik WJ, Paez X, Ulivi M (1989) A transduction mechanism for GABA$_B$ receptors In: Barnard EA, Costa E (eds) Allosteric modulation of amino acid receptors: therapeutic implications. Raven, New York, pp 173–193

Wong EHF, Leeb-Lundberg LMF, Teichberg VI, Olsen RW (1984) γ-Aminobutyric acid activation of ^{36}Cl$^-$ flux in rat hippocampal slices and its potentiation by barbiturates. Brain Res 303:267–275

Wright DJ (1985) Biological activity and mode of actions of avermectins. In: Ford MG, Lunt GG, Reay RC, Usherwood PNR (eds) Neuropharmacology and pesticide action. Horwood, Chichester, pp 174–202

Yang JSJ, Olsen RW (1987) γ-Aminobutyric acid receptor-regulated ^{36}Cl flux in mouse cortical slices. J Pharmacol Exp Ther 241:677–685

Yokoyama N, Ritter B, Neubert A (1982) 2-Arylpyrazolo [4,3-c]-quinolin-3-ones: novel agonist, partial agonist and antagonist of benzodiazepines. J Med Chem 25:337–339

CHAPTER 15

Convulsants Acting at the Inhibitory Glycine Receptor*

C.-M. Becker

A. The Glycinergic System of the Mammalian Central Nervous System

Inhibitory neurotransmission in the CNS is predominantly mediated by γ-aminobutyric acid (GABA) and glycine. Whereas GABAergic synapses are abundant in the cortex and cerebellum, glycine predominates in the spinal cord and brain stem (reviewed by Betz and Becker 1988; Langosch et al. 1990a,b). Glycine-mediated synaptic inhibition was first demonstrated in the spinal cord where, at the segmental level, neuronal pathways regulate the tonus of skeletal muscle (reviewed by Krnjevic 1981; Aprison 1990). Glycinergic synapses prevail in the spinal sensory, auditory, and visual system and in other parts of the CNS.

I. Postsynaptic Receptors

At the postsynaptic membrane, glycine binds to neurotransmitter receptors associated with an intrinsic anion channel. The resulting chloride current hyperpolarizes the membrane potential and depresses neuronal firing (Curtis et al. 1968a,b). Inhibitory miniature potentials recorded from rat moto-neurons are consistent with a presynaptic vesicular release of glycine (Takahashi 1984). Glycine-mediated inhibition is efficiently antagonized by the convulsant alkaloid strychnine, a high-affinity ligand of postsynaptic glycine receptors (reviewed by Curtis and Johnston 1970; Betz and Becker 1988; Langosch et al. 1990a,b). Only at higher concentrations does strychnine detectably interfere with nonglycinergic receptors and channels (reviewed by Barron and Guth 1987; but see Becker and Betz 1987). Conversely, [³H]strychnine was introduced as a reliable and, at nanomolar concentrations, highly specific probe of the glycine receptor (Young and Snyder 1973, 1974a). Affinity chromatography on 2-aminostrychnine columns has been used to purify glycine receptor from mammalian spinal cord. Immunological and cDNA cloning experiments characterize the glycine receptor as a typical member of the ligand-gated ion channel family.

---

*This work was supported by Deutsche Forschungsgemeinschaft (SFB 317 and Heisenberg-Programm).

Glycine receptors exist in developmentally regulated isoforms composed of distinct subunit variants (reviewed by Betz and Becker 1988; Langosch et al. 1990a,b; Betz 1990).

The inhibitory glycine receptor is distinct from 'strychnine-insensitive' [³H]glycine binding sites which represent a regulatory domain of the NMDA (N-methyl-D-aspartate) subtype of the glutamate receptor. Potentiation by glycine of NMDA receptor activity is mimicked by D-serine and D-alanine (reviewed by Becker and Betz 1987; Monaghan et al. 1989). At micromolar concentrations, however, strychnine exerts a voltage-dependent block on the cation channel associated with the NMDA receptor (Bertolino and Vicini 1988).

II. Glycine Receptors and Convulsants

Strychnine intoxication resembles the loss of glycinergic inhibition in the CNS. Consistent with the physiology of glycinergic synapses, symptoms of strychnine toxicity include severe motor disturbances, convulsions, and altered sensory, visual, and acoustic perception (Poulsson 1920; Dusser de Barenne 1933; Gosselin et al. 1976). The convulsant mechanism of various toxins and drugs has also been attributed to glycine receptor antagonism. Distribution, structure, and pharmacological profile of the inhibitory glycine receptor will be reviewed in relation to the convulsant mechanism of strychnine and functionally related compounds in the mammalian CNS.

B. Toxicology: Strychnine as a Prototypic Glycine Receptor Antagonist

Strychnine is an indole-alkaloid contained in the seeds (nux vomica), bark, and roots of the Indian tree *Strychnos nux vomica* L. and other members of the *Strychnos* plant family (Loganiaceae). In 1818, strychnine was first isolated by Pelletier and Caventou from St. Ignatius beans, the seeds of a climbing shrub native to the Philippines (*Strychnos ignatii* Bergius) (Poulsson 1920; Teuscher and Lindequist 1988). Following elucidation of its chemical structure, it was completely synthesized by Woodward et al. (1954).

I. Symptoms of Strychnine Intoxication

Only recently was therapeutical application of strychnine as a roborant (Franz 1985) eliminated from the US drug index (Perper 1985), whereas it is still used in European countries (Rote Liste 1991). Strychnine at higher doses is highly toxic; the lethal dose is about 1 mg/kg in man (Gosselin et al. 1976). Use of the alkaloid as a rodenticide has contributed to suicidal, criminal, and, mostly in children, accidental poisoning. During the prodromal phase of intoxication, or at subconvulsive doses, strychnine causes hyperreflexia and, although less prominent, alters tactile, acoustic, and visual perception. Sensory stimuli initiate convulsions which last seconds to minutes, affecting all skeletal muscles. During the tonic phase of the seizure, tetanic muscle contraction leads to a painful, backward posture (opisthotonus). Arrest of respiration may lead to immediate death (GOSSELIN et al. 1976; PERPER 1985). Strychnine toxicity is thoroughly described in an earlier contribution (POULSSON 1920) and by DUSSER DE BARENNE (1933). Successful treatment of strychnine intoxication with diazepam (HEIDRICH et al. 1969; JACKSON et al. 1971) indicates that loss of glycinergic inhibition is therapeutically compensated by benzodiazepine potentiation of $GABA_A$ receptors, which are far less sensitive to strychnine (CURTIS et al. 1968b; HOUAMED et al. 1984; HORIKOSHI et al. 1988a,b). In the following, the toxicity of strychnine will be correlated to the distribution and function of its neuronal target molecule, the inhibitory glycine receptor.

II. Strychnine's Effects on Motor Regulation and the Somatosensory System

Strychnine-induced motor disturbances are attributed to a loss of glycinergic inhibition of spinal motoneurons mediated by small interneurons including the Renshaw cells (KRNJEVIC 1981; DENGLER 1990). Substantial decreases in glycinergic transmission are tolerated before "spinal seizures" are triggered within segmental neuronal circuits (TRIBBLE et al. 1983). Supraspinal disinhibition by strychnine is not essential as convulsions also occur in decerebrate animals. However, transsection of dorsal roots, i.e., sensory afferents, prevents seizures, indicating that they are reactive to external stimuli (POULSSON 1920). In the isolated spinal cord, strychnine block of Renshaw inhibition exaggerates motoneuron responses to dorsal root stimulation (BISCOE and DUCHEN 1986; LONG et al. 1989; AL ZAMIL et al. 1990).

Strychnine-sensitive glycine receptors are present in motor nuclei of various CNS areas. [³H]Strychnine autoradiography (ZARBIN et al. 1981; ROTTER et al. 1984; FROSTHOLM and ROTTER 1985; GILLBERG and AQUILONIUS 1985; BRISTOW et al. 1986; PROBST et al. 1986; BRUNING et al. 1990; WHITE et al. 1990) and immunohistology (TRILLER et al. 1985, 1987; ARAKI et al. 1988; VAN DEN POL and GORCS 1988) reveal a high receptor density in the spinal cord anterior horn and cranial nuclei including the oculomotor, trochlear, trigeminal motor, abducens, facial, hypoglossal, and ambiguus nucleus. Glycine receptors are expressed in motoneurons identified by retrogradely transported dyes injected into the target muscle (ST. JOHN et al. 1986; SCHAFFNER et al. 1987) and the marker enzyme choline acetyl transferase (GEYER et al. 1987). Interestingly, mouse (WHITE 1985; BECKER et al. 1986; BECKER 1990) and bovine (GUNDLACH 1990) mutants suffering from a hereditary glycine receptor deficit display motor symptoms indistinguishable from strychnine intoxication.

Strychnine-induced changes in sensory perception range from increased tactile discrimination and hyperesthesia to intense pain (Poulsson 1920; Dusser de Barenne 1933; Gosselin et al. 1976; Perper et al. 1985). Tactile stimuli, not painful to the untreated animal, result in agitation and autonomic responses after administration of strychnine. Apparently, strychnine causes hyperesthesia by interference with glycinergic modulation of non-nociceptive signals but does not lower the pain threshold of nociceptive somatic and visceral inputs (Beyer et al. 1985; Yaksh 1989).

Although the cellular physiology of sensory modulation remains to be elucidated, it is known that glycine inhibits and strychnine facilitates discharge of sensory neurons in the spinal cord dorsal horn (Zieglgänsberger and Herz 1971). Glycine receptors have been demonstrated in the dorsal horn by [^{3}H]strychnine autoradiography (Zarbin et al. 1981; Frostholm and Rotter 1985; Gillberg and Aquilonius 1985; Bristow et al. 1986; Probst et al. 1986; Bruning et al. 1990; White et al. 1990) and immunohistology (Araki et al. 1988; Basbaum 1988; van den Pol and Gorcs 1988) using antibodies against a cytoplasmic component (93-kDa protein) of the glycine receptor complex (Pfeiffer et al. 1984; Schmitt et al. 1987). Although both receptor markers codistribute in other CNS regions, a mismatch of intense [^{3}H]strychnine binding (Zarbin et al. 1981; Gillberg and Aquilonius 1985; Bruning et al. 1990; White et al. 1990) and low immunoreactivity (Basbaum 1988; van den Pol and Gorcs 1988) was been noted for the substantia gelatinosa (laminae I and II of Rexed), an area implicated in nociception (Basbaum 1988; van den Pol and Gorcs 1988). This discrepancy is not understood but may relate to the differential expression of the glycine receptor strychnine binding and 93-kDa protein (Becker et al. 1988). The reticular formation, a supraspinal structure containing high levels of glycine receptors (Zarbin et al. 1981; Araki et al. 1988; White et al. 1990), is thought to trigger sensory seizures during the prodromal phase of strychnine intoxication (Faingold et al. 1985).

III. Audition and Vision

Subconvulsive doses of strychnine induce hyperacusis and a prominent startle response that appear to have a physiological basis in glycine-mediated neuronal circuits of the central auditory pathway (Kehne and Davis 1984; Hirsch and Oertel 1988). Glycine receptors are detected in auditory centers by [^{3}H]strychnine binding (Sanes and Wooten 1987; Sanes et al. 1987; Glendenning and Baker 1988; Frostholm and Rotter 1986) and immunocytochemistry. In the cochlear nucleus, the superior olivary complex, and the lateral lemniscal neurons, receptor immunoreactivity has been located to neuronal cell bodies (Altschuler et al. 1986; Wenthold et al. 1988). Glycine receptors in the vestibular nuclei correlate to strychnine-induced disturbances in equilibrium (Zarbin et al. 1981; Araki et al. 1988; White et al. 1990).

Strychnine increases visual acuity as impressively described by the ophthalmologist VON HIPPEL (1873), who exposed himself to a carefully chosen dose of the convulsant (see also POULSSON 1920). In the retina, strychnine blocks light responses by antagonizing amacrine cell output onto a subset of ganglion cells (MÜLLER et al. 1988) and alters the receptive field surround by interference with glycinergic bipolar cells (KARSCHIN and WÄSSLE 1990). Inhibitory glycine receptors are located in the inner plexiform layer (SCHAEFFER and ANDERSON 1981; JÄGER and WÄSSLE 1987), and retinal mRNA induces glycine receptors in *Xenopus laevis* oocytes (PARKER et al. 1985). Glycine receptors are present in the central visual pathway (ZARBIN et al. 1981; ARAKI et al. 1988; VAN DEN POL and GORCS 1988; WHITE et al. 1990) where strychnine-sensitive synapses also modulate motor responses to optic stimuli. Strychnine application to the superior colliculus results in convulsions triggered by visual stimuli (ASCHER and GACHELIN 1967).

IV. Other Central Nervous System Functions

Although topical administration of strychnine to the cerebral cortex induces convulsions, the role of glycinergic synapses in cortical inhibition remains to be elucidated (BISCOE and CURTIS 1967). Glycine receptor levels are low in the cerebral cortex of the adult rodent (ZARBIN et al. 1981; ARAKI et al. 1988; VAN DEN PGL and GORCS 1988; BECKER et al. 1989; NAAS et al. 1991) and man (PROBST et al. 1986; NAAS et al. 1991) but high in the rabbit (FERENCI et al. 1984). Expression studies employing *Xenopus laevis* oocytes indicate that the cortical glycine receptors are developmentally regulated. In contrast to RNA extracted from adult rodent cortex (ASANUMA et al. 1987; AKAGI and MILEDI 1988a), samples from neonatal rodent (AKAGI and MILEDI 1988a,b; CARPENTER et al. 1988) and human fetal cortex (GUNDERSEN et al. 1984) efficiently induce glycine-gated current responses. In addition, Northern blot and *in situ*-hybridization data reveal the expression of glycine receptor subunit mRNAs in rodent cortex (GRENNINGLOH et al. 1990b; AKAGI et al. 1991; MALOSIO et al. 1991b). The pharmacological profile and the function of these receptors are still uncharacterized. Strychnine at submicromolar concentrations appears to protect rat cortical cultures against kainate- and NMDA-induced neurotoxicity. These observations suggest that cortical inhibitory glycine receptors contribute to neuronal vulnerability (ERDÖ, 1990; McNAMARA and DINGLEDINE, 1990).

Strychnine toxicity extends to autonomic functions of the CNS. Extreme increases in blood pressure induced by strychnine apparently result from alkaloid effects on vasomotor centers including the nucleus tractus solitarii (TALMAN and ROBERTSON 1989) and the vagal nucleus (TALMAN 1988), which contain high levels of glycine receptors (ZARBIN et al. 1981; PROBST et al. 1986; ARAKI et al. 1988; VAN DEN POL and GORCS 1988; WHITE et al. 1990). Hormonal alterations by strychnine certainly are less obvious than convulsant effects. Strychnine effectively antagonizes pituitary prolactin release

Table 1. Ligand affinities at the inhibitory glycine receptor

Ligand	K_d or K_i[a]	Species	References
Amino acid agonists			
Glycine	$2-40\,\mu M$ ⎫	rat, mouse,	1,3,7–9,
β-Alanine	$4-57\,\mu M$ ⎬	pig, pigeon	11–17
Taurine	$29-135\,\mu M$ ⎭		
β-Aminoisobutyric acid	$200\,\mu M$	rat	8,17
Strychnine and derivatives			
Strychnine	$2-14\,nM$[b] ⎫	rat, mouse,	3,8–11,13–14,
Pseudostrychnine	$6-10\,nM$ ⎪	pig, pigeon	See also
α-Colubrine	$12-18\,nM$ ⎬		1,2,7,12–15
2-Aminostrychnine	$8-23\,nM$ ⎪		for
Brucine	$15-180\,nM$ ⎪		[³H]strychnine
2-Nitrostrychnine	$53-500\,nM$ ⎭		binding
Cacotheline	$240\,nM$	rat	13
N-Methylstrychnine	$500\,nM$	rat	8
	$31\,nM$	pigeon	11
Isostrychnine	$37-2000\,nM$	rat	3, 10
	$>700\,nM$	pigeon	11
THIP and analogues			
Iso-THAZ	$1.7-6.6\,\mu M$	rat	3,8,13,14
4,5-TAZA	$2.8-67\,\mu M$	rat	3,8
THAZ	$20-48\,\mu M$	rat	3,8
THIP	$39-120\,\mu M$	rat	3,8
β-Spiro[pyrrolidinoindolines]	$>20\,\mu M$	rat	7
Steroids			
RU 5135	$3.6-9.8\,nM$	rat	3,13,14
GABA$_A$ receptor ligands			
Pitrazepin	$70-580\,nM$	rat	2,3
(+) Bicuculline	$2-7\,\mu M$	rat	3,4,8,14
Bicuculline MeCl/MeI	$55-77\,\mu M$	rat, mouse	3,4,16
Flunitrazepam	$9\,\mu M$	rat	17
Bromazepam, nitrazepam	$10\,\mu M$	rat	17
Diazepam	$13\,\mu M$	rat	17
PK 8165	$4-14\,\mu M$	rat	8,14
Norhamane	$52\,\mu M$	rat	14
Avermectin B$_{1a}$	$800-1300\,nM$	rat	6,8,14
Opioids			
Thebaine	$900\,nM$	rat	5,13
Sinomenine	$23\,\mu M$	rat	5
L-Methadone	$30\,\mu M$	rat	5
Alkaloids			
Boldine	$67\,\mu M$	rat	13
Gelsemine	$18\,\mu M$	rat	13
Hydrastine	$13\,\mu M$	rat	13
Laudanosine	$16\,\mu M$	rat	13
Vincamine	$31\,\mu M$	rat	13
Miscellaneous agents			
1,5-Diphenyl-3,7-diazaadamantan-9-ol	$27\,nM$	rat	10
Imipramine	$17\,\mu M$	pigeon	11

Table 1. (continued)

[a] Ligand affinities were derived from published displacement data of [³H]strychnine binding as described (BETZ and BECKER 1988). This procedure could not be applied to the IC$_{50}$ values of [³H]strychnine displacement from ref. [13] included in this table. Modified from BETZ and BECKER (1988).

[b] A neonatal glycine receptor isoform containing the α2* subunit is characterized by low-affinity strychnine binding ($K_i \approx 10\,\mu M$) (1).

1, BECKER et al. (1986, 1988); HOCH et al. (1989); *2*, BRAESTRUP and NIELSEN (1985); *3*, BRAESTRUR et al. (1986); *4*, GOLDINGER and MÜLLER (1980); *5*, GOLDINGER et al. (1981); *6*, GRAHAM et al. (1982); *7*, HERSHENSON et al. (1977); *8*, MARVIZON et al. (1986a,b); *9*, HARVIZON et al. (1986b); *10*, MACKERER et al. (1977); *11*, LEFORT et al. (1978); *12*, PFEIFFER and BETZ (1981); *13*, PHELAN et al. (1989); *14*, RUIZ-GOMEZ et al. (1989b); *15*, SCHAEFFER and ANDERSON (1981); *16*, WHITE (1985); *17*, YOUNG et al. (1974).

THIP, 4,5,6,7-tetrahydroisoxazolo-[5,4-c]pyridin-3-ol; THAZ, 5,6,7,8,-tetrahydro-4H-isoxazolo-[4,5-d]azepin-3-ol; 4,5-TAZA, 2,3,6,7-tetrahydro-1H-azepine-4-carboxylic acid.

(ONDO and BUCKHOLTZ 1983) and alters the membrane potential of hypo-thalamic neurosecretory neurons (AKAIKE and KANEDA 1989). Outside the CNS, high affinity [^{3}H]strychnine binding sites have been detected in the adrenal medulla (YADID et al. 1990).

C. The Inhibitory Glycine Receptor

I. Receptor Structure and Isoforms

Glycine-displaceable [^{3}H]strychnine binding is a reliable probe of the in-hibitory glycine receptor. In synaptosomal membranes from spinal cord and other CNS regions of most species, [^{3}H]strychnine recognizes a single class of binding sites with affinity constants of $3-10\,nM$. The agonistic amino acids glycine, β-alanine, and taurine display affinities at these sites in the range of $10-100\,\mu M$ (YOUNG et al. 1973, 1974a; PFEIFFER and BETZ 1981; PFEIFFER et al. 1982; GALLI et al. 1983; LLOYD et al. 1983a,b; GRAHAM et al. 1985; BECKER et al. 1986; see also Table 1). Two classes of [^{3}H]strychnine binding sites have been reported for rabbit cortex (FERENCI et al. 1984).

The glycine receptor from the spinal cord of adult mammals behaves as a complex glycoprotein of >250 kD molecular weight (PFEIFFER and BETZ 1981; PFEIFFER et al. 1982; GRAHAM et al. 1985; HOCH et al. 1989). After affinity purification on a 2-aminostrychnine agarose resin, it contains three polypeptides of 48, 58, and 93 kDa (Fig. 1) (PFEIFFER et al. 1982; GRAHAM et al. 1985; BECKER et al. 1986). [^{3}H]Strychnine is a natural photoaffinity ligand which, upon UV illumination, is irreversibly incorporated into the 48 kDa (α_1) polypeptide (Fig. 1) (GRAHAM et al. 1981, 1983, 1985; PFEIFFER et al. 1982; BECKER et al. 1986). The ligand binding α_1 subunit and the 58 kDa (β) polypeptide are integral membrane proteins which assemble into a pentameric chloride channel structure (LANGOSCH et al. 1988). Upon reconstitution in phospholipid vesicles, the isolated receptor mediates glycine-enhanced, strychnine-sensitive anion translocation (GARCIA-CALVO et al. 1989; RIQUELME et al. 1990).

A distinct neonatal isoform of the glycine receptor is expressed in the spinal cord of newborn rodents (BECKER et al. 1988). This protein is charac-terized by low strychnine binding affinity, molecular weight (49 kDa), and antigenic epitopes of its ligand binding subunit α_2^* (low affinity strychnine binding is indicated by*) (BECKER et al. 1988; HOCH et al. 1989; KUHSE et al. 1990a; SCHRÖDER et al. 1991). The neonatal isoform accounts for about 70% of the total glycine receptor in the spinal cord of the newborn rat (BECKER et al. 1988) and is fully replaced by the adult receptor within 2 weeks after birth (BECKER et al. 1988; compare ZUKIN et al. 1975; BENAVIDES et al. 1981; BRUNING et al. 1990). Developmental changes in strychnine sensitivity of rodents have been attributed to alterations in glycine receptor isoform expression (BECKER et al. 1988; KUHSE et al.

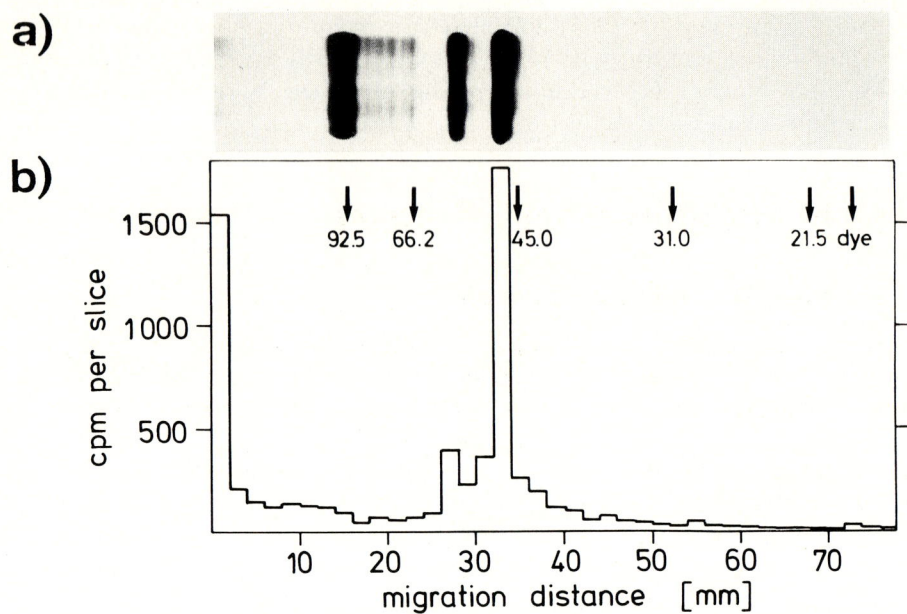

Fig. 1a,b. Glycine receptor polypeptide composition. **a** SDS/polyacrylamide gel electrophoresis of affinity-purified glycine receptor from adult rat spinal cord. **b** UV-light-induced incorporation of [³H]strychnine into glycine receptor polypeptides. After electrophoresis, level of incorporated radioactivity was determined in individual gel slices. Radioactivity label exhibits a peak at the position of the α subunit (48 kDa). (From Pfeiffer et al. 1982)

1990a). Primary cultures of spinal cord predominantly express the neonatal isoform of the glycine receptor protein (Hoch et al. 1989). Strychnine-resistant, glycine-activated chloride currents have been recorded from cultured neurons (Lewis et al. 1989).

Molecular cloning has identified variants of the glycine receptor ligand binding subunit that are highly homologous to each other (Grenningloh et al. 1987a, 1990a; Kuhse et al. 1990a,b, 1991; Akagi et al. 1991) and to the glycine receptor β subunit (Grenningloh et al. 1990b) with respect to deduced primary structure and transmembrane topology (Fig. 2). In addition, alternative splicing contributes to glycine receptor α subunit heterogeneity (Kuhse et al. 1991; Malosio et al. 1991a). For each subunit, hydropathy profiles predict four hydrophobic segments (M1 to M4) spanning the postsynaptic membrane (Grenningloh et al. 1987a). Transmembrane region M2 is thought to form the inner wall of the receptor channel, while the ligand binding site(s) of the α subunits have been tentatively located to a large extracellular N-terminal domain (Fig. 2a). Homologies of the deduced primary and predicted transmembrane structures to other ligand-

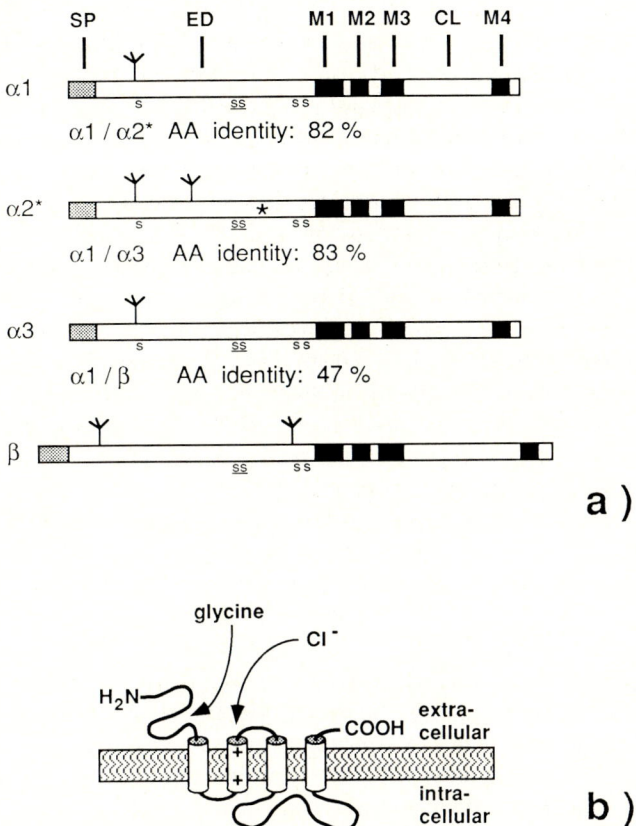

Fig. 2a,b. Structure of glycine receptor polypeptides. **a** Organization of rat glycine receptor subunits. The putative signal peptides (*SP*) are indicated by *shaded bars*, transmembrane segments (*M1-M4*) by *black bars*, extracellular domains (*ED*) and cytoplasmic loop regions (*CL*) by *open bars*, cysteine residues by *S*, the conserved cysteine domain by *SS*, and potential glycosylation sites by *branch symbols*. (*) denotes the position of an isolated amino acid exchange in the rat α_2^* subunit which leads to low-affinity strychnine binding. Amino acid identities to the rat α_1 subunit are indicated. (Modified from BETZ 1990) **b** Predicted transmembrane topology of the glycine receptor α subunit. (+) indicates charged amino acids adjacent to the M2 transmembrane segment thought to line the inner wall of the receptor anion pore

gated ion channels including the $GABA_A$ and nicotinic acetylcholine receptor are also apparent. These homologies indicate the existence of a superfamily of receptor subunit genes which presumably evolved from a common ancestor (GRENNINGLOH et al. 1987b; BETZ 1990).

Glycine receptor isoforms are developmentally regulated at the transcriptional level. Size fractionation and expression in *Xenopus laevis* oocytes of poly(A)$^+$ mRNA result in the separation of two classes of mRNAs encoding glycine-gated chloride channels. A rapidly sedimenting mRNA prevails in the spinal cord of adult rats, whereas preparations of neonatal spinal cord contain a light glycine receptor mRNA species. The light mRNA is also detected in adult cortex although at minute concentrations (Akagi and Miledi 1988a,b; Akagi et al. 1989). Developmental analysis of defined glycine receptor subunit transcripts reveals that spinal cord levels of both splice variants known of the α_1 subunit show a similar postnatal increase (Malosio et al. 1991a). Likewise, α_3 subunit mRNAs increase postnatally (Kuhse et al. 1990b). At all times, mRNA expression of the α_1 subunit exceeds the levels of α_3 transcripts (Malosio et al. 1991b). In contrast, α_2* transcripts are abundant around birth and display a drastic postnatal decrease (Kuhse et al. 1990a), as expected for the ligand binding subunit of the neonatal isoform (Becker et al. 1988). Expression in *Xenopus laevis* oocytes of the α_2* polypeptide induces glycine-gated chloride channels with low strychnine binding affinity (Kuhse et al. 1990a), consistent with biochemical studies on the neonatal receptor protein (Becker et al. 1988; Hoch et al. 1989). The α_2* glycine receptor polypeptide of rodent receptor is highly homologous to the human and rat α_2 subunit (Grenningloh et al. 1990a; Akagi et al. 1991; Kuhse et al. 1991) which, however, is highly sensitive to strychnine.

Whereas the α and β polypeptides are transmembrane components of the glycine receptor anion channel (Schmitt et al. 1987), the 93-kDa polypeptide resides at the cytoplasmic face of the postsynaptic membrane (Triller et al. 1985, 1987). As a peripheral membrane protein, it reversibly associates with the transmembrane core of the receptor and has been implicated in its synaptic anchoring and cluster formation (Schmitt et al. 1987; Becker et al. 1989; Triller et al. 1990). By photobleach recovery employing fluorescent strychnine derivatives, a laterally mobile and an immobile glycine receptor population are distinguished on cultured neurons (Srinivasan et al. 1990). It is not known whether receptor lateral mobility is restricted by interaction with the 93-kDa protein.

II. Ligand Binding Domains

Glycine receptor α-subunits carry binding sites for glycine, the antagonist strychnine, and functional determinants of the ion pore. Heterologous expression of α-subunit variants leads to the formation of functional, homooligomeric receptors. The pharmacological profile of recombinant glycine-gated ion channels is indistinguishable from native receptors, i.e., dose-response relationships for glycine, β-alanine, and taurine as well as strychnine sensitivity are conserved (Schmieden et al. 1989; Sontheimer et al. 1989; Grenningloh et al. 1990a; Kuhse et al. 1990a,b, 1991). These studies provide the final proof that agonist and antagonist binding sites are contained in the α-subunits.

Upon UV illumination of membrane fractions from adult rat spinal cord, [^{3}H]strychnine is irreversibly incorporated into the 48-kDa (α_1) subunit (Fig. 1) (GRAHAM et al. 1981, 1983, 1985; PFEIFFER et al. 1982; BECKER et al. 1986). This covalent incorporation is specifically antagonized by glycine (GRAHAM et al. 1983). Trypsinization of photolabelled synaptosomes removes a proteolytic fragment of about 11 kDa from the extracellular part of the α-subunit but does not extract the incorporated radioligand from the membrane-bound receptor protein (GRAHAM et al. 1983). As noted by GRENNINGLOH et al. (1987a), this observation excludes the first ≈100 of 223 amino acids forming the N-terminal extracellular domain of the α_1-subunit as the site of strychnine incorporation. In addition, protein modification experiments show that tyrosine residues are essential for [^{3}H]strychnine binding (YOUNG and SNYDER 1974a). In fact, a region of the extracellular domain immediately preceding the first transmembrane segment (M1) of the α-subunit is characterized by tyrosine residues, two neighboring cysteines, and charged amino acid residues thought to provide functional determinants of ligand recognition (Fig. 2) (GRENNINGLOH et al. 1987a; BETZ 1990).

In the α-subunit of the nicotinic acetylcholine receptor, acetylcholine binding depends on a stretch of amino acids including two neighboring cysteines that precede the first transmembrane segment. A sequence comparison shows that this region of the acetylcholine receptor corresponds to the pre-M1 domain of the highly homologous glycine receptor (BETZ 1990). Based on these considerations, the glycine receptor pre-M1 domain has been proposed by GRENNINGLOH et al. (1987a) to be the site of strychnine binding. In fact, analysis of proteolytic fragments of the glycine receptor has shown that UV illumination leads to incorporation of [^{3}H]strychnine into peptide fragments of this α_1-subunit region (RUIZ-GOMEZ et al. 1990).

The rat α_2* subunit variant of the neonatal glycine receptor isoform, upon *Xenopus laevis* oocyte expression, induces chloride channels characterized by low affinities for glycine and strychnine (KUHSE et al. 1990a). Glycine receptors formed by recombinant expression of the highly homologous human and rat α_2-subunit, in contrast, are highly sensitive to strychnine (GRENNINGLOH et al. 1990a; AKAGI et al. 1991; KUHSE et al. 1991). Within the large N-terminal domain, the glycine residue at position 167 of the convulsant-sensitive α_2-subunit is exchanged for glutamate in the more resistant α_2* variant (Fig. 2). Reversing this exchange by site-directed mutagenesis restores strychnine sensitivity in the α_2* 167 Glu-Gly variant (KUHSE et al. 1990a). Although distant from the UV-stimulated incorporation site of [^{3}H]strychnine, position 167 appears to be an important determinant of high-affinity strychnine binding (BETZ 1990). This position is close to two cysteines (Fig. 2, denoted by *SS*, positions 145 and 159 of the α_2* subunit variant) which are conserved and exactly 13 amino acid residues apart in all known α-subunit variants of glycine, GABA$_A$, and nicotinic acetylcholine receptors. These cysteines are thought to form a disulfide bridge essential for receptor conformation (BETZ 1990).

The lack of high-affinity agonists has precluded mapping of the α-subunit agonist recognition site. Several lines of evidence nevertheless indicate that glycine and strychnine recognize overlapping, though not identical, binding domains (see below). Protein modification differently affects receptor binding of glycine and strychnine. Treatment of spinal cord membranes by diazonium tetrazole (modification of histidyl and tryptophanyl residues) or the amino-reactive agent acetic anhydride deletes displacement of [^{3}H]strychnine binding by glycine but not by unlabelled strychnine (Young and Snyder 1974a; Marvizón et al. 1986a). Modification by fluorescein 5'-isothiocyanate (FITC) of lysine residues likewise affects glycine but not strychnine binding, indicating that agonist and antagonist are recognized by distinct amino acid residues. Attempts to map the incorporation site of fluorescent label within the α-subunit sequence were not successful (Ruiz-Gomez et al. 1989a). Analysis of ion effects on the glycine receptor further supports the concept of nonidentical agonist and antagonist binding determinants, as sodium chloride increases strychnine but lowers glycine binding affinity (see below). Taken together, biochemical and molecular cloning studies have permitted a regional localization of α-subunit domains involved in ligand binding, whereas determinants of agonist recognition are not defined.

Little is known about the functional importance of cytosolic glycine receptor subunit domains. The α$_1$ polypeptide is rapidly phosphorylated by protein kinase C. Analysis of the phosphoprotein suggests that serine 391 is phosphorylated which resides within the major intracellular loop preceding the M4 transmembrane segment. Homologous positions of phosphorylation sites have been identified in other ligand-gated ion channels (Ruiz-Gómez et al. 1991a). In spinal trigeminal neurons, cAMP-dependent protein kinase dramatically increases the opening probability of glycine-gated chloride channels (Song and Huang 1990).

III. Interaction of Ligand Binding Domains

Glycine-induced chloride current responses recorded from spinal motoneurons in situ (Werman et al. 1968), cultured neurons (Tokutomi et al. 1989; Lewis et al. 1989), and *Xenopus laevis* oocytes after injection of brain mRNA (Gundersen et al. 1984) show positive cooperativity characterized by Hill coefficients (n$_H$) of 1.7–2.7. Recombinant expression of glycine receptor α-subunit variants in *Xenopus laevis* oocytes (Schmieden et al. 1989; Grenningloh et al. 1990a; Kuhse et al. 1990a,b) or transfected eukaryotic cells (Sontheimer et al. 1989) also induces cooperative glycine current responses (n$_H$ ≈ 3.0). These data suggest that the occupation of 2–3 glycine binding sites per receptor complex is necessary to induce channel opening. For unknown reasons receptor activation by β-alanine and taurine lacks cooperativity (n$_H$ ≈ 1.0) (Schmieden et al. 1989).

Reciprocal antagonism of receptor binding by glycine and strychnine is well established, but the mode of inhibition is still debated. Radioligand studies conducted under different conditions of pH (YOUNG and SNYDER 1974a), ion activities (MARVIZON et al. 1986a,b), redox states (RUIZ-GÓMEZ et al. 1991b), temperature (RUIZ-GÒMEZ et al. 1989b), and binding assay used (O'CONNOR 1989) have led to diverse models of glycine receptor ligand binding:

1. Receptor binding of strychnine and glycine are mutually exclusive, consistent with Hill coefficients of ≈ 1.0 for glycine displacement of [^{3}H]strychnine binding (GUNDLACH and BEART 1981; FRY and PHELAN 1985; WHITE 1985; BECKER et al. 1986; O'CONNOR 1989). Strychnine inhibition of glycine currents induced in heterologous expression systems also lacks cooperativity ($n_H \approx 1.0$) (SCHMIEDEN et al. 1989; SONTHEIMER et al. 1989). These results comply with a fully competitive inhibition and are compatible with overlapping but conformationally distinct binding domains for glycine and strychnine (GRENNINGLOH et al. 1987a; O'CONNOR 1989; see also APRISON et al. 1987; APRISON 1990).

2. Hill coefficients above unity reported for the complete displacement of [^{3}H]strychnine binding by glycine are compatible with a negative allosteric influence of the convulsant on receptor agonist binding (YOUNG and SNYDER 1974a). A kinetic study showing that receptor-ligand dissociation of [^{3}H]strychnine is accelerated by glycine provides further evidence for an allosteric agonist influence on antagonist sites. Agonist enhancement of [^{3}H]strychnine dissociation is diminished by chemical modification of tyrosine residues (MAKSAY 1990) essential for glycine binding (YOUNG and SNYDER 1974a; MARVIZON et al. 1986a). Furthermore, the interaction of ligand binding sites is modulated by the redox state of glycine receptor sulfhydryl groups. Reduction of thiol groups increases, and oxidation decreases, Hill coefficients for glycine-induced displacement of [^{3}H]strychnine binding without affecting the affinity for either ligand. This observation indicates that sulfhydryl groups contribute to an allosteric interaction between different glycine sites for displacing bound [^{3}H]strychnine (RUIZ-GÒMEZ et al. 1991b).

3. Glycine only partially inhibits [^{3}H]strychnine binding, suggesting that receptor, agonist, and radiolabelled antagonist form ternary complexes (MARVIZON et al. 1986a). Binding of [^{35}S]tertiary-butylcyclophosphorothionate ([^{35}S]TBPS) and [^{3}H]strychnine to GABA$_A$ and glycine receptors, respectively, shows an equivalent anion enhancement. This similarity has been taken as evidence for a functional homology of both convulsants (MARVIZON and SKOLNICK 1988a). As a "cage" convulsant, TBPS is thought to block the intrinsic ion channel of the GABA$_A$ receptor complex (SQUIRES et al. 1983). This noncompetitive mechanism of inhibition is compatible with the formation of ternary receptor-ligand-inhibitor com-

plexes. However, strychnine does not meet the criteria of a "cage" convulsant, since it does not induce changes in conductance or average open-time of glycine-gated channels (Barker et al. 1983).

Glycine receptor agonist binding induces changes in protein conformation that lead to the opening of an anion-selective pore. Thermodynamic parameters have been identified which are thought to reflect corresponding energy changes of the receptor complex. Higher temperature increases affinities for glycine, β-alanine, and taurine and favors formation of the receptor-agonist complex. In contrast, affinities for strychnine and other antagonists are significantly lowered. Furthermore, thermodynamic analyses indicate that glycine receptor agonist binding is driven by increases and antagonist binding by decreases in enthalpy ($\Delta H°$) (Ruiz-Gomez et al. 1989b). It appears that thermodynamic parameters are of predictive value to characterize agonist and antagonist binding at the inhibitory glycine receptor.

IV. Ion Effects on Ligand Binding

The ion pore induced by glycine binding to the pentameric receptor complex is selective for monovalent anions which, in turn, modulate ligand recognition. The neurophysiological activity of monovalent anions correlates to their ion channel permeability (Eccles 1964; Bormann et al. 1987). In a patch-clamp study on cultured neurons, a permeability sequence has been established of $SCN^- > I^- > Br^- > Cl^- > F^-$ which is different from ion mobilities in free solution. Thus, the receptor ion pore does not merely function as a molecular sieve, its selectivity may rather be explained by an electrostatic interaction of ions with functional sites of the channel (Bormann et al. 1987).

Glycine receptor ligand binding is strongly influenced by anions permeable to the intrinsic ion channel. Ammonium salts of a series of anions reduce [^{3}H]strychnine binding. The extent of inhibition closely correlates to channel permeability as smaller anions (Br^-, Cl^-) are more effective than larger ones (I^-, SCN^-). Ammonium chloride inhibition of [^{3}H]strychnine binding is described by Hill coefficients of 2.3–2.7 (Young and Snyder 1974b; Müller and Snyder 1978; Marvizón et al. 1986b). Counter-cations, however, modulate anion effects on glycine receptor binding (Marvizon et al. 1986b). Sodium salts of these regulatory anions exert a biphasic effect on [^{3}H]strychnine binding. Low salt concentrations enhance (Young and Snyder 1974b; Müller and Snyder 1978; Braestrup et al. 1986; Marvizon et al. 1986b; Marvizon and Skolnick 1988), higher concentrations inhibit binding of the radioligand (Müller and Snyder 1978; Marvizon and Skolnick 1988b). When both cations are present, ammonium overrides low-concentration enhancement by sodium salts and diminishes [^{3}H]strychnine binding (Marvizon et al. 1986b). Whereas sodium-dependent, anion-induced enhancement of [^{3}H]strychnine binding is sensitive to protein modification, ammonium-dependent anion inhibition of [^{3}H]strychnine binding is not affected (Marvizon et al. 1986b), suggesting that the receptor domains modulated by ionic conditions may not be the same.

The biphasic, sodium-dependent anion regulation of antagonist [³H]strychnine binding is consistent with mathematical models predicting two anion binding sites (Marvizon and Skolnick 1989). Patch-clamp analysis of glycine receptor channels indicates that ion pore selectivity is also achieved by two anion binding sites (Bormann et al. 1987). It is, however, not known whether channel permeability and ligand binding are regulated by the same anion domains. Within the predicted transmembrane regions of glycine receptor subunits, segment M2 is thought to form the inner wall of the anion channel. A ring of positively charged amino acids at the presumptive mouth of the channel has been suggested to be the structural correlate of the channel anion binding sites (Fig. 2) (Langosch et al. 1990a,b; 1991).

Ionic conditions differently affect receptor binding of glycine and strychnine (Young and Snyder 1974b; Braestrup et al. 1986; Marvizon et al. 1986b). Sodium chloride increases the affinity for [³H]strychnine but reduces affinities for glycine and other agonistic amino acids. In contrast to thermodynamic analysis (Ruiz-Gomez et al. 1989b), however, the resulting shift in association constants is not of predictive value for agonistic or antagonistic properties of glycine receptor ligands (Braestrup et al. 1986). Ion effects on ligand binding are not readily reconciled with caesium chloride activation of strychnine-sensitive glycine receptor currents in cultured neurons (Lewis et al. 1989).

D. Glycine Receptor Pharmacology

Only a few high-affinity ligands of the inhibitory glycine receptor have been identified, all of which are convulsants including strychnine, its derivatives, the agent 1,5-diphenyl-3,7-diazaadamantan-9-ol (Mackerer et al. 1977), and the steroid RU 5135 (Hunt and Clemens-Jewery 1981; Braestrup et al. 1986). The agonist profile of the receptor is limited to a few ligands outside glycine and related β-amino acids. On the other hand, several drugs with principle modes of action at other neurotransmitter receptors display micromolar affinities to the glycine receptor. Common structural properties characterizing glycine receptor agonists or antagonists have not been identified.

I. Agonist Pharmacology of the Glycine Receptor

Glycine is the smallest amino acid, and one might imagine that merely the size of its receptor binding pocket imposes rigid steric restrictions on potential agonists and thus drastically limits their number. Nevertheless, the depressant effect of glycine on motoneuron activity in the spinal cord is mimicked by iontophoretic application of a series of amino acids. In the spinal cord (Curtis et al. 1968a) and isolated hypothalamic neurons (Tokutomi et al. 1989), the relative potency at the glycine receptor channel of "glycine-like" amino acids decreases in the order: glycine > β-alanine > taurine > L- and D-α-alanine > L-serine ≫ D-serine. The relative potencies of agonistic amino acids are conserved in recombinant α_1-subunit glycine receptors expressed in *Xenopus laevis* oocytes (SCHMIEDEN et al. 1989; GRENNINGLOH et al. 1990a; KUHSE et al. 1990a,b). Strychnine effectively antagonizes neuronal inhibition by "glycine-like" amino acids (CURTIS et al. 1968b; KRISHTAL et al. 1988; TOKUTOMI et al. 1989; SCHMIEDEN et al. 1989; GRENNINGLOH et al. 1990; KUHSE et al. 1990a,b).

Although inhibitory glycine and GABA$_A$ receptors are highly selective for the respective neurotransmitters glycine and GABA, both receptor channels appear to respond to β-amino acids. In isolated rat optic nerve and cuneate nucleus (SIMMONDS 1983), cultured chick spinal neurons (CHOQUET and KORN 1988), and *Xenopus laevis* oocytes injected with brain mRNA (PARKER et al. 1988; HORIKOSHI et al. 1988a), β-alanine and taurine display pharmacological and cross-desensitization properties compatible with a dual action on both receptors. Taurine and β-alanine responses are sensitive to both the GABA$_A$ receptor antagonist bicuculline and the glycine receptor antagonist strychnine (SIMMONDS 1986; HORIKOSHI et al. 1988a). This observation is somewhat qualified by a decrease of receptor selectivity of these convulsants at the micromolar concentrations employed (DAVIDOFF and APRISON 1969; see below). Further analysis of chloride currents induced by brain mRNA in *Xenopus laevis* oocytes shows that taurine elicits a smaller maximum response than glycine (SCHMIEDEN et al. 1989). Upon coapplication of both amino acids, taurine partially reduces the glycine response (HORIKOSHI et al. 1988b). These effects are explained by a difference in ligand intrinsic activities, with taurine being a partial agonist at the glycine receptor.

The inhibitory efficacy of "glycine-like" amino acids parallels their ability to displace [^{3}H]strychnine binding from synaptic membranes of mammalian spinal cord (Table 1) (YOUNG and SNYDER, 1973, 1974a,b). Glycine displays the highest ($K_i \approx 10-30 \mu M$) and β-amino acids lower affinities decreasing in the order: β-alanine > β-aminoisobutyric acid > taurine. GABA and higher ω-amino acids are without significant effect (YOUNG and SNYDER 1973, 1974a; PFEIFFER and BETZ 1981; PFEIFFER et al. 1982; GRAHAM et al. 1985; BECKER et al. 1986).

Substitutions to the glycine molecule result in a loss of glycine receptor agonistic action. TOKUTOMI et al. (1989) have noted that the agonists β-alanine and taurine exist in extended and folded conformations which might permit the amino and acidic groups to assume spatial positions resembling glycine. It may be tempting to speculate that the dual effect of these β-amino acids on inhibitory glycine and GABA$_A$ receptors (SIMMONDS 1983; CHOQUET and KORN 1988; HORIKOSHI et al. 1988a,b) results from a transition between these conformations. Despite close structural similarities to glycine, larger compounds such as substituted heterocycles including aminohydroxyisoxazoles and pyrazoles are inactive at the glycine receptor (DRUMMOND et al. 1989).

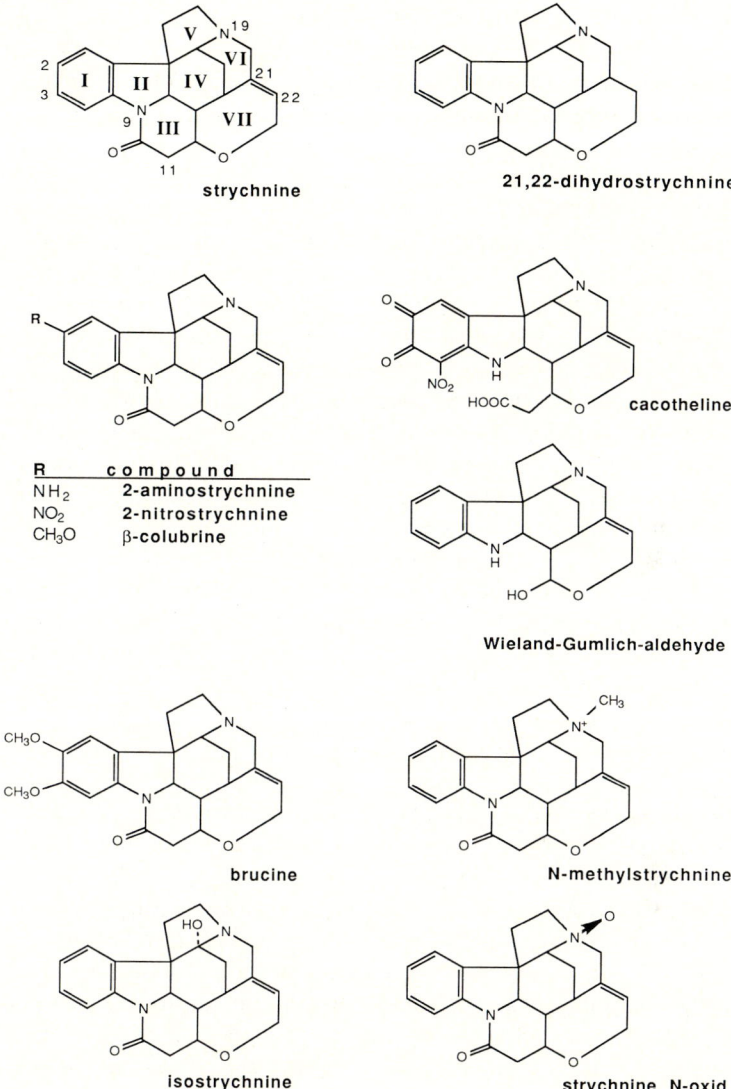

Fig. 3. Structures of strychnine and related alkaloids

II. Strychnine, Derivatives, and Analogues

1. Structure-Activity Relationship of Strychnine-Like Alkaloids

Strychnine, the principle antagonist of the inhibitory glycine receptor, has been subjected to a number of structure-activity studies (Fig. 3). However, none of the modifications of its chemical structure has increased receptor affinity or yielded an agonist or positive allosteric activator. Pseudostrychnine, α-colubrine, and 2-aminostrychnine represent a class of high-affinity antagonists with K_i values in the range of 10–20 nM. These agents either possess an unsubstituted aromatic ring or small substitutents like amino- or methoxygroups in position 2 (Mackerer et al. 1977; Betz 1985; Phelan et al. 1989). The following modifications of the aromatic ring yield medium-affinity derivatives ($K_i \approx$ 50–500 nM): a nitrogroup in position 2, substitution in position 3 or in both positions as in brucine (Mackerer et al. 1977; Marvizon et al. 1986; Betz 1985; Phelan et al. 1989). Apparently, benzene ring substitutions do not delete receptor binding of strychnine derivatives. Distortions in the carbon skeleton of strychnine, e.g., loss of the double bond in 21,22-dihydrostrychnine ($K_i \approx$ 130 nM; Fig. 3), moderately reduce binding affinity (Mackerer et al. 1977; Betz 1987; Phelan et al. 1989). Reduction of the 21,22-double bond together with benzene ring substitution drastically affects receptor binding as in pseudobrucine and 21,22-dihydrobrucine (Mackerer et al. 1977). Secoderivatives of strychnine obtained by reductive fission and saturation of the 21,22-bond exert a tranquillizing, imipramine-like effect on animals but do not act as convulsants (Rees and Smith 1967). However, glycine receptor binding of these secoderivatives has not been analyzed. Interestingly, the tricyclic antidepressant imipramine itself binds to pigeon CNS glycine receptors with micromolar affinity ($K_i \approx$ 17 μM; Table 1) (LeFort et al. 1978).

The lactam ring of strychnine is an important determinant of glycine receptor binding that in topology and electronic charge distribution closely resembles glycine (Fig. 3). With small deviations, the carboxylate and nitrogen of glycine map onto the amide moiety of the lactam and the carbon in position 11, respectively, of strychnine (Aprison et al. 1987; Aprison 1990). Opening of the lactam in cacotheline moderately reduces (Phelan et al. 1989) and loss of the lactam carbonyl group in Wieland-Gumlich aldehyde ($K_i \approx$ 10 μM) dramatically lowers glycine receptor affinity (Fig. 3) (Phelan et al. 1989). Strychnine carries a basic nitrogen in position 19 that is largely protonated at physiological pH (Mackerer et al. 1977). Accordingly, quaternization of this nitrogen in N-methylstrychnine (Fig. 3) only moderately depresses glycine receptor binding (LeFort et al. 1978; Marvizon et al. 1986a). In contrast, the N-oxide in position 19 of strychnine-N-oxide ($K_i \approx$ 2.3 μM; Marvizon et al. 1986a; Phelan et al. 1989), brucine N-oxide ($K_i \approx$ 1.8 mM; Phelan et al. 1989) or N-oxystrychnic acid ($K_i \approx$ 10 μM; Phelan et al. 1989) dramatically diminishes receptor affinity compared with the structural analogues strychnine ($K_i \approx$ 4.0 nM; Marvizon et al. 1986a,b), brucine ($K_i \approx$ 180 nM; Phelan et al. 1989), and cacotheline ($K_i \approx$ 240 nM; Phelan et al. 1989), respectively. In the pigeon CNS, modifications of the lactam ring (Wieland-Gumlich aldehyde) or the nitrogen in position 19 (strychnine-N-oxide) are only of minor effect on glycine receptor binding (LeFort 1978).

β-Spiropyrrolidinoindolines

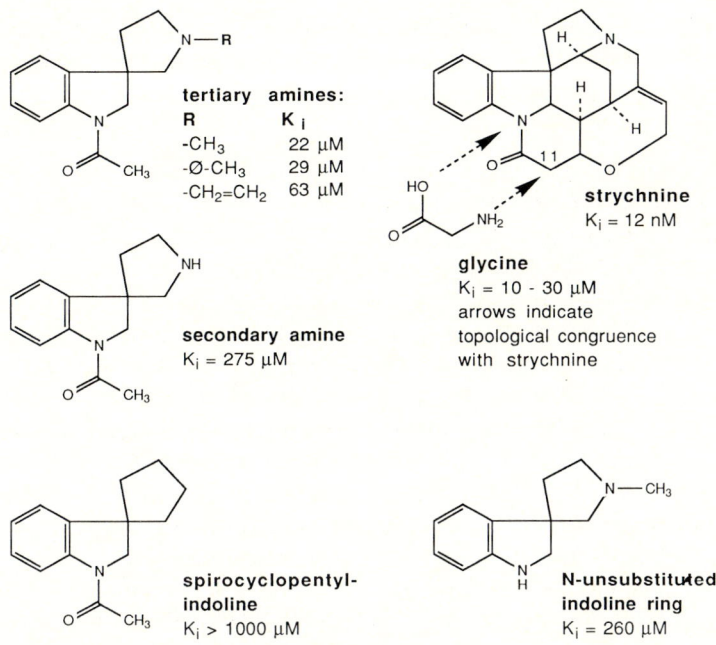

tertiary amines:

R	K_i
-CH₃	22 μM
-Ø-CH₃	29 μM
-CH₂=CH₂	63 μM

secondary amine
K_i = 275 μM

glycine
K_i = 10 - 30 μM
arrows indicate
topological congruence
with strychnine

strychnine
K_i = 12 nM

spirocyclopentyl-
indoline
K_i > 1000 μM

N-unsubstituted
indoline ring
K_i = 260 μM

Fig. 4. Structures of convulsant β-spiropyrrolidinoindolines representing fragments of the strychnine molecule. K_i values indicating glycine receptor affinities of these compounds were derived from published displacement data of [³H]strychnine binding (HERSHENSON et al. 1977; MACKERER et al. 1977) as described (BETZ and BECKER 1988)

2. Synthetic Fragments of Strychnine: β-Spiropyrrolidinoindolines

A group of β-spiropyrrolidinoindolines represent structural fragments of the strychnine molecule (Fig. 4). Convulsant activity and glycine receptor affinity (K_i > 20 μM) of these compounds closely correlate; agonistic β-spiropyrrolidinoindolines are not known (HERSHENSON et al. 1977). The following structural characteristics contribute to glycine receptor binding: (a) Receptor binding critically depends on the nitrogen atom of the β-spiropyrrolidino ring which corresponds to position 19 of strychnine. The cyclopentyl analogue is virtually inactive. Substitution to the pyrrolidino ring nitrogen increases the affinity about tenfold, as evident from a comparison of tertiary amines and the secondary amine (Fig. 4) (HERSHENSON et al. 1977). (b) Convulsant activity of β-spiropyrrolodinoindolines is promoted by an acetyl group on the indoline ring nitrogen corresponding to position 9 of the strychnine lactam ring. Loss of this acetyl indoline moiety causes a pronounced loss in receptor binding affinity (HERSHENSON et al. 1977).

III. Nonstrychnine Convulsants as Glycine Receptor Ligands

A number of convulsant alkaloids and synthetic drugs has been identified which affect glycine receptor function, but none resembles strychnine in affinity and specificity. The most powerful glycine receptor ligands among this group are 1,5-diphenyl-3,7-diazaadamantan-9-ol and the steroid derivative RU 5135 (see below).

1. Synthetic Glycine Receptor Antagonists

The amidine derivative 1,5-diphenyl-3,7-diazaadamantan-9-ol (Fig. 5) is a potent convulsant (Longo et al. 1959; Chiavarelli et al. 1962), efficiently antagonizing neuronal inhibition mediated by glycine but not GABA (Curtis et al. 1968b). The agent also appears to interfere with central nicotinic cholinergic synapses (Curtis et al. 1968b). 1,5-Diphenyl-3,7-diazaadamantan-9-ol is a high-affinity inhibitor of [^{3}H]strychnine binding (Mackerer et al. 1977). Its chemical structure is reminiscent of the "cage" convulsant at the GABA$_A$ receptor, TBPS. Whereas TBPS is thought to block the ion pore of the GABA$_A$ receptor channel complex (Squires et al. 1983; Rienitz et al. 1987), the mechanism of glycine receptor inhibition by 1,5-diphenyl-3,7-diazaadamantan-9-ol is not known. The synthetic agents 4-phenyl-4-formyl-N-methylpiperidine (1762 I.S.) and hexahydro-2'-methylspiro(cyclohexane-1,8'-(6H)-oxazino-(3,4-A)-pyrazine likewise alleviate glycine-induced depression of neuronal inhibition (Curtis et al. 1968b).

2. Alkaloid Antagonists and Opioids

Glycine receptors are also blocked by alkaloids structurally unrelated to strychnine (Table 1). Gelsemine is a convulsant occurring in the American yellow jasmine that antagonizes glycine receptor function with micromolar affinity. Similar to the structure of strychnine, however, gelsemine carries a lactam ring situated next to an aromatic ring but distant from a tertiary amine (Curtis et al. 1971b; Phelan et al. 1989). Other alkaloids binding to the glycine receptor are vincamine and boldine. The glycine but not GABA antagonistic agents hydrastine (Phelan et al. 1989) and laudanosine (Curtis et al. 1971b; Phelan et al. 1989) are structurally related to bicuculline, an inhibitor at the GABA$_A$ receptor agonist site (Curtis et al. 1970a,b, 1971; Olsen and Tobin 1990). Dendrobine, which structurally resembles the GABA$_A$ receptor channel blocker, picrotoxin, likewise suppresses currents induced by glycine but not by GABA (Curtis et al. 1970b).

Opioids find a wide therapeutical use as analgesics. However, high intrathecal concentrations of morphine provoke hyperesthesia reminiscent of strychnine-induced misperceptions instead of analgesia (Yaksh and Harty 1988). In addition, high doses of opioids exert a proconvulsant effect (Jaffe and Martin 1990). Morphine and other opioids are ligands of micromolar

Synthetic agents

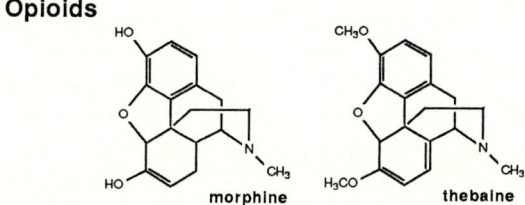

RU 5135
3α-hydroxy-16-imino-
5β-17-aza-androstan-11-on

**1,5-diphenyl-
3,7-diaza-
adamantan-9-ol**

Nonselective GABA$_A$ receptor ligands

bicuculline

ivermectin B$_{1a}$

picrotoxinin

pitrazepine

Opioids

morphine

thebaine

Fig. 5. Nonselective glycine receptor ligands

affinity at the inhibitory glycine receptor. The K$_i$ value at the glycine receptor of thebaine ($\approx 1\,\mu M$) exceeds more than tenfold the affinities of L-methadone, levorphanol, codeine, and morphine (GOLDINGER et al. 1981; PHELAN et al. 1989). Conversely, thebaine depresses glycine-mediated, postsynaptic inhibition of cat spinal cord neurons (CURTIS et al. 1968b). These observations suggest that glycine receptor antagonism by opioids contributes to their proconvulsant and hyperesthetic effect.

IV. Nonselective GABA$_A$ Receptor Ligands

Structurally, the inhibitory glycine receptor is highly homologous to the GABA$_A$ receptor. This homology extends to functional characteristics including conductance properties of the intrinsic ion channels (Bormann et al. 1987), ionic modulation (Marvizon and Skolnick 1989), and kinetics (Maksay 1990) of ligand binding. In addition, a wide spectrum of ligands recognizes both receptors.

1. Muscimol Analogues

Both inhibitory glycine and GABA$_A$ receptors apparently respond to β-amino acids (Simmonds 1983; Choquet and Korn 1988; Horikoshi et al. 1988a,b). A similar overlap in binding specificity exists for structural analogues of muscimol, a selective high-affinity (K$_i$ ≈ 5 nM; Krogsgaard-Larsen et al. 1982) agonist at the GABA$_A$ receptor. In a series of structure-activity studies on heterocyclic muscimol derivatives, Krogsgaard-Larsen et al. (1982) have identified structural determinants of ligand specificity for glycine and GABA$_A$ receptors. The amino group of muscimol correlates to the 4-amino group of GABA; incorporation into a piperidine ring of this nitrogen generates the heterocyclic zwitterion, 4,5,6,7-tetrahydroisoxazolo-[5,4-c] pyridin-3-ol (THIP) (Fig. 6; Table 1). This compound antagonizes glycine receptor binding with medium affinity (K$_i$ ≈ 40 μM) but nevertheless remains a high-affinity agonist at the GABA$_A$ receptor (K$_i$ ≈ 100 nM). In physiological experiments, however, glycine antagonism by THIP is concealed by the GABA-like action, which accounts for the antispastic effect of this agent (Krogsgaard-Larsen et al. 1982; Braestrup et al. 1986; Brehm et al. 1986). The secondary amine of the piperidine ring is essential for a GABA$_A$ receptor agonistic but not the glycine receptor antagonistic effect. The tertiary amine N-methyl-THIP inhibits glycine but not GABA responses of cat spinal neurons (Krogsgaard-Larsen et al. 1982; Braestrup et al. 1986). Exchange of the piperidine ring of THIP for the seven-membered azepine rings of 5,6,7,8-tetrahydro-4H-isoxazolo-[5,4-c]azepin-3-ol (THIA) and 5,6,7,8-tetrahydro-4H-isoxazolo-[4,5-d]azepin-3-ol (THAZ) only slight affects glycine receptor binding but generates GABA$_A$ receptor antagonists of micromolar affinities (Fig. 6) (Krogsgaard-larsen et al. 1975, 1982; Braestrup et al. 1986). Similar to THIP, inclusion into a pyrrolidine ring of the amino function of the selective GABA$_A$ receptor agonist homomuscimol generates an antagonist (5-(3-pyrrolidinyl)-3-isoxazol, 3-PYOL) at both GABA$_A$ and glycine receptors (K$_i$ ≈ 100 μM) (Krogsgaard-Larsen et al. 1982; Braestrup et al. 1986).

The isoxazolol moiety of THIP analogues carries another determinant of glycine receptor binding, as affinity is weakened by opening of the 3-isoxazolol ring of THAZ and THIA, producing the amino acids, 4,5-TAZA and 3,4-TAZA, respectively (Fig. 6). Apparently, constraining the carboxy function and delocalizing its electrons within the 3-isoxazolol ring elevates receptor

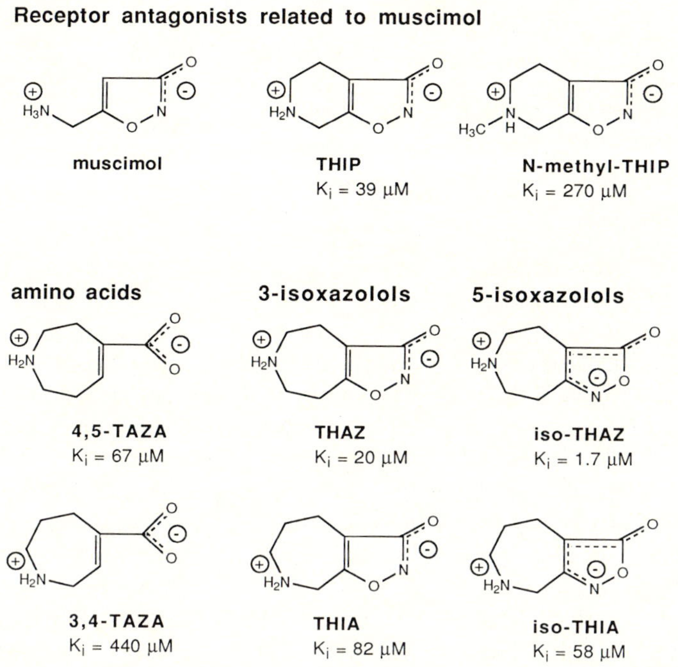

Receptor antagonists related to muscimol

muscimol

THIP
$K_i = 39 \mu M$

N-methyl-THIP
$K_i = 270 \mu M$

amino acids

3-isoxazolols

5-isoxazolols

4,5-TAZA
$K_i = 67 \mu M$

THAZ
$K_i = 20 \mu M$

iso-THAZ
$K_i = 1.7 \mu M$

3,4-TAZA
$K_i = 440 \mu M$

THIA
$K_i = 82 \mu M$

iso-THIA
$K_i = 58 \mu M$

Fig. 6. Structures of glycine receptor antagonists related to muscimol and 4,5,6,7-tetrahydroisoxazolo-[5,4-c]pyridin-3-ol (THIP). *THIA*, 5,6,7,8-tetrahydro-4*H*-isoxazolo-[5,4-c]azepin-3-ol; *3,4-TAZA*, 2,5,6,7-tetrahydro-1*H*-azepine-4-carboxylic acid; *4,5-TAZA*, 2,3,6,7-tetrahydro-1*H*-azepine-4-carboxylic acid; *THAZ*, 5,6,7,8-tetrahydro-4*H*-isoxazolo-[4,5-d]azepin-3-ol

affinity (KROGSGAARD-LARSEN et al. 1982; BRAESTRUP et al. 1986). The position of the isoxazolol nitrogen further contributes to glycine receptor potency of THIP derivatives. 5-Isoxazolols, e.g., iso-THAZ and iso-THIA, possess higher affinites than the corresponding 3-isoxazolol (THAZ, THIA). Iso-THAZ ($K_i \approx 1.7 \mu M$) even exceeds all other THIP derivatives in affinity. The crystal structures indicate that the negative charge is more delocalized in the 5-isoxazolol (Fig. 6), which may influence hydrogen bonding to corresponding receptor determinants (KROGSGAARD-LARSEN et al. 1982; BRAESTRUP et al. 1986; BREHM et al. 1986).

Although heterocyclic isoxazolol zwitterions antagonize glycine-induced inhibition and thus resemble strychnine, two lines of evidence indicate that THIP derivatives share recognition of receptor domains with glycine rather than strychnine: (a) Sodium chloride decreases glycine and THIP but favors strychnine binding to the receptor (BRAESTRUP et al. 1986). (b) Protein modification attenuates receptor binding of glycine and isoxazolol zwitterions (iso-THAZ, THAZ, and 4,5-TAZA) but not of strychnine (MARVIZON et al. 1986a). However, glycine receptor binding of the isoxazolol derivative iso-THAZ is diminished by increasing temperature, i.e., is enthalpy driven, which is distinctive for antagonists (Ruiz-Gomez et al. 1989b). Apparently, THIP and other isoxazolols recognize glycine binding determinants of the receptor complex but lack the intrinsic activity to induce an open channel conformation.

The convulsant alkaloid bicuculline (Fig. 5) antagonizes binding of GABA to the GABA$_A$ receptor (Curtis et al. 1971b; Barker et al. 1983; Enna and Möhler 1987). Electrophysiological observations on glycine receptor sensitivity to bicuculline are contradictory. Inhibition has been reported for cat cuneate neurons (Hill et al. 1973, 1976) while glycine responses of spinal cord neurons (Curtis et al. 1971b), channels induced by brain mRNA in *Xenopus laevis* oocytes (Horikoshi et al. 1988a,b; Parker et al. 1988), and recombinant glycine receptors expressed in a human cell line (Sontheimer et al. 1989) appear to be resistant. Surprisingly, the alkaloid efficiently antagonizes [^{3}H]strychnine binding at concentrations which do not affect current responses to glycine. The more active (+)-isomer of bicuculline displays micromolar K_i values at the glycine receptor (Goldinger and Müller 1980; Braestrup et al. 1986; Marvizon et al. 1986a). Resistance to protein modification, enhancement of binding by sodium chloride, and decreases by elevated temperature characterize bicuculline as a strychnine-like glycine receptor antagonist (Braestrup et al. 1986; Marvizon et al. 1986a; Ruiz-Gomez et al. 1989b).

The synthetic agent pitrazepine (Fig. 5) is an antagonist of nanomolar affinity at the agonist binding domain of the GABA$_A$ receptor (Gähwiler et al. 1984; Braestrup and Nielsen 1985). At the glycine receptor, pitrazepin displays K_i values in the range of 70–580 nM and a strychnine-like ion shift of binding affinities (Braestrup and Nielsen 1985; Braestrup et al. 1986). The pyridazinyl-GABA derivative SR 95103, a GABA$_A$ receptor antagonist less potent than bicuculline, also inhibits glycine-induced neuronal responses in the rat cuneate nucleus (Michaud et al. 1986).

2. Ligands of the GABA$_A$ Receptor Modulatory Domain

Benzodiazepines enhance GABAergic neurotransmission by allosteric modulation of the GABA$_A$ receptor agonist site (Enna and Möhler 1987; Olsen and Tobin 1990). Despite its high degree of homology to the GABA$_A$ receptor, similar modulatory domains have not been identified for the glycine receptor. However, flunitrazepam, diazepam, clonazepam, and other benzodiazepines bind to the glycine receptor with K_i values of $\approx 10\,\mu M$ (Young et al. 1974; Marvizon et al. 1986a). These benzodiazepine affinities are about 1000-fold lower than those at the GABA$_A$ receptor (Enna and Möhler 1987). It is unlikely that glycine receptor binding of benzodiazepines is related to the anxiolytic effect of these agents.

Other ligands of the GABA$_A$ receptor modulatory site likewise recognize the inhibitory glycine receptor. The quinoline derivative, PK 8165, is a partial agonist of submicromolar affinity at the benzodiazepine site

(BENAVIDES et al. 1984). At the $GABA_A$ receptor, β-carbolines produce effects opposite to anxiolytic benzodiazepines. Both the β-carboline nor-hamane and PK 8165 are strychnine-like ligands displaying micromolar glycine receptor affinities (MARVIZON et al. 1986a; RUIZ-GOMEZ et al. 1989b). Barbiturates are powerful modulators of $GABA_A$ receptor function (ENNA and MÖHLER 1987). Whereas nonconvulsant barbiturates do not influence the glycine receptor (GUNDERSEN et al. 1984; PARKER et al. 1988; NICHOLSON et al. 1988; SCHMIEDEN et al. 1989), some convulsant barbiturates antagonize neuronal glycine responses in isolated spinal cord (NICHOLSON et al. 1988). Receptor binding of these agents has not been explored.

3. Avermectin B_{1a}

Loss of neuronal inhibition by drugs is a well-established cause of convulsions. On the contrary, excessive activation of inhibitory receptor channels appears to account for the selective toxicity of the anthelmintic drug avermectin B_{1a} (Fig. 5), a macrocyclic lactone disaccharide derived from *Streptomyces avermectilis*, and its dihydroderivative ivermectin B_{1a}. In invertebrate muscle, avermectin B_{1a} eliminates postsynaptic potentials by increasing the chloride conductance of peripheral GABA receptor channels (FRITZ et al. 1979). In the vertebrate CNS, it binds to $GABA_A$ receptors (DREXLER and SIEGHART 1984), where it stimulates ligand binding (WILLIAMS and RISLEY 1982) and leads to a long-lasting channel activation (SIGEL and BAUR 1988). It appears to define an as yet uncharacterized regulatory domain on the $GABA_A$ receptor complex.

Avermectin B_{1a} also acts as a noncompetitive inhibitor of [^{3}H]strychnine binding to the inhibitory glycine receptor (apparent $K_i \approx 1-3\,\mu M$) (GRAHAM et al. 1982; MARVIZON et al. 1986a; RUIZ-GOMEZ et al. 1989b). Interestingly, the anthelmintic agent reduces the apparent number of binding sites but not the affinity for [^{3}H]strychnine (MARVIZON et al. 1986a). A Hill coefficient close to 0.5 characterizes the complex interaction of strychnine and avermectin B_{1a} binding (GRAHAM et al. 1982; MARVIZON et al. 1986a). Thermodynamic analysis shows that glycine receptor binding of ivermectin is characterized by substantial increases in entropy, as otherwise seen with agonistic ligands (RUIZ-GOMEZ et al. 1989b). Although the influence of avermectin B_{1a} on glycine receptor channel opening has not been explored, these observations suggest that the anthelmintic agent interferes with an as yet unidentified modulatory domain of the receptor.

4. Convulsant Steroids

The convulsant amidine steroid RU 5135 (3α-hydroxy-16-imino-5β-17-azaandrostan-11-one) (Fig. 5; Table 1) is a powerful antagonist of glycine-mediated inhibition in spinal cord neurons (CURTIS and MALIK 1985). It is characterized by the highest affinity known ($K_i \approx 4-10\,nM$) of any glycine receptor ligand (Braestrup et al. 1986; Marvizon et al. 1986a; Ruiz-Gomez et al. 1989b). It is, however, not entirely specific for glycine receptors and also antagonizes GABA$_A$ receptor ligand binding (Hunt and Clemens-Jewery 1981). RU 5135 and strychnine share some characteristics, e.g., receptor binding is enthalpy driven (Ruiz-Gomez et al. 1990) and enhanced by sodium chloride (Braestrup et al. 1986), suggesting that the agent recognizes the strychnine binding site at the glycine receptor. This conclusion is not readily reconciled with observations on other steroids including progesterone, deoxycorticosterone, and pregnenolone sodium sulfate which, at micromolar concentrations, likewise suppress neuronal glycine responses. Further analysis suggests that these steroids act through a glycine receptor domain distinct from the strychnine binding site (Wu and Oertel 1990). It thus remains to be shown whether RU 5135 and progesterone-like steroids recognize identical or distinct glycine receptor domains. At the GABA$_A$ receptor, progesterone and other steroids are positive modulators (Wu and Oertel 1990).

5. Channel Blockers

Channel blockers of the GABA$_A$ receptor complex including TBPS and picrotoxin are noncompetitive inhibitors that antagonize permeability of the anion pore. The "cage" convulsant TBPS does not bind to inhibitory glycine receptors (Rienitz et al. 1987; see above). Picrotoxin, which readily separates into equimolar amounts of the pharmacologically active component picrotoxinin (Fig. 5) and the inactive picrotin (Ticku et al. 1978), in contrast, poorly inhibits [^{3}H]strychnine binding but efficiently antagonizes glycine-gated chloride currents. Current responses induced in heterologous expression systems are almost completely abolished by picrotoxin concentrations (Akagi and Miledi 1988b; Schmieden et al. 1989; Sontheimer et al. 1989) which fail to displace [^{3}H]strychnine binding (Graham et al. 1982; F. Holzinger et al., unpublished data). This discrepancy is best explained by picrotoxin affecting two functional domains of the glycine receptor complex: Occupation of a high-affinity site related to the ion channel leads to noncompetitive inhibition, while the antagonist binding domain is antagonized only by higher picrotoxin concentrations. However, this explanation does not account for contrasting reports about picrotoxin effects on glycine-induced neuronal currents. Whereas glycine-mediated inhibition of spinal interneurons (Davidoff and Aprison 1969) and cuneate neurons (Hill et al. 1976) appears to be susceptible to picrotoxin, no sensitivity is seen in the oculomotor nucleus (Obata and Highstein 1970), isolated spinal cord of immature rats (Evans 1978), or cultured neurons (Barker et al. 1983). Developmental or regional heterogeneity of pharmacologically distinct glycine receptor isoforms (Akagi and Miledi 1988b; Becker et al. 1988; Kuhse et al. 1990a,b, 1991; Akagi et al. 1991) may contribute to this puzzling situation.

V. Various Effects on Glycine Receptor Function

In addition to identified receptor ligands, convulsant activity of several toxins has been related to the inhibitory glycine receptor. In rodents, the insecticides dichlorodiphenyl trichloroethane (DDT) and allethrin generate myoclonus which closely resembles strychnine intoxication (CHUNG and VAN WOERT 1984). Urea-induced myoclonus has likewise been attributed to glycine receptor dysfunction. Whereas other receptors are not affected, urea ($100\,\mu M$) drastically reduces the number, but not the affinity, of [^{3}H]strychnine binding sites in rat medulla (CHUNG et al. 1985; CHUNG and VAN WOERT 1986). Enhanced glycine receptor activity has been postulated to contribute to the depressant effect on CNS function of anesthetic gases (SMITH et al. 1984) and ethanol (CELENTANO et al. 1988). However, further studies are required to characterize glycine receptor responses to depressant drugs.

E. Conclusions and Perspectives

The distribution of inhibitory glycine receptors within the CNS represents the basis of the selective neurotoxicity of strychnine and related convulsants. At the glycine receptor complex, convulsants and agonistic amino acids recognize distinct ligand binding determinants. Structural identification of these receptor domains should contribute to our understanding of ligand-gated ion channels. In particular, the action of noncompetitive receptor antagonists is not fully understood and warrants further research. Positive modulators of the glycine receptor resembling benzodiazepine enhancement of GABA$_A$ receptor function have not been identified, with the possible exception of avermectin B$_{1a}$. Thus, it remains a rewarding goal of glycine receptor research to develop agonistic drugs which may have a therapeutical potential as antispastic or anticonvulsive agents.

Acknowledgement. I thank H. Betz for encouragement and valuable comments and H. Lueddens for a critical reading of the manuscript.

References

Akagi H, Miledi R (1988a) Expression of glycine and other amino acid receptors by rat spinal cord mRNA in *Xenopus* oocytes. Neurosci Lett 95:262–268

Akagi H, Miledi R (1988b) Heterogeneity of glycine receptors and their messenger RNAs in rat brain and spinal cord. Science 242:270–273

Akagi H, Patton DE, Miledi R (1989) Discrimination of heterogenous mRNAs encoding strychnine-sensitive glycine receptors in *Xenopus* oocytes by antisense oligonucleotides. Proc Natl Acad Sci USA 86:8103–8107

Akagi H, Hirai K, Hishinuma F (1991) Cloning of a glycine receptor subtype expressed in rat brain and spinal cord during a specific period of neuronal development. FEBS Lett 281:160–166

Akaike N, Kaneda M (1989) Glycine-gated chloride current in acutely isolated rat hypothalamic neurons. J Neurophysiol 62:1400–1409

Altschuler RA, Betz H, Parakkal MH, Reeks KA, Wenthold RJ (1986) Identification of glycinergic synapses in the cochlear nucleus through immunocytochemical localization of the postsynaptic receptor. Brain Res 369:316–320

Al Zamil Z, Bagust J, Kerkut GA (1990) Tubocurarine and strychnine block Renshaw cell inhibition in the isolated mammalian spinal cord. Gen Pharmacol 21:499–509

Aprison MH (1990) The discovery of the neurotransmitter role of glycine. In: Ottersen O, Storm-Mathisen V (eds) Glycine neurotransmission. Wiley, Chichester, pp 1–23

Aprison MH, Lipkowitz KB, Simon JR (1987) Identification of a glycine-like fragment on the strychnine molecule. J Neurosci Res 17:209–213

Araki T, Yamano M, Murakami T, Wanaka A, Betz H, Tohyama M (1988) Localization of glycine receptors in the rat central nervous system: an immunocytochemical analysis using monoclonal antibody. Neuroscience 25:613–624

Asanuma A, Horikoshi T, Yanagisawa K, Anzai K, Goto S (1987) The distribution of GABA and glycine response in the mouse brain using *Xenopus* oocytes. Neurosci Lett 76:87–90

Ascher P, Gachelin G (1967) Rôle du colliculus supérieur dans l'élaboration de réponse motrices à des stimulations visuelles. Brain Res 3:327–342

Barker JL, McBurney RN, Mathers DA (1983) Convulsant-induced depression of amino acid responses in cultured mouse spinal neurones studied under voltage clamp. Br J Pharmacol 80:619–629

Barron SE, Guth PS (1987) Use and limitations of strychnine as a probe in neurotransmission. Trends Pharmacol Sci 8:204–206

Basbaum AI (1988) Distribution of glycine receptor immunoreactivity in the spinal cord of the rat: cytochemical evidence for a differential glycinergic control of lamina I and V nociceptive neurons. J Comp Neurol 278:330–336

Becker C-M (1990) Disorders of the inhibitory glycine receptor: the spastic mouse. FASEB J 4:2767–2774

Becker C-M, Betz H (1987) Strychnine: useful probe or not? Trends Pharmacol Sci 8:379–380

Becker C-M, Hermans-Borgmeyer I, Schmitt B, Betz H (1986) The glycine receptor deficiency of the mutant mouse spastic: evidence for normal glycine receptor structure and localization. J Neurosci 6:1358–1364

Becker C-M, Hoch W, Betz H (1988) Glycine receptor heterogeneity in rat spinal cord during postnatal development. EMBO J 7:3717–3726

Becker C-M, Hoch W, Betz H (1989) Sensitive immunoassay shows selective association of peripheral and integral membrane proteins of the inhibitory glycine receptor complex. J Neurochem 53:124–131

Benavides J, Lopez-Lahoya J, Valdivieso F, Ugarte M (1981) Postnatal development of synaptic glycine receptors in normal and hyperglycinergic rats. J Neurochem 37:315–320

Benavides J, Malgouris A, Flamier A, Tur C, Quarteronet F, Begasset F, Camelin JC, Uzan C, Gueremy C, LeFur G (1984) Biochemical evidence that 2-phenyl-4[2-(4-piperidinyl)ethyl]quinoline, a quinoline derivative with pure anticonflict properties, is a partial agonist of benzodiazepine receptors. Neuropharmacology 23:1129–1136

Bertolino M, Vicini S (1988) Voltage-dependent block by strychnine of N-methyl-D-aspartic acid-activated cationic channels in rat cortical neurons in culture. Mol Pharmacol 34:98–103

Betz H (1985) The glycine receptor of rat spinàl cord: exploring the site of action of the plant alkaloid strychnine. Angew Chem (Engl) 24:365–370

Betz H (1990) Ligand-gated ion channels in the brain: the amino acid receptor family. Neuron 5:383–392

Betz H, Becker C-M (1988) The mammalian glycine receptor: biology and structure of a neuronal chloride channel protein. Neurochem Int 13:137–146

Beyer C, Roberts LA, Komisaruk BR (1985) Hyperalgesia induced by altered glycinergic activity at the spinal cord. Life Sci 37:875–882

Biscoe TJ, Curtis DR (1967) Strychnine and cortical inhibition. Nature 214:914–915

Biscoe TJ, Duchen MR (1986) Synaptic physiology of spinal motoneurones of normal and spastic mice: an in vitro study. J Physiol (Lond) 379:275–292

Bormann J, Hamill OP, Sakmann B (1987) Mechanism of anion permeation through channels gated by glycine and γ-aminobutyric acid in mouse cultured spinal cord neurons. J Physiol (Lond) 385:243–286

Braestrup C, Nielsen M (1985) Interaction of pitrazepin with the GABA/benzodiazepine receptor complex and with glycine receptors. Eur J Pharmacol 118:115–121

Braestrup C, Nielsen M, Krogsgaard-Larsen P (1986) Glycine antagonists structurally related to 4,5,6,7-tetrahydroisoxazolo-[5,4-c]pyridin-3-ol inhibit binding of [³H]strychnine to rat brain membranes. J Neurochem 47:691–696

Brehm L, Krogsgaard-Larsen P, Schaumburg K, Johansen JS, Falch E (1986) Glycine antagonists. Synthesis, structure, and biological effects of some bicyclic 5-isoxazolol zwitterions. J Med Chem 29:224–229

Bristow DR, Bowery NG, Woodruff GN (1986) Light microscopic autoradiographic localisation of [³H]glycine and [³H]strychnine binding sites in rat brain. Eur J Pharmacol 126:303–307

Bruning G, Bauer R, Baumgarten HG (1990) Postnatal development of [³H]flunitrazepam and [³H]strychnine binding sites in rat spinal cord localized by quantitative autoradiography. Neurosci Lett 110:6–10

Carpenter MK, Parker I, Miledi R (1988) Expression of GABA and glycine receptors by messenger RNAs from the developing rat cerebral cortex. Proc R Soc Lond [Biol] 234:159–170

Celentano JJ, Gibbs TT, Farb DH (1988) Ethanol potentiates GABA- and glycine-induced chloride currents in chick spinal cord neurons. Brain Res 455:377–380

Chiavarelli S, Fennoy LV, Settimj G, DeBarran L (1962) The effect of methoxyphenyl substitutions on the strychnine-like activity of arylazaadamantanes and aryldiazaadamantanols. J Med Chem 5:1293–1297

Choquet D, Korn H (1988) Does β-alanine activate more than one chloride channel associated receptor? Neurosci Lett 84:329–334

Chung E, van Woert MH (1984) DDT myoclonus: sites and mechanism of action. Exp Neurol 85:273–282

Chung EY, van Woert MH (1986) Urea myoclonus: possible involvement of glycine. Adv Neurol 43:565–568

Chung E, Yocca F, van Woert MH (1985) Urea-induced myoclonus: medullary glycine antagonism as mechanism of action. Life Sci 36:1051–1058

Curtis DR, Johnston GAR (1970) Amino acid transitters in the mammalian central nervous system. Ergeb Physiol 69:98–188

Curtis DR, Malik R (1985) Glycine antagonism by RU 5135. Eur J Pharmacol 110:383–384

Curtis DR, Hösli L, Johnston GAR, Johnston IH (1968a) The hyperpolarization of spinal motoneurones by glycine and related amino acids. Exp Brain Res 5:235–258

Curtis DR, Hösli L, Johnston GAR (1968b) A pharmacological study of the depression of spinal neurones by glycine and related amino acids. Exp Brain Res 6:1–18

Curtis DR, Dugan AW, Johnston GAR (1971a) The specificity of strychnine as a glycine antagonist in the mammalian spinal cord. Exp Brain Res 12:547–565

Curtis DR, Duggan AW, Felix D, Johnston, GAR (1971b) Bicuculline, an antagonist of GABA and synaptic inhibition in the spinal cord of the cat. Brain Res 32:69–96

Davidoff RA, Aprison MH (1969) Picrotoxin antagonism of the inhibition of interneurons by glycine. Life Sci 8:107–112

Dengler R. (1990) The motor unit: physiology, diseases, regeneration. Urban and Schwarzenberg, Munich

Drexler G, Sieghart W (1984) Properties of a high affinity binding site for [^{3}H]avermectin B$_{1a}$. Eur J Pharmacol 99:269–277

Drummond J, Johnson G, Nickell DG, Ortwine DF, Bruns RF, Welbaum B (1989) Evaluation and synthesis of aminohydroxyisoxazoles and pyrazoles as potential glycine agonists. J Med Chem 32:2116–2128

Dusser de Barenne, JG (1933) The mode and site of action of strychnine on the nervous system. Physiol Rev 13:325–335

Eccles JC (1964) The physiology of synapses. Springer, Berlin Heidelberg New York

Enna SJ, Möhler H (1987) γ-Aminobutyric acid (GABA) receptors and their association with benzodiazepine recognition sites. In: Meltzer HY (ed) Psychopharmacology: the third generation of progress. Raven, New York pp 265–272

Erdö SL (1990) Strychnine protection against excitotoxic cell death in primary cultures of rat cerebral cortex. Neurosci Lett 115:341–344

Evans RH (1978) The effect of amino acids and antagonists on the isolated hemisected spinal cord of the immature rat. Br J Pharmacol 62:171–176

Faingold CL, Hoffmann, WE, Caspary, D.M. (1985) Mechanisms of sensory seizures: brain-stem neuronal response changes and convulsant drugs. Fed Proc 44:2436–2441

Ferenci P, Pappas SC, Munson PJ, Heenson K, Jones EA (1984) Changes in the status of neurotransmitter receptors in a rabbit model of hepatic encephalopathy. Hepatology 4:186–191

Franz DN (1985) Central nervous system stimulants. In: Goodman Gilman A, Goodman LS, Rall TW, Murad F (eds) The pharmacological basis of therapeutics, 7th edn. MacMillan, New York, pp 582–588

Fritz LC, Wang CC, Gorio A (1979) Avermectin B$_{1a}$ irreversibly blocks postsynaptic potentials at the lobster neuromuscular junction by reducing muscle membrane resistance. Proc Natl Acad Sci U.S.A. 76:2062–2066

Frostholm A, Rotter A (1985) Glycine receptor distribution in mouse CNS: autoradiographic localization of [^{3}H]strychnine binding sites. Brain Res Bull 15:473–486

Frostholm A, Rotter A (1986) Autoradiographic localization of receptors in the cochlear nucleus of the mouse. Brain Res Bull 16:189–203

Fry JP, Phelan PP (1985) Interaction of glycine receptor ligands in the normal mouse spinal cord. J Physiol 373:21

Gähwiler BH, Maurer R, Wuthrich HJ (1984) Pitrazepin, a novel GABA$_A$ antagonist. Neurosci Lett 45:311–316

Galli A, Nocchi M, Sciarra P (1983) Evidence of enrichment in glycine receptors of crude synaptic membranes from rat spinal cord following Triton X-100 treatment. Biochem Biophys Res Commun 112:809–816

Garcia-Calvo M, Ruiz-Gomez A, Vazquez J, Morato E, Valdivieso F, Mayor F Jr (1989) Functional reconstitution of the glycine receptor. Biochemistry 28:6405–6409

Geyer SW, Gudden W, Betz H, Gnahn H, Weindl A (1987) Co-localization of choline acetyltransferase and postsynaptic glycine receptors in motoneurons of rat spinal cord demonstrated by immunocytochemistry. Neurosci Lett 82:11–15

Gillberg PG, Aquilonius SM (1985) Cholinergic, opioid and glycine receptor binding sites localized in human spinal cord by in vitro autoradiography. Changes in amyotrophic lateral sclerosis. Acta Neurol Scand 72:299–306

Glendenning KK, Baker BN (1988) Neuroanatomical distribution of receptors for three potential inhibitory neurotransmitters in the brainstem auditory nuclei of the cat. J Comp Neurol 275:288–308

Goldinger A, Müller WE (1980) Stereospecific interaction of bicucculline with specific [^{3}H]strychnine binding to the glycine receptor of rat spinal cord. Neurosci Lett 16:91–95

Goldinger A, Müller WE, Wollert U (1981) Inhibition of glycine and GABA receptor binding by several opiate agonists and antagonists. Gen Pharmacol 12:477–479

Gosselin RE, Hodge HC, Smith RP, Gleason MN (1976) Clinical toxicology of commercial products, 4th edn. Williams and Wilkins, Baltimore, pp 303–307

Graham D, Pfeiffer F, Betz H (1981) UV-light induced cross-linking of strychnine to the glycine receptor of rat spinal cord membranes. Biochem Biophys Res Commun 102:1330–1335

Graham D, Pfeiffer F, Betz H (1982) Avermectin B_{1a} inhibits the binding of strychnine to the glycine receptor of rat spinal cord. Neurosci Lett 29:173–176

Graham D, Pfeiffer F, Betz H (1983) Photoaffinity-labelling of the glycine receptor of rat spinal cord. Eur J Biochem 131:519–525

Graham D, Pfeiffer F, Simler R, Betz H (1985) Purification and characterization of the glycine receptor of pig spinal cord. Biochemistry 24:990–994

Grenningloh G, Rienitz A, Schmitt B, Methfessel C, Zensen M, Beyreuther K, Gundelfinger ED, Betz H (1987a) The strychnine-binding subunit of the glycine receptor shows homology with nicotinic acetylcholine receptors. Nature 328:215–220

Grenningloh G, Gundelfinger E, Schmitt B, Betz H, Darlison MG, Barnard EA, Schofield PR, Seeburg PH (1987b) Glycine vs GABA receptors (Letter). Nature 330:25–26

Grenningloh G, Schmieden V, Schofield P, Seeburg PH, Siddique T, Mohandas T K, Becker C-M, Betz H (1990a) Alpha subunit variants of the human glycine receptor: primary structures, functional expression and chromosomal localization of the corresponding genes. EMBO J 9:771–776

Grenningloh G, Pribilla I, Prior P, Multhaup G, Beyreuther K, Taleb O, Betz H (1990b) Cloning and expression of the 58 kd beta subunit of the inhibitory glycine receptor. Neuron 4:963–970

Gundersen CB, Miledi R, Parker I (1984) Properties of human brain glycine receptors expressed in Xenopus oocytes. Proc R Soc Lond [Biol] 221:235–244

Gundlach, AL (1990) Disorder of the inhibitory glycine receptor: inherited myoclonus in Hereford calves. FASEB J 4:2761–2766

Gundlach AL, Beart PM (1981) [^{3}H]Strychnine binding suggests glycine receptors in the ventral tegmental area of rat brain. Neurosci Lett 22:289–294

Heidrich H, Ibe K, Klinge D (1969) Akute Vergiftung mit Strychnin-N-oxydhydrochlorid (Movellan-Tabletten®) und ihre Behandlung mit Diazepam. Arch Toxicol 24:188–200

Hershenson MF, Prodan KA, Kochman RL, Bloss JL, Mackerer CR (1977) Synthesis of β-spiro[pyrrolodinoindolines], their binding to the glycine receptor, and in vivo biological activity. J Med Chem 20:1448–1451

Hill RG, Simmonds MA, Straughan DW (1973) Amino acid antagonists and the depression of cuneate neurones by γ-aminobutyric acid (GABA) and glycine. Br J Pharmacol 47:642–643

Hill RG, Simmonds MA, Straughan DW (1976) Antagonism of γ-aminobutyric acid and glycine by convulsants in the cuneate nucleus of cat. Br J Pharmacol 56:9–19

Hirsch JA, Oertel D (1988) Synaptic connections in the dorsal cochlear nucleus of mice, in vitro. J Physiol (Lond) 396:549–562

Hoch H, Betz H, Becker C-M (1989) Primary cultures of mouse spinal cord express the neonatal isoform of the inhibitory glycine receptor. Neuron 3:339–348

Horikoshi T, Asanuma A, Yanagisawa K, Anzai K, Goto S (1988a) Taurine and β-alanine act on both GABA and glycine receptors in Xenopus oocytes injected with mouse brain messenger RNA. Brain Res 464:97–105

Horikoshi T, Asanuma A, Yanagisawa K, Goto S (1988b) Taurine modulates glycine response in Xenopus oocytes injected with messenger RNA from mouse brain. Brain Res 464:243–246

Houamed KM, Bilbe G, Smart TG, Constanti A, Brown DA, Barnard EA, Richards BM (1984) Expression of functional GABA, glycine and glutamate receptors in Xenopus oocytes injected with rat brain mRNA. Nature 310:318–321

Hunt P, Clemens-Jewery S (1981) A steroid derivative, R 5135, antagonizes the GABA/benzodiazepine receptor interaction. Neuropharmacology 20:357–361

Jackson G, Diggle GE, Bourke IG (1971) Strychnine poisoning treated successfully with diazepam. Br Med J 3:519–520

Jaffe JH, Martin WR (1990) Opioid analgesics and antagonists. In: Goodman Gilman A, Rall TW, Nies AS, Taylor P (eds) The pharmacological basis of therapeutics, 8th edn. Pergamon, New York, pp 485–521

Jäger J, Wässle H (1987) Localization of glycine uptake and receptors in the cat retina. Neurosci Lett 75:147–151

Karschin A, Wässle H (1990) Voltage and transmitter-gated currents in isolated rod bipolar cells of the retina. J Neurophysiol 63:860–876

Kehne JH, Davis M (1984) Strychnine increases acoustic startle amplitude but does not alter short-term or long-term habituation. Behav Neurosci 98:955–968

Krishtal OA, Osipchuk YV, Vrublevsky SV (1988) Properties of glycine-activated conductances in rat brain neurones. Neurosci Lett 84:271–276

Krnjevic K (1981) Transmitters in motor systems. In: Geiger SR (ed) Handbook of physiology. Am Physiol Soc, Baltimore, pp 107–154

Krogsgaard-Larsen P, Johnston GAR, Curtis DR, Game CJA, McCulloch RM (1975) Structure and biological activity of a series of conformationally restricted analogues of GABA. J Neurochem 25:803–809

Krogsgaard-Larsen P, Hjeds H, Curtis DR, Leah JD, Peet MJ (1982) Glycine antagonists structurally related to muscimol, THIP, or isoguvacine. J Neurochem 39:1319–1324

Kuhse J, Schmieden V, Betz H (1990a) A single amino acid exchange alters the pharmacology of neonatal rat glycine receptor subunit. Neuron 5:867–873

Kuhse J, Schmieden V, Betz H (1990b) Identification and functional expression of a novel ligand binding subunit of the inhibitory glycine receptor. J Biol Chem 265:22317–22320

Kuhse J, Kuryatov A, Maulet Y, Malosio M-L, Schmieden V, Betz H (1991) Alternative splicing generates two isoforms of the inhibitory glycine receptor. FEBS Lett 283:73–77

Langosch D, Thomas L, Betz H (1988) Conserved quaternary structure of ligand-gated ion channels: the postsynaptic glycine receptor is a pentamer. Proc Natl Acad Sci USA 85:7394–7398

Langosch D, Betz H, Becker C-M (1990a) Molecular structure and developmental regulation of the inhibitory glycine receptor. In: Ottersen O, Storm-Mathisen V (eds) Glycine neurotransmission. Wiley, Chichester, pp 67–82

Langosch D, Becker C-M, Betz H (1990b) The inhibitory glycine receptor: ligand-gated chloride channel of the central nervous system. Eur J Biochem 194:1–8

Langosch D, Hartung K, Grell E, Bamberg E, Betz H (1991) Ion channel formation by synthetic transmembrane segments of the inhibitory glycine receptor-a model study. Biochim Biophys Acta 1063:36–44

LeFort D, Henke H, Cuenod M (1978) Glycine specific [3H]strychnine binding in the pigeon CNS. J Neurochem 30:1287–1291

Lewis CA, Ahmed Z, Faber DS (1989) Characteristics of glycine-activated conductances in cultured medullary neurons from embryonic rat. Neurosci Lett 96:185–190

Lloyd KG, de Montis G, Javoy-Agid F, Beaumont K, Lowenthal A, Constantinidis J, Agid Y (1983a) Glycine receptors in the human brain: characterization of ^{3}H-strychnine binding and status in pathological conditions. Adv Biochem Psychopharmacol 36:233–238

Lloyd KG, de Montis G, Broekkamp CL, Thuret F, Worms P (1983b) Neurochemical and neuropharmacological indications for the involvement of GABA and glycine receptors in neuropsychiatric disorders. Adv Biochem Psychopharmacol 37:137–148

Long SK, Evans RH, Krijzer F (1989) Effects of depressant amino acids and antagonists on an in vitro spinal cord preparation from the adult rat. Neuropharmacology 28:683–688

Longo VG, Silvestrini B, Bovert D (1959) An investigation of convulsant properities of the 5-7-diphenyl-1-3-diazaadamantan-6-ol (1757 I.S.) J Pharmacol Exp Ther 126:41–49

Mackerer CR, Kochman RL, Shen TF, Hershenson FM (1977) The binding of strychnine and strychnine analogs to synaptic membranes of rat brainstem and spinal cord. J Pharmacol Exp Ther 201:326–331

Maksay G (1990) Dissociation of muscimol, SR 95531, and strychnine from GABA A and glycine receptors, respectively, suggests similar cooperative interactions. J Neurochem 54:1961–1966

Malosio M-L, Grenningloh G, Kuhse J, Schmieden V, Schmitt B, Prior P, Betz H (1991a) Alternative splicing generates two variants of the alpha 1 subunit of the inhibitory glycine receptor. J Biol Chem 266:2048–2053

Malosio M-L, Marquèze-Poney B, Kuhse J, Betz H (1991b) Widespread expression of glycine receptor subunit mRNAs in the adult and developing rat brain. EMBO J 10:2401–2409

Marvizón JC, Skolnick P (1988a) Enhancement of t-[^{35}S]butylbicyclophosphoro thionate and [^{3}H]strychnine binding by monovalent anions reveals similarities between γ-aminobutyric acid- and glycine-gated chloride channels. J Neurochem 50:1632–1639

Marvizón JC, Skolnick P (1988b) Anion regulation of [^{3}H]strychnine binding to glycine-gated chloride channels is explained by the presence of two anion binding sites. Mol Pharmacol 34:806–813

Marvizón JC, Vazquez J, Garcia-Calvo M, Mayor F Jr, Ruiz-Gomez A, Valdivieso F, Benavides J (1986a) The glycine receptor: pharmacological studies and mathematical modeling of the allosteric interaction between the glycine- and strychnine-binding sites. Mol Pharmacol 30:590–597

Marvizón JC, Garcia-Calvo M, Vazquez J, Mayor F Jr, Ruiz-Gomez A, Valdiviesco F, Benavides J (1986b) Activation and inhibition of ^{3}H-strychnine binding to the glycine receptor by Eccles' anions: modulatory effects of cations. Mol Pharmacol 30:598–602

McNamara D, Dingledine R (1990) Dual effect of glycine on NMDA-induced neurotoxicity in rat cortical cultures. J Neurosci 10:3970–3976

Michaud JC, Mienville JM, Chambon JP, Biziere K (1986) Interactions between three pyridazinyl-GABA derivatives and central GABA and glycine receptors in the rat, an in vivo microiontophoretic study. Neuropharmacology 25:1197–1203

Monaghan DT, Bridges RJ, Cotman CW (1989) The excitatory amino acid receptors: their classes, pharmacology, and distinct properties in the function of the central nervous system. Annu Rev Pharmacol Toxicol 29:365–402

Müller WE, Snyder SH (1978) Strychnine binding associated with synaptic glycine receptors in rat spinal cord membranes: ionic influences. Brain Res 147:107–116

Müller F, Wässle H, Voigt T (1988) Pharmacological modulation of the rod pathway in the cat retina. J Neurophysiol 59:1657–1672

Naas E, Zilles K, Gnahn H, Betz H, Becker C-M, Schröder H (1991) Glycine receptor immunoreactivity in rat and human cerebral cortex. Brain Res 561:139–146

Nicholson GM, Spence I, Johnston GAR (1988) Differing actions of convulsant and nonconvulsant barbiturates: an electrophysiological study in the isolated spinal cord of the rat. Neuropharmacology 27:459–465

Obata K, Highstein SM (1970) Blocking by picrotoxin of both vestibular inhibition and GABA action on rabbit oculomotor neurons. Brain Res 18:538–541

O'Connor VM (1989) Chemical modification of overlapping but conformationally distinct recognition sites for glycine and strychnine in isolated spinal cord membranes. J Physiol (Lond) 415:49 P

Olsen RW, Tobin AJ (1990) Molecular biology of GABA$_A$ receptors. FASEB J 4:1469–1480

Ondo JG, Buckholtz NS (1983) The central actions of glycine and strychnine on prolactin and LH secretion. Brain Res Bull 11:7–10

Parker I, Sumidawa K, Miledi R (1985) Messenger RNA from bovine retina induces kainate and glycine receptors in Xenopus oocytes. Proc R Soc Lond [Biol] 225:99–106

Parker I, Sumikawa K, Miledi R (1988) Responses to GABA, glycine and β-alanine induced in Xenopus oocytes by messenger RNA from chick and rat brain. Proc R Soc Lond [Biol] 233:201–216

Perper JA (1985) Fatal strychnine poisoning – a case report and review of the literature. J Forensic Sci 30:1248–1255

Pfeiffer F, Betz H (1981) Solubilization of the glycine receptor from rat spinal cord. Brain Res 226:273–279

Pfeiffer F, Graham D, Betz H (1982) Purification by affinity chromatography of the glycine receptor of rat spinal cord. J Biol Chem 257:9389–9393

Pfeiffer F, Simler R, Grenningloh G, Betz H (1984) Monoclonal antibodies and peptide mapping reveal structural similarities between the subunits of the glycine receptor of rat spinal cord. Proc Natl Acad Sci USA 81:7224–7227

Phelan PP, Fry JP, Martin IL, Johnston GAR (1989) Polyclonal antibodies to the glycine receptor antagonist strychnine. J Neurochem 52:1481–1486

Poulsson E (1920) Die Strychningruppe. In: Hefftner A (ed) Handbuch der Experimentellen Pharmakologie, vol 2. Springer, Berlin

Probst A, Cortes R, Palacios JM (1986) The distribution of glycine receptors in the human brain. A light microscopic autoradiographic study using [^{3}H]strychnine. Neuroscience. 17:11–35

Rees R, Smith H (1967) Structure and biological activity of some reduction products of strychnine, brucine, and their congeners. J Med Chem 10:624–627

Rienitz A, Becker C-M, Betz H, Schmitt B (1987) The chloride channel blocking agent, t-butyl bicyclophosphorothionate, binds to the gamma-aminobutyric acid-benzodiazepine, but not to the glycine receptor in rodents. Neurosci Lett 76:91–95

Riquelme G, Morato E, Lopez E, Ruiz-Gomez A, Ferragut JA, Gonzalez-Ros JM, Mayor F Jr (1990) Agonist binding to purified glycine receptor reconstituted into giant liposomes elicits two types of chloride currents. FEBS Lett 276:54–58

Rote Liste (1991) Cantor, Aulendorf

Rotter A, Schultz CM, Frostholm A (1984) Regulation of glycine receptor binding in the mouse hypoglossal nucleus in response to axotomy. Brain Res Bull 13:487–492

Ruiz-Gómez A, Fernandez-Shaw C, Valdivieso F, Mayor F Jr (1989a) Chemical modification of the glycine receptor with fluorescein isothiocyanate specifically affects the interaction of glycine with its binding site. Biochem Biophys Res Commun 160:374–381

Ruiz-Gómez A, Garcia-Calvo M, Vazquez J, Marvizòn JC, Valdivieso F, Mayor F Jr (1989b) Thermodynamics of agonist and antagonist interaction with the strychnine-sensitive glycine receptor. J Neurochem 52:1775–1780

Ruiz-Gómez A, Morato E, Garcia-Calvo M, Valdivieso F, Mayor F Jr (1990) Localization of the strychnine binding site on the 48-kilodalton subunit of the glycine receptor. Biochemistry 29:7033–7040

Ruiz-Gómez A, Vaello ML, Valdivieso F, Mayor F Jr (1991a) Phosphorylation of the 48-kDa subunit of the glycine receptor by protein kinase C. J Biol Chem 266:559–566

Ruiz-Gómez A, Fernandez-Shaw C, Morato E, Marvizòn JC, Vazquez J, Valdivieso F, Mayor F Jr (1991b) Sulfhydryl groups modulate the allosteric interaction between glycine binding sites at the inhibitory glycine receptor. J Neurochem 56:1690–1697

Ryan GP, Hackman JC, Davidoff, RA (1984) Spinal seizures and excitatory amino acid-mediated synaptic transmission. Neurosci Lett 44:161–166

Sanes DH, Wooten GF (1987) Development of glycine receptor distribution in the lateral superior olive of the gerbil. J Neurosci 7:3803–3811

Sanes DH, Geary WA, Wooten GF, Rubel EW (1987) Quantitative distribution of the glycine receptor in the auditory brain stem of the gerbil. J Neurosci 7:3793–3802

Schaeffer JM, Anderson SM (1981) Identification of strychnine binding sites in the rat retina. J Neurochem 36:1597–1600

Schaffner AE, St John PA, Barker JL (1987) Fluorescence-activated cell sorting of embryonic mouse and rat motoneurons and their long-term survival in vitro. J Neurosci 7:3088–3104

Schmieden V, Grenningloh G, Schofield PR, Betz H (1989) Functional expression in *Xenopus* oocytes of the strychnine binding 48 kd subunit of the glycine receptor. EMBO J 8:695–700

Schmitt B, Knaus P, Becker C-M, Betz H (1987) The M_r 93 000 polypeptide of the postsynaptic glycine receptor is a peripheral membrane protein. Biochemistry 26:805–811

Schröder S, Hoch W, Becker C-M, Grenningloh G, Betz H (1991) Mapping of antigenic epitopes on the α_1 subunit of the inhibitory glycine receptor. Biochemistry 30:42–47

Sigel E, Baur R (1988) Effect of avermectin B_{1a} on chick neuronal γ-aminobutyrate receptor channels expressed in *Xenopus* oocytes. Mol Parmacol 32:749–752

Simmonds MA (1983) Depolarizing responses to glycine, beta-alanine and muscimol in isolated optic nerve and cuneate nucleus. Br J Pharmacol 79:799–806

Simmonds MA (1986) Classification of inhibitory amino acid receptors in the mammalian nervous system. Med Biol 64:301–311

Smith EB, Bowser-Riley F, Daniels S, Dunbar IT, Harrison CB, Paton WB (1984) Species variation and the mechanism of pressure-anaesthetic interactions. Nature 311:56–57

Song YM, Huang LY (1990) Modulation of glycine receptor chloride channels by cAMP-dependent protein kinase in spinal trigeminal neurons. Nature 348:242–245

Sontheimer H, Becker C-M, Pritchett DP, Schofield PR, Grenningloh G, Kettenmann H, Betz H, Seeburg PH (1989) Functional chloride channels by mammalian cell expression of rat glycine receptor subunit. Neuron 2:1491–1497

Squires RF, Casida JE, Richardson M, Saedrup E (1983) [^{35}S]t-Butylcyclophosphorothionate binds with high affinity to brain specific sites coupled to γ-aminobutyric acid-A and ion recognition sites. Mol Parmacol 23:326–336

Srinivasan Y, Guzikowski AP, Haugland RP, Angelides KJ (1990) Distribution and lateral mobility of glycine receptors on cultured spinal cord neurons. J Neurosci 10:985–995

St John PA, Wayne MK, Mazetta JS, Lange GD, Barker JL (1986) Analysis and isolation of embryonic mammalian neurons by fluorescence-activated cell sorting. J Neurosci 6:1492–1512

Takahashi T (1984) Inhibitory miniature synaptic potentials in rat motoneurons. Proc R Soc Lond [Biol] 221:103–109

Talman WT (1988) Glycine microinjected in the rat dorsal vagal nucleus increases arterial pressure. Hypertension 11:664–667

Talman WT, Robertson SC (1989) Glycine, like glutamate, microinjected into the nucleus tractus solitarii of rat decreases arterial pressure and heart rate. Brain Res 16(477):7–13

Teuscher E, Lindequist U (1988) Biogene Gifte. Akademie, Berlin

Ticku MK, Ban M, Olsen RW (1978) Binding of [³H]α-dihydropicrotoxinin, a γ-aminobutyric acid synaptic antagonist, to rat brain membranes. Mol Pharmacol 14:391–402

Tokutomi N, Kaneda M, Akaike N (1989) What confers specificity on glycine for its receptor site? Br J Pharmacol 97:353–360

Tribble GL, Schwindt PC, Crill WE (1983) Reduction of postsynaptic inhibition tolerated before seizure initiation: spinal cord. Exp Neurol 80:288–303

Triller A, Cluzeaud F, Pfeiffer F, Betz H, Korn H (1985) Distribution of glycine receptors at central syhnapses: an immunoelectron microscopy study. J Cell Biol 101:683–688

Triller A, Cluzeaud F, Korn H (1987) Gamma-aminobutyric acid-containing terminals can be apposed to glycine receptors at central synapses. J Cell Biol 104:947–956

Triller A, Seitanidon T, Franksson O, Korn H (1990) Size and shape of glycine receptor clusters in a central neuron exhibit a somato-dendritic gradient. New Biol 2:637–641

Van den Pol AN, Gorcs T (1988) Glycine and glycine receptor immunoreactivity in brain and spinal cord. J Neurosci 8:472–492

Von Hippel A (1873) Über die Wirkungen des Strychnins auf das normale und kranke Auge. Springer, Berlin

Wenthold RJ, Parakkal MH, Oberdorfer MD, Altschuler RA (1988) Glycine receptor immunoreactivity in the ventral cochlear nucleus of the guinea pig. J Comp Neurol 276:423–435

Werman R, Davidoff RA, Aprison MH (1968) Inhibitory action of glycine on spinal neurons in the cat. Nature 214:681–683

White WF (1985) The glycine receptor in the mutant mouse spastic (spa): strychnine binding characteristics and pharmacology. Brain Res 329:1–6

White WF, O'Gorman S, Roe AW (1990) Three-dimensional autoradiographic localization of quench-corrected glycine receptor specific activity in the mouse brain using ³H-strychnine as the ligand. J Neurosci 10:795–813

Williams M, Risley EA (1982) Interaction of avermectins with [³H]β-carboline-3-carboxylate ethyl ester and [³H]diazepam binding sites in rat brain cortical membranes. Eur J Pharmacol 77:307–315

Woodward RB, Cava MP, Ollis WD, Hunger A, Daeniker HU, Schenker K (1954) The total synthesis of strychnine. J Am Soc Chem 76:4749–4751

Wu SH, Oertel D (1990) Inhibitory circuitry in the ventral cochlear nucleus is probably mediated by glycine. J Neurosci 6:2691–2706

Yadid G, Youdim MB, Zinder O (1990) High-affinity strychnine binding to adrenal medulla chromaffin cell membranes. Eur J Pharmacol 175:365–366

Yaksh TL (1989) Behavioral and autonomic correlates of the tactile evoked allodynia produced by spinal glycine inhibition: effects of modulatory receptor systems and excitatory amino acid antagonists. Pain 37:111–123

Yaksh TL, Harty GJ (1988) Pharmacology of the allodynia in rats evoked by high dose intrathecal morphine. J Pharmacol Exp Ther 244:501–507

Young AB, Snyder SH (1973) Strychnine binding associated with glycine receptors of the central nervous system. Proc Natl Acad Sci USA 70:2832–2836

Young AB, Snyder SH (1974a) Strychnine binding in rat spinal cord membranes associated with the synaptic glycine receptor: cooperativity of glycine interactions. Mol Pharmacol 10:790–809

Young AB, Snyder SH (1974b) The glycine synaptic receptor: evidence that strychnine binding is associated with the ionic conductance mechanism. Proc Natl Acad Sci USA 71:4002–4005

Young AB, Zukin SR, Snyder SH (1974) Interaction of benzodiazepines with central nervous glycine receptors: possible mechanism of action. Proc Natl Acad Sci USA 71:2246–2250

Zarbin MA, Wamsley JK, Kuhar MJ (1981) Glycine receptor: light microscopic autoradiographic localization with [^{3}H] strychnine. J Neurosci 1:532–547

Zieglgänsberger W, Herz A (1971) Changes of cutaneous receptive fields of spinocervical tract neurones and other dorsal horn neurons by microelectrophoretically administered amino acids. Exp Brain Res 13:111–126

Zukin SR, Young AB, Snyder SH (1975) Development of the synaptic glycine receptor in chick embryo spinal cord. Brain Res 83:525–530

CHAPTER 16

Peptide Toxins Acting on the Nicotinic Acetylcholine Receptor*

F. Hucho

A. Introduction

Ever since Claude Bernard identified the neuromuscular junction as the site of action of the arrow poison curare (BERNARD 1857), neurotoxins have been at the focus of interest for researchers investigating the mechanisms of propagation and transmission of nerve impulses. Neurotoxins turned out to be very valuable tools in these investigations (MEBS and HUCHO 1990; HUCHO and OVCHINNIKOV 1983). They are currently used for elucidating the proteins of nerve and muscle membranes with atomic resolution.

I. Toxins as Tools in Neurochemical Research

The cholinergic synapse is among the most successful objects of application of this approach: Neurotoxins are used as tools for the identification, isolation, and structural analysis of key components involved in presynaptic transmitter release and postsynaptic transmitter reception. This and the following chapter focus on the postsynaptic membrane. For merely technical reasons, this treatise is subdivided into two parts, the first chapter summarizing polypeptide toxins and the second one concentrating on the low molecular weight toxins.

Nature apparently developed neurotoxins as an evolutionary advantage for defending or predatory organisms: A very small amount of substance can prevent an enemy from attacking or a prey from escaping if this substance is targeted at a few sites of central importance. Animal venoms contain a variety of toxic substances, but overall cytotoxicity, for example, would not be rapid enough as a means to exert an effect. What use would it be for a snake, if its prey escapes and then dies slowly? A neurotoxin can stop all vital functions (including escape mechanisms) immediately.

Therefore, an important characteristic of neurotoxic substances is their selectivity. No material is wasted if toxic molecules act specifically at sites of special importance. Selectivity and specificity apparently correlate with high affinity of binding of a toxic molecule to its molecular target.

* Work from the lab included in this review was supported by the Deutsche Forschungsgemeinschaft (SfB 312) and the Fonds der Chemischen Industrie.

For the organism this means that the toxic molecule sticks selectively to the spot at which it is supposed to act. For the researcher this means that a toxin can be used to guide him to a site of special importance.

A high-affinity toxin binding site can be assumed to play a central role in a cellular system. Because of this, toxins that bind with high affinity can be used to extract molecular components of such sites and, subsequently, to elucidate their structures and functional mechanisms.

II. Toxins vs. Antibodies

For the neurochemist, neurotoxins have several features in common with antibodies. The immune system can be harmful or even lethal to a nervous system – severe autoimmune diseases like myasthenia gravis are convincing examples (Fuchs 1979; Lindstrom and Dau 1980; Drachman 1987). Like neurotoxins, antibodies have been widely used as tools for investigating important molecules of the nervous system, ranging from receptors and ion channels to G proteins. Antibodies, especially monoclonal antibodies, are more versatile as tools than toxins are, because researchers can design and produce them according to their needs. Their affinity for their antigen, on the other hand, is often lower than that of neurotoxins. This fact limits their selectivity and may cause ambiguous results. A major advantage of neurotoxins as compared with antibodies lies in the fact that nature designed them to be directed against functional epitopes of the target molecule. A monoclonal antibody, on the other hand, is directed against an antigenic determinant which may be selected to be a functional site, but only with special effort.

B. Scope

This chapter's topic is the postsynaptic membrane of nicotinic cholinergic synapses as targets of peptide neurotoxins. As a matter of fact, the main and best-known target in this membrane is the nicotinic acetylcholine receptor (AChR). For the past 2 decades this protein has served as a model for several similar systems. I shall discuss it here under the two viewpoints outlined above: The receptor is described as a target for various selective neurotoxins, and the neurotoxins are described as tools for elucidating the structure and functional mechanism of a model receptor. I shall start with a brief review of the biochemistry of the AChR, focusing on the relevant aspects. For a more detailed account, the reader is referred to special review articles on receptor biochemistry (Changeux 1981, 1990; Maelicke 1984, 1988; Hucho 1986; Karlin 1991), immunology (Fuchs 1979; Lindstrom et al. 1988; Lindstrom 1980), electrophysiology (Sakmann and Neher 1984; Colquhoun 1986), developmental and molecular biology (Claudio 1989), regulation (Changeux 1990), pharmacology (Albuquerque et al. 1988; Taylor 1990a,b), toxicology (Mebs and Hucho 1990), and ultrastructure (Popot and Changeux 1984; Unwin et al. 1988; Kunath et al. 1989; Stroud et al. 1990).

C. The Biochemistry of the Acetylcholine Receptor

I. Assay and Isolation

CHANG and LEE (1963) were the first to show that a purified α-neurotoxin (see definition below) blocks synaptic transmission at the neuromuscular synapse of higher vertebrates, and LESTER et al. (1975) extended this finding to the cholinergic synapse at the electroplax of the electric ray. (Electric tissue of *Electrophorus* and *Torpedo* species was introduced previously into neurobiological research by NACHMANSOHN, as reviewed in his book published in 1959.) The electric tissue is extremely rich in cholinergic synapses and in connection with the blocking effect of α-bungarotoxin was extensively exploited in a pioneering work by Jean-Pierre Changeux and his coworkers, which led to the first purifications of the AChR from the electric tissue of the electric eel (*Electrophorus electricus*) (OLSEN et al. 1972; KARLIN and COWBURN 1973; LINDSTROM and PATRICK 1974; MEUNIER et al. 1974) and electric ray (*Torpedo marmorata*) (SCHMIDT and RAFTERY 1972; ELDEFRAWI and ELDEFRAWI 1973). The α-neurotoxin from the krait *Bungarus multicinctus*, α-bungarotoxin, was used in these studies, especially in its iodinated or tritiated form, as a ligand for a binding assay essential in determining receptor concentrations in various fractions of detergent-solubilized tissue. Using this binding assay (MEUNIER et al. 1972; later modified by other authors) and in combination with an appropriate affinity chromatography, it was shown for the first time that the AChR is a molecular entity distinct from, e.g., acetylcholinesterase (OLSEN et al. 1972). Furthermore, the toxin-binding assay was the basis for the purification of the receptor protein to homogeneity from electric fish and other sources.

I shall come back to the AChR as a target of α-neurotoxins below. I start by describing the receptor protein, primarily of the *Torpedo* receptor for which the greatest quantity of molecular data is available, extending the description as necessary to the homologues from other tissues, e.g., from muscle of higher vertebrates and from nervous tissue. This description will be largely biochemical. Biophysical, electron microscopical, developmental, molecular biological, and immunological aspects are touched upon only in passing. For further details the reader is referred to the many excellent and comprehensive reviews (e.g., those quoted above).

II. Structure

1. Primary and Quaternary Structures

The AChR from *Torpedo* electric tissue is a heteropentameric protein with an apparent M_r of 290 000. It is composed of two identical polypeptide chains (α) and three different chains (β, γ, δ), resulting in the quaternary structure $α_2βγδ$ (REYNOLDS and KARLIN 1978). The cDNAs of the precursor molecules have been cloned and sequenced (NODA et al. 1982, 1983a,b; CLAUDIO et al. 1983), yielding calculated relative molecular masses for the individual chains of α = 50 116 (437 amino acids; apparent M_r on SDS-PAGE 40 000); β = 53 681 (469; 50 000); γ = 56 279 (489; 60 000) and δ = 57 565 (501; 65 000). These calculated relative molecular masses do not include the various posttranslational modifications which account, in part, for the discrepancies between the calculated and the apparent values.

The AChR of higher vertebrate muscle has been shown to differ from the *Torpedo* receptor in having an ϵ-subunit instead of the γ-subunit (Takai et al. 1985; for sequences of fetal and adult rat AChR subunits, see Witzemann et al. 1990). Only the fetal AChR seems to be composed of the α-, β-, γ-, and δ-subunits which are also found in electric tissue. This difference between fetal and adult AChR seems to be correlated with some functional properties, e.g., with the kinetics and conductivity of the receptor channel (Mishina et al. 1986).

The primary structures of all receptor subunits from *Torpedo* and more than twenty other species was deduced from cloned and sequenced cDNAs (see Table 1). Obviously, the four different subunits are coded for by different genes. The great similarity of their amino acid sequences indicates that they evolved from a common ancestral gene (Hunkapiller et al. 1979). A similar homology is observed among the various subunits prevailing among AChRs of species as different as electric ray and man. Homology also exists among AChR from peripheral and CNS tissue, although the quaternary structure of the latter seems to be significantly different (Deneris et al. 1991): So far, only α- and β-subunits have been identified in brain receptors (see below).

2. Posttranslational Modifications

The main posttranslational modification contributing about 20 000 Dalton to the receptor's total molecular weight of about 290 000 Dalton is glycosylation. All receptor subunits are glycoproteins (Vandlen et al. 1979); one glycosylation site was localized in position αN141 (143 in consensus numbering) of the primary structure of the α-subunit. The structures of six major oligosaccharides were elucidated (Nomoto et al. 1986; Poulter et al. 1989); their function is unknown. During biosynthesis no assembly of functional AChR occurs when subunit glycosylation is inhibited (Fambrough 1979, 1983).

Another posttranslational modification is phosphorylation. The receptor protein from *Torpedo californica* contains 7–8 phosphate groups (Vandlen et al. 1979). Phosphorylation at γS353 and δS361 is catalyzed by protein kinase A (Huganir and Greengard 1983), at αS333, δS377, and δS362 by protein kinase C (Safran et al. 1987; Schröder et al. 1990, 1991), and at βY355, γY364, and δY372 by a tyrosine kinase (Huganir et al. 1984). Phosphorylation was implied in short-term regulatory events like receptor "desensitization" (Huganir and Greengard 1987) (see below) and long-term developmental events (Changeux 1981). Various observations including

Table 1. A Selection of the known primary structures of AChR α-subunits (squares indicate conserved residues). (Sequences taken from: Noda et al. 1982, 1983; Fornasari et al. 1990; Schoepfer et al. 1990)

```
                      Signal sequence                    1■   ■   ■   ■
Torpedo α1    MILCSYWHVGLVLLLLFSCC--GLVL-------------------G│SEHETRLVANLLENY   015
Human α1      M------EPWPLLLLLFSLCSAGLVL-------------------G│SEHETRLVAKLFKDY
Human α3      MAL----------------AVSLPLACRARLLLLLLLSLLPVARA │SEAEHRLFERLFEDY
Chick BgtBP1  MGLRA-----LMLWLLAAA--GLVRES----------------LQ│GEFQRKLYKELLKNY


                   ■■■                    ■                ■  ■ ■■    ■■
Torpedo α1    NKVIRPVEHHTHFVDITVGLQLIQLISVDEVNQIVETNVRLRQQWIDVRLRWNPADYGGI   075
Human α1      SSVVRPVEDHRQVVEVTVGLQLIQLINVDEVNQIVTTNVRLKQQWVDYNLKWNPDDYGGV
Human α3      NEIIRPVANVSDPVIIHFEVSMSQLVKVDEVNQIMETNLWLKQIWNDYKLKWNPSGYGGA
Chick BgtBP1  NPLERPVANDSQPLTVYFTLSLMQIMDVDEKNQVLTTNIWLQMYWTDHYLQWNVSEYPGV


                  ■   ■ ■■       ■      ■           ■■      ■■■■■  ■ ■
Torpedo α1    KKIRLPSDDVWLPDLVLYNNADGDFAIVHMTKLLLDYTGKIMWTPPAIFKSYCEIIVTHF   135
Human α1      KKIHIPSEKIWRPDLVLYNNADGDFAIVKFTKVLLQYTGHITWTPPAIFKSYCEIIVTHF
Human α3      EFMRVPAQKIWKPDIVLYNNAVGDFQVTTKTKALLKYTGEVTWIPPAIFKSSCKIDVTYF
Chick BgtBP1  KNVRFPDGLIWKPDILLYNSADERFDATFHTNVLVNSSGHCQYLPPGIFKSSCYIDVRWF


                ■■■■ ■ ■■■■                           ■■      ■
Torpedo α1    PFDQQNCTMKLGIWTYDGTKVSISPESDRPDLSTFMESGEWVMKDYRGWKHWVYYTCCPD   195
Human α1      PFDEQNCSMKLGTWTYDGSVVAINPESDQPDLSNFMESGEWVIKESRGWKHSVTYSCCPD
Human α3      PFDYQNCTMKFGSWSYDKAKIDLVLIGSSMNLKDYWESGEWAIIKAPGYKHDIKYNCCEE
Chick BgtBP1  PFDVQKCNLKFGSWTYGGWSLDLQWQEADISGYISN--GEWDLVGIPGKRTESFYECCKE


                ■ ■■      ■ ■     ■      ■                         ■
Torpedo α1    TPYLDITYHFIMQRIPLYFVVNVIIPCLLFSFLTGLVFYLPTDSGEKMTLSISVLLSLTV   255
Human α1      TPYLDITYHFVMQRLPLYFIVNVIIPCLLFSFLTGLVFYLPTDSGEKMTLSISVLLSLTV
Human α3      I-YPDITYSLYSRRLPLFYTINLIIPCLLISFLTVLVFYLPSDCGEKVTLCISVLLSLTV
Chick BgtBP1  -PYPDITFTVTMRRRTLYYGLNLLIPCVLISALALLVFLLPADSGEKISLGITVLLSLTV


                ■■ ■      ■■                  ■                        ■
Torpedo α1    FLLVIVELIPSTSSAVPLIGKYMLFTMIFVISSIIITVVVINTHHRSPSTHTMPQWVRKI   315
Human α1      FLLVIVELIPSTSSAVPLIGKYMLFTMVFVIASIIITVIVINTHHRSPSTHVMPNWVRKV
Human α3      FLLVITETIPSTSLVIPLIGEYLLFTMIFVTLSIVITVFVLNVHYRTPTTHTMPSWVKTV
Chick BgtBP1  FMLLVAEIMPATSDSVPLIAGYFASTMIIVGLSVVVTVIVLQYHHHDPDGGKMPKWTRVI


Torpedo α1    FIDTIPNVMFFSTMK---------------------------------------------   330
Human α1      FIDTIPNIMFFSTMK---------------------------------------------
Human α3      FLNLLPRVMFMTRPTSNEGNAQKPRPLYGAELSNLNCFSRAESKGCKEGYPCQDGMCGYC
Chick BgtBP1  LLNWCA---WFLRMK---------------------------------------------


                                 ■                                       ■
Torpedo α1    ----------------RASKEKQENKIFADDIDISDISGKQVTGEVIFQTPLIKNPDVKSA   375
Human α1      ----------------RPSREKQDKKIFTEDIDISDISGKPGPPPMGFHSPLIKHPEVKSA
Human α3      HHRRIKISNFSANLTRSSSSESVDAVVS----LSALS-----------------PEIKEA
Chick BgtBP1  ----------------RPGEDK------------------------------------VRPA


Torpedo α1    ------------------------------------------------------------
Human α1      ------------------------------------------------------------
Human α3      ------------------------------------------------------------
Chick BgtBP1  CQHKQRRCSLSSMEMNTVSGQQCSNGNMLYIGFGRGLDGVHCTPTTDSGVICGRMTCSPTE


                                  ■■          ■      ■■ ■■■ ■■ ■
Torpedo α1    --------------------IEGVKYIAEHMKSDEESSNAAEEWKYVAMVIDHILLCVFM   415
Human α1      --------------------IEGIKYIAETMKSDQESNNAAAEWKYVAMVMDHILLGVFM
Human α3      --------------------IQSVKYIAENMKAQNEAKEIQDDWKYVAMVIDRIFLWVFT
Chick BgtBP1  EENLLHSGHPSEGDPDLAKILEEVRYIANRFRDQDEEEAICNEWKFAASVVDRLCLMAFS


                  ■ ■
Torpedo α1    LICIIGTVSV-------------FAGRLIELSQEG   437 COOH
Human α1      LVCIIGTLAV-------------FAGRLIELNQQG
Human α3      LVCILGTAGL------FLQPLMAREDA
Chick BgtBP1  FVTIICTIGILMSAPNFVEAVSKDFA
```

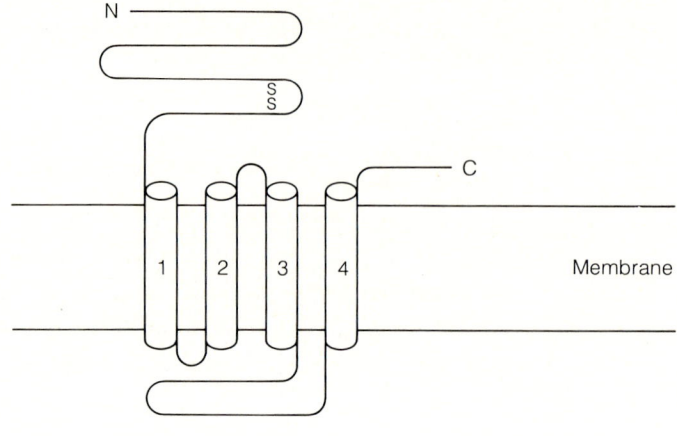

Fig. 1. Predicted transmembrane folding of ligand-gated ion channels like acetylcholine receptor (nAChR), glycine, and γ-aminobutyric acid (GABA_A) receptors

stability of the incorporated phosphate groups were interpreted as being in favor of the latter (Schröder et al. 1991).

A third type of posttranslational modification found in AChR from some species is esterification with fatty acids (shown to occur on the β-subunit). Again, nothing is known of the function of this feature.

3. Secondary Structure

The AChR is a member of the type-A receptor superfamily. Receptors of this class are so-called ligand-gated ion channels (Betz 1990; Stroud et al. 1990). Presumably, each of their subunits spans the membrane with four transmembrane helices (Claudio et al. 1983), although other models with three (Claudio 1989), five (Finer-Moore and Stroud 1984), or even seven (Criado et al. 1985) transmembrane sequences have been proposed. Contradictory models are mainly due to ambiguous immunological studies. The four-helix model as predicted from the occurrence of four hydrophobic stretches of about 20 amino acids each implies that both the N- and C-terminus of the receptor subunits are oriented toward the extracellular side of the postsynaptic membrane (Fig. 1).

Evaluation of the secondary structure by spectroscopic methods yielded 34% α- and 29% β-structure for the receptor solubilized in detergent (Moore et al. 1974). In the native membrane-bound protein the β-structure seems to predominate (NAUMANN et al. 1990). The true location of these secondary structure elements and the exact tertiary structure have to await X-ray analysis of the crystallized receptor (which is not yet available).

4. Neuronal Acetylcholine Receptors and Brain α-Bungarotoxin Binding Protein

With the primary structures and the corresponding cDNA of *Torpedo* and muscle AChR at hand, low stringency screening of cDNA libraries obtained from neuronal sources produced several further AChR subunits (DENERIS et al. 1991) expressed to different extents in tissue from various CNS areas and from peripheral ganglia. Six neuronal α-subunits ($\alpha_2-\alpha_7$) and three neuronal β-subunits ($\beta_2-\beta_4$) were identified to date. Entities corresponding to the γ-, ε-, and δ-subunits were not detected in central and peripheral neuronal tissue. All those sequences are defined as α-subunits which possess the two tandem cysteine residues corresponding to position 192/193 in the *Torpedo* homologue (according to this nomenclature *Torpedo* and muscle subunits are called α_1 and β_1). The principal functional distinction of the neuronal α-subunits is their inability to bind α-neurotoxins; however, see below, COUTURIER et al. (1990b), while the other pharmacological properties seem to be similar. Neuronal AChR binds and is blocked by κ-bungarotoxin (also called neuronal bungarotoxin or toxin 3.1) instead (CHIAPPINELLI 1983) (see below).

Finally, one further AChR species should be mentioned: α-bungarotoxin was shown to bind specifically to sites in the CNS, albeit without blocking cholinergic function. Two α-bungarotoxin binding proteins (α-BgBPα1 and α-BgBPα2) were identified and their sequences deduced from cloned cDNA (SCHOEPFER et al. 1990). They are similar to the other receptor α-subunits (Table 1), and α-BgBPα1 is identical to α_7 (COUTURIER et al. 1990a). Their function is unclear except for α_7; α_7, expressed in *Xenopus* oocytes, assembles into a homooligomeric channel responding to acetylcholine and nicotine and is blocked by α-bungarotoxin (COUTURIER et al. 1990a).

III. Function

As mentioned above, the AChR is a ligand-gated ion channel. This classification indicates a dual function of this membrane protein: It recognizes a specific molecular signal, i.e., the neurotransmitter acetylcholine, arriving at the extracellular side of the postsynaptic cell, and it converts this signal into an intracellular effect. The effect is a depolarization of the postsynaptic membrane caused by an influx of cations through an ion channel, whose properties are regulated by the transmitter-protein interaction.

The two α-chains contain the binding sites for the physiological transmitter acetylcholine and for a variety of effectors of the cholinergic synapse, including the α-neurotoxins (see below). The functions of the other chains

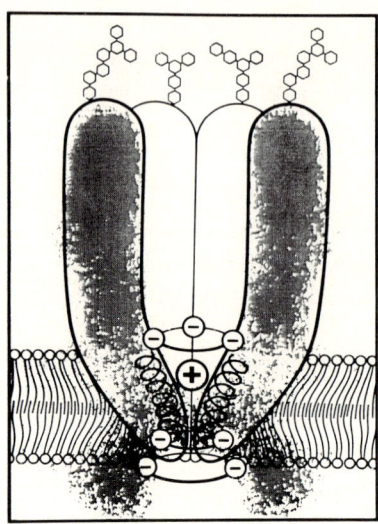

Fig. 2. Helix-M2 model of acetylcholine receptor and its ion channel composed from biochemical (Hucho et al. 1986; Guy and Hucho 1987), electron microscopical (Toyoshima and Unwin 1989; Kunath, et al. 1989), and electrophysiological (Imoto et al. 1988) data (Hucho and Hilgenfeld 1989). The *shaded areas* represent the image elucidated by electron microscopy

are not yet understood comprehensively. As concluded from reconstitution experiments with purified AChR, they form, together with the α-chains, the transmembrane cation channel regulated by the transmitter (Schiebler and Hucho 1978; Huganir et al. 1979; Popot et al. 1981).

The ion channel is the target of low molecular weight toxins and, furthermore, contains the binding site for a variety of noncompetitive antagonists (Changeux 1990) like chlorpromazine (Schofield et al. 1981; Heidmann et al. 1983), phencyclidine (Albuquerque et al. 1980), various local anesthetics (Grünhagen and Changeux 1976), and the organic cation triphenylmethylphosphonium (Lauffer and Hucho 1982). Its structure has been partially elucidated by a combination of chemical (Hucho et al. 1986; Hucho and Hilgenfeld 1989), electron microscopical (Toyoshima and Unwin 1988), and molecular biological (Imoto et al. 1986, 1988) methods. It appears to consist of a funnel (Fig. 2) formed by homologous helical sequences contributed by each of the receptor subunits: helix-M2 (helix-M2 model, Hucho et al. 1986; Hucho and Hilgenfeld 1989) and helix-M1 (DiPaola et al. 1990). Its extracellular opening is about 3 nm wide; this narrows down to about 1.15 nm at the local anesthetic binding site and to 0.65 nm at the narrowest part, which seems to represent the selectivity filter

Table 2. Lethal doses of some selected snake venom neurotoxins

	LD_{50} (mg/kg)
Short-chain neurotoxins	
Naja naja atra, cobrotoxin	0.09 subcutaneous
Naja nigricollis, a-toxin	0.036 subcutaneous
Laticauda semifasciata, erabutoxin b	0.15 intramuscular injection
Long-chain neurotoxins	
Naja naja siamensis, toxin 3	0.1 intravenous
Naja naja, toxin α	0.15 subcutaneous
Bungarus multicinctus, α-bungarotoxin	0.3 subcutaneous

(Endo and Tamiya 1987).

formed by a ring of negatively charged amino acid side chains (Imoto et al. 1988).

Opening and closing (gating) of the channel is induced by sequential binding of two ACh molecules (Prinz and Maelicke 1983a,b). Transmitter binding sites and the channel seem to interact as do allosteric proteins (Changeux et al. 1984). Channel opening times are on the order of milliseconds, during which about 10^4 cations permeate. Prolonged presence of transmitter causes receptor "desensitization", a state of high affinity for agonists and of inactivated (closed) channel. Desensitization is promoted among others by noncompetitive antagonists and calcium ions (reviewed in Changeux et al. 1984; Changeux 1990).

D. α-Neurotoxins from Snake Venoms

I. General Properties and Classification

Snake venoms of the families *Elapidae* and *Hydrophiidae* contain, among others, postsynaptically acting peptide toxins called α-neurotoxins (Endo and Tamiya 1987; Mebs and Hucho 1990). Because of their functional similarity to the arrow poison curare, they are called curare-mimetic. Like D-tubocurarine they block the binding site for the cholinergic transmitter ACh on the α-subunits of the receptor. Their primary target is the peripheral AChR of the neuromuscular endplate in striated muscle; snake venom neurotoxins blocking neuronal AChR have been described too (see below).

Some 10%–40% of a venom's dry weight may be neurotoxin. Venom from one specimen of the black mamba is sufficient to kill ten men (Dufton and Hider 1980). The toxicity of snake venom neurotoxins is 15–40 times higher than that of D-tubocurarine. LD_{50} values of 0.07–0.15 mg/kg have been observed (Table 2). Toxicity is grossly correlated with the affinity towards the receptor which ranges from K_d 10^{-11} M (α-bungarotoxin) to K_d 10^{-9} M

Table 3. Amino Acid Sequences of Neurotoxins and Neurotoxin Homologs. (1–45) Short Neurotoxins; (46–80) Long Neurotoxins; (81) κ-Bungarotoxin That Blocks Specifically the Neuronal Nicotinic AChR (17) and (18) are Identical with (16), (20) Identical with (19), and (34) and (35) Identical with (33)

Neurotoxin	Amino acid residue

```
                  ----+----1----+----2----+----3----+----4----+----5----+----6----+----7----+--
 1 Lc II     RRCYNQQSSQPKTTKSCPPGENSCYNKQWRD H  RGSITERGC  GCPTVKPGIKLRCCESEDCNN
 2 Lc c      RRCYNQQSSQPKTTKSCPPGENSCYNKQWRD H  RGSITERGC  GCPKVKPGIKLRCCESEDCNN
 3 Lc d      RRCFNQQSSQPKTTKSCPPGENSCYNKQWRD H  RGTIIERGC  GCPTVKPGIKLTCCQSEDCNN
 4 Ll a (Am)ᵃ RRCFNHPSSQPQTNKSCPPGENSCYNKQWRD H  RGTITERGC  GCPTVKPGIKLTCCQSEDCNN
 5 Ll a (NC)ᵃ RRCFNHPSSQPQTNKSCPPGENSCYNKQWRD H  RGTITERGC  GCPTVKPGIKLTCCQSEDCNN
 6 Ll b (Am)ᵃ RRCFNHPSSQPQTNKSCPPGENSCYNKQWRD H  RGTITERGC  GCPQVKSGIKLTCCQSDDCNN
 7 Ll c (Am)ᵃ RRCFNQQSSQPQTNKSCPPGENSCYNKQWRD H  RGTITERGC  GCPTVKPGVKLRCCQSEDCNN
 8 Lcr aᵃ    RRCFNHPSSQPQTNKSCPPGENSCYNKQWRD H  RGTITERGC  GCPTVKPGIKLTCCQSDDCNN
 9 Lcr bᵇ    RRCFNHPSSQPQTNKSCPPGENSCYNKQWRD H  RGTIIERGC  GCPQVKSGIKLTCCQSDDCNN
10 Lcr cᵃ    RRCFNQQSSQPQTNKSCPPGENSCYRKQWRD H  RGTIIERGC  GCPTVKPGIKLRCCQSEDCNN
11 Ea        RICFNHQSSQPQTTKTCSPGESSCYHKQWSD F  RGTIIERGC  GCPTVKPGIKLSCCESEVCNN
12 Eb        RICFNHQSSQPQTTKTCSPGESSCYHKQWSD F  RGTIIERGC  GCPTVKPGINLSCCESEVCNN
13 Ec        RICFNHQSSQPQTTKTCSPGESSCYHKQWSD F  RGTIIERGC  GCPTVKPGINLSCCESEVCNN
14 As a      MTCCNQQSSQPKTTTNCA GNSCYKKTWSD H   RGTIIERGC  GCPQVKKGIRLECCHTNECNN
15 Hc a      MTCCNQQSSQPKTTTNCA ESSCYKKTWSD H   RGTRIERGC  GCPQVKSGIKLECCHTNECNN
16 Hc b      MTCCNQQSSQPKTTTNCA ESSCYKKTWSD H   RGTRIERGC  GCPQVKSGIKLECCHTNECNN
17 Es 5
18 Ptx a
19 Es 4      MTCCNQQSSQPKTTTNCA ESSCYKKTWSD H   RGTRIERGC  GCPQVKPGIKLECCHTNECNN
20 Lh
21 Hl a      MTCCNQQSSQPKTTTNCA ESSCYKKTWRD F   RGTRIERGC  GCPQVKPGIKLECCHTNECNN
22 Al a      LTCCNQQSSQPKTTTDCA DNSCYKKTWQD H   RGTRIERGC  GCPQVKPGIKLECCKTNECNN
23 Al b      LTCCNQQSSQPKTTTDCA DNSCYKMTWRD H   RGTRIERGC  GCPQVKPGIKLECCKTNECNN
24 Al c      LTCCNQQSSQPKTTTDCA DNSCYKKTWKD H   RGTRIERGC  GCPHVKPGIKLTCCKTDECNN
25 Pa a      MTCCNQQSSQPKTTTICAGGESSCYKKTWRD H  RGSRTERGC  GCPRVKPGIRLICCKTDECNN
26 Au c      MQCCNQQSSQPKTTTCPGGVSSCYKKTWRD H   RGTIIERGC  GCPKVKQGIHLHCCQSDKCNN
27 Djk Vn-II RICYNHQSTTPATTKSC GENSCYKKTWSD H  RGTIIERGC  GCPKVKRGVHLHCCQSDKCNN
28 Dv 4.11.3 RICYNHQSTTPATTKSC GENSCYKKTWSD H  RGTIIERGC  GCPKVKPGVGIHCCQSDKCNY
29 Dpp α     RICYNHQSTTRATTKSC EENSCYKKTWRD H  RGTIIERGC  GCPSVKKGVKINCCTTDRCNN
30 Nm d      MECHNQQSSQPPTTKTCP GETNCYKKQWSD H  RGTIIERGC  GCPTVKPGIKLNCCTTDKCNN
31 Tx α      LECHNQQSSQRPTIKTCP GETNCYKKRWRD H  RGTIIERGC  GCPSVKKGVGIYCCKTDKCNR
32 Nnv β     MICHNQQSSQRPTIKTCP GETNCYKKRWRD H  RGTIIERGC  GCPSVKKGVGIYCCKTDKCNR
33 Nnv δ     LECHNQQSSQPPTTKTCP GETNCYKKRWRD H  RGSITERGC  GCPSVKKGIEINCCTTDKCNN
34 Nhh 6
35 Nha α
36 Nhh 10a   MICHNQQSSQPPTIKTCP GETNCYKKQWRD H  RGTIIERGC  GCPSVKKGVGIYCCKTDKCNR
37 Hh II     LECHNQQSSQPPTIKTCP GDTNCYNKRWRD H  RGTIIERGC  GCPTVKPGIKLKCCTTDRCNK
38 Hh IV     LECHNQQSSQTPTTQTCP GETNCYKKQWSD H  RGSRTERGC  GCPSVKNGIEINCCTTDRCNK
39 Cbt       LECHNQQSSQPTTTGCSGGETNCYKKWWSD H  RGTIIERGC  GCPKVKPGVKLNCCRTDRCNN
40 Nnp       LECHNQQSSQAPTTKTCS GETNCYKKWWSD H  RGTIIERGC  GCPKVKPGVNLNCCRTDRCNN
41 Nno II    LECHNQQSSEPPTTKTCS GETNCYKKWRD H   RGYRTERGC  GCPTVKKGIELNCCTTDRCNN
42 Nmm I     RICYNHQSTTPATTRCS GETNCYKKRWRD H  RGYRTERGC  GCPTVKKGIQLHCCTSDNCNN
43 Nmm II    LNCHNQMSAQPPTTTRCSRWETNCYKKRWRD H RGYKTERGC  GCPTVKKGIQLHCCTSDNCNN
44 Nha 10    MICYKQQSLQFPITTVCP GEKNCYKKQWSG H  RGTIIERGC  GCPSVKKGIEINCCTTDKCNR
45 Nha 14    MICHNQQSSQPPTIKTCP GETNCYKKWWRD H  RGTIIERGC  GCPSVKKGVGIYCCKTNKCNR
46 Lc a      RICY  LAP RDT QICAPGQEICYLKSWDDGTGFLKGNRLEFGCAATCPTVKPGIKLRCCESTDKCNPHPKLA
47 Lc b      RICY  LAP RDT QICAPGQEICYLKSWDDGTGSIRGNRLEFGCAATCPTVKRGIHIKCCSTDKCNPHPKLA
48 Ls III    RECY  LNP HDT QTCPSGQEICYVKSWCNAWCSSRGKVLEFGCAATCPSVNTGIEIKCCSADKCNTYP
49 As bᵇ     LSCY  LGY K HSQTCPPGENVCFVKTWCDGFCNTRGERIIMGCAATCPTAKSGVHIACCSTDNCNIYTKWGSGR
50 As cᵇ     LSCY  LGY K HSQTCPPGENVCFVKTWCDAFCSTRGERIVMGCAATCPTAKSYNEVKCCSTDNCNPFPVRPRRPP
51 Aa b      VICY  RGY NN PQTCPPGENVCFTRTWCDAFCSSRGKVVELGCAATCPIAKSYEDVTCCSTDNCNPFPVRPRHPP
52 Nss III4  LICY  MGP K TPRTCPRGQNLCYTKTWCDAFCSSRGKRVELGCAATCPKVKAGVGIKCCSTDNCNPFPVWNPRG
53 Dv I      RTCY  KTP SVKPETCPHGENICYTETWCDAWCSQRGKRVELGCAATCPKVKAGVGIKCCSTDNCNPFPVWNPR
54 Dv V      RICY  KTP SVKPETCPHGENICYTETWCDAWCSQRGKRELGCAATCPKVKAGVGIKCCSTDNCDPFPVKNPR
55 Dv 4.9.3  RTCY  KTP SVKPETCPHGENICYTETWCDAWCSQRGKREELGCAATCPKVKAGVGIKCCSTDNCDPFPVKNPR
56 Dv 4.7.3ᶜ RTCN  KTF SDQSKICPPGENICYTKTWCDAWCSQRGKIVELGCAATCPKVKAGVEIKCCSTDNCNLKFGKPR
57 Dpp Vn2   RTCN  KTF SDQSKICPPGENICYTKTWCDAWCSQRGKIVELGCAATCPKVKAGVEIKCCSTDDCDKFQFGKPR
58 Dpp γ     RTCN  KTF SDQSKICPPGENICYTKTWCDAWCSQRGKIVELGCAATCPKVKAGVEIKCCSTDNCNKFKFGKPR
59 Dpp δ     RTCN  KTF SDQSKICPPGENICYTKTWCDAWCSQRGKIVELGCAATCPKVKAGVEIKCCSTDDCKFQFGKPR
60 Djk Vn-III1 RTCY KTY SDKSKTCPRGENICYTKTWCDGFCSQRGKRVELGCAATCPKVKTGVEIKCCSTDYCNPFPVWNPR
61 Oh a      TKCY  VTP DVKSQTCPAGQDICYTETWCDAWCFSSRGKRVNLGCAATCPIVKPGVEIKCCSTDNCNPFPTWRKRP
62 Oh b      TKCY  VTP DATSQTCPDGQDICYTETWCDGFCSSRGKRIDLGCAATCPVKPGVDIKCCSTDNCNPFPTWRKRH
63 Oh 9      TKCY  VTP DVKSETCPAGQDLCYTDTWCVAWCTVAWCAAWCKRVSLGCAAICPIVPPKVSIKCCSTDACGPFPTWPNVR
64 Nha III   IRCF  ITP DVTSQACPDGQNICYTKTWCDNFCGMRGKRVDLGCAATCPTVKPGVDIKCCSTDNCNPFPTRKRS
65 Nhh 5     IRCF  ITP DVTSQACPDGH VCYTKTWCDNFCASRGKRVDLGCAATCPTVKPGVNIKCCSTDNCNPFPTRKRS
66 Nm b      IRCF  ITP DVTSQACPDGH VCYTKTWCDNFCASRGKRVDLGCAATCPTVKPGVNIKCCSRDNCNPFPTRKRS
67 Nm 3.9.4  KRCY  RTP DLKSQTCPPGEDLCYTKTWCDAWCTSRGKVIELGCVATCPKVKPYEQITCCSTDNCNPHPKMKP
68 Nhh D-VII IRCF  ITP DVTSQACPDGH VCYTKTWCDNFCGMRGKRVDLGCAATCPTVKPGVNIKCCSRDNCNPFPTRKRS
69 Nnv α     IRCF  ITP DITSKDCPNGH VCYTKTWCDAFCSIRGKRVDLGCAATCPTVRTGVDIQCCSTDNCNPFPTRKRP
70 α-Cbt     IRCF  ITP DITSKDCPNGH VCYTKTWCDGFCSIRGKRVDLGCAATCPTVRTGVDIQCCSTDNCNPFPTKKRP
71 Nnn 3     IRCF  ITP DITSKDCPNGH VCYTKTWCDGFCSIRGRVDLGCAATCPTVRTGVDIQCCSTDNCNPFPTRKRP
72 Nnn 4     IRCF  ITP DITSKDCPNGH VCYTKTWCDGFCSSRGKRVDLGCAATCPTVRTGVDIQCCSTDNCNPFPTRKRP
73 Nnn 3WP   IRCF  ITP DITSKDCPNGH VCYTKTWCDGFCSIRGKRVDLGCAATCPTVRTGVDIQCCSTDDCDPFPTRKRP
74 Tx A      IRCF  ITP DITSKDCPNGH VCYTKTWCDGFCSIRGKRVDLGCAATCPTVRTGVDIQCCSTDNCNPFPTRKRP
75 Tx B      IRCF  ITP DITSKDCPNGH VCYTKTWCDGFCSIRGKRVDLGCAATCPTVRTGVDIQCCSTDDCDPFPTRKRP
76 Tx C      IRCF  ITP DITSKDCPNGH VCYTKTWCDGFCSIRGKRVDLGCAATCPTVRTGVDIQCCSTDDCDPFPTRKRP
77 Tx D      IRCF  ITP DITSKDCPNGH VCYTKTWCDGFCRIRGKRVDLGCAATCPTVRTGVDIQCCSTDDCDPFPTRKRP
78 Tx E      IRCF  ITP DITSKDCPNGH VCYTKTWCDGFCSRGERVDLGCAATCPTVESYQDIKCCSTDNCNPHPKQKP
79 Nno I     ITCY  KTPIPITSETCAPGQNLCYTKTWCDAWCGSRGKVIELGCAATCPTVESYQDIKCCSTDNCNPHPKQKP
80 α-Bgt     IVCHTTAT IPSSAVTCPPGENLCYRKMWCDAFCSSRGKVVELGCAATCPSKKPYEEVTCCSTDKCNHPPKRQPG

                  ----+----1----+----2----+----3----+----4----+----5----+----6----+----7
81 κ-Bgtᵈ    RTCL  ISP SSTPQTCPNGQDICFLKAQCDKFCSIRGPVIEQGCVATCPQFRSNYRSLLCCTT DNCNH
```

ᵃ Mixture of two homologous toxins with Thr-41 or Ile-41. The sequence of the major fraction is shown in the table. ᵇ C-terminal is amidated. ᶜ Trp 29 is oxidized. ᵈ One amino acid residue is inserted between Cys 49 and Cys 60. ᵉ One amino acid residue is inserted between Cys 61 and Cys 66. ᶠ From Japan and Philippines. ᵍ From the Solomon Islands and Fiji. ʰ From New Caledonia. ⁱ From Amami Island, Kagoshima, Japan. ʲ From the Solomon Islands. ᵏ Quoted by Karlsson (1979).

Table 3 (continued)

	Species	Neurotoxin	Abbreviation	Reference
1	*Laticauda colubrina (J.P.)*[f]	Laticauda colubrina II	Lc II	Tamiya et al. (1983b)
2	*Laticauda colubrina (S.F.)*[g]	Laticauda colubrina c	Lc c	Tamiya et al. (1983b)
3	*Laticauda colubrina (NC)*[h]	Laticauda colubrina d	Lc d	Tamiya et al. (1983b)
4	*Laticauda laticaudata (Am)*[i]	Laticauda laticaudata a	Ll a (Am)	Tamiya et al. (1983b)
5	*Laticauda laticaudata (NC)*[h]	Laticauda laticaudata a	Ll a (NC)	Tamiya et al. (unpublished)
6	*Laticauda laticaudata (Am)*[i]	Laticauda laticaudata b	Ll b (Am)	Tamiya et al. (1983b)
7	*Laticauda laticaudata (Am)*[i]	Laticauda laticaudata c	Ll c (Am)	Tamiya et al. (unpublished)
8	*Laticauda crockeri*	Laticauda crockeri a	Lcr a	Tamiya et al. (1983b)
9	*Laticauda crockeri*	Laticauda crockeri b	Lcr b	Tamiya et al. (1983b)
10	*Laticauda crockeri*	Laticauda crockeri c	Lcr c	Tamiya et al. (unpublished)
11	*Laticauda semifasciata*	Erabutoxin a	Ea	Sato and Tamiya (1971)
12	*Laticauda semifasciata*	Erabutoxin b	Eb	Sato and Tamiya (1971)
13	*Laticauda semifasciata*	Erabutoxin c	Ec	Tamiya and Abe (1972)
14	*Astrotia stokesii*	Astrotia stokesii a	As a	Maeda and Tamiya (1978)
15	*Hydrophis cyanocinctus*	Hc Hydrophitoxin a	Hc a	Liu and Blackwell (1974)
16	*Hydrophis cyanocinctus*	Hc Hydrophitoxin b	Hc b	Liu and Blackwell (1974)
17	*Enhydrina schistosa*	Es Toxin 5	Es 5	Fryklund et al. (1972)
18	*Pelamis platurus*	Pelamitoxin a	Ptx a	Wang et al. (1976)
19	*Enhydrina schistosa*	Es Toxin 4	Es 4	Fryklund et al. (1972)
20	*Lapemis hardwickii*	Lh Neurotoxin	Lh	Fox et al. (1977)
21	*Hydrophis lapemoides*	Hydrophis lapemoides a	Hl a	Tamiya et al. (1983a)
22	*Aipysurus laevis*	Aipysurus laevis a	Al a	Maeda and Tamiya (1976)
23	*Aipysurus laevis*	Aipysurus laevis b	Al b	Maeda and Tamiya (1976)
24	*Aipysurus laevis*	Aipysurus laevis c	Al c	Maeda and Tamiya (1976)
25	*Pseudechis australis*	Pseudechis australis a	Pa c	Takasaki and Tamiya (unpublished)
26	*Acanthophis antarcticus*	Acanthophis antarcticus c	Aa c	Kim and Tamiya (1981b)
27	*Dendroaspis jamesoni kaimosae*	Djk Toxin Vn-II	Djk Vn-II	Strydom (1973a)
28	*Dendroaspis viridis*	Dv Toxin 4.11.3	Dv 4.11.3	Banks et al. (1974)
29	*Dendroaspis polylepis polylepis*	Dpp Toxin α	Dpp α	Strydom (1972)
30	*Naja melanoleuca*	Nm Toxin d	Nm d	Botes (1972)
31	*Naja nigricollis*	Nn Toxin α	Tx α	Eaker and Porath (1967)
32	*Naja nivea*	Nn Toxin β	Nnv β	Botes (1971)
33	*Naja nivea*	Nn Toxin δ	Nnv δ	Botes et al. (1971)
34	*Naja haje haje*	Nhh Toxin CM-6	Nhh 6	Joubert and Taljaard (1978)
35	*Naja haje annulifera*	Na Toxin α	Nha α	Botes and Strydom (1969)
36	*Naja haje haje*	Nhh Toxin CM-10a	Nhh 10a	Joubert and Taljaard (1978)
37	*Hemachatus haemachatus*	Hh Toxin II	Hh II	Strydom and Botes (1971)
38	*Hemachatus haemachatus*	Hh Toxin IV	Hh IV	Strydom and Botes (1971)
39	*Naja naja atra*	Nna Cobrotoxin	Cbt	Yang et al. (1969)
40	*Naja naja philippinensis*	Nnp Toxin	Nnp	Hauert et al. (1974)
41	*Naja naja oxiana*	Nno Neurotoxin II	Nno II	Grishin et al. (1973)
42	*Naja mossambica mossambica*	Nmm Neurotoxin I	Nmm I	Gregoire and Rochat (1977)
43	*Naja mossambica mossambica*	Nmm Neurotoxin III	Nmm III	Gregoire and Rochat (1977)
44	*Naja haje annulifera*	Nha CM-10	Nha 10	Joubert (1975)
45	*Naja haje annulifera*	Nha CM-14	Nha 14	Joubert (1975)
46	*Laticauda colubrina (S)*[i]	Laticauda colubrina a	Lc a	Kim and Tamiya (1982)
47	*Laticauda colubrina (J.P)*[f]	Laticauda colubrina b	Lc b	Kim and Tamiya (1982)
48	*Laticauda semifasciata*	Laticauda semifasciata III	Ls III	Maeda and Tamiya (1974)
49	*Astrotia stokesii*	Astrotia stokesii b	As b	Maeda and Tamiya (1978)
50	*Astrotia stokesii*	Astrotia stokesii c	As c	Maeda and Tamiya (1978)
51	*Acanthophis antarcticus*	Acanthophis antarcticus b	Aa b	Kim and Tamiya (1981a)
52	*Notechis scutatus scutatus*	Nss Toxin III-4	Nss III4	Halpert et al. (1979)
53	*Dendroaspis viridis*	Dv Toxin I	Dv I	Bechis et al. (1976)
54	*Dendroaspis viridis*	Dv Toxin V	Dv V	Bechis et al. (1976)
55	*Dendroaspis viridis*	Dv Toxin 4.9.3	Dv 4.9.3	Banks et al. (1974)
56	*Dendroaspis viridis*	Dv Toxin 4.7.3	Dv 4.7.3	Banks et al. (1974)
57	*Dendroaspis polylepis polylepis*	Dpp Toxin Vn2	Dpp Vn2	Strydom and Haylett (1977)
58	*Dendroaspis polylepis polylepis*	Dpp Toxin γ/Vn1	Dpp γ	Strydom (1973)
59	*Dendroaspis polylepis polylepis*	Dpp Toxin δ	Dpp δ	Strydom (1973b)
60	*Dendroaspis jamesoni kaimosae*	Djk Toxin Vn-IIII	Djk Vn-IIII	Strydom (1973a)
61	*Ophiophagus hannah*	Oh Toxin a	Oh a	Joubert (1973)
62	*Ophiophagus hannah*	Oh Toxin b	Oh b	Joubert (1973)
63	*Ophiophagus hannah*	Oh Toxin CM-9	Oh 9	Nan-Qin et al. (1984)
64	*Naja haje annulifera*	Nha Toxin III	Nha III	Kopeyan et al. (1975)
65	*Naja haje haje*	Nhh Toxin CM-5	Nhh 5	Joubert and Taljaard (1978)
66	*Naja melanoleuca*	Nm Toxin b	Nm b	Botes (1972)
67	*Naja melanoleuca*	Nm Toxin 3.9.4	Nm 3.9.4	Shipolini et al. (1974)
68	*Naja haje haje, Desert*	Nhh Desert Toxin VII	Nhh D-VII	Shimazu et al. (unpublished)
69	*Naja nivea*	Nh Toxin α	Nnv α	Botes (1971)
70	*Naja naja siamensis*	Nns 3/α-Cobratoxin	α-Cbt	Karlsson et al. (1972)
71	*Naja naja naja*	Nnn Toxin 3	Nnn 3	Arnberg et al. (unpublished)[k]
72	*Naja naja naja*	Nnn Toxin 4	Nnn 4	Arnberg et al. (unpublished)[k]
73	*Naja naja naja*	Nnn Toxin 3WP	Nnn 3WP	Ryden et al. (1973)
74	*Naja naja (Indian)*	Nn Toxin A	Tx A	Nakai et al. (1971)
75	*Naja naja (Indian)*	Nn Toxin B	Tx B	Ohta et al. (1976)
76	*Naja naja (Indian)*	Nn Toxin C	Tx C	Ohta et al. (1981a)
77	*Naja naja (Indian)*	Nn Toxin D	Tx D	Ohta et al. (1981b)
78	*Naja naja (Indian)*	Nn Toxin E	Tx E	Hayashi (private communication)
79	*Naja naja oxiana*	Nno Neurotoxin I	Nno I	Grishin et al. (1974)
80	*Bungarus multicinctus*	α-Bungarotoxin	α-Bgt	Mebs et al. (1972)
81	*Bungarus multicinctus*	κ-Bungarotoxin	κ-Bgt	Grant and Chiappinelli (1985)

for the cobra toxins. α-Neurotoxins were among the main tools used in isolating and characterizing receptors and ion channels in general and AChRs in particular (Hucho and Ovchinnikov 1983). They were used in binding assays for identifying, localizing, and quantifying AChR, for isolating the receptor, and for mapping its agonist and antagonist binding domains (see below).

Because of this manifold interest α-neurotoxins have been thoroughly investigated in the past 3 decades. The primary structures of more than 80 α-neurotoxins have been elucidated (see Table 3). The tertiary structures of several of them are known from X-ray analysis of the crystallized molecules and from MR spectroscopy in solution. The mechanism of their toxic action is being investigated and will be described in the foreseeable future with atomic resolution. One of the α-neurotoxins (neurotoxin II from *Naja naja oxiana*) has been produced by chemical synthesis (Deigin et al. 1982) and was shown to have full toxic activity. The structure of the gene coding for the short (see below) neurotoxin erabutoxin c was elucidated (Fuse et al. 1990); it contains two introns, three exons, and a promotor region (TATA box) 29–33 nucleotides upstream.

α-Neurotoxins may be classified into two groups of different molecular size. Short toxins are polypeptides of 60–62 amino acids (about 7000 relative molecular mass) with four disulfide bridges stabilizing their tertiary structures. Most long toxins contain 70–74 amino acids (M_r about 8000) and a fifth disulfide bridge. Reducing all of the disulfide bridges reversibly renders the toxins inactive (Yang 1965; Menez et al. 1980), although a residual binding affinity of $K_d = 3 \times 10^{-6} M$ was observed (Martin et al. 1983a).

II. Primary Structure

As seen in Table 3 the primary structures of short and long toxins are very similar. Allowing for a few deletions alignment of sequences is possible, indicating the conserved and mutated regions. Especially the cysteine residues involved in disulfide formation are conserved in all α-neurotoxins investigated. Some amino acid residues are present in both neurotoxins and the structurally related cardiotoxins. They may be involved in the stabilization of the tertiary structures. Others are conserved within the neurotoxin family only. They may be important for the specific neurotoxic function. These residues comprise among others (numbering in the aligned sequences): Lys 27, Trp 29, Asp 31, His/Phe/Trp 33, Arg/Lys 37, Gly 38, Glu/Asp 42, Val/Ala 52, and Lys/Arg 53. I will come back to some of these when discussing the functional domains of the neurotoxins.

A primary structure deduced from cloned and sequenced cDNA of *Laticauda semifasciata* indicates an N-terminal hydrophobic peptide which may serve as a signal peptide involved in the secretion of the toxin (Tamiya et al. 1985).

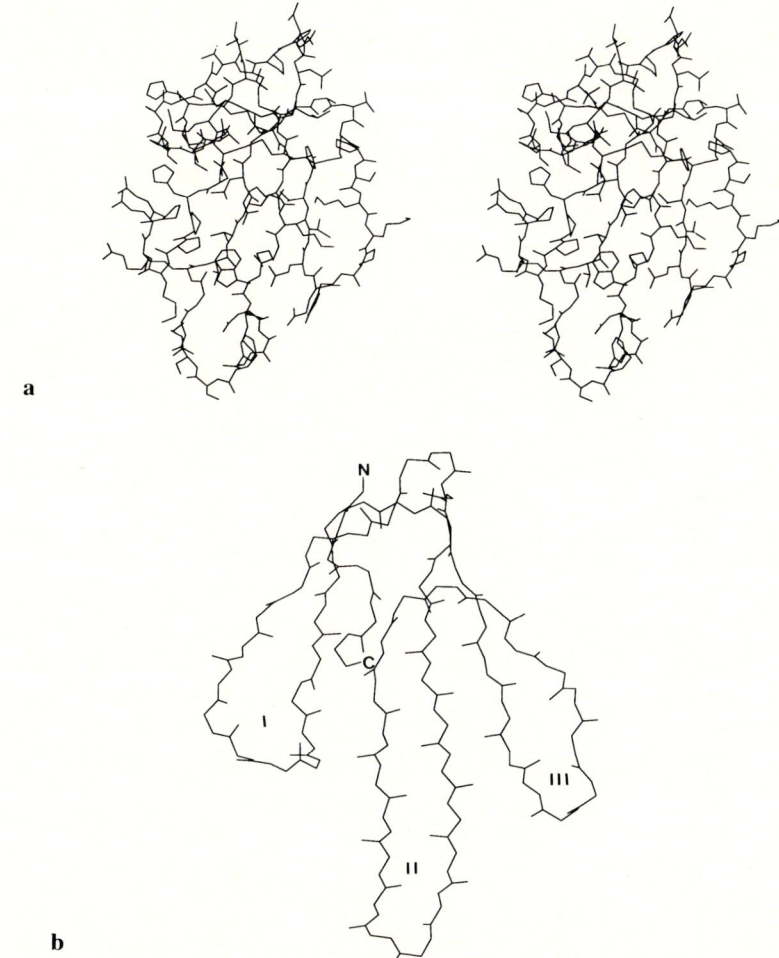

Fig. 3a,b. Crystal structure of erabutoxin b (a short α-neurotoxin) from *Laticauda semifasciata* (Low et al. 1976); **a** stereo view; **b** peptide backbone showing the three protruding loops (*I, II, III*)

III. Tertiary Structure

The first tertiary structure of an α-neurotoxin in the crystalline state was elucidated by X-ray analysis of crystals of the short neurotoxin erabutoxin b from the sea snake *Laticauda semifasciata* (Fig. 3; Low et al. 1976; TSERNOGLOU and PETSKO 1976). Crystal structures of the long neurotoxins α-cobrotoxin from the cobra *Naja naja siamensis* (Fig. 4; WALKINSHAW et al. 1980) and of α-bungarotoxin (Fig. 5) from the krait *Bungarus multicinctus* are also available (AGARD and STROUD 1982; LOVE and STROUD 1986). MR

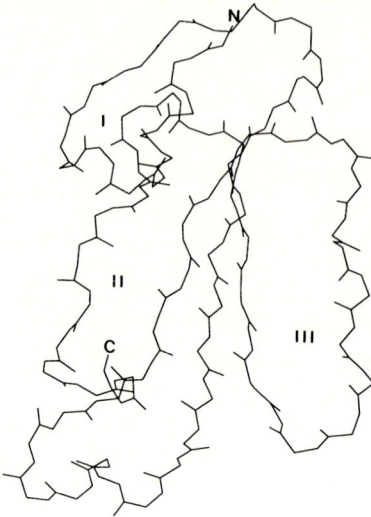

Fig. 4. Crystal structure of α-cobrotoxin (a long α-neurotoxin) from the cobra *Naja naja siamensis* (WALKINSHAW et al. 1980)

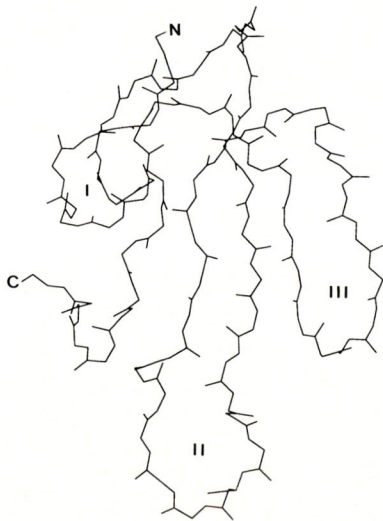

Fig. 5. Crystal structure of α-bungarotoxin (a long α-neurotoxin) from *Bungarus multicinctus* (LOVE and STROUD 1986)

spectroscopical investigations revealed that the gross structure of these molecules in solution is similar to the crystal structure (ENDO et al. 1981; HIDER et al. 1982; BYSTROV et al. 1983; YU et al. 1990). The following general features result from both types of conformational analysis.

The conservation of structure through evolution is even more conspicuous with the tertiary structures. They all consist of a core containing the four cystine bridges from which three loops protrude. This latter part of the molecule contains a characteristic three-stranded antiparallel β-sheet structure stabilized by hydrogen bonds. With the long toxins, the central loop contains the fifth disulfide bridge.

IV. Functional Domain

Knowing the tertiary structure of typical representatives of snake peptide toxins, one may ask: Which is the "lethal conformation" (DUFTON and HIDER 1980)? The tertiary structures were screened for features they have in common and which may be comparable to configurations in other well-known nonpeptide toxins as, for example, D-tubocurarine. A picture of the architecture of the domains interacting with the receptor is beginning to emerge, but it is far from complete. In particular, defining structural analogies in, e.g., D-tubocurarine and peptide neurotoxins (DUFTON and HIDER 1980) by comparing distances of polar moieties, charges, or hydrophobic groups on the basis of known crystal structures seems premature, because both the ligand (D-tubocurarine and neurotoxin; BYSTROV et al. 1983) and the receptor (MAELICKE et al. 1977; ENDO et al. 1986) are flexible and have been shown to undergo conformational changes upon binding. (However, GRÜNHAGEN and CHANGEUX reached different results in 1976.) Ideally, only a crystal structure obtained by X-ray analysis of a crystallized receptor-toxin complex would give results equally reliable as those achieved with, e.g., trypsin/trypsin inhibitor complexes. The following discussion therefore should be considered preliminary.

The overall structure of α-neurotoxin molecules is a flat, slightly bent disk, with the antigenic determinants located on its convex rear surface. On the concave side, several invariant amino acid residues such as Trp 29, Asp 31, and Arg 37 (aligned sequences) point to the same direction and may be involved directly in binding to the AChR (Fig. 6).

Several methods have been applied in attempting to determine which residues of the toxin are involved in binding to the receptor. One already discussed is comparing primary structures and spotting conserved residues. Such a comparison suggests that loops II and III are part of the receptor binding domain. The invariant "toxic" residues all are located near the tip of these loops and may interact by entering a cleft on the surface of the receptor. The positively charged side chains may bind to the binding site for the cholinium moiety of the transmitter ACh, and burying the hydrophobic side chain of Trp 29 in a hydrophobic environment within the receptor cleft may contribute to the binding force.

Interestingly, sequence comparison has shown that parts of the glycoprotein of the neurotropic rabies virus has considerable homology to the entire sequence of long neurotoxins (LENTZ et al. 1982). This may point to a possible target of this virus.

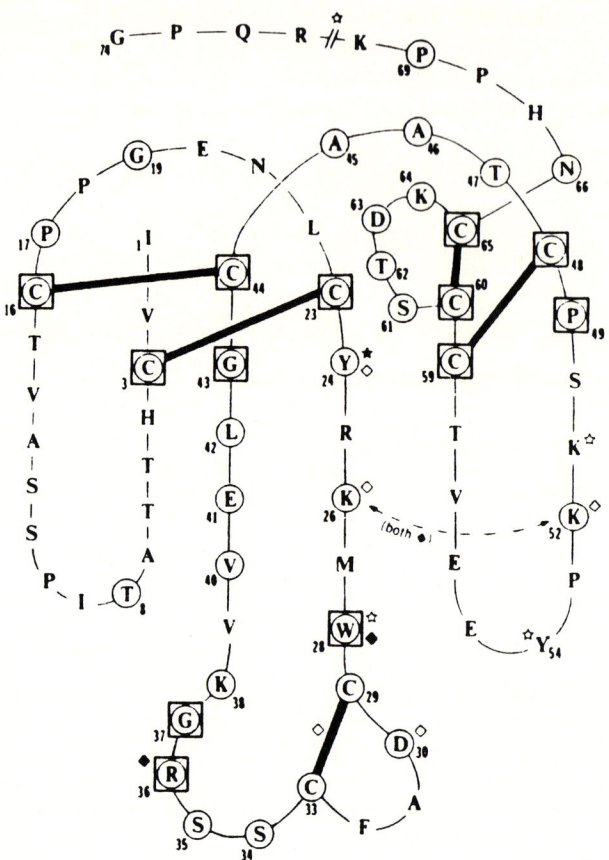

Fig. 6. Schematic representation of the bungarotoxin resolution structure. Residues which are highly conserved among long neurotoxins are *circled*, those invariant among all neurotoxins are enclosed by a *square*. A summary of consequences of chemical modification is indicated by the symbols: *stars* appear next to residues which have been modified in bungarotoxin; *diamonds* next to those modified in other toxins. A *solid symbol* means that the modification led to a large loss of toxicity (more than 90%) or binding affinity (increase of dissociation constant by at least 100 times). An *open symbol* indicates little effect on toxicity or affinity. (From Love and Stroud 1986)

Chemical modification experiments have made clear that there is no single amino acid residue essential for toxin binding (Karlsson 1979). Rather, a multipoint attachment has to be assumed. Even oxidation of the invariant Trp 29 does not render the toxin inactive (Chang et al. 1990). Moreover, the effect of such modification varies in magnitude with different toxins. The oxidation product N-formylkynurenine-α-bungarotoxin is almost as toxic as the native peptide, while with ˙cobrotoxin the same modification leads to considerable reduction in lethal toxicity (Chang et al. 1990).

From spin labeling and fluorescence labeling experiments (TSETLIN et al. 1979, 1982; ROUSSELET et al. 1984) it was concluded that in the case of the short toxin from *Naja naja oxiana* among others Lys 26 and Lys 46 (Lys 27 and Lys 53 in the aligned sequences), as predicted, may be in contact with the receptor.

Finally, ATASSI et al. (1988) proposed mapping the binding surface of the toxin by synthesizing partial sequences of, for example, α-bungarotoxin. Using this approach it was confirmed that there are three binding regions, comprising the N-terminal sequence, the tip of loop II, and part of loop III (see also JUILLERAT et al. 1982; MARTIN et al. 1983a,b).

V. Binding Domain on the Acetylcholine Receptor

1. Quaternary Structure

By a similar approach, the elucidation of the receptor domains which are in contact with the neurotoxins has been attempted. As mentioned above, the toxin binding site is not identical but overlaps with the binding sites for the low molecular weight agonists and competitive antagonists and is located primarily on the two α-subunits (POPOT and CHANGEUX 1984). Isolated α-subunits in SDS (HAGGERTY and FROEHNER 1981; TZARTOS and CHANGEUX 1983a,b, 1984) and protein blotted by transferring the subunits separated by SDS polyacrylamide gel electrophoresis to a foil (GERSHONI et al. 1983) have been shown to bind α-bungarotoxin.

Energy transfer measurements with two differently labeled neurotoxins have shown that the distance of the two toxin binding sites within one receptor molecule is larger than that between neighboring receptor molecules (JOHNSON et al. 1984). This means that the toxin-binding sites are located at the receptor's periphery rather than in its center. The distance between the two toxin molecules bound to the same receptor complex was estimated to be 67 Å.

By similar energy transfer measurements it was calculated that the toxin binding sites are 22–40 Å apart (BOLGER et al. 1984) from the binding site for noncompetitive inhibitors which are assumed to bind within the ion channel (NEHER and STEINBACH 1978; CHANGEUX 1990).

Covalent labeling of AChR with photoactive derivatives of α-neurotoxins presented proof for the localization of the binding site on the α-subunits of the AChR. However, the same type of photoaffinity labeling yielded also evidence for the close proximity of other subunits to the α-neurotoxin binding sites. Nitroazidophenyl derivatives of α-bungarotoxin were shown to cross-link upon irradiation to the α-subunit of the receptor; the same derivatives of the toxin from cobra venom surprisingly reacted with the δ-subunit as well (HUCHO 1979). Similar observations were presented by WITZEMANN et al. (1979), who found varying labeling patterns with toxin-receptor cross-links of different chain lengths. With a 3.3-nm cross-linker, label was found exclusively in the δ-subunit. More precise were investigations by TSETLIN and his coworkers (1983, 1984, 1987) who prepared photoaffinity derivatives of the short toxin of *Naja naja oxiana* with the cross-linking photoactive group in defined positions of the toxin primary structure. With the cross-linker in position Lys 46, most of the covalent labeling occurred in the α-subunit; with the Lys 26 analogue predominant labeling of the γ- and δ-subunit was observed (KREIENKAMP et al. unpublished observation). Interestingly, significantly different labeling patterns were obtained with AChR in the membrane-bound state or detergent-solubilized state (TSETLIN et al. 1987). More recently, a detailed mapping of the receptor surface with photoactive cobra toxin derivatives was attempted (CHATRENET et al. 1990). With α-neurotoxin from *Naja nigricollis*, carrying a photoactive group in positions Lys 47, Lys 15, and Lys 51, all four subunits of the receptor were photo-labeled under certain conditions.

From all these investigations it is obvious that the receptor's surface area binding peptide neurotoxins is rather extended and that other subunits besides the α-subunit are in contact with the curare-mimetic α-neurotoxins. The same seems to be true for D-tubocurarine (PEDERSEN and COHEN 1988, 1990), which can be photo-cross-linked to α, γ, and δ. One should point out that the two α-subunits of the AChR are not equivalent because of the different environment within the quaternary structure (KARLIN 1980). The nonequivalence is evidenced, e.g., by two binding sites for D-tubocurarine, having significantly different affinities (NEUBIG and COHEN 1979) and different susceptibility to affinity labeling by [^{3}H]-MBTA [4-(*N*-maleimido)benzyltrimethyl ammonium; DAMLE and KARLIN 1978]. It may have consequences for the fine structure of the toxin-receptor interaction, which may vary at the two binding sites. Monoclonal antibodies have been produced which block only half of the toxin binding sites (WATTERS and MAELICKE 1983; DOWDING and HALL 1987).

2. Primary Structure

Photolabeling, to date, using photoactive toxins indicated the role of the receptor's quaternary structure but not yet the amino acid sequences involved in peptide toxin binding. The binding site for small cholinergic ligands has been partially elucidated by affinity labeling experiments. Cys 192–Cys 193 was identified as part of this domain with the help of MBTA (KAO et al. 1984; KAO and KARLIN 1986) and Tyr 190 with lopho-toxin (ABRAMSON et al. 1988, 1989); the diazonium compound p-(N,N-dimethyl)amino benzene diazonium fluoroborate (DDF) (DENNIS et al. 1988; GALZI et al. 1990) specified Tyr 93, Trp 149, and again Tyr 190, Cys 192, and Cys 193 as being part of the agonist/antagonist binding pocket located on the α-subunit.

As expected for a competitive antagonist, mostly the same domains seem to be involved in α-neurotoxin binding. First of all, several attempts have been made to map the α-neurotoxin binding site with antisera or monoclonal antibodies raised against synthetic peptides comprising partial sequences of the receptor α-subunit (MULAC-JERICEVIC and ATASSI 1986; MAELICKE et al. 1989; CONTI-TRONCONI et al. 1990; ATASSI 1991). Two sequences were considered: (a) peptides including the two disulfide-bridge forming cysteine residues 192–193 (WILSON et al. 1984; NEUMANN et al. 1986) and (b) sequences including Cys 128 and Cys 142 (McCORMICK and ATASSI 1984; CRIADO et al. 1986). While considerable controversy exists as to the involvement of the latter (SMART et al. 1984; MULAC-JERICEVIC et al. 1987; COCKCROFT et al. 1990 on the one side, and CRIADO et al. 1986; GRIESMANN et al. 1990 on the other side), there is general agreement that the α-subunit sequence in the range of α185–196 participates in toxin binding and that possibly other sequences are involved as well, the toxin binding site being composed of a continuous surface formed by discontinuous receptor sequences (ATASSI 1991).

Epitope mapping with antibodies was supplemented by other methods. Binding studies investigating the affinity of synthetic receptor peptides towards α-bungarotoxin showed that sequences around Cys 192–Cys 193 possess a high affinity for the toxin (see below). As an example, a K_d of $3 \times 10^{-7} M$ was observed for the 17 amino acid sequence αW184–αD200 of *Torpedo* receptor (ARONSHEIM et al. 1988). This is almost four orders of magnitude lower than for α-bungarotoxin binding to the native *Torpedo* receptor, but it is much too high to be called 'non-specific'. Again, several authors found other sequences having comparable affinities, but this chapter is not the place to describe in too much detail the experimental discrepancies.

In an elegant investigation, the characterization of the hypothetical ligand-binding epitope was performed by measuring the interaction of genetically engineered receptor peptides with various receptor ligands by selective T1 MR relaxation (FRAENKEL et al. 1990a,b). This technique allows precise measurements of interactions in the mM K_d range. The results were in agreement with the known pharmacology of the receptor, albeit affinities of the engineered product were shifted to much higher values as compared with native receptor.

Energy transfer calculations exploiting the intrinsic fluorescence of tryptophan residues of native α-bungarotoxin and of the synthetic dodecamer α185–196 showed that these residues must be located at a distance of about 12 Å (PEARCE and HAWROT 1990), yielding another piece of information as to the architecture of the toxin binding site.

ATASSI and coworkers (RUAN et al. 1990) carried the mapping of the receptor-toxin interface one step further. A binding assay was designed allowing the measurement of the affinities between three synthetic peptides representing the three loops of the neurotoxin (see above) and the various receptor peptides under consideration. Based on the known three-dimensional structure of the toxin, the receptor peptides showing significant affinity were assembled by computer modeling into a model of the receptor's

Table 4. Comparison of sequences around the tandem cysteine residues of acetylcholine receptor (AChR) α-chains from various organisms (10 residues each upstream and downstream of C-C)

Muscle/electric organ AChR:

	182	204	
Torpedo cal. α	RGWK HWVYYT CCPDT PYLDITYH		Noda et al. (1982)
Bovine muscle α	RGWK HWVF YACCPS TPYLDITYH		Noda et al. (1983)
Human muscle α	RGWK HS VTYS CCPDTPYLDITYH		Noda et al. (1983c)
Mouse muscle $α_1$	RGWK HWVF YS CCPT TPYLDITYH		Isenberg et al. (1986)
Chicken $α_1$	RGWK HWVYYACCPDTPYLDITYH		Nef et al. (1988)
Xenopus α	RCWK HWVYYT CCPDKPYLDITYH		Baldwin et al. (1988)
Naja naja/Natte	RGF WHS VNYS CCLDTPYLDITYH		Neumann et al. (1989)

Neuronal AChR:

Rat $α_2$	T GTYNS KKYDCCA EI	YP DVTYY	Wada et al. (1988)
Rat $α_3$	P GYKHEI KYNCCE EI	YQDI TYS	Boulter et al. (1986)
Rat $α_4$	V GTYNT RKYECCA EI	YP DI TYA	Goldman et al. (1987)
Rat $α_5$	181-MGS KGNRTDS CCW	YP YI TYS -200	Boulter et al. (1990)
Chicken $α_2$	189-I GRYNS KKYDCCT EI	YP DI TF Y-210	Nef et al. (1988)
Chicken $α_3$	181-P GYKHDI KYNCCE EI	YT DI TYS -202	Nef et al. (1988)
Chicken $α_4$	187-V GNYNS KKYECCT EI	YP DI TYS -208	Nef et al. (1988)
Chicken $α_5$	152-T GS KGNRTDGCCW	YP F VTYS -171	Couturier et al. (1990b)
Locust $αL_1$	191-P AERHE KYYP CCA EP	YP DI FFN-212	Marshall et al. (1990)
Droso SAD	P AERHE KYYP CCA EP	YP DI FFN	Sawruk et al. (1990)
Droso ALS	191-P AVRNE KFI S CCE EP	YL DI VFN-212	Bossy et al. (1988)
Human $α_3$	P GYKHDI KYNCCE EI	YP DI TYS	Fornasari et al. (1990)
Chicken $αBgtBPα_1$	180-P GKRTE S F YECCK EP	YP DI TFT-201	Schoepfer et al. (1990)
Chicken $αBgtBPα_2$	180-G GKRNEL YYECCK EP	YP DVTYT-201	Schoepfer et al. (1990)
Goldfish $α_3$	P GYKHDI KYNCCE EI	YP DI TYS	Hieber et al. (1990)

toxin binding pocket. Since some of the receptor peptides were not found by others to bind specifically to α-neurotoxins this model needs confirmation. The solid-phase binding assay used in these studies should be considered for similar investigations.

From the description presented so far it is obvious that the sequence around αCys 192−Cys 193 is thought by most investigators to be part of the receptor domain binding cholinergic ligands including α-neurotoxins. Initially identified by blotting proteolytic α-subunit fragments which bound radio-active α-bungarotoxin (Wilson et al. 1984, 1985), the amino acid sequence involved was narrowed down to residues α185−196 (*Torpedo*) with the help of synthetic peptides and antibodies raised against them (Neumann et al. 1986). The slightly longer sequence α184-WKHWVYYTCCPDTPYLDα200 binds [^{125}I]-α-bungarotoxin with K_d of $3 \times 10^{-7}\,M$ (Aronsheim et al. 1988; Pearce and Hawrot 1990). This affinity drops only slightly (to $K_d = 4.7 \times 10^{-7}\,M$) with the shorter peptide α186−198. Removing further amino acids from the C- or N-terminus diminishes the affinity to a large extent (Fraenkel et al. 1990).

An evolutionary survey (Table 4) shows that the respective sequence is well conserved among AChR from *Torpedo*, *Xenopus*, chick, mouse, calf, and human. This might be taken as an argument in favor of the importance of this sequence for receptor function. An argument against it should be seen in

Table 5. K_d values (mM) for ligands binding to recombinant peptides representing the putative toxin binding sequence surrounding the tandem cysteine residues of the receptor α-subunit

Sequence	AChR	Nicotine	TC	GA	BTX
Torpedo	1.9	3.6	0.24	0.15	6.3×10^{-5}
Chick	1.9	3.8	0.17	0.11	1.5×10^{-4}
Xenopus	2.1	3.1	0.18	0.21	5.4×10^{-4}
Drosophila	3.9	12.0	0.46	0.39	1.7×10^{-3}
Mouse	3.3	3.25	0.34	0.21	3.2×10^{-3}
Calf	2.5	3.4	0.40	0.32	6.2×10^{-3}
Human	2.7	2.5	0.12	0.11	6.5×10^{-3}

(FRAENKEL et al. 1990a,b).
AChR, acetylcholine receptor; TC, D-tubocurarine; GA, Gallamine; BTX, α-bungarotoxin.

the observation (Table 5) that the affinity of the human sequence for α-bungarotoxin is insignificant ($K_d = 6.5 \times 10^{-3} M$; FRAENKEL et al. 1990a,b), and affinities of the mouse and calf peptides are not much better.

It is tempting to extract from the many known sequences a "minimum toxin-binding sequence." Such an attempt should be supported by including AChR α-subunit sequences of a receptor which does not bind α-neurotoxins, as for example the neuronal subunits α2, α3, α4, and the α-subunits of AChR from poisonous and nonpoisonous snakes which seem to be protected against α-neurotoxin poisoning (NEUMANN et al. 1989). Furthermore, one should include the two α-bungarotoxin binding subunits from brain recently cloned and sequenced (SCHOEPFER et al. 1990) which seem to represent no classic AChR. As mentioned above, the putative ligand binding sequence can be easily identified by the vicinal cysteine residues in the consensus position 192/193. Four major substitutions around these might be responsible for the inability of the snake receptor to bind snake-peptide neurotoxins: As compared with the *Torpedo* sequence, Trp 184 became Phe 184, Lys 185 is Trp 185 in the snakes, Trp 187 is Ser 187, and Pro 194 mutated to Leu 194.

If one continues these sequence comparisons, no obvious pattern strikes the eye. Not even the tandem cysteine residues are necessary for toxin binding. Mutants obtained by site-directed mutagenesis in which Cys 192 or Cys 193 was converted to serine showed when expressed in *Xenopus* oocytes no response to Ach, had little affinity for small ligands, but bound α-bungarotoxin almost normally (MISHINA et al. 1985).

E. Other Peptide Neurotoxins

Of the other peptide neurotoxins binding to the AChR, I have already mentioned κ-bungarotoxin (neuronal bungarotoxin, toxin 3.1) which is specific for AChR from peripheral and CNS neuronal tissue. Its primary

Table 6. Amino acid sequences of α-conotoxins

GI	ECCNPACGRHYSC[a]
GIA	ECCNPACGRHYSCGK[a]
GII	ECCHPACGKHFSC[a]
MI	GRCCHPACGKNYSC[a]

[a] C-terminus is amidated.
(MEBS and HUCHO 1990).

structure shows considerable homology to the long α-neurotoxins (see Table 3).

A series of curare-mimetic peptide neurotoxins have been isolated (CRUZ et al. 1978; McINTOSH et al. 1982) from *Conus geographus* and *Conus magus* (Table 6). These α-conotoxins are peptides with 13–15 amino acids containing two disulfide bridges (GRAY et al. 1983) and an amidated C-terminus. They do not block neuronal AChR, while at muscle AChR they compete with D-tubocurarine and α-bungarotoxin (CRUZ et al. 1985; OLIVERA et al. 1985). Their structure-activity relationships are discussed elsewhere (HIDER 1985; HASHIMOTO et al. 1985).

F. Conclusions and Outlook

From this discussion it is obvious that we do not have a clear picture of the α-neurotoxin binding site and thus of the mechanism of action of snake neurotoxins on the AChR. In the far future, X-ray analysis of crystalline receptor protein, perhaps at least of the hydrophilic extracellular portion or of the ligand-binding domain, will answer most of our questions. More realistic for the forseeable future are hopes for a model of the architecture of the binding pocket emerging from the various techniques outlined in this chapter, ranging from spectroscopic methods to the mapping of the receptor surface with antibodies or toxin derivatives.

Acknowledgements. I wish to thank Dr. Wolfram Saenger for providing the X-ray structures, Dr. Christoph Weise for his help with aligning the sequences, and Mary Wurm for preparing the manuscript.

References

Abramson SN, Culver P, Klines T (1988) Lophotoxin and related coral toxins covalently label the α-subunit of the nicotinic acetylcholine receptor. J Biol Chem 263:18.568–18.573
Abramson SN, Li Y, Culver P, Taylor P (1989) An analog of lophotoxin reacts covalently with Tyr 190 in the α-subunit of the nicotinic acetylcholine receptor. J Biol Chem 253:12666–12672
Agard DA, Stroud RM (1982) α-Bungarotoxin structure revealed by a rapid method for averaging electron density of noncrystallographically translationally related molecules. Acta Crystallogr [A]38:186–194
Albuquerque EX, Adler M, Spivak CE, Aguayo LG (1980) Mechanism of nicotinic channel activation and blockade. Ann NY Acad Sci 358:204–238

Albuquerque EX, Daly JW, Warnick JE (1988) Macromolecular sites for specific neurotoxins and drugs on chemosensitive synapses and electrical excitation in biological membranes. In: Narahashi T (ed) Jon channels, p 95

Aronsheim A, Eshel Y, Mosckowitz R, Gershoni JM (1988) Characterization of the binding of α-bungarotoxin to bacterially expressed cholinergic binding sites. J Biol Chem 263:9933–9937

Atassi MZ (1991) Postsynaptic neurotoxin-acetylcholine receptor interaction and the binding sites on the two molecules. In: Tu A (ed) Reptile and amphibian venoms. Dekker, New York, pp 53–83 (Handbook of natural toxins, vol 5)

Atassi MZ, McDaniel CS, Manshouri T (1988) Mapping by synthetic peptides of the binding sites for acetylcholine receptor on α-bungarotoxin. J Protein Chem 7:655–666

Baldwin TJ, Yoshihara CM, Blackmer, Kikiner CR, Burden SJ (1988) Regulation of acetylcholine receptor transcript expression during development in *Xenopus laevis* J Cell Biol 106:469–478

Banks BEC, Miledi R, Shipolini RA (1974) The primary sequences and neuromuscular effects of three neurotoxic polypeptides from the venom of *Dendroaspis viridis*. Eur J Biochem 45:457–468

Bechis G, Granier C, van Rietschoten J, Jover E, Rochat H, Miranda F (1976) Purification of six neurotoxins from the venom of *Dendroaspis viridis*: primary structure of two long toxins. Eur J Biochem 68:445–456

Bernard MC (1857) Leçon sur les effets des substances toxiques et médicamenteuses. Baillière, Paris, pp 238–306

Betz H (1990) Ligand-gated ion channels in the brain: the amino acid receptor superfamily. Review. Neuron 5:383–392

Bolger MB, Dionne V, Chrivia J, Johnson DA, Taylor P (1984) Interaction of a fluorescent acyldicholine with the nicotinic acetylcholine receptor and acetylcholinesterase. Mol Pharmacol 26:57–69

Bossy B, Ballivet M, Spierer P (1988) Conservation of neural nicotinic acetylcholine receptor from *Drosophila* to vertebrate central nervous systems. EMBO J 7:611–618

Botes DP (1971) The amino acid sequences of toxin α and β from *Naja nivea* venom and the disulfide bonds of toxin α. J Biol Chem 246:7383–7391

Botes DP (1972) Snake venom toxins: the amino acid sequences of toxins b and d from *Naja melanoleuca* venom. J Biol Chem 247:2866–2871

Botes DP, Strydom DJ (1969) A neurotoxin, toxin α, from Egyptian cobra (*Naja naja haje*) venom: purification, properties, and complete amino acid sequence. J Biol Chem 244:4147–4157

Botes DP, Strydom DJ, Anderson CG, Christensen PA (1971) Snake venom toxins: purification and properties of three toxins from *Naja nicea* (linnaeus) (Cape cobra) venom and the amino acid sequence of toxin α. J Biol Chem 246:3132–3139

Boulter J, Evans K, Goldman D, Martin G, Heinemann S Patrick J (1986) Isolation of a cDNA clone coding for a possible neuronal nicotinic acetylcholine receptor alpha-subunit. Nature 319:368–374

Boulter J, O'Shea-Greenfield A, Duvoisin R, Connolly JG, Wada E, Jensen A, Gardner PD, Ballivet M, Deneris ES, McKinnon D, Heinemann S, Patrick J (1990) α3, α5 and β4: three members of the rat neuronal nicotinic acetylcholine receptor-related gene family form a gene cluster. J Biol Chem 265:4472–4482

Bystrov VF, Tsetlin VI, Karlsson E, Pashkov VS, Utkin Y, Kondakov VI, Pluzhnikov KA, Arseniev AS, Ivanov VT, Ovchinnikov YA (1983) Magnetic resonance evaluation of snake neurotoxin structure-function relationship. In: Hucho F, Ovchinnikov YA (eds) Toxins as tools in neurochemistry. de Gruyter, Berlin, p 193

Chang CC, Lee CY (1963) Isolation of neurotoxins from the venom of *Bungarus multicinctus* and their modes of neuromuscular blocking action. Arch Int Pharmacodyn 144:316–332

Chang CC, Kawata Y, Sakiyama F, Hayashi K (1990) The role of an invariant tryptophan residue in α-bungarotoxin and cobrotoxin – investigation of active derivatives with the invariant tryptophan replaced by kynurenine. Eur J Biochem 193:567–572

Changeux JP (1965) Sur les propriétés allostériques de la L-thréonine désaminases de biosynthèse. VI. Discussion générale. Bull Soc Chim Biol 47:281–300

Changeux JP (1981) The acetylcholine receptor: an "allosteric" membrane protein. Harvey Lect 75(85):254

Changeux JP (1990) Functional architecture and dynamics of the nicotinic acetylcholine receptor: an allosteric ligand-gated ion channel. In: Fidia research foundation neuroscience award lectures, vol 4. Raven, New York, pp 21–168

Changeux JP, Devillers-Thiéry A, Chemouilli P (1984) Acetylcholine receptor: an allosteric protein. Science 225:1335–1345

Chatrenet B, Trémeau O, Bontems F, Goeldner MP, Hirth CG, Ménez A (1990) Topography of toxin-acetylcholine receptor complexes by using photoactivatable toxin derivatives. Proc Natl Acad Sci USA 87:3378–3382

Chiappinelli VA (1983) Kappa toxin: a probe for neuronal nicotinic receptor in the avian ciliary ganglion. Brain Res 277:9–21

Claudio T (1989) Molecular genetics of acetylcholine receptor-channels. In: Glover DM, Hames BC (eds) Frontiers in molecular biology: molecular neurobiology volume. IRL, Oxford, pp 63–142

Claudio T (1991) cAMP stimulation of acetylcholine receptor expression is mediated through posttranslational mechanisms. Proc Natl Acad Sci USA 88:854–858

Claudio T, Ballivet M, Patrick J, Heinemann S (1983) Nucleotide and deduced amino acid sequences of *Torpedo californica* acetylcholine receptor gamma-subunit. Proc Natl Acad Sci USA 80:1111–1115

Cockcroft VB, Osguthorpe DJ, Barnard EA, Lunt GG (1990) Modeling of agonist binding to the ligand-gated ion channel superfamily of receptors. Proteins Struct Funct Genet 8:386–397

Colquhoun D (1986) On the principles of postsynaptic action of neuromuscular blocking agents. In: Kharkevich DA (ed) Handbook of experimental pharmacology, vol 79. Springer, Berlin Heidelberg New York, pp 59–113

Conti-Tronconi BM, Tang F, Diethelm BM, Spencer SR, Reinhardt-Maelicke S, Maelicke A (1990) Mapping of a cholinergic binding site by means of synthetic peptides, monoclonal antibodies and α-bungarotoxin. Biochemistry 29:6221–6230

Couturier S, Bertrand D, Matter J-M, Hernandez M-C, Bertrand S, Millar N, Valera S, Barkas T, Ballivet M (1990a) A neuronal nicotinic acetyclholine receptor subunit (α7) is developmentally regulated and forms a homo-oligomeric channel blocked by α-BTX. Neuron 5:847–856

Couturier S, Erkman L, Valera S, Rungger D, Bertrand S, Boulter J, Ballivet M, Bertrand D (1990b) α5, α3 and non-α3 – three clustered avian genes encoding neuronal nicotinic acetylcholine receptor-related subunits. J Biol Chem 265:17560–17567

Criado M, Hochschwender S, Sarin V, Fox JL, Lindstrom J (1985) Evidence for unpredicted transmembrane domains in acetylcholine receptor subunits. Proc Natl Acad Sci USA 82:2004–2008

Criado M, Sarin V, Fox JL, Lindstrom J (1986) Evidence that the acetylcholine binding site is not formed by the sequence alpha 127–143 of the acetyclholine receptor. Biochemistry 25:2839–2846

Cruz LJ, Gray WR, Olivera BM (1978) Purification and properties of a myotoxin from *Conus geographus* venom. Arch Biochem Biophys 190:539

Cruz LJ, Gray WR, Yoshikami E, Olivera BM (1985) Conus venoms: a rich source of neuroactive peptides. J Toxicol, Toxin Rev 4:107

Damle VN, Karlin A (1978) Affinity labeling of one of two α-neurotoxin binding sites in acetylcholine receptor from *Torpedo californica*. Biochemistry 18:2039–2045

Deigin V, Ulyashin V, Mikhaleva N, Ivanov V (1982) Total synthesis of neurotoxin II from the Central Asian cobra (*Naja naja oxiana*) venom. In: Blaha K, Malon P (eds) Peptides 1982. de Gruyter, Berlin, p 276

Deneris ES, Connolly J, Rogers SW, Duvoisin R (1991) Pharmacological and functional diversity of neuronal nicotinic acetylcholine receptors. TIPS 12: 34–40

Dennis M, Giraudat J, Kotzyba-Hibert F, Goeldner M, Hirth C, Chang JY, Lazure C, Chretién M, Changeux JP (1988) Amino acids of the *Torpedo marmorata* acetylcholine receptor α-subunit labeled by a photoaffinity ligand for the acetycholine binding site. Biochemistry 27:2346–2357

DiPaola M, Czajkowski D, Karlin A (1989) The sidedness of the COOH terminus of the acetylcholine receptor δ subunit. J Biol Chem 264:15457–15463

DiPaola M, Kao PN, Karlin A (1990) Mapping the α-subunit photolabeled by the noncompetitive inhibitor [^{3}H]quinacrine azide in the active state of the nicotinic acetylcholine receptor. J Biol Chem 265:11017–11029

Dowding AJ, Hall ZW (1987) Monoclonal antibodies specific for each of the two toxin-binding sites of *Torpedo* acetylcholine receptor. Biochemistry 26:6372–6381

Drachman DB (ed) (1987) Myasthenia gravis: biology and treatment. Ann NY Acad Sci 505

Dufton MJ, Hider RC (1980) Lethal protein conformations. TIBS 5:52–56

Eaker D, Porath J (1967) The amino acid sequence of a neurotoxin from *Naja nigricollis* venom. Jpn J Microbiol 11:353–355

Eldefrawi ME, Eldefrawi AT (1973) Purification and molecular properties of the acetylcholine receptor from *Torpedo* electroplax. Arch Biochem Biophys 159:362

Endo T, Tamiya N (1987) Current view on the structure-function relationship of postsynaptic neurotoxins from snake venoms. Pharmacol Ther 34:403–451

Endo T, Inagaki F, Hayashi K, Miyazawa T (1981) Local conformational transition of toxin B from *Naja naja* as studied by nuclear magnetic resonance and circular dichroism. Eur J Biochem 122:541–547

Endo T, Nakanishi M, Furukawa S, Joubert FJ, Tamiya N, Hayashi K (1986) Stopped-flow fluorescence studies on binding kinetics of neurotoxins with acetylcholine receptor. Biochemistry 25:395–404

Fambrough DM (1979) Control of acetylcholine receptors in skeletal muscle. Physiol Rev 59:165–227

Fambrough DM (1983) Biosynthesis and intracellular transport of acetylcholine receptors. Methods Enzymol 96:331–352

Finer-Moore J, Stroud RM (1984) Amphipathic analysis and possible formation of the ion channel in an acetylcholine receptor. Proc Natl Acad Sci USA 81:155–159

Fornasari D, Chini B, Tarroni P, Clementi F (1990) Molecular cloning of human neuronal nicotinic receptor α3-subunit. Neurosci Lett 111:351–356

Fox JW, Elzinga M, Tu AT (1977) Amino acid sequence of a snake neurotoxin from the venom of *Lapemis hardwickii* and the detection of a sulfhydryl group by laser Raman spectroscopy. FEBS Lett 80:217–220

Fraenkel Y, Navon G, Aronheim A, Gershoni JM (1990a) Direct measurement of agonist binding to genetically engineered peptides of the acetylcholine receptor by selective T_1 NMR relaxation. Biochemistry 29:2617–2622

Fraenkel Y, Ohana B, Mosckovitz R, Gershoni J, Navon G (1990b) NMR studies of specific binding of acetylcholine, nicotine, gallamine and D-tubocurare to recombinant active site peptides of the nAChR. In: Gershoni J, Hucho F, Silman I (eds) International symposium: The cholinergic synapse, Berlin

Fryklund L, Eaker D, Karlsson E (1972) Amino acid sequences of the two principal neurotoxins of *Enhydrina schistosa* venom. Biochemistry 11:4633–4640

Fuchs S (1979) Immunological analysis of acetylcholine receptor. Adv Cytopharmacol 3:279–286

Fuse N, Tsuchiya T, Nonomura Y, Menez A, Tamiya T (1990) Structure of the snake short-chain neurotoxin, erabutoxin c-precursor gene. Eur J Biochem 193:629–633

Galzi JL, Revah F, Black D, Goeldner M, Hirth C, Changeux JP (1990) Identification of a novel amino acid α Tyr 93 within the active site of the acetylcholine receptor by photoaffinity labeling: additional evidence for a three-loop model of the acetylcholine binding site. J Biol Chem 265:10430–10437

Gershoni JM, Hawrot E, Lentz TL (1983) Binding of α-bungarotoxin to isolated α-subunit of the acetylcholine receptor of *Torpedo californica*: quantitative analysis with protein blots. Proc Natl Acad Sci USA 80:4973–4977

Goldman D, Deneris E, Luyten W, Kochhar A, Patrick J, Heinemann S (1987) Members of a nicotinic acetylcholine receptor gene family are expressed in different regions of the mammalian central nervous system. Cell 48:965–973

Grant GA, Chiappinelli VA (1985) K-bungarotoxin: complete amino acid sequence of a neuronal nicotinic receptor probe. Biochemistry 24:1532–1537

Gray WR, Luque A, Olivera BM, Barrett J, Cruz LJ (1983) Conotoxin MI: disulfide bonding and conformational states. J Biol Chem 258:12247–12251

Green WN, Ross AF, Claudio T (1991) cAMP stimulation of acetylcholine receptor expression is mediated through posttranslational mechanisms. Proc Natl Acad Sci USA 88:854–858

Gregoire J, Rochat H (1977) Amino acid sequences of neurotoxin I and III of the elapidae snake *Naja mossambica mossambica*. Eur J Biochem 80:283–293

Griesmann GE, McCormick DJ, de Aizpurua HJ, Lennon VA (1990) α-Bungarotoxin binds to human acetylcholine receptor α-subunit peptide 185–199 in solution and solid phase but not to peptides 125–147 and 389–409. J Neurochem 54:1541–1547

Grishin EV, Sukhikh, AP, Lukyanchuk NN, Slobodyan LN, Lipkin VM, Ovchinnikov YA, Sorokin VM (1973) Amino acid sequence of neurotoxin II from *Naja naja oxiana* venom. FEBS Lett 36:77–78

Grishin EV, Sukhikh AP, Slobodyan LN, Ovchinnikov YA, Sorokin VM (1974) Amino acid sequence of neurotoxin I from *Naja naja oxiana* venom. FEBS Lett 45:118–121

Grünhagen HH, Changeux JP (1976) Studies on the electrogenic action of acetylcholine with *Torpedo marmorata* electric organ. Quinacrine: a fluorescent probe for the conformational transitions of the cholinergic receptor protein in its membrane bound state. J Mol Biol 106:497–516

Guy HR, Hucho F (1987) The ion channel of the nicotinic acetylcholine receptor. TINS 10:318–321

Haggerty JG, Froehner SC (1981) Restoration of ^{125}I-α-bungarotoxin binding activity to the α-subunit of *Torpedo* acetylcholine receptor isolated by gel electrophoresis in sodium dodecyl sulfate. J Biol Chem 256:8294–8297

Halpert J, Fohlman J, Eaker D (1979) Amino acid sequence of a postsynaptic neurotoxin from the venom of the Australian tiger snake *Notechis scutatus scutatus*. Biochimie 61:719–723

Hashimoto K, Uchida S, Yoshida H, Nishiuchi Y, Sakakibara S, Yukari K (1985) Structure-activity relations of conotoxins at the neuromuscular junction. Eur J Pharmacol 118:351

Hauert J, Maire M, Sussmann A, Bargetzi JP (1974) The major lethal neurotoxin of the venom of *Naja naja philippinensis*. Int J Pept Protein Res 6:201–222

Heidmann T, Oswald RE, Changeux JP (1983) Multiple sites of action for non-competitive blockers on acetylcholine receptor-rich membrane fragments from *Torpedo marmorata*. Biochemistry 22:3112–3127

Hider RC (1985) A proposal for the structure of conotoxin: a potent antagonist of the nicotinic acetylcholine receptor. FEBS Lett 184:181–184

Hider RC, Drake AF, Inagaki F, Williams RJP, Endo T, Miyazawa T (1982) Molecular conformation of α-cobratoxin as studied by nuclear magnetic resonance and circular dichroism. J Mol Biol 158:275–291

Hieber V, Bouchey J, Agranoff B, Goldman D (1990) Nucleotide and deduced amino acid sequence of the goldfish neural nicotinic acetylcholine receptor α-3 subunit. Nuclcic Acids Res 18:5293

Hucho F (1979) Photoaffinity derivatives of α-bungarotoxin and α-*Naja naja siamensis* toxin. FEBS Lett 103:27–32

Hucho F (1986) The nicotinic acetylcholine receptor and its ion channel. Review. Eur J Biochem 158:211–226

Hucho F, Hilgenfeld R (1989) The selectivity filter of a ligand-gated ion channel – the helix-M2 model of the ion channel of the nicotinic acetylcholine receptor. FEBS Lett 257:17–23

Hucho F, Ovchinnikov YA (1983) Toxins as tools in neurochemistry. de Gruyter, Berlin

Hucho F, Oberthür W, Lottspeich F (1986) The ion channel of the nicotinic acetylcholine receptor is formed by the homologous helices M II of the receptor subunits. FEBS Lett 205:137–142

Huganir RL, Greengard P (1983) cAMP-dependent protein kinase phosphorylates the nicotinic acetylcholine receptor. Proc Natl Acad Sci USA 80:1130–1134

Huganir RL, Greengard P (1987) Regulation of receptor function by protein phosphorylation. TIPS 8:472–477

Huganir RL, Schell MA, Racker E (1979) Reconstitution of the purified acetylcholine receptor from *Torpedo californica*. FEBS Lett 108:155–160

Huganir RL, Miles K, Greengard P (1984) Phosphorylation of the nicotinic acetylcholine receptor by an endogenous tyrosine-specific protein kinase. Proc Natl Acad Sci USA 81:6968–6972

Hunkapiller MW, Strader CD, Hood L, Raftery MA (1979) Amino terminal amino acid sequence of the major polypeptide subunit of *Torpedo californica* acetylcholine receptor. Biochem Biophys Res Commun 91:164–169

Imoto K, Methfessel C, Sakmann B, Mishina M, Konno T, Nakai J, Bujo H, Fujita Y, Numa S (1986) Location of a delta-subunit region determining transport through the acetylcholine receptor channel. Nature 324:670–674

Imoto K, Busch C, Sakmann B, Mishina M, Konno T, Nakai J, Bujo H, Mori Y, Fukuda K, Numa S (1988) Rings of negatively-charged amino acids determine the acetylcholine receptor channel conductance. Nature 335:645–648

Isenberg KE, Mudd J, Shah V, Merlie JP (1986) Nucleotide sequence of the mouse muscle nicotinic acetylcholine receptor α subunit. Nucleic Acids Res 14:5111

Johnson DA, Voet JG, Taylor P (1984) Fluorescence energy transfer between cobra α-toxin molecules bound to the acetylcholine receptor. J Biol Chem 259:5717–5725

Joubert FJ (1973) The amino acid sequences of two toxins from *Ophiophagus hannah* (King cobra) venom. Biochim Biophys Acta 317:85–98

Joubert FJ (1975) The amino acid sequences of three toxins (CM-10, CM-12, and CM-14) from *Naja haje annulifera* (Egyptian cobra) venom. Hoppe Seylers Z Physiol Chem 356:53–72

Joubert FJ, Taljaard N (1978) Purification, some properties and the primary structures of three reduced and S-carboxymethylated toxins (CM-5, CM–6 and CM-10a) from *Naja haje haje* (Egyptian cobra) venom. Biochim Biophys Acta 579:1–8

Juillerat MA, Schwendimann B, Hauert J, Fulpius BW, Bargetzi JP (1982) Specific binding to isolated acetylcholine receptor of a synthetic peptide duplicating the sequence of the presumed active center of a lethal toxin from snake venom. J Biol Chem 257:2901–2907

Kao PN, Karlin A (1986) Acetylcholine receptor binding site contains a disulfide cross-link between adjacent half-cystinyl residues. J Biol Chem 261:8085–8088

Kao PN, Dwork AJ, Kaldany RRJ, Silver ML, Wideman J, Stein S, Karlin A (1984) Identification of the alpha-subunit half-cystine specifically labeled by an affinity reagent for the acetylcholine receptor binding site. J Biol Chem 259:11662–11665

Karlin A (1980) Molecular properties of nicotinic acetylcholine receptors. Cell Surf Rev 6:191–260

Karlin A (1991) Explorations of the nicotinic acetylcholine receptor. Harvey Lect 86

Karlin A, Cowburn DA (1973) The affinity labelling of partially purified acetylcholine receptor from electric tissue of *Electrophorus*. Proc Natl Acad Sci USA 70:3636–3640

Karlsson E (1979) Chemistry of protein toxins in snake venoms. In: Lee CY (ed) Snake venoms. Springer, Berlin Heidelberg New York, pp 159–212 (Handbook of experimental pharmacology, vol 2)

Karlsson E, Eaker DL, Ponterius G (1972) Modification of amino groups in *Naja naja* neurotoxin and the preparation of radioactive derivatives. Biochim Biophys Acta 257:235–248

Kim HS, Tamiya N (1981a) Isolation, properties and amino acid sequence of a long-chain neurotoxin, *Acanthophis antarcticus b*, from the venom of an Australian snake (the common-death-adder, *Acanthophis antarcticus*). Biochem J 193:899–906

Kim HS, Tamiya N (1981b) The amino acid sequence and position of the free thiol group of a short-chain neurotoxin from common-death-adder (*Acanthophis antarcticus*) venom. Biochem J 199:211–219

Kim HS, Tamiya N (1982) Amino acid sequences of two novel long-chain neurotoxins from the venom of the sea snake *Laticauda colubrina*. Biochem J 207:215–223

Kopeyan C, Miranda F, Rochat H (1975) Amino-acid sequence of toxin III of *Naja haje*. Eur J Biochem 58:117–122

Kunath W, Giersig M, Hucho F (1989) The electron microscopy of the nicotinic acetylcholine receptor. Electron Microsc Rev 2:349–366

Lauffer L, Hucho F (1982) Triphenylmethylphosphonium is an ion channel ligand of the nicotinic acetylcholine receptor. Proc Natl Acad Sci USA 79:2406–2409

Lentz TL, Burrage TG, Smith AL, Crick J, Tignor GH (1982) Is the acetylcholine receptor a rabies virus receptor? Science 215:182–184

Lester H, Changeux JP, Sheridan RE (1975) Conductance increases produced by bath application of cholinergic agonists to *Electrophorus* electroplaques. J Gen Physiol 65:797–816

Lindstrom J (1980) Probing nicotinic acetylcholine receptors with monoclonal antibodies. TINS 9:401–407

Lindstrom J, Shelton D, Fujii Y (1988) Myasthenia gravis. Adv Immunol 42:233–284

Lindstrom J, Patrick J (1974) Purification of the acetylcholine receptor by affinity chromatography. In: Bennett MVL (ed) Synaptic transmission and neuronal interaction. Raven, New York, pp 191–216

Liu CS, Blackwell RQ (1974) Hydrophitoxin b from *Hydrophis cyanocinctus* venom. Toxicon 12:543–546

Love RA, Stroud RM (1986) The crystal structure of α-bungarotoxin at 2.5 Å resolution: relation to solution structure and binding to acetylcholine receptor. Protein Eng 1:37–46

Low BW, Preston HS, Sato A, Rosen LS, Searl JE, Rudko AD, Richardson JR (1976) Three-dimensional structure of erabutoxin b neurotoxic protein: inhibitor of acetylcholine receptor. Proc Natl Acad Sci USA 73:2991–2994

Maeda N, Tamiya N (1974) The primary structure of the toxin *Laticauda semifasciata* III, a weak and reversibly acting neurotoxin from the venom of a sea snake, *Laticauda semifasciata*. Biochem J 141:389–400

Maeda N, Tamiya N (1976) Isolation, properties and amino acid sequences of three neurotoxins from the venom of a sea snake. *Aipysurus laevis*. Biochem J 153:79–87

Naeda N, Tamiya N (1978) Three neurotoxins from the venom of a sea snake, *Astrotia stokesii*, including two long-chain neurotoxic proteins with amidated C-terminal. Biochem J 175:507–517

Maelicke A (1984) Biochemical aspects of cholinergic excitation. Angew Chem Int Ed Engl 23:195–221

Maelicke A (1988) Structure and function of the nicotinic acetylcholine receptor. In: Whittaker VP (ed) Handbook of experimental pharmacology, vol 86. Springer, Berlin Heidelberg New York, pp 267–313

Maelicke A, Fulpius BW, Klett RP, Reich E (1977) Acetylcholine receptor: responses to drug binding. J Biol Chem 252:4811–4830

Maelicke A, Plümer-Wilk R, Fels G, Spencer SR, Engelhard M, Veltel D, Conti-Tronconi BM (1989) The limited sequence specificity of anti-peptide antibodies may introduce ambiguity in topological studies. In: Maelicke A (ed) Molecular biology of neuroreceptors and ion channels. Springer, Berlin Heidelberg New York, pp 321–326 (NATO ASI series, vol 32)

Marshall J, Buckingham SD, Shingia R, Lunt GG, Goosey MW, Darlison MG, Sattelle DB, Barnard EA (1990) Sequence and functional expression of a single α-subunit of an insect nAChR. EMBO J 9:4391–4398

Martin BM, Chibber BA, Maelicke A (1983a) The sites of neurotoxicity in α-cobratoxin. J Biol Chem 258:8714–8722

Martin BM, Chibber BA, Maelicke A (1983b) Toxic peptides obtained by enzymatic cleavage of α-cobratoxin. Toxicon [Suppl]3:273–276

McCormick DJ, Atassi MZ (1984) Localization and synthesis of the acetylcholine-binding site in the alpha-chain of the Torpedo californica acetylcholine receptor. Biochem J 224:995–1000

McIntosh M, Cruz LJ, Hunkapiller MW, Gray WR, Olivera BM (1982) Isolation and structure of a peptide toxin from the marine snail Conus magus. Arch Biochem Biophys 218:329

Mebs D Hucho F (1990) Toxins acting on ion channels and synapses. In: Shier WT, Mebs D (eds) Handbook of toxicology. Dekker, New York, pp 493–600

Mebs D, Narita K, Iwanaga S, Samejima Y, Lee CY (1972) Purification, properties and amino acid sequence of α-bungarotoxin from the venom of Bungarus multicinctus. Hoppe Seylers Z Physiol Chem 353:243–262

Ménez A, Bouet F, Guschlbauer W, Fromageot P (1980) Refolding of reduced short neurotoxins: circular dichroism analysis. Biochemistry 19:4166–4172

Meunier JC, Olsen RW, Menez A, Fromageot P, Boquet P, Changeux JP (1972) Studies on the cholinergic receptor protein of Electrophorus electricus. II. Some physical properties of the receptor protein revealed by a tritiated alpha-toxin from Naja nigricollis venom. Biochemistry 11:1200–1210

Meunier JC, Sealock R, Olsen R, Changeux JP (1974) Purification and properties of the cholinergic receptor from Electrophorus electricus electric tissue. Eur J Biochem 45:371–394

Mishina M, Tobimatsu T, Imoto K, Tanaka K, Fujita Y, Fukuda K, Kurasaki M, Takahashi H, Morimoto Y, Hirose T, Inayama S, Takahashi T, Kuno M, Numa S (1985) Location of functional regions of acetylcholine receptor α-subunit by site-directed mutagenesis. Nature 313:364–369

Mishina M, Takai T, Imoto K, Noda M, Takahashi T, Numa S, Methfessel C, Sakmann B (1986) Molecular distinction between fetal and adult forms of muscle acetylcholine receptor. Nature 321:406–411

Moore WM, Holladay LA, Puett D, Brady RN (1974) On the conformation of the acetylcholine receptor protein from Torpedo nobiliana. FEBS Lett 45:145–149

Mulac-Hericevic B, Atassi MZ (1986) Segment alpha-182–198 Torpedo californica acetylcholine receptor contains a second toxin-binding region and binds anti-receptor antibodies. FEBS Lett 199:68–74

Nachmansohn D (1959) Chemical and molecular basis of nerve activity. Academic, New York, p 235

Nakai K, Sasaki T, Hayashi K (1971) Amino acid sequence of toxin A from the venom of the Indian cobra (Naja naja). Biochem Biophys Res Commun 44:893–897

Nan-Qin L, Yao-Shi Z, Jian-Feng M, Wan-Yü W, Chang-Jiu Y, Zu-Liang X (1984) Amino acid sequence of the neurotoxin (CM-9) from the snake venom of Quangxi kind cobra (in Chinese). Acta Biochim Biophys Sin 16:592–596

Naumann D, Schultz C, Hucho F (1990) Probing acetylcholine receptor secondary structure by Fourier transform infrared spectroscopy. In: Synaptic channels and membrane receptors. Abstracts of the 10th International Biophysics Congress "Biophysics for the 90's", July/August, Vancouver

Nef P, Oneyser C, Alliod C, Couturier S, Ballivet M (1988) Genes expressed in the brain define three distinct neuronal nicotinic acetylcholine receptors. EMBO J 7:595–601

Neher E, Steinbach JH (1978) Local anaesthetics transiently block currents through single acetylcholine receptor channels. J Physiol (Lond) 277:153–176

Neubig RR, Cohen JB (1979) Equilibrium binding of (^{3}H) D-tubocurarine and]^{3}H] acetylcholine by Torpedo postsynaptic membranes: stoichiometry and ligand interactions. Biochemistry 18:5464–5475

Neumann D, Gershoni JM, Fridkin M, Fuchs S (1985) Antibodies to synthetic peptides as probes for the binding site on the alpha-subunit of the acetylcholine receptor. Proc Natl Acad Sci USA 82:3490–3493

Neumann D, Barchan D, Safran A, Gershoni JM, Fuchs S (1986) Mapping of the alpha-bungarotoxin binding site within the α-subunit of the acetylcholine receptor. Proc Natl Acad Sci USA 83:3008–3011

Neumann D, Barchan D, Horowitz M, Kochva E, Fuchs S (1989) Snake acetylcholine receptor: cloning of the domain containing the four extracellular cysteines of the α subunit. Proc Natl Acad Sci USA 86:7255–7259

Noda M, Takahashi H, Tanabe T, Toyosato M, Furutani Y, Hirose T, Asai M, Inayama S, Miyata T, Numa S (1982) Primary structure of alpha-subunit precursor of Torpedo californica acetylcholine receptor deduced from cDNA sequence. Nature 299:793–797

Noda M, Takahashi H, Tanabe T, Toyosato M, Kikyotani S, Hirose T, Asai M, Takashima H, Inayama S, Miyata T, Numa S (1983a) Primary structures of beta and delta-subunit precursors of Torpedo californica acetylcholine receptor deduced from cDNA sequences. Nature 301:251–255

Noda M, Takahashi H, Tanabe T, Toyosato M, Kikyotani S, Furutani Y, Hirose T, Takashima H, Inayama S, Miyata T, Numa S (1983b) Structural homology of Torpedo californica acetylcholine receptor subunits. Nature 302:528–532

Noda M, Furutani Y, Takahashi H, Toyosato M, Tanabe T, Shimizu S, Kikyotani S, Kayano T, Hirose T, Inayama S, Numa S (1983c) Cloning and sequence analysis of calf cDNA and human genomic cDNA encoding α-subunit precursor of muscle AChR subunits. Nature 305:818–823

Nomoto H, Takahashi N, Nagaki Y, Endo S, Arata Y, Hayashi K (1986) Carbohydrate structures of acetylcholine receptor from Torpedo californica and distribution of oligosaccharides among the subunits. Eur J Biochem 157:133–142

Ohta M, Sasaki T, Hayashi K (1976) The primary structure of toxin B from the venom of the Indian cobra Naja naja. FEBS Lett 71:161–166

Ohta M, Sasaki T, Hayashi K (1981a) The primary structure of toxin C from the venom of the Indian cobra (Naja naja). Chem Pharm Bull 29:1458–1462

Ohta M, Sasaki T, Hayashi K (1981b) The amino acid sequence of toxin D isolated from the venom of Indian cobra (Naja naja). Chem Pharm Bull (Tokyo) 29:1458–1462

Olivera BM, Gray WR, Zeikus R, McIntosh JM, Varga J, Rivier J, deSantos V, Cruz LJ (1985) Peptide neurotoxins from fish-hunting cone snails. Science 230:1338–1343

Olsen R, Meunier JC, Changeux JP (1972) Progress in purification of the cholinergic receptor protein from Electrophorus electricus by affinity chromatography. FEBS Lett 28:96–100

Pearce SF, Hawrot E (1990) Intrinsic fluorescence of binding-site fragments of the nicotinic acetylcholine receptor: perturbations produced upon binding α-bungarotoxin. Biochemistry 29:10649–10659

Pedersen SE, Cohen JB (1988) Photoaffinity labelling of the high and low affinity *d*-tubocurare binding sites of the nicotinic acetylcholine receptor (AChR) by [^{3}H]*d*-tubocurare (*d*-Tc). Biophys J 53:351a

Pedersen SE, Cohen JB (1990) *d*-Tubocurarine binding sites are located at α–gamma and α–β subunit interfaces of the nicotinic acetylcholine receptor. Proc Natl Acad Sci USA 87:2785–2789

Popot JL, Changeux JP (1984) Nicotinic receptor of acetylcholine: structure of an oligomeric integral membrane protein. Physiol Rev 64:1162–1239

Popot JL, Cartaud J, Changeux JP (1981) Reconstitution of a functional acetylcholine receptor: incorporation into artificial lipid vesicles and pharmacology of the agonist-controlled permeability changes. Eur J Biochem 118:213–214

Poulter L, Earnest JP, Stroud RM, Burlingame AL (1989) Structure, oligosaccharide structures and posttranslationally modified sites of the nicotinic acetylcholine receptor. Proc Natl Acad Sci USA 86:6645–6649

Prinz H, Maelicke A (1983a) Interaction of cholinergic ligands with the purified acetylcholine receptor protein. I. Equilibrium binding studies. J Biol Chem 258:10263–10271

Prinz H, Maelicke A (1983b) Interaction of cholinergic ligands with the purified acetylcholine receptor protein. II. Kinetic studies. J Biol Chem 258:10273–10282

Reynolds JA, Karlin A (1978) Molecular weight in detergent solution of acetylcholine receptor from *Torpedo californica*. Biochemistry 17:2035–2038

Rousselet A, Fauer G, Boulain J-C, Ménez A (1984) The interaction of neurotoxin derivatives with either acetylcholine receptor or a monoclonal antibody: an electron-spin-resonance study. Eur J Biochem 140:31–37

Ruan K-H, Spurling J, Quiocho FA, Atassi MZ (1990) Acetylcholine receptor-α-bungarotoxin interactions: determination of the region-to-region contacts by peptide-peptide interactions and molecular modeling of the receptor cavity. Proc Natl Acad Sci USA 87:6156–6160

Rydén L, Gabel D, Eaker D (1973) A model of the three-dimensional structure of snake venom neurotoxins based on chemical evidence. Int J Pept Protein Res 5:261–273

Safran A, Sagi-Eisenberg R, Neumann D, Fuchs S (1987) Phosphorylation of the acetylcholine receptor by protein kinase C and identification of the phosphorylation site within the receptor delta subunit. J Biol Chem 262:10506–10510

Sakmann B, Neher E (1984) Patch-clamp techniques for studying ionic channels in excitable membranes. Annu Rev Physiol 46:455–472

Sato S, Tamiya N (1971) The amino acid sequences of erabutoxins, neurotoxic proteins of sea-snake (*Laticauda semifasciata*) venom. Biochem J 122:453–461

Sawruk E, Schloss P, Betz H, Schmitt B (1990) Heterogeneity of *Drosophila* nicotinic acetylcholine receptors: SAD, a novel developmentally regulated α-subunit. EMBO J 9:2671–2677

Schiebler W, Hucho F (1978) Membranes rich in acetylcholine receptor: characterization and reconstitution to excitable membranes from exogenous lipids. Eur J Biochem 88:55–63

Schmidt TJ, Raftery MA (1972) Use of affinity chromatography for acetylcholine receptor purification. Biochem Biophys Res Commun 49:572

Schmidt TJ, Raftery MA (1973) Purification of acetylcholine receptors from *Torpedo californica* electroplax by affinity chromatography. Biochemistry 49:572

Schoepfer R, Conroy W, Whiting P, Gore M, Lindstrom J (1990) Brain α-bungarotoxin binding protein cDNAs and MAbs reveal subtypes of this branch of the ligand-gated ion channel gene superfamily. Neuron 5:35–48

Schofield GG, Witkop B, Warnick JE, Albuquerque EX (1981) Differentiation of the open and closed states of the ionic channels of nicotinic acetylcholine receptors by tricyclic antidepressants. Proc Natl Acad Sci USA 89:5240–5244

Schröder W, Covey T, Hucho F (1990) Identification of phosphopeptides by mass spectrometry. FEBS Lett 273:31–35

Schröder W, Meyer HE, Buchner K, Bayer H, Hucho F (1991) Phosphorylation sites of the nicotinic acetylcholine receptor – a novel site detected in position delta-S362. Biochemistry (in press)

Shipolini RA, Bailey GS, Banks BEC (1974) The separation of a neurotoxin from the venom of *Naja melanoleuca* and the primary sequence determination. Eur J Biochem 42:203–211

Smart L, Meyers H-W, Hilgenfeld R, Saenger W, Maelicke A (1984) A structural model for the ligand-binding sites at the nicotinic acetylcholine receptor. FEBS Lett 178:64–68

Stroud RM, McCarthy MP, Schuster M (1990) Nicotinic acetylcholine receptor superfamily of ligand-gated ion channels. Biochemistry 29:11010–11023

Strydom AJC (1972) Snake venom toxins: the amino acid sequences of two toxins from *Dendroaspis polylepis polylepis* (black mamba) venom. J Biol Chem 247:4029–4042

Strydom AJC (1973a) Snake venom toxins: the amino acid sequences of two toxins from *Dendroaspis jamesoni kaimosae* (Jameson's mamba) venom. Biochim Biophys Acta 328:491–509

Strydom AJC (1973b) Studies on the toxins of *Dendroaspis polylepis* (black mamba) venom. Dissertation, University of South Africa, Pretoria

Strydom AJC, Botes DP (1971) Snake venom toxins: purification, properties and complete amino acid sequences of two toxins from ringhals (*Hemachatus haemachatus*) venom. J Biol Chem 246:1341–1349

Strydom DJ, Haylett T (1977) Snake venom toxins: the amino acid sequence of toxin V_n2 of Dendroaspis polylepis polylepis (black mamba) venom. S Afr J Chem 30:40–48

Takai T, Noda M, Mishina M, Shimizu S, Furutani Y, Kayano T, Ikeda T, Kubo T, Takahashi H, Takahashi T, Kuno M, Numa S (1985) Cloning, sequencing and expression of cDNA for a novel subunit of acetylcholine receptor from calf muscle. Nature 315:761–764

Tamiya N, Abe H (1972) The isolation, properties and amino acid sequences of erabutoxin c, a minor neurotoxic component of the venom of a sea snake *Laticauda semifasciata*. Biochem J 130:547–555

Tamiya N, Maeda N, Cogger HG (1983a) Neurotoxins from the venoms of the sea snakes *Hydrophis ornatus* and *Hydrophis lapemoides*. Biochem J 213:31–38

Tamiya N, Sato A, Kim HS, Teruuchi T, Tkasaki C, Ishikawa Y, Guinea ML, McCoy M, Heatwole H, Cogger HG (1983b) Neurotoxins of sea snakes of the genus *Laticauda*. Toxicon [Suppl]3:445–447

Tamiya T, Lamouroux A, Julien J-F, Grima B, Mallet J, Fromageot P, Mènez A (1985) Cloning and sequence analysis of the cDNA encoding a snake neurotoxin precursor. Biochimie 67:185–189

Taylor P (1990a) Cholinergic agonists. In: Goodman Gilman A, Rall TW, Nies AS, Taylor P (eds) The pharmacological basis of therapeutics, 8th edn. Pergamon, New York, p 122

Taylor P (1990b) Agents acting at the neuromuscular junction and autonomic ganglia. In: Goodman Gilman A, Rall TW, Nies AS, Taylor P (eds) The pharmacological basis of therapeutics, 8th edn. Pergamon, New York, p 166

Toyoshima C, Unwin N (1988) Ion channel of acetylcholine receptor reconstructed from images of postsynaptic membranes. Nature 336:214–251

Tsernoglou D, Petsko GA (1976) The crystal structure of a postsynaptic neurotoxin from sea snake at 2.2 Å resolution. FEBS Lett 68:1–4

Tsetlin VI, Karlsson E, Arseniev AS, Utkin YN, Surin AM, Pashkov VS, Pluzhnikov KA, Ivanov VT, Bystrov VF, Ovchinnikov YA (1979) EPR and fluorescence study of interaction of *Naja naja oxiana* neurotoxin II and its derivatives with acetylcholine receptor protein from *Torpedo marmorata*. FEBS Lett 106:47–52

Tsetlin VI, Karlsson E, Utkin YN, Pluzhnikov KA, Arseniev AS, Surin AM, Kondakov VV, Bystrov VF, Ivanov VT, Ovchinnikov YA (1982) Interacting surfaces of neurotoxins and acetylcholine receptor. Toxicon 20:83–93

Tsetlin VI, Pluzhnikov KA, Karelin A, Ivanov V (1983) Acetylcholine receptor interaction with the neurotoxin II photoactivable derivatives. In: Hucho F, Ovchinnikov YA (eds) Toxins as tools in neurochemistry. de Gruyter, Berlin

Tsetlin VI, Pluzhnikov KA, Karelin AA, Karlsson E, Ivanov VT (1984) Mutual disposition of the bound neurotoxins and acetylcholine receptor subunits (in Russian). Bioorg Khim 10:176–187

Tsetlin VI, Alyonycheva TN, Kuryatov AB, Pluzhnikov KA (1987) Selective labeling study on topography of acetylcholine receptor and bacteriorhodopsin. In: Ovchinnikov YA, Hucho F (eds) Receptors and ion channels. de Gruyter, Berlin

Tzartos SJ, Changeux JP (1983a) High affinity binding of alpha-bungarotoxin to the purified alpha-subunit and its 27.000 dalton proteolytic peptide from *Torpedo marmorata* acetylcholine receptor. Requirement for sodium dodecyl sulfate. EMBO J 2:381–387

Tzartos SJ, Changeux JP (1983b) Lipid-dependent recovery of alpha-bungarotoxin and monoclonal antibody binding to the purified alpha-subunit from *Torpedo marmorata* acetylcholine receptor. J Biol Chem 259:11512–11519

Unwin N, Toyoshima C, Kubalek E (1988) Arrangement of the acetylcholine receptor subunits in the resting and desensitized states, determined by cryoelectron microscopy of crystallized *Torpedo* postsynaptic membranes. J Cell Biol 107:1123–1138

Vandlen RL, Wu WC-S, Eisenach JC, Raftery MA (1979) Studies of the composition of purified *Torpedo californica* acetylcholine receptor and of its subunits. Biochemistry 10:1845–1854

Vijayaraghavan S, Schmid HA, Halvorsen SW, Berg DK (1990) Cyclic AMP-dependent phosphorylation of a neuronal acetylcholine receptor α-type subunit. J Neurosci 10:3255–3262

Wada K, Ballivet M, Boulter J, Connolly J, Wada E, Deneris ES, Swanson LW, Heinemann S, Patrick J (1988) Functional expression of a new pharmacological subtype of brain nicotinic acetylcholine receptor. Science 240:330–334

Walkinshaw MD, Saenger W, Maelicke A (1980) Three-dimensional structure of the "long" *Torpedo californica* acetylcholine receptor in reconstituted membranes. Biochemistry 21:5384–5389

Wang CL, Liu CS, Hung YO, Blackwell RQ (1976) Amino acid sequence of pelamitoxin a, the main neurotoxin of the sea snake, *Pelamis platurus*. Toxicon 14:459–466

Watters D, Maelicke A (1983) Organization of ligand binding sites at the acetylcholine receptor: a study with monoclonal antibodies. Biochemistry 22:1811–1819

Wilson PT, Gershoni JM, Hawrot E, Lentz TL (1984) Binding of alpha-bungarotoxin to proteolytic fragments of the alpha-subunit of *Torpedo acetylcholine* receptor analyzed by protein transfer on positively charged membrane filters. Proc Natl Acad Sci USA 81:2553–2557

Wilson PT, Lentz TL, Hawrot E (1985) Determination of the primary amino acid sequence specifying the α-bungarotoxin binding site on the α-subunit of the acetylcholine receptor from *Torpedo californica*. Proc Natl Acad Sci USA 82:8790–8794

Witzemann V, Muchmore D, Raftery MA (1979) Affinity-directed cross-linking of membrane-bound acetylcholine receptor polypeptides with photolabile α-bungarotoxin derivatives. Biochemistry 18:5511–5518

Witzemann V, Stein E, Barg B, Konno T, Koenen M, Kues W, Criado M, Hofmann M, Sakmann B (1990) Primary structure and functional expression of the α-, β-, γ- and ε-subunits of the acetylcholine receptor from rat muscle. Eur J Biochem 194:437–448

Yang CC (1965) Crystallization and properties of cobrotoxin from Formosan cobra venom. J Biol Chem 240:1616–1618

Yang CC, Yang HJ, Huang JS (1969) The amino acid sequence of cobrotoxin. Biochim Biophys Acta 188:65–77

Yu C, Lee C-S, Chuang L-C, Shei Y-R, Wang CY (1990) Two-dimensional NMR studies and secondary structure of cobrotoxin in aqueous solution. Eur J Biochem 193:789–799

CHAPTER 17

Nicotinic Acetylcholine Receptors and Low Molecular Weight Toxins*

K.L. SWANSON and E.X. ALBUQUERQUE[1]

A. Introduction

The study of the nicotinic acetylcholine receptor has since its very inception been associated with the use of toxins. The work by LANGLEY in 1905 demonstrated (amongst many important and lasting findings) the presence of the "receptive substance" in striated muscle and sympathetic ganglia. Indeed, the receptor was named for its responsiveness to the tobacco alkaloid nicotine. Although the physiological transmitter in these and other tissues was later identified as acetylcholine (ACh), this receptor has been classically distinguished from other cholinergic receptors by their responses to nicotine and muscarine (DALE 1914). Langley also offered a precocious description of multiple sites or multiple ways in which a toxin might produce both agonist and antagonist effects:

> Both stimulation and paralysis imply a combination of the poison with some substance, so that an action of whatever nature implies some similarity in the substance acted on. The different physiological effects produced by the combination may be regarded as due to minor differences, such as the presence of different radicles in (or the existence of different molecular associations of) the substances primarily affected. (LANGLEY 1905)

Although many studies supported the existence of allosteric sites, a broad spectrum of electrophysiological studies with the antagonist histrionicotoxin (HTX) led to the description of the ion conductance modulator of the nicotinic acetylcholine receptor-ion channel macromolecule (AChR) (ALBUQUERQUE et al. 1973a) via its action at a high-affinity allosteric site on the AChR (reviewed in ALBUQUERQUE et al. 1988d).

Low molecular weight (nonpeptide) toxins have remained a focus of nicotinic receptor studies for several reasons. Perhaps above all others, the most persuasive reason to study such toxins is that agonists or partial agonists are found in this class. Ganglionic and skeletal muscle receptor types were distinguished by their relatively selective stimulation by hexamethonium

---

*This work was supported by NIH Grant NS 25296 and U.S. Army Medical Research and Development Command Contract DAMD17-88-C-8819.
[1] To whom reprint requests should be sent.

and decamethonium, respectively (Paton and Zaimis 1952). Further, the majority of nicotinic toxins are alkaloids, and many are secondary or tertiary amines that can penetrate nonpolar cell membranes and the BBB in their equilibrium uncharged form. Such toxins can be administered systemically to affect the brain. In contrast to the polypeptide snake toxins, smaller dissociable toxins have reversible actions which facilitate their use in experiments and as anesthetic adjuncts.

Our objective here is to present, from the viewpoint of the small toxins, the essence of the various nicotinic receptor recognition sites and their physiological and pharmacological roles in receptor function. While the tremendous contribution of experiments using snake toxin peptides, particularly α-bungarotoxin, to the understanding and isolation of the AChR is readily acknowledged, this topic has been discussed earlier (Hucho, this volume). Some historical comments are a consequence of the century of research and vastly differing methodologies used to derive conclusions and standard classification of receptor types; relevant topic reviews will be cited. The process of evolution has dictated several receptor sites which interact with the physiological transmitter ACh. The principal ones are nicotinic receptors (peripheral, ganglionic, and central), muscarinic receptors, and cholinesterases. In addition, presynaptic sites react with choline during the transmitter reuptake process. A basic project of nicotinic pharmacology is the development of selective ligands for the nicotinic receptor. To complicate the task, along with agonist sites for channel activation which respond to mimics of ACh, the AChR has several other target areas which interact with agonists and agonist analogues such that many toxins and drugs have mixed agonist-antagonist actions. The recent work with potent and selective ligands identifies target sites of the AChR most clearly. Therefore, we will address such examples, recognizing the unfortunate consequence that many interesting toxins will be neglected. Nonpolar sites probably exist at the interface of the channel proteins and the lipid membrane; the effects of general anesthetics and alcohols in this regard will not be discussed. In this sense, this chapter may be more appropriately described as an update, rather than a comprehensive review. We will illustrate how the continued use of low molecular weight toxins to investigate the properties of the AChR promises to be fruitful in further distinguishing subtypes of neuronal nicotinic receptors, in characterizing discrete physiological functions, and in clinical manipulation of the nicotinic systems.

B. Nicotinic Acetylcholine Receptor Ion Channel Macromolecules

The early biochemical characterization of the peripheral receptor was greatly facilitated by the presence of a high density of AChR in the electric organ of fishes. These receptors were biochemically and pharmacologically very similar to the AChR found in vertebrate skeletal muscle (ALBUQUERQUE et al. 1974; BARNARD et al. 1975; FERTUCK and SALPETER 1974). These peripheral AChRs are made up of five subunit proteins of four types with the stoichiometry $\alpha_2\beta\gamma\delta$ (the γ is substituted by ε in some cases depending upon cell maturity). The assembly of five glycoprotein subunits into a ring of trans-membrane units around a central ion channel pore yields a functional receptor (excellent reviews are available: KARLIN 1980, 1983; STROUD 1983; HUCHO 1986; OCHOA et al. 1989; CHANGEUX 1990). In neuronal forms of the nicotinic receptor, both α- and β-subunits were necessary and sufficient to assemble a functional receptor (based on gene cloning and synthesis of subunits in oocytes). Several different α-subunit sequences have been identified, each presumably bearing a distinct ACh recognition site. Like the peripheral receptor, α- and non-α-subunits have been proposed to comprise a pentameric neuronal receptor containing at least two ACh binding sites (WHITING et al. 1987; COOPER 1990).

Some confusion initially arose during the identification of central neuronal receptors: both sites for (-)nicotine and α-bungarotoxin were identified. It is now considered that high-affinity (-)-nicotine binding sites in the brain are correlated with high-affinity ACh recognition sites and that α-bungarotoxin sites in brain tissue represent a separate category of receptor (CLARKE 1987; WONNACOTT 1986). The high-affinity receptors for (-)-nicotine appear to have excitatory roles on postsynaptic membranes and facilitatory roles on presynaptic transmitter release (RAPIER et al. 1988; WESSLER 1989). Physiological functions are implicated for the nicotinic AChR in respiration (BRADLEY and LUCY 1983) and hearing (KNAPPE et al. 1988), in vision at the retina (LIPTON et al. 1987) and visual cortex (PARKINSON et al. 1988), at the brain stem prolactin release (GIBLIN et al. 1988), and seizure induction at unlocalized receptors (probably associated with the hippocampus or cortex) (COLLINS et al. 1986; MEYERHOFF and BATES 1985; MINER et al. 1986). Alterations of the AChR may disturb mnemonic function in Alzheimer's disease (WHITEHOUSE et al. 1986; NILSSON et al. 1987; ROBERTS and LAZARENO 1989; LONDON et al. 1989; KELLAR and WONNACOTT 1990).

A fundamental question of nicotinic receptor physiology is "how does interaction of the receptor with a transmitter induce a conformational shift resulting in ion conductance?" Various theories have been put forth to explain the efficacy of certain drugs or toxins to enhance channel activation. Two extremes shall be considered. The occupancy theory suggests that drug efficacy (activation) is proportional to binding affinity. However, this does not explain the observation that competitive antagonists bind to the same site on the receptor while only rarely initiating channel currents. In response, rate theory suggests that the activation is proportional to the binding rate and that antagonism is a result of high-affinity binding with a low dissociation rate. Electrophysiological studies, including single-channel recordings and analyses developed in the last decade, sought to define the kinetics of the AChR at the molecular level. The receptor was proposed to undergo agonist binding and channel opening in sequence (del Castillo and Katz 1957; Magleby and Stevens 1972). Later studies showing cooperativity led to a model which incorporated two agonists per receptor:

$$2A + R \rightleftharpoons A + AR \rightleftharpoons A_2R \rightleftharpoons A_2R^*$$

where A is the agonist, R is the receptor, * indicates the open conformation. The intrinsic efficacy, i.e., the efficacy of agonists after normalizing for their binding affinity, would be determined by the rate of transitions between agonist-bound resting and open conformations, while the single-channel conductance remains constant. Intrinsic efficacy was indeed partially related to the mean channel open duration or lifetime (Auerbach et al. 1983). In structural terms, there may be a correlation of agonist "ionicity" with channel opening and closing rates (Papke et al. 1988). Thus, ion channels that remain open for a longer period pass a larger net current and produce a greater response. However, time constants for open-channel lifetime for almost all agonists that have been tested fell in a range from about 1 to 50 ms, depending upon recording conditions. This is much narrower than the scope of agonist potencies. An accumulation of data from many agonists reveals widely varying potencies, thus suggesting that the binding affinities, which also encompassed several orders of magnitude, are the overwhelming contributor to agonist potency. This strong correlation between potency and affinity suggests that as the second agonist molecule (including semirigid agonists described below) approaches the receptor, it may conform the agonist binding site to itself with sufficient forces to trigger opening of the ion channel. However, a confounding problem is that the affinity of agonists, which is measured in binding competition assays, was apparently a high affinity associated with a desensitized (nonconducting) conformation of the receptor and a lower affinity was associated with the non-desensitized (activatable) conformation of the receptor. (Rang and Ritter 1969; Colquhoun and Rang 1976; Ochoa et al. 1989).

Antagonists come in several forms. Antagonists that bind to the ACh site apparently (in some cases undoubtedly due to lack of data to the contrary) do not elicit measurable ion fluxes or single-channel, whole-cell or end plate currents, i.e., the intrinsic efficacy at the ACh site is negligible. These compounds vary in the degree of reversibility but are generally larger molecules than agonists, and several dissociate very slowly or bind irreversibly. The peptide toxins including snake toxins have been described as competitive antagonist (Hucho, this volume). Allosteric "noncompetitive" antagonists are far more common. [The term noncompetitive antagonists is used in nicotinic pharmacology to describe an allosteric interaction with the receptor (Monod et al. 1965). This term will be avoided here because of the confusion which surrounds its usage. This is not limited to the irreversible biochemical interactions which result in a decrease in the maximum activity of an enzyme.] The number of allosteric sites has not been precisely determined. A site for ion channel blocking drugs, which have comparatively low specificity, may be formed by the cooperative efforts of the channel domains of each subunit protein. This or other sites enhance receptor activation or desensitize receptors. Target areas for allosteric modification, including additional agonist sites, are especially in need of characterization.

There are some aspects of the receptor that can be addressed by examining the actions of various toxins. What chemical features make a ligand specific for a target on the nicotinic receptor? What are the features for each type of interaction with the receptor target? How does the toxin-receptor interaction determine the response? What relationship exists among agonist binding, channel activation and desensitization? What distinguishes an agonist from an antagonist?

C. Toxins Affecting the ACh Recognition Site for Channel Activation

Several α-protein subtypes bear the putative agonist recognition domain, each with a distinct amino acid sequence containing vicinal cysteines. These ACh targets from various tissues and species can be distinguished by their primary amino acid sequence (Table 1). The receptor site may also incorporate other specific regions on the α-subunit N-terminal sequence (see review by Changeux 1990). Furthermore, the two agonist sites of the peripheral receptor, which are on α-subunits with identical sequences, can be distinguished from each other by large molecules such as antibodies and α-bungarotoxin and apparently also by (+)-tubocurarine and lophotoxin. Thus, these toxins revealed inequalities in the final receptor which arise from subunit interactions. For the nicotinic receptor in the brain, the sensitivity to agonists depended not only on the composition of the α-subunit but also on its interaction with the β-subunit, e.g., the α_3 and β_2 pair was less sensitive to nicotine than was α_3 and β_4 pair (Luetje and Patrick 1990). Probably this result was due to low affinity of nicotine for the ACh site, which in this case was *not* specific for the α-subunit unit alone but resulted from an interaction between the α- and β-subunits. These findings are particularly interesting in comparison with the effects of the antagonist tubocurarine which acted on α-subunit sites and on neighboring subunits (see below in III).

I. Sulfhydryl Modification of the ACh Target

Relatively early studies of the AChR showed that the receptor protein could be reduced and reoxidized using dithiothreitol and dithiobis-2-nitrobenzoate, respectively, and furthermore, the reduced form of the receptor was alkylated by the addition of a maleimide compound (Fig. 1; Karlin 1969). Chemical reduction of the receptor decreased the potency of many agonists at the AChRs from several sources including *Torpedo* electric organ (Barrantes

Table 1. Amino acid sequences of various α-subunit proteins in the region of the proposed acetylcholine (ACh) recognition site (Connolly 1989; Peterson 1989)

Species	Type	183	184	185	186	187	188	189	190	191	192	193	194	195	196	197	198	199	200	201	202	203	204
Xenopus		gly	trp	lys	his	trp	val	tyr	tyr	thr	cys-cys		pro	asp	lys	pro	tyr	lys	asp	ile	thr	tyr	his
Torpedo		gly	trp	lys	his	trp	val	tyr	tyr	thr	cys-cys		pro	asp	thr	pro	tyr	lys	asp	ile	thr	tyr	his
Chick	muscle	gly	trp	lys	his	trp	val	tyr	tyr	ala	cys-cys		pro	asp	thr	pro	tyr	lys	asp	ile	thr	tyr	his
Mouse	muscle	gly	trp	lys	his	trp	val	phe	tyr	ser	cys-cys		pro	thr	thr	pro	tyr	lys	asp	ile	thr	tyr	his
Calf	muscle	gly	trp	lys	his	trp	val	phe	tyr	ala	cys-cys		pro	ser	thr	pro	tyr	lys	asp	ile	thr	tyr	his
Human	muscle	gly	trp	lys	his	ser	val	tyr	tyr	ser	cys-cys		pro	asp	thr	pro	tyr	lys	asp	ile	thr	tyr	his
Drosophila		ala	val	arg	asn	glu	lys	phe	tyr	ser	cys-cys		glu	glu	–	pro	tyr	lys	asp	ile	val	phe	asn
Locust	αBGT	ala	glu	arg	his	glu	lys	tyr	tyr	pro	cys-cys		ala	glu	–	pro	tyr	pro	asp	ile	phe	pro	asn
Chick brain	αBGT	gly	lys	arg	thr	glu	ser	phe	tyr	glu	cys-cys		lys	glu	–	pro	tyr	pro	asp	ile	thr	pro	asn
Chicken	α₂	gly	arg	tyr	asn	ser	lys	lys	tyr	asp	cys-cys		thr	glu	–	ile	tyr	pro	asp	ile	thr	phe	tyr
Rat brain	α₂	gly	thr	tyr	asn	ser	lys	lys	tyr	asp	cys-cys		ala	glu	–	ile	tyr	pro	asp	val	thr	tyr	ser
Rat	α₄	gly	thr	tyr	asn	thr	arg	lys	tyr	glu	cys-cys		ala	glu	–	ile	tyr	pro	asp	ile	thr	tyr	ala
Chicken	α₄	gly	asn	tyr	asn	ser	lys	lys	tyr	glu	cys-cys		thr	glu	–	ile	tyr	pro	asp	ile	thr	tyr	ser
Rat PC12	α₃	gly	tyr	lys	his	glu	ile	lys	tyr	asn	cys-cys		glu	glu	–	ile	tyr	gln	asp	ile	thr	tyr	ser
Chicken	α₃	gly	tyr	lys	his	asp	ile	lys	tyr	asn	cys-cys		glu	glu	–	ile	tyr	thr	asp	ile	thr	tyr	ser

Putative ACh binding site amino acid sequence

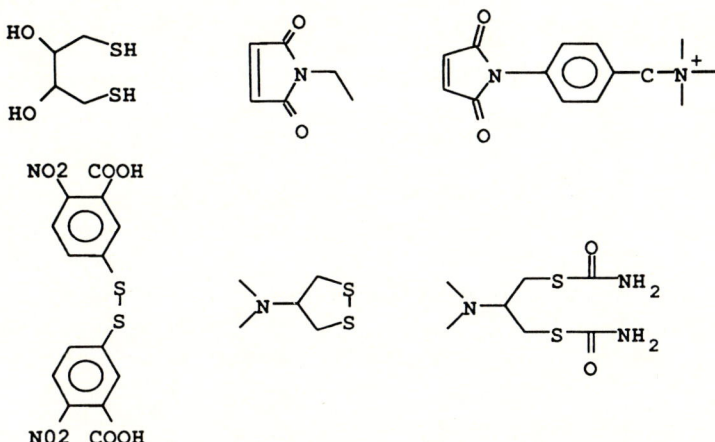

Fig. 1. Compounds interacting with the nicotinic receptor at or near the ACh binding site. *Above* dithiothreitol, *N*-ethylmaleimide (NEM), 4-(-N-maleimido) benzyltrimethylammonium; *below*, 5,5'-dithiobis-(2-nitrobenzoic acid) (DTNB), nereistoxin, and (5,5'-[2-(dimethylamino)-trimethylene]thiocarbamate (Cartap). Normal saturation of carbons are assumed and hydrogens are not shown throughout; thus, peripheral bonds represent attached carbons unless otherwise indicated

1980), chick muscle (RANG and RITTER 1971), PC12 cells (LEPRINCE 1983) and mammalian brain (STITZEL et al. 1988). In contrast, the potency of hexamethonium and decamethonium increased (RANG and RITTER 1971; BARRANTES 1980). Sulfhydryl reduction of AChR has been reported to either decrease (ADAMS 1983; BEN-HAIM et al. 1975) or increase (TERRAR 1978) conductance, to increase the endplate decay rate (TERRAR 1978) and to decrease channel lifetime (BEN-HAIM et al. 1975) precise. However, more methods should be used to review these points. Structure-activity studies of the maleimides showed a specific relationship of size to alkylating strength, thus suggesting that a specific dimension (on the order of 8 Å) exists between the point of alkylation and the anionic site of the receptor (KARLIN 1969). Thus, these early studies predicted the presence of vicinal cysteines in the ACh binding region of the receptor.

Nereistoxin (Fig. 1), from the marine annelid *Lumbriconereis*, is a dithiolane partial agonist of the AChR that appeared to bind to the ACh site of the receptor and to induce cation currents and then, more slowly, underwent a covalent reaction to produce irreversible inhibition (DEGUCHI et al. 1971; ELDEFRAWI et al. 1980). This toxin was selective for the postjunctional type of AChR of vertebrates, while having only mild presynaptic effects. Its toxicity to insects formed the basis for the development of pesticides including Cartap (Fig. 1). It seems probable that nereistoxin and the pesticide metabolites may interact with the disulfide bond on the receptor.

II. Agonists

The simplest compounds which stimulate the AChR are charged ammonium compounds. (These same compounds also can block the nicotinic receptor and the voltage-dependent K^+ channel.) [The nitrogen of the alkaloids is present in two pharmacologically important forms, charged and uncharged. The quaternary nitrogens retains a permanent positive charge in all conditions, whereas tertiary, secondary, and primary amines bear a positive charge at physiological pH. The latter are able to penetrate cell membranes and the BBB in their equilibrium uncharged form, and yet the charged center is extremely important to interaction with the nicotinic receptor.] Complete methylation of the amine seemed to be necessary for agonist potency, but ethylation was excessive. Although the choline moiety is well suited as antagonist it is not essential that the charged nucleus be in this form. Thus, many trimethylammonium analogue series have been tested to evaluate other aspects of their structures. The length of the methylene chain between the bis-trimethyl moieties of hexamethonium and decamethonium has been used to distinguish the ganglionic from neuromuscular receptors (PATON and ZAIMIS 1952). [Both hexamethonium and decamethonium are also non-competitive antagonists (ADAMS and SAKMANN 1978; MILNE and BYRNE 1981).]

The addition of a carbonyl component (Fig. 2) contributes potency to agonists and stability to the open conformation of the agonist-bound receptor (HEY 1952; AUERBACH et al. 1983). Attempts to describe the agonist pharmacophore have relied upon either solution or crystalline conformations of the molecules (see reviews: TRIGGLE and TRIGGLE 1976; SKOK et al. 1989). Thus, ACh meets two characteristics with chemical simplicity: a positively charged nucleus for electrostatic interaction and a carbonyl presumed to H-bond with the receptor. Proceeding toward selective agents, the most commonly recognized difference between the nicotinic and muscarinic agonists corresponds to the distance between the two sites on the agonists, with the nicotinic receptor requiring a 4.8 Å separation (5.9 Å to the hydrogen bonding point) and the muscarinic receptor preferring about 4.5 Å (BEERS and REICH 1970; SHERIDAN et al. 1986). These modest descriptions of agonists were formulated from nicotinic agonists with relatively low selective potential, prior to the discovery of more potent and stereoselective agonist toxins. Early attempts to synthesize rigid analogues with greater selectivity based on this model were not especially productive (TRIGGLE and TRIGGLE 1976). In contrast to this method, BEHLING et al. (1988) recently assessed the conformation of ACh while bound to the AChR; unfortunately, their results describe ACh associated with the high-affinity and presumably desensitized conformation of the AChR receptor. We have turned to analysis of semirigid analogues for definition of the ACh recognition site.

Recognition of the potential for stereoselectivity by the receptor relationship focused initially on nicotine. A modest stereoselectivity (6- to 8-fold) of the agonist target site of the peripheral AChR was observed (BARLOW and

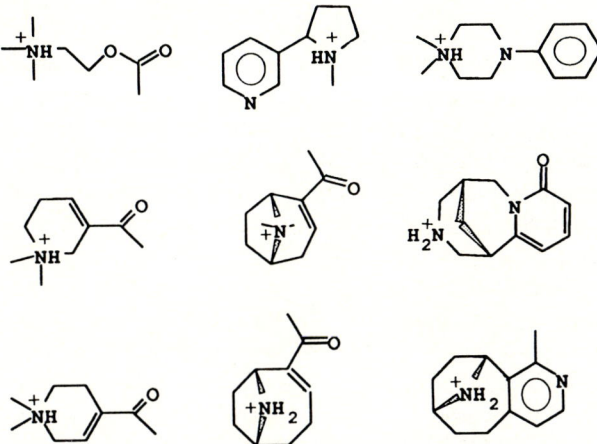

Fig. 2. Structures of potent agonists. *Above*, acetylcholine ACh, nicotine, 1,1-dimethyl-4-phenyl-piperazine (DMPP); *middle*, methylarecolone, methylferruginine, cytisine; *below*, methylisoarecolone, anatoxin-a (AnTX), pyridohomotropane. These compounds are protonated at physiological pH

HAMILTON 1965), while the ion channel site which results in blockade did not exhibit a similar steric sensitivity to nicotine enantiomers (WANG and NARAHASHI 1972; ROZENTAL et al. 1989). Although the peripheral receptor had rather low stereoselectivity for the natural (−)-isomer, it was found that the receptor in vertebrate brain demonstrated a great preference (63-fold) for the natural alkaloid (ROMANO and GOLDSTEIN 1980). [³H]-(−)-Nicotine has since become extremely useful in the identification and localization of AChR in vertebrate brain (CLARKE et al. 1985).

The discovery, isolation, and characterization of anatoxin-a, a natural, semirigid alkaloid toxin produced by blooms of the blue-green algae *Anabaena flosaquae* expanded the field for description of stereochemical aspects of the nicotinic receptors' ACh recognition sites. The toxicity was attributed to depolarizing blockade of neuromuscular transmission (CARMICHAEL et al. 1975; SPIVAK et al. 1980) and (+)-anatoxin-a is, indeed, the most potent agonistic, natural, small toxin at the peripheral nicotinic receptor. Anatoxin-a conformed to classical models for agonists (Fig. 2) and, thus, reaffirmed many of the early observations about structure-activity relationships at the nicotinic receptor. In addition, studies of the molecular actions of synthetic enantiomers showed a 150-fold difference in potency at the peripheral nicotinic receptor (SWANSON et al. 1986) and high stereoselectivity at neuronal receptors (MACALLAN et al. 1988). In order to describe the additional active sites of the receptor-toxin interaction that confer the stereospecificity and to test previous structure-activity hypotheses of binding mechanisms, a series of (+)-anatoxin-a analogues were synthesized. Initial results (summarized below) are promising for the development of drugs that discriminate between peripheral and brain nicotinic receptors.

Considerable attention has been given to an apparent need for nicotinic agonists to have steric bulk around the nitrogen (discussed in Spivak and Albuquerque 1982). However, rigid agonist structures were an exception; steric bulk was unfavorable. [Similar observations have been made for muscarinic agonists (Hanin et al. 1966; Brimblecombe and Rowsell 1969; Ehlert and Jenden 1984).] Strong intramolecular interaction between carbonyl and amine moieties could reduce flexibility so that receptor binding conformations became energetically disfavored; therefore, methylation limited intramolecular interaction and was advantageous for flexible agonists, but intramolecular interaction was not considered to be an important factor for rigid molecules (Triggle and Triggle 1976) such as cytisine, for which methylation caused a relatively small (10-fold) decrease in potency (compare across the row of Table 2). This generalization no longer appears to hold true. The dramatic contrast in the effect of methylation on ferruginine and anatoxin series deserves some notice: upon N,N-dimethylation, the potency of nor ferruginine increased 30-fold but the potency of anatoxin decreased 10 000-fold (Table 2). It therefore becomes very important to learn why these agonist sezies changed potencies in opposite directions, even though the chemical structures appeared similar (Fig. 2). This underlines the importance of the high stereospecificity of anatoxin and the role of a major portion of the molecule in its interaction with the agonist recognition site. Much remains to be explained about anatoxin and ferruginine and two chemically related synthetic compounds (Fig. 2) with similar potency, methylisoarecolone (Waters et al. 1988) and pyridohomotropane (Kanne and Abood 1988) (Fig. 2). [Pyridohomotropane has essentially the same affinity as anatoxin at neuronal [^{3}H]-(−)-nicotine sites; agonist activity has not been proven.]

The second functional moiety due for reconsideration is the carbonyl region. For the action of ACh at the muscarinic receptor, the ether oxygen (of the ester) was proposed to be important. However, the opposite appeared to be true for the nicotinic receptor. For example, arecolone was 20 times more potent than arecoline, methylarecolone (Fig. 2) was 7 times more potent than methylarecoline (Albuquerque and Spivak 1984), and anatoxin was similarly more potent than the corresponding synthetic analogue anatoxinic acid methyl ester (Swanson et al. 1990; see Table 3). [It was necessary to eliminate the possibility that allosteric antagonist sites might contribute to the difference in potency. The data for anatoxin and its analogues show that the recognition site itself is responsible for the effect: ligand binding studies show a decrease in the inhibition of [^{125}I]-α-bungarotoxin binding.] The addition of an ether oxygen increases the electron withdrawing effect of the carbonyl and, if hydrogen bonding strength is the most important factor, should increase the potency. The fact that this repeatedly did not occur points toward other factors for agonist-pharmacophore interaction: (a) modification of the agonist conformation as a whole, or (b) interaction of another specific

Table 2. Effect of N-methylation on the relative potency of various semirigid nicotinic agonists at peripheral nicotinic receptors

Secondary	Degree of nitrogen methylation Tertiary	Quaternary	Stereo-specificity
		Acetylcholine 14	
		Carbamylcholine 1.0	
Nornicotine <0.1	Nicotine 0.77	Methylnicotine 0.7	7.0
	Arecolone 0.17	Methylarecolone 1.3	
Norferruginine 0.09	(−)-Ferruginine 0.04	Methylferruginine 3.3	
(−)-Cytisine 1.2	Caulophylline 0.28	Methylcaulophylline ≪0.2	
(+)-Anatoxin 110	Methylanatoxin ≈0.1	Dimethylanatoxin <0.01	150

Each row lists the potencies for a single type of nicotinic agonist with the only difference being the degree of nitrogen methylation. All potencies are related to carbamylcholine for the practical reason that carbamylcholine has been the most frequently used as a standard for comparison, except as stated below. Most potencies are relative potencies for inducing contracture in frog rectus abdominis muscle (ALBUQUERQUE and SPIVAK 1984; COSTA et al. 1990; ROZENTAL et al. 1989; SPIVAK et al. 1983b; SWANSON et al. 1986). The potency of ACh was determined after irreversible cholinesterase inhibition. Values for cytisine (2°) and caulophylline (3°) were calculated from potencies relative to (−)-nicotine; chick muscle indicates that caulophylline methiodide (4°) is very weak (BARLOW and McLEOD 1969). A direct comparison of racemic anatoxin to (−)-cytisine (SPIVAK et al. 1983b) also gave a similar value. Suggestion for nornicotine was based on the 25-fold lower potency of (±)-nornicotine relative to (±)-nicotine, measured as inhibition of [^{3}H]-(−) nicotine binding to *Torpedo* membranes (KANNE and ABOOD 1988). The stereospecificities are for nicotine and anatoxin.

Table 3. Comparison of relative potency of agonists with acetyl moieties to methyl ester analogs

Ketone form			Ester form
Arecolone	0.17	0.009	Arecoline
Methylarecolone	8.7	1.3	Methylarecoline
Methyisoarecolone	50	−	
Anatoxin	110	1	Anatoxin-methylester
Methylanatoxin	0.2	<0.02	Methylanatoxin-methylester

Potencies are relative to carbamylcholine. (ALBUQUERQUE and SPIVAK 1984; SWANSON et al. 1991; SWANSON unpublished data).

Table 4. Selectivity of anatoxin analogs for neuronal receptor

	Electric organ [^{125}I]αBGT IC$_{50}$	Rat brain [^{3}H] (−)-nicotine K_i	Selectivity ratio (−)-nicotine αBGT
Anatoxin	2.5×10^{-8}	3.5×10^{-9}	7.1
Anatoxinic acid isoxazolidide	3.2×10^{-5}	3.6×10^{-8}	888.
Anatoxinic acid methoxyamide	8.9×10^{-5}	4.1×10^{-8}	2170.
Anatoxinic acid methyl ester	2.4×10^{-6}	3.7×10^{-8}	65.
αhydroxy-anatoxin	7.2×10^{-6}	2.6×10^{-6}	2.7
N-methyl anatoxin	1.7×10^{-5}	2.6×10^{-6}	6.5

The data for inhibition of [^{125}I]αBGT are from Swanson et al., 1991 and the data for inhibition of [^{3}H] (−)-nicotine binding are from Wonnacott et al., 1991.

point in the pharmacophore with this portion of the agonist. (c) altering of the hydrophobic binding properties of the ligands. To address these issues, we have examined the affinities, potencies, and efficacies of an expanded series of acetyl side-chain analogues of anatoxin. Additional hydrophobic components of the side chain are certainly important to the interaction of anatoxin analogues with the receptor. Throughout a large series, the potencies at the peripheral receptor are well correlated with binding affinities (Swanson et al. 1991). Extension of the chain to become a proprionyl analogue had the least effect on potency (Wonnacott et al., in press). All other modifications had varying detrimental impacts (Table 4). Simple α-hydroxylation radically diminished the agonist activity. Overall, steric factors play a significant role and in addition the more active compounds at neuronal receptors seem to retain the potential for an extended planar configuration (see below).

Anatoxin is also a very effective and selective stimulant of neuronal receptors; the K_i for inhibition of [^{3}H]-(−)-nicotine binding to rat brain membranes was 0.3 nM (MacAllan et al. 1988). The closely related semirigid agonists (Fig. 2) also bound to the neuronal receptor: [^{3}H]cytisine, K_d < 1 nM (Pabreza et al. 1991); pyridohomotropane, K_i = 5 nM for [^{3}H]-(−)-nicotine sites (Kanne and Abood 1988). Although poisoning by pure anatoxin produced profound weakness, which may be due to neuromuscular blockade as described above, rats also display tremor, and furthermore, clonic convulsions occurred immediately prior to death (Swanson and Albuquerque, unpublished observations). Anatoxin applied to motor nerve terminals increased the rate of spontaneous miniature endplate potentials (Spivak et al. 1980) and transiently potentiated twitch (50-400 nM; Swanson et al., unpublished observations). In the autonomic system, anatoxin elicited a neurally mediated contraction of the guinea-pig ileum that could be blocked by hexamethonium or tetrodotoxin (Carmichael et al. 1979). Systemic administration of anatoxin to rats greatly elevated blood pressure, via

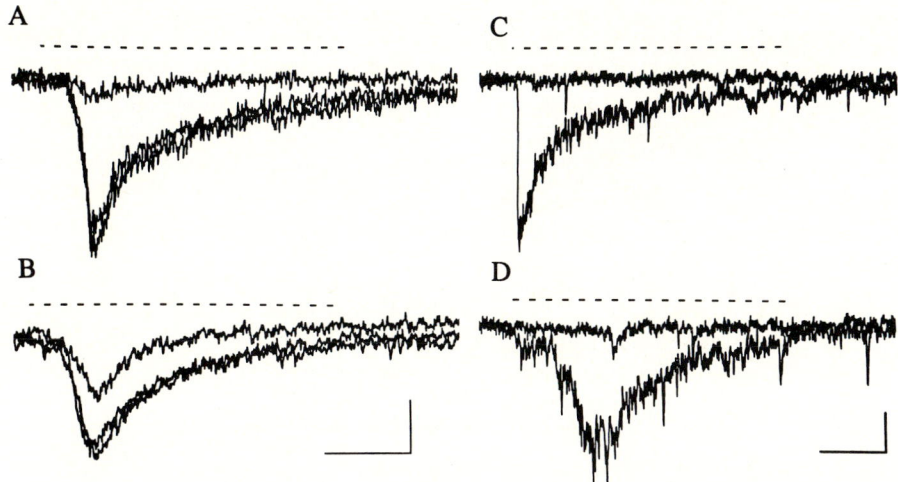

Fig. 3A–D. Whole-cell currents elicited by anatoxin ($10 \mu M$) in cultured hippocampal neurons grown for 17 to 42 days. Nicotinic currents were inhibited by simultaneous application of D-tubocurarine ($50 \mu M$) (**A**) or mecamylamine ($10 \mu M$) (**B**), and after bungarotoxins, 12 min after $0.02 \mu M$ κ-BGT (**C**) or 50 min after $0.3 \mu M$ α-BGT (**D**). H.P. = -50 mV. Calibrations: (**A**) 500 ms, 50 pA (**B**) 500 ms 25 pA (**C,D**) 250 ms 50 pA

actions at autonomic ganglia and secondarily through the stimulation of catecholamine release (SIREN and FEUERSTEIN 1990). At the CNS nicotinic receptor, agonist binding studies found a high (1000-fold) stereospecificity of (+)-anatoxin-a for the [^{3}H]-(−)-nicotine site (MACALLAN et al. 1988) as well as a high selectivity (100-fold) over the muscarinic receptor (ARONSTAM and WITKOP 1981). Anatoxin ($1 \mu M$) stimulated nicotinic receptor responses in cultured hippocampal cells (ARACAVA et al. 1987b; RAMOA et al. 1990) that could be blocked by α-cobratoxin (ALKONDON and ALBUQUERQUE 1990), κ-bungarotoxin, and D-tubocurarine (Fig. 3).

Although other high-affinity agonists for the neuronal AChR are known, including [^{3}H]N-methylcarbamylcholine with K_d = 11 nM (BOKSA and QUIRION 1987), these more flexible molecules have less to offer for identification of the molecular interaction with the agonist site and development of neuroselective agonists. In the same series of anatoxin analogues mentioned above, peripheral potencies and affinities of many analogs were well correlated with the affinities for the brain (−)-nicotine site (WONNACOTT et al. 1991). The intriguing exceptions to the correlation were an isoxazolidide and a methoxyamide that were rejected by the peripheral receptor (SWANSON et al. 1991), while they maintained significant affinity for the neuronal receptor (Table 4; Fig. 4; WONNACOTT et al. 1991). Thus, the semirigid nicotinic agonists (Fig. 2) provide a strong basis for further studies of nicotinic molecular pharmacology, particularly in the brain where many nicotinic

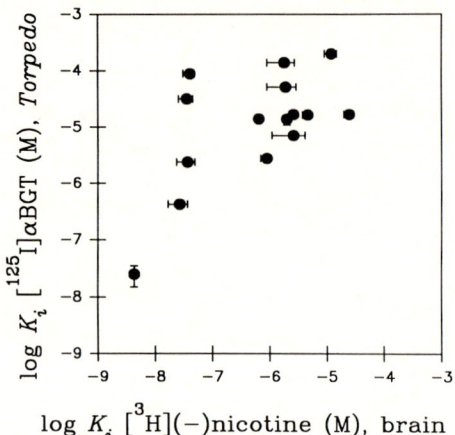

log K_i [^{3}H](−)nicotine (M), brain

Fig. 4. Relationship between competition of anatoxin analogues for acetylcholine sites of rat brain and *Torpedo* electroplax receptors determined using [^{3}H]-(−)-nicotine and α-bungarotoxin ([^{125}I]αBGT), respectively Standard error bars are drawn for both parameters, although they are often smaller than the symbols

receptor subtypes are present but relatively little detail is known about their physiology or pharmacology.

III. Antagonists

Antagonists which act at the ACh site are generally larger than agonists and as a result probably interact with a larger receptor region. Several also have covalent reactions which have made feasible to disassemble the receptor after the bond is formed and determine the location of the antagonist binding site. A selection of antagonists which are promising for characterizing various receptor subtypes will be considered: curare, lophotoxin, neosurugatoxin, methyllycaconitine, and dihydro-β-erythroidine (Fig. 5).

D-Tubocurarine is a reversible antagonist at the ACh target site, and it has the further interesting property that it readily distinguished between the two ACh recognition sites on the peripheral AChR with K_d of 35 nM and 1.2 µM (see Pedersen and Cohen 1990). At these sites, (+)-dimethyl-tubocurarine (DMT) also exhibited a high and a low affinity which differ about 100-fold (Sine and Taylor 1981). When the action of DMT was examined using single-channel current analysis (Sine and Steinbach 1986), it was found that only higher concentrations of toxin, which bound to both agonist sites, actually activated ion currents albeit at an intrinsically low frequency. (Ion channel blocking also occurs at high concentrations; Manalis 1977; Katz and Miledi 1978; Lambert et al. 1980; Shaker et al. 1982; Karpen and Hess 1986b.) Selective expression of subunits showed that α-γ and α-δ pairs formed specific sites for D-tubocurarine (Blount and Merlie

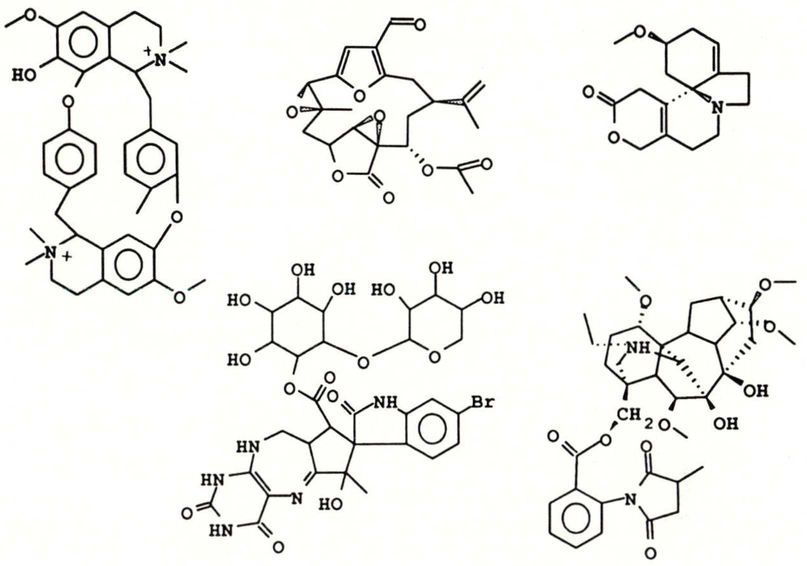

Fig. 5. Structure of competitive antagonists at the ACh site. *Above*, D-tubocurarine, lophotoxin, dihydro-β-erythroidine; *below*, neosurugatoxin, methyllycaconitine

1989). When the toxin was covalently linked to the receptor by using ultraviolet irradiation, sites on α-subunit proteins were labelled, as expected, along with sites on the γ- and δ-subunits (PEDERSEN and COHEN 1990). The site with high affinity for toxin was associated with α- and γ-subunits, and the lower affinity site was associated with α- and δ-subunits. This suggests that the recognition site is near the interface of the subunits and that a sufficiently large molecule may bind to more than one subunit. It is not difficult to imagine that this toxin might stabilize the receptor in a manner to inhibit motion between subunits and to oppose channel activation.

Lophotoxin is found in several Pacific horny corals. This cyclic diterpene toxin is devoid of any alkaloid function but nonetheless produced neuromuscular blockade at a low concentration (80 nM, CULVER and JACOBS 1981; FENICAL et al. 1981) via reduction of endplate current amplitude (ATCHISON et al. 1984). The toxin acted as a competitive irreversible inhibitor of agonist binding to the recognition site; just one α-subunit was labelled at low toxin concentrations (CULVER et al. 1984). Fortunately for the scientists' tool box, this toxin preferred the site which had a low affinity for tubocurarine: Only a high concentration (300 μM) of D-tubocurarine blocked both α-subunit sites and protected against irreversible binding of lophotoxin (3 μM). In contrast, the lophotoxin analog-1 did not distinguish between the two α-subunit sites (CULVER et al. 1985). Lophotoxin bound covalently to tyrosine 190 (ABRAMSON et al. 1989) that is conserved in all known nicotinic receptor α-subunit sequences and is adjacent to the vicinal cysteine (Table 1). Because the toxin exhibited high selectivity between α-subunits, depending upon oligomeric structure when the macromolecule is imbedded in the membrane, its actions on neuronal receptors were also investigated. At the autonomic ganglion, lophotoxin also produced a block of nicotinic transmission (frog, LANGDON and JACOBS 1985; chick ciliary ganglion, SORENSON et al. 1987). [In contrast, the muscarinic and GABAergic electric responses of chick sympathetic ganglia (SORENSON et al. 1987) and muscarinic contractile responses of guinea-pig and rabbit ileum were insensitive to lophotoxin (LANGDON and JACOBS 1985).] For the neuronal α_2- and α_3-subunits expressed in oocytes, lophotoxin produced less than complete inhibition of ACh-induced currents (LUETJE et al. 1990); therefore the toxin seems to have additional interactions that depend upon the nicotinic receptor components beyond the tyrosine 190.

Neosurugatoxin was found in the Japanese ivory mollusc, but as with other shellfish poisonings (e.g., saxitoxin poisoning), the toxin levels varied and likely originated from another source and were accumulated by the mollusc. The source has not been identified, and the toxin was named after Suruga Bay, where this form of shellfish poisoning was first characterized. Initially, the toxicity of extracts was attributed to surugatoxin, and studies demonstrated a selective postsynaptic ganglionic blockade (HAYASHI and YAMADA 1975). The mydriatic effects in mice and the inhibition of the contractile response of the ileum to nicotine were used for the bioassay during the isolation of neosurugatoxin (KOSUGE et al. 1982). This toxin also inhibited the carbachol-induced influx of cations and secretion of catecholamines from adrenal medullary cells (WADA et al. 1989). In the mammalian brain, neosurugatoxin inhibited nicotine binding to two sites, which were distinguished by their affinities (3 nM and 2 μM, YAMADA et al. 1985). Comparison of related compounds showed that binding was dependent upon the bromine moiety rather than the sugar residues (YAMADA et al. 1987). The toxin displayed an irreversible or slow dissociation from the receptor site and an allosteric reduction of nicotine binding (YAMADA et al. 1985, 1987). The similar inhibition by neosurugatoxin analogues of guinea pig ileum responses and forebrain [^{3}H]nicotine binding reinforced the similarity of the ganglionic and CNS nicotinic receptors (YAMADA et al. 1987). Neosurugatoxin also proved to be an extremely potent inhibitor of neurotransmitter release from synaptosomal preparations, via an action on the presynaptic nicotinic receptor (RAPIER et al. 1990). While it seems clear that the neosurugatoxin is a selective antagonist of neuronal nicotinic receptors, this toxin merits detailed analysis of its molecular actions.

The wild flower larkspur, *Delphinium*, produces a toxic syndrome characterized by muscle weakness which has most often been seen in cattle. The aconite alkaloid toxin methyllycaconitine (MLA) isolated from the seeds of *Delphinium*, produced blockade of muscle twitch in rats and frogs (NAMBI AIYAR et al. 1979). This toxin has been found to be an impressively potent insecticide (JENNINGS et al. 1986). Competitive inhibition of α-bungarotoxin binding in fly heads demonstrated a high affinity (2.5×10^{-10} M) of MLA for the cholinergic receptor site (JENNINGS et al. 1986) and thus also suggests postsynaptic action at the nicotinic receptor. By comparison of binding sites in locust ganglion and rat brain, MLA demonstrated 1000-fold greater affinity for the α-bungarotoxin sites ($K_i = 1.4 \times 10^{-9} M$) than for the nicotine sites (MacALLAN et al. 1988). In cultured fetal rat hippocampal neurons, whole cell current evoked by ACh or anatoxin and single channel currents evoked by anatoxin were inhibited by picomolar to femtomolar concentrations of MLA (ALKONDON and ALBUQUERQUE 1991; PEREIRA et al. 1991a). Interestingly, the potency order for aconitine and MLA at vertebrate muscle and insect neurons was reversed (JENNINGS et al. 1986). During sciatic nerve stimulation, $10^{-8} M$ of MLA caused inhibition of twitch in frog sartorius muscle (NAMBI AIYAR et al. 1979), despite the observed K_i of $10^{-5} M$ (measured against $[^{125}I]$-α-bungarotoxin) in frog muscle extract (WARD et al. 1990); thus MLA may have allosteric antagonistic actions at the peripheral AChR. The competitive action of MLA seen at very low concentrations of toxin with binding sites in mammalian and insect neurons could provide a very important tool for receptor discrimination.

Dihydro-β-erythroidine (DHβE) and other alkaloids from beanlike seeds of *Erythrina* trees and shrubs were found to be curaremimetic. Despite relatively little available detail on their molecular actions, DHβE is perhaps worthy of a short note here because of its remarkably smaller molecular size than most antagonists and because the chemical structure reveals carbonyl and amine components typical of agonists. Although DHβE is not a cholinesterase inhibitor, the neuromuscular block produced by it was reversed by neostigmine but not by physostigmine (UNNA et al. 1944); both of the latter also acted directly on the nicotinic receptor (see below in F). DHβE had an anticonvulsant effect (UNNA et al. 1944) and bound to two sites in the brain with high affinity (K_d's of 2 and 22 nM; WILLIAMS and ROBINSON 1984). Several electrophysiological studies demonstrated that DHβE reversed general cholinergic or nicotinic effects on neurons (BIRD and AGHAJANIAN 1976; ROVIRA et al. 1983; VIDAL and CHANGEUX 1989), although the high (μM) concentration of DHβE used in these and other studies allowed that relatively nonspecific actions could be involved.

D. Allosteric Agonist Sites

Several carbamates, including pyridostigmine, (−)physostigmine (Fig. 15), and neostigmine, and the quaternary amine edrophonium, in addition to causing inhibition of cholinesterase and being allosteric antagonists of the peripheral AChR, have also been shown to activate AChR channel currents in single channel-recording experiments (PASCUZZO et al. 1984; AKAIKE et al. 1984; SHAW et al. 1985; SHERBY et al. 1985; ARACAVA et al. 1987a; ALBUQUERQUE et al. 1988c; for discussion of the role of carbamates in poisoning by organophosphorus compounds see below in F). (−)-Physostigmine was a weak agonist at frog muscle (Shaw et al., 1985). Indeed, recent evidence from the laboratory of A. Maelicke has raised the possibility that such channel activation occurs via an action at a site other than the normal agonist site which recognizes ACh: 1) radiolabelled binding studies showed a $K_d = 70\,\mu M$ for (−)physostigmine and a $K_i = 3.9\,mM$ for ACh, clearly not the high affinity ACh-site associated with the normal channel activation; 2) after membrane vesicles from *Torpedo marmorata* were preincubated with α-bungarotoxin or tubocurarine (−)-physostigmine was still able to induce ion fluxes, and 3) these currents were antagonized by benzoquinonium and dibucaine (Okonjo et al. 1991). Furthermore, prolonged treatment with ACh, which led to desensitization of the AChR, was overcome by (−)-physostigmine (Kuhlmann et al. 1991). These authors suggested that "the conformational state of the receptor corresponding to desensitization is confined to the transmitter binding region, leaving the channel fully activatable – albeit only from other than the transmitter binding site(s)". In a manner similar to the above results, stimulation of the nicotinic receptors of cultured rat hippocampal neurons by (−)-physostigmine was noted in single channel studies; the (−)-physostigmine-induced currents were not blocked by the antagonist MLA, while benzoquinonium was effective in blocking the agonistic action of (−)-physostigmine (Albuquerque et al. 1991). Since the channel activating property of (−)-physostigmine is not associated with the carbamate moiety that accounts for its anticholinesterase properties (Okonjo et al. 1991), the stimulation of single channel currents by organophosphorus compounds (Albuquerque et al. 1984) must still be evaluated to determine their mechanism.

E. Allosteric Antagonist Toxin Binding Sites

Toxins which inhibit AChR function through allosteric sites have exhibited a multitude of phenomena when examined at the molecular level. Several discrete antagonist effects have been distinguished electrophysiologically: voltage dependence, time dependence, or voltage- and time dependence and also agonist dependence. An allosteric increase in agonist affinity and efficacy that secondarily results in desensitization at high concentrations has also been noted. Consideration of these different effects suggests that several sites or modes of drug-receptor interaction are possible. Both the open and the closed conformation states of the AChR are altered by allosteric ligands. The type of allosteric action may be determined by the specific structure of the ligand. The importance of studying allosteric mechanisms is emphasized by recent findings of homology among neurotransmitter receptors. The homology among ion channel peptide sequences of nicotinic, γ-aminobutyric acid (GABA$_A$), and glycine receptors led to the suggestion of a genetic relationship or "superfamily" of homologous cation and anion channels that may operate under similar principles (SCHOFIELD et al. 1987; GRENNINGLOH et al. 1987). Pharmacological homology between the GABA and *N*-methyl-D-aspartate (NMDA) receptors has also been reported (SCHWARTZ and MINDLIN 1988; NMDA homology is also described below) and anticonvulsant antagonists of the NMDA receptor were found to inhibit the nicotinic currents. There may be an entire class of such compounds which are in general less polar than those which interact with the Na^+ channel. Here, we will consider a spectrum of toxins (Fig. 6) which illustrate the capacity of drugs to modulate AChR function in selective ways.

There are several conflicting fundamental issues regarding allosteric sites and their effects on AChR activity. (a) Ion channel blockade was described as a reversible concentration- and voltage-dependent blockade of the open conformation of the AChR (STEINBACH 1968). The fact that many nicotinic ion channel blockers are positively charged at physiological pH was suggested as significant to the voltage dependence of the ligand binding to the receptor, almost as though cationic drugs were drawn into the channel along with the cationic current, which is greater at hyperpolarized potentials. For simple ion channel blockers, the voltage-dependent kinetic parameters have been used to estimate the position of the binding site in the membrane electric field (WOODHULL 1973) and authors have suggested a variety of transmembrane locations (see summary in SWANSON and ALBUQUERQUE 1987). Using different techniques, measurement of ion permeation determined that one main binding site is located in the smallest point of the pore, although others may be present in wider regions (DANI 1989). (Secondary sites for the voltage-independent drug actions have been proposed; SANCHEZ et al. 1986.) However, these locations conflict with the observation of competitive inhibition at a single binding site (KARPEN and HESS 1986a). Furthermore, procaine exhibited the same voltage dependence for blockade when in the uncharged form (high pH) or when applied intracellularly (GAGE et al. 1983). (b) Biochemical interactions between drugs with distinct electrophysiological modes of action have been observed. For example, bupivacaine competed for [^{3}H]phencyclicine (PCP) and [^{3}H]perhydrohistrionicotoxin (H_{12}-HTX) binding sites (Fig. 7), and carbamylcholine stimulation intensified these competitive effects. Cocaine, which produces similar ion channel blocking phenomena as bupivacaine, competes for the [^{3}H]-PCP binding site (KARPEN and HESS 1986a). Thus, these voltage-dependent open channel blockers (local anesthetics) and voltage- and time-dependent closed channel blockers (PCP and HTX) apparently compete for regions of the AChR substantially close to each other. Chemical modification of PCP and HTX did not produce parallel effects on binding site interactions (see below). Thus, the comparison of antagonists by competition binding assays may be hazardous because it may not be sufficiently sensitive to distinguish among numerous allosteric actions that electrophysiological methods can distinguish. (c) The amphipathic α-helices of several subunits formed a high-affinity binding site for

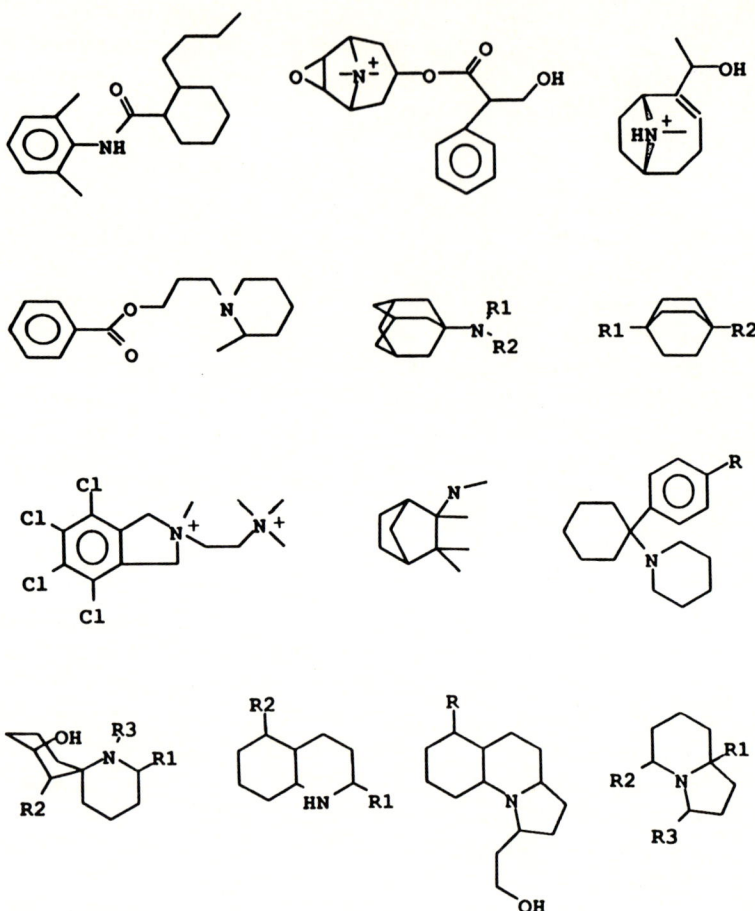

Fig. 6. Allosteric antagonists. *Top row*, bupivacaine, scopolamine, *N*-methyl-anatoxinol; *second row*, piperocaine, adamantanamine (amantadine R1 and R2 are hydrogens), bicyclooctane analogues (R1 is methyl propyl or cyclohexyl and R2 is carboxyl amino, amino methyl, or dipropylaminomethyl groups); *third row*, chlorisondamine, mecamylamine, phencyclidine (R is hydrogen in PCP and is substituted in analogues); *bottom row*, several dendrobatid alkaloids: histrionicotoxin (HTX, a spiropiperidine alkaloid, R1, R2, and R3 are −C−C=C−C≡C, −C=C−C=C, and −H, respectively), pumiliotoxin C (a *cis*-decahydroquinoline alkaloid, R1 is *n*-propyl and R2 methyl), gephyrotoxin (R is −C−C=C−C≡C), and indolizidines (for compound 223$_{AB}$ R1, R2, and R3 are hydrogen, propyl, and butyl moieties, respectively)

chlorpromazine and were proposed to line the ion channel (GIRAUDAT et al. 1987, 1989; OBERTHÜR et al. 1986). However, the electrophysiological characteristics of chlorpromazine-induced blockade were incongruous with a voltage-dependent channel blocking effect mediated by a site located in the electric field of the membrane; rather, the data suggested that this drug

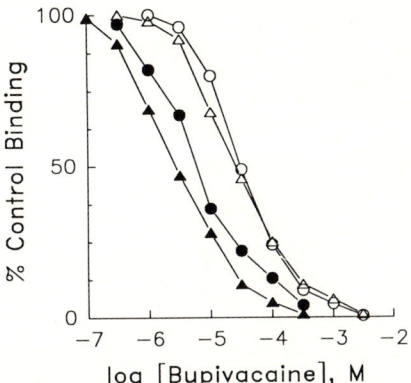

Fig. 7. Binding interactions of allosteric antagonists bupivacaine, phencyclidine, and perhydrohistrionicotoxin (H_{12}-HTX). The competition of bupivacaine for phencyclidine (*circles*) and H_{12}-HTX sites (*triangles*) are enhanced in the presence of carbamylcholine (*filled symbols*) (IKEDA et al. 1984)

allosterically modified desensitization of the AChR (CARP et al. 1983). In summary, the number, specificity, location, and functional effects of allosteric sites are unclear. While a high-affinity site dependent upon van der Waals interactions with a specific drug binding domain could predict sensitivity to minor chemical differences, it does not readily suggest the observed variety of electrophysiological responses. Thus, the mechanism by which binding site interactions can generate different electrophysiological phenomena needs further explanation. How many sites are there? Are various drugs bound by electrostatic or hydrophobic interaction? Do these drugs bind to resting, closed, or open channels? We will consider those properties which distinguish the actions of toxins and then discuss structure-activity relationships.

I. Electrophysiological Characteristics of Allosteric Toxin Actions

1. Open Channel Blockade: Agonist and Voltage Dependence

Many alkaloids interact with several ion channel types. Simple quaternary ammonium ions with large radicals, the smallest being tetraethylammonium, are capable of interaction with the ion channel of nicotinic receptors and also K^+ channels (MEVES this volume). This interaction relies first upon opening of the channel, which allows access of the cation to the pore (STEINBACH 1968; COLQUHOUN and HAWKES 1983). Two groups of alkaloids deserve special mention because of their nonspecific actions: Both antimuscarinics and local anesthetics have dissociable reactions with the nicotinic channel and the molecules are sufficiently large to block the usual cation currents through the ion channel pore. Micromolar concentrations of the drugs are usually necess-

Endplate Currents

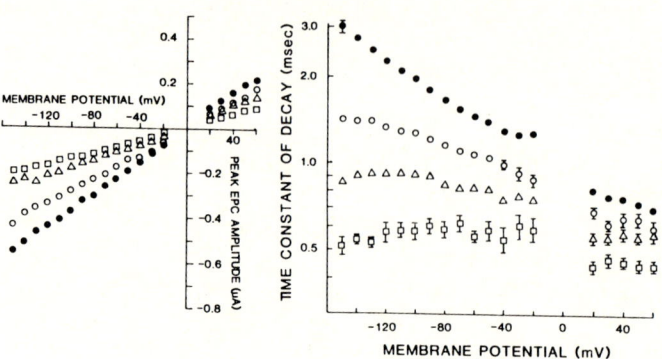

Spontaneous Miniature Endplate Currents

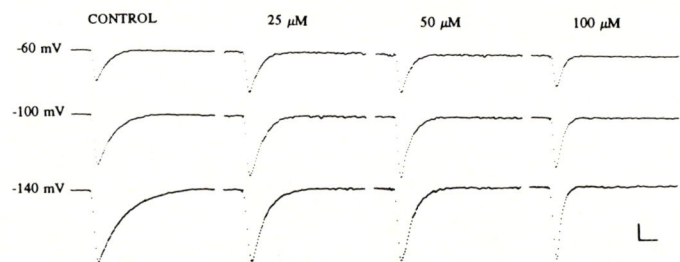

Single Channel Currents

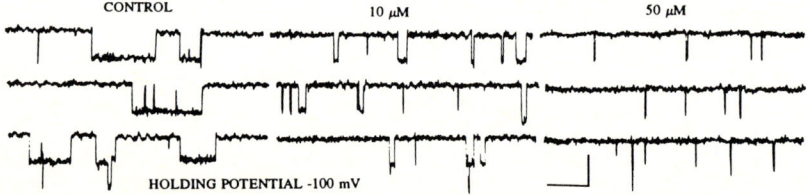

Fig. 8. Voltage-dependent open channel blockade characterized by bupivacaine. *Top*: Endplate currents recorded from *Rana pipiens* sciatic nerve-sartorius muscle preparation using two-electrode voltage-clamp technique. The concentration-dependent decrease of peak amplitude and voltage sensitivity of the τ_{EPC} under control conditions (●) and after exposure to bupivacaine 25 (○), 50 (△), or 100 (□) μM. Each symbol represents the mean of 8–24 fibers from 2 to 3 muscles (Ikeda et al. 1984). *Center*: Families of averaged digitized miniature endplate currents recorded from frog cutaneous pectoris muscle. Each trace represents the signal average of 16–45 miniature endplate currents from a single endplate. Calibrations are 1 ms horizontally and 1.58, 0.91, 0.95, and 0.73 nA vertically for control and bupivacaine concentrations of 25, 50, and 100 μM, respectively (Ikeda et al. 1984). *Bottom*: Samples of ACh-activated single-channel currents recorded from cultured rat myoballs; ACh alone, or ACh with 10, or 50 μM bupivacaine. Holding potential, -100 mV; calibration: 100 ms horizontally and 2 pA vertically (Aracava et al. 1984)

ary to produce ion channel blockade. Potent antimuscarinics, with K_d for the muscarinic receptor in the nanomolar range, can be used systemically in lower concentrations without producing neuromuscular block as a side effect. In contrast, local anesthetics have more similar dose-response relationships for antagonism of the Na^+ channel (BECKER and GORDON, this volume) and nicotinic receptors, and therefore appropriate doses of drugs such as procaine and cocaine can block neuromuscular transmission by a postsynaptic action (ADAMS 1977; SWANSON and ALBUQUERQUE 1987).

Drugs which appeared to act primarily through open channel blocking are, for example, bupivacaine (ARACAVA et al. 1984, IKEDA et al. 1984), procaine (ADAMS 1977), cocaine (SWANSON and ALBUQUERQUE 1987), quinacrine (TSAI et al. 1979), and quinuclidinylbenzilate (QNB) (SCHOFIELD et al. 1981). As shown by bupivacaine (Fig. 8), the open time of the channel was determined by the rate of drug binding; the decay of the (EPC) reflected this open channel blockade directly. In addition, scopolamine and atropine illustrated a fundamental kinetic principle in receptor ion channel blockade (ADLER et al. 1978). These muscarinic receptor antagonists differ structurally only by the presence of an epoxide radical on scopolamine (Fig. 6). Scopolamine produced a reversible block of the receptor when in its open channel conformation. Atropine acted similarly but dissociated more slowly from the blocked-open ion channel. When observing endplate currents, scopolamine produced a double-exponential decay from the peak current, whereas atropine produced a rapid, single-exponential decay (Fig. 9). The corollary in single-channel recordings was a pattern of multiple channel openings in bursts versus a predominantly single-channel opening pattern, respectively. In the case of atropine, the total open duration of the channel was greatly reduced to a similar extent as the endplate current was shortened (CUNS et al. 1990). A similar comparison of channel blocking modes has been made with the local anesthetics QX-314 and QX-222, (quaternary N-ethyl- and N-methyllidocaine derivatives), respectively (NEHER and STEINBACH 1978; NEHER 1983). Further specificity of the open channel blocking reaction may be present: The diastereomeric pair (S)- and (R)-N-methyl-anatoxinol both exhibited reversible ion channel blockade, but their actions displayed voltage-dependent and voltage-independent kinetics, respectively (see Fig. 14 in ARACAVA et al. 1988; SWANSON et al. 1989). Thus, the dissociation of the ligands from the blocked-open channel had a specific chemical basis. The possibility for the creation or discovery of more specific toxins with additional steric dependence is suggested by those already observed.

Amantadine represents a class of weak agonists for which the EPC decay time constant appeared to be reversed relative to the control voltage dependence (TSAI et al. 1978). The decay phases are single exponentials, thus the longer decay time could not result from an erroneous measurement of a fragmented decay phase. Figure 10 shows that amantadine in addition to decreasing the decay time constant (τ_{EPC}) at hyperpolarized potentials, increased the channel decay time constant occurred at depolarized potentials.

CONTROL

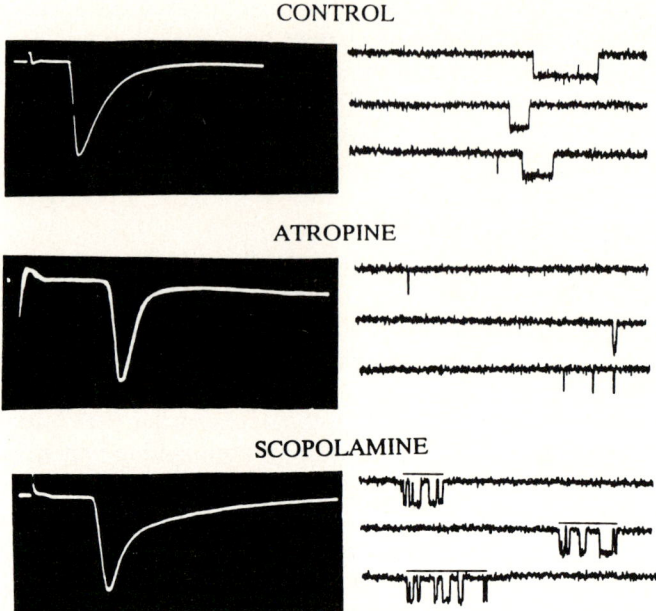

ATROPINE

SCOPOLAMINE

Fig. 9. Comparison of the effects of rapidly and slowly dissociating ion channel blockers on endplate currents and single-channel currents. The endplate currents shown were recorded at a membrane potential of $-100\,mV$. Single-channel currents were activated by acetylcholine ($0.4\,\mu M$) and are shown at $-100\,mV$; atropine ($20\,\mu M$) produced a fast single-exponential decay and well-spaced single-channel currents. With both techniques the total current was greatly decreased. Scopolamine ($10\,\mu M$) induced double-exponential decay, and the single-channel currents appeared as bursts of numerous brief openings; the total currents were conserved relative to control (Adler et al. 1978; Cuns et al. 1990)

As verified with the patch-clamp technique, amantadine did have open channel blocking actions which were greater at the hyperpolarized potentials (Cintra and Albuquerque, unpublished results). However, it is most unlikely for open channel blocking (in the sense that drug binding prematurely obstructs the ion conductance path) to prolong the currents, and therefore alternative mechanisms have been considered. The kinetics for conformational shifts may be allosterically prolonged to yield a slower channel closing rate, particularly at depolarized potentials, or a large allosteric increase of the agonist affinity would prolong the agonist effect. Further complications were that amantadine had a weak agonistic action, which resulted in relatively prolonged channel openings (Cintra and Albuquerque, unpublished results), and it inhibited cholinesterase. Any of these conditions in combination with a simultaneous voltage-dependent ion channel blockade at hyperpolarized potentials have appeared as a reversal of the normal voltage dependence of the EPC decay. The case of amantadine was also useful to emphasize the important principle that the effect on τ_{EPC} can be

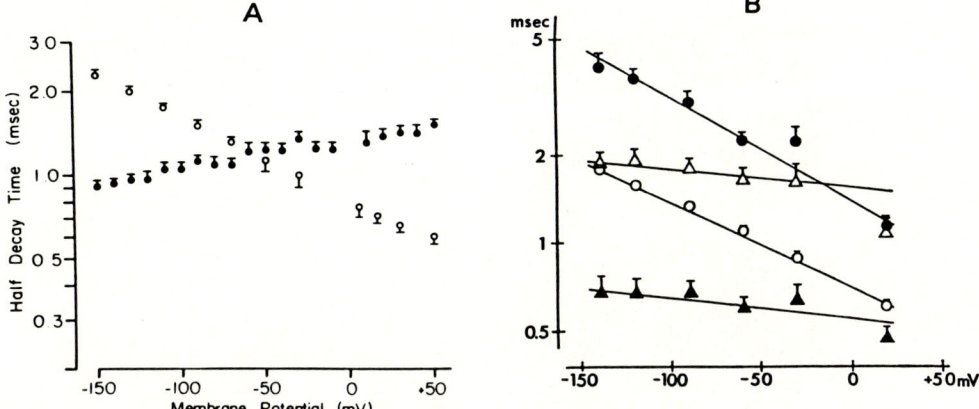

Fig. 10A,B. Effects of amantadine and diisopropylfluorophosphate (DFP) on endplate current (EPC) decay. **A** The relation between the half decay time of the EPC and membrane potential in control (○) and 200 μM amantadine (●). Each symbol represents the mean ±SEM of 20 fibers from five sartorius muscles (Tsai et al. 1978). **B** The same relationship in control (○), during exposure to 1.1 mM DFP (▲), after washing out the excess DFP (●), and after treatment with 9 mM of the oxime 2-pralidoxime (△). Each point represents the mean half decay time ±SEM of 6–38 cells (Kuba et al. 1974)

discrete from the effects on EPC amplitude; e.g., at −50 mV the τ_{EPC} was unchanged by amantadine, but the amplitude of the EPC was drastically reduced.

2. Closed Channel Blockade: Voltage Dependence or Voltage and Time Dependence

Several conditions have been recognized for the depression of peak synaptic current amplitude and, therefore, blockade of transmission. The peak amplitude of the current largely reflects the states of the receptors prior to agonist release, and linear depression of the plot of peak amplitude versus current reflects a decreased number of receptors available for activation. Interactions at the agonist site, be they competitive or irreversible binding, will reduce the population of accessible agonist sites and hence the number of activatable receptors. By accumulating receptors into any state other than the resting condition, the amplitude of the response may also be decreased. The open channel blocking mechanism, already described above, has reduced the EPC rise time and along with it the peak amplitude of the current via a rapid onset of channel block at extremely high concentrations of blocker (Swanson and Albuquerque 1987). Persistent (slowly reversible) open channel blockade keeps the receptors in the blocked-open channel state. Closed channel blockade and desensitization can also reduce the pool of activatable receptors.

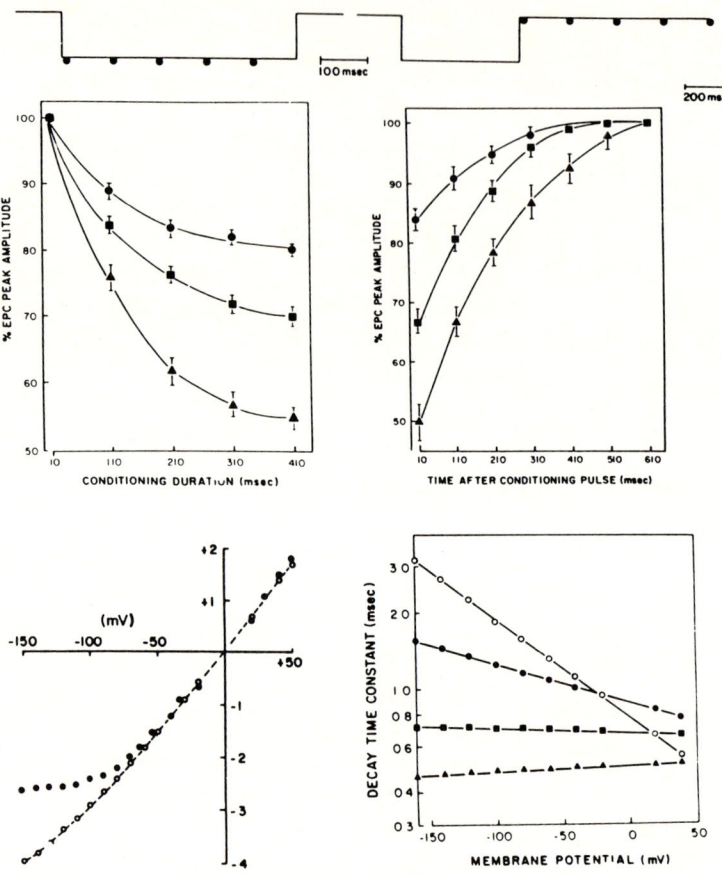

Fig. 11. Voltage- and time-dependent blockade of endplate currents (EPCs) characterized by piperocaine. Development of blockade following a voltage jump from −50 mV to the potential indicated (*upper left*) and recovery after voltage conditioning for 500 ms and returning to −50 mV (*upper right*): Effects on peak amplitude at various holding potentials, −75 mV (●), −100 mV (■), and −130 mV (▲) in the presence of 50 μM piperocaine. *Vertical bars* represent ±SE in 5 experiments. The peak current-voltage relationships in 50 μM piperocaine (*lower left*) was dependent upon the conditioning duration of 25 (○) or 500 (●) ms. The relationship of decay time contant (τ_{EPC}, *lower right*) and membrane potential in control (○) or piperocaine 25 (●), 50 (■), or 75 (▲) μM was also voltage-dependent (TIEDT et al. 1979)

When neuromuscular junctions were exposed to tetraethylammonium (TEA), the relationship of maximum current to clamp potential (I-V plot) was nonlinear; the amplitude was only decreased at hyperpolarized potentials (ADLER et al. 1979). This was independent of the conditioning duration from 5 ms to 2.5 s. Whereas with TEA no brevity of voltage conditioning revealed a linear I-V relationship, for some antagonists another form of blockade develops during an interstimulus interval (in the absence of transmitter) that depends upon the time of conditioning voltage. With piperocaine, the response measured after varying conditioning times at a hyperpolarized potential decreased progressively, and the responses measured shortly after subsequent return to a less hyperpolarized potential also recovered progressively and the responses measured shortly after subsequent return to a less hyperpolarized potential also recovered progressively (Fig. 11). Curvature of the I-V plot was only present with long voltage conditioning durations (2.5 s) but not with short conditioning (10 ms) (TIEDT et al. 1979). Hence, this type of blockade is called voltage- and time-dependent. This blockade developed at hyperpolarized potentials during the absence of agonist and is, therefore, a form of closed channel blockade. The same voltage- and time-dependent effect was seen for HTX, PCP, and meproadifen, and, in addition, an even greater change occurred. In a related voltage- and time-dependent paradigm, the macroscopic conductance portrayed a clockwise hysteresis in the third quadrant of the I-V plot (Fig. 12).

The curvature or hysteresis of the I-V plot associated with closed channel blockade was dissociated from effects on the decay time constant. The actions on τ_{EPC} were not related to time of conditioning. The effects of TEA on EPC amplitude are also independent of effects on τ_{EPC}: The reduction of peak amplitudes continued above $100\,\mu M$ TEA, although there was no further effect on the decay time constant. Although piperocaine shortened the decay phase of EPCs, no difference in decay time was noted with the voltage sequence which produced marked differences in amplitude of responses (Fig. 11; TIEDT et al. 1979).

3. Desensitization: Agonist and Time Dependence

Although desensitization does not normally develop during a single neuromuscular synaptic event, repetitive stimulation or cholinesterase inhibition may induce desensitization. The receptor in this state has greater than normal affinity for agonists but does not undergo transformation into the open channel conformation. Prolonged or repetitive stimulation by agonist led to a decrement of response by transition to nonconducting, agonist-bound states (BARRANTES 1978) which only slowly reverted to the activatable state. These desensitized states were an intrinsic characteristic of the receptor and did not depend upon its modification by processes such as phosphorylation (see below). Two phases of desensitization and recovery were apparent when agonists are applied for seconds or minutes (FELTZ and TRAUTMANN 1982). The rate of desensitization was also voltage-sensitive, being more rapid at hyperpolarized potentials (NASTUK and PARSONS 1970; FIEKERS et al. 1980).

Enhancement of desensitization also occurred through allosteric mechanisms. The binding of ligands such as HTX or PCP enhanced the affinity of the agonists. Phenothiazines are among the most selective drugs in this regard: chlorpromazine (CPZ) had neither an effect on τ_{EPC} nor a voltage- and time-dependent effect on the amplitude of the EPC, yet macroscopic

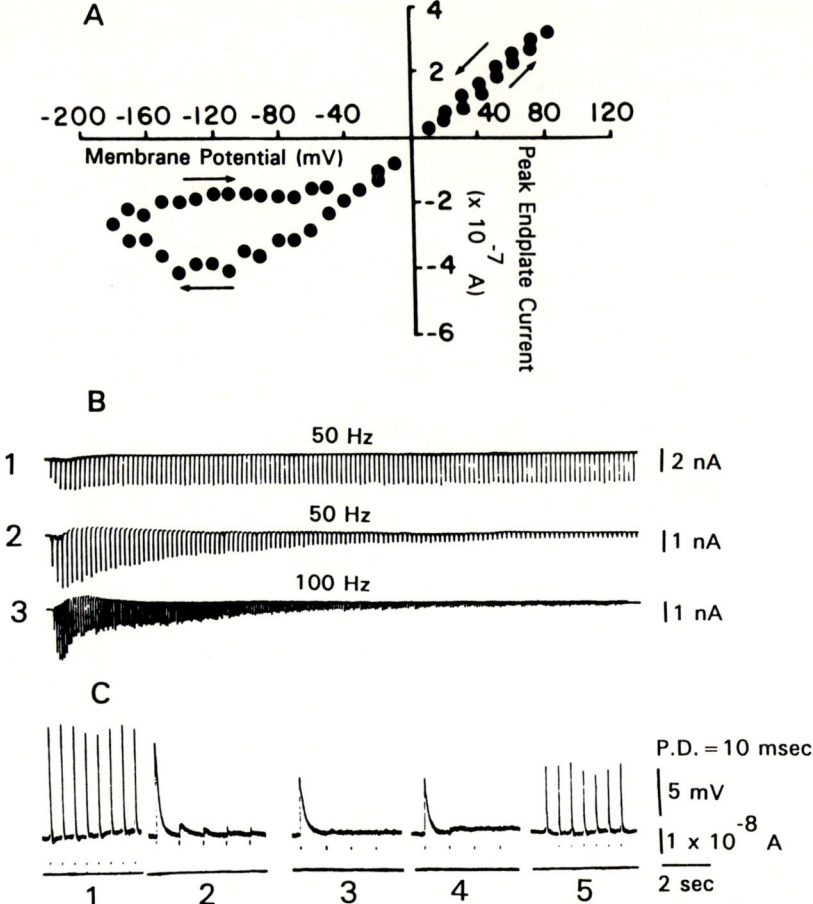

Fig. 12. A Significant hysteresis in the current-voltage (I-V) relationship of peak endplate current (EPC) amplitude in the presence of histrionicotoxin (30 μ*M*). The endplate region of a representative fiber was conditioned for 3 s at each potential level by voltage clamp. EPCs were evoked by nerve stimulation at the end of each step. The sequence of steps began at −50 mV and progressed first toward positive values following the direction of the *arrows*. At initial depolarized potentials the response was stable, but hyperpolarization attenuated the response, producing a concave upward curvature of the I-V plot, and when the holding potentials were finally depolarized by 10 mV steps, the response amplitude recovered (MASUKAWA and ALBUQUERQUE 1978). **B** EPC amplitudes during repetitive stimulation under control conditions (*1*) and in the presence of perhydrohistrionicotoxin (H_{12}-HTX; *2* and *3*). H_{12}-HTX (20 μ*M*) first enhanced facilitation, probably by prolonging the action potential decay phase in the presynaptic nerve terminal. The relative depression of peak amplitude at the end of the train in the presence of H_{12}-HTX was profound. The membrane potential was −90 mV (SPIVAK et al. 1982). **C** Effect on the sensitivity of the 10-day chronically denervated soleus muscles of the rat to acetylcholine (ACh) microiontophoretically applied at the extrajunctional region. Pulse duration (*P.D.*) of microiontophoretic current was 10 ms. The frames show *1* control potentials, *2* after 15 min exposure to H_{12}-HTX (17.5 μ*M*), *3* after 30 min, *4* 30 s after the previous frame, and *5* recovery after washing for 30 min. The first potential response of ACh was always elicited, followed by complete blockade of the potentials during subsequent applications of ACh. In order for the first ACh potential in sequential trains of impulses to be observed in the presence of HTXs, about 10–15 s must have elapsed between trains (ALBUQUERQUE et al. 1973b)

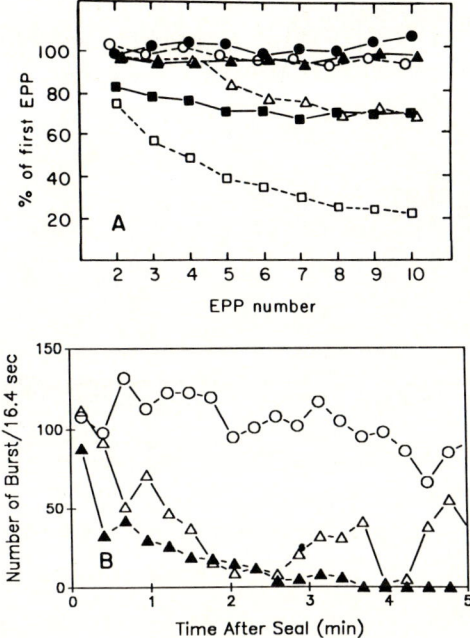

Fig. 13. Desensitization. *Top*, Effect chlorpromazine on endplate potentials (EPPs) evoked by repetitive microiontophoretic application of acetylcholine (ACh). Trains of EPPs were evoked at frequencies of 0.5 Hz (*solid line with closed symbols*) and 5.0 Hz (*broken lines with open symbols*). Each point represents the mean peak amplitude ($n = 3$) of the second to the tenth EPP expressed on the ordinate as a percentage of the peak amplitude of the first EPP for controls ($\bigcirc$, $\bullet$), $1 \mu M$ ($\triangle$, $\blacktriangle$), and $5 \mu M$ ($\square$, $\blacksquare$) chlorpromazine. The SEM for each point is <10% of the mean (Carp et al. 1983). *Bottom*, Effect of metaphit on the activation of single-channel currents in isolated frog muscle fiber. The graph shows the frequency of channels activated by ACh ($0.4 \mu M$) after obtaining the seal at -120 mV membrane potential at 0, 0.5, and 1.0 μM metaphit ($\bigcirc$, $\triangle$, and $\blacktriangle$, respectively) (Tano et al. 1990)

conductance decreased at all potentials. More importantly, the EPC amplitude was decreased by repetitive stimulation, more so at 5 Hz than 0.5 Hz, due to desensitization (Fig. 13; Carp et al. 1983). Similarly, metaphit, an antagonist of PCP-induced behaviors (Contreras et al. 1985; French et al. 1987), was recently shown to have an extremely selective action to decrease channel activation (Fig. 13; Tano et al. 1990) without changing lifetime in single-channel studies. Meproadifen, PCP, and HTX also elicited desensitization, but less selectively: Meproadifen had voltage- and time-dependent effects on the EPC amplitude, and PCP and HTX had both voltage- and time-dependent effects and open channel blocking effects (already described above). Still, HTX and its analogues prominently display the agonist-dependent phenomenon at high (tetanic) frequencies. In control solution, EPCs were stable up to tetanic frequencies. In the presence of HTX, high-frequency stimulation of EPCs, at a single voltage-clamp potential (after the

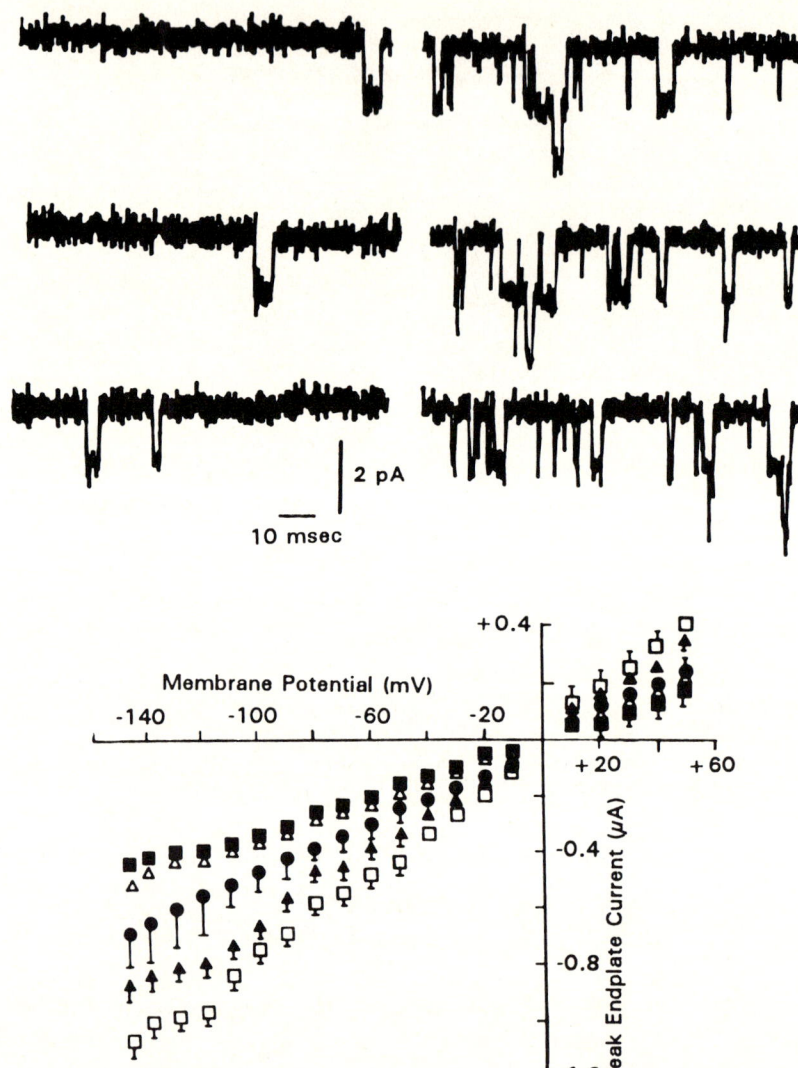

Fig. 14. Allosteric activation. *Top*, Single-channel records from the same perfused outside outpatch excised from cultured rat myoballs showing the excitatory effect of naltrexone. Data were recorded at 1 kHz band width, temperature 10°C, holding potential −40 mV. *Left*, 0.2 μM acetylcholine (ACh); *right*, 0.2 μM ACh plus 0.2 μM naltrexone. Downward deflections represent positive inward currents (Madsen and Albuquerque 1985). *Bottom*, The potentiation of the peak endplate current (EPC) amplitude at nanomolar concentrations of naltrexone. The relationship between peak EPC amplitude and membrane potential at low concentrations of naltrexone: control (●), 250 nM (□), 500 nM (▲), 1 μM (△), and 2 μM (■). At least 30-min exposure to naltrexone was allowed before recording EPCs (Oliveira et al. 1987)

voltage- and time-dependent effect responsible for hysteresis, (described above, had reached equilibrium), demonstrated a further loss of response amplitude (Fig. 12). Microiontophoresis of ACh also had a similar inhibitory action in HTX-treated preparations (Fig. 12).

At minimal concentrations of some allosteric antagonists, i.e., antagonists which increased desensitization at higher concentrations, an increase in the activation response to agonist was seen. Naltrexone is one such drug (Fig. 14): submicromolar concentrations greatly enhanced the rate of single-channel activation (MADSEN and ALBUQUERQUE 1985) and doubled the amplitude of the EPC (OLIVEIRA et al. 1987). (The decay time constant was also decreased by micromolar concentrations of naltrexone.)

The physiological mechanism of desensitization and the role of phosphorylation in producing desensitization of nicotinic receptors are as yet unclear (HUGANIR and GREENGARD 1986; STEINBACH and ZEMPLE 1987; OCHOA et al. 1989). Phosphorylation of several sites on various AChR protein subunits have been observed (HUGANIR et al. 1986), but which of these contributes to physiological effects is uncertain. Stimulation of cAMP-dependent cyclase by forskolin did not change the resting sensitivity of the receptor but did increase the rate of desensitization produced by repetitive stimulation with agonists (ALBUQUERQUE et al. 1986; MIDDLETON et al. 1986).

II. Antagonist Actions at Homologous Receptors

In the central nervous system as in muscle, H_{12}-HTX was an antagonist: The release of neurotransmitters from rat brain synaptosomes stimulated by nicotine was inhibited in the presence of $5\,\mu M$ HTX, while the release due to K^+-induced depolarization was unchanged (RAPIER et al. 1987). However, this action in the brain is not selective. Due to the homology of the nicotinic receptor and other neurotransmitter receptor ion channels, H_{12}-HTX was also an antagonist of the NMDA receptor. The NMDA-evoked single channel currents in cultured neurons in the presence of H_{12}-HTX, measured with the single-channel recording technique, showed characteristics of ion channel blockade and channel activation was reduced (LIMA-LANDMAN and ALBUQUERQUE 1988). Similarly, PCP was shown to antagonize the NMDA receptor in cultured neurons; however, its mode of action here seemed to differ quantitatively from that at the peripheral AChR: the predominant action of PCP on NMDA receptors was a drastic reduction in the frequency of channel activation at hyperpolarized potentials (RAMOA and ALBUQUERQUE 1988). Surprisingly, the m-NO_2 analogue of PCP stimulated NMDA channel currents. In the reciprocal homologous relationship, antagonists initially thought to specifically block the NMDA receptor are also now recognized as nicotinic receptor antagonists: for example, MK-801 inhibited transmission in skeletal muscle, blocked AChR ion currents of retinal ganglion cells, and inhibited transmitter release in synaptosomal preparations from rat forebrain (RAMOA et al. 1990). Thus, the allosteric drug binding sites present on receptors with ion channel sequence homology have produced related (though not necessarily identical) antagonist effects.

III. Structure-Activity Relationships at Allosteric Antagonist Sites

Many studies have indicated that the action of allosteric antagonists is related to the hydrophobicity of the compounds including PCP derivatives and dendrobatid alkaloids such as HTX and gephyrotoxin analogues. For example, substitutions on the phenyl ring of PCP yielding more polar compounds (p-NO$_2$-PCP and p-CH$_3$O-PCP) reduced the potency, but less polar compounds (p-Cl-PCP, p-F-PCP, or p-CH$_3$-PCP) were more potent (Albuquerque et al. 1988a). However, this potency was unrelated to the mode of action, i.e., the chloro, fluoro, and nitro substitutions reduced voltage- and time-dependent effects (hysteresis) while methyl and methoxy substitution maintained these effects. Replacement of the phenyl ring with thienyl ring also increased potency, without changing the mode of action. Comparison of the potency of various PCP analogues revealed that, apparently, the actions of PCP and its analogues on the AChR were unrelated to the receptors that are the target for the hallucinogenic properties of PCP. Instead, the studies of alternating behaviors elicited by these compounds strongly supported the theory that alteration of K$^+$ conductance is significant for the behavioral effects of PCP. The effect on NMDA is most likely secondary (Ramoa and Albuquerque 1988), and the presynaptic effect on g_k is probably the key effect (Albuquerque et al. 1988a).

Amine moieties are prevalent among allosteric antagonists. Using the bicyclooctane analogues of amantadine, only those with amine moieties (R$_2$) were active, whereas carboxyl substitution (R$_2$ = COOH) eliminated the efficacy (Warnick et al. 1984). N-alkylation of amantadine (R$_1$ and R$_2$ = H) to yield methyl, ethyl, propyl, butyl, and dimethyl derivatives increased the potency of antagonism of muscle twitch; although N-butyl amantadine was stronger than amantadine, the shorter chain substituted analogues were even more potent. The inhibition of H$_{12}$-HTX binding was similarly increased by N-alkylated amantadine derivatives. The addition of larger alkyl groups apparently reduced the neuromuscular blocking effects: ID-16 (1-amine-4-hexylbicyclooctane) was ineffective (Warnick et al. 1982c, 1984). Four pumiliotoxin C variations from 0 to 4 carbon side chain length indicated that di-n-propyl substitutions yielded the most potent compound (Warnick et al. 1982b). A series of N-alkyl substitutions to PCP also demonstrated a general increase in blocking potency with chain length up to 4 carbons (Warnick et al. 1982a). The replacement of the piperidine moiety of PCP by pyrrolidine (removing one carbon from the heterocyclic ring) produced a compound with similar actions to PCP. However, upon replacement with a morpholino group (oxygen substitution for carbon in the *para* position of the heterocyclic ring) the voltage- and time-dependent actions on the AChR were eliminated (Aguayo and Albuquerque 1986).

Several of the allosteric antagonists have hydroxyl moieties, and for some of those for which structure-activity comparisons can be made, the stereochemistry of the hydroxyl may be relevant. For indolizidine alkaloids (a dendrobatid alkaloid), the addition of hydroxyl groups reduced the affinity for the H_{12}-HTX binding site, with little specificity as to hydroxyl orientation (ARONSTAM et al. 1986). However, note that both of the natural frog toxins HTX and gephyrotoxin already have hydroxyl moieties present in their natural structures. Removal of the hydroxyl from HTX had little or no effect on HTX binding but decreased potency against PCP binding three- to four-fold (ARONSTAM et al. 1985). Although the hydroxyl appeared to interact with the N in the crystalline HTX (SPIVAK et al. 1983a), alteration of the hydroxyl did not have remarkable effects on the activity of the analogues (ARONSTAM et al. 1985). The presence of a hydroxyl on indolizidine analogues was detrimental to the inhibition of $[^3H]$-H_{12}-HTX or $[^3H]$-PCP binding. In contrast to these mild effects, the orientation of the secondary alcohol in the N-methylanatoxinols was considerably more important: The R isomer inhibited H_{12}-HTX binding more effectively (K_i against $[^3H]$-H_{12}-HTX was several-fold lower) than the S isomer, and although the open channel blockade by the R isomer was slower, it was also considerably more stable than for the S configuration (SWANSON et al. 1989).

The degree of saturation of the two side chains of HTX derivatives was related to the time for inhibition of muscle twitch (SPIVAK et al. 1982; MALEQUE et al. 1984): analogues with greater saturation required longer to develop blockade. This assay incorporated the combined effects on the AChR, Na^+, and K^+ channels. The specific effects of these compounds on the AChR produced a less clear pattern. The potency to inhibit the specific binding to $[^3H]$-H_{12}-HTX and $[^3H]$-PCP sites was increased by reduction of several different positions in the five-carbon side chain, indicating that the side chains in natural HTX are not optimal for the allosteric binding site. Alteration of $[^3H]$-H_{12}-HTX binding did not parallel the alterations of $[^3H]$-PCP binding. Changes in either conformation or electronic configuration could have been responsible for these effects. Depentyl-HTX was less potent but had relatively greater effects on the decay time constant than did HTX or H_{12}-HTX and retained the closed channel blocking properties (MALEQUE et al. 1984). Depentyl-HTX (= desamyl-HTX) had a lower affinity for $[^3H]$-H_{12}-HTX and $[^3H]$-PCP sites (ARONSTAM et al. 1985). Azaspiro-HTX (both side chains removed) was particularly weak (SPIVAK et al. 1982).

Stereospecificity has not generally been reported for the noncompetitive blocking actions of nicotinic antagonists. The optical antipodes of H_{12}-HTX were tested at $10 \mu M$ (the 50% inhibition of amplitude) by EPC assay. In no case were the effects of natural $(-)$- and synthetic $(+)$-H_{12}-HTX on EPC amplitude or decay time constants different (SPIVAK et al. 1983a). Further, both H_{12}-HTX enantiomers were equally effective at inhibiting the binding of $[^3H]$-$(-)$-H_{12}-HTX to the ion channel, in the presence and absence of carbamylcholine (Aronstam et al. 1985). An important implication arose from these observations on the ion channels of the AChR, i.e., the ion channel does not distinguish stereoisomers of noncompetitive ligands which blocked the channel in either the open or the closed conformation. The binding of natural physostigmine (-) and its optical enantiomer (+) both produced very similar blockade of the nicotinic AChR ion channel, even though they were dissimilar in regard to anticholinesterase actions. Gephyrotoxin enantiomers were not different with respect to inhibition of binding of either $[^3H]$-PCP or $[^3H]$-H_{12}-HTX (Aronstam et al. 1986). Both (+)- and (−)-nicotine produced similar noncompetitive blockade of the ion channel in a variety of electrophysiological studies (Rozental et al. 1989). One exception describes a weakly enantio-specific interaction: the (−)-H_{12}-HTX was three fold more potent than the (+)-enantiomer for inhibition of $[^3H]$-PCP binding (K_i of 25 and 85 nM, respectively; Aronstam et al. 1985). The general lack of stereospecificity of the allosteric blocking agents has become useful in separating the various roles of agents with multiple modes of action such as agonist, antagonist, and anticholinesterase properties, e.g., the stereospecific anticholinesterase actions of (−)-physostigmine were eliminated in studies of the enantiomer (+)-physostigmine so that the toxicological significance of noncompetitive blockade could be determined (see below in F).

F. Poisoning by Anticholinesterase Agents and Antidotal Therapy Mediated by AChR Actions

Most anticholinesterase agents certainly produce many effects, i.e., blockade of acetylcholinesterase and a variety of effects on AChRs. Their effects on the CNS nicotinic receptors are largely unknown. Attention has focused on their cholinesterase inhibiting properties, largely because this was the topic of early research using biochemical techniques. The classic concept of anticholinesterase poisoning is that after inhibition of acetylcholinesterase, there is an excess of ACh in the synapse, and activation of AChR leads to an accumulation of receptor in the desensitized state, i.e., depolarizing blockade, the signs of which include flaccid (except in birds) paralysis and respiratory failure. Three components of respiratory failure were identified as the peripheral neuromuscular blockade through nicotinic mechanisms, increased resistance to lung inflation, and inhibition of respiratory centers through muscarinic mechanisms (Wright 1954). However, it has been shown repeatedly that anticholinesterase agents also produced profound, reversible actions by acting directly on the AChR. Diisopropylfluorophosphate (DFP) examplifies the effect of many such agents (Fig. 10). Treatment of skeletal muscle with DFP causes a blockade of the ion channel, and thereby shortened the time constant for EPC decay (Kuba et al. 1973, 1974). Only after washing was the irreversible anticholinesterase effect seen, and the decay time was prolonged because a high concentration of ACh is maintained in

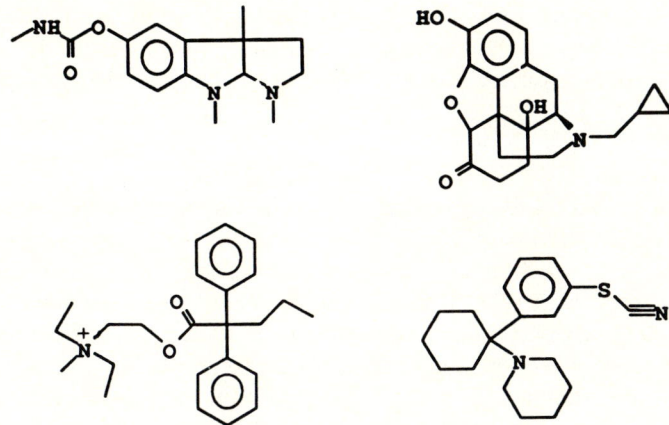

Fig. 15. Agents which modify acetylcholine receptor activation. *Above*, physostigmine, naltrexone; *below*, meproadifen, metaphit

the synaptic cleft so that ACh is able to reassociate with the receptor and reactivate ion currents. Besides being cholinesterase inhibitors, triphenylmethylphosphonium TPMP (LAUFFER and HUCHO 1982; SPIVAK and ALBUQUERQUE 1985), the nerve gas agent VX (0-ethyl S-2-diiosopropylaminoethyl methyl phosphonothiolate; RAO et al. 1987), and several carbamates (ALBUQUERQUE et al. 1988c) also had ion channel blocking actions on the nicotinic receptor. Indeed, the photoaffinity labeling of the AChR with TPMP has been useful in describing a channel site for blockade (HUCHO et al. 1986; HUCHO 1986). Recall that the agonist effects of carbamates and nerve agents were described above. These allosteric effects of pesticides and chemical warfare agents must be considered in prophylactic and antidotal treatment. The presently accepted treatments with atropine and cholinesterase reactivators are not optimal. The multiple effects of anticholinesterase agents apply not only to peripheral AChR but also to the CNS receptors, as is evident from the greater protection against nerve agent (sarin) poisoning afforded by physostigmine than by quaternary anticholinesterases, which do not cross the BBB, even after pretreatment with atropine which does cross the BBB (DESHPANDE et al. 1986).

The significance of noncholinesterase actions in both the development of toxicity and prophylactic treatment was highlighted by the use of a stereoselective agent. The use of reversible anticholinesterases (carbamates) in the prophylaxis of organophosphate poisoning was based on the concept that reversible association of a carbamate with acetylcholinesterase prevents the covalent binding of the irreversible organophosphorus compounds to the enzyme. (−)-Physostigmine (Fig. 15) in combination with antimuscarinic and ganglionic blockers (atropine and mecamylamine, respectively) did indeed reduce the lethality of VX (ALBUQUERQUE et al. 1985). In contrast to the natural isomer, synthetic (+)-physostigmine had very weak anticholinesterase activity (ALBUQUERQUE et al. 1988c) yet had antidotal effects against sarin poisoning (ALBUQUERQUE et al. 1987; KAWABUCHI et al. 1988). Therefore, the protective effects of physostigmine enantiomers may be more appropriately assigned to both their actions on acetylcholinesterase and their direct effects on the AChR.

Prolonged exposure of acetylcholinesterase to organophosphate leads to a secondary chemical reaction known as aging. Desire to remove organophosphates from acetylcholinesterase was the basis for the development of cholinesterase reactivators such as 2-pralidoxime (2-PAM). Some reactivators successfully provided some prophylactic or therapeutic effect, perhaps only coincidentally, and thereby reinforced the hypothesis. The properties of the oximes 2-PAM and HI-6 [1-(2-hydroxyiminomethyl-1-pyridino)-3-(4-carbamoyl-1-pyridino)-2-oxapropane] at the AChR have been examined in fine detail (ALKONDON et al. 1988). Two allosteric characteristics may contribute to recovery from organophosphate poisoning: while rapidly reversible ion channel blockade and an increased desensitization were seen at higher concentrations, low concentrations of oximes tended to enhance the rate of channel activation. Evidence from in vitro studies of tetanic stimulation of neuromuscular transmission indicated that the relative potency of various allosteric actions is important in the selection of the optimal therapeutic depending on the nerve agent exposure; i.e., HI-6 was preferable for antagonism of soman effects, but 2-PAM was preferable for antagonism of tabun effects (ALBUQUERQUE et al. 1988b; REDDY et al. 1991). There is a paradox associated with the allosteric AChR effector as an antidote: How does one allosteric blocker reduce the toxicity of another? The extremely rapid association and dissociation kinetics of the channel blockade by the oximes may be an important factor, such that oximes can inhibit the action of other noncompetitive antagonists without themselves producing a prolonged receptor blockade. Alternatively, the mild agonist effect produced by lower concentrations of these agents may be very important (see above in D).

Stimulation of the motoneuron presynaptic terminal and antidromic conduction of action potentials was another symptom elicited by anticholinesterase agents (BLABER and BOWMAN 1963; HUBBARD et al. 1965; LASKOWSKI and DETTBARN 1975); this phenomenon might follow from ephaptic transmission or presynaptic stimulation by an excess of synaptic ACh. However, terminals of other types (different neurotransmitter systems) were also affected. For example, at the glutamatergic neuromuscular synapse of the locust, exposure to physostigmine ($20\,\mu M$), DFP ($1\,mM$), and nerve gas agents (VX, $10\,\mu M$; tabun, $20\,\mu M$) apparently evoked presynaptic action potentials; excitatory postsynaptic currents (not miniatures) were also observed (IDRISS et al. 1986). The frequency of synaptic potentials was also increased when VX was applied to hippocampal cell cultures (ALKONDON and ALBUQUERQUE, unpublished observations; PEREIRA et al. 1991b). Following electrical stimulation of the locust motor nerve, the amplitudes of excitatory postsynaptic currents were reduced by DFP, VX, and physostigmine; the reduction in amplitude was compatible with both a presynaptic action and enhanced glutamate-induced desensitization. The possible role of a direct stimulatory action via presynaptic receptors (perhaps nicotinic) has not been explored. Thus, it has become evident that an excess of synaptic ACh does not account for all presynaptic actions of the cholinesterase agents, and much more detailed analysis of the activity of these agents on the CNS with the aim of determining their molecular targets and possible antidotal therapies, remains to be resolved.

A noncholinergic receptor mechanism was certainly involved in the postsynaptic actions at the locust neuromuscular synapse, where the receptor is a quisqualate-type glutamate receptor. After its exposure to VX and DFP, the exponential decay time constant and channel lifetime (measured by noise analysis) of the postsynaptic currents were shortened (IDRISS et al. 1986); this effect of anticholinesterases was compatible with an open channel blocking mechanism at the quisqualate receptor, similar to that described for the nicotinic receptor. Structural homology of the nicotinic receptor ion channel and the glutamatergic ion channel may account for the toxicologically similar sensitivities of endplate (cholinergic) and postsynaptic glutamatergic responses.

G. Conclusions

It seems to be a natural consequence of the vital nature of nicotinic systems in animals that the AChR has fallen target to many natural toxins throughout the phylogenetic hierarchy. Many nicotinic agonists and antagonists are found in algae (anatoxin), plants (curare dihydro-β-erythroidine, nicotine, methylly-caconitine), molluscs (neosurugatoxin), insects (ACh), amphibians (batrachotoxin, gephyrotoxin, HTX), and reptiles (bungarotoxins). In addition, the plant toxins known to inhibit cholinesterase have been the prototype for pesticides and organophosphates, most of which also affect the nicotinic receptor. Using the variety of specific ligands described, the pharmacologist is equipped to differentiate neuronal nicotinic receptor subtypes in various brain regions and to delineate their physiological roles. The addiction liability associated with nicotine and the chronic abuse of nicotine in tobacco are areas which demand attention. We are only now beginning to realize the magnitude of long-term changes in the AChR population that results after chronic exposure. This rapidly expanding field of medical research is on the threshold for understanding the development of nicotinic receptors and their relationship to normal brain development and abnormalities in brain disease. The loss of nicotinic receptors, as a component of Alzheimer's disease, demonstrates their importance in higher brain functions. Through investigating the properties of nicotinic toxins on these receptors, we hope to learn how to use them for a better benefit for life.

Acknowledgement. We thank Manickavasagom Alkondon for the figure of anatoxin stimulation and various antagonist effects on cultured hippocampal neurons, Tania Tano for the figure of metaphit-enhanced desensitization, and Edna de Fatima Rezende Pereira for the information from their experimental studies in our laboratories on physostigmine, MLA and benzoquinonium.

References

Abramson SN, Li Y, Culver P, Taylor P (1989) An analog of lophotoxin reacts covalently with Tyr[190] in the α-subunit of the nicotinic acetylcholine receptor. J Biol Chem 264:12666–12672

Adams DJ (1983) Chemical modification of endplate channels in frog skeletal muscle. J Physiol (Lond) 343:29P

Adams PR (1972) Voltage jump analysis of procaine action at frog end-plate. J Physiol (Lond) 268:291–318

Adams PR, Sakmann B (1978) Decamethonium both opens and blocks endplate channels. Proc Natl Acad Sci USA 75:2994–2998

Adler M, Albuquerque EX, Lebeda FJ (1978) Kinetic analysis of endplate currents altered by atropine and scopolamine. Mol Pharmacol 14:514–529

Adler M, Oliveira AC, Albuquerque EX, Mansour NA, Eldefrawi AT (1979) Reaction of tetraethylammonium with the open and closed conformations of the acetylcholine receptor ionic channel complex. J Gen Physiol 74:129–152

Aguayo LG, Albuquerque EX (1986) Effects of phencyclidine and its analogs on the end-plate current of the neuromuscular junction. J Pharmacol Exp Ther 239:15–24

Akaike A, Ikeda SR, Brookes N, Pascuzzo GJ, Rickett DL, Albuquerque EX (1984) The nature of the interaction of pyridostigmine with the nicotinic acetylcholine receptor-ionic channel complex. II. Patch clamp studies. Mol Pharmacol 25:102–112

Albuquerque EX, Spivak CE (1984) Natural toxins and their analogues that activate and block the ionic channel of the nicotinic acetylcholine receptor. In: Krogsgaard-Larsen P, Brøgger Christenses S, Kofod H (eds) Natural products and drug development. Munksgaard, Copenhagen, pp 301–323 (Alfred Benzon symposium, vol 20

Albuquerque EX, Barnard EA, Chiu TH, Lapa AJ, Dolly JO, Jansson S-E, Witkop B (1973a) Acetylcholine receptor and ion conductance modulator sites at the murine neuromuscular junction: evidence from specific toxin reactions. Proc Natl Acad Sci USA 70:949–953

Albuquerque EX, Kuba K, Lapa AJ, Daly JW, Witkop B (1973b) Acetylcholine receptor and ionic conductance modulator of innervated and denervated muscle membranes. Effect of histrionicotoxins. Excerpta Med Int Cong Ser 333:585–597

Albuquerque EX, Barnard EA, Porter CW, Warnick JE (1974) The density of acetylcholine receptors and their sensitivity in the postsynaptic membrane of muscle endplates. Proc Natl Acad Sci USA 71:2818–2822

Albuquerque EX, Akaike A, Shaw K-P, Rickett DL (1984) The interaction of anticholinesterase agents with the acetylcholine receptor-ionic channel complex. Fund Appl Toxicol 4:S27–S33

Albuquerque EX, Deshpande SS, Kawabuchi M, Aracava Y, Idriss M, Rickett DL, Boyne AF (1985) Multiple actions of anticholinesterase agents on chemosensitive synapses: molecular basis for prophylaxis and treatment of organophosphate poisoning. Fundam Appl Toxicol 5:S182–S203

Albuquerque EX, Deshpande SS, Aracava Y, Alkondon M, Daly JW (1986) A possible involvement of cyclic AMP in the expression of desensitization of the nicotinic acetylcholine receptor. A study with forskolin and its analogs. FEBS Lett 199:113–120

Albuquerque EX, Swanson KL, Deshpande SS, Aracava Y, Cintra WM, Kawabuchi M, Alkondon M (1987) The direct interaction of cholinesterase inhibitors with the acetylcholine receptor and their involvement with cholinergic autoregulatory mechanisms. In: Sixth Medical Chemical Defense Bioscience Review US Army Medical Research Institute of Chemical Defense, Aberdeen Proving Ground, Maryland, pp 27–34

Albuquerque EX, Aguayo L, Swanson KL, Idriss M, Warnick JE (1988a) Multiple interactions of phencyclidine at central and peripheral sites. In: Domino EF, Kamenka J-M (eds) Sigma and phencyclidine-like compounds as molecular probes in biology. NPP, Ann Arbor, pp 425–438

Albuquerque EX, Alkondon M, Deshpande SS, Cintra WM, Aracava Y, Brossi A (1988b) The role of carbamates and oximes in reversing toxicity of organophosphorus compounds: a perspective into mechanisms. In: Lunt GG (ed) Neurotox '88: molecular basis of drug and pesticide action. Elsevier, Cambridge, pp 349–373

Albuquerque EX, Aracava Y, Cintra WM, Brossi A, Schönenberger B, Deshpande SS (1988c) Structure-activity relationship of reversible cholinesterase inhibitors: activation, channel blockade and stereospecificity of the nicotinic acetylcholine receptor-ion channel complex. Braz J Med Biol Res 21:1173–1196

Albuquerque EX, Daly JW, Warnick JE (1988d) Macromolecular sites for specific neurotoxins and drugs on chemosensitive synapses and electrical excitation in biological membranes. Ion Channels 1:95–162

Albuquerque EX, Maelicke A, Pereira EFR (1991) Single channel currents activated by physostigmine in hippocampal neurons are blocked by benzoquinonium but not by methyllycaconitine. Soc Neurosci Abs 17:585

Alkondon M, Albuquerque EX (1990) α-Cobratoxin blocks the nicotinic acetylcholine receptor in rat hippocampal neurons. Eur J Pharmacol 191:505–506

Alkondon M, Albuquerque EX (1991) Initial characterization of the nicotinic acetylcholine receptors in rat hippocampal neurons. J Receptor Res 11:1001–1022

Alkondon M, Rao KS, Albuquerque EX (1988) Acetylcholinesterase reactivators modify the properties of nicotinic acetylcholine receptor ion channels. J Pharmacol Exp Ther 245:543–556

Aracava Y, Ikeda SR, Daly JW, Brookes N, Albuquerque EX (1984) Interactions of bupivacaine with ionic channels of the nicotinic receptor: analysis of single channel currents. Mol Pharmacol 26:304–313

Aracava Y, Deshpande SS, Rickett DL, Brossi A, Schönenberger B, Albuquerque EX (1987a) The molecular basis of anticholinesterase actions on nicotinic and glutamatergic synapses. Ann NY Acad Sci 505:226–255

Aracava Y, Deshpande SS, Swanson KL, Rapoport H, Wonnacott S, Lunt G, Albuquerque EX (1987b) Nicotinic acetylcholine receptor in cultured neurons from the hippocampus and brain stem of the rat characterized by single channel recording. FEBS Lett 222:63–70

Aracava Y, Swanson KL, Rozental R, Albuquerque EX (1988) Structure-activity relationships of (+)anatoxin-a derivatives and enantiomers of nicotine on the peripheral and central nicotinic acetylcholine receptor subtypes. In: Lunt GG (ed) Neurotox '88: molecular basis of drug and pesticide action. Elsevier, Cambridge, pp 157–184

Aronstam RS, Witkop B (1981) Anatoxin-a interactions with cholinergic synaptic molecules. Proc Natl Acad Sci USA 78:4639–4643

Aronstam RS, King CT Jr, Albuquerque EX, Daly JW, Feigl DM (1985) Binding of [3H]perhydrohistrionicotoxin and [3H]phencyclidine to the nicotinic receptor-ion channel complex of *Torpedo* electroplax. Inhibition by histrionicotoxins and derivatives. Biochem Pharmacol 34:3037–3047

Aronstam RS, Daly JW, Spande TF, Narayanan TK, Albuquerque EX (1986) Interaction of gephyrotoxin and indolizidine alkaloids with the nicotinic acetylcholine receptor-ion channel complex of *Torpedo* electroplax. Neurochem Res 11:1227–1240

Atchison WD, Narahashi T, Vogel SM (1984) Endplate blocking actions of lophotoxin. Br J Pharmacol 82:667–672

Auerbach A, del Castillo J, Specht PC, Titmus M (1983) Correlation of agonist structure with acetylcholine receptor kinetics: studies on the frog end-plate and on chick embryo muscle. J Physiol (Lond.) 343:551–568

Barlow RB, Hamilton JT (1965) The stereospecificity of nicotine. Br J Pharmacol 25:206–212

Barlow RB, McLeod LJ (1969) Some studies on cytisine and its methylated derivatives. Br J Pharmacol 35:161–174

Barnard EA, Dolly JO, Porter CW, Albuquerque EX (1975) The acetylcholine receptor and ionic conductance modulation system of skeletal muscle. Exp Neurol 48:1028

Barrantes FJ (1978) Agonist-mediated changes of the acetylcholine receptor in its membrane environment. J Mol Biol 124:1–26

Barrantes FJ (1980) Modulation of acetylcholine receptor states by thiol modification. Biochemistry 19:2957–2965

Beers WH, Reich E (1970) Structure and activity of acetylcholine. Nature 288: 917–922

Behling RW, Yamane T, Navon G, Jelinski LW (1988) Conformation of acetylcholine bound to the nicotinic acetylcholine receptor. Proc Natl Acad Sci USA 85:6721–6725

Ben-Haim D, Dreyer F, Peper K (1975) Acetylcholine receptor: modification of synaptic gating mechanism after treatment with a disulfide bond reducing agent. Pflugers Arch 355:19–26

Bird SJ, Aghajanian JK (1976) The cholinergic pharmacology of hippocampal pyramidal cells: a microiontophoretic study. Neuropharmacology 15:273–282

Blaber LC, Bowman WC (1963) The effects of some drugs on the repetitive discharges produced in nerve and muscle by anticholinesterases. Int J Neuropharmacol 2:1–16

Blount P, Merlie JP (1989) Molecular basis of the two nonequivalent ligand binding sites of the muscle nicotinic acetylcholine receptor. Neuron 3:349–357

Boksa P, Quirion R (1987) [^{3}H]N-Methyl-carbamylcholine, a new radioligand sepcific for nicotinic acetylcholine receptors in brain. Eur J Pharmacol 139:323–333

Bradley PB, Lucy AP (1983) Cholinoceptive properties of respiratory neurones in the rat medulla. Neuropharmacology 22:853–858

Brimblecombe RW, Rowsell DG (1969) A comparison of the pharmacological activities of tertiary bases and their quaternary ammonium derivatives. Int J Neuropharmacol 8:131–141

Carmichael WM, Biggs DF, Gorham PR (1975) Toxicology and pharmacological action of *Anabaena flos-aquae* toxin. Science 187:542–544

Carmichael WW, Biggs DF, Peterson MA (1979) Pharmacology of anatoxin-a produced by the freshwater cyanophyte *Anabaena flos-aquae* NRC-44-1. Toxicon 17:229–236

Carp JS, Aronstam RS, Witkop B, Albuquerque EX (1983) Electrophysiological and biochemical studies on enhancement of desensitization by phenothiazine neuroleptics. Proc Natl Acad Sci USA 80:310–314

Changeux J-P (1990) Functional architecture and dynamics of the nicotinic acetylcholine receptor: an allosteric ligand-gated ion channel. Fidia Res Found Neurosci Award Lect 4:21–168

Clarke PBS (1987) Recent progress in identifying nicotinic cholinoceptors in mammalian brain. TINS 8:32–35

Clarke PBS, Schwartz RD, Paul SM, Pert CB, Pert A (1985) Nicotinic binding in rat brain: autoradiographic comparison of [^{3}H]acetylcholine, [^{3}H]nicotine, and [^{125}I]-α-bungarotoxin. J Neurosci 5:1307–1315

Collins AC, Evans CB, Miner LL, Marks MJ (1986) Mecamylamine blockade of nicotine responses: evidence for two brain nicotinic receptors. Pharmacol Biochem Behav 24:1767–1773

Colquhoun D, Hawkes AG (1983) The principles of the stochastic interpretation of ion-channel mechanisms. In: Sakmann B, Neher E (eds) Single-channel recording. Plenum, New York. pp 135–175

Colquhoun D, Rang JP (1976) Effects of inhibitors on the binding of iodinated α-bungarotoxin to acetylcholine receptors in rat muscle. Mol Pharmacol 12:519–535

Connolly JG (1989) Structure-function relationships in nicotinic acetylcholine receptors. Comp Biochem Physiol [A] 93:221–231

Contreras PC, Rafferty MF, Lessor RA, Rice KC, Jacobson AE, O'Donohue TL (1985) A specific alkylating ligand for phencyclidine (PCP) receptors antagonizes PCP behavioral effects. Eur J Pharmacol 111:405–406

Cooper E (1990) Evidence that functional neuronal nicotinic AChRs have pentameric structures. Soc Neurosci Abstr 16:10

Costa ACS, Swanson KL, Aracava Y, Aronstam RS, Albuquerque EX (1990) Molecular effects of dimethylanatoxin on the peripheral nicotinic acetylcholine receptor. J Pharmacol Exp Ther 252:507–516

Culver P, Jacobs RS (1981) Lophotoxin: a neuromuscular acting toxin from the sea whip (*Lophorgorgia rigida*). Toxicon 19:825–830

Culver P, Fenical W, Taylor P (1984) Lophotoxin irreversibly inactivates the nicotinic acetylcholine receptor by preferential association at one of the two primary agonist sites. J Biol Chem 259:3763–3770

Culver P, Burch M, Potenza C, Wasserman L, Fenical W, Taylor P (1985) Structure-activity relationships for the irreversible blockade of nicotinic receptor agonist sites by lophotoxin and congeneric diterpene lactones. Mol Pharmacol 28:436–444

Cuns JCT, Aracava Y, Albuquerque EX (1990) Atropine actions on nicotinic receptors: single channel analysis. Biophys J 57:123a

Dale HH (1914) The action of certain esters and ethers of choline, and their relation to muscarine. J Pharmacol Exp Ther 6:147–190

Dani JA (1989) Open channel structure and ion binding sites of the nicotinic acetylcholine receptor channel. J Neurosci 9:884–892

Deguchi R, Narahashi T, Haas HG (1971) Mode of action of nereistoxin on the neuromuscular transmission in the frog. Pestic Biochem Physiol 1:196–204

Del Castillo J, Katz B (1957) Interaction at end-plate receptors between different choline derivatives. Proc R Soc Lond 146:369–381

Deshpande SS, Viana GB, Kauffman FC, Rickett DL, Albuquerque EX (1986) Effectiveness of physostigmine as a pretreatment drug for protection of rats from organophosphate poisoning. Fundam Appl Toxicol 6:566–577

Ehlert FJ, Jenden DJ (1984) Comparison of the muscarinic receptor binding activity of some tertiary amines and their quaternary ammonium analogues. Mol Pharmacol 25:46–50

Eldefrawi A, Bakry NM, Eldefrawi NE, Tsai M-C, Albuquerque EX (1980) Nereistoxin interaction with the acetylcholine receptor-ionic channel complex. Mol Pharmacol 17:172–179

Feltz A, Trautmann A (1982) Desensitization at the frog neuromuscular junction: a biphasic process. J Physiol (Lond) 322:257–272

Fenical W, Okuda RK, Bandurraga MM, Culver P, Jacobs RS (1981) Lophotoxin: a novel neuromuscular toxin from pacific sea whips of the genus *Lophogorgia*. Science 212:1512–1514

Fertuck HC, Salpeter MM (1974) Localization of acetylcholine receptor by ^{125}I-labeled α-bungartoxin binding at mouse motor endplates. Proc Nat Acad Sci USA 71:1376–1378

Fiekers JF, Spannbauer PM, Scubon-Mulieri B, Parsons RL (1980) Voltage dependence of desensitization: influence of calcium and activation kinetics. J Gen Physiol 75:511–529

French ED, Jacobson AE, Rice KC (1987) Metaphit, a proposed phencyclidine (PCP) antagonist, prevents PCP-induced locomotor behavior through mechanisms unrelated to specific blockade of PCP receptors. Eur J Pharmacol 140:267–274

Fróes-Ferrão MM, Rozental R, Albuquerque EX (1991) Single channel currents activated by physostigmine at junctional nicotinic acetylcholine receptor of mammalian and amphibian. Soc Neurosci Abs 17:751

Gage PW, Hamill OP, Wachtel RE (1983) Sites of action of procaine at the motor end-plate. J Physiol (Lond) 335:123–137

Giblin BA, Lumpkin MD, Kellar KJ (1988) Repeated administration of nicotine results in long-term decrease in prolactin release by acute nicotine in rats. Soc Neurosci Abstr 14:1328

Giraudat J, Dennis M, Heidmann T, Haumont P-Y, Lederer F, Changeux J-P (1987) Structure of the high-affinity binding site for noncompetitive blockers of the acetylcholine receptor: [^{3}H]chlorpromazine labels homologous residues in the β and γ chains. Biochemistry 26:4210–4218

Giraudat J, Galzi J-L, Revah F, Changeux J-P, Haumont P-Y, Lederer F (1989) The noncompetitive blocker [^{3}H]chlorpromazine labels segment M2 but not segment M1 of the nicotinic acetylcholine receptor a-subunit. FEBS Lett 253:190–198

Grenningloh G, Reinitz A, Schmitt C, Methfessel C, Zensen M, Beyreuther K, Gundelfinger ED, Betz H (1987) The strychnine-binding subunit of the glycine receptor shows homology with nicotinic acetylcholine receptors. Nature 328:215–220

Hanin I, Jenden DJ, Cho AK (1966) The influence of pH on the muscarinic action of oxotremorine, arecoline, pilocarpine, and their quaternary ammonium analogs. Mol Pharmacol 2:352–359

Hayashi E, Yamada S (1975) Pharmacological studies on surugatoxin, the toxic principle from Japanese ivory mollusc (*Babylonia japonica*). Br J Pharmacol 53:206–215

Hey P (1952) On relationships between structure and nicotine-like stimulant activity in choline esters and ethers. Br J Pharmacol 7:117–129

Hubbard JI, Schmidt RF, Yokota T (1965) The effect of acetycholine upon mammalian motor nerve terminals. J Physiol (Lond) 181:810–829

Hucho F (1986) The nicotinic acetylcholine receptor and its ion channel. Eur J Biochem 158:211–226

Hucho FL, Oberthür W, Lottspeich F (1986) The ion channel of the nicotinic acetylcholine receptor is formed by the homologous helices MII of the receptor subunits. FEBS LETT 205:137–142

Huganir RL, Delcour AH, Greengard P, Hess GP (1986) Phosphorylation of the nicotinic acetylcholine receptor regulates its rate of desensitization. Nature 321:744–776

Huganir RL, Greengard P (1987) Regulation of receptor function by protein phosphorylation. Trends Pharmacol Sci 8:472–477

Idriss MK, Aguayo LG, Rickett D, Albuquerque EX (1986) Organophosphate and carbamate compounds have pre- and postjunctional effects at the insect glutamatergic synapse. J Pharmacol Exp Ther 239:279–285

Ikeda SR, Aronstam RS, Daly JW, Aracava Y, Albuquerque EX (1984) Interactions of bupivacaine with ionic channels of the nicotinic receptor: electrophysiological and biochemical studies. Mol Pharmacol 26:293–303

Jennings KR, Brown DG, Wright DP Jr (1986) Methyllycaconitine, a naturally occurring insecticide with a high affinity for the insect cholinergic receptor. Experientia 42:611–613

Kanne DB, Abood LG (1988) Synthesis and biological characterization of pyrido-homotropanes. Structure-activity relationships of conformationally restricted nicotinoids. J Med Chem 31:506–509

Karlin A (1969) Chemical modification of the active site of the acetylcholine receptor. J Gen Physiol 54:245s–264s

Karlin A (1980) Molecular properties of nicotinic acetylcholine receptors In: Cotman CW, Poste G, Nicolson GL (eds) The Cell Surface and Neuronal Function. Elsevier Amsterdam, pp 191–206

Karlin A (1983) The anatomy of a receptor. Neuroscience Commentaries 1:111–123

Karpen JW, Hess GP (1986a) Cocaine, phencyclidine, and procaine inhibition of the acetylcholine receptor: characterization of the binding site by stopped-flow measurements of receptor-controlled ion flux in membrane vesicles. Biochemistry 25:1777–1785

Karpen JW, Hess GP (1986b) Acetylcholine inhibition by D-tubocurarine involves both a competitive and a noncompetitive binding site as determined by stopped-flow measurements of receptor-controlled ion flux in membrane vesicles. Biochemistry 25:1786–1792

Katz B, Miledi R (1978) A re-examination of curare action at the motor endplate. Proc R Soc Lond [Biol] 203:119–133

Kawabuchi M, Boyne AF, Deshpande SS, Cintra WM, Brossi S, Albuquerque EX (1988) Enantiomer (+)physostigmine prevents organophosphate-induced subjunctional damage at the neuromuscular synapse by a mechanism not related to cholinesterase carbamylation. Synapse 2:139–147

Kellar KJ, Wonnacott S (1990) Nicotinic cholinergic receptors in Alzheimer's disease. In: Wonnacott S, Russell MAH, Stolerman IP (eds) Nicotine psychopharmacology. Oxford University Press, Oxford, pp 341–373

Knappe U, Wirtz-Brugger F, Cornfeldt M, Fielding S (1988) Effects of hemicholinium-3 on brainstem auditory evoked potentials in the rat. Soc Neurosci Abstr 14:800

Kosuge T, Tsuji K, Hirai K (1982) Isolation of neosurugatoxin from the Japanese ivory shell, *Babylonia japonica*. Chem Pharm Bull (Tokyo) 9:3255–3259

Kuba K, Albuquerque EX, Barnard EA (1973) Diisopropylfluorophosphate: suppression of ionic conductance of the cholinergic receptor. Science 181:853–856

Kuba K, Albuquerque EX, Daly J, Barnard EA (1974) A study of the irreversible cholinesterase inhibitor, diisopropylfluorophosphate, on time course of endplate currents in frog sartorius muscle. J Pharmacol Exp Ther 189:499–512

Kuhlmann J, Okonjo KO, Maelicke A (1991) Desensitization is a property of the cholinergic binding region of the nicotinic acetylcholine receptor, not of the receptor-integral ion channel. FEBS Lett 279:216–218

Lambert JJ, Volle RL, Henderson EG (1980) An attempt to distinguish between the actions of neuromuscular blocking drugs on the acetylcholine receptor and on its associated ionic channel. Proc Natl Acad Sci USA 77:5003–5007

Langdon RB, Jacobs RS (1985) Irreversible autonomic actions by lophotoxin suggest utility as a probe for both C6 and C10 nicotinic receptors. Brain Res 359:233–238

Langley JN (1905) On the reaction of cells and of nerve-endings to certain poisons, chiefly as regards the reaction of striated muscle to nicotine and to curari. J Physiol (Lond) 33:374–413

Laskowski MB, Dettbarn W-D (1975) Presynaptic effects of neuromuscular cholinesterase inhibition. J Pharmacol Exp Ther 194:351–361

Lauffer L, Hucho F (1982) Triphenylmethylphosphonium is an ion channel ligand of the nicotinic acetylcholine receptor. Proc Natl Acad Sci USA 79:2406–2409

Leprince P (1983) Chemical modification of the nicotinic cholinergic receptor of PC-12 nerve cell. Biochemistry 22:5551–5556

Lima-Landman MT, Albuquerque EX (1988) The novel neurotoxin H_{12}-histrionicotoxin blocks the N-methyl-D-aspartate receptor of cultured hippocampus of the rat. Soc Neurosci Abstr 14:96

Lipton SA, Aizeman E, Loring RH (1987) Neural nicotinic acetylcholine responses in solitary mammalian retinal ganglion cell. Pflugers Arch 410:37–43

London ED, Ball MJ, Waller SB (1989) Nicotinic binding sites in cerebral cortex and hippocampus in Alzheimer's dementia. Neurochem Res 14:745–750

Luetje CW, Patrick J (1991) Both alpha and beta subunits contribute to the agonist sensitivity of neuronal nicotinic acetylcholine receptors. J Neurosci 11:837–845

Luetje CW, Wada K, Rogers S, Abramson SN, Tsuji K, Heinemann S, Patrick J (1990) Neurotoxins distinguish between different neuronal nicotinic acetylcholine receptor subunit combinations. J Neurochem 55:632–640

MacAllan DRE, Lunt GG, Wonnacott S, Swanson KL, Rapoport H, Albuquerque EX (1988) Methyllycaconitine and (+)anatoxin-a differentiate between nicotinic receptors in vertebrate and invertebrate nervous systems. FEBS Lett 226:357–363

Madsen BW, Albuquerque EX (1985) The narcotic antagonist naltrexone has a biphasic effect on the nicotinic acetylcholine receptor. FEBS Lett 182:20–24

Magleby KL, Stevens CF (1972) A quantitative description of end-plate currents. J Physiol (Lond.) 223:173–197

Maleque MA, Takahashi K, Witkop B, Brossi A, Albuquerque EX (1984) A study of the novel synthetic analog (±)-depentylperhydrohistrionicotoxin on the nicotinic receptor-ion channel. J Pharmacol Exp Ther 230:619–626

Manalis RS (1977) Voltage-dependent effect of curare at the frog neuromuscular junction. Nature 267:366–367

Masukawa LM, Albuquerque EX (1978) Voltage- and time-dependent action of histrionicotoxin on the endplate current of the frog muscle. J Gen Physiol 72:351–367

Meyerhoff JL, Bates VE (1985) Combined treatment with muscarinic and nicotinic cholinergic antagonists slows development of kindled seizures. Brain Res 339:386–389

Middleton P, Jamarillo F, Schuetze SM (1986) Forskolin increases the rate of acetylcholine receptor desensitization at rat soleus endplates. Proc Natl Acad Sci USA 83:4967–4971

Milne RJ, Byrne JH (1981) Effects of hexamethonium and decamethonium on end-plate current parameters. Mol Pharmacol 19:276–281

Miner LL, Marks MJ, Collins AC (1986) Genetic analysis of nicotine-induced seizures and hippocampal nicotinic receptors in the mouse. J Pharmacol Exp Ther 239:853–860

Monod J, Wyman J, Changeux J-P (1965) On the nature of allosteric transitions: a plausible model. J Mol Biol 12:88–118

Nambi Aiyar V, Benn MH, Hanna T, Jacyno J, Roth SH, Wilkens JL (1979) The principal toxin of *Delphinium brownii* Rydb. and its mode of action. Experentia 35:1367–1368

Nastuk WL, Parsons RL (1970) Factors in the inactivation of postjunctional membrane receptors of frog skeletal muscle. J Gen Physiol 56:218–249

Neher E (1983) The charge carried by single channel currents of rat cultured muscle cells in the presence of local anesthetics. J Physiol (Lond) 339:663–678

Neher E, Steinbach JH (1978) Local anaesthetics transiently block currents through single acetylcholine-receptor channels. J Physiol (Lond) 277:173–176

Nilsson L, Adem A, Hardy J, Winblad B, Nordberg A (1987) Do tetrahydroaminoacridine (THA) and physostigmine restore acetylcholine release in Alzheimer brains via nicotinic receptors? J Neural Transm 70:357–368

Oberthür W, Muhn P, Baumann H, Lottspeich F, Wittmann-Liebold B, Hucho F (1986) The reaction site of a non-competitive antagonist in the δ-subunit of the nicotinic acetylcholine receptor. EMBO J 5:1815–1519

Ochoa ELM, Chattopadhyay A, McNamee MG (1989) Desensitization of the nicotinic acetylcholine receptor: molecular mechanisms and effect of modulators. Cell Mol Neurobiol 9:141–178

Okonjo KO, Kuhlmann J, Maelicke A (1991) A second pathway of activation of the nicotinic acetylcholine receptor ion channel. Eur J Biochem 200:671–677

Oliveira L, Madsen BW, Kapai N, Sherby SM, Swanson KL, Eldefrawi ME, Albuquerque EX (1987) Interaction of narcotic antagonist naltrexone with nicotinic acetylcholine receptor. Eur J Pharmacol 140:331–342

Pabreza LA, Dhawan S, Kellar K (1991) [^{3}H]Cytisine binding to nicotinic cholinergic receptors in brain. Mol Pharmacol 39:9–12

Papke RL, Millhauser G, Lieberman Z, Oswald RE (1988) Relationships of agonist properties to the single channel kinetics of nicotinic acetylcholine receptors. Biophys J 53:1–10

Parkinson D, Kratz KE, Daw NW (1988) Evidence for a nicotinic component to the actions of acetylcholine in cat visual cortex. Exp Brain Res 73:553–568

Pascuzzo GJ, Akaike A, Maleque MA, Shaw K-P, Aronstam RS, Rickett DL, Albuquerque EX (1984) The nature of the interactions of pyridostigmine with the nicotinic acetylcholine receptor-ionic channel complex. I. Agonist, desensitizing and binding properties. Mol Pharmacol 25:92–101

Paton WDM, Zaimis EJ (1952) The methonium compounds. Pharmacol Rev 4:219–253

Pedersen SE, Cohen JB (1990) D-Tubocurarine binding sites are located at $\alpha-\gamma$ and $\alpha-\delta$ subunit interfaces of the nicotinic acetylcholine receptor. Proc Natl Acad Sci USA 87:2785–2789

Pereira EFR, Wonnacott S, Albuquerque EX (1991a) Methyllycaconitine is a potent antagonist of nicotinic acetylcholine receptors on rat hippocampal neurons: single channel studies. Soc Neurosci Abs 17:960

Pereira EFR, Alkondon M, Albuquerque EX (1991b) Effects of organophosphate (OP) compounds and physostigmine (PHY) on nicotinic acetylcholine receptors (AChR) in the mammalian central nervous system (CNS). Proceedings of the 1991 Medical Defense Bioscience Review. US Army Medical Research Institute of Chemical Defense, Aberdeen Proving Ground, MD. pp 229–233

Peterson GL (1989) Consensus residues at the acetylcholine binding site of cholinergic proteins. J Neurosci Res 22:488–503

Ramoa AS, Albuquerque EX (1988) Phencyclidine and some of its analogues have distinct effects on NMDA receptors of rat hippocampal neurons. FEBS Lett 235:156–162

Ramoa AS, Alkondon M, Aracava Y, Irons J, Lunt GG, Deshpande SS, Wonnacott S, Aronstam RS, Albuquerque EX (1990) The anticonvulsant MK-801 interacts with peripheral and central nicotinic acetylcholine receptor ion channels. J Pharmacol Exp Ther 254:71–82

Rang HP, Ritter JM (1969) The relationship between desensitization and the metaphilic effect at cholinergic receptors. Mol Pharmacol 6:383–390

Rang HP, Ritter JM (1971) The effect of disulfide bond reduction on the properties of cholinergic receptors in chick muscle. Mol Pharmacol 7:620–631

Rao KS, Aracava Y, Rickett DL, Albuquerque EX (1987) Noncompetitive blockade of the nicotinic acetylcholine receptor-ion channel complex by an irreversible cholinesterase inhibitor. J Pharmacol Exp Ther 240:337–344

Rapier C, Wonnacott S, Lunt GG, Albuquerque EX (1987) The neurotoxin histrionicotoxin interacts with the putative ion channel of the nicotinic acetylcholine receptors in the central nervous system. FEBS Lett 212:292–296

Rapier C, Lunt GG, Wonnacott S (1988) Stereoselective nicotine-induced release of dopamine from striatal synaptosomes: concentration dependence and repetitive stimulation. J Neurochem 50:1123–1130

Rapier C, Lunt GG, Wonnacott S (1990) Nicotinic modulation of [^{3}H]dopamine release from striatal synaptosomes: pharmacological characterisation. J Neurochem 54:937–945

Reddy VK, Deshpande SS, Cintra WM, Scoble GT, Albuquerque EX (1991) Effectiveness of oximes 2-PAM and HI-6 in recovery of muscle function depressed by organophosphate agents in the rat hemidiaphragm: an in vitro study. Fundam Appl Toxicol 17:746–760

Roberts F, Lazareno S (1989) Cholinergic treatments for Alzheimer's disease. Biochem Soc Trans 17:76–79

Romano C, Goldstein A (1980) Stereospecific nicotine receptors on rat brain membranes. Science 210:647–650

Rovira C, Ben-Ari Y, Cherubini E, Krnjevic K, Roper N (1983) Pharmacology of the dendritic action of acetylcholine and further observations on the somatic disinhibition in the rat hippocampus in situ. Neuroscience 8:97–106

Rozental R, Aracava Y, Scoble GT, Swanson KL, Wonnacott S, Albuquerque EX (1989) The agonist recognition site of the peripheral acetylcholine receptor ion channel complex differentiates the enantiomers of nicotine. J Pharmacol Exp Ther 251:395–404

Sanchez JA, Dani JA, Siemen D, Hille B (1986) Slow permeation of organic cations in acetylcholine receptor channels. J Gen Physiol 87:985–1001

Schofield GG, Warnick JE, Albuquerque EX (1981) Elucidation of the mechanism and site of action of quinuclidinyl benzilate (QNB) on the electrical excitability and chemosensitivity of the frog sartorius muscle. Cell Mol Neurobiol 1:209–230

Schofield PR, Darlison MG, Fujita N, Burt DR, Stephenson FA, Rodriguez H, Rhee LM, Ramachandran J, Reale V, Glencorse TA, Seeburg PH, Barnard EA (1987) Sequence and functional expression of the $GABA_A$ receptor shows a ligand-gated receptor super-family. Nature 328:221–227

Schwartz RD, Mindlin MC (1988) Inhibition of the GABA receptor-gated ion channel in brain by noncompetitive inhibitors of the nicotinic receptor-gated cation channel. J Pharmacol Exp Ther 244:963–970

Shaker N, Eldefrawi AT, Aguayo LG, Warnick JE, Albuquerque EX, Eldefrawi ME (1982) Interactions of D-tubocurarine with the nicotinic acetylcholine receptor/channel molecule. J Pharmacol Exp Ther 220:172–177

Shaw K-P, Aracava Y, Akaike A, Daly JW, Rickett DL, Albuquerque EX (1985) The reversible cholinesterase inhibitor physostigmine has channel-blocking and agonist effects on the acetylcholine receptor-ion channel complex. Mol Pharmacol 28:527–538

Sherby SM, Eldefrawi AT, Albuquerque EX, Eldefrawi ME (1985) Comparison of the actions of carbamate anticholinesterases on the nicotinic acetylcholine receptor. Mol Pharmacol 27:343–348

Sheridan RP, Nilakantan R, Dixon JS, Venkataraghavan R (1986) The ensemble approach to distance geometry: application to the nicotinic pharmacophore. J Med Chem 29:899–906

Sine SM, Steinbach JH (1986) Acetylcholine receptor activation by a site-selective ligand: nature of brief open and closed states in BC3H-1 cells. J Physiol (Lond) 370:357–379

Sine SM, Taylor P (1981) Relationship between reversible antagonist occupancy and the functional capacity of the acetylcholine receptor. J Biol Chem 256:6692–6699

Siren A-L, Feuerstein G (1990) Cardiovascular effects of AnTX-a in the conscious rat. Toxicol Appl Pharmacol 102:91–100

Skok VI, Selyanko AA, Derkach BA (1989) Neuronal acetylcholine receptors. Plenum, New York

Sorenson EM, Culver P, Chiappinelli VA (1987) Lophotoxin: selective blockade of nicotinic transmission in autonomic ganglia by a coral neurotoxin. Neuroscience 20:875–884

Spivak CE, Albuquerque EX (1982) Dynamic properties of the nicotinic acetylcholine receptor ionic channel complex: activation and blockade. In: Hanin I, Goldberg AM (eds) Progress in cholinergic biology: model cholinergic synapses. Raven, New York, pp 323–357

Spivak CE, Albuquerque EX (1985) Triphenylmethylphosphonium blocks the nicotinic acetylcholine receptor noncompetitively. Mol Pharmacol 27:246–255

Spivak CE, Witkop B, Albuquerque EX (1980) Anatoxin-a: a novel, potent agonist at the nicotinic receptor. Mol Pharmacol 18:384–394

Spivak CE, Maleque MA, Oliveira AC, Masukawa LM, Tokuyama T, Daly JW, Albuquerque EX (1982) Actions of histrionicotoxins at the ion channel of the nicotinic acetylcholine receptor and at the voltage-sensitive ion channels of muscle membranes. Mol Pharmacol 21:351–361

Spivak CE, Maleque MA, Takahashi K, Brossi A, Albuquerque EX (1983a) The ionic channel of the nicotinic acetylcholine receptor is unable to differentiate between the optical antipodes of perhydrohistrionicotoxin. FEBS Lett 163:189–193

Spivak CE, Waters J, Witkop B, Albuquerque EX (1983b) Potencies and channel properties induced by semirigid agonists at frog nicotinic acetylcholine receptors. Mol Pharmacol 23:337–343

Steinbach AB (1968) Alteration by xylocaine (lidocaine) and its derivatives of the time course of the end-plate potential. J Gen Physiol 52:144–161

Steinbach JH, Zemple J (1987) What does phosphorylation do for the nicotinic acetylcholine receptor. Trends Neurosci 10:61–64

Stitzel JA, Campbell SM, Collins AC, Marks MJ (1988) Sulfhydryl modification of two nicotinic binding sites in mouse brain. J Neurochem 50:920–928

Stroud RM (1980) Acetylcholine receptor structure In: Cotman CW, Poste G, Nicolson GL (eds) The Cell Surface and Neuronal Function. Elsevier/North-Holland Biomedical Press, pp 124–138

Swanson KL, Albuquerque EX (1987) Nicotinic acetylcholine receptor ion channel blockade by cocaine: the mechanism of synatpic action. J Pharmacol Exp Ther 243:1202–1210

Swanson KL, Allen CN, Aronstam RS, Rapoport H, Albuquerque EX (1986) Molecular mechanisms of the potent and stereospecific nicotinic receptor agonist (+)-anatoxin-a. Mol Pharmacol 29:250–257

Swanson KL, Aracava Y, Sardina FJ, Aronstam RS, Rapoport H, Albuquerque EX (1989) N-methylanatoxinol isomers: derivatives of the agonist anatoxin-a block the nicotinic acetylcholine receptor ion channel. Mol Pharmacol 35:223–231

Swanson KL, Rapoport H, Aronstam RS, Albuquerque EX (1990) Nicotinic acetylcholine receptor function studied with synthetic (+)-anatoxin-a and derivatives. ACS Symp Ser 418:107–118

Swanson KL, Aronstam RS, Wonnacott S, Rapoport H, Albuquerque EX (1991) Nicotinic pharmacology of anatoxin analogs. I. Side chain structure-activity relationships at peripheral agonist and noncompetitive antagonist sites. J Pharmacol Exp Therap 259:377–386

Tano T, Rice K, Aronstam RS, Oliveira AC, Aracava Y, Albuquerque EX (1990) Effect of metaphit on the peripheral nicotinic acetylcholine receptor. Soc Neurosci Abstr 16:207

Terrar DA (1978) Effects of dithiothreitol on end-plate currents. J Physiol (Lond) 26:403–417

Tiedt TN, Albuquerque EX, Bakry NM, Eldefrawi ME, Eldefrawi AT (1979) Voltage- and time-dependent actions of piperocaine on the ion channel of the acetylcholine receptor. Mol Pharmacol 16:909–921

Triggle DJ, Triggle CR (1976) Chemical pharmacology of the synapse. Academic, London

Tsai M-C, Mansour NA, Eldefrawi AT, Eldefrawi ME, Albuquerque EX (1978) Mechanism of action of amantadine on neuromuscular transmission. Mol. Pharmacol 14:787–803

Tsai M-C, Oliveira AC, Albuquerque EX, Eldefrawi ME, Eldefrawi AT (1979) Mode of action of quinacrine on the acetylcholine receptor ionic channel complex. Mol Pharmacol 16:382–392

Unna K, Kniazuk M, Greslin JG (1944) Pharmacologic action of *Erythrina* alkaloids. I. β-erythroidine and substances derived from it. J Pharmacol 80:39–53

Vidal C, Changeux J-P (1989) Pharmacological profile of nicotinic acetylcholine receptors in the rat prefrontal cortex: an electrophysiological study in a slice preparation. Neuroscience 29:261–270

Wada A, Uezono Y, Arita M, Tsuji K, Yanagihara N, Kobayashi H, Izumi F (1989) High-affinity and selectivity of neosurugatoxin for the inhibition of ^{22}Na influx via nicotinic receptor-ion channel in cultured bovine adrenal medullary cells: comparative study with histrionicotoxin. Neuroscience 33:333–339

Wang CM, Narahashi T (1972) Mechanisms of dual action of nicotine on end-plate membranes. J Pharmacol Exp Ther 182:427–441

Ward JM, Cockcroft VB, Lunt GG, Smillie FS, Wonnacott S (1990) Methyl-lycaconitine: a selective probe for neuronal αbungarotoxin binding sites. FEBS Lett 270:45–48

Warnick JE, Aguayo LG, Maleque MA, Albuquerque EX (1982a) N-Alkyl analogs of phencyclidine on twitch and endplate currents. Fed Proc 41:1333

Warnick JE, Jessup PS, Overman LE, Eldefrawi ME, Nimit Y, Daly JW, Albuquerque JW (1982b) Pumiliotoxin-C and synthetic analogues. A new class of nicotinic antagonists. Mol Pharmacol 22:565–573

Warnick JE, Maleque MA, Bakry N, Eldefrawi AT, Albuquerque EX (1982c) Structure-activity relationships of amantadine. I. Interaction of the N-alkyl analogues with the ionic channels of the nicotinic acetylcholine receptor and electrically excitable membrane. Mol Pharmacol 22:82–93

Warnick JE, Maleque MA, Albuquerque EX (1984) Interaction of bicyclo-octane analogs of amantadine with ionic channels of the nicotinic acetylcholine receptor and electrically excitable membrane. J Pharmacol Exp Ther 228:73–79

Waters JA, Spivak CE, Hermsmeier M, Yadav JS, Liang RF, Gund TM (1988) Synthesis, pharmacology, and molecular modeling studies of semirigid, nicotinic agonists. J Med Chem 31:545–554

Wessler I (1989) Control of transmitter release from the motor nerve by presynaptic nicotinic and muscarinic autoreceptors. TIPS 10:110–114

Whitehouse PJ, Martino AM, Antuono PH, Lowenstein PR, Coyle JT, Price DL, Kellar KJ (1986) Nicotinic acetylcholine binding sites in Alzheimer's disease. Brain Res 371:146–151

Whiting P, Esch F, Shimasake S, Lindstrom J (1987) Neuronal nicotinic acetylcholine receptor β-subunit is coded for by the cDNA clone α4. FEBS Lett 219:459–464

Williams H, Robinson JL (1984) Binding of the nicotinic cholinergic antagonist, dihydro-β-erythroidine, to rat brain tissue. J Neurosci 4:2906–2911

Woodhull AM (1973) Ionic blockage of sodium channels in nerve. J Gen Physiol 61:687–708

Wonnacott S (1986) α-Bungarotoxin binds to low-affinity nicotine binding sites in rat brain. J Neurochem 47:1706–1712

Wonnacott S, Jackman S, Swanson KL, Rapoport H, Albuquerque EX (1991) Nicotinic pharmacology of anatoxin analogs. II. Side chain structure-activity relationships at neuronal nicotinic ligand binding sites. J Pharmacol Exp Therap 259:387–391

Wonnacott S, Swanson KL, Albuquerque EX, Huby NJS, Thompson P, Gallagher, T. Honoanatoxin: a potent analogue of anatoxin-a. Biochem Pharmacol (in press)

Wright PG (1954) An analysis of the central and peripheral components of respiratory failure produced by anticholinesterase poisoning in the rabbit. J Physiol (Lond.) 126:52–70

Yamada S, Iosgai M, Magawa Y, Takayanagi N, Hayashi E, Tsuji K, Kosuge T (1985) Brain nicotinic acetylcholine receptors: biochemical characterization by neosurugatoxin. Mol Pharmacol 28:120–127

Yamada S, Kagawa Y, Takayangi N, Nakayama K, Tsuji K, Kosuge T, Hayashi E, Okada K, Inoue S (1987) Comparison of antinicotinic activity by neosurugatoxin and the structurally related compounds. J Pharmacol Exp Ther 243:1153–1158

CHAPTER 18

Neurotoxic Agents Interacting with the Muscarinic Acetylcholine Receptor

J. Järv

A. Introduction

The neurotransmitter acetylcholine (ACh) exerts two types of effects, differing in their location and in the nature of the associated physiological responses and pharmacology (DALE 1914). These observations have led to the definition of two major types of ACh receptors, specified as nicotinic and muscarinic because their characteristic responses can be evoked by nicotine and muscarine, respectively (MICHAELSON and ZEIMAL 1973).

Muscarinic receptors can be identified in both the central (YAMAMURA and SNYDER 1974) and peripherial nervous systems, including smooth muscle (BOLTON 1979), cardiac muscle (LÖFFELHOLZ and PAPPANO 1985), and glands (GALLACHER and MORRIS 1986), but also on circulating cells such as erythrocytes (ARONSTAM et al. 1977) and T lymphocytes (LOPKER et al. 1980). Such a wide distribution of these receptors reflects the involvement of muscarinic mechanisms in processes as various as cognitive, emotional, associative, autonomic, and motor behavior. For the same reason the defects in the muscarinic synaptic transmission can lead to a variety of pathological states and neurological and mental disorders.

Firstly, these defects can be caused by exogenous compounds, which interact directly with the muscarinic receptor and thus can be treated as primary muscarinic neurotoxic agents. Secondly, they may arise from the excess of endogenous neurotransmitter ACh in the synaptic cleft due to the inhibition of synaptic acetylcholinesterase by exogenous inhibitors. The latter compounds may be classified as secondary muscarinic neurotoxins. Thirdly, the defects connected with muscarinic responses can be caused by toxicants interfering with the biochemical and biophysical events which mediate those responses. The targets of these toxicants are other than the muscarinic receptor.

The present chapter is focused on neurotoxic agents of the first type, and an attempt is made to outline the recent developments in understanding the molecular mechanism and specificity of their interaction with the muscarinic receptor. Although a large number of reviews on muscarinic receptors exist, these aspects have been neglected in spite of their fundamental significance for receptor studies.

B. Toxicants Acting on the Muscarinic Receptor

I. Agonists and Antagonists

Several poisonous herbs and plants (*Hyoscyamus niger*, *Atropa belladonna*, *Mandagora officinalis*, *Datura stramonium*, *Scopolia carniolica*) and fungae (*Amanita muscaria*), whose hallucinogenic and narcotic effects were realized by our ancestors, contain compounds acting on the muscarinic receptor as exogenous agonists and antagonists.

Muscarinic agonists mimic the action of ACh on muscarinic nerve junctions, and their peripheral effects include increased salivation, hypotension, and diarrhea. Their central activities lead to restlessness, mania, and hallucination. These effects can be treated by muscarinic antagonists, for instance, atropine (BECKER et al. 1979; TAYLOR 1980).

The exogenous agonist muscarine, which gives its name to the appropriate type of acetylcholine receptors, was isolated from the fungus *Amanita muscaria* by Oswald Schmiedeberg in Tartu University more than 120 years ago (SCHMIEDEBERG and KOPPE 1869). In this classic study, the nerve endings were identified as the place for action of muscarine and the antagonizing influence of atropine against this acute poisoning was discovered. It should be emphasized that these conclusions were made considerably earlier than the genesis of the synaptic theory of nerve transmission.

The intoxication with muscarinic antagonists or the "cholinergic syndrome" arises from an overdose of these agents, especially of the phenothiazine class, which possess high affinity against the receptor in the peripheral and central nervous systems. The "cholinergic syndrome" presents symptoms such as confusion, delirium, and coma (HEISER and GILLIN 1971; RUMACK 1973).

The muscarinic antagonist quinuclidinyl benzilate was originally developed as a hallucinogenic chemical warfare agent with the code name BZ (Merck Index 8005). However, its extremely high activity seems to be rather an exception, as many substances used as pharmaceutical agents possess moderate affinities for the muscarinic receptor (JÄRV and BARTFAI 1988).

Beginning in the early 1970s, when the methods of radioligand analysis became available, more than 500 drugs have been investigated in respect of their binding with the muscarinic receptor. These binding data together with the specification of tissue preparation and other experimental conditions have been collected in "The Data Bank of Muscarinic Ligands," compiled in the Laboratory of Bioorganic Chemistry at Tartu University. The list of references to this Data Bank involves over 1700 publications, and their number is continuously increasing.

It should be emphasized that many of these compounds whose binding with the muscarinic receptor has been studied are not recognized by pharmacologists as true neurotoxins (see BOWMAN and BOWMAN 1986), as they do not cause acute muscarinic disorders under the conditions applied, mostly due to their moderate affinity for this receptor. At the same time, the application of these drugs may lead to muscarinic side-effects in both the central and peripherial nervous systems, like dryness of mouth, blurred vision, and confusion.

II. Partial Agonists

A group of compounds can be identified which are able to initiate weak effects compared with the "true" agonists, even if all the receptor sites are occupied by these drugs (STEPHENSON 1956; ABRAMSON et al. 1963). Such an action of these partial agonists has been explained by their low intrinsic activity or low efficacy compared with agonists. This concept removes the absolute distinction between agonists and antagonists; however, it does not explain the molecular basis of the phenomenon. Moreover, in some cases the bell-shaped dose-response curves were obtained for the partial agonists by analogy to the phenomenon of substrate inhibition (STEPHENSON 1956), formally showing the dependence of the properties of the ligand upon its concentration. The latter fact cannot be fitted with the explanation referred to above.

In general, a question should be formulated about the mechanism of recognition of agonists, partial agonists, and antagonists by the receptor. All these ligands of different physiological activity may reveal a rather similar binding affinity, but some minor alterations in their structure, like the addition of one methylene group into the alkyl substituent of alkyltrimethylammonium salts, may alter the type of their activity (ABRAMSON et al. 1963).

III. Snake Venom Toxins

Recently, two toxins have been isolated from green mamba (*Dendroapsis angusticeps*) venom which interfere with binding of [^{3}H]quinuclidinyl benzilate to rat cerebral cortex synaptsososomal membranes, pointing to the specificity of these toxins for the muscarinic receptor (ADEM et al. 1988). They are compounds of peptide nature, and the amino acid sequence of toxin 2, consisting of 65 amino acid residues, was recently determined (KARLSSON et al. 1990). This sequence reveals several homologies with other short-chain neurotoxins acting on the nicotinic receptor (KARLSSON 1979). There are several highly conserved sites, including several positively charged groups and eight cysteine residues in analogous positions. Moreover, common features can be seen in the ultraviolet circular dichroism spectra of these toxins (DUCANCEL et al. 1991). Despite these sequential homologies and spectral similarities, the two toxins have different specificities, recognizing muscarinic and nicotinic receptors, respectively.

The pharmacological properties of these muscarinic toxins as individual compounds have not been studied yet, but on the basis of the analogy with other neurotoxins interacting with the nicotinic receptor, their antagonistic effects can be predicted on the muscarinic receptor.

C. Mechanism of Antagonist Interaction

I. Muscarinic Antagonists as Radioligands

The first attempt of direct determination of muscarinic binding sites was made with radioactive atropine (PATON and RANG 1965). However, the use of this ligand was limited due to its low specific radioactivity. Presently, the list of radioactively labelled muscarinic antagonists used in binding studies includes, besides atropine, [³H]dexetimide (LADURON et al. 1979), [³H]N-methyl-4-piperidinyl benzilate (KLOOG and SOKOLOVSKY 1978), [³H]N-methylatropine (BIRDSALL et al. 1983), [³H]N-methylscopolamine (BIRDSALL et al. 1983), [³H]pirenzepine (WATSON et al. 1982). [³H]N-propylbenzilylcholine mustard (HULME et al. 1978), [³H]3-quinuclidinyl benzilate (YAMAMURA and SNYDER 1974), [³H]scopolamine (KLOOG and SOKOLOVSKY 1978), and [¹²⁵I]3-quinuclidinyl benzilate (GIBSON et al. 1984).

Determination using these radioligands is based on their stoichiometric interaction with the receptor sites and the subsequent separation of the formed complex from the excess of radioligand. In the case of a membrane-bound receptor, filtration on glass-fiber filters (YAMAMURA and SNYDER 1974; EHLERT et al. 1984) or rapid centrifugation (EHLERT et al. 1984; HRDINA 1986) are generally employed. However, all these procedures are accompanied by extensive dilution of the reaction mixture while washing the filters or pellet to minimize the amount of nonspecifically bound radioligand. The same happens in the case of the solubilized radioligand-receptor complex during its separation from the excess of free radioligand by gel filtration (HURKO 1978). As all these procedures shift the binding equilibrium, only those ligands which dissociate sufficiently slowly from the receptor-ligand complex can be used in this assay.

On the other hand, however, the slowness of the dissociation process may hamper the experiments of radioligand displacement from the receptor complex by nonradioactive ligands. For this kinetic reason quinuclidinyl benzilate, which is incontestably the most widely used muscarinic radioligand, cannot be applied in displacement experiments which require the achievement of the true equilibrium state of the system (JÄRV and ELLER 1988). This fact clearly shows that the knowledge of the kinetic aspects of radioligand interaction with the receptor is a prerequisite for correct analysis of the binding data, especially if this approach is taken for the quantitative study of receptor-ligand interactions.

II. Kinetic Studies with Radioactive Antagonists

In the classic receptor theory it has been considered that the interaction of antagonists with the muscarinic receptor occurs as a single-step reversible process (CLARK 1937; STEPHENSON 1956):

$$R + A \underset{}{\overset{K_d}{\rightleftharpoons}} RA \qquad (1)$$

characterized by the dissociation constant K_d. The simplicity of this binding scheme has led to its general use in receptorology, including ligand binding studies with the muscarinic receptor (HEILBRONN and BARTFAI 1978; EHLERT et al. 1984; SOKOLOVSKY 1984; HRDINA 1986; JÄRV and BARTFAI 1988).

Nevertheless, the kinetic evidence for a more complex mechanism of antagonist binding was obtained in the case of [³H]quinuclidinyl benzilate (JÄRV et al. 1979; SCHIMERLIK and SEARLES 1980; SILLARD et al. 1985), [³H]N-methylquinuclidinyl benzilate (ELLER et al. 1988), [³H]N-methylpiperidinyl benzilate (JÄRV et al. 1979; 1987), [³H]N-methylscopolamine (ELLER and JÄRV 1989), and [³H]pirenzepine (LUTHIN and WOLFE 1984). In all these studies the observed rate constants k_{obs} of radioligand binding with the muscarinic receptor were determined given an excess of ligand over the receptor concentration:

$$B_t = B_{nonsp} + B_{sp} \left(1 - \exp\left(-k_{obs}\, t\right)\right) \qquad (2)$$

where B_t stands for the bound radioligand at time t and B_{nonsp} and B_{sp} for nonspecifically and specifically bound radioligand, respectively. Further, the plots of k_{obs} versus ligand concentration were analyzed as generally accepted in physical organic chemistry and enzymology (JENCKS 1969).

For the radioactive antagonists listed above the plots of k_{obs} versus ligand concentration were not linear, as follows from the reaction scheme (1), but showed a more complex form which regularly includes a hyperbolic part. The latter fact points to an at least two-step binding mechanism which should involve a fast reversible step of complex formation followed by a slower step of complex isomerization:

$$R + A \underset{k_{-i}}{\overset{K_a}{\rightleftharpoons}} RA \overset{k_i}{\rightleftharpoons} (RA) \qquad (3)$$

The rate equation for this reaction scheme is described as follows:

$$k_{obs} = \frac{k_i\,[A]}{K_A + [A]} + k_{-i} \qquad (4)$$

and kinetic parameters calculated from this plot for some radioactive antagonists are listed in Table 1.

An important conclusion from these data is that only the "isomerized" complex (RA) can be experimentally measured, while the rapidly dissociat-

Table 1. Kinetic data for radioactive antagonist binding with muscarinic receptor from the rat brain, 25°C, pH 7.40. (Data from JÄRV et al. 1987; ELLER et al. 1988; SILLARD et al. 1985; ELLER and JÄRV 1989)

Antagonist	K_A (nM)	$10^2 k_i$ (s^{-1})	$10^4 k_{-i}$ (s^{-1})	K_d (nM)	K_i
[³H]N-Methylpiperidinyl benzilate					
Ph HOĊC(O)O⟨N-Me⟩ Ph	5.5 ± 3.1	1.4 ± 0.3	6.7 ± 0.8	0.40 ± 0.03	0.05
[³H]N-Methylquinuclidinyl benzilate					
Ph HOĊC(O)O ⟨ ⟩ Ph N$^+$–Me	5.0 ± 2.3	4.6 ± 0.5	14 ± 4	0.13 ± 0.03	0.03
L-[³H]Quinuclidinyl benzilate					
Ph HOĊC(O)O ⟨ ⟩ Ph N	1.3 ± 0.5	1.2 ± 0.3	1.3 ± 0.4	–	–
[³H]N-Methylscopolamine					
Ph HOCH₂ĊHC(O)⟨ NMe₂ ⟩	9.7 ± 2.1	11.9 ± 0.8	9 ± 4	0.082 ± 0.008	0.008

ing complex RA escapes determination by the assay procedures used (JÄRV et al. 1979). Thus, the existence of the isomerization step seems to be the main criterion for the selection of compounds which can be used as radioactive "tools" in studies of the muscarinic receptor.

Secondly, it should be taken into consideration that the isomerization process involves monomolecular steps, and therefore this equilibrium cannot be shifted by addition of an excess of antagonist (JÄRV et al. 1979). As a result, a portion of the receptor sites, presented by the complex RA escapes detection independently of the antagonist concentration used. This circumstance is important to consider when measuring the number of receptor binding sites by radioactive antagonists as the results may depend upon the equilibrium constant of the isomerization step.

As both steps of the binding scheme (3) are reversible, the overall process can also be characterized by the dissociation constant K_d, which is, however, a combination of the constants K_A and $K_i = k_{-i}/k_i$, $K_d = K_A K_i$. This fact explains the particularly high potency of several muscarinic antagonists, reaching even the picomolar concentration range (JÄRV and ELLER 1988).

Under certain experimental conditions biphasic kinetic curves have been obtained for the dissociation of [³H]quinuclidinyl benzilate (GALPER et al. 1977; ELLER and JÄRV 1988), [³H]N-methylpiperidinyl benzilate (KLOOG and SOKOLOVSKY 1978), and [³H]N-methylscopolamine (ELLER and JÄRV 1989) from their complexes with the muscarinic receptor, suggesting that a heterogeneous population of binding sites exists. A more thorough analysis of these data has revealed the possibility of a multistep "isomerization" mechanism for these antagonists (ELLER and JÄRV 1988, 1989):

$$R + A \underset{k_{-i}}{\overset{K_A}{\rightleftharpoons}} RA \underset{k_{-i}}{\overset{k_i}{\rightleftharpoons}} (RA) \underset{k'_{-i}}{\overset{k'_i}{\rightleftharpoons}} ((RA)) \tag{5}$$

Both of the isomerized complexes (RA) and ((RA)) dissociate slowly and can be detected by common assay procedures. Due to that the transformation of (RA) into ((RA)) cannot be followed through the binding experiments. On the other hand, this extra step of complex isomerization may additionally decrease the apparent K_d value if $K_{i2} \ll 1$, as $K_d = K_A K_{i1} K_{i2}$ (JÄRV and ELLER 1988).

Moreover, in the case of [^{3}H]quinuclidinyl benzilate (SILLARD et al. 1985) and [^{3}H]N-methylpiperidinyl benzilate (JÄRV et al. 1987), cooperative regulation of the isomerization rate by an excess of antagonist was discovered, pointing to an even more complex reaction scheme.

III. Kinetic Studies with Nonradioactive Ligands

Until recently, kinetic studies have found only limited usefulness for investigation into the receptor-ligand interactions, obviously due to the traditional thinking based on the concept of ligand-receptor equilibria. On the other hand, this is certainly connected with the small number of radioactively labelled ligands available for receptor studies. To overcome the latter difficulty, an experimental procedure was proposed to investigate the kinetic mechanism of interaction of nonradioactive ligands with the muscarinic receptor by making use of one radioactive "reporter ligand" (SCHREIBER et al. 1985; ELLER et al. 1988, 1989). This approach is based on analysis of the influence of nonradioactive ligand A on the kinetics of binding of a radioactive "reporter ligand" A* to the receptor.

In principle, for a nonradioactive ligand both binding mechanisms (1) and (3) should be taken into consideration, which leads to two different kinetic models. The first model involves the case in which the nonradioactive compound A does not initiate isomerization of the receptor-ligand complex:

$$R \begin{cases} + A^* \underset{k_{-i}^*}{\overset{K_A^*}{\rightleftharpoons}} RA^* \underset{k_{-i}^*}{\overset{k_i^*}{\rightleftharpoons}} (RA^*) \\ + A \overset{K_A}{\rightleftharpoons} RA \end{cases} \tag{6}$$

The second model takes into consideration the possibility that the nonradioactive ligand A also initiates the isomerization process:

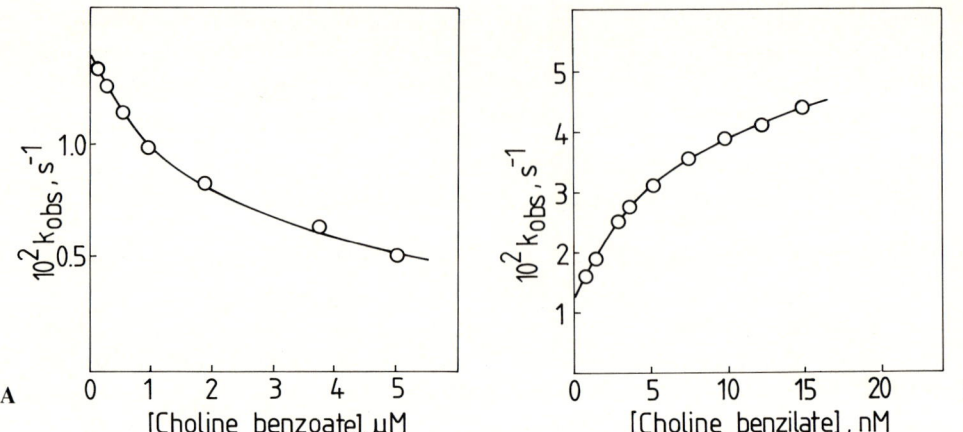

Fig. 1A,B. Effect of choline benzoate (*left*) (data from ELLER et al. 1989) and choline benzilate (*right*) (our unpublished data) on kinetics of [³H] *N*-methylscopolamine binding with the muscarinic receptor from rat brain, 25°C, 0.05 *M* K-phosphate buffer, pH 7.4

$$
R \quad
\begin{array}{l}
+ A^* \underset{}{\overset{K_A{}^*}{\rightleftharpoons}} RA^* \underset{k_{-i}{}^*}{\overset{k_i{}^*}{\rightleftharpoons}} (RA^*) \\[2ex]
+ A \underset{k_{-i}}{\overset{K_A}{\rightleftharpoons}} RA \underset{}{\overset{k_i}{\rightleftharpoons}} (RA)
\end{array}
\tag{7}
$$

The formation of the complex (RA*) can be followed experimentally by common assay procedures, i.e., the filtration method, and the observed rate constant k_{obs} for the "reporter ligand" A* can be calculated from these data as described above. Further more, the plot of k_{obs} versus the concentration of nonradioactive ligand can be analyzed to differentiate between the first and second kinetic models (ELLER et al. 1988).

In the first case, the rate of association of the "reporter ligand" A* is inhibited by the nonradioactive compound A which acts as a reversible inhibitor of the radioligand binding:

$$
k_{obs} = \frac{k_i{}^*[A^*]}{(1 + [A]/K_d)\,K_A{}^* + [A^*]}
\tag{8}
$$

In the second case, acceleration of the binding of the "reporter" ligand can be observed in the presence of ligand A:

$$
k_{obs} = \frac{k_i\,[A]}{K_A + [A]} + \text{constant}
\tag{9}
$$

On the basis of these k_{obs} versus [A] plots the kinetic parameters were calculated for association of several nonradioactive ligands with the

Table 2. Kinetic data for nonradioactive antagonist binding with the muscarinic receptor from rat brain, 25°C, pH 7.4, measured by using [^{3}H]N-methylscopolamine as "reporter ligand". (From ELLER et al. 1989)

Antagonist	K_A (nM)	$10^2 k_i$ (s^{-1})	$10^4 k_{-i}$ (s^{-1})	K_d (nM)	K_i
N,N-Dimethylaminoethyl benzilate					
Ph HOĊC(O)OC$_2$H$_4$N(CH$_3$)$_2$ Ph	38 ± 13	3.4 ± 0.6	68 ± 10	9.6 ± 1.7	0.20
Choline benzilate					
Ph HOĊC(O)OC$_2$H$_4$N$^+$(CH$_3$)$_3$ Ph	9.2 ± 2.2	3.3 ± 0.5	50 ± 9	2.3 ± 0.2	0.15
Atropine					
Ph HOCH$_2$CHC(O)O ⟨ NMe ⟩	6.3 ± 2.0	6.0 ± 1.0	10 ± 1	0.20 ± 0.03	0.017
N-Methylatropine					
Ph HOCH$_2$CHC(O)O ⟨ N$^+$Me$_2$ ⟩	3.0 ± 1.0	6.1 ± 1.0	21 ± 3	0.11 ± 0.02	0.034
Scopolamine					
Ph HOCH$_2$CHC(O)O ⟨ NMe ⟩O	3.7 ± 1.2	3.0 ± 0.5	30 ± 7	0.14 ± 0.03	0.05

muscarinic receptor (ELLER et al. 1988, 1989), and some of these data are listed in Table 2.

IV. Subtypes of Muscarinic Antagonists

Kinetic studies on the binding of antagonists with the muscarinic receptor from rat brain gave evidence for both mechanisms (6) and (7) (ELLER et al. 1988, 1989). The dose-dependent decrease of the apparent rate constant k_{obs} for [^{3}H]N-methylscopolamine binding in the presence of choline benzoate points to a reversible complex formation between the latter ligand and the muscarinic receptor (Fig. 1, left). On the other hand, the increase in the k_{obs} values for [^{3}H]N-methylscopolamine binding in the presence of choline benzilate (Fig. 1, right) gives evidence for the slow isomerization step in the case of this nonradioactive ligand. The kinetic constants K_A and k_{obs}, calculated for choline benzilate as well as for some other nonradioactive antagonists, are listed in Table 2.

Thus, at least two different subtypes can be differentiated among the classic set of muscarinic antagonists on the basis of the kinetic mechanism of their interaction with the receptor. These varied binding mechanisms seem to be directly connected to the potency and dynamics of the effects exerted by muscarinic antagonists; in the case of the isomerization processes, the apparent K_d value is a combination of the dissociation constant K_A and the isomerization constant K_i. This increases the potency of the antagonist without increasing the number and intensity of the interactions between the ligand and receptor. This fact is important in the case of relatively small ligand molecules providing limited contacts with the receptor site. Therefore, it is not surprising that the most potent muscarinic toxicants belong to the class of benzilic or tropic esters which are able to initiate the "isomerization" of the ligand-receptor complex.

The presence or absence of the isomerization step determines obviously the dynamics of the physiological effects of muscarinic antagonists, as both the rate of onset and recovery of the receptor depend upon the rate constants of the ligand association and dissociation. Such variable behavior of muscarinic antagonists was found in experiments with an ileum preparation, in which some potent antagonists were slow in onset and very slow to wear off after washing (ABRAMSON et al. 1969). With some other compounds both onset and recovery were rapid (ABRAMSON et al. 1969), pointing to the absence of the "isomerization" step according to the kinetic theory of ligand-receptor interaction.

Finally, the different mechanisms of antagonist binding with the muscarinic receptor must be taken into consideration while analyzing the structure-activity relationships, because the dissociation constants measured under equilibrium conditions have a different physical meaning for different types of antagonists (JÄRV and ELLER 1989).

D. Mechanism of Agonist Interaction

I. Agonist-Antagonist Competition Studies

The binding of agonists with the muscarinic receptor is most commonly studied by displacement of radioactive antagonists from their complexes with the receptor. These experiments have revealed the existence of a heterogeneous population of receptor sites, which have different affinities and agonist binding capacities (BIRDSALL et al. 1978; FIELDS et al. 1978; JIM et al. 1982; KLOOG et al. 1979). Besides the low- and high-affinity receptor sites the existence of binding sites with superhigh affinity was reported (BIRDSALL et al. 1980), but in most studies it was sufficient to postulate only the two former sites. As the heterogenity of agonist binding sites has been noted in the brain (BIRDSALL et al. 1978; KLOOG et al. 1979), heart (FIELDS et al. 1978; NATHANSON 1983), and smooth muscle (JIM et al. 1982), it seems to be characteristic of different subtypes of muscarinic receptors, including the classic M1 and M2 receptors (WATSON et al. 1986).

The high- and low-affinity sites are obviously related to two different physical states of the muscarinic receptor. It has been proposed that these states are generated by complexing of the receptor with GTP-binding protein (EHLERT 1985). The elegant reconstitution experiments with purified muscarinic receptor and GTP-binding protein revealed the conversion of the low-affinity agonist binding sites into the high-affinity sites (HAGA et al. 1985, 1986).

On the other hand, however, the radioactive antagonists which are used in the displacement studies themselves initiate the heterogenity of the receptor states in the assay system due to the isomerization of the receptor-antagonist complex. Unfortunately, the influence of this process on the results of the displacement studies has never been discussed, although the appearance of the low-affinity binding sites may well be related to the agonist interaction with the "isomerized" antagonist-receptor complex.

II. Binding Studies with Radioactive Agonists

The amount of receptor sites traced by radioactive agonists, such as [³H]-ACh (GURWITZ et al. 1984), [³H]N-methyloxotremorine (BEVAN 1984; BIRDSALL et al. 1980; HARDEN et al. 1983), [³H]pilocarpine (HEDLUND and BARTFAI 1981), and [³H]cis-methyldioxolane (CLOSSE et al. 1987; EHLERT et al. 1980; VICKROY et al. 1984), is considerably less compared with the antagonist binding data. At the same time, the dissociation constants calculated for agonists from these direct binding data are in reasonably good agreement with the dissociation constants for the high-affinity binding sites obtained from the displacement experiments. Therefore, it was proposed that only the high-affinity receptor sites can be detected by agonists as radioligands.

On the other hand, considerable differences are found in the binding of various agonists with the receptor (GURWITZ et al. 1985). It has been shown that [³H]-ACh binds in most tissues with considerably more binding sites than [³H]N-methyloxotremorine and [³H]cis-methyldioxolane (EHLERT et al. 1980; VICKROY et al. 1984). The Scatchard plots for these ligands also varied, representing ACh binding with a homogeneous population of receptor sites, while the two latter agonists revealed heterogeneity of their binding sites. The latter fact was explained by their interaction with both high- and superhigh-affinity sites. Nevertheless, these data clearly show that agonist binding with the muscarinic receptor should be quite incomplete even within the group of high- and superhigh-affinity states of the receptor, because the total amount of these sites cannot be less than the number of the ACh binding sites.

Such situations, as well as the strong influence of the assay conditions on the results of the direct binding studies (CLOSSE et al. 1987), can be explained by the rapid dissociation of the agonist-receptor complex, which interferes with the applicability of the filtration or centrifugation methods.

III. Kinetic Studies on Agonist–Antagonist Competition

Due to the complications in the case of direct agonist binding studies, competition between a nonradioactive agonist and a radioactive antagonist was used to investigate the agonist-receptor interaction mechanism (Järv et al. 1980). It has been found that in the presence of carbachol and oxotremorine the rate constant k_{obs} for the binding of [^{3}H]N-methyl-piperidinyl benzilate with the rat brain muscarinic receptor decreased in a dose-dependent manner. This means that there is no slow isomerization of the agonist-receptor complex. Later, the same conclusion was made by Schreiber et al. (1985).

The kinetic approach provides the unique possibility to study separately the agonist interaction with the free receptor R, the fast antagonist-receptor complex RA, and the "isomerized" complex (RA) as denoted in Eqn. (3). It has been found that the dissociation constant K_A for antagonist binding was not affected by the presence of carbachol and oxotremorine (Järv et al. 1980), while these agonists inhibited the rate of isomerization of the antagonist-receptor complex. This means that [^{3}H]N-methylpiperidinyl benzilate binds equally well with the free receptor and the agonist-receptor complex, giving evidence for the formation of the ternary complex ABR, where B stands for agonist, A for antagonist, and R for receptor:

$$
\begin{array}{ccccc}
 & K_A & & k_i & \\
R & \rightleftharpoons & RA & \underset{k_{-i}}{\overset{}{\rightleftharpoons}} & (RA) \\
\Big\updownarrow K_B & & \Big\updownarrow K_B & & \\
 & K_A & & & \\
RB & \rightleftharpoons & BRA & &
\end{array}
\tag{10}
$$

At the same time, the isomerization of the antagonist-receptor complex is inhibited by agonist in a dose-dependent manner. For this reason, the displacement of antagonists by agonists can be followed in the binding experiments, leaving the impression of a competitive mechanism of the agonist-antagonist interaction on the muscarinic receptor.

E. Structure of Muscarinic Neurotoxic Agents

I. Agonists and Antagonists: Qualitative Aspects

Proceeding from the structure of ACh, the important role of the structural fragment NCCOCC has been postulated for muscarinic drugs (Abramson et al. 1969, 1974; Baker et al. 1971; Barlow et al. 1963; Ing 1949), and on the basis of this hypothesis several series of effective ligands were designed (Table 3). These compounds differ remarkably in the rigidity of their NCCOCC backbone that was used for deducing the geometry of the "muscarinic pharmacophore" (Schulman et al. 1983; Takemura 1984) and

Table 3. Some basic structures of muscarinic drugs

Type of ligand	Principle structure
Esters	$-N^+-C-C-O-\overset{\overset{O}{\|\|}}{C}-C-$
Ethers	$-N^+-C-C-O-C-C-$
Tetrahydrofuran derivatives	
Furfuryl derivatives	
1,3-Dioxolane derivatives	
1,3-Oxathiolane derivatives	
Pyridine derivatives	
1,3-Dioxane derivatives	
Benzylammonium ions	
Alkylammonium ions	$-N^+-R$
Butynyl derivatives	$\supset N-C-C \equiv C-C-N\subset$

to rationalize the stereoselectivity of the agonist binding site (SCHULMAN et al. 1983).

The overlapping patterns for active conformations of potent agonists led to the conclusion that the atoms of the NCCOCC backbone must be arranged nearly in a plane to exhibit muscarinic activity, pointing to a flat shape of the complementary agonist binding site (TAKEMURA 1984). On the other hand, this site recognizes chirality centers of agonist molecules, most sensitively at the second and fourth carbon atoms of the backbone, NCC*OCC. and NCCOCC*, respectively. In the former case, the S-configuration of agonists is preferred as shown for muscarine (BECKETT 1967), acetyl-β-methylcholine (BECKETT et al. 1963), and cis-2-methyl-4-trimethylammoniummethyl-1,3-dioxolane (CHANG and TRIGGLE 1973).

Muscarinic antagonists revealed much weaker stereospecificity compared with agonists (ELLENBROEK et al. 1965; TRIGGLE and TRIGGLE 1976).

Moreover, the more potent enantiomers of 3-quinuclidinyl acetate and 3-quinuclidinyl benzilate, which possess agonistic and antagonistic properties, respectively, have opposite absolute configurations (LAMBRECHT 1976; REHAVI et al. 1977). Thus, the stereospecificity pattern of muscarinic agonists and antagonists is rather varied.

The oxygen atom of the NCCOCC group can be replaced by a tertiary nitrogen atom. The presence of these basic atoms in this fragment increases the binding effectiveness of agonists, pointing to the possibility of a donor-acceptor interaction between the ligand and receptor at this point (PRATESI et al. 1984). But this interaction is not crucial for agonist binding as simple alkylammonium ions reveal quite a strong agonistic activity. This conclusion is also valid for muscarinic antagonists (ABRAMSON et al. 1969).

The variation of substituents at the C-end of the NCCOCC fragment of the agonist molecule can lead to two different results. Firstly, a moderate increase in the substituent size decreases the affinity of ligands, while still retaining agonistic properties, especially in the case of flexible (open chain) compounds (PRATESI et al. 1984). Secondly, the introduction of large bulky groups to this C-atom transforms these ligands into partial agonists or antagonists, and the binding affinity of the latter is, as a rule, rather high (ABRAMSON et al. 1969, 1974; BARLOW et al. 1963; BAUMGOLD et al. 1975; BURGEN 1965).

Similar effects can be observed at the N-atom of the NCCOCC fragment. For example, the replacement of ACh methyl groups with ethyl groups (BIRDSALL et al. 1983) or the substitution of the whole choline moiety by quinuclidinole or piperidinole (MAAYANI et al. 1973) led to a sharp decrease in the affinity of these ligands, which, however, still remained agonistic. Variation of the ammonium group structure in the 1,3-dioxolane series, however, transformed these ligands into antagonists (ANGELI et al. 1984). Similar changes occurred in the case of pentyltrimethylammonium and ethoxyethyltrimethylammonium ions, and the replacement of the methyl groups by larger alkyl substituents yielded partial agonists and then antagonists (ABRAMSON et al. 1969).

The oxygen atom of the NCCOCC fragment can be replaced with a triple bond, yielding 2-butynylamine derivatives (Table 3), among which oxotremorine is the most effective muscarinic agonist known to date (BEBBINGTON et al. 1966; DAHLBOM et al. 1982; RINGDAHL and JENDEN 1983). The stereoelectronic properties of the acetylenic group are absolutely vital for the biological activity, as the replacement of the triple bond by a double bond or a phenylenic group yielded inactive derivatives (BEBBINGTON et al. 1966).

Variation of the oxotremorine structure at both ends of the $\rangle NCH_2C\equiv CCH_2N\langle$ fragment yielded either weak agonists or transformed the compounds into antagonists (RINGDAHL and JENDEN 1983). For example, the replacement of the β-lactam ring with an N-methylethylamide group retained the agonistic properties of the compound, while the introduction of an N-ethylethylamide group yielded an antagonist (RESUL et al. 1979). The size of substituents at the other end of the oxotremorine molecule is also critical, and the replacement of the pyrrolidine ring with other cyclic amines of larger ring size, or by a diethylamino group, reduced their affinity and yielded antagonistic properties (RINGDAHL and JENDEN 1983).

In summary, quite similar trends in the structure-activity relations can be observed for different types of compounds in their interaction with the muscarinic receptor. In virtually all cases, the size of the drug molecules seems to be critical for distinction between agonists and antagonists. The opposing effects of this structural factor on the binding effectiveness of these ligands can be easily understood on the basis of two, discrete binding sites for agonists and antagonists, as was proved by the kinetic studies.

II. Agonists and Antagonists: Quantitative Aspects

Some further insight into the mechanism of discrimination between muscarinic agonists and antagonists can be provided by structure-activity relationships, in which the "bulkiness' of the drug molecules is quantified by means of the molar refraction constant MR (HANSCH and LEO 1979). In these plots, the affinity of muscarinic antagonists was characterized by the pK_d values calculated from Eqn. (1), while in the case of agonists the dissociation constants, calculated for the high-affinity site, were used to obtain at least an interim solution.

For different groups of muscarinic antagonists of NCCOCC type, including esters, ethers, and alkylammonium ions, a common linear correlation between the pK_d and MR values was found (JÄRV and ELLER 1989; ELLER et al. 1989). This relationship has a positive slope, as larger ligands give more stable complexes. It can be seen in Fig. 2 that rather "bulky" compounds, like phencyclidine derivatives (MAAYANI et al. 1973; WEINSTEIN et al. 1983) fit this correlation. In the case of benzilic esters, which initiate the isomerization of the receptor-ligand complex, the constants pK_A fit the correlation, pointing to the common nature of the complex RA in Eqns. (1) and (3).

It should be mentioned that the relationship between pK_d and MR is rather distinct from the results of correlation of the same binding data with hydrophobicity parameters logP (JÄRV and BARTFAI 1982; JÄRV and ELLER 1989). In the latter case, separate linear dependences were found for esters, ethers, and alkylammonium ions that can be explained by differences in the donor-acceptor interactions between these ligands and the receptor site, if compared with octanol, as this solvent is used as the reference system for determining the hydrophobicity parameters.

Data for two sets of muscarinic agonists, including choline esters and alkyltrimethylammonium ions, are presented in Fig. 2. In both cases an increase in ligand size led to a decrease of binding affinity. However, the intensity of these effects is very different for esters and alkylammonium

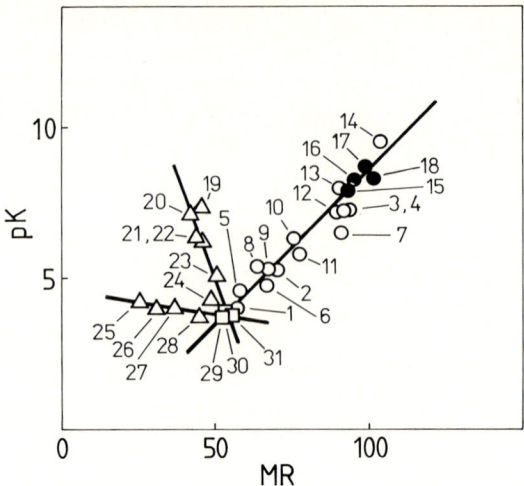

Fig. 2. Correlation of binding data for muscarinic drugs with the molar refraction constants MR. The points correspond to the following compounds:

Antagonists (pK_d ○, pK_A ●):
1, Pent-N$^+$Et$_3$; *2*, Ph (CH$_2$)$_4$N$^+$Me$_3$; *3*, Ph$_2$CH(CH$_2$)$_4$N$^+$Me$_3$ (Abramson et al. 1969);

4, ⬡⬡—N⬡ (Gabrielevitz et al. 1980); *5*, EtOC$_2$H$_4$N$^+$E;

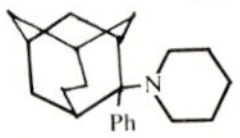

6, PhC$_2$H$_4$OC$_2$H$_4$N$^+$Me$_3$; *7*, Ph$_2$CHCH$_2$OC$_2$H$_4$N$^+$Me$_3$ (Abramson et al. 1969);
8, PhCOOC$_2$H$_4$N$^+$Me$_3$; *9*, PhCH$_2$COOC$_2$H$_4$N$^+$Me$_3$ (Eller et al. 1988);

10, PhCH$_2$COO⬡N$^+$Me2 Me$_2$; *11*, PhCH$_2$COO⬡N$^+$–Me ;

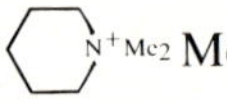

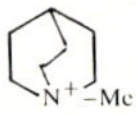

12, Ph$_2$CHCOOC$_2$H$_4$N$^+$Me$_3$ (Hulme et al. 1978);

   cyPent
   |
13, HO–CCOOC$_2$H$_4$N$^+$Me$_3$ (Eller et al. 1989);
   |
   Ph

   cyHex
   |
14, HO–COOC$_2$H$_4$N$^+$Me$_3$ (Ensing and de Zeeuw 1986);
   |
   Ph

15, Ph$_2$C(OH)COOC$_2$H$_4$N$^+$Me$_3$ (Eller et al. 1989);

16, Ph$_2$C(OH)COO⬡NMe (Järv et al. 1987); *17*, Ph$_2$C(OH)COO⬡N

(Sillard et al. 1985); *18*, Ph$_2$C(OH)COO⬡N$^+$–Me (Eller et al. 1988).

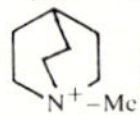

Agonists (pK_h △):
19, Me$^-$⬡O CH$_2$N$^+$Me$_3$; *20*, MeCOOC$_2$H$_4$N$^+$Me$_3$; *21*, NH$_2$COOC$_2$H$_4$N$^+$Me$_3$; *22*,
MeOC(O)⬡N-Me ; *23*, Me⬡HO–CH$_2$N$^+$Me$_3$; *24*, EtCOOC$_2$H$_4$N$^+$Me$_3$; *25*,
Me–N$^+$Me$_3$; *26*, Et–N$^+$Me$_3$; *27*, Pr–N$^+$Me$_3$ (Birdsall et al. 1978); *28*, Pent-
N$^+$Me$_3$ (Abramson et al. 1969).

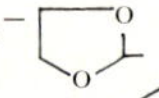

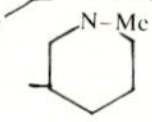

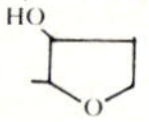

Partial agonists (pK_d □):
29, EtOC$_2$H$_4$N$^+$Me$_2$Et; *30*, EtOC$_2$H$_4$N$^+$Et$_3$; *31*, EtOC$_2$H$_4$N$^+$–N⬡Et

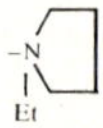

Et, ethyl, C$_2$H$_5$; Me, methyl, CH$_3$; Pent, penthyl, C$_5$H$_{11}$; Ph, phenyl, C$_6$H$_5$; Pr, propyl, C$_3$H$_5$; CyPent, cyclopenthyl, C$_5$Hg; CyHex, cyclo-hexyl, C$_6$H$_{11}$.

salts, pointing to a sepcific interaction between the O-atom of the NCCOCC backbone and the receptor binding site. However, if this interaction is hampered due to the variation of the acyl group structure, as seems to happen in the case of butyrylcholine, esters and simple alkylammonium ions exhibit a similar binding affinity (BIRDSALL et al. 1978). Thus, the tolerance by the receptor for the "bulkiness' of agonists seems not to play a crucial role in their binding unless the alterations in agonist structure hamper some other specificity-determining factors.

The common intercept of the linear pK versus MR plots refers to a hypothetical structure which has an equal affinity for both agonist and antagonist binding sites. It is remarkable that the data for affinity of partial agonists lie close to this intersection point, supporting the hypothesis that these ligands have a similar affinity for both sites. Thus, the quantitative structure-activity relationships provide a quite clear criterion towards the definition of the type of activity of a toxicant acting on the muscarinic receptor.

F. Two-Site Model of the Muscarinic Receptor

It is obvious from the results above that at least two different types of binding sites can be found on the muscarinic receptor, which exert specificity for either agonists or antagonists. Therefore, it can be assumed that the functions of these sites are related to the physiological and biochemical responses brought about by muscarinic agonists and antagonists. At the same time the varying specificity of these sites can be the key point for understanding the molecular mechanisms used by the receptor to distinguish between agonist and antagonist structures.

The recognition of these compounds by the muscarinic receptor seems to be based chiefly on quantitative variations in interactions between the receptor and ligand molecule, because no principal differences can be found in the basic framework of agonist and antagonist structures. Thus, these sites cannot be absolutely specific for agonists and antagonists. This means that muscarinic antagonists bind more effectively at the site which blocks the physiological response and less effectively at the site which triggers this response. As the former site is occupied at lower ligand concentrations, the inhibitory effect prevails in the case of antagonists. On the other hand, muscarinic agonists show preference for the site which governs the appearance of the response, but their binding with the inhibitory site is also possible at higher ligand concentrations. Therefore, the model proposed predicts inhibition of muscarinic responses at high agonist concentrations.

Such a two-site receptor model provides a simple explanation for the mechanism of action of partial agonists. It is obvious that between the typical sets of agonists and antagonists some compounds may possess a similar affinity for both types of binding sites. Therefore, these ligands simultaneously evoke and inhibit the physiological activity of the receptor.

G. Conclusions

The questions concerning the mechanism and specificity of action of muscarinic toxicants have been the subject of considerable debate for several decades. While the conventional pharmacological methods of receptor study, including structure-activity relationships, have provided a substantial base for understanding several features of the muscarinic receptor, the methods of radioligand analysis have proved to be extremly useful in revealing the events that underlie the initial steps of the response of nerve cell membranes to agonists and antagonists. This approach has been improved by complex kinetic studies, which have supplied additional information concerning the mechanism of interaction and evidence that these ligands occupy discrete binding sites. The latter fact will substantially contribute to the further understanding of the physicochemical factors which are used by the muscarinic receptor to discriminate between agonists and antagonists and thus determine the type of action of neurotoxic agents.

References

Abramson FB, Barlow RB, Mustafa MG, Stephenson RP (1969) Relationships between chemical structure and affinity for acetylcholine receptors. Br J Pharmacol 37:207–233

Abramson FB, Barlow RB, Franks FM, Pearson FBM (1974) Relationships between chemical structure and affinity for postganglionic acetylcholine receptors of the guinea-pig ileum. Br J Pharmacol 51:81–93

Adem A, Asblom A, Johansson G, Mbugua PM, Karlsson E (1988) Toxins from the venom of the green mamba *Dendroaspis angusticeps* that inhibit the binding of quinuclidinyl benzilate to muscarinic acetylcholine receptors. Biochim Biophys Acta 968:340–345

Angeli P, Gianella M, Pigini M (1984) Size of muscarinic receptor anionic binding site related to onium group of ligands with a 1,3-oxathiolane nucleus. Eur J Med Chem 19:495–500

Aronstam RS, Abood LG, MacNeil MK (1977) Muscarinic cholinergic binding in human erythrocyte membranes. Life Sci 20:1175–1180

Baker RW, Chotia CH, Pauling P, Petcher TJ (1971) Structure and activity of muscarinic stimulants. Nature 230:439–445

Barlow B, Scott KA, Stephenson RP (1963) An attempt to study the effects of chemical structure on the affinity and efficacy of compounds related to acetylcholine. Br J Pharmacol 21:509–522

Baumgold J, Abood LG, Hoss WP (1975) Chemical factors influencing the psychomimetic potency of glycolate esters. Life Sci 17:603–612

Bebbington A, Brimblecombe RW, Shekeshaft D (1966) The central and peripheral activity of acetylenic amines related to oxotremorine. Br J Pharmacol 26:56–67

Becker CE, Tong TG, Boerner U, Roe RL, Scott RAT, MacQuarrie MB, Bartter F (1979) Diagnosis and treatment of *Amanita phalloides* type mushroom poisoning. West J Med 125:100–109

Beckett AH (1967) Stereospecificity in reactions of cholinesterases and the cholinergic receptor. Ann NY Acad Sci 144:675–688

Beckett AH, Harper NJ, Cliterow JW (1963) The absolute configurations of the alfa- and beta-methylcholine isomers and their acetyl and succinyl esters. J Pharm Pharmacol 15:349–355

Bevan P (1984) [³H]Oxotremorine-M binding to membrances prepared from rat brain and heart: evidence for subtypes of muscarinic receptors. J Eur Pharmacol 101:101–110

Birdsall NJM, Burgen ASV, Hulme EC (1978) Binding of agonists to brain muscarinic receptors. Mol Pharmacol 14:723–736

Birdsall NJM, Hulme EC, Burgen ASV (1980) The character of muscarinic receptors in different regions of the rat brain. Proc R Soc Lond [Biol] 207(1166):1–12

Birdsall NJM, Burgen ASV, Hulme EC, Wong EHF (1983) The effect of p-chloromercuribenzoate on structure-binding relationships of muscarinic receptors on the rat cerebral cortex. Br J Pharmacol 80:197–204

Bolton TB (1979) The mechanisms of action of transmitters and other substances on smooth muscle. Physiol Rev 59:606–718

Bowman WC, Bowman A (1986) Dictionary of pharmacology. Blackwell, Oxford

Burgen ASV (1965) The role of ionic interaction at the muscarinic receptor. Br J Pharmacol Chemother 25:4–17

Chang KJ, Triggle DJ (1973) Stereoselectivity of cholinergic activity in a series of 1,3-dioxolanes. J Med Chem 16:718–720

Clark AJ (1937) General pharmakology. Springer, Berlin (Heffter's Handbuch der experimentellen Pharmakologie, vol 4)

Closse A, Bittiger H, Langenegger D, Wanner A (1987) Binding studies with [³H]cis-methyldioxolane in different tissues. Naunyn Schmiedebergs Arch Pharmacol 335:372–377

Dahlbom R, Jenden DJ, Resul B, Ringdahl B (1982) Stereochemical requirements for central and peripheral muscarinic and antimuscarinic activity of some acetylenic compounds related to oxotremorine. Br J Pharmaccol 76:299–304

Dale HH (1914) The action of certain esters and ethers of choline and their relation to muscarine. J Pharmacol 6:147–190

Ducanel F, Rowan EG, Cassart E, Harvey AL, Menez A, Boulain J-C (1991) Aminoacid sequence of a muscarinic toxin deduced from the cDNA nucleotide sequence. Toxicon 29:No, 4/5, 516–520

Ehlert FJ (1985) The relationship between muscarinic receptor occupancy and adenylate cyclase inhibition in the rabbit myocardium. Mol Pharmacol 28:410–421

Ehlert FJ, Dumont Y, Roeske WR, Yamamura HI (1980) Muscarinic receptor binding in rat brain using the agonist [³H]cis-methyldioxolane. Life Sci 26:961–967

Ehlert FJ, Roeske WR, Yamamura HI (1984) Muscarinic receptor [³H]ligand binding methods. In: Marangos PJ, Campbell IC, Cohen RM (eds) Amines and acetylcholine. Academic, Orlando, pp 339–355 (Brain receptor methodology, part A)

Ellenbroek BWJ, Nivard RJF, van Rossum JM, Ariens EJ (1965) Absolute configuration and parasympathetic action: pharmacodynamics of enantiomorphic and diastereoisomeric esters of beta-methylcholine. J Pharm Pharmacol 17:393–404

Eller M, Järv J (1988) Two-step isomerization of quinuclidinyl benzilate-muscarinic receptor complex. Neurochem Int 12:285–289

Eller M, Järv J (1989) Kinetics of N-methylscopolamine interaction with muscarinic receptor from rat cerebral cortex. Neurochem Int 15:301–305

Eller M, Järv J, Palumaa P (1988) Influence of non-radioactive ligands on kinetics of N-methyl-[³H]scopolamine binding to muscarinic receptor. Org React 25:372–386

Eller M, Järv J, Loodmaa E (1989) Kinetic analysis of interaction of ester antagonists with muscarinic receptor. Org React 26:199–210

Ensing K, de Zeeuw RA (1986) Differential behavior toward muscarinic receptor binding between quaternary anticholinergics and their tertiary analogues. Pharmacol Res 3:327–332

Fields JZ, Roeske WR, Morkin E, Yamamura HI (1978) Cardiac muscarinic cholinergic receptors. Biochemical identification and characterization. J Biol Chem 253:3251–3258

Gabrielevitz A, Kloog Y, Kalir A, Balderman D, Sokolovsky M (1980) Interaction of phencyclidine and its new adamantyl derivatives with muscarinic receptors. Life Sci 26:89–95

Gallacher DV, Morris PA (1986) A patch-clamp study of potassium currents in resting and acetylcholine-stimulated mouse submandibular acinar cells. J Physiol (Lond) 373:379–395

Galper JB, Klein N, Catterall WA (1977) Muscarinic acetylcholine receptors in developing chick heart. J Biol Chem 252:8692–8699

Gibson RE, Rzeszotarski WJ, Jagoda EM, Francis BE, Reba RC, Eckelman WC (1984) [^{125}I]3-quinuclidinyl-4-iodobenzilate: a high affinity high specific activity radioligand for the M1 and M2 acetylcholine receptors. Life Sci 34:2287–2296

Gurwitz D, Kloog Y, Sokolovsky M (1984) Recognition of the muscarinic receptor by its endogenous transmitter: binding of [^{3}H]acetylcholine and its modulation by transition metal ions and guanine nucleotide. Proc Natl Acad Sci USA 81:3650–3654

Gurwitz D, Kloog Y, Sokolovsky M (1985) High affinity binding of [^{3}H]acetylcholine to muscarinic receptors: regional distribution and modulation by guanine nucleotides. Mol Pharmacol 28:295–297

Haga K, Haga T, Ichiyama A, Katada T, Kurose H, Ui M (1985) Functional reconstruction of purified muscarinic receptors and inhibitory guanine nucleotide regulatory protein. Nature 316:731–733

Haga K, Haga T, Ichiyama A (1986) Reconstitution of the muscarinic acetylcholine receptor: guanine nucleotide-sensitive high affinity binding of agonists to purified muscarinic receptors reconstituted with GTP-binding proteins (Gi and Go). J Biol Chem 261:10133–10140

Hansch C, Leo A (1979) Substituent constants for correlation analysis in chemistry and biology. Wiley, New York

Harden TK, Meeker RB, Martin MW (1983) Interaction of radiolabelled agonist with cardiac muscarinic cholinergic receptors. Pharmacol Exp Ther 227:570–577

Hedlund B, Bartfai T (1981) Binding of ^{3}H-pilocarpine to membranes from rat cerebral cortex. Naunyn Scmiedebergs Arch Pharmacol 317:126–130

Heilbronn E, Bartfai T (1978) Muscarinic acetylcholine receptor. Prog Neurobiol 11:171–188

Heiser JF, Hillin JC (1971) The reversal of anticholinergic drug induced delirium and coma with physiostigmine. Am J Psychiatry 127:1050–1052

Hrdina PD (1986) General principles of receptor binding. In: Boulton AA, Baker GB, Hrdina PD (eds) Neuromethods. Humana, Clifton, pp 1–21

Hulme EC, Birdsall NJM, Burgen ASV, Mehta P (1978) The binding of antagonists to brain muscarinic receptors. Mol Pharmacol 14:737–750

Hurko O (1978) Specific [^{3}H]-quinuclidinyl benzilate binding activity in digitonin-solubilized preparations from bovine brain. Arch Biochem Biophys 190:434–445

Ing HR (1949) The structure-action relationships of the choline group. Science 109:264–266

Järv J, Bartfai T (1982) The importance of hydrophobic interactions in the antagonist binding to the muscarinic acetylcholine receptor. Acta Chem Scand [B] 36: 489–498

Järv J, Bartfai T (1988) Muscarinic acetylcholine receptor. In: Whittaker VP (ed) The cholinergic synapse. Springer, Berlin Heidelberg New York, pp 315–345 (Handbook of experimental pharmacology, vol 86)

Järv J, Eller M (1988) Kinetic aspects of L-quinuclidinyl benzilate interaction with muscarinic receptor. Neurochem Int 13:419–428

Järv J, Eller M (1989) Effect of ligand volume on affinity of muscarinic antagonists. Org React 26:188–198

Järv J, Hedlund B, Bartfai T (1979) Isomerization of the muscarinic receptor-antagonist complex. J Biol Chem 254:5595–5598

Järv J, Hedlund B, Bartfai T (1980) Kinetic studies on muscarinic antagonist-agonist competition. J Biol Chem 255:2649–2651

Järv J, Sillard R, Bartfai T (1987) Influence of temperature on cooperative binding of N-methylpiperidinyl benzilate with muscarinic receptor from rat cerebral cortex. Chemistry 36:172–180

Jencks WP (1969) Catalysis in chemistry and enzymology. McGraw-Hill, New York

Jim K, Bolger GT, Triggle DJ (1982) Muscarinic receptors in guinea pig ileum. A study by agonist-[^{3}H]labelled antagonist competition. Can J Physiol Pharmacol 60:1707–1714

Karlsson E (1979) Chemistry of protein toxins in snake venoms. In: Lee CY (ed) Snake venoms. Springer, Berlin Heidelberg New York, pp 259–276 (Handbook of experimental pharmacology, vol 52)

Karlsson E, Risinger C, Jolkkonen M, Wernstedt C, Adem A (1990) Amino acid sequence of a snake venom toxin that binds to the muscarinic acetylcholine receptor. Toxicon 29:521–528

Kier LB, Hall LH (1978) Molecular study of muscarinic receptor affinity of acetylcholine antagonists. J Pharm Sci 67:1408–1412

Kloog Y, Sokolovsky M (1978) Studies on muscarinic AChR from mouse brain: characterization of the interaction with antagonists. Brain Res 144:31–48

Kloog Y, Egozi Y, Sokolovsky M (1979) Characterization of muscarinic acetylcholine receptors from mouse brain: evidence for regional heterogeneity and isomerization. Mol Pharmacol 15:545–558

Laduron PM, Verwimp M, Leysen JE (1979) Stereospecific in vitro binding of [^{3}H]dexetimide to brain muscarinic receptors. J Neurochem 32:421–427

Lambrecht G (1976) Struktur- und Konformationswirkungs-Beziehungen heterocyclischer Acetylcholinanaloge. I Muscarinwirkung enantiomerer 3-Acetoxy-chinuclidine und 3-Acetooxypiperidine. Eur J Med Chem 11:461–466

Löffelholz K, Pappano AJ (1985) The parasympathetic neuroeffector junction of the heart. Pharmacol Rev 37:1–24

Lopker A, Abood LG, Hoss W, Lionetti FJ (1980) Stereoselective muscarinic acetylcholine and opiate receptors in human phagocytic leucocytes. Biochem Pharmacol 29:1361–1365

Luthin GR, Wolfe BB (1984) [^{3}H]pirenzepine and [^{3}H]quinuclidinylbenzilate binding to brain muscarinic cholinergic receptors. Differences in measured receptor density are not explained by differences in receptor isomerization. Mol Pharmacol 26:164–169

Maayani S, Weinstein H, Cohen S, Sokolovsky M (1973) Acetylcholine-like molecular arrangement in psychomimetic anticholinergic drugs. Proc Natl Acad Sci USA 70:3103–3107

Michaelson MJ, Zeimal EV (1973) Acetylcholine. Pergamon, New York

Nathanson NM (1983) Binding of agonists and antagonists to muscarinic acetylcholine receptor on intact cultured heart cells. J Neurochem 41:1545–1549

Paton WDM, Rang HP (1965) The uptake of atropine and related drugs by intestinal smooth muscle of guinea pig in relation to acetylcholine receptors. Proc R Soc Lond [Biol] 163:1–44

Pratesi P, Villa L, Ferri V, de Micheli C, Grana E, Silipo C, Vittoria A (1984) Recent advances in the study of muscarinic receptor by QSAR. In: Melchirre C, Gianella M (eds) Highlights in receptor chemistry. Elsevier, Amsterdam, pp 225–250

Rehavi M, Maayani S, Sokolovsky M (1977) Enzymatic resolution and cholinergic properties of 3-quinuclidinyl derivates. Life Sci 21:1293–1302

Resul B, Ringdahl B, Hacksell U, Svensson U, Dahlbom R (1979) Acetylenic compounds of potential pharmacological value. XXXI. Studies on N-(4-tert-amino-2-butynyl)carboxamides as muscarinic agonists. Acta Pharm Suec 16:225–232

Ringdahl B, Jenden DJ (1983) Pharmacological properties of oxotremorine and its analogs. Life Sci 32:2401–2413

Rumack BH (1973) Anticholinergic poisoning: treatment with physostigmine. Pediatrics 52:449–451

Schimerlik ML, Searles RP (1980) Ligand interactions with membrane bound porcine atrial muscarinic receptor(s). Biochemistry 19:3407–3413

Schmiedeberg O, Koppe R (1869) Das Muscarin, das giftige Alkaloid des Fliegenpilzes (*Agaricus muscarius* L.), seine Darstellung, chemischen Eigenschaften, physiologischen Wirkungen, toxikologische Bedeutung und sein Verhältnis zu Pilzvergiftungen im Allgemeinen. Vogel, Leipzig

Schreiber G, Henis YI, Sokolovsky M (1985) Rate constants of agonist binding to muscarinic receptors in rat brain medulla: evaluation by competition kinetics. J Biol Chem 260:8795–8802

Schulman JM, Sabio ML, Disch RL (1983) Recognition of cholinergic agonists by the muscarinic receptor. I. Acetylcholine and other agonists with the NCCOCC backbone. J Med Chem 26:817–823

Sillard RG, Järv JL, Bartfai T (1985) Kinetic evidence for cooperativity of quinuclidinyl benzilate interaction with muscarinic receptor from rat brain. Biol Membr 2:426–431

Sokolovsky M (1984) Muscarinic receptors. Int Rev Neurobiol 25:139–183

Stephenson RP (1956) A modification of receptor theory. Br J Pharmacol 11:379–393

Takemura S (1984) Geometry of muscarinic agonists. J Pharmacobiodyn 7:436–444

Taylor P (1980) Cholinergic agonists. In: Goodman LS, Gilman AG (eds) The pharmacological basis of therapeutics. Macmillan, New York, pp 91–99

Triggle DJ, Triggle CR (1976) Chemical pharmacology of the synapse. Academic, New York, pp 233–430

Vickroy TW, Roeske WR, Yamamura HI (1984) Pharmacological differences between high-affinity muscarinic agonist binding sites of the rat heart and cerebral cortex labelled with $(+)$-[^{3}H]*cis*-methyldioxolane. J Pharmacol Exp Ther 229:747–755

Watson M, Roeske WR, Yamamura HI (1982) [^{3}H]pirenzepine selectively identifies a high affinity population of muscarinic cholinergic receptors in the rat cerebral cortex. Life Sci 31:2019–2023

Watson M, Roeske WR, Bickroy TW, Smith TL, Akyiama K, Gulya K, Duckles SP, Serra M, Adam A, Nordberg A, Gehlert DR, Wamsley JK, Yamamura HI (1986) Biochemical and functional basis of putative muscarinic receptor subtype and its implications. In: Levine RR, Birdsall NJM, Giachetti A, Hammer R, Iversen LL, Jenden DJ, North RA (eds) Subtypes of muscarinic receptors II. Supplement to Trends in Pharmacological Sciences. Elsevier, Amsterdam, pp 46–55

Weinstein H, Maayani S, Pazhenchevsky B (1983) Multiple actions of phencyclidine: discriminant structure-activity relationships from molecular conformations and assay conditions. Fed Proc 42:2574–2578

Yamamura HI, Snyder SH (1974) Muscarinic cholinergic binding in rat brain. Proc Natl Acad Sci USA 71:1725–1729

CHAPTER 19

Peptide Toxins that Alter Neurotransmitter Release*

J.O. DOLLY

A. Introduction

Despite the fundamental importance of quantal transmitter release in neural communication, there is only limited information available on the detailed mechanism(s) involved (VALTORTA et al. 1990; TAUC and POULAIN 1991). Clearly, progress with this challenging problem ought to be achieved by identifying molecular components in nerve terminals concerned with the process or its regulation. As several neurotoxins have been characterized with abilities to perturb transmitter release in very selective ways (Table 1), using them as probes to discover such functional entities is proving to be a successful experimental strategy (reviewed by DOLLY 1990, 1991). Of course, antibodies to particular proteins involved in synaptic vesicle trafficking (e.g. synaptophysin; VALTORTA et al. 1988) or thought to be concerned with acetylcholine (ACh) release (e.g. mediatophore; ISRAEL et al. 1990) can also provide useful tools, but few such macromolecules have been identified. Moreover, when antibodies are raised, they often bind to non-essential epitopes (or their activity cannot be detected in screening due to the lack of suitable assays). In contrast, neurotoxins are by definition designed to cause malfunction of their targets and, fortunately for research purposes, these can represent a series of proteins (e.g. Fig. 1) that function in the multi-step process of synaptic transmission. In other cases, a family of homologous toxins (e.g. dendrotoxins) exhibit subtle differences in specificities for variants of voltage-dependent, fast-activating K^+ channels (cf. AWAN and DOLLY 1991). Thus, diligent searches for unrecognised naturally occurring neurotoxins, particularly those with selectivity for novel macromolecules, as well as complete characterization (in terms of specificity) of established probes, are fully warranted.

Most of the toxins (Table 1) change transmitter release indirectly by affecting a cation channel (for Na^+, Ca^{2+} or K^+) on the plasma membrane, whilst only a few (i.e. the *Clostridial* proteins) act at a point after Ca^{2+} entry into the nerve terminal; additionally, others exert multiple effects (i.e. phospholipase A_2-containing proteins and α-latrotoxin). This overview will

*The work reported herein from my laboratory is supported by MRC, Wellcome Trust and USAMRIID (Contract No. DAMD 17-88-C-8008).

Table 1. Naturally occurring polypeptide toxins that affect (indirectly or directly) the release of transmitters

Toxin	Source	M_r (kDa)	Toxicity in rodents (wt/g body wt intraperitoneal injection)	Effects on transmitter release *Indirect*
α, β and γ Scorpion toxins	Various species of scorpions		see BECKER and GORDON (this volume)	Facilitation by binding to distinct sites on voltage-activated Na^+ channels and altering activation and/or inactivation
ω-Conotoxins	*Conus* snails		see STRIESSNIG and GLOSSMAN (this volume)	Inhibition due to blockade of certain voltage-sensitive Ca^{2+} channels
Toxin I	*Dendroaspis polylepis*	7		Facilitation at motor, parasympathetic and sympathetic nerves; elevation of neuronal excitability, often resulting in repetitive firing in CNS and PNS, by inhibiting fast-activating, aminopyridine-sensitive, voltage-dependent K^+ channels
α-Dendrotoxin	*D. angusticeps*	7	>50 µg	
Dendrotoxins: β γ				Thought to block non-inactivating, voltage-dependent K^+ channels in brain synaptosomes
δ	*D. angusticeps*			Action seems to be similar to that of α, though not studied extensively; binds to most, but not all, of the K^+ channels in CNS labelled by α-dendrotoxin
Noxiustoxin	*Centruroides noxius*	4.2	20 µg	Facilitation: inhibits K^+ permeability of brain synaptosomes; also, blocks delayed rectifier K^+ channel in squid axon and a Ca^{2+}-activated K^+ current in skeletal muscle

Toxin	Source			Description
Charybdotoxin	*Leiurus quinquestriatus*	4.3		Facilitation in CNS (not at motor nerve endings); could result from inhibition of α-dendrotoxin-susceptible K$^+$ current(s) (seen in sensory neurons) and, possibly, from blockade of Ca^{2+}-activated, large and intermediate K$^+$ conductances (observed in various cells)
β-Bungarotoxin	*Bungarus multicinctus*	20.5	10 ng	*Multiple* Initially depress, then transiently increase and eventually abolish ACh release; block certain K$^+$ currents. (In CNS, β-bungarotoxin affects the release of several transmitters)
Notexin	*Notechis scutatus*	13.5	17 ng	
Crotoxin	*Crotalus durissus terrificus*	41.0	35 ng	
Taipoxin	*Oxyuranus scutellatus*	46.0	2 ng	
α-Latrotoxin	Black widow spiders	130	20 ng	Increases the release of transmitters in many preparations forms membrane pores resulting in massive influx of divalent cations
Tetanus toxin	*Clostridium tetani*	150	~0.5 pg	*Direct* Inhibition primarily at central inhibitory synapses; much less effective at motor nerve terminals. Active inside all reason types tailed
Botulinum neurotoxins	*C. botulinum*	150	~0.5 pg	Blockade at peripheral cholinergic nerves; when applied intracellularly inhibits release of numerous transmitters

PNS, peripheral nervous system; CNS, central nervous system.

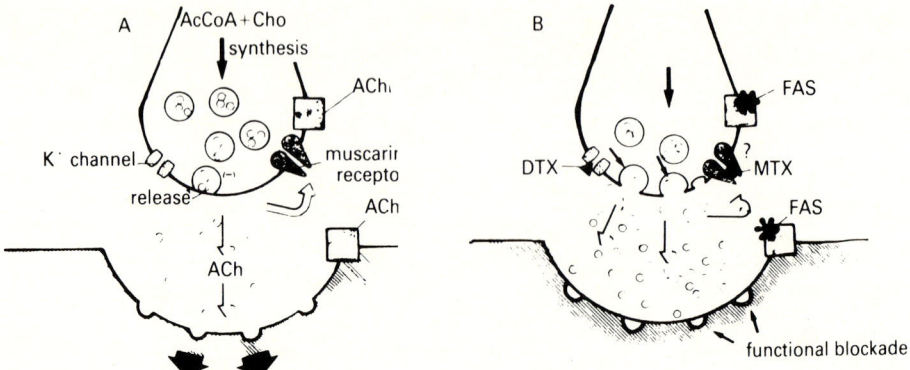

Fig. 1A,B. Hypothetical cholinergic synapse showing acetylcholin esterase (*AChE*), presynaptic K^+ channels and muscarinic receptors as targets for protein toxins from mamba venoms [fasciculins (*FAS*), dendrotoxins (▲, *DTX*) and muscarinic toxin (●, *MTX*)] (**A**). By increasing the release of ACh (DTX, MTX) and inhibiting its hydrolysis (FAS), the result of this synergistic activity of the three toxins would be a net increase in ACh concentration in the synaptic cleft and an eventual functional blockade of neurotransmission (**B**). *Ac*, acetyl; *Cho*, choline (From DAJAS et al. 1988)

focus primarily on botulinum (BoNT) and tetanus (TeTX) neurotoxins because of the apparent direct involvement of their targets in Ca^{2+}-dependent secretion. As K^+-channel toxins are discussed in detail in another chapter, only their effects on transmitter release are discussed herein, whilst those active on other channels (Table 1) are dealt with elsewhere in this volume. Emphasis is also placed on β-bungarotoxin (β-BuTX) and α-latrotoxin because of the intriguing actions of the latter and the exclusive presynaptic selectivity of the former for peripheral cholinergic nerve terminals.

B. Toxins that Facilitate Indirectly the Neuronal Release of Transmitters by Blocking K^+ Channels

I. Dendrotoxins

These are single-chain homologous polypeptides, with sequence similarities to Kunitz-type protease inhibitors, that have been purified from the venoms of mamba snakes [α, β, γ and δ dendrotoxin (DTX) from *Dendroaspis angusticeps*; toxins I and K from *D. polylepis*] and contain 59–61 amino acids with 3 intra-chain disulphide bonds (STRYDOM 1973; JOUBERT and TALJAARD 1980; HARVEY and KARLSSON 1982; BENISHIN et al. 1988).

1. Effects in the Peripheral Nervous System

α-DTX, $C_{13}S_1C_3$ (δ-DTX), toxins I and K were first found to augment nerve-evoked twitch tension in chick biventer cervicis nerve-muscle prep-

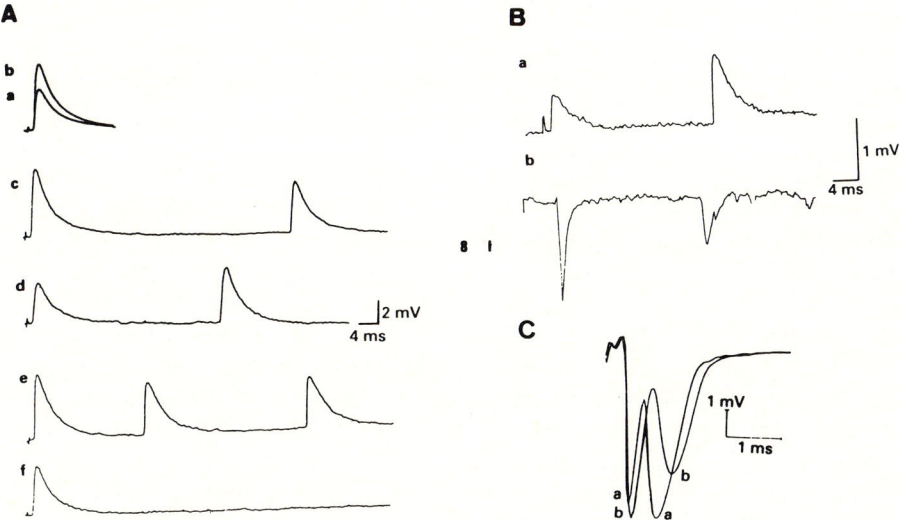

Fig. 2. Effects of toxin I (**A, B**) and β-bungarotoxin (**C**) on endplate potentials (EPPs) and perineural waveform in frog cutaneous pectoris (**A, B**) and mouse triangularis sterni (**C**) preparations. **A** EPPs were recorded at 20°C in medium containing 0.3 mM Ca^{2+} and 2 mM Mg^{2+} before (*a*) and at the same endplate after 5(*b*), 20(*c*), 25 (*d, e*) and 30(*f*) min after exposure to 2.6 μM toxin I. The motor nerve was stimulated at a frequency of 0.5 (*a–d*), 0.1 (*e*) or l(*f*)Hz. Averages of 20–30 EPPs are shown in (*a*) and (*b*); (*c–f*) typical responses. The average amplitude of MEPPs (0.4 mV) and muscle fibre resting membrane potential (−80 mV) remained constant during the recording period. **B**, EPPs (*a*) and perineural waveforms (*b*) following single stimuli after 65 min of treatment with 25 nM toxin I at 23°C; the bathing solution contained 0.1 mM Ca^{2+} and 1.9 mM Mg^{2+}. **C** Control waveform (*a*) and that after 60 min exposure to 150 nM β-bungarotoxin. Note the large reduction in the second negative component of the waveform. (**A, B** from ANDERSON and HARVEY 1988; **C** from ROWAN and HARVEY 1988)

arations without affecting responses (except at relatively high concentrations) to exogenous ACh or KCl, thereby indicating a presynaptic facilitation of ACh release (HARVEY and KARLSSON 1980, 1982; HARVEY et al. 1984). This conclusion was later confirmed in a fraction of frog (4 of 7) and mouse (10 of 31) neuromuscular preparations examined by demonstrating that α-DTX or toxin I within 2–15 min of application cause a *transient* (usually) but significant increase (20%–50%) in the amplitude of Epps (Fig. 2A), whilst leaving their time course unchanged (ANDERSON and HARVEY 1988). Curiously, at murine endplates this effect was only pronounced at 37°C, and in each tissue a relatively high toxin concentration was necessary. As α-DTX did not increase the frequency or alter the amplitude or kinetics of miniature endplate potentials (MEPPs), it was deduced that these toxins can produce an initial increase in quantal content (the number of quanta released per stimulus).

Interestingly, after longer periods of exposure (15–60 min normally) these two toxins produced multiple EPPs following a single stimulation of the motor nerve in frog (Fig. 2A,B) or mouse preparations (24 of 31) studied. This was accompanied by coincident repetitive activity of the motor nerve (Fig. 2B), as observed by recording the current flow through the nerve terminal membrane following stimulation (with an extra-cellular electrode placed inside the perineural sheath). In some cases, the toxins reduced to a variable extent the amplitude of the second negative deflection of the perineural waveform (thought to represent presynaptic K^+ currents), whereas on other occasions there was little or no change (as reported by DREYER and PENNER 1987), but repetitive firing appeared (ANDERSON and HARVEY 1988). In view of the demonstrated abilities of α-DTX and toxin I to inhibit certain fast-activating, voltage-dependent K^+ currents (reviewed by DOLLY et al. 1987a; STRONG 1990; DREYER 1990; DOLLY 1991; MEVES, this volume) and their recognised importance in controlling neuron excitability, it was suggested by the authors that these toxins block at least one K^+ current in each terminal and, probably, two in others. As one of these was assumed to be only a minor fraction of the total K^+ component measured, its alteration by toxin would not have been detected; nevertheless, blockade of this current could destabilise the neuronal membrane (as noted below for central neurons), resulting in repetitive firing. The relative importance of the larger K^+ current (clearly seen to be attenuated by the toxins but only in some experiments) in α-DTX-induced repetitive firing and/or the *transient* elevation of quantal content remains unclear. For example, when the second negative deflection was abolished by 3,4-diaminopyridine in mouse motor nerve terminals, subsequent addition of α-DTX caused repetitive firing (ANDERSON and HARVEY 1988). In any case, despite the practical difficulties of recording presynaptic currents and associated complexities in interpreting the resultant data, it can be concluded that toxin-sensitive K^+ channel(s) exist in motor nerve terminals where they, apparently, serve important roles; however, their complete definition and relation to those in other neurons (Meves, this volume) is a topic of great interest and importance that warrants further study. Finally, it is noteworthy that α-DTX and toxin I also facilitate release of transmitters in the autonomic nervous system, although their effects there are more rapid in onset than in motor nerves, not sustained and readily reversible (HARVEY et al. 1984).

2. Actions of α-Dendrotoxin and Its Homologues in the Central Nervous System

a) Neurophysiological Demonstration of Toxin-Induced Increased Neuronal Excitability, Facilitated Release of Transmitters and Blockade of an I_A-like K^+ Current

In contrast to the peripheral nervous system where α-DTX or toxin I have limited toxicity in whole animals (Table 1), they are several orders of magnitude more toxic centrally (2.5 and 0.5 ng/g body wt, respectively) (BLACK et al. 1986); when injected intraventricularly into rat brain, α-, β-, γ-, δ-DTX or toxin I induce pronounced convulsive symptoms, resulting in death. Silveira et al. (1988) have characterised a series of behavioural effects (e.g. postural asymmetry and stereotypes) induced by intrastriatal injection of α-DTX, as well as increased turnover of monoamines. By means of electrophysiological recording in CA1 pyramidal cells of rodent hippocampal slices, a basis for the toxins' facilitatory actions was unveiled (detailed in Dolly et al. 1984a, 1986; Halliwell et al. 1986). Synaptic transmission and cell excitability in the hippocampal slice are enhanced by α-DTX. These changes do not result from its depression of inhibitory transmission; instead, the toxin potentiates the level of inhibitory activity, as revealed by enhanced, spontaneous, picrotoxin-sensitive, inhibitory, postsynaptic potentials. This is assumed to arise from increased excitatory input to inhibitory neurons or potentiated output of an inhibitory transmitter, most likely γ-aminobutyrate. In summary, the complex effects of α-DTX can be ascribed to an overall elevation of neuronal activity encompassing facilitated release of both excitatory and inhibitory transmitters. These can be explained, at least in part, by the selective ability of α-DTX or toxin I (Dolly et al. 1984a; Halliwell et al. 1986) to block a fast-activating, voltage-dependent, transient, outward K^+ current (resembling I_A). Also, their inhibitory action on the related but slowly inactivating K^+ current (Stansfeld et al. 1986, 1987; Penner et al. 1986a; Benoit and Dubois 1986) in peripheral neurons raises the likelihood that attenuation by the toxins of similar variants in the CNS could contribute to the epileptiform activity they induce. However, such a postulation has yet to be documented, although a slowly decaying K^+ current (of unknown sensitivity to α-DTX) has been detected in hippocampal neurons (Storm 1988), and some K^+ channels cloned from mammalian brain yield slowly inactivating K^+ currents susceptible to α-DTX or toxin I when expressed in oocytes (detailed by Meves, this volume).

b) Dendrotoxins Elicit Resting Efflux of Transmitters from Synaptosomes and Elevate Cytosolic Ca^{2+} Concentration

In accordance with the aforementioned electrophysiological data from hippocampus, α-DTX and toxin I were found to facilitate the release of both excitatory and inhibitory transmitters from isolated nerve terminals. Notably, α-DTX (or 4-aminopyridine at very much higher concentrations) increases the Ca^{2+}-dependent efflux of glutamate from guinea-pig cerebrocortical synaptosomes; its proportional reduction in the amount of glutamate released by subsequent K^+-depolarisation (Fig. 3A) suggests that a common exocytotic pool of transmitter is affected (Tibbs et al. 1989a). A qualitatively similar result has been obtained with glutamate release from striatal synaptosomes, even when pre-exposure to D-aspartate was used to deplete the cytoplasmic amino-acid pool (Barbeito et al. 1990). Also, α-DTX is known to raise the Ca^{2+}-dependent efflux of $[^3H]$-γ-aminobutyrate from a

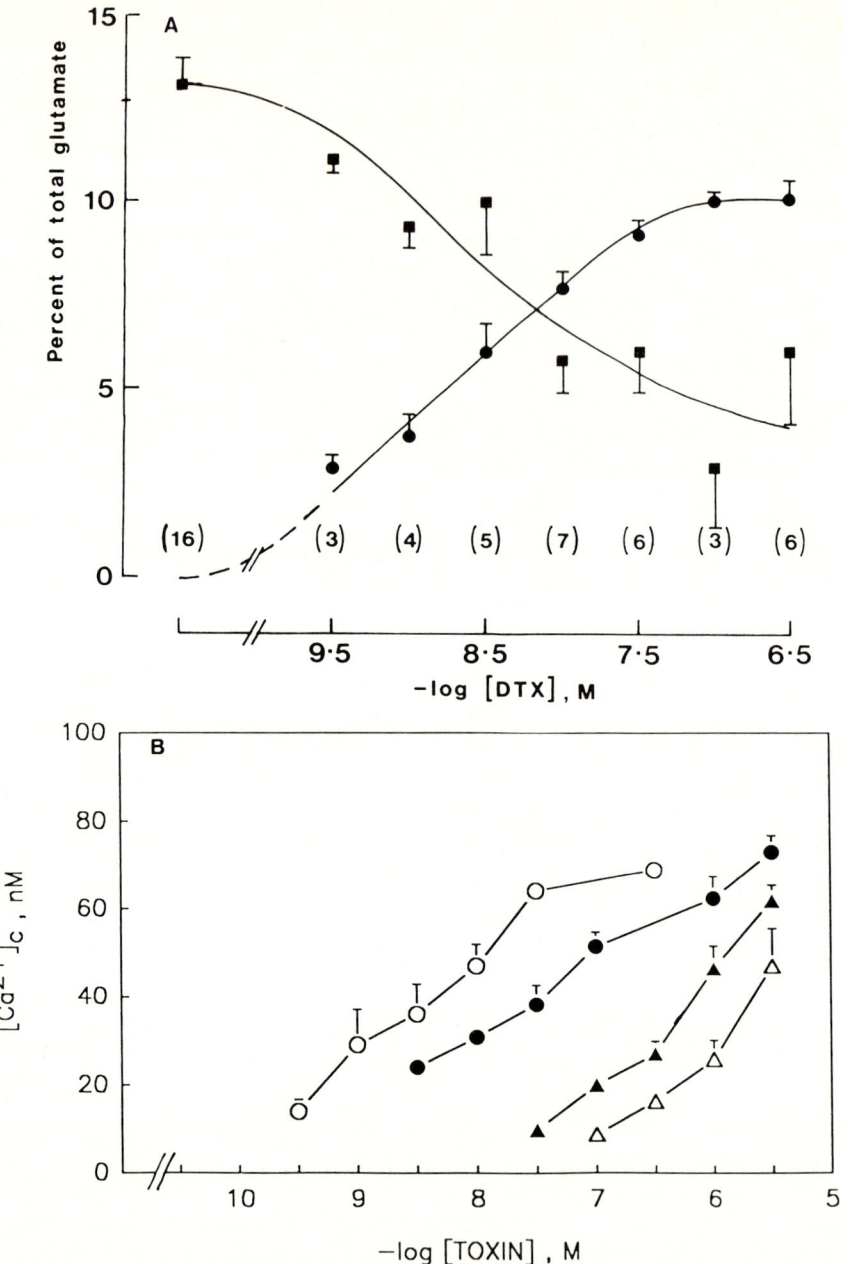

Fig. 3A,B. Elevation of synaptosomal glutamate release (**A**) and cytosolic Ca^{2+} concentration ($[Ca^{2+}]_c$) (**B**) by dendrotoxin (α-DTX) or its homologues. **A** Ca^{2+}-dependent efflux of endogenous glutamate (●) in the 10 min following toxin addition to guinea-pig cerebrocortical synaptosomes and (■) that evoked by 30 mM KCl added for 10 min after α-Dtx. Values (mean±SEM) are expressed as a percentage of the total released by Triton X-100; *numbers in parentheses* represent the number of independent experiments. **B** $[Ca^{2+}]_c$ was measured in Fura-2-loaded synaptosomes equilibrated in 1.3 mM Ca^{2+}; toxins (o, α-; ●, δ; ▲, γ-; △, β-DTX) were added 10 min after Ca^{2+}. Each value represents the mean ± SEM of at least 4 separate determinations; where a bar is absent, it encompasses the symbol. (**A, B** from TIBBS et al. 1989a and MUNIZ et al. 1990a, respectively)

brain particulate preparation (WELLER et al. 1985). These consistent stimulatory effects of α-DTX on the *resting* efflux of transmitters from central
nerve terminals differs from those observed at the neuromuscular junction,
at which a transient reduction is caused in the frequency of MEPPs (in
addition to facilitating neurally evoked ACh release) (ANDERSON and HARVEY
1988). Thus, it is reasonable to deduce that α-DTX-sensitive, outward K^+
currents are more important in controlling the excitability of polarised
synaptosomes compared with the situation at motor nerve terminals.

Consistent with its blockade of nerve terminal K^+ currents, α-DTX
elevates the intrasynaptosomal levels of free Ca^{2+} as measured by the Fura-
2 technique (TIBBS et al. 1989b); importantly, its dose-dependence (Fig. 3B)
matches that for facilitation of synaptosomal glutamate release (Fig. 3A)
and saturable binding of ^{125}I-labelled α-DTX to K^+ channels in these same
synaptosomes (TIBBS et al. 1989a). Homologues of α-DTX can also raise the
cytoplasmic concentration of Ca^{2+} in synaptosomes (Fig. 3B); however,
with the exception of δ-DTX, these toxins are relatively impotent. Also, no
additivity of the effects of these toxins on Ca^{2+} levels was detected (MUNIZ
et al. 1990a), although using a ^{86}Rb flux assay α- and δ-DTX were deemed
to block preferentially an inactivating K^+ current in synaptosomes, whilst
β- and γ-DTX seemed to inhibit non-inactivating K^+ channels (BENISHIN et
al. 1988). In this regard, it is noteworthy that the overall distribution in rat
brain of binding sites for α- and δ-DTX is similar, and mutually exclusive
binding is observed, except for K^+ channels in the Purkinje cell layer of the
cerebellum and the lacunosum moleculare layer of the hippocampus which
are labelled by α- but not δ-DTX (AWAN and DOLLY 1991). Despite the
observed abilities of these homologues to inhibit (in a "non-competitive-
like" manner, at least in the case of α- and δ-DTX) the binding of $[^{125}I]$-
labelled α- and δ-DTX to synaptosomes (MUNIZ et al. 1990b) or brain
sections (AWAN and DOLLY 1991), their apparent K_i values are relatively
high. Clearly, further research, especially electrophysiological measurements, are needed to establish their precise specificities for K^+ channel
subtypes; this will undoubtedly improve their usefulness as much-needed
probes.

II. Noxiustoxin and Charybdotoxin

The Mexican scorpion (*Noxius centruroides*) is the source of noxiustoxin, a
polypeptide containing 39 amino acids and cross-linked by 3 disulphide
bonds (POSSANI et al. 1982). Although reported initially to block the delayed
rectifier K^+ current in squid giant axons (CARBONE et al. 1982), it was found
recently to facilitate nerve-evoked ACh release in chick biventer cervices
nerve-muscle preparation (though directly stimulated twitch tension was
also raised albeit to a lesser extent) (HARVEY et al. 1990). In this context,
it is pertinent that the toxin increases the synaptosomal release of γ-
aminobutyrate (SITGES et al. 1986) with an EC_{50} value (2.9 n*M*) which is very

much lower than that reported for its aforementioned action on squid axon
(300 nM). Also, a single concentration was shown to inhibit ^{86}Rb efflux from
the synaptosomes. Hence, considering these collective findings, especially
its ability to antagonise the binding of [^{125}I]-labelled toxin I to synaptosomal
membranes (Harvey et al. 1990), noxiustoxin clearly interacts with a number
of K[+] channel types including certain Ca[2+]-activated varieties (see Meres,
this volume).

Charybdotoxin, a basic single-chain polypeptide with 37 amino acids and
3 disulphide bonds, was isolated from another scorpion (Table 1) and shows
homology to noxiustoxin but not to dendrotoxins (for details on its sequence
and action on Ca[2+]-activated and voltage-sensitive K[+] channels, see Meves,
this volume). Although charybdotoxin, unlike α-DTX, is ineffective on ACh
release at chick neuromuscular junction (Anderson et al. 1988) it increases
cytosolic free Ca[2+] concentration in guinea-pig cerebrocortical synaptosomes
in a manner similar to α-DTX (Tibbs et al. 1989b). Importantly, little additivity was seen in the elevation of Ca[2+] levels by α-DTX and charybdotoxin,
suggesting that both block the same voltage-activated K[+] channel. Indeed,
Schweitz et al. (1989) demonstrated electrophysiologically that it blocks an
α-DTX-sensitive, Ca[2+]-independent K[+] current in rat dorsal root ganglion
cells; measurement of ^{86}Rb flux from rat brain synaptosomes showed that
it inhibits a voltage-activated, non-Ca[2+]-dependent K[+] channel but with
~ threefold lower potency than its blockade of a Ca[2+]-dependent K[+]
conductance (Schneider et al. 1989). From the changes induced in intrasynaptosomal Ca[2+] levels, it appears that voltage-operated K[+] channels

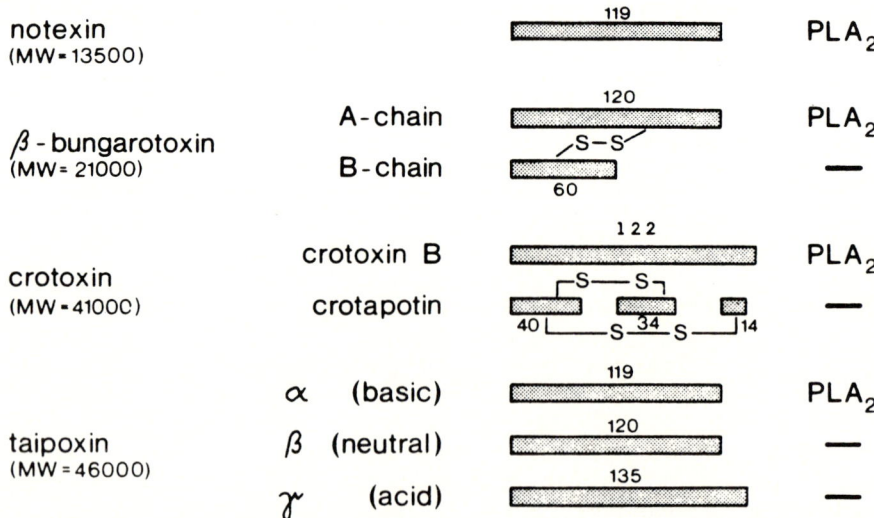

Fig. 4. Gross structures of the most intensely studied phospholipase A$_2$ (PLA$_2$)-
containing toxins showing their molecular weights (*MW*), number of amino acids in
each chain and which ones possess PLA$_2$ activity (Adapted from Dreyer 1990)

sensitive to α-DTX and charybdotoxin make a more significant contribution to maintaining the polarised potential of synaptosomes than do those activated by Ca^{2+} that are susceptible to charybdotoxin.

C. Polypeptide Toxins with Multiple Effects on the Neuronal Release of Transmitters

I. β-Bungarotoxin and Related Phospholipase A$_2$-Containing Toxins

1. General Properties of β-Bungarotoxin, Notexin, Crotoxin and Taipoxin

These toxins, obtained from various snake venoms, have different structural properties (Fig. 4), but all possess Ca^{2+}-dependent phospholipase A$_2$ (PLA$_2$) activity and share the ability to affect (evoked and spontaneous) ACh release at motor nerve terminals in a triphasic manner: an initial depression followed by a period of facilitation that precedes an irreversible blockade (recently reviewed in STRONG 1990; DREYER 1990; HARVEY 1990). Although the same pattern of toxin-induced changes is seen with EPPs and MEPPs, the time course of events for nerve-evoked release is more rapid. Members of this family of toxins include (Table 1; Fig. 4) (a) β-BuTX, from the Formosan banded krait, whose larger A chain ($M_r \approx 13.5$) has considerable sequence similarity to PLA$_2$ enzymes and is disulphide-linked to a B chain ($M_r = 7$) that shows homology with dendrotoxins and Kunitz-type protease inhibitors; in fact, 5 isotoxins (β$_{1-5}$) have been isolated and shown from their sequences to arise from pairing of 3 A and 2 B chains (ROSENBERG 1990). (b) Notexin, from the Australian tiger snake, is a basic single-chain protein. (c) Crotoxin, from the South American rattlesnake, consists of a basic chain (B) displaying PLA$_2$ activity that is non-covalently attached to an acidic protein (crotapotin or crotoxin A); the latter contains 3 disulphide-linked polypeptides. Combination of crotoxin B (low toxicity) with crotapotin (non-toxic) greatly increases the level of toxicity; crotapotin appears to act as a chaperone preventing non-specific binding of crotoxin B but dissociates on interaction of the latter with its target (BON et al. 1988). (d) Taipoxin, from the Australian taipan, comprises three non-covalently associated proteins that are highly homologous to PLA$_2$ enzymes, but only the basic α-chain exhibits significant enzymic activity; the complex is the most toxic of all snake proteins (Table 1).

2. Triphasic Effects of Phospholipase A$_2$ Toxins on Acetylcholine Release: Importance of the Blockade of a Presynaptic K$^+$ Current

Although all these toxins act preferentially on the presynaptic membrane, with the exception of β-BuTX (CHANG et al. 1973), they also exert postsynaptic effects [e.g. action of crotoxin (BON et al. 1988) on nicotinic receptors and myotoxicity of notexin (Rosenberg 1990)]. The demonstrated selectivity of β-BuTX for cholinergic nerves in the periphery is striking and contrasts with its lack of transmitter specificity in the brain (Rosenberg 1990). Some interesting differences are also apparent in the action of this group of toxins at amphibian and mammalian motor nerves (detailed in Strong 1990). For example, the triphasic pattern of effects is clear-cut at the frog neuromuscular junction, with the first being independent of PLA_2 activity, unlike the two subsequent phases; structural damage to the nerve terminals results from exposure to toxin, at least in the case of β-BuTX. On the other hand, in rodent nerve-muscle preparations, the initial suppression of ACh release is minimal relative to the dominant facilitatory phase, and both are independent of the toxin's enzymic activity. Although the final irreversible blockade depends on PLA_2 action, apparently, mammalian nerve endings are not disrupted irreversibly by β-BuTX. This species variation was further documented and insight gained into the toxins' action by extracellular recording of currents in the presynaptic membrane at mouse and frog endplates (Rowan and Harvey 1988). With the murine prep-aration, β-BuTX (Fig. 2C) and the other three toxins (notexin, crotoxin and taipoxin) produced a similar effect – a selective depression of the second negative deflection of the perineural waveform assumed to represent partial blockade of the terminal's K^+ current (the remainder was abolished by 3,4-diaminopyridine). Notably, the result remained unchanged when record-ings were made in the absence of extracellular Ca^{2+}, excluding a direct effect on Ca^{2+}-dependent K^+ currents or Ca^{2+} currents and indicating that PLA_2 activity was not required; however, residual activity may be excluded in this way. From a rather complex series of electrophysiological measure-ments of presynaptic currents at mouse nerve terminals, indirect evidence was obtained for β-BuTX, notexin, crotoxin and taipoxin augmenting ACh release due to inhibition of a slowly activating, voltage-sensitive K^+ current, though no change was seen in the perineural wave form (Dreyer and Penner 1987). As in the case of α-DTX, such an attenuation of a K^+ current in the nerve terminal should slow repolarisation after an action potential, allowing voltage-activated Ca^{2+} channels to open for longer and, thereby, raise the amount of ACh released. Although this has been offered as a plausible mechanism/basis for the facilitatory phase of the toxin's action by both groups, no explanation has been put forward for the initial sup-pression. Moreover, β-BuTX failed to alter the perineural waveform at frog motor nerve endings despite exerting its normal pattern of effects on ACh release (Rowan and Harvey 1988); again as noted for α-DTX, β-BuTX may inhibit a minor K^+ current that remains undetected, and this could underlie its transient facilitation. Lastly, a basis for the final and irreversible blockade of ACh release by this group of toxins at motor endplates has not been documented, although it is likely to be mediated by phospholipolysis, as PLA_2-induced changes in a number of parameters have been implicated from studies on central preparations (see below).

The overall similarity in the pharmacological effects of the four toxins on motor nerves indicates an oversimplification, possibly arising from the inability of the electrophysiological techniques used to resolve their subtle mechanistic differences, because complete mutual interaction is not observed with these toxins (CHANG and SU 1980; HARVEY and KARLSSON 1982) (though this has not yet been investigated fully using the perineural recording technique). For example, blockade of distinct, though neighbouring, K^+ channel variants by the different toxins could result in a similar alteration in the composite presynaptic currents. Also, because of their different structures, it has been speculated that these toxins bind to distinct sites (BON et al. 1988). Resolving this important discrepancy will undoubtedly shed light on their precise binding sites and/or neuronal targets, together with the physiological functions they serve.

3. Insights into the Action of β-Bungarotoxin Gained from Studies on Central Neuronal Preparations

Relative to the vast literature available on β-BuTX, few reports (reviewed in ROSENBERG 1990) have appeared on the central effects of notexin, crotoxin or taipoxin; thus, coverage herein is restricted to β-BuTX.

a) Electrophysiological Recordings in Brain Slices Showing that β-Bungarotoxin Preferentially Blocks Transmitter Release

Using slices of rat olfactory cortex, β-BuTX was shown to irreversibly block neurotransmission by primarily inhibiting transmitter release; in addition, it reduced the amplitude of the presynaptic action potential but at a slower rate (HALLIWELL et al. 1982; DOLLY et al. 1980). Minimizing the toxin's PLA_2 activity only attenuated the inhibitory effect on transmitter release but abolished its reduction of the action potential. β-BuTX-induced synaptic depression in brain slices occurs without detectable change in the sensitivity of the postsynaptic neuron to applied transmitters (HALLIWELL and DOLLY 1982a). Collectively, these results are taken as evidence that β-BuTX initially suppresses transmitter release, a deduction supported by the observed loss of spontaneous synaptic activity in hippocampal cells; indeed, in the latter, the toxin exerts a preferential action on presynaptic terminal regions (HALLIWELL and DOLLY 1982b). Prolonged exposure of the slices to the toxin induced less specific changes in neuronal excitability which were much more dependent on PLA_2 activity. Thus, these electrophysiological data offer no evidence of a triphasic action of β-BuTX on transmitter release, as observed at the neuromuscular junction; this could be due to an inability to detect the other phases or a difference in action at central and peripheral synapses.

b) Biochemical Evidence for Acceptor Binding Underlying the Preferential
Action of β-Bungarotoxin on Nerve Terminals

Extensive early studies showing inhibitory effects of β-BuTX on synaptosomal
uptake and release (basal plus evoked) of transmitters and blockade of
active transport of ACh into vesicles are difficult to interpret in terms of a
unified model for its action because relatively high toxin concentration and
varied conditions were employed (reviewed by ROSENBERG 1990; HARVEY
1990). However, several facts emerged that aided future investigations. (a)
β-BuTX bound to agarose beads (and thus unable to be internalized)
retained its ability to decrease γ-aminobutyrate uptake into synaptosomes,
indicating that internalisation of the toxin is not a prerequisite for its action
(HOWARD and WU 1976). (b) Oxidative phosphorylation could be inhibited
by β-BuTX (or fatty acids), and this caused efflux of deoxyglucose or γ-
aminobutyrate from synaptosomes (WERNICKE et al. 1975). (c) β-BuTX could
liberate fatty acids from synaptic membranes and depolarise synaptosomes,
thereby promoting transmitter efflux, inhibiting uptake and leading to
eventual blockade of transmission (SEN and COOPER 1978). Subsequently,
low concentrations of β-BuTX ($2\,nM$) were shown to cause an uncoupling of
mitochondria in synaptosomes, due to fatty acids released from its enzymic
action on the plasma membrane, and a depolarisation of the synaptosomal
plasma membrane when fatty acids are complexed by albumin; both effects
required high-affinity saturable binding of toxin that was antagonised by
α-DTX (NICHOLLS et al. 1985; RUGOLO et al. 1986). Although these measurements do not preclude initial effects resulting from β-BuTX binding
to its membrane acceptors (i.e. K^+ channel proteins) such as blockade
of a K^+ current, it is clear that in central nerve terminals site-directed,
PLA_2-induced permeabilisation (as reflected by efflux of cytoplasmic lactate
dehydrogenase) of the plasma membrane occurs, leading to depolarisation
and fatty-acid-mediated uncoupling of mitochondria. The pronounced efflux
of endogenous glutamate evoked from cerebrocortical synaptosomes by β-
BuTX [$EC_{50} \approx 0.4\,nM$ in accord with K_d of $0.5\,nM$ for binding to putative
K^+ channels (OTHMAN et al. 1982; BREEZE and DOLLY 1989)] that occurs in
the presence of Ca^{2+} but not Sr^{2+} (which affords release but not PLA_2
activity) and is antagonised by α-DTX is consistent with a generalised
permeabilisation following targetting to the presynaptic membrane via
acceptor binding (TIBBS et al. 1989a). The selective and characteristic distribution patterns observed for β-BuTX binding sites in rat brain (OTHMAN
et al. 1983; PELCHEN-MATTHEWS and DOLLY 1988) lends credence to its
specific acceptor recognition. A simplified outline of the sequence of multiple
effects produced by β-BuTX at central nerve terminals, and by analogy
to the dominant phases of its action at peripheral motor nerve endings,
could include: (a) a fast-onset, transient augmentation of release (not seen
electrophysiologically) resulting from a combination of K^+ channel block,
initial plasma membrane depolarisation and mitochondrial uncoupling and

(b) an abolition of transmission due to collapse of potentials across plasma and mitochondrial membranes with the associated de-energisation and failure to accumulate transmitters into the terminals or vesicles. This final stage is not observed with isolated synaptosomes, presumably because of the limited duration of experimental measurements. A variation of this scheme has recently been postulated from elegant fluorescence and electron-spin resonance measurements of β-BuTX-mediated fusion of liposomes (Rufini et al. 1990); it is envisaged that the facilitatory phase arises from toxin-induced fusion of synaptic vesicles with the plasma membrane (Strong 1990), whilst permeabilisation accounts for the blockade.

These electrophysiological and biochemical studies establish conclusively that the documented cholinergic specificity of β-BuTX in the periphery is not maintained in the CNS, where numerous transmitters are affected. In view of the emerging evidence for β-BuTX blocking voltage-activated K^+ channels (Petersen et al. 1986; Dreyer 1990), such a difference may arise from susceptible cholinergic nerve endings possessing an unique β-BuTX-sensitive K^+ channel not shared with others in the periphery.

II. α-Latrotoxin

A number of protein toxins have been isolated from black widow spider venom that potentiate secretion of neuromediators in PC-12 cells and at various synapses in vertebrates or insects (Grasso 1988). α-Latrotoxin, the most extensively studied, acts at the neuromuscular junction in normal Ringer's or in Ca^{2+}-free medium (containing $1\,mM$ EGTA and $4\,mM\,Mg^{2+}$) to induce massive release of ACh; initially, there is a gradual increase in the frequency of MEPPs followed by a decrease and, eventually, a blockade of evoked release (Ceccarelli et al. 1988). It depletes the small clear synaptic vesicles but does not alter the number of dense core vesicles containing calcitonin gene related peptide (Matteoli et al. 1988). Depletion of the clear vesicles occurs with α-latrotoxin in Ca^{2+}-free medium, provided other divalent cations are present; this apparently results from its pronounced stimulation of small synaptic vesicle exocytosis and blockade of the recycling mechanism concerned with maintaining a constant population of the latter under normal conditions (Matteoli et al. 1988). Based on such a striking observation with these clearly distinct vesicle populations, it was deduced that two secretory pathways exist in the same nerve terminal; apparently, they differ in some way and probably are differentially regulated by physiological stimuli. Saturable binding sites for α-latrotoxin have been localised on the presynaptic membrane of motor nerve terminals (Valtorta et al. 1984) and their distribution in brain found to coincide with the nerve terminal protein, synapsin I, though not in all areas (Malgaroli et al. 1989). Its acceptor protein when purified from bovine brain membranes consists of multiple subunits with M_r (kDa) of 200 (α), 160 (α), 79 (β) and 43 (γ) in a ratio of 1:1:2:2 (Petrenko et al. 1990). In PC-12 cells, Ca^{2+}-insensitive binding of the toxin leads to activation or generation of a non-specific cation channel and, thus, depolarisation associated with Ca^{2+} entry. The toxin-induced Ca^{2+} influx and elevation of Ca^{2+}-dependent and -independent release of dopamine are antagonised by La^{3+} (ROSENTHAL et al. 1990), an effect first observed in synaptosomes. The reported hydrolysis of phosphoinositides by α-latrotoxin appears to arise from activation of a Ca^{2+}-sensitive phospholipase C, following a persistent elevation of intracellular Ca^{2+} concentration by toxin in Ca^{2+}-containing medium, rather than from direct coupling of the α-latrotoxin acceptor to the enzyme. Although this spider toxin has proved useful as a tool for investigating transmitter release (noted above) particularly at the neuromuscular junction, its full potential can only be realised upon deciphering the molecular basis of the many effects it exerts (particularly on synaptosomes; cf. MCMAHON et al. 1990). This is a worthwhile task.

D. Botulinum Neurotoxins and Tetanus Toxin: Unique Probes for Studying Neurotransmitter Release

I. Clostridial Proteins that Block Ca^{2+}-Dependent Secretion by Analogous Multi-Step Mechanisms

At least seven immunologically distinguishable types of BoNT are each produced by different strains of *Clostridium botulinum*, whereas only a single serotype of TeTX has been isolated from *Clostridium tetani*. These

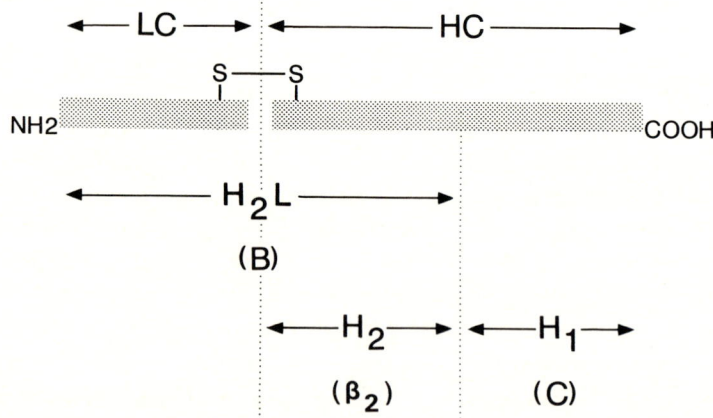

Fig. 5. Generalised structures of botulinum neurotoxin, its chains (*LC*, hight chain; *HC*, heavy chain) and proteolytic fragments; the equivalent moieties of tetanus toxin are shown in parentheses (Adapted from POULAIN et al. 1991)

are large proteins ($M_r \approx 150000$) which, after conversion to the di-chain forms by proteolytic "nicking", display amazing potencies in causing botulism and tetanus, respectively. The flaccid neuromuscular paralysis caused by BoNT is due to selective (MACKENZIE et al. 1982) and near-irreversible inhibition of ACh release in the peripheral nervous system (DOLLY et al. 1988). On the other hand, the spastic paralysis characteristic of tetanus results from a primary blockade of transmitter release from inhibitory (e.g. glycinergic) inter-neurons in the spinal cord; apparently, it also affects γ-aminobutyrate-containing cells which mediate descending inhibition from the brainstem (reviewed by WELLHÖNER 1989). Despite this difference in specificity for *intact* neurons, there are striking analogies in the gross structures (Fig. 5), multiphasic intoxication process and, particularly, the intracellular actions of certain BoNT types and TeTX (outlined below). Although it is, thus, pertinent to compare certain properties of the latter two herein, TeTX is otherwise discussed in detail by Wellhöner (this volume).

1. Structural Features Common to Botulinum Neurotoxins and Tetanus Toxin

Each BoNT (and TeTX) consists of a heavy chain (HC; $M_r \approx 100000$) attached by at least one disulphide bond to a light chain (LC; $M_r \approx 55000$); HC and LC can be isolated chromatographically in the presence of a reducing agent and urea (DASGUPTA 1989). Although the individual, separately renatured polypeptides are relatively non-toxic in the whole animal, reconstitution of the chains (from the same serotype) results in the formation of covalently linked di-chain species and a recovery of toxicity (MAISEY et al. 1988). Assignment of roles to domains in BoNT (and TeTX) has been aided by the isolation of proteolytic fragments: H_2L (toxin minus the C-terminal half of HC), H_1 and H_2 (C- and N-terminal halves of HC) (POULAIN et al. 1989a,b; KOZAKI et al. 1989). Also, the sequences of the genes for BoNT A, C and D should prove helpful in identifying functional residues (THOMPSON et al. 1990; BINZ et al. 1990).

2. Experimental Support for Triphasic Mechanisms of Toxin Action

A striking feature of BoNT is the remarkable specificity with which it virtually abolishes evoked release of ACh at motor nerve terminals and, also, diminishes the spontaneous efflux (reviewed by SIMPSON 1989; DOLLY 1990). This is highlighted by the absence of any other *direct* effects (e.g. on synthesis of ACh or its vesicular packaging), presynaptic Ca^{2+} (or cation) currents or propagation of action potentials to the nerve terminals. In fact, BoNT appears to act at some point after Ca^{2+} entry (see below); hence, it is a highly selective and very valuable probe for studying the fundamental process of transmitter release. Based on extensive pharmacological studies at the mammalian neuromuscular junction, a multiphasic intoxication has been proposed involving: (a) toxin binding to ecto-acceptors on cholinergic nerves; (b) acceptor-mediated, temperature- and energy-dependent uptake and (c) an internal action (Simpson 1981, 1989). Direct evidence for the first two phases has been obtained by electron microscope autoradiographic analysis (Dolly et al. 1984b; Black and Dolly 1986a,b) of mouse endplates incubated with biologically active [^{125}I]-labelled BoNT A or B (Williams et al. 1983; Evans et al. 1986). This revealed saturable interaction of the latter with acceptors on the unmyelinated plasma membrane of motor nerve terminals. Moreover, acceptor-mediated internalisation was observed; most importantly, such uptake seems to be a prerequisite for intoxication (reviewed by Dolly et al. 1986, 1987b). Furthermore, the unique ability of cholinergic nerves in the periphery to recognise and internalise the toxins (Dolly et al. 1984a,b; Black and Dolly 1987) accords with their susceptibility to BoNT. Hence, at least some of these observed binding sites appear to be implicated in the intoxication. Moreover, the remarkable ability of BoNT to target to motor nerve endings, together with its long-lasting inhibition of ACh release, form the basis of its very successful application in the clinical treatment of several movement disorders (Elston 1990).

The triphasic action of BoNT proposed by Simpson (1981) has been validated in *Aplysia* neurons. Saturable interaction of toxin with cholinergic cells of the buccal ganglion has been observed, and this is relatively insensitive to temperature (i.e. detectable at 10°C); blockade of ACh release did not ensue until a temperature-dependent uptake step had occurred (Poulain et al. 1989b). As in the case of mammalian nerves, non-cholinergic transmission was found to be very much less sensitive to BoNT (Fig. 6A,C).

As regards the final phase, conclusive evidence has been obtained recently (detailed below) for inhibition by BoNT (and TeTX) of Ca^{2+}-dependent secretion of a variety of transmitters when administered intracellularly, its nominal selectivity for cholinergic nerves being created by the ecto-acceptors located exclusively thereon. By analogy, the above-noted contrasting preference of TeTX for inhibitory nerve endings must arise from their possession of acceptors that can mediate its internalisation into the cytoplasm and, thus, its localised action. However, intoxication with TeTX is more complicated in that this local scheme of events is preceded by its uptake at peripheral nerve terminals, followed by long-distance transport to inhibitory inter-neurons in the spinal cord. This multiphasic process apparently encompasses binding to ecto-acceptors on motor nerve terminals, internalisation, retrograde axonal transport and subsequent transfer into presynaptic inhibitory neurons (via specific acceptors) where it produces dysinhibition (Lazarovici et al. 1988). It will be intriguing to establish the differences between these internalisation processes that are responsible for local delivery of TeTX and its distant transport; clearly, both BoNT and TeTX provide useful tools to allow such generally important aspects of protein traffic to be investigated.

3. Subtile Differences in the Pharmacological Effects of Botulinum Neurotoxin Types A and B

Despite all BoNT types having similar gross structures, details of the inhibition of transmitter release differ with certain serotypes (reviewed in DOLLY et al. 1986; SIMPSON 1989; DOLLY 1990). Neurally evoked release at endplates paralyzed with BoNT A can be reversed, to variable extents, by raising the intracellular Ca^{2+} concentration, e.g. with Ca^{2+} ionophores or K^+ channel blockers. Curiously, the blockade caused by other types of BoNT (e.g. B) is usually less complete, maintained over a shorter period and can be reversed much less readily with 4-aminopyridine. Moreover, asynchronous release can be detected upon high-frequency stimulation of motor nerves treated with type B; in contrast, this is never seen with type A. Intriguingly, the pattern of inhibition of ACh release produced by TeTX (which can block ACh release at the neuromuscular junction though with much lower potency) is identical in most respects to that of BoNT B, including the shared ability to desynchronise release (GANSEL et al. 1987). On the basis of such pharmacological data, it was envisaged that BoNT A lowers the Ca^{2+}-responsiveness of an unidentified macromolecule in the release process (CULL-CANDY et al. 1976), whilst BoNT B (plus apparently other types) and TeTX probably alter the interaction of transmitter vesicles with some component of the cytoskeleton, thereby perturbing synchronised release.

II. Conclusive Evidence that Botulinum Neurotoxins Act Intracellularly to Block Ca^{2+}-Dependent Release of Transmitters from Various Neurons and Exocrine Cells

To demonstrate directly an intracellular target for BoNT, advantage was taken of large cholinergic and non-cholinergic neurons in the buccal and cerebral ganglia, respectively, of *Aplysia* (POULAIN et al. 1988a,b). These allow micro-injection of the toxin into the presynaptic member of identified pairs of cells forming characterized synapses; voltage-clamp recording can be performed in both neuron types, thereby facilitating the quantitation of transmitter release. Intracellular application of BoNT A to cholinergic and non-cholinergic neurons inhibited evoked transmitter release (Fig. 6B,D). This accords with the ability of BoNT (when high concentrations were used to ensure uptake) to block Ca^{2+}-dependent release of numerous transmitters from brain synaptosomes (DOLLY et al. 1987b; ASHTON and DOLLY 1988) or other preparations (BIGALKE et al. 1981a,b; HABERMANN 1989). Similarly, BoNT A reduces Ca^{2+}-elicited release of catecholamines from chromaffin cells when applied intracellularly (PENNER et al. 1986b); the same result was found with the latter or PC-12 cells when the internalisation step is obviated by permeabilisation of the plasma membrane (BITTNER et al. 1989a; STECHER et al. 1989a; McINNES and DOLLY 1990). It is noteworthy that similar IC_{50}

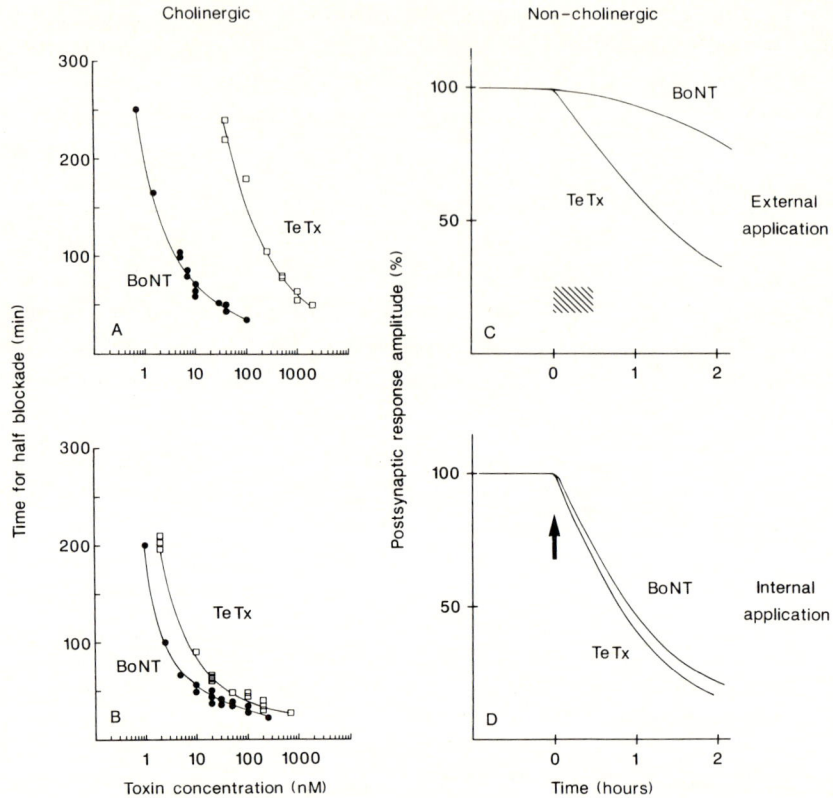

Fig. 6A–D. Relative potencies of botulinum neurotoxin (BoNT) and tetanus toxin (TeTX) in blocking transmitter release at cholinergic and non-cholinergic synapses of *Aplysia* when applied extra- and intraneuronally. **A, B** BoNT or TeTX was either bath applied (**A**) to the buccal ganglion for 30 min or air-pressure injected into the presynaptic cell body (**B**). In each case, the time taken for a 50% reduction in neurally evoked release was determined (by recording the amplitude of the postsynaptic response) and plotted against the extracellular concentration of toxin (**A**) or the calculated intrasomatic concentration (**B**). **C, D** An equivalent set of measurements for non-cholinergic transmission was made using the cerebral ganglion after bath application (**C**) of 10 nM TeTX or 100 nM BoNT A for the time shown (*hatched area*) or following intraneuronal administration (**D**) (*arrow*) of the same concentration of each toxin. In these experiments, the time courses of blockade of transmitter release were monitored and the average points plotted, after making allowance for the extent of run-down of the postsynaptic response seen in control samples. (Adapted from POULAIN *et al.* (1991)

values (2–5 n*M*) were obtained for BoNT blocking catecholamine release from the two types of permeabilised cells studied; also, cholinergic and non-cholinergic transmissions in *Aplysia* neurons seem equally sensitive to the toxin when applied intracellularly (detailed below). At present, it is unclear if this general level of potency matches that for mammalian motor nerves; though liposomally delivered BoNT A was recently shown to act there, the concentration effective intraneuronally could not be determined (de PAIVA and DOLLY 1990). Notwithstanding the desirability of such a quantitative comparison, it is clear that BoNT acts intracellularly, with similar efficacy, on some component(s) of the release process that is common to more than one neuron or cell type. As TeTX, also, acts intracellularly (see Fig. 6B,D) in the model systems tested (MOCHIDA et al. 1989; POULAIN et al. 1990; BITTNER et al. 1989b; STECHER et al. 1989a), it is emerging that both toxins are universal probes for Ca^{2+}-dependent secretion.

III. Light Chain of Botulinum Neurotoxin Blocks Ca^{2+}-Dependent Secretion when Applied Inside Motor Nerve Endings and Other Mammalian Cells: A Vestigial Intracellular Role for Its Heavy Chain

For convenience, permeabilisation of chromaffin or PC-12 cells with digitonin (BITTNER et al. 1989a,b) or streptolysin 0 (STECHER et al. 1989a,b) was used initially for structure-activity studies because the individual toxin chains can gain access to the cytosol, whilst the cells retain their ability to perform Ca^{2+}-elicited exocytosis. The separated, renatured LC of BoNT A proved to be as equally effective as the parent toxin in reducing Ca^{2+}-dependent [³H]noradrenaline release from digitonin-treated PC-12 cells; in contrast, HC alone was ineffective and did not augment the activity of LC (McINNES and DOLLY 1990), in agreement with results reported for chromaffin cells permeabilised with digitonin (BITTNER et al. 1989a) or streptolysin (STECHER et al. 1989a). Consistently, LC of TeTX is also active in both types of permeabilised chromaffin cells (AHNERT-HILGER et al. 1989; BITTNER et al. 1989b).

Due to concern about the action of the toxin's chains on secretion from such artificial model systems being representative of that on quantal trans-mitter release from nerve terminals, *Aplysia* neurons were studied (POULAIN et al. 1988a,b). Micro-injection (or bath application) of LC (or HC) from BoNT A or B into cells in the buccal ganglion proved ineffective in altering evoked quantal release of ACh, in contrast to its action in permeabilised cultured cells. However, intracellular administration of both the individually renatured chains at low concentration (~1 n*M*) produced a blockade; pre-injection of LC followed by external application of HC (which can be internalized alone) was also effective. Moreover, combinations of the respective chains from types A and B were active together, reaffirming that, at least in *Aplysia* neurons, HC and LC participate in the toxin's intra-cellular action. Recently, both chains of BoNT A were also found to be necessary for intracellular blockade of transmitter release at non-cholinergic synapses in the cerebral ganglion of *Aplysia* (POULAIN et al. 1991). It has been deduced that H_1 contributes to such activity because of the failure of injected (or bath applied) H_2L to block ACh release until HC was included (POULAIN et al. 1989a). Notably, LC alone of TeTX blocked ACh release when applied intracellularly, with HC failing to alter its potency (MOCHIDA et al. 1989; Poulain et al. 1990). Thus, the intracellular requirement of both chains of BoNT relates to some characteristic of the toxin or its target in *Aplysia* rather than to an artefact of the experimental system.

The question thus arose as to whether results obtained with intact invertebrate ganglionic neurons or *permeabilised* exocrine and undifferentiated PC-12 cells were typical of mammalian motor nerve endings. This has been addressed recently using liposomal entrapment of LC from BoNT A to achieve delivery into the nerve terminals of the mouse hemidiaphragm, together with measurement of effects on ACh release by monitoring nerve-evoked muscle-twitch tension (De Paiva and Dolly 1990). LC was added to a suspension of phosphatidyl choline/cholesterol and phosphatidyl serine (7:2:1 w/w) in 118 mM NaCl/10 mM Hepes buffer, pH 7.4, vortexed and sonicated; the resultant liposomes were separated by gel filtration (Fig. 7A). By including a trace amount of ^{125}I-labelled LC, it was revealed that this chain became encapsulated into the liposomes, because its addition after sonication resulted in negligible amounts eluting with the liposomal peak in the void volume of the Sephacryl column (Fig. 7A). Application of such liposomes to mouse phrenic nerve hemidiaphragm at 37°C in Krebs solution resulted in their fusion with nerve and muscle membranes and delivery of the contents, as established by quantitation of uptake of an inert radioactive tracer into endplate-enriched and non-endplate areas of diaphragm. When LC-containing liposomes were used, a time- and concentration-dependent diminution of neuromuscular transmission occurred (Fig. 7B). Such an effect was not seen with liposomal encapsulated HC or control liposomes devoid of toxin and made with the buffer used in the twitch experiments; likewise, a contribution from any contaminating intact toxin was excluded. Evidence for a presynaptic inhibition of ACh release was provided by the temporary, though incomplete, reversal caused by 4-aminopyridine, a blocker of certain voltage-activated K^+ channels in the nerve membrane that facilitates transmitter release as a result of increasing Ca^{2+} influx. Because bath application of the same concentration of LC in the absence of liposomes failed to affect synaptic transmission, it is concluded that LC alone mimics the toxin's action inside mammalian motor nerve terminals, the toxin's prime target. This represents a major advance because it consolidates the proposed scheme of intoxication and, also, establishes that LC can be used as a safer and more selective probe for future studies on its intracellular target.

In the light of these collective findings, the involvement of HC in the toxin's action inside *Aplysia* neurons, a feature lost through evolution in mammals and for the action of TeTX in all cells examined, would seem to be secondary. For example, it could be speculated that interaction of HC (probably via its hydrophobic H_1 domain) with the toxins' intracellular target is a prerequisite for recognition of LC. This scenario is more likely than HC associating with LC because HC (processed) of type A can enable single-chain BoNT E (or H_2L) to block ACh release in *Aplysia* neurons

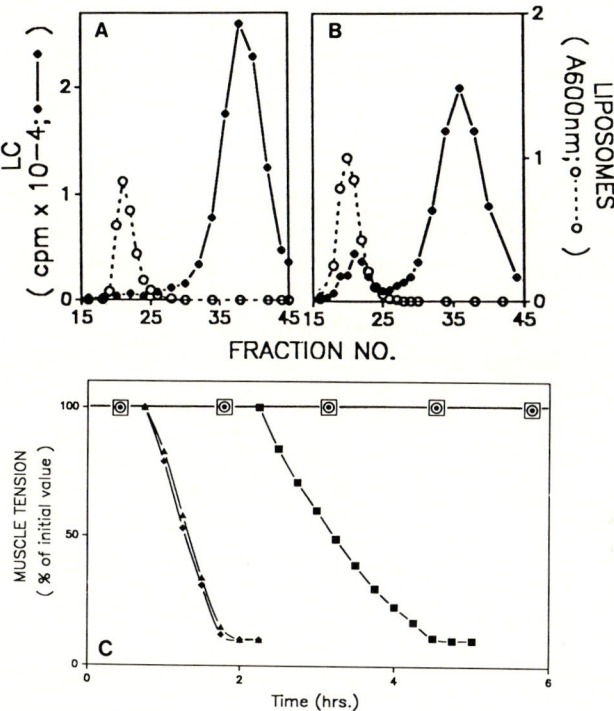

Fig. 7A–C. Liposomally delivered botulinum neurotoxin light chain (BoNT LC) blocks neuromuscular transmission at mouse endplates. **A, B** Gel filtration profiles on a Sephacryl S200 HR column of LC (containing a trace of [^{125}I]-labelled chain) mixed with *preformed* liposomes (**A**) or after encapsulation (**B**) by sonication of the lipids in the presence of LC. **C,** Muscle twitch measured in phrenic nerve hemidiaphragm preparation after adding liposomes encapsulating LC [final approximate concentrations (nM): ◆, 20; ▲, 15; and ■, 9], heavy chain (75 nM, ○), buffer only (●) or BoNT (□) at the minute concentration (0.2 pM) that could be contaminating the LC. (Taken from De Paiva and Dolly (1990)

(Poulain et al. 1989b). Deciphering these possibilities as well as establishing the importance of effects produced in neuronal preparations by HC (Weller et al. 1989) may yield further insight into related facets of the toxin's action.

IV. Optimal Targetting/Internalisation of Botulinum Neurotoxin at Mammalian Nerve Terminals Requires the Intact Di-Chain Form: Functional Domains in the Heavy Chain

Bath application of H$_2$L to mouse hemidiaphragm does not alter nerve-evoked twitch tension, whilst addition of a mixture of separately renatured HC and LC of BoNT A gives only a feeble blockade (~300-fold less potent than BoNT) of neuromuscular transmission (WADSWORTH et al. 1988). Moreover, a 100-fold excess of H_2L or HC is unable to antagonise the action thereon of $0.3\,nM$ BoNT A, even when the conditions used maximise the possibility of competition in the binding to ecto-acceptors (MAISEY et al. 1988; POULAIN et al. 1989b). Accordingly, [^{125}I]-labelled H_2L does not show saturable binding to motor nerves, revealing that removal of H_1 destroys both binding and the consequent toxicity. Clearly, the di-chain species is required for productive binding to ecto-acceptors; presence of the inter-disulphide bond seems essential outside since free LC is active inside these nerve endings (see above). In contrast, HC (Fig. 8A) or indeed H_2 can interact saturably with the membrane of *Aplysia* cholinergic neurons and execute the uptake of LC (POULAIN et al. 1989a,b). Thus, notwithstanding the greater structural stringency for BoNT binding in mammalian nerve membranes, it is reasonable to assume that the role of H_2 documented in *Aplysia* (see later) is maintained.

V. Activities of Hybrid Mixtures of the Chains of Botulinum Neurotoxins and Tetanus Toxin: Binding Via the Heavy Chain to Distinct Ecto-Acceptors Underlies their Neuronal Specificities

When a comparison was made (POULAIN et al. 1991) of the action of BoNT A and TeTX on the buccal ganglion of *Aplysia*, BoNT proved ~100-fold more potent in inhibiting ACh release when bath applied (Fig. 6A); in striking contrast, the reverse was true at a non-cholinergic synapse in the cerebral ganglion (Fig. 6C), where the identity of the transmitter remains unclear although it is not ACh. Yet, when injected inside the cell body of either the cholinergic or non-cholinergic neurons, both toxins were near-equally potent (Fig. 6B,D). These findings highlighted the possibility of the toxins' neuron specificites being due to differences in targetting/uptake; this postulation has been validated by recent investigations on *Aplysia* neurons using heterologous mixtures of the chains from BoNT and TeTX (POULAIN et al. 1990, 1991). Bath application of equimolar amounts ($40\,nM$) of sep-arately renatured HC and LC of BoNT A to the buccal ganglion resulted in a blockade of ACh release (Fig. 8A); note that such a mixture was much less effective at the murine neuromuscular junction at which intact BoNT is required for full toxicity (see above). As noted above with intact TeTX, a combination of its renatured chains at the same concentration was very much less effective on cholinergic transmission in the time period examined (Fig. 8A). However, substitution of HC of TeTX with its counterpart from BoNT produced a dramatic increase in the extent of blockade of ACh release (Fig. 8B). This observation is supported by the recent report that HC of BoNT A plus LC of TeTX is appreciably more potent than TeTX in blocking neuromuscular transmission (WELLER et al. 1991). When an equivalent series of experiments were performed on non-cholinergic synapses in the cerebral ganglion, adding HC and LC of TeTX together diminished

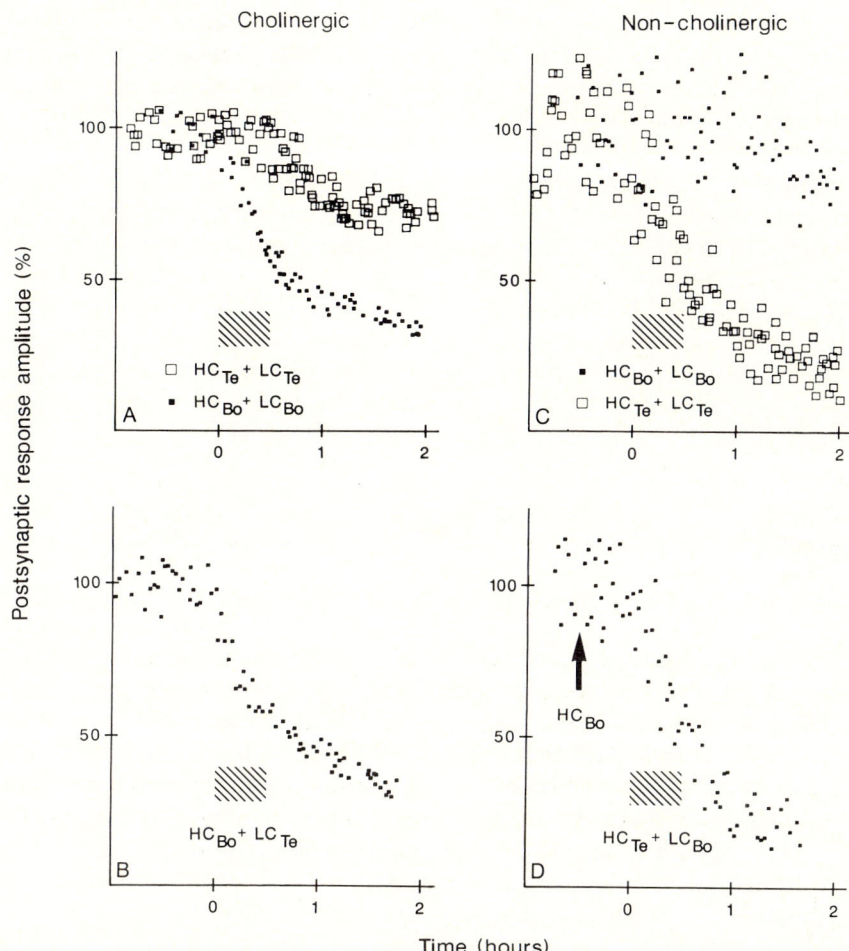

Fig. 8A–D. Effects of hybrid mixtures of the chains of botulinum neurotoxin (Bo) and tetanus toxin (Te) on neurotransmitter release at cholinergic and non-cholinergic synapses in *Aplysia*. These experiments were performed as outlined in Fig. 6 except mixtures of the individually renatured heavy and light chains (HC and LC) of the two toxins (40 nM) were bath applied (*hatched area*); in **D**, HC_{Bo} was injected intraneuronally (see text). Note that HC_{Bo} made LC_{Te} more effective in blocking ACh release (**B**), whereas the reverse occurred in non-cholinergic cells (**D**). (Adapted from POULAIN et al. 1991)

transmission, whilst only a slight reduction resulted from the constituent chains of BoNT (Fig. 8C), again in accord with the respective potencies of the parent toxins (Fig. 6A,C). Intriguingly, by using HC of TeTX together with LC of BoNT transmission was greatly reduced (Fig. 8D), provided HC of BoNT was pre-injected into the neuron [as a source of H_1 (see above) that cannot be provided by HC of TeTX (POULAIN et al. 1990, 1991)]. It can reasonably be deduced from these collective results that HC of BoNT and TeTX mediate the uptake of *either* LC into cholinergic and non-cholinergic nerves, respectively; in fact, the respective potencies of the heterologous combinations match those of the intact toxins from which each HC was derived. Clearly, the two neuron types tested possess distinct ecto-acceptors with which the appropriate HC can specifically interact. If it can be assumed that BoNT and TeTX can act with similar potencies inside *mammalian* motor and central nerve terminals, their preferential actions in the peripheral and central nervous systems, respectively, would be ascribable to their selective uptake at these sites. Furthermore, the high toxicity of TeTX in the whole animal denotes that it is internalised very efficiently by peripheral nerve endings and transported to its site of action in the spinal cord with presumably little "escaping" locally; this would explain its "apparent" low potency in blocking neurotransmission peripherally.

VI. Clues to the Molecular Basis of the Intracellular Action of Botulinum Neurotoxin

Because of the generally accepted view that an enzymically amplified mechanism underlies the toxin's unique potency, a thorough search was made for BoNT-A-induced covalent modification of nerve terminal components (reviewed in DOLLY et al. 1987b, 1988, 1990). Although phosphorylation/dephosphorylation has been implicated in the control of transmitter release, BoNT A seems unable to alter the phosphorylation state of proteins in intact or broken cerebrocortical synaptosomes or in synaptic vesicles poisoned with the toxin (DOLLY et al. 1987b; ASHTON et al. 1988a). Likewise, using synaptosomes, no endogenous ADP-ribosyltransferase activity was detected in type A or B toxin (ASHTON et al. 1988b), although a commercial impure preparation of type D did label small G proteins ($M_r \approx 24\,000$–$26\,000$). As back-titration experiments revealed that this modification could not be antagonised by types A, B or D under conditions in which they all blocked transmitter release, it does not underlie the toxins' action (ASHTON et al. 1990) and has been shown to be contributed by C_3, a contaminating ADP-ribosyltransferase (RÖSENER et al. 1987). Nevertheless, it ought to be stressed that purified synaptic vesicles were found to contain such low M_r G proteins (MATSUOKA and DOLLY 1990) that may well function in secretion, but they are not ADP-ribosylated by BoNT types A or B. In a more global approach, altering intrasynaptosomal levels of second messengers (e.g. cAMP, cGMP) were without effect on the inhibition of transmitter release by BoNT A or B (ASHTON and DOLLY 1991). Raising the extracellular Ca^{2+} concentration was ineffective in antagonising the toxins' action unless the Ca^{2+} ionophore A23187 was included. Interestingly, this treatment caused a more pronounced reversal of the intoxication seen with type A than B, reminiscent of findings at motor endplates (see above). Ca^{2+}-induced changes in protein kinase C activity do not appear to be involved in this reversal because an active phorbol ester could not mimic the ionophore (Dolly et al. 1990).

Due to the implication of the cytoskeleton in transmitter release (reviewed in Linstedt and Kelly 1987), an influence on intoxication of disassembling its elements was examined (Dolly et al. 1990; Ashton and Dolly 1991). Breakdown of microfilaments in synaptosomes with cytochalasin D did not change the blockade of K^+-evoked noradrenaline by BoNT A or B. In contrast, prior depolymerisation of microtubules with colchicine attenuated the reduction in transmitter release normally produced by BoNT B, whilst type A remained unaffected (Fig. 9A,B). To exclude secondary effects of colchicine underlying this interesting effect, several other microtubule-perturbing drugs having different structures were investigated. Pretreatment of synaptosomes with nocodazole or griseofulvin produced a similar alleviation of the inhibition by BoNT B of Ca^{2+}-dependent, K^+-evoked noradrenaline release (Fig. 9C–F) whilst being ineffective towards type A. Additivity was not observed in the effects of griseofulvin and nocodazole or griseofulvin plus colchicine in reducing the intoxication with type B; also, none of the drug treatments alone, or in combination, modified synaptosomal noradrenaline release or uptake (Ashton and Dolly 1991). Even more convincing evidence for the involvement of microtubules was provided by the observation that stabilisation of this cytoskeletal component with taxol counteracted the normal effect of colchicine on intoxication with type B (Fig. 10). Taxol alone caused no change in noradrenaline release from synaptosomes. Thus, the consistent results seen with the four drugs excludes the possibility of side effects and clearly implicates the microtubules; moreover, the absence of an alteration in the action of type A provides a very reliable control and accords with the difference (noted above) in the precise modes of action of BoNT A and B. Also, it is important to note that available evidence indicates that colchicine does not affect the *measurable* binding or internalisation of BoNT B into synaptosomes but rather exerts its effect intrasynaptosomally. Using specially devised conditions (e.g. temperatures of the various incubations and the appropriate washing procedures), colchicine was found to be effective when added after the toxin is internalised but before it acts (Ashton and Dolly 1991). One can, therefore, deduce that intact microtubules are needed for the action of BoNT B and, by inference, for transmitter release. With regard to the latter, the tubulin-based cytoskeleton is known to be involved in axonal transport of synaptic vesicles (Llinas et al. 1989), and it has been speculated that they may be delivered by microtubules to the region of the active zones/release sites on the plasma membranes (Smith 1971; Gray 1975). In fact, synaptic vesicles attached to microtubules in nerve terminals have been visualised by quick-freeze deep-etch electron microscopy (Hirokawa et al. 1989). Although direct evidence for participation of microtubules in the release of transmitters is difficult to obtain (e.g. synaptosomal transmitter release seen in presence of colchicine may be

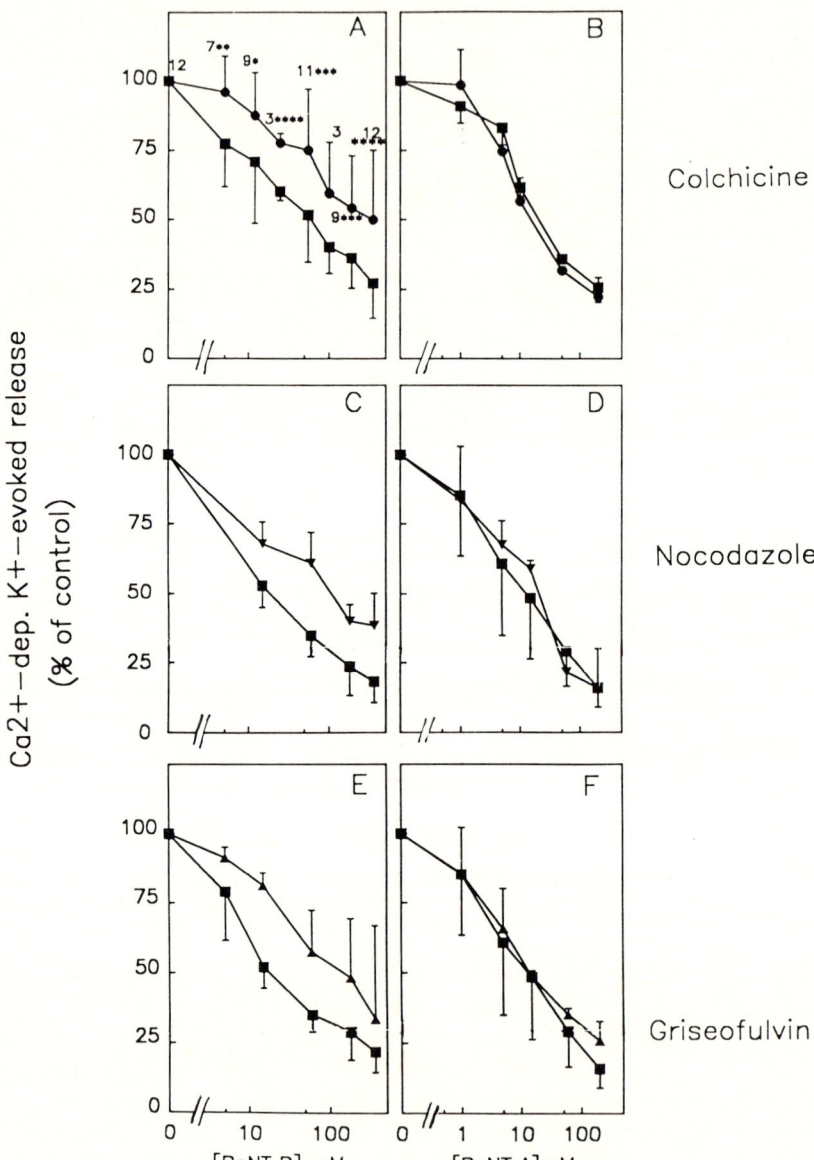

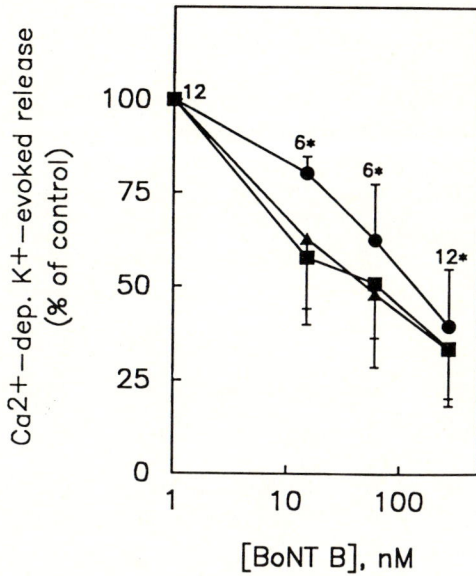

Fig. 10. Stabilisation of microtubules with taxol counteracts colchicine-induced antagonism of the inhibition of synaptosomal transmitter release by BoNT B. This experiment was carried out as described in Fig. 9 except the synaptosomes were pre-treated at 37°C with 10 µM taxol for 1 h, then simultaneously with 1 mM colchicine for another 30 min (▲). Taxol was present during the washing steps and whilst measuring noradrenaline release; alone, it had no effect on the latter. (■), Control without added drug; (●), colchicine alone. The number of independent experiments for each value (mean ± SD) is shown; the significance of difference between taxol/colchicine-treated samples and those exposed to colchicine alone is indicated by an asterisk above the paired values: (* 10%). (From Ashton and Dolly 1991)

Fig. 9A–F. Microtubule-dissociating agents antagonise the inhibitory action of botulinum neurotoxin (BoNT) B (but not A) on Ca^{2+}-dependent, K^+-evoked [^{3}H]noradrenaline release from synaptosomes. Rat cerebrocortical synaptosomes were incubated at 37°C for 30 min with 1 mM colchicine (**A, B**), nocodazole 10 µg/ml (**C, D**) or griseofulvin 200 µg/ml (**E, F**). Various concentrations of BoNT were then added and incubation continued for 90 min. Control samples were treated similarly except for the omission of the microtubule drugs (■). Synaptosomes were subsequently washed, and Ca^{2+}-dependent, K^+-evoked [^{3}H]noradrenaline release was measured over a 5 min period. The observed amounts of Ca^{2+}-independent [^{3}H]noradrenaline efflux have been subtracted from the total: resultant values are expressed relative to that for the non-toxin-treated control. The values presented for release are the mean for 3 (± SD) (**B**) or 2 (± range) (**C–F**). For **A**, the number of independent experiments for each value is indicated by the number shown; the significance of difference between colchicine-treated and non-treated samples is indicated by *asterisks* above the paired points: * 10%; ** 5%, *** 2%; **** 1%. (From Ashton and Dolly 1991)

due to documented drug-resistant microtubules and/or arise from vesicles already located near the release sites), the results with BoNT B extend the circumstantial evidence available in support of their involvement. Further experimentation is warranted to evaluate this possibility, as has already been achieved for an actin-based cytoskeleton that has been shown to form a barrier preventing vesicle/granule fusion (CHEEK and BURGOYNE 1986; LELKES et al. 1986). Nevertheless, it is tempting to propose that BoNT B prevents the detachment of synaptic vesicles from microtubules assumed to be associated with the overall process of transmitter release; on the other hand, as suggested previously, type A may modify a site concerned with the Ca^{2+}-sensing mechanism, thereby preventing vesicle fusion/emptying. A major attraction of our hypothesis is that it would explain the asynchronous ACh release detectable after treatment of motor nerves with BoNT B (but not A) and could be applicable to both vesicular and non-vesicular release mechanisms, provided that vesicles serve a role in the latter (e.g. buffering of Ca^{2+} that enters the nerve terminal upon stimulation), as reviewed recently (TAUC and POULAIN 1991). Although it is unclear at present if BoNT B acts directly on the microtubules/associated protein or whether their disassembly results in some reaction/process that counteracts the ability of BoNT B to block transmitter release, this first mechanistic clue should prove helpful in deciphering the molecular basis of botulism and, also, in gaining insight into Ca^{2+}-dependent secretion – both worthwhile and challenging goals.

Acknowledgements. I am grateful to the researchers in my group and all our collaborators who have contributed to the data presented, especially those (Dr. B. Poulain, Dr. G. Tibbs, A. de Paiva and A. Ashton) who provided figures. Also, I thank Profs. F. Dajas, A. Harvey and F. Dreyer for allowing inclusion of illustrations from their publications.

References

Ahnert-Hilger G, Weller U, Dauzenroth M-E, Habermann E, Gratzl M (1989) The tetanus toxin light chain inhibits exocytosis. FEBS Lett 242:245–248
Anderson AJ, Harvey AL (1988) Effects of the potassium channel blocking dendrotoxins on acetylcholine release and motor nerve terminal activity. Br J Pharmacol 93:215–221
Anderson AJ, Harvey AL, Rowan EG, Strong PN (1988) Effect of charybdotoxin, a blocker of Ca^{2+}-activated K^+ channels, on motor nerve terminals. Br J Pharmacol 95:1329–1335
Ashton AC, Dolly JO (1988) Characterization of the inhibitory action of botulinum neurotoxin type A on the release of several transmitters from rat cerebrocortical synaptosomes. J Neurochem 50:1808–1816
Ashton AC, Dolly JO (1991) Microtubule-dissociating drugs and A23187 reveal differences in the inhibition of synaptosomal transmitter release by botulinum neurotoxins types A and B. J Neurochem 56:827–835
Ashton AC, Edwards K, Dolly JO (1988a) Action of botulinum neurotoxin A on protein phosphorylation in relation to blockade of transmitter release. Biochem Soc Trans 16:885–886

Ashton AC, Edwards K, Dolly JO (1988b) Lack of detectable ADP-ribosylation in synaptosomes associated with inhibition of transmitter release by botulinum neurotoxins A and B. Biochem Soc Trans 16:883–884

Ashton AC, Edwards K, Dolly JO (1990) ADP-ribosylation of cerebrocortical synaptosomal proteins by cholera, pertussis and botulinum toxins. Toxicon 28:963–973

Awan KA, Dolly JO (1991) K^+ channel sub-types in rat brain: characteristic locations revealed using β-bungarotoxin, α- and δ-dendrotoxins. Neuroscience 40:29–39

Barbeito L, Siciliano J, Dajas F (1990) Depletion of the Ca^{++}-dependent releasable pool of glutamate in striatal synaptosomes associated with dendrotoxin-induced potassium channel blockade. J Neural Transm 80:167–179

Benishin CG, Sorensen RG, Brown WE, Krueger BK, Blaustein MP (1988) Four polypeptide components of green mamba venom selectively block certain potassium channels in rat brain synaptosomes. Mol Pharmacol 34:152–159

Benoit E, Dubois J-M (1986) Toxin 1 from the snake *Dendroaspis polylepis polylepis*: a highly specific blocker of one type of potassium channel in myelinated nerve fiber. Brain Res 377:374–377

Bigalke H, Ahnert-Hilger G, Habermann E (1981a) Tetanus toxin and botulinum A toxin inhibit acetylcholine release from but not calcium uptake into brain tissue. Naunyn Schmiedebergs Arch Pharmacol 316:143–148

Bigalke H, Heller I, Bizzini B, Habermann E (1981b) Tetanus toxin and botulinum A toxin inhibit release and uptake of various transmitters, as studied with particulate preparations from rat brain and spinal cord. Naunyn Schmiedebergs Arch Pharmacol 316:244–251

Binz T, Kurazono H, Wille M, Frevert J, Wernars K, Niemann H (1990) The complete sequence of botulinum neurotoxin type A and comparison with other clostridial neurotoxins. J Biol Chem 265:9153–9158

Bittner MA, DasGupta BR, Holz RW (1989a) Isolated light chains of botulinum neurotoxins inhibit exocytosis: studies in digitonin-permeabilized chromaffin cells. J Biol Chem 264:10354–10360

Bittner MA, Habig WH, Holz RW (1989b) Isolated light chain of tetanus toxin inhibits exocytosis: studies in digitonin-permeabilized cells. J Neurochem 53:966–968

Black JD, Dolly JO (1986a) Interaction of ^{125}I-labeled botulinum neurotoxins with nerve terminals. I. Ultrastructural autoradiographic localization and quantitation of distinct membrane acceptors for types A and B on motor nerves. J Cell Biol 103:521–534

Black JD, Dolly JO (1986b) Interaction of ^{125}I-labeled botulinum neurotoxins with nerve terminals. II. Autoradiographic evidence for its uptake into motor nerves by acceptor-mediated endocytosis. J Cell Biol 103:535–544

Black JD, Dolly JO (1987) Selective location of acceptors for botulinum neurotoxin A in the central and peripheral nervous systems. Neuroscience 23:767–779

Black AR, Breeze AL, Othman IB, Dolly JO (1986) Involvement of neuronal acceptors for dendrotoxin in its convulsive action in rat brain. Biochem J 237:397–404

Bon C, Choumet V, Faure G, Jiang M-S, Lembezat M-P, Radvanyi F, Saliou B (1988) Biochemical analysis of the mechanism of action of crotoxin, a phospholipase A_2 neurotoxin from snake venom. In: Dolly JO (ed) Neurotoxins in neurochemistry. Horwood, Chichester, pp 53–63

Breeze AL, Dolly JO (1989) Interactions between discrete neuronal membrane binding sites for the K^+-channel ligands β-bungarotoxin, dendrotoxin and mast-cell-degranulating peptide. Eur J Biochem 178:771–778

Carbone E, Wanke E, Prestifino G, Possani LD, Maelicke A (1982) Selective blockage of voltage-dependent K^+ channels by a novel scorpion toxin. Nature 296:90–91

Ceccarelli B, Hurlbut WP, Iezzi N (1988) Effect of α-latrotoxin on the frog neuromuscular junction at low temperature. J Physiol (Lond) 402:195–217

Chang CC, Su MJ (1980) Mutual potentiation, at nerve terminals, between toxins from snake venoms which contain phospholipase A_2 activity: β-bungarotoxin, crotoxin, taipoxin. Toxicon 18:641–648

Chang CC, Chen TF, Lee CY (1973) Studies of the presynaptic effect of β-bungarotoxin on neuromuscular transmission. J Pharmacol Exp Ther 184:339–345

Cheek TR, Burgoyne RD (1986) Nicotine-evoked disassembly of cortical actin filaments in adrenal chromaffin cells. FEBS Lett 207:110–114

Cull-Candy SG, Lundh H, Thesleff S (1976) Effects of botulinum toxin on neuromuscular transmission in the rat. J Physiol (Lond) 260:177–203

Dajas F, Cervenansky C, Silveira R, Barbeito L (1988) Fasciculins: some aspects of their anticholinesterase activity in the central nervous system. In: Dolly JO (ed) Neurotoxins in neurochemistry. Horwood, Chichester, pp 241–251

DasGupta BR (1989) The structure of botulinum neurotoxin. In: Simpson LL (ed) Botulinum neurotoxin and tetanus toxin. Academic, New York, pp 53–67

De Paiva A, Dolly JO (1990) Light chain of botulinum neurotoxin is active in mammalian motor nerve terminals when delivered via liposomes. FEBS Lett 277:171–174

Dolly JO (1990) Functional components at nerve terminals revealed by neurotoxins. In: Vincent A, Wray D (eds) Neuromuscular transmission: basic and applied aspects. Manchester University Press, Manchester, pp 107–131

Dolly JO (1991) Components involved in neurotransmission probed with toxins: voltage-dependent K^+ channels. In: Tipton K, Iversen L.L. (eds) Probes for neurochemical target sites. Royal Irish Academy Press, Dublin, pp 127–140

Dolly JO, Halliwell JV, Spokes JW (1980) Interaction of β-bungarotoxin with synapses in the mammalian central nervous system. In: Eaker D, Wadstrom T (eds) Natural toxins. Pergamon, Oxford, pp 549–559

Dolly JO, Halliwell JV, Black JD, Williams RS, Pelchen-Matthews A, Breeze AL, Mehraban F, Othman IB, Black AR (1984a) Botulinum neurotoxin and dendrotoxin as probes for studies on transmitter release. J Physiol (Paris) 79:280–303

Dolly JO, Black J, Williams RS, Melling J (1984b) Acceptors for botulinum neurotoxin reside on motor nerve terminals and mediate its internalization. Nature 307:457–460

Dolly JO, Black JD, Black AR, Pelchen-Matthews A, Halliwell JV (1986) Novel role of neural acceptors for inhibitory and facilitatory toxins. In: Harris JB (ed) Natural toxins – animal, plant and microbial. Oxford University Press, (Oxford) pp 237–264

Dolly JO, Stansfeld CE, Breeze A, Pelchen-Matthews A, Marsh SJ, Brown DA (1987a) Neuronal acceptor sub-types for dendrotoxin and their relation to K^+ channels. In: Jenner P (ed) Neurotoxins and their pharmacological implications. Raven, New York, pp 81–96

Dolly JO, Ashton AC, Evans DM, Richardson PJ, Black JD, Melling J (1987b) Molecular action of botulinum neurotoxins: role of acceptors in targetting to cholinergic nerves and in the inhibition of the release of several transmitters. In: Dowdall MJ, Hawthorne JN (eds) Cellular and molecular basis for cholinergic function. Horwood, Chichester, pp 517–533

Dolly JO, Poulain B, Maisey EA, Breeze AL, Wadsworth JD, Ashton AC, Tauc L (1988) Neurotransmitter release and K^+ channels probed with botulinum neurotoxins and dendrotoxin. In: Dolly JO (ed) Neurotoxins in neurochemistry. Horwood, Chichester, pp 79–99

Dolly JO, Ashton AC, McInnes C, Wadsworth JDF, Poulain B, Tauc L, Shone CC, Melling J (1990) Clues to the multi-phasic inhibitory action of botulinum neurotoxins on release of transmitters. J Physiol (Paris) 84:237–246

Dreyer F (1990) Peptide toxins and potassium channels. Rev Physiol Biochem Pharmacol 115:93–136

Dreyer F, Penner R (1987) The action of presynaptic snake toxins on membrane currents of mouse motor nerve terminals. J Physiol (Lond) 386:455–463

Elston JS (1990) Botulinum toxin A in clinical medicine J Physiol (Paris) 84:285–289

Evans DM, Williams RS, Shone CC, Hambleton P, Melling J, Dolly JO (1986) Botulinum neurotoxin type B. Its purification, radioiodination and interaction with rat brain synaptosomal membranes. Eur J Biochem 154:409–416

Gansel M, Penner R, Dreyer F (1987) Distinct sites of action of clostridial neurotoxins revealed by double-poisoning of mouse motor nerve terminals. Pflugers Arch 409:533–539

Grasso A (1988) α-Latrotoxin as a tool for studying ion channels and transmitter release process. In: Dolly JO (ed) Neurotoxins in neurochemistry. Horwood, Chichester, pp 67–78

Gray EG (1975) Presynaptic microtubules and their association with synaptic vesicles. Proc R Soc Lond [Biol] 190:369–372

Habermann E (1989) Clostridial neurotoxins and the central nervous system: functional studies on isolated preparations. In: Simpson LL (ed) Botulinum neurotoxin and tetanus toxin. Academic, New York, pp 255–279

Halliwell JV, Dolly JO (1982a) Electrophysiological analysis of the presynaptic action of β-bungarotoxin in the central nervous system. Toxicon 20:121–127

Halliwell JV, Dolly JO (1982b) Preferential action of β-bungarotoxin at nerve terminal regions in the hippocampus. Neurosci Lett 30:321–327

Halliwell JV, Tse CK, Spokes JW, Othman I, Dolly JO (1982) Biochemical and electrophysiological demonstrations of the actions of β-bungarotoxin on synapses in brain. J Neurochem 39:543–550

Halliwell JV, Othman IB, Pelchen-Matthews A, Dolly JO (1986) Central action of dendrotoxin: selective reduction of a transient K^+ conductance in hippocampus and binding to localized acceptors. Proc Natl Acad Sci USA 83:493–497

Harvey AL (1990) Presynaptic effects of toxins. Int Rev Neurobiol 32:201–239

Harvey AL, Karlsson E (1980) Dendrotoxin from venom of the green mamba, Dendroaspis angusticeps, a neurotoxin that enhances acetylcholine release at neuromuscular junctions. Naunyn Schmiedebergs Arch Pharmacol 312:1–6

Harvey AL, Karlsson E (1982) Protease inhibitor homologues from mamba venoms: facilitation of acetylcholine release and interactions with prejunctional blocking toxins. Br J Pharmacol 77:153–161

Harvey AL, Anderson AJ, Karlsson E (1984) Facilitation of transmitter release by neurotoxins from snake venoms. J Physiol (Paris) 79:222–227

Harvey AL, Anderson AJ, Marshall DL, Pemberton KE, Rowan EG (1990) Facilitatory neurotoxins and transmitter release. J Toxicol Toxin Rev 9:225–242

Hirokawa N, Sobue K, Kanda K, Maruda A, Yorifugi H (1989) The cytoskeletal architecture of the presynaptic terminal and molecular structure of synapsin I. J Cell Biol 108:111–126

Howard BD, Wu WCS (1976) Evidence that β-bungarotoxin acts at the exterior of nerve terminals. Brain Res 103:190–192

Israel M, Lesbats B, Suzuki A (1990) Characterisation of a polyclonal antiserum raised against mediatophore: a protein that translocates acetylcholine. In: Clementi F, Meldolesi J (eds) Neurotransmitter release: the neuromuscular junction. Academic New York, pp 117–127

Joubert FJ, Taljaard N (1980) Snake venoms. The amino acid sequences of two proteinase inhibitors from Dendroaspis angusticeps venom. Hoppe Seylers Z Physiol Chem 361:661–674

Kozaki S, Miki A, Kamata Y, Ogasawara J, Sakaguchi G (1989) Immunological characterisation of papain-induced fragments of Clostridium botulinum type A neurotoxin and interaction of the fragments with brain synaptosomes. Infect Immun 57:2634–2639

Lazarovici P, Yavin E, Bizzini B, Fedinec A (1988) Retrograde transport in sciatic nerves of ganglioside-affinity-purified tetanus toxin. In: Dolly JO (ed) Neurotoxins in neurochemistry. Horwood, Chichester, pp 100–108

Lelkes PI, Friedman JE, Rosenheck K, Oplatka A (1986) Destabilisation of actin filaments as a requirement for the secretion of catecholomines from permeabilised chromaffin cells. FEBS Lett 208:357–363

Linstedt AD, Kelly RB (1987) Overcoming barriers to exocytosis. Trends Neurosci 10:446–448

Llinas R, Sugimori M, Lin J-W, Leopold PL, Brady ST (1989) ATP-dependent directional movement of rat synaptic vesicles injected into the presynaptic terminal of squid giant synapse. Proc Natl Acad Sci USA 86:5656–5660

MacKenzie I, Burnstock G, Dolly JO (1982) The effects of purified botulinum neurotoxin type A on cholinergic, adrenergic and non-adrenergic, atropine-resistant autonomic neuromuscular transmission. Neuroscience 7:997–1006

Maisey EA, Wadsworth JDF, Poulain B, Shone CC, Melling J, Gibbs P, Tauc L, Dolly JO (1988) Involvement of the constituent chains of botulinum neurotoxins A and B in the blockade of neurotransmitter release. Eur J Biochem 177:683–691

Malgaroli A, de Camilli P, Meldolesi J (1989) Distribution of α-latrotoxin receptor in the rat brain by quantitative autoradiography: comparison with the nerve terminal protein, synapsin I. Neuroscience 32:393–404

Matsuoka I, Dolly JO (1990) Identification and localization of low-molecular-mass GTP-binding proteins associated with synaptic vesicles and other membranes. Biochim Biophys Acta 1026:99–104

Matteoli M, Haimann C, Torri-Tarelli F, Polak JM, Ceccarelli B, de Camilli P (1988) Differential effect of α-latrotoxin on exocytosis from small synaptic vesicles and from large dense-core vesicles containing calcitonin gene-related peptide at the frog neuromuscular junction. Proc Natl Acad Sci USA 85:7366–7370

McInnes C, Dolly JO (1990) Ca^{2+}-dependent noradrenaline release from permeabilised PC12 cells is blocked by botulinum neurotoxin A or its light chain. FEBS Lett 261:323–326

McMahon HT, Rosenthal L, Meldolesi J, Nicholls DG (1990) α-Latrotoxin releases both vesicular and cytoplasmic glutamate from isolated nerve terminals. J Neurochem 55:2039–2047

Mochida S, Poulain B, Weller U, Habermann E, Tauc L (1989) Light chain of tetanus toxin intracellularly inhibits acetylcholine release at neuro-neuronal synapses, and its internalization is mediated by heavy chain. FEBS Lett 253:47–51

Muniz ZM, Tibbs GR, Maschot P, Bougis P, Nicholls DG, Dolly JO (1990a) Homologues of a K^+ channel blocker α-dendrotoxin: characterization of synaptosomal binding sites and their coupling to elevation of cytosolic free calcium concentration. Neurochem Int 16:105–112

Muniz ZM, Diniz CR, Dolly JO (1990b) Characterisation of binding sites for δ-dendrotoxin in guinea-pig synaptosomes: relationship to acceptors for the K^+-channel probe α-dendrotoxin. J Neurochem 54:343–346

Nicholls D, Snelling R, Dolly JO (1985) Bioenergetic actions of β-bungarotoxin, dendrotoxin and bee-venom phospholipase A_2 on guinea-pig synaptosomes. Biochem J 229:653–662

Othman IB, Spokes JW, Dolly JO (1982) Preparation of neurotoxic ^{3}H-β-bungarotoxin: demonstration of saturable binding to brain synapses and its inhibition by toxin I. Eur J Biochem 128:267–276

Othman IB, Wilkin GP, Dolly JO (1983) Synaptic binding sites in brain for [^{3}H]β-bungarotoxin – a specific probe that perturbs transmitter release. Neurochem Int 5:487–496

Pelchen-Matthews A, Dolly JO (1988) Distribution of acceptors for β-bungarotoxin in the central nervous system of the rat. Brain Res 441:127–138

Penner R, Petersen M, Pierau F-K, Dreyer F (1986a) Dendrotoxin: a selective blocker of a non-inactivating potassium current in guinea-pig dorsal root ganglion neurones. Pflugers Arch 407:365–369

Penner R, Neher E, Dreyer F (1986b) Intracellularly injected tetanus toxin inhibits exocytosis in bovine adrenal chromaffin cells. Nature 324:76–78

Petersen M, Penner R, Pierau F-K, Dreyer F (1986) β-Bungarotoxin inhibits a non-inactivating potassium current in guinea pig dorsal root ganglion neurones. Neurosci Lett 68:141–145

Petrenko AG, Kovalenko VA, Shamotienko OG, Surkova IN, Tarasyuk TA, Ushkaryov YA, Grishin EV (1990) Isolation and properties of the α-latrotoxin receptor. EMBO J 9:2023–2027

Possani LD, Martin BM, Svendsen I (1982) The primary structure of noxiustoxin: a K^+ channel blocking peptide purified from the venom of the scorpion *Centruroides noxius*. Carlsberg Res Commun 47:285–289

Poulain B, Tauc L, Maisey EA, Wadsworth JDF, Mohan PM, Dolly JO (1988a) Neurotransmitter release is blocked intracellularly by botulinum neurotoxin, and this requires uptake of both toxin polypeptides by a process mediated by the larger chain. Proc Natl Acad Sci USA 85:4090–4094

Poulain B, Tauc L, Maisey EA, Dolly JO (1988b) Ganglionic synapses of *Aplysia* as model for the study of the mechanism of action of botulinum neurotoxins. CR Acad Sci [III] 306:483–488

Poulain B, Wadsworth JDF, Maisey EA, Shone CC, Melling J, Tauc L, Dolly JO (1989a) Inhibition of transmitter release by botulinum neurotoxin A. Eur J Biochem 185:197–203

Poulain B, Wadsworth JDF, Shone CC, Mochida S, Lande S, Melling J, Dolly JO, Tauc L (1989b) Multiple domains of botulinum neurotoxin contribute to its inhibition of transmitter release in *Aplysia* neurons. J Biol Chem 264:21928–21933

Poulain B, Mochida S, Wadsworth JDF, Weller U, Habermann E, Dolly JO, Tauc L (1990) Inhibition of neurotransmitter release by botulinum neurotoxins and tetanus toxin at *Aplysia* synapses: role of the constituent chains. J Physiol (Paris) 84:247–261

Poulain B, Mochida S, Weller M, Högy B, Habermann E, Wadsworth JDF, Shone CC, Dolly JO, Tauc L (1991) Heterologous combinations of heavy and light chains from botulinum neurotoxin A and tetanus toxin inhibit neurotransmitter release in *Aplysia*. J Biol Chem 266:9580–9585

Rosenberg P (1990) Phospholipases. In: Shier WT, Mebs D (eds) Handbook of toxicology. Dekker, New York, pp 67–277

Rösener S, Chhatwal GS, Aktories K (1987) Botulinum ADP-ribosyltransferase C3 but not botulinum neurotoxins C1 and D ADP-ribosylates low molecular mass GTP-binding proteins. FEBS Lett 224:38–42

Rosenthal L, Zacchetti D, Madeddu L, Meldolesi J (1990) Mode of action of α-latrotoxin: role of divalent cations in Ca^{2+}-dependent and Ca^{2+}-independent effects mediated by the toxin. Mol Pharmacol 38:917–923

Rowan EG, Harvey AL (1988) Potassium channel blocking actions of β-bungarotoxin and related toxins on mouse and frog motor nerve terminals. Br J Pharmacol 94:839–847

Rufini S, Pedersen Jens Z, Desideri A, Luly P (1990) β-Bungarotoxin-mediated liposome fusion: spectroscopic characterization by fluorescence and ESR. Biochemistry 29:9644–9651

Rugolo M, Dolly JO, Nicholls DG (1986) The mechanism of action of β-bungarotoxin at the presynaptic plasma membrane. Biochem J 233:519–523

Schneider MJ, Rogawski RS, Krueger BK, Blaustein MP (1989) Charybdotoxin blocks both Ca-activated K channels and Ca-independent voltage-gated K channels in rat brain synaptosomes. FEBS Lett 250:433–436

Schweitz H, Stansfeld CE, Bidard J-N, Fagni L, Maes P, Lazdunski M (1989) Charybdotoxin blocks dendrotoxin-sensitive voltage-activated K^+ channels. FEBS Lett 250:519–522

Sen I, Cooper JR (1978) Similarities of β-bungarotoxin and phospholipase A_2 and their mechanism of action. J Neurochem 30:1369–1375

Silveira R, Siciliano J, Abo V, Viera L, Dajas F (1988) Intrastriatal dendrotoxin injection: behavioral and neurochemical effects. Toxicon 26:1009–1015

Simpson LL (1981) The origin, structure, and pharmacological activity of botulinum toxin. Pharmacol Rev 33:155–188

Simpson LL (1989) Peripheral actions of the botulinum toxins. In: Simpson LL (ed) Botulinum neurotoxin and tetanus toxin. Academic, New York pp 153–178

Sitges M, Possani LD, Bayon A (1986) Noxius toxin, a short-chain toxin from the Mexican scorpion Centruroides noxius induces transmitter release by blocking K^+ permeability. J Neurosci 6:1570–1574

Smith DS (1971) On the significance of cross-bridges between microtubules and synaptic vesicles. Philos Trans R Soc Lond [Biol] 261:395–405

Stansfeld CE, Marsh SJ, Halliwell JV, Brown DA (1986) 4-Aminopyridine and dendrotoxin induce repetitive firing in rat visceral sensory neurones by blocking a slowly inactivating outward current. Neurosci Lett 64:299–304

Stansfeld CE, Marsh SJ, Parcej DN, Dolly JO, Brown DA (1987) Mast cell degranulating peptide and dendrotoxin selectively inhibit a fast-activating potassium current and bind to common neuronal proteins. Neuroscience 23:893–902

Stecher B, Gratzl M, Ahnert-Hilger G (1989a) Reductive chain separation of botulinum A toxin – a prerequisite to its inhibitory action on exocytosis in chromaffin cells. FEBS Lett 248:23–27

Stecher B, Weller U, Habermann E, Gratzl M, Ahnert-Hilger G (1989b) The light chain but not the heavy chain of botulinum A toxin inhibits exocytosis from permeabilised adrenal chromaffin cells. FEBS Lett 255:391–394

Storm JF (1988) Temporal integration by a slowly inactivating K^+ current in hippocampal neurons. Nature 336:379–381 (698)

Strong PN (1990) Potassium channel toxins. Pharmacol Ther 46:137–162

Strydom DJ (1973) Protease inhibitors as sn(ake)ake venom toxins. Nature [New Biol] 243:88–89

Tauc L, Poulain B (1991) Vesigate hypothesis of neurotransmitter release explains the formation of quanta by a non-vesicular mechanism. Physiol Res 40:279–291

Thompson DE, Brehm JK, Oultram JD Swinfield TJ, Shone CC, Atkinson T, Melling J, Minton NP (1990) The complete amino acid sequence of the Clostridium botulinum type A neurotoxin, deduced by nucleotide sequence analysis of the encoding gene. Eur J Biochem 189:73–91

Tibbs GR, Dolly JO, Nicholls DG (1989a) Dendrotoxin, 4-aminopyridine and β-bungarotoxin act at common loci but by two distinct mechanisms to induce Ca^{2+}-dependent release of glutamate from guinea-pig cerebrocortical synaptosomes. J Neurochem 52:201–206

Tibbs GR, Nicholls DG, Dolly JO (1989b) Dendrotoxin and charybdotoxin increase the cytosolic concentration of free Ca^{2+} in cerebrocortical synaptosomes: an effect not shared by apamin. FEBS Lett 255:159–162

Valtorta F, Madeddu L, Meldolesi J, Ceccarelli B (1984) Specific localisation of the α-latrotoxin receptor in the nerve terminal plasma membrane. J Cell Biol 99:124–132

Valtorta F, Fesce R, Greeengard P, Ceccarelli B (1988) Synaptophysin (p38) at the frog neuromuscular junction: its incorporation into the axolemma and recycling after intense quantal secretion. J Cell Biol 107:2717–2727

Valtorta F, Fesce R, Grohovaz F, Haimann C, Hurlbut WP, Iezzi N, Torri Tarelli F, Villa A, Ceccarelli B (1990) Neurotransmitter release and synaptic vesicle recycling. Neuroscience 35:477–489

Wadsworth JDF, Shone CC, Melling J, Dolly JO (1988) Roles of the constituent chains of botulinum neurotoxin type A in the blockade of neuromuscular transmission in mice. Biochem Soc Trans 26:886–887

Weller U, Bernhardt U, Siemen D, Dreyer F, Vogel W, Habermann E (1985) Electrophysiological and neurobiochemical evidence for the blockade of a potassium channel by dendrotoxin. Naunyn Schmiedebergs Arch Pharmacol 330:77–83

Weller U, Dauzenroth M-E, Heringdorf MD, Habermann E (1989) Chains and fragments of tetanus toxin: separation, reassociation and pharmacological properties. Eur J Biochem 182:649–656

Weller U, Dauzenroth M-E, Gansel M, Dreyer F (1991) Cooperative action of the light chain of tetanus toxin and the heavy chain of botulinum toxin type A on the transmitter release of mammalian motor endplates. Neurosci Lett 122:132–134

Wellhöner HH (1989) Clostridal toxins and the central nervous system: studies on in situ tissues. In: Simpson LL (ed) Botulinum neurotoxin and tetanus toxin. Academic, New York pp 231–253

Wernicke JF, Vanker AD, Howard BD (1975) The mechanism of action of β-bungarotoxin. J Neurochem 25:483–496

Williams RS, Tse C-K, Dolly JO, Hambleton P, Melling J (1983) Radioiodination of botulinum neurotoxin type A with retention of biological activity and its binding to brain synaptosomes. Eur J Biochem 131:437–445

Sodium Channel Specific Neurotoxins: Recent Advances in the Understanding of Their Molecular Mechanisms

S. BECKER and R.D. GORDON

A. Introduction

The voltage-dependent sodium channel is a large transmembrane glyco-protein that mediates the sodium current during the action potential, a characteristic property of excitable cells (HODGKIN and HUXLEY 1952). A number of small molecules (<8 kDa) of diverse structure have quite specific effects on the physiological properties of the sodium channel by directly binding to the protein. These toxins, which have been extensively used for the molecular characterization of the sodium channel, include (a) those that inhibit ion transport, e.g., tetrodotoxin (TTX), (b) lipid-soluble activators of the sodium channel, e.g., batrachotoxin (BTX), (c) the North African α-scorpion toxins, e.g., *Leiurus quinquestriatus* toxin, (d) the American β-scorpion toxins, e.g., *Tityus* toxin, (e) ciguatoxin and brevetoxin (see Table 1 and Fig. 1).

This review will concentrate on the recently described biochemical properties of these toxins which have given new insight into the mechanism of sodium channel action in the light of current molecular models. This field of study has been the subject of many previous reviews (CATTERALL 1980; LAZDUNSKI and RENAUD 1982; HILLE 1984; KAO and LEVINSON 1988; ALBUQUERQUE et al. 1988).

B. Current Status of the Molecular Structure of the Sodium Channel

The molecular characterization of the sodium channel has progressed rapidly over the past 10 years and has transformed our understanding of how the toxins interact with it. The sodium channel has been purified using radiolabelled toxins [TTX, saxitoxin (*STX*), and *Tityus*] as markers. The common finding is that the pure sodium channel from *Electrophorus electricus* (AGNEW 1984; NORMAN et al. 1983), rat brain (CATTERALL 1986; BARHANIN et al. 1983), rat skeletal muscle (KRANER et al. 1985; CASADEI et al. 1986), and chick heart (LOMBET and LAZDUNSKI 1984) consists of a 260–270-kDa α-subunit together with one or two 32–39-kDa β-subunits in the mammalian species.

Table 1. Toxins acting on the sodium channel

Toxin group	Examples	Functional effect	Reference
1. Blockers	Tetrodotoxin Saxitoxin Conotoxin GIIIA Conotoxin GIIIB Conotoxin GS	Blocks ion transport	Catterall (1980), Gray et al. (1988)
2. Lipophilic activators	Batrachotoxin Veratridine Aconitine Grayanotoxin	Causes persistent activation	Albuquerque et al. (1988), Hille (1984)
3. North African α-scorpion toxins, Sea anemone toxins	*Leiurus quinquestriatus Androcotonus Anemonia, Anthropleura, Radianthus*	Slow inactivation	Catterall (1980, 1988), Lazdunski et al. (1986)
4. American β-scorpion toxins	*Titys* and *Centruroides* toxins	Shift inactivation Persistent activation	Lazdunski et al. (1986)
5. Other toxins	Brevetoxin Ciguatoxin	Causes repetitive firing	Huang et al. (1984) Bidard et al. (1984)

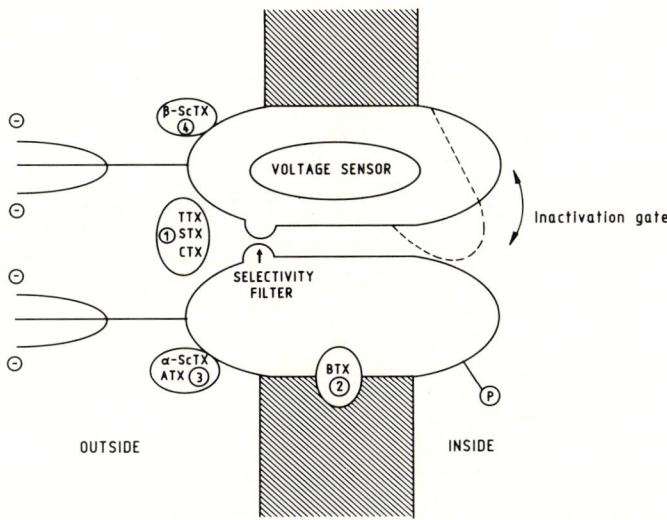

Fig. 1. A working model for the sodium channel and its toxin binding sites. The sodium channel is depicted as a large transmembrane glycoprotein. The external surface is glycosylated (see negative charges), while the inside surface is phosphorylated (*P*). The binding sites for the toxins are depicted as (*1*) tetrodotoxin (*TTX*), saxitoxin (*STX*), and μ-conotoxin (*CTX*) site, (2) the lipophilic activator site [e.g., batrochotoxin (*BTX*)], (3) the North Aftican α-scorpion toxin (*a-SCTX*) and sea anemone (*ATX*) site, and (4) the American β-scorpion toxin (*β-ScTX*) site. The diagram also shows a working hypothesis for the location of the voltage sensor, inactivation gate, and selectivity filter

These initial studies resulted in the purification of a toxin binding site; the full identification of the sodium channel was done using reconstitution experiments which extensively relied on the use of toxins to activate and block sodium fluxes through the channel into phospholipid vesicles (AGNEW 1984; TANAKA et al. 1986; CATTERALL 1988). In a final approach, pure sodium channels in phospholipid vesicles were fused with artifical lipid bilayers. Using voltage alone or in conjunction with toxin activators, sodium currents were obtained of the correct conductance and opening time, which could be blocked by TTX (AGNEW 1984; TANAKA et al. 1986; CATTERALL 1988).

Thus, the purified sodium channel preparations were shown to have the properties of the sodium channel as described in vivo. More to the point, the *E. electricus* sodium channel consisting of only a preparation enriched in a 260-kDa protein possessed all these properties, which strongly implied that the sodium channel requires only the large subunit for full functional activity (AGNEW 1984; TRIMMER and AGNEW 1989).

Molecular biological studies have revealed the primary amino acid sequence of the 260-kDa protein from *E. electricus* (NODA et al. 1984), rat

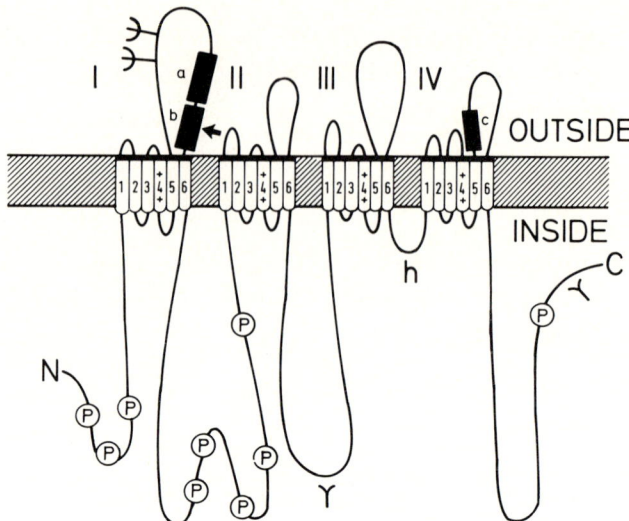

Fig. 2. Model of the sodium channel transmembrane helices (*1–6*) organised into four domains (*I-IV*). Phosphorylation sites (*P*) are shown, h is the inactivation gate, the solid boxes (*a–c*) are the binding sites for *Leiurus quinqestriatus* toxin, and the arrow indicates the position of glutamate 387, the proposed tetrodotoxin/saxitoxin binding site. The "*Y*" symbol shows the regional topology identified with antibodies. The ψ symbol in domain I shows the putative glycosylation sites in the skeletal muscle sodium channel. The + sign indicated in transmembrane domain 4 shows the presence of positively charged amino acid residues in the segment

brain (Noda et al. 1986a; Auld et al. 1988), rat heart (Rogart et al. 1989), rat skeletal muscle (Trimmer et al. 1989; Kallen et al. 1990), and *Drosophila* (Salkoff et al. 1987). An important outcome of this work was the opportunity it presented to propose molecular models of how the sodium channel folded through the membrane (Noda et al. 1984, 1986a; Kosower 1985; Greenblatt et al. 1985; Guy and Seetharamulu 1986). It was initally clear that the sodium channel consists of four repeating domains, each consisting of six transmembrane helices. These models have been tested using antibodies, scorpion toxins, and phosphorylation studies (Catterall 1988; Gordon et al. 1987, 1988; Emerick and Agnew 1989; Meiri et al. 1987), and a broad agreement has been obtained for the proposals of Noda et al. (1986a) (see Fig. 2).

The cDNA encoding the 260-kDa subunit has been expressed in oocytes (Noda et al. 1986b; Stühmer et al. 1989; Trimmer et al. 1989) and in cell culture (Scheuer et al. 1990). There are some differences in the electrophysiological properties, but the overall result is that the cDNA encodes a protein possessing the activation, voltage dependence, TTX/STX sensitivity, and inactivation properties of the sodium channel as described in vivo (Noda et al. 1986b).

Site-directed mutagenesis studies have been used to alter the sodium channel sequence and then to examine the biophysical properties of the result (STÜHMER et al. 1989). This approach has its limitations as the three-dimensional structure of the sodium channel is yet to be determined; nevertheless it has provided evidence for the involvement of the polypeptide loop between domains III and IV in the fast inactivation (STÜHMER et al. 1989). This work confirmed the initial findings of VASSILEV and CATTERALL (1986), who showed that antibodies directed against this region also prevent inactivation. The same approach has revealed that the fourth transmembrane segments in each domain (S4) are involved in activation (STÜHMER et al. 1989).

C. Description of Toxins by Class

I. Toxins that Inhibit Ion Transport

1. Heterocyclic Guanidines

TTX and STX are heterocylic guanidines that block the inward sodium current in a reversible manner. TTX was originally extracted from the ovaries and liver of the puffer fish (*Tetraodon*), but other tissue sources have now been found (LAZDUNSKI and RENAUD 1982). Saxitoxin is produced by dinoflagellates of genus *Gonyaulax* and is, of course, found in the shellfish that feed on them, a potentially serious public health problem during dinoflagellate blooms (CATTERALL 1980).

The mechanism of block of sodium currents by both toxins is similar. Extensive voltage-clamp studies have shown that the block of sodium currents by TTX/STX is reversible, with a binding constant from 1 to $10\,nM$, a range of values which correlates well with the results obtained with binding studies (HILLE 1984). The exceptions to this rule are the TTX-insensitive sodium channels observed in adult heart tissue, during skeletal muscle development, and after denervation of adult skeletal muscle (BARCHI 1988). The sodium channels of the nerves and muscles of the puffer fish also exhibit TTX-insensitivity (ALBUQUERQUE et al. 1988).

TTX and STX inhibit ion flux through sodium channels without altering the rate of voltage dependence of the activation or inactivation processes. The binding of these toxins is not inhibited by direct competitive inhibition by other toxins except for µ-conotoxins (see next section).

The simplest mechanism by which these two toxins act is by direct occlusion of the pore. An early hypothesis for TTX blockade placed the guanidinium group, a common feature of both toxins, in a intrinsic region of the sodium conduction pathway known as the selectivity filter, where an ion pair formed between the guanidinium group and a nearby carboxyl group on the sodium channel (HILLE 1984). There are many lines of evidence in support of this, including the block of sodium channels by guanidinium ions (HILLE 1984) and the well-documented block of TTX binding by trimethyloxonium ions, a carboxyl group modifing reagent; however, channels made TTX-insensitive by chemical reagents have been shown to have all the properties (including ion selectivity) of the in vivo sodium channel (LAZDUNSKI and RENAUD 1982).

In contrast to this initial hypothesis, a more recent proposal has placed this toxin binding site on the outside of the pore on the surface of the mouth, well away from the selectivity filter (MOCZYDLOWSKI et al. 1984a, b). Thus, there are two concepts, a "deep" and a "shallow" binding site with respect to the pore of the sodium channel.

Recent evidence from NUMA and coworkers has shed new light on our concept of a TTX binding site (NODA et al. 1989). Using their site-directed mutagenesis technique, they were able to construct a mutant rat brain sodium channel and examine TTX/STX sensitivity. Substitution of glutamic acid residue 387 by a glutamine residue resulted in a brain sodium channel of drastically reduced sensitivity to both toxins. This site is on the loop formed between transmembrane segments 5 and 6 on domain 1 (see Fig. 2) and is located in the corresponding region that is involved in charybdotoxin binding in the voltage-dependent potassium channel (MCKINNON and MILLER 1989). This glutamate residue lies in a region currently thought to be the α-scorpion toxin binding site (THOMSEN and CATTERALL 1989; see below). Could this be the glutamate residue that contains the carboxyl group long hypothesized (HILLE 1984) to be a key feature of the binding site of TTX and STX?

A simple test of this hypothesis would be to examine the corresponding sequence of the TTX-insensitive channels from heart and skeletal muscle and to look for conservation of this glutamic residue. These sequences are both available (ROGART et al. 1989; KALLEN et al. 1990) and reveal that although this glutamate residue is conserved, an adjacent aparagine residue is mutated to arginine. It has been proposed that this arginine neutralizes the charge on the glutamic acid residue, thus causing TTX insensitivity (NODA et al. 1989). Does this imply that there is more than one mechanism of TTX insensitivity? There is chemical mechanism by which the glutamic acid residue on the sodium channel becomes modified by reagents like trimethyloxonium, and a mutation mechanism occurring in TTX-insensitive sodium channels in which the binding site is modulated by the presence of positive charges on adjacent arginines.

2. μ-Conotoxins

A series of polypeptide toxins (generically termed μ-conotoxins) that act on the sodium channel have been isolated from the Pacific gastropod *Conus geographus* (GRAY et al. 1988): conotoxin GIIIA (also called geographutoxin I), conotoxin GIIIB (also called geographutoxin II), and conotoxin GS.

Other conotoxins of the GIII class have been described, but their physiological properties have not been fully elucidated (Cruz et al. 1985). The conotoxins GIIIA and GIIIB are 22 amino acid residues long, while conotoxin GS is 34 amino acids long (Sato et al. 1983; Cruz et al. 1985; Yanagawa et al. 1988). They are all basic, crosslinked by three disulfide bonds, and all possess the unusual amino acid *trans*-4-hydroxyproline.

Conotoxin GS also possesses the unusual amino acid 4-carboxyglutamic acid, usually found in calcium binding proteins (Yanagawa et al. 1988). It has very little sequence homology to the conotoxin GIIIA or GIIIB. Yanagawa et al. (1988) were able to show that this toxin can displace the binding of [^{3}H]-lysyl-TTX and [^{3}H]-propionyl-conotoxin GIIIA to *E. electricus* membranes with a K_i of 34 nM and 24 nM, respectively, but with little effect on the binding to brain membranes. Interestingly, the displacement of [^{3}H]-lysyl-TTX binding by conotoxin GS to rat skeletal muscle membranes occurred at concentrations an order of magnitude greater than that necessary to displace [^{3}H]-lysyl-TTX from *E. electricus* membranes. Nevertheless, it is clear that this toxin binds to the TTX/conotoxin GIIIA binding site (Yanagawa et al. 1988).

There is abundant evidence in the literature to show that conotoxins GIIIA and GIIIB can block sodium currents in muscle with a similar behavior to TTX but have little effect on the sodium channels from nervous tissue (Cruz et al. 1985; Gray et al. 1988). Direct ligand binding experiments with [^{3}H]-STX, [^{3}H]-lysyl-TTX, or [^{3}H]-propionyl-conotoxin GIIIA indicate that conotoxins GIIIA and GIIIB bind to skeletal muscle and *E. electricus* membranes with a K_d of 1–50 nM, but with very little effect on nervous tissues (Ohizumi et al. 1986; Moczydlowski et al. 1986; Yanagawa et al. 1986). This strongly suggests that the binding site for these toxins on the sodium channel in nerves is different to that in skeletal muscles. Conotoxins are also able to distinguish between TTX-sensitive and -insensitive sodium channels during the development of rat skeletal muscle (Gonoi et al. 1987).

Conotoxins are very attractive tools to study the molecular nature of the TTX/STX binding site. We, like others, have worked out methods of synthesizing conotoxin GIIIA (Becker et al. 1989; Cruz et al. 1989), thus facilitating its derivatization. Also, the geometry of the disulfide bonds is now known, and some idea can be deduced about the structure of the toxin (Hidaka et al. 1990). We have substituted nearly every amino acid in this toxin synthetically, thus making a series of toxin analogues, in order to obtain a structure-activity relationship for conotoxin GIIIA. We found that the lysine, hydroxyproline, glutamine, and aspartate residues are not necessary for activity as measured by their ability to displace [^{3}H]-STX binding. The three arginine residues were also individually substituted; however, we found that only an analogue in which arginine 13 (N-terminal amino acid = 1) was substituted for a glutamine residue was unable to displace [^{3}H]-STX binding. This strongly suggests that an arginine guanidino

Table 2. Conotoxins acting on the sodium channel

Conotoxin	Sequence	Reference
Conotoxin GS	ACSGRGSRCHHQCCMGLRLGRGNPQKCIGA MXEV	Yanagawa et al. (1988)
Conotoxin GIIIA	RDCCTHHKKC KDRQCK HQRCCA	Sato et al. (1983), Cruz et al. (1985)
Conotoxin GIIIB	RDCCTHHRKC KDRRCK HMKCCA	Sato et al. (1983), Cruz et al. (1985)
Conotoxin GIIIC	RDCCTHHKKC KDRRCK HLKCCA	Cruz et al. (1985)

* Conserved residues
H, 4-trans-hydroxyproline; X, 4-carboxyglutamic acid.

group is the key group necessary for binding (BECKER et al., unpublished observations). Inspection of the sequences of conotoxins GIIIA, GIIIB, GIIIC, and GS shows that this arginine 13 is conserved across all four (see Table 2). We have investigated the solution structure of u-conotoxin GIIIA by 2-D NMR methods (OTT et al. 1991). The derived structure indicates that arginine 13 is on an exposed loop. One may speculate that this loop enters the pore of the sodium channel during u-conotoxin GIIIA block of the sodium current.

We also used our synthetic approach to derivatize conotoxin GIIIA with the photoactive group 4-azidosalicyl (BECKER et al. 1990). Photoactive conotoxin GIIIA derivatives have been described before (HATANAKA et al. 1990), but we were able to iodinate our derivative and to show specific binding to the sodium channel. In addition to these findings, we photo-activated this compound in the presence of *E. electricus* membranes and showed using SDS-PAGE together with autoradiography analysis that it binds to a 260-kDa protein. Taken together, these results demonstrate that conotoxin GIIIA can be derivatized and potentially can be used bio-chemically to probe the molecular nature of the TTX/STX binding site on the sodium channel.

II. Lipid-Soluble Toxic Activators of the Sodium Channel

These neurotoxins are all lipid-soluble compounds obtained from plants and various species of tropical frogs. The main toxins in this class include batrachotoxin (BTX), veratridine, aconitine, and grayanotoxin (CATTERALL 1980). Veratridine is steroidal and is the most abundant toxin obtained from the mixture of toxins extracted from the liliaceous plant genera *Schoenocaulon*, *Zygadenus*, *Stenanthium*, and *Veratum* (LAZDUNSKI and RENAUD 1982). Aconitine is a plant alkaloid produced by the plant *Aconitum napellus*, and not resembling BTX or veratridine in structure. Grayanotoxin is a group of three diterpenoid toxins (grayanotoxin I, II, and III) obtained from the plant genera *Rhododendron*, *Kalmia*, and *Leucothoe* (family *Ericacae*). Grayanotoxin II and III are derivatives of I. They all have a very similar action, causing depolarization of excitable cells by activating the sodium channel to various degrees, thus evoking sodium currents. The hydrophobic nature of these toxins suggests that the receptor lies in a strongly hydrophobic region in the membrane, interacting with the trans-membrane helices of the sodium channel. This review will focus on the best-studied activator, BTX.

1. Batrachotoxin

BTX is a steroidal alkaloid isolated from the skin of frogs (genus *Phyllobates*) from South and Central America (ALBUQUERQUE et al. 1988). It is the most potent and most selective toxin acting on >95% of sodium channels. It is a most toxic substance, its lethal effect on mammals being mainly due to its cardiac arrhythmogenic action (Lazdunski and Renaud 1982). It is active on almost all sodium channels from skeletal muscle, nerve, and heart. It is well-known to cause depolarization of membranes due to an increase in sodium permeability. Sodium channels are considered to be persistantly activated by this toxin. Detailed voltage-clamp studies have shown that the toxin has two effects on the sodium channel. It blocks the voltage-dependence of inactivation and shifts the voltage-dependence of activation to more negative potentials. Also, single-channel conductances and peak macroscopic currents are smaller than in control (untreated) channels. BTX additionally causes the characteristic ion selectivity of the sodium channel to be decreased. One function of sodium channels remains unaltered: TTX still blocks the channel even after BTX treatment (Hille 1984).

As with the case of animals that produce TTX, the motor nerves and skeletal muscles of *Phyllobate* frogs are insensitive to BTX even when exposed to concentrations which cause depolarization in other frogs (Albuquerque et al. 1988). The action of BTX on sodium channels is greatly enhanced by α-scorpion and sea anemone toxins (see Catterall 1980). The action is long-lasting, and it has been proposed that BTX interacts with the sodium channel allosterically (Catterall 1980). Radioligand binding studies with [^{3}H]-BTX have demonstrated that this toxin binds to synaptosomal membranes with a K_d of $40-160\,nM$, showing that BTX is a high-affinity ligand for the sodium channel and providing direct evidence that the mechanism of BTX binding is allosteric. To summarize, BTX has its own specific binding site on the sodium channel distinct from the heterocyclic guanidines, sea anemone, and scorpion toxins.

III. North African α-Scorpion and Sea Anemone Toxins

1. α-Scorpion Toxins

These are small, basic, polypeptide toxins of molecular weight approximately 7 kDa (Catterall 1980). They have been purified from the venoms of scorpions of the general *Leiurus*, *Buthus*, and *Androctonus* (Lazdunski and Renaud 1982). This section will concentrate on the best studied α-scorpion toxin, that from *L. quinquestriatus*. Every venom from each scorpion species contains multiple toxins which have extensive sequence homology. Substantial sequence homology is also observed between α- and β-scorpion toxins. The purified toxins possess at least four disulfide bonds (Catterall 1980).

These toxins exert their lethal effects by causing the secretion of neurotransmitters, arrhythymia in the heart, and repetitive firing and depolarization in nerves (Catterall 1980). It has long been known that this last effect is due to an increase in sodium permeability. Analysis of the effect of *L. quinquestriatus* toxin V (LqV) on sodium channels under voltage-clamp conditions showed that the voltage dependence of inactivation was slowed and incomplete. The binding of LqV to excitable tissues was studied with a radiolabelled derivative, and specific high-affinity binding was demonstrated with a K_d of $0.5-15\,nM$, which is unaffected by STX or TTX (CATTERALL 1986). It has been shown that the binding of this toxin to the sodium channel is voltage-dependent (CATTERALL et al. 1976). The voltage dependence of binding closely parallels that of activation of sodium channels measured in voltage-clamp experiments.

Considerable progress has been made by CATTERALL (1988) in the localization of the binding site of LqV on the sodium channel protein. These workers first showed that an LqV azido derivative covalently attached to a 270-kDa and 37-kDa protein in solubilized extracts of synaptosomal proteins (BENESKI and CATTERALL 1980; SHARKEY et al. 1984). These proteins represent the α- and β-subunits of the brain sodium channel. CATTERALL and coworkers later went on to demonstrate using a $[^3H]$-STX probe that it was possible to purify the sodium channel from rat brain by conventional chromatography (CATTERALL 1986, 1988). They showed that the respective subunits of this complex were 270, 39, and 37 kDa. In order to check the functional integrity of the complex, the pure sodium channel was reconstituted into phospholipid vesicles, and the functional properties characteristic of sodium channels as described in vivo were examined. Radiolabelled LqV binding study was performed in order to examine the functional integrity of the purified sodium channel (FELLER et al. 1985). Very little binding was observed when the sodium channel was reconstituted into phosphatidylcholine (PC) vesicles or phosphatidylethanolamine (PE) vesicles. Optimal binding was only obtained when at least two combinations of PC, PE, and phosphatidylserine were used. This provided the first evidence of the lipid dependance of the binding of the LqV toxin to the voltage-dependent sodium channel.

In a series of elegant experiments CATTERALL and coworkers localized a radiolabelled photoactivate derivative of LqV in peptide maps of the α-subunit of the brain sodium channel (TEJEDOR et al. 1988). They found that the toxin bound to a 14-kDa fragment after sequential cyanogen bromide treatment and trypsinization of the sodium channel. The peptide was not identified by peptide sequencing but by a series of peptide site-specific antibodies. The results showed specific binding of LqV to a series of sites between transmembrane segments 5 and 6 on domains I and IV (THOMSEN and CATTERALL 1989) (see Fig. 2). One of these sites (amino acids 382–400) overlaps glutamic acid residue 387, which is currently thought to be the TTX/STX binding site (NODA et al. 1989), a surprising finding in view of the fact that α-scorpion toxins are not known to compete with TTX or STX binding (CATTERALL 1980).

These findings helped the current understanding of the topology of the sodium channels, as LqV was a toxin well-known to act on the outside of the channel. It had been proposed that there were additional transmembrane helices between segments S5 and S6 (Greenblatt et al. 1985; Guy and Seetharamulu 1986); it is now clear that these findings are not consistent with data showing the localization of the LqV binding site (Thomsen and Catterall 1989). Thus, the model of Numa and coworkers fits the topological data most correctly (Noda et al. 1986a) (see Fig. 2).

The binding site information also gave considerable insight into how activation is coupled to inactivation. LqV binds to an extracellular loop between segments 5 and 6 near segment 4, the proposed voltage sensor (Stühmer et al. 1989), and the inactivation gate is probably the segment between domains III and IV (Vassilev and Catterall 1988; Stühmer et al. 1989). It may well be that the toxin interacts with the segment 4 gating charges to slow or prevent the coupling of activation gating charges to the inactivation site between domains III and IV, thus effecting inactivation. It's a rather speculative explanation, and it may well have to be revised when the three-dimensional structure of the sodium channel is known, but it is the best explanation of the toxicity of LqV.

2. Sea Anemone Toxins

A group of sodium-channel-specific polypeptide neurotoxins have been purified from the nematocysts of sea anemones of the following genera: *Anemonia*, *Anthopleura*, and *Radianthus* (Lazdunski and Renaud 1982; Schweitz et al. 1985). They are 27–51 amino acid residues in length, with at least four disulfide bonds (Catterall 1980; Lazdunski et al. 1986). A most important recent advance within this toxin group has been the structure determination of an *Anemonia sulcata* (Widmer et al. 1989) and an *Anthropleura* (Torda et al. 1988) toxin using magnetic resonance imaging, distance geometry, and molecular dynamics.

It should be pointed out that the toxins from all three sea anemone species possess considerable sequence homology with each other; however, there is very little sequence homology with the α-scorpion toxins (*Androcotonus australis* or *Leiurus quinquestriatus*). Chemical modification of arginine, lysine, or glutamic acid residues has been shown to abolish activity, indicating a complicated interaction with the channel.

These toxins have been well described electrophysiologically. Initial observations in *Crustacea* showed that they prolong action potentials (Catterall 1980). Subsequent detailed voltage-clamp analysis revealed that this was due to a slowing of the rate of inactivation (Lazdunski et al. 1988).

Sea anemone toxins are considered to bind to the same site as the North African α-scorpion toxins on the basis of competitive binding studies (Catterall 1980). It should be pointed out, however, that there is a discrepancy in the number of sea anemone toxin binding sites to the number of the *Androcotonus* (an α-scorpion toxin) binding sites (Lazdunski et al. 1986). There are ten times more of the former than of the latter.

As far as the molecular action of these toxins is concerned, the sea anemone toxins remain rather obscure in relation to the fine details now available for LqV. It has been shown that sea anemone toxins act on purifed reconstituted brain sodium channels which in their hands only contain the 270-kDa subunit (HANKE et al. 1984). This evidence crudely localizes the receptor site to the 270-kDa protein; however, reservations remain about the properties of this reconstituted sodium channel preparation as this preparation does not have all the properties of the sodium channel as described in vivo.

One of the most interesting properties of the sodium channel is the existence of subtypes that are observed during skeletal muscle development and after denervation (BARCHI 1988). These subtypes are clearly distinguished by their sensitivity to TTX; furthermore, the mammalian cardiac sodium channel is a TTX-insensitive channel. *Anemonia sulcata*, but not *Androctonus australis*, toxins are able to bind to TTX-insensitive sodium channels in neuroblastoma cells and rat myoblasts ($K_d = 0.15-150\,nM$), while they bind weakly to TTX-sensitive channels in chick myotubules and fibroblasts ($K_d = 60-1000\,nM$) (LAZDUNSKI et al. 1986). Clearly, sea anemone toxins have a role as molecular probes in the examination of the α-scorpion toxin/TTX cluster of binding sites in the sodium channel.

IV. American β-Scorpion Toxins

American β-scorpion toxins are polypeptide toxins extracted from the venoms of the American scorpion genera *Tityus* and *Centruroides* (LAZDUNSKI et al. 1988; CATTERALL 1988).

This section will concentrate on the description of *Tityus* gamma-toxin from *Tityus serulatus*, which is the best characterized American scorpion toxin. It is 61 amino acid residues in length, with four disulfide bridges (LAZDUNSKI et al. 1988). Its structure is unknown, but one has been derived from crystallographic studies of a *Centruroides* toxin (FONTECILLA-CAMPS et al. 1980), to which *Tityus* toxin is highly homologous.

Tityus toxin has been well characterized by LAZDUNSKI and coworkers in preference to the *Centruroides* toxin which has a lower affinity for the sodium channel ($4\,pM$ versus $1\,nM$) (BARHANIN et al. 1983). Detailed voltage-clamp studies have shown that this toxin acts on both the voltage dependence of activation and of inactivation. The voltage dependence of activation is shifted to a more negative potential, while the rate of inactivation is affected, albeit in a minor way (BARHANIN et al. 1983).

Tityus toxin has proved on account of its high affinity and slow off-rate to be an excellent biochemical marker for the sodium channel from a wide range of tissues, including electric eel, synaptosomes, and skeletal muscle. *Tityus* toxin has a high affinity for cardiac muscle where it binds with high affinity ($K_d = 3-750\,pM$) (BARHANIN et al. 1983).

LAZDUNSKI and coworkers have suceeded in using radiolabelled *Tityus* toxin as a biochemical marker to purify sodium channel-*Tityus* toxin complexes from electric eel and rat brain (NORMAN et al. 1983; BARHANIN et al. 1983). The results, using broadly similiar chromatographic procedures, were essentially the same. The *Tityus* binding protein was found to consist of a single subunit protein of molecular weight 270 kDa. These investigators went on to reconstitute it into phospholipid bilayers and were able to demonstrate that it had a considerable number, although not all, of the properties of the brain sodium channel (Hanke et al. 1984). The results from electric eel studies confirmed the findings of Agnew (1984) who previously reported that the electric eel sodium channel consists only of a 260-kDa protein.

These results imply that the brain sodium channel consists of a 260-kDa protein without any small subunits of 32–38 kDa. This is in contrast to the results of Catterall (1986) but consistent with the results of Elmer et al. (1985).

Catterall's group have consistently emphasized that in a addition to the large subunit small subunits of 32–39 kDa are always found in their preparations (Catterall 1986, 1988). A important point must be stressed here: Numa and others have clearly shown by expression of cDNA encoding the 260-kDa protein in oocytes and in cell culture that all of the functions of the sodium channel as described in vivio, namely, voltage-dependent activation and inactivation, sodium current, and all the toxin binding sites, can be ascribed to the 260-kDa protein (Noda et al. 1986a,b; Scheuer et al. 1990). This almost certainly means that the small subunits of the brain sodium channel described by Catterall (1988) do not play a physiological role, even though this group has shown that the α-scorpion toxin binding site is on both the 260- and the 39-kDa proteins (Beneski and Catterall 1980).

How can these results be reconciled? Purified preparations of the voltage-dependent calcium and potassium channels also consist of the subunit that effects the physiological function in addition to smaller proteins whose role is as yet unknown (Catterall 1988; Rehm and Lazdunski 1988). Could it be that different investigators of the sodium channel from rat brain are managing to copurify associated proteins in different proportions?

Less controversially, *Tityus* toxin has been used as a biochemical marker to purify the sodium channel from chick cardiac muscle (Lombet and Lazdunski 1984). Here, it was shown that the channel consists of a 230–260-kDa protein with no small subunits. *Tityus* toxin remains the only effective biochemical probe for cardiac muscle.

V. Other Toxins

1. Ciguatoxin

This is a toxin of unknown structure extracted from the dinoflagellate *Gymnothorax javanicus*. Ciguatoxin induces membrane depolarization and can stimulate sodium flux through sodium channels in neuroblastoma cells or skeletal muscle myoblasts when used in synergy with veratridine, BTX, pyrethroids, sea anemone or scorpion toxins. Ciguatoxin is unable to block the binding of any *Tityus* gamma-toxin or sea anemone toxins (*Anemonia sulcata* and *Androctonus austrailis*). Thus, it is a poorly characterized toxin, which in preliminary studies showed an effect on the sodium channel at a site distinct from the other toxin groups (BIDARD et al. 1984).

2. Brevetoxin

This toxin is from a lipophilic series of toxins extracted from dinoflagellate blooms from *Ptychodiscus brevis* (formerly *Gymnodinium breve*). It causes depolarization of squid axonal membranes, which is reversed by TTX. Detailed voltage-clamp studies indicate that the voltage-dependence of activation is shifted to more negative potentials with prevention of inactivation. There is a synergistic effect between these toxins and those from the sea anemone *Anthropleura xanthogrammica*. Its actions bear a close resemblance to the effects of BTX and grayanotoxin. As this toxin potentiates the action of veratridine, it is not normally put in the same class as the lipophilic activators of the sodium channel (HUANG et al. 1984).

D. Conclusions

There are at least five different classes of toxin binding sites on the sodium channel. Site 1 is for TTX, STX, and the μ-conotoxins, which all block sodium transport. The location of this site is unknown but is clearly associated with the ion pore. Site 2 is hydrophobic and binds BTX, veratridine, grayanotoxin, and aconitine. Site 3 binds North African α-scorpion and sea anemone toxins, which slow inactivation. This site lies on the outside of the sodium channel at a position which undergoes a large conformational change upon depolarization. Site 4 is also on the outside of the sodium channel and binds the American β-scorpion toxins. Site 5 is for ciguatoxin and brevetoxin (see Table 1 and Fig. 1).

References

Agnew WS (1984) Voltage-regulated sodium channel molecules. Annu Rev Physiol 46:517–530

Albuquerque EX, Daly JW, Warnick JE (1988) Macromolecular sites for specific neurotoxins and drugs on chemosensitive synapses and electrical excitation in biological membranes. In: Narahashi T (ed) Ion channels, vol 1. Plenum, New York, pp 95–150

Auld VJ, Goldin AL, Krafte DS, Marshall J, Dunn JM, Catterall WA, Lester HA, Davidson N, Neuron Dunn RJ (1988) A rat sodium channel α subunit with novel gating properties. Neuron 1:449–461

Barchi RL (1988) Probing the structure of the voltage-dependent sodium channel. Annu Rev Neurosci 11:455–495

Barhanin J, Pauron D, Lombert A, Norman R, Vijverberg HP, Giglio J, Lazdunski M (1983) Electrophysiological characterization, solubilization and purification of *Tityus* toxin. EMBO J 2:915–920

734 S. Becker and R.D. Gordon

Becker S, Atherton E, Gordon RD (1989) Synthesis and characterization of μ-conotoxin IIIA. Eur J Biochem 185:79–84

Becker S, Liebe R. Gordon RD (1990) Synthesis and characterization of a N-terminal-specific 125 I-photoaffinity derivative of μ-conotoxin GIIIA which binds to the voltage-dependent sodium channel. FEBS Lett 272:152–154

Beneski DA, Catterall WA (1980) Covalent labelling of the protein components of the sodium channel with a photoactivable derivative of scorpion toxin. Proc Natl Acad Sci USA 77:639–643

Bidard J-N, Vijverberg HPM, Frelin C, Changue E, Legrand A-M, Bagnis R, Lazdunski M (1984) Ciguatoxin is a novel type of sodium channel toxin. J Biol Chem 259:8353–8357

Casadei JM, Gordon RD, Barchi RL (1986) Immunoaffinity isolation of sodium channels from rat skeletal muscle. Analysis of subunits. J Biol Chem 261:4318–4323

Catterall WA (1980) Neurotoxins that act on the voltage sensitive sodium channels in excitable membranes. Annu Rev Pharmacol Toxicol 20:15–43

Catterall WA (1986) Molecular properties of voltage-sensitive sodium channels. Annu Rev Biochem 55:953–985

Catterall WA (1988) Structure and function of voltage sensitive ion channels. Science 242:50–61

Cruz LJ, Gray WE, Olivera BM, Zeikus RD, Kerr L, Yoshikami D, Moczydlowski E (1985) Conus geographus toxins that discriminate between neuronal and muscle sodium channels. J Biol Chem 260:9280–9288

Cruz LJ, Kupryszewski G, LeCheminant GW, Gray BW, Olivera BM, Rivier J (1989) μ-Conotoxin IIIA a peptide ligand for muscle sodium channels: chemical synthesis, radiolabelling and receptor characterization. Biochemistry 28:3437–3442

Elmer LW, O'Brien BJ, Nutter TJ, Angelides KJ (1985) Physiochemical characterization of the α-peptide of the sodium channel from rat brain. Biochemistry 24:8128–8137

Emerick ME, Agnew WS (1989) Identification of phosphorylation sites for adenosine 3′,5′ cyclic phosphate dependent protein kinase on the voltage dependent sodium channel from E. electricus. Biochemistry 28:8367–8380

Feller DJ, Talvenheimo JA, Catterall WA (1985) The sodium channel from rat brain. Reconstitution of the voltage-dependent scorpion toxin binding in vesicles of defined lipid composition. J Biol Chem 260:11 542–11 547

Fontecilla-Camps JC, Almassy RJ, Duddath FL, Watt DD, Bugg CE (1980) Three dimensional structure of a protein from scorpion venom: a new structural class of neurotoxins. Proc Natl Acad Sci USA 77:6496–6500

Gonoi T, Ohizumi Y, Nakamura H, Kobayashi J, Catterall WA (1987) Conus toxin geographutoxin II distinguishes two functional sodium channel subtypes in rat muscle cells developing in vitro. J Neurosci 7:1726–1731

Gordon RD, Fieles WE, Schotland DL, Angeletti RA, Barchi RL (1987) Topological localization of a C-terminal region of the voltage dependent sodium channel from E. electricus using antibodies against a synthetic peptide. Proc Natl Acad Sci USA 84:308–313

Gordon RD, Li Y, Fieles WE, Schotland DL, Barchi RL (1988) Topological localization of a segment of the eel voltage-dependent sodium channel primary sequence (AA 927-938) that discriminates between models of tertiary structure. J Neurosci 8:3742–3749

Gray WR, Olivera BM, Cruz LJ (1988) Peptide toxins from venomous Conus snails. Annu Rev Biochem 57:665–700

Greenblatt RE, Blatt Y, Montal M (1985) Structure of the voltage sensitive sodium channel. Inferences derived from computer-aided analysis of Electrophorus electricus channel primary structure. FEBS Lett 193:125–134

Guy HR Seetharamulu P (1986) A molecular model of the action potential sodium channel. Proc Natl Acad Sci USA 83:508–513

Hanke W, Boheim G, Barhanin J, Pauron D, Lazdunshi M (1984) Reconstitution of highly purified saxitoxin sensitive Na$^+$ channels into planar lipid bilayers. EMBO J 3:509–515

Hatanaka Y, Yoshida E, Nakayama H, Abe T, Satake M, Kanaoka Y (1990) Photolabile μ-conotoxins with a chromogenic phenyldizirine. A novel probe for the muscle type sodium channels. FEBS Lett 260:27–30

Hidaka Y, Sato K, Nakamura H, Kaboyashi J, Ohizumi Y, Simonishi Y (1990) Disulphide pairings in geographutoxin, a peptide neurotoxin from *Conus geographus*. FEBS Lett 264:29–32

Hille (1984) Ion channels of excitable membranes. Sinauer, Sunderland

Hodgkin AL, Huxley AF (1952) The components of membrane conductance in the giant axon of *Loligo*. J Physiol (Lond) 116:473–496

Huang JMC, Wu CH, Baden DG (1984) Depolarizing action of a red tide dinoflagellate brevetoxin on axonal membranes. J Pharmacol Exp Ther 229:615–621

Kallen RG, Sheng ZH, Yang J, Chen L, Rogart RB, Barchi RL (1990) Primary structure and expression of a sodium channel characteristic of denervated and immature rat skeletal muscle. Neuron 4:233–242

Kao CY, Levinson SR (eds) (1988) Tetrodotoxin, saxitoxin and the molecular biology of the sodium channel. Ann NY Acad Sci 479

Kosower E (1985) A structural and dynamic molecular model for the sodium channel of *Electrophorus electricus*. FEBS Lett 182:234–242

Kraner SD, Tanaka JC, Barchi RL (1985) Purification and functional reconstitution of the voltage sensitive sodium channel from rabbit "T" tubular membranes. J Biol Chem 260:6342–6347

Lazdunski M, Renaud JF (1982) The action of cardiotoxins on the cardiac plasma membranes. Annu Rev Physiol 44:463–473

Lazdunski M, Frelin C, Barhanin J, Lombet A, Meiri H, Pauron D, Romey G, Schmid A, Schweitz H, Vigue P, Vijverberg HPM (1986) Polypeptide toxins as tools to study voltage-sensitive Na$^+$ channels. Ann NY Acad Sci 479:204–221

Lombet A, Lazdunski M (1984) Characterization, solubilization and affinity labelling of the cardiac Na$^+$ channel using *Tityus* gamma toxin. Eur J Biochem 141:651–660

Mckinnon R, Miller C (1989) Mutant potassium channels with altered binding of charybdotoxin, a pore-blocking peptide inhibitor. Science 245:1382–1385

Meiri H, Spira G, Sammar M, Namir M, Schwartz A, Komoriya A, Kosower EM, Palti Y (1987) Mapping a region associated with the sodium channel inactivation using antibodies to a synthetic peptide corresponding to a part of the channel. Proc Natl Acad Sci USA 84:5058–5062

Moczydlowski E, Hall S, Garber SS, Strichartz GR, Miller C (1984a) Voltage-dependent blockade of muscle Na$^+$ channels by guanidinium ions. Effect of toxin charge. J Gen Physiol 84:687–704

Moczydlowski E, Garber SS, Miller C (1984b) Batrachotoxin-activated Na$^+$ channels in planar lipid bilayers. Competition of tetrodotoxin block by Na$^+$ channels. J Gen Physiol 84:665–686

Moczydlowski E, Olivera BM, Gray WR, Strichartz GR (1986) Discrimination of muscle and neuronal Na-channel subtypes by binding competition between [^{3}H]saxitoxin and μ-conotoxins. Proc Natl Acad Sci USA 83:5321–5325

Noda M, Shimizu S, Tanabe T, Takai T, Kayano T, Ikeda T, Takahashi H, Nakayama H, Kanoka Y, Minamino N, Kangawa K, Matsuo H, Raftery MA, Hirose T, Inayama S, Hayashida H, Miyata T, Numa S (1984) Primary structure of *Electrophorus electricus* sodium channel deduced from cDNA sequence. Nature 312:121–127

Noda M, Ikeda T, Kayano T, Suzuki H, Takeshima H, Kurasaki M, Takahashi H, Numa S (1986a) Existence of distinct sodium channel messenger RNAs in rat brain. Nature 320:188–192

Noda M, Ikeda T, Suzuki H, Takeshima H, Takahashi T, Kuno M, Numa S (1986b)
    Expression of functional sodium channels from cDNA. Nature 322:826–828
Noda M, Suzuki H, Numa S, Stühmer W (1989) A single point mutation confers
    tetrodotoxin insensitivity on the sodium channel II. FEBS Lett 259:213–216
Norman RI, Schmid A, Lombert A, Barhanin J, Lazdunski M (1983) Purification of
    the binding protein for *Tityus* gamma toxin identified with a gating component
    of the voltage-dependent sodium channel. Proc Natl Acad Sci USA 80:4164–
    4168
Ohizumi Y, Nakamura H, Kobayashi J, Catterall WA (1986) Specific inhibition of
    [^{3}H]saxitoxin binding to skeletal muscle sodium channels by geographutoxin II,
    a polypeptide channel blocker. J Biol Chem 261:6149–6152
Ott K-H, Becker S, Gordon RD, Rueterjans H (1991) FEBS letts. Solution structure
    of μ-conotoxin GIIIA analysed by 2D-NMR and distance geometry calculations.
    FEBS Lett 278:160–166
Rehm H, Lazdunski M (1988) Purification and subunit structure of a putative
    K$^+$-channel protein identified by its binding properties for dendrotoxin. Proc
    Natl Acad Sci USA 85:4919–4923
Rogart RB, Cribbs LL, Muglia LK, Kephart DD, Kaiser MW (1989) Molecular
    cloning of a putative tetrodotoxin-resistant rat heart Na$^+$ channel isoform. Proc
    Natl Acad Sci USA 86:8170–8174
Salkoff L, Butler A, Wei A, Scavarda N, Giffen K, Ifune C, Goodman R, Mandel G
    (1987) Genomic organization and deduced amino acid sequence of a putative
    sodium channel in *Drosophila*. Science 237:744–749
Sato S, Nakamura H, Ohizumi Y, Kaboyashi J, Hirata Y (1983) The amino acid
    sequences of homologous hydroxyproline containing myotoxins from the snail
    Conus geographus venom. FEBS Lett 155:277–280
Scheuer T, Auld V, Boyd S, Offord J, Dunn R, Catterall WA (1990) Functional
    properties of the rat brain sodium channel expressed in a somatic cell line.
    Science 247:854–858
Schweitz H, Bidard J-N, Frelin C, Pauron D, Vijverberg HPM, Mahashneh DH,
    Lazdunski M (1985) Purification, sequence and pharmacological properties of
    sea anemone toxins from *Radianthus paumotensis*. A new class of sea anemone
    toxin acting on the sodium channel. Biochemistry 24:3554–3561
Sharkey RG, Beneski DA, Catterall WA (1984) Differential labelling of the α and
    the ß subunits of the sodium channel by photoactivable derivatives of scorpion
    toxin. Biochemistry 23:6078–6086
Stühmer W, Conti F, Suzuki H, Wong X, Noda M, Yahagi N, Kubo H, Numa S
    (1989) Structural parts involved in the activation and inactivation of the sodium
    channel. Nature 339:597–603
Tanaka JC, Furman RE, Barchi RL (1986) Skeletal muscle sodium channels:
    isolation and reconstitution. In: Miller C (ed) Ion channel reconstitution.
    Plenum, New York, pp 307–336
Tejedor FJ, Catterall WA (1988) Site of covalent attachment of α-scorpion toxin
    derivatives in the domain I of the sodium channel α-subunit. Proc Natl Acad Sci
    USA 85:8742–8746
Thomsen WJ, Catterall WA (1989) Localization of the receptor site for α-scorpion
    toxin by antibody binding mapping: implications for the sodium channel
    topology. Proc Natl Acad Sci USA 86:10 161–10 165
Torda AE, Mabbut BC, van Gunsteren WI, Norton RS (1988) Backbone folding of
    the polypeptide cardiac stimulant *Anthropleura A* determined by nuclear
    magnetic resonance, distance geometry and molecular dynamics. FEBS Lett
    239:266–270
Trimmer JS, Agnew WS (1989) Molecular diversity of voltage sensitive sodium
    channels. Annu Rev Physiol 51:401–418
Trimmer JS, Cooperman SS, Tomiko SA, Zhou J, Crean SM, Boyle MB, Kallen
    RG, Sheng Z, Barchi RL, Sigworth FJ, Goodman R, Agnew WS, Mandel G

(1989) Primary structure and functional expression of a mammalian skeletal muscle sodium channel. Neuron 3:33–49

Vassilev PM, Scheuer T, Catterall WA (1988) Identification of an intracelluler peptide seqment involved in sodium channel inactivation. Science 241:1658–1661

Widmer H, Billeter M, Wüthrich K (1989) Three dimensional structure of the neurotoxin ATX Ia from *Anemonia sulcata* in aqueous solution determined by nuclear resonance spectroscopy. Proteins Struct Funct Genet 6:357–371

Yanagawa Y, Abe T, Satake M (1986) Blockade of [³H] lysine-tetrodotoxin binding to sodium channel proteins by conotoxin GIIIA. Neurosci Lett 64:7–12

Yanagawa Y, Abe T, Satake M, Odani S, Suzuki J, Ishikawa K (1988) A novel sodium channel inhibitor from *Conus geographus*: purification, structure and pharmacological properties. Biochemistry 27:6256–6262

CHAPTER 21

Potassium Channel Toxins

H. Meves

A. Introduction

The past decade has seen rapid progress in our knowledge about K channels and their pharmacology. Various types have been discovered: voltage-gated (separated into delayed rectifier, inward rectifier, transiently activating channels); Ca-activated [subdivided into high conductance (150–250 pS) and low conductance (10–14 pS) channels]; Na-activated; ATP-sensitive; M channels (inhibited by acetylcholine acting through muscarinic receptors); and S channels (inhibited by a cAMP-dependent protein kinase through serotonin receptors). Twelve years ago, the first report that an animal toxin (the bee venom toxin apamin) blocks a certain type of K channel (the Ca-activated K channel of hepatocytes) appeared (Banks et al. 1979). Three years later, a fraction of the venom of the scorpion *Centruroides noxius* was found to depress the permeability of voltage-dependent K channels (but not of Na channels) (Carbone et al. 1982). A further milestone in the toxinology of K channels was the discovery that dendrotoxin, a potent convulsant purified from the venom of mamba snakes (Harvey and Karlsson 1980), selectively blocks a transient K conductance in hippocampal neurons (Dolly et al. 1984). Great interest was also attracted by charybdotoxin, a fraction of the venom of the scorpion *Leiurus quinquestriatus*; it has a high affinity for Ca-activated K channels of high conductance (Miller et al. 1985) but is also a potent blocker of voltage-sensitive K channels.

Several reviews on the diversity of K channels and their pharmacology have appeared (Moczydlowski et al. 1988; Rudy 1988; Castle et al. 1989; Cook and Quast 1990; Dreyer 1990; Strong 1990). The present one deals with K channel toxins and will concentrate on the literature published in the years 1988–1990.

B. Chemistry

Table 1 shows the amino acid sequence of several K channel toxins. The venom of the honey-bee, *Apis mellifera*, contains the 18-residue peptide apamin and the structurally similar 22-residue mast cell degranulating peptide (MCDP). Both are strongly basic at neutral pH. The presence of 5

Table 1. Amino acid sequences of potassium channel toxins from bee venom (A), scorpion venoms (B), and snake venoms (C, D)

(A)   Apamin (AP) (Habermann 1972) and mast cell degranulating peptide (MCDP) (Hider and Ragnarsson 1981)

```
            1     5     1 0    1 5    2 0
AP:      — CNCK– – APETALCARRCQQH-NH₂
MCDP:  IKCNCKRHVI KPHI CRKI CGKN-NH₂
```

(B)   Charybdotoxin (ChTX) (Gimenez-Gallego et al. 1988), iberiotoxin (IbTX) (Galvez et al. 1990), noxiustoxin (NTX) (Possani et al. 1982; Gurrola et al. 1989), and leiurotoxin I(LTX) (Chicchi et al. 1988)

```
         1    5    1 0   1 5   2 0   2 5   3 0    3 5    4 0
ChTX:  ZFTNVS CTTS– KE CWS V–CQRLHNTS RG– KCMNKKCRCYS
IbTX:  ZFTDVDCSVS– KE CWS V–CKDLF GVDRG– KCMGKKCRCYQ
NTX:   TI I NVKC–TSPKQCS KP–CKELYGS S AGAKCMNGKCKCYNN
LTX:      AF C–NL–RMCQLS –CRS L– GLL– G– KCI GDKCECVKH
```

(C)   Dendrotoxin (DTX) (Rehm 1984; Dufton 1985; Harvey and Anderson 1985; Moczydlowski et al. 1988)

```
            1    5    10    1 5   2 0   2 5    3 0    3 5    4 0    4 5    5 0    5 5    6 0
DTX      ZPRRKLCI LHRNP GRCYDKIPAFYYNQ KKKQCERFDWSGCGGNS NRFKTIEE–CRRTCIG
C13S1C3  ––AAKYCKLP VRYGPCKKKIPS FYYKWKAKQCLP FDY SGCGGNANRFKTIEE–CRRTCVG
Toxin I  ZPL RKLCI LHRNP GRCYQKIPAFYYNQ KKKQCEGFT WSGCGGNS NRFKTIEE–CRRTCIRK
Toxin K  ––AAKYCKLP L RI GPCKRKIPS FYYKWKAKQCLP FDY SGCGGNANRFKTIEE–CRRTCVG
DV14  .  ––AAKYCKLP VRYGPCKK KIPS FYYKWKAKQCL YFDY SGCGGNANRFKTIEE–CRRTCVG
```

(D)   Subunits of β-bungarotoxin (Rehm 1984; Dufton 1985; Harvey and Anderson 1985; Moczydlowski et al. 1988)

```
     1    5    1 0    1 5    2 0    2 5    3 0    3 5    4 0    4 5    5 0    5 5    6 0
B   RQRHRDCDKP P DKGNC– GP VRAF YYDTRL KTCKAF QYRGCDGDHGNFKTETL– CRCECLVYP
A   GAGGS GRP I DAL DRCCYVHDNC YGDAEKKHKCNRKTS QI CYGAAGGTCRI VCDCDRTAALCF
```

The sequences are aligned by the cysteine residues and are arranged for maximum homology by inserting gaps denoted by a *hyphen*. The single amino acid code is: (A) ala, (E) glu, (Q) gln, (D) asp, (N) asn, (L) leu, (G) gly, (K) lys, (S) ser, (V) val, (R) arg, (T) thr, (p) pro, (I) ile, (M) met, (F) phe, (Y) tyr, (c) cys, (w) trp, (H) his, (Z) pyroglutamic acid
Subunits of ß-bungarotoxin: B = complete sequence of subunit B, A= residues 30–101 of subunit A.

lysine, 2 arginine, and 2 histidine residues and the absence of any carboxyl groups result in a net charge greater than +8 per molecule of MCDP. There is little similarity between the bee venom toxins and the scorpion toxins charybdotoxin (ChTX, from *Leiurus quinquestriatus hebraeus*), iberiotoxin (IbTX, from *Buthus tamulus*), noxiustoxin (NTX, from *Centruroides noxius*), and leiurotoxin I (LTX, also from *L. quinquestriatus*). The four scorpion toxins which act on potassium channels are much smaller than the 57–66 amino acid scorpion toxins which affect sodium channels. The molecular weights of ChTX, IbTX, NTX, and LTX are 4.3, 4.3, 4.2, and 3.4 kDa, respectively. Gimenez-Gallego et al. (1988) emphasize the great similarity between ChTX, NTX, and a number of other neurotoxins from marine worms and snakes (e.g., α-bungarotoxin from *Bungarus multicinctus*). They also stress the lack of similarity between ChTX and the dendrotoxins. The latter are a group of five homologous, strongly basic peptides of 57–61 residues from snake venoms: dendrotoxin (DTX) and C13S1C3 from *Dendroaspis angusticeps*, toxin I and K from *D. polylepis*, and toxin DV14 from *D. viridis*. The differences between these five homologues are very small: DV14 is identical to C13S1C3 except for one amino acid, toxin I differs from DTX only in six amino acids. BENISHIN et al. (1988) isolated four polypeptides from the venom of the green mamba *D.angusticeps*, designated α-, β-, γ-, and δ-DaTX; α-DaTX is identical to DTX, δ-DaTX nearly identical to toxin K. According to SCHWEITZ et al. (1990) the venom of *D. polylepis* contains 28 different peptides; the 14 most cationic ones form a structurally and functionally homogeneous group of analogues of the most abundant toxin, I. As Table 1 shows, the dendrotoxins have many structural similarities with subunit B (and A) of β-bungarotoxin (β-BuTX) from the venom of the snake *Bungarus multicinctus* and, interestingly, with protease inhibitors like the bovine pancreatic trypsin inhibitor (BPTI). Like the dendrotoxins, ChTX constitutes probably a group of homologues (LUCCHESI et al., 1989), which may explain why different ChTX preparations have different toxicities (MACKINNON et al. 1990; KIRSCH et al. 1991).

For none of the K channel toxins has it been feasible to determine the three-dimensional structure by X-ray crystallography. However, predictions from statistical rules and data from spectroscopic studies are available. Based on earlier work, FREEMAN et al. (1986) arrived at a structural model of apamin: An α-helix core surrounded by two β-turns and stabilized by two disulfide bonds, altogether a very compact conformation. PEASE and WEMMER (1988) used two-dimensional proton MR data with distance geometry to obtain an improved picture of apamin's solution structure. Further MR measurements suggested that the secondary structure of MCDP is very similar to that of apamin (KUMAR et al. 1988). From the assignment of the disulfide bridging in native and synthetic ChTX, SUGG et al. (1990) predict a highly folded tertiary structure. MR measurements on ChTX (MASSEFSKI et al. 1990) show a roughly ellipsoidal, very tightly folded structure consisting mainly of three antiparallel β-strands without an α-helix. By contrast, more recent MR measurements on synthetic ChTX (LAMBERT et al. 1990) and natural ChTX (BONTEMS et al. 1991) reveal a β-sheet linked to an α-helix by two disulfide bridges. MARTINS et al. (1990) report a similar structure for LTX 1. According to MASSEFSKI et al. (1990), nine of the ten charged groups of ChTX are located on one side of the ellipsoid, with seven of the eight positive residues lying in a stripe 2.5 nm in length. Based on circular dichroic measurements, HOLLECKER and LARCHER (1989) came to the conclusion that the dendrotoxin homologues I and K have a BPTI-like backbone conformation and follow essentially similar ways of folding.

C. Binding Studies

Measurements of the binding of the radioiodinated or tritiated K channel toxins apamin, MCDP, DTX, β-BuTX and ChTX to cell membranes gave K_d values ranging between 0.02 and 0.7 nM (see Table 2). In some preparations a second population of receptors with a much higher K_d value

Table 2. K_d and B_{max} values from equilibrium binding experiments with radioiodinated toxins

Toxin	Preparation	K_d (nM)	B_{max} (pmoles/ mg of protein)	Reference
Apamin	Mouse neuroblastoma cell (N1E 115)	0.022	0.012	For original references, see Schmid-Antomarchi et al. (1986)
	Rat skeletal muscle	0.060	0.0035	
	Guinea-pig hepatocyte	0.350	0.001	
	Rat neuron	0.098	0.003–0.008	
	Rat pheochromocytoma cell	0.350	0.600	
Mast cell	Rat synaptosomes	0.15	0.200	Taylor et al. (1984)
degranulating peptide (MCDP)	Rat brain membranes	0.158	0.163	Bidard et al. (1989)
Dendrotoxin (DTX)	Rat synaptosomes	0.4	1.1	Dolly et al. (1984)
	Chick synaptic membrane	0.5 and 15	0.09 and 0.4	Black and Dolly (1986)
	Rat synaptic membrane	0.5	1.3	Black et al. (1988)
	Rat synaptic membrane	0.69	1.8	Sorensen and Blaustein (1989)
	Xenopus nerve	0.022	0.023	Bräu et al. (1990)
	Rat nerve	0.078 and 8.8	0.705 and 6.784	Corrette et al. (1991)
Toxin I	Rat brain membranes	0.041	0.847	Bidard et al. (1989)
β-Bungarotoxin	Rat synaptosomes	0.6	0.15	Othman et al. (1982)
	Chick synaptic membrane	0.47	0.05	Rehm and Betz (1982)
	Rat synaptic membrane	0.7 and 16	0.3 and 1.0	Black et al. (1988)
	Rat central nervous system	0.6–2.1	0.04–0.44	Pelchen-Matthews and Dolly (1988)
Charybdotoxin (ChTX)	Bovine aortic smooth muscle	0.1	0.5	Vázquez et al. (1989)
	Rat brain	0.73	0.076	Schweitz et al. (1989a)
		0.025–0.03	0.3–0.5	Vázquez et al. (1990)

is discernible. As Table 2 shows, the maximum binding capacity B_{max} is generally small (<0.01 pmoles/mg of protein) for apamin (for further data, see HABERMANN 1984) and larger (>0.3 pmoles/mg of protein) for DTX and β-BuTX. Notable exceptions are the apamin receptor-rich rat pheochromocytoma cells and the relatively DTX receptor-poor *Xenopus* nerve fibers.

As an example, the saturable binding of [^{125}I]-DTX to nerve fiber membranes is shown in Fig. 1A. The figure also illustrates the important finding that the K_d value for [^{125}I]-DTX binding (but not the B_{max} value) is much larger in high K$^+$ solution (105 mM) than in Ringer's solution (2.5 mM K), indicating that K$^+$ ions interfere with its binding.

DTX, toxin I, MCDP, β-BuTX, and ChTX bind to interacting (not necessarily identical) sites on the same membrane protein. This is shown by inhibition experiments. Saturable [^{125}I]-DTX binding is inhibited by the DTX homologue toxin I and, less effectively, by MCDP; partial or complete (see below) inhibition is observed with β-BuTX. Conversely, toxin I inhibits the binding of [^{3}H]- or [^{125}I]-β-BuTX and [^{125}I]-MCDP. Inhibition experiments on chick and rat synaptic membrane have been done by OTHMAN et al. (1982), DOLLY et al. (1984), BLACK and DOLLY (1986), BIDARD et al. (1987), STANSFELD et al. (1987), SCHMIDT et al. (1988), and BREEZE and DOLLY (1989). Figure 1B illustrates an inhibition experiment on peripheral nerves. In high K$^+$ solution the inhibition curves are shifted to the right, indicating that K$^+$ ions reduce the affinities of the toxins for their binding sites. For a quantitative analysis, the ligand concentration causing half-maximal inhibition of [^{125}I]-DTX binding (IC$_{50}$) and the apparent equilibrium dissociation constant for the unlabelled toxins (K$_I$) were determined (see legend of Fig. 1). As shown by the K$_I$ values in the legend of Fig. 1, the affinity of *Xenopus* nerve fibers for DTX is 47 times higher than for MCDP.

Inhibition is of a noncompetitive type, compatible with the idea of different but interacting binding sites on the same protein. This was demonstrated by BIDARD et al. (1987), SCHMIDT et al. (1988), and BREEZE and DOLLY (1989). They measured toxin binding (e.g., binding of [^{125}I]-MCDP) in the presence of different concentrations of the inhibitor (e.g., toxin I). In all cases, increasing concentrations of the inhibitor did not change the slope of the Scatchard plot and only affected B_{max} values. Interestingly, the inhibition of [^{125}I]-δ-DaTX binding by α-DaTX is also noncompetitive, indicating that the two toxins from *D.angusticeps* bind at distinct sites (MUNIZ et al. 1990). The distribution of δ-DaTX binding sites in sections of rat brain is less widespread than that seen for α-DaTX and resembles the distribution of MCDP acceptors (AWAN and DOLLY 1991).

BREEZE and DOLLY (1989) discovered that the directly measured binding of β-BuTX and its ability to inhibit the binding of [^{125}I]-DTX are markedly altered by changes in buffer composition. When measured in Krebs/phosphate buffer, the binding appeared monotonic, at least at low

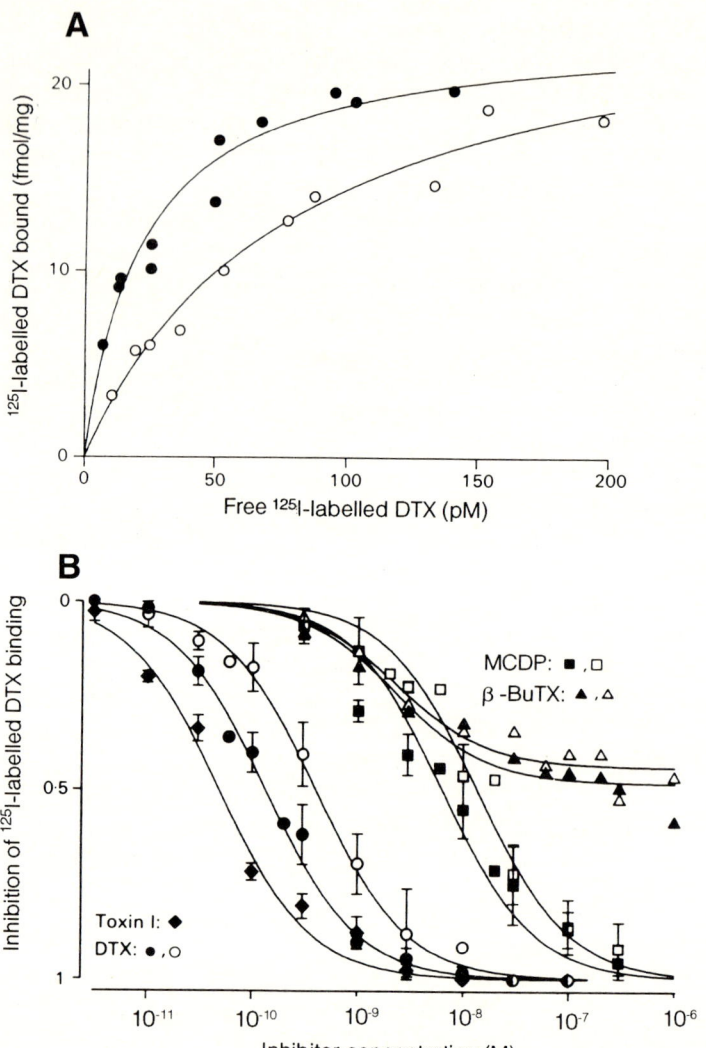

Fig. 1A,B. Labelled dendrotoxin ($[^{125}I]$-DTX) binding to nerve fiber membranes. **A** Saturable binding of $[^{125}I]$-DTX to *Xenopus* nerve fiber membrane as a function of $[^{125}I]$-DTX concentration in Ringer's solution (*filled circles*) and in high K^+ solution (*open circles*); the values are the difference between total and nonspecific binding and are fitted with the function $B = B_{max} L/(L + K_d)$ where B is the amount of bound $[^{125}I]$-DTX, B_{max} the maximum binding capacity, L the concentration of free $[^{125}I]$-DTX, and K_d the equilibrium dissociation constant. **B** Inhibition of $[^{125}I]$-DTX binding to nerve fiber membranes by native toxin I (*diamonds*), DTX (*circles*), mast cell degranulating peptide (*MCDP; squares*), and β-BuTX (*triangles*) in Ringer's solution (*filled symbols*) and in high K^+ solution (*open symbols*). The concentration of labelled DTX was $60 \pm 4\,pM$. Points with error bars are means $\pm$SEM from three experiments; points without error bars are means of two determinations. To estimate the ligand concentration causing half-maximal inhibition of $[^{125}I]$-DTX binding (IC_{50}) the data were fitted with the equation $IB = IB_{max}/(1 + IC_{50}/I)$, where I is the concentrations of $[^{125}I]$-β-BuTX. By contrast, equilibrium binding experiments conducted in imidazole- and HEPES-based buffers yielded distinctly biphasic Scatchard plots, indicating two populations of sites with high and low affinity, respectively. Consistent with this, β-BuTX was considerably more potent an inhibitor of the binding of $[^{125}I]$-DTX in imidazole buffer than in Krebs buffer. In the latter, β-BuTX acted with only very low efficacy (see Fig. 1B). In imidazole medium, however, the efficacy of inhibition was markedly increased.

Binding experiments consistently indicate fewer high-affinity binding sites for β-BuTX and MCDP than for DTX. An autoradiographic analysis of $[^{125}I]$-DTX and $[^{125}I]$-β-BuTX binding to a large number of rat brain areas revealed that all those examined specifically bound higher levels of the former (PELCHEN-MATTHEWS and DOLLY 1989). BREEZE and DOLLY (1989) determined the relative number of binding sites for DTX, β-BuTX, and MCDP in rat brain synaptic membranes. As illustrated in Fig. 2, they postulated two subpopulations of acceptors, A and B. A is 3 times more frequent than B and contains DTX sites, low-affinity β-BuTX sites, and low-affinity MCDP sites. B contains in addition high-affinity β-BuTX sites. This is based on B_{max} values from equilibrium binding experiments, which indicate that the number of DTX binding sites approximately equals the number of low-affinity β-BuTX binding sites but is much larger than that of high-affinity binding sites. This observation was confirmed by SCOTT et al. (1990).

REHM and LAZDUNSKI (1988a) used affinity chromatography of detergent extracts of rat brain to provide direct evidence for the existence of two subtypes of toxin I-binding proteins: those with high affinity for β-BuTX (30%–40% of total) and those with low affinity for β-BuTX (60%–70%). The latter subtype contains most (85%–90%) of the binding sites for MCDP. The two subtypes differ by a factor of 40–50 in their affinity for β-BuTX.

---

inhibitor concentration and IB_{max} is the maximal inhibition ($IB_{max} = 1$, except for β-BuTX where it is 0.48 in Ringer's and 0.44 in high K^+). The apparent equilibrium dissociation constant for the unlabelled toxins (K_I) was calculated from the equation $K_I = IC_{50}/(1 + L/K_d)$, in which L is the concentration of the free $[^{125}I]$-DTX and K_d the equilibrium dissociation constant for labelled DTX. The following values were obtained:

	Ringer's		High K^+	
	$IC_{50}(nM)$	$K_I(nM)$	$IC_{50}(nM)$	$K_I(nM)$
DTX	0.14	0.038	0.43	0.25
MCDP	6.8	1.8	14.5	8.4
Toxin I	0.052	0.014		
β-BuTX	2.4	0.64	2.4	1.4

(From BRÄU et al. 1990)

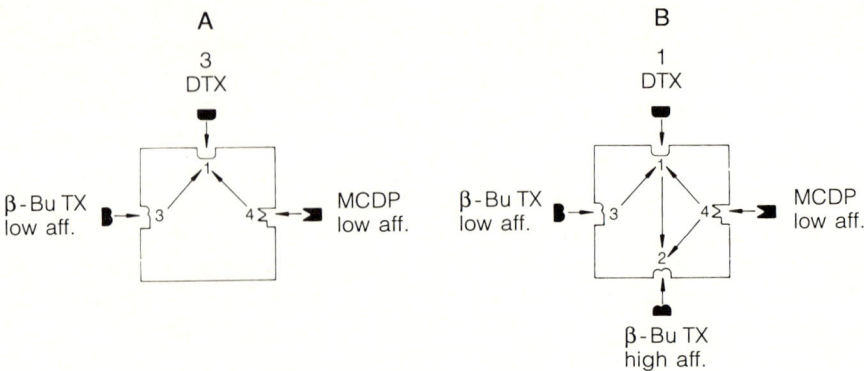

Fig. 2. Model of the dendrotoxin/β-bungarotoxin/mast cell degranulating peptide (*DTX/β-BuTX/MCDP*) binding component from rat brain synaptic membranes. Two subtypes of the acceptor are shown, one of which possesses a high-affinity binding site for β-BuTX in addition to the low-affinity site present on both subtypes; 3:1 represents their molar ratio. *Arrows* between binding sites represent noncompetitive interactions; *arrows* between toxins and binding sites represent competitive interactions. The high-affinity MCDP site is not included because its location and stoichiometry relative to the others remain unclear. (From BREEZE and DOLLY 1989)

In addition to the two subtypes of acceptor postulated by BREEZE and DOLLY (1989) and REHM and LAZDUNSKI (1988a) there exists a third type with a high-affinity MCDP site. BIDARD et al. (1989) found that the number of high-affinity ($K_d = 0.158\,nM$) binding sites for MCDP is about five times lower than the number of high-affinity ($K_d = 0.041\,nM$) binding sites for the DTX homologue toxin I (see Table 2).

The situation is further complicated by the discovery that the DTX/MCDP acceptor possesses ChTX binding sites of high and low affinity. ChTX (which blocks large conductance Ca-activated K channels but also voltage-activated K channels) inhibits the binding of radioiodinated toxin I or MCDP to rat synaptic membrane with an IC_{50} of about $30\,nM$ (SCHWEITZ et al. 1989a,b) or a K_I of $3\,nM$ (HARVEY et al. 1989). Also, in *Xenopus* and rat nerve fibers ChTX displaces [^{125}I]-DTX from its binding sites in nanomolar concentrations (BRÄU et al. 1990; CORRETTE et al. 1991). Conversely, binding of [^{125}I]-ChTX to voltage-dependent K channels in rat brain membranes is completely inhibited by the DTX homologue toxin I and MCDP (SCHWEITZ et al. 1989a). As can be seen for the rat brain in Table 2, there are many fewer high-affinity binding sites for MCDP (163 fmol/mg of protein) and ChTX (76 fmol/mg of protein) than for toxin I (847 fmol/mg of protein). Nevertheless, both MCDP and ChTX inhibit completely toxin I binding at its high-affinity site. To explain this, it is assumed that high-affinity toxin I binding sites are associated not only with high-affinity binding sites for MCDP and ChTX but also with low-affinity binding sites for MCDP and ChTX.

Unfortunately, some doubt has arisen whether the K_d and B_{max} values of SCHWEITZ et al. (1989a) for $[^{125}I]$-ChTX binding to rat brain membranes are correct; VÁZQUEZ et al. (1990) claim that the site density is 5 times higher and the K_d 30 times smaller than estimated. They confirm, however, that $[^{125}I]$-ChTX binding in the brain is completely inhibited by α-DTX. They show furthermore that it is inhibited by NTX, but not by iberiotoxin (a selective inhibitor of Ca-activated K channels, see below) and not by tetraethylammonium (TEA; which inhibits $[^{125}I]$-ChTX binding to Ca-activated K channels, see below). Also, $[^{125}I]$-ChTX binding in brain is stimulated by low Na and K concentrations and inhibited by high ones. There is agreement that the high-affinity ChTX binding sites in brain are associated with voltage-activated (rather than Ca-activated) K channels.

The classic inhibitors of K channels, TEA, 4-aminopyridine, and 3,4-diaminopyridine, do not affect $[^{125}I]$-β-BuTX binding to rat and chick brain membranes (SCHMIDT et al. 1988; BREEZE and DOLLY 1989) nor $[^{125}I]$-DTX binding to rat brain and *Xenopus* nerve fiber membranes (BLACK et al. 1986; BRÄU et al. 1990). Also, apamin is unable to displace $[^{125}I]$-MCDP from its binding sites in rat brain even at $10\,\mu M$ concentrations (TAYLOR et al. 1984). Thus, the two bee venom toxins bind to different receptors.

$[^{125}I]$-apamin binding is inhibited by high concentrations of K^+, as is binding of $[^{125}I]$-DTX (see Fig. 1) and $[^{125}I]$-ChTX (see above). The affinity of apamin for its receptor has an optimum at about $5\,mM$, i.e., in the range of the normal $[K]_0$ (HABERMANN and FISCHER 1979; HUGUES et al. 1982c; SEAGAR et al. 1987). Interestingly, AUGUSTE et al. (1990) observed a similar optimum for the binding of LTX (= scyllatoxin). This minor component from the venom of the scorpion *L. quinquestriatus* inhibits $[^{125}I]$-apamin binding to rat brain synaptosomal membranes ($K_i = 75\,pM$) (CHICCHI et al. 1988; AUGUSTE et al. 1990). According to CHICCHI et al. (1988), it increases the K_d and decreases B_{max} for $[^{125}I]$-apamin binding. In the later study (AUGUSTE et al. 1990), it increased K_d without changing B_{max}, suggesting competitive inhibition. A toxin (LQ-VIII) which CASTLE and STRONG (1986) isolated from *L. quinquestriatus* venom inhibits $[^{125}I]$-apamin binding to guinea-pig hepatocytes and is presumably identical with LTX. LTX has little structural homology to apamin but some homology to other scorpion toxins like ChTX and NTX (Table 1). ChTX also inhibits $[^{125}I]$-apamin binding to rat brain synaptosomal membranes but is 2000-fold less potent than LTX (CHICCHI et al. 1988).

Binding of $[^{125}I]$-ChTX to Ca-activated K channels has been studied by VÁZQUEZ et al. (1989) on sarcolemmal membrane vesicles derived from bovine aortic smooth muscle, a particularly rich source of high conductance Ca-activated K channels (see Table 2). Binding of $[^{125}I]$-ChTX to this preparation was inhibited by mono- and divalent cations and TEA (see Fig. 3B). IbTX, a minor component of the venom of the scorpion *Buthus tamulus*, is a noncompetitive inhibitor of $[^{125}I]$-ChTX binding (GALVEZ et al. 1990).

D. Cross-Linking Experiments

Cross-linking experiments with electrophoretic analysis of the cross-linked membrane proteins yield the molecular weight of the binding peptides. As emphasized by SCHMIDT and BETZ (1989), molecular weight estimates after cross-linking are not very reliable. Nevertheless, it seems safe to conclude that the binding peptides for DTX, MCDP, and β-BuTX consist of larger and smaller ones with molecular weights of 65–80 and 28–35 kDa, respectively (Table 3). This is consistent with the data obtained from gel analysis of the solubilized and purified binding protein (see below). It is tempting to speculate that the larger peptide corresponds to the 55–77 kDa proteins identified by cDNA cloning and/or purification of *Drosophila* A-type and mammalian brain K channels (see below). The function of the smaller polypeptides is unclear.

Cross-linking experiments with [^{125}I]-ChTX on rat brain identified a polypeptide component of $M_r = 79$, very similar to $M_r = 76–77$ for the toxin I and MCDP binding subunits (see Table 3). Cross-linking of [^{125}I]-toxin I to its target was prevented by ChTX, consistent with the view that the α-subunit of the DTX/MCDP-sensitive, voltage-dependent K channel also contains ChTX binding sites (see above). Unfortunately, VÁZQUEZ et al. (1990) were not able to confirm this observation.

Cross-linking studies with [^{125}I]-ChTX on sarcolemmal membrane vesicles derived from bovine aortic smooth muscle showed that ChTX binds to a 35-kDa protein, presumably a subunit of the high conductance Ca-activated K channel (GARCIA-CALVO et al. 1991).

Affinity labeling of the apamin receptor with a variety of cross-linking reagents has identified receptor-associated polypeptides of 22, 30 or 33, 45, 59, and 86 kDa (Table 3). According to LEVEQUE et al. (1990), the 45- and 59-kDa components result from cleavage of the 86-kDa polypeptide. Thus, the apamin binding site, like the binding sites for DTX, MCDP, and β-BuTX (see above), consists of a larger and a smaller peptide ($M_r = 86$ and 30, respectively). Consistent with this view, the M_r of the whole apamin-channel complex has been estimated as 84–115 by radiation-inactivation technique (SEAGAR et al. 1986).

In cross-linking experiments with LTX, an inhibitor of apamin binding (see above), two polypeptides of 27 and 57 kDa were identified (Table 3).

E. Solubilized and Purified Binding Protein

The low number of apamin binding sites renders the purification of the apamin receptor difficult. Nevertheless, SEAGAR et al. (1987) succeeded in solubilizing the [^{125}I]-apamin receptor of rat brain membranes. Under optimum conditions, the receptor retained high affinity for [^{125}I]-apamin with $K_d = 40 \, pM$ and a binding capacity of 0.017 pmoles/mg protein.

Table 3. Molecular weights (M_r) of the binding peptides estimated from cross-linking experiments (C) and purification studies (P)

Toxin	Method	Preparation	M_r	Reference
Apamin	C	Cultured neurons and rat brain	22, 33, 59, 86	SEAGAR et al. (1985, 1986)
	C	Rat brain and pheochromocytoma cells	30, 45, 58, 86	AUGUSTE et al. (1989)
Leiurotoxin	C	Rat brain	27, 57	AUGUSTE et al. (1990)
Mast cell degranulating peptide	C	Rat brain	77	REHM et al. (1988)
Charybdotoxin	C	Rat brain	79	SCHWEITZ et al. (1989a)
	C, P	Bovine smooth muscle	35	GARCIA-CALVO et al. (1991)
Noxiustoxin	C, P	Squid axon	60, 160, 220	PRESTIPINO et al. (1989)
α-Dendrotoxin	C	Rat brain	65	MEHRABAN et al. (1984)
	C	Rat brain	65	SORENSEN and BLAUSTEIN (1989)
	C	Chick brain	75, 69	BLACK and DOLLY (1986)
	C	Guinea-pig brain	69	MUNIZ et al. (1990)
	C, P	Bovine brain	78, 39	SCOTT et al. (1990)
δ-Dendrotoxin	C	Guinea-pig brain	82, 69	MUNIZ et al. (1990)
Toxin I	C	Rat brain	76	REHM et al. (1988)
	P	Rat brain	76–80, 38, 35	REHM and LAZDUNSKI (1988b)
	P	Bovine brain	75, 37	PARCEJ and DOLLY (1989)
β-Bungarotoxin	C	Chick brain	95	REHM and BETZ (1983)
	C	Chick brain	75, 28	SCHMIDT and BETZ (1989)

The common binding protein which binds DTX, toxin I, MCDP, and β-BuTX has been solubilized and purified in the laboratories of H. BETZ, J.O. DOLLY, M. LAZDUNSKI, and H. REHM. REHM and BETZ (1984) reported the solubilization and characterization of the β-BuTX-binding protein from chick brain membranes. They estimated the molecular weight of the solubilized protein part as 431 kDa. In support of its presumed K channel nature, the protein exhibited a marked K^+ dependence for conformational stability upon solubilization. As shown by Schmidt and Betz (1988), binding of $[^{125}I]$-β-BuTX to the solubilized and partially purified binding protein is inhibited by toxin I and MCDP, as it is in binding studies on membrane fractions. The affinity for all three toxins is unaltered or even increased upon solubilization and purification. The observation that toxin I- and MCDP-binding sites copurify with the β-BuTX-binding protein supports the notion that binding of these toxins occurs at allosteric sites of the same membrane protein. Rehm and Lazdunski (1988b) described the complete purification of the toxin I/MCDP-binding protein from rat brain. It has a molecular weight of 450 kDa and, according to chromatography, is a mixture of polypeptide chains of 76–80, 38, and 35 kDa (see Table 3). A combination of, for instance, four copies of the 76–80-kDa subunit and multiple copies of the 38-kDa polypeptide would result in a protein complex of the order of 450 kDa.

In the laboratory of J.O. Dolly, the acceptor for DTX and β-BuTX was extracted from rat cortex by Mehraban et al. (1985). Black et al. (1988) estimated the molecular weight of the solubilized acceptors as 405–465 kDa with a Stokes radius of 8.6 nm. Parcej and Dolly (1989) described the detergent extraction and complete purification of the toxin I-acceptor from bovine brain; according to affinity chromatography it contains two major subunits with M_r values of 75 and 37 (see Table 3). For the α-DTX acceptor of bovine brain two very similar M_r values (78 and 39, see Table 3) were obtained (Scott et al. 1990).

Sorensen and Blaustein (1989) solubilized the rat brain receptor for α-DaTX and β-DaTX, two toxins from the venom of *D. angusticeps* (see above). The molecular weight of the solubilized α-DaTX receptor was estimated to be about 270 kDa by sucrose density gradient centrifugation, similar to earlier estimates from radiation inactivation experiments (240–265 kDa , see Dolly et al. 1984) but much smaller than the values (405–465 kDa) estimated by Black et al. (1988).

The toxin-binding peptides are sialated and N-glycosylated glycoproteins (see arrowheads in Fig. 6A). Treatment with neuraminidase and glycopeptidase F reduces the molecular mass of the larger toxin-binding subunit (Rehm 1989; Rehm et al. 1989a). The reduction by neuraminidase is from 90 to 70 kDa in rat toxin-binding peptide and from 83 to 66 kDa in bovine toxin-binding peptide. Additional treatment with glycopeptidase F produces a further reduction of the M_r of the rat toxin-binding peptide (to 65) but has no effect on the bovine toxin-binding peptide. The M_r value of 65 for the deglycosylated larger subunit of the α-DTX acceptor of bovine brain was confirmed by Scott et al. (1990).

Conclusive evidence for the toxin-acceptor being a K channel constituent requires reconstitution of the purified protein into lipid membranes and demonstration of its electrical activity. Prestipino et al. (1989) isolated NTX-binding proteins from a detergent extract of squid axon membranes by affinity column (for M_r values see Table 3) and reconstituted them into planar lipid bilayers. A main conductance of 11 pS (in 300/100 mM KCl) and two larger conductances (22 and 32 pS) were observed, similar to the single-channel conductances of 10, 20, and 40 pS reported for K channels of squid axon by LLANO et al. (1988). Reconstitution of the purified DTX-binding protein was achieved by REHM et al. (1989b) and resulted in multiple conductance levels of elementary current. The most frequently observed conductance (>65% of all openings) was 21 pS (with symmetrical 150 mM K). The activity of the reconstituted channel was greatly enhanced by both cAMP-dependent and endogenous phosphorylation of the protein on its cytoplasmic side. The activity of the phosphorylated 21 pS channel was drastically reduced by the DTX homologue toxin I.

Further evidence for the toxin-binding protein being a K channel constituent comes from the observation that the binding protein for DTX, MCDP, and β-BuTX has a strong immunological relationship to a mouse homologue (MBK1, see below) of the *Drosophila* A channel (REHM et al. 1989a). Antibodies raised against two synthetic peptides from the N-terminal and the C-terminal end of the MBK1 K channel recognize the toxin-binding subunit of the DTX-binding protein from rat and bovine brain.

Final proof that the α-DTX acceptor is synonymous with a K channel protein was obtained by microsequencing the N-terminal of the 78-kDa subunit from bovine brain (SCOTT et al. 1990): The first 27 amino acids were identical (except for one conserved substitution) to the N-terminal sequence deduced from cDNA of RCK5, a K channel protein from rat brain known to be susceptible to α-DTX (see below and Table 4). Also, the M_r of the 78-kDa subunit (65 after deglycosylation) is similar to that of the RCK5 protein (57, see Table 4).

Reconstitution of the purified ChTX receptor into liposomes which are then incorporated into artificial phospholipid bilayers has been reported in abstract form (GARCIA-KALVO et al. 1991).

F. Electrophysiological Experiments

I. Apamin

The bee venom toxin apamin blocks Ca-activated K channels of low conductance in neuroblastoma cells (HUGUES et al. 1982a), skeletal muscle (HUGUES et al. 1982b), and hepatocytes (BANKS et al. 1979; CAPIOD and OGDEN 1989; for review, see MARTY 1989). In nerve cells, the apamin-sensitive Ca-activated K current, $I_{K(Ca)}$, is responsible for long-lasting after-hyperpolarizations that follow action potentials. In hepatocytes, $I_{K(Ca)}$ is induced by hormone action. Figure 3A shows single-channel currents of a hepatocyte activated by 1.5 μM internal Ca and the reversible inhibition by a very brief apamin application. With longer application times, the block becomes irreversible. The unitary conductance of 20 pS found for the channel in guinea-pig hepatocytes at high (150 mM) K concentrations is similar to

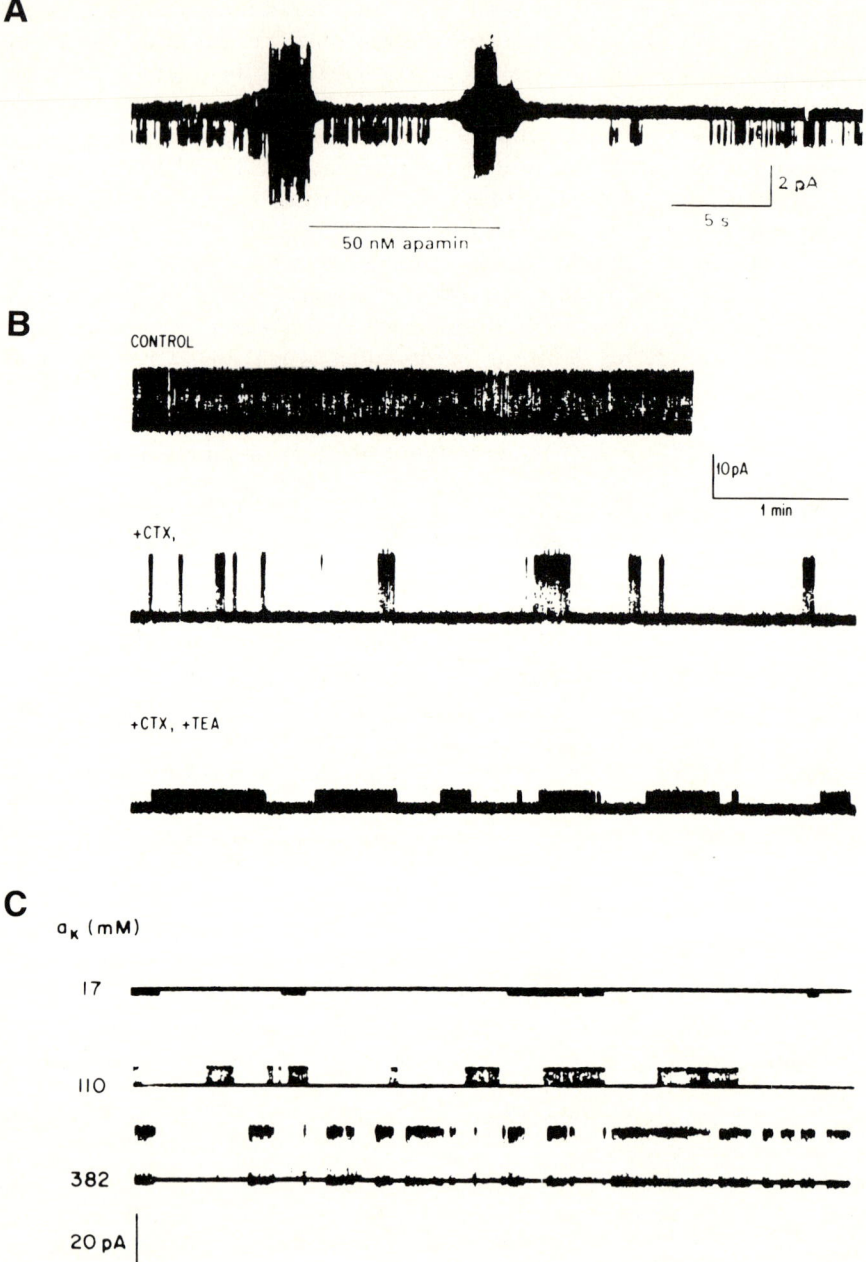

Fig. 3A–C. Effects of apamin (**A**) and charybdotoxin (*CTX*, **B**, **C**) on single-channel currents. **A** Outside-out patch from a guinea-pig hepatocyte showing reversible block of Ca-activated K channels by a brief exposure to apamin. Note large tap-turning artifacts. (From CAPIOD and OGDEN 1989). **B**, **C**, Single Ca-activated K channel from rat muscle inserted into a phospholipid bilayer. **B** Records

the value of 18 pS reported for the apamin-sensitive Ca-activated K channel of rat myotubes with symmetrical K solutions (BLATZ and MAGLEBY 1986). A larger unitary conductance of 42 pS in symmetrical KCl solutions has been reported for the apamin-sensitive $I_{K(Ca)}$ of rabbit proximal tubule cells (MEROT et al. 1989).

Apamin does not exclusively act on K channels. It inhibits the electrogenic effect of the Na/K pump and the activity of the Na/K-ATPase at nanomolar concentrations (ZEMKOVÁ et al. 1988). In frog cardiac fibers, apamin in a concentration of 10^{-9}–$10^{-8} M$ depresses the slow Ca inward current by 40%–50% but has no effect at all on the Na and K currents (FILIPPOV et al. 1988). Likewise, in cultured chick embryonic heart muscles, apamin was reported to be a specific Ca channel blocker (BKAILY et al. 1985).

II. Charybdotoxin, Leiurotoxin, and Iberiotoxin

ChTX, a minor component of the venom of the scorpion *Leiurus quinquestriatus*, was first shown to block Ca-activated K channels of high conductance (~200 pS) from rat skeletal muscle at nanomolar concentrations (MILLER et al. 1985). Venoms of other Old World scorpions (*Buthus occitanus*, *Buthacus arenicola*) and a New World scorpion (*Centruroides sculpturatus*) also contain ChTX-like activity (MOCZYDLOWSKI et al. 1988). Its effect was studied in detail by ANDERSON et al. (1988), MILLER (1988, 1990), and MACKINNON and MILLER (1988), using high-conductance Ca-activated K channels from skeletal muscle plasma membranes reconstituted into planar lipid bilayers. Figure 3B illustrates the block by ChTX and the competition between ChTX and TEA. Under control conditions the channel is active almost all the time; because of the slow time scale individual opening and closing events cannot be discerned. Addition of ChTX to the external solution induces the appearance of long-lived (~15 s) nonconducting intervals (see VÁZQUEZ et al. 1989 for similar records from bovine aortic smooth muscle). Each of them represents a randomly distributed interval of time during which a single ChTX molecule is bound to the single channel under observation. Addition of 0.6 mM TEA shortens the nonconducting intervals, i.e., weakens the ChTX effect, suggesting competition between ChTX and TEA. In addition TEA reduces the current size, as expected for

---

at a potential of 35 mV in control, in the presence of 30 nM CTX and in the presence of 30 nM CTX + 0.6 mM tetraethylammonium (*TEA*). CTX and TEA are added to the external side of the membrane. Note the long nonconducting intervals in 30 nM CTX, their shortening by TEA, and the reduction in current size by TEA. (From MILLER 1988). **C** Records at a potential of 30 mV. The solution on the external side of the membrane contained 100 mM KCl and 5 nM CTX, while KCl activity on the inside was increased from 17 to 110 to 382 mM. As internal K concentration rises, the blocked intervals become shorter. i.e., CTX becomes a less effective blocker; also, the K current increases in size. (From MACKINNON and MILLER 1988)

a rapidly (on the time scale of microseconds) reversible blocking reaction. Further experiments showed that both the open and the closed states of the channel bind the toxin, that dissociation of ChTX from the channel is enhanced by depolarization, and that the toxin association rate is strongly dependent on the ionic strength of the external medium. The rate-determining step for ChTX binding and unbinding is diffusion of toxin up to or away from the channel, as shown by the strictly inverse relation between solution viscosity and the rate constants of both association and dissociation of ChTX with the channel mouth. Presumably, ChTX physically plugs the pore of the Ca-activated K channel by binding to its external mouth. Interestingly, K^+ and Rb^+ ions (but not Na^+ and Cs^+) relieve toxin block when added to the opposite (i.e., internal) side of the membrane (Fig. 3C). This is due to an increase in the ChTX dissociation rate, while the association rate remains unchanged.

The channel has several carboxyl groups close to its externally facing mouth which electrostatically attract ChTX with its net charge of about +5. Consequently, the affinity for ChTX can be reduced in two ways: By replacing 1 of the 8 positively charged residues (in particular lysine 27) in the ChTX molecule by the neutral glutamine (Park and Miller 1991) or by treating the channels with the carboxyl group modifying reagent trimethyloxonium (MacKinnon and Miller 1989a; MacKinnon et al. 1989). In addition to the carboxyl groups, a negatively charged glutamate residue is located close to the ChTX binding site (see below).

ChTX blocks also Ca-activated K channels of smaller conductance: 127 pS channels in cultured kidney cells (Guggino et al. 1987), 35 pS channels in *Aplysia* neurons (Hermann and Erxleben 1987), and 25 pS channels in human erythrocytes (Castle and Strong 1986; Beech et al. 1987). [The toxin used on erythrocytes was originally called LQ-X/2 and later shown to be identical with ChTX, see Strong et al. (1989).] By incorporating rat brain plasma membrane vesicles into planar lipid bilayers, Farley and Rudy (1988) and Reinhart et al. (1989) found and characterized different types of Ca-activated K channels. Farley and Rudy (1988) described two types, the large type I (200–250 pS in symmetrical KCl solutions) and the intermediate-sized type II (65–145 pS) which are both blocked by ChTX. Reinhart et al. (1989) described four types with unitary conductances of 242, 236, 135, and 76 pS in symmetrical 150 mM KCl buffers. They are all activated by depolarizing voltages and micromolar concentrations of internal Ca. Three of the four channels are blocked by nanomolar concentrations of ChTX. One of the high-conductance channels is novel in that it is not blocked by ChTX and exhibits long closed and open times. Presumably, this channel differs from the other three in the structure of the external mouth. The various Ca-activated K channels of the rat brain give rise to the fast Ca-activated after-hyperpolarization which in rat hippocampal slices is blocked by 25 nM ChTX (Lancaster and Nicoll 1987). In human red cells, ChTX inhibits partially (72%) but with high affinity ($IC_{50} = 0.88 nM$) the Ca-activated K efflux into K-free media (Wolff et al. 1988).

Recent studies have shown that ChTX is not selective for Ca-dependent K channels. CHRISTIE et al. (1989) found a slowly inactivating rat brain K channel expressed in oocytes sensitive to ChTX (EC_{50} = 5.5 nM). SANDS et al. (1989) reported that ChTX blocks some voltage-dependent K channels (which are not activated by intracellular Ca^{2+} ions) in human and murine T lymphocytes with a K_d value between 0.5 and 1.5 nM. ChTX also inhibits proliferation and interleukin-2 production in human lymphocytes, effects probably mediated by the blockade of the voltage-gated K channels (PRICE et al. 1989). GRINSTEIN and SMITH (1990) describe two ChTX-sensitive types of K conductance in human lymphocytes, a Ca-activated and a Ca-independent. In human platelets, too, ChTX inhibits voltage-gated K channels (MAHAUT-SMITH et al. 1990). In rat dorsal root ganglion cells, ChTX blocks a voltage-activated, non-Ca-activated, DTX-sensitive K current ($IC_{50} \sim$ 30 nM) (SCHWEITZ et al. 1989b), consistent with the observation that ChTX inhibits binding of the DTX homologue toxin I (see above). Likewise, in *Xenopus* nerve fibers, ChTX blocks the K current component I_{Kf1} which is specificly inhibited by DTX (see below). In rat brain synaptosomes, ChTX inhibits the ^{86}Rb efflux through Ca-activated K channels ($IC_{50} \sim$ 15 nM) and through Ca-independent voltage-gated K channels ($IC_{50} \sim$ 40 nM) (SCHNEIDER et al. 1989), whereas the delayed rectifier channel which FRECH et al. (1989) isolated from rat brain by expression cloning is insensitive to ChTX (up to 1 μM). Differences in ChTX sensitivity may arise from different processing of the primary translational products: ChTX blocks neither native nor transformant A-current channels in *Drosophila* muscle (see MACKINNON et al. 1988; ZAGOTTA et al. 1989a,b) but blocks the transient, Ca-insensitive A-current produced by injection of *Drosophila* Shaker gene transcripts into *Xenopus* oocytes (MACKINNON et al. 1988; ZAGOTTA et al. 1989a) or mammalian cells (LEONARD et al. 1989). A possible explanation is a difference in post-translational processing (e.g., glycosylation) between *Drosophila* and vertebrate cells.

ChTX is not the only K channel toxin present in *L. quinquestriatus* venom. LTX has been mentioned above as an inhibitor of apamin binding. It blocks the long-lasting after-hyperpolarization which follows the action potential of rat myotubes, as does apamin (AUGUSTE et al. 1990). CASTLE and STRONG (1986) report block of apamin-sensitive K fluxes in guinea-pig hepatocytes by LTX (originally named LQ-VIII).

While ChTX is no longer considered selective for Ca-activated K channels, a new toxin, named IbTX, from the venom of the scorpion *Buthus tamulus* is reported to be a potent, selective, and reversible blocker of the large-conductance, Ca-activated K channel in vascular smooth muscle (GALVEZ et al. 1990). Block of channel activity appears markedly different from that of ChTX since IbTX decreases both the probability of channel opening as well as the channel mean open time. This (together with the fact that it is a noncompetitive inhibitor of [^{125}I]-ChTX binding, see above) suggests that it interacts with the channel at a site distinct from the ChTX receptor. It is a selective inhibitor of Ca-activated K channels and does not affect the voltage-gated K channels which are sensitive to ChTX.

III. Noxiustoxin and Other Scorpion Toxins

NTX, fraction II-11 of the venom of the scorpion *Centruroides noxius*, blocks the K current of the squid giant axon without affecting the Na current (CARBONE et al. 1982, 1987). Outward K currents are reduced to 50% at $0.5\,\mu M$ and to about 25% at $3\,\mu M$. At concentrations greater than $1.5\,\mu M$ the block becomes voltage- and frequency-dependent: It increases with increasing negativity of the holding potential and decreases during repetitive pulsing, especially when large pulses are used. Inhibition of K channels by NTX is never complete, suggesting that either the toxin is unable to block the channel fully or that only a fraction of the channels possess a receptor site for the toxin. As also observed with other K channel blockers, the aminopyridines (YEH et al. 1976; MEVES and PICHON 1977), and barium (ÅRHEM 1980; ARMSTRONG et al. 1982), the blocking action of NTX is reduced at high external [K].

NTX has also been studied in other preparations. It blocks voltage-dependent K channels of T lymphocytes with high affinity ($K_d = 0.2\,nM$) (SANDS et al. 1989). A much higher concentration ($0.2\,\mu M$) is required to block the efflux of ^{86}Rb from mouse brain synaptosomes (SITGES et al. 1986). ChTX-sensitive, Ca-activated K channels from skeletal muscle, incorporated into lipid bilayers, are blocked by NTX with a K_d of $450\,nM$ (VALDIVIA et al. 1988). The latter observation demonstrates that the (external) toxin binding site of Ca-activated and non-Ca-activated K channels is similar, a conclusion supported by the finding that ChTX also acts on both types of channels (see above).

NTX is not the only scorpion toxin which acts exclusively on K channels. A similar toxin has been isolated from the venom of the Brazilian scorpion *Tityus serrulatus* (CARBONE et al. 1983). The venom of the scorpion *Pandinus imperator* selectively reduces the K currents of frog myelinated nerve fibers without affecting the Na currents (PAPPONE and CAHALAN 1987). It also blocks voltage-gated potassium channels in cultured clonal rat anterior pituitary cells (GH_3 cells), but in this preparation the Na current is also somewhat reduced (PAPPONE and LUCERO 1988). *Pandinus* venom differs from other known K channel scorpion toxins in that its effects are highly voltage-dependent and irreversible. Block of the delayed rectifier channels is greater at negative potentials than at positive ones; even high concentrations of venom fail to block K currents completely at positive potentials. The toxin which is responsible for the effect on K channels has not yet been isolated from the crude venom; it can, however, be estimated that it is effective in the nanomolar range of concentrations ($K_d < 200\,nM$).

IV. Dendrotoxin, Mast Cell Degranulating Peptide, β-Bungarotoxin

DOLLY et al. (1984) and HALLIWELL et al. (1986) showed that DTX (50–300 nM) and its more toxic homologue toxin I (10–150 nM) inhibit the A current, a transient voltage-dependent K current, in hippocampal neurons of rat or guinea-pig. The effect was selective, i.e., other currents (e.g., $I_{K(Ca)}$, I_M) were not affected. Interestingly, the A current of neurons from the nodose or superior cervical ganglion of the rat is not affected by DTX or toxin I (STANSFELD et al. 1987), an illustration of the enormous heterogeneity of K channels. In these cells, instead, DTX blocks a slowly inactivating K outward current, probably part of the delayed rectifier. Similarly, DTX selectively inhibits a portion of the classic non-inactivating, delayed outward current in dorsal root ganglion cells (PENNER et al. 1986); most (but not all) of the DTX-sensitive current is also abolished by TEA (30 mM). A complete block of the delayed rectifier by externally or internally applied DTX (with a K_d of 176 and 150 nM, respectively) is observed in *Myxicola* giant axons (SCHAUF 1987); interestingly, in this preparation DTX also significantly slows Na current inactivation. DREYER (1990) emphasizes that DTX acts only on neuronal K channels and gives a list of cell preparations in which DTX is ineffective.

In the frog node of Ranvier, DTX doses as low as 0.1 nM suppress the K conductance considerably, whereas Na currents are not affected even at concentrations as high as 850 nM (WELLER et al. 1985). BENOIT and DUBOIS (1986) made the important discovery that toxin I specifically blocks one component of the K current (I_{Kf1}) in the frog node of Ranvier with high affinity ($K_d = 0.4$ nM) and has a much smaller affinity for the components I_{Kf2} ($K_d = 0.4$ μM) and I_{Ks} ($K_d = 1.2$ μM) and no effect on the Na current. Their observation is illustrated in Fig. 4. More recently, BRÄU et al. (1990) found with DTX, also in the frog node of Ranvier, $K_d = 68$ nM for I_{Kf1} and $K_d = 1.8$ μM for I_{Kf2}, i.e., a difference of almost two orders of magnitude, and concluded that DTX can be regarded as a "rather specific blocker of the component f_1". In rat nodes of Ranvier, by contrast, DTX blocks about 50% of the f_2 component with the same K_d as for f_1 (CORRETTE et al. 1991).

As can be expected from the binding studies, MCDP and β-BuTX have the same electrophysiological effects as DTX and toxin I. This was shown for MCDP on neurons of the nodose ganglion (STANSFELD et al. 1987), for β-BuTX on dorsal root ganglion cells (PETERSEN et al. 1986), and for MCDP and β-BuTX on the frog node of Ranvier (BRÄU et al. 1990) (see Fig. 5). STANSFELD et al. (1987) emphasized the fact that the outward current suppression by MCDP and DTX is nonadditive, indicating that the two toxins act on the same conductance. In all studies on peripheral nerves and ganglia, MCDP and β-BuTX were less effective in blocking the K current than DTX and toxin I. In the nodose ganglion, half-maximal inhibition occurred at 2.1 nM DTX and 37 nM MCDP, giving a ratio of 18. For the component f_1

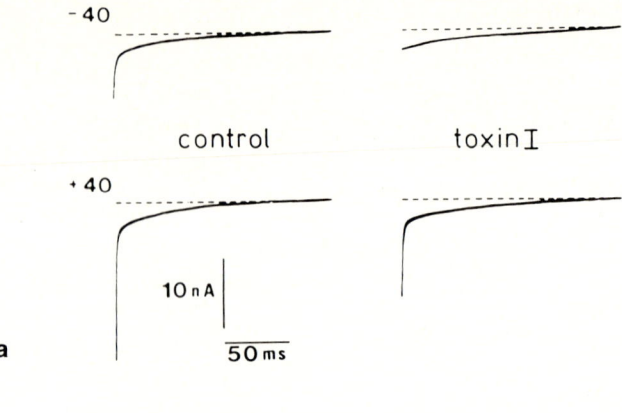

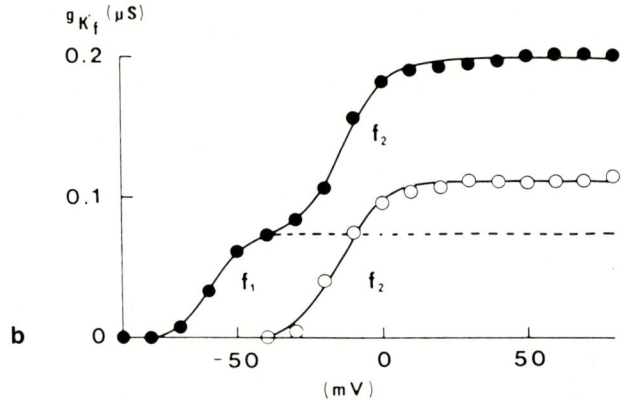

Fig. 4a,b. Effects of toxin I on the fast and slow K conductances of the frog node of Ranvier. **a** Tail currents upon repolarization to -90 mV (following a 100-ms pulse to 40 mV) in high K solution before and after adding 5.3 nM toxin I to the external solution. The tail current consists of a fast and a slow component, and only the former is decreased by toxin I. **b** Fast conductance vs. pulse potential in the absence (*filled circles*) and in the presence (*open circles*) of toxin I. The fast conductance, g_{kf}, was calculated from the fast current (= total instantaneous current minus extrapolated instantaneous slow current). In control, g_{kf} consists of two components, g_{kf1} (turning on between -80 and -30 mV) and g_{kf2} (activating between -40 and $+30$ mV). In toxin I, only g_{kf2} remains. The curves are fits of the equation:

$$g_{kf} = g_{kf1} + g_{kf2}$$
$$\text{with } g_{kf1} = g_{kf1max}/(1 + \exp((\overline{V}_{f1} - V)/k_{f1}))$$
$$\text{and } g_{kf2} = g_{kf2max}/(1 + \exp((\overline{V}_{f2} - V)/k_{f2}))$$

where g_{kfmax} is the maximum conductance, $\overline{V}$ is the potential for half-maximum conductance, and k is the steepness factor. The parameters are:

	g_{kf1max}	$\overline{V}_{f1}$	k_{f1}	g_{kf2max}	$\overline{V}_{f2}$	k_{f2}
Control	0.075 μS	-59 mV	5.2 mV	0.124 μS	-14 mV	6.8 mV
Toxin I	0	–	–	0.111 μS	-15 mV	7.1 mV

(From BENOIT and DUBOIS 1986)

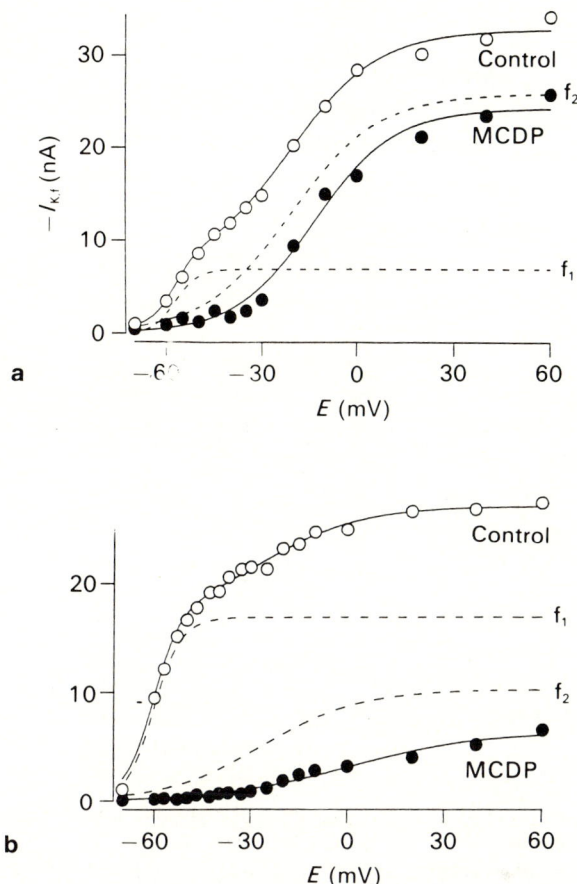

Fig. 5a,b. Voltage-dependence of the fast component of tail current, I_{kf}, in a sensory (**a**) and a motor (**b**) fiber before (*open circles*) and after (*filled circles*) application of $1\,\mu M$ mast cell degranulating peptide (*MCDP*). Tail currents were recorded in high K^+ solution at $-110\,mV$ (**a**) and at $-90\,mV$ (**b**) after 100-ms activating pulses to various potentials (E). The component I_{kf1} is much larger in motor fibers (**b**) than in sensory fibers (**a**). The points were fitted by an equation corresponding to the Eqn. in the legend of Fig. 4 with the following parameters:

	I_{kf1max}	$\bar{V}_{f1}$	k_{f1}	I_{kf2max}	$\bar{V}_{f2}$	k_{f2}
Sensory fiber:						
Control	6.9 nA	−56.5 mV	3.4 mV	26.0 nA	−20.2 mV	13.9 mV
MCDP	0	−	−	24.3 nA	−12.6 mV	12.7 mV
Motor fiber:						
Control	16.9 nA	−59.6 mV	4.2 mV	10.2 nA	−25.6 mV	15.2 mV
MCDP	0	−	−	6.3 nA	1.6 mV	20.3 mV

(From Bräu et al. 1990)

of the node of Ranvier, IC_{50} was 68 nM for DTX and 99 nM for MCDP; for component f_2 the corresponding values were 1.8 µM and 7.6 µM. β-BuTX was considerably less potent than MCDP on the node of Ranvier.

A comparison of [^{125}I]-DTX binding and electrophysiological DTX effect *on the same preparation* was made by Bräu et al. (1990) and Corrette et al. (1991), the former on *Xenopus* nerve, the latter on rat nerve. Corrette et al. (1991) estimated IC_{50} = 8 nM for the block of the total fast K current in 160 mM KCl solution, in satisfactory agreement with the K_d of the low-affinity binding site, which is 14.6 nM in 160 mM KCl (and 8.8 nM in Ringer's, see Table 2). The high-affinity binding site (K_d = 0.166 nM in 160 mM KCl and 0.078 nM in Ringer's, see Table 2) could be localized on the thin or medium-sized fibers (which are not used for electrophysiological experiments) or on structures which only become accessible after homogenization. In *Xenopus* nerve, Bräu et al. (1990) found only a high-affinity binding site, with K_d = 0.081 nM in 105 mM KCl (and 0.022 nM in Ringer's, see Table 2). They did not find a low-affinity binding site which could be connected with the DTX block of the fast K current (IC_{50} = 107 and 11 nM in 105 mM KCl and in Ringer's, respectively), presumably because the density of this site is too low.

Jonas et al. (1989) succeeded in recording single-channel currents from Na and K channels in the nodal and paranodal membrane of a frog nerve fiber demyelinated by treatment with enzymes. They found three types of K channels, called I (intermediate), F (fast), and S (slow) according to their deactivation kinetics. The I channel is the most frequent type. It has a single-channel conductance of 23 pS (with 105 mM K on both sides of the membrane), activates between −60 and −30 mV (similar to I_{Kf1} in Figs. 4 and 5), deactivates more slowly than I_{Kf1}, and is completely blocked by DTX (50–500 nM). The F and S channels are not affected by DTX.

The single-channel conductance of the I channel (23 pS) is remarkably similar to the main conductance level of the reconstituted DTX-binding protein (21 pS). It is also remarkably similar to the DTX-sensitive low-conductance channel of dorsal root ganglion cells (Stansfeld and Feltz 1988) and to the delayed rectifier channel expressed in *Xenopus* oocytes after injection of rat brain RCK1 cRNA (Stühmer et al. 1988; see below). The former has a conductance of 18–33 pS (in symmetrical K concentrations), the latter has multiple conductance states with a main one at 22 pS (with 100 mM K in the extracellular fluid).

Central K channels (as opposed to K channels in peripheral nerves and ganglia) can be studied by measuring the different components of the ^{86}Rb efflux from rat brain synaptosomes. With this method Benishin et al. (1988) were able to show that the four toxins from the venom of *D. angusticeps* (α-, β-, γ-, and δ-DaTX, see above) act on different types of K channels: α- and δ-DaTX preferentially block an inactivating channel, β- and γ-DaTX preferentially block a noninactivating channel. The noninactivating (α-DaTX-insensitive) channel is selectively blocked by β-BuTX (Benishin 1990).

Thus, there exist central K channels which have a higher sensitivity for β-BuTX than for α-DaTX, a behavior not observed on peripheral K channels. As further shown by BENISHIN (1990), the channel blocking activity is associated only with the B-subunit of β-BuTX and not with the A-subunit (see Table 1). The B-subunit is 10 times more effective in blocking the noninactivating channel than the native toxin and, at higher concentrations, also inhibits the inactivating channel.

Recently, LUCCHESI and MOCZYDLOWSKI (1990) reported an effect of toxin I on maxi Ca-activated K channels from rat skeletal muscle. Intracellular application of toxin I reversibly induces a long-lived, inwardly rectifying, subconductance state. Its dwell time corresponds to the actual time that a single toxin I molecule occupies a particular site at the internal mouth of the channel. Thus, there is a shared sensitivity not only to ChTX but also to DTX between voltage-gated and Ca-activated K channels.

V. Other Toxins

The most potent K^+ channel blockers so far described are phalloidin and antamanide, two cyclic peptide toxins from the death-cup mushroom, *Amanita phalloides*, which reversibly block part ($\sim$50%) of the fast K outward current of frog skeletal muscle fibers with an ED_{50} of $2 \cdot 10^{-12}$ and 10^{-11} M, respectively (COGNARD et al. 1986; RAYMOND et al. 1987). However, delayed rectifier K^+ currents in rat alveolar epithelial cells (which can be blocked by nanomolar concentrations of ChTX) are not affected by $10\,\mu M$ phalloidin (JACOBS and DECOURSEY 1990).

Chiriquitoxin, a naturally occurring analogue of TTX from the skin and eggs of the Costa Rican frog *Atelopus chiriquensis*, has been reported to affect both Na and K channels of the frog muscle fiber at nanomolar concentrations (KAO et al. 1981; YANG and KAO 1990). The same seems to occur with TTX in a small proportion of fibers (C.Y. KAO, personal communication).

The venom of the marine snail *Conus striatus* contains toxins which affect the size and kinetics of the delayed rectifier K currents in *Aplysia* neurons (CHESNUT et al. 1987).

G. The Presumed Site of Action

Knowledge about the molecular structure of voltage-dependent K channels has been considerably advanced by the analysis of cloned cDNAs encoding K-channel-forming proteins from the Shaker gene complex of *Drosophila melanogaster* (for review, see JAN and JAN 1989) and from rat and mouse brain (BAUMANN et al. 1988; TEMPEL et al. 1988; CHRISTIE et al. 1989; STÜHMER et al. 1989).

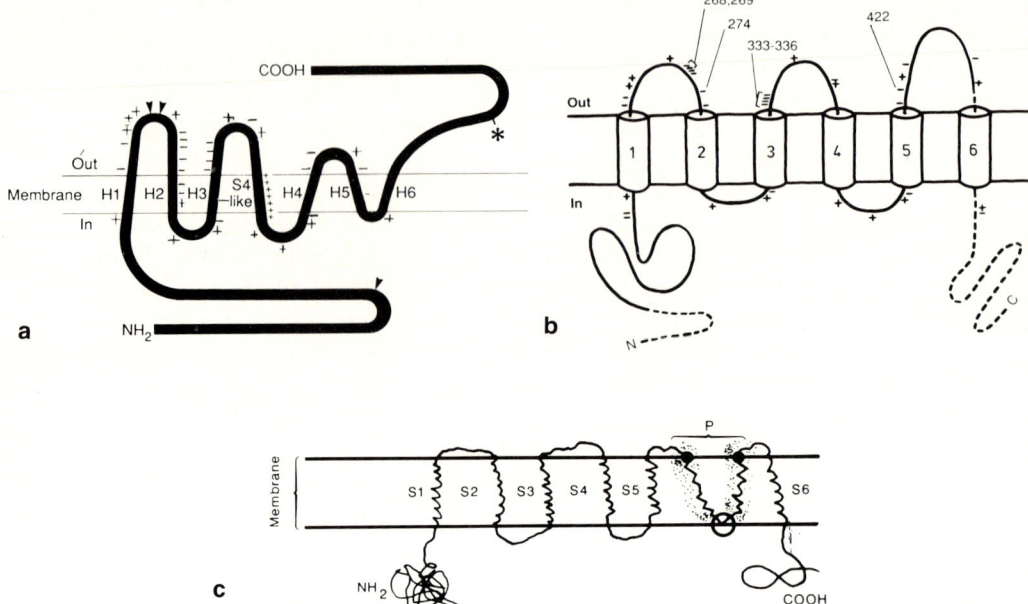

Fig. 6a–c. Three protein folding models for the K channel: **a** and **b** for *Drosophila* A-current K channel from TEMPEL et al. (1987) and MACKINNON and MILLER (1989b). **c** from STEVENS (1991). In **a**, the potential sites for N-glycosylation (*arrowheads*) and cAMP-mediated phosphorylation (*) are shown; the basic and acidic residues found in or near the hydrophobic regions are indicated. In **b**, six membrane-spanning regions are assumed, and the termini are thought to be cytoplasmic; the approximate position of every charged residue in the extracellular loops together with some residue numbers are indicated. In **c**, the pore-forming region P (= H5 region) is indicated; *closed* and *open circles* are external and internal binding sites for tetraethylammonium, respectively

Figure 6A shows the proposed transmembrane topography of an A-current K channel protein of *Drosophila*. This particular variant (ShA1) consists of 616 amino acids with a molecular mass of 70 kDa, similar in size to the toxin-binding subunits identified by cross-linking studies (see above). The predicted protein is homologous to the vertebrate Na channel in the S4 region, an arginine-rich sequence that has been proposed to be involved in voltage-dependent gating of the Na channel. Negatively charged residues predominate in the extracellular loops between (and especially near) proposed membrane-spanning regions. It is natural to suppose that these loops form the binding site for basic channel blockers like DTX, toxin I, MCDP, β-BuTX, or ChTX. The whole K channel is probably a homo- or heteromultimeric aggregate in which the protein of Fig. 6A combines with other proteins.

MACKINNON and MILLER (1989b) used site-directed mutagenesis to localize the binding site of ChTX. They measured A currents in *Xenopus* oocytes injected with mRNA coding for either wild-type *Drosophila* K channels or for mutants in which the negatively charged glutamate residue 422 (see Fig. 6B) was changed to a neutral glutamine, a positive lysine, or a negative aspartate. Addition of 8 nM ChTX inhibited wild-type currents and aspartate mutants by 70%, but reduced the glutamine and lysine mutants by only 40% and 15%, respectively. This indicates that Glu 422 stabilizes the binding of ChTX, probably by exerting an electrostatic influence on the cationic toxin. Mutations of residue numbers 268, 269, 274, and 333–336 (Fig. 6B) did not alter the ChTX effect. In further experiments MacKinnon et al. (1990) found that mutations at positions 427, 431, 449, 451, and 452 also affect the ChTX block. The amino acid residues that alter toxin inhibition are located at both ends of the S5–S6 linker (see Fig. 6B), separated by the pore-forming H5 region (see Fig. 6C). Interestingly, the pharmacological properties of the K channels depend also on the amino acid sequence of the amino-terminus. Stocker et al. (1990) found that two members of the Shaker K channel family (variants ShA2 and ShD2) which differ only in the amino-terminus have very different rates of inactivation and, in addition, very different ChTX sensitivity (IC_{50} = 4 and 105 nM for ShA2 and ShD2, respectively). One has to assume that a change in the amino-terminus (which is presumably located intracellularly, see Fig. 6) triggers a change of the ChTX binding site on the extracellular side.

The predicted size and amino acid sequence of the rat cortex K channel proteins RCK are very similar to those of the A channel protein of *Drosophila*. More than 60% of the amino acids are identical, although sequences facing the extracellular side (and probably containing the toxin binding sites) are not well conserved between *Drosophila* and rat (Baumann et al. 1988). Table 4 summarizes the electrical and pharmacological properties of five different types of K channels expressed in *Xenopus* oocytes after injection of RNA specific for different RCK-forming proteins. Several of the RCK channels are sensitive to DTX and MCDP (which the *Drosophila* Shaker K channels are not, see Stocker et al. 1990; Lichtinghagen et al. 1990). The RCK4 channels are different from the other channels: They inactivate quickly and have a larger molecular weight, smaller single-channel conductance, and lower 4-aminopyridine (4-AP) sensitivity than the others. Also, the conductance-voltage curve of the RCK4 channels is much flatter than that of the other channels (Stühmer et al. 1989). It is clear that the two very slowly inactivating channels (RCK1 and RCK5) have the highest DTX sensitivity. Heteromultimeric channels composed of a channel protein with high DTX sensitivity (RCK1) and a channel protein with low DTX sensitivity (RCK4) have intermediate DTX sensitivity (Ruppersberg et al. 1990). As pointed out by Stühmer et al. (1989), the reduced DTX sensitivity of RCK3 and RCK4 channels may be due to a replacement of glutamic acid residue 353 (RCK1) and of aspartic acid residue 354 (RCK5) in the extracellular loop S5–S6 by uncharged amino acids. In addition to the S5-S6 loop, the S3-S4 loop may play a role in toxin binding: The number of acidic

Table 4. Rat cortex K channel forming proteins (RCK) and pharmacological characteristics of the K channels expressed in *Xenopus* oocytes after injection of RCK-specific RNA[a]. Functional characteristics of the human brain K channel 2 (HBK2) and its rat homologue RCK2 are also included[b]. Parameters of the K channels encoded by mouse K channel gene MK3 are given in the last line[c]

	Amino acids (n)	M_r	γ (pS)	Inactivation	$I_{3.2}/I_p$	4-AP (mM)	TEA (mM)	DTX (nM)	MCDP (nM)	ChTX (nM)
RCK1	495	56	8.7	Very slow	0.83	1.0	0.6	12	45	22
RCK3	525	58	9.6	Intermediate	0.14	1.5	50	>600	>1000	1
RCK4	655	73	4.7	Fast	<0.02	12.5	>100	>200	>2000	>40
RCK5	498	57	10.2	Very slow	0.60	0.8	129	4	175	6
HBK2, RCK2	529	59	8.9	Very slow	0.74	1.5	7	20	10	1
MK3	530		13	Intermediate		0.4	11			0.5–2

4-AP, 4-aminopyridine; TEA, tetraethylammonium; DTX, dendrotoxin; MCDP, mast cell degranulating peptide; ChTX, charybdotoxin.
[a] Single-channel conductances γ from single-channel current-voltage relations measured in cell-attached patches with normal frog Ringer's solution on the extracellular side. Inactivation time course refers to currents recorded from macro-patches; $I_{3.2}/I_p$ gives current at end of 3.2-s pulse to 0 mV relative to peak current. Concentrations refer to ID_{50} values (50% inhibition of peak current), measured at 20 mV pulse potential with whole-cell current recording. (From BAUMANN et al. 1988; STÜHMER et al. 1988, 1989).
[b] From GRUPE et al. (1990).
[c] From GRISSMER et al. (1990).

amino acid residues in the extracellular loop S3–S4 correlates with the affinity of the RCK channels to ChTX. According to a recent report by BUSCH et al. (1991), the highest DTX sensitivity ($K_d \sim 1\,nM$) is observed in the RBK1 channels (= rat brain channels), which CHRISTIE et al. (1990) expressed in *Xenopus* oocytes.

GRUPE et al. (1990) have isolated and sequenced a human cDNA from which a novel K-channel-forming protein was derived (HBK2 = human brain K channel 2). The corresponding rat cDNA (RCK2) was also isolated and characterized for comparison. As shown in Table 4, the electrical and pharmacological properties of the channels formed by the HBK2 and RCK2 proteins after injection of their respective cRNAs into *Xenopus* oocytes are similar to those of the RCK1 and RCK5 channels. SWANSON et al. (1990) and KIRSCH et al. (1991) studied the same rat brain channel RCK2 (also called KV2) as GRUPE et al. (1990) but found somewhat different ID_{50} values. Particularly striking is the difference for the ID_{50} of ChTX: $1\,nM$ (GRUPE et al. 1990; see Table 4) vs. $>0.2\,\mu M$ (SWANSON et al. 1990) and $3.3\,\mu M$ (KIRSCH et al. 1991). The enormous difference is probably due to the use of different methods for the preparation of ChTX, yielding different ChTX isoforms (see above). Interestingly, both SWANSON et al. (1990) and KIRSCH et al. (1991) report blockage of the RCK2 channel by NTX ($ID_{50} = 0.2$–$0.4\,\mu M$).

The mouse K channel gene MK3 closely resembles the amino acid sequence of the RCK3 cDNA (CHANDY et al. 1990). Not surprisingly, MK3 currents are remarkably similar to those of RCK3 (see Table 4). The biophysical and pharmacological properties of MK3 channels are indistinguishable from those of voltage-gated type n K channels in T lymphocytes (GRISSMER et al. 1990).

Interestingly, none of the K channels in Table 4 seems to be sensitive to β-BuTX. By contrast, binding studies with radioiodinated toxins suggest the existence of DTX acceptors with (low- or high-affinity) β-BuTX binding sites (Fig. 2). Apparently, K channels with high-affinity β-BuTX binding sites have not yet been cloned.

Acknowledgement. My thanks to Dr. J.P. Ruppersberg (Heidelberg) for discussions about the molecular biology of the potassium channel.

References

Anderson CS, MacKinnon R, Smith C, Miller C (1988) Charybdotoxin block of single Ca^{2+} activated K^+ channels. J Gen Physiol 91:317–333

Århem P (1980) Effects of rubidium, caesium, strontium, barium and lanthanum on ionic currents in myelinated nerve fibres from *Xenopus laevis*. Acta Physiol Scand 108:7–16

Armstrong CM, Swenson RP Jr, Taylor SR (1982) Block of squid axon K channels by internally and externally applied barium ions. J Gen Physiol 80:663–682

Auguste P, Hugues M, Lazdunski M (1989) Polypeptide constitution of receptors for apamin, a neurotoxin which blocks a class of Ca^{2+}-activated K^+ channels. FEBS Lett 248:150–154

Auguste P, Hugues M, Gravé B, Gesquière J-C, Maes P, Tartar A, Romey G, Schweitz H, Lazdunski M (1990) Leiurotoxin I (Scyllatoxin), a peptide ligand for Ca^{2+}-activated K^+ channels. Chemical synthesis, radiolabeling, and receptor characterization. J Biol Chem 265:4753–4759

Awan KA, Dolly JO (1991) K^+ channel sub-types in rat brain: characteristic locations revealed using β-bungarotoxin, α- and δ-dendrotoxins. Neuroscience 40:29–39

Banks BEC, Brown C, Burgess GM, Burnstock G, Claret M, Cocks TM, Jenkinson DH (1979) Apamin blocks certain neurotransmitter-induced increases in potassium permeability. Nature 282:415–417

Baumann A, Grupe A, Ackermann A, Pongs O (1988) Structure of the voltage-dependent potassium channel is highly conserved from *Drosophila* to vertebrate central nervous systems. EMBO J 7:2457–2463

Beech DJ, Bolton TB, Castle NA, Strong PN (1987) Characterization of a toxin from scorpion (*Leiururus quinquestriatus*) venom that blocks in vitro both large (BK) K^+ channels in rabbit vascular smooth muscle and intermediate (IK) conductance Ca^{2+}-activated K^+ channels in human red cells. J Physiol (Lond) 387:32 P

Benishin CG (1990) Potassium channel blockade by the B subunit of β-bungarotoxin. Mol Pharmacology 38:164–169

Benishin CG, Sorensen RG, Brown WE, Krueger BK, Blaustein MP (1988) Four polypeptide components of green mamba venom selectively block certain potassium channels in rat brain synaptosomes. Mol Pharmacol 34:152–159

Benoit E, Dubois J-M (1986) Toxin 1 from the snake *Dendroaspis polylepis polylepis*: a highly specific blocker of one type of potassium channel in myelinated nerve fiber. Brain Res 377:374–377

Bidard J-N, Mourre C, Lazdunski M (1987) Two potent central convulsant peptides, a bee venom toxin, the MCD peptide, and a snake venom toxin, dendrotoxin I, known to block K^+ channels, have interacting receptor sites. Biochem Biophys Res Commun 143:383–389

Bidard J-N, Mourre C, Gandolfo G, Schweitz H, Widmann C, Gottesmann C, Lazdunski M (1989) Analogies and differences in the mode of action and properties of binding sites (localization and mutual interactions) of two K^+ channel toxins, MCD peptide and dendrotoxin I. Brain Res 495:45–57

Bkaily G, Sperelakis N, Renaud JF, Payet MD (1985) Apamin, a highly specific Ca^{2+} blocking agent in heart muscle. Am J Physiol 248:H961–H965

Black AR, Dolly JO (1986) Two acceptor sub-types for dendrotoxin in chick synaptic membranes distinguishable by β-bungarotoxin. Eur J Biochem 156:609–617

Black AR, Breeze AL, Othman IB, Dolly JO (1986) Involvement of neuronal acceptors for dendrotoxin in its convulsive action in rat brain. Biochem J 237:397–404

Black AR, Donegan CM, Denny BJ, Dolly JO (1988) Solubilization and physical characterization of acceptors for dendrotoxin and β-bungarotoxin from synaptic membranes of rat brain. Biochemistry 27:6814–6820

Blatz AL, Magleby K (1986) Single apamin-blocked Ca-activated K channels of small conductance in rat cultured skeletal muscle. Nature 323:718–720

Bontems F, Roumestand C, Boyot P, Gilquin B, Doljansky Y, Menez A, Toma F (1991) Three-dimensional structure of natural charybdotoxin in aqueous solution by ^{1}H-NMR. Charybdotoxin possesses a structural motif found in other scorpion toxins. Eur J Biochem 196:19–28

Bräu ME, Dreyer F, Jonas P, Repp H, Vogel W (1990) A K^+ channel in *Xenopus* nerve fibres selectively blocked by bee and snake toxins: binding and voltage-clamp experiments. J Physiol (Lond) 420:365–385

Breeze AL, Dolly JO (1989) Interactions between discrete neuronal membrane binding sites for the putative K^+-channel ligands β-bungarotoxin, dendrotoxin and mast-cell-degranulating peptide. Eur J Biochem 178:771–778

Busch A, Hurst R, Kavanaugh MP, Christie MJ, Osborne PB, Adelman JP, North RA (1991) The site of binding of α-dendrotoxin to voltage-dependent potassium channels. J Physiol 434:32 P

Capiod T, Ogden DC (1989) The properties of calcium-activated potassium ion channels in guinea-pig isolated hepatocytes. J Physiol (Lond) 409:285–295

Carbone E, Wanke E, Prestipino G, Possani LD, Maelicke A (1982) Selective blockage of voltage-dependent K^+ channels by a novel scorpion toxin. Nature 296:90–91

Carbone E, Prestipino G, Wanke E, Possani LD, Maelicke A (1983) Selective action of scorpion neurotoxins on the ionic currents of the squid giant axon. Toxicon 3:57–60

Carbone E, Prestipino G, Spadavecchia L, Franciolini F, Possani LD (1987) Blocking of the squid axon K^+ channel by noxiustoxin: a toxin from the venom of the scorpion *Centruroides noxius*. Pflügers Arch 408:423–431

Castle NA, Strong PN (1986) Identification of two toxins from scorpion (*Leiurus quinquestriatus*) venom which block distinct classes of calcium-activated potassium channel. FEBS Lett 209:117–121

Castle NA, Haylett DG, Jenkinson DH (1989) Toxins in the characterization of potassium channels. Trends Neurosci 12:59–65

Catterall WA (1988) Structure and function of voltage-sensitive ion channels. Science 242:50–61

Chandy KG, Williams CB, Spencer RH, Aguilar BA, Ghanshani S, Tempel BL, Gutman GA (1990) A family of three mouse potassium channel genes with intronless coding regions. Science 247:973–975

Chesnut TJ, Carpenter DO, Strichartz GR (1987) Effects of venom from *Conus striatus* on the delayed rectifier potassium current of molluscan neurons. Toxicon 25:267–278

Chicchi GG, Gimenez-Gallego G, Ber E, Garcia ML, Winquist R, Cascieri MA (1988) Purification and characterization of a unique, potent inhibitor of apamin binding from *Leiurus quinquestriatus hebraeus* venom. J Biol Chem 263:10192–10197

Christie MJ, Adelman JP, Douglass J, North RA (1989) Expression of a cloned rat brain potassium channel in *Xenopus* oocytes. Science 244:221–224

Cognard C, Ewane-Nyambi G, Potreau D, Raymond G (1986) The voltage-dependent blocking effect of phalloidin on the delayed potassium current of voltage-clamped frog skeletal muscle fibres. Eur J Pharmacol 120:209–216

Cook NS, Quast U (1990) Potassium channel pharmacology. In: "Potassium channels. Structure, classification, function and therapeutic potential". (N.S. Cook, ed.) Ellis Horwood Ltd, Chichester

Corrette BJ, Repp H, Dreyer F, Schwarz JR (1991) Two types of fast K^+ channels in rat myelinated nerve fibres and their sensitivity to dendrotoxin. Pflügers Arch 418:408–416

Dolly JO, Halliwell JV, Black JD, William RS, Pelchen-Matthews A, Breeze AL, Mehraban F, Othman IB, Black AR (1984) Botulinum neurotoxin and dendrotoxin as probes for studies on transmitter release. J Physiol (Paris) 79:280–303

Dreyer F (1990) Peptide toxins and potassium channels. Rev Physiol Biochem Pharmacol 115:93–136

Dufton MJ (1985) Proteinase inhibitors and dendrotoxins. Sequence classification, structural prediction and structure/activity. Eur J Biochem 153:647–654

Farley J, Rudy B (1988) Multiple types of voltage-dependent Ca^{2+}-activated K^+ channels of large conductance in rat brain synaptosomal membranes. Biophys J 53:919–934

Filippov AK, Skatova GE, Porotikov VI, Orlov BN, Asafova NN (1988) Effect of the neurotoxin apamin on ion currents in membranes of heart fibers. Biofizika 33:109–112

Frech GC, van Dongen AMJ, Schuster G, Brown AM, Joho RH (1989) A novel potassium channel with delayed rectifier properties isolated from rat brain by expression cloning. Nature 340:642–645

Freeman CM, Catlow CRA, Hemmings AM, Hider RC (1986) The conformation of apamin. FEBS Lett 197:289–296

Galvez A, Gimenez-Gallego G, Reuben JP, Roy-Contancin L, Feigenbaum P, Kaczorowski GJ, Garcia ML (1990) Purification and characterization of a unique, potent, peptidyl probe for the high conductance calcium-activated potassium channel from venom of the scorpion *Buthus tamulus*. J Biol Chem 265:11083–11090

Garcia-Calvo M, Smith MM, Kaczorowski GJ, Garcia ML (1991) Purification of the charybdotoxin receptor from bovine aortic smooth muscle. Biophys J 59: 78a

Gimenez-Gallego G, Navia MA, Reuben JP, Katz GM, Kaczorowski GJ, Garcia ML (1988) Purification, sequence, and model structure of charybdotoxin, a potent selective inhibitor of calcium-activated potassium channels. Proc Natl Acad Sci USA 85:3329–3333

Grinstein S, Smith JD (1990) Calcium-independent cell volume regulation in human lymphocytes. Inhibition by charybdotoxin. J Gen Physiol 95:97–120

Grissmer S, Dethlefs B, Wasmuth JJ, Goldin AL, Gutman GA, Cahalan MD, Chandy KG (1990) Expression and chromosomal localization of a lymphocyte K^+ channel gene. Proc Natl Acad Sci USA 87:9411–9415

Grupe A, Schröter KH, Ruppersberg JP, Stocker M, Drewes T, Beckh S, Pongs O (1990) Cloning and expression of a human voltage-gated potassium channel. A novel member of the RCK potassium channel family. EMBO J 9:1749–1756

Guggino SE, Guggino WB, Green N, Sacktor B (1987) Blocking agents of Ca^{2+}-

activated K$^+$ channels in cultured medullary thick ascending limb cells. Am J Physiol 252:C128–C137

Gurrola GB, Molinar-Rode R, Sitges M, Bayon A, Possani LD (1989) Synthetic peptides corresponding to the sequence of noxiustoxin indicate that the active site of this K$^+$ channel blocker is located on its amino-terminal portion. J Neural Transm 77:11–20

Habermann E (1972) Bee and wasp venom. Science 177:314–322

Habermann E (1984) Apamin. Pharmacol Ther 25:255–270

Habermann E, Fischer K (1979) Bee venom neurotoxin (apamin): iodine labeling and characterization of binding sites. Eur J Biochem 94:355–364

Halliwell JV, Othman IB, Pelchen-Matthews A, Dolly JO (1986) Central action of dendrotoxin: selective reduction of a transient K conductance in hippocampus and binding to localized acceptors. Proc Natl Acad Sci USA 83:493–497

Harvey AL, Anderson AJ (1985) Dendrotoxins: snake toxins that block potassium channels and facilitate neurotransmitter release. Pharmacol Ther 31:33–55

Harvey AL, Karlsson E (1980) Dendrotoxin from the venom of the green mamba, *Dendroaspis angusticeps*. Naunyn Schmiedebergs Arch Pharmacol 312:1–6

Harvey AL, Marshall DL, De-Allie FA, Strong PN (1989) Interactions between dendrotoxin, a blocker of voltage-dependent potassium channels, and charybdotoxin, a blocker of calcium-activated potassium channels, at binding sites on neuronal membranes. Biochem Biophys Res Commun 163:394–397

Hermann A, Erxleben C (1987) Charybdotoxin selectively blocks small Ca-activated K channels in *Aplysia* neurons. J Gen Physiol 90:27–47

Hider RC, Ragnarsson U (1981) A comparative structural study of apamin and related bee venom peptides. Biochim Biophys Acta 667:197–208

Hollecker M, Larcher D (1989) Conformational forces affecting the folding pathways of dendrotoxins I and K from black mamba venom. Eur J Biochem 179:87–94

Hugues M, Romey G, Duval D, Vincent JP, Lazdunski M (1982a) Apamin as a selective blocker of the calcium-dependent potassium channel in neuroblastoma cells: voltage-clamp and biochemical characterization of the toxin receptor. Proc Natl Acad Sci USA 79:1308–1312

Hugues M, Schmid H, Romey G, Duval D, Frelin C, Lazdunski M (1982b) The Ca^{2+}-dependent slow K$^+$ conductance in cultured rat muscle cells: characterization with apamin. EMBO J 1:1039–1042

Hugues M, Duval D, Kitabgi P, Lazdunski M, Vincent J-P (1982c) Preparation of a pure monoiodo derivative of the bee venom neurotoxin apamin and its binding properties to rat brain synaptosomes. J Biol Chem 257:2762–2769

Jacobs ER, DeCoursey TE (1990) Mechanisms of potassium channel block in rat alveolar epithelial cells. J Pharm Exp Therap 255:459–472

Jan LY, Jan YN (1989) Voltage-sensitive ion channels. Cell 56:13–25

Jonas P, Bräu ME, Hermsteiner M, Vogel W (1989) Single-channel recording in myelinated nerve fibers reveals one type of Na channel but different K channels. Proc Natl Acad Sci USA 86:7238–7242

Kao CY, Yeoh PN, Goldfinger MD, Fuhrman FA, Mosher HS (1981) Chiriquitoxin, a new tool for mapping ionic channels. J Pharmacol Exp Ther 217:416–429

Kirsch GE, Drewe JA, Verma S, Brown AB, Joho RH (1991) Electrophysiological characterization of a new member of the RCK family of rat brain K$^+$ channels. FEBS Lett 278:55–60

Kumar NV, Wemmer DE, Kallenbach NR (1988) Structure of P401 (mast cell degranulating peptide) in solution. Biophys Chem 31:113–119

Lambert P, Kuroda H, Chino N, Watanabe TX, Kimura T, Sakakibara S (1990) Solution synthesis of charybdotoxin (ChTX), a K$^+$ channel blocker. Biochem Biophys Res Commun 170:684–690

Lancaster B, Nicoll RA (1987) Properties of two calcium-activated hyperpolarizations in rat hippocampal neurones. J Physiol (Lond) 389:187–203

Leonard RJ, Karschin A, Jayashree-Aiyar S, Davidson N, Tanouye MA, Thomas L, Thomas G, Lester HA (1989) Expression of *Drosophila* Shaker potassium channels in mammalian cells infected with recombinant vaccinia virus. Proc Natl Acad Sci USA 86:7629–7633

Leveque C, Marqueze B, Couraud F, Seagar M (1990) Polypeptide components of the apamin receptor associated with a calcium activated potassium channel. FEBS Lett 275:185–189

Lichtinghagen R, Stocker M, Wittka R, Boheim G, Stühmer W, Ferrus A, Pongs O (1990) Molecular basis of altered excitability in *Shaker* mutants of *Drosophila melanogaster*. EMBO J 9:4399–4407

Llano I, Webb CK, Bezanilla F (1988) Potassium conductance of the squid giant axon. J Gen Physiol 92:179–196

Lucchesi K, Moczydlowski E (1990) Subconductance behavior in a maxi Ca^{2+}-activated K^+ channel induced by dendrotoxin I. Neuron 4:141–148

Lucchesi K, Ravindran A, Young H, Moczydlowski E (1989) Analysis of the blocking activity of charybdotoxin homologs and iodinated derivatives against Ca^{2+}-activated K^+ channels. J Membrane Biol 109:269–281

MacKinnon R, Miller C (1988) Mechanism of charybdotoxin block of the high-conductance, Ca^{2+}-activated K^+ channel. J Gen Physiol 91:335–349

MacKinnon R, Miller C (1989a) Functional modification of a Ca^{2+}-activated K^+ channel by trimethyloxonium. Biochemistry 28:8087–8092

MacKinnon R, Miller C (1989b) Mutant potassium channels with altered binding of charybdotoxin, a pore-blocking peptide inhibitor. Science 245:1382–1385

MacKinnon R, Reinhart PH, White MM (1988) Charybdotoxin block of *Shaker* K^+ channels suggests that different types of K^+ channels share common structural features. Neuron 1:997–1001

MacKinnon R, Latorre R, Miller C (1989) Role of surface electrostatics in the operation of a high-conductance Ca^{2+}-activated K^+ channel. Biochemistry 28:8092–8099

MacKinnon R, Heginbotham L, Abramson T (1990) Mapping the receptor site for charybdotoxin, a pore-blocking potassium channel inhibitor. Neuron 5:767–771

Mahaut-Smith MP, Rink TJ, Collins SC, Sage SO (1990) Voltage-gated potassium channels and the control of membrane potential in human platelets. J Physiol 428:723–735

Martins JC, Zhang W, Tartar A, Lazdunski M, Borremans FAM (1990) Solution conformation of leiurotoxin I (scyllatoxin) by ^1H nuclear magnetic resonance. Resonance assignment and secondary structure. FEBS Lett 260:249–253

Marty A (1989) The physiological role of calcium-dependent channels. Trends Neurosci 12:420–424

Massefski W, Redfield AG, Hare DR, Miller C (1990) Molecular structure of charybdotoxin, a pore-directed inhibitor of potassium ion channels. Science 249:521–524

Mehraban F, Breeze AL, Dolly JO (1984) Identification by cross-linking of a neuronal acceptor protein for dendrotoxin, a convulsant polypeptide. FEBS Lett 174:116–122

Mehraban F, Black AR, Breeze AL, Green DG, Dolly JO (1985) A functional membranous acceptor for dendrotoxin in rat brain: solubilization of the binding component. Biochem Soc Trans 13:507–508

Merot J, Bidet M, Le Maout S, Tauc M, Poujeol P (1989) Two types of K^+ channels in the apical membrane of rabbit proximal tubule in primary culture. Biochim Biophys Acta 978:134–144

Meves H, Pichon Y (1977) The effect of internal and external 4-aminopyridine on the potassium currents in intracellularly perfused squid giant axons. J Physiol (Lond) 268:511–532

Miller C (1988) Competition for block of a Ca^{2+}-activated K^+ channel by charybdotoxin and tetraethylammonium. Neuron 1:1003–1006

Miller C (1990) Diffusion-controlled binding of a peptide neurotoxin to its K^+ channel receptor. Biochemistry 29:5320–5325

Miller C, Moczydlowski E, Latorre R, Phillips M (1985) Charybdotoxin, a protein inhibitor of single Ca^{2+}-activated K^+ channels from mammalian skeletal muscle. Nature 313:316–318

Moczydlowski E, Lucchesi K, Ravindran A (1988) An emerging pharmacology of peptide toxins targeted against potassium channels. J Membr Biol 105:95–111

Muniz ZM, Diniz CR, Dolly JO (1990) Characterisation of binding sites for δ-dendrotoxin in guinea-pig synaptosomes: relationship to acceptors for the K^+-channel probe α-dendrotoxin. J Neurochem 54:343–346

Othman IB, Spokes JW, Dolly JO (1982) Preparation of neurotoxic ^3H-β-bungarotoxin: demonstration of saturable binding to brain synapses and its inhibition by toxin I. Eur J Biochem 128:267–276

Pappone PA, Cahalan MD (1987) *Pandinus imperator* scorpion venom blocks voltage-gated potassium channels in nerve fibers. J Neurosci 7:3300–3305

Pappone PA, Lucero MT (1988) *Pandinus imperator* scorpion venom blocks voltage-gated potassium channels in GH_3 cells. J Gen Physiol 91:817–833

Parcej DN, Dolly JO (1989) Dendrotoxin acceptor from bovine synaptic plasma membranes. Binding properties, purification and subunit composition of a putative constituent of certain voltage-activated K^+ channels. Biochem J 257:899–903

Park C-S, Miller C (1991) High-level expression of charybdotoxin in *E. coli*. Biophys J 59:197a

Pease JH, Wemmer DE (1988) Solution structure of apamin determined by nuclear magnetic resonance and distance geometry. Biochemistry 27:8491–8498

Pelchen-Matthews A, Dolly JO (1988) Distribution of acceptors for β-bungarotoxin in the central nervous system of the rat. Brain Res 441:127–138

Pelchen-Matthews A, Dolly JO (1989) Distribution in the rat central nervous system of acceptor sub-types for dendrotoxin, a K^+ channel probe. Neuroscience 29:347–361

Penner R, Petersen M, Pierau F-K, Dreyer F (1986) Dendrotoxin: a selective blocker of non-inactivating potassium current in guinea-pig dorsal root ganglion neurones. Pflügers Arch 407:365–369

Petersen M, Penner R, Pierau F-K, Dreyer F (1986) β-bungarotoxin inhibits a non-inactivating potassium current in guinea pig dorsal root ganglion neurones. Neurosci Lett 68:141–145

Possani LD, Martin BM, Svendsen I (1982) The primary structure of noxiustoxin: a K^+ channel blocking peptide, purified from the venom of the scorpion *Centruroides noxius* Hoffmann. Carlsberg Res Commun 47:285–289

Prestipino G, Valdivia HH, Liévano A, Darszon A, Ramírez AN, Possani LD (1989) Purification and reconstitution of potassium channel proteins from squid axon membranes, FEBS Lett 250:570–574

Price M, Lee SC, Deutsch C (1989) Charybdotoxin inhibits proliferation and interleukin 2 production in human peripheral blood lymphocytes. Proc Natl Acad Sci USA 86:10171–10175

Raymond G, Potreau D, Cognard C, Jahn W, Wieland T (1987) Antamanide antagonizes the phalloidin-induced negative inotropic effect and blocks voltage dependently the fast outward K^+ current in voltage-clamped frog muscle fibres. Eur J Pharmacol 138:21–27

Rehm H (1984) News about the neuronal membrane protein which binds the presynaptic neurotoxin β-bungarotoxin. J Physiol (Paris) 79:265–268

Rehm H (1989) Enzymatic deglycosylation of the dendrotoxin-binding protein. FEBS Lett 247:28–30

Rehm H, Betz H (1982) Binding of β-bungarotoxin to synaptic membrane fractions of chick brain. J Biol Chem 257:10015–10022

Rehm H, Betz H (1983) Identification by cross-linking of a β-bungarotoxin binding polypeptide in chick brain membranes. EMBO J 2:1119–1122

Rehm H, Betz H (1984) Solubilization and characterization of the β-bungarotoxin-binding protein of chick brain membranes. J Biol Chem 259:6865–6869

Rehm H, Lazdunski M (1988a) Existence of different populations of the dendrotoxin

I binding protein associated with neuronal K$^+$ channels. Biochem Biophys Res Commun 153:231–240

Rehm H, Lazdunski M (1988b) Purification and subunit structure of a putative K$^+$ channel protein identified by its binding properties for dendrotoxin I. Proc Natl Acad Sci USA 85:4919–4923

Rehm H, Bidard J-N, Schweitz H, Lazdunski M (1988) The receptor site for the bee venom mast cell degranulating peptide. Affinity labeling and evidence for a common molecular target for mast cell degranulating peptide and dendrotoxin I, a snake toxin active on K$^+$ channels. Biochemistry 27:1827–1832

Rehm H, Newitt RA, Tempel BL (1989a) Immunological evidence for a relationship between the dendrotoxin-binding protein and the mammalian homologue of the *Drosophila* Shaker K$^+$ channel. FEBS Lett 249:224–228

Rehm H, Pelzer S, Cochet C, Chambaz E, Tempel BL, Trautwein W, Pelzer D, Lazdunski M (1989b) Dendrotoxin-binding brain membrane protein displays a K$^+$ channel activity that is stimulated by both cAMP-dependent and endogenous phosphorylations. Biochemistry 28:6455–6460

Reinhart PH, Chung S, Levitan IB (1989) A family of calcium-dependent potassium channels from rat brain. Neuron 2:1031–1041

Rudy B (1988) Diversity and ubiquity of K channels. Neuroscience 25:729–749

Ruppersberg JP, Schröter KH, Sakmann B, Stocker M, Sewing S, Pongs O (1990) Heteromultimeric channels formed by rat brain potassium-channel proteins. Nature 345:535–537

Sands SB, Lewis RS, Cahalan MD (1989) Charybdotoxin blocks voltage-gated K$^+$ channels in human and murine T lymphocytes. J Gen Physiol 93:1061–1074

Schauf CL (1987) Dendrotoxin blocks potassium channels and slows sodium inactivation in *Myxicola* giant axons. J Pharmacol Exp Ther 241:793–796

Schmid-Antomarchi H, Hugues M, Lazdunski M (1986) Properties of the apamin-sensitive Ca^{2+}-activated K$^+$ channel in PC12 pheochromocytoma cells which hyper-produce the apamin receptor. J Biol Chem 261:8633–8637

Schmidt RR, Betz H (1988) The β-bungarotoxin-binding protein from chick brain: binding sites for different neuronal K$^+$ channel ligands co-fractionate upon partial purification. FEBS Lett 240:65–70

Schmidt RR, Betz H (1989) Cross-linking of β-bungarotoxin to chick brain membranes. Identification of subunits of a putative voltage-gated K$^+$ channel. Biochemistry 28:8346–8350

Schmidt RR, Betz H, Rehm H (1988) Inhibition of β-bungarotoxin binding to brain membranes by mast cell degranulating peptide, toxin I, and ethylene glycol bis(β-aminoethyl ether)-N,N,N',N'-tetraacetic acid. Biochemistry 27:963–967

Schneider MJ, Rogowski RS, Krueger BK, Blaustein MP (1989) Charybdotoxin blocks both Ca-activated K channels and Ca-independent voltage-gated K channels in rat brain synaptosomes. FEBS Lett 250:433–436

Schweitz H, Bidard J-N, Maes P, Lazdunski M (1989a) Charybdotoxin is a new member of the K$^+$ channel toxin family that includes dendrotoxin I and mast cell degranulating peptide. Biochemistry 28:9708–9714

Schweitz H, Stansfeld CE, Bidard J-N, Fagni L, Maes P, Lazdunski M (1989b) Charybdotoxin blocks dendrotoxin-sensitive voltage-activated K$^+$ channels. FEBS Lett 250:519–522

Schweitz H, Bidard JN, Lazdunski M (1990) Purification and pharmacological characterization of peptide toxins from the black mamba (*Dendroaspis polylepis*) venom. Toxicon 28:847–856

Scott VES, Parcej DN, Keen JN, Findlay JBC, Dolly JO (1990) α-Dendrotoxin acceptor from bovine brain is a K$^+$ channel protein. Evidence from the N-terminal sequence on its larger subunit. J Biol Chem 265:20094–20097

Seagar MJ, Labbé-Jullié C, Granier C, van Rietschoten J, Couraud F (1985) Photoaffinity labeling of components of the apamin sensitive K$^+$ channel in neuronal membranes. J Biol Chem 260:3895–3898

Seagar MJ, Labbé-Jullié C, Granier C, Goll A, Glossmann H, van Rietschoten J,

Couraud F (1986) Molecular structure of the rat brain apamin receptor: differential photoaffinity labeling of putative K$^+$ channel subunits and target size analysis. Biochemistry 25:4051–4057

Seagar MJ, Marqueze B, Couraud F (1987) Solubilization of the apamin receptor associated with a calcium-activated potassium channel from rat brain. J Neurosci 7:565–570

Sitges M, Possani LD, Bayon A (1986) Noxiustoxin, a short-chain toxin from the Mexican scorpion Centruroides noxius, induces transmitter release by blocking K$^+$ permeability. J Neurosci 6:1570–1574

Sorensen RG, Blaustein MP (1989) Rat brain dendrotoxin receptors associated with voltage-gated potassium channels: dendrotoxin binding and receptor solubilization. Mol Pharmacol 36:689–698

Stansfeld CE, Feltz A (1988) Dendrotoxin-sensitive K$^+$ channels in dorsal root ganglion cells. Neurosci Lett 93:49–55

Stansfeld CE, Marsh SJ, Parcej DN, Dolly JO, Brown DA (1987) Mast cell degranulating peptide and dendrotoxin selectively inhibit a fast-activating potassium current and bind to common neuronal proteins. Neuroscience 23:893–902

Stevens CF (1991) Ion channels. Making a submicroscopic hole in one. Nature 349:657–658

Stocker M, Stühmer W, Wittka R, Wang X, Müller R, Ferrus A, Pongs O (1990) Alternative Shaker transcripts express either rapidly inactivating or noninactivating K$^+$ channels. Proc Natl Acad Sci USA 87:8903–8907

Strong PN (1990) Potassium channel toxins. Pharmacol Ther 46:137–162

Strong PN, Weir SW, Beech DJ, Hiestand P, Kocher HP (1989) Effects of potassium channel toxins from Leiurus quinquestriatus hebraeus venom on responses to cromakalim in rabbit blood vessels. Br J Pharmacol 98:817–826

Stühmer W, Stocker M, Sakmann B, Seeburg P, Baumann A, Grupe A, Pongs O (1988) Potassium channels expressed from rat brain cDNA have delayed rectifier properties. FEBS Lett 242:199–206

Stühmer W, Ruppersberg JP, Schröter KH, Sakmann B, Stocker M, Giese KP, Perschke A, Baumann A, Pongs O (1989) Molecular basis of functional diversity of voltage-gated potassium channels in mammalian brain. EMBO J 8:3235–3244

Sugg EE, Garcia ML, Reuben JP, Patchett AA, Kaczorowski GJ (1990) Synthesis and structural characterization of charybdotoxin, a potent peptidyl inhibitor of the high conductance Ca^{2+}-activated K$^+$ channel. J Biol Chem 265:18745–18748

Swanson R, Marshall J, Smith JS, Williams JB, Boyle MB, Folander K, Luneau CJ, Antanavage J, Oliva C, Buhrow SA, Bennett C, Stein RB, Kaczmarek LK (1990) Cloning and expression of cDNA and genomic clones encoding three delayed rectifier potassium channels in rat brain. Neuron 4:929–939

Taylor JW, Bidard JN, Lazdunski M (1984) The characterization of high affinity binding sites in rat brain for the mast cell degranulating peptide from bee venom using the purified monoiodinated peptide. J Biol Chem 259:13957–13967

Tempel BL, Papazian DM, Schwarz TL, Jan YN, Jan LY (1987) Sequence of a probable potassium channel component encoded at Shaker locus of Drosophila. Science 237:770–775

Tempel BL, Jan YN, Jan LY (1988) Cloning of a probable potassium channel gene from mouse brain. Nature 332:837–839

Valdivia HH, Smith JS, Martin BM, Coronado R, Possani LD (1988) Charybdotoxin and noxiustoxin, two homologous peptide inhibitors of the K$^+$ (Ca^{2+}) channel. FEBS Lett 226:280–284

Vázquez J, Feigenbaum P, Katz G, King VF, Reuben JP, Roy-Contancin L, Slaughter RS, Kaczorowski GJ, Garcia ML (1989) Characterization of high affinity binding sites for charybdotoxin in sarcolemmal membranes from bovine aortic smooth muscle. Evidence for a direct association with the high conductance calcium-activated potassium channel. J Biol Chem 264:20902–20909

Vázquez J, Feigenbaum P, King VF, Kaczorowski GJ, Garcia ML (1990) Characterization of high affinity binding sites for charybdotoxin in synaptic plasma membranes from rat brain. Evidence for a direct association with an inactivating, voltage-dependent, potassium channel. J Biol Chem 265:15564–15571

Weller U, Bernhardt U, Siemen D, Dreyer F, Vogel W, Habermann E (1985) Electrophysiological and neurobiochemical evidence for the blockade of a potassium channel by dendrotoxin. Naunyn-Schmiedebergs Arch Pharmacol 330:77–83

Wolff D, Cecchi X, Spalvins A, Canessa M (1988) Charybdotoxin blocks with high affinity the Ca-activated K^+ channel of Hb A and Hb S red cells: individual differences in the number of channels. J Membr Biol 106:243–252

Yang L, Kao CY (1990) Chiriquitoxin and ionic currents in frog skeletal muscle fibers. Biophys J 57:104a

Yeh JZ, Oxford GS, Wu CH, Narahashi T (1976) Dynamics of aminopyridine block of potassium channels in squid axon membrane. J Gen Physiol 68:519–535

Zagotta WN, Hoshi T, Aldrich RW (1989a) Gating of single Shaker potassium channels in *Drosophila* muscle and in *Xenopus* oocytes injected with Shaker mRNA. Proc Natl Acad Sci USA 86:7243–7247

Zagotta WN, Germeraad S, Garber SS, Hoshi T, Aldrich RW (1989b) Properties of ShB A-type potassium channels expressed in Shaker mutant *Drosophila* by germline transformation. Neuron 3:773–782

Zemková H, Teisinger J, Vyskočil F (1988) Inhibition of the electrogenic Na, K pump and Na, K-ATPase activity by tetraethylammonium, tetrabutylammonium, and apamin. J Neurosci Res 19:497–503

Note Added in Proof

Several further papers on potassium channel toxins have appeared:

Apamin:

Labbé-Jullié C, Granier C, Albericio F, Defendini M-L, Ceard B, Rochat H, Van Rietschoten J (1991) Binding and toxicity of apamin. Characterisation of the active site. Eur J Biochem 196:639–645

Messier C, Mourre C, Bontempi B, Sif J, Lazdunski M, Destrade C (1991) Effect of apamin, a toxin that inhibits Ca^{2+}-dependent K^+ channels, on learning and memory processes. Brain Res 551:322–326

Charybdotoxin:

Deutsch C, Price M, Lee S, King VF, Garcia ML (1991) Characterisation of high affinity binding sites for charybdotoxin in human T lymphocytes. J biol Chem 266:3668–3674

Park C-S, Hausdorff SF, Miller C (1991) Design, synthesis, and functional expression of a gene for charybdotoxin, a peptide blocker of K^+ channels. Proc Natl Acad Sci USA 88:2046–2050

Murray MA, Berry JL, Cook SJ, Foster RW, Green KA, Small RC (1991) Guinea-pig isolated trachealis: the effects of charybdotoxin on mechanical activity, membrane potential changes and the activity of plasmalemmal K^+-channels. Br J Pharmacol 103:1814–1818

Leiurotoxin:

Gossen D, Gesquière J-C, Tastenoy M, De Neef P, Waelbroeck M, Christophe J (1991) Characterisation and regulation of the expression of scyllatoxin (leiurotoxin I) receptors in the human neuroblastoma cell line NB-OK 1. FEBS Letters 285:271−274

Dendrotoxin:

Lucchesi KJ, Moczydlowski E (1991) On the interaction of bovine pancreatic trypsin inhibitor with maxi Ca^{2+}-activated K^+ channels. J gen Physiol 97:1295−1319
Stocker M, Pongs O, Hoth M, Heinemann SH, Stühmer W, Schröter K-H, Ruppersberg JP (1991) Swapping of functional domains in voltage-gated K^+ channels. Proc R Soc Lond B 245:101−107
Hurst RS, Busch AE, Kavanaugh MP, Osborne PB, North RA, Adelman JP (1991) Identification of amino acid residues involved in dendrotoxin block of rat voltage-dependent potassium channels. Molec Pharmacol 40:572−576

Mast Cell Degranulating Peptide:

Heurteaux C, Lazdunski M (1991) MCD peptide and dendrotoxin I activate *c-fos* and *c-jun* expression by acting on two different types of K^+ channels. A discrimination using the K^+ channel opener lemakalim. Brain Res 554:22−29
Kobayashi Y, Sato A, Takashima H, Tamaoki H, Nishimura S, Kyogoku Y, Ikenaka K, Kondo T, Mikoshiba K, Hojo H, Aimoto S, Moroder L (1991) A new α-helical motif in membrane active peptides. Neurochem Int 18:525−534

CHAPTER 22

Calcium Channel Toxins*

J. Striessnig and H. Glossmann

A. Introduction

Electrical excitability of neurons is based on the ability to maintain a resting membrane potential and to respond to potential changes with alterations in the plasma membrane permeability for ions on a time scale of milliseconds. The Na^+/K^+-ATPase provides the ionic imbalance (low intracellular Na^+ and high intracellular K^+ concentrations relative to the extracellular fluid). In combination with a high permeability for K^+ under resting conditions, the (inside negative) membrane potential is generated. Upon excitation, an electrical signal can be propagated along the plasmalemma in the form of an action potential to allow rapid information transfer resulting finally in neurotransmitter release. The properties of the ionic currents underlying an action potential were first sorted out in voltage-clamp experiments (Hodgkin and Huxley 1952). Opening and closing of voltage-dependent ion channels, mainly Na^+, K^+, and Ca^{2+} channels, cause these discrete changes in membrane permeability. Improved electrophysiological methods, i.e., the patch-clamp technique, allowed the characterization even of subtypes of these channels on the single-channel level and provided insight into their functional properties. Complementation by biochemical data eventually led to the cloning and functional expression of the genes of several Na^+, K^+, and Ca^{2+} channels. They all share common structural features (e.g., a putative voltage-sensing transmembrane region, for review see Catterall 1986, 1988) and are therefore considered as members of a gene family originating from a common ancestor.

In case of the voltage-dependent Na^+ channel, functional as well as biochemical studies have been greatly facilitated by the discovery of a number of neurotoxins which bind with high affinity to distinct receptor sites of the channel pore. In contrast, synthetic radiolabeled and unlabeled Ca^{2+} channel drugs ("Ca^{2+} antagonists", like dihydropyridines, phenyl-alkylamines, benzothiazepines; for reviews, see Glossmann and Striessnig 1988, 1990; Glossmann et al. 1988) rather than naturally occurring com-

* H.G.'s work is supported by Fonds zur Förderung der wissenschaftlichen Forschung (Schwerpunktprogramm S 4501/2-Med), Oesterreichische Nationalbank, Bundes-ministerium fur wissenschaft und Forschung, and Dr. Legerlotz-Foundation. J.S. is a recipient of a Max-Kade-Foundation Inc. research fellowship.

Table 1. Criteria for calcium channel toxins

1. The toxin should alter Ca^{2+} channel function and/or the binding of channel-specific ligands. Alterations in channel function should preferably be measured by the patch-clamp method. Modulation of ligand binding to L-type channels should be probed for on at least three of the L-type Ca^{2+}-channel-associated drug receptors (dihydropyridines, phenylalkylamines, and benzothiazepines; for review, see Glossmann and Striessnig 1988, 1990; Glossmann et al. 1988). For N-type channels, radiolabelled ω-conotoxin GVIA and neuronal tissue from different species may be employed.
2. The toxin should bind directly to the Ca^{2+}-channel . . . complex [i.e. $α_1$-subunit of L-, N- or P-type channels]. An indirect effect on channel function via a second messenger system must be excluded.
3. The toxin should be devoid of enzymatic – in particular phospholipase or protease – activity.

(Adapted and modified from Hamilton and Perez 1987).

pounds (including toxins) have been indispensable tools for Ca^{2+} channel research. Several toxins proposed to act on voltage-dependent Ca^{2+} channels have been described in the past, but most of them cannot be classified as Ca^{2+} channel-selective toxins based on the criteria formulated by Hamilton and Perez (1987) (summarized in Table 1). The best characterized Ca^{2+}-channel toxins to date are peptides isolated from marine snails of the genus *Conus*, termed ω-conotoxins. They are the only known blockers so far of a subtype of neuronal Ca^{2+} channels (N-type channels), which play an eminent role for neurotransmitter release from central and peripheral neurons. The structure and toxicology of toxins from venomous snails (including the ω-conotoxins) as well as of other putative Ca^{2+}-channel toxins have been covered in recent reviews (Gray et al. 1986; Wu and Narahashi 1988). We therefore focus on current studies with ω-conotoxins and several newly discovered spider toxins. The importance of these toxins for Ca^{2+}-channel classification will be discussed. This review covers work available to us in published form or as preprints up to September 1990.

B. Calcium Channel Toxins from Venomous Snails: ω-Conotoxins

Conus is the best-known genus of venomous snails containing more than 300 species. Especially the fish-hunting ones paralyze their prey very rapidly, usually within a few seconds by injecting their venom through a disposable, harpoon-like tooth. Their venom is also toxic for other vertebrates including man, and several human fatalities from *Conus* stings have been reported (see references in Gray et al. 1986). The typical *Conus* venom is a milky fluid, containing small insoluble granules, soluble proteins,

Table 2. Cysteine arrangements in different toxins

Arrangement	Toxin	Number of peptides found
CCC**C(**)	α-Conotoxins	>4
CCC**C**CC**	μ-Conotoxins	10
C**C**CC**C**C	ω-Conotoxins	16
CC**CC**C**C**C**	μ-Agatoxins	>6
CC**C**C**C**C**C**C**	ω-Agatoxin IA	–
CC**CC**C–C**C–C**	Curtatoxins	3

Asteriks between the cysteine (C) residues denote a variable number of intercysteine residues [**, $n > 1$; (**), $n = 0$ or $n > 1$; –, $n = 1$]. Note that arrangements are highly conserved within a single class of toxins. Once this arrangement in evolution was found to confer high affinity for a certain receptor type, it was used as a matrix for other toxins perhaps even for targets having no relevance for the snail or the spider.

C K S P̲ G S S C S P̲ T S Y N C C R + S C N P̲ Y T K R C Y * GVIA
C K G̲ K G A K C S̲ R L M Y D C C T G S C̲ R + + S G K C *MVIIA

Fig. 1. Comparison of the amino acid sequence of the calcium channel blockers ω-conotoxin GVIA and MVIIA. Disulfide bridges are formed between Cys residues 1 and 16, 8 and 19, and 15 and 26. *, Amidated C-terminal; P, hydroxyproline; +, gap for alignment

peptides, and low molecular weight compounds. The term "conotoxins" was first assigned to paralytic, relatively small peptides (13–22 amino acid residues) from the venom of *Conus geographus*. Paralysis of vertebrates is very effectively induced by several different conotoxins in the venom which synergistically block three types of ion channels in the neuromuscular junction: α-Conotoxins block the nicotinic acetylcholine receptor of the post-synaptic membrane, whereas μ- and ω-conotoxins block voltage-dependent Na^+ and Ca^{2+} channels on the presynaptic membrane, respectively (for a systematic nomenclature of conotoxins, see GRAY et al. 1988). Conotoxins are rich in cysteine residues, representing perhaps the most disulfide-cross-linked peptides known (WOODWARD et al. 1990). Within the different classes of conotoxins the position of the cysteine residues is highly conserved, and different motifs can be described (Table 2). ω-Conotoxins, for example, are 25–30-mers containing 6 cysteine residues able to form three disulfide bonds to stabilize the molecule's structure (Fig. 1). Although there are many more different possibilities of disulfide bond formation, only this biologically most active form is synthesized. How is the correct folding of the toxin managed in vivo? Cloning of the cDNA of several conotoxins allows a first insight into this mechanism (WOODWARD et al. 1990). The amino acid sequences derived from several cloned cDNAs indicate the synthesis of propeptides more than 70 amino acids in length. They contain a N-terminal region which is highly conserved among various toxins and a hypervariable C-terminal portion. The latter corresponds to the toxic peptide, which is released from the propeptide by proteolytic processing. It is very likely that the conserved region is essential for the correct folding of the propeptide. This could then result in the formation of the biologically active, disulfide-bonded conformation of the toxin. Several ω-conotoxins have so far been isolated from *Conus geographus* (GVIA, GVIB, GVIC, GVIIA, GVIIB) and *Conus majus* (MVIIA, MVIIB) (Gray et al. 1988; Olivera et al. 1990). All of them are highly basic peptides. Some contain an amidated C-terminus and the unusual amino acid hydroxyproline. The structure of the most investigated ω-conotoxins, GVIA and MVIIA, is shown in Fig.1. Their length allows synthesis by solid phase methods. Synthetic ω-conotoxin GVIA and its monoiodinated derivative, [^{125}I-tyrosyl]-ω-conotoxin GVIA, are commercially available and widely used to study neuronal Ca^{2+} channels. An overview of radioligand binding studies with the iodinated toxin is given elsewhere (Gray et al. 1986; Glossmann and Striessnig 1990). Intra-peritoneal (i.p.) injection of ω-conotoxin GVIA causes death in fish, frogs, and birds by blocking neuromuscular transmission. In mammals, no effect on the neuromuscular junction is observed. Intraperitoneal injection (3.5 nmol/g) into neonatal mice causes a decrease in the respiration rate and leads to death, whereas 6-day-old mice showed marked depression of respiration but survived. As a possible target for peripheral actions of ω-conotoxin GVIA the carotid body was suggested (Myers et al. 1991). Intracerebroventricular (i.c.v.) injection causes a persistent shaking syn-drome, suggesting that Ca^{2+} channels of central neurons are toxin-sensitive. The onset and persistence of this tremor is dose-dependent, lasting from a few minutes after injection to up to 3 days (Olivera et al. 1985), and was therefore used as a bioassay to follow the purification of ω-conotoxin GVIA, the "shaker peptide", over several chromatographic steps. Rats appear to be more sensitive to the toxin. A complex pattern of hemodynamic changes (tachycardia and a transient increase of mean arterial pressure), neurological motor deficits, and disturbances of the thermoregulation are observed after i.c.v. injection. None of these could be reproduced by extreme doses of L-type channel blockers [(+)-*cis*-diltiazem, nifedipine] but were abolished by the i.c.v. application of muscimol (a γ-aminobutyric acid agonist) or combined treatment with a β-adrenoceptor antagonist and a muscarinic acetylcholine receptor blocker (Shapira et al. 1990a). After unilateral i.c.v. administration the CA3 pyramidal cells in the hippocampus of the injected hemisphere are necrotic, suggesting that the Ca^{2+} channels in these cells are essential for viability (Shapira et al. 1990b). Kerr and Yoshikami (1985) first established neuronal voltage-dependent Ca^{2+} channels as the target for ω-conotoxins. They observed an irreversible block of nerve stimulus-evoked release of transmitter at the frog skeletal muscle junction by the prevention of Ca^{2+} entry into nerve terminals during depolarization. They also demonstrated an irreversible inhibition of Ca^{2+} currents in chick dorsal root ganglion cells. However, ω-conotoxins were without effect on Ca^{2+} currents in smooth, heart, and skeletal muscle (KERR and YOSHIKAMI 1985; REYNOLDS et al. 1986). ω-Conotoxin MVIIA causes death in fish and birds upon i.p. injection but, in contrast to ω-conotoxin GVIA, not in amphibians. Species- and synapse-dependent toxin effects indicate that ω-conotoxin-sensitive Ca^{2+} channels can be further subdivided into ω-conotoxin MVIIA-sensitive and -insensitive species. This was also confirmed in radioreceptor assays (OLIVERA et al. 1987). Unfortunately, ω-conotoxin MVIIA is not commercially available, so detailed information about its effects in the mammalian nervous system are missing.

I. ω-Conotoxin GVIA is a Selective Blocker of N-Type, Dihydropyridine-Insensitive Calcium Channels

Voltage-clamp and single-channel studies allowed the discrimination of different types of calcium channels in excitable cells. Initially, three types were described in chick dorsal root ganglions (CARBONE and LUX 1984; NOWYCZKY et al. 1985; Fox et al. 1987a,b), which were termed T-, L-, and N-type. T-type channels (I_{fast}, low-voltage-activated channel) are activated at relatively negative membrane potentials and show rapidly transient channel kinetics. In all tissues studied (e.g., heart, brain, skeletal and smooth muscle) they possess a relatively small unitary conductance. Specific ligands are still unclear, but drugs which interact with other Ca^{2+}-channel subtypes or with Na^+ channels and Na^+ transporters also affect T-type currents (e.g., gallopamil, nicardipine, flunarizine, amiloride, phenytoin) (TAKAHASHI et al. 1989; HESS 1990; PELZER et al. 1990; GLOSSMANN and STRIESSNIG 1990).

L-type Ca^{2+} channels (long lasting-channel, I_{slow}, high-threshold activated channel) are modulated by chemically different classes of drugs ("Ca^{2+} antagonists"), which bind to distinct drug receptors on the channel α_1-subunit (for review, see GLOSSMANN and STRIESSNIG 1988, 1990). These highly specific tools eventually led to the first isolation of a L-type-like Ca^{2+} channel from skeletal muscle transverse tubules and the cloning and sequencing of its cDNA (for review, see CATTERALL et al. 1988; HOSEY and LAZDUNSKI 1988; CAMPBELL et al. 1988; GLOSSMANN and STRIESSNIG 1990). In addition to their dihydropyridine-sensitivity, L-type Ca^{2+} channels are characterized by an activation range at positive test potentials, large unitary conductances, slow inactivation kinetics (if Ca^{2+} is not the charge carrier) and modulation by cAMP-dependent protein kinase (in heart, skeletal muscle, and GH_3 cells; HOFMANN et al. 1987). Like T-type channels, L-type Ca^{2+} channels are functionally expressed in neuronal as well as nonneuronal cells, like muscle cells.

N-type Ca^{2+} channels (non-T, non-L channels) are also high-voltage-activated channels, but they are neuron-specific and, generally speaking, dihydropyridine-insensitive (see, however, TAKAHASHI et al. 1989). First described by Nowyczky et al. (1985) in dorsal root ganglion cells, they were reported to be activated at more positive membrane potentials than T-type channels. They possess an intermediate single-channel conductance and inactivate rapidly and completely during strong depolarization. Therefore, it was believed that rapidly decaying components of the whole cell Ca^{2+} (or Ba^{2+}) current represents flux through N-type channels, whereas the maintained, slowly incativating component occurs through L-type currents (Fox et al. 1987a,b). Earlier studies describing the effects of ω-conotoxin GVIA on neurotransmitter release, whole-cell Ca^{2+} currents, and single-channel currents have been summarized by Gray et al. (1988) and Tsien et al. (1988). All three types of Ca^{2+} channels were reported to be blocked by the toxin in peripheral (chick dorsal root ganglion cells, sympathetic neurons) as well as central (rat hippocampus) neurons. N- and L-type channels were blocked irreversibly, but T-type channels only weakly and in a reversible manner. The toxin also inhibited K^+-induced or electrically evoked neurotransmitter release (e.g., Dooley et al. 1987; Reynolds et al. 1986; Rivier et al. 1987). As dihydropyridines (e.g., nitrendipine, PN200-110) and other L-type Ca^{2+}-channel active drugs were ineffective, it was concluded that N-type Ca^{2+} channels must be responsible for neurotransmitter release from peripheral and central neurons. Hirning et al. (1988) provided direct evidence that only Ca^{2+} transients (measured by fura-2 fluorescence) elicited by Ca^{2+} influx through N-type, but not L-type, channels resulted in norepinephrine (NE) release from rat sympathetic ganglia. Taken together, these results explained the previously observed insensitivity of neurotransmitter release against L-type Ca^{2+} channel blockers in differentiated neuronal cells (Miller 1987).

More recent studies confirm ω-conotoxin GVIA block of N-type Ca^{2+} channels and inhibition of neurotransmitter release in a variety of neuronal cells and are summarized in Tables 3 and 4. However, the initial conclusion that ω-conotoxin GVIA also persistently blocks L-type channels is no longer supported. Plummer et al. (1989) found that ω-conotoxin GVIA indeed inhibits a rapidly decaying as well as slowly inactivating component of neuronal Ca^{2+} currents in PC12 cells and rat sympathetic ganglia. In contrast to the earlier view, both components were carried by dihydropyridine-insensitive N-type Ca^{2+} channels. Only a minor portion of the whole cell current was carried by L-type channels ($<20\%$), which, like their counterparts in nonneuronal cells, were unaffected by ω-conotoxin GVIA but always stimulated by the dihydropyridine agonist S-202-791. Similar findings were recently reported for chick and rat sensory neurons (Aosaki and Kasai 1989; Carbone et al. 1990) and rat hippocampus neurons (Toselli and Taglietti 1990). ω-Conotoxin GVIA was either a reversible and partial blocker of L-currents or did not block them at all. It was therefore concluded that the effects of ω-conotoxin GVIA on L-type currents (which are, using pharmacological criteria for definition, dihydropyridine-sensitive) can be more reliably evaluated after selective blockade of the dihydropyridine-

Table 3. Effect of ω-conotoxin GVIA on neurotransmitter release, Ca^{2+} influx, and excitatory postsynaptic potentials of neuronal Ca^{2+} channels in various neuronal cells[a]

Neuronal cell	Effect/concentration	Ca^{2+} antagonist sensitivity	Reference	Comment
Guinea-pig sympathetic nerve terminals	Irreversible block of EE NTR (10 nM)	–	Brock et al. (1990)	
Rat cultured superior cervical ganglia neurons	70% inhibition of K^+-induced [^{3}H] norepinephrine release	(yes)	Hirning et al. (1988)	Small stimulation of release by BAY K-8644 after prolonged depolarization, no effect of nitrendipine
Rat submandibular ganglia	No effect on synaptic transmission	yes	Seabrook and Adams (1989)	$(+)$-cis-diltiazem (30 μM) induces 80% block of e.p.s.p. 90% inhibition of the whole cell Ca^{2+} current from cultured cardiac parasympathetic neurons
Rat myenteric plexus	70% block of EE [^{3}H]Ach release (EC_{50} = 1 nM)	no	Wessler et al. (1990)	Increase in extracellular Ca^{2+} from 0.9 to 3.5 mM decreases potency of ω-conotoxin GVIA
Rat phrenic nerve	No effect at 100 nM and 1.8 mM extracellular Ca^{2+}	no	Wessler et al. (1990)	Inhibitory effect seen after prolonged preincubation with 100 nM before stimulation and low extracellular Ca^{2+} (0.9 mM)

Table 3 (continued)

Neuronal cell	Effect/concentration	Ca^{2+} antagonist sensitivity	Reference	Comment
Rat cortical/hippocampal slices	Partial block of NMDA and kainate-induced NT release	no	KEITH et al. (1989)	
Rat neocortex slices	90% block of EE [^{3}H]Ach release (EC$_{50}$ = 10 nM)	–	WESSLER et al. (1990)	
Human neocortex slices	60%–70% inhibition of [^{3}H]Ach and [^{3}H]norepinephrine release	–	FEUERSTEIN et al. (1990)	[^{3}H]Ach release was inhibited more effectively (70% inhibition, EC$_{50}$ = 3 nM) than [^{3}H] norepinephrine release (60% inhibition, EC$_{50}$ = 14 nM)
Rat neostriatum slices	48% reduction of K$^+$-evoked (25 mM) [^{3}H]-GABA release at 100 pM	–	THATE and MEYER (1989)	Indirect evidence for inhibition of somatostatin release by ω-conotoxin
Rat hypothalamic slices	80% inhibition of K$^+$-evoked histamine release (EC$_{50}$ = 1–10 nM)	no	TAKEMURA et al. (1989)	Effect of the agonist BAY K-8644 was not tested
Guinea-pig hippocampus transverse sections	65% block of synaptic transmission from mossy fibers to CA1 neurons (100 nM)	no	KAMIYA et al. (1988).	

Cultured hypothalamic mouse neurons	45% inhibition of K^+-evoked TRH release (100 nM)	yes	LOUDES et al. (1988)	TPA stimulates basal TRH release; BAY K-8644 enhances basal and K^+-evoked release; TPA plus BAY K-8644 gives a more than additive response; PN200-100 does not inhibit basal or K^+-evoked release
Hippocampal CA1 pyramidal neurons	Decrease of Ca^{2+} spike (5 µM)	–	HIGASHI et al. (1990)	
Mouse neurohypophysis	50% block of neuropeptide release (5 µM)	no	OBAID et al. (1989)	Changes in light scattering due to neuropeptide secretion was the optical signal
Chick brain homogenates	Nearly complete inhibition of K^+-depolarization-induced InsP$_3$ production and NT release	no	HOFMANN and HABERMANN (1990)	Moderate inhibition of InsP$_3$ production but large inhibition of NT release upon stimulation with carbachol; only weak effects of the toxin in rat brain
Cultured bovine adrenal chromaffin cells	Partial inhibition of $^{45}Ca^{2+}$ uptake (250 nM)	yes	BALLESTA et al. (1989)	1 µM Nitrendipine inhibits 60% of K^+-evoked Ca^{2+} uptake; conotoxin effect was only seen when cells were hyperpolarized before depolarization

Table 3 (continued)

Neuronal cell	Effect/concentration	Ca²⁺ antagonist sensitivity	Reference	Comment
Isolated bovine adrenal chromaffin cells	No effect on BAY K-8644 evoked Ca²⁺ transients	yes	JAN et al. (1990)	
Isolated bovine adrenal chromaffin cells	No effect on K⁺ depolarization-induced ⁴⁵Ca²⁺ uptake (20 nM); no effect on K⁺ bradykinin or prostaglandin E₁-induced norepinephrine release	yes	OWEN et al. (1989) (⁴⁵Ca²⁺ uptake)	1 μM Nitrendipine inhibits and BAY K-8644 stimulates
Rat cerebrocortical synaptosomes	40% block of Ca²⁺ influx (20 μM)	no	SUSZKIW et al. (1989)	
Rat neostriatum synaptosomes	Inhibition of K⁺-stimulated dopamine release (25% at 10 nM, 60% at 10 μM) and Ca²⁺ influx	–	WOODWARD et al. (1988)	Biphasic effect (high- and low-affinity site)

[a] Papers already reviewed in GRAY et al. (1988) are not listed

Ach, acetylcholine; Ca²⁺ antagonist sensitivity, sensitivity to classic L-type Ca²⁺ channel drugs (e.g. dihydropyridine), EE, electrically evoked; e.p.s.p., excitatory postsynaptic potentials; NTR, neurotransmitter release; irr, irreversible block; NT, neurotransmitter; Bay K-8644 is a L-type Ca²⁺-channel-activating dihydropyridine; TPA, 10-tetradecanoylphorbol-13-acetate; GABA, γ-aminobutyric acid; InsP₃, inositol 1,4,5-triphosphate; TRH, thyrotropin releasing hormone; NMDA, N-methyl-D-aspartate.

Table 4. Effects of ω-conotoxin GVIA: voltage-clamp and single-channel studies

Cell	Channel	Inhibition	Concentration (μM)	DHP sensitivity	Reference	Comment
Cultured rat hippocampal neurons	"N"	irrev	4	no	TOSELLI and TAGLIETTI (1990)	
	"L"	no	4	yes		
Cultured chick sensory neurons	"N"	irrev	0.5–5	no	AOSAKI and KASAI (1989)	The single-channel conductance of the toxin- and dihydropyridine-sensitive components was 13 and 25 pS (100 mM Ba^{2+}), respectively; 50% of the dihydropyridine-sensitive current was reversibly blocked by the toxin; only 67/175 cells displayed a dihydropyridine-sensitive current
	"L"	partial rev	0.5–5	yes		
Rat sensory neurons	T	rev, weak	6.4	no	CARBONE et al. (1990)	Few differences were found between toxin- and dihydropyridine-sensitive currents with respect to inactivation rate and activation range; the existence of toxin- and dihydropyridine-insensitive channels was not ruled out
	"N"	irrev	6.4	no		
	"L"	no	6.4	yes		
Rat dorsal root ganglion × mouse hybrid (F11) cells	"N"	irrev	10	yes	BOLAND and DINGLEDINE (1990)	27% block of N-type, 81% block of L-type; evidence for multiple components of transient and sustained currents
	"L"	irrev	10	yes		

Table 4 (continued)

Cell	Channel	Inhibition	Concentration (μM)	DHP sensitivity	Reference	Comment
Differentiated rat PC12 cells and rat superior cervical ganglia	N L	irrev/rev no	16 16	no yes	Plummer et al. (1989)	Two N-type components which were blocked reversibly and irreversibly, respectively. N-type currents gave also rise to a non-inactivating current and have a single-channel conductance similar, to L-type channels
Rat superior cervical ganglion neurons	N L	irrev irrev	10 10	no yes	Hirning et al. (1988)	Single-channel conductances were 11 and 27 pS (110 mM Ba^{2+}) for N- and L-type channels, respectively; toxin-resistant high-voltage-activated currents were not probed with BAY K-8644
Retinal ganglion cells	T L	rev rev		no yes	Karschin and Lipton (1989)	Single-channel conductances were 8 and 20 pS (96–110 mM Ba^{2+}) for the T- and L-component, respectively

Channel types are placed in quotation marks when they were not designated by the authors but assigned from context DHP, dihydropyridine; rev, reversible; irrev, irreversible.

sensitive current component. This was not the case in earlier studies in which current components were defined by biophysical rather than pharmacological criteria. Insensitivity of L-type Ca^{2+} channels towards ω-conotoxin GVIA was also found in adrenal chromaffin cells: K^+ depolarization induced Ca^{2+} entry, and Ca^{2+} transients or neurotransmitter release is modulated by agonistic (e.g., BAY K-8644) or antagonistic dihydropyridines but not by ω-conotoxin GVIA. MINTZ et al. (1991) determined the relative contribution of L-, N-, and T-type Ca^{2+} channels to the whole cell Ca^{2+} current in peripheral and central neurons. In addition to omega-conotoxin they also used the spider toxin omega-agatoxin-IIIA, an inhibitor of N- and L-type Ca^{2+} channels (see section C). A fraction of (high treshold) Ca^{2+} current left unblocked by Ca^{2+} antagonists drugs and these toxins (PLUMMER et al., 1989; AOSAKI and KASAI, 1989; MINTZ et al., 1991) is believed to be mediated, at least in part, by P-type Ca^{2+} channels (see section C.II.).

The recent observation of a tetrodotoxin-sensitive Na^+ channel in hippocampus neurons which is permeable to Ca^{2+} (and is blocked by low concentrations of flunarizine; AKAIKE, personal communication) adds an additional complicating factor. It seems that the contradictions which exist between classification of channels by biophysical parameters and by pharmacological sensitivities can only be resolved by additional specific pharmacological tools and cloning and expression of neuronal Ca^{2+} channels.

With the help of ω-conotoxin GVIA attempts have also been made to determine the relative contribution to neurotransmitter release of Ca^{2+} influx versus Ca^{2+} release from intracellular stores (HOFMANN and HABERMANN 1990). In chick brain, ω-conotoxin GVIA inhibits K^+- as well as carbachol-induced NE release. However, inositolphosphate production due to muscarinic receptor activation is only slightly inhibited by the toxin or by omission of Ca^{2+} from the extracellular medium. Therefore, in contrast to Ca^{2+} entry, carbachol-induced release of intracellularly sequestered Ca^{2+} (via inositolphosphate production) is apparently not involved in transmitter release. Muscarinic receptor activation not only directly stimulates inositolphosphate production but indirectly causes depolarization of the cell membrane due to the closing of receptor-coupled (M) K^+ channels, thereby promoting Ca^{2+} ion influx through voltage-dependent Ca^{2+} channels. It is the latter pathway which explains the effectiveness of ω-conotoxin GVIA on carbachol-induced neurotransmitter release (HOFMANN and HABERMANN 1990).

A consistent finding with ω-conotoxin GVIA is the relative insensitivity (IC_{50} values $>1\,\mu M$) of acetylcholine release from peripheral motoneurons of mammals (WESSLER et al. 1990), whereas noradrenergic (PRUNEAU and ANGUS 1990) and cholinergic autonomic neurotransmitter release (WESSLER et al. 1990) is inhibited at nanomolar concentrations. Taken together with the effects of the toxin in CNS neurotransmitter release paradigms, these observations are best explained with a mosaic building block model of Ca^{2+} channels: The narrow specificity of the toxin is of utmost importance to the snail (see OLIVERA et al. 1990) to poison the prey (i.e., fish) effectively. The target amino acids on presynaptic Ca^{2+} channels of fish (which form the extracellular docking domain for the toxin) may or may not be present on the subtypes of channels that developed later in an apparently tissue- and cell-function-adapted manner. In mammals, some CNS neurons and perhaps most vegetative peripheral neurons retained the high-affinity docking domain. Others (i.e., the motoneurons) did not. It is already emerging (see ω-agatoxin below) that this interpretation is the most appropriate until the entire master plan of all Ca^{2+} channels is unveiled by molecular biology methods. Furthermore, we caution against a classification of neuronal Ca^{2+} channels based on findings with *one* toxin only.

ω-Conotoxin GVIA also discriminates between different channel conformations of one channel. Ca^{2+} channels become permeable to Na^+ when the external Ca^{2+} concentration is lowered to submicromolar levels. This is explained by removal of Ca^{2+} from some sort of selectivity filter (for review, see HESS 1990). CARBONE and LUX (1988) found that the toxin is able to discriminate between the Na^+- and Ca^{2+}-conducting state of the channel. They blocked Ca^{2+} currents through high-voltage-activated Ca^{2+} channels (mainly carried by N-type channels) irreversibly with ω-conotoxin GVIA in chick dorsal root ganglion neurons. Washing the cells with a toxin-free, low Ca^{2+} solution resulted in the appearance of Na^+ currents. When Ca^{2+} was added back to the solution, the block of Ca^{2+} currents was rapidly re-established, suggesting that the Na^+ current occurred through ω-conotoxin GVIA-modified Ca^{2+} channels. Relief of the inhibition in the presence of Na^+ and absence of Ca^{2+} ions cannot be explained by a selective block of the toxin-receptor interaction. Although [^{125}I-tyrosyl]-ω-conotoxin GVIA binding to mammalian brain membranes is inhibited in the presence of millimolar concentrations of Na^+, Ca^{2+}, and other cations, these Na^+ ions are unable to dissociate the [^{125}I-tyrosyl]-ω-conotoxin GVIA-channel complex once formed (CRUZ and OLIVERA 1986; KNAUS et al. 1987; GLOSSMANN and STRIESSNIG 1990). Therefore, the most likely explanation for this observation is that different permeating ions induce different Ca^{2+}-channel conformations, and ω-conotoxin GVIA persistently depresses only the Ca^{2+}-permeable state.

II. ω-Conotoxin GVIA Reveals Functional Associations Between Excitatory Amino Acid Receptors and Neuronal Calcium Channels

The main neurotransmitter receptors mediating synaptic excitation in the mammalian CNS are those for the excitatory amino acids, like L-glutamate or L-aspartate. At least 5 different types can be distinguished. The NMDA (*N*-methyl-D-aspartate), AMPA (α-amino-3-hydroxy-5-methyl-4-isoxazole-propionic acid), and kainate receptors act as cation-selective ionophores (for review, see WATKINS et al. 1990), causing an increase in the permeability of the postsynaptic membrane for cations like Na^+, K^+, and Ca^{2+}. NMDA receptors are permeable to all three cations, whereas the Ca^{2+} permeability for kainic acid receptors is controversial (MURPHY and MILLER 1989). Their activation results in membrane depolarization, which triggers the opening of voltage-dependent Ca^{2+} channels, providing additional Ca^{2+} influx. Both L-and N-type channels seem to be involved in this process. Inhibition of NMDA- and kainate-induced Ca^{2+} influx by L-type channel blockers has been described (DINGLEDINE 1983; MURPHY and MILLER 1989; WEISS et al. 1990). On the other hand, RIVEROS and ORREGO (1986) observed activation of a Cd^{2+}-sensitive but dihydropyridine-insensitive Ca^{2+} influx by excitatory amino acids. With ω-conotoxin GVIA as a new pharmacological tool for N-type channels, KEITH et al. (1989) studied the effect of the toxin on NMDA- and kainate-induced neurotransmitter ($[^3H]$-NE) release from rat cortical and hippocampal brain slices. $[^3H]$-NE release induced by both agonists was similarly inhibited by $100\,nM$ ω-conotoxin GVIA (by about 30%) whereas the dihydropyridine PN200-110 was without significant effect. This means that the majority of the Ca^{2+} influx which triggers transmitter release seems to be mediated by the NMDA- or kainate-gated ionophores (KEITH et al. 1989; MURPHY and MILLER 1990) but that a significant amount may be provided by N-type, voltage-dependent Ca^{2+} channels.

Other modes of functional coupling between excitatory amino acids receptors and Ca^{2+} channels were found in rat hippocampus. Here the excitatory synaptic input to pyramidal CA1 neurons is operated by glutaminergic mechanisms, which are sensitive to non-NMDA blockers during low frequency stimulation but become NMDA blocker-sensitive during intensive stimulation. The transient activation of NMDA receptors is accompanied by an elevation of intracellular Ca^{2+} concentrations, which is believed to be a necessary step in the generation of long-term potentiation (LTP, long lasting enhancement of synaptic transmission at synapses after high frequency stimulation; for a review of LTP, see KENNEDY 1989). KRISHTAL et al. (1989) stimulated the appropriate input pathway to evoke postsynaptic stimulatory potentials (PSPs) on hippocampus CA1 neurons. As expected these PSPs were insensitive to NMDA antagonists, but they were blocked by ω-conotoxin GVIA to 20%–30% of the control levels. PSPs were greatly suppressed after bath application of L-glutamate or L-aspartate but largely recovered after maintaining perfusion with the excitatory amino acids. The recovered responses after L-glutamate treatment displayed clearly altered pharmacological properties. First, they were NMDA antagonist-sensitive, and second, they lost their sensitivity towards ω-conotoxin GVIA. Upon subsequent cycles of washout and treatment with excitatory amino acids, long-lasting ("remembered") pharmacological changes in the sensitivity of the recovered responses towards NMDA antagonists as well as ω-conotoxin GVIA were observed. It was concluded that this form of hippocampal plasticity involves transitions between distinct states of synaptic functioning, which include (but are not limited to) changes in the functioning of ω-conotoxin GVIA-sensitive Ca^{2+} channels.

III. Probing the Structure and Location of N-Type Calcium Channels with ω-Conotoxin GVIA Derivatives

[^{125}I-tyrosyl]-ω-conotoxin GVIA has been widely used as a radioligand (for review, see Gray et al. 1988; Glossmann and Striessnig 1990) to identify putative N-type Ca^{2+}-channel-associated components in the vertebrate CNS. Binding of [^{125}I-tyrosyl]-ω-conotoxin GVIA to its receptor, in accordance with its toxic effect, occurs in an apparently irreversible manner and with very high affinity. The binding is not modulated by L-type Ca^{2+}-channel drugs. In contrast, [^{3}H-proprionyl]-ω-conotoxin GVIA, synthesized by Yamaguchi et al. (1988), exhibits partially reversible binding which was stereoselectively inhibited by the diastereoisomers of *cis*-diltiazem, which are L-type Ca^{2+}-channel blockers. [^{125}I-tyrosyl]-ω-conotoxin GVIA has also been covalently(photoaffinity) cross-linked to the receptor polypeptides (e.g., as its azidobenzoate derivative). Apparent molecular weights of its targets could be determined by SDS-PAGE and autoradiography. Several specifically labeled polypeptides with apparent molecular weights between 135 000 and 310 000 have been reported (Cruz and Olivera 1986; Barhanin et al. 1988; Glossmann et al. 1988; Myers et al. 1991). Despite the fact that cross-linking or photoaffinity labeling experiments provide a very sensitive tool for the detection of putative Ca^{2+}-channel-associated components, . . . it took several years to develop procedures for the purification of the N-type channel complex (McEnery et al. 1991). It consists of a 230 kDa alpha1 subunit (already identified by photoaffinity labeling) associated with a neuronal alpha2-delta and beta subunits (Sakamoto et al., 1991; McEnery et al., 1991; Ahlijanian et al., 1991). At least a subset of N-type Ca^{2+} channel-associated alpha1 subunits is phosphorylated in vitro by the protein kinases A and C (Ahlijanian et al., 1991).

For the immunological characterization of these N-type channel subunits, monoclonal (Sakamoto et al., 1991; Ahlijanian et al., 1991) as well as polyclonal, sequence-directed antibodies (Ahlijanian et al., 1991) have been employed. In addition, sera from patients with Lambert-Eaton-myasthenic syndrome (LEMS) immunoprecipitate [^{125}I-tyrosyl]-omega-conotoxin GVIA labeled N-type channels (Vincent et al., 1989). These antibodies inhibit channel function, which explains the clinical symptoms of LEMS patients (general signs of a deficit in neurotransmitter release, like e.g. reduced tendon reflexes, muscle weakness, constipation). In contrast to the vertebrate CNS membrane system, ω-conotoxin GVIA binds with low affinity and reversibly (($K_{0.5} \approx 0.6\,\mu M$) to membranes from the electric organ of the ray *Omata discorpyge* and inhibits transmitter release from cholinergic synaptosomes (YEAGER et al. 1987). It was recently reported that this low-affinity binding copurified with the nicotinic acetylcholine receptor (HORNE et al. 1991).

If N- and L-type Ca^{2+} channels are different entities, one would also expect a varied distribution of ω-conotoxin and L-type channel-associated drug receptors within the nervous system. Autoradiography of [^{125}I-tyrosyl]-ω-conotoxin GVIA-labeled brain sections demonstrated the highest density of binding sites in the cerebral cortex, hippocampus, nucleus caudatus-putamen, nucleus of the solitary tract, and glomerular layer of the olfactory bulb (MAEDA et al. 1989). The distribution of [^{125}I-tyrosyl]-ω-conotoxin GVIA receptors differed in many respects from the distribution of the Ca^{2+} antagonist sites for, e.g., (+)-[^{3}H]PN200-110, [^{3}H]nitrendipine, or (−)-[^{3}H] desmethoxyverapamil (for review, see GLOSSMANN and STRIESSNIG 1988). In the hippocampus, the highest density of [^{125}I-tyrosyl]-ω-conotoxin GVIA was in the stratum oriens and the stratum radiatum, and moderate densities were seen at the pyramidal cell layer, the stratum moleculare and lacunosum, as well as the molecular layer of the gyrus dentatus. In contrast, the highest density of L-type Ca^{2+}-channel drug receptors are present in the molecular layer of the gyrus dentatus, where they are presumably located on the dendrites of granule cells (CORTES et al. 1983). Different distributions were also observed in the olfactory bulb and cerebellum. Accordingly, binding studies comparing the densities of receptors for [^{125}I-tyrosyl]-ω-conotoxin GVIA and the L-type Ca^{2+}-channel ligand (+)-[^{3}H]PN200-110 in membranes of several brain regions also lead to the conclusion that the binding sites for the two ligands are nonidentical, are localized to CNS regions exhibiting high densities of synaptic connections, and are relatively stable during the aging process, when studied in brains of rats of different ages (DOOLEY et al. 1988). JONES et al. (1989) synthesized tetramethyl-rhodamine-ω-conotoxin GVIA (a fluorescent derivative) and biotin-ω-conotoxin GVIA to study the distribution and lateral mobility of ω-conotoxin GVIA on hippocampal CA1 neurons. The affinity of the respective monomodified probes was 20–40 times lower than for the native toxin. Colloidal gold decoration of biotin-ω-conotoxin GVIA labeled neurons suggested binding to cell bodies and processes. The latter were identified as dendrites. Fluorescent staining revealed the localization of putative N-type Ca^{2+} channels in multiple clusters within restricted areas. Presumably, these clusters were located on cell bodies, although no electron microscopic evidence was provided for this assumption. After innervation, labeling was organized into a single cluster that coincided with the site of synaptic contact. It was reduced by 80% in the presence of an excess of the native toxin. The majority of the Ca^{2+} channels were immobile. Most likely, association of the channel molecules with, e.g., cytoskeletal proteins limit their mobility and concentrate them regionally. The formation of arrays of Ca^{2+} channels may serve to initiate very large focal cytoplasmic Ca^{2+} transients. One model predicts that from these "hot spots" Ca^{2+} could reach even more distant targets by diffusion, thus mediating fast and synchronous triggering of neurotransmitter release (Smith and Augustine 1988). An uneven distribution on neuronal somata has also been reported for L-type Ca^{2+} channels, which are preferentially located at the base of major dendrites (Ahlijanian et al. 1990). Clustering of voltage-dependent Ca^{2+} channels and the formation of intramembrane particles was found in squid nerve terminals, mammalian motorneuron terminals, parallel-fiber-Purkinje synapses (Smith and Augustine 1988), and even in skeletal muscle transverse tubules where Ca^{2+} influx is not required to trigger the ryanodine-sensitive Ca^{2+} release channel from sarcoplasmic reticulum (Block et al. 1988; Hymel et al. 1988).

C. Calcium Channel Toxins from Spider Venoms

I. Peptide Toxins

1. ω-Agatoxins

Several classes of toxins isolated from the venom of the funnel web spider *Agelenopsis aperta* have been recently characterized (Skinner et al. 1989). Like the *Conus* venoms described above, they act synergistically by interfering with normal neuromuscular transmission at pre- as well as postsynaptic sites. α-Agatoxins are acylpolyamines (Quistad et al. 1990) and are postsynaptic blockers of glutamate-receptor controlled channels in the insect neuromuscular junction, causing transitory paralysis. μ-Agatoxins are peptides which induce repetitive firing and excessive neurotransmitter release at insect neuromuscular junctions and are presumably, like α-scorpion toxins, activators of presynaptic voltage-dependent Na^+ channels. A third class of toxins produces a long-lasting suppression of neurotransmitter release from the presynaptic membrane (Adams et al. 1990). Responsible for this effect are at least two different classes of peptides, termed ω-agatoxins I, II, and III. The type I toxins are of similar size (7.5 kDa) and display high sequence similarity. ω-Agatoxin IA has been completely sequenced. It consists of 66 amino acids with 5 tryptophan and 9 cysteine residues, respectively. The structure of these molecules is also stabilized by intramolecular disulfide bonds. The type II toxin is larger (9 kDa) and shares only low sequence homology with the type I toxins. ω-Agatoxin IA and IIIA are the most abundant agatoxins in the venom (0.6% and 0.4% of the total venom protein, respectively). If applied to the neuromuscular junction of prepupal house flies (*Musca domestica*) ω-agatoxin IA at a concentration of 18 nM reduces neurally evoked excitatory junctional potentials (EJPs, recorded in muscle, evoked by presynaptic neurotransmitter relase) by 90% but does not affect potentials elicited by the direct application of glutamate to the postsynaptic membrane. However, the block is completely abolished by increasing the extracellular Ca^{2+} ion concentration from 0.75 mM to

$5\,mM$. This indicates that ω-agatoxin IA blocks neuromuscular transmission on presynaptic Ca^{2+} channels. An effect on presynaptic voltage-dependent Na^+ channels was ruled out, as antidromic nerve spikes could still be elicited by focal stimulation of nerve terminals. ω-Agatoxin IA also suppresses high-threshold, presumably N-type, Ca^{2+} currents in cultured neonatal dorsal root ganglion (DRG) neurons (ADAMS et al. 1990) and blocks transmission at the frog cutaneous pectoris nerve-muscle junction (BINDOKAS and ADAMS 1989). Unfortunately, no data concerning mammalian neuromuscular junctions (where conotoxins are without effect) are as yet available. ω-Agatoxin IIA and IIIA (but not ω-agatoxin IA) are able to inhibit [^{125}I-tyrosyl]-ω-conotoxin GVIA binding to chick synaptosomal membranes (IC_{50} = 9 and $15\,nM$, respectively) and K^+-depolarization-induced $^{45}Ca^{2+}$ uptake (EC_{50} = 1.6 and $1.1\,nM$, respectively). The mechanism of binding inhibition (competitive or allosteric) is not known. It will also be interesting to see if the class I agatoxins displace [^{125}I-tyrosyl]-ω-conotoxin GVIA binding to rat neuronal tissue since ω-conotoxin GVIA and ω-agatoxin IA efficiently block N-type currents in rat DRG neurons. ω-Agatoxin IIIA blocks (in a voltage-independent manner and reversibly) L-type channel currents ($IC_{50} \approx 1\,nM$) in guinea-pig atrial myocytes (LEIBOWITZ et al. 1990) and L- and N-type Ca^{2+} channels in peripheral and central neurons of frogs and rats with similar potencies ($IC_{50} \sim 1.5\,nM$) (MINTZ et al., 1991).

Thus, it appears that these new classes of toxins have less channel subtype selectivity and broader species activity than the conotoxins. They are promising candidates for the further subclassification of channels and as structural probes. ω-Agatoxin IIIA appears to be the first peptide ligand for cardiac L-type channels. Identification of its binding domain within the known primary structure of the α_1-subunit could reveal insights into channel architecture (e.g., channel lining segments) and function as recently demonstrated for the classic L-type channel blockers of the verapamil and dihydropyridine type (STRIESSNIG et al. 1990, STRIESSNIG et al. 1991; NAKAYAMA et al., 1991).

2. *Hololena curta* Toxins

BOWERS et al. (1987) screened the venom of the hunting spider *Hololena curta* for potent toxins that inhibit neuronal Ca^{2+} channels in *Drosophila*. They isolated a peptide toxin fraction by size-exclusion and high performance liquid chromatography which was composed of two peptides of 7000 and 9000 dalton. This fraction produced a complete and long-lasting inhibition of neuromuscular transmission in *Drosophila* larvae due to interaction with presynaptic Ca^{2+} channels. Recently, three polypeptide neurotoxins (termed curtatoxin I, II, and III) were sequenced (STAPLETON et al. 1990). Their molecular weight is smaller (4.1 kDa, 36–38 amino acid residues); all three are sequence-related, cysteine-rich, and C-terminal amidated. The purified peptides produced a rapid and irreversible flaccid paralysis in crickets (*Acheta domestica*), suggesting a presynaptic blockade affecting neuromuscular function. Effects of these toxins on vertebrate neurons are unknown. As long as more detailed studies are not available, these toxins should be classified as putative Ca^{2+} channel toxins.

3. Plectreurys tristes Toxins

The venom of the spider *Plectreurys tristes* contains excitatory and inhibitory neurotoxins which irreversibly act on *Drosophila* larval neuromuscular junctions (BRANTON et al. 1987). Inhibition was consistent, with a specific, irreversible block of presynaptic Ca^{2+} channels, and confined to fractions with molecular weights of 6000–7000 upon size exclusion chromatography. The most abundant (0.1% of venom protein) component, a single polypeptide (molecular weight 7000) termed α-PLTX II, completely blocks Ca^{2+} currents in *Drosophila* but is inactive at the frog neuromuscular junction. The crude venom inhibits the binding of [^{125}I-tyrosyl]-ω-conotoxin GVIA to rat brain membranes in a noncompetitive manner (FEIGENBAUM et al. 1988). The venom component responsible for this effect has not yet been identified.

4. *Agelena opulenta* Toxins

A 35-amino acid toxin (agelenin) has been isolated and sequenced from the venom of the spider *Agelena pulenta* (HAGIWARA et al. 1990). The toxin contains six cysteine residues having homologies to the μ-agatoxins and to ω-agatoxin I. It blocks neuromuscular transmission at lobster neuromuscular synapses in an irreversible manner. It is suggested that agelenin is an inhibitor of presynaptic Ca^{2+} channels and can be classified among the putative Ca^{2+} channel toxin candidates.

II. Nonpeptide Toxins

In addition to T-, L-, and N-type Ca^{2+} channels, a fourth type has recently been postulated to exist in the dendrites of cerebellum Purkinje cells, termed "P-type". P-type channels apparently mediate the depolarization-induced repetitive Ca^{2+}-dependent spikes (TANK et al. 1988; LLINÁS et al. 1989a,b,). P-type channels are insensitive to L-type Ca^{2+}-channel drugs and ω-conotoxin GVIA and ω-ogatoxin IIIA, but are blocked by inorganic Ca^{2+} antagonists like Cd^{2+} and Co^{2+}. The only known organic blocker for these channels is a fraction (FTX) isolated from the venom of funnel web spiders (*Agelenopsis aperta*, *Hololena curta*, and *Calilena*). The toxic fraction induces death in mice after i.p. injection by central respiratory failure (LLINÁS et al. 1989a). The toxic principle(s), whose structure is not yet unequivocally clarified, affects central neurons but not mammalian motoneuron Ca^{2+} channels. Because voltage-clamping of Purkinje cells is technically not feasible, the channel was studied after purification from guinea-pig cerebella on a FTX affinity matrix and reconstitution into lipid membranes. The observed conductances exhibited voltage-dependent open probabilities and kinetics. Similar FTX-sensitive P channels were also isolated from squid optic lobe. In the squid giant synapse, whose action potentials and ion currents can be easily recorded, the FTX completely blocked synaptic transmission without affecting the presynaptic action potential. The selective effect of the toxin on presynatic Ca^{2+} current was shown in voltage-clamp experiments (LLINAS et al. 1989a; CHARLTON and AUGUSTINE 1990).

P channels are not very abundant in the mammalian brain. ω-Conotoxin GVIA and dihydropyridine-insensitive Ca^{2+} currents have been observed in different neurons (PLUMMER et al. 1989; AOSAKI and KASAI 1989), but their relationship to P channels has not yet been investigated. When rat brain mRNA or cRNA of a cloned neuronal Ca^{2+} channel α_1-subunit gene (MORI et al., 1991) is injected into *Xenopus* oocytes, calcium currents are induced that are insensitive to classic L-type channel blockers (dihydropyridines) and ω-conotoxin GVIA: Crude spider venom or partially purified toxic fractions (from *Agelenopsis apperta*) partially blocked the current, depending on Ba^{2+} concentrations in the extracellular medium (LIN et al. 1990, MORI et al., 1991).

D. Calcium Channel Toxins from Snake Venoms: Taicatoxin

The toxin purified from the freshly collected venom of the Australian Taipan snake (*Oxyuranus scutellatus*) is a basic, highly charged polypeptide of about 65 amino acids. Its name originates from *Tai*pan and *ca*lcium (BROWN et al. 1987). Taicatoxin was reported to block calcium currents in the skeletal muscle BH_3C_1 line and in mammalian smooth muscle and to inhibit (in a noncompetitive manner) the binding of $[^3H]$-(+)-PN200-110 to dihydropyridine receptors linked to L-type channels in isolated cardiac membranes.

Whole-cell patch-clamp analysis of guinea-pig ventricular cells revealed a decrease of the Ca^{2+} current by taicatoxin. The effect was reversible upon washout. K^+ and Na^+ currents were not changed by taicatoxin. Injection of taicatoxin produced no inhibition of Ca^{2+} channels. Inclusion of the toxin in the pipette using the cell-attached, single-channel mode led to blockade, whereas in the same experimental setup, application outside of the patch was ineffective. It was concluded that the extracellular mouth of the Ca^{2+} channel was the target.

The IC_{50} value of the toxin at a holding potential of $-30\,mV$ was $10\,nM$, and the block was complete at saturating taicatoxin concentrations. At $-80\,mV$, the block was incomplete and the apparent affinity of the toxin reduced, suggesting block and binding to be voltage-dependent. Further evidence that the toxin did not act via second messenger systems but directly on the high-threshold (L-type) channel came from outside-out patch-clamp experiments with ventricular cells from neonatal rat heart. Taicatoxin suppressed channel activity without changing single-channel conductance by reducing (re)opening and increasing the frequency of records in which the channel was silent.

In summary, taicatoxin appears to be a candidate for further research on L-type Ca^{2+} channels. Its usefulness for differentiation of subtypes, as a labelled ligand, and for purification remains to be established.

E. Toxins with Claimed but Unproven Action on Calcium Channels

I. Maitotoxin

Maitotoxin was originally described as a Ca^{2+}-channel activator in rat pheochromocytoma cells by Takahashi et al. (1983). In a recent review, we summarized all published evidence against this proposed mechanism of action (Glossmann and Striessnig 1990). Perhaps it acts like palytoxin to activate the Na^+-Ca^{2+} exchange mechanism (Sauviat 1989) by forming pores with membrane proteins (Chatwal et al. 1983).

II. Leptinotoxin-h

This toxin is a constituent of the hemolymph of the Colorado potato beetle, *Leptinotarsa haldemani*. The acidic 57-kDa protein (Crosland et al. 1984) stimulates acetylcholine release in the peripheral and central nervous systems in a Ca^{2+}-dependent manner (McClure et al. 1980). It depolarizes guinea-pig synaptosomes and neurosecretory (PC12) cells, is not antagonized by tetrodotoxin or verapamil, and induces $^{45}Ca^{2+}$ influx as well as a rise in the cytosolic free Ca^{2+} concentration.

Depolarization due to the toxin required the presence of external calcium, suggesting that the divalent cation was needed for binding (Madeddu et al. 1985). Leptinotoxin-h induced Ca^{2+}-dependent ATP release from resting synaptosomes from electroplax of the ray, *Ommata discopyge*. The release was not blocked by ω-conotoxin GVIA (as was depolarization-induced release), indicating different loci of action (Yeager et al. 1987). There is no direct proof that leptinotoxin-h acts directly on Ca^{2+} channels (see, however, Miljanich et al. 1988).

III. *Goniopora* Toxin

This polypeptide toxin was isolated from the *Goniopora coral* by Qar et al. (1986). It migrated with a M_r of 19000 as a single band on SDS-gel electrophoresis. The toxin stimulated $^{45}Ca^{2+}$ influx into chick cardiac cell cultures with an EC_{50} value of $5.3\,\mu M$. In the guinea-pig ileum system contractions were induced ($EC_{50} = 1.7\,\mu M$). The toxin's effects were blocked by the L-type channel-selective drugs nitrendipine and devapamil. The toxin also inhibited the binding of $[^3H]$-(+)-PN200-110 to rabbit skeletal transverse tubule membranes with an IC_{50} value of $5.3\,\mu M$.

IV. Apamin

Apamin, a potent bee venom toxin (for review, see HABERMANN 1984), was claimed to be a highly specific Ca^{2+} blocking agent in heart muscle. The peptide at picomolar concentrations blocked naturally occurring slow action potentials in cultured heart cell aggregates from chick and noncultured chick hearts (BKAILY et al. 1985). The effects depended on extracellular K^+ concentration and resisted washing. The slow action potential could be restored by the application of quinidine. Although the authors suggested that apamin might be used as a tool to study Ca^{2+} channels from heart tissue, no further data supporting this claim have been forwarded. Apamin is known to block certain types of Ca^{2+}-dependent K^+ channels (LAZDUNSKI 1983) and binds in a K^+-dependent manner to its receptor sites (HABERMANN and FISCHER 1979).

V. Atrotoxin

Atrotoxin is a protein fraction ($>15\,kDa$) isolated from the rattle snake, *Crotalus atrox*, venom by gel filtration and ion exchange chromatography. It increases the Ca^{2+} currents in mammalian heart cells (HAMILTON et al. 1985). The effects were reversible upon washout and not seen when toxin was applied intracellularly. Atrotoxin is reported to lose activity during further purification (HAMILTON and PEREZ 1987). It fulfills some of the criteria of a direct Ca^{2+} channel activating toxin but remains a doubtful candidate, as reported before (GLOSSMANN and STRIESSNIG 1990).

F. Future Prospects

In contrast to voltage-dependent Na^+ channels, no natural toxins were available for the characterization of voltage-dependent Ca^{2+} channels until recently. L-type Ca^{2+} channels were studied and purified by means of their high affinity for Ca^{2+} antagonists (for review, see CATTERALL 1988; HOSEY and LAZDUNSKI 1988; GLOSSMANN and STRIESSNIG 1988, 1990) designed by pharmaceutical companies for the therapy of cardiovascular disorders like hypertension, angina, or cardiac arrhythmias. Natural Ca^{2+} antagonists were isolated from plants, e.g., (+)-tetrandrine from a Chinese medical herb (KING et al. 1988) which binds to the benzothiazepine receptor on L-type Ca^{2+} channels and (with very high affinity) to Ca^{2+} channels in the insect nervous system (GLOSSMANN and STRIESSNIG 1990; GLOSSMANN et al. 1991). However, (+)-tetrandrine seems to be only of therapeutical but no toxicological relevance, with the possible exception of insects. Ca^{2+} antagonist receptors are widely distributed in the mammalian nervous system, but pharmacological effects of the drugs in differentiated neurons were either small or absent (MILLER 1987). Thus, the subset of voltage-dependent Ca^{2+} channels responsible for major neuronal functions, e.g., neurotransmitter release or synaptic plasticity was pharmacologically nearly inaccessible.

The isolation of toxins from the venom of animals which kill their prey by blocking neuromuscular transmission constitutes a major breakthrough in neuronal Ca^{2+}-channel research. The preterminal Ca^{2+} channels of the neuromuscular junction which were the original target, e.g., for conotoxins in fish are not toxin-sensitive in higher vertebrates, despite the presence of toxin-sensitive channels in their nervous systems. Preservation of the toxin binding site throughout evolution indicates that it may be part of a highly conserved motif, which is probably essential for channel function. Such highly conserved regions between fish and rabbit skeletal muscle (L-type-like) Ca^{2+} channels have been described (GRABNER et al. 1991).

Recently characterized spider toxins block neuronal Ca^{2+} channels in invertebrates and vertebrates, but effects on mammalian neuromuscular transmission have not yet been described.

With the help of the toxins reviewed in this article at least 5 types of voltage-dependent Ca^{2+} channels can be distinguished in neurons: T, N, L, and P channels and an ω-conotoxin GVIA-, dihydropyridine-, and FTX-insensitive channel in mammalian motoneurons. This diversity is reflected in the existence of at least four different classes of homologous, putative Ca^{2+}-channel mRNAs in brain (SNUTCH et al. 1990, TSIEN et al., 1991). Each represents a distinct hybridization pattern when probed in Northern blots with an oligonucleotide corresponding to a portion of skeletal muscle (L-type-like) Ca^{2+} channels. Expression studies in *Xenopus* oocytes showed that one of the isolated cDNAs most likely encodes a ω-conotoxin GVIA- and dihydropyridine-insensitive but FTX-sensitive channel (SNUTCH et al. 1990; LIN et al. 1990). As purification proved to be very difficult (see, e.g., HORNE et al. 1990), cloning and functional expression will be a powerful method to reveal the structure of neuronal Ca^{2+} channels (as already shown for a putative P-type Ca^{2+} channel (MORI et al., 1991)). Ca^{2+} channels can be characterized by biophysical or pharmacological criteria. So far (with a few exceptions), both classifications are in good agreement. With the increasing number of toxins available and with a conservative estimate of two dozen Ca^{2+} channel subtypes (some of which originate by alternative splicing), the current classification will be revised in the near future, based on sequence and toxin or drug data. Isolation of the genes for the toxins and expression in suitable systems will have a bright future. The ingenious way by which, e.g., *Conus* can generate (from a relatively simple building block scheme) numerous high-affinity ligands for receptors may be duplicated in the laboratory. We could then obtain easy access to novel tools. These will help to classify and isolate Ca^{2+} channels and to develop novel drugs for CNS disorders.

Well-characterized toxins recognizing muscle T- and L-type channels (with the possible exception of ω-agatoxin IIIA) are not yet available. These Ca^{2+} channels are of relatively low abundance in smooth and cardiac muscle and therefore biochemically difficult to characterize. Here, toxins would also be of great help for future research.

Acknowledgements. One of the authors (H.G.) would like to thank Drs. B. Olivera, M. Adams, N. Akaike, and H. Nakayama for fruitful discussions and supply of preprints.

References

Adams ME, Bindokas VP, Hasegawa L, Venema VJ (1990) Omega-agatoxins: novel calcium channel antagonists of two subtypes from funnel web spider (*Agelenopsis apaerta*) venom. J Biol Chem 265:861–867

Ahlijanian MK, Westenbroek RE, Catterall WA (1990) Subunit structure and localization of dihydropyridine-sensitive calcium channels in mammalian brain, spinal cord, and retina. Neuron 4:819–832

Ahlijanian MK, Striessnig J, Catterall WA (1991) Phosphorylation of an α_1-like subunit of an ω-conotoxin-sensitive brain calcium channel by cAMP-dependent protein kinase and protein kinase C. J. Biol. Chem. 266:20192–20197.

Aosaki T, Kasai H (1989) Characterization of two kinds of high-voltage-activated Ca-channel currents in chick sensory neurons. Pflugers Arch 414:150–156

Ballesta JJ, Palermo M, Hidalgo MJ, Gutierrez LM, Reig JA, Viniegra S, Garcia AG (1989) Separate binding and functional sites for omega-conotoxin and nitrendipine suggest two types of calcium channels in bovine chromaffin cells. J Neurochem 53:1050–1056

Barhanin J, Coppola T, Schmid A, Borsotto M, Lazdunski M (1988) Properties of structure and interaction of the receptor for omega-conotoxin, a polypeptide active in calcium channels. Biochem Biophys Res Commun 150:1051–1062

Bindokas VP, Adams ME (1989) Omega-aga-I: a presynaptic calcium channel antagonist from venom of the funnel web spider, *Agelenopsis aperta*. J Neurobiol 20:171–188

Bkaily G, Sperelakis N, Renaud JF, Payet MD (1985) Apamin, a highly specific calcium blocking agent in heart muscle. Am J Physiol 248:H961–H965

Block BA, Imagawa T, Campbell KP, Franzini-Armstrong C (1988) Structural evidence for direct interaction between the molecular components of the transverse-tubule/sarcoplasmic reticulum junction in skeletal muscle. J Cell Biol 107:2587–2600

Boland LM, Dingledine R (1990) Multiple components of both transient and sustained barium currents in a rat dorsal root ganglion cell line. J Physiol (Lond) 420:223–245

Bowers CW, Philips HS, Lee P, Jan YN, Jan LY (1987) Identification and purification of an irreversible presynaptic neurotoxin from the venom of the spider *Hololena curta*. Proc Natl Acad Sci USA 84:3506–3510

Branton WD, Kotton L, Jan YN, Jan LY (1987) Neurotoxins from Plectreuris spider venom are potent presynaptic blockers in Drosophila. J Neurosci 7:4195–4200

Brock JA, Cunnane TC, Ziogas J (1990) Local application of omega-conotoxin GVIA to sympathetic nerve terminals in the guinea-pig isolated vas deferens. Br J Pharmacol 98:775P

Brown AM, Yatani A, Lacerda AE, Gurrola GB, Possani LD (1987) Neurotoxins that act selectively on voltage-dependent cardiac calcium channels. Circ Res 61:6–9

Campbell KP, Leung AT, Sharp AH (1988) The biochemistry and molecular biology of the dihydropyridine-sensitive calcium channel. Trends Neurosci 11:425–430

Carbone E, Lux HD (1984) A low-voltage-activated, fully inactivating Ca^{2+}-channel in vertebrate sensory neurons. Nature 310:501–502

Carbone E, Lux HD (1988) Omega-conotoxin blockade distinguishes Ca from Na permeable states in neuronal calcium channels. Pflugers Arch 413:14–22

Carbone E, Formenti A, Pollo A (1990) Multiple actions of BAY K 8644 on high-threshold Ca channels in adult rat sensory neurons. Neurosci Lett 111:315–320

Catterall WA (1986) Molecular properties of voltage-sensitive sodium channels. Annu Rev Biochem 55:953–985

Catterall WA (1988) Structure and function of voltage-sensitive ion channels. Science 242:50–61

Catterall WA, Seagar MJ, Takahashi M (1988) Molecular properties of dihydropyridine-sensitive calcium channels. J Biol Chem 263:3535–3538

Charlton MP, Augustine GJ (1990) Classification of presynaptic calcium channels at the squid giant synapse: neither T-, L- nor N-type. Brain Res 525:133–139

Chatwal GS, Hessler HJ, Habermann E (1983) The action of palytoxin on erythrocytes and resealed ghosts. Formation of small nonselective pores linked with Na,K ATPase. Naunyn Schmiedebergs Arch Pharmacol 323:261–268

Cortes R, Supavilai P, Karobath M, Palacios JM (1983) The effects of lesions in the rat hippocampus suggest the association of calcium channel blocker binding sites with a specific neuronal population. Neurosci Lett 42:249–254

Crosland RD, Hsaio TH, McClure WO (1984) Purification and characterization of beta-leptinotarsin-h, an activator of presynaptic calcium channels. Biochemistry 23:734–741

Cruz LJ, Olivera BM (1986) Calcium channel antagonists. Omega-conotoxin defines a new high affinity site. J Biol Chem 261:6230–6233

Dingledine R (1983) N-Methyl-aspartate activates voltage-dependent calcium conductance in rat hippocampal pyramidal cells. J Physiol (Lond) 343:385–405

Dooley DJ, Lupp A, Hertting G (1987) Inhibition of central neurotransmitter release by omega-conotoxin GVIA, a peptide modulator of the N-type voltage-sensitive calcium channel. Naunyn Schmiedebergs Arch Pharmacol 336:467–470

Dooley DJ, Lickert M, Lupp A, Osswald H (1988) Distribution of [^{125}I]omega-conotoxin GVIA and [^{3}H]isradipine binding sites in the central nervous system of rats of different ages. Neurosci Lett 93:318–323

Feigenbaum P, Garcia ML, Kaczorowski GJ (1988) Evidence for distinct sites coupled to high affinity ω-conotoxin receptors in rat brain synaptic plasma membrane vesicles. Biochem Biophys Res Commun 154:298–305

Feuerstein TJ, Dooley DJ, Seeger W (1990) Inhibition of norepinephrine and acetylcholine release from human neocortex by omega-conotoxin GVIA. J Pharmacol Exp Ther 252:778–785

Fox AP, Nowycky MC, Tsien RW (1987a) Kinetic and pharmacological properties distinguishing three types of calcium currents in chick sensory neurones. J Physiol (Lond) 394:149–172

Fox AP, Nowycky MC, Tsien RW (1987b) Single channel recordings of three types of calcium channels in chick sensory neurones. J Physiol (Lond) 394:173–200

Glossmann H, Striessnig J (1988) Calcium channels. Vitam Horm 44:155–328

Glossmann H, Striessnig J (1990) Molecular properties of calcium channels. Rev Physiol Biochem Pharmacol 114:1–105

Glossmann H, Striessnig J, Hymel L, Zernig G, Knaus HG, Schindler H (1988) The structure of the calcium channel: photoaffinity labeling and tissue distribution. In: Morad M, Nayler WG, Kazda S, Schramm M (eds) The calcium channel: structure, function and implications. Springer, Berlin Heidelberg New York, pp 168–192

Glossmann H, Zech C, Striessnig J, Staudinger R, Hall L, Greenberg R, Armah BI (1991) Very high affinity interaction of DPI 201-106 and BDF 8784 enantiomers with the phenylalkylamine-sensitive Ca^{2+}-channel in *Drosophila* head membranes. Br J Pharmacol 102:446–452

Grabner M, Friedrich K, Knaus HG, Striessnig J, Scheffauer F, Staudinger R, Koch WJ, Schwartz A, Glossmann H (1991) Calcium channels from *Cyprinus carpio* skeletal muscle. Proc Natl Acad Sci USA 88:727–731

Gray WR, Luque A, Olivera BM, Barrett J, Cruz LJ (1986) Peptide toxins from *Conus geographus* venom. J Biol Chem 256:4734–4740

Gray WR, Olivera BM, Cruz LJ (1988) Peptide toxins from venomous conus snails. Annu Rev Biochem 57:665–700

Habermann E (1984) Apamin. Pharmacol Ther 25:255–270

Habermann E, Fischer K (1979) Bee venom neurotoxin (apamin): iodine labeling and characterization of binding sites. Eur J Biochem 94:355–364

Hagiwara K, Sakai T, Miwa A, Kawai N, Nakajima T (1990) Complete amino acid sequence of a new type of neurotoxin from the venom of the spider *Agelena opulenta*. Biomed Res 11:181–186

Hamilton SL, Perez M (1987) Toxins that affect voltage-dependent calcium channels. Biochem Pharmacol 36:3325–3329

Hamilton SL, Yatani A, Hawkes MJ, Redding K, Brown AM (1985) Atrotoxin: a specific agonist for calcium currents in heart. Science 229:182–184

Hess P (1990) Calcium channels in vertebrate cells. Annu Rev Neurosci 13:337–356

Higashi H, Sugita S, Matsunari S, Nishi S (1990) Calcium-dependent potentials with different sensitivities to calcium agonists and antagonists in guinea-pig hippocampal neurons. Neuroscience 34:35–47

Hirning LD, Fox AP, McCleskey EW, Olivera BM, Thayer SA, Miller RJ, Tsien RW (1988) Dominant role of N-type calcium channels in evoked release of norepinephrine from sympathetic neurons. Science 239:57–60

Hodgkin AL, Huxley AF (1952) A quantitative description of membrane current and its application to conduction and excitation in nerve. J Physiol (Lond) 117:500–544

Hofmann F, Habermann E (1990) Role of omega-conotoxin-sensitive calcium channels in inositolphosphate production and noradrenaline release due to potassium depolarization or stimulation with carbachol. Naunyn Schmiedebergs Arch Pharmacol 341:200–205

Hofmann F, Nastainczyk W, Roehrkasten A, Schneider T, Sieber M (1987) Regulation of L-type calcium channels. Trends Pharmacol Sci 8:393–398

Horne WA, Hawrot E, Tsien RW (1991) ω-Conotoxin GVIA receptors of Discopyge electric organ. J. Biol. Chem. 266:13719–13725.

Hosey MM, Lazdunski M (1988) Calcium channels: molecular pharmacology, structure and regulation. J Membr Biol 104:81–106

Hymel L, Striessnig J, Glossmann H, Schindler H (1988) Purified skeletal muscle 1,4-dihydropyridine receptor forms phosphorylation-dependent oligomeric calcium channels in planar lipid bilayers. Proc Natl Acad Sci USA 85:4290–4294

Jan C-R, Titeler M, Schneider AS (1990) Identification of omega-conotoxin binding sites on adrenal medullary membranes: possibility of multiple calcium channels in chromaffin cells. J Neurochem 54:355–358

Jones OT, Kunze DL, Angelides KJ (1989) Localization and mobility of omega-conotoxin sensitive Ca^{2+} channels in hippocampal CA1 neurons. Science 244:1189–1193

Kamiya H, Sawada S, Yamamoto C (1988) Synthetic omega-conotoxin blocks synaptic transmission in the hippocampus in vitro. Neurosci Lett 91:84–88

Karschin A, Lipton SA (1989) Calcium channels in solitary retinal ganglion cells from post-natal rat. J Physiol (Lond) 418:379–396

Kasai H, Aosaki T, Fukuda J (1987) Presynaptic Ca-antagonist omega-conotoxin irreversibly blocks N-type Ca-channels in chick sensory neurons. Neurosci Res 4:228–235

Keith RA, Mangano TJ, Salama AI (1989) Inhibition of *N*-methyl-D-aspartate- and kainic acid-induced neurotransmitter release by omega-conotoxin GVIA. Br J Pharmacol 98:767–772

Kennedy MB (1989) Regulation of synaptic transmission in the central nervous system: long-term potentiation. Cell 59:777–787

Kerr LM, Yoshikami D (1985) A venom peptide with novel presynaptic blocking action. Nature 308:282–284

King VF, Garcia ML, Himmel D, Reuben JP, Lam YK, Pan JX, Han GQ, Kaczorowski GJ (1988) Interaction of tetrandrine with slowly inactivating calcium channels. Characterization of calcium channel modulation by an alkaloid of Chinese medical herb origin. J Biol Chem 263:2238–2244

Knaus HG, Striessnig J, Koza A, Glossmann H (1987) Neurotoxic aminoglycoside antibiotics are potent inhibitors of [^{125}I]omega-conotoxin binding to guinea-pig cerebral cortex membranes. Naunyn Schmiedebergs Arch Pharmacol 336:583–586

Krishtal OA, Petrov AV, Smirnov SV, Nowycky MC (1989) Hippocampal synaptic plasticity induced by excitatory amino acids includes changes in sensitivity to the calcium channel blocker, omega-conotoxin. Neurosci Lett 102:197–204

Lazdunski M (1983) Apamin, a neurotoxin specific for one class of calcium-dependent K$^+$-channels. Cell Calcium 4:421–428

Leibowitz MD, Balc T, Adams ME, Venema VJ, Cohen CJ (1990) Selective block of atrial L-type Ca channels by the spider toxin ω-aga-IIIA. FASEB J 16:956

Lin J-W, Rudy B, Llinas RR (1990) Funnel-web spider venom and a toxin fraction block calcium current expressed from rat brain mRNA in *Xenopus* oocytes. Proc Natl Acad Sci USA 87:4538–4542

Llinas RR, Sugimori M, Lin J-W, Cherksey B (1989a) Blocking and isolation of a calcium channel from neurons in mammals and cephalopods utilizing a toxin fraction (FTX) from funnel-web spider poison. Proc Natl Acad Sci USA 86:1689–1693

Llinas RR, Sugimori M, Cherksey B (1989b) Voltage-dependent calcium conductances in mammalian neurons. Ann NY Acad Sci 560:103–111

Loudes C, Faivre-Bauman A, Patte C, Tixier-Vidal A (1988) Involvement of DHP voltage-sensitive calcium channels and protein kinase C in thyroliberin (TRH) release by developing hypothalamic neurons in culture. Brain Res 456:324–332

Madeddu L, Pozzan T, Robello T, Rolandi R, Hsiao TH, Meldolesi J (1985) Leptinotoxin-h action in synaptosomes, neurosecretory cells, and artificial membranes: stimulation of ion fluxes. J Neurochem 45:1708–1718

Maeda N, Wada K, Yuzaki M, Mikoshiba K (1989) Autoradiographic visualization of a calcium channel antagonist, [^{125}I]omega-conotoxin GVIA, binding site in the brains of normal and cerebellar mutant mice (pcd and weaver). Brain Res 489:21–30

McClure WO, Abbott BC, Baxter DE, Hsaio TH, Satin LS, Siger A, Yoshino JE (1980) Leptinotarsin: a presynaptic neurotoxin that stimulates release of acetylcholine. Proc Natl Acad Sci USA 77:1219–1223

McEnery MW, Snowman A, Sharp AH, Adams ME, Snyder SH (1991) The purified ω-conotoxin GVIA receptor of rat brain resembles a dihydropyridine-sensitive L-type calcium channel. Proc. Natl. Acad. Sci (USA): in press

Miljanich GP, Yeager RE, Hsiao TH (1988) Leptinotarsin-d, a neurotoxic protein, evokes neurotransmitter release from, and calcium flux into, isolated electric organ nerve terminals. J Neurobiol 19:373–386

Miller RJ (1987) Multiple calcium channels and neuronal function. Science 235:46–52

Mintz I, Venema VJ, Adams ME, Bean BP (1991) Inhibition of N- and L-type Ca^{2+} channels by the spider venom toxin ω-Aga-IIIA. Proc. Natl. Acad. Sci (USA) 88:6628–6631

Mori Y, Friedrich T, Kim MS, Mikami A, Nakai J, Ruth P, Bosse E, Hofmann F, Flockerzi V, Furuichi T, Mikoshiba K, Imoto K, Tanabe T, Numa S (1991) Primary structure and functional expression from complementary DNA of a brain calcium channel. Nature 350:398–402

Murphy SN, Miller RJ (1989) Regulation of Ca^{++} influx into striatal neurons by kainic acid. J Pharmacol Exp Ther 249:184–193

Myers RA, McIntosh JM, Imperial J, Williams RW, Oas T, Haack JA, Hernandes J-F, Rivier J, Cruz LJ, Olivera BM (1991) Peptides from *Conus* venoms which affect Ca^{2+} entry into neurons. J Toxin Toxin Revs (in press)

Nakayama H, Taki M, Striessnig J, Glossmann H, Catterall WA, Kanaoka Y (1991) Identification of 1,4-dihydropyrioline binding regions within the α_1 subunit of skeletal muscle calcium channels by photo affinity labeling with diazipine. Proc. Natl. Acad. Sci (USA) 88:9203–9207

Nowyczky MC, Fox AP, Tsien RW (1985) Three types of neuronal calcium channel with different calcium agonist sensitivity. Nature 316:440–443

Obaid AL, Flores R, Salzberg BM (1989) Calcium channels that are required for secretion from intact nerve terminals of vertebrates are sensitive to omega-conotoxin and relatively insensitive to dihydropyridines. J Gen Physiol 93:715–729

Olivera BM, Gray WR, Zeikus R, McIntosh JM, Varga J, Rivier J, de Santos V, Cruz LJ (1985) Peptide neurotoxins from fish-hunting cone snails. Science 230:1338–1343

Olivera BM, Cruz LJ, de Santos V, LeCheminant GW, Griffin D, Zeikus R, McIntosh JM, Galyean R, Varga J Gray WR, Rivier J (1987) Neuronal calcium channel antagonists. Discrimination between calcium channel subtypes using omega-conotoxin from Conus *magnus* venom. Biochemistry 26:2086–2090

Olivera BM, Rivier J, Clark C, Ramilo CA, Corpuz GP, Abogadie C, Mena EE, Woodward SR, Hillyard DR, Cruz LJ (1990) Diversity of *Conus* neuropeptides. Science 249:257–263

Owen PJ, Marriott DB, Boarder MR (1989) Evidence for a dihydropyridine-sensitive and conotoxin-insensitive release of noradrenaline and uptake of calcium in adrenal chromaffin cells. Br J. Pharmacol 97:133–138

Pelzer D, Pelzer S, McDonald TF (1990) Properties and regulation of calcium channels in muscle cells. Rev Physiol Biochem Pharmacol 114:107–207

Plummer MR, Logothetis DE, Hess P (1989) Elementary properties and pharmacological sensitivities of calcium channels in mammalian peripheral neurons. Neuron 2:1453–1463

Pruneau D, Angus JA (1990) ω-Conotoxin GVIA is a potent inhibitor of sympathetic neurogenic responses in rat small mesenteric arteries. Br J Pharmacol 100:180–184

Qar J, Schweitz H, Schmid A, Lazdunski M (1986) A polypeptide toxin from the coral *Goniopora*. Purification and action on calcium channels. FEBS Lett 202:331–336

Quistad GB, Suwanrumpha S, Jarema MA, Shapiro MJ, Skinner WS, Jamieson GC, Lui A, Fu EW (1990) Structures of paralytic acylpolyamines from the spider *Agelenopsis aperta*. Biochem Biophys Res Commun 169:51–56

Reynolds IJ, Wagner JA, Snyder SH, Thayer SA, Olivera BM, Miller RJ (1986) Brain voltage-sensitive calcium channel subtypes differentiated by omega-conotoxin fraction GVIA. Proc Natl Acad Sci USA 83:8804–8807

Riveros N, Orrego F (1986) N-methylaspartate-activated calcium channels in rat brain cortex slices. Effect of calcium channel blockers and of inhibitory and depressant substances. Neuroscience 17:541–546

Rivier J, Galyean R, Gray WR, Azimi-Zonooz A, McIntosh JM, Cruz LJ, Olivera BM (1987) Neuronal calcium channel inhibitors. Synthesis of omega-conotoxin

GVIA and effects on ^{45}Ca uptake by synaptosomes. J Biol Chem 262:1194–1198

Sano K, Enomoto K, Maeno T (1987) Effects of synthetic omega-conotoxin, a new type Ca^{2+} antagonist, on frog and mouse neuromuscular transmission. Eur J Pharmacol 141:1283–1286

Sauviat M-P (1989) Effect of palytoxin on the calcium current and the mechanical activity of frog heart muscle. Br J Pharmacol 98:773–780

Seabrook GR, Adams DJ (1989) Inhibition of neurally-evoked transmitter release by calcium channel antagonists in rat parasympathetic ganglia. Br J Pharmacol 97:1125–1136

Shapira S, Adeyemo OM, Feuerstein G (1990a) Integrated autonomic and behavioral responses to L/N Ca^{2+}-channel blocker x-conotoxin in conscious rats. Am J Physiol 259:R427–R438

Shapira S, Kadar T, Adeyemo OM, Feuerstein G (1990b) Selective hippocampal lesion following ω-conotoxin administration in rats. Brain Res 523:291–294

Sher E, Biancardi E, Passaforo M, Clemendi F (1991) Physiopathology of neuronal voltage-operated calcium channels. FASEB J. 5:2677–2683

Skinner WS, Adams ME, Quistad GB, Kataoka H, Cesarin BJ, Enderlin FE, Schooley DA (1989) Purification and characterization of two classes of neurotoxins from the funnel web spider, *Agelenopsis aperta*. J Biol Chem 264:2150–2155

Smith SJ, Augustine GJ (1988) Calcium ions, active zones and synaptic transmitter release. Trends Neurosci 11:458–464

Snutch TP, Leonard JP, Gilbert MM, Lester HA, Davidson N (1990) Rat brain expresses a heterogeneous family of calcium channels. Proc Natl Acad Sci USA 87:3391–3395

Stapleton A, Blankenship DT, Ackermann BL, Chen T-M, Gorder GW, Manley GD, Palfreyman MG, Coutant JE, Cardin AD (1990) Curtatoxins. Neurotoxic insecticidal polypeptides isolated from the funnel-web spider *Hololena curta*. J Biol Chem 265:2054–2059

Striessnig J, Glossmann H, Catterall WA (1990) Identification of a phenylalkylamine binding region within the α_1 subunit of skeletal muscle Ca^{2+} channels. Proc Natl Acad Sci USA 87:9108–9112

Striessnig J, Murphy BJ, Catterall WA (1991) Dihydropyridine receptor of L-type Ca^{2+} channels: Identification of binding domains for [^{3}H](+)-PN200-110 and [^{3}H]azidopine within the alpha$_1$ subunit. Proc. Natl. Acad. Sci. 88: in press

Suszkiw JB, Murawsky MM, Shi M (1989) Further characterization of phasic calcium influx in rat cerebrocortical synaptosomes: inferences regarding calcium channel type(s) in nerve endings. J Neurochem 52:1260–1269

Takahashi M, Tatsumi M, Ohizumi Y, Yasumoto T (1983) Calcium channel-activating function of maitotoxin, the most potent marine toxin known, in clonal rat pheochromocytoma cells. J Biol Chem 258:10944–10949

Takahashi K, Wakamori M, Akaike N (1989) Hippocampal CA1 pyramidal cells of rats have four voltage-dependent calcium conductances. Neurosci Lett 104:229–234

Takemura M, Kishino J, Yamatodani A, Wada H (1989) Inhibition of histamine release from rat hypothalamic slices by omega-conotoxin GVIA, but not by nilvadipine, a dihydropyridine derivative. Brain Res 496:351–356

Tank DW, Sugimori M, Connor JA, Llinas RR (1988) Spatially resolved calcium dynamics of mammalian Purkinje cells in cerebellar slice. Science 242:773–777

Thate A, Meyer DK (1989) Effect of omega-conotoxin on release of ^{3}H-gamma-aminobutyric acid from slices of rat neostriatum. Naunyn Schmiedebergs Arch Pharmacol 339:359–361

Toselli M, Taglietti V (1990) Pharmacological characterization of voltage-dependent calcium currents in rat hippocampal neurons. Neurosci Lett 112:70–75

Tsien RW, Lipscombe D, Madison DV, Bley KR, Fox AP (1988) Multiple types of neuronal calcium channels and their selective modulation. Trends Neurosci 11:431–438

Tsien RW, Ellinor PT, Horne WA (1991) Molecular diversity of voltage-dependent Ca^{2+}-channels. Trends Pharmacol. Sci. 12:349–354

Vincent A, Lang B, Newsom-Davis J (1989) Autoimmunity to the voltage-gated calcium channel underlies the Lambert-Eaton myasthenic syndrome, a paraneoplastic disorder. TINS 12:496–502

Watkins JC, Krogsgaard-Larsen P, Honore T (1990) Structure-activity relationships in the development of excitatory amino acid receptor agonists and competitive antagonists. Trends Pharmacol Sci 11:25–33

Weiss JH, Hartley DM, Koh J, Choi DW (1990) The calcium channel blocker nifedipine attenuates slow excitatory amino acid neurotoxicity. Science 247:1474–1477

Wessler I, Dooley DJ, Werhand J, Schlemmer F (1990) Differential effects of calcium channel antagonists (ω-conotoxin GVIA, nifedipine, verapamil) on the electrically-evoked release of [³H]acetylcholine from the myenteric plexus, phrenic nerve and neocortex of rats. Naunyn Schmiedebergs Arch Pharmacol 341:288–294

Woodward JJ, Rezazadeh M, Leslie SW (1988) Differential sensitivity of synaptosomal calcium entry and endogenous dopamine release to omega-conotoxin. Brain Res 475:141–145

Woodward SR, Cruz LJ, Olivera BM, Hillyard DR (1990) Constant and hypervariable regions in conotoxin propeptides. EMBO J 9:1015–1020

Wu CT, Narahashi T (1988) Mechanism of action of novel marine neurotoxins on ion channels. Annu Rev Pharmacol Toxicol 28:141–161

Yamaguchi T, Saisu H, Mitsui H, Abe T (1988) Solubilization of the omega-conotoxin receptor associated with voltage-sensitive calcium channels from bovine brain. J Biol Chem 263:9491–9498

Yeager RE, Yoshikami D, Rivier J, Cruz LJ, Miljanich GP (1987) Transmitter release from presynaptic terminals of electric organ: inhibition by the calcium channel antagonist omega *Conus* toxin. J Neurosci 7:2390–2396

ADP-Ribosylation of Signal-Transducing Guanine Nucleotide Binding Proteins by Cholera and Pertussis Toxin

P. Gierschik and K.H. Jakobs

A. Introduction

Heterotrimeric guanine nucleotide binding proteins (G proteins) are key elements of many transmembrane signalling systems. They function at the level of the plasma membrane and couple a large number (>100) of receptors for extracellular mediators (hormones, neurotransmitters, growth factors, chemoattractants, pheromones, odorants) or even physical signals such as light to effector moieties, which generate a "second message" inside the cell. The list of G-protein-regulated effectors includes enzymes [e.g., adenylyl cyclase, various phospholipases, and phosphodiesterase(s)], a great number of ion channels (e.g., K^+ and Ca^{2+} channels), and possibly various types of pumps, transporters, and antiporters as well (see Birnbaumer 1990; Birnbaumer et al. 1990; Gierschik et al. 1990a,b for recent reviews and references).

G proteins are heterotrimeric proteins, consisting of an α-subunit (≈ 39–$52\,kDa$) that reversibly associates with a tight complex composed of a β- and a γ-subunit (≈ 36 and $8\,kDa$, respectively). The α-subunit carries the binding site for guanine nucleotides and is capable of hydrolyzing GTP. Biochemical and molecular biological studies have revealed that all three G-protein sub-units are members of large protein families. Thus, at least 18 different α-subunits have been identified in mammalian cells to date (Lochrie and Simon 1988; Hsu et al. 1990; Strathmann et al. 1989; Strathmann and Simon 1990). Furthermore, there is a minimum of four distinct β-subunits (Birnbaumer 1990) and seven γ-subunits (Gautam et al. 1990). In some cases, the molecular diversity of the α-subunits is due to alternative splicing of a single precursor RNA. For example, there are four variants of the α-subunit of G_s (the stimulatory G protein of adenylyl cyclase) and two variants of α_o (an α-subunit abundant in brain), which are derived from a single α_s or α_o gene, respectively (Bray et al. 1986; Hsu et al. 1990). On the other hand, three genes exist in the mammalian genome which encode α_i proteins. G_i proteins were initially believed to be specifically involved in inhibition of adenylyl cyclase (and hence termed G_i) but are now known to be involved in regulating other effectors as well 1990; (see Birnbaumer 1990; Birnbaumer et al. 1990; Gierschik et al. 1990b for recent reviews and references).

The G-protein-mediated activation or inhibition of an effector moiety by an agonist-activated receptor is thought to result from a series of sequential interactions of receptor, G protein, and effector. First, the activated receptor interacts with the heterotrimeric G protein and facilitates an exchange of GTP for the GDP bound to its α-subunit. Activation of the GTP-bound form of the G protein appears to coincide with its dissociation from the receptor as well as its own dissociation into a free α-subunit and the βγ-dimer. In certain cases (e.g., stimulation of adenylyl cyclase), the free GTP-bound α-subunit is responsible for effector regulation. In others, however, free βγ-dimers may also be directly involved in controlling effector activity (see Neer and Clapham 1990 for further information). Deactivation of the G protein is thought to result mainly from the hydrolysis of bound GTP to GDP and the subsequent association of the GDP-bound α-subunit with the βγ-dimer.

An additional important property of many G protein α-subunits is their ability to undergo ADP-ribosylation by certain bacterial toxins. The most important of these toxins are cholera and pertussis. The latter is also referred to as "islet activating protein" (IAP). A third, less commonly used toxin which is also capable of ADP-ribosylating G protein α-subunits and shares many properties with cholera toxin is the heat-labile enterotoxin of *Escherichia coli* (see Vaughan and Moss 1981; Moss and Vaughan 1988 for further information). Although cholera and pertussis toxins are – strictly speaking – not considered neurotoxins, both have proved to be extremely useful tools for the neuroscientist. This is due to the fact that a large number of receptors for neurotransmitters is coupled to their effector moieties via G

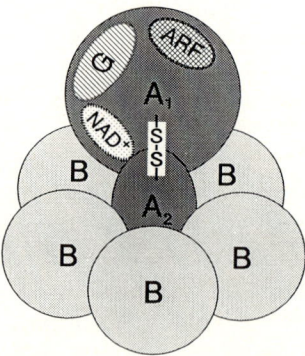

Fig. 1. Subunit structure of cholera toxin. The mature toxin is composed of five noncovalently linked B (binding) subunits and the disulfide-linked A_1A_2-heterodimer. A_1A_2 is derived from the monomeric A subunit by proteolytic cleavage. A_1 is the enzymatically active subunit and contains the binding sites for nicotinamide adenine dinucleotide (NAD^+), the G-protein substrate (G), and – most likely – ADP-ribosylating factor (ARF). Reduction of the intermolecular disulfide bond is required for activation of the toxin

proteins sensitive to cholera and/or pertussis toxin. In these cases, the investigator can use cholera and/or pertussis toxin to (1) radiolabel selectively and thus identify G protein α-subunits in crude membrane preparations (using [^{32}P]-ADP-ribosylation followed by analysis on SDS-PAGE) and (2) alter selectively the function of specific G proteins by treatment of intact cells or membrane preparations with the toxins and nonradioactive nicotinamide adenine dinucleotide (NAD$^+$).

Four aspects of the structure and function of cholera and pertussis toxin are reviewed here: (1) their molecular properties, (2) the structure and function of the ADP-ribosylating factors (ARFs), (3) the molecular mechanism of G-protein ADP-ribosylation by these toxins, and (4) the functional consequences elicited by these modifications. Given the limits in space, we have not been able to provide a complete overview of the literature on this topic. The reader is referred to several excellent previous reviews for more complete information on early work (FINKELSTEIN 1973; MOSS and VAUGHAN 1979; VAUGHAN and MOSS 1981; UI 1984, 1986).

B. Structure-Function Relationships of Cholera and Pertussis Toxins

I. Cholera Toxin

Cholera toxin is produced by the gram-negative bacterium *Vibrio cholerae* and is composed of six noncovalently linked subunits, one A-subunit and five identical B-subunits (Fig. 1) (GILL 1976). The genes encoding the A- and B-subunits are arranged in a single transcriptional unit with one A cistron preceding one B cistron (MEKALANOS et al. 1983). The mature products of the A and B cistrons consist of 240 and 103 amino acids with calculated molecular weights of 27 215 and 11 677, respectively. The A-subunit contains two half-cysteines (Cys 187 and Cys 199). The region between these disulfide-bonded residues is highly sensitive to proteolytic cleavage. Proteolytic cleavage, which is presumably catalyzed by a protease produced by the bacterium and occurs between residues 194 and 195 (MENDEZ et al. 1975; MEKALANOS et al. 1983), generates two disulfide-linked polypeptides designated A$_1$ and A$_2$. The A$_1$-subunit is the enzymatically active component and displays both ADP-ribosyltransferase and NAD$^+$-glycohydrolase activity (MOSS et al. 1979). The A$_2$-subunit probably serves as an adaptor piece that allows the A$_1$ fragment to properly associate with the B pentamer (GILL 1976; MEKALANOS et al. 1979). The B monomers are noncovalently linked to each other to form a ringlike structure with a binding site for the A$_2$ fragment located in its axis (GILL 1976). The B pentamer is responsible for the interaction of the holotoxin with the surface of the target cell (see MOSS and VAUGHAN 1988 for further details and references).

Both proteolytic cleavage and reduction of the disulfide bond of the A subunit are required for the A_1 fragment to become enzymatically active (Gill 1976; Mekalanos et al. 1979). Disulfide reduction is accomplished in vitro by treating the toxin with dithiothreitol (see Ribeiro-Neto et al. 1985; Gill and Woolkalis 1988 for experimental details) and in vivo by cellular glutathione (Gill 1976). The latter process may be enhanced by a thiol:protein disulfide oxidoreductase that is present in many cells and lowers and concentration of glutathione required to activate the toxin (Moss et al. 1980). Treatment of the reduced holoprotein or the A_1A_2-heterodimer with detergents like SDS has been reported to give rise to a further enhancement of the enzymatic activities of A_1 (Gill 1976). This observation led to the suggestion that the A_1 fragment expresses its maximal activity only when released from the rest of the toxin molecule (Gill 1976). Subsequent experimentation showed, however, that the free A_1 fragment was at best only twofold more active than the thiol-treated nicked toxin, indicating that the ADP-ribosylation catalyzed by the A_1 chain might occur while A_1 is still associated with the membrane-bound B and A_2 components (Mekalanos et al. 1979). It is important to note in this context that detergents may exert multiple effects on the activity of the A_1 chain, including enhancing its stability at low protein concentration (Moss et al. 1979; Gill and Woolkalis 1988) and facilitating its interaction with ARF (Bobak et al. 1990; Noda et al. 1990). Permeabilization of sealed membrane vesicles containing the toxin substrates at their inner surface has also been suggested to be a mechanism by which detergents may enhance G-protein ADP-ribosylation by A_1 (Neer et al. 1987).

Little information is available on the organization of the A_1 fragment of cholera toxin into functional domains. The N-terminal region of A_1 is likely to be involved in the interaction with NAD^+. This suggestion is based on the striking N-terminal amino acid sequence homology between the A_1 chain of cholera toxin and the S1-subunit of pertussis toxin (see Locht and Keith 1986; Nicosia et al. 1986; Cortina and Barbieri 1989 for amino acid sequence comparison), together with the increasing body of evidence that this region of S1 is crucial for its interaction with NAD^+ (see below) The C-terminal region of A_1 has been suggested to be important for the interaction with the G-protein substrate, since immobilization of A_1 through Cys 187 did not block ADP-ribosylation of small molecular weight ADP-ribose acceptors like agmatin or prevent the regulation of A_1 by ARF but did impair the ADP-ribosylation of α_s (Noda et al. 1989).

II. Pertussis Toxin

Pertussis toxin is a 105-kDa hexameric protein produced by the gram-negative bacterium *Bordetella pertussis* and composed of five distinct, noncovalently linked polypeptides designated (in the order of decreasing molecular weight) S1 through S5 (Fig. 2). Similar to cholera toxin, pertussis

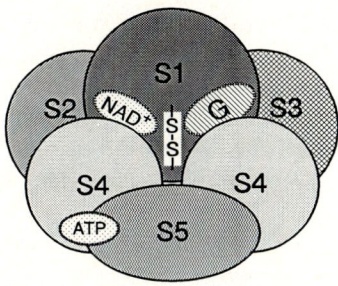

Fig. 2. Subunit structure of pertussis toxin. Pertussis toxin is composed of six subunits. S1 is the enzymatically active subunit and contains the binding site for nicotinamide adenine dinucleotide (NAD^+) and the G-protein substrate (G). The remainder of the molecule (the two dimers, *S4-S3* and *S4-S2*, and the connecting subunit *S5*) is responsible for binding of the toxin to the target cell (B-oligomer). Reduction of the intramolecular disulfide bond of S1 and binding of ATP to the B-oligomer leads to release of S1 from the B-oligomer and S1 activation

toxin can be divided into two distinct functional units, the enzymatically active A-protomer consisting of a single polypeptide (S1) and the pentameric B-oligomer (S2, S3, two copies of S4, and S5). The B-oligomer is involved in binding to the surface of eukaryotic target cells and presumably in translocation of the toxin across the plasma membrane (Tamura et al. 1982, 1983; Burns et al. 1988; Witvliet et al. 1989). The two S4 polypeptides form two distinct heterodimers with S2 and S3 (S2-S4 and S3-S4), which are in turn held together by S5 (Tamura et al. 1982). The site(s) of interaction of S1 with the B-oligomer is unknown.

The genes coding for the pertussis toxin subunits have been cloned and sequenced (Nicosia et al. 1986; Locht and Keith 1986). The five genes are arranged in tandem and form a single operon. All subunits contain signal peptides of variable length. The calculated molecular weights of the mature subunits are 26 220 (235 aa) for S1, 21 920 (199 aa) for S2, 21 860 (199 aa) for S3, 12 060 (110 aa) for S4, and 10 940 (99 aa) for S5 (Nicosia et al. 1986). Two distinct sequences have been published for S1, differing in the region from amino acids 190–200. As the S1 sequence published by Nicosia et al. (1986) has been independently confirmed by others (M.A. Schmidt, personal communication), we will use this sequence for further reference.

The S1-subunit of pertussis toxin harbors two enzymatic activities: (1) an ADP-ribosyltransferase activity that catalyzes the transfer of an ADP-ribose moiety from NAD^+ to a cysteine localized three amino acids upstream of the C-terminal residue of the α-subunits of various heterotrimeric G proteins (see below) and (2) a NAD^+-glycohydrolase activity that hydrolyzes NAD^+ to ADP-ribose and nicotinamide and is operative in the absence of a suitable acceptor (Katada et al. 1983; Moss et al. 1983). The enzymatic activity of S1 is silent in both the native A-protomer (Katada et al. 1983) and in the holotoxin (Moss et al. 1983) and is synergistically activated in vitro by a variety of substances, including sulfhydryl agents, adenine nucleotides, certain phospholipids and detergents (Katada et al. 1983; Moss et al. 1983, 1986; Kaslow et al. 1987). Cellular glutathione (possibly in conjunction with a thiol:protein disulfide oxidoreductase), ATP, and membrane lipids of the toxin target cells may be the relevant activators in vivo (Kaslow et al. 1987).

S1 contains two cysteine residues in positions 41 and 201 which are disulfide bonded in the native toxin (Sekura et al. 1983). Reduction of this disulfide bond by dithiothreitol leads to a marked stimulation of S1 activity, which coincides with a release of S1 from the B-oligomer (Burns and Manclark 1989). Alkylation or deletion by site-directed mutagenesis of Cys 41 (but not of Cys 201) markedly reduces the NAD^+-glycohydrolase and ADP-ribosyltransferase activity of S1 (Kaslow and Lesikar 1987; Kaslow et al. 1989; Burns and Manclark 1989; Locht et al. 1990). On the other hand, replacement of Cys 41 by a variety of amino acids reveals that the side chain of Cys 41 is not absolutely necessary for the catalytic functions of S1 (Kaslow et al. 1989; Locht et al. 1990). Thus, Cys 41 is very likely to be located close to the NAD^+ binding site of S1 and may be an important determinant of the affinity of the $S1-NAD^+$ interaction. The detrimental effect of replacing Cys 41 with negatively charged residues suggests that Cys 41 is located close to at least one of the phosphates of NAD^+ (Locht et al. 1990).

Adenine nucleotides facilitate activation of pertussis toxin by interacting with the B-oligomer (Moss et al. 1986; Burns and Manclark 1986; Burns et al. 1988). This activation is probably due to the ATP-induced release of S1 from the B-oligomer (Burns and manclark 1986). The ability of ATP to facilitate dithiothreitol-mediated toxin activation is markedly enhanced in the presence of phospholipids and certain ionic and nonionic detergents (Moss et al. 1986; Kaslow et al. 1987). Interestingly, the ability of certain detergents (e.g., 3-[(3-cholamidopropyl)dimethylammonio]-1-propanesulfonate CHAPS) to stimulate NAD^+-glycohydrolysis does not correlate with their effect on ADP-ribosyltransferase activity (Moss et al. 1986). Presumably, these detergents have additional effects on the ADP-ribose acceptor. The site of interaction of lipophilic substances with the toxin has been mapped to S1 (Moss et al. 1986).

In addition to Cys 41, at least four other residues have been identified to be critical for S1 function (Arg 9, Arg 13, Trp 26, and Glu 129). Substitution of Arg 9 with Lys or of Arg 13 with either Lue or Lys leads to a marked reduction of ADP-ribosyltransferase activity (Burnette et al. 1988; Pizza et al. 1989). An S1 protein carrying mutations in both positions displays ADP-ribosyltransferase activity which is at least 10 000-fold lower than that of the native S1 protein (Pizza et al. 1989). Interestingly, both Arg 9 and Arg 13 lie within a stretch of eight amino acids which are nearly identical in sequence to similarly located regions of cholera and *E. coli* heat-labile toxin.

This suggests that all three toxins may interact with NAD$^+$ via this very homologous region (Locht and Keith 1986).

A Glu-X-X-X-X-Trp motif has been known for quite some time to be important for the interaction of both diphteria toxin and *Pseudomonas aeruginosa* exotoxin A (toxins which ADP-ribosylate elongation factor 2) with the nicotinamide moiety of NAD$^+$ (see Carroll and Collier 1987 for further references). Recent evidence suggests that Glu 129 and Trp 26 fulfill similar functions in the S1-subunit of pertussis toxin and that the protein folding provides for the appropriate spacing of the two residues in the native S1 protein (Pizza et al. 1988; Barbieri and Cortina 1988; Locht et al. 1989; Cortina and Barbieri 1989). Of interest in this regard, Glu 129 has recently been shown to serve as a substrate for photolabeling with NAD$^+$ (Barbieri et al. 1989). By analogy to similar studies performed with diphteria toxin and *P. aeruginosa* exotoxin A, these results suggest that the number 6 carbon atom of the NAD$^+$ nicotinamide moiety comes into close proximity of the γ-methylene group of Glu 129 (Carrol et al. 1985; Carrol and Collier 1987). Thus, Glu 129 is located within or at least very close to the catalytic center of the S1 subunit.

C. Structure and Function of ADP-Ribosylation Factors

The requirement of both an additional protein factor and GTP for the cholera-toxin-mediated ADP-ribosylation of G_s was noted as early as 1979 (Enomoto and Gill 1979). However, detailed information on the nature of this additional factor, initially referred to as cytosolic factor (Enomoto and Gill 1979, 1980; Le Vine and Cuatrecasas 1981) or "second site" (Gill and Meren 1983) and now known as ADP-ribosylation factor (ARF) (Schleifer et al. 1982), and on its mechanism(s) of action was not available until the first ARF polypeptide was purified in 1984 (Kahn and Gilman 1984a).

It is now clear that there are several ARF polypeptides which are highly homologous members of a family of proteins. ARFs are expressed in virtually all eukaryotic organisms examined thus far, including slime mold, yeast, fruit fly, and even plants (Kahn et al. 1988; Stearns et al. 1990). In those tissues and cell types which have been analyzed to date, ARFs are localized predominantly (≥50%, often about 90%) in the cytosol but are also present in very clean membrane preparations (Kahn et al. 1988). Mammalian brain appears to be a particularly abundant source of ARF polypeptides. In this tissue, ARFs account for up to 1% of the total cellular protein (Kahn et al. 1988). ARF polypeptides have been purified from rabbit liver, bovine brain, turkey erythrocyte membranes (Kahn and Gilman 1984a, 1986; Tsai et al. 1987; Kahn et al. 1988), and bovine brain cytosol (Tsai et al. 1988). Two forms each of membrane and soluble ARF can be distinguished upon protein purification (Kahn and Gilman 1984a; Tsai et al. 1988). It is unclear whether these four forms are modified versions of the same polypeptide or products of distinct genes. One of the mARFs is myristoylated at the N-terminal glycine (Kahn et al. 1988; Weiss et al. 1989). The N-terminus of one of the soluble ARFs is blocked upon amino acid sequence determination, suggesting that this form carries a N-terminal modification as well (Price et al. 1988). These findings clearly indicate that differences in N-terminal myristoylation are unlikely to be the sole reason for the distinct behavior of the various ARFs upon subcellular fractionation and protein purification.

ARF polypeptides specifically bind guanine nucleotides with high affinity (Kahn and Gilman 1986; Weiss et al. 1989; Bobak et al. 1990). As purified ARF polypeptides contain equimolar amounts of noncovalently bound GDP, binding of radiolabeled exogeneous guanine nucleotides to ARF is complex, since it involves both the release of bound GDP and the binding of radiolabeled nucleotide. A few general statements can, however, be made. Binding of radiolabeled GTP or GTP[S] to purified membranous and soluble ARF polypeptides is dependent on Mg^{2+} and is enhanced by phospholipids like dimyristoylphosphatidylcholine (DMPC) and certain detergents like sodium cholate (Kahn and Gilman 1986; Weiss et al. 1989; Bobak et al. 1990). High ionic strength has been reported to be essential for binding of GTP[S] to purified membranous ARF (Kahn and Gilman 1986). However, a high salt concentration appears to have no effect on GTP binding to purified soluble ARF (Bobak et al. 1990) and actually inhibits GTP[S] binding to recombinant bARF1 (Weiss et al. 1989). Purified membranous ARF displays an extremely low GTPase activity, even when tested under conditions optimal for binding of exogeneous guanine nucleotides (Kahn and Gilman 1986). This raises the question of whether an ARF-specific GTPase activating protein (GAP) similar to the *ras* GAP (see Bourne et al. 1990 for review) exists to support GTP hydrolysis by ARF in membrane preparations or intact cells. The fact that binding of GTP is required for ARF to be capable of supporting ADP-ribosylation of α_s (see below) together with the finding that ARFs are purified with bound GDP clearly suggests that GTP hydrolysis takes place in crude systems. Interestingly, preliminary evidence has been reported for the existence of an ARF GAP in yeast (Stearns et al. 1990). Future experimentation will show whether ARF GAPs exist in mammalian cells as well.

Purification of ARF polypeptides from abundant sources such as the brain paved the way for amino acid sequence determination and isolation of cDNAs coding for ARF polypeptides. Two bovine, three human, and two yeast ARF cDNAs have been characterized to date (Sewell and Kahn 1988; Price et al. 1988; Bobak et al. 1989; Monaco et al. 1990; Stearns et al. 1990). The predicted amino acid sequence of one of the three human cDNAs (hARF1) is identical to that predicted for one of the bovine ARF cDNAs (bARF1; Sewell and Kahn 1988; Bobak et al. 1989). The other three mammalian cDNAs (bARF2, hARF3, and hARF4) and the two yeast cDNAs (yARF1, yARF2) are highly homologous to each other and to bARF1 and hARF1 but are distinct entities. Thus, at least four mRNAs coding for ARF polypeptides appear to exist in mammalian cells. Discrepancies between amino acid sequence information obtained from purified ARFs and the sequence predicted by the known ARF cDNAs (KAHN et al. 1988; BOBAK et al. 1989; MONACO et al. 1990), together with complex hybridization patterns obtained upon Northern analysis of poly(A)$^+$ RNA from various mammalian tissues and species probed with these cDNAs (TSUCHIYA et al. 1989), strongly suggest that more members of the ARF family are yet to be discovered.

ARF cDNAs encode proteins consisting of $\approx$180 residues with predicted molecular masses of about 20 kDa. Several features of the ARF structure are worth noting. First, similar to α-subunits of heterotrimeric G proteins and small molecular weight GTP binding proteins related to the *ras* (proto)oncogene products, ARF polypeptides contain regions homologous to all four of the conserved regions identified by HALLIDAY (1984) for GTP binding proteins. One of these regions, however, carries an important difference between the ARFs and all other GTP binding proteins. A glycine, which corresponds to Gly 12 of the cellular *ras* proteins, is replaced by an aspartic acid in position 26 of all ARF polypeptides. A glycine in this position is essential for GTP hydrolysis by *ras* and other GTP binding proteins (BARBACID 1987). Replacement by Asp thus explains the extremely low GTPase activity of purified ARFs (KAHN and GILMAN 1986). The second important feature relates to their modification by lipid moieties. None of the ARFs contains a cysteine among the four C-terminal amino acids, which is found farnesylated in many other small molecular weight GTP binding proteins (HANCOCK et al. 1989; MALTESE et al. 1990; see MALTESE 1990 for review). On the other hand, all ARFs identified to date contain a glycine next to the initiating methionine, which is part of the consensus sequence for N-terminal myristoylation (TOWLER et al. 1988). However, only hARF4 carries a serine in position 6, which is known to be very favorable to N-myristoylation. ARF1 and hARF3 contain a less favorable Ala 6 or Gly 6, respectively. As bARF2 contains a glutamic acid in position 6, it is expected to be a poor substrate for N-myristoylation. Thus, the differences in N-myristoylation of purified membranous ARFs (see above) may be due to distinct primary structures but may also reflect partial modification of a single polypeptide.

It is very likely that the major biological function of ARF polypeptides is distinct from assisting the cholera toxin-dependent ADP-ribosylation of G$_s$. This assumption is based in part on the wide distribution of ARF proteins, both among the various tissues of a given organism and among the various divisions of the plant and animal kingdom (KAHN et al. 1988; TSUCHIYA et al. 1989; STEARNS et al. 1990). Microinjection of ARF has been reported to inhibit progesterone- or insulin-elicited maturation of *Xenopus*

oocytes (Bahnson et al. 1989). These findings led to the suggestion that
ARFs may be involved in the signal transduction mechanisms elicited by
these two mediators or in modulating the subsequent response. A recent
report suggests that ARFs may be involved in regulating intracellular protein transport (Stearns et al. 1990). This proposal is based on the finding
that deletion of one of the two known ARF genes in yeast leads to defective
protein secretion. Furthermore, ARFs are localized to the Golgi apparatus
upon immunocytochemical analysis of various mammalian cells. Thus,
ARFs may very well join the rapidly growing subfamily of the small molecular weight GTP binding proteins involved in regulating vesicle transport
and organelle function (see Balch 1990 for a recent review).

D. Molecular Mechanisms of G-Protein
ADP-Ribosylation by Cholera and Pertussis Toxins

I. Cholera Toxin

In the presence of ARF, the A_1 fragment of cholera toxin is capable of
ADP-ribosylating a variety of proteins, including phosphorylase b, bovine
serum albumin, α-lactalbumin, and even the β-subunit of retinal transducin
(Tsai et al. 1987). In addition, the A_1 fragment also ADP-ribosylates ARF
and itself. The natural and most readily modified substrates of the toxin,
however, are the α-subunits of the stimulatory G protein of adenylyl cyclase,
G_s. It is very likely that all four variants of α_s are ADP-ribosylated by
cholera toxin. In addition, an α_s-related protein found in olfactory epithelium (termed α_{olf}) is also modified by cholera toxin (D.T. Jones and
Reed 1989; D.T. Jones et al. 1990). As alluded to earlier, maximal ADP-
ribosylation of α_s and α_{olf} requires the presence of GTP in addition to ARF
and the A_1 fragment of the toxin. Using somewhat different conditions
(see below), the α-subunits of several other G proteins, including retinal
transducin and members of the G_i/G_o subfamily, can also be shown to serve
as specific substrates for ADP-ribosylation by cholera toxin.

1. ADP-Ribosylation of G_s

It is now well established that ARF needs to be liganded with GTP or one of
its poorly hydrolyzable analogues in order to support ADP-ribosylation of α_s
(Kahn and Gilman 1986). GDP-liganded ARF is ineffective in this regard.
In contrast, binding of GTP[S] or other poorly hydrolyzable analogues of
GTP to α_s renders the protein a poor substrate for the toxin (Kahn and
Gilman 1984b). This explains why GTP[S] is less effective than GTP in
supporting ADP-ribosylation of purified G_s (Kahn and Gilman 1984b). As
ARFs are poor GTPases, GTP is expected to activate ARF preferentially,
whereas GTP[S] simultaneously activates ARF and reduces the substrate
quality of G_s. The reason why GTP[S]-activated α_s is such a poor cholera

toxin substrate is unknown. The most logical explanation would be that GTP[S] promotes dissociation of G_s into βγ-complexes and $α_s$, which would then be resistant to modification by the toxin. The observation that exogeneous βγ-subunits markedly enhance the cholera-toxin-dependent ADP-ribosylation of bacterially expressed $α_s$ is certainly very much in favor of such a hypothesis (GRAZIANO et al. 1987; 1989). However, two findings are not easily consistent with this concept. First, $α_s$ purified from rabbit liver and separated from βγ-subunits was just as well ADP-ribosylated by cholera toxin as the holoprotein (KAHN and GILMAN 1984b). Second, AlF_4^-, another agent known to elicit both activation and subunit dissociation of G_s (NORTHUP et al. 1983b), did not affect ADP-ribosylation of $α_s$ by cholera toxin (KAHN and GILMAN 1984b). Thus, the mechanism(s) by which binding of poorly hydrolyzable GTP analogues to $α_s$ reduces its ADP-ribosylation by cholera toxin have yet to be elucidated. Likewise, the question whether or not βγ-subunits are required for ADP-ribosylation of nonrecombinant $α_s$ proteins in their native plasma membrane environment will have to be addressed by future experimentation.

As will be discussed below, the amino acid ADP-ribosylated by cholera toxin in $α_t$ has been identified as Arg 174. An arginine is conserved in an equivalent position in all of the known α-subunits cDNAs (see LOCHRIE and SIMON 1988 for amino acid sequence information). It is thus very likely, albeit not proven, that this arginine (Arg 187, Arg 188, Arg 201, or Arg 202, depending on the splice variant of $α_s$) serves as the major ADP-ribose acceptor in $α_s$ as well. It is important to note, however, that replacement of this Arg by either Ala, Glu, Lys, His, or Cys through in vitro or in vivo mutagenesis of $α_s$ markedly but not completely reduced the ADP-ribosylation by cholera toxin (FREISSMUTH and GILMAN 1989; LANDIS et al. 1989). As cholera toxin specifically modifies arginine residues (MOSS and VAUGHAN 1977, 1979; HSIA et al. 1985), it is likely that an arginine(s) in addition to the one corresponding to Arg 174 of $α_t$ is also ADP-ribosylated by cholera toxin. The observation of multiple, distinct, ADP-ribosylated $α_s$ polypeptides upon fractionation by isoelectric focussing may be a reflection of this phenomenon (SCHLEIFER et al. 1980; OWENS et al. 1985).

An important question that has not been adequately answered to date is whether the stimulatory effect of ARF on cholera-toxin-mediated ADP-ribosylation is due entirely to a direct interaction of ARF with A_1 or whether an ARF-G_s interaction contributes as well. There is ample evidence to support the first hypothesis (TSAI et al. 1987, 1988), but both experimental data and convincing arguments supporting a direct ARF-G_s interaction have also been presented (KAHN 1990).

2. ADP-Ribosylation of Retinal Transducin

In 1981, COOPER and colleagues reported that the retinal G protein transducin is ADP-ribosylated by cholera toxin. A subsequent report demonstrated that cholera-toxin-mediated ADP-ribosylation of transducin was markedly enhanced by light and the poorly hydrolyzable GTP analogue guanosine-5'-(β,γ-imino)triphosphate (GppNHp) (Abood et al. 1982). Under optimal conditions, cholera toxin led to a high level of incorporation of ADP-ribose into α_t (0.7 mol/mol), which was accompanied by an about 70% inhibition of light-stimulated GTPase activity. Navon and Fung (1984) were able to reproduce and extend these findings in a reconstituted system. Their results demonstrated that maximal incorporation of ADP-ribose into α_t was achieved when GppNHp and $\beta\gamma_t$ were present at concentrations equal to that of α_t and when rhodopsin was continuously irradiated with visible light, suggesting that a transient protein complex consisting of α_t-GppNHp, $\beta\gamma_t$, and light-activated rhodopsin was the actual cholera toxin substrate. Purified heterotrimeric transducin was only poorly ADP-ribosylated by the toxin. Whether or not ARF-like protein is present in rod outer segment membranes to support light-stimulated ADP-ribosylation of α_t has not been clarified up to the present time. It is unlikely, however, that GppNHp is required in the ADP-ribosylation reaction to activate a putative retinal ARF protein, since ADP-ribosylation of α_t preloaded with GppNHp was not further enhanced by exogeneous GppNHp (Navon and Fung 1984).

The peptide ADP-ribosylated by cholera toxin in α_t has been identified by direct sequencing as Ser-Arg-Val-Lys, with Arg being the modified amino acid residue (Van Dop et al. 1984a). Comparison of this peptide sequence with the complete amino acid sequence of α_t predicted by its cDNA sequence (see Lochrie and Simon 1988 for references) revealed that cholera toxin ADP-ribosylates an arginine residue in position 174.

3. ADP-Ribosylation of G_i/G_o-Like G Proteins

Although it is generally believed that the various forms of G_s are the only cholera-toxin-sensitive G proteins in non-retinal systems, there is an increasing body of evidence suggesting that members of the pertussis-toxin-sensitive G-protein subfamilies G_i and G_o are also ADP-ribosylated by cholera toxin, at least under certain conditions. A 40-kDa protein present in rat adipocyte plasma membranes was the first non-G_s G protein found ADP-ribosylated by cholera toxin in nonretinal tissues (Graves et al. 1983). Interestingly, in contrast to the ADP-ribosylation of α_s the modification of the 40-kDa substrate was suppressed by GTP. This finding not only indicated that the 40-kDa protein was in fact a GTP-binding protein but also demonstrated that guanine nucleotides differentially affect the ADP-ribosylation of different G proteins by cholera toxin. Owens et al. (1985) subsequently extended these findings and showed that the 40-kDa cholera toxin substrate could be resolved into a M_r 39 000/41 000 doublet, which comigrated with the main pertussis toxin substrates present in these membranes on both one- and two-dimensional SDS polyacrylamide gels. Interestingly, incubation of fat cell membranes with cholera toxin in the absence of GTP led to a marked attenuation of both receptor- and GTP-mediated inhibition of adenylyl cyclase, indicating that at least one of the ADP-ribosylated proteins was the inhibitory G protein of adenylyl cyclase, G_i (see below). Similar to adipocytes, many other cell types contain $\approx$40-kDa poteins that serve as specific cholera toxin substrates in the absence of GTP. These include mouse macrophages (AKSAMIT et al. 1985), human neutrophils, monocytes, and cultured phagocytes (VERGHESE et al. 1986; GIERSCHIK and JAKOBS 1987; GIERSCHIK et al. 1989a), NG108-15 cells (MILLIGAN and McKENZIE 1988; KLINZ and COSTA 1989, 1990), and C6 glioma cells (MILLIGAN 1987).

Similar to the cholera-toxin-mediated ADP-ribosylation of transducin, the modification of other non-G_s G proteins is markedly enhanced by agonist-activated receptors. This phenomenon was first described for human HL-60 granulocytes, in which labelling of both α_{i2} and α_{i3} by cholera toxin was markedly enhanced by the chemotactic peptide fMet-Leu-Phe (GIERSCHIK and JAKOBS 1987; GIERSCHIK et al. 1989a). It is very likely that cholera and pertussis toxin modify the α_i proteins of HL-60 cells at two distinct sites, most likely at an arginine residue corresponding to Arg 174 of α_t and at a Cys four residues from the C-terminus, respectively (IIRI et al. 1989). Both residues are present in the appropriate position in both α_{i2} and α_{i3} (LOCHRIE and SIMON 1988). In NG108-15 neuroblastoma glioma hybrids and C6 glioma cells a 40-kDa membrane protein(s) was reported to be ADP-ribosylated by cholera toxin in response to opioid peptides and an unidentified receptor agonist present in fetal calf serum, respectively (MILLIGAN and McKENZIE 1988; KLINZ and COSTA 1989; MILLIGAN 1989). Note that – at least under certain conditions – even the ADP-ribosylation of G_s by cholera toxin appears to be enhanced by agonist occupancy of G_s-coupled receptors (ENOMOTO and GILL 1980).

Using membranes prepared from NG108-15 cells, KLINZ and COSTA (1989) made the interesting observation that in contrast to opioid peptide agonists, certain antagonists (those that inhibit basal GTPase activity in these membranes and are hence termed "negative antagonists") reduced the basal labeling of the 40-kDa substrate by cholera toxin. (–)Naloxone, an "inert" antagonist, did not affect the ADP-ribosylation. This finding, together with the fact that the basal ADP-ribosylation was also reduced in cells from which opioid receptors had been functionally or physically removed by desensitization led the authors to suggest that, in certain systems, G proteins may be coupled to and activated by receptors even in the absence of agonists, a concept that is supported by certain effects of G-protein ADP-ribosylation to be discussed below.

Whether or not an ARF-like protein is involved to support ADP-ribosylation of α-subunits of the G_i/G_o family is uncertain. In one report, addition of cytosolic protein (which may contain an ARF-like activity) was shown to enhance ADP-ribosylation of the 40-kDa substrate (GRAVES et al. 1983). IIRI et al. (1989) demonstrated that partially purified bovine brain sARFII did not support ADP-ribosylation of purified G_{i2}. N-formyl-Met-Leu-Phe receptor-mediated ADP-ribosylation of the 40-kDa substrate was lost when HL-60 membranes were solubilized and reconstituted into phospholipid vesicles, despite the fact that exogenous sARFII was added and that there was functional receptor-G-protein coupling in these reconstituted vesicles. These findings were taken by the authors to suggest that the receptor-mediated ADP-ribosylation of G_i/G_o proteins is dependent on a labile component distinct from sARFII which is lost upon membrane solubilization. Identification of this factor(s) awaits further experimentation.

II. Pertussis Toxin

1. ADP-Ribosylation of Retinal Transducin

Retinal transducin is an excellent substrate for ADP-ribosylation by pertussis toxin, both in its native membrane environment and when the purified protein is treated with the toxin in solution (Van Dop et al. 1984b; Watkins et al. 1984, 1985). The latter finding demonstrates that ARF-like proteins are not involved in pertussis-toxin-mediated ADP-ribosylation of G proteins. Pertussis toxin specifically ADP-ribosylates the α-subunit of transducin. The ADP-ribosylated amino acid residue is a cysteine in position 347, which corresponds to the fourth amino acid from the C-terminus (West et al. 1985). Thus, cholera and pertussis toxin ADP-ribosylate $α_t$ at distinct sites. It is very likely that a cysteine present in position -4 from the C-terminus in all other pertussis toxin substrates (cf. Lochrie and Simon 1988) is modified by the toxin in these proteins as well (Hsia et al. 1985; Hoshino et al. 1990).

The ability of $α_t$ to serve as as a pertussis toxin substrate is markedly dependent on the association of the protein with $βγ_t$ and guanine nucleotides and on the presence of the light receptor rhodopsin in the ADP-ribosylation mixture. βγ-subunits markedly enhance the ADP-ribosylation of purified $α_t$ (Watkins et al. 1985). Whether there is an absolute requirement of $βγ_t$ for ADP-ribosylation of $α_t$ is uncertain. In two reports, ADP-ribosylation of $α_t$ was only reduced but not completely abolished in the absence of $βγ_t$, in particular in the presence of ATP and centain phospholipids (Watkins et al. 1985; Moss et al. 1986). Whether this "basal" ADP-ribosylation of $α_t$ was due to the presence of $βγ_t$ contaminating the $α_t$ preparations remains to be examined. Of note, βγ-subunits have been reported to support catalytically ADP-ribosylation of α-subunits by pertussis toxin (Casey et al. 1989). Thus, a minute quantity of $βγ_t$ could have enhanced the radiolabelling of "pure" $α_t$ in the above studies.

In native rod outer segments, ADP-ribosylation of transducin was reported to be reduced by about 95% when bleached rather than dark-adapted membranes were used in the absence of exogenous guanine nucleotides (Van Dop et al. 1984b). This interesting observation was taken by the authors to suggest that the receptor-coupled (i.e., guanine-nucleotide-free) form of transducin is not a pertussis toxin substrate (possibly because the

ADP-ribosylation site on α_t is masked by the activated receptor). Interestingly, GDP[S] markedly stimulated the ADP-ribosylation of α_t complexed by light-activated rhodopsin, suggesting that binding of guanine nucleoside diphosphates to the G protein relaxes – at least to some extent – the receptor-G-protein interaction, which would then lead to an exposure of the ADP-ribosylation site to the toxin (VAN DOP et al. 1984b).

2. ADP-Ribosylation of G_i/G_o-Like G Proteins

The molecular mechanisms involved in pertussis-toxin-mediated ADP-ribosylation of G proteins of the G_i/G_o subfamily are very similar to those observed with retinal transducin. This is particularly true when the G-protein ADP-ribosylation is studied following the reconstitution of the G protein with a receptor protein, as done by TSAI et al. (1984), who reconstituted partially purified rabbit liver α_i- and $\beta\gamma$-subunits with retinal rhodopsin. Retinal rhodopsin is capable of replacing inhibitory receptors of adenylyl cylase in reconstituted systems, as it is fully capable of interacting with and activating G_i (KANAHO et al. 1984; CERIONE et al. 1985). As observed with α_t (see above), $\beta\gamma$-subunits markedly enhanced ADP-ribosylation of α_i. This effect was particularly prominent in the absence of ATP. Interestingly, photolyzed rhodopsin markedly inhibited the ADP-ribosylation of the G_i holoprotein in the absence of guanine nucleotides, which is very reminscent of the effects observed by VAN DOP et al. (1984b) for retinal transducin. In the presence of photolyzed rhodopsin, addition of a variety of guanine nucleotides markedly enhanced ADP-ribosylation of G_i. Maximal stimulation was observed with GDP, GDP[S], and GTP. Surprisingly, the GTP analogues guanosine-5'-(α,β-methylene)triphosphate [Gpp(CH$_2$)p], GppNHp, and GTP[S] also stimulated ADP-ribosylation, albeit to a lesser extent than the other guanine nucleotides (see below).

Similar to transducin, purified G_i and G_o proteins are excellent substrates for pertussis toxin even in the absence of any additional protein factor and a lipid bilayer. All variants of α_i and α_o so far examined require $\beta\gamma$-subunits in order to be ADP-ribosylated by pertussis toxin (NEER et al. 1984; HUFF and NEER 1986; KATADA et al. 1986a; GIERSCHIK et al. 1987; CASEY et al. 1989). This requirement also holds true for recombinant α_i- and α_o-subunits produced in E. coli, in reticulocyte lysates by in vitro translation, or in mammalian cells by DNA transfection (LINDER et al. 1990; HSU et al. 1990; T.L.Z. JONES et al. 1990). When the α-subunits were expressed in E. coli, less that 2% of the protein that could be modified in the presence of $\beta\gamma$ was ADP-ribosylated in its absence (LINDER et al. 1990).

The precise mechanism(s) by which $\beta\gamma$-subunits enhance ADP-ribosylation of pertussis-toxin-sensitive α-subunits is unknown. Protection of a-subunits from denaturation may contribute to a considerable extent, because heterotrimeric G proteins may dissociate under certain conditions (e.g., low protein concentration) even in the absence of activating ligands

(Smigel et al. 1982). The free α-subunit has a considerably lower affinity for GDP than the heterotrimeric protein and is thus more likely to lose bound GDP (Higashijima et al. 1987). Unliganded α-subunits are exceptionally labile and thus rapidly lost by denaturation (Smigel et al. 1982; Northup et al. 1983a,b). Addition of βγ-dimers and/or exogenous guanine nucleotides counteracts subunit dissociation and/or loss of guanine nucleotide and thus markedly enhances the amount of viable α-subunits available for toxin modification (Katada et al. 1986a; Mattera et al. 1986). A second, possibly even more important mechanism for the stimulatory effect of βγ-subunits on ADP-ribosylation of α-subunits by pertussis toxin is an βγ-dimer-induced conformational change of the GDP-liganded α-subunit that is unrelated to α-subunit stability. This suggestion is based on the observation that βγ-subunits led to a marked ($\approx$10-fold) stimulation of ADP-ribosylation of α_i and α_o even after care was taken to prevent denaturation of the α-subunits by inclusion of GTP in the incubation mixture (Katada et al. 1986a).

The effects of guanine nucleotides and Mg^{2+} on pertussis-toxin-mediated ADP-ribosylation of G proteins are complex and, at least in part, still poorly understood. This is mainly due to the fact that both agents probably exert multiple effects in the toxin-catalyzed ADP-ribosylation reaction. For example, magnesium ions have been reported by one group to inhibit activation of the toxin by ATP (Moss et al. 1986). Another group observed a marked inhibition of the toxin-mediated modification of human erythrocyte G_i by Mg^{2+}, Mn^{2+}, and Ca^{2+} but was unable to observe an effect of Mg^{2+} on the ability of ATP to activate the toxin (Mattera et al. 1986).

Similar to Mg^{2+}, guanine nucleotides (which are either added on purpose or unintentionally supplied as a contaminant with the ATP used to activate the toxin) affect G-protein ADP-ribosylation by acting at serveral levels. First, GTP, GDP, GDP[S], and even GTP[S] have been reported to interact directly with and facilitate the activation of pertussis toxin just like ATP (or AppNHp) (Mattera et al. 1986, 1987). This effect was particularly prominent when ATP was absent from the incubation medium. Second, guanine nucleotides may act as G-protein stabilizers and thereby enhance ADP-ribosylation (Mattera et al. 1986; Katada et al. 1986a). Third, binding of stable GTP analogues like GTP[S] to the G protein induces a conformation that is only poorly or not at all ADP-ribosylated by the toxin (Bokoch et al. 1984; Huff and Neer 1986; Navon and Fung 1987; Mattera et al. 1987; Jones et al. 1990). Interestingly, subunit dissociation does not appear to be required for GTP[S]-mediated inhibition of G_i ADP-ribosylation (Mattera et al. 1987). Finally, as suggested above, both guanosine nucleoside di- and triphosphates as well as their analogues may loosen the interaction of the G protein with an activated receptor and thus expose the ADP-ribosylation site (Van Dop et al. 1984a; Tsai et al. 1984). Only if one considers each of these actions is one able to rationalize some of the seemingly paradoxical effects of guanine nucleotides on pertussis-toxin-mediated G-protein ADP-ribosylation. For example, the observation that poorly hydrolyzable GTP analogues stimulated ADP-ribosylation of G_i when reconstituted with photolyzed rhodopsin (Tsai et al. 1984) is certainly rather puzzling in light of the fact that these nucleotides are inhibitory in most other cases. However, binding of these analogues to a receptor-bound G protein is well-known to disrupt the receptor-G-protein interaction, which could easily give rise, albeit only transiently, to a G protein that is a better substrate for the toxin than the one complexed to the receptor.

A very important question concerning the molecular mechanism of pertussis-toxin-mediated ADP-ribosylation of G proteins is which of the structural elements of a given α-subunit (aside from the cysteine in position -4 from the C-terminus) are absolutely required for the toxin-mediated modification. Although this question is still subject to further investigation, serveral initial conclusions can already be drawn from the literature. First, it is clear that the ≈20 amino terminal residues of the α-subunit are absolutely required for pertussis-toxin-mediated ADP-ribosylation (Watkins et al. 1985; Navon and Fung 1987; Neer et al. 1988). As the same region is also required for the association of the α-subunit with βγ-dimers (Navon and Fung 1987; Neer et al. 1988), it is very likely that an N-terminal truncation precludes the ADP-ribosylation of the α-subunit by preventing the formation of the αβγ-heterotrimer. Modification of the N-terminal residue by myristoylation, found with many pertussis toxin-sensitive α-subunits (Buss et al. 1987; Schultz et al. 1987), does not appear to be required for ADP-ribosylation (T.L.Z. Jones et al. 1990). Integrity of the α-subunit C-terminus is important for ADP-ribosylation by pertussis toxin. This is based on the observation that treatment of α_o with carboxypeptidase A (which most likely removes the two C-terminal residues, Leu 353 and Tyr 354, but not the substrate amino acid Cys 351) prevented its ADP-ribosylation (Neer et al. 1988). The ability of the protein to interact with βγ appeared to be unaltered by this treatment. Also pertinent to this issue, replacement of the tripeptide Gln 390-Tyr 391-Glu 392 of α_{s1} (394 residues) by Asp 390-Cys 391-Gly 392 (a sequence corresponding to that found in all three forms of α_i; Lochrie and Simon 1988) generated an α-subunit that was ADP-ribosylated by pertussis toxin only poorly, despite the fact that the mutant protein contained a Cys in position -4 from the C-terminus and appeared to interact with the βγ-dimer (Freissmuth and Gilman 1989). A similar finding was obtained by Osawa et al. (1990), who showed that replacement of the C-terminal 38 residues of α_{s1} by the 36 C-terminal residues of α_{i2} did not suffice to make the chimeric protein pertussis-toxin-sensitive. Results presented in the latter study also demonstrate that two N-terminal domains of α_{i2} are required for ADP-ribosylation of the α_{s1}/α_{i2} fusion protein (in addition to the 36 amino acid α_{i2} C-terminus). One domain is expected to lie between residues 1 through 64, the other between residues 65 through 212. It has been suggested that both domains are necessary for the interaction of α-subunits with the βγ-dimer (OSAWA et al. 1990). Although more experimentation is required to substantiate this hypothesis, the concept is consistent with the earlier observation that sulfhydryl alkylation of Cys 108 of $α_o$ by N-ethylmaleimide blocks ADP-ribosylation of the protein (WINSLOW et al. 1987), presumably by preventing the association of $α_o$ with βγ-subunits (HOSHINO et al. 1990).

E. Functional Consequences of G-Protein ADP-Ribosylation by Cholera and Pertussis Toxins

I. ADP-Ribosylation of G_s by Cholera Toxin

1. Short-Term Effects

ADP-ribosylation of $α_s$ by cholera toxin has long been known to activate adenylyl cyclase persistently in intact cells and membrane preparations (see FINKELSTEIN 1973; MOSS and VAUGHAN 1979; ROSS and GILMAN 1980 for early reviews). In 1977, CASSEL and SELINGER discovered that cholera toxin treatment of turkey erythrocyte membranes markedly reduces the receptor-stimulated increase in the steady-state rate of GTP hydrolysis. The toxin treatment specifically prolonged the lifetime of the agonist plus GTP-activated G_s, but did not alter the rate of the receptor-mediated activation of G_s. These observation led to the concept that the toxin-induced modification of G_s specifically reduces the $k_{cat,GTP}$ of the overall GTPase reaction and thereby markedly enhances the fraction of G_s in the GTP-bound state capable of activating adenylyl cyclase (CASSEL and SELINGER 1977; CASSEL et al. 1977).

Several lines of recent evidence support the validity of such a concept. First, somatic mutations of the $α_s$ gene replacing the arginine serving as the substrate amino acid for cholera toxin in the wild-type protein by either cysteine or histidine have been identified in growth hormone-secreting human pituitary tumors (LANDIS et al. 1989). The mutated G_s proteins constitutively activate adenylyl cyclase (VALLAR et al. 1987). Analysis of the $k_{cat,GTP}$ of the G_s proteins present in the tumor tissues revealed a mutation-dependent decrease by a factor of about 30 (LANDIS et al. 1989). Second, intentional replacement of the same arginine residue in $α_{s3}$ (Arg 187) by either Ala, Glu, or Lys led to an about 100-fold reduction of the $k_{cat,GTP}$ when the wild-type and mutant proteins were expressed in $E.$ $coli$ and functionally compared with each other (FREISSMUTH and GILMAN 1989). In contrast, the rate of GTP[S] binding (which is limited by and thus reflects the release of bound GDP) was not greatly altered by these mutations. In the presence of GTP, the fractional occupancy of the mutated $α_s$ proteins by GTP was approximately 40-fold higher (80% vs. 2%) than that of the wild-type protein.

Despite the fact that these recent results strongly favor the hypothesis that a reduction of the $k_{cat,GTP}$ is the major consequence of the cholera-toxin-mediated ADP-ribosylation of α_s, it is important to note that this concept is still based on indirect evidence. Thus, an inhibition of GTPase activity by cholera-toxin-mediated ADP-ribosylation has so far not been demonstrated for purified or recombinant G_s. In one report, ADP-ribosylation of purified G_s by cholera toxin failed to reduce the basal GTPase activity of G_s, and even a slight stimulation was observed (KAHN and GILMAN 1984b). It should be noted, however, that the value reported in this study for the control GTPase activity of unmodified G_s was about 100-fold lower than the activity reported subsequently when the experimental conditions for determining GTP hydrolysis by purified G_s had been optimized (BRANDT and Ross 1985). Thus, the question whether purified G_s has reduced GTPase activity when ADP-ribosylated by cholera toxin may have to be reexamined under the latter conditions.

A further important point to be mentioned in this context is that a reduction of the $k_{cat,GTP}$ of α_s does not necessarily have to be the only functional consequence of G_s ADP-ribosylation by cholera toxin; for example, there are several lines of evidence suggesting that it results in a decreased affinity of G_s for guanine nucleotides. Specifically, cholera toxin treatment of turkey erythrocyte or rat adipocyte membranes has been shown to enhance the release of G_s-bound GDP from these membranes (BURNS et al. 1982, 1983; MURAYAMA and UI 1984). In turkey erythrocyte membranes, cholera toxin treatment shortened the lag time normally observed for activation of adenylyl cyclase by the poorly hydrolyzable GTP analogue GppNHp (LAD et al. 1980). In rat liver membranes, cholera toxin decreased the apparent affinity of guanine nucleotides to stimulate adenylyl cyclase and to regulate glucagon receptor binding (LIN et al. 1978; ROJAS and BIRNBAUMER 1985). Of interest in this regard, ADP-ribosylated α_s appears to have a lower affinity for $\beta\gamma$-subunits (KAHN and GILMAN 1984b). As a free α-subunit binds GDP with a lower affinity than the $\alpha\beta\gamma$-trimer (HIGASHIJIMA et al. 1987), a reduced subuhnit interaction could very well explain the altered guanine nucleotide binding properties of ADP-ribosylated G_s observed in membrane preparations.

2. Long-Term Effects

In addition to modifying GTP hydrolysis, α-$\beta\gamma$ interaction, and possibly the guanine nucleotide binding properties of G_s, cholera-toxin-mediated ADP-ribosylation renders G_s extremely sensitive to denaturation and/or degradation. This phenomenon is observed in membrane preparations and in intact cells. In membrane preparations, GTP markedly and specifically stabilized the ADP-ribosylated α_s protein (NAKAYA et al. 1980, 1981). Whether the decreased stability of α_s in cholera-toxin-treated membranes is secondary to a toxin-induced release of GDP and generation of an unoccupied and thus very labile species of α_s is unknown. It is clear, however, that GTP serves two purposes when G_s is ADP-ribosylated by cholera toxin in membrane preparations. One is activation of ARF, and the other is stabilization of the modified toxin substrate.

Treatment of intact cells for extended periods (>1 h) with relatively high concentrations (0.1–10 µg/ml) of cholera toxin led to a 95% decrease in the level of immunoreactive α_s present in membrane preparations of a variety of cell types (Chang and Bourne 1989; Klinz and Costa 1989; Milligan et al. 1989; Macleod and Milligan 1990; Carr et al. 1990; Hermouet et al. 1990). Interestingly, in most cases the α_s protein lost from the membrane was not recovered in soluble fractions, suggesting that α_s was degraded rather than translocated. ADP-ribosylation of α_s preceded its removal from the membrane, indicating that ADP-ribosylation was required for down-regulation (Chang and bourne 1989; Milligan et al. 1989). Surprisingly, the decrease in α_s seen upon cholera toxin treatment did not appear to be due to an increase in cellular cAMP (Chang and Bourne 1989; Milligan et al. 1989). In most studies, the toxin-induced down-regulation was limited to α_s with no change in β-subunits, α_i or α_o proteins (Chang and Bourne 1989; Milligan et al. 1989; Carr et al. 1990; Macleod and Milligan 1990; Hermouet et al. 1990). In one report, however, there was a marked toxin-dependent reduction in both α_o and β_{36} in addition to the decrease in α_s, although higher concentrations of cholera toxin were required to down-regulated G-protein subunits other than α_s (Klinz and Costa 1990). The molecular mechanisms involved in cholera-toxin-mediated down-regulation of α_s (and possibly other G-protein subunits) are unknown.

The functional consequences of long-term cholera toxin treatment of intact cells leading to α_s down-regulation on adenylyl cyclase activity are somewhat variable and presumably depend on the amount of functional α_s protein remaining in the membrane after toxin treatment and the excess of α_s over the adenylyl cyclase catalyst present in a particular cell type. Thus, the long-term changes range from a persistence of elevated GTP-dependent adenylyl cyclase activity despite a substantial loss of α_s (Chang and Bourne 1989) to a biphasic response of adenylyl cyclase activity with an inhibitory phase following stimulation when cells are incubated with the toxin for extended periods and/or at high toxin concentrations (Macleod and Milligan 1990).

II. ADP-Ribosylation of G Proteins Other than G_s by Cholera Toxin

The functional consequences of cholera-toxin-mediated ADP-ribosylation of G proteins other than G_s are very similar to the effects observed with G_s, at least with respect to their GTPase activity. Thus, ADP-ribosylation of retinal transducin by cholera toxin has been shown to reduce markedly light-stimulated GTP hydrolysis by transducin in rod outer segment membranes (Abood et al. 1982). Determination of the single turnover rate of GTP hydrolysis revealed that ADP-ribosylation of α_t by cholera toxin slowed the hydrolysis of bound GTP about 20-fold but did not alter the rate of $[^3H]$-GppNHp binding (NAVON and FUNG 1984).

Cholera toxin treatment of membranes prepared from HL-60 granu-locytes led to a marked reduction of GTPase activity in response to various chemotactic stimuli, which bind to receptors coupled via G_{i2} and/or G_{i3} to stimulate phospholipase C (GIERSCHIK and JAKOBS 1987; MCLEISH et al. 1989; GIERSCHIK et al. 1989a). Treatment of intact NG108-15 cells with high concentrations (10 µg/ml) of cholera toxin has been reported to abolish the $[^{32}P]$-ADP-ribosylation of α_{i2} and/or α_o when plasma membranes were treated with activated cholera toxin, opioid peptide agonist, and $[^{32}P]$-NAD$^+$, suggesting that α_{i2} and/or α_o were ADP-ribosylated by cholera toxin even in intact cells (KLINZ and COSTA 1990). Interestingly, the same intact cell toxin treatment also led to a marked reduction of opioid-stimulated GTPase activity in membranes, indicating that cholera-toxin-mediated ADP-ribosylation of G_{i2} and/or G_o led to reduced GTP hydrolysis by these G proteins (KLINZ and COSTA 1989). It is important to keep in mind, however, that cholera toxin treatment of NG108-15 cells also led to a reduction of membrane-bound α_o, α_{i2}, and β_{36} (KLINZ and COSTA 1990). It is uncertain to what extent these changes in protein levels contribute to the cholera-toxin-induced loss in opioid-receptor-stimulated GTPase activity.

Very little is known about the effects of cholera-toxin-mediated ADP-ribosylation of G proteins other than G_s on effector regulation by these G proteins. ADP-ribosylation of retinal transducin has been reported to lead to either no change (ABOOD et al. 1982) or even to a marked reduction of light-stimulated cGMP phosphodiesterase (PDE) activity (JELSEMA 1987). Both studies, however, have to be interpreted with great caution. In the former study, the effect of cholera-toxin-mediated ADP-ribosylation on PDE activation was examined under conditions where α_t-GTP was in excess of PDE. These conditions do not allow for detection of a toxin-mediated activation of α_t. In the latter study, cholera toxin treatment of rod outer segments was performed using low concentrations of the toxin (0.01–1 µg/ml) under conditions which are very unfavorable to ADP-ribosylation of α_t (i.e., absence of light, lack of GppNHp). No correlation was shown between the observed effects and ADP-ribosylation of α_t. Thus, the effects of cholera-toxin-mediated ADP-ribosylation of α_t on regulation of cGMP PDE remain to be examined under more appropriate conditions.

To what extent cholera-toxin-mediated ADP-ribosylation of other G proteins, e.g., G_i proteins, affects their ability to regulate effector moieties and whether the functional consequence is stimulation or inhibition is essen-tially unknown. This is in part due to the fact that we still do not know which effector(s) are regulated by a given G_i or G_o protein. In one report, however, opioid-peptide-dependent inhibition of adenylyl cyclase was found to be reduced in membranes prepared from cholera-toxin-treated NG108-15 cells (KLEE et al. 1984). Similarly, cholera toxin treatment of adipocyte membranes led to a reduction of both receptor- and GTP-mediated adenylyl cyclase inhibition (Owens et al. 1985). Chemotaxis of mouse macrophages has been reported to be inhibited by cholera toxin via a mechanism(s) not involving an increase in cellular cAMP (Aksamit et al. 1985). Furthermore, in at least two other systems, there was a marked, cAMP-independent inhibition by cholera toxin of phosphoinositide hydrolysis (Imboden et al. 1986; Lo and Hughes 1987). Very recently, mutated α_{i2} genes (designated *gip*2) have been identified in a significant proportion of human adrenal cortical and ovarian tumors (Lyons et al. 1990). Interestingly, in all cases the mutation replaced Arg 179 (which corresponds to Arg 174 of α_t) with either cysteine or histidine. The mutation was confined to the tumor cells and was not found in other cells or normal tissue of the affected patients, suggesting that the mutations played a direct causal role in development of the tumors. It is thus tempting to speculate that Arg 179 plays an important role in regulating the function of α_{i2} and that modifying this residue by either mutation or cholera-toxin-mediated ADP-ribosylation may indeed have profound consequences on the protein's function. Future experimentation will have to clarify whether these consequences involve changes in GTP hydrolysis and whether they lead to activation or inhibition. It is expected that these studies will also allow us to establish in which signaling pathway the normal α_{i2} protein participates.

III. G-Proteins ADP-Ribosylation by Pertussis Toxin

The main functional consequence of pertussis-toxin-mediated G-protein ADP-ribosylation is an uncoupling of the receptor from effector regulation. The number of transmembrane signalling systems influenced by pertussis toxin in this way is considerable and still increasing. It includes the pathways leading to inhibition of adenylyl cyclase, stimulation of phospholipases C, D, and A_2, stimulation of retinal cGMP PDE, opening or closing of various ion channels, and probably many more (see Gierschik et al. 1990b; Birnbaumer et al. 1990 for recent reviews and further references).

The molecular mechanisms leading to the toxin-mediated inhibition of receptor-effector interaction have been analyzed in detail using membrane preparations, reconstituted systems consisting of receptors and/or G proteins implanted into artificial lipid bilayers, and purified G proteins in solution. As a result of these studies, it is now generally agreed that ADP-ribosylation by pertussis toxin prevents the interaction of the agonist-activated receptor with the GDP-liganded G protein. Evidence in support of this concept is threefold. First, G-protein ADP-ribosylation prevents formation of a receptor-G-protein complex that binds agonists with high affinity. This phenomenon has been demonstrated in many membrane systems as well as in phospholipid vesicles (see Moss and Vaughan 1988 for references). Second, pertussis toxin treatment of hamster adipocyte membranes led to a marked decrease in the α_2-adrenoceptor-mediated release of [^{3}H]-GDP from membrane bound G_i (which had been prelabelled with [^{3}H]-GTP prior to the release assay) (Murayama and Ui 1984). Third, retinal transducin ADP-ribosylated by pertussis toxin bound less tightly to photolyzed rhodopsin than did the unmodified G protein. Thus, ADP-ribosylated transducin detached from rod outer segment membranes even at physiologic ionic strength and in the absence of guanine nucleotides (Van Dop et al. 1984b). It is important to note that none of the "intrinsic" G-protein functions (GDP release, GTP[S] binding, GTPase activity, subunit dissociation) were altered when purified G_i proteins were examined that had been treated with pertussis toxin (Haga et al. 1985; Enomoto and Asakawa 1986; Huff and Neer 1986; Katada et al. 1986b; Sunyer et al. 1989; Casey et al. 1989). Thus, there is strong evidence that the major consequence of the toxin-mediated modification is in fact a reduced ability of the G protein to interact with the receptor.

Despite the compelling nature of the concept just developed, there are certain findings in the literature that are not easily explained by it. First, treatment of S49 cyc⁻ lymphoma cells with pertussis toxin blocked inhibition of adenylyl cyclase by GTP even in the absence of inhibitory hormones (Hildebrandt et al. 1983) and markedly increased the time required to observe maximal adenylyl cyclase inhibition by GTP[S] (Jakobs et al. 1984). Second, pertussis toxin treatment of hamster fibroblasts markedly reduced the stimulation of inositol phosphate production by AlF_4^- (Paris and Pouyssegur 1987). These findings are rather puzzling since inhibition of adenylyl cyclase by GTP or GTP[S] and stimulation of phospholipase C by AlF_4^- (in the absence of a hormonal stimulus) are generally believed to be receptor-independent processes, which should be neither inhibited nor delayed by ADP-ribosylation of G_i. Also pertinent to this issue, pertussis toxin treatment of HL-60 cells markedly reduced basal GTPase activity and basal [^{35}S]-GTP[S] binding to their membranes (Gierschik et al. 1989b, 1991). Note that both activities were assayed in the absence of a receptor agonist and are thus expected to be receptor-independent.

There are two possibilities that might explain these findings. First, it is conceivable that inhibition of the receptor-G-protein interaction is not the only functional consequence of G-protein ADP-ribosylation by pertussis toxin. This possibility may appear unlikely since ADP-ribosylation of purified G proteins by pertussis toxin has no effect on their various functions (see above). However, it is entirely possible that the functions of membrane-bound G-proteins differ significantly from those observed in detergent-containing solutions or even in synthetic lipid vesicles (cf. Asano and Ross 1984; Florio and Sternweis 1985). The second possibility is that unoccupied receptors are not completely silent with regard to G-protein activation. This concept has recently gained considerable support from several laboratories, which can be summarized as follows: (1) Reconstitution of the purified β-adrenoceptor into lipid vesicles containing purified G_s significantly enhances GTPase activity of G_s, even in the absence of adrenoceptor agonist (Cerione et al. 1984). (2) It has been suggested that activation of G proteins by GTP (as opposed to the nonhydrolyzable analogues GppNHp or GTP[S]) does not proceed in the absence of a receptor protein and may not be receptor-independent but dependent on unoccupied, but nevertheless active receptors (SUNYER et al. 1989). (3) Recent evidence suggests that opioid receptors present in the membranes of NG108-15 cells spontaneously associate with G proteins and give rise to increased GTPase activity even in the absence of receptor agonists (COSTA et al. 1990). Interestingly, the increased "basal" GTPase activity is inhibited by certain antagonists (negative antagonists) in a stereo-specific fashion (COSTA and HERZ 1989). In contrast to what is known for opioid agonists, the affinity of negative antagonists for opioid receptors is enhanced when G proteins have been ADP-ribosylated by pertussis toxin.

F. Future Perspectives

Despite the tremendous recent progress in both analyzing the molecular mechanism of cholera- and pertussis-toxin-mediated G-protein ADP-ribosylation and characterizing the functional alterations elicited by these modifications, there is a considerable number of questions that remain to be addressed by future experimentation. For example, it is still not clear whether the main functional consequences of G-protein ADP-ribosylation (inhibition of GTP hydrolysis by cholera toxin and inhibition of receptor-G-protein interaction by pertussis toxin) are in fact the only sequelae of toxin action. Another line of investigation will be to analyze which of the structural elements of the various G-protein subunits are required for toxin-mediated ADP-ribosylation. Results obtained in these studies will provide important insights into the organization of α, β, and γ-subunits into functional domains. Finally, it is important to find out whether eukaryotes contain endogenous ADP-ribosyltransferases capable of ADP-ribosylating G proteins just like their prokaryotic counterparts. Evidence in support of this concept is emerging (FELDMAN et al. 1987; TANUMA et al. 1988; JACOBSON et al. 1990).

References

Abood ME, Hurley JB, Pappone M-C, Bourne HR, Stryer L (1982) Functional homology between signal-coupling proteins: cholera toxin inactivates the GTPase activity of transducin. J Biol Chem 257:10540–10543
Aksamit RR, Backlund PS, Cantoni GL (1985) Cholera toxin inhibits chemotaxis by a cAMP-independent mechanism. Proc Natl Acad Sci USA 82:7475–7479
Asano T, Ross EM (1984) Catecholamine-stimulated guanosine 5'-O-(3-thiotriphosphate) binding to the stimulatory GTP-binding protein of adenylate cyclase: kinetic analysis in reconstituted phospholipid vesicles. Biochemistry 23:5467–5471
Bahnson TD, Tsai S-C, Adamik R, Moss J, Vaughan M (1989) Microinjection of a 19-kDa guanine nucleotide-binding protein inhibits maturation of Xenopus oocytes. J Biol Chem 264:14824–14828

Balch WE (1990) Small GTP-binding proteins in vesicular transport. Trends Biochem Sci 15:473–477

Barbacid M (1987) *ras genes*. Ann Rev Biochem 56:779–827

Barbieri JT, Cortina G (1988) ADP-ribosyltransferase mutations in the catalytic S-1 subunit of pertussis toxin. Infect Immun 56:1934–1941

Barbieri JT, Mende-Mueller LM, Rappuoli R, Collier RJ (1989) Photolabelling of Glu-129 of the S-1 subunit of pertussis toxin with NAD. Infect Immun 57:3549–3554

Birnbaumer L (1990) G proteins in signal transduction. Annu Rev Pharmacol Toxicol 30:675–705

Birnbaumer L, Abramowitz J, Brown AM (1990) Receptor-effector coupling by G proteins. Biochim Biophys Acta 1031:163–224

Bobak DA, Nightingale MS, Murtagh JJ, Price SR, Moss J, Vaughan M (1989) Molecular cloning, characterization, and expression of human ADP-ribosylation factors: two guanine nucleotide-dependent activators of cholera toxin. Proc Natl Acad Sci USA 86:6101–6105

Bobak DA, Bliziotes MM, Noda M, Tsai S-C, Adamik R, Moss J (1990) Mechanism of activation of cholera toxin by ADP-ribosylation factor (ARF): both low- and high-affinity interactions of ARF with guanine nucleotides promote toxin activation. Biochemistry 29:855–861

Bokoch GM, Katada T, Northup JK, Ui M, Gilman AG (1984) Purification and properties of the inhibitory guanine nucleotide-binding regulatory component of adenylate cyclase. J Biol Chem 259:3560–3567

Bourne HR, Sanders DA, McCormick F (1990) The GTPase superfamily: a conserved switch for diverse cell functions. Nature 348:125–132

Brandt DR, Ross EM (1985) GTPase activity of the stimulatory GTP-binding regulatory protein of adenylate cyclase, G_s. Accumulation and turnover of enzyme-nucleotide intermediates. J Biol Chem 260:266–272

Bray P, Carter A, Simons C, Guo V, Puckett C, Kamholz J, Spiegel A, Nirenberg M (1986) Human cDNA clones for four species of $G\alpha_s$ signal transduction protein. Proc Natl Acad Sci USA 83:8893–8897

Burnette WN, Cieplak W, Mar VL, Kaljot KT, Sato, H, Keith JM (1988) Pertussis toxin S1 mutant with reduced enzyme activity and a conserved protective epitope. Science 242:72–74

Burns DL, Manclark CR (1986) Adenine nucleotides promote dissociation of pertussis toxin subunits. J Biol Chem 261:4324–4327

Burns DL, Manclark CR (1989) Role of cysteine 41 of the A subunit of pertussis toxin. J Biol Chem 264:564–568

Burns DL, Moss J, Vaughan M (1982) Choleragen-stimulated release of guanyl nucleotides from erythrocyte membranes. J Biol Chem 257:32–34

Burns DL, Moss J, Vaughan M (1983) Release of guanyl nucleotides from the regulatory subunit of adenylate cyclase. J Biol Chem 258:1116–1120

Burns DL, Hausman SZ, Witvliet MH, Brennan MJ, Poolman JT, Manclark CR (1988) Biochemical properties of pertussis toxin. Tokai J Exp Clin Med 13 [Suppl]:181–185

Buss JE, Mumby S, Casey PJ, Gilman AG, Sefton BM (1987) Myristoylated α subunits of guanine nucleotide-binding regulatory proteins. Proc Natl Acad Sci USA 84:7493–7497

Carr C, Loney C, Unson C, Knowler J, Milligan G (1990) Chronic exposure of rat glioma C6 cells to cholera toxin induces loss of the α-subunit of the stimulatory guanine nucleotide-binding protein (G_s). Eur J Pharmacol 188:203–209

Carroll SF, Collier RJ (1987) Active site of *Pseudomonas aeruginosa* exotoxin A: glutamic acid 553 is photolabeled by NAD and shows functional homology with glutamic acid 148 of diphteria toxin. J Biol Chem 262:8707–8711

Carroll SF, McCloskey JA, Crain PF, Oppenheimer NJ, Marschner TM, Collier RJ (1985) Photoaffinity labeling of diphteria toxin fragment A with NAD: structure of the photoproduct at position 148. Proc Natl Acad Sci USA 82:7237–7241

Casey PJ, Graziano MP, Gilman AG (1989) G protein βγ subunits from bovine brain and retina: equivalent catalytic support of ADP-ribosylation of α subunits by pertussis toxin but differential interactions with $G_s\alpha$. Biochemistry 28:611–616

Cassel D, Selinger Z (1977) Mechanism of adenylate cyclase activation by cholera toxin: inhibition of GTP hydrolysis at the regulatory site. Proc Natl Acad Sci USA 74:3307–3311

Cassel D, Levkovitz H, Selinger Z (1977) The regulatory GTPase cyclase of turkey erythrocyte adenylate cyclase. J Cyclic Nucleotide Res 3:393–406

Cerione RA, Codina J, Benovic JL, Lefkowitz RJ, Birnbaumer L, Caron MG (1984) The mammalian β_2-adrenergic receptor: reconstitution of functional interactions between pure receptor and pure stimulatory nucleotide binding protein of the adenylate cyclase system. Biochemistry 23:4519–4525

Cerione RA, Staniszewski C, Benovic JL, Lefkowitz RJ, Caron MG, Gierschik P, Somers RL, Spiegel AM, Codina J, Birnbaumer L (1985) Specificity of the functional interactions of the β-adrenergic receptor and rhodopsin with guanine nucleotide regulatory proteins reconstituted in phospholipid vesicles. J Biol Chem 260:1493–1500

Chang F-H, Bourne HR (1989) Cholera toxin induces cAMP-independent degradation of G_s. J Biol Chem 264:5352–5357

Cooper DMF, Jagus R, Somers RL, Rodbell M (1981) Cholera toxin modifies diverse GTP-modulated regulatory proteins. Biochem Biophys Res Commun 101:1179–1185

Cortina G, Barbieri JT (1989) Role of tryptophan 26 in the NAD glycohydrolase reaction of the S-1 subunit of pertussis toxin. J Biol Chem 264:17322–17328

Costa T, Herz A (1989) Antagonists with negative intrinsic activity at δ opioid receptors coupled to GTP-binding proteins. Proc Natl Acad Sci USA 86:7321–7325

Costa T, Lang J, Gless C, Herz A (1990) Spontaneous association between opioid receptors and GTP-binding regulatory proteins in native membranes: specific regulation by antagonists and sodium ions. Mol Pharmacol 37:383–394

Enomoto K, Asakawa T (1986) Inhibition of catalytic unit of adenylate cyclase and activation of GTPase of Ni protein by βγ-subunits of GTP-binding proteins. FEBS Lett 202:63–68

Enomoto K, Gill DM (1979) Requirement for guanosine triphosphate in the activation of adenylate cyclase by cholera toxin. J Supramol Struct 10:51–60

Enomoto K, Gill DM (1980) Cholera toxin activation of adenylate cyclase: roles of nucleoside triphosphates and a macromolecular factor in the ADP ribosylation of the GTP-dependent regulatory component. J Biol Chem 255:1252–1258

Feldman AM, Levine MA, Baughman KL, Van Dop C (1987) NAD^+-mediated stimulation of adenylate cyclase in cardiac membranes. Biochem Biophys Res Commun 142:631–637

Finkelstein RA (1973) Cholera. CRC Crit Rev Microbiol 2:553–623

Florio VA, Sternweis PC (1985) Reconstitution of resolved muscarinic cholinergic receptors with purified GTP-binding proteins. J Biol Chem 260:3477–3483

Freissmuth M, Gilman AG (1989) Mutations of $G_s\alpha$ designed to alter the reactivity of the protein with bacterial toxins: substitutions at Arg^{187} result in loss of GTPase activity. J Biol Chem 264:21907–21914

Gautam N, Northup J, Tamir H, Simon MI (1990) G protein diversity is increased by associations with a variety of γ subunits. Proc Natl Acad Sci USA 87:7973–7977

Gierschik P, Sidiropoulos D, Spiegel AM, Jakobs KH (1987) Purification and immunochemical characterization of the major pertussis-toxin-sensitive guanine-nucleotide-binding protein of bovine-neutrophil membranes. Eur J Biochem 165:185–194

Gierschik P, Sidiropoulos D, Jakobs KH (1989a) Two distinct G_i-proteins mediate formyl peptide receptor signal transduction in human leukemia (HL-60) cells. J Biol Chem 264:21470–21473

Gierschik P, Sidiropoulos D, Steisslinger M, Jakobs KH (1989b) Na$^+$ regulation of formyl peptide-receptor-mediated signal transduction in HL-60 cells: evidence that the cation prevents activation of the G-protein by unoccupied receptors. Eur J Pharmacol 172:481–492

Gierschik P, Sidiropoulos D, Dieterich K, Jakobs KH (1990a) Structure and function of signal-transducing heterotrimeric guanosine triphosphate binding proteins. In: Habenicht A (ed) Growth factors, differentiation factors, and cytokines. Springer, Berlin Heidelberg New York, p 395

Gierschik P, Sidiropoulos D, Dieterich K, Jakobs KH (1990b) Transmembrane signalling by G_i-proteins. In: Nahorski SR (ed) Transmembrane signalling, intracellular messengers and implications for drug development. Wiley, Chichester, p 73

Gierschik P, Moghtader R, Straub C, Dieterich K, Jakobs KH (1991) Signal amplification in HL-60 granulocytes: evidence that the chemotactic peptide receptor catalytically activates G-proteins in native plasma membranes. Eur J Biochem 197:725–732

Gill DM (1976) The arrangement of subunits in cholera toxin. Biochemistry 15:1242–1248

Gill DM, Meren R (1983) A second guanyl nucleotide-binding site associated with adenylate cyclase: distinct nucleotides activate adenylate cyclase and permit ADP-ribosylation by cholera toxin. J Biol Chem 258:11908–11914

Gill DM, Woolkalis M (1988) [^{32}P]ADP-ribosylation of proteins catalyzed by cholera toxin and related heat-labile enterotoxins. Methods Enzymol 165:235–245

Graves CB, Klaven NB, McDonald JM (1983) Effects of guanine nucleotides on cholera toxin catalyzed ADP-ribosylation in rat adipocyte plasma membranes. Biochemistry 22:6291–6296

Graziano MP, Gilman AG (1989) Synthesis in *Escherichia coli* of GTPase-deficient mutants of $G_s\alpha$. J Biol Chem 264:15475–15482

Graziano MP, Casey PJ, Gilman AG (1987) Expression of cDNAs for G proteins in *Escherichia coli*: two forms of $G_s\alpha$ stimulate adenylate cyclase cyclase. J Biol Chem 262:11375–11381

Graziano MP, Freissmuth M, Gilman AG (1989) Expression of $G_{s\alpha}$ in *Escherichia coli*: purification and properties of the two forms of the protein. J Biol Chem 264:409–418

Haga K, Haga T, Ichiyama A, Katada T, Kurose H, Ui M (1985) Functional reconstitution of purified muscarinic receptors and inhibitory guanine nucleotide regulatory protein. Nature 316:731–733

Halliday KR (1984) Regional homology in GTP-binding protooncogene products and elongation factors. J Cyclic Nucleotide Protein Phosphorylation Res 9:435–448

Hancock JF, Magee AI, Childs JE, Marshall CJ (1989) All *ras* proteins are polyisoprenylated but only some are palmitylated. Cell 57:1167–1177

Hermouet S, Milligan G, Lanotte M (1990) Reduction of adenylyl cyclase activity by cholera toxin in myeloid cells: long-term down-regulation of $G_s\alpha$ subunits by cholera toxin treatment. FEBS Lett 267:221–225

Higashijima T, Ferguson KM, Sternweis PC, Smigel MD, Gilman AG (1987) Effects of Mg^{2+} and the βγ-subunit complex on the interactions of guanine nucleotides with G proteins. J Biol Chem 262:762–766

Hildebrandt JD, Sekura RD, Codina J, Iyengar R, Manclark CR, Birnbaumer L (1983) Stimulation and inhibition of adenylyl cyclases mediated by distinct regulatory proteins. Nature 302:706–709

Hoshino S, Kikkawa S, Takahashi K, Itoh H, Kaziro Y, Kawasaki H, Suzuki K, Katada T, Ui M (1990) Identification of sites for alkylation by *N*-ethylamaleimide and pertussis toxin-catalyzed ADP-ribosylation on GTP-binding proteins. FEBS Lett 276:227–231

Hsia JA, Tsai S-C, Adamik R, Yost DA, Hewlett EL, Moss J (1985) Amino acid-specific ADP-ribosylation: sensitivity to hydroxylamine of [cysteine(ADP-ribose)]protein and [arginine(ADP-ribose)]protein linkages. J Biol Chem 260:16187–16191

Hsu WH, Rudolph U, Sanford J, Bertrand P, Olate J, Nelson C, Moss LG, Boyd III AE, Codina J, Birnbaumer L (1990) Molecular cloning of a novel splice variant of the α subunit of the mammalian G_o protein. J Biol Chem 265:11220–11226

Huff RM, Neer EJ (1986) Subunit interactions of native and ADP-ribosylated α_{39} and α_{41}, two guanine nucleotide-binding proteins from bovine cerebral cortex. J Biol Chem 261:1105–1110

Iiri T, Tohkin M, Morishima N, Ohoka Y, Ui M, Katada T (1989) Chemotactic peptide receptor-supported ADP-ribosylation of a pertussis toxin substrate GTP-binding protein by cholera toxin in neutrophil-type HL-60 cells. J Biol Chem 264:21394–21400

Imboden JB, Shoback DM, Pattison G, Stobo JD (1986) Cholera toxin inhibits the T-cell antigen receptor-mediated increases in inositol trisphosphate and cytoplasmic free calcium. Proc Natl Acad Sci USA 83:5673–5677

Jacobson MK, Loflin PT, Aboul-Ela N, Mingmuang M, Moss J, Jacobson EL (1990) Modification of plasma membrane protein cysteine residues by ADP-ribose in vivo. J Biol Chem 265:10825–10828

Jakobs KH, Aktories K, Schultz G (1984) Mechanism of pertussis toxin action on the adenylate cyclase system: inhibition of the turn-on reaction of the inhibitory regulatory site. Eur J Biochem 140:177–181

Jelsema CL (1987) Light activation of phospholipase A_2 in rod outer segments of bovine retina and its modulation by GTP-binding proteins. J Biol Chem 262:163–168

Jones DT, Reed RR (1989) G_{olf}: an olfactory neuron specific-G protein involved in odorant signal transduction. Science 244:790–795

Jones DT, Masters SB, Bourne HR, Reed RR (1990) Biochemical characterization of three stimulatory GTP-binding proteins: the large and small forms of G_s and the olfactory-specific G-protein G_{olf}. J Biol Chem 265:2671–2676

Jones TLZ, Simonds WF, Merendino JJ Jr, Brann MR, Spiegel AM (1990) Myristoylation of an inhibitory GTP-binding protein α subunit is essential for its membrane attachment. Proc Natl Acad Sci USA 87:568–572

Kahn RA (1990) ADP-ribosylation factor of adenylyl cyclase: a 21-kDa GTP-binding protein. In: Iyengar R, Birnbaumer L (eds) G proteins. Academic Press, San Diego, p 201

Kahn RA, Gilman AG (1984a) Purification of a protein cofactor required for ADP-ribosylation of the stimulatory regulatory component of adenylate cyclase by cholera toxin. J Biol Chem 259:6228–6234

Kahn RA, Gilman AG (1984b) ADP-ribosylation of G_s promotes the dissociation of its α and β subunits. J Biol Chem 259:6235–6240

Kahn RA, Gilman AG (1986) The protein cofactor necessary for ADP-ribosylation of G_s by cholera toxin is itself a GTP binding protein. J Biol Chem 261:7906–7911

Kahn RA, Goddard C, Newkirk M (1988) Chemical and immunological characterization of the 21-kDa ADP-ribosylation factor of adenylate cyclase. J Biol Chem 263:8282–8287

Kanaho Y, Tsai S-C, Adamik R, Hewlett EL, Moss J, Vaughan M (1984) Rhodopsin-enhanced GTPase activity of the inhibitory GTP-binding protein of adenylate cyclase. J Biol Chem 259:7378–7381

Kaslow HR, Lesikar DD (1987) Sulfhydryl-alkylating reagents inactivate the NAD glycohydrolase activity of pertussis toxin. Biochemistry 26:4397–4402

Kaslow HR, Lim, LL, Moss J, Lesikar DD (1987) Structure-activity analysis of the activation of pertussis toxin. Biochemistry 26, 123–127

Kaslow HR, Schlotterbeck JD, Mar VL, Burnette WN (1989) Alkylation of cysteine 41, but not cysteine 200, decreases the ADP-ribosyltransferase activity of the S1 subunit of pertussis toxin. J Biol Chem 264:6386–6390

Katada T, Tamura M, Ui M (1983) The A protomer of islet-activating protein, pertussis toxin, as an active peptide catalyzing ADP-ribosylation of a membrane protein. Arch Biochem Biophys 224:290–298

Katada T, Oinuma M, Ui M (1986a) Two guanine nucleotide-binding proteins in rat brain serving as the specific substrate of islet-activating protein, pertussis toxin: interactions of the α-subunits with βγ-subunits in development of their biological activities. J Biol Chem 261:8182–8191

Katada T, Oinuma M, Ui M (1986b) Mechanisms for inhibition of the catalytic activity of adenylate cyclase by the guanine nucleotide-binding proteins serving as the substrate of islet-activating protein, pertussis toxin. J Biol Chem 261:5215–5221

Klee WA, Koski G, Tocque B, Simonds WF (1984) On the mechanism of receptor-mediated inhibition of adenylate cyclase. Adv Cyclic Nucleotide Protein Phosphorylation Res 17:153–159

Klinz F-J, Costa T (1989) Cholera toxin ADP-ribosylates the receptor-coupled form of pertussis toxin-sensitive G-proteins. Biochem Biophys Res Commun 165:554–560

Klinz F-J, Costa T (1990) Cholera toxin differentially decreases membrane levels of α and β subunits of G proteins in NG108-15 cells. Eur J Biochem 188:567–576

Lad PM, Nielsen TB, Preston MS, Rodbell M (1980) The role of the guanine nucleotide exchange reaction in the regulation of the β-adrenergic receptor and in the actions of catecholamines and cholera toxin on adenylate cyclase in turkey erythrocyte membranes. J Biol Chem 255:988–995

Landis CA, Masters SB, Spada A, Pace AM, Bourne HR, Vallar L (1989) GTPase inhibiting mutations activate the α chain of G_s and stimulate adenylyl cyclase in human pituitary tumours. Nature 340:692–696

Le Vine H III, Cuatrecasas P (1981) Activation of pigeon erythrocyte adenylate cyclase by cholera toxin: partial purification of an essential macromolecular factor from horse erythrocyte cytosol. Biochim Biophys Acta 672:248–261

Lin MC, Welton AF, Berman MF (1978) Essential role of GTP in the expression of adenylate cyclase activity after cholera toxin treatment. J Cyclic Nucleotide Res 4:159–168

Linder ME, Ewald DA, Miller RJ, Gilman AG (1990) Purification and characterization of $G_o\alpha$ and three types of $G_i\alpha$ after expression in *Escherichia coli*. J Biol Chem 265:8243–8251

Lo WWY, Hughes J (1987) A novel cholera toxin-sensitive G protein (G_c) regulating receptor-mediated phosphoinositide signalling in human pituitary clonal cells. FEBS Lett 220:327–331

Lochrie MA, Simon MI (1988) G protein multiplicity in eukaryotic signal transduction systems. Biochemistry 27:4957–4965

Locht C, Keith JM (1986) Pertussis toxin gene: nucleotide sequence and genetic organization. Science 232:1258–1264

Locht C, Capiau C, Feron C (1989) Identification of amino acid residues essential for the enzymatic activities of pertussis toxin. Proc Natl Acad Sci USA 86:3075–3079

Locht C, Lobet Y, Feron C, Cieplak W, Keith JM (1990) The role of cysteine 41 in the enzymatic activities of the pertussis toxin S1 subunit as investigated by site-directed mutagenesis. J Biol Chem 265:4552–4559

Lyons J, Landis CA, Harsh G, Vallar L, Grünewald K, Feichtinger H, Duh Q-Y, Clark OH, Kawasaki E, Bourne HR, McCormick F (1990) Two G protein oncogenes in human endocrine tumors. Science 249:655–659

Macleod KG, Milligan G (1990) Biphasic regulation of adenylate cyclase by cholera toxin in neuroblastoma x glioma hybrid cells is due to the activation and subsequent loss of the α subunit of the stimulatory GTP binding proteins (G_s). Cell Signal 2:139–151

Maltese WA (1990) Posttranslational modification of proteins by isoprenoids in mammalian cells. FASEB J 4:3319–3328

Maltese WA, Sheridan KM, Repko EM, Erdman RA (1990) Posttranslational
    modification of low molecular mass GTP-binding proteins by isoprenoid. J Biol
    Chem 265:2148–2155
Mattera R, Codina J, Sekura RD, Birnbaumer L (1986) The interaction of
    nucleotides with pertussis toxin: direct evidence for a nucleotide binding site on
    the toxin regulating the rate of ADP-ribosylation of N_i, the inhibitory regulatory
    component of adenylyl cyclase. J Biol Chem 261:11173–11179
Mattera R, Codina J, Sekura RD, Birnbaumer L (1987) Guanosine 5'-O-(3-
    thiotriphosphate)reduces ADP-ribosylation of the inhibitory guanine nucleotide-
    binding regulatory protein of adenylyl cyclase (N_i) by pertussis boxin without
    causing dissociation of the subunits of N_i: evidence of existence of
    heterotrimeric pt^+ and pt^+ conformations of N_i. J Biol Chem 262:11247–11251
McLeish KR, Gierschik P, Schepers T, Sidiropoulos D, Jakobs KH (1989) Evidence
    that activation of a common G-protein by receptors for leukotriene B_4 and
    N-formylmethionyl-leucyl-phenylalanine in HL-60 cells occurs by different
    mechanisms. Biochem J 260:427–434
Mekalanos JJ, Collier RJ, Romig WR (1979) Enzymic activity of cholera toxin:
    relationship to proteolytic processing, disulfide bond reduction, and subunit
    composition. J Biol Chem 254:5855–5861
Mekalanos JJ, Swartz DJ, Pearson GDN, Harford N, Groyne F, De Wilde M (1983)
    Cholera toxin genes: nucleotide sequence, deletion analysis and vaccine
    development. Nature 306:551–557
Mendez E, Lai CY, Wodnar-Filipowicz A (1975) Location and the primary structure
    around the disulfide bonds in cholera toxin. Biochem Biophys Res Commun
    67:1435–1443
Milligan G (1987) Guanine nucleotide regulation of the pertussis and cholera toxin
    substrates of rat glioma C6 BU1 cells. Biochim Biophys Acta 929:197–202
Milligan G (1989) Foetal calf serum enhances cholera toxin-catalyzed ADP-
    ribosylation of the pertussis toxin-sensitive guanine nucleotide binding protein,
    G_i2, in rat glioma C6BU1 cells. Cell Signal 1:65–74
Milligan G, McKenzie FR (1988) Opioid peptides promote cholera-toxin-catalysed
    ADP-ribosylation of the inhibitory guanine-nucleotide-binding protein (G_i) in
    membranes of neuroblastoma x glioma hybrid cells. Biochem J 252:369–373
Milligan G, Unson CG, Wakelam MJO (1989) Cholera toxin treatment produces
    down-regulation of the α-subunit of the stimulatory guanine-nucleotide-binding
    protein (G_s). Biochem J 262:643–649
Monaco L, Murtagh JJ, Newman KB, Tsai S-C, Moss J, Vaughan M (1990) Selective
    amplification of an mRNA and related pseudogene for a human ADP-
    ribosylation factor, a guanine nucleotide-dependent protein activator of cholera
    toxin. Proc Natl Acad Sci USA 87:2206–2210
Moss J, Vaughan M (1977) Mechanism of action of choleragen: evidence for ADP-
    ribosyltransferase activity with arginine as an acceptor. J Biol Chem 252:2455–
    2457
Moss J, Vaughan M (1979) Activation of adenylate cyclase by choleragen. Annu Rev
    Biochem 48:581–600
Moss J, Vaughan M (1988) ADP-ribosylation of guanyl nucleotide-binding
    regulatory proteins by bacterial toxins. Adv Enzymol 61:303–379
Moss J, Stanley SJ, Lin MC (1979) NAD glycohydrolase and ADP-ribosyltransferase
    activities are intrinsic to the A_1 peptide of choleragen. J Biol Chem 254:11993–
    11996
Moss J, Stanley SJ, Morin JE, Dixon JE (1980) Activation of choleragen by
    thiol:protein disulfide oxidoreductase. J Biol Chem 255:11085–11087
Moss J, Stanley SJ, Burns DL, Hsia JA, Yost DA, Myers GA, Hewlett EL (1983)
    Activation by thiol of the latent NAD glycohydrolase and ADP-
    ribosyltransferase activities of Bordetella pertussis toxin (islet-activating protein).
    J Biol Chem 158:11879–11882

Moss J, Stanley SJ, Watkins PA, Burns DL, Manclark CR, Kaslow, HR, Hewlett EL (1986) Stimulation of the thiol-dependent ADP-ribosyltransferase and NAD glycohydrolase activities of *Bordetella pertussis* toxin by adenine nucleotides, phospholipids, and detergents. Biochemistry 25:2720–2725

Murayama T, Ui M (1984) [^{3}H]GDP release from rat and hamster adipocyte membranes independently linked to receptors involved in activation or inhibition of adenylate cyclase. J Biol Chem 259:761–769

Nakaya S, Moss J, Vaughan M (1980) Effects of nucleoside triphosphates on choleragen-activated brain adenylate cyclase. Biochemistry 19:4871–4874

Nakaya S, Watkins PA, Bitonti AJ, Hjelmeland LM, Moss J, Vaughan M (1981) GTP stabilization of adenylate cyclase activated and ADP-ribosylated by choleragen. Biochem Biophys Res Commun 102:66–75

Navon SE, Fung BK-K (1984) Characterization of transducin from bovine retinal rod outer segments: mechanism and effects of cholera toxin-catalyzed ADP-ribosylation. J Biol Chem 259:6686–6693

Navon SE, Fung BK-K (1987) Characterization of transducin from bovine retinal rod outer segments; participation of the amino-terminal region of Tα in subunit interaction. J Biol Chem 262:15746–15751

Neer EJ, Clapham DE (1990) Structure and function of G-protein βγ subunit. In: Iyengar R, Birnbaumer L (eds) G proteins. Academic, San Diego, p 41

Neer EJ, Lok JM, Wolf LG (1984) Purification and properties of the inhibitory guanine nucleotide regulatory unit of brain adenylate cyclase. J Biol Chem 259:14222–14229

Neer EJ, Wolf LG, Gill DM (1987) The stimulatory guanine-nucleotide regulatory unit of adenylate cyclase from bovine cerebral cortex: ADP-ribosylation and purification. Biochem J 241:325–336

Neer EJ, Pulsifer L, Wolf LG (1988) The amino terminus of G-protein α-subunits is required for interaction with βγ. J Biol Chem 263:8996–9000

Nicosia, A, Perugini M, Franzini C, Casagli MC, Borri MG, Almoni M, Neri P, Ratti G, Rappuoli R (1986) Cloning and sequencing of the pertussis toxin genes: operon structure and gene duplication. Proc Natl Acad USA 83:4631–4635

Noda M, Tsai S-C, Adamik R, Bobak DA, Moss J, Vaughan M (1989) Activation of immobilized, biotinylated choleragen A$_1$ protein by a 19-kilodalton guanine nucleotide-binding protein. Biochemistry 28:7936–7940

Noda M, Tsai S-C, Adamik R, Moss J, Vaughan M (1990) Mechanism of cholera toxin activation by a guanine nucleotide-dependent 19 kDa protein. Biochim Biophys Acta 1034:195–199

Northup JK, Sternweis PC, Gilman AG (1983a) The subunits of the stimulatory regulatory component of adenylate cyclase: resolution, activity, and properties of the 35 000-dalton (β) subunit. J Biol Chem 258:11361–11368

Northup JK, Smigel MD, Sternweis PC, Gilman AG (1983b) The subunits of the stimulatory regulatory component of adenylate cyclase: resolution of the activated 45 000-dalton (α) subunit. J Biol Chem 258:11369–11376

Osawa S, Dhanasekaran N, Woon CW, Johnson GL (1990) Gα$_i$-Gα$_s$ chimeras define the function of α chain domains in control of G protein activation and βγ subunit complex interaction. Cell 63:697–706

Owens JR, Frame LT, Ui M, Cooper DMF (1985) Cholera toxin ADP-ribosylates the islet-activating protein substrate in adipocyte membranes and alters its function. J Biol Chem 260:15946–15952

Paris S, Pouyssegur J (1987) Further evidence for a phospholipase C-coupled G protein in hamster fibroblasts: induction of inositol phosphate formation by fluoroaluminate and vanadate and inhibition by pertussis toxin. J Biol Chem 262:1970–1976

Pizza M, Bartoloni A, Prugnola A, Silvstri S, Rappuoli R (1988) Subunit S1 of pertussis toxin: mapping of the regions essential for ADP-ribosyltransferase activity. Proc Natl Acad Sci USA 85:7521–7525

Pizza M, Covacci A, Bartoloni A, Perugini M, Nencioni L De Magistris MT, Villa L, Nucci D, Manetti R, Bugnoli M, Giovannoni F, Olivieri R, Barbieri JT, Sato H, Rappuoli R (1989) Mutants of pertussis toxin suitable for vaccine development. Science 246:497–500

Price SR, Nightingale MS, Tsai S-C, Williamson KC, Adamik R, Chen H-C, Moss J, Vaughan M (1988) Guanine nucleotide-binding proteins that enhance choleragen ADP-ribosyltransferase activity: nucleotide and deduced amino acid sequence of an ADP-ribosylation factor cDNA. Proc Natl Acad Sci USA 85:5488–5491

Ribeiro-Neto FAP, Mattera R, Hildebrandt JD, Codina J, Field JB, Birnbaumer L, Sekura RD (1985) ADP-ribosylation of membrane components by pertussis and cholera toxin. Methods Enzymol 109:566–573

Rojas FJ, Birnbaumer L (1985) Regulation of glucagon receptor binding: lack of effect of Mg and preferential role for GDP. J Biol Chem 260:7829–7835

Ross EM, Gilman AG (1980) Biochemical properties of hormone-sensitive adenylate cyclase. Annu Rev Biochem 49:533–564

Schleifer LS, Garrison JC, Sternweis PC, Northup JK, Gilman AG (1980) The regulatory component of adenylate cyclase from uncoupled S49 lymphoma cells differs in charge from the wild type protein. J Biol Chem 255:2641–2644

Schleifer LS, Kahn RA, Hanski E, Northup JK, Sternweis PC, Gilman AG (1982) Requirement for cholera toxin-dependent ADP-ribosylation of the purified regulatory component of adenylate cyclase. J Biol Chem 257:20–23

Schultz AM, Tsai S-C, Kung H-F, Oroszlan S, Moss J, Vaughan M (1987) Hydroxylamine-stable covalent linkage of myristic acid in $G_o\alpha$, a guanine nucleotide-binding protein of bovine brain. Biochem Biophys Res Commun 146:1234–1239

Sekura RD, Fish F, Manclark CR, Meade B (1983) Pertussis toxin: affinity purification of a new ADP-ribosyltransferase. J Biol Chem 258:14647–14651

Sewell JL, Kahn RA (1988) Sequences of the bovine and yeast ADP-ribosylation factor and comparison to other GTP-binding proteins. Proc Natl Acad Sci USA 85:4620–4624

Smigel MD, Northup JK, Gilman AG (1982) Characteristics of the guanine nucleotide-binding regulatory component of adenylate cyclase. Recent Progr Horm Res 38:601–624

Stearns T, Willingham MC, Botstein D, Kahn RA (1990) ADP-ribosylation factor is functionally and physically associated with the Golgi complex. Proc Natl Acad Sci USA 87:1238–1242

Strathmann M, Simon MI (1990) G protein diversity: a distinct class of α subunits is present in vertebrates and invertebrates. Proc Natl Acad Sci USA 87:9113–9117

Strathmann M, Wilkie TM, Simon MI (1989) Diversity of the G-protein family: sequences from five additional α subunits in the mouse. Proc Natl Acad Sci USA 86:7407–7409

Sunyer T, Monastirsky B, Codina J, Birnbaumer L (1989) Studies on nucleotide and receptor regulation of G_i proteins: effects of pertussis toxin. Mol Endocrinol 3:1115–1124

Tamura M, Nogimori K, Murai S, Yajima M, Ito K, Katada T, Ui M, Ishii S (1982) Subunit structure of islet-activating protein, pertussis toxin, in conformity with the A-B model. Biochemistry 21:5516–5522

Tamura M, Nogimori K, Yajima M, Ase K, Ui M (1983) A role of the B-oligomer moiety of islet-activating protein, pertussis toxin, in development of the biological effects on intact cells. J Biol Chem 258:6756–6761

Tanuma S, Kawashima K, Endo H (1988) Eukaryotic mono(ADP-ribosyl)transferase that ADP-ribosylates GTP-binding regulatory G_i protein. J Biol Chem 263:5485–5489

Towler DA, Gordon JI, Adams SP, Glaser L (1988) The biology and enzymology of eukaryotic protein acylation. Annu Rev Biochem 57:69–99

Tsai S-C, Adamik R, Kanaho Y, Hewlett EL, Moss J (1984) Effects of guanyl nucleotides and rhodopsin on ADP-ribosylation of the inhibitory GTP-binding component of adenylate cyclase by pertussis toxin. J Biol Chem 259:15320–15323

Tsai S-C, Noda M, Adamik R, Moss J, Vaughan M (1987) Enhancement of choleragen ADP-ribosyltransferase activities by guanyl nucleotides and a 19-kDa membrane protein. Proc Natl Acad Sci USA 84:5139–5142

Tsai S-C, Noda M, Adamik R, Chang PP, Chen H-C, Moss J, Vaughan M (1988) Stimulation of choleragen enzymatic activities by GTP and two soluble proteins purified from bovine brain. J Biol Chem 263:1768–1772

Tsuchiya M, Price SR, Nightingale MS, Moss J, Vaughan M (1989) Tissue and species distribution of mRNA encoding two ADP-ribosylation factors, 20-kDa guanine nucleotide binding proteins. Biochemistry 28:9668–9673

Ui M (1984) Islet-activating protein, pertussis toxin: a probe for functions of the inhibitory guanine nucleotide regulatory component of adenylate cyclase. Trends Pharmacol Sci 5:277–279

Ui M (1986) Pertussis toxin as a probe of receptor coupling to inositol lipid metabolism. In: Putney JW (ed) Phosphoinositides and receptor mechanisms. Liss, New York, p 163

Vallar L, Spada A, Giannattasio G (1987) Altered G_s and adenylate cyclase activity in human GH-secreting pituitary adenomas. Nature 330:566–568

Van Dop C, Tsubokawa M, Bourne HR, Ramachandran J (1984a) Amino acid sequence of retinal transducin at the site ADP-ribosylated by cholera toxin. J Biol Chem 259:696–698

Van Dop C, Yamanaka G, Steinberg F, Sekura RD, Manclark CR, Stryer L, Bourne HR (1984b) ADP-ribosylation of transducin by pertussis toxin blocks the light-stimulated hydrolysis of GTP and cGMP in retinal photoreceptors. J Biol Chem 259:23–26

Vaughan M, Moss J (1981) Mono(ADP-ribosyl)transferases and their effects on cellular metabolism. Curr Top Cell Regul 20:205–246

Verghese M, Uhing RJ, Snyderman R (1986) A pertussis/cholera toxin-sensitive N protein may mediate chemoattractant receptor signal transduction. Biochem Biophys Res Commun 138:887–894

Watkins PA, Moss J, Burns DL, Hewlett EL, Vaughan M (1984) Inhibition of bovine rod outer segement GTPase by *Bordetella pertussis* toxin. J Biol Chem 259:1378–1381

Watkins PA, Burns DL, Kanaho Y, Liu T-Y, Hewlett EL, Moss J (1985) ADP-ribosylation of transducin by pertussis toxin. J Biol Chem 260:13478–13482

Weiss O, Holden J, Rulka C, Kahn RA (1989) Nucleotide binding and cofactor activities of purified bovine brain and bacterially expressed ADP-ribosylation factor. J Biol Chem 264:21066–21072

West RE Jr, Moss J, Vaughan M, Liu T, Liu T-Y (1985) Pertussis toxin-catalyzed ADP-ribosylation of transducin: cysteine 347 is the ADP-ribose acceptor site. J Biol Chem 260:14428–14430

Winslow JW, Bradley JD, Smith JA, Neer EJ (1987) Reactive sulfhydryl groups of α_{39}, a guanine nucleotide-binding protein from brain: location and function. J Biol Chem 262:4501–4507

Witvliet MH, Burns DL, Brennan MJ, Poolman JT, Manclark CR (1989) Binding of pertussis toxin to eucaryotic cells and glycoproteins. Infect Immun 57:3324–3330

CHAPTER 24

Clostridium botulinum C2 Toxin and *C. botulinum* C3 ADP-Ribosyltransferase

K. Aktories

A. Introduction

At least eight botulinum toxins (A, B, C1, C2, D, E, F, and G) are produced by the different types of *Clostridium botulinum* (Habermann and Dreyer 1986). With the exception of C2 toxin, seven are neurotoxins which presynaptically inhibit neurotransmitter release. To date, the molecular mechanism of the action of these most potent neurotoxins is unknown. *C. botulinum* C2 toxin is not a neurotoxin but acts on various peripheral cells (Simpson 1982; Ohishi et al. 1984). It belongs to a novel family of ADP-ribosylating toxins which modify actin (Aktories et al. 1986a,b). Besides C2 toxin, various strains of *C. botulinum* type C and D produce another ADP-ribosyltransferase, which was called C3 (Aktories et al. 1987a). This exoenzyme ADP-ribosylates low molecular mass GTP binding proteins. In this chapter, both clostridial ADP-ribosyltransferases are described in more detail.

B. *Clostridium botulinum* C2 Toxin

I. Origin and Structure

Many strains of *C. botulinum* which produce the neurotoxins C1 and D and the exoenzyme C3 have also been shown to synthesize C2 toxin. Whereas the production of C3 and the synthesis of the neurotoxins C1 and D depend on bacteriophages (Eklund et al. 1972), C2 toxin most likely is chromosomally encoded (Rubin et al. 1988).

Various ADP-ribosylating protein toxins such as diphtheria, cholera, and pertussis toxins are constructed according to the A-B model and are comprised of an enzymatically active component A and a binding component B, which transfers the enzyme component into the target cell (Collier and Cole 1969; Collier and Kandel 1971; Gill 1977; Tamura et al. 1982). Also, C2 toxin consists of two components (C2I and C2II) (Ohishi et al. 1980). However, in contrast to the above-mentioned ADP-ribosylating toxins, both components of C2 toxin are separate proteins. Thus, C2 toxin is a real binary toxin and comparable with anthrax toxin (Leppla 1982) or leukocidin (Noda 1981).

II. The Binding Component

The binding component (C2II) of C2 toxin (M_r 100 000) has to be activated by trypsin treatment to gain full activity (Eklund and Poysky 1972; Ohishi 1987). The active proteolytic fragment has a molecular weight of about 74 000. This component binds to the cell surface, thereby inducing a binding site for C2I (Ohishi and Miyake 1985). Since the transport of the toxin is inhibited at low temperatures (4°C) and by ammonium chloride or methylamine, it has been concluded that receptor-mediated endocytosis is involved in this process (Simpson et al. 1989).

III. The Enzyme Component

The enzyme component C2I has a molecular weight of about 45 000 and possesses mono-ADP-ribosyltransferase activity (Simpson 1984). The K_m value of the reaction for nicotinamide adenine dinucleotide (NAD) is about 5 μM. C2I ADP-ribosylates monomeric G-actin but not polymerized F-actin (Aktories et al. 1986a). Therefore, phalloidin, which induces polymerization of actin, inhibits the toxin-catalyzed ADP-ribosylation of actin (Aktories et al. 1986b). C2 toxin modifies actin at arginine 177 (Vandekerckhove et al. 1988). Although all vertebrate actin isoforms have arginine at this identical position (Vandekerckhove and Weber 1979), the toxin modifies cytoplasmic β/γ-actin and γ-smooth muscle actin but not the α-actin isoforms from skeletal, cardiac, and smooth muscle (Mauss et al. 1990). Because smooth muscle α-actin and γ-actin are different only in the N-terminal region, it has been suggested that this region defines the substrate specificity of the C2 toxin-catalyzed ADP-ribosylation (Vandekerckhove and Weber 1979; Mauss et al. 1990).

IV. Functional Consequences of the ADP-Ribosylation of Actin

The C2 toxin-induced ADP-ribosylation drastically affects the functional properties of actin. The modified actin loses its ability to polymerize (Aktories et al. 1986a,b). Furthermore, ADP-ribosylated actin acts like a capping protein and binds to the fast growing (barbed) end of actin filaments (Wegner and Aktories 1988; Weigt et al. 1989). The interaction of ADP-ribosylated actin with the actin filament blocks further polymerization of nonmodified actin at this site. In contrast, ADP-ribosylated actin does not interact with the slow-growing (pointed) end of actin filaments. Actin is an ATP-binding protein and possesses ATPase activity (Pollard and Cooper 1986) which is largely stimulated concomitantly with polymerization. ADP-ribosylation by C2 toxin inhibits the actin ATPase activity (Geipel et al. 1989, 1990). This effect is not simply caused by inhibition of polymerization, because the blockade of ATP hydrolysis is observed at low concentrations of Mg^{2+} (50 μM), below the critical concentration of actin, and even with actin, which is bound in the monomeric actin-DNAse complex, indicating that the G-actin ATPase is inhibited by ADP-ribosylation (GEIPEL et al. 1990).

ADP-ribosylation of actin is a reversible process. Reversal of ADP-ribosylation of previously modified actin occurs at a high concentration of nicotinamide and in the absence of NAD (JUST et al. 1990). It is accompanied by restoration of the ability of actin to polymerize and by an increase in actin's ATPase activity. ADP-ribosylation and its reversal are not observed after treatment of actin with ethylenediaminetetraacetic acid (EDTA), which denatures actin, indicating that the native structure of actin is necessary for the forward and reverse reaction of ADP-ribosylation.

V. Cytopathic Effects

C2 is a cytotoxin which induces rounding up of a large variety of cells (OHISHI et al. 1984; REUNER et al. 1987; ZEPEDA et al. 1988). The cytopathic effects occur with a delay of 0.5–1 h. This interval is most likely necessary for the toxin to enter the cells. Also, in intact cells actin is ADP-ribosylated by C2 toxin (REUNER et al. 1987). This was demonstrated by reverse ADP-ribosylation of actin in the cell lysate of previously toxin-treated cells in the presence of [^{32}P]-NAD. Furthermore, after loading of cells with ^{32}P$_i$, the toxins caused labelling of a 43-kDa protein which was identified as actin. Under these conditions, no other proteins are labelled by the toxin. Rounding up of cells correlates with the ADP-ribosylation of actin in intact cells in a time- and concentration-dependent manner.

Treatment of cells with C2 toxin causes drastic alteration of the microfilament network stained with fluorescein-conjugated phalloidin (REUNER et al. 1988; WIEGERS et al. 1991). In toxin-treated fibroblasts the actin filaments are broken, condensed, and/or rarefied and finally disappear. Concomitantly, the cellular amount of G-actin increases (AKTORIES et al. 1989a). The following model has been suggested for the cytotoxic action of C2 toxin (Fig. 1) (AKTORIES and WEGNER 1989): The binding component C2II facilitates the transfer of the enzyme component C2I into the cell, where monomeric actin is ADP-ribosylated. The modified actin acts like a capping protein, binds to the barbed end of actin filaments, and inhibits further polymerization. At the pointed end of filaments, monomeric actin is still released and becomes a substrate of C2I. The ADP-ribosylated actin is not able to polymerize and is therefore trapped in its monomeric form. Thus, the modified actin is withdrawn from the treadmilling pool of actin. Both processes ("capping" and "trapping") finally cause destruction of the microfilament network and induce subsequently the redistribution of intermediate filaments.

VI. Pharmacological Actions

In rats, C2 toxin induces hypotension, hemorrhage, and fluid accumulation around the lungs (SIMPSON 1982). These effects can be explained by an

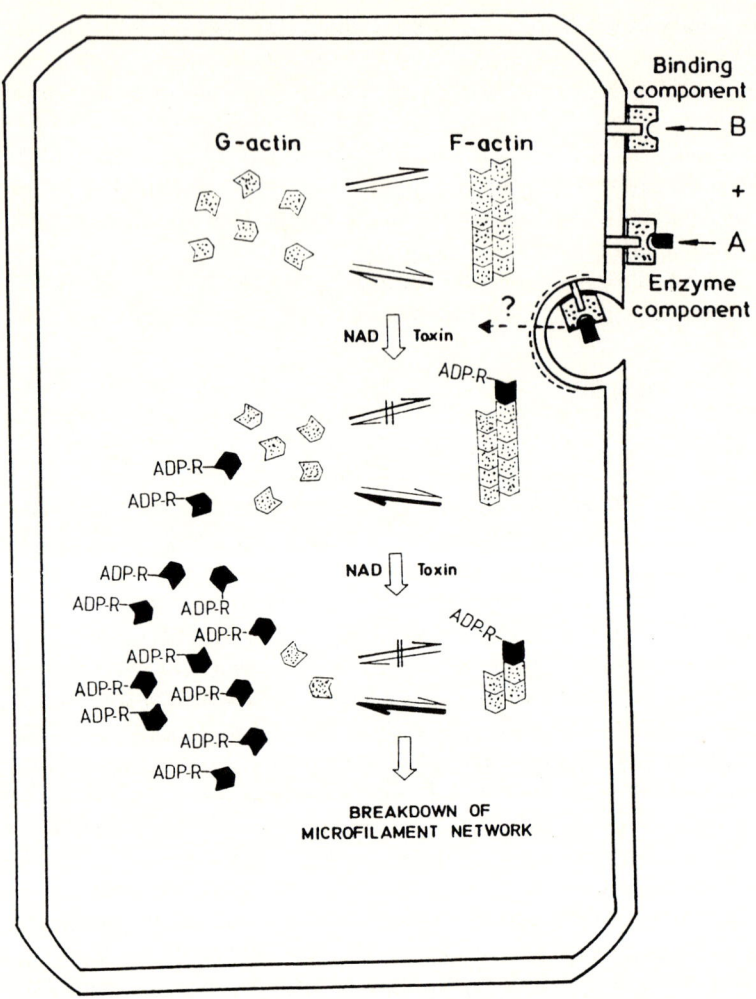

Fig. 1. Model of the pathophysiological mechanism of actin-ADP-ribosylating toxins. The binary toxins consist of a binding component (*B*) and an enzyme component (*A*). The binding component binds to the cell surface receptor of the target cell, thereby inducing a binding site for the enzyme component. Most probably mediated by endocytosis the enzyme component is transferred into the cell where the toxin-catalyzed ADP-ribosylation disturbs the equilibrium between polymerization and depolymerization of actin filaments. The toxins ADP-ribosylate monomeric G-actin, which behaves like a capping protein and binds to the barbed ends of actin filaments, thereby inhibiting actin polymerization. The pointed ends are not affected by ADP-ribosylated toxins. At this end, actin filaments depolymerize. The released monomeric G-actin is ADP-ribosylated by the toxins. Since ADP-ribosylated actin has lost its ability to polymerize, it is trapped in the monomeric form and withdrawn from the treadmilling pool of actin. Both phenomena, "capping" of F-actin and "trapping" of G-actin, are probably responsible for the destruction of the microfilament network by actin-ADP-ribosylating toxins, (Modified from Aktories and Wegner 1989)

increase in vascular permeability induced by the toxin (OHISHI 1983b). Accordingly, C2 toxin largely increases the permeability of endothelial cell monolayers (SUTTORP et al. 1991). In mouse intestinal loop, the toxin causes fluid accumulation (OHISHI 1983a), a phenomenon which was accompanied by damage of the intestinal mucosa (OHISHI and ODAGIRI 1984).

Actin is one of the most important components of the cell architecture. The protein is involved in cell movement and in a variety of other motile processes like phagocytosis, vesicle transport, and exocytosis (BERSHADSKY and VASILIEV 1988). The actin-ADP-ribosylating toxins are novel tools to study the physiological functions of actin. In human neutrophil leukocytes, C2 toxin increases superoxide anion production and secretion of the vitamin B12 binding protein as well as *N*-acetylglucosamine from specific and azurophilic granules stimulated by chemotactic agents (NORGAUER et al. 1988; AL-MOHANNA et al. 1987). In contrast, the random and stimulated migration of leukocytes are inhibited by the toxin (NORGAUER et al. 1988). These effects are paralleled by ADP-ribosylation of actin and its depolymerization in neutrophils (NORGAUER et al. 1989). In PC12 cells, C2 toxin increases and decreases the carbachol-stimulated neurotransmitter release dose-dependently in a biphasic manner (MATTER et al. 1989). Whereas in mast cells the secretion of histamine is inhibited by toxin treatment (BÖTTINGER et al. 1987), the steroid release from Y1 adrenal cells is increased by C2 toxin (ZEPEDA et al. 1989). All these studies indicate the involvement of actin in the described physiological processes.

Furthermore, C2 toxin was shown to inhibit smooth muscle contraction of a guinea-pig ileum longitudinal smooth muscle preparation, most likely by a direct action of the toxin on the smooth muscle cell (MAUSS et al. 1989). In contrast, the toxin has no effect on the contraction of the aortic smooth muscle (SIMPSON 1982). This discrepancy is most likely due to variation in the actin isoform composition of the smooth muscle preparations studied (SKALLI et al. 1987). Moreover, in consideration of the fact that C2 toxin ADP-ribosylates G- but not F-actin (AKTORIES 1986a,b), inhibition of smooth muscle contractility by the toxin suggests that a G/F-actin transition plays a role.

VII. Other Actin-ADP-Ribosylating Toxins

Recently, additional actin-ADP-ribosylating toxins have been described. These include *C. perfringens* iota toxin (SCHERING et al. 1988; SIMPSON et al. 1987), *C. spiroforme* toxin (POPOFF and BOQUET 1988; SIMPSON ct al. 1989), and an ADP-ribosyltransferase produced by *C. difficile* (POPOFF et al. 1988). These toxins are binary in structure and immunologically related (POPOFF and BOQUET 1988; POPOFF et al. 1988; STILES and WILKINS 1986). However, no immunological cross-reactivity has been shown between these toxins and C2 toxin (POPOFF and BOQUET 1988). Furthermore, the binding components of the iota-like toxins can replace each other to transport the enzyme component into the target cell (POPOFF and BOQUET 1988; SIMPSON et al. 1989). In contrast, the components of C2 toxin are not interchangeable with components of iota-like toxins. The iota-like toxins apparently ADP-ribosylate actin also at arginine 177 (VANDEKERCKHOVE et al. 1987; SCHERING et al. 1988). There exists, however, one major difference in the substrate specificity of C2 toxin and iota toxin, because the latter was shown to ADP-ribosylate all six vertebrate actin isoforms including skeletal muscle, cardiac muscle, and smooth muscle α-actin (SCHERING et al. 1988; MAUSS et al. 1990).

C. *Clostridium botulinum* C3 ADP-Ribosyltransferase

I. Origin and Structure

C3 ADP-ribosyltransferase is produced by type C and D strains of *C. botulinum* (AKTORIES et al. 1987; RUBIN et al. 1988). The exoenzyme is encoded by bacteriophages which also encode neurotoxins C1 and D. Because C3 is a contamination of commercially available neurotoxin preparations, its effects have been mixed up with those of the neurotoxins. However, it is now generally accepted that C3 and neurotoxins are structurally, immunologically, and functionally unrelated (AKTORIES and FREVERT 1987; RÖSENER et al. 1987; ADAM-VIZI et al. 1988; RUBIN et al. 1988). C3 has been purified to apparent homogeneity, having a molecular weight of about 25 kDa, which is evidently different from the 150-kDa neurotoxins (AKTORIES et al. 1988a; RUBIN et al. 1988). Recently, the cDNA sequence of C3 has been reported (POPOFF et al. 1990). From this work it is clear that C3 is related to neither other bacterial ADP-ribosylating toxins nor botulinum neurotoxins. In support of this view, we observed that a strain of *C. limosum*, which synthesizes no neurotoxins, produces an ADP-ribosyltransferase largely related to C3 (JUST and SCHALLEHN 1991). Analysis of the amino acid sequences of peptides derived from the *C. limosum* exoenzyme shows about 70% homology with the respective sequences of C3 (VANDEKERCKHOVE and AKTORIES, unpublished results). *C. limosum* exoenzyme and C3 share the same protein substrates (*rho* proteins) and apparently modify the identical acceptor amino acid (asparagine 41). Antibodies raised against C3 cross-reacted with the *C. limosum* exoenzyme and blocked the ADP-ribosylation by this transferase (JUST and SCHALLEHN 1991).

II. Enzyme Activity

C3 mono-ADP-ribosylates 21–26-kDa GTP binding proteins in all cells and tissues studied so far (AKTORIES and HALL 1989). The K_m value of the reaction for NAD is about $1 \mu M$. The ADP-ribosylation by C3 is reversible in the presence of high concentrations of nicotinamide ($>1\,\text{m}M$) with the pH optima at 7.5 and 5.5 for the forward and reverse reaction, respectively (HABERMANN et al. 1991). Again, as for C2 toxin, C3 possesses NAD-glycohydrolase activity and splits NAD in ADP-ribose and nicotinamide (AKTORIES et al. 1988a).

III. Substrates

Recently, the protein substrates of C3 have been identified to be the GTP-binding proteins *rho* (KIKUCHI et al. 1988; BRAUN et al. 1989; AKTORIES et al. 1989; NARUMIYA et al. 1988) and *rac* (DIDSBURY et al. 1989). These proteins are encoded by a superfamily of *ras*-related genes. The *ras* gene family (Ki-, Ha-, N-*ras*) has attracted special attention because mutations of these genes are frequently found in certain types of human tumors and are suggested to contribute to the malignancy of the cells (BARBACID 1987; HALL 1989). Other genes belonging to this multigene superfamily are for example *rap* (1A, 1B, 2) (PIZON et al. 1988), *rab* (1–6) (TOUCHOT et al. 1987), *ral* (CHARDIN and TAVITIAN 1986), and *YPT* (HAUBRUCK et al. 1987). All these genes encode the GTP-binding proteins of $21-24\,\text{kDa}$, which are regulated by associated GTPase activities (BOURNE et al. 1991). The gene products are active in the GTP form and inactive with GDP bound. Recently, different GTPase activating proteins (GAP) have been described, which largely increase the hydrolysis of protein-bound GTP, thereby inactivating the GTP-binding proteins (ADARI et al. 1988; GARRETT et al. 1989). Furthermore, it appears that the active state of the small GTP-binding proteins is additionally regulated by nucleotide exchange promoting (Downward et al. 1990; Wolfman and Macara) and exchange inhibiting factors (UEDA et al. 1990).

The *rho* (= ras *ho*mologues) genes were first described in *Aplysia* (MADAULE and AXEL 1985). In yeast (*Saccharomyces cerevisiae*) two *rho* genes (*RHO*-1 and *RHO*-2) have been detected; deletion analysis revealed that *RHO*-1 but not *RHO*-2 is essential for yeast viability (MADAULE et al. 1987). In human cells, three different *rho* genes (A, B, C) have been found which encode proteins with about 90% homology between each other (CHARDIN et al. 1988; YERAMIAN et al. 1987). The human *rho* genes are about 35% homologous to *ras* genes. All three *rho* proteins (A, B, C) are substrates of C3.

The second family of C3 substrates are the *rac* 1 and *rac* 2 proteins (*rac* = *ras*-related *C*3-substrate) (DIDSBURY et al. 1989). The *rac* 1 and *rac* 2 genes are together 92% homologous and share 58% homology with the human *rho* genes. The homology of the *rac* genes with human *ras* is about 30%. Whereas transcript of *rac* 1 was detected in various tissues, *rac* 2 transcript was predominantly observed in myeloid cells.

No effect of ADP-ribosylation has been reported on the binding of guanine nucleotides to the *rho* proteins or on its basal or GTPase-activating protein- (GAP)-stimulated GTPase activity (Braun et al. 1989; Paterson et al. 1990). However, divalent cations and guanine nucleotides appear to regulate the ADP-ribosylation by C3 (Aktories et al. 1988a). Whereas at low concentrations of Mg^{2+} ($100\,\mu M$) the ADP-ribosylation is increased, high concentrations ($1\,mM$) decrease it. When *rho* A protein has been loaded with various guanine nucleotides after "opening" the nucleotide binding site by treatment with EDTA, the favored substrate of C3 is the GDP-bound form of the *rho* protein (Habermann et al. 1991). By analogy, the pertussis toxin-catalyzed ADP-ribosylation of the heterotrimeric G-proteins (e.g., G_i, transducin) is largely reduced with the active GTP- or GTP[S]-bound forms of the proteins (Katada et al. 1984; Van Dop et al. 1984). Accordingly, the activation of transducion e.g., by interaction with the stimulated light-receptor rhodopsin, decreases the ADP-ribosylation of the GTP-binding protein by pertussis toxin (Van Dop et al. 1984). Apparently, the same holds for the C3-induced ADP-ribosylation. Wieland et al. (1990) showed that in bovine rod outer segment membranes, the ADP-ribosylation by C3 was largely reduced after activation of rhodopsin by light. These findings were interpreted to indicate that the C3 substrates also interact with rhodopsin and to propose a role of these GTP-binding proteins in membrane signal transduction.

IV. ADP-Ribosylation of Asparagine

C3 ADP-ribosylates *rho* (probably all C3 substrates) at asparagine 41 (Sekine et al. 1989). The ADP-ribose-asparagine bond has been shown to be particularly stable towards NH_2OH and mercury ions (Aktories et al. 1988b). These agents are able to cleave the ADP-ribose-arginine and ADP-ribose-cysteine-linkage formed by cholera toxin, *C. botulinum* C2 toxin, and pertussis toxin, respectively (Hsia et al. 1985; Mayer et al. 1988). Mainly based on the proposed structural homologies between *rho/rac* and *ras* proteins, it has been suggested that asparagine 41 is located in the so-called effector region of the GTP-binding proteins (Pai et al. 1990; Nishiki et al. 1990).

V. Functional Consequences

The functional consequences of the ADP-ribosylation by C3 are still unclear. So far, only a few studies have been performed with highly purified C3 preparations, which exclude effects elicited by the extremely potent botulinum neurotoxins or other clostridial contaminants. Rubin et al. (1988) reported that C3, which was introduced into NIH 3T3 cells by the osmotic shock method, caused changes in the cell morphology with rounding up and formation of binucleated cells. In PC12 cells, C3 induces the formation of short neurites, inhibition of cell growth, and stimulation of acetylcholinesterase (Rubin et al. 1987; Nishiki et al. 1990). C3 treatment of Vero cells leads to rounding up and destruction of the microfilament network (CHARDIN et al. 1989). Similarly, in rat hepatoma FAO cells the microfilament network is largely altered by C3 treatment (WIEGERS et al. 1991). In addition, the intermediate filaments stained with vimentin-specific antibody redistribute and aggregate near the nucleus. In contrast, the microtubule system is almost unaffected and shows still typical filaments. From the latter studies it has been suggested that the C3 substrates are somehow involved in the regulation of cytoskeletal proteins. More direct evidence for an involvement of *rho* proteins in the regulation of the cytoskeletal architecture comes from studies with recombinant valine-14 *rho* A protein. This mutated *rho* protein is inhibited in its endogeneous GTPase activity and therefore constitutively active. Microinjection of valine-14 *rho* protein into Swiss 3T3 cells caused dramatic morphological changes. Interestingly, the ADP-ribosylation of the protein by C3 largely prevented the effects on cell morphology. These findings were interpreted to indicate that the C3-induced modification of *rho* renders the protein inactive (PATERSON et al. 1990).

D. Concluding Remarks

The ADP-ribosylating toxins have been shown to be valuable tools in studying the functions of their target proteins. The same holds true for the clostridial ADP-ribosyltransferases C2 and C3. While C2 appears to be highly instrumental for studies on actin, C3 may help to elucidate the role of their eukaryotic substrate proteins *rho* and *rac*. Recently, it was suggested that bacterial toxins mimic endogenous ADP-ribosylating enzymes. For instance, evidence has been presented that eukaryotic endogenous enzymes catalyze the ADP-ribosylaton of elongation factor 2 (LEE and IGLEWSKI 1984) or of the heterotrimeric GTP-binding protein G_i (TANUMA et al. 1988). Further studies are necessary to elucidate whether actin and the low molecular mass GTP-binding proteins *rho* and *rac* are also substrates of endogenous enzymes.

References

Adam-Vizi V, Rösener S, Aktories K, Knight DE (1988) Botulinum toxin-induced ADP-ribosylation and inhibition of exocytosis are unrelated events. FEBS Lett 238:277–280

Adari H, Lowy DR, Willumsen BM, der Channing J, McCormick F (1988) Guanosine triphosphate-activating protein (GAP) interacts with the p21 *ras* effector binding domain. Science 240:518–521

Aktories K, Frevert J (1987) ADP-ribosylation of a 21–24 kDa eukaryotic protein (s) by C3, a novel botulinum ADP-ribosyltransferase, is regulated by guanine nucleotide. Biochem J 247:363–368

Aktories K, Hall A (1989) Botulinum ADP-ribosyltransferase C3: a new tool to study low molecular weight GTP-binding proteins. Trends Pharmacol Sci 10:415–418

Aktories K, Wegner A (1989) ADP-ribosylation of actin by clostridial toxins. J Cell Biol 109:1385–1387

Aktories K, Bärmann M, Ohishi I, Tsuyama S, Jakobs KH, Habermann E (1986a) Botulinum C2 toxin ADP-ribosylates actin. Nature 322:390–392

Aktories K, Ankenbauer T, Schering B, Jakobs KH (1986b) ADP-ribosylation of platelet actin by botulinum C2 toxin. Eur J Biochem 161:155–162

Aktories K, Weller U, Chhatwal GS (1987) *Clostridium botulinum* type C produces a novel ADP-ribosyltransferase distinct from botulinum C2 toxin. FEBS Lett 212:109–113

Aktories K, Rösener S, Blaschke U, Chhatwal GS (1988a) Botulinum ADP-ribosyltransferase C3. Purification of the enzyme and characterization of the ADP-ribosylation reaction in platelet membranes. Eur J Biochem 172:445–450

Aktories K, Just I, Rosenthal W (1988b) Different types of ADP-ribose protein bonds formed by botulinum C2 toxin, botulinum ADP-ribosyltransferase C3 and pertussis toxin. Biochem Biophys Res Commun 156:361–367

Aktories K, Reuner K-H, Presek P, Bärmann M (1989a) Botulinum C2 toxin treatment increases the G-actin pool in intact chicken cells: a model for the cytopathic action of actin-ADP-ribosylating toxins. Toxicon 27:989–993

Aktories K, Braun U, Rösener S, Just I, Hall A (1989b) The *rho* gene product expressed in *E. coli* is a substrate of botulinum ADP-ribosyltransferase C3. Biochem Biophys Res Commun 158:209–213

Al-Mohanna FA, Ohishi I, Hallett MB (1987) Botulinum C_2 toxin potentiates activation of the neutrophil oxidase. FEBS Lett 219:40–44

Barbacid M (1987) *ras* Genes. Annu Rev Biochem 56:779–827

Bershadsky AD, Vasiliev JM (1988) Cytoskeleton. Plenum, New York

Böttinger H, Reuner KH, Aktories K (1987) Inhibition of histamine release from rat mast cells by botulinum C2 toxin. Int Arch Allergy Appl Immunol 84:380–384

Bourne HR, Sanders DA, McCormick F (1991) The GTPase superfamily: conserved structure and molecular nmechanism. Nature 349:117–127

Braun U, Habermann B, Just I, Aktories K, Vandekerckhove, J (1989) Purification of the 22 kDa protein substrate of botulinum ADP-ribosyltransferase C3 from porcine brain cytosol and its characterization as a GTP-binding protein highly homologous to the *rho* gene product. FEBS Lett 243:70–76

Chardin P, Tavitian A (1986) The *ral* gene: a new *ras* related gene isolated by the use of a synthetic probe. EMBO J 5:2203–2208

Chardin P, Madaule P, Tavitian A (1988) Coding sequence of human rho cDNAs clone 6 and clone 9. Nucl Acids Res 16:2717

Chardin P, Boquet P, Madaule P, Popoff MR, Rubin EJ, Gill DM (1989) The mammalian G protein rho C is ADP-ribosylated by *Clostridium botulinum* exoenzyme C3 and affects actin microfilaments in Vero cells. EMBO J 8:1087–1092

Collier RJ, Cole HA (1969) Diphtheria toxin subunit active in vitro. Science 164:1179–1181

Collier RJ, Kandel J (1971) Structure and activity of diphtheria toxin. I. Thiol-dependent dissociation of a fraction of toxin into enzymatically active and inactive fragments. J Biol Chem 246:1496–1503

Didsbury J, Weber RF, Bokoch GM, Evans T, Snyderman R (1989) Rac, a novel *ras*-related family of proteins that are botulinum toxin substrates. J Biol Chem 264:16378–16382

Downward J, Riehl R, Wu L, Weinberg RA (1990) Identification of a nucleotide exchange-promoting activity for p21[ras]. Proc Natl Acad Sci USA 87:5988–6002

Eklund MW, Poysky FT (1972) Activation of a toxic component of *Clostridium* types C and D by trypsin. Appl Microbiol 24:108–113

Eklund MW, Poysky FT, Reed SM, Smith CA (1971) Bacteriophage and the toxigenicity of *Clostridium botulinum* type C. Science 172:480–482

Garrett MD, Self AJ, von Oers C, Hall A (1989) Identification of distinct cytoplasmic targets for *ras*, R-*ras* and *rho* regulatory proteins. J Biol Chem 264:10–13

Geipel U, Just I, Schering B, Haas D, Aktories K (1989) ADP-ribosylation of actin causes increase in the rate of ATP exchange and inhibition of ATP hydrolysis. Eur J Biochem 179:229–232

Geipel U, Just I, Aktories K (1990) Inhibition of cytochalasin D-stimulated G-actin ATPase by ADP-ribosylation with *Clostridium perfringens* iota toxin. Biochem J 266:335–339

Gill DM (1977) Mechanism of action of cholera toxin. Adv Cyclic Nucleotide Res 8:85–118

Habermann E, Dreyer F (1986) Clostridial neurotoxins: handling and action at the cellular and molecular level. Curr Top Microbiol Immunol 129:93–179

Habermann B, Mohr C, Just I, Aktories K (1991) ADP-ribosylation and de-ADP-ribosylation of the *rho* protein by *Clostridium botulinum* exoenzyme C3. Regulation by EDTA, guanine nucleotide and pH. Biochim Biophys Acta 1077:253–258

Hall A (1989) The cellular functions of small GTP-binding proteins. Science 249:635–640

Haubruck H, Disela C, Wagner P, Gallwitz D (1987) The *ras*-related *ypt* protein is an ubiquitous eukaryotic protein: isolation and sequence analysis of mouse cDNA clones highly homologous to the yeast *YPT1* gene. EMBO J 6:4049–4053

Hsia JA, Tsai S-C, Adamik R, Yost DA, Hewlett EL, Moss J (1985) Amino acid-specific ADP-ribosylation. J Biol Chem 260:16187–16191

Just I, Schallehn G (1991) A novel C3-like ADP-ribosyltransferase produced by *Clostridium limosum*. Naunyn Schmiedeberg Arch Pharmacol [Suppl]343:R38

Just I, Geipel U, Wegener A, Aktories K (1990) De-ADP-ribosylation of actin by *Clostridium perfringens* iota toxin and *Clostridium botulinum* C2 toxin. Eur J Biochem 192:723–727

Katada T, Northup JK, Bokoch GM, Ui M, Gilman AG (1984) The inhibitory guanine nucleotide-binding regulatory component of adenylate cyclase: subunit dissociation and guanine nucleotide-dependent hormonal inhibition. J Biol Chem 259:3578–3585

Kikuchi A, Yamamoto K, Fujita T, Takai Y (1988) ADP-ribosylation of the bovine brain *rho* protein by botulinum toxin type C1. J Biol Chem 263:16303–16308

Lee H, Iglewski WJ (1984) Cellular ADP-ribosylation with the same mechanism of action as diphtheria toxin and *Pseudomonas* toxin A. Proc Natl Acad Sci USA 81:2703–2707

Leppla SH (1982) Anthrax toxin edema factor: bacterial adenylate cyclase that increases cyclic AMP concentrations in eukaryotic cells. Proc Natl Acad Sci USA 79:3162–3166

Madaule P, Axel R (1985) A novel *ras*-related gene family. Cell 41:31–40

Madaule P, Axel R, Myers AM (1987) Characterization of two members of the *rho* gene family from the yeast *Saccharomyces cerevisiae*. Proc Natl Acad Sci USA 84:779–783

Matter K, Dreyer F, Aktories K (1989) Actin involvement in exocytosis from PC12 cells: studies on the influence of botulinum C2 toxin on stimulated noradrenaline release. J Neurochem 52:370–376

Mauss S, Koch G, Kreye VAW, Aktories K (1989) Inhibition of the contraction of the isolated longitudinal muscle of the guinea-pig ileum by botulinum C2 toxin: evidence for a role of G/F-actin transition in smooth muscle contraction. Naunyn Schmiedebergs Arch Pharmacol 340:345–351

Mauss S, Chaponnier C, Just I, Aktories K, Gabbiani G (1990) ADP-ribosylation of actin isoforms by *Clostridium botulinum* C2 toxin and *Clostridium perfringens* iota toxin. Eur J Biochem 194:237–241

Meyer T, Koch R, Fanick W, Hilz H (1988) ADP-ribosyl proteins formed by pertussis toxin are specifically cleaved by mercury ions. Biol Chem Hoppe Seyler 369:579–583

Narumiya S, Sekine A, Fujiwara M (1988) Substrate for botulinum ADP-ribosyltransferase, Gb, has an amino acid sequence homologous to a putative *rho* gene product. J Biol Chem 263:17255–17257

Nishiki T, Narumiya S, Morii N, Yamamoto M, Fujiwara M, Kamata Y, Sakaguchi G, Kozacki S (1990) ADP-ribosylation of the rho/rac proteins induces growth inhibition, neurite outgrowth and acetylcholine esterase in cultured PC-12 cells. Biochem Biophys Res Commun 167:265–272

Noda M, Kato I, Matsuda F, Hirayama T (1981) Mode of action of staphylococcal leukocidin: relationship between binding of ^{125}I-labeled S and F components of leukocidin to rabbit polymorphnuclear leukocytes and leukocidin activity. Infect Immun 34:362–367

Norgauer J, Kownatzki E, Seifert R, Aktories K (1988) Botulinum C2 toxin ADP-ribosylates actin and enhances O_2^--production and secretion but inhibits migration of activated human neutrophils. J Clin Invest 82:1376–1382

Norgauer J, Just I, Aktories K, Sklar LA (1989) Influence of botulinum C2 toxin on F-actin and N-formyl peptide receptor dynamics in human neutrophils. J Cell Biol 109:1133–1140

Ohishi I (1983a) Response of mouse intestinal loop to botulinum C2 toxin: enterotoxic activity induced by cooperation of nonlinked protein components. Infect Immun 40:691–695

Ohishi I (1983b) Lethal and vascular permeability activities of botulinum C2 toxin induced by separate injection of the two toxin components. Infect Immun 40:336–339

Ohishi I (1987) Activation of botulinum C_2 toxin by trypsin. Infect Immun 55:1461–1465

Ohishi I, Miyake M (1985) Binding of the two components of C2 toxin to epithelial cells and brush borders of mouse intestine. Infect Immun 48:769–775

Ohishi I, Odagiri Y (1984) Histopathological effect of botulinum C2 toxin on mouse intestines. Infect Immun 43:54–58

Ohishi I, Iwasaki M, Sakaguchi G (1980) Purification and characterization of two components of botulinum C2 toxin. Infect Immun 30:668–673

Ohishi I, Miyake M, Ogura K, Nakamura S (1984) Cytopathic effect of botulinum C2 toxin on tissue-culture cell lines. FEMS Lett Microbiol 23:281–284

Pai EF, Kabsch W, Krengel U, Holmes KC, John J, Wittinghofer A (1989) Structure of the guanine-nucleotide-binding domain of the Ha-*ras* oncogene product p21 in the triphosphate conformation. Nature 341:209–214

Paterson HF, Self AJ, Garrett MD, Just I, Aktories K, Hall A (1990) Microinjection of recombinant p21[rho] induces rapid changes in cell morphology. J Cell Biol 111:1001–1007

Pizon V, Chardin P, Lerosey I, Olofson B, Tavitian A (1988) Human cDNAs rap 1 and rap 2 homologous to the *Drosophila* gene Dras3 encode proteins closely related to *ras* in the "effector" region. Ocogene 3:201–204

Pollard T, Cooper JA (1986) Actin and actin-binding Proteins. A critical evaluation of mechanism and functions. Annu Rev Biochem 55:987–1035

Popoff MR, Boquet P (1988) *Clostridium spiroforme* toxin is a binary toxin which ADP-ribosylates cellular actin. Biochem Biophys Res Commun 152:1361–1368

Popoff MR, Rubin EJ, Gill DM, Boquet P (1988) Actin-specific ADP-ribosyltransferase produced by a *Clostridium difficile* strain. Infect Immun 56:2299–2306

Popoff MR, Boquet P, Gill DM, Eklund MW (1990) DNA sequence of exoenzyme C3, an ADP-ribosyltransferase encoded by *Clostridium botulinum* C and D phages. Nucleic Acids Res 18:1291

Reddy E, Reynolds R, Santos E, Barbacid (1982) A point mutation is responsible for the acquisition of transforming properties by the T24 human bladder carcinoma oncogene. Nature 300:149–152

Reuner KH, Presek P, Boschek CB, Aktories K (1987) Botulinum C2 toxin ADP-ribosylates actin and disorganizes the microfilament network in intact cells. Eur J Cell Biol 43:134–140

Rösener S, Chhatwal GS, Aktories K (1987) Botulinum ADP-ribosyltransferase C3 but not botulinum neurotoxins C1 and D ADP-ribosylates low molecular mass GTP-binding proteins. 1987. FEBS Lett 224:38–42

Rubin EJ, Gill DM, Boquet P, Popoff MR (1988) Functional modification of a 21-kilodalton G protein when ADP-ribosylated by exoenzyme C3 of *Clostridium botulinum*. Mol Cell Biol 8:418–426

Schering B, Bärmann M, Chhatwal GS, Geipel U, Aktories K (1988) ADP-ribosylation of skeletal muscle and non-muscle actin by *Clostridium perfringens* iota toxin. Eur J Biochem 171:225–229

Sekine A, Fujiwara M, Narumiya S (1989) Asparagine residue in the *rho* gene product is the modification site for botulinum ADP-ribosyltransferase. J Biol Chem 264:8602–8605

Simpson LL (1982) A comparison of the pharmacological properties of *Clostridium botulinum* type C1 and C2 toxins. J Pharmacol Exp Ther 223:695–701

Simpson LL (1984) Molecular basis for the pharmacological actions of *Clostridium botulinum* type C2 toxin. J Pharmacol Exp Ther 230:665–669

Simpson LL (1989) The binary toxin produced by *Clostridium botulinum* enters cells by receptor-mediated endocytosis to exert its pharmacologic effects. J Pharmacol Exp Ther 251:1223–1228

Simpson LL, Stiles BG, Zapeda HH, Wilkins TD (1987) Molecular basis for the pathological actions of *Clostridium perfringens* iota toxin. Infect Immun 55:118–122

Simpson LL, Stiles BG, Zepeda H, Wilkins TD (1989) Production by *Clostridium spiroforme* of an iotalike toxin that possesses mono(ADP-ribosyl)transferase activity: identification of a novel class of ADP-ribosyltransferases. Infect Immun 57:255–261

Skalli O, Vandekerckhove J, Gabbiani G (1987) Actin-isoform pattern as a marker of normal or pathological smooth-muscle and fibroblastic tissues. Differentiation 33:232–238

Stiles BG, Wilkins TD (1986) Purification and characterization of *Clostridium perfringens* iota toxin: dependence on two nonlinked proteins for biological activity. Infect Immun 54:6783–688

Tamura M, Nogimuri K, Murai S, Yajima M, Ito K, Katada T, Ui M, Ishii S (1982) Subunit structure of islet-activating protein, pertussis toxin, in conformity with the A-B model. Biochemistry 21:5516–5522

Tanuma S, Kawashima K, Endo N (1988) Eukaryotic mono(ADP-ribosyl)transferase that ADP-ribosylates GTP-binding regulatory G_i protein. J Biol Chem 263:5485–5489

Touchot N, Chardin P, Tavitian A (1987) Four additional members of the *ras* gene superfamily isolated by an oligonucleotide strategy: molecular cloning of YPT-related cDNAs from a rat brain library. Proc Natl Acad Sci USA 84:8210–8214

Trahey M, McCormick F (1987) A cytoplasmic protein stimulates normal N-*ras* p21 GTPase but does not affect oncogenic mutants. Science 238:542–545

Ueda T, Kikuchi A, Ohga N, Yamamoto J, Takai Y (1990) Purification and characterization from bovine brain cytosol of a novel regulatory protein inhibiting the dissociation of GDP from and the subsequent binding of GTP to *rho*B p20, a *ras*-like GTP-binding protein. J Biol Chem 265:9373–9380

Vandekerckhove J, Weber K (1979) The complete amino acid sequence of actins from bovine aorta, bovine heart, bovine fast skeletal muscle and rabbit slow skeletal muscle. Differentiation 14:123–133

Vandekerckhove J, Schering B, Bärmann M, Aktories K (1987) *Clostridium perfringens* iota toxin ADP-ribosylates skeletal muscle actin in Arg-177. FEBS Lett 225:48–52

Vandekerckhove J, Schering B, Bärmann M, Aktories K (1988) Botulinum C2 toxin ADP-ribosylates cytoplasmic β/γ-actin in arginine 177. J Biol Chem 263:696–700

Van Dop C, Yamanaka G, Steinberg F, Sekura RD, Manclark CR, Stryer L, Bourne HR (1984) ADP-ribosylation of transducin by pertussis toxin blocks the light-stimulated hydrolysis of GTP and cGMP in retinal photoreceptors. J Biol Chem 259:23–26

Wegner A, Aktories K (1988) ADP-ribosylated actin caps the barbed ends of actin filaments. J Biol Chem 263:13739–13742

Weigt C, Just I, Wegner A, Aktories K (1989) Nonmuscle actin ADP-ribosylated by botulinum C2 toxin caps actin filaments. FEBS Lett 246:181–184

Wiegers W, Just I, Müller H, Hellwig A, Traub P, Aktories K (1991) Alteration of the cytoskeleton of mammalian cells cultured in vitro by *Clostridium botulinum* C2 toxin and C3 ADP-ribosyltransferase. Eur J Cell Biol 54:237–245

Wieland T, Ulibarri I, Aktories K, Gierschik P, Jakobs KH (1990) Interaction of small G proteins with photoexcited rhodopsin. FEBS Lett 263:195–198

Wolfman A, Macara IG (1990) A cytosolic protein catalyzes the release of GDP from p21^{ras}. Science 248:67–69

Yeramian P, Chardin P, Madaule P, Tavitian A (1987) Nucleotide sequence of human rho cDNA clone 12. Nucleic Acids Res 15:1869

Zepeda H, Considine RV, Smith HL, Sherwin JA, Ohishi I, Simpson LL (1988) Actions of the *Clostridium botulinum* binary toxin on the structure and function of Y-1 adrenal cells. J Pharmacol Exp Ther 246:1183–1189

Subject Index

Printing: Saladruck, Berlin
Binding: Buchbinderei Lüderitz & Bauer, Berlin